The Encyclopedia of Housing

Board of Consulting Editors

The Encyclopedia of Housing

Willem van Vliet--

EDITOR

SAGE Publications
International Educational and Professional Publisher
Thousand Oaks   London   New Delhi

For information:

SAGE Publications, Inc.
2455 Teller Road
Thousand Oaks, California 91320
E-mail: order@sagepub.com

SAGE Publications Ltd.
6 Bonhill Street
London EC2A 4PU
United Kingdom

SAGE Publications India Pvt. Ltd.
M-32 Market
Greater Kailash I
New Delhi 110 048 India

Printed in the United States of America

Library of Congress Cataloging-in-Publication Data

The encyclopedia of housing / edited by Willem van Vliet--.
   p. cm.
   Includes bibliographical references and indexes.
   ISBN 0-7619-1332-7 (cloth)
   1. Housing—Encyclopedias.   2. Housing—United
States—Encyclopedias.   I. Van Vliet--, Willem, 1952–
HD7287 .E53 1998
363.5′03—-ddc21
                     98-8949

Visit the Sage website at www.sagepub.com and the special website for this volume at www.sagepub.com/sagepage/encyclopedia_of_housing.htm

98 99 00 01 02 03 04 7 6 5 4 3 2 1

Acquiring Editor:	Catherine Rossbach
Editorial Assistants:	Kathleen Derby/Heidi Van Middlesworth
Production Editor:	Diana E. Axelsen
Production Assistants:	Karen Wiley/Lynn Miyata
Copy Editor:	Linda Gray
Typesetter/Designer:	Christina M. Hill
Cover Designer:	Ravi Balasuriya
Indexer:	Rosi Hauber
Print Buyer:	Anna Chin

Contents

List of Tables

List of Figures

Foreword

ROBERT B. BECHTEL

This is a time when housing studies and programs are proliferating and diversifying into a bewildering array. Government programs have been cut, and this has stimulated private efforts. Meanwhile, postindustrial culture and electronic technology have produced smart houses and work-home environments. *The Encyclopedia of Housing* comes at an appropriate time and fills a long-felt need.

When I looked over the contents, I found myself wishing I had had such a reference when I started out more than 30 years ago. It would have saved many hours of searching in the library and playing phone tag with colleagues. It is a great benefit to those of us who work in this area to have so many basic references handy and in one place.

But this is not a simple historic volume; it is conceptual in nature so that the issues are dealt with in a long-term manner. Any reader will be able to scan the history and come up with the central problems and prospects of each area covered long into the future. It saves a lot of time in trying to understand what is and will be happening.

An unfortunate modern trend has been to treat social problems from a narrow, academic point of view. Universities are the principal offenders because they have a highly compartmentalized and fragmented system of departments organized along disciplinary boundaries that are not easily crossed. Thus, we have a sociology of housing, a psychology of housing, an economics of housing, an anthropology of housing, a geography of housing, a political science of housing, an architectural point of view, and a legal perspective. Each of these is found in a separate department and often the twain never meet or even tangle. The *Encyclope-*

dia seeks to bridge these gaps and presents many points of view. Its language aims across the disciplines.

It was a delight to look over the roster of contributors and see so many old friends and new collaborators, too many to mention them all, some of them an encyclopedia of housing themselves.

This volume should be on the shelves of every library in the world; at every office of professionals who design, manage, or research housing; and in the office of every politician from Congress down to the county and city level. This is the basic reference, and it will keep that status for some time to come. I can't help reiterating that I wish I had had something like it over the years.

It is often said that those who don't learn history are doomed to repeat it. Although the stance of the *Encyclopedia* is not solely historical, it chronicles the major programs, such as Section 8 and many others, so that it should be required reading for everyone concerned with housing. There are so many lessons to be learned here about the failures, the successes, and the sterling examples. If only the policymakers would make themselves familiar with these basics, we would have an easier world to live in. Better still, we would have a world in which we don't keep repeating the same mistakes.

I was pleased to see Habitat for Humanity listed among the many entries. This nongovernmental effort has had great success throughout the world, not only in the United States but in more than 50 countries. Over 40,000 homes have been built worldwide with the organization's no-interest loan policy and self-help format that brings poor and middle class together. Here is another light to the future that may help make up for government withdrawal.

We have an incredible amount of knowledge about many problems. In their separate disciplines, political scientists, psychologists, sociologists, anthropologists, geographers, architects, and even maintenance crews have accumulated great treasures of information about what is needed to improve the quality of life. The main difficulty seems to be the inability to apply that knowledge to the problems at hand. *The Encyclopedia of Housing* is an important step in the process of using the knowledge we have accumulated to address this persistent and worldwide problem of how to provide decent and affordable shelter for everyone. On every page, there is hope that at the very least we don't have to repeat a mistake again or at the very best, we can take programs that have already been tried and apply them to new and challenging situations.

—*Robert B. Bechtel*
Professor of Psychology
University of Arizona, Tucson

Foreword

John M. Goering

One of the most critical and yet typically undervalued tasks in social science research and policy is to translate concepts, principles, and policy ideas into terms that are understandable across disciplines as well as across national boundaries. *The Encyclopedia of Housing* makes an enormous contribution to this field of improved international understanding of key issues in housing research and policy-making. It makes real a promise that social scientists have frequently made over the last half century—that interdisciplinary research and analysis should be central in the process of doing applied housing research. Nowhere has this been less true, to date, than in many aspects of housing research—which often have strong disciplinary traditions deeply rooted in a single country's policies and programs.

The Encyclopedia of Housing includes entries from many of the most eminent researchers and analysts of housing issues in the United States and indeed in other parts of the world. It includes a wealth of information and analyses covering key concepts and issues in the field of housing, many of which apply without regard to national context. There are, for example, a superb series of entries addressing a cross-cutting set of race segregation, redlining, and civil rights issues that affect housing—material that would often be neglected when only a "bricks-and-mortar" approach to housing is taken.

The reader will be challenged by authors whose views are clear and who do not shy away from offering strong recommendations for policy and action. In some instances, readers may find entries more argumentative than they are used to seeing in a standard encyclopedia, but the advantage of this collection is that the contributions are written with conviction drawn from years of research. This is not a timid volume but, rather, one that includes principled arguments for many types of housing reform, much debated over the last decade. There are also a number of instances in which various authors find common ground in addressing large policy and program changes so that the reader sees an issue from more than one perspective.

Readers need not agree with all of the arguments and perspectives expressed, and indeed debate should be a natural outgrowth of this *Encyclopedia,* but they will welcome the care and clarity with which large fields of policy analysis and research have been synthesized in a nontechnical manner.

This volume includes entries by clearly recognized leaders in their fields who have assembled concise and often self-critical assessments of the research, policies, and programs in an area—no small feat for areas typically filled with shelves of books on these subjects. The lists of organizations, journals, and legislation are added services the volume provides, making it a useful reference resource beyond that normally offered in encyclopedias.

In this volume, readers will find a lack of academic jargon accompanied by concise, relevant assessments of state-of-the-art information about research, legislation, and programs—a substantial contribution to the field of housing. They will find themselves challenged by the broad sweep of issues summarized in the individual entries but helped by each author's conscientious indication of other readings and background material of relevance.

An additional advantage is that many of the authors make reference to policies, programs, or research from other countries, especially European housing policy research. Such assessments help clarify for the reader the strengths as well as some of the current or inevitable limits of current U.S. policy options. Editor Willem van Vliet--deserves great credit for assembling such an eminent group of scholars and selecting a wide-ranging and enticing list of topics and reference materials for use—even to those long familiar with the field of housing research.

—John M. Goering
U.S. Department of Housing and
Urban Development, Washington, D.C.

Foreword

James A. Johnson

Today, 66% of Americans own their own homes, an all-time high. This is a monumental achievement, owing in large part to the tremendous commitment of individuals, many of whom make enormous sacrifices to purchase a home. But for Americans, homeownership goes beyond a mere investment in property. It's a concept rooted in the founding ideals of the United States, a symbol of security, hope, and community. The quality of housing and access to it help define a nation. The importance of home has been an enduring value for me, as well. Growing up in Benson, Minnesota, I cherished stories about my forebears who emigrated from Norway to the northern plains to make a new life. The small log cabin they built now stands in the Swift County Historical Museum. Their homestead is a testament not to a lost era but to a dream of homeownership that is still very much alive.

Today, community building is no longer a matter of land, wood, bricks, and mortar. Now technology, demographics, financing alternatives, and affordability are critical factors. Public policies on crime prevention, drugs, the environment, and transportation influence the growth and development of neighborhoods. Accommodations for people with AIDS, the homeless, the disabled, and others with special needs are a vital part of the housing discussion. Jobs, historic preservation, and access to child care have an impact. Fighting discrimination in housing and educating immigrants and other new consumers about the home-ownership process are two of the numerous challenges the housing industry addresses. A huge amount of data is available to those with an interest in real estate, housing finance, construction, and the social and cultural implications of housing. However, much of this information has been disseminated without adequate perspective on closely related topics.

It gives me pleasure to write this foreword to *The Encyclopedia of Housing*. This exceptional resource provides what is certain to become a much appreciated reference tool for students, teachers, and professionals interested in housing. Professor Van Vliet-- has taken great care to create a convenient, comprehensive, substantive resource.

By compiling the work of many of the nation's experts on housing issues, he has provided a compendium that takes a multidisciplinary approach to housing. With subjects ranging from sweat equity to mortgage-backed securities, *The Encyclopedia of Housing* highlights significant topics about contemporary issues that affect housing. As such, this book can help scholars and lay people alike become better informed about the broad spectrum of factors that make up the housing industry.

—James A. Johnson
Chairman and CEO, Fannie Mae,
Washington, D.C.

Foreword

DANIEL STOKOLS

Studies of environment and behavior examine the dynamic transactions between people and their everyday sociophysical environments. Research in the environment–behavior field encompasses the perspectives of several disciplines, including environmental psychology, environmental sociology, behavioral geography, environmental design, urban planning, natural resources management, and cultural anthropology (Moore and Marans 1997; Stokols and Altman 1987; Zube and Moore 1991). Regardless of the particular theories emphasized in these studies, all are fundamentally concerned with the effects of sociophysical environments on people's behavior and well-being and with the reciprocal efforts made by individuals and groups to modify their surroundings in accord with their everyday goals and activities. This process of striving to enhance the fit between people and their surroundings is referred to as "human environment optimization" (Stokols 1977).

Of all environmental settings encountered by individuals during their daily routines, housing is in many respects the most pivotal, in view of its psychological and social significance. Housing environments, under the best of circumstances, provide a haven of security and a comfortable, supportive "anchor point" from which individuals organize their daily plans and activities (see Michelson 1998 [this volume]; Wapner 1981). When the congruence between people and their surroundings is impaired, however, emotional disturbances, physical health problems, and social disorders often occur (see Fried 1963). Thus, to improve the fit between people and their residential settings, it is crucial that we understand those aspects of housing that support or constrain occupants' goals and activities.

A fact of life in the late 20th century is that people's physical and social environments are changing rapidly because of demographic and technological shifts (Stokols 1995). That our population is aging, for example, suggests that the diversity and quantity of housing stock designed for the elderly will increase in the coming years (Parmelee and Lawton 1990). Moreover, the growing prevalence of desktop computing and our greater reliance on telecommunications will alter the design and locational pattern of housing. Examples of these changes include the increasing need for dedicated office space within residential units and the emerging trend toward telecommuting—working at home or at some other location away from one's primary workplace via electronic mail, telefax, and phone (Handy and Mokhtarian 1995).

As environmental, technological, and demographic changes accelerate and become more pervasive, efforts to ensure a fit between people and their housing arrangements will become more challenging. These challenges are attributable, in part, to the diversity of physical forms, household structures, behavior patterns, and regulatory issues associated with housing. Housing designs have become more varied, and the economic and legal contexts surrounding these designs have become more complex. At the same time, "nontraditional" household groups have become more prevalent because of recent demographic shifts (Franck and Ahrentzen 1989). Thus, both the physical forms and social arrangements associated with housing have become more diverse. Consequently, the task of optimizing the fit between housing environments and the diverse needs of prospective residents has become more complex.

The *Encyclopedia of Housing* makes several important contributions that will enable housing researchers and practitioners to better grasp the determinants of fit between residential settings and the needs of their occupants. First, the entries reflect the variety of physical and social housing types that currently exist—encompassing alternative construction and energy conservation techniques, diverse residential designs and neighborhood plans (e.g., loft arrangements, high-rise apartment buildings, new towns, and planned communities), and a host of social innovations and management strategies (such as cooperative housing, self-help housing, cohousing, community-based housing, resident management, and tenant organizing strategies).

Second, the *Encyclopedia* suggests certain overarching dimensions to understand the diverse, and sometimes conflicting, needs of multiple occupant groups. For instance, the entries focusing on children, adolescents, and the elderly show the value of a life span developmental perspective in housing research and practice. These entries remind us that the design features and site plans of housing should be matched to support the developmental needs and activity patterns of those age groups for whom the residences are designed.

Similarly, another set of entries—those focusing on feminist housing design, Asian Americans, feng-shui, African Americans, Hispanic Americans, Native Americans, and household composition more generally—highlight other dimensions of occupant diversity (based on gender, ethnicity, and household structure) that must be considered in any effort to improve the fit between residents and their housing arrangements.

Third, the *Encyclopedia* outlines several types of psychological distress, physical health problems, and social disorders that often arise when housing and neighborhood arrangements are dysfunctional—that is, when they undermine rather than support residents' goals and activities. These behavioral and social disorders are examined in the entries on problems of crowding stress; on drug abuse, crime, and abandonment that have often plagued high-rise apartment buildings and low-income neighborhoods; and on community-wide conditions of housing discrimination, segregation, redlining, and homelessness that have become all too common in many U.S. cities.

Fourth, the *Encyclopedia* identifies several powerful economic, organizational, and legal strategies that can be used by housing practitioners—including environmental designers, real estate developers, urban planners, and policymakers—to reduce or eliminate the dysfunctional outcomes of inferior housing noted above. These strategies include gentrification processes to revitalize urban areas that have undergone disinvestment and economic decline; principles of "new urbanism" aimed at revitalizing public spaces and accommodating the needs of diverse population groups; and fair housing audits and inclusionary zoning practices intended to reverse discriminatory housing practices. Entries also identify legal and policy approaches to redress or prevent housing dysfunctions. These regulatory and public policy strategies include federally funded affordable housing initiatives, rent supplements, and housing subsidies for low-income residents; the Community Development Block Grant program; and the implementation of housing codes and space standards to ensure the healthfulness of residential environments.

Finally, the *Encyclopedia* presents the diverse governmental agencies, corporate entities, and nonprofit organizations that will play important roles in designing, constructing, and managing housing environments in the 21st century. These organizations include federal and state governments, city councils, housing developers, the Neighborhood Reinvestment Corporation, Fannie Mae, and community development corporations, among many others. To the extent that these different entities can find ways to coordinate their respective efforts, the goal of providing high-quality, supportive housing to all segments of the population is more likely to be achieved.

In sum, *The Encyclopedia of Housing* is unprecedented in its multidisciplinary scope and its integration of major research and practice issues that are central to the field of housing. This volume will be enormously useful to housing researchers and students, as well as to environmental designers, residential developers, urban planners, and elected officials. It is an impressive scholarly achievement that will stand as a major contribution to both environment-behavior research and housing practice.

—Daniel Stokols
Dean, School of Social Ecology,
University of California, Irvine

References

Franck, K. A. and S. Ahrentzen, eds. 1989. *New Households,* New Housing. New York: Van Nostrand Reinhold.

Fried, M. 1963. "Grieving for a Lost Home." In *The Urban Condition,* edited by L. Duhl. New York: Basic Books.

Handy, S. L. and P. L. Mokhtarian. 1995. "Planning for Telecommuting: Measurement and Policy Issues." *Journal of the American Planning Association* 61:99-111.

Michelson, W. H. 1998. "Behavioral Aspects." In *The Encyclopedia of Housing,* edited by W. van Vliet--. Thousand Oaks, CA: Sage.

Moore, G. T. and R. W. Marans, eds. 1997. *Advances in Environment, Behavior, and Design: Vol. 4. Toward the Integration of Theory, Methods, Research, and Utilization.* New York: Plenum.

Parmelee, P. A. and M. P. Lawton. 1990. "The Design of Special Environments for the Aged." Pp. 464-88 in *Handbook of the Psychology of Aging,* edited by J. E. Birren and K. W. Schaie. 3d ed. New York: Academic Press.

Stokols, D. 1977. *Perspectives on Environment and Behavior: Theory, Research, and Applications.* New York: Plenum.

Stokols, D. 1995. "The Paradox of Environmental Psychology." *American Psychologist* 50:821-37.

Stokols, D. and I. Altman, eds. 1987. *Handbook of Environmental Psychology.* Vols. 1 and 2. New York: John Wiley.

Wapner, S. 1981. "Transactions of Persons-in-Environments: Some Critical Transitions." *Journal of Environmental Psychology* 1:223-39.

Zube, E. H. and G. T. Moore, eds. 1991. *Advances in Environment, Behavior, and Design.* Vol. 3. New York: Plenum.

Acknowledgments

This *Encyclopedia* owes its existence to the contributions of many people. The members of the Board of Consulting Editors provided much appreciated assistance during various stages of the project, including the development of the list of entries, the selection of authors, and the review of draft entries. It was a rare opportunity being able to benefit from their collective wisdom. Most of them also shared their expertise by writing one or more entries.

Each entry underwent careful review. Aside from the board members, colleagues from around the country served as reviewers, offering useful feedback and suggestions for revision. These reviewers include the following: S. Aitken, Irwin Altman, Bob Applebaum, Larry Bourne, Satya Brink, Terry Clark, William Clark, Hemalata Dandekar, David E. Dowall, Joan Draper, Joe Feagin, Herbert Gans, Stephen Golant, William Grigsby, Jan Gudmand-Høyer, Michael Harloe, Charlene Harrington, Richard Harris, Karen Hill, Steve Kendall, Mark La Gory, M. Powell Lawton, John Logan, Robert Marans, Guido Francescato, S. M. Keigher, Jane Lillydahl, John Miron, Iouri Moisseev, Daniel Monti, John O'Loughlin, Zev Paiss, David Popenoe, Walter Rybeck, David Satterthwaite, Michael Schill, Hilary Silver, Susan Smith, Marcia Steinberg, Karl Taeuber, Lawrence Vale, Greg Van Ryzin, Allan Wallis, Peter Ward, Irving Welfeld, Kioe Sheng Yap, and John Zeisel.

In addition, there were others who were helpful in a variety of ways, including Maria N. Briones, Steve Brobeck, Laurie De Freese, Gary Kubis, Jeff Lisher, Roger Montgomery, Mary Fran Myers, Karen Pace, Jim Ratzenberger, Michael Siewert, Cathie Sullivan, and Jean Whelan.

Special thanks are due to the authors who wrote the entries, revised them, and generously gave of their time to furnish informative descriptions, insightful assessments, and otherwise authoritative coverage in concise form. Working with them was a unique privilege and an exceptional learning experience.

Caroline Nagel, Laurel Phoenix, and Mara Sidney spent more hours than they will care to remember providing competent assistance. Ray Studer, former dean of the then College of Environmental Design at the University of Colorado at Boulder, made available essential support from the project's inception. Ursi Chappelle, Bill Henry, Loree Kaleth, Mary MacKay, Meghann Ormond, and Lisa Perry lent a helping hand in bringing it to a conclusion. Lynn Lickteig and Carole Cardon of the Visual Resource Center of the College of Architecture and Planning skillfully produced original artwork for many of the illustrations. Regina Ahram, Elizabeth Gould, and Lynnette Westerlund of Interlibrary Loan Services, University of Colorado, Boulder, went beyond the call of duty in tracking down and making available relevant material, no matter how arcane. From the beginning to close to the end, Linda Stevens provided excellent secretarial support, flawlessly developing and maintaining a complex information base with superlative expertise and invariable good cheer. Rosi Hauber deserves a lifetime supply of chocolate chip bagels for her meticulous processing of seemingly interminable indices.

I would be remiss in not acknowledging two colleagues, who shall remain unnamed, whose threats of legal action (one for being excluded from the *Encyclopedia,* the other for being included) amply made up for the occasional dull moment during the project's seven-year span.

In a bizarre twist, when all was done, the original publisher decided to cancel production because of business considerations. At first disappointed and infuriated, I came to appreciate this as a happy turn of events after successfully reclaiming the rights to the manuscript. During this difficult stage, Peter Marcuse offered sound advice and board members expressed welcome support, while Michael Pollet, Esq., was indispensable in the drawn-out negotiations. In retrospect, I would like to thank Garland Publishing, Inc., for not publishing the project. Susan Clarke proposed it would be of interest to Sage Publications, Inc., where Catherine Rossbach, Acquisitions Editor, at once recognized she had no dog by the tail. She was a perfect ambassadress for the publishing industry in general and Sage in particular. Thanks to her professionalism and dedication and the enthusiastic backing of Nancy Hammerman, Senior Vice President of the Books Division, the abrupt cancellation proved to be a blessing in disguise.

Diana Axelsen once again proved her incomparable mastery of the production process, with a firm grasp of the overall schedule and keen attention to important details. Tina Hill gave the project the benefit of her typesetting expertise, untiringly making adjustments resulting from late revisions. Linda Gray vanquished anguished English in her copyediting of more than 4,000 manuscript pages and astutely identified points needing further clarification. Nancy Lambert caught gremlins undetected by anyone else. I am grateful to Ravi Balasuriya for being receptive to a child's drawing when using his graphic artistry to create the front cover. Having been involved in the project from up close for a long time, more than anyone I am aware of its limitations. Skepticism about whether the goal merited the enormous collaborative effort sometimes beset me. Aside from other things, Yvonne Könneker deserves credit for

having challenged my doubts. Others at Sage who made invaluable contributions include Kathleen Derby, Lynn Miyata, Karen Wiley, Heidi van Middlesworth, Jennifer Morgan, Jacklyn Paciulan, Tracy Tomakin, Shelly Crane, Wendy Westgate, Denise Santoyo, Kristi White, Elaine Benditson, and Jill Willman.

Never last, nor least, the sun's terpsichorean rays were instrumental in sustaining the ever-present inspiration.

Willem van Vliet--
University of Colorado

Editor's Introduction

WILLEM VAN VLIET--

Every day is a journey,
and the journey itself is home.

—Matsuo Bashō
(1644–1694)

Functions and Conceptualizations of Housing

Housing addresses basic human needs. At its most elemental level, it acts as shelter, offering protection against excessive cold and heat, rain, high winds, and other intemperate weather situations threatening people's well-being. By the same token, dampness, lead paint, vermin, overcrowding, and other substandard conditions undermine residents' physical and mental health. Housing also protects people against the risk of victimization by street crime.

Housing fulfills other important functions as well. At the household level, it provides a physical enclosure for domestic behavior—a place of relative privacy for daily activities, where people can cook, eat, socialize, and rest, away from the public realm and a place where, in many cultures, they are born and die. It is also a setting, removed from external scrutiny, where child beating and spousal abuse often go undetected. At the same time, through its location, housing forms the basis for activities in the community and larger outside world, such as interactions with neighbors, work, school, and shopping.

For most residents, however, housing is more than just a structure of "bricks-and-mortar." Usually, it is a place that people want to make into a home, a place to which they tend to hold emotional attachments resulting from its as-

sociation with accumulated life experience. Psychologically and socially significant aspects of housing are also evident in people's desire to personalize the interior and exterior space. People attempt to express their individual or group identity through housing. Research has found, for example, that the design of housing may reflect the occupational values of the residents, with self-made businessmen choosing somewhat ostentatious display homes and service professionals opting for more inward-looking designs. Furthermore, from an economic perspective, housing represents the largest financial investment most households will make during their lifetimes. Housing cost burdens for those with low incomes often leave insufficient resources for other necessities, such as food, clothing, medical care, and transportation.

In a wider community context, the design and location of housing can denote a household's affiliation with a particular cultural or religious group, serving to reinforce the social bonds among its members or making possible the carrying out of certain ritual activities. On the other hand, these same housing characteristics also reflect segregation from other population groups and can reinforce unequal access to day care, education, jobs, and life chances generally. In this sense, housing is inextricably connected to questions of redistributive justice and, thereby, to political and economic processes in the wider community and society at large.

It is not only to its occupants that housing is important. Aside from the users, there are the producers: land developers, builders, lenders, realtors, investors, architects, planners, construction unions, contractors, and many specialized professions and trades. Each of these groups has its own particular interests. In market societies, these interests revolve around financial gain because, in them, housing is treated foremost as a commodity, to be produced and traded for profit. In these systems, access to housing is a function of ability to pay. This perspective contrasts with a view of housing as a right, similar to entitlements to elementary education and basic health care. After years of back-and-forth negotiation among opponents and proponents, at the 2nd U.N. Conference on Human Settlements (Habitat II), held in Istanbul in June 1996, governments officially affirmed a commitment to the full and progressive realization of the right to adequate housing. However, in practice, this remains an elusive goal.

At the policy level, governments use housing to attain various other objectives. Chief among them are economic ones. Internationally, housing investments constitute between 2% and 8% of GNP, between 10% and 30% of gross capital formation, between 20% and 50% of accumulated wealth, and between 10% and 40% of household expenditures. Residential construction has numerous forward linkages (e.g., furniture, household appliances) and backward linkages (e.g., building components). Using this multiplier effect, governments often stimulate new construction to boost employment. Alternatively, when inflation is high, governments may seek to slow building by, for example, restricting credit supply. Housing can also be used as a tool to attain population-related objectives. Countries such as Singapore and Israel have attempted to achieve integration of their populations by means of housing, whereas others, such as Great Britain and the Netherlands, have used housing as part of population redistribution efforts. In an earlier era, the Republic of South Africa exploited housing as an instrument to implement apartheid.

Just as housing can be used to advance other policy goals, so also can nonhousing policies have significant effects on housing. A case in point was the U.S. transportation policy during the 1950s, when the interstate highway construction program played a key role in promoting low-density suburban sprawl. Other nonhousing policies affecting housing in major ways are those concerned with public finance, trade, employment, and social welfare.

Turner (1972) distinguished between housing as a noun and housing as a verb. According to this distinction, housing can be viewed as a *product,* ranging from individual artifacts (i.e., a house) to a collection of units (e.g., an apartment building, the national housing stock). Another conceptualization sees housing as a *process:* the provision and maintenance of housing by means of informal activities (e.g., self-help) and formal programs, embedded in public policies. The latter perspective is less oriented to what housing *is* (e.g., in terms of its physical standards) and more to what it *does* (e.g., provides security and access). Although these two are very different conceptualizations, both make it possible to link different levels of analysis and practice. Studies and programs concerned with homelessness, for example, can (and should) go beyond a narrow focus on the personal characteristics of those who are homeless to include also consideration of the broader contexts with which these individual experiences are twined, such as local housing and labor markets, national economic and social policies, and the global mobility of investment capital. Because it implicates different levels of action, housing can provide important insights into the workings of society. Thus, there exist studies of how, where, and how much money enters into and leaves the housing system on behalf of households, large corporations, and government entities. There are also many studies of housing interest group politics vis-à-vis local and national government and of how public policies are the outcome of a shifting balance of power among contending forces.

The Multidisciplinarity of the Field

Considering the many functions and different conceptualizations of housing, it is not surprising that, as a field of research, housing draws on many disciplines, including political science, sociology, economics, geography, anthropology, and psychology. Housing practitioners likewise have diverse origins: planning, architecture, law, social work, the policy sciences, and public administration, among others. As a result, much of the work in housing has been scattered rather than cumulative. For example, those concerned with legal aspects of housing and those involved with its financial aspects tend to publish in different journals, and they generally go to different conferences. Planners similarly have their own arena of activities, often more directed to physical aspects of housing yet quite distinct from architects, who are more specifically oriented to issues of design.

Organizational structure (e.g., separate professional associations) and communication impediments (e.g., jargon) further reinforce these disciplinary and professional boundaries.

This rather amorphous situation is unfortunate because most housing issues do not divide up neatly according to these contrived boundaries. Combining insights from one area with those from others produces a broader-based and more effective approach to housing problems. There is a growing recognition that the resolution of practical housing issues increasingly demands approaches based on several disciplines and professions. There is a rising demand for individuals whose thinking is informed by expertise from multiple domains.

The Encyclopedia of Housing responds to this greater interest in integration. Its coverage is multidisciplinary. Its authors represent more than a dozen disciplines and work in widely different settings: a range of academic departments, a variety of public agencies at the federal and local level, nonprofit organizations, and several private firms. This broad spectrum results from a deliberate decision to cast a wide net to capture the diversity and scope of the housing field. By consequence, readers will find entries written by contributors favoring demand-side economics as well as others by proponents of a supply-side perspective. The list of authors includes long-time champions of a progressive housing agenda as well as leading advocates of a more conservative political paradigm. The field does not speak in unison. My purpose was to reflect this diversity and to bring together the many different voices in a balanced mix. In doing so, I have tried not to impose my personal viewpoints, which are available in more appropriate places elsewhere (e.g., Van Vliet-- 1992, 1996). An alternative approach would have been to remain within a narrowly defined, single disciplinary or political framework. This would have meant more in-depth treatment but of fewer subjects. It would have been a different project. It also would have prevented the work from bringing into focus the many and important connections between the different disciplines in the field.

Other Reference Works

A number of other reference works related to housing are available. To begin with, there are several quite voluminous dictionaries of housing-related terms. For example, Sayegh's (1987) effort lists over 28,000 entries. Not all of these terms have a direct bearing on housing (e.g., *self-esteem, spinster,* and *schizophrenia,* among others, seem only marginally relevant), but the comprehensive scope ensures that users with a question about any aspect of housing will find a listing. For instance, in case one seeks a definition of *door, window, roof,* or *driveway,* Sayegh provides an answer. However, considering the exhaustive listing, by necessity these answers tend to be brief, generally limited to one or two lines. Rostron (1997) has compiled a similar dictionary of housing.

Several other works also feature brevity but with a stronger concern for aspects of application. Examples are Dumouchel's (1985) *The Commissioner's Dictionary of Housing and Community Development Terminology* and Moskowitz and Lindbloom's (1993) *New Illustrated Book of Development Definitions.* A much more specialized effort along these lines is *The Automated Builder* by Carlson (1995), which concerns itself strictly with industrialized housing. However, the choice of some terms is a bit curious (e.g., *potable water:* safe drinking water; *PR:* public relations; *PC:* personal computer), whereas many others have a rather technical orientation, describing certain construction-related activities, defining certain tools or building components, at times with reference to their commercial manufacturer.

A number of discipline-based reference works also exist. Morrow's (1987) *A Dictionary of Landscape Architecture* contains an extensive list of brief definitions, although few of them are directly relevant to housing. On the business side, there are several reference works on real estate investment, development, and management (e.g., Blankenship 1989; Tosh 1990). Packard and Korab (1995) offer a lavishly illustrated *Encyclopedia of American Architecture* whose eclectic selection of 234 entries includes diverse topics such as adhesives, computer, fountain, housing, L. Mies van der Rohe, I. M. Pei, seismic design, and wood frame structure. A somewhat dated work by Heyer (1978) focuses exclusively on individual architects, as does

Placzek's (1982) four-volume encyclopedia. Emanuel's (1994) *Contemporary Architects* is a similar resource of more recent vintage. Detailed information and visual documentation of cultural, geographic, environmental, and climatic aspects of traditional architecture in some 80 countries around the world can be found in Oliver's (1997) recent three-volume opus. Whittick (1974) provides a not quite current but still interesting international survey of urban planning, including some aspects of housing. Finally, there also exist several practice- oriented publications—for example, *The Subdivision and Site Plan Handbook,* by Listokin and Walker (1989), and *Time-Saver Standards for Housing and Residential Development,* by De Chiara, Panero, and Zelnik (1995). Additional publications in this genre include Harris (1988) and Colley (1993).

All these reference works differ in various ways from the present volume. To begin with, this *Encyclopedia* is strictly oriented to housing, albeit broadly viewed. Its focus and scope are defined in substantive terms from a multidisciplinary perspective. This approach contrasts with one that would adopt a narrower, sectoral perspective delineated by a particular discipline or profession. Furthermore, in this volume, treatment of subjects goes well beyond providing merely a definition or brief description, whether the topic be Abandonment or Zoning. For example, to the latter subject Morrow (1987) devotes two brief paragraphs, Dumouchel (1985) gives it one short paragraph, and it gets three lines from Moskowitz and Lindbloom (1993) and one sentence from Carlson (1995). This *Encyclopedia* provides more detailed coverage: In addition to a full essay on Zoning, it also includes related entries on, for example, Exclusionary Zoning, Inclusionary Zoning, and Subdivision Controls. Although the aforementioned sources are useful for purposes of quick reference, they serve a different function from this volume, which, aside from offering initial definitions, provides expanded treatment.

About This Volume

This book, then, is intended to address a gap in the literature. It offers a brief description of each topic, further elaboration of selected aspects, and critical assessment. Virtually all entries are also accompanied by a brief bibliography, listing classic works, recent authoritative studies, or other key publications. These references serve less as source documentation for the entries than as suggestions for readers who want to study a subject in greater depth. Inevitably, they are selections from much larger compilations. In some cases, opinions may differ over what should be included. However, the intent in each instance was to provide a reasonable representation of the relevant literature on any given topic. The entries also include brief descriptions on housing organizations and periodical publications.

The length of entries varies. Short entries typically furnish concise reviews of relevant research findings, whereas longer, more detailed entries usually include historical background, a review of legislative context, or an analysis of policy trends. Key topics get more coverage than more peripheral ones. A strategy of "triangulation" ensured that major topics are discussed under more than one title. In most of those instances, one of these titles has been chosen as a "nodal" entry (see below), and related aspects are discussed in additional entries, identified by cross-references. For example, the first entry, Abandonment, presents a compact and informative discussion of the topic, and cross-references identify related aspects, such as Blight, Code Enforcement, Redlining, and Urban Redevelopment, which are discussed in other entries.

With respect to housing-related organizations and publications, coverage was sometimes restricted by the availability of information. We considered eliminating those cases or imposing the same word limit on all these entries. However, accommodating a certain unevenness in length appeared to be preferable over truncating and impoverishing a substantial number of entries in favor of uniformity. Furthermore, all these entries include address information letting readers know where to obtain further details.

The cross-references are an important feature of this *Encyclopedia*. Often, a number of entries form a cluster of interrelated topics. Cross-references at the end of entries help identify these clusters and facilitate an understanding of the connections between different subjects. Their organization follows a nodal pattern; that is, in each case, one entry will serve as a central point with a relatively extensive list of cross-references to related "satellite" entries. Each of these satellites will have fewer cross-references, one of which always leads back to the nodal entry and via it to other, related satellites. The entries on Affordability, Discrimination, and the Elderly are examples of such nodal entries. A full list is provided in Appendix A. For the sake of convenience, nodal cross- references are printed in boldface roman type to set them apart from the other cross-references.

The *Encyclopedia* has four appendices. The first comprises the nodal entries. The second lists the housing organizations included as entries. The third does the same for housing publications. These lists provide convenient overviews. The fourth appendix is a chronology of major U.S. housing legislation. Most of these acts and their major components are also covered in separate entries.

There are three indices at the end of this volume. The first index lists the authors of this encyclopedia and enables readers to find the contribution(s) each has written. The second index comprises the names of the authors whose works are cited in the brief bibliographies that conclude most entries under the brief heading "Further Reading." The third index includes terms appearing in the text of the entries. Whenever a term in this index refers to a topic to which an entire entry is devoted, the pagination inclusive of coverage is printed in boldface; additional mention of that topic, incidental to other entries, is included as well.

The Audience

Scientific reference works have a reputation for being as dull as dishwater. Some unwritten law appears to proscribe plain language. Although the use of jargon can be a parsimonious way of communicating within a particular discipline or profession, it hinders effective cross-boundary communication. In line with the intent to foster multidisciplinary integration, most contributors to this *Encyclo-*

pedia purposely tried to make their writing accessible to others outside their specialization. Hence, this is a reference work meant to serve several user groups, including students, teachers, researchers, housing professionals, and government officials. A prepublication version proved to be a useful teaching tool; students found the succinct coverage informative and indicated that the system of cross-references helped them to develop a more coherent grasp of multifaceted issues.

A number of entries were written by housing practitioners, and people in the field have reacted positively to parts of the manuscript while it was in progress. Nevertheless, this *Encyclopedia* is not primarily oriented to immediate applications in practice. It is not a "how-to-do-it" manual. Thus, readers will find few specific guidelines for planning and design, no hands-on architectural programs, and no specification of the detailed steps borrowers go through when seeking a mortgage loan. However, the entries do assist in the conceptualization of housing problems and outline methods for studying and resolving them. For example, the entry on Discrimination identifies domains of housing where discrimination occurs, the forms it takes, the reasons behind it, and its effects. Related, cross-referenced entries go into greater depth on how to detect housing discrimination, the extent to which it occurs, and how to prevent it. Other entries provide information on organizations actively working to counter discriminatory practices.

Similarly, in addition to the entry on Affordability, related entries address questions of how affordability can be measured, the advantages and disadvantages of alternative measurement methods, the characteristics of a variety of subsidy approaches, and the activities of organizations whose mission it is to promote affordability. Here, again, the cross-references are a useful tool.

Limitations of Scope

An acknowledgement of humbleness: Its title notwithstanding, this volume cannot be truly encyclopedic in the sense of being completely comprehensive. It is extensive in its coverage. However, it is not exhaustive. Although it includes many important housing topics in several disciplines, the choice for multidisciplinarity has also meant that no single discipline could be fully covered. Some readers may find that a favorite topic is not included or not discussed as fully as they would like. Limits on the project necessitated a selection of topics and required authors to put strict boundaries on write-ups of subjects meriting book-length discussion. However, throughout, an attempt was made to arrive at decisions that would produce results representative of pertinent research and practice.

It is important to be aware that this volume is not a compilation of empirical research. The entries were not written to report detailed findings of specific studies. Instead, they offer more broad-ranging coverage that summarizes and synthesizes the current state of knowledge. Furthermore, although every effort was made to include the latest information, the lag between collection and availability of data inevitably results in the description of circumstances at least several years in the past. However, many housing issues have historical continuity and fundamental characteristics that endure. The entries are meant to provide understanding and insights not radically changed by tomorrow's developments. For those interested in housing statistics, Simmons (1997) has provided a convenient organization in a single source, containing some 200 tables with detailed data on housing demand, housing starts, housing investment, household trends, ownership rates and other aspects of the U.S. housing stock and market dynamics.

Having described what this *Encyclopedia* includes, it is appropriate as well to be clear about what it does not include. Space restrictions and the wish to avoid duplication produced a decision not to include biographical entries. Instead, many persons who have been prominent in housing in the United States are mentioned in the context of the situations in which they made their presence felt. Readers interested in more extensive coverage of particular individuals are referred to various Who's Who publications and existing biographies (e.g., Macfadyen 1970; Davis 1973; Richards 1977; Placzek 1982; Novak 1988; Doumato 1989; Tafel 1993; Tominaga 1993; Emanuel 1994).

Nor does this *Encyclopedia* provide a comprehensive history of housing. Its orientation is more to the present. Selected historical topics have been included, however, when they offer a useful background for contemporary issues. For example, the essay on the housing of slaves, while of interest in itself, informs about antecedents to the subsequent housing situation of African Americans, discussed in a separate essay. Likewise, the Civil Rights Act of 1866 is covered as an important precursor to the Civil Rights Act of 1968. In other cases, essays include a historical dimension insofar as it sets the context for a better understanding of a current topic. Examples are the entries on Lending Institutions, Cooperative Housing, Federal Government, and Housing Codes. Readers interested in more in-depth treatment of historical aspects of housing are referred to specialized works for specific countries (e.g., Bullock and Read 1985; Daunton 1990; Doan 1997; Gauldie 1974; Mason 1982; Wright 1983) or cities (Bowly 1978; Fairbanks 1988; Plunz 1990; Shapiro 1985).

Full treatment of past and current housing legislation would constitute an encyclopedia in its own right. This volume contains a chronological listing of major housing legislation in the United States (Appendix D) as well as selective coverage of individual acts. Many sections give brief descriptions that state, for example, a law's intent and its eligibility requirements. A number of housing acts, generally viewed as having been more important, get more extensive discussion. Sometimes this occurs under a legislative heading for all or part of a given act (for example, Section 8, Community Development Block Grant). Other times, cross-references point to more extensive, separate discussion of major elements of a certain act or topics related to it. Examples are Wagner-Steagall Housing Act (Public Housing), Housing Act of 1949 (Urban Redevelopment), Housing and Urban Development Act of 1968 (Section 235, Section 236), the Civil Rights Act of 1968–Title VIII (Discrimination), and the Stewart B. McKinney Homeless Assistance Act (Homelessness). Readers will find detailed coverage of U.S. housing legislation in U.S. House of Representatives (1991) and Milgram (1994). More general treatment of legal aspects of housing is provided by, for example, Burnet (1996) and in periodical publications such as *Housing Law Bulletin, Housing Affairs Letter,* and *Housing and Development Reporter.* A comprehensive, ongoing information service on housing law and practice in Britain is the multiple-volume set edited by Arden, Hunter, and Pramall (1972), supplemented by regular looseleaf additions.

Finally, the primary frame of reference for this *Encyclopedia* is North America—in particular, the United States. Housing legislation and institutions, for example, are discussed predominantly within this context.

Lack of space would not permit equal coverage of other countries. To be sure, there are a number of international entries, but they do not focus on any single country in particular. Instead, they concern themselves with a broad geographical region or a topic that is more generally relevant, irrespective of national boundaries. They were included because housing in the United States can be better understood when placed on a wider spectrum of perspectives (Van Vliet-- 1990). Public housing, privatization, self-help, and community-based initiatives, to name but a few, all have origins or parallels elsewhere. Therefore, readers will find entries on housing in Canada, Western and Eastern Europe, the Third World, cross-national housing analysis, the Global Strategy for Shelter, the U.N. Centre for Human Settlements, and the World Bank, among others. These international entries help set out a framework for evolutionary policy-making, which enables U.S. housing policies and practices to be informed by the experiences from other countries. In a generic sense, many housing issues transcend the peculiarities of the U.S. housing system. Therefore, I hope that also the reading public elsewhere will find the *Encyclopedia*'s treatment of these issues to be of interest.

References

Arden, A., C. Hunter, and S. A. Pramall, eds. 1972. *Encyclopedia of Housing Law and Practice.* 5 Vols., updates. London: Sweet & Maxwell.

Blankenship, Frank J. 1989. *The Prentice Hall Real Estate Investor's Encyclopedia.* Englewood Cliffs, NJ: Prentice Hall.

Bowly, Devereux. 1978. *The Poorhouse: Subsidized Housing in Chicago, 1895-1976.* Carbondale: Southern Illinois University Press.

Bullock, Nicholas and James Read. 1985. *The Movement for Housing Reform in Germany and France, 1840-1914.* New York: Cambridge University Press.

Burnet, David. 1996. *Introduction to Housing Law.* Homes Beach, FL: Gaunt.

Carlson, Don O., ed. 1995. *Automated Builder: Dictionary/Encyclopedia of Industrialized Housing.* Carpenteria, CA: CMN Associates.

Colley, B. 1993. *Practical Manual of Land Development.* 2d ed. New York: McGraw-Hill.

Daunton, M. J., ed. 1990. *Housing the Workers, 1850-1915: A Comparative Perspective.* London: Leicester University Press.

Davis, A. F. 1973. *American Heroine: The Life and Legend of Jane Addams.* New York: Oxford University Press.

De Chiara, Joseph, Julius Panero, and Martin Zelnik, eds. 1995. *Time-Saver Standards for Housing and Residential Development.* 2d ed. New York: McGraw-Hill.

Doan, Mason C. 1997. *American Housing Production, 1880-2000: A Concise History.* Lanham, MD: University Press of America.

Doumato, Lamia. 1989. *Catherine Bauer, 1905-1964: A Bibliography*. Monticello, IL: Vance Bibliographies.

Dumouchel, Robert. 1985. *The Commissioner's Dictionary of Housing and Community Development Terminology*. Washington, DC: National Association of Housing and Redevelopment Officials.

Emanuel, M. 1994. *Contemporary Architects*. 3d ed. New York: St. James Press.

Fairbanks, Robert B. 1988. *Making Better Citizens: Housing Reform and the Community Development Strategy in Cincinnati, 1890-1960*. Urbana: University of Illinois Press.

Gauldie, Enid. 1974. Cruel Habitations: A History of Working-Class Housing 1780-1918. London: Allen & Unwin.

Harris, C. M. 1988. Time-Saver Standards for Landscape Architecture. New York: McGraw-Hill.

Heyer, Paul. 1978. *Architects on Architecture*. New York: Walker.

Listokin, David and Carole Walker. 1989. *The Subdivision and Site Plan Handbook*. New Brunswick: Rutgers, State University of New Jersey, Center for Urban Policy Research.

Macfadyen, D. 1970. *Sir Ebenezer Howard and the Town Planning Movement*. Cambridge: MIT Press.

Mason, Joseph B. 1982. *History of Housing in the U.S., 1930-1980*. Houston, TX: Gulf.

Milgram, Grace. 1994. *A Chronology of Housing Legislation and Selected Executive Actions, 1892-1992*. Congressional Research Service Report. Washington, DC: Government Printing Office.

Morrow, Baker H. 1987. *A Dictionary of Landscape Architecture*. Albuquerque: University of New Mexico Press.

Moskowitz, Harvey S. and Carl G. Lindbloom. 1993. *New Illustrated Book of Development Definitions*. New Brunswick: Rutgers, State University of New Jersey, Center for Urban Policy Research.

Novak, F. G. 1988. *The Autobiographical Writings of Lewis Mumford*. Honolulu: University of Hawaii Press.

Oliver, Paul, ed. 1997. *The Encyclopedia of Vernacular Architecture of the World*. Cambridge, UK: Cambridge University Press.

Packard, Robert and Balthazar Korab. 1995. *Encyclopedia of American Architecture*. 2d ed. New York: McGraw-Hill.

Placzek, A. K. 1982. *Macmillan Encyclopedia of Architects*. New York/London: Free Press.

Plunz, Richard. 1990. *A History of Housing in New York City: Dwelling Type and Social Change in the American Metropolis*. New York: Columbia University Press.

Richards, T. M., ed. 1977. *Who's Who in Architecture: From 1400 to the Present Day*. London: Weidenfeld & Nicolson.

Rostron, Jack. 1997. *Dictionary of Housing*. Brookfield, MA: Ashgate.

Sayegh, Kamal S. 1987. *Housing: A Multidisciplinary Dictionary*. Ottawa: Academy Book.

Shapiro, Ann-Louise. 1985. *Housing the Poor of Paris, 1850-1902*. Madison: University of Wisconsin Press.

Simmons, P. A., ed. 1997. *Housing Statistics of the United States*. Lanham, MD: Bernan.

Tafel, Edgar. 1993. *About Wright: An Album of Recollections by Those Who Knew Frank Lloyd Wright*. New York: John Wiley.

Tominaga, Yuzuru. 1993. *Essays on Residential Masterpieces: Le Corbusier 2*. Tokyo: A. D. A. Edita.

Tosh, Dennis S. 1990. *Handbook of Real Estate Terms*. Englewood Cliffs, NJ: Prentice Hall.

Turner, J. F. C. 1976. *Housing by People: Towards Autonomy in Building Environments*. New York: Pantheon.

U.S. House of Representatives. Subcommittee on Housing and Community Development. 1991. *Basic Laws on Housing and Community Development*. Revised through September 30, 1991. Washington, DC: Committee on Banking, Finance, and Urban Affairs.

Van Vliet--, Willem, ed. 1990. *International Handbook of Housing Policies and Practices*. Westport, CT: Greenwood/Praeger.

Van Vliet--, Willem. 1992. "A House Is Not an Elephant: Centering the Marginal." *The Meaning and Use of Housing*, edited by Ernesto Arias. London: Gower.

Van Vliet--, Willem, 1997. "Learning from Experience: The Ingredients and Transferability of Success." Pp. 247-76 in *Affordable Housing and Urban Development in the United States*, edited by Willem van Vliet--. Thousand Oaks, CA: Sage.

Whittick, A., ed. 1974. *Encyclopedia of Urban Planning*. New York: McGraw-Hill.

A

▶ Abandonment

Among the distinctive characteristics of housing are its permanence and its sensitivity to the economic and demographic fate of a given location—its block, neighborhood, town, and region. Housing is built in response to human needs, mediated as they are through markets, political programs, and personal actions. Housing characteristics are thus functions of the combined political, economic, and demographic forces that affect households. This is particularly the case with the pushes and pulls affecting migration. Cities, especially older manufacturing centers, arose in large part in the United States as a response to those factors. Many of these cities have experienced a reversal of fortunes in the past several decades as people have opted for less dense communities where jobs have been more plentiful. Thus, the abandonment of the built environment signals two events: (a) A given location was once important enough that a community's housing needs were substantially satisfied, and (b) that same community subsequently declined in importance so that there remained little in the existing location to retain those households.

The streetscapes of many older cities in the United States, particularly outside their office and commercial centers, are dominated by neighborhoods that have thinned out in population. In many cases, this deconcentration left empty and crumbling houses, many eventually replaced by vacant lots. Because most paradigms of urban development assumed relatively constant growth in urban areas, explanations of urban downsizing and abandonment have tended to focus on "irrational" factors affecting the housing market, such as overly greedy land speculators or landlords "milking" properties of their last possible dollar, the overregulation of low-income housing, or racial issues. In addition, some have argued that a fundamental systemic factor—the change in the dominant economic base of cities—has been responsible.

Housing abandonment began to emerge as a concern in the United States in the middle to late 1960s, as a renewed sense of concern about urban America accompanied the riots and civil disturbances of that era. During this period, the theme of the greedy slumlord dominated many discus-

sions. However, ethnographic research of landlords and tenants indicated a more complex relationship. Many landlords had inherited their inner-city properties or had been unable to sell them when their households moved from inner-city areas. Many felt constrained by the economic realities of very poor tenants and by large-scale political and economic forces suggesting the impossibility of making a regular profit from their buildings. Although some landlords undoubtedly saw this as an opportunity for property accumulation and milking, this was hardly a universal picture of the low-income rental market.

This theme of the absentee, greedy, and speculative landlord remains a common thread in many discussions of abandonment. In this view, abandonment is argued to be a final stage in the ownership of rental housing. Maintenance budgets decline, as do rent levels and collections; tax payments are foregone and vacancies increase. The landlord is reduced to treating the building itself as a write-off, because there is no viable market for either the building or the land under it. As a property holder in tax delinquency, the landlord is essentially treating the underlying land as the sole remaining investment, waiting for social or economic conditions to change or for the city to seize the property. In either case, the "endgame" approach to housing appears as the sole economically rational choice available to such landlords.

Three important components of the endgame process have contributed to abandonment. First, professional landlords, holding a varied portfolio of buildings, can effectively milk the last bits of income from emptying buildings by purposely reducing maintenance and services on buildings. Second, there have been several recorded instances, in Boston, New York, and Chicago, in which property owners have been able to "kite" the property value of an emptying building by rapidly selling the same property within a closely linked group of investors (sometimes shell companies), covering the property with insurance, and then committing arson on the property. Third, in earlier periods of the federal tax law, it was easier to use poor-quality, nearly abandoned properties as components of real estate investment trusts (REITS) or to qualify for historic renewal or low-income housing tax credits when little in the way of renovation was actually taking place. An argument can be advanced that historic renewal tax credits encouraged displacement of populations from neighborhoods, creating an interim form of abandonment (which could become more permanent if gentrification did not follow the displacement process).

Landlords themselves cite government regulation of housing as a contributing component in the abandonment process. In a few cities, rent control is specifically called into account, particularly in New York City, the largest single rental housing market in the United States. There is substantial evidence that rent controls affect the maximum theoretical rent that could be obtained from a housing unit, thus limiting profits and inhibiting the passing on of real costs to the tenant. Landlords extend this argument to maintain that limiting the income stream from rental properties encourages a cutback on reinvestment in properties and must, therefore, lead to abandonment. This argument has been shown to be limited by the extent of low-income households in New York (creating a problem of affordability) and by New York's atypically low vacancy rate (usually below 3%), which create a monopoly rent situation for landlords (raising their expectations for rental income beyond a market return). Nonetheless, this argument achieved great currency during the hyperinflation years of the late 1960s and during the energy crises of the late 1970s (when the costs associated with rental property, such as interest and fuel, could not easily be passed along in a rent-controlled environment).

More generically, some analysts have argued that the costs of local taxes and regulations—for rent control certification, for licensing and inspection, for maintenance of minimal living standards, for the use of copper wiring and pipes rather than aluminum and PVC, respectively—are pivotal in driving landlords of low-income tenants into abandonment. Although rent control, taxes, and housing regulations carry hidden costs that cannot be easily justified in market terms, analyses that focus on the abandonment of buildings have recurrently found that it is the low income of a significant component of the market that accounts for the inadequate cash flow for landlords. Although the costs of regulation affect the abandonment process, their direct effect is marginal, at best.

The disproportionate amount of abandonment concentrated in minority neighborhoods—particularly those identified as "underclass" communities—has led some analysts to consider the role of race and racial discrimination in the housing market in the abandonment process. Although abandonment occurred in close historical proximity to the riots affecting many U.S. cities, there is some danger of overstating the impact of these disturbances on white withdrawal from large central cities during the 1970s and 1980s. Suburbanization had been a normal part of the U.S. urban landscape throughout the 20th century, but the operation of the U.S. housing market in the aftermath of World War II provided both an inducement and a subsidy for white households to leave the city behind, through the introduction of large-scale subsidies in the FHA and VA programs. These same programs openly blocked African Americans

from sharing in ownership of the vast numbers of new, single-family homes made available in the new suburban communities. Paradoxically, the migration of southern blacks during the early part of the century in search of economic and political opportunities led to a limited set of housing choices within the inner city and older communities associated with early manufacturing. These communities were essentially locked into place during the subsequent deindustrialization during the 1960s and 1970s, while new migrants to metropolitan areas were both disproportionately white and increasingly drawn to urban areas.

In addition to these macrosocial forces, racially biased realtors encouraged blockbusting to encourage the dumping of properties on already glutted inner-city housing markets that also felt the effects of widespread mortgage redlining. People looking to finance the sale of their homes found that financing for the purchaser was not available, simply because banks assumed that racial transition would necessarily lead to the devaluation of property. The effect of this process was both to create a group of absentee landlords who rented their former homes out when they could not be sold and to help concentrate poor and abandoned housing in African American communities.

In a postmanufacturing era, U.S. cities have been coping with increased suburbanization, movement from older manufacturing cities in the Northeast and Midwest (Frostbelt) of the country to the newer, decentralized, and service-oriented cities of the South and West—the Sunbelt. More recently, the offshore placement of many basic industries has further disrupted the traditional economic base of the densely settled, housing-intensive industrial city. The great urban era of the late 19th and early 20th centuries has been supplanted by the decentralized, lower-density metropolitan areas associated with the new production and service economies. In many cities, both the presence of a once-vital industry and the shift in population away from the city can be seen in the empty factories and warehouses and abandoned houses and lots of older neighborhoods. Even the jobs that remain are typically lower waged, more often part-time and less available, yielding a lower income per capita and creating an aggregate problem for the low-income housing market.

Thus, a combination of demographics (e.g., population shifts and declines), very low income groups, and local market irrationalities (such as racial discrimination) creates a set of events in which a part of the housing stock is neglected, empties out, and eventually deteriorates. Usually, but not always, this occurs in the inner core of declining manufacturing cities and differentially affects African American communities. Although no common operational definition of abandonment exists such that all cities measure the same thing when they discuss abandonment, a sense of the comparative differences between cities can be gained by a careful examination of the housing vacancy statistics of the decennial census. In 1990, for instance, Albuquerque—a Sunbelt growth center—had one-tenth of 1% of its housing stock boarded up and only about 5.6% of its nonmarket vacancies boarded up. Atlantic City, however—a Frostbelt city that lost both its traditional manufacturing and tourism industries—had more than 2% of its standing stock boarded up and over 45% of its nonmarket vacancies boarded. In a small sample of cities examined, Baltimore, Houston, Philadelphia, and St. Louis each demonstrated a much higher than average percentage of housing vacant and not on the market, as well as boarded up. Cities such as Los Angeles, Anaheim, San Francisco, and Colorado Springs demonstrated a much lower percentage of such housing than did other cities. Most cities average about 10% of their stock standing vacant, between 1% and 2% of the stock vacant and not on the market, and around .75% of vacancies boarded up.

Other problems of definition remain. Although it is easy to verify that a crumbling, unoccupied structure is in fact abandoned, there is considerable debate over the origins of the abandonment process. In particular, does abandonment begin at the point at which landlords skimp on maintenance? When they skip a real estate or corporate tax payment to a municipality or state? When apartments lie vacant for long periods of time with no effort to fill the vacancies? When landlords stop collecting rent?

Regardless of the answers to these questions, abandonment is a process linked to the uneven development of urban neighborhoods and regions. As the economic base of a region declines, housing abandonment is one of the likely consequences of such changes. Policy concerning abandonment should be oriented toward issues of appropriate neighborhood planning, creation of land banking for these properties, and concerns for revitalizing the economic base of a neighborhood or city.

Although no massive intervention has been introduced to counter abandonment, neighborhood-specific strategies abound. Strategies that have proven effective involve either intervention before abandonment, such as in the case of New York City's in rem, early property turnover program. Tax delinquent properties are seized after one year's tax delinquencies and often transferred to community groups. Many times, new housing units are substituted in heavily abandoned neighborhoods, at a lower density than previously. Some neighborhoods have used creative approaches to community-based renewal efforts, incorporating nontraditional financial sources from both the public and private sectors (such as linkage programs, housing trust funds, ethical investment funds, and funding). Most of these efforts reflect the continued devolution of federally financed and administered programs, and virtually all revitalization now leverages public sector funds with private sector investments and, to a variable extent, some combination of "sweat equity" and "urban homesteading." (SEE ALSO: *Blight; Code Enforcement; Gentrification; In Rem Housing; Obsolescence; Power of Eminent Domain; Redlining;* **Urban Redevelopment**)

—*David W. Bartelt*

Further Reading

Bartelt, David W. and Ron Lawson. 1982. "Rent Control and Abandonment: A Second Look at the Evidence." *Journal of Urban Affairs* 4:49-64.

Bartelt, David W. and George Leon. 1986. "Differential Decline: The Neighborhood Context of Abandonment." *Housing and Society* 13:81-106.

Bratt, Rachel. 1989. *Rebuilding a Low-Income Housing Strategy*. Philadelphia: Temple University Press.

Bratt, Rachel, Chester Hartman, and Ann Meyerson, eds. 1986. *Critical Perspectives on Housing*. Philadelphia: Temple University Press.

Ketchum, James. 1994. "Spatial Aspects of Housing Abandonment in the 1990s: The Cleveland Experience." *Housing Studies* 9(4):493-510.

Salins, Peter. 1980. *The Ecology of Housing Destruction*. New York: New York University Press.

Smith, Neil. 1984. *Uneven Development*. New York: Basil Blackwell.

Stegman, Michael A. 1986. *Housing Finance and Public Policy: Cases and Supplementary Readings*. New York: Van Nostrand Reinhold.

Sternlieb, George. 1965. *The Tenement Landlord*. New Brunswick: Rutgers, State University of New Jersey, Center for Urban Policy Research.

Sternlieb, George. 1969. *The Urban Housing Dilemma*. New Brunswick: Rutgers, State University of New Jersey, Center for Urban Policy Research. ◀

▶ Accessory Dwelling Units

Apartments created in the surplus space of single-family homes make up the most common form of accessory dwelling units. They are complete and separate. They have their own kitchen and bath and almost always their own separate entrance. They are also known as "mother-in-law" units and single-family conversions.

The terms *second units,* in California, and *ohana* units, in Hawaii, cover both accessory apartments and accessory cottages. Accessory cottages are typically separate from the main home or may be attached to it but not structurally part of it. The term *granny flats* is also often applied in the media to both accessory apartments and accessory cottages. It originally referred only to elderly cottage housing opportunity (ECHO) housing, a third form of accessory dwelling unit. The distinction between accessory apartments and accessory cottages is important. Accessory apartments cost much less to construct than accessory cottages, because they typically use existing space within a home. They are, and will probably continue to be, the more common of the two accessory housing types.

Potential for Providing Housing

The baby boom left behind an empty-nester boom. In addition, young couples are having fewer children. Data from the American Housing Survey indicate that roughly one-third of the single-family housing stock consists of homes of five rooms or more—not counting basements, garages, or attics—occupied by households of two persons or less. Theoretically, about one-third of the single-family homes in the United States have the potential for an accessory apartment.

These apartments could be built for about one-third of the cost of constructing a conventional rental unit. The cost will vary dramatically with the house type, and the amount spent often depends on whether the apartment is intended for rental income or a relative. In 1996, a split-level ranch house with a walk-out basement "rec room" and full bath converted for little more than the cost of installing a kitchen, often under $5,000.

Benefits

Accessory apartments provide added income, security, companionship if wanted, and the opportunity to trade rent reductions for needed services. Older homeowners, in particular, benefit because many are house rich and cash poor, and many are frail enough to need added security and service exchanges. Typically, young homebuyers benefit most from the added income to help meet mortgage payments until their income and family size grow. Single parents can use accessory apartments to retain the family home in the wake of divorce so that the children do not lose their neighborhood at the same time they are losing the presence of one of their parents in the home. Disabled people may like accessory apartments because they provide privacy in close proximity to support. Transportation planners also see accessory units as a good way to increase the number of people living close to transit stops.

Studies show that about half of all tenants are related to the homeowner. Often, the tenant is either the aging parent of an adult child or an adult child who has "returned to the nest." Regardless of whether an apartment was constructed for related or unrelated tenants, over time, it will typically be used for both.

Rents for unrelated tenants are often "deep subsidies"— that is, no rent at all. In addition, U.S. and Canadian studies have shown that accessory apartments rent at below-market rents for comparable conventional apartments, without rent control or subsidies. The anecdotal explanations for low rents in accessory apartments are that they cost less to create, that homeowners are more afraid of vacancies than typically landlords are, and that homeowners also charge less so that they can attract and keep tenants they like.

An important advantage of accessory apartments is that they take a house type designed to serve one stage in the life cycle and make it flexible enough to serve almost any stage.

Community Resistance and Zoning Provisions

In the past, opposition to accessory apartments on the part of civic associations has been strong. Increasingly, however, it is possible to find prestigious communities with long-standing and effective ordinances covering accessory housing, such as Westport and Weston, Connecticut; many towns in Westchester County, New York; Boulder, Colorado; Marin County, California; and Montgomery County, Maryland. Vancouver, British Columbia, is an outstanding Canadian example, where 40% of the homes in one neighborhood have accessory apartments.

Zoning for accessory apartments must calm neighborhood fears without hampering homeowners who want to install apartments. Common provisions include making accessory apartments a special-exception use, requiring a resident homeowner, and permitting little or no exterior change. Common zoning problems that inhibit installation

include long approval periods, public hearings on applications, and high permit fees.

California has a state law, the "Mello Bill," that requires most communities to zone for accessory apartments, but it sets no standards. As a result, in most communities the zoning for accessory apartments is so restrictive that few units are permitted. In 1995, Ontario, Canada, was considering legislation with standards requiring all local governments to permit accessory apartments. Hawaii had such legislation but has made it voluntary, apparently in reaction sales of accessory cottages as condominiums, something rarely, if ever, permitted elsewhere.

Contribution to the Housing Stock: Installation Rate

A 1989 study of installation rates in communities permitting accessory apartments indicated that when zoning provisions are not a burden for homeowners, approximately one accessory apartment will be installed each year for each 1,000 single-family homes in a community. The national rate of construction is about 10 new homes per 1,000 existing homes per year. In other words, accessory apartments appear to be capable of increasing housing production by about 10% a year. Almost all of that housing would be affordable. It should also be emphasized that none of it would require any subsidy.

Despite their potential usefulness, accessory apartments suffer from several drawbacks. Although, the concept was very attractive in the early 1980s, two problems emerged. First, advocates underestimated the problems of installation for homeowners, including the burden of altering zoning provisions requiring public hearings. Second, no one had defined success. There were no clear expectations, and successful ordinances received no attention. However, where built, accessory apartments appear to be quite successful. (SEE ALSO: *ECHO Housing; Efficiency Apartment; Elderly*)

—*Patrick H. Hare*

Further Reading

Gellen, Martin. 1986. *Accessory Apartments in Single Family Housing.* New Brunswick: Rutgers, State University of New Jersey, Center for Urban Policy Research.

Hare, Patrick H. and Jolene Ostler. 1987. *Creating an Accessory Apartment.* New York: McGraw-Hill. (This book is now available only through Patrick H. Hare Planning and Design, Washington, DC.)

Hare, Patrick H. Continually updated. *Accessory Units: The State of the Art* (Four separately published reports). Report 1. Summary of Experience with Accessory Units in the U.S. and Canada; Report 2. Resource Guide to Accessory Units; Report 3. Model Zoning Ordinances for Accessory Units; Report 4. Installations of Accessory Units in Communities Where They Are Legal. Washington, DC: Patrick H. Hare Planning and Design.

Ministry of Housing and Ministry of Municipal Affairs, Ontario. 1992. *Apartments in Houses.* Toronto, Ontario, Canada: Ministry of Housing and Ministry of Municipal Affairs.

Rudel, Thomas K. 1984. "Household Change, Accessory Apartments, and Low Income Housing in Suburbs." *Professional Geographer* 36(2):174-81. ◄

► Acquired Immune Deficiency Syndrome

Individuals with acquired immune deficiency syndrome (AIDS) are confronted with significant challenges for both securing and maintaining housing. At the same time, changes in the character of housing across communities have important implications for the diffusion of the AIDS virus. This entry will elaborate both of these dilemmas and suggest a number of areas in which remedies are needed.

Housing Destruction

In communities where preexisting poverty and overcrowding exist, the destruction and decline of housing units contributes to the spread of the AIDS virus. Housing destruction in poor communities is fueled by policies of "planned shrinkage" whereby municipal services such as fire and police protection or street and sewer maintenance are reduced. The resultant increase in housing burnout or decay contributes to an out-migration of residents who are able to leave and overcrowding among residents who remain.

The effect of these dynamics is an intensification of deviant behavior (e.g., drug abuse, homicide, and suicide) that promotes the spread of AIDS. The effects of these problems are exacerbated by the severing of social networks wrought by the loss of community. Thus, changes in the character of housing in poor communities can have a very real impact on the spread of the AIDS virus.

Financial Disaster

People with AIDS suffer severe financial setbacks as a result of the disease. AIDS is very expensive to treat, and insurance may not cover the array of costs related to the loss of employment and treatment of illness. Given the prevalence of AIDS in poor communities, many individuals with AIDS do not have insurance. More important, AIDS impairs the ability to support oneself. Persons suspected of being infected with the AIDS virus may be fired from their jobs or pressured to quit even if they are still healthy. Inevitably, those with full-blown AIDS symptoms will become too sick to work.

The financial circumstance of people with AIDS is compounded by the fact that they are less likely than other people to have connections to family members who could offer financial support. This may occur because of the stigma attached to the disease itself or to certain lifestyles, such as intravenous drug use and homosexuality, associated with its transmission. Thus, unless individuals infected with AIDS have substantial savings, they eventually end up on public assistance. This, of course, severely restricts the housing options available to them.

Health

Besides affecting one's employment potential, deteriorating health may also make certain living arrangements infeasible. Mobility problems for a person with AIDS can make a formerly suitable living arrangement inaccessible. For example, an apartment located up several flights of stairs or a house that cannot accommodate a wheelchair may make it necessary for the resident to move. As health deteriorates and greater assistance is needed, independent living may become impossible. Group homes, nursing homes, or hospice living arrangements may be the only adequate housing options possible.

Zoning Ordinances

The availability of nontraditional housing, such as group homes or hospice care for people with AIDS, is affected by local zoning ordinances. These ordinances may limit the number of people who can live together as an unrelated family or restrict the number of special-use dwellings in residential areas. Typically, zoning restrictions encompass any type of group use function, requiring AIDS-related housing to compete with myriad other housing programs (such as developmentally disabled group homes or drug halfway houses) for locations.

Zoning regulations usually allow for residential review and opposition to the establishment or continuation of special-use facilities, providing an institutionalized outlet for potential discrimination. This can effectively close off many neighborhoods, particularly in more well-to-do areas, restricting nontraditional housing options to poorer communities with substandard housing. However, a number of state courts have invalidated zoning that prohibits group homes by limiting the number of unrelated persons who can live together.

Discrimination

In addition to discrimination in the workplace and the potential for veiled discrimination through residential review of nontraditional housing options, people with AIDS can face discrimination from landlords who refuse to rent to them or evict them from housing they occupy. Often, discrimination is fueled by fears of contagion or loss of property value and by loathing of particular lifestyles associated with AIDS transmission. This more direct form of housing discrimination can take place in legitimate fashion in the form of month-to-month leases for which no reason for termination need be given or in evictions for which certain lifestyles (drug use, homosexual behavior) have no legal protection.

Besides legitimate avenues of discrimination, landlords can simply refuse to rent to persons suspected of having AIDS or evict such persons, ostensibly for other reasons. Proving discriminatory practices is a difficult and potentially expensive matter. Governmental bodies mandated to investigate unfair housing practices are capable of investigating only a fraction of cases they receive. Thus, most people with AIDS simply move on and try to disguise behavior or symptoms that might get them evicted.

Remedies

Resolution of the housing dilemmas faced by people with AIDS calls for a threefold approach. First, the destruction of housing and community in low-income neighborhoods needs to be reversed to help stem the diffusion of the disease. Municipal funding for basic infrastructural services is essential for maintaining existing housing. Incentives for replenishing the decimated low-cost housing stock, whether through tax breaks, low-interest loans, or subsidized rent, is imperative for restoring community in low-income areas. Subtitle D of Title VIII of the Cranston-Gonzales National Affordable Housing Act (1990) offers grants to help states and local governments provide housing for those with AIDS.

Second, legislation prohibiting discrimination in housing for people with AIDS needs to be enforced. Legal protection has expanded from local ordinances to national policy with the passage of the Cranston-Gonzales National Affordable Housing Act, but absent the resources to investigate complaints, legal protection has little practical value. Given the pessimistic outlook for increasing expenditures for such protection. AIDS advocacy organizations could play an important role in exposing discriminatory practices by landlords.

Finally, alternative housing models, particularly at a medium-care level, are needed. Many people with AIDS end up in nursing homes or hospices when only moderate care or periodic intensive care is required. Independent group-living models, residence hotels, and board-and-care facilities can meet the needs of those who are only moderately ill at a fraction of the cost of nursing homes and hospices while providing a greater sense of personal independence for the resident. (SEE ALSO: **Health;** *Hospice Care*)

—*Daniel M. Cress*

Further Reading

Arno, P. S. et al. 1996. "The Impact of Housing Status on Health Care Utilization among Persons with HIV Disease." *Journal of Health Care for the Poor and Underserved* 7(1): 3-14.

Foner, Geoffrey. 1988. "AIDS and Homelessness." *Journal of Psychoactive Drugs* 20:197-202.

Mandelker, Daniel R. 1987. "Housing Issues." Pp. 142-52 in *AIDS and the Law: A Guide for the Public,* edited by Harlon L. Dalton and Scott Burris. New Haven, CT: Yale University Press.

Wallace, Rodrick. 1990. "Urban Desertification, Public Health and Public Order: Planned Shrinkage, Violent Death, Substance Abuse and AIDS in the Bronx." *Social Science and Medicine* 31:801-13. ◀

▶ Adjustable-Rate Mortgages

Ever since the early 1930s, the standard, residential mortgage instrument used in the United States was a fixed-rate mortgage (FRM). Although it was not often called an FRM, this instrument had a fixed interest rate, established at the beginning of the loan period; a constant debt service payment, consisting of both interest and principal; and a fully amortizing repayment schedule. By the end of the repayment schedule, the entire principal was scheduled to be systematically repaid to the lending institution with level payments at a constant interest rate.

During periods of anticipated inflation, however, it is cumbersome to use the FRM as a viable mortgage instrument. This stems from what is called the *tilt effect*. This phenomenon refers to the fact that the actual (real) burden of the loan increases dramatically at the beginning of the loan during periods of anticipated inflation. During certain high-interest-rate periods of the 1960s and 1970s, it be-

came clear that the FRM proved deficient as the primary mortgage instrument in the U.S. financial system.

In 1978, a state-chartered savings and loan association in California introduced what was initially called a variable-rate mortgage (VRM) as an alternative to the FRM. The VRM became the predecessor for the adjustable-rate mortgage (ARM). By 1982, the Federal Home Loan Bank Board had adopted the ARM as its alternative mortgage instrument (AMI) of choice.

Both the original VRM and the ARM were aimed at alleviating the burden of the tilt effect for household borrowers. Because the ARM did not require the lender to estimate the effects of future interest rate changes as required under the FRM, the lender could offer the ARM at a lower interest rate. This is because the borrower bears the risk associated with future interest rate changes when an ARM is used.

ARMs require specification of some additional parameters, including interest rate floors, ceilings (called "caps," limiting both the total adjustment allowable over the life of the loan and the maximum adjustment per period), and a specified mechanism for adjusting the interest rate on the note (or the choice of an "index"). Over time, these parameters have become somewhat standardized according to market experience. For example, the One Year Treasury Bill average serves as the most common index chosen in residential mortgage markets. There has also been research about the optimal design of ARMs (and other AMIs) in recent years, including the effects of different parameters. Although these instruments are more complicated than FRMs, it is safe to say that much more is known now, after years of experience, than when they were first introduced nationally in 1982.

Perhaps the best way to think about the difference between an FRM and an ARM is to consider what the borrower (lender) receives (gives) at the time the mortgage instrument is originated. Regardless of interest rate trends, the borrower using an FRM has the right to repay the loan at a *fixed* rate of interest. The borrower using an ARM does not acquire this right. Thus, one could think of an FRM consisting of an ARM and a financial option to repay the note at a fixed rate. This option is quite valuable, especially when mortgage rates are volatile. Therefore, the borrower should expect FRM interest rates to be higher; the lender will not offer this right (i.e., the financial option) at a zero price.

During the past decade, the success of the ARM is testimonial to the preferences and choices that households make in many mortgage markets throughout the United States. For households with short planning horizons, it may not be necessary to shift the risk of future increases in mortgage interest rates to the lender. For households with variable income, the choice of an ARM is like taking a hedged position with respect to future interest rates; adopting an FRM requires the borrower to pay a premium to shift the risk to the lender. Some households might decide that the option price to shift the risk to the financial institution is too high (i.e., they would rather assume the interest rate risk themselves).

One of the standard assumptions in the analysis of American financial institutions during much of the 20th century has been that regulated financial institutions such as banks and savings and loan associations were better suited to bear interest rate risk than were individuals or households. For a majority of cases during many years, this assumption was no doubt correct. However, in recent years, this assumption no longer seems to be valid. The ARM has become a permanent feature of mortgage financing in the United States. Households now rely on the ARM for permanent mortgage financing in as much as 30% of U.S. mortgages. (SEE ALSO: *Alternative Mortgage Instruments*)

—*Austin J. Jaffe*

Further Reading

Asay, Michael R. 1984. "Pricing and Analysis of Adjustable Rate Mortgages." *Mortgage Banking* (December): 60-72.

Brueggeman, William B. and Jeffrey D. Fisher. 1993. *Real Estate Finance and Investments*. 9th ed. Homewood, IL: Irwin.

Clauretie, Terrence M. and James R. Webb. 1993. *The Theory and Practice of Real Estate Finance*. Fort Worth, TX: Dryden.

Dhillon, J., J. Sa-Aadu, and J. D. Shilling. 1996. "Choosing between Fixed- and Adjustable-Rate Mortgages: The Case of Commercial Mortgages." *Journal of Real Estate Finance and Economics* 12(3):265-78.

Fabozzi, Frank J. and Franco Modigliani. 1992. *Mortgage and Mortgage-Backed Securities Markets*. Boston: Harvard Business School Press.

McNulty, James E. 1988. "Mortgage Securities: Cash Flows and Prepayment Risk." *Real Estate Issues* 13 (Spring/Summer):10-17.

Peek, Joe. 1990. "A Call to ARMs: Adjustable Rate Mortgages in the 1980s." Federal Reserve Bank of Boston. *New England Economic Review* (March/April): 47-61.

Sirmans, C. F. 1989. *Real Estate Finance*. 2d ed. New York: McGraw-Hill. ◄

► Adolescents

Research on adolescence and housing forms an emerging area of study within the interdisciplinary field of environmental and behavior research that relates age-specific and developmental issues of adolescence to sociophysical aspects of housing and neighborhoods.

Attention to this topic was limited until the 1970s. Since that time, a growing body of work has been spurred on by the development of fields such as environmental psychology, urban sociology, environmental education, and planning and architectural studies. Some of the work has been directed toward understanding basic relationships between adolescents and their physical surroundings. Other work has sought to influence housing and planning policies to benefit children and adolescents. In some cases, attempts have been made to demonstrate effective ways in which adolescents can participate in planning and modifying settings, and the potentially beneficial consequences of doing so.

Adolescence loosely corresponds with the teenage years, although the beginning and end of adolescence, and its progression, vary among individuals and among cultures. Some researchers have defined adolescence as beginning as early as age 10 and ending as late as in one's 20s. Adolescence is generally defined as a transitional period between childhood and adulthood, characterized by the biological changes of puberty and physical growth, psychological changes in cognitive abilities, emotional development, identity formation, social shifts toward an increased focus on involvement with age peers, decreased dependence on adults, and sociocultural shifts toward attaining adult status and preparing for adult roles.

Housing and Near-Home Environments as Settings for Adolescents

During early childhood, the interior of the home is a focal point of daily activities and of time spent. Movement to places outside the home is constrained or facilitated to a large degree by caretakers. During adolescence, there is a shift toward activities, exploration, and time spent outside the home. Independent movement is much more possible. Nonetheless, under certain conditions, the interior of the home may still remain an important setting for adolescents.

Some research has shown that the ability to secure privacy within the home by appropriating a room or section of the home can serve an important psychological function. The preference among children for having a "room of one's own" begins early and is likely to intensify during adolescence when issues of identity formation and independence come to the fore. Having this place within the home can allow solitude and retreat, which can be helpful in regulating one's emotional life. In addition, this place can be personalized and used for keeping valued possessions. Several studies suggest that places in or immediately around suburban, single-family homes are valued by adolescents who live there, although some other more public settings (e.g., parks, natural environments, commercial areas, "free spaces" in schools, such as cafeterias) may be valued even more highly for interaction with peers if they are not subject to direct adult control. Institutional settings, which are highly controlled by adults, are least favored by adolescents. The focus on, and preference for, housing interiors as an important place may reflect the availability of space and other resources within the home and a lack of easy access to community resources. Some research also suggests that females are more likely to spend time in, and focus more on, interiors of homes than are males because of socialization and constraints placed on females by adults.

If the bedroom is shared, and especially if the dwelling is overcrowded, adolescents may find ways to extend their territories into semipublic and public spaces, and the interior of the home will be viewed as less important. In addition, the form of housing (e.g., multiple-unit apartment dwelling versus suburban tract house) and the availability of community resources within easy reach of the home (or with transportation supporting access) seems to influence the perceived importance of the home to adolescents.

One "resource" that a community can have is a relatively high density of age peers for adolescents in the neighborhood. Higher age peer densities are associated with adolescents' engaging in more activities with friends and having fewer complaints about a lack of friends in the neighborhood; in turn, this may have positive effects on adolescents' social adjustment and peer relations. Some authors have suggested, however, that increases in child densities in neighborhoods should also be matched by other community resources to help avoid the emergence of social problems and to provide a positive focus for activities. The limited research on this topic suggests that community resources such as organized, out-of-school programs and activities are more prevalent and varied in affluent suburbs than in low-income, inner-city areas. Resources may be least available where most needed.

Access and Participation

Although research is limited on the issues, a few studies have suggested that the availability of, and access to, settings within reach of the home environment may have important developmental consequences for adolescents. The term *access* implies that physical entry into a setting is possible, that information about the characteristics of the setting is available, that there is a perception of the setting as being open for entry, and that involvement and participation within the setting are possible. Adolescents' perception of the availability of, and potential for access to, settings is an area that deserves more study because it may help to explain how and why teens use their environments the way they do. For example, adults' perceptions that adolescents "like to hang out" around housing may in part reflect teens' efforts to spend time with age peers freely, without adult interference, and may also reflect adolescents' perceptions of not having suitable alternative settings in which to meet this goal. The issue of access may also prove important because it may affect adolescents' developing understanding of their role, or lack of a role, in their communities and in the adult social world. In addition, the kinds of settings available are likely to have an impact on adolescents' opportunities to develop skills and abilities useful in adulthood. Housing is related to this issue to the extent that location determines proximity, that housing and community design are integrated, and that the housing setting itself contains a variety of possibilities for adolescents.

The participation of adolescents in the design or modification of housing and community settings is relatively rare, although some projects have demonstrated the potential of participation as a way of improving settings and of enhancing adolescents' skills and development. Advocates of participation define it as the process of sharing decision making in a way that affects both one's community and one's own life. Participation is viewed by these advocates as an avenue by which democracy is built, and it is seen as a fundamental right of citizenship. When adults and adolescents work together in participatory projects, they learn about their rights and responsibilities as citizens. All parties are also confronted with issues of power, collaboration, and negotiation. Documented examples of successful participatory projects suggest that children and adolescents do learn a great deal about their community and develop skills and motivation that prepare them for future projects.

Summary

This emerging area of study has yielded some interesting research results and some practical suggestions for involv-

ing adolescents in the design of their housing and community. Current research is focusing more closely on the developmental dynamics involved in adolescents' relationships with aspects of their housing and near-home environments. The ways in which class, race, gender, and culture are related to these dynamics are also being investigated more closely. In addition, advocates of participatory planning and design are attempting to better understand how family and cultural contexts may influence the process of participation. (SEE ALSO: *Children; Cohousing*)

—*Michael K. Conn*

Further Reading

Chawla, Louise. 1991. "Homes for Children in a Changing Society." Pp. 187-228 in *Advances in Environment, Behavior, and Design*, Vol. 3. edited by Ervin H. Zube and Gary T. Moore. New York: Plenum.

Csikszentmihalyi, Mihaly and Reed Larson. 1984. *Being Adolescent: Conflict and Growth in the Teenage Years*. New York: Basic Books.

Hart, Roger A. 1992. *Children's Participation: From Tokenism to Citizenship*. Florence, Italy: UNICEF International Child Development Centre.

Littell, Julia and Joan Wynn. 1989. *The Availability and Use of Community Resources for Young Adolescents in an Inner-City and a Suburban Community*. Chicago: Chapin Hall Center for Children at the University of Chicago.

Lynch, Kevin. 1977. *Growing Up in Cities: Studies of the Spatial Environment of Adolescence in Cracow, Melbourne, Mexico City, Salta, Toluca, and Warszawa*. Cambridge: MIT Press.

Van Vliet--, Willem. 1985. "The Role of Housing Type, Household Density, and Neighborhood Density in Peer Interaction and Social Adjustment." Pp. 165-200 in *Habitats for Children: The Impacts of Density*, edited by Joachim F. Wohlwill and Willem van Vliet--. Hillsdale, NJ: Lawrence Erlbaum. ◀

▶ Advisory Commission on Regulatory Barriers to Affordable Housing

During President Bush's administration and under the guidance of then-HUD (Department of Housing and Urban Development) Secretary Jack Kemp, a blue-ribbon-panel was established with former New Jersey Governor Thomas Kean as chairman. This panel was asked to evaluate the causes of the growing housing affordability crisis in many cities in the United States. Specifically, the panel was asked to evaluate legal barriers common in many locales throughout the country. Land use controls as state and local legislative mechanisms were the primary focus in terms of their impacts on affordability.

The 1991 study has come to be known as the NIMBY (or "Not in My Back Yard") Report. The report identified the pervasiveness of the "NIMBY syndrome" among homeowners in U.S. cities and suburbs. This phenomenon refers to the actions of local residents who support various types of land use policies, including inclusionary zoning and environmental legislation *as long as* these programs do not affect themselves. For example, the setting aside of sites for halfway houses for recovering drug addicts would be of interest to many members of the community so long as the site is located in a different neighborhood. Environmental protection is applauded by some so long as it affects only other people's properties. Land use controls appear to be helpful so long as it is someone else's land who bears the impacts.

Although NIMBY as a U.S. urban phenomenon was not invented by this distinguished panel, this commission emphasized the representative nature of this syndrome throughout the urban United States in the late 20th century. In addition, the study evaluated the impact of NIMBY on affordability and reported that affordability problems were exacerbated by the ability of local residents to shift social policies elsewhere. Because many residents share the same view, the supply of affordable sites has been greatly reduced.

In addition, the study emphasized the substantial regulatory barriers that have developed extensively over the past half century in many U.S. cities. Regulations such as growth controls, exclusionary zoning, subdivision controls, and impact fees were cited as barriers for low- and middle-income households who seek better housing. Considerable academic research in recent years has shown that some land use controls result in restrictions that limit the housing choices of these moderate-income Americans.

One of the major legislative devices identified as a culprit was rent control. Although not a new finding, the report emphasized the detrimental effects associated with rent control as a housing policy. Rent control remains in place in certain cities, despite evidence of its detrimental effect on the housing stock, its benefits to undeserving tenants, and its failure to solve housing affordability problems.

Environmental laws were also identified as one of the critical barriers. Controversies over wetlands legislation and the Endangered Species Act and their relationship to affordable housing were discussed. It is clear that the panel sought to link social objectives associated with the environment to the cost of supplying housing to the populace.

Whether the report of the Advisory Commission on Regulatory Barriers to Affordable Housing will have a major impact on housing remains to be seen. The focus differs from previous affordability studies in emphasizing that policy-making decisions and mechanisms are local in nature. Thus, it is appropriate to use NIMBY as a symbol for the claims made in the report and for the focus of the recommendations. So long as local governments remain active in land use issues, the NIMBY syndrome is likely to remain relevant. If nothing else, this report focused attention on local land use regulations and their impacts on affordability. (SEE ALSO: *Not in My Back Yard*)

—*Austin J. Jaffe*

Further Reading

Advisory Commission on Regulatory Barriers to Affordable Housing. 1991. *"Not in My Back Yard": Removing Barriers to Affordable Housing.* Washington, DC: Department of Housing and Urban Development. ◄

► Affordability

Housing affordability is the relation of a consumer's housing costs to his or her available resources. Both facts and standards are involved, often in conflicting fashion. Affordability is more likely to be a perceived and real problem among consumers with fewer available resources and in those areas where housing costs are high and rising rapidly.

Rents generally rise faster than incomes, and renters as a class have lower incomes than homeowners. For example, in 1989 constant dollars, gross rents (median contract rent plus fuel, utilities, and some other costs) rose from $363 in 1970 to $402 in 1993. During this same period, renters' median income fell from $18,915 to $15,618. Consequently, the median percentage of renters' income devoted to housing rose from 20.5% in 1970 to 26.8% in 1993.

Homeowners' expenses, and their relationship to incomes, are far more complex to calculate, because they vary enormously with (a) credit status (whether there is a mortgage or not—about one-third of all homeowners have no mortgage), (b) the downpayment/mortgage split, (c) opportunity cost of the downpayment, (d) historical differences in mortgage rates depending on when the loan was assumed, (e) widely differential individual tax savings attributable to the homeowner deduction, (f) whether and how value appreciation is taken into account, and (g) other factors. As a result, data on affordability among homeowners are far less reliable and meaningful. There is general consensus that housing cost burdens have been rising for homeowners as well, although less severely and uniformly than is true for renters.

American Housing Survey data for 1993 indicated that 6.9 million renter households were paying 50% or more of their income for housing, of whom 4.1 million were paying 70% or more. Such extreme housing cost burdens often result in eviction and eventual homelessness—the ultimate housing affordability problem. There are, however, no national data on evictions (not all of which actions are due to nonpayment, however) and little in the way of local data; part of the difficulty is definitional: What constitutes an actual eviction, as opposed to a "voluntary" departure before formal legal proceedings are instituted or carried out? Involuntary displacement of tenants, from a variety of causes, is a major phenomenon, and much, if not most, of this is attributable to affordability problems.

With respect to homeowners, The Mortgage Bankers Association collects national data on mortgage delinquency and foreclosures; at any given time, about 1% of all residential mortgages are in foreclosure proceedings, although only half of these advance to the point of eviction of the occupant homeowner. Delinquency rates (being 30 or more days behind in payments) are considerably higher: At any given time, usually 4% to 5% of all homeowners with mortgages are in this category.

One other common way people deal with the financial squeeze caused by high housing costs relative to incomes is overcrowding or doubling up. Because rents are a function of housing quality, location, and amount of space, leasing a smaller unit, and sleeping several people per bedroom or using nonbedroom space for sleeping, can reduce housing affordability problems but at the cost of the various psychological stresses and health and safety dangers that extreme overcrowding can produce. The more extreme forms of overcrowding come from doubling up—more than one household sharing a single space, to share housing costs or avoid homelessness. (This is to be distinguished from more positively motivated space sharing by extended families or households of unrelated people who wish to live together.) Such situations often are short-term in nature, because the strains can be unbearable, and usually cause considerable negative effects. Data on overcrowding and doubling up will understate the true extent of these conditions, because residents are understandably reluctant to report these facts, for fear that landlords, code inspectors, immigration agents, or welfare officials will find out.

Large regional variations in affordability exist within the United States, because housing prices and costs range widely and exhibit a wider spread than do incomes. Median sales prices of existing single-family homes (third quarter, 1994), for example, ranged from lows of $50,000 to $60,000 in housing markets such as Amarillo, Texas; Davenport, Iowa/Moline-Rock Island, Illinois; Ocala, Florida; Saginaw-Bay City-Midland, Michigan; Topeka, Kansas; Waterloo-Cedar Falls, Iowa; and Youngstown-Warren, Ohio, to highs of $180,000 to $215,000 in the Orange County, California; Boston, Massachusetts; Los Angeles, California; Bergen/Passaic, New Jersey; and Newark, New Jersey, housing markets, with San Francisco, California ($250,000), and Honolulu, Hawaii ($368,000), topping the list.

Significant racial variations in housing affordability exist as well, with black and Hispanic households paying higher proportions of their income for housing than is true for white households. Lower incomes among the former, as well as discriminatory pricing systems, account for this. This racial disparity is more pronounced among renters than among homeowners.

Affordability Standards

What people "ought" to pay for housing has always been a somewhat murky matter. The usual term applied to such standards, *rules of thumb,* captures the imprecise, ad hoc quality of the concept and measure. What proportion of income people pay for housing is subject to factors well beyond their control—what they can earn, what the market charges for housing—and, where income levels or the housing market permit, personal preference. As is to be expected, the data consistently show that the higher one's income, the lower the proportion (although not the absolute amount) of one's income is actually spent on housing and on basic needs generally.

Such standards as are promulgated may be used (a) as a guideline to personal expenditures, (b) as a factor determining whether a lender will grant a mortgage or whether a landlord will rent an apartment to a given applicant, or (c) as the basis, under various government programs, for determining what rent to charge or the amount of subsidy that will be granted.

Because, generally speaking, housing cost increases have outpaced incomes, it is no surprise that affordability has become the principal U.S. housing problem, replacing slum conditions, which motivated earlier government efforts in urban areas, principally the public housing and urban re-

newal programs. Along with this changing reality has gone the ratcheting upward of the government's "should" or "must" standard with respect to its own programs. Thus, the federal government's low-rent public housing program began in the 1930s with a 20% standard as the portion of income each tenant would be charged. (Capital costs for construction were totally subsidized by the government, and tenant rents were to cover operating costs.) As tenants' income-based rents increasingly were unable to cover sharply rising operating costs, local housing authorities were forced to increase rents, which in some cases rose to 50% or more of the tenant's income. In response, Congress in 1969-71 introduced a series of amendments to the public housing financing program, one of which placed a 25% cap (in effect, a floor as well for most families) on the amount of a tenant's income that could be charged for rent. Following that, as the gap between operating costs and tenants' incomes increasingly widened, the Reagan administration's Office of Management and Budget in 1981 raised the percentage to 30%, a "standard" later codified by Congress. Thus, the government's solution to its fiscal concerns was simply to change the standard—in essence, extracting more money from poor tenants to reduce the need for government subsidies. (Actual rent:income ratios for public housing tenants are in fact somewhat lower than the mandated percentage, because the income figure to which the percentage is applied is in many cases reduced by various allowable deductions.)

Definitional problems are also evident. Both income and rent can be defined in many ways. The former can be before- or after-tax income and can include or exclude in-kind income. The latter can be contract rent (the amount paid to the landlord, which may or may not include some or all utilities), gross rent, or total housing expenditures. (A particular problem pertains to income from lodgers or subtenants: whether to treat it as additional income or as lower rent, each of which yields very different results for the rent-to-income ratio.) Using 1950 data, housing economist Chester Rapkin showed that for the lowest income range, where variations in definition can have the greatest influence, "a twist of the tongue can shrink the housing expenditure ratio from 50.9 to 19.3%" (p. 9).

An important conceptual and policy critique of the affordability standard, whatever the actual figure used, has been raised by housing economist Michael Stone, who has argued that such ratios realistically are and ought to be a function both of household size and household income. That is, the larger the number of people in the household, the more that household must spend for food, clothing, transportation, medical care, and other nonshelter necessities; therefore, at any given income level, the proportion of income devoted to housing should decrease as household size increases. (Because of the fixed nature of housing costs—compared with food, the other major expense item for most households, expenditures for which can be rapidly and substantially changed by altering the amount, frequency, quality, and composition of meals—lower-income households especially seek to keep housing costs to a minimum, because rent or mortgage payments represent the overriding claim on disposable income and the consequences of not meeting that claim—eviction, foreclosure—are drastic.)

According to Stone, the household's income level must similarly be considered in arriving at an acceptable housing expenditure-to-income ratio. The higher one's income, the more one has left to devote to housing once other nonshelter basics have been taken care of; conversely, lower-income households (of a given size) can afford to devote less of their income to housing if they are to satisfy the household's basic needs for food, clothing, and so on.

In place of traditional fixed rules of thumb, Stone has suggested a sliding scale of affordability related to both household size and income level. Applying this basic analytic concept—which he has termed *shelter poverty*—to U.S. Bureau of the Census data on actual housing expenditures and to minimum family budgets prepared by the Bureau of Labor Statistics, Stone calculated that in 1991, 29 million U.S. households (owners as well as renters)—30% of all households—were "shelter poor"; that is, they were paying more than they could afford for housing if they were to have enough left to purchase the minimum nonshelter basics they needed, according to the same Bureau of Labor Statistics budgets. The data also showed that the number of such households had increased by 54% since 1970. Most striking, Stone estimated that half the shelter-poor households could not afford a single cent for housing if they were to have sufficient income left for the nonshelter basics the government said were the minimum needed. (SEE ALSO: *Affordability Indicators; Affordable Housing Indices; American Affordable Housing Institute; Cooperative Housing; Council for Affordable and Rural Housing [CARH]; Cranston-Gonzalez National Affordable Housing Act of 1990; Delinquency on Loan; Emergency Low-Income Housing Preservation Act of 1987; Employer-Assisted Housing; Fair Market Rent; Flat-Rent Concept; Foreclosures;* **Homelessness;** *Housing Allowances; Housing Costs; Housing-Income Ratios; Housing Price; Low-Income Housing Preservation and Resident Homeownership Act of 1990; Military-Related Housing; National Foundation for Affordable Housing Solutions, Inc.; National Low Income Housing Coalition; National Low Income Housing Preservation Commission; Rent Burden; Resolution Trust Corporation's Affordable Housing Disposition Program;* Shelterforce Magazine; *Shimberg Center for Affordable Housing*)

—*Chester Hartman*

Further Reading

Hartman, Chester. 1988. "The Affordability of Housing." Pp. 111-29 in *Handbook of Housing and the Built Environment in the United States,* edited by Elizabeth Huttman and Willem van Vliet--. New York: Greenwood.

Hulchanski, J. David. 1995. "The Concept of Housing Affordability: Six Contemporary Uses of the Housing Expenditure-to-Income Ratio." *Housing Studies* 10:471-92.

Joint Center for Housing Studies of Harvard University. 1994. *The State of the Nation's Housing 1994.* Cambridge, MA: Author.

Rapkin, Chester. 1957. "Rent-Income Ratio." *Journal of Housing* 14:8-12.

Stone, Michael E. 1993. *Shelter Poverty: New Ideas on Housing Affordability.* Philadelphia: Temple University Press.

Van Vliet--, Willem, ed. 1997. *Affordable Housing and Urban Redevelopment in the United States.* Thousand Oaks, CA: Sage.

Worst Case Needs for Housing Assistance in the United States. 1994. Report to Congress, prepared for the U.S. Department of Housing and Urban Development, Office of Policy Development and Research. ◄

► Affordability Indicators

Housing affordability indicators act as "measuring sticks" for determining people's ability to pay for housing. Perhaps the best known affordability indicator is the rule of thumb that suggests a household should pay no more than a certain percentage of income (e.g., 30%) for housing. There are several types of indicators, including housing-income ratios, constant quality comparisons, and market basket methods. No single standard of affordability is accurate for all situations. Policy analysts and scholars often devise housing affordability indices based on a mix of indicators, assumptions, and analytical methods. When it comes to assessing housing affordability, scholars need to determine which indicators and methods best suit their research needs.

No assessment of housing need would be complete without defining affordability for the situation at hand. The use of appropriate indicators is the key to determining the abilities of certain groups of people to pay for housing in particular places at given times. Housing-income ratios, which show relationships between a family's income and housing costs, are the most widely used type of affordability indicators. Constant-quality comparison is a method that can be used to control for quality standards to determine the relative affordability of housing. For example, a study of housing costs in a given area would compare homes that are of similar size, age, and condition; contain the same number of bedrooms; and have similar amenities. Other approaches, including market basket or shelter poverty methods, consider nonhousing costs that compete for a family's housing dollars.

Most measures of housing affordability use some form of housing-income ratio. The home loan industry, for example, uses a percentage of gross income to determine how much borrowers can afford for mortgage payments and other long-term debt. Similarly, resident contributions toward rent in federally assisted housing are set at a percentage (in 1996, 30%) of the renter's adjusted income. Ratios provide a practical way to assess affordability because data on incomes, rent, and mortgage costs are relatively reliable, stable, and easy to obtain compared with other kinds of household information. The percentage-of-income approach presents difficulties, however, when making comparisons across income groups. To illustrate, a family earning $42,000 annually might easily pay 30% of its income, or $1,050 per month, for housing; it would, however, be more difficult for a comparable household with only $12,000 annual income to pay 30% of income ($300 per month) because the lower-income family would have less discretionary income remaining for food, clothing, health care, and other basic needs. Also, a family of five or six people would need more nonhousing goods and services than would a family of two or three. Generally, the 30% rule tends to underestimate affordability problems for large households compared with standards that consider total demands on household income.

Some housing advocates have recommended the establishment of separate standards for different income groups and for owners and renters. Consumer economists generally advise using a market basket approach by establishing costs for food, apparel, medical care, and items such as those listed in the U.S. Department of Labor Consumer Price Index (CPI). The National Low Income Housing Coalition estimates a family's ability to pay for housing by subtracting the cost of nonhousing essentials from net income—if there is nothing left over, a family cannot afford a single dollar for housing, regardless of income. In reality, however, rent or mortgage payments generally make first claim on income. So households may sacrifice adequate nutrition, child care, or other basics to pay for housing—a condition called "shelter poverty." Michael Stone, in a 1990 study, developed shelter poverty standards of affordability for both renter and homeowner households based on minimum adequacy levels for nonhousing costs. Stone found that many low-income households and large families paid less than 25% of income for housing but were shelter poor; moreover, not all households paying excessive percentage-of-income amounts were shelter poor.

Although simple ratio-based indicators of affordability are often used arbitrarily, established models for assessing housing affordability can be used or adapted for particular situations. Or indices can be constructed for the purpose of a specific needs assessment. Indices for tracking affordability that are national in scope include (a) the National Association of Home Builders Housing Opportunity Index, (b) the National Low Income Coalition Rental Housing Index, (c) the Federal Home Loan Bank of Atlanta Lower-Income Housing Affordability Index, (d) the National Association of Realtors Affordability Index, (e) the Census Bureau Price Index of New One-Family Homes Sold, (f) the American Chamber of Commerce Researchers Association Cost of Living Index, and (g) the U.S. Department of Agriculture Cost of Food at Home. (SEE ALSO: **Affordability**; *Housing-Income Ratios; Rent Burden*)

—*Marjorie E. Jensen*

Further Reading

Dolbeare, Cushing N. 1989. *Low Income Housing Needs*. Washington, DC: National Low Income Housing Coalition.

———. 1989. *Out of Reach: Why Everyday People Can't Find Affordable Housing*. 2d printing. Washington, DC: Low Income Housing Information Service.

Feins, Judith D. and Terry Saunders Lane. 1983. "Defining the Affordability Issue." In *Housing Supply and Affordability*, edited by Frank Schnidman and Jane A. Silverman. Washington, DC: Urban Land Institute.

Fronczek, Peter J. and Howard A. Savage. 1991. *Who Can Afford to Buy a House? Survey of Income and Program Participation*. Current Housing Reports H121/91-1. Washington DC: U.S. Department of Commerce, Economics and Statistics Administration, Bureau of the Census.

Stone, Michael E. 1990. *One-Third of a Nation: A New Look at Housing Affordability in America*. Washington, DC: Economic Policy Institute.

———. 1993. *Shelter Poverty: New Ideas on Housing Affordability*. Philadelphia: Temple University Press. ◀

▶ Affordable Housing Indices

Several indices have been developed by government, private, and nonprofit organizations to measure housing affordability on the national and regional scales. These indices provide a basis from which to compare the status of housing affordability in various jurisdictions, and they are frequently used to formulate and evaluate local housing policies. Following are examples of affordable housing indices.

Housing Opportunity Index (HOI)

Developed by the National Association of Home Builders, the HOI ranks 169 metropolitan areas by the relative ease with which a family of median income can purchase a home. This index is based on median household income, as estimated by the Department of Housing and Urban Development (HUD), and on the prices of new and existing housing units. Using an allowable-purchase multiplier derived from interest rates, the index measures the percentage share of homes on the market that a median-income household can afford. The higher the index value, the greater the degree of affordability on the market. For example, an index of 85 means that a family of median income can afford 85% of the homes sold in a certain region, whereas a value of 15 signifies that such a family can afford only 15% of the homes sold.

The National Low Income Coalition Rental Housing Index (RHI)

The RHI measures the affordability of rental housing using HUD's Fair Market Rent data. Assuming an affordable monthly rent set at 30% of household income, the index calculates the income necessary to rent housing in each metropolitan area.

National Association of Realtors (NAR) Affordability Index

The NAR index is the ratio of the median income to the minimum income required to qualify for a conventional loan covering 80% of the median price of existing single-family homes. An index value of 110 means that a family earning the median family income of an area has 110% of the income necessary to qualify for the 80% loan. Twenty-two metropolitan areas are covered by the index.

Price Index of New One-Family Sold

This price index was designed by the U.S. Bureau of the Census to measure changes over time in the sale price of new single-family houses with the same characteristics. Index scores are derived from five separate price models, four of them for detached houses in each census region (Northeast, South, West, and Midwest) and one for attached houses throughout the United States. Seven components are included in all five price models: geographic division within region, location in relation to a metropolitan statistical area, floor area, number of fireplaces, number of bathrooms, type of parking facilities, and type of foundation. The sale price used in the index covers not only cost of labor and materials but also land cost, direct and indirect selling expenses, and seller's profit. (SEE ALSO: **Affordability;** *Housing-Income Ratios: Rent Burden*)

—*Caroline Nagel* ◄

► African Americans

Housing is both a consumer and a durable good—we value it for its services in the present and as an asset in the future. In their ability to acquire housing as either a consumer or a durable good, throughout their history in this nation, African Americans have faced limited choices—both of type and location. These limits have resulted, first, from their status as slaves and, later, from zoning ordinances, federal statutes and programs of housing assistance, and court rulings, all of which have combined to relegate the majority of African Americans to segregated neighborhoods with a limited range of housing options that are decent, safe, sanitary, and appropriate to their needs. Racial discrimination in the processes of renting and purchasing homes, another infringement on the housing choices of African Americans, continues today.

Before 1950
Between 1619 and 1863 (when the Emancipation Proclamation was signed), Africans brought to this country were slave laborers on the plantations in the South. Their housing was usually log cabins, with few of the comforts found in the "Big House," occupied by their owners or masters. Slave quarters usually protected their inhabitants from the elements but did not afford separate bedrooms or privacy. Slaves were not allowed to own property, so they could not enjoy the "durable" aspects of housing that come through homeownership.

After the end of the Civil War in 1865, the federal government took a series of steps to provide rights to the descendants of slaves, comparable to those enjoyed by all other U.S. citizens. The first of these actions was passage of the Civil Rights Act of 1866, which established the same rights to purchase, sell, and hold property for African Americans as for white Americans. Although the Fourteenth and Fifteenth Amendments to the Constitution were adopted to buttress the property rights of African Americans, shortly after the Constitution had been amended, Supreme Court rulings in the civil rights cases of 1883 began to erode these rights. The Court ruled that, under the Fourteenth Amendment, although states could not limit where African Americans may buy homes or refuse to sell properties to them on terms comparable to those offered white Americans, private actors and individuals could.

The exclusion of private actors from the mandate of the Civil Rights Act of 1866 set the stage for the "separate-but-equal" doctrine, established in 1896. In *Plessy v. Ferguson* (163 U.S. 537), the Supreme Court established the separate-but-equal doctrine for railway travel. This doctrine, however, soon became the guidepost for relationships between the races and was used by real estate professionals throughout the nation to justify steering home seekers to communities in a manner that would establish and maintain neighborhoods that were completely segregated by race.

While these legislative and judicial actions were taking place, the descendants of slaves were leaving the agricultural South and moving to the industrializing North and West. Despite *Plessy v. Ferguson,* in the early years of the 20th century, African Americans were able to acquire decent homes in racially integrated neighborhoods in cities such as Chicago, Columbus, Minneapolis, New York City, Pittsburgh, Portland, San Francisco, and San Diego. This limited period of freedom of choice and open access to housing for African Americans was turned around by local events that fell into line with the separate-but-equal doctrine. First, race riots sparked by competition between African and European Americans seeking jobs in northern cities caused many African Americans to choose to live in neighborhoods with other African Americans for self-protection. Second, many southern municipalities passed residential segregation (or racial zoning) ordinances that designated the residential blocks that could be occupied by African Americans and by white Americans. Although racial zoning ordinances were outlawed by the 1917 Supreme Court decision in *Buchanan v. Warley* (245 U.S. 60), they were replaced by racially restrictive covenants that achieved the same purpose. Racially restrictive covenants were attached to deeds and dictated the racial or ethnic groups to whom a given property could not be sold; they were not ruled unenforceable until the 1948 Supreme Court decision in *Shelley v. Kraemer* (334 U.S. 1).

Thus, it took the nation from the end of the Civil War until the middle of the 20th century before its legal and judicial framework conformed with the ideal of equal opportunity in housing—that all persons of any racial or ethnic group should be able to seek out and acquire housing they can afford in any locality. Over this same period, according to Gries and Ford, the housing of the growing numbers of African Americans who were leaving the South to settle in other parts of the land could be characterized to an increasing degree as ghettoized, and in areas with a high rate of delinquency, a high rate of mortality, and a distorted standard of living. The phrase "distorted standard of living" reflects factors such as (a) high rents because landlords knew the demand for housing by African Americans far exceeded the supply; (b) the large proportion of income that African Americans were required to spend for housing (because of low wages and high rents); (c) the taking in of lodgers to help meet rental payments, which overcrowds units and leads to their rapid deterioration; and

(d) limited homeownership and, thereby, limited wealth acquisition.

Federal Housing Assistance Programs

Housing assistance programs funded by the federal government also contributed to the segregation of the races and the access to a limited set of housing options for African Americans. These programs were initiated in the 1930s (such as the low-rent public housing program and the mortgage insurance programs of the Federal Housing Administration) and were expanded on in the 1940s (such as the loan guarantee program of the Veterans Administration, now the Department of Veterans Affairs). The federal Urban Renewal Program, established in 1949, may have had the greatest influence on the patterns of residential location of African Americans throughout the nation. Although funded by the federal government, many of these programs were administered by the same states and localities that had sanctioned the use of racial zoning to determine where the two races should reside within their boundaries.

Although African American families are half of all families living in low-rent public housing today, court suits provide historical evidence of the following barriers to equal access to quality housing for African Americans in this program: (a) exclusion from participation, (b) delay or prohibition of project construction, (c) concentration of projects in central cities, and (d) complete segregation by race in the projects. In *Vann v. Toledo Metropolitan Housing Authority,* 1953, OH (113 F. Supp. 210), the District Court of Ohio ruled that a municipality charged with managing low-rent public housing projects erected with public funds could not exclude persons of the "colored race," as the Toledo Metropolitan Housing Authority had been doing, because exclusion violated the Fourteenth Amendment of the Constitution. In 1982, in *Atkins v. Robinson,* 1982, VA (545 F. Supp. 852), the federal court found that Greenville City, Virginia, had vetoed plans for low-rent public housing for racially discriminatory reasons, thereby violating the Civil Rights Act of 1866. Although many jurisdictions have been found culpable of selecting sites for low-rent public housing in ways that concentrate these projects in inner-city neighborhoods and perpetuate segregation of the races, the most famous court cases filed about this were the Gautreaux cases. The remedy for the Gautreaux cases, filed by the late Dorothy Gautreaux of Chicago—*Gautreaux v. Romney,* 1971, IL (448 F. 2d 731) and *Hills v. Gautreaux,* 1976, IL (425 U.S. 284)—was the establishment of a program to provide funds to tenants of low-rent public housing (through the Federal Section 8 program) to enable them to rent units in the suburbs of Chicago. The ultimate goal of the program was to reduce the concentration of low-rent public housing projects and low-income tenants on the South Side of Chicago. Because tenant selection to maintain racial segregation in low-rent public housing projects was found as recently as the 1980s in two cases titled *Young v. Pierce* [(544 F. Supp. 1010) and (628 F. Supp. 1037)], a court-ordered reassignment of tenants was mandated.

Although the mortgage insurance program of the Federal Housing Administration (FHA) and the loan guarantee program of the Department of Veterans Affairs (DVA) have helped many white Americans follow jobs and new homes to the suburbs, these programs have helped few African Americans move to the suburbs. In fact, prior to February 15, 1950, both agencies had provided insurance and guarantees for properties with racially restrictive covenants on their deeds. Although the FHA today is a major source of home financing for minority households, its history includes redlining and initially starving these same households and inner-city neighborhoods of needed mortgage loan funds. The operation of these programs has contributed to the current differentials in wealth accumulation between African and white Americans.

The Urban Renewal Program, established in 1949 and active during the 1950s and 1960s, is dubbed the "Negro Removal" program because often its impact was to destroy neighborhoods in which African Americans lived. The objective of the program was to lower the cost of constructing new rental and owner projects in urban areas by lowering the cost of land acquisition. Land cost was lowered because localities were able to use their power of eminent domain to acquire properties, which would then be razed. New structures would be built on these same sites. Although households dislocated by the Urban Renewal Program were entitled both to payments for their housing and land and to relocation assistance, many stable African American communities with standard quality housing were demolished, and, in some cases, luxury housing was built in its place.

1950 to the Present

Although the civil rights movement (1954-1956) did not have a strong housing mandate, as a result of this movement, two statutes were passed in the 1960s that targeted discrimination on the basis of race in both the private sector and in federal programs of housing assistance. The Civil Rights Act of 1964 (Title VI) prohibits discrimination on the basis of race, color, religion, or national origin against persons eligible to participate in and receive the benefits of any program receiving federal financial assistance. The Civil Rights Act of 1968 (Title VIII) established federal fair housing policy for the nation. To strengthen the enforcement authority provided to the U.S. Department of Housing and Urban Development (HUD) under the 1968 Act, 20 years later, the 1988 Fair Housing Amendments Act was passed.

The enactment and enforcement of these three pieces of civil rights legislation have not been able to eradicate the residential patterns that had developed during the era of "separate but equal" or to guarantee fair housing and equal access to mortgage funds throughout the United States. Neither have these statutes, nor the passage of time since their enactment, been sufficient to close the gap in ownership and wealth accumulation between African and white Americans.

As early as the 1950s, census data revealed out-migration of white Americans and in-migration of and rates of increase among African Americans, the combination of which have left the populations of major metropolitan areas increasingly nonwhite and increasingly segregated. In cities such as Baltimore, Cleveland, Houston, and Richmond, in 1950, over 90% of either the African American or white American population would have had to move for there to be the same percentage of African Americans and white Americans per city block as in the entire metropolitan area. By 1970, although segregation had decreased in these four places, in three of them (Cleveland, Houston, and Richmond), over 90% of either the African American or white population still would have had to relocate to achieve the metrowide racial distribution per block. By 1980, although the trend in segregation had continued to decline, 91% of Clevelanders, 86% of Baltimoreans, 81% of Houstonians, and 79% of Richmond residents of either race were still required to move to achieve the desired racial distribution per block. Between 1980 and 1990, for most of the segregation indices used to determine the percentages who need to move to eliminate the spatial segregation of the races, the mean values remained constant or declined slightly (generally less than 10%).

Discrimination in the home purchase and home rental markets persists and further limits the options for African Americans to find residences of their choosing, which could become the cornerstones of their family estates. Since the late 1970s, the use of testers (paired white and African American persons given nearly identical socioeconomic characteristics and requirements for residences to use when visiting apartment management and real estate offices) has revealed significant differences in the experiences of potential African American and white home purchasers and renters. Discrimination against African American purchasers includes (a) limited amounts of time spent with the potential client, (b) limited information provided on whether any houses are available in the requested price range and neighborhoods, (c) few houses offered as serious possibilities, (d) few houses shown, (e) unwillingness to discuss financing, and (f) discourteous or rude treatment, such as canceling or coming late to appointments for viewing houses. For rental units, instances of racial discrimination are revealed in differential treatment with respect to (a) acknowledging the availability of an advertised unit and of similar units, (b) the number of units shown, and (c) the number of units recommended but not shown.

When differential ownership rates by race are examined, the pattern evident in 1950 remained evident in 1997—ownership rates are much higher for white Americans than for African Americans. In 1950, 57% of whites were homeowners, whereas 35% of African Americans were. By 1970, the ownership rates had risen for both groups—to 42% for African Americans and 65% for whites. In 1997, although more than 45% of African American were owners, over two-thirds of whites (69.8%) were. The gap of at least 22 percentage points has remained for more than 40 years.

The ownership disparity is compounded when net worth is compared by race in 1988. Although an owned home constitutes 68% of the total net worth of African Americans, their mean equity from owning a home is only $36,770.

An owned home constitutes only 42% of the net worth of whites, but mean equity from homeownership among whites is $64,164. Thus, even though the value of an owned home constitutes a larger share of the net worth of African Americans than of whites, the equity in this home is considerably less for African Americans than for white Americans. These findings contribute to the marked disparity between median net worth of white and African Americans. Median net worth for white Americans ($45,740) was more than 10 times the median net worth for African Americans ($4,418) in 1993. (SEE ALSO: **Discrimination;** *Segregation; Slaves, Housing of)*

—*Wilhelmina A. Leigh*

Further Reading

Abrams, C. 1955. *Forbidden Neighbors; A Study of Prejudice in Housing.* New York: Harper.

Gries, J. and J. Ford. 1932. *Negro Housing, VI, Report on the President's Conference on Home Building and Home Ownership.* Washington, DC: National Capital Press.

Grigsby, J. Eugene and Mary L. Hruby. 1991. "Recent Changes in the Housing Status of Blacks in Los Angeles." *Review of Black Political Economy* (Winter-Spring):211-40.

Hirsch, Arnold. 1983. *Making the Second Ghetto: Race and Housing in Chicago.* New York: Cambridge University Press.

Leigh, Wilhelmina A. 1991. "Civil Rights Legislation and the Housing Status of Black Americans: An Overview." *Review of Black Political Economy* (Winter-Spring):5-28.

Leigh, Wilhelmina A. and James B. Stewart, eds. 1992. *The Housing Status of Black Americans.* New Brunswick, NJ: Transaction Publishers.

Taeuber, Karl E. and Alma F. Taeuber. 1965. *Negroes in Cities: Residential Segregation and Neighborhood Change.* Chicago, IL: Aldine.

U.S. Bureau of the Census. 1995. *What We're Worth—Asset Ownership of Households: 1993.* Statistical Brief SB/95-26. Washington, DC: U.S. Department of Commerce.

Wienk, Ronald et al. 1979. *Measuring Racial Discrimination in American Housing Markets: The Housing Market Practices Survey.* Washington, DC: U.S. Department of Housing and Urban Development. ◀

▶ Alternative Mortgage Instruments

During the 1930s, the Federal Housing Administration (FHA) created a standard mortgage instrument. This mortgage came to be known as the fixed-rate mortgage (FRM). The FRM had a number of attractive characteristics: (a) The required debt service included both an interest and an amortization (principal repayment) portion with each payment; (b) the periodic debt service was constant so that the repayment schedule was level; (c) at the end of the repayment period, the original principal had been completely repaid, or fully amortized; and (d) the interest rate on the mortgage note was fixed throughout the life of the loan. This algorithm enabled loans to extend to as much as 20, 25, or 30 years. Prior to the development of this instrument, mortgages tended to be 5 to 7 years without amortization.

Lenders liked the FRM because their risk exposure was reduced with each successive payment. In addition, collection procedures were streamlined, and over time, the valuation of their mortgages became easier. Borrowers liked the FRM because by extending the repayment period, affordability was enhanced a great deal, and for the next 50 years, mortgage financing grew at an unprecedented rate. Indeed, the FRM revolutionized the mortgage financing system in the United States and elsewhere.

But during the periods of volatile interest rates in the 1960s and 1970s, the FRM came under pressure as a consumer product. During periods of anticipated inflation, a small change in mortgage rates translated into a larger change in the required debt service payment. This is because the debt service payment covers both interest and principal payments for the fixed term of the loan. As interest rates rise, the burden of making a constant payment over a fixed term increases on the borrower. This factor (called the *tilt effect*) becomes severe during periods of high expected inflation.

Beginning in the late 1970s and continuing in the 1980s, financial institutions developed alternative mortgages to the FRM. This class of mortgages is called alternative mortgage instruments (AMIs). They are frequently viewed as competing choices for the standard FRM and have many forms, objectives, and methods.

The more common AMIs include the adjustable-rate mortgage (ARM), the graduated-payment mortgage (GPM), the growing-equity mortgage (GEM), the price-level-adjusted mortgage (PLAM), the shared appreciation mortgage (SAM), the reverse-annuity mortgage (RAM), and many others. Each of these mortgage instruments attempts to solve one or more deficiencies with the FRM. Despite their brief lives, there are some success stories, along with some tales of failure.

For example, to offset the impact of the tilt effect, the ARM, GPM, and PLAM are intended to moderate the effect of anticipated inflation. The GEM and SAM are designed to share appreciation potential with the lender. The RAM tries to provide liquidity for property owners who are "income poor," but "property rich."

In the beginning of the 1980s, it was often feared that AMIs would drive FRMs out of the mortgage market. Some consumer advocates even argued that AMIs were nasty, strategic devices unleashed by financial institutions to take advantage of household consumers. In the beginning, there was strong sentiment in some quarters to regulate AMIs out of existence.

With the experience of the past several years, it has become quite apparent that AMIs and the traditional FRM can coexist. In fact, most financial institutions offer quotations for both types of mortgages with various provisions at the same time. In addition, there are examples of hybrid instruments, using elements of both types of loans. The issue about financial institutions taking advantage of consumers has tended to die down because borrowers can usually choose to stay with the FRM tradition or elect one of the newer AMIs. It is also apparent that the prohibition of AMIs would have had a tremendous impact on the millions of consumers who chose, for example, ARMs, during the 1990s. Such restrictions would have limited consumer choices regarding the bearing of interest rate risk in the mortgage market.

Some AMIs have not worked out very well (e.g., the GPM has proven to be too difficult for lenders to price). Others have required some modification from their initial definitions for use in actual markets. The development of AMIs in the 1980s has produced a major addition to the American financial system. No doubt, other instruments will be created, experimented with, and used in the future. (SEE ALSO: *Adjustable-Rate Mortgages; Graduated-Payment Mortgage; Growing-Equity Mortgage;* **Mortgage Finance;** *Price-Level-Adjusted Mortgage; Reverse-Equity Mortgage; Shared-Appreciation Mortgage*)

—*Austin J. Jaffe*

Further Reading

Clauretie, Terrence M. and James R. Webb. 1993. *The Theory and Practice of Real Estate Finance.* Fort Worth, TX: Dryden.

Cohn, Richard A. and Stanley Fischer. 1975. "Alternative Mortgage Designs." *New Mortgage Designs for an Inflationary Environment.* Federal Reserve Bank of Boston Conference Series No. 14.

Guttentag, Jack M. 1984. "Recent Changes in the Primary Home Mortgage." *Housing Finance Review* 3:221-55.

Lessard, Donald and Franco Modigliani. 1975. "Inflation and the Housing Market: Problems and Potential Solutions." *New Mortgage Designs for an Inflationary Environment.* Federal Reserve Bank of Boston Conference Series No. 14.

Seiders, David F. 1981. "Changing Patterns of Housing Finance." *Federal Reserve Bulletin* 67 (June): 461-72. ◀

▶ American Affordable Housing Institute

The American Affordable Housing Institute (AAHI) is an applied policy research center located in the School of Planning and Public Policy at Rutgers, the State University of New Jersey. The institute's principal objective is to develop a new knowledge base for action that will enhance the ability of all Americans to live in decent, safe, and affordable housing. AAHI conducts research designed to yield concrete innovative public and private solutions to housing affordability and availability problems in local, state, and national settings. The institute undertakes projects with the private sector, nonprofit housing agencies, foundations, and governments.

Its research emphasizes innovation in (a) preventing homelessness, (b) revitalizing urban neighborhoods, (c) increasing the housing affordability by expanding the pool of private capital for investment in low- and moderate-income housing—especially via employer-assisted housing programs, (d) meeting the health-related housing needs among the growing number of frail elderly in the United States, and (e) assisting the development of housing policy at the national, state, and local levels.

AAHI publishes books, articles, and technical reports and conducts conferences and seminars for government officials and for-profit and nonprofit shelter industries. The institute also conducts outreach, technical assistance services, and demonstration projects. AAHI draws on the multidisciplinary resources of Rutgers University, using expertise in areas such as planning, political science, labor studies, business, finance, and administration. AAHI is also a member of the National Association of Homebuilders' University Research Centers Consortium. Contact: AAHI, RUTGERS, 33 Livingston Ave., Suite 160, P.O. Box 118, New Brunswick, NJ 08903. Phone: (732) 932-6812. Fax: (732) 932-7974. (SEE ALSO: **Affordability**)

—Daniel Hoffman ◄

► American Association of Homes and Services for the Aging

The American Association of Homes and Services for the Aging (AAHSA) is a national organization of over 5,000 nonprofit senior housing facilities, retirement communities, nursing homes, and community agencies servicing the elderly. AAHSA serves its members by representing the concerns of nonprofit elder care organizations before Congress and federal agencies. It also strives to enhance the professionalism of elder care practitioners and facilities through the Certification Program for Retirement Housing Professionals, the Continuing Care Accreditation Commission, conferences, educational services, and publications. Founded in 1961, in 1996, the AAHSA operated with 95 staff members on an annual budget of $9.5 million. Contact person: Larry McNickle, Director of Housing Policy. Address: 901 E Street, NW, Suite 500, Washington, DC 20004-2037. Phone: (202) 783-2242. Fax: (202) 783-2255. (SEE ALSO: **Elderly**)

—Caroline Nagel ◄

► American Association of Housing Educators

The American Association of Housing Educators (AAHE) was founded in 1965 to strengthen and promote the interdisciplinary study of housing as a vital part of both consumer education and professional preparation for housing careers in the private and public sectors. The association publishes a journal, *Housing and Society,* and a newsletter, *In House* (both published three times a year); holds an annual conference; provides an educational materials service; and sponsors a student competition. Members are encouraged to become active in the various committees through which the majority of the association's work is accomplished. Members of the AAHE are involved in resident instruction, research, and extension in colleges and universities, as well as in government, industry, and elementary and secondary education. The interests of members cover a diversity of housing-related topics, including architecture, behavioral sciences, consumer education, energy conservation, engineering, historic preservation, home furnishings and equipment, human factors, interior and environmental design, political and legal aspects, real estate and finance, socioeconomic factors, and urban planning and community development. Executive Director: Jean Memken. Address: 5060 FCS Dep't., Illinois State University, Normal, IL 61790-5000. Phone: (309) 438-5802. Fax: (309) 438-5037.

—Laurel Phoenix ◄

► American Housing Survey

Purpose
The purpose of the American Housing Survey (AHS, formerly the Annual Housing Survey) is to provide a current and continuous series of data on selected housing and demographic characteristics. The AHS is the largest regular national sample that describes people and their homes in the United States. It is sponsored by the Department of Housing and Urban Development (HUD) and conducted by the U.S. Bureau of the Census, which uses interviews to gather information on over 200,000 housing units and the households that occupy them.

The AHS contains a wealth of information that can be used by professionals in nearly every housing field, whether for planning, decision making, market research, or various kinds of program development. The survey collects data on apartments, single-family homes, and mobile homes; vacant housing units; age, sex, and race of householders; income; housing and neighborhood quality; housing costs; equipment and fuels; and size of housing units. The survey also collects data on homeowners' repairs and mortgages, rent control, rent subsidies, previous units of recent movers, and reasons for moving.

Survey Design and Sample Size
The AHS is actually two surveys. There is a national survey and a survey of metropolitan areas.

The national survey is conducted biennially in housing units selected from the 1980 Census of Housing and from housing units added since the census (new construction) obtained from a sample of building permits. The 1993 national survey used a combination of computer-assisted telephone interviewing (CATI), telephone interviewing with conventional questionnaires, and personal visit interviewing involving a sample of approximately 61,000 units. The sample consisted of units included in the 1991 survey plus a sample of building permits issued since 1991. The 1993 sample also included a supplemental sample of 9,000 additional "neighbor sample" units, about 6,600 of which were last interviewed in 1989. The neighbor sample in 1985 and 1989 consisted of about 600 "kernel" units and the 10 closest neighbors to each kernel; thus, the AHS in 1985 and 1989 contained data for 600 "mini-neighborhoods." In 1987 and 1991, the AHS sample in rural areas was increased by 50% to improve the reliability of estimates of rural housing characteristics.

The metropolitan survey (MS) is conducted in 44 metropolitan areas interviewed on a rotating basis. For budgetary reasons, the 1993 survey includes only seven metropolitan areas, with a sample size of 4,700 units in each MS. The original metropolitan sample was selected from the 1970 census and updated with a sample of building permits to include housing units added since the census (new construction). The 1993 sample consists of units interviewed in previous years and supplemented by a new construction sample.

Historical Background

Interviewing for the first national survey was done in 1973, with a sample size of 60,000 housing units. In 1974, the sample size was increased by 16,000 rural units; these units were dropped from the 1981 sample because of budget constraints but were reinstated for the 1983 survey. The survey was conducted annually from 1973 to 1981; then it became biennial because of budget constraints. The national sample was redesigned in 1985 based on data from the 1980 census, with a base sample size of approximately 47,000 units and rotating supplemental samples of approximately 6,000 to 9,000 units.

The MS survey was called the standard metropolitan statistical areas (SMSAs) survey before changes were made in the definitions and composition of such areas in 1984. The original SMSA survey consisted of 60 SMSAs divided into three groups of 20 each, which were interviewed on a rotating basis beginning in 1974. Each group had a total sample size of 140,000. Budget constraints forced a change to four groups of 15 SMSAs beginning in 1978. Further budget constraints in 1982 required a reduction in the number of SMSAs to be interviewed yearly and a reduction of approximately 50% in the sample size. Plans project future interviews in 45 areas once every four years. The name of the survey was changed to the American Housing Survey starting with the 1984 metropolitan sample.

Respondents

For occupied units, the respondent must be a knowledgeable household member 16 years of age or over. The respondent provides the information on the unit and the household.

For vacant units, the respondent can be the landlord, owner, real estate agent, or if these people are not available, a knowledgeable neighbor.

Special Features

The AHS uses longitudinal interviewing. The interviewers go back to the same housing units each interview period, recording changes in the characteristics of the units and their occupants. This approach gives a picture of the homes and households as they change over time. Since 1987, the national survey has used computer-assisted telephone interviewing (CATI).

Data Products

The results of the national survey are published in series H150 and H151. These reports present data for the United States and for the four census regions. Statistics are also provided for subgroups using the following distinctions: central city/suburban/nonmetropolitan, urban/rural, owner/renter, black, Hispanic, over 65 years of age, recent movers, and households below poverty level.

The results of the MS surveys are published in series H170 and H171. Statistics are disaggregated as above for the national survey. Each report also shows statistics for the three largest jurisdictions in the metropolitan area.

Microfiche copies of both the national (H150 and H151) and metropolitan (H170 and H171) reports are also available.

The Bureau of the Census occasionally issues special analytical reports using data from the AHS. These reports are published in series H121 and H123. Examples of these reports are *Home Alone in 1989*, *Housing Characteristics of Recent Movers*, *Housing Characteristics of Selected Races and Hispanic Origin Households in the United States: 1987*, *Housing Characteristics of One-Parent Households: 1989*, and *Homeowners and Home Improvements: 1987*.

For the data user whose needs are not met by the reports, there are magnetic computer tapes and CD-ROMs with copies of each respondent's answers (microdata), so these answers can be tabulated by computer programs in any way desired. To protect the confidentiality of the respondents, all names and addresses are removed and geographic areas with fewer than 100,000 people are not identified. The sample design generally will not support analysis for areas smaller than those shown in the reports.

The *American Housing Briefs* are a series of short, nontechnical fact sheets presenting the latest housing data complete with bar charts and graphs for each metropolitan area beginning with the 1988 group. Also available are *Statistical Briefs* that give data for the United States as a whole on topics of current interest (such as first-time homeowners, new homes, and recent movers).

The Bureau of the Census also publishes a one-page data chart (24 × 36 inches). The data chart presents a complete distribution of data for each characteristic collected in the AHS for the United States. Separate charts are available for each national survey starting with the 1987 survey.

Product Availability

The American Housing Survey products are available from a variety of sources:

HUD User (800-245-2691 or 301-241-5141)
 Box 6091
 Rockville, Maryland 20850
 (301-763-8551)
Bureau of the Census
Housing and Household Economic Statistics
 Division
 Washington, DC 20233-3300
Bureau of the Census
 Customer Services (301-763-4100)
 Washington, DC 20233-5300
Superintendent of Documents (202-783-3238)
 Washington, DC 20402-9325.
(SEE ALSO: *U.S. Bureau of the Census*)

—*Edward D. Montfort*

Further Reading

U.S. Department of Housing and Urban Development. Biennial. Current Housing Reports, H-150. *American Housing Survey*. Washington DC: U.S. Department of Commerce, Bureau of the Census. ◀

► The American Institute of Architects

The American Institute of Architects (AIA) is an organization founded in 1857 to unite the nation's architects, to promote the architecture profession, to advance research in the field, and to coordinate the building and design industries. The AIA circulates a monthly newsletter to its members and maintains a 30,000-volume lending library, including slide and video collections. Promotional materials and books on a variety of design topics are available through the AIA Bookstore.

For over 100 years, AIA committees have been addressing professional issues. The AIA launched the Professional Interest Areas (PIAs) program in 1993 to promote increased participation by all AIA members. PIAs form the knowledge base of the AIA as the principal sources of information, expertise, research, development, and education on specific practices and careers in architecture. PIAs provide resources, forums, and opportunities for the collection of information to benefit all members of the PIAs, the membership at large, the construction industry, and the public.

The Housing Professional Interest Area (Housing PIA) is an alliance of architects involved in all facets of housing: custom housing, affordable housing, codes and regulations, community preservation, and environmentally sensitive housing. Developments designed by members include both multifamily and single-family housing, as well as shared housing arrangements for special needs populations. Members act as the AIA's ambassadors to other organizations, including the National Association of Homebuilders, the American Planning Association, Urban Land Institute, and the Construction Specifications Institute. AIA members and the general public are invited to attend workshops and conferences sponsored by the Housing PIA. For complete information on the Housing PIA's activities (and all other PIAs), call 1-800-242-3837.

In the area of affordable housing, the AIA's reach extends beyond the Housing PIA. In 1986, the AIA, in partnership with the American Institute of Architectural Students (AIAS) sponsored the "Search for Shelter Program." Search for Shelter contributed to the national dialogue on solving the crisis of homelessness and the critical lack of affordable housing. From 1987 to 1991, 83 Search for Shelter projects (i.e., designs for decent, safe, affordable housing and economic development program models for homeless and very low income people) were presented in public exhibits across the nation. AIA local chapters and components played a pivotal role in sharing this work with the public.

In 1995, at the AIA's National Convention in Atlanta, the AIA (national) and AIA Atlanta collaborated with Habitat for Humanity to build six houses on-site. Called the "Legacy Project," this new program's mission had a goal of leaving behind an architectural residual of lasting value and service to the host community of an AIA National Convention. The Legacy Program benefited from an expanded vision in 1997 when the New Orleans AIA chapter created a multifaceted residential service center and park for homeless people in New Orleans. Legacy Projects are also planned for (1998) San Francisco and (1999) Dallas conventions.

A standing AIA Task Force on Affordable Housing is appointed by the board of directors. The AIA is one of 57 organizations in partnership with the White House and the U.S. Department of Housing and Urban Development to promote homeownership in America. The AIA has approximately 60,000 members and operates on an annual budget of $135 million. Staff size: 150. Contact: The American Institute of Architects. Address: 1735 New York Avenue NW, Washington, DC 20006. Phone: (202) 626-7300. Fax: (202) 626-7426. (SEE ALSO: *Architectural Review Boards*)

—Cheryl P. Derricotte ◄

► American Land Title Association

The American Land Title Association (ALTA) is a national organization founded in 1907 to encourage the safe and efficient transfer of property ownership by informing and educating consumers, regulators, legislators, and members. ALTA publishes *Title News,* a bimonthly magazine, as well as a variety of educational materials, including correspondence courses. Membership: 2,000 corporations and individuals nationwide. Staff size: 17. Contact person: James R. Maher, Executive Vice President. Address: 1828 L Street NW, Suite 705, Washington, DC 20036. Phone: (202) 296-3671. Fax: (202) 223-5843.

—Caroline Nagel ◄

► American National Standards Institute (and Other) Accessibility Standards and Housing

The American National Standards Institute (ANSI) is a national coordinating body charged with establishing design standards for construction or manufacturing in a wide range of public and private industries. The ANSI accessibility standard for people with disabilities is one of more than 10,000 standards approved by ANSI. ANSI standard for accessibility A117.1 dates back to the early 1960s when an ANSI committee arrived at a national consensus for minimum accessibility guidelines. For public facilities, the early ANSI standard did not address access to housing. Today, ANSI, through periodic revisions, has evolved into a comprehensive set of accessibility guidelines that have been adopted by federal agencies as well as for most state and local building codes and that set an important context for residential environments. The recent Americans with Disabilities Act (ADA) Accessibility Guidelines (1991) and the ADA Standards for Accessible Design were in large part based on the 1986 revision of the ANSI accessibility standard.

Inception of the ANSI Accessibility Standard
In 1961, the first ANSI standard for accessibility (ANSI A117.1, dubbed "American National Standard Specifications for Making Buildings and Facilities Accessible to and Usable by the Physically Handicapped") was approved by a joint committee composed of 35 representatives from

professional and trade associations, government agencies, associations representing people with disabilities, and other health-related organizations.

It marked a turning point. All or part of the original or revised standards have since become the foundation for federal guidelines and for state and local building codes in all 50 states and the District of Columbia. Compliance with the standard is voluntary unless it has been adopted by or referenced in legislation or regulations that control design and construction of facilities.

Revisions of the 1961 Standard

Since 1961, ANSI A117.1 has been reviewed and revised periodically. By the mid 1970s, there were approximately 65 different codes and regulations for accessible design in the United States, and most of them required very different solutions to similar accessibility problems.

In response to the need for uniformity, in 1974 the secretariat for the ANSI A117.1 standard, funded by the Department of Housing and Urban Development (HUD), began research necessary to evaluate and revise the 1961 standard. The review added the issue of accessible housing for people with disabilities and folded in additional technical specifications. These included requirements for kitchens and bathing facilities, the environmental needs of people with hearing and visual impairments, and more detailed illustrations.

What began in 1961 as a 6-page discussion of very basic issues—toilet stalls, parking spaces, water fountains, entrances, elevators, and ramping—emerged in 1990 as a 60-page document addressing site planning as well as building and interior design with detailed specifications gleaned from studies of people with disabilities performing daily living tasks.

In addition, the 1980 and 1986 ANSI A117.1 revisions introduced the concepts of accessible routes and accessible elements. These issues need to be addressed during the initial design stages of a construction project. Basically, an accessible route is a continuous unobstructed path connecting accessible spaces and elements. An element may be a telephone, fire extinguisher, water fountain, or other item required by code to be accessible.

In 1984, the ANSI (1980) technical specifications, including those for housing, were largely incorporated into the new Uniform Federal Accessibility Standards (UFAS), the current standard for all federally funded construction subject to the Architectural Barriers Act of 1968 and one of the applicable standards under the program accessibility requirement of Section 509 of the Rehabilitation Act of 1973 and the local and state requirements of Title III of the ADA of 1990. The major difference between ANSI A117.1 and UFAS are the "scoping" provisions added to the UFAS, which detail the method for determining the required number and location of accessible bathrooms, parking spaces, theater seats, and so on, based on the size and type of building project. During the regular five-year review for ANSI A117.1 in 1985, the ANSI committee adopted some of the UFAS refinements to ensure increased uniformity in access requirements in both federal and private construction. In 1992, the Council of American Building Officials

(CABO), as the new secretariat for ANSI 117.1, brought refinements from the ADA guidelines in the revised CABO/ANSI117.1, titled "American National Standard Accessible and Usable Buildings and Facilities."

ANSI and Accessible Housing

Prior to 1980 and the updated ANSI standard, national and federal standards for accessibility did not include housing. During this period, many states and federal agencies had developed their own specifications for fixed accessible housing units. Many of these requirements focused on features for wheelchair users only and mandated that a small percentage of new, multifamily units must comply. However, there were significant problems with these housing regulations. To begin with, they excluded all of the existing housing stock. Furthermore, the disabled people for whom the new accessible units were intended often could not rent them because the rent was too high. Also, if only one bedroom units were accessible (as was often the case), families with a disabled individual or a disabled person with an attendant could not use them. If provisions forbade the rental of these units to nondisabled persons, landlords lost money, thus discouraging the private sector from complying with the law.

To remedy these problems, both the ANSI standard research team and the disability community have evolved a more flexible standard for accessible housing—namely, adaptable housing. An adaptable housing unit includes all of the accessibility features required by ANSI (1986) and UFAS (1984), such as wider doors, clear floor space, and an accessible route, while allowing a choice of certain adjustable features or fixed accessible features, such as adjustable height countertops, closet rods, and closet shelves. Cosmetically no different from a standard unit, an adaptable housing unit is thus marketable to everyone.

Standards for accessible housing have kept pace with the recent disability nondiscrimination laws. The Fair Housing Act Amendment of 1988 prohibits discrimination in housing against people with disabilities. Its fair housing design guidelines, based on earlier ANSI A117.1 specifications, require a low level of accessibility and some adaptability for most multifamily apartments on ground floors and all apartments in buildings with elevators. The Americans with Disabilities Act of 1990 (ADA) prohibits discrimination in a broader range of activities. The law interprets inaccessible buildings as one form of discrimination and establishes the ADA Standard for Accessible Design as mandatory nationwide requirements. The ADA requires some forms of housing, including temporary facilities such as hotels, motels, and dormitories, to be accessible but does not address single- and multifamily dwelling units. The ADA standard is based on ANSI A117.1 and UFAS. The Fair Housing guidelines also derived from early ANSI specifications directly reference ANSI A117.1 (1996) for public and common use areas of multifamily projects. Today, UFAS is largely the standard applicable to housing for which federal money is used and when housing is perceived as part of a federal, local, or state program. (SEE ALSO: *Barrier-Free Design; Physical Disabilities, Housing of Persons With*)

—Ronald L. Mace

Further Reading

American National Standard Accessible and Usable Buildings and Facilities (CABO/ANSI A117.1-1992). 1992. New York: American National Standards Institute.

American National Standard Specifications for Making Buildings and Facilities Accessible to and Usable by Physically Handicapped People. 1980, 1986. ANSI A117.1-1980, 1986. New York: American National Standards Institute.

The Americans with Disabilities Act (ADA) Accessibility Guidelines for Buildings and Facilities. 1991. In "Standards for Accessible Design." *Federal Register* 56(144): 35605-35691.

Barrier Free Environments. 1987. *Adaptable Housing: A Technical Manual for Implementing Adaptable Dwelling Unit Specifications.* Prepared for the U.S. Department of Housing and Urban Development. Raleigh, NC: Author.

———. 1996. *Fair Housing Act Design Manual: A Manual to Assist Designers and Builders in Meeting the Accessibility Requirements of the Fair Housing Act.* Prepared for the U.S. Department of Housing and Urban Development. Raleigh, NC: Author.

"Final Fair Housing Accessibility Guidelines." 1991. *Federal Register* 56(44):9472-95115.

"Uniform Federal Accessibility Standards." 1985. *Federal Register* 49. ◀

▶ American Planning Association

The American Planning Association (APA) is a nonprofit public interest and research organization established to advance the art and science of planning, and to foster physical, economic, and social planning at the local, regional, and national levels. It resulted from a merger between the American Institute of Planners, founded in 1917, and the American Society of Planning Officials, established in 1934. Its 30,000 members—including practicing planners, elected and appointed officials, and citizens—share a commitment to the use of sound planning to meet the nation's economic and community development needs, to conserve resources, and to preserve the environment.

The organization is composed of 46 regional chapters and 16 divisions of specialized planning interests. The Housing and Human Services Division serves those concerned with the social issues of planning and focuses on improving the policy framework and practical implementation of planning as it relates to housing and human services. This division's interests include housing, neighborhoods, demographics, the elderly, minorities, children, and the homeless. It publishes the *Housing and Human Services Quarterly Newsletter.*

APA offers housing-related information through all its services, including training, conferences and seminars, policy and legislative research, and publications. It publishes *Planning,* a monthly magazine; *Zoning News,* a monthly newsletter on local land use control; the *Journal of the American Planning Association,* published quarterly; and *Land Use Law,* a monthly publication highlighting land use regulations.

It is funded through membership dues, journal subscriptions, and conference and research fees. It has offices in Chicago and in Washington, D.C., with 75 employees. Contact Frank S. So, AICP, Executive Director, 122 South Michigan Avenue, Suite 1600, Chicago, IL 60603. Phone:

(312) 431-9985. Fax: (312) 431-9100. Or Jeffrey L. Soule, AICP, Policy Director, 1776 Massachusetts Ave., NW, Suite 400, Washington, DC 20036. Phone: (202) 872-0611. Fax: (202) 872-0643.

—Sarah E. Polster ◀

▶ American Real Estate and Urban Economics Association

The association publishes a quarterly journal on scholarly research in real estate issues, including housing, to facilitate communication among academic researchers and industry professionals. Submission Editors: Joe Gyourko and Susan M. Wachter. Address: Wharton Real Estate Dept., University of Pennsylvania, Lauder-Fischer Hall; 256 S. 37th St., Philadelphia, PA 19104-6330. Phone: (215) 898-2841. Fax: (215) 573-4062. (SEE ALSO: *Real Estate Developers* and *Housing*)

—Laurel Phoenix ◀

▶ American Seniors Housing Association

The American Seniors Housing Association (ASHA) is a national association of corporations and agencies in the seniors' housing sector founded in 1992. Affiliated with the National Multi Housing Council. ASHA is dedicated to (a) promoting policies and legislation favorable to the development and preservation of seniors' housing; (b) disseminating information on tax and finance laws, building codes, and regulations that affect seniors' housing; and (c) fostering the exchange of information and ideas in the industry through clearinghouse services, publications, a comprehensive library, and annual meetings. It publishes *Seniors Housing Update,* a bimonthly newsletter alerting members to trends, legislation, and regulatory developments in the seniors' housing industry; an annual seniors' housing report; and periodic reports and advocacy pieces on issues of concern. ASHA operates on an annual budget of $750,000 and is funded by membership dues. Membership: 260 corporate members nationwide (1995). Staff size: 18. Executive Director: David S. Schless. Address: 1850 M Street, NW, Suite 540, Washington, DC 20036. Phone: (202) 659-3381. Fax: (202) 775-0112. (SEE ALSO: Elderly)

—Caroline Nagel and Laurel Phoenix ◀

▶ Amortization

Using the standard fixed-rate mortgage (FRM), each debt service payment is divided into two portions: the interest payment (i.e., the payment *on* the mortgage balance) and the principal payment (i.e., the payment *of* the mortgage balance). The principal payment is sometimes called the amortization payment. The systematic repayment of an FRM is termed the *amortization* of the mortgage loan.

Although the algorithm used to amortize loans was developed in the 19th century, not until 1934 did the Federal Housing Administration (FHA) create what has come to be known as the FRM. Only then did amortizing loans became popular and important. Because the FRM payment can be divided into interest and a sinking fund payment used to repay the loan (principal), lenders are more comfortable with this type of self-amortizing mechanism compared with other instruments. This is because with each payment, a portion of the outstanding balance of the loan is repaid and the amortization portion directly reduces the lender's exposure to losses.

The FRM has a constant total debt service payment each period. The sum of interest and principal payments will always equal the debt service payment. However, each payment has a different percentage of interest and principal. This is because the amortization procedure requires that the debt service payment be applied first to satisfy the interest owed for the period; any money remaining can be applied directly to the principal balance. Therefore, as the outstanding balance of the loan is reduced, the amount of interest due each period falls. With a constant payment, the amount of principal (amortization payment) increases.

Some observers have relied on the "cross-over" point (i.e., the date at which the percentage of debt service used to pay interest is equal to the percentage allocated to principal) to signal the time to refinance. Modern analysis has shown that this is not a good refinancing rule; the time when principal is equal to or greater than interest is simply determined by the parameter values of the level of interest and repayment term. It is now known that the decision to refinance must come from some other type of analysis. (SEE ALSO: **Mortgage Finance**)

—*Austin J. Jaffe*

Further Reading

Brueggeman, William B. and Jeffrey D. Fisher. 1993. *Real Estate Finance and Investments.* 9th ed. Homewood, IL: Irwin.
Jaffe, Austin J. and C. F. Sirmans. 1995. *Fundamentals of Real Estate Investment.* 3d ed. Englewood Cliffs, NJ: Prentice Hall.
McNulty, James E. 1988. "Mortgage Securities: Cash Flows and Prepayment Risk." *Real Estate Issues* 13 (Spring/Summer):10-17.
Sirmans, C. F. 1989. *Real Estate Finance.* 2d ed. New York: McGraw-Hill. ◀

▶ Appraisal

An appraisal is a systematic estimate of the value of real property, usually conducted for a fee by a professional valuator working in the private sector. It differs from an assessment, which is an estimate of property valuation conducted by a government official. Appraisals are most often conducted in preparation for the sale, loan, or other transfer of real property, such as eminent domain takings. In addition, property appraisals may be conducted to determine property value for purposes of insurance, improvement, or liquidation.

An appraisal may be calculated in several ways, depending on the need and reason for conducting the appraisal. The calculations may include a determination of original cost, replacement cost, market comparison, or income generation. (SEE ALSO: *Appraisal Institute* [AI]; *Carrying Costs; Closing Costs; Real Estate Agency*)

—*John S. Klemanski and C. Michelle Piskulich*

Further Reading

American Institute of Real Estate Appraisers. 1987. *The Appraisal of Real Estate.* 9th ed. Chicago: Author.
American Institute of Real Estate Appraisers. 1989. *The Dictionary of Real Estate Appraisal.* 2d ed. Chicago: Author.
Kinnard, Jr., William N., ed. 1986. *1984 Real Estate Colloquium: A Redefinition of Real Estate Appraisal Receipts and Processes.* Cambridge, MA: Lincoln Institute of Land Policy. ◀

▶ Appraisal Institute

Founded in 1991 with the unification of the American Institute of Real Estate Appraisers and the Society of Real Estate Appraisers, the Appraisal Institute (AI) serves the public by conferring membership designations to properly qualified real estate appraisers and by enforcing a code of standards and ethics on its members. The AI also represents its members before Congress and government agencies and promotes research in the appraisal field. It publishes several periodicals, as well: the *Appraisal Journal,* which contains peer-reviewed articles on commercial and residential appraisal, appraisal law, and international finance; *Quarterly Byte,* which presents information on applied computer technology in the field; and *MarketSource,* which features real estate market data and forecasts. In 1993, the Appraisal Institute had 32,000 members in North America and operated on an annual budget of $23 million with a staff of 170. Address: 875 North Michigan Avenue, Suite 2400, Chicago, IL 60611. Phone: (312) 335-4100. Fax: (312) 335-4400. (SEE ALSO: *Appraisal*)

—*Caroline Nagel* ◀

▶ Architectural Review Boards

Architectural review boards establish design regulations to control development in a community's built environment. Unbridled development will produce disparities between design quality and solutions by builders motivated by profit seeking. Many developments in the United States are not regulated by planning or zoning boards, leaving architectural review boards to provide additional checks and balances for design control.

Public and private architectural review boards differ in their goals, authority, and power. Public architectural review boards generally fall under the direct authority of either planning or zoning departments within a local or municipal government's structure, demanding mandatory compliance to their decisions. Private architectural review boards gain their authority and power from a contractual agreement attached to the covenants in a parcel of land's deed of trust.

Public Architectural Review Boards

Governmental power to regulate building, zoning, and development generally stems from a municipal government's structure as adopted and enforced by the planning, building, and zoning departments. State codes and zoning laws do exist in several regions of the United States, but the power to adopt and enforce those codes lies with the municipalities.

Small municipalities with limited resources may not possess the means to create and enforce planning regulations or zoning laws, allowing development to proceed without proper controls. In these cases, design issues such as density, use, height, materials, parking, and access are the responsibility of the developers and builders. Even where departments adopt regulations, these generally focus on broad issues, such as the type of land use and general density or parking requirements.

In these communities, specific design quality issues, such as sustainability, a building's bulk and/or size, solar rights, materials, facade, quality, or historic preservation value, remain at the discretion of the developers. Progressive communities with a long-range vision or comprehensive plan supported by financial resources will address these specific design questions through public design review boards or historical commissions.

Public architectural reviews, conducted in public hearings, encourage and solicit adjacent property owners' and community members' input. Should the board disapprove the project, it may request reworking of the design, or an appeals board may review the architectural review board's decision at the developer's request.

Private Architectural Review Boards

The gap created by the lack of publicly enforced control is often filled by private architectural review boards, enforced by a homeowners' association. Their power lies within the covenants that are attached to the deed of trust for the land. Purchasing land or in the case of condominiums air rights within that plat of land, binds the purchaser to the provisions of the covenants. This is a contractual agreement and, as such, the purchaser of the land agrees to a review by the architectural review board. Often, these private review boards enforce far more restrictive codes and aesthetic details than the public boards.

Private review boards can exist in conjunction with public planning and zoning reviews. In this case, the most restrictive requirements govern. The agendas of private review boards may extend far beyond public requirements. Private boards generally examine aspects of open space, landscaping design, facade design, floor area, entries, windows, rooflines, massing, bulk plane, building materials, and access, demanding compliance to specific requirements. The only recourse for disagreement is through the courts. Although private review boards' methods may be strict, the results often create carefully designed communities.

(SEE ALSO: *American Institute of Architects]; Custom-Built Homes; Subdivision Controls*)

—Fred Andreas

Further Reading
Calthorpe, P. 1993. *The Next American Metropolis: Ecology, Community, and the American Dream.* New York: Princeton Architectural Press.
Katz, P. 1993. *The New Urbanism: Towards an Architecture of Community.* New York: McGraw-Hill.
Krieger, Alex. 1991. *Towns and Town-Making Principles: Andres Duany and Elizabeth Platter-Zybeck, Architects.* New York: Rizolli.
Scheer, B. C. and W. F. E. Preiser, eds. 1994. *Design Review: Challenging Urban Aesthetic Control.* New York: Chapman & Hall.
U.S. General Accounting Office. 1977. *Procedures Used for Holding Architects and Engineers Accountable for the Quality of Their Design Work.* Washington, DC: Author. ◄

► Architecture and Behavior/Architecture et Comportement

Architecture and Behavior/Architecture et Comportement is an English and French language journal concerned with the interests of architects, planners, and scholars of environmental design. The journal is devoted to the dissemination of multidisciplinary and usable research on the relationships between human beings and their built environment, including housing. It was first published in 1980. Circulation: 700 internationally; up to 3,000 for special issues. Price: 150 Swiss Francs per year for institutions; 70 Swiss Francs per year for individuals. Editor: Kaj Noschis, Ph.D., Department of Architecture, Swiss Federal Institute of Technology, P.O. Box 555, 1001 Lausanne, Switzerland. Phone: 41 21-693 4225. Fax: 41 21-617 6317. (SEE ALSO: Behavioral Aspects)

—Caroline Nagel ◄

► Asian Americans

Overall Patterns

Housing conditions of Asian Americans are generally inferior to those found among white households in the United States. Asian American households have lower homeownership rates, live in smaller and more crowded units, are less concentrated in suburban areas, and pay more to own or rent housing. In 1990, just over 62% of white households owned homes, but only 53% of Asians American households were homeowners. Nearly a quarter of Asian American households had more than one person per room, compared with only 3% of whites. Although nearly two-thirds of metropolitan whites live in the suburbs, less than half (47%) of Asian Americans in metropolitan areas are suburbanites. At the same time, housing costs incurred by Asian Americans are high, especially among those who own homes. A median of 23% of Asian American homeowners' household incomes is spent on housing (compared with only 18% for white homeowners), and the median percentage of income expended on housing by Asian renters is 28% (2% higher than among whites).

Explaining the Patterns

This general portrait of the housing of Asian Americans reflects several factors that characterize the social position of this population in U.S. society—most important, immigration and the regional and metropolitan concentration of this group. Asian Americans are largely a recent immigrant group with the largest portion of all immigration having taken place since 1965. Immigrants often live in more crowded housing; they double up because of strong familial and ethnic ties used in the migration process and because of limited economic resources. Weak knowledge of the operation of the housing market, limited financial assets, and a desire to use one's resources for other types of socioeconomic mobility such as entrepreneurship also make homeownership and suburban residence less likely for immigrants.

Characteristics of immigrants explain much of the inferior housing of the Asian American population. Improved English language ability and increased economic resources, both of which occur with longer residency in the United States, have large effects on increasing levels of homeownership within Asian American populations. In 1990, only 29.6% of Asian immigrants who arrived in the United States within the prior 10 years owned a home. However, nearly 70% of Asians arriving between 1970 and 1979 were homeowners by 1990. Crowding is also substantially reduced with the improved economic position that accrues with a longer stay in the United States. And suburbanization levels increase as Asian households become more assimilated and improve their socioeconomic status.

These national patterns for Asian Americans mostly reflect housing conditions in large metropolitan areas in the West, New York, and Hawaii. The largest portion of all Asian American households reside in metropolitan areas (94% in 1990), and a large majority live in metropolitan areas with a population of over 1 million (67% in 1990). Also, three-quarters of the Asian American population lives in the West (39% in California alone in 1990), New York, and Hawaii. The extraordinarily high housing costs and more modest homeownership rates in large metropolitan areas where a disproportionate share of Asian Americans live, such as Honolulu, New York, Los Angeles-Long Beach, and the San Francisco Bay area, increase the prices these households pay and make it more difficult to rapidly improve their housing status.

Residential Segregation

Despite the conditions described above, the residential segregation of Asians from whites is relatively modest. In the typical metropolitan area in which Asians resided in 1990, approximately 40% of either Asians or whites would have had to move to a different neighborhood for all areas to have had the same percentage of Asians. This is substantially less segregation than found among African Americans. In 1990, nearly 70% of metropolitan blacks (or whites) would have had to change their neighborhood of residence to achieve an even distribution. Areas with sizable and growing Asian populations, such as Los Angeles, have somewhat higher than average segregation (46% of Asians or whites there would have to move to eliminate Asian/white segregation), but these levels are never very high anywhere in the country. Although Asian/white residential segregation increased between 1980 and 1990, this trend is explained by the large flows of recent Asian immigrants to U.S. metropolitan areas. Coethnic ties among immigrants increase segregation because they channel group members into similar neighborhood locations. The socioeconomic mobility that generally occurs with continued duration of residence in the United States leads to consistent declines in Asian/white residential segregation.

Asian Subpopulations

Much of what we know about Asian Americans, especially in terms of segregation and housing conditions, derives from sources that examine this population as a single group. However, the Asian American population comprises diverse national origin groups, the largest of which are Chinese, Filipino, Japanese, Asian Indian, Korean, and Vietnamese Americans. The diverse current and historical experiences of these groups have significant implications for their residential conditions. In particular, Japanese Americans have a long history in the United States and low levels of immigration at present. Japanese also have higher socioeconomic status. Taken together, these factors lead to much better housing outcomes than among other groups. Compared with all of the other Asian American populations, Japanese Americans have a higher rate of homeownership (60%), have a lower level of crowding (only 5% of households having more than one person per room), and are the least segregated from whites. Sizable immigration over the last two decades has produced much more crowding and somewhat more segregation among Chinese, Filipinos, Asians Indians, and Koreans. But these and other residential conditions, like homeownership and suburbanization, are strongly influenced by economic factors (e.g., income) and immigrant characteristics (e.g., English language ability) in each of these populations.

Vietnamese Americans provide a contrast. Homeownership rates among Vietnamese are relatively low and only partly explained by economic, familial, and immigrant characteristics. Vietnamese Americans also are more segregated from whites and from other Asian American populations. The most disadvantaged Asian American groups are the small Cambodian, Laotian, and Hmong populations. These groups increased at astounding rates between 1980 and 1990 (819%, 210%, and 1,700%, respectively), but there were still only 147,000 Cambodians, 149,000 Laotians, and 90,000 Hmongs in the United States in 1990. The combination of rapid growth through immigration with low economic and educational attainment and sometimes extremely difficult paths of immigration (for example, the Cambodian experience) resulted in obviously depressed housing conditions. For example, in 1990, only 12% of Hmong households were homeowners and the vast majority (78%) lived in units with more than one person per room. Among Cambodian and Laotian households, respectively, 24% and 30% owned homes and 61% and 55% were crowded (more than one person per room). Given the recency of their immigration and their small size, it is difficult to predict how the housing situation of these groups will change.

Conclusion

In sum, although the current picture for Asian Americans is mixed, most of the evidence suggests that assimilation to ever-better housing conditions is the likely path for most groups within this population. However, continuing immigration may produce poorer housing for immigrants themselves and also may dampen the housing mobility of native Asian Americans. One study of Hispanic housing found just this pattern: Latino households obtained better housing as they became more assimilated; but living in a metropolitan area with more non-English speakers and recent immigrants made it more difficult for natives or immigrants to improve their housing. This may occur because of greater stereotyping and discrimination against all Latinos in areas with large influxes of immigrants or because of greater pressure on the housing stock. The same processes could operate for Asian Americans. However, any pattern is unlikely to be identical for all Asian populations. It is critical that future analysts explore the variation in detail to provide a more complete understanding of the role of immigration, ethnicity, and race in the housing market experiences of immigrant and minority groups in the United States. (SEE ALSO: **Discrimination;** *Racial Integration; Segregation*)

—*Lauren J. Krivo*

Further Reading

Alba, Richard D. and John R. Logan. 1992. "Assimilation and Stratification in the Homeownership Patterns of Racial and Ethnic Groups." *International Migration Review* 26:1314-41.

———. 1991. "Variations on Two Themes: Racial and Ethnic Patterns in the Attainment of Suburban Residence." *Demography* 28:431-54.

Frey, William H. and Reynolds Farley. 1996. "Latino, Asian, and Black Segregation in U.S. Metropolitan Areas: Are Multiethnic Metros Different?" *Demography* 33:35-50.

Myers, Dowell and Seong Woo Lee. 1996. "Immigration Cohorts and Residential Overcrowding in Southern California." *Demography* 33:51-65.

Woodward, Jeanne M. 1994. America's Racial and Ethnic Groups: Their Housing in the Early Nineties. *Current Housing Reports* (H121/94-3). Washington, DC: U.S. Department of Housing and Urban Development. ◄

► Assisted Living

Assisted living is a planned retirement housing option in the United States offering shelter, protective oversight, personalized assistance, and health care for physically or mentally vulnerable older people. Although professionally managed, it usually offers a less institutionalized and less medically oriented environment than found in nursing homes. Ideally, it seeks to promote the maximum independence and personal autonomy of its residents in a homelike environment. Facilities are variously owned by public, nonprofit, or proprietary entities and may be part of a chain operation. Federally insured mortgage insurance financing (under the Section 232 program) became available in late 1992 for developers of these facilities. Although some observers argue that frail elderly persons in the United States have been accommodated in such facilities since the 1930s, the term *assisted living* is itself new and emerged in the 1980s. The private sector has increasingly favored the assisted-living label to distinguish its facilities from the mom-and-pop "board-and-care" option that is often associated with a lower-income and less upscale elderly consumer market.

The assisted-living option is identified by a confusing array of labels, usually a function of a state's licensing or regulatory requirements. Among the most common labels: residential care facilities, personal care homes, catered-living facilities, retirement homes, home for adults, board-and-care homes, domiciliary care homes, rest homes, community residences, and sheltered care. An increasing number of state governments, however, use the term assisted living in their regulatory language. Private developers, nonprofit sponsors, management firms, service providers, and policymakers are currently struggling to agree on a uniform definition of this living alternative. Confusion also exists because experts often do not carefully distinguish between "ideal" (what it should look like) and "actual" (what's actually out there) models of assisted living. The varied names of these facilities, the lack of agreement as to how they look and what services they provide, and uncertainty as to what percentage of their residents are elderly (as opposed to nonelderly mentally or physically disabled) result in varying estimates of the number of assisted-living facilities found in the United States (from 35,000 to 45,000) and the number of their elderly occupants (from 500,000 to 1,000,000).

Facilities usually consist of multi-unit buildings that accommodate from 20 to 100 older persons, but some are occupied by as many as 300 persons. They range from one- to three-story buildings to high-rise buildings greater than nine stories. The building may have been specially created for assisted care or adaptively remodeled from another use. The assisted-care facility sometimes is located in a separate wing or floor of a congregate housing facility or of a nursing home or housed in a building that stands alone. They are found in all such ways on the campus grounds of continuing-care retirement communities. Assisted-care services are also provided in some low-rent elderly housing projects funded by federal and state housing programs.

Accommodations in assisted-living facilities range from single rooms to small studio and one-bedroom apartments with toilet and bathing facilities usually available in each unit. These units differ as to whether they have kitchen facilities, are occupied by more than one resident, and whether residents can lock their own doors. Common living areas for resident use may be similar to those found in adult congregate-living facilities and include living rooms, libraries, social halls, and television lounges. Special design features (e.g., wider doorways, lever handles, grab-bars, emergency call systems) are often found both in individual units and in the public areas. Some facilities have more of a residential feel about them, whereas others more resemble nursing homes.

Most assisted-living facilities offer services such as housekeeping, personal laundry, maintenance, transportation services, and structured leisure activities. They will also offer two or three daily meals in a central dining room. The types and level of care offered to residents, however, varies widely among facilities, reflecting the regulatory environment of their respective state agencies. In some states, assisted living involves the delivery of services to residents with needs for light care; in others, it is seen as almost replacing the nursing home level of care. Lower-end services include personal care, such as help with bathing, dressing, eating, getting in and out of bed and chairs, walking, going outdoors, using the toilet, and assistance with medications. Higher-end services include nursing services such as skin care, dressing changes, health assessment and monitoring, specialized programs for incontinence, programs for memory impairment, and the monitoring of less stable medical conditions. Some facilities use permanent staff to provide these services, whereas others depend on third-party providers (e.g., physical and speech therapists from home health agencies). Nearly all facilities, however, will have at least one staff member awake and on duty all night to meet unscheduled, unpredictable needs.

Typically, older persons enter these facilities in their late 70s and early 80s and are disproportionately women living alone. Their occupants have a median age of about 83. Older persons enter these facilities because they have disabilities making it difficult or impossible for them to function safely in their own homes. Candidates for this option may suffer from chronic health problems (such as arthritis, hypertension, diabetes, stroke-related paralysis, incontinence, musculoskeletal disorders) that result in their needing personal care and health assistance. They may depend on a walker or be confined to a wheelchair, but they are not permanently bedridden or infirm and do not require the 24-hour skilled care of the nursing home. Alternatively, they may have few physical impairments but may be suffering from dementia, such as Alzheimer's disease. Thus, they may be confused, have memory lapses, have difficulty comprehending their environment, and subsequently experience restlessness, irritability, and display behavioral problems. They often need reminders to take their medications and supervision when doing so. The extent of frailty of the population admitted to and retained in these facilities will depend much on the types and level of assisted care provided. Residents are discharged from these facilities for three major reasons: They die, enter a hospital, or relocate to a nursing home.

Most assisted-living facilities currently cater to private-pay, middle-class and higher residents who can afford monthly fees ranging from $700 to over $3,000 a month. The cost will depend on the size of unit, single or double occupancy, extensiveness of offered services, and luxuriousness of facility. A small number of assisted-living facilities also accommodate lower-income elderly residents. They subsidize the cost of their services with funds from federally supported Medicaid waivers or state-funded long-term care programs.

The growth of this housing alternative is attributed to several factors: (a) a larger frail older population needing help with their activities of daily living, (b) the practical difficulties of providing personal care in older people's homes, (c) the increased inability of families to serve as caregivers (e.g., because of workplace obligations), (d) the absence of stringent federal regulations dictating the level or quality of care offered in this alternative, (e) the desire by state governments to encourage less costly alternatives to Medicaid-subsidized nursing home beds, and (f) the perception by families, older people, and professionals that assisted-care facilities offer a more humane and autonomous living environment than nursing homes. A major growth stimulus would be the widespread availability of Medicaid subsidies lowering the elderly consumer's cost of this alternative. Although these factors are likely to fuel the expansion of this alternative, some experts identify countervailing forces. They point to the possible introduction of federal or state regulations that could make this shelter-and-care option more administratively difficult and expensive to operate and make less available state Medicaid funding. They suggest that demand could be dampened if older people (and their families) cope with their frailties in their own homes and move only when they require the more intense level of care now offered in nursing homes. They also worry that very frail residents will receive inadequate care in these less regulated facilities.

The Assisted Living Facilities Association of America (ALFAA), a membership organization located in Fairfax, Virginia, promotes the assisted-living industry by working with government and elected officials at the national, state, and local levels on regulatory concerns. (SEE ALSO: *Assisted Living Facilities Association of America; Congregate Housing; Continuing Care Retirement Communities;* Elderly; *Section 232*)

—*Stephen M. Golant*

Further Reading

Assisted Living Facilities Association of America and Coopers & Lybrand, Inc. 1993. *An Overview of the Assisted Living Industry.* Fairfax, VA: Assisted Living Facilities Association of America.

Brummet, W. 1997. *The Essence of Home: Design Solutions for Assisted Living Housing.* New York: Van Nostrad Reinhold.

Kalymun, M. 1990. Toward a definition of assisted-living. *Journal of Housing for the Elderly,* 7:97-132.

———. 1992. Board and care versus assisted living: Ascertaining the similarities and differences. *Adult Residential Care,* 6(1):35-44.

Newcomer, R. J. and L. A. Grant. 1990. "Residential Care Facilities: Understanding Their Role and Improving Their Effectiveness." Pp. 101-24 in *Aging in Place,* edited by D. Tilson. Glenview, IL: Scott Foresman.

Regnier, V. 1994. *Assisted Living Housing for the Elderly: Design Innovations from the U.S. and Europe.* New York: Van Nostrand Reinhold.

Regnier, V., J. Hamilton, and S. Yatabe. 1995. *Assisted Living for the Aged and Frail.* New York: Columbia University Press.

Wilson, Keren Brown. 1996. *Assisted Living: Reconceptualizing Regulation to Meet Consumers' Needs & Preferences.* Washington, DC: American Association of Retired Persons. ◄

► Assisted Living Federation of America

The Assisted Living Federation of America (ALFA) is a national organization of residential care providers, who offer health and daily living assistance to those who cannot or choose not to live alone. ALFA represents its members before legislative bodies and regulatory agencies, monitors legislation that affects the assisted-living industry, and develops research, position papers, and policy statements to advance the industry. The association also sponsors conventions, meetings, and educational programs for its members. In addition, ALFA publishes a variety of educational materials and periodicals, including the Assisted Living Training System for providers; a comprehensive manual on assisted living; a monthly newsletter titled *The ALFA Advisor*; a quarterly journal, *The ALFA Treasure Chest,* and a bimonthly magazine titled *Assisted Living Today,* which reports the latest industry news and trends. Founded in 1991, ALFA has acquired a membership of 5,000 assisted-living companies and industry partners nationwide. In 1993, the association operated with a staff of 18 on an annual budget of $3 million. President and CEO: Karen A. Wayne. Address: 10300 Eaton Place, Suite 400, Fairfax, VA 22030. Phone: (703) 691-8100. Fax: (703) 691-8106. E-mail: info@alfa.org. (SEE ALSO: **Elderly**)

—Caroline Nagel ◄

► Association of Community Organizations for Reform Now

The Association of Community Organizations for Reform Now (ACORN) is a national association founded in 1970 representing low- and moderate-income families through community organizations dealing with issues of nationhood, banking, housing, insurance, jobs, and education. The association produces several publications, including *United States of ACORN,* a bimonthly, for $20 per year, and *Homesteader,* a quarterly newsletter about housing, for $10 per year. ACORN operates on an annual budget of $5 million, funded through membership dues and fundraising. Current membership: 70,000. Staff size: 150. Chief organizer: Wade Rathke. Address: 1024 Elysian Fields Ave., New Orleans, LA 70117. Phone: (504) 943-0044. Fax: (504) 944-7078.

—Laurel Phoenix

Further Reading
Borgas, Seth. 1986. "Low-Income Homeownership and the ACORN Squatters Campaign." Pp. 428-46 in *Critical Perspectives on Housing,* edited by Rachel Bratt, Chester Hartman, and Ann Meyerson. Philadelphia: Temple University Press. ◄

► Association of Local Housing Finance Agencies

The Association of Local Housing Finance Agencies (ALHFA) is a national, nonprofit organization of public and private sector agencies responsible for developing and financing affordable housing. ALHFA serves both as an advocate before Congress and regulatory agencies on issues affecting affordable housing and as a forum in which members can interact with others in the field. It has lobbied extensively for the preservation of governmental incentives for the production of affordable housing, including the Mortgage Revenue Bond and the Low-Income Housing Tax Credit. ALHFA also publishes *Housing Finance Report,* a bimonthly newsletter that focuses on developments in affordable housing finance, and periodic legislative news bulletins. Founded in 1982, the association has 180 member agencies nationwide. In 1993, it operated with four staff members on an annual budget of $400,000. Contact person: John C. Murphy, Executive Director. Address: 1101 Connecticut Avenue, NW, Suite 700, Washington, DC 20036. Phone: (202) 857-1197. Fax: (202) 857-1111.

—Caroline Nagel ◄

► Assumption of Loan

The assumption of a loan is a technique in real estate finance wherein a buyer pays the seller a sum of money equal to the seller's equity in the property and then takes responsibility for (or assumes) the seller's mortgage payments on the same terms specified in the mortgage. In particular, when interest rates are high, this practice is very advantageous for both buyers and sellers. Because the financing is already in place, loan assumption aids in the quick sale of a property, a condition that aids the seller. And the buyer benefits because this financing is available on terms more favorable than those prevailing under market conditions characterized by high interest rates. Thus, assumable loans were a crucial factor supporting real estate markets in the late 1970s and early 1980s in the United States.

On the other hand, the practice of assuming loans works to the detriment of lending institutions. Because loan assumption decreases the amount of new mortgage loans, lenders cannot generate fees associated with the origination of such mortgages. In addition, it prevents lenders from reinvesting their capital at prevailing interest rates, a factor that also erodes their profits. But most critical, lenders' loan portfolios are burdened by mortgages that carry interest rates below market. As a result, the value of those portfolios is eroded, and this condition can precipitate the ultimate insolvency of the lending institution; this was a contributing factor in the collapse of many such U.S. institutions during the 1980s. Not surprisingly, steps have been taken to severely restrict the assumability of mortgage loans, although a few notable exceptions remain: Federal Housing Administration (FHA) and Department of Veterans Affairs

(DVA) loans and some adjustable-rate mortgages (ARMs) can still be assumed, although this, too, could change.

The inclusion of a due-on-sale (or alienation) clause in the mortgage contract discontinues the assumability of real estate loans. By the mid-1970s, most mortgages began to include this provision; it enables the lender to demand payment of the mortgage balance if the property is sold. However, the clause was often ignored and only sporadically enforced. By the mid-1980s, congressional legislation (the Garn-St. Germain Depository Institutions Act) and litigation decided by the U.S. Supreme Court upheld the strict enforcement of the due-on-sale clause. As a result, before a loan may be assumed, the lender has the right to ascertain if the buyer is creditworthy, to determine what assumption fees will apply, and to renegotiate the terms of the loan with respect to perceptions of present and future interest rates. (SEE ALSO: **Mortgage Finance**)

—*Daniel J. Garr*

Further Reading

Lowry, Ira S. 1983. *Creative Financing in California: The Morning After.* Santa Monica, CA: RAND.

B

► Balloon Mortgage

One of the more well-known creative financing techniques has been the use of the balloon mortgage. This instrument requires a large payment, often the final payment due on the mortgage note, after a series of "interest-only" payments. Sometimes, the debt service payment is the same as under a fixed-rate mortgage (FRM), but the outstanding balance is due as of a specific date. On that date, the balloon payment is made. It was the principal instrument for homeownership until 1933.

The balloon mortgage is helpful, often as supplementary financing, because the borrower saves money by not having to make the principal payments, as with the FRM, and the lender receives interest payments (often at a slightly higher interest rate). Also, at the balloon payment date, the lender receives the remainder of the balance due. Alternatively, when property is transacted between buyer and seller and some seller financing is involved, a balloon mortgage could be used to increase the rate of interest paid to the seller for a few years and then the balance is "due in full" earlier than 30 years. As a result, balloon mortgages are often used as second mortgages because all parties can gain in that everyone's objectives are achieved.

However, nothing could be this good for free. The primary cost is that there are some risks involved with balloon mortgages. Borrowers need to obtain financing to make the balloon payment (almost always, this involves a refinancing rather than a payment in cash). Lenders worry that conditions in financial markets might be such that refinancing of the balance at the time the balloon payment is due might be difficult. Sellers who agree to finance the sale with the balloon mortgage may become concerned about whether the buyer will complete the final payment.

Balloon mortgages permit flexibility in the scheduling of repayment schedules. Because they are so flexible, the schedules can be adjusted to the needs of the parties involved. Another feature is that the repayment schedules are very easy to structure; in fact, the mechanics of the balloon mortgage are very easy to work with. Finally, balloon mortgages do not require principal payments, which in some cases, is attractive because the rate of return can be en-

hanced by using the cash more profitability than to build equity in the property. (SEE ALSO: **Mortgage Finance**)

—*Austin J. Jaffe*

Further Reading

Fabozzi, Frank J. and Franco Modigliani. 1992. *Mortgage and Mortgage-Backed Securities Markets*. Boston: Harvard Business School Press.

Jacobus, Charles J. 1996. *Real Estate Principles*. Upper Saddle River, NJ: Prentice Hall.

Shim, Jae K. 1996. *Dictionary of Real Estate*. New York: John Wiley.

Tosh, Dennis S. 1990. *Handbook of Real Estate Terms*. Englewood Cliffs, NJ: Prentice Hall. ◄

► Barrier-Free Design

In the United States, the concept of barrier-free design took shape soon after World War II, in part to address the needs of returning disabled veterans facing reemployment and education obstacles. Efforts to remove workplace barriers, or "architectural" barriers, coincided with government encouragement to hire "handicapped" people, including veterans. Although the disabled population faced many types of barriers, including attitude barriers, the most limiting were physical obstacles, such as curbs, entrances, stairs, and transportation. As more people recognized the benefits of removing barriers for the population as a whole, the term *barrier-free design* evolved, a more positive and inclusive designation.

In the past 30 years, the need for a barrier-free environment resulted in numerous federal and state laws addressing physical accessibility and participation in public programs. More recently, laws have addressed discrimination as well. Standards for barrier-free design were initially developed by the American National Standards Institute, Inc. (ANSI) to test and document minimum space and element measurements to accommodate people with disabilities. The standard was intended to ensure consistency among the various codes. These "accessibility standards" were adopted and amended in 1984 by federal agencies as the Uniform Federal Accessibility Standard (UFAS). UFAS was subsequently used as the basis for the technical standards appended to the Americans with Disabilities Act (ADA) of 1990.

The broad, national jurisdiction of the ADA, in contrast to former program-specific laws, prohibits discrimination against people with disabilities. It applies to all public and commercial buildings and facilities, requiring new and renovated buildings, as well as most existing buildings and facilities, to be accessible. The act addresses barriers impeding people with mobility, visual, and hearing impairments—barriers such as parking, stairs, toilet stalls, protruding objects, and elevator controls. It also addresses barriers faced by persons with less visible disabilities, such as cognitive dysfunction and chemical intolerance. The corresponding design standards, Americans with Disabilities Act Standards for Accessible Design, as adopted by the U.S. Department of Justice, are now the minimum standards for barrier-free design.

The concept of "universal design" emerged when it was recognized that many of the features required for accessibility make life easier and more convenient for everyone. Thus, the practice of barrier-free design entered the mainstream of construction design. For example, wider toilet stalls for wheelchair users could be enjoyed by people with children in strollers and travelers carrying bags. Curb ramps could aid delivery people using hand carts and children on bicycles. Traffic signals that are both audible and visual could increase safety for even the most distracted pedestrian.

The universal design approach argues for accessible features that are no longer identified solely for use by the "handicapped," "disabled," or "elderly." Accessibility features that are not labeled as "special" mean that no stigma will be attached to their use. In addition, mass production and innovation in many cases can keep costs down to match the costs of inaccessible alternatives. Barrier-free design has come a long way from the architectural-barrier removal practices of the past, and current approaches to universal design go yet further beyond the minimum accessibility standards. (SEE ALSO: *American National Standards Institute (and Other) Accessibility Standards and Housing; Home; Physical Disabilities, Housing of Persons With*)

—Ronald L. Mace

Further Reading

"Access: Special Universal Design Report." 1992. *Metropolis* November, pp. 39-67.

Barrier Free Environments, Inc. 1991. *Accessible Housing Design File.* Florence, KY: Van Nostrand Reinhold.

Canestaro, N. and R. Null, eds. 1995. "Universal Design." *Housing and Society* 22(1&2 [Special issue]).

Center for Universal Design. 1996. *The Seven Principles of Universal Design.* Raleigh, NC: North Carolina State University, Center for Univeral Design.

Covington, G. A. and B. Hannah. 1997. *Access by Design.* New York: Van Nostrand.

Liebrock, Cynthia, with Susan Behar. 1993. *Beautiful Barrier-Free: A Visual Guide to Accessibility.* Florence, KY: Van Nostrand Reinhold.

Mace, Ron, Graeme Hardie, and Jaine Place. 1990. *Accessible Environments: Toward Universal Design.* Raleigh, NC: Center for Accessible Housing.

Moore, Robin, Susan Goltsman, and Daniel Iacofano, eds. 1992. *Play for All Guidelines: Planning, Design and Management of Outdoor Play Settings for All Children.* Berkeley, CA: MIG Communications.

Peloquin, Albert. 1993. *Barrier-Free Residential Design.* New York: McGraw-Hill.

Plae, Inc., USDA Forest Service, et al. 1993. *Universal Access to Outdoor Recreation: A Design Guide.* Berkeley, CA: MIG Communications.

Universal Design Newsletter. Rockville, MD: Universal Designers and Consultants. ◀

▶ Behavioral Aspects

How people's behavior reflects the nature of their home environment is a question that many housing researchers have sought to answer. They have documented many aspects of housing and behavior but have not reached a consensus on the conception or importance of behavior, on theory and types of explanation or on research methods.

When behavior is linked to housing, it can take on any number of meanings. The most common type of housing behavior is how the consumer deals with it in a market or administrative setting. For example, housing choice is a behavior. Similarly, the actions of persons and interest groups who influence the nature, cost, and supply of housing are also behavior—in this case the behavior of major actors in the housing market. These kinds of behavior are important to our understanding of housing but are dealt with in other entries (e.g., Federal Government, State Government, Local Government, Growth Machines, Lending Institutions, Real Estate Developers, Residential Development).

Behavior is often viewed in terms of specific, fragmented actions. But one can also think of behavior as an integrated flow of activity: the totality of what goes on in and around housing. Behavior thus conceived is a dynamic flow of activity. Although housing is something to build, sell, buy, and rent, to have as an asset, to protect one from the wet and cold, it is also a setting in which behavior occurs. It is a locus for self-maintenance activities, such as sleep, hygiene, eating, and recuperation. Child rearing is traditionally home centered. Many forms of recreation occur in one's home. Housing also generates activities, because of the need for cleaning and maintenance; it provides a setting in which to deal with commodities such as food, dishes, and clothing. Thus, although researchers are often interested in specific actions, much of what occurs in and around housing is a rich combination of largely repetitive activities, the behavior of living one's everyday life.

Housing takes on social relevance to the extent that its design influences the performance of specific behaviors. To what extent is a particular behavior likely to occur because an individual is in a certain housing context? To what extent is this behavior unlikely to be found elsewhere?

Theory, Explanation, Causality

Addressing the above questions involves the issue of causality. Does housing cause certain behaviors to occur? Is it in some way partly responsible for behavior? Behavior and housing can be causally linked, although not in a strictly deterministic way. Causation can take a number of forms, with different degrees of strength. By and large, studies of the housing-behavior relationship show that housing design can provide the opportunity for certain behaviors to occur. But design does not make behavior happen; people first must want to behave in certain ways. When people want to do so, housing design can influence the extent and the location of the behavior. By the same token, housing design can make particular behaviors more difficult or even impossible to carry out.

A number of theoretical formulations have been employed in housing-behavior research, beginning with Kurt Lewin's *field theory*. Each addresses delimited situations, without becoming a general theory. For example, Hall assessed housing and other built environments in terms of how their designs reflected culture-bound definitions of the interpersonal distances people found appropriate for specific kinds of behavior. Sommer studied how individuals maintain personal territories in their choices of where to sit, given the presence of others, and hence in the arrangements of sitting in congregate areas. Ankerl rooted explanation of these phenomena in communications theory. Most applications of these popular theories have been at the microlevel, relevant to room design within residential contexts, but less applicable to the linkage and interaction of the separate rooms and spaces making up the whole housing environment.

A more recent theoretical insight has steered attention away from a view of behavior as purely a function of environmental conditions. Daniel Stokols has fostered the growth of what he calls *interactive theories,* in which people and environments simultaneously influence one another. This view of theory reconciles two competing explanations of behavioral patterns found in different kinds of housing. One perspective, called *self-selection,* holds that people assess their own lifestyle and then choose housing to optimize their ability to live it; environment does not determine behavior because people bring a predetermined behavior pattern (their lifestyle) to a setting hospitable for its realization. The other perspective holds that housing design, through its opportunities and constraints, creates circumstances in which people typically adopt common habits.

A different theoretical approach to the relationship of environment and housing aims to reveal what people regard as meaningful in such contexts and therefore act on. Design and spatial arrangement according to this view are secondary to the symbolic message that housing conveys to a particular person or group. Lawrence wrote a significant treatise spotlighting this approach while integrating it with several others.

Research Methods

The multiplicity of methods used to study housing and behavior reflects the diversity of information needed to understand the complex subject, as well as the strengths and weaknesses of various methods.

Behavior, for example, may pertain to individuals, primary groups, and/or collectivities. Behaviors of interest may be specific movements or activities or, at the other end of a continuum, rich mixtures of everyday activities. The housing context involved should be analyzed explicitly. Beyond documentation of behavior and housing, data on attitudes and effects help to provide a more secure basis for understanding and practical application. On the one hand, we have to know what people think about their in situ behavior. Is it desired or undesired? Voluntary or involuntary? Pleasant or unpleasant? High or low priority? What meanings does it have for them? On the other hand, we need to know what residual effects occur, if any, as a result of the housing-behavior match: to people and groups, to the housing itself, and to the organizations and institutions involved in the planning and provision of housing.

Research options include asking questions, observation, simulation, and archival and epidemiological approaches.

Asking questions of individuals and groups is unsurpassed for gaining information about what people think about a situation, for routine background and factual information, for intentions and motives, for attitudes and opinions, for interpretation and expression of meaning, and for expressions of personal evaluation. The time use survey is particularly useful because it grounds activities during a tangible period in their social, spatial, and subjective contexts in a naturally quantitative way.

Observation is useful for assessing the details of actual behavior in groups, at least when an observer can gain entrance and is capable of observing all that goes on. This generally rules out personal situations, very large-scale settings, and behavior that occurs across many settings.

Simulation, usually involving modeling of environments, allows observation of people's reactions to and behaviors in prototypes of planned housing. When simulated housing can be adjusted easily, successive episodes of feedback are possible. Game playing is another form of simulation, in which people are forced to make trade-offs and then see and react to the results of their priorities.

Despite the usefulness of simulations and prospective surveys, most housing research is postoccupancy, to assess behavioral aspects of experience over time in actual environments.

Archival and epidemiological techniques are intended to use systematic information on outcomes of particular situations. The former may contain insights into local situations over a period of time, to serve as a basis for analysis and inferences. But such techniques depend on the quality of records and the adequacy of reporting, and they are seldom contextually explicit.

Some Aspects Illuminated

Despite the formidable theoretical and methodological complexities of carrying out definitive research on behavioral aspects of housing, a number of areas of interest have been illuminated.

One subject of particular interest in the 1960s and 1970s was behavior in high-rise apartments, then relatively novel forms of housing. Many studies criticized their impact on families with young children. They typically found that parents could not control the conditions under which their children played outside nor the safety and noise levels throughout the building. Such criticisms did not necessarily extend to the experiences of groups other than families with children or to the impacts of high-rise buildings with increased on-site resources. The focus on children was used to assess other forms of housing as well, because the play of younger children is visible, localized, and an important criterion for the adequacy of family housing.

Another area of housing-behavior research has been the design and planning of features that foster or constrain criminal behaviors such as mugging and vandalism.

The later 1980s and 1990s have been marked by active research on the relationship of gender to housing and urban environments. Particular interest in women arises from the major societal changes in women's roles, daily activities, and household structures. The starting point of such research is that the design of traditional family housing does not correspond to the everyday behavior and needs of contemporary women, particularly in suburban locations. This research interest spills over to housing for nontraditional household structures, whose needs may not be satisfied within normally available housing. The emerging literature examines not only the pros and cons of the status quo but also emerging solutions such as cohousing.

The thrust of research on behavioral aspects of housing is that housing does matter for people's behavior. This seldom occurs in a determinative way, but housing, through its design, can either facilitate or frustrate what people might wish to do, and it sets up the conditions within which residents' everyday routines are established. (SEE ALSO: *Architecture and Behavior/Architecture et Comportement; Cultural Aspects; Environment and Behavior; Home; Postoccupancy Evaluation; Residential Autobiographies*)

—William Michelson

Further Reading

Ankerl, Guy. 1981. *Experimental Sociology of Architecture.* The Hague: Mouton.

Barker, Roger. 1968. *Ecological Psychology.* Stanford, CA: Stanford University Press.

Bechtel, Robert, Robert Marans, and William Michelson, eds. 1990. *Methods in Environmental and Behavioral Research.* Malibar, FL: Robert Krieger.

Franck, Karen and Sherry Ahrentzen, eds. 1989. *New Households, New Housing.* New York: Van Nostrand Reinhold.

Hall, Edward. 1966. *The Hidden Dimension.* Garden City, New York: Doubleday.

Lawrence, Roderick. 1987. *Housing, Dwellings, and Homes.* New York: John Wiley.

Michelson, William. 1977. *Environmental Choice, Human Behavior, and Residential Satisfaction.* New York: Oxford University Press.

Pollowy, Anne-Marie. 1977. *The Urban Nest.* Stroudsburg, PA: Dowden, Hutchinson & Ross.

Sommer, Robert. 1969. *Personal Space: The Behavioral Basis of Design.* Englewood Cliffs, NJ: Prentice Hall.

Zeisel, John. 1981. *Inquiry by Design.* Monterey, CA: Brooks/Cole. ◀

▶ Blight

Blight is a condition of an area often defined in economic terms—as a pattern of deterioration and decay of economic vitality in cities or neighborhoods. As a public health issue, most of the national housing acts (1937, 1949, 1974) note the problem of urban blight. The Housing Act of 1949 provided for a large-scale urban renewal program targeted at "blighted areas." The determination of a blighted area that may qualify for federal funding is left largely to state statutory language and to local interpretation and implementation. Operationally, local governments may use a combination of indices to indicate blight. Relevant factors may include deteriorating economic, physical, and social conditions. Often, factors such as low housing occupancy rates, substandard housing, high crime rates, and below-average public health statistics are used. In most cases, local governments have been given latitude in defining a blighted area so that locally relevant factors can combine to determine blight. In this way, it is hoped that remedial action can be taken to prevent the early stages of blight from becoming more severe.

Although *blight* and *slum* are often used together or interchangeably, most literature considers *blight* an economic term and *slum* a social one; moreover, a blighted condition can lead to the end result of a slum. With the elimination of the federal urban renewal program in 1974, a neighborhood's or local area's blighted condition is still used to determine which areas might qualify for community development or redevelopment funding. (SEE ALSO: **Abandonment;** *Displacement; Slums;* **Urban Redevelopment**)

—*John S. Klemanski and John W. Smith*

Further Reading

Housing Act of 1949 (42 U.S.C. 1401 et seq.).

Jacobs, Jane. 1963. *The Death and Life of Great American Cities.* New York: Vintage.

Walker, Mabel L. 1938. *Urban Blight and Slums: Economic and Legal Factors in the Origin, Reclamation, and Prevention.* Cambridge, MA: Harvard University Press. ◀

▶ Blockbusting

Blockbusting is a now-illegal tactic used by real estate speculators to acquire property from owners in segregated neighborhoods and manipulate a decline in housing values. One house, normally acquired at market value, would be sold to a minority family. Following this sale, other homeowners in the neighborhood would be contacted and pressured to sell their homes at below-market value. Tactics used emotionally charged suggestions—for example, that children would be "mingling" with black children, that daughters or sons could be raped by black youths, and that property values would plummet. Pressure was maintained until panic-stricken homeowners sold. After a number of such homes were acquired, the real estate agents would sell the homes at higher prices to blacks and keep the panic differential as profit.

Blockbusting was rendered an illegal practice under Title VIII, Section 804, of the federal Civil Rights Act of 1968, which prohibited (among other discriminatory practices in housing) any profit-oriented behavior intended "to induce or attempt to induce any person to sell or rent any dwelling by representations regarding the entry or prospective entry into the neighborhood of a person or persons of a particular race, color, religion, or national origin." Categories protected from discrimination by Title VIII of the Civil Rights Act of 1968 (the Fair Housing Act) have been expanded over time. The Fair Housing Amendments of 1988, Section 3604, include the prohibition to discrimination on the basis of sex, handicap, and familial status in addition to race, color, religion, and national origin.

Aside from prohibiting the behavior, Congress included procedures for redress of alleged discriminatory housing practices. Individuals who believe they are the victim of discrimination must file a written complaint within one year of the alleged violation with the secretary of Housing and Urban Development who is charged with investigating such complaints. The secretary notifies the respondent within 10 days and allows an additional 10 days for the accused individual or group to respond to the charges. If an investigation reveals that "reasonable cause" exists that discrimination occurred, the secretary can refer the case to the attorney general or can refer the case for civil action in state or local court. Here, complainants can sue for damages.

Although blockbusting may be less well-known than redlining, in fact, the two can be complementary practices. For example, in the late 1960s, in an infamous case of blockbusting in Boston, the Boston Banks Urban Renewal Group (B-BURG) decided to use federally subsidized loans to move blacks into a Jewish neighborhood. Although the loans were intended to help any low-income person to purchase a home, the B-BURG group determined that only blacks would receive the loans and only in this neighborhood. Real estate agents used blockbusting techniques to persuade the longtime residents of the community to sell their homes so that the loans could be made to black clients.

Blockbusting has several negative effects. First, homeowners are harmed because they sell at below-market rates and must relocate. In the Boston case, the homeowners had often paid their mortgages in full and were strongly attached to the neighborhood. Second, blockbusting causes rapid neighborhood transition leading to instability and hostility between the entering and exiting groups. Finally, blockbusting prolongs the segregation in housing patterns that leads to de facto segregation in the nation's schools.

The link between blockbusting and federal housing loan subsidy programs cannot be overlooked. The Federal Home Administration (FHA) expanded its low downpayment or no downpayment loan programs to cover existing housing stock in addition to new construction units. Low-income individuals are less capable of maintaining older homes because they are unlikely to have the financial ability to make the likely repairs without missing mortgage payments. Mortgage companies profit because they are protected from the risks associated with selling to low-income families. As a result, during the 1950s and 1960s, as real estate agents fueled blockbusting, the mortgage companies made loans to low-income blacks, repossessed the property if a mortgage payment was missed, and either abandoned the property or resold it to another low-income client without experiencing a loss. The result was highly unstable, isolated, and blighted neighborhoods that still negatively affect cities such as Detroit and Chicago nearly 30 years after the practice of blockbusting was outlawed. (SEE ALSO: **Discrimination;** *Community Reinvestment Act;* **Fair Housing Amendments Act of 1988**)

—*C. Michelle Piskulich and John S. Klemanski*

Further Reading

Hirsch, Arnold R. 1983. *Making the Second Ghetto: Race and Housing in Chicago, 1940-1960.* New York: Cambridge University Press.

Lake, Robert W. 1981. *The New Suburbanites: Race and Housing in the Suburbs.* New Brunswick: Rutgers, State University of New Jersey, Center for Urban Policy Research.

Levine, Hillel and Lawrence Harmon. 1992. *The Death of an American Jewish Community: A Tragedy of Good Intentions.* New York: Free Press.

Orser, W. Edward. 1994. *Blockbusting in Baltimore.* Lexington: University Press of Kentucky.

U.S. Congress. 1990. Senate Subcommittee on Consumer and Regulatory Affairs on the Reports from Thrift Regulatory Agencies and the Department of HUD on the Evil of Racial Discrimination in the Home Mortgage Lending and the Inadequate Regulatory Response to the Situation. 1990. *Mortgage Discrimination.* Hearing. 101st Cong., 2d sess. ◄

► Board-and-Care Homes

Board-and-care homes are generally chosen by people who need some services but cannot afford to have them provided in their own home, or services provided to the home are not available in their community. Most residents are women who have lost their spouses or were never married and who do not have family available to provide needed services. The average age of residents is between 60 and 75 years old. Most residents are able to walk. Residents whose health deteriorates so that extensive medical services are required may be asked to leave the board-and-care home and move to a nursing home.

Board-and-care homes are often located in residential neighborhoods and may serve those who have previously lived in the neighborhood, thus providing the person with the possibility of continued contact with friends living nearby. Most were originally built as private, single-family dwellings. Board-and-care homes are usually mom-and-pop enterprises in which the owner is also the major caretaker. The number of facilities varies greatly by state, with some states having less than 100 and others having several thousand.

Board-and-care homes vary in size, ranging from the household that uses the spare bedroom to accommodate one or two persons, to a home that might serve a number of people. Most have 30 or fewer residents. Facilities that accommodate more than 30 people are likely to be called assisted-living or residential care facilities. These larger facilities differ from board-and-care homes in that they are staffed by paid professionals. Likewise, board-and-care homes generally do not have self-contained apartment units that are professionally managed as found in congregate housing.

Although the costs of board-and-care homes can range from several hundred dollars a month to several thousand dollars a month, in general, the residents of board-and-care homes tend to have lower incomes. About one-half pay for the cost of the accommodation from their own private funds with the other half receiving assistance to pay for their costs. Financial assistance for those who qualify is provided by federal Supplemental Security Income (SSI), which is supplemented in some states by state supplemental payments.

Most board-and-care homes are for-profit facilities operated by women, although some are nonprofit. Others are operated by the state. The state facilities generally provide for the mentally ill.

A current concern regarding board-and-care homes is establishing a mechanism and procedure for ensuring safe, healthy, and sanitary conditions, while at the same time, maintaining a low-cost housing alternative. It is estimated that nearly one-half of the facilities are not licensed either because it is not required (the home serves just a few persons) or because they should be licensed but are not. Agencies lack the resources to require conformity from nonconforming facilities. Court action is often unsuccessful and very time-consuming. Some argue that too many regulations may frustrate and discourage current and potential providers, thus decreasing the supply of board-and-care homes. Operators and staff often experience burnout from attempting to meet the physical and emotional needs of frail residents. They are often frustrated because the funds received from people on SSI are insufficient to provide the level of services needed. Because of low profit margins and demanding schedules, there is a high turnover of staff in board-and-care facilities. (SEE ALSO: *Elderly; Section 232; Single-Room Occupancy Housing*)

—*E. Raedene Combs*

Further Reading

Haske, Margaret. 1986. *A Home Away from Home. Consumer Information on Board and Care Homes.* Washington DC: American Association of Retired Persons.

Hawes, Catherine, Judith B. Wildfire, and Linda J. Lux. 1993. *The Regulation of Board and Care Homes: Results of a Survey in the 50 States and the District of Columbia.* Washington, DC: American Association of Retired Persons.

Lewin/ICF, Inc., and James Bell Associates. 1990. *Descriptions of and Supplemental Information on Board and Care Homes Included in the Update of the National Health Provider Inventory.* Washington, DC: U.S. Department of Health and Human Services, Office of the Assistant Secretary for Planning and Evaluation.

U.S. Department of Health and Human Services, Office of the Inspector General. 1990. *Board and Care.* New York: U.S. Department of Health and Human Services, Regional Inspector General of Region II, Office of Evaluation and Inspections. ◄

► Brooke Amendments

The Brooke Amendments are a series of amendments to the Housing and Urban Development Acts of 1969, 1970, and 1971. These amendments, introduced to Congress by Republican Senator Edward Brooke of Massachusetts, represent a shift in national public housing policies from a predominance of low-income housing construction to a combination of rent supplements and capital cost subsidies. The initial Brooke Amendment placed an upper limit of 25% on the proportion of income that a family could be charged for rent in a public housing unit. In addition, it redefined income to exclude (a) nonrecurring income, (b) 5% of gross income, (c) $300 for each dependent and each secondary wage earner, and (d) extraordinary medical expenses. The resultant loss of rental income for local housing authorities (estimated at $60 million-$100 million annually) was to be offset by the federal government. Subsequent Brooke Amendments authorized additional federal funds for local housing authorities to upgrade and maintain existing housing projects.

Despite passage in Congress, some provisions of the Brooke Amendments were obstructed by the Office of Management and Budget (OMB) and the Department of Housing and Urban Development (HUD) under the Nixon administration. Opposition to the amendments stemmed from the charge that the subsidization of local housing authorities was precipitating a fiscal crisis within the Department of Housing and Urban Development. Belatedly realizing the costliness of the amendments, the Nixon administration declared a freeze on noncash assistance proposals and refused to continue allocating federal funds to local housing authorities as mandated by the amendments. Critics of Nixon's housing strategies asserted that the withholding of funds by HUD and the OMB forced already beleaguered local housing authorities into bankruptcy, thus creating even greater chaos and stress within the public housing system. Embroiled in controversy, the Nixon administration articulated a national housing policy that would eventually be embodied in the Housing and Community Development Act of 1974. (SEE ALSO: *Housing-Income Ratios*)

—*Caroline Nagel*

Further Reading

Hartman, Chester W. 1975. *Housing and Social Policy.* Englewood Cliffs, NJ: Prentice Hall.

Hays, R. Allen. 1995. *The Federal Government and Urban Housing: Ideology and Change in Public Policy.* Albany: State University of New York Press.

Mitchell, J. Paul. 1985. *Federal Housing Policy and Programs: Past and Present.* New Brunswick: Rutgers, State University of New Jersey, Center for Urban Policy Research. ◄

► Brownfields

The decline of the U.S. manufacturing sector includes abandoned factories. The facilities and their surrounding areas are frequently contaminated by hazardous wastes. These "brownfields" constitute a large obstacle to the redevelopment of many urban areas by lowering property values and tax bases and adversely affecting health and employment opportunities. They are a problem in communities across the United States. The General Accounting Office (GAO) estimates that there are between 130,000 and 425,000 sites nationwide. However, there is no national list to track the actual number of sites. Instead, it is left to the states, where different standards for reporting make it impossible to pinpoint a precise number. Adding to the difficulty of quantifying the problem is that contamination is continually being discovered at new sites.

The two largest hindrances to the redevelopment of brownfields are cost and liability concerns. The GAO estimates total cleanup cost at $650 billion. The federal Environmental Protection Agency (EPA) estimates the costs of the first phase of a site assessment at between $1,000 and $10,000. If that assessment shows that it is necessary to examine what is buried underground, the costs can climb to $70,000. If environmental hazards are then located, the cleanup costs will further spiral upward, sometimes exceeding the value of the property itself.

Lenders are hesitant to finance redevelopment of these sites because they fear that if they assume responsibility for these properties, they will be liable as potentially responsible parties (PRPs) and forced to finance the cleanup before they can sell the site and reclaim their investment. This fear is a direct result of the *United States v. Fleet Factors Corp.* decision in 1990, which allowed a financial institution to be held liable as a PRP if "its involvement with a facility's management is sufficiently broad to support the inference that it could affect hazardous waste disposal decisions if it so chose." The decision also stated that this was true even if the creditor was not responsible for the daily operations of such a facility. This has led to "greenlining"—lenders refusing loans based solely on a location's potential for contamination.

The issue of PRPs was created through the Comprehensive Environmental Response, Compensation, and Liability Act (CERCLA). CERCLA is the federal government's attempt to address the environmental issues raised by the 1980 Love Canal disaster. It provides guidelines for public and private parties involved in cleanups; established the National Priority List (NPL) for the worst, most contaminated, highest-priority sites requiring cleanup; and created the Hazardous Substances Response Fund (also known as Superfund) to pay for the cleanups. Thousands of acres of slightly contaminated land, much of it slated for new development (industrial, commercial, and residential), are frozen by the inability of owners, developers, or lenders to gain certainty with respect to the costs and liabilities that may be incurred. These problems are exacerbated in urban centers and old industrial areas, where a weak real estate market typically cannot command the sizable financial investment needed to undertake the required environmental assessments and cleanups. Although less contaminated sites often do not pose an imminent hazard, they are nonetheless governed by regulations designed to handle highly hazardous sites. The result is that any level of actual or suspected contamination has a stifling effect on investment and development.

Many property owners have found it easier to abandon properties than to clean them up. Developers have found it easier to construct new buildings in "greenfields"— undeveloped suburban and rural areas—even when they have to furnish new infrastructures. However, despite these obstacles, brownfields remain assets that, if properly re-developed, could help bring affordable housing, new industries, and businesses to some of the most depressed urban areas. Remediating these sites would also help provide jobs for residents of surrounding communities, broaden the tax base, and restore the physical landscape. (SEE ALSO: *Environment and Housing; Environmental Contamination: Toxic Waste; Land Value Taxation; Locally Unwanted Land Use; Not in My Back Yard; Temporarily Obsolete Abandoned Derelict Sites;* **Urban Redevelopment**)
—*Cassandra M. Hanley*

Further Reading
Bartsch, Charles and Elizabeth Collaton. 1994. *Industrial Site Reuse, Contamination, and Urban Redevelopment: Coping with the Challenge of Brownfields* (Report). Washington, DC: Northeast-Midwest Institute.
Bartsch, Charles et al. 1991. *New Life for Old Buildings: Confronting Environmental and Economic Issues to Industrial Reuse* (Report). Washington, DC: Northeast-Midwest Institute.
Hanley, R., ed., "Focus Issue on Recycling Brownfields." 1995. *Journal of Urban Technology* 2(2). ◄

► Building Codes

The earliest known building code was contained in the Code of Hammurabi, the 18th-century B.C. Babylonian ruler. This code specified that the builder would be slain if the house fell in and killed the head of the household. Professional liability was also required of engineers in ancient Rome who were expected to stand below the scaffolding when it was removed from an arch they had designed.

Today's builders do not have to lay their lives on the line, but they must comply with strict building construction standards that are intended to ensure health and safety and the protection of property.

Although building codes have evolved over time to cover every facet of residential, commercial, and industrial construction, many jurisdictions do not have such codes in place. Unfortunately, it often takes a tragedy before there is enough support for enforceable construction standards. New York State allowed local governments to adopt building codes if they wanted to do so; the state even provided a model code for consideration. When 27 executives died in a motel fire in 1980, the state responded by developing a uniform fire prevention and building code that became mandatory in 1984 for all jurisdictions besides New York City (the city has its own building code).

Builders and those in allied industries have been in the forefront of the push for the adoption of one of three regional model codes: the uniform, national, and standard building codes. These are sponsored by organizations that have the capacity to do the research necessary to ensure that the models are up-to-date. Consistency and predictability in regulations among jurisdictions simplifies the challenges inherent in building construction. Appropriate standards and adequate enforcement also protects the industry from unscrupulous and shoddy construction practices.

Code amendments reflect innovations in building construction technology as well as compliance with mandates such as the federal 1990 Americans with Disabilities Act. Some changes are in response to problems that have been encountered in the field. When Hurricane Andrew devastated southern Florida in 1992, researchers and building officials closely examined the resulting damage to determine the extent to which it was related to the building codes. They discovered, in a number of cases, that damage was due to poor construction techniques coupled with lax code enforcement. They were also able to observe which code requirements were effective and which needed to be strengthened.

Building codes are often criticized for their complexity, their associated bureaucracy, and their tendency to be protective of unions. Chicago's code prohibits the use of polyvinyl chloride pipe for plumbing and flexible wire for electrical work, both of which are laborsaving alternatives. The increasing use of the model codes, training, and certification opportunities for inspectors; an emphasis on performance as opposed to prescriptive standards; and a more open decision-making process for the amendment of the three models is alleviating some of the problems that have been historically associated with building codes.

In accordance with building code requirements, almost any type of construction, no matter how minor, needs a building permit. This includes housing renovation involving structural changes such as moving a wall or installing a window where none previously existed. Construction activity that includes electrical and plumbing work may require separate permits if this work involves addressing standards that are not incorporated in the building code. Permit fees are set so as to cover the actual costs of inspections, the number of which will vary depending on the size and complexity of the project. A reroofing may involve just a single, drive-by inspection; a new single-family house will require five inspections at different stages of construction.

A change in use for a given piece of property will often involve the application of more stringent building code standards. For example, the conversion of a single-family house into a bed and breakfast or a place for wedding receptions may require the installation of fire escapes and fire doors as well as bathrooms that are accessible to the disabled.

Despite the legitimacy of building codes, a considerable amount of construction activity takes place without proper permits, particularly by do-it-yourself remodelers who either do not know about the need for permits or, more likely, do not want to pay the added costs or deal with the inconveniences associated with inspections. In many respects, construction codes are geared toward those who make a living in the building industry. Permits specify deadlines by which a particular activity is to be completed. The code fails to give allowance for the fact that the process for those who do their own remodeling is often sporadic, piecemeal, and executed over a relatively long time. Work may also be done primarily on weekends when inspections are not feasible.

Some owner-builders as well as others concerned with building affordable housing raise questions about the extent to which code requirements actually go beyond their explicit purpose of protecting public health and safety. For instance, in very low-density, rural areas, it can be argued that indoor toilets and plumbing are an amenity and not a need, yet they are typically required by building codes, thus driving housing costs beyond the means of certain households. New York State attempted to address this issue by enabling building officials to give owner-occupants the right not to have interior plumbing as long as a potable source of water is nearby.

Although owner-occupants may claim that they should have a right to assume risks that can be avoided with code compliance, this perspective begs the question of what protection will be given to future owners of the property. And although there is clearly room for improvement in building codes and the manner in which they are enforced, it can be said without qualification that they have contributed to a vast improvement in construction quality over the past 100 years. (SEE ALSO: *Code Enforcement; Health Codes; Housing Codes; HUD Minimum Property Standards; Model Codes; National Commission on Urban Problems*)

—*Deborah A. Howe*

Further Reading

Colling, R. C. and Colling, Hal. 1950. "Five Thousand Years of Building Regulation." In *Modern Building Inspection*, edited by R. C. Colling and Hal Colling. Los Angeles: Building Standards Monthly Publishing.

Hageman, Jack M. 1990. *Contractor's Guide to the Building Code Revisited*. Carlsbad, CA: Craftsman.

Kern, Ken, Ted Kogan, and Rob Thallon. 1976. *The Owner-Builder and the Code: The Politics of Building Your Home*. Oakhurst, CA: Owner-Builder Publications.

Lurz, William H. 1989. "The Drive for Uniform Statewide Building Codes Gains Momentum." *Professional Builders* 54(14):69.

New York State Uniform Fire Prevention and Building Code. Updated regularly. New York: Division of Housing and Community Renewal. ◀

▶ Building Cycle

Building cycle is a period of time over which the level of building activity is thought to systematically oscillate about a normal level. Cycles differ in length and arise for different reasons.

In temperate climates, an annual (seasonal) cycle is commonplace because of the difficulties that arise in undertaking construction outdoors during the winter months. Table 1 presents the indices used by the U.S. Department of Commerce to deseasonalize monthly housing starts. Housing starts tend to peak during April through August and to trough in November through February. Not surprisingly, seasonal variations are greater in the Northeast and Midwest and smaller in the South and West.

A longer cycle (variable in length but averaging four years) is associated with the business cycle, wherein new building activity increases with the expansion of income during an upswing and declines during a recession (see Table 2). Some proponents argue that (a) an increase in demand for new buildings causes developers to begin much new

TABLE 1 Seasonal Indexes Used to Adjust Housing Units Started, United States, 1996 (in percentages)

Month	All Starts	Northeast	Midwest	South	West
January	75	59	53	85	83
February	76	57	53	88	85
March	97	88	91	105	97
April	116	115	118	117	114
May	117	118	124	114	112
June	111	128	125	104	114
July	111	115	117	107	111
August	114	119	126	107	118
September	105	110	110	100	103
October	112	120	123	104	107
November	90	94	96	90	80
December	83	82	68	87	80

SOURCE: U.S. Department of Commerce, Bureau of the Census (1997), *Housing Starts*, selections from January to March, 1997.
NOTE: This index is a ratio of unadjusted housing starts to seasonally adjusted starts expressed as a percentage.

TABLE 2 Business Cycles in the United States since 1945

Postwar Business Cycle	Trough	Peak	Trough	Trough to Peak	Peak to Trough
				Elapsed time (months)	
1	Oct 45	Nov 48	Oct 49	37	11
2	Oct 49	Jul 53	May 54	45	10
3	May 54	Aug 57	Apr 58	39	8
4	Apr 58	Apr 60	Feb 61	24	10
5	Feb 61	Dec 69	Nov 70	106	11
6	Nov 70	Nov 73	Mar 75	36	16
7	Mar 75	Jan 80	Jul 80	58	6
8	Jul 80	Jul 81	Nov 82	12	16
9	Nov 82	Jul 90	Mar 91	92	8

SOURCE:, Updated by author from *McGraw-Hill Encyclopedia of Economics.* 2d ed. pp. 98-99.

construction and (b) initially construction may be very profitable; but (c) new construction takes many months or years to complete, (d) developers are myopic, and (e) new construction eventually overshoots the increase in demand, leading to a phase of sluggish construction. Other proponents argue that the building cycle occurs because of a rationing of mortgage credit that they say occurs near the peak of the business cycle.

The economist Simon Kuznets found evidence of a still-longer cycle in the U.S. building industry in the 19th and early 20th centuries, averaging about 15 to 20 years in length. New building construction in the United States was found to have peaked around 1870, 1890, 1910, and 1925. Proponents argue that demographic factors, particularly immigration into the United States, were primarily responsible for these building booms: waves of new immigrants leading to waves of new building and other capital investment. If so, the relevance of the Kuznets cycle in the present day is not clear at a time when the volume of immigration is relatively smaller. In other countries, evidence of a Kuznets cycle is mixed. In Canada, for example, there was also a boom in housing construction in the early 1870s, but the next major boom was not until around 1910.

An even longer cycle (over 50 years in length) was proposed by the Russian economist Nikolai Kondratieff, writing in the 1920s. He found evidence of a long swing (upswing followed by downswing in growth rates) from the 1780s to 1840s, another from the 1840s to 1890s, and a third starting from the 1890s (proponents suggest that it ended in the 1940s). Some proponents argue that Kondratieff waves result from major innovations in technology: the first wave from the application of steam to textile manufacturing, the second wave from the expansion of iron and steel production, and the third from the emergence of electrical power, automobiles, and trucking. In each case, technological innovation led to bursts of new construction early on that petered out as the innovation

became widely adopted. (SEE ALSO: *Housing Completions; Housing Starts*)

—*John R. Miron*

Further Reading

Eatwell, J. M. Milgate and P. Newman, eds. 1987. *The New Palgrave: A Dictionary of Economics.* Vol. 3. London: Macmillan (see pp. 60-62, 70-71, 235-36, 241-42).

Greenwald, D., ed. 1994. *The McGraw-Hill Encyclopedia of Economics.* 2d ed. New York: McGraw-Hill (see pp. 96-104, 593-97, 607-14).

Henderson, D. R. 1993. *The Fortune Encyclopedia of Economics.* New York: Warner (see pp. 173-77).

Pozdena, R. J. 1988. *The Modern Economics of Housing: A Guide to Theory and Policy for Finance and Real Estate Professionals.* New York: Quorum (see chap. 10). ◀

▶ Building Permit

Construction of a new building requires prior administrative approval from a local enforcement agency, usually a building or engineering department. The permit is issued after a site plan review is granted and if the drawn and written construction documents conform to building codes and ordinances.

Local code standards cover the structure itself as well as the internal workings of a building (plumbing, mechanical and electrical, fire and life safety, and so on). Model codes have been promulgated by national professional associations such as the Building Officials and Code Administrators (BOCA) and the International Conference of Building Officials (ICBO).

The building permit requires payment of a fee to cover some of the costs associated with site inspection by the enforcement agency. In some cases, impact fees may be levied to help offset costs connected with additionally needed public infrastructure. Building permits are issued for a limited time period, usually six months, and may be renewed for a minimum fee. (SEE ALSO: *Certificate of Occupancy*)

—John S. Klemanski and John W. Smith

Further Reading

Blankenship, Frank J. 1989. *The Prentice Hall Real Estate Investor's Encyclopedia*. Englewood Cliffs, NJ: Prentice Hall.

Building Officials and Code Administrators International, Inc. 1993. *National Building Code*. Country Club, IL: BOCA. Annual Supplements.

International Conference of Building Officials (ICBO). 1991. *Uniform Building Code*. Whittier, CA: ICBO. Annual Supplements.

Jacobus, Charles J. 1996. *Real Estate Principles*. Upper Saddle River, NJ: Prentice Hall.

Peiser, Richard B. 1992. *Professional Real Estate Development: The ULI Guide to the Business*. Washington, DC: Urban Land Institute.

Shim, Jae K. 1996. *Dictionary of Real Estate*. New York: John Wiley.

Tosh, Dennis S. 1990. *Handbook of Real Estate Terms*. Englewood Cliffs, NJ: Prentice Hall. ◀

▶ Building Systems Councils of the National Association of Home Builders

The Building Systems Councils of the National Association of Home Builders (BSC) is a national organization composed of four councils representing distinct segments of the housing manufacturing industry: the Modular Building Systems Council, the Panelized Building Systems Council, the Log Homes Council, and an Associate Members Council (product and service suppliers in the industry). The councils represent their members on legislative and regulatory issues; promote the architectural and engineering excellence of building systems through conventions and trade expositions; and offer quality control programs, technology seminars, and informational services. The BSC also publishes industry studies, technical manuals, consumer guides, and *Building Systems Review,* a monthly newsletter for BSC members. Founded in 1981, the BSC has 250 corporate members in the United States and Canada. Staff size: 4. Executive Director: Barbara K. Martin. Address: 1201 15th Street, NW, Washington, DC 20005. Phone: (202) 822-0576. Fax: (202) 861-2141.

—Caroline Nagel ◀

▶ Built Environment

Built Environment is a quarterly journal concerned with the research interests of urban, regional, and environmental planners, geographers, and architects. First published in the 1930s under the name *Official Architecture and Planning*, it has been published under the current name since 1974. Each issue is devoted to a single topic. Past topics include inner-city revitalization, transportation planning, and alternative housing arrangements. Circulation: approximately 600 worldwide. Price: £70 per year; £77 for overseas yearly; £19.50 per issue. Submission policy: Most papers specially commissioned, but all submitted papers are considered for publication. Editors: Professor Peter Hall and Dr. David Banister, c/o Alexandrine Press, P.O. Box 15, 51 Cornmarket Street, Oxford OX1 3EB England. Phone: 01865-724627. Fax: 01865-792309.

—Caroline Nagel

C

▶ Canada Mortgage and Housing Corporation

Canada Mortgage and Housing Corporation (CMHC) is the crown corporation, wholly owned by the federal government, that administers the National Housing Act (NHA). Originally established in 1946 as the Central Mortgage and Housing Corporation, CMHC was renamed in 1979. Currently, the principal objectives of federal involvement in housing are to improve quality of life for all Canadians through better living environments and equal access to affordable, suitable, and adequate housing. The NHA includes provisions for both market housing (owner occupied and for-profit rental) and social housing (public housing, nonprofit housing, and nonequity cooperatives). In 1996, CMHC administered operating grants, contributions, and subsidies totaling CAN$1.9 billion; almost all of this constituted operating subsidies for social housing. In the same year, an additional CAN$0.8 billion was committed in direct loans, with much of this going to public housing.

At CMHC's inception, the housing stock of Canada was in poor shape due to an early-postwar shortage of building materials, an illiquid mortgage market, little social housing, and an absence of local planning expertise. Although housing is a provincial responsibility under Canada's constitution, federal-provincial cooperation remains commonplace. In public housing, for example, typically, the federal government provides loan guarantees, provides capital and operating subsidies, and sets national standards; the provincial (and municipal) governments choose sites and designs, share funding, and manage projects. From the early postwar years, activities of CMHC have included the following: (a) joint mortgage loans (and direct lending); (b) property appraisal and building inspection related to this lending; (c) research into the social, economic, and technical aspects of housing and related fields; (d) architectural innovation; (e) building, site planning, and neighborhood standards; (f) land banking; (g) municipal infrastructure subsidies; (h) urban renewal and neighborhood revitalization; (i) data gathering and information dissemination; and

(j) coordination of provincial housing activities. In 1954, the joint loan provisions of the NHA were replaced by a cost-recovering mortgage insurance scheme that continues to the present day. At the end of 1996, there was about CAN$131 billion in outstanding NHA-insured mortgage loans.

To improve efficiency and adequacy in the construction of market housing, CMHC has sponsored innovations in building materials, methods of erection, residential building standards, and streamlined regulation; developed better designs for moderate-cost housing; and used mortgage approval as a stick to get smaller builders to improve building standards and site design in an era when municipal planning was still in its infancy. CMHC is also widely credited with improving access to long-term, low-down payment residential mortgage lending in the 1940s and 1950s. In 1984, CMHC introduced mortgage-backed securities (MBS) that further improved the liquidity of mortgages and thus the supply of mortgage funds. In MBS, monthly payment of principal and interest by a pool of borrowers, net of an administrative charge, is passed through to security holders. Because MBS pools are similar in terms of risk, standardized debt can be created and distributed among many holders, rendering liquid the secondary market in MBS.

Low-income households were supported primarily in the rental sector, at first through the public housing programs, rent supplements, and limited dividend housing and other subsidies to for-profit landlords, and later through subsidies to nonprofit housing corporations and nonequity cooperatives. However, CMHC also experimented with improvements in access to homeownership among low-income families during the 1960s and 1970s—for example, the Assisted Home Ownership Plan that operated from 1973 to 1978 subsidized payments in the early years of a mortgage.

CMHC has also done much to encourage research and technical expertise. It sponsored the establishment of the Community Planning Association of Canada, the Canadian Housing Design Council, the Canadian Council on Urban and Regional Research, and the Intergovernmental Committee on Urban and Regional Research and encouraged the establishment of departments of urban and regional planning in universities. CMHC also conducts important surveys. Their *Starts and Completions Survey* (SCS) collects building permit summary statistics: monthly, from all urban areas with more than 10,000 people, and quarterly, elsewhere in the nation. Twice a year, CMHC also conducts a survey of rents and vacancies in every conventional apartment building of six dwellings or more in Canada's larger urban centers. Altogether, CMHC spent about CAN$30 million in 1996 on housing research and information.

Today, CMHC is at a crossroads. That annual investment in renovations and repairs to the existing stock are now at a level comparable to investment in new residential construction suggests that housing policy be more focused on rehabilitation. The residential mortgage market today in Canada is efficient, liquid, and sophisticated. Direct lending by CMHC, which peaked at almost CAN$700 million in 1975, is rarely used anymore. Canada now has a not-insubstantial stock of social housing. CMHC now faces competition in the form of a private mortgage insurer. And provincial housing departments now have housing expertise of their own and undertake much research and planning. Much of the original impetus for CMHC has disappeared. It has been argued that the NHA mortgage insurance program should be privatized. Similarly, much of the data gathering and information dissemination could also be undertaken by private enterprise, although there continues to be a need for some mechanism by which Canadians can have standardization and avoid duplication. (SEE ALSO: *Housing Abroad: Canada*)

—*John R. Miron*

Further Reading

Canada Mortgage and Housing Corporation. 1991. *Strategic Plan*. Ottawa: The Corporation (see p. 32).

Canada Mortgage and Housing Corporation. 1994. *Canadian Housing Statistics*. Ottawa: The Corporation.

Miron, J. R. 1993. *House, Home, and Community: Progress in Housing Canadians: 1945-1986*. Montreal: McGill-Queen's University Press. ◀

▶ *Canadian Housing/Habitation Canadienne*

First published in 1983, *Canadian Housing* is published by the Canadian Housing and Renewal Association (CHRA) to promote a discussion of affordable housing policies and alternatives, understanding of housing, community renewal, rehabilitation, and property standards affecting the urban environment. It recognizes adequate and affordable housing as a fundamental human right. A quarterly journal, its price is $40 per year in Canada and $50 per year outside Canada. Managing Editor: Sharon Chisholm. Address: Canadian Housing and Renewal Association, 401—251 Laurier Ave. West, Ottawa, Ontario, Canada, K1P 5J6. Phone: (613) 594-3007. Fax: (613) 594-9596. (SEE ALSO: *Housing Abroad: Canada*)

—*Laurel Phoenix* ◀

▶ **Capital Gain**

Capital gain (or loss) is the amount by which the selling price of an asset on disposition exceeds (or falls short of) the original purchase price paid. Realized nominal capital gains have been considered part of normal income in the United States since the 1986 Tax Reform Act; however, some carrying costs are deductible, and special provisions apply in the case of death, gifts, and sale of principal residence. Until recently, U.S. taxpayers who sold their primary home and used the proceeds to buy or build another home within two years could defer payment of a capital gains tax on the sale for as long as they continued to own their primary home. Taxpayers aged 55 or over at the time of the sale could exclude up to $125,000 of gain from primary home sale from their taxes. With revisions that came into effect in August 1997, federal tax law now allows taxpayers to exempt $250,000 of gain ($500,000 for married couples filing a joint return) on the sale of a principal residence. Unlike the one-time-only exemption provided under prior law, the exemption may be claimed each time a taxpayer sells a principal residence, although the exemption generally may be claimed not more than once every two years. Also unlike prior law, the taxpayer need not reinvest the sales proceeds in a new residence to claim the exemption. To be eligible, the residence must have been used as the taxpayer's principal residence for at least two of the five years prior to sale.

Because they choose when to sell assets, taxpayers can be expected to realize a capital loss now but to defer a capital gain; hence, tax legislation in most countries limits the ability of the taxpayer to declare a capital loss or to offset it against other income. The rationales for lower taxation of capital gain are fourfold: (a) to offset the fact that capital losses are not fully deductible, (b) to encourage risk taking and venture capital formation that is thought to be essential to economic growth, (c) to simplify accounting by ignoring the carrying costs of the asset, and (d) to account for the effect of inflation generally on asset pricing.

Critics argue, however, that the capital gains rate should be the same as the rate of taxation for normal income and that a lower taxation rate for capital gains constitutes an unwarranted tax expenditure. The capital gain on owner-occupied housing alone in the United States is about $200 billion in an average year. The income tax foregone on this would exceed all public spending on rental housing assistance. However, were capital gains to be taken into normal income, allowance would have to be given for carrying costs incurred by homeowners, hence reducing the net income arising from a capital gain. (SEE ALSO: *Tax Expenditures*)

—*John R. Miron*

Further Reading

Dale Johnson, D. and G. M. Phillips. 1984. "Housing Attributes Associated with Capital Gain." *American Real Estate and Urban Economics Association Journal* 12(2):162-75.

Feenberg, D. and L. Summers. 1990. "Who Benefits from Tax Reductions?" *Tax Policy and the Economy* 4:1-24.

Geiger, M. A. and H. G. Hunt, III. 1989. "Taxation: A Critical Analysis of Historical and Current Issues." *Advances in Taxation* 2:21-39.

Ling, D. C. 1992. "Real Estate Values, Federal Income Taxation, and the Importance of Local Market Conditions." *American Real Estate and Urban Economics Association Journal* 20(1): 125-39.

Poterba, J. M. 1989. 1987. "How Burdensome Are Taxes? Evidence from the United States." *Journal of Public Economics* 33(2): 157-72.

———. "Venture Capital and Capital Gains Taxation." *Tax Policy and Capital Gains Taxation* 3:47-67. ◄

▶ Carrying Costs

Carrying costs are the nominal pretax costs to the owner of holding an asset for a period of time. These costs vary widely depending on the nature and value of the asset and on the mode of ownership. When the asset is raw land held prior to development, for example, carrying costs include the financing charges and property taxes paid by a developer. When the asset is a rental building operated by a landlord, carrying costs might include mortgage expense, property taxes, heat, light, utilities, insurance, grounds-keeping, building maintenance and depreciation, and super-intendency. When the asset is an owner-occupied dwelling, carrying costs are normally thought to include mortgage, installment or contract payments (principal plus interest), and real estate taxes (including taxes on mobile homes or owned trailer sites). In measuring housing expense, the American Housing Survey (AHS) also includes property insurance, homeowners association fees, cooperative or condominium fees, mobile home park fees, land rent, utilities (electricity, gas, water, and sewage disposal), fuels, and garbage and trash collection. An important use of the carrying cost concept is in measurement of the affordability of ownership to prospective homebuyers. Here, carrying cost is estimated by assuming a particular combination of dwelling purchase price, down payment, mortgage interest rate, and other homeownership expense.

Sometimes, the expenses incurred by renters are also considered to be carrying costs, although the only asset involved is a lease. Here, carrying costs are taken to include contract rent, utilities (electricity, gas, water, and sewage disposal), fuels, property insurance, mobile home land rent, and garbage and trash collection.

Carrying costs differ from *user cost;* the latter is the economic cost of asset holding. In the case of homeownership, user cost ignores the component of mortgage payments that constitutes repayment of principal but includes the imputed return on homeowner equity and nets out capital gains and income tax deductions. User cost is a measure of the opportunity cost, as opposed to the out-of-pocket cost, to an asset holder. (SEE ALSO: *Appraisal; Real Estate Agency*)

—*John R. Miron*

Further Reading

Blankenship, Frank J. 1989. *The Prentice Hall Real Estate Investor's Encyclopedia.* Englewood Cliffs, NJ: Prentice Hall.

Jacobus, Charles J. 1996. *Real Estate Principles.* Upper Saddle River, NJ: Prentice Hall.

McKenzie, Dennis J. 1996. *Essentials of Real Estate Economics.* Upper Saddle River, NJ: Prentice Hall.

Peiser, Richard B. 1992. *Professional Real Estate Development: The ULI Guide to the Business.* Washington, DC: Urban Land Institute.

Shim, Jae K. 1996. *Dictionary of Real Estate.* New York: John Wiley.

Tosh, Dennis S. 1989. *Handbook of Real Estate Terms.* Englewood Cliffs, NJ: Prentice Hall. ◄

▶ Center for Universal Design

Formerly named the Center for Accessible Housing, the center was founded in 1989 to develop a variety of affordable housing options and innovative approaches to investigating designing, financing, and managing models of adaptable housing and applications of universal design in the home environment. Publications include *UD Newsline,* a quarterly newsletter; *Infopacks,* various resource lists; *Techpacks,* technical assistance booklets; monographs, and media products. The center offers information and referral services (free basic information and referral about universal design and accessible housing) and technical design assistance services (plan review, design consultation, Americans With Disabilities Act [ADA] consultation, fair housing/multifamily consultation, and technical design materials), which are offered for a fee. Staff: 20. Budget sources: National Institute on Disability and Rehabilitation Research. Director of Outreach and Dissemination: Jan Reagan. Address: North Carolina State University, Box 8613, Raleigh, NC 27695-8613. Phone: (800) 647-6777, (919) 515-3082. Fax: (919) 515-3023. (SEE ALSO: *Physical Disabilities, Housing of Persons with*)

—*Laurel Phoenix* ◄

▶ Center for Urban Policy Research, Rutgers University

The Center for Urban Policy Research (CUPR) provides Rutgers University with an interdisciplinary focal point for studies of urban and regional policy. Its mission is to serve the university, the state, and the nation by conducting research, by disseminating research results, and by educating the researchers and practitioners of the future. Publications include *CUPReport Newsletter,* a quarterly with a circulation of 11,000; CUPR Working Papers; CUPR Policy Reports; monographs on urban policy topics; and a book catalog with full listings of publications, printed twice a year, with a circulation of 50,000. In addition to publications and research, CUPR teaches graduate courses, trains students in research projects, conducts conferences and seminars, and provides expert advice. Founded in 1968, it has a staff of 25 people and a budget of $1.2 million, from university support and public and privately funded grants. Director: Dr. Norman J. Glickman. Address: Rutgers University, 33 Livingston Avenue, Suite 400, New Brunswick, NJ 08901-1982. Phone: (732) 932-3133. Fax: (732) 932-2363.

—*Laurel Phoenix* ◄

TABLE 3 Tenure Forms in Selected East–Central European Countries, 1988 to 1989

	Albania	Bulgaria	Czechoslovakia	Hungary	Poland	Romania	Soviet Union
Public rental	40.0	15.2	34.1	24.0	27.2	33.0	23.0
Private rental		2.8	0.3	2.0	4.1		
Cooperative			18.7		35.4		3.7
Owner occupied	60.0	80.9	46.9	74.0	21.4	67.0	39.0
Other		1.1[a]			11.9[b]		31.7[c]
Other							2.3[d]
Total	100.0	100.0	100.0	100.0	100.0	100.0	100.0

SOURCE: Turner, Hegedüs, and Tosics (1992).

a. Owners and tenants
b. State firms
c. Employers (state firms, ministries)
d. Collective farms

▶ Centrally Planned Housing Systems

Housing systems usually show a mixture of market processes and state intervention. An extreme version of this mixture is the centrally planned housing system in which state regulation is dominant and market mechanisms play an insignificant role. With state control over property rights (limiting individual ownership and eliminating private rental tenure) and concentration of financial means in state institutions (with control over incomes and expenditures of individuals, enterprises, and local governments), the central planning system takes over the determination of housing investments and the shaping of housing policy.

This kind of housing policy was introduced in 1917 in the Soviet Union with the nationalization of all urban land as a first step. After World War II, the countries of Soviet-type political regimes in Central and Eastern Europe had to adopt the logic of this system. The different "stakeholders" of the housing system tried to control the funds allocated to the housing system within the limits of the centrally planned economy. The sectoral ministries, big state-owned companies, "loan" institutions, and cooperatives under the supervision of regional councils also wanted to influence the allocation of the units. For example, the National Saving Bank in Hungary controlled allocation of a huge number of units for sale, most of them with high subsidies. The same was true for the state-owned companies, which tried to increase their control over the subsidies distributed in the system and not to maximize their "profit."

The specific forms of the East European housing systems were different—regarding nationalization and tenure structure, for example—but a common logic determined the "rules" for behavior of the state and private sector and for the different social and economic groups. Based on the analysis of similarities and differences in the housing systems of these countries, the theory of the East European housing model was developed.

Centrally planned housing systems dissolved in Europe around 1989-1990, parallel to the demise of the socialist political and economic system. The then prevailing housing tenures are shown in Table 3. In some countries (notably in Hungary), changes had started earlier with significant modifications in property rights system and central planning.

The Economic Background

After World War II, East-Central European (and some Asian) countries were brought into the sphere of influence of the Soviet Union and under the reign of communist parties. An economic development system was introduced that aimed to restructure the economy, to increase economic development, and to strengthen the economic and military potential. The primary aim was to increase investment in the production sphere, even at the expense of holding back internal consumption, including housing. One of the most important elements of this system was income regulation. Wages did not include the cost of housing, education, health care, and infrastructure. The costs for these services were covered by taxes on companies and then redistributed through the budget. In this way the determination of the level of collective consumption was transferred to the state through the central planning system, excluding market relations. The centrally planned system is based on a permanent shortage situation in the economy. This shortage means that there is a tendency toward a long-term underinvestment in infrastructure, including housing.

Two important rules determined the development of housing policies:

▶ *Limitation of individual consumption:* One family could possess only one flat (in addition, a weekend house or plot was allowed)
▶ *Concentration of the budget on selected forms of new construction:* State rental flats were built and allocated for free or very cheaply.

The Original Setup of the Model

The socialist housing model has never been realized in its pure form. One of the reasons was that the "stock" was very large compared with the "flow." Each country inherited an existing housing system and was not able to carry out a complete redistribution of flats overnight. Both redistribution of the housing stock and control over the private transactions of citizens entailed extreme administrative costs. In the absence of continuous strict control, either the complete prohibition of private transactions or a more liberal approach were the only alternatives. The latter gave private transactions an increasing role in the allocation process, with merely nominal state control. In many countries, cooperative or state-owned flats were also subject of private transactions.

Housing policy relied on indirect means for regulating housing construction (for example, the control over building material supply, land policy, etc.) As a consequence, not only the allocation of housing but also its construction included a sphere that was not centrally controlled. Although officially prohibited, there was an "underground economy," missing from the official statistics.

Consequently, the realities of the East European housing model were mixed scenarios, in which the state sphere and the private sphere were working in different combinations. However, the privately controlled sphere of socialist housing had no effect on the production process. The fact that demand exceeded supply did not increase supply. Similarly, within the state sector, no feedback mechanism was formed based on the interaction of price, demand, and supply.

The state sector means all institutions controlled by the Communist Party. The aim of the institutions in the housing system was to control the distribution process—that is, the allocation of advantages. It was the task of the central state organizations to establish consensus among the institutions and to ensure that the individual institutional endeavors would not threaten macroeconomic targets.

Tensions in the Model

Individuals in the centrally planned housing systems had two theoretical possibilities. One was "voice": realizing housing needs in the state sphere. The other possibility was "exit": stepping out of the state sphere and searching for a solution in the private sphere. The voice option conformed to the East European housing model and was the driving force of organizational changes. This option took various forms, including the housing application, corruption, and different forms of patronage. State housing and regional policy had an important effect on the exit option—for example, through the regulation of migration (in almost all countries up to the late 1980s, the state limited flats for families moving into cities). The exit option included transactions based on agreements between private persons and on self-help construction. Sometimes, the exit option was excluded by the state—for example in elite areas of cities, where strict control was a basic state interest.

In the history of East European housing systems, many institutions (e.g., local councils, companies, ministries) distributed flats and provided subsidies according to preferential principles and procedures, producing clear inequalities. Social groups controlling power used their positional advantages.

Secondary Housing Market

In the shortage economies of East European countries, systematic forced savings characterized household consumption behavior. These accumulated savings represented a potential demand in the housing system. However, the secondary market prices were not the result of open housing market transactions; in many cases, these were illegal or informal transactions. Private rental and subtenancy

prices were much higher than in the state sector. This was illegal because of rent control, but nobody could check the real prices paid.

Self-Built Housing

One of the most typical forms of exit was private housing construction. It existed in each East European housing system, although not always and everywhere to the same extent. The informal economy provided the conditions for private housing construction. Besides private transactions and state building site assignments, building plots were sometimes secured by "self reorganization" from relatives or by building without permission. To overcome the shortage of building materials people used materials from demolition and those "overflowing" or "leaking" from state housing construction. Self-help provided the necessary human resources. Supply could be solved in a way that avoided the real human resource market and state intervention. Financing of private housing construction was sometimes supplemented on a self-help basis. Finally, administrative control of private housing construction could be circumvented by construction without permission or by implementing plans different from the officially approved version. Thus, the state could restrict private housing construction but could not prevent it completely.

The Economic Crisis and the Housing Sector

Since the end of the 1970s, the economic crisis, which began in the capitalist countries with the escalation of oil prices, also spread to the socialist countries. Three different scenarios resulted from these increasing economic difficulties.

In the first, politics did not allow the crisis to affect the housing sector; society maintained the same output of housing construction and did not change its housing regulations. This was the case in the German Democratic Republic (and to a lesser extent in Bulgaria) where growth lasted until the mid-80s, much longer than in the other socialist countries. In the second scenario, the output of housing construction was reduced and its internal structure changed, but no general reforms were initiated (e.g., Romania, Czechoslovakia, and Poland). The third type of reaction to the economic crisis experimented with housing reform to stop the declining housing supply by restructuring the financial burdens and subsidy system (e.g., Hungary). The change in Eastern Europe in 1989 created a completely new political and economic situation in the housing systems. The process of radical change and total transformation of the economic and political socialist system had just begun.

Until 1991, Russia represented perhaps the most extreme form of the East European housing model—state control and a very limited role for the private sector. Housing reform in Russia started with a series of changes in legislation regarding property rights and with a massive privatization. More than 50% of the housing stock had become private at the end of 1994. Banking reform laid the basis for a housing finance system that made possible privatization of the state-owned enterprises. Removing administrative control over private investment was also an important step toward a market-oriented system and reform of municipal housing management. However, by the mid-1990s, housing reform was hindered by worsening macro-economic conditions (recession, inflation), which slowed down institutional restructuring and behavioral adjustment by housing consumers.

Evaluation

The economic performance of the centrally planned housing system has been as questionable as the social character of its allocation mechanism. The basic problem was that housing was separated from the economy and the economic performance of households was only very indirectly tied to their housing situation. According to the original socialist model, housing was to be regarded as a social good, but no society really could afford this. The system generated its own reward system by allocating the best units to individuals in "strategic" positions. This was a very inefficient linkage of job performance and housing, much less effective than in market economies. The majority of state allocation or state-controlled allocation had nothing or very little to do with the economy. As a result, consumer feedback on quality was very ineffective. The cost of housing was low because of price control (grants and subsidies, rent control, subsidized utilities fees, loan repayment, and interest rate). The housing shortage created a secondary, informal high-price market. This market was a quasi market, which influenced only the allocation of the existing stock but not new production because there was no mechanism for feedback to the job market, construction material, and housing finance sector. (SEE ALSO: **Cross-National Housing Research;** *Housing Distribution Mechanisms; Privatization; Welfare State Housing*)

—*Jozsef Hegedüs and Ivan Tosics*

Further Reading

DáNiel, Zsuzsa. 1989. "Housing Demand in a Shortage Economy: Results of a Hungarian Survey." *Acta Economica* 41(1-2):157-80.

Hegedüs, J. and I. Tosics. Forthcoming. "Disintegration of the East-European Housing Model." In *Chasing the Market: Housing Privatization in Europe,* edited by D. Clapham, J. Hegedüs, J. Kintrea, and I. Tosics. Wesport, CT: Greenwood.

Kosareva, N. B., Puzanov, A. S., and Tikhomirova, M. V. 1996. "Housing Sector Reform 1991-1995." In *Economic Restructuring of the Former Soviet Block: The Case of Housing,* edited by Raymond J. Struyk. Washington DC: Urban Institute Press.

Struyk, R., ed. 1996. *Economic Restructuring of the Former Soviet Bloc: The Case of Housing.* Washington, DC: Urban Institute Press.

Szelényi, Iván. 1983. *Urban Inequalities Under State Socialism.* Oxford, UK: Oxford University Press.

Turner, B., J. Hegedüs, and I. Tosics, eds. 1992. *The Reform of Housing in Europe and the Soviet Union.* New York: Routledge (see pp. 151-79). ◄

► Centre for Urban and Community Studies

Founded in 1964, the Centre for Urban and Community Studies facilitates and supports multidisciplinary studies by housing and administering research projects, publishing papers, and improving communications among research-

ers and institutions in Canada and elsewhere. It promotes urban research in developing countries as well as in Canada. Its members are predominantly academic, with strong links to policy and practitioner communities. The center also publishes a newsletter, and organizes seminars and conferences. Publications include research papers, bibliographies, and books. Their circulation varies from 100 to 500. Major publications are also released through commercial publishers. The center operates on a base budget from the University of Toronto, with additional funding from research grants. Administrative staff size: 4. Information Officer: Judith Bell. Address: Room 426, 455 Spadina Ave., University of Toronto, Toronto, Canada, M5S 2G8. Phone: (416) 978-2072. Fax: (416) 978-7162.

—*Laurel Phoenix* ◄

► Certificate of Occupancy

Construction and use of new buildings requires building permits and certificates of occupancy issued by a local government code enforcement agency. That local agency, usually a building or engineering department, conducts inspections of the building at various phases of construction. The building code applies minimum standards to skeletal structure and to the mechanical, plumbing, and electrical systems. A certificate of occupancy, also called an occupancy permit, is the final issuance granting legal permission to use a structure.

Most local code enforcement agencies apply standards recommended by state building codes, which in turn are derived from standards established by the federal government and the various model codes established by national professional organizations. These model codes include the Uniform Building Code, established by the International Conference of Building Officials; the BOCA National Building Code, set by the Building Officials and Code Administrators International, Inc.; and the Standard Building Code, established by the Southern Building Code Congress International. The certificate of occupancy is issued for a specific use. With change of use, the owner-user must apply for a new certificate. Should a building without a certificate of occupancy incur structural damage or fault, the owner likely would be unable to collect insurance proceeds. Occupancy cannot legally occur without a certificate of occupancy; a temporary certificate may be issued in some cases. Unlike commercial properties that are constructing a new addition, residential additions do not require a new certificate of occupancy. (SEE ALSO: *Building Permit; Housing Occupancy Codes*)

—*John S. Klemanski and John W. Smith*

Further Reading

Building Officials and Code Administrators International, Inc. 1993. *National Building Code.* Country Club, IL: Author. Annual Supplements.

Council of American Building Officials. 1991. *CABO One and Two Family Dwelling Code.* Falls Church, VA: Author. Annual Supplements.

International Conference of Building Officials. 1991. *Uniform Building Code*. Whittier, CA: Author. Annual Supplements.
Southern Building Code Congress International. 1991. *Standard Building Code*. Birmingham, AL: Author. Annual Supplements. ◄

► Chattel Mortgage

A chattel mortgage is an agreement in which an item of personal property is pledged to secure a loan. The adoption of Article 9 of the Uniform Commercial Code (UCC) by each state legislature led to the replacement of most chattel mortgages with the modern security agreement. Under the UCC, items of personal property that are to become real property fixtures may be pledged as collateral through the use of a security agreement. A common example is a security interest in a permanently installed air-conditioning unit. A security agreement in fixtures must be perfected by filing a financing statement (called a "fixture filing") with the government office designated within the county in which the real property is situated. In some cases, this may be the same office where a mortgage of the real property would be recorded. If the secured party perfects the security interest by filing at the proper office, the secured party will have a priority over all subsequent interests and claims relating to the fixture, subject to certain exceptions set forth in the UCC. Despite the changes brought about by the adoption of the UCC, security agreements covering fixtures often are labeled "chattel mortgages." (SEE ALSO: **Mortgage Finance**)

—*Jeffery M. Sharp*

Further Reading

French, William B. and Harold F. Lusk. 1979. *The Law of the Real Estate Business*. 4th ed. Homewood, IL: Irwin.
Seidel, George J. 1993. *Real Estate Law*. 3d. ed. Minneapolis/St. Paul: West.
Uniform Commercial Code § 9-313. ◄

► Children

Creating housing in which children do not merely live, but thrive, requires attention to basic physical, cognitive, emotional, and social needs. These basic needs are constant, whereas the housing context is changing rapidly—family structures, community networks, technologies. Detached single-family housing has been marketed as the ideal Western child-rearing environment, whereas the contemporary reality for many children is urban environments, small families, single parents, mothers in the paid labor force, poverty, or migration: conditions that often make conventional ideals unattainable or dysfunctional. Good housing can facilitate positive family functioning and child development, just as bad housing can be an impediment. Although children are affected by their housing conditions, they have little or no voice in housing decisions. Therefore, the following sections review important considerations in designing and building for them, with a focus on the period from birth through age 12 and with attention to urban as well as suburban and rural conditions.

Health and Safety

Healthy physical growth, a basic need, requires housing that is safe from injury and pollution. A problem, however, is that standards of acceptable risk for exposure to environmental contaminants are currently normed for adults, whereas children are far more vulnerable. Children have an underdeveloped immune system, lesser mass, more active metabolism, faster breathing rate, less sanitary habits (playing on the floor, eating dirt, sucking their thumb, scratching surfaces), and a longer future in which to develop long-term diseases such as cancer. Therefore, every effort should be made to minimize indoor and outdoor contamination beyond guidelines for adults.

In terms of siting decisions, family housing should not be located downwind from toxic industries or near airports, heavy traffic, waste sites, abandoned military sites, sewage treatment plants, incinerators, or garage vents. In renovations, asbestos and lead-based paint should be professionally encapsulated or removed. In new construction and interior furnishings, natural materials should replace particle boards, vinyls, and plastics, which outgas formaldehyde, benzene, styrene, and other suspected carcinogens. New paint should be left to cure before a child is moved into a bedroom. To prevent accumulations of indoor pollution, ventilation by means of windows, heat exchangers, and stove exhausts should be a priority beyond the minimum regulated by legislation.

Sensory and Social Stimulation

The largest body of research regarding children's housing experience relates cognitive development to indoor stimulus levels in the form of noise, density, and opportunities for sensorimotor play. Toddlers and preschoolers have been observed to spend 80% to 90% of their waking time interacting with their physical rather than their social environment. Hence, it is not surprising that young children score higher in sensory coordination and cognition when they have changing home decorations and views out windows and when they can move freely around rooms and explore a variety of complex, responsive objects.

Unit-per-acre and person-per-room densities need to strike a balance between too low densities associated with social isolation and too high densities associated with high levels of noise and distractions. Both high residential density and high noise levels relate to poor language development in infants and impaired information processing and language achievement in older children. High person-per-room densities are negatively related to cognitive performance in both infants and older children; and among preschool and grade school children, they are positively associated with aggression, anger, and acting out. Thresholds issued by the Environmental Protection Agency's Office of Noise Abatement and Control are set with adults in mind, but children of language-learning age are particularly at risk.

These findings suggest several design recommendations. To facilitate play, through which children discover the prop-

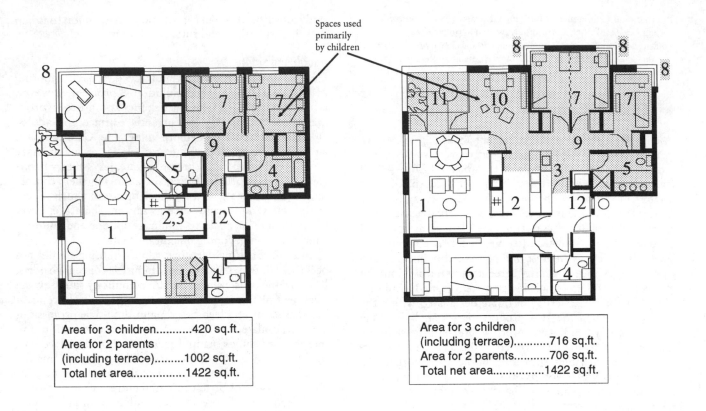

Spaces used primarily by children

Area for 3 children...........420 sq.ft.
Area for 2 parents
(including terrace).........1002 sq.ft.
Total net area................1422 sq.ft.

Area for 3 children
(including terrace)...........716 sq.ft.
Area for 2 parents............706 sq.ft.
Total net area................1422 sq.ft.

APARTMENT IN FAVOR OF ADULTS

1. Large living/dining area primarily for adults' entertaining.
2. Compact one-person kitchen
3. Kitchen for adults' use only
4. 2½ or 3 bathrooms
5. Custom-designed fixtures and finishes for parents' bathroom
6. Spacious master bedroom
7. Bedrooms too small for 2 children, relatively large for 1 child
8. Distinctive corner window allotted to master bedroom
9. No "wasted" corridor space
10. Play restricted to child's bedroom and a corner of living room
11. Direct access from adult space to terrace
12. Entrance designed for maximum privacy and noninvolvement

APARTMENT IN FAVOR OF CHILDREN

1. Smaller living/dining area for adults' entertaining
2. Larger kitchen, centrally located
3. Low children's sink in counter
4. Smaller budget for bathrooms
5. Children's bathroom designed for their access, safety, and play
6. Flexible-use parents' bedroom
7. Small autonomous bedroom space guaranteed for each child
8. Distinctive corner windows allotted for children's rooms
9. Bedroom corridor usable for play
10. Supervisable indoor play area of at least 35 sq.ft. per child
11. Direct access from indoor play are to outdoor play space
12. Vision panel at entry overlooking public hallway

Figure 1. Apartments of Equal Area but with Different Impact on Children's Home Life.
Christine Benglia Bevington, Architect.

erties of things and their own abilities and limits, the primary needs are sufficient floor space, where children can be in sight or hearing but not underfoot, and accessible storage space for play materials. Given budget constraints, when parents have participated in design or in postoccupancy evaluations, they have preferred to have a large kitchen/family room, where children can play in sight of parents' work, at the expense of living room or bedroom size. Figure 1 illustrates some design options to achieve these ends within the limits of an apartment.

Negative associations with noise and density indicate a need for what has been termed a "stimulus shelter": a room or corner where children can withdraw for rest or quiet.

This research also indicates a need for multiple areas for family activities, within or beyond the dwelling, to reduce room densities, and effective soundproofing between rooms and between the residence and the exterior. Closets, hallways, exterior plantings, construction methods, and siting can be designed to buffer noise.

Although overcrowding may result in antisocial behavior, low densities decrease opportunities for social learning. Canadian recommendations for family housing call attention to the social danger of suburban or rural environments where preschoolers often have to play alone. Added to this issue is the need for adequate density to facilitate child care arrangements.

Security and Identity

As well as a place for physical growth and cognitive learning, the dwelling is a center for the development of emotional attachments and personal identity. When it forms a reliable, familiar base for shelter, nourishment, social relations, and sleep, the home fosters a basic trust in the world. This essential function is suggested by high measures of emotional disturbance among homeless children.

As the location where children control personal possessions and where they may be allowed to arrange and decorate personal space, the home—particularly the child's room—contributes to the development of a sense of individual rights and self-identity. Through spatial negotiations with other family members, children learn basic concepts of autonomy and privacy.

As early as the age of three, children associate a room of their own with personal control and identity. In interviews, young subjects of all ages commonly name their bedroom as the interior space where they most like to be, where they keep treasured possessions, and where they go when they are upset or do not want to be disturbed. However, they are less likely to name their room as a favorite place or where they go to play, read, or do homework if they share it with siblings; and they are less likely to consider it personal territory if they share a small rather than a large room. Parents—who, unlike children, have control over the rest of the house—tend to rank their own bedroom low in importance.

These results contrast with the custom of designing a large master bedroom for parents. A more useful practice is flexible design that can change functions in time with the family life cycle. If the placement of windows, doors, closets, and bathrooms allows rooms to be divided or turned into detachable units, the dwelling can be adjusted to changing family needs: playroom, workroom, guest room, study, office, granny flat, apartment to rent, or a live-in area for a caregiver (see Figure 1).

Children's drawings, as well as residential autobiographies in which adults recall childhood homes, show an attraction for special places such as attics, landings, window seats, the underside of stairs, fireplaces, alcoves, balconies, porches, and roomy convivial kitchens. Here, children can balance on the fringes of where the action is, poised between inside and out, togetherness and solitude. In the name of cost-effectiveness, these "frills" are often eliminated from modern housing. Despite the trend to smaller families and smaller dwellings, these associations suggest that each home should include at least one such "special place" that can be claimed by its children.

Access to Community Resources

According to parents' reports, children engage in more quiet, passive play than active, creative play in the home, and active play decreases along with dwelling size. When the outdoors is both safe and challenging, it offers an outlet for gross motor exercise and creative play and is a refuge from household noise, density, and stress. Out of doors, young people encounter the social world and natural world, extending their spheres of knowledge, attachment, and competence. Thus, access to outdoor resources needs to be an integral goal of housing policy and design, in multifamily as well as single-family schemes.

Children can negotiate with caregivers to gradually extend their free range when residences and sites include safe spaces that can be overseen from dwellings and a graduated series of transitional spaces that can be used for play: hallways, stairways, porches, paths, yards, courtyards, and adjoining streets. Young people are observed to be the heaviest users of the housing site, and these close-to-home spaces compose the main part of children's habitual range of outdoor activity.

High-rise housing forces families into "all or nothing" decisions in which parents must either restrict children to the apartment, supervise them outside, or relinquish contact when they go out alone. Studies show that preschool and school-age children spend less time outdoors when they live on high floors than when they live at or near ground level. This problem can be addressed by low-rise multifamily housing with play areas in sight and call of parents' work areas, by reserving the lower three floors of high-rise housing for families with children, or by constructing rooftop playgrounds or terrace gardens.

Observations show that children play all over sites. Paved areas—sidewalks and streets (not playgrounds)—receive the heaviest use. Acknowledging this fact, some design schemes have created protected internal pathways within developments (e.g., Radburn, New Jersey, by Clarence Stein and Henry Wright; and Village Homes in Davis, California, by Michael Corbett). Children, however, like to be wherever the action is. A Dutch solution that aims to preserve outdoor access in urban areas is the *woonerf*—a residential street designated primarily for pedestrian use, where traffic must slowly navigate bumps, bends, parked cars, plantings, street furniture, and play equipment. (Boundary streets carry through traffic.)

When child densities exceed the capacity of legitimate site attractions, vandalism can be expected. Whenever densities exceed 30 children per acre, supervised on-site recreation should be implemented.

Access to Social Diversity

Through the economic, racial, cultural, and age diversity of neighbors, children can grow to understand the wider society, become more tolerant of various lifestyles, and be better prepared for the future. Segregation by age, ethnic group, and income fails to prepare children for the contemporary global level of social interactions. Single-use, low-density suburbs, which have been marketed as ideal settings for raising children, offer limited activity settings and opportunities for this needed socialization.

Within communities lacking social diversity, scattered-site affordable or subsidized housing can introduce a range of incomes and ethnic groups. Mixed-use zoning increases activity choices by allowing educational, cultural, commercial, and other nonharmful activities within residential neighborhoods, making it easier for children to connect home experience with school and the world of work, travel independently, find adult role models, and envision their place in society. For working parents and single parents, mixed-use zoning can reduce or eliminate commuting time and bring services and child care in proximity of the home.

An innovative example of mixed use is Laguna West near Sacramento, California, conceived by Peter Calthorpe. Not only is it pedestrian oriented, with all housing within walking distance of public transportation, but it also is zoned to integrate homes, schools, work, shopping, recreation, natural areas, and even agriculture. Children thus recapture some of the advantages of living in a preautomobile era village.

A means to accommodate a mix of ages is cohousing, introduced to North America from Scandinavia by Charles Durrett and Kathy McCamant. Cohousing groups, which are self-selected, deliberately try to create a socially integrated environment for themselves and their children. Children in a cohousing group have as much access to their parents as children in conventional housing, but they also have access to shared indoor and outdoor spaces where they can meet other people of all ages. This arrangement facilitates supervision, child care, and the availability of playmates.

Access to Nature

Although children use porches, stoops, sidewalks, streets, and yards most heavily, when they are questioned regarding their favorite outdoor place, they primarily name natural areas. Fields, woodlots, earth, and water allow children to create their own imaginary worlds with nature's loose parts and to learn about the natural world. Therefore, site plans should preserve distinctive landscape features and include natural areas for play. Gardening and recycling also teach natural cycles.

In projects such as the Danièle Casanova community in Ivry on the outskirts of Paris, innovative ways to incorporate green space within high-rise living have been pioneered by the late French architect Jean Renaudie and his followers (see Figure 2). Patterned after dense Mediterranean hill towns, these projects form artificial hills where dwellings are grouped around private turf-covered terraces and linked by paths, ramps, and steps winding their way along elevated gardens and semipublic spaces. This intricate and challenging high-density environment preserves urban children's connection to vegetation and reduces heavy reliance on expensive-to-maintain elevators.

Conclusion: Child-Sensitive Policy and Design

There is a need to reevaluate housing policy and design in the United States. Housing codes, zoning laws, Federal Housing Administration guidelines, Department of Housing and Urban Development minimum standards, and common management and maintenance practices that inadequately support child development need to be amended. Table 4 proposes the creation of a category of "childhood dwellings" with a child-responsive set of codes and regulations.

Because children are primary users of the home and its surroundings but usually have no direct voice in selecting or shaping their place, housing providers have a special

Figure 2. At the Danièle Casanova housing project in Ivry-sur-Seine, designed by Jean Renaudie, each unit has its own turfed terrace for play, and children also share common paths and green spaces.
Photograph by Christine Benglia Bevington.

responsibility to address children's needs. One approach to this goal is to review and conduct relevant research and to derive design guidelines from it.

A second approach is for planners, designers, and policymakers to compose residential autobiographies, or written and graphic records of personal home experiences, beginning with earliest memories. This method can sensitize adults to children's perspectives and values.

A third approach, which may be combined effectively with the preceding two, is to involve the young in participatory research, design, and planning. Involving young people in environmental planning and management not only yields significant information but prepares them to take democratic responsibility for the quality of their homes and communities. At its best, good housing for children today provides a model for good building and caretaking in the future. (SEE ALSO: *Adolescents; Cohousing; Home; High-Rise Housing; Multifamily Housing*)

—*Christine Benglia Bevington and Louise Chawla*

TABLE 4 Childhood Dwellings—Proposed Codes and Regulations

CHILDHOOD DWELLINGS	HEALTH AND SAFETY	VENTILATED KITCHENS
• Create a housing category for dwellings specifically designed to raise children 12 years old and under. • In this category relax standards and regulations which, although of advantage to some adults, are of little value to most children (master bedrooms, large living rooms, high ceilings, wheelchair access, parking requirements . . .) in exchange for provisions known to support their healthy physical, mental, and social growth. • Require a minimum of 20% of new housing to qualify as "childhood dwellings." **1**	THE SURGEON GENERAL HAS DETERMINED THAT THIS HOME IS NOT HARMFUL TO A CHILD'S HEALTHY GROWTH • Adopt more conservative environmental health standards for children than for adults. • Offer periodic testing of noise, materials, and air and water quality in spaces where children live **2**	 • Allow full-sized kitchens (not just kitchenettes) to be ventilated mechanically so as to encourage affordable large kitchens best located for child supervision. • Require a ventilated hood over the cooking range whether or not a kitchen has a window. **3**
SAFE, LOW WINDOWS	**ACCESSIBLE DESIGN**	**CHILD-SIZED BEDROOMS**
• Allow low window sills so that small children may look out. • Prevent accidents through alternative safety measures—fixed sash, window guards, screen, special hardware, reinforced glass, wire mesh, child-proofed parapet—not lack of view. **4**	• Recognize children's boundless energy as significant a design factor as elder frailty. • Within "childhood dwellings," consider wheelchair access requirements secondary to those of children: Allow loft spaces, low headroom, steps, ladders, tunnels, nooks, and crannies in most cases, although not all. • Give developers and parents freedom of choice as to controls or height of kitchen equipment, counters, bathroom fixtures, light switches, door handles, etc. • Encourage flexible solutions that can adapt to growth. **5**	 • Reduce minimum bedroom area to 45 sq.ft. per child, minimum width to 6 feet per child. • Guarantee a window, a door, and a closet for each child even when a room is shared. • Require an indoor play area in addition to above minimums. **6**
INDOOR PLAY AREA	**OUTDOOR ACCESS**	**NEIGHBORING OPTIONS**
• Designate on all floor plans an indoor play area of no less than 35 sq.ft. per child and within 50 feet of a point of adult supervision. • The indoor play area may be within children's rooms, within the dwelling, or adjacent to it. • Apply same standards of natural ventilation, light, and fire egress for the play area as for other habitable rooms. • Its access, use, maintenance, supervision, or decoration may be shared with immediate neighbors, if desired, following a precise agreement between the families concerned and the management. **7**	 • Require an outdoor or indoor/outdoor space easily accessible to children of one or more dwellings. • Specify safety regulations when such spaces occur on high floors. • Require adult participation and supervision over all play areas. **8**	 • Remove barriers to the possibility of sharing spaces or services with adjacent dwellings. • Allow communicating doors, accessory units, semiprivate play spaces, swing rooms, and sharing maintenance, storage, or work spaces if desired by residents. **9**
FRIENDLY HALLWAYS	**NEIGHBORHOOD FITNESS**	**FAMILY CARE HOMES**
• Allow alcoves for "doorstep play" and some furniture, toys, plants, and decoration in public hallways used primarily by families with children. • Allow vision panels from inside dwellings to public hallways. • Protect children from elevator doors with gates or other means. • Require a high level of acoustic protection in corridors and between dwellings with children. • Accept alternatives to hard-surfaced, bare corridors to achieve fire safety: sprinklers, automatic shutters, fire drills, mandatory child supervision, etc. **10**	 • Declare buildings, projects, streets, playgrounds, or parks where crime, drugs, pollution, or other major dangers grossly interfere with a child's use of the residential setting as "no kids" areas. • Offer families with young children housing alternatives where a normal childhood is possible. • Mandate immediate and effective actions to restore physical, mental, and social health to "no kids" areas, to remove the severe violation, and to return access to its children. **11**	• Confer commercial value to "childhood dwellings" equipped with sufficient indoor and outdoor play space for 6 children by certifying them as potential family day care homes. • In high-density areas suffering from lack of child care, reward project submissions that integrate child care and home design with zoning bonuses, speedy reviews, and helpful guidance toward approval. Family day care homes, group family care homes, play groups, intergenerational day care, and housing-based child care centers are some examples. **12**

Further Reading

Altman, Irwin and Joachim F. Wohlwill, eds. 1978. *Children and the Environment.* New York: Plenum.

Bartlett, Sheridan. 1997. "Housing as a Factor in the Socialization of Children." *Merrill-Palmer Quarterly* 43(2):169-98.

Chawla, Louise. 1991. "Homes for Children in a Changing Society." Pp. 187-228 in *Advances in Environment, Behavior, and Design.* Vol. 3, edited by E. H. Zube & G. T. Moore. New York: Plenum.

Cooper Marcus, Clare and Wendy Sarkissian. 1986. *Housing as If People Mattered.* Berkeley: University of California Press.

Pollowy, Anne-Marie. 1977. *The Urban Nest.* Stroudsburg, PA: Dowden, Hutchinson & Ross.

Schoemaker, Rosemary and Charity Vitale. 1991. *Healthy Homes, Healthy Kids.* Covelo, CA: Island Press.

Van Vliet--, Willem. 1983. "Families in Apartment Buildings." *Environment and Behavior* 15:211-34.

Weinstein, Carol and Thomas David, eds. 1987. *Spaces for Children.* New York: Plenum.

Wohlwill, Joachim F. and Willem van Vliet--, eds. 1985. *Habitats for Children.* Hillsdale, NJ: Lawrence Erlbaum. ◄

► City Limits

City Limits is an urban affairs magazine covering community action and government policy as it affects low-income neighborhoods in New York City and elsewhere. It includes news and investigative reporting on housing, economic development, homelessness, public health, the urban environment, and related topics, as well as commentary essays by practitioners. Subscribers are community leaders, nonprofit developers, activists, planners, politicians, city officials, academics, journalists, and other makers of opinion and policy. First published in 1976, its circulation is 5,000 nationwide, about 80% located in the New York region. Subscriptions are $25 per year for individuals and nonprofits, $35 per year for government and business. It is published 10 times a year, monthly except June/July and August/September. Submissions policy: Query first by letter or phone. Editors like to see a writer's clips and résumé before assigning an article. Most *City Limits* writers are professional journalists, except for commentary authors. Submissions editor: Carl Vogel. Address: 120 Wall St., New York, NY 10005. Phone: (212) 479-3344. Fax: (212) 344-6457.

—Andrew White ◄

► Civil Rights Act of 1866

Be it enacted by the Senate and House of Representatives of the United States of America in Congress assembled, That all persons born in the United States and not subject to any foreign power, excluding Indians not taxed, are hereby declared to be citizens of the United States; and *such citizens, of every race and color,* without regard to any previous condition of slavery or involuntary servitude, except as a punishment for crime whereof the party shall have been duly convicted, *shall have the same right, in every State and Territory in the United States, to make and enforce con-*tracts, to sue, be parties, and give evidence, to inherit, purchase, lease, sell, hold, and convey real and personal property, and to full and equal benefit of all laws and proceedings for the security of person and property, as is enjoyed by white citizens, and shall be subject to like punishment, pains, and penalties, and to none other, any law, statute, ordinance, regulation, or custom, to the contrary not withstanding.
—April 9, 1866, An Act to protect all Persons in the United States in their Civil Rights, and furnish the Means of their Vindication (emphasis added)

In the wake of the Civil War, Congress passed the Civil Rights Act of 1866. The two main elements of the act were these: (a) Racial discrimination could not be a factor in decisions to engage in a contract of any sort, and (b) racial discrimination could not enter into decisions concerning the acquisition and subsequent use of housing or other property. In effect, this statute was the first U.S. fair housing act and is still used either independently or in conjunction with Title VIII of the 1968 Civil Rights Act to adjudicate housing discrimination cases. (Title VIII, with amendments, is known as the Fair Housing Act [FHA].) Although the two civil rights acts overlap in many areas of jurisdiction, each has significant unique provisions.

The Civil Rights Act of 1866 is codified in the United States Code in Title 24, the Public Health and Welfare, § 1981 and § 1982 (the Fair Housing Act is also in U.S.C. Title 42). Section 1981, "Equal rights under the law," guarantees the right to contract to all citizens regardless of race. Although this right was originally made law in 1866, there were dissenting opinions concerning Congress's authority to enact these antidiscriminatory measures. In 1870, Congress reenacted the right of all people to contract (May 31, 1870, chap. 114, § 16 Stat.144). The second provision, § 1982, more specifically deals with housing; it is subtitled "Property Rights of Citizens."

This entry discusses § 1982, the provision legislating that all citizens have the same right to property as white citizens. (For enabling case law, refer to the bibliography below.) The 1866 act had lain nearly dormant as a remedy for private housing discrimination until its resurgence in April 1968, with *Jones v. Alfred H. Mayer Co.* (392 U.S. 409). Here, a developer refused to sell a home to Mr. Jones, only because he was black. The Supreme Court's ruling in Mr. Jones' favor came just two days before the enactment of Title VIII of the Civil Rights Act and only weeks after the assassination of Martin Luther King, Jr. Previously, § 1982 had been generally held to bar discrimination only in state actions.

Title VIII makes it illegal to discriminate in the rental, sale, or advertising of a dwelling based on a person's race, color, religion, sex, familial status, national origin, or handicap (42 U.S.C. § 3604). This broad definition of federally protected categories of people contrasts markedly with the earlier § 1982, which protects only people who have been discriminated against because of their race or color.

However, the 1866 Civil Rights Act covers a larger population than might be expected by its restriction to race and color. Courts construe "race" to cover much of what is now categorized as ethnicity; Jewish, Arab, and Latino plaintiffs have successfully invoked § 1982. The Court reasons that Congress's intentions when drafting the act was to include all groups that, in 1866, were considered racial groups.

Section 1982 also covers some situations not included in the FHA. Most notably, owner-occupied multifamily dwellings with four or fewer separate units are exempt from the FHA [§ 3603(b)(2)]. Thus, when a person is refused a unit in a small multifamily dwelling because of race, his or her recourse in law is § 1982—the Civil Rights Act of 1866. Cases brought under § 1982, even when simultaneously brought under Title VIII, may enjoy a longer statute of limitations than pure Title VIII cases.

Further benefits of § 1982 are that it covers discrimination related to any "real or personal property," not just dwellings, and any race-based interference with their use. For instance, if a recreational club is nominally open to everyone in a predominately white area, yet membership is refused to an eligible nonwhite family because of color, the family members are protected by § 1982. Public housing tenants have used the statute to fight racially segregated public housing. And in 1948, the Supreme Court found racially restrictive covenants to be unenforceable under § 1982 (*Shelley v. Kraemer*, 334 U.S. 1).

The focus of the older act is on providing individuals with specific property rights, whereas the FHA defines what practices constitute discrimination in housing. Thus, although only the FHA covers discriminatory advertising, the two acts are often used in a complementary manner. For instance, both cover *steering*, the creation of dual housing markets, and protect white people who have been denied housing or have been evicted on account of their association with people of other races or ethnicities, be the association through entertaining in one's home or through marriage.

The U.S. Department of Justice is not authorized to bring housing discrimination cases under the 1866 Civil Rights Act; it relies on Title VIII, the Fair Housing Act. Private attorneys, however, use § 1981 and § 1982 in response to discriminatory practices.

The right to contract and enjoy property continues to be hindered by racial and ethnic discrimination. It is noteworthy that an act designed during the Reconstruction era, with the explicit purpose to outlaw such discrimination, is still invoked so often. (SEE ALSO: **Fair Housing Amendments Act of 1988**)

—*Ellen-J. Pader*

Further Reading

Schwemm, Robert G. 1990. *Housing Discrimination: Law and Litigation.* New York: Clark Boardman.

United States Code Annotated (U.S.C.A.). Title 42. The Public Health and Welfare, Chapter 21—Civil Rights. West. 1995. ◀

▶ Civil Rights Act of 1968, Title VIII (Fair Housing Act)

Title VIII of the 1968 Civil Rights Act (P.L. 90-284; 42 U.S.C. 3601-3631), known as the Fair Housing Act, was the first national legislation to address broadly the problem of housing discrimination. Specifically, Title VIII sought "to provide, within constitutional limitations, for fair housing throughout the United States" by prohibiting discrimination based on race, religion, or national origin in the sale and rental of housing. The Department of Housing and Urban Development (HUD) shared enforcement duties with the Department of Justice (DOJ), although a compromise struck in Congress to pass the controversial bill provided weak enforcement mechanisms relying primarily on voluntary compliance and offering few sanctions for discriminatory behavior. Mounting criticism of the legislation's weak enforcement mechanisms and of HUD's slow and incomplete implementation of Title VIII, combined with evidence that discriminatory housing practices persisted, resulted in the 1988 amendments to the Fair Housing Act. These amendments broadened the scope of the act and changed its enforcement procedures.

The Fair Housing Act was the last in a series of 1960s civil rights legislation that included the prohibition of discrimination in public accommodations and employment and the guarantee of voting rights. Although, in 1962, President Kennedy had issued Executive Order 11063 prohibiting discrimination in new government-owned or -operated housing, this order applied to only about 1% of the nation's housing stock. An attempt to pass a fair housing bill in 1966 had failed because of opposition from Southern Democrats in Congress.

Although the U.S. Commission on Civil Rights, scholars, and advocacy groups had been studying the problems of housing discrimination and residential segregation for many years, in 1967, riots in black ghetto neighborhoods across the United States focused the nation's attention on these issues. Investigations into the riots cited blacks' poor housing conditions and their difficulty finding jobs as motivating factors of the urban unrest. In addition, documentation was accumulating of discriminatory practices in the real estate industry that were contributing to segregated living patterns.

The Johnson administration and a group of liberal senators renewed the fight for national fair housing legislation in 1967. Opponents, mostly Southern Democrats, claimed it would infringe on property rights and states' rights, calling the proposed bill "forced housing" legislation. In early March 1968, during Senate debate of the bill, the Kerner Commission, a body appointed by the president, released its report on civil disorders. It warned that the United States was becoming "two societies, one black, one white—separate and unequal." This highly publicized report, combined with the appeals by the president and some senators to patriotism and justice, prompted the Senate to pass a compromise bill. The compromise bill covered most of the nation's housing but weakened HUD's enforcement powers. In April 1968, Martin Luther King's assassination prompted passage of the bill in the House of Representatives.

The Fair Housing Act outlines seven types of practices that constitute discrimination when based on race, religion, or national origin:

1. Refusing to sell or rent to an individual after he or she has made an offer on a dwelling, refusing to

negotiate the sale or rental of a unit, making a unit unavailable, or denying the dwelling

2. Discrimination regarding the terms, conditions, or privileges of sale or rental
3. Indicating racial or ethnic preference in real estate advertisements
4. Lying to an individual about the availability of a unit
5. Blockbusting, or attempting, for profit, to induce individuals to sell or leave their homes
6. Denying a loan or giving different terms to an individual
7. Denying realtors access to multiple-listing services or real estate organizations.

The Fair Housing Act applies to nearly all housing in the United States, with two exceptions: (a) single-family homes rented or sold by the owner without the use of a broker or of advertisements and (b) multifamily dwellings with up to four units, one of which is occupied by the owner. In addition, religious organizations and private clubs that provide lodging for purposes other than commercial gain may give preference to members of their religions or clubs.

Under the original act, HUD and the DOJ shared responsibility for implementing the fair housing policy. Individuals could file complaints with HUD, whose staff would investigate and engage in a conciliation process designed to achieve voluntary compliance. Although HUD lacked the authority to enforce resolutions, it could refer cases to the DOJ. HUD was also charged with ensuring that federal housing programs were administered in a nondiscriminatory fashion. The DOJ could prosecute cases in which a "pattern and practice" of discrimination was evident or in which the discrimination was of "general public importance." In addition, individuals could file civil suits seeking actual damages and up to $1,000 in punitive damages.

Despite the Fair Housing Act, residential segregation and discrimination in real estate and mortgage lending has persisted. The statute's inherent weaknesses along with its inadequate implementation limited its effectiveness. The act relied heavily on individuals who suffered discrimination to file complaints or suits, yet the fines established in the law were relatively small, and plaintiffs were responsible for lawyer and court fees. To file a suit or complaint, an individual had to be aware that discrimination had occurred, which can be difficult in light of realtors' and leasing agents' subtle actions. Thus, fair housing advocacy groups have played an important role in working with individuals to identify discrimination and to file lawsuits. Advocacy groups, sometimes sponsored by HUD, also have conducted fair housing audits to identify housing discrimination.

Congress's failure to articulate its intentions clearly in several parts of the original act contributed to its weakness. Vague statutory language left the act open to a variety of interpretations. For example, Section 808 charges executive agencies to administer their programs "in a manner affirmatively to further the purposes of this title." Some advocates and courts have interpreted this phrase to mean that the government must work to achieve integrated housing patterns, whereas others believe it charges government only to ensure that discriminatory actions do not occur.

HUD's implementation of the Fair Housing Act was slow and spotty during the 20 years leading up to the 1988 amendments. Evaluations published during this time portray an agency reluctant to issue regulations, coordinate its activities with other agencies affected by the legislation, and integrate fair housing enforcement into its agenda. HUD did not issue rules interpreting what would constitute a violation of the Fair Housing Act or what standard of proof would be used. The agency was not able to reach conciliation in most of the complaints it handled. Although developers receiving federal money were required to submit affirmative marketing plans, HUD did not monitor their compliance.

The DOJ responded more vigorously than HUD to the Fair Housing Act, although limited resources and a changing political climate constrained its capacity. During the 1970s, for instance, the DOJ won most of the fair housing suits it initiated, although the number of cases it pursued declined during the 1980s. In addition, the U.S. Commission on Civil Rights praised the department in the late 1970s for the consent decrees negotiated with offenders. Although these decrees secured monetary relief for individual victims, many also included affirmative remedies, procedures to be followed by leasing agents that would guard against discrimination.

At the same time, however, the DOJ defended HUD in privately initiated discrimination suits, notably in the *Gautreaux* litigation, which began with a 1966 class action suit in which black public housing tenants charged HUD and the Chicago Housing Authority (CHA) with intentionally segregating Chicago's public housing. The Supreme Court held in *Hills v. Gautreaux* (425 U.S. 284, 1976) that the two agencies had violated federal statutes and the Fourteenth Amendment. The DOJ has also initiated suits that some observers find antithetical to the spirit of the Fair Housing Act. In the 1980s, *United States v. Starrett City Associates* (488 U.S. 946, 1988), the department successfully challenged a subsidized housing development's integration maintenance program.

In the absence of HUD regulations, the courts have acted as primary interpreters of the Fair Housing Act. Case law emerged from DOJ-initiated and private litigation regarding who had standing to file suit under the act, the standard of proof required, and which actions were covered by the law. The courts have also spurred HUD to halt its own discriminatory practices and to remedy their effects. Yet litigation has proved an inefficient method of implementation, requiring much time and many resources and providing no guarantee that an individual's immediate housing needs will be met. In addition, most remedies have been narrow in scope, focusing on the individual (or group of individuals in the case of class actions) bringing the suit. Remedies typically require an "offender," for example, a real estate agent or owner of rental housing, to compensate victims and to take affirmative measures to prevent the individual's company from discriminating in the future. Such remedies rarely address the broader conditions and dynamics that give rise to discrimination and segregation. Litigation also provides geographically uneven implementation. Not every city and state, for example, has an active fair housing group; these are the groups that have been responsible for initiating much of the private fair housing litigation and, in many cases, for prompting publicly initiated litigation.

Although attempts to amend the Fair Housing Act began in 1977, Congress finally acknowledged its weaknesses in 1988 when it passed the Fair Housing Amendments Act. This legislation strengthens HUD's enforcement powers and extends the act's coverage to disabled people and to families with children. (A 1974 amendment had extended the act to cover discrimination on the basis of sex.) The amended act also prohibits discrimination in home repair and improvement loans and in the secondary mortgage market.

Over the years, Congress has passed other legislation that directly or indirectly seeks to combat discrimination in housing. For example, the 1974 Housing and Community Development Act required cities to prepare "housing assistance plans" that described low-income housing needs and stated compliance with the Fair Housing Act. Also that year, the Equal Credit Opportunity Act prohibited discrimination in mortgage lending and required banks to collect information on the race of loan applicants. The 1975 Home Mortgage Disclosure Act and the 1977 Community Reinvestment Act aim to fight redlining by requiring banks to disclose which neighborhoods receive their mortgages and home improvement loans and to document lending to low-income neighborhoods.

Policy related to the dispersal of low-income housing throughout a metro area can work to decrease racial discrimination in housing because of the disproportionate number of minorities who are poor. Thus, fair-share housing efforts stemming from the Mt. Laurel court cases have provided some opportunity for minorities to move into white neighborhoods. Indeed, because of this convergence of race and class, community objections to low-income housing can be interpreted as discrimination not only against the poor but also against minorities. (SEE ALSO: *Fair-Share Housing;* **Fair Housing Amendments Act of 1988;** *Mt. Laurel*)

—*Mara Sidney*

Further Reading

Lief, Beth J. and Susan Goering. 1987. "The Implementation of the Federal Mandate for Fair Housing." Pp. 227-67 in *Divided Neighborhoods: Changing Patterns of Racial Segregation,* edited by Gary A. Tobin. Newbury Park, CA: Sage.

Massey, Douglas S. and Nancy A. Denton. 1933. *American Apartheid: Segregation and the Making of the Underclass* Cambridge, MA: Harvard University Press.

Schwemm, Robert G., ed. 1989. *The Fair Housing Act After Twenty Years: A Conference at Yale Law School, March 1988.* New Haven, CT: Yale Law School.

Schwemm, Robert G. 1990. *Housing Discrimination: Law and Litigation.* New York: C. Boardman. ◀

▶ Closing Costs

Closing costs are the charges and fees incurred in transferring ownership of a home, including charges by the lender for loan processing and by an attorney to examine the title.

Items that may be included in settlement charges are title search, title insurance, attorney's fees, a property survey, a credit report, points, an appraisal fee, recording fee, state and local transfer taxes, escrow fees, and mortgage insurance. They are paid at a meeting between the buyer and seller, representatives of the lender, the real estate broker if one was involved, and attorneys hired by any and all parties. Closing charges may be 3% of the sales price but vary by local laws, customs, and lending institution practices. They are payable to someone other than the seller. The Real Estate Settlement Procedures Act applies to the settlement of all federally related residential mortgage loans. Closing adjustments are made between the buyer and seller for items such as property taxes, which are charged on a pro rata basis. (SEE ALSO: *Appraisal; Points; Real Estate Settlement Procedures Act*)

—*Carol B. Meeks*

Further Reading

ABA Standing Committee on Lawyers' Title Guaranty Funds. 1991. *Buying or Selling Your Home: Your Guide to: Contracts, Titles, Brokers, Financing, Closings.* Chicago: American Bar Association.

Jacobus, Charles J. 1996. *Real Estate Principles.* Upper Saddle River, NJ: Prentice Hall.

McKenzie, Dennis J. 1996. *Essentials of Estate Economics.* Upper Saddle River, NJ: Prentice Hall.

Peiser, Richard B. 1992. *Professional Real Estate Development: The ULI Guide to the Business.* Washington, DC: Urban Land Institute. ◀

▶ Code Enforcement

Local governments seek to protect public health and safety by setting standards for construction and remodeling as well as for housing maintenance. These standards are reflected in various building, plumbing, electrical, and housing codes. Given the intent of these codes, it is ironic that their strict enforcement can lead to an overall loss in the housing stock. This is because housing improvements cost money. If the costs of making required improvements exceeds what an investor can realize in rental income or a homeowner can hope to recoup on sale of a property, the property may be abandoned or withdrawn from the market. This can happen when landowners decide to invest elsewhere or a building official orders that the property be vacated (or not reoccupied after vacancy) until improvements are made. The low-income renter is the most seriously affected by this worst-case scenario because code enforcement can eliminate low-cost housing alternatives throughout a community.

Theorists have argued that in the absence of deep subsidies, public policy must allow for the deterioration of portions of a metropolitan region to provide housing for very-low and no-income households. At the same time, it is deemed desirable to prevent the intrusion of low-quality housing within middle- and upper-middle-income areas. Rather than explicitly establishing different construction and maintenance standards for low-, middle-, and upper-income areas, municipalities accomplish the same goal through differential enforcement of their building and housing codes. Thus, inspectors look the other way in low-income areas and have a more visible and persistent presence in areas that have housing that is already in better condition.

Some municipalities (such as Baltimore, Memphis, Dallas, and Cincinnati) have found that flexible code enforcement can be a valuable tool for maintaining if not improving overall housing quality. This involves using the codes to identify needed repairs and at the same time providing information and access to public and private funding alternatives for the repairs. In some cases, technical as well as social service assistance are provided.

Flexible enforcement requires an emphasis on cooperation rather than heavy-handed enforcement. This is particularly true for programs targeted to owner-occupants. It is also the case in working with landlords who may be inclined to take punitive action against tenants. Inspectors in some programs are allowed to overlook nonhazardous violations, depending on the circumstances of the landowner. Cincinnati takes into consideration current residents and their use of the property. If an elderly couple has peeling paint in a room and there are no children in the house, this violation would not be cited. The correction would be required, however, with a change in occupancy.

At the same time that efforts are made to develop personalized solutions to the problems that are encountered, municipalities must be willing to go to court to force compliance if landowners resist making the required improvements.

Flexible enforcement programs are most effective in neighborhoods that are experiencing the initial phases of deterioration. If this has been allowed to proceed too far, the costs of repair will be such that demolition will be the only recourse.

Historic preservation is another instance in which vigorous code enforcement can have detrimental effects. Imposition of modern standards can be physically impossible to accomplish, cost prohibitive, or in violation of the historic integrity of the structure. As a result, some preservation efforts are simply not feasible and the structure may have to be demolished.

The problems arise most frequently with respect to fire and egress requirements. Some structures were built as single-family homes with no notion that at a future point they would serve as a meeting place, reception hall, museum, or the like. Codes may require enclosure of open stairways, widening of exits, the construction of a second stairway, or the reconstruction of single-board partitions to make them fire resistant. Various successful historic preservation efforts have involved the close collaboration of building officials and architects who are willing to define alternatives that merge the intent of the codes with the requirements of meaningful preservation. The end result may involve restrictions on open flames within the structure or construction of a new stairway that works well with the original design of the structure. In New Orleans, exit requirements for historic buildings were modified so that conversions to higher-density uses would be feasible.

Ultimately, effective code enforcement will happen only if there is a clear understanding of the relationship between the codes and other community objectives, such as affordable housing and historic preservation. (SEE ALSO: **Abandonment;** *Building Codes; Gentrification; Health Codes; Historic Preservation; Housing Codes*)

—*Deborah A. Howe*

Further Reading

Ahlbrandt, Roger S., Jr. 1976. *Flexible Code Enforcement: A Key Ingredient in Neighborhood Preservation Programming.* Washington, DC: National Association of Housing and Redevelopment Officials.

Downs, Anthony. 1981. *Neighborhoods and Urban Development.* Washington, DC: Brookings Institution.

Preservation and Building Codes. 1975. Washington, DC: Preservation Press. ◄

► Cohousing

Cohousing describes a community of individual households with private dwellings and shared common facilities. Cohousing communities are designed to enhance social contact; most have been developed with the residents; and all are managed and maintained by residents with no set ideology but that of creating a supportive living environment.

Cohousing was pioneered in Denmark in the early 1960s by the architect Jan Gudmand-Høyer. A growing dissatisfaction with existing housing options and a realization that the quality of life can be improved by creating a sense of community has popularized cohousing in Europe and transferred the idea to the United States where the first cohousing was built in 1991.

Background

In Northern Europe, by 1998, more than 500 cohousing communities had been built under a variety of tenures and building types. In Denmark, the first communities (1972) were under a type of condominium ownership, privately developed by those residents who could afford the initially high costs of development. The first rental cohousing, Tinggarden (1978), a 79-unit development, was initiated by government and nonprofit housing organizations. Today, a type of limited-equity tenure accounts for most Danish cohousing.

In The Netherlands, cohousing communities are called *centraal wonen* (central living). The first rental community opened in 1977. About 90% of *centraal wonen* units have been developed by nonprofit housing corporations. To increase a sense of community, larger Dutch developments have divided the residents into subgroups with their own common facilities—a cluster kitchen and dining room—in addition to the amenities shared by all.

Swedish *kollektivhus* (collective housing) are predominantly multistoried rental units planned by public housing authorities in conjunction with future residents. Although greater interest in socializing spurred the creation of Danish cohousing, the Swedes placed an emphasis on task cooperation among residents. The first cohousing development, Stacken, a high-rise renovation, opened in 1980.

Finnish *asuinyhteiso* (dwelling community), similar to the Danish cohousing model, consist of privately owned units with some nonprofit rentals. Cohousing in Norway, generally resident owned, typically has three to nine households.

Within Northern Europe, senior cohousing is a growing movement. Senior cohousing is limited to residents 55 years and older with an emphasis on mutual support and care. Begun in the 1980s, more than 100 senior cohousing developments exist, mostly in Holland and Denmark, predominantly developed as nonprofit rentals or limited-equity housing. A few cohousing communities are for singles, single parents, and other specialized groups.

Development

Cohousing is tailored to a specific group's requirements. Development usually begins when interested people meet—through their social network or a newspaper ad—and together decide on basic issues, such as size, type of location, level of collectivity, and price range.

The core group may change because development often takes two or more years. Future residents take an active role in the design programming. Most groups also choose consultants and obtain part or all of the financing. Differences are often resolved through consensus building. Members learn to switch from an individual mode of thinking to one that respects the needs of the group. Prospective members also use these managing skills, learned collaboratively, when the project is completed and they move in. The importance of creating a core group first has led nonprofit organizations, communities, and government housing authorities in Denmark, Sweden, and Holland to promote their formation as the first step in building subsidized cohousing, then helping the group along through the purchase of land and providing financing.

Extensive Common Facilities

An average "Danish style" cohousing development has between 20 and 36 households, small enough so that residents know each other well and large enough to allow for some diversity and to avoid instability resulting from turnover. Cohousing can range from fewer than 10 to as many as 80 households.

The shared indoor facilities are located in one or more common houses (or on a separate floor in a multistory building) and include a commercial-sized kitchen and a dining area. Other shared facilities can include a workshop, guest room, playroom, laundry, TV room, room for teenagers, darkroom, craft room, exercise room, office, sauna, or music room. As residents pool their resources, they can afford a variety of outdoor spaces: common parking, gardens, paved areas for gatherings and dancing, play areas, greenhouses, and occasionally, an orchard. Facilities used in common vary, depending on the requirements of the residents, and alter over time as needs change.

The common house has evolved over the past two decades from a place to be used once a week to one used almost daily, from seating a smaller group of residents to seating all the residents together at one time, from a relatively small facility with basic amenities to an impressive structure that is often the focal point of the site.

The first Danish cohousing communities had smaller common facilities (300-400 m^2) and large private homes (110-220 m^2), about 10 to 23 square meters common area per unit. By the early 1980s, as housing prices increased and residents realized the advantages of shared facilities, the common spaces had grown (400-800 m^2) and the private units had shrunk (65-116 m^2), about 20 to 35 square meters common area per unit. In Denmark, the desire to get the most use out of common space year round has led to the creation of some cohousing under one roof with glass-covered streets connecting units to the common house, increasing the covered common area (up to 1,300 m^2), 30 to 60 square meters common space per unit. By the mid-1980s, Danish cohousing began to be included as a part of larger planned unit developments, often increasing outdoor amenities.

Private Units

To build the common facilities, the floor area of each residence is about 15% to 20% smaller than conventional units. The cost savings of this reduced floor area make possible the construction of the common facilities. The private units are self-sufficient dwellings complete with kitchen, living/dining room, one or more bedrooms, and baths. The largest space reductions have occurred in the kitchen, dining room, hallway, and living room because the common facility has taken over some of these functions and now houses the bulky freezer, washer and dryer, the hobby room, and TV room.

The smaller, tighter unit has a layout that reflects community priorities. The kitchen and dining areas are usually on the front of the dwelling, visually connected to the common outdoor areas. Residents can work in their kitchens and see who is passing by or keep an eye on young children playing in the commons. Bedrooms and the living room are generally oriented toward the back of the house for privacy, along with a small private outdoor area.

The two basic types of units are low-rise dwellings (row houses or walk-up apartments) around a common house, and multistory cohousing, typically new construction, although there are also renovations that include warehouses, schools, and old farms. Units are owned as condominiums, cooperatives, or as rentals developed by a nonprofit housing association.

Layout and Design

Cohousing design is noted for the clustering of units, the creation of semipublic spaces, the design of the common house, and the layout of pedestrian circulation away from cars.

Low-rise units are clustered to create a sense of community, with part of the site remaining open space for common use. Parking is located at the periphery of the site, creating a safe and quiet car-free interior. Site plans have evolved over the past two decades from housing loosely placed around shared areas to housing ordered around streets, squares, and plazas.

Figure 3. Shared Open Space in South Park Cohousing, Sacramento, California
Photograph by Dorit Fromm.

Figure 4. Pathways Promoting Neighborly Interaction, Nyland Cohousing, Lafayette, Colorado
Photograph by Dorit Fromm.

The layout is designed to bring residents in daily contact with one another. Walking to their homes, residents have a chance to stop and chat with others, and children can freely visit the homes of friends to play. There is often a variety of common areas for different-sized groups to meet—from two households to the entire community.

In addition, social opportunities have been increased through gradual transitions between private, common, and public areas. For example, by creating soft edges (places to sit, see, and linger) between the private unit and the common areas, residents can spend time on their porches, balconies, or terraces while at the same time keeping an eye on the common areas, watching children play or greeting those walking by. A main pedestrian path, with housing on each side or through a courtyard, increases these chance encounters.

In multistory buildings, with or without elevators, units have often been arranged around one or more stairwells. The stairway is the link between apartments and the common spaces, similar to a pedestrian street along a cluster of houses.

The common facilities are the heart of the community and often centrally located. They can be difficult to design because they balance functionality—part restaurant, part entertainment center, part hotel, part garage workshop—with the need to be at least partly as intimate as home. Residents pass them on their way daily to see what events are occurring. Directly linked to the facility are the outdoor commons, allowing easy access between interior and exterior use (see Figures 3 and 4).

The greatest strengths of cohousing design have also been its drawbacks. Dwellings clustered around a shared central area have often resulted in housing turned inward. The car-free zone invariably results in parking located nearest the road, which may not enhance the surrounding neighborhood.

Creating a distinct community within the larger neighborhood has sometimes caused neighbors to view cohousing with suspicion, concerned not only about the added traffic and noise but also about the kind of people who might live there. A number of cohousing developments have faced strong neighborhood opposition. Some developments have changed the structure of the area by introducing a large number of people with different outlooks. Some have roused the envy of neighbors who do not have the open space or common facilities of cohousing. Recent developments recognize the need to create both a community within and to present a good orientation to the public.

Management and Maintenance

Members are expected to participate in management and tasks but not forced to do so. However, to ensure high levels of participation, the active approval of residents in all areas of running the community is sought. Each resident has an equal voice at regular meetings where concerns are discussed and problems solved. Attendance is usually high. The week-to-week issues and details are delegated to committees or working groups (such as Gardening, Common House, Social Issues) in which most residents participate. Residents volunteer anywhere from 0 to 60 hours a month; about 8 hours a month work is average.

Decision making is generally accomplished through a consensus process. Although more time-consuming than majority vote, the results better guarantee that each member is heard and shares in the responsibility of the decision.

New people interested in moving into an existing cohousing community learn the requirements of this different way of living and attend a few meetings. Their names are placed on a list, and they are contacted when a resident moves out. In owner-occupied cohousing, resale has usually been at market rate and open to any interested person. Equity has been comparable to that of dwellings in the surrounding neighborhood.

In rental cohousing, the tenants negotiate with the nonprofit sponsor on management and maintenance and on choosing new tenants. This has often resulted in cost savings as well as greater satisfaction with the housing.

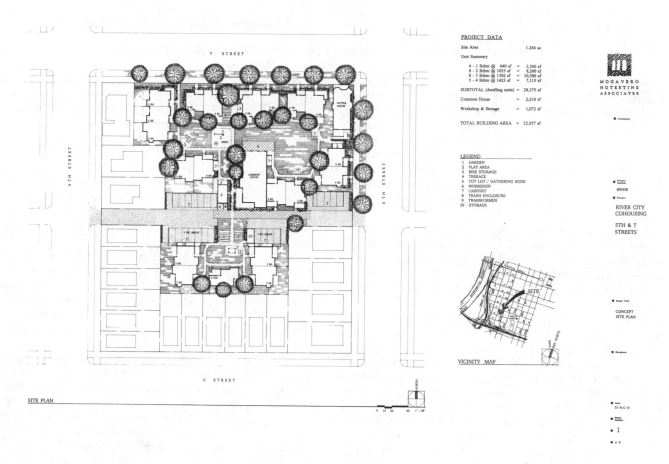

Figure 5. Site plan for South Park, Sacramento, California, shows clustered housing typical of cohousing, placed within a typical block plan.
Mogavero Notestine Associates.

Task Cooperation

The essence of cohousing is that community is created by meeting everyday needs in a common way. The most utilitarian chores—cooking, watching children, sweeping the walkway—provide an opportunity to talk with neighbors and develop relationships. Preparation of shared evening meals at the common house is a routine that saves each household shopping, cooking, and cleaning for its own evening meal. Depending on the size of the scheme or the agreement on meal arrangements, meal preparation may vary from once or twice a month to six times a week, each meal providing food for 10 to more than 100 residents. Typically, a resident can eat a prepared meal in the common house three times a week in exchange for participating in the meal preparation about once a month. Meals are inexpensive. Although common dining is voluntary, cooking is usually a requirement. The food preparation is organized in a variety of methods, often around cooking crews made up of two to six residents.

Cohousing in the United States

Transplanting the Danish cohousing concepts to the United States has not been easy. Whereas Europeans interested in cohousing are now supported by their governments through technical assistance, start-up financing and low-interest loans, people starting cohousing in the United States have struggled in their search for a site, in obtaining city agency approvals, and in securing financing, not unlike the first Danish cohousing residents. Also, reaching consensus among 30 to 60 individuals on the numerous decisions involved in multifamily housing is not a skill that comes easily to many Americans.

In the United States, unlike Europe, professional consultants and developers have taken the lead in developing cohousing communities with emerging groups. The first new American cohousing, Muir Commons in Davis, California, opened in August 1991 with 26 low-rise units and a 350-square meter common house, under condominium ownership. The second, Winslow Cohousing near Seattle (1992), is under cooperative ownership. The third, Doyle Street Cohousing (1992), a renovated warehouse in Emeryville, across the bay from San Francisco, is held in condominium ownership. The 12 units and 200 square meters of common space are all under one roof. Nyland (1993), in Lafayette, Colorado, has 42 units (43 acres) with 800 square meters of built common space (see Figure 4). Cohousing has also been developed by households living in existing housing, where one house has been added or altered to be a common house. Seven years after Muir Commons, 50 communities have been built or are nearing completion, in North America. More than twice that number of groups are in the planning process.

A number of the developments have emphasized conservation through the siting and orientation of buildings, the preservation of open space, greater energy efficiency buildings, and the recycling of goods and building materials. Diversity of households, a goal of cohousing communities, has often been achieved in regard to age and household type. Less successful has been a diversity of income, class, ethnicity, and racial backgrounds. As more varied and affordable models are built, a wider range of households may be served.

Some hybrids of the European cohousing model are evolving, pushed by financing realities and by Americans' preference for individualized dwellings secured in a less time-consuming process. The question of how much the development process can be streamlined and still create successful strong cohousing poses difficult challenges.

Aside from differences in development, tenure in Europe has favored cooperative and nonprofit ownership reflecting government support. In the United States, condominium ownership predominates. Nevertheless, both European and U.S. cohousing have provided the benefits of close and caring neighbors, the availability of shared evening meals, play areas for children, and a spacious outdoor area. Living costs have often been reduced by sharing tasks such as maintenance and child care, sharing tools and transportation, growing vegetables in the common garden, or buying household necessities in bulk. Residents have created a strong sense of security and a network of support outside the family.

In return, living in cohousing requires time, effort, and more tolerance and responsibility toward one's neighbors. Residents need more collaborative communication skills than individuals in a typical neighborhood. Individual sovereignty can also be restricted because some decisions usually made within the family move into the sphere of the group. Group restrictions on children's behavior in the common spaces, pets, fencing, and the color of one's home can be found in some cohousing communities. Although not for everyone, cohousing is an important alternative to the isolation and autonomy of the single-family home. (SEE ALSO: *Adolescents; Children;* **Community-Based Housing**)

—*Dorit Fromm*

Further Reading

Cohousing, the journal of the Cohousing Network. Berkeley, CA, and Boulder, CO. P.O. Box 2584, Berkeley, CA 94702.

Fromm, Dorit. 1991. *Collaborative Communities, Cohousing, Central Living, and Other New Forms of Housing with Shared Facilities.* New York: Van Nostrand Reinhold.

Hanson, Chris. 1996. *The Cohousing Handbook.* Vancouver-Pt. Roberts, WA: Hartley & Marks.

McCamant Kathryn and Charles Durrett. 1994. *Cohousing: A Contemporary Approach to Housing Ourselves.* Rev. ed. Berkeley, CA: Ten Speed Press.

Woodward, A. 1989. "Communal Housing in Sweden." Pp. 71-94 in *New Households, New Housing,* edited by Karen Franck and Sherry Ahrentzen. New York: Van Nostrand Reinhold. ◄

► Collateral Mortgage Obligation

One of the more successful types of mortgage-backed security (MBS) is called a collateral mortgage obligation (CMO). CMOs were introduced as a specialized MBS in June 1983 by the Federal Home Loan Mortgage Corporation (Freddie Mac). Since then, more than $175 billion of all types of CMOs have been issued by Freddie Mac, Wall Street firms, savings and loan associations, life insurance companies, and others. They have proven to be popular because the structure divides up the cash flow into classes, thus permitting the mortgage originator to be able to sell mortgages more easily by dividing them into parts. Investors can choose from a wider range of vehicles because each class is available with various risk-and-return characteristics.

The CMO is a debt instrument that has been collateralized by mortgage pass-throughs or by individual mortgages. However, there are differences between the prepayment risk exposure facing holders of mortgage pass-throughs and holders of CMOs. In the case of CMOs, investors do not share all of the prepayment risk equally because of the structure of the security.

Specifically, a CMO is created with a number of classes for the receipt of cash flow payments from the borrowers. These classes are called "tranches." As payments are made, they are divided into various tranches according to the structure of the CMO. The first class ("Class A") receives interest and principal payments up to a certain limit; after that limit, all payments go to the next class ("Class B") and so on. "Class Z" would be the final class, and investors would not receive any interest or principal until all of the other classes have received their structured payments.

Investors choose which tranche matches their investment needs. Research shows that certain types of institutional investors prefer early repayment schedules (such as commercial banks), and they tend to invest in Class A and Class B CMOs. Other institutions (such as insurance companies and pension funds) allocate more of the investment resources to slow-paying tranches.

The investment advantages of CMOs stem from the structuring of the cash flow payments into these tranches. Each class has different maturities and average repayment lives. Each class has different risk characteristics associated with its repayments. Compared with traditional pass-through securities, CMOs permit investors to broaden their range of investment choices.

CMOs have been successful in providing options in the MBS market that have traditionally been unavailable. As is often noted, CMOs are not really new mortgage instruments; rather, they are innovative mechanisms for redirecting mortgage cash flows into smaller investments to better match the needs of institutional and other investors. (SEE ALSO: **Housing Finance;** *Real Estate Mortgage Investment Conduits*)

—*Austin J. Jaffe*

Further Reading

Angell, Robert J. 1991. "Evaluating Investments in CMOs." *Real Estate Review* (Summer):41-48.

Fabozzi, Frank J., ed. 1992. *The Handbook of Mortgage-Backed Securities*. 3d ed. Chicago: Probus.

Fabozzi, Frank J. and Franco Modigliani. 1992. *Mortgage and Mortgage-Backed Securities Markets.* Boston: Harvard Business School Press.

McNulty, James E. 1988. "Mortgage Securities: Cash Flows and Prepayment Risk." *Real Estate Issues* 13 (Spring/Summer):10-17.

Sellon, Gordon H., Jr. and Deana VanHahman. 1988. "The Securitization of Housing Finance." Federal Reserve Bank of Kansas City. *Economic Review* (July/August):3-20.

Smith, Donald J. and Mark D'Annolfo. 1986. "Collateralized Mortgage Obligations: An Introduction." *Real Estate Review* 16:30-42.

Winger, Alan R. 1987. "The Securitization of Real Estate Finance." Federal Home Loan Bank of Cincinnati. *Review* (August) 1:2-5. ◄

► Colonias

In Spanish, the word *colonia* refers to a district or residential area of a city. In the United States, colonias are generally defined as rural settlements on the U.S. side of the U.S.-Mexico border, housing the poorest of the poor—new immigrants (both legal and illegal), with lower educational attainment levels and lesser degrees of proficiency to speak and read the English language. They are unincorporated subdivisions characterized by substandard housing, inadequate plumbing and sewage disposal systems, and inadequate access to clean water. They are highly concentrated poverty pockets that are physically and legally isolated from neighboring cities.

Colonias first began to appear in earnest in the 1950s as developers expanded into rural areas in pursuit of cheap land. In 1998, there were approximately 1,600 colonias, populated by approximately 400,000 persons. An overwhelming majority of colonias (approximately 1,400) are in Texas. In the 1990s, about 60% of all residences now constructed along the U.S.-Mexico border were in colonias. Often, colonias are analogous to squatter settlements in developing nations. The primary difference is that land and homes in colonias are purchased by legal means, primarily under a contract-for-sale arrangement.

Colonias exist for two major reasons. First, colonia households are very poor; in 1995, about 40% had an income of less than $15,000 per year and 70% of households had an income of less than $30,000. Because colonia lots lack water, sewer, paved streets, and other basic services, such as fire and police protection (along with the attendant taxes to finance such services), they are considerably more affordable. In other words, colonia settlements meet a significant demand for low-income housing. Second, colonias help satisfy a strong cultural demand of residents to be a part of the "landed." In the Spanish culture, there are three broad cultural status categories: the landless, the tenants, and the landed. Owning a colonia lot, beginning the process of constructing a family residence, and inhabiting the residence, even as construction continues over a number of years, moves residents from landless or tenant status to landed, or landowner, status. For colonia residents, the desire to own a part of the American dream and to provide an inheritable tenure is a strong push in colonia creation.

The proliferation of colonia settlements presents major economic, environmental, and health problems. Because of the absence of basic infrastructure, drinking water is imported from private ad hoc vendors and often stored in 55 gallon drums once used for other purposes. In some areas, shallow wells are dug even though the groundwater may be tainted with agricultural chemicals or made otherwise undrinkable because of its high saline or mineral content. In many cases, outdoor pit-privies exist in the absence of a sewer system. In other areas, often inadequate septic systems exist. Garbage collection is spotty at best and in most colonias, nonexistent. With the absence of a drainage system, periodic flooding is a major problem, particularly for colonia settlements in the Lower Rio Grande Valley of Texas. When floods or heavy rains occur, the absence of paved streets makes transportation by car or bus almost impossible; school buses cannot get children to school, and the employed have difficulty getting to work. Furthermore, all of these conditions contribute to significant health problems in colonias, where dysentery and hepatitis A, among other endemic diseases, occur at alarming rates.

To further complicate these problems, colonia residents are most often left without access to credit. Under a contract-for-sale arrangement, title to the property does not pass from the developer (seller) to the buyer, and the buyer has few true rights of ownership until the debt is completely retired. Until that time, the property and all of its improvements can be repossessed by the developer even if only one payment is missed. Because of the contract-for-sale arrangement, coupled with the low income levels of colonia residents, it is very difficult, if not impossible, to finance improvements through traditional financial institutions.

In Texas, the State stepped in to finance the installations of water and sewer systems and provide regulatory power to counties to stop the proliferation of colonias. Prior to 1989, counties in Texas had authority neither to control land use nor to establish and enforce subdivision regulations. Consequently, developers were free to subdivide land outside incorporated city limits and without the requirement to provide basic infrastructure. The Texas state legislature also enacted the Economically Depressed Areas Program (EDAP) to bring water and sewer systems to colonia settlements through a $700 million state-financed bond program. To participate in the program, counties along the Texas-Mexico border were required to adopt model subdivision regulations. Another problem arose when residents could not connect to the water and sewer systems because their structures did not meet building codes required by the county's subdivision regulations. Because of their income levels, colonia residents could not qualify for loans to make the needed improvements and, even if they did, few traditional lending institutions would be willing to lend money to a homeowner who has purchased under a contract-for-sale arrangement and, consequently, cannot offer the property as collateral.

Slow progress is being made. A pilot program has begun to convert contract-for-sale agreements into deeds of trust. Under Department of Housing and Urban Development

Figures 6 and 7. Typical housing structures found in colonias along the border between Texas and Mexico. Although some colonia houses have electricity, few have water and sewer service.
Photographs by David R. Ellis, Center for Housing and Urban Development, College of Architecture, Texas A&M University.

Title I loan provisions, loans for home improvements can be made without placing the home as collateral, thereby permitting loans to residents without a deed of trust. Finally, in Texas, the State has prohibited the creation of new subdivisions in unincorporated areas without adequate water and sewer infrastructure.

Still, much is left to be done. Even if no new colonias are created, their population is likely to double because only half of the lots in already existing colonias are occupied. Furthermore, the State will need to remain vigilant as potentially unscrupulous developers find new ways to meet a legitimate demand for low-cost, low-income housing. (SEE ALSO: *Substandard Housing;* **Third World Housing**)

—*David R. Ellis and A. Kermit Black*

Further Reading

Betts, Dianne C. and Daniel J. Slottje. 1994. *Crisis on the Rio Grande.* Boulder, CO: Westview.

Copeland, Claudia and Mira Courpas. 1987. *Border State "Colonias": Background and Options for Federal Assistance.* Washington, DC: Congressional Research Service.

Davies, Christopher S. 1995. "Colonia Settlements: Working-Class Refuge Stations along the Texas-Mexico Border." *Planning Forum* 1:33-54.

Davies, C. S. and R. K. Holz. 1992. "Settlement Evolution of 'Colonias' along the US-Mexico Border: The Case of the Lower Rio Grande Valley of Texas." *Habitat International* 16(4):119-42.

Maril, Robert Lee. 1989. *Poorest of Americans: The Mexican Americans of the Lower Rio Grande Valley of Texas.* Notre Dame, IN: University of Notre Dame Press. ◄

► Columbia Point

Boston's Columbia Point has been the site of two significant housing experiments. Taken together, these have enabled the place to be viewed as a symbol of both the worst and the best thinking about U.S. public housing. Once a notorious example of public housing gone awry, Columbia Point was transformed into an award-winning, mixed-income, privately run rental community now known as Harbor Point.

The first Columbia Point experiment began with the completion in 1954 of New England's largest public housing project, located in prominent isolation on a harborfront site next to the city dump, south of Boston's downtown. Beginning in the late 1960s, as the economic circumstances of its residents declined and the management and maintenance efforts of the Boston Housing Authority (BHA) faltered, Columbia Point gained widespread notoriety as one of the most devastated and dangerous housing developments in the United States. A cycle of neglect and abandonment accelerated, and by 1979, only 350 of the original 1,502 apartments remained legally occupied.

At the same time, the future of low-income housing at the site faced other threats, as the rest of Columbia Point peninsula became more attractive to developers. A series of construction ventures on tracts of land near the housing project, including a new Boston campus for the University of Massachusetts and the John F. Kennedy Library, drew attention to the scenic amenities of the site and underscored the anomaly of the substantially boarded-up housing project that remained.

After more than a decade of false starts and indecision on the part of the BHA and others, Columbia Point's tenants themselves took the lead in forging a viable redevelopment plan. Rather than accept piecemeal renovations or risk wholesale demolition, the Columbia Point Community Task Force (CPCTF), led mostly by African American women, lobbied for a large-scale redevelopment effort that would preserve low-income housing in perpetuity. In 1983, the Massachusetts Housing Finance Agency organized a multi-agency effort and solicited redevelopment proposals from developers willing to guarantee that existing residents of Columbia Point could be rehoused. Out of this effort began the second housing experiment at Columbia Point: its $250 million transformation into the mixed-income community now called Harbor Point.

In 1986, the Harbor Point Apartments Company was formed as an equal partnership between the CPCTF and a private development team led by Corcoran, Mullin, Jennison, Inc. The extraordinarily complex financing of Harbor Point involved a high degree of coordination between public sector agencies at all levels of government working together with a wide variety of private investors. Overall, approximately $175 million of the financing came from diverse government loans and grant programs, and $75 million was raised from private syndications, made feasible only when Congress, in 1987, passed special rules to preserve tax breaks for investors in Harbor Point. An attempt to obtain public funding to subsidize moderate-income housing to complement the low-income and market-rate units did not come through, however.

The redevelopment involved the demolition of approximately half of the buildings on the site, and the wholesale rehabilitation of all of the others. There was also substantial new construction of both townhouses and midrise buildings. The architects, Goody, Clancy & Associates and Mintz Associates, working with the landscape architecture firm of Carol R. Johnson & Associates, thoroughly reconfigured the site plan to make better use of water views, while creating a street grid reminiscent of Boston's Back Bay neighborhood. The Harbor Point program also provided for new amenities, such as tennis courts, swimming pools, and various other recreational opportunities, plus an on-site system of support services, including day care, health services, and senior citizen programs.

After the first buildings were opened in 1987, 335 of the 350 tenant families remaining at Columbia Point when the redevelopment began moved into new apartments at Harbor Point. The redevelopment was completed in 1991 and, as of 1998, Harbor Point was operating at more than 90% occupancy. That said, it operated at a loss during its first years, due primarily to a soft real estate market that necessitated lower than expected rents.

With 400 of Harbor Point's 1,283 units reserved for low-income families for the length of a 99-year lease, some heavily subsidized housing is guaranteed to remain. At Harbor Point, there is no difference in the design and level of amenities provided for subsidized and market rate apartments, and there is a formal agreement to make sure that no building houses more than 50% subsidized tenants, thereby ensuring a mixture of incomes throughout the development.

Both during and after the redevelopment process, Harbor Point was subject to considerable scrutiny in the local press, with criticism centered on the high level of public investment and on doubts about the prospects for social mixing between market rate tenants (most of whom are single individuals or couples and are highly transient) and the low-income, former public housing residents (most of whom are from larger families intending to remain for the long term). Other critics bemoaned the loss of subsidized

units, noting that Harbor Point would house only about one quarter of the low-income families that Columbia Point did when it was fully occupied.

Overall, however, praise for the Harbor Point experiment has more than balanced the voices of its detractors. Despite its high public costs, many point to the inestimable societal value of transforming the single most stigmatized place in Boston into an amenity. Co-winner of the 1993 Rudy Bruner Award for Excellence in the Urban Environment, Harbor Point has been lauded not only for the physical and socioeconomic transformation of a dangerous, largely abandoned public housing project into a safe and attractive mixed-income and mixed-race community but it has also been commended as a model for resident involvement in all stages of a redevelopment process. Although the labyrinthine financing of this project may never again be replicated, Harbor Point may yet be an inspiration for other mixed-income public-private housing ventures using tenant-developer partnerships. (SEE ALSO: **Urban Redevelopment**)

—*Lawrence J. Vale*

Further Reading

DiMambro, Antonio. 1986. "Restoration or Liquidation? Two American Experiments in Public Housing in Boston, Massachusetts." *Spazio e Società/Space & Society* 33(Winter):6-19.

Down the Project. 1981. Film. Producer: Richard Boardman. Distributor: Cine Research, 170 Garden St., Cambridge, MA 02138. Phone: (617) 442-9756.

Farbstein, Jay and Richard Wener. 1993. "Harbor Point, Boston, Massachusetts." Pp. 36-57 in *Report to the Selection Committee on the Five Finalists.* Prepared for the 1993 Rudy Bruner Award for Excellence in the Urban Environment. New York: Bruner Foundation.

Pader, Ellen-J. and Myrna Margulies Breitbart. 1993. "The Transformation of a Public Housing Project." *Places* 8(4):34-41.

Swartz, Linda. *Columbia Point: Life in the Ghetto, USA.* 1989. Video slide transfer, with accompanying essay by Marie Kennedy. Minneapolis: Intermedia Arts Minnesota, 425 Ontario St. S.E., Minneapolis, MN 55414. ◄

► Common Interest Development

Common interest development (CID) is any real estate development in which land or structures or both have been partitioned for sale to individual investors (typically landlords or homeowners) but wherein shared property rights (and sometimes financial responsibilities) are also assigned, whether explicitly or implicitly, with respect to tangible or intangible elements. Commonly, this term is used in reference to a "gated estate": a walled-off cluster of dwellings built on private roads and accessed through a security checkpoint. Here, residents typically have exclusive use of their home and lot, a deeded right to use roadways and other community facilities, and an obligation to share in the maintenance of community facilities. In principle, the term could also be used to describe a condominium or cooperative housing project because these, too, involve elements of both shared property rights and exclusive use and often also have a concierge or doorman who controls access to the site. Consumers are attracted to such development in part because of the secure environment and in part because these developments provide services and amenities simply not possible elsewhere.

When the CID involves more than just a few owners, a system of governance is required. As with condominiums and cooperatives, many CIDs are governed by a special-purpose nonprofit corporation. Such corporations have two main purposes in practice. One is to maintain and operate the elements that compose the CID. The corporation does this typically by levying fees on owners and then spending this revenue on maintenance and repairs. Typically, fees can vary from one owner to the next but have to conform to the contractual agreement by which the CID was created (e.g., a legal declaration in the case of a condominium). The second main purpose of the corporation is to develop and enforce rules for access to, and use of, the common elements and to resolve disputes among owners arising therefrom.

Arguably, the sharing of property rights among neighboring landowners has its legal foundations in a device, originating in common law, known as deed restrictions (alternatively known as restrictive covenants). In land subdivision, deed restrictions are used to prevent purchasers from subsequently using their property in a nonconforming way—for example, building a store in a residential subdivision or storing scrap metal on-site. Early in this century, U.S. courts ruled that each property owner in a land subdivision has the right to see enforced the deed restrictions on other owners.

It can be argued that CIDs are also rooted in the volunteerist homeowner associations that sprang up in many neighborhoods over the last half century. Homeowner associations generally form to protect and promote the interests of existing residents. They give local residents greater scope for participating in community decision making as well as a united voice in dealing with politicians, bureaucrats, and real estate developers. However, there is a major difference. Unlike many CIDs, homeowner associations do not own neighborhood assets (e.g., local roads and parks) and cannot prevent nonresidents from entering the neighborhood.

In some respects, a CID is the private sector equivalent of a local government. A CID may maintain roads, sidewalks, parks, streetlights, and water and sewage systems; police traffic; and perhaps even have its own firefighters. However, a CID typically gives each dwelling a vote in community decision making rather than one vote per adult as in local government, and votes typically go just to homeowners (that is, members). In these respects, a CID is more like a club than a municipality. Inherent to a CID, as with almost any club, is some degree of exclusivity, self-segregation, and hence homogeneity. A CID attracts people with similar tastes and incomes and excludes those who cannot afford or do not want that particular bundle of shared property rights. In contrast, a democratic government is thought to be responsive to the needs of all citizens living in the community. (SEE ALSO: *Condominum; Cooperative Housing*)

—*John R. Miron*

Further Reading

Dillon, D. 1994. "Fortress America." *Planning* 60(6):8-12.
McKenzie, E. 1994. *Privatopia: Homeowner Associations and the Rise of Residential Private Government.* New Haven, CT: Yale University Press. ◄

► Community Associations Institute

The Community Associations Institute (CAI) is a national organization of condominium, cooperative, and homeowner associations; builders; developers; insurance and real estate brokers; and individual homeowners. Founded in 1973, the institute is dedicated to developing and distributing the most advanced and effective guidelines for creating, financing, operating, and maintaining community facilities and services. It provides information and referral services, tracks legislation concerned with community association and housing issues, and hosts two national conferences and two law seminars per year. CAI also distributes more than 70 moderately priced publications and newsletters covering legal, financial, and management issues. Membership: 13,000 nationwide (1993). Staff size: 31. Contact person: Barbara Beach, staff vice president, Public Affairs Department. Address: 1630 Duke Street, Alexandria, VA 22314. Phone: (703) 548-8600. Fax: (703) 684-1581.

—*Caroline Nagel* ◄

► Community-Based Housing

Community-based housing refers to a range of activities that attempt to provide the residents of low-income neighborhoods with control over housing development and management in their communities. Community-based housing approaches include nonprofit development corporations rehabilitating and operating affordable housing, tenant ownership and management, self-help housing strategies, and land trusts for housing affordability.

The objectives behind a community-based approach to housing center on the development of affordable housing that is sensitive to neighborhood and community needs, that is developed through a democratic process that includes all elements of the community, and that puts ownership and control of housing development in the hands of community members.

The neighborhood movement of the 1970s in the United States saw the first development of a self-conscious, neighborhood-based agenda for housing policy. Groups from across the country coalesced in the early part of the 1970s to discuss methods of combating neighborhood decline, disinvestment, and rapid racial turnover in inner-city neighborhoods. In addition, a number of local groups continued to act to keep urban neighborhoods integrated through efforts to stop the blockbusting, redlining, and steering practices of lenders and real estate brokers. As community-based groups attempted to expand their neighborhood stabilization efforts, they began to focus more on methods to control and participate in development and to retain ownership of neighborhood assets. Community-based housing became a significant trend in the United States during the 1980s, when the federal government slashed its housing assistance budget and local communities were forced to fill in the gap.

Community Development Corporation
Perhaps the most important element of a community-based approach to housing is the community development corporation (CDC). The CDC is a nonprofit housing development corporation, committed to the creation, rehabilitation, or management of affordable housing. Typically, CDC boards of directors are dominated by community residents, which allows local control over development strategies. CDCs provide community activists with a vehicle for engaging in the development process in a proactive manner and a means of securing common ownership of housing and land assets. CDCs, unlike for-profit companies that participate in affordable housing programs, provide for continued affordability of the housing they develop. CDCs are also seen as more in touch with community interests and therefore more able to create a housing strategy that is responsive to residents' needs than are government agencies or for-profit developers.

Community Land Trusts
Community land trusts are nonprofit organizations that own land in a trust and lease the improvements on the land for community purposes such as affordable housing. Although leaseholders may own the houses that rest on the land, the trust maintains ownership of the land. This removes land inflation from the cost of housing, thereby reducing housing costs and eliminating land speculation. Community land trusts are run by local residents and are another means of more democratic ownership and control over land use. Landownership remains in the community, while leaseholders maintain equity in their homes.

Cooperative Housing
In this category are lumped a number of housing arrangements that attempt to provide residents with collective ownership or control over their living environment. Cohousing, for example, is a community living environment of typically 15 to 30 households, organized, planned, and managed by residents. Practiced most extensively in Denmark, cohousing combines the autonomy of private living space with shared facilities, such as cooking, day care, play spaces, workshops, and laundry. Each household has a private living space that is physically and functionally independent, whereas other aspects of life are carried out in common spaces.

Forms of cooperative ownership such as *limited-equity cooperatives,* and *mutual housing associations* allow for democratic (resident controlled) housing management, collective ownership of a building or buildings by residents, and continued affordability through limitations on equity appreciation.

Tenant management and "leasehold" cooperatives provide for user control in a rental setting. In these models, tenants assume management responsibilities for the properties in which they live, although ownership resides in another entity—usually a nonprofit organization or limited partnership.

Community Lending Programs
Other community-based housing efforts focus on making private lenders more active in affordable housing efforts. The primary tools for these efforts are the Home Mortgage Disclosure Act (HMDA), which requires lending institutions to disclose information on the geographic location of their lending, and the Community Reinvestment Act (CRA), which provides a means for community groups to challenge the mortgage-lending practices of federally chartered banks and savings and loans. Communities across the United States have organized in attempts to stop disinvestment in inner-city neighborhoods. A number of national organizations, such as Association of Community Organizations for Reform Now (ACORN), the Center for Community Change, and the National Training and Information Center (NTIC), provide technical assistance to local community groups across the United States in their efforts to force banks to establish community lending programs. Successful programs that have channeled millions of dollars in mortgage loans to inner-city neighborhoods have been established in many large cities across the country, including Chicago, Atlanta, Detroit, Minneapolis, and Boston.

Responsive lending campaigns have also given rise to the development of various financial institutions oriented toward the provision of capital for community revitalization. Community credit unions, development banks, and community development loan funds contribute to local economic strength by keeping financial resources in the community and recycling them through loans for housing, small business development, and other local social and physical infrastructure.

Self-Help Housing
Community-based efforts at producing affordable housing have also included various self-help efforts. Perhaps the best known in the United States is the Atlanta-based "Habitat for Humanity" program. This program and others like it provide the means by which low-income families can build or rehabilitate their own home. Using donated materials, volunteer workers, and the "sweat equity" of the residents themselves, self-help programs produce inexpensive housing for low-income families. The sweat equity approach is also designed to enhance homeowners' feelings of commitment to the community and increase their ability to maintain and improve the dwelling units in future years.

The self-help model is extensively used in developing countries. Resident participation in community building and self-help housing is a significant means by which low-income settlements are created and upgraded in developing countries. Governments in these countries have begun to support self-help settlements through the provision of basic services.

Summary

These disparate efforts share a number of characteristics that identify them as community-based housing initiatives. These techniques generally provide for greater responsiveness of housing management to the needs of residents, more democratic forms of housing ownership and management, a supportive community-based living environment and sharing of resources, continued affordability of housing for residents, means of interest organization for tenants and residents, and greater control over private investment in neighborhoods. Community-based housing emphasizes the use of the nonprofit sector and other alternatives to private or public sector development and landownership. (SEE ALSO: *Cohousing; Community Development Corporations; Community Land Trust; Cooperative Housing; Limited-Equity Cooperatives; Mutual Housing; Nonprofit Housing; Participatory Design and Planning; Resident Management; Self-Help Housing;* **Tenure Sectors**)

—*Edward G. Goetz*

Further Reading

Bruyn, Severyn T. and James Meehan, eds. 1987. *Beyond the Market and the State.* Philadelphia: Temple University Press.

Gilbert, Alan and Peter M. Ward. 1986. *Housing, the State and the Poor.* Cambridge, UK: Cambridge University Press.

Gunn, Christopher and Hazel Dayton Gunn. 1991. *Reclaiming Capital: Democratic Initiatives and Community Development.* Ithaca, NY: Cornell University Press.

Mathéy, Kosta, ed. 1992. *Beyond Self-Help Housing.* London: Mansell.

Nelson, Joan M. 1979. *Access to Power: Politics and the Urban Poor in Developing Nations.* Princeton, NJ: Princeton University Press.

Saltman, Juliet. 1990. *A Fragile Movement: The Struggle for Neighborhood Stabilization.* Westport, CT: Greenwood.

Squires, Gregory D., ed. 1992. *From Redlining to Reinvestment: Community Responses to Urban Disinvestment.* Philadelphia: Temple University Press.

Taub, Richard P. 1988. *Community Capitalism.* Boston: Harvard Business School Press. ◀

▶ Community Development Block Grant

The Community Development Block Grant (CDBG) (P.L. 93-383) was signed into law August 22, 1974. CDBG is a *block grant,* which means that most of its funds are awarded in entitlements determined by formula to nearly 900 eligible jurisdictions (cities and suburbs of 50,000 or more population and urban counties) throughout the nation. These jurisdictions have a relatively high degree of discretion in spending the funds. CDBG replaced eight *categorical* programs that required jurisdictions to compete for funding and that were very explicit about how funds could be spent: Urban Renewal; Model Cities; Open Space, Urban Beautification, and Historic Preservation Grants; Public Facility Loans; and Water and Sewer and Neighborhood Facilities Grants. Since FY 1982, 70% of funds has gone to entitlement communities and 30% of funds has gone to small cities. Also since 1982, the small cities component of CDBG has been administered by the states, with only New York and Hawaii not directly administering Small Cities grants. An average of $3.2 billion has

been appropriated annually for the CDBG program. CDBG recipients are also entitled to retain program income when CDBG funds are repaid to them. Program income was slightly more than $0.5 billion in FY 1991.

The primary objective of CDBG is to develop "viable urban communities, by providing decent housing and suitable living environments and expanding economic opportunities principally for persons of low and moderate income." In addition, all CDBG projects must meet one of three national objectives:

1. Principally benefit low- and moderate-income persons
2. Aid in the prevention or elimination of slums or blight
3. Meet other urgent community needs

Democratic and Republic administrations have had divergent views on whether the low- and moderate-income benefit was more important than the other two benefits. Finally, in 1983, Congress responded by stipulating that CDBG funds principally benefit low- and moderate-income persons by requiring that at least 51% of CDBG funds be spent to benefit low- and moderate-income persons. Since then, this percentage has been raised so that 70% of CDBG funds must be used for low- and moderate-income persons (households earning 80% or less of area median income).

Eligible activities that can be funded with CDBG include acquisition, disposition, or retention of real property; provision of public works; rehabilitation of residential and nonresidential buildings; public services (social services); and economic development, including assistance to for-profit businesses. Generally, any activity not authorized by statute is ineligible for CDBG funding. The Department of Housing and Urban Development (HUD) regulations specify ineligible activities, including support for general government buildings and political activities. New housing construction is not generally an eligible activity unless it is housing of last resort (in the case of relocation where comparable replacement housing cannot be found) or when the housing is developed by CDBG subrecipients as part of a neighborhood revitalization, community economic development, or energy conservation project.

Over time, CDBG allocations by entitlement communities have largely stabilized, with nearly 40% spent for housing, most of which goes to housing rehabilitation. In the early years of the CDBG program, a large share of funds went to public works, in many cases completing urban renewal projects that had begun prior to 1974. More recently, public works have consumed about one-fifth of entitlement jurisdiction's CDBG funds. Administration and economic development each consume about 13% of all CDBG entitlement expenditures, and public services (social services) consume about 10% of CDBG expenditures. For CDBG Small Cities jurisdictions, a little more than half of all funds go for public works, and about one-fourth go to housing activities.

Section 108 is the loan guarantee component of the CDBG program. This provision allows HUD to guarantee local taxable bonds for community development. Loans are guaranteed for up to 20 years and for a maximum loan amount that cannot exceed five times the amount of the most recent CDBG grant. The purpose of the Section 108 loan guarantee is to permit jurisdictions to commit to large community development projects that could not be done with a single year's CDBG allocation.

Under the CDBG entitlement program, cities of 50,000 or more and central cities of metropolitan areas are eligible to receive an annual amount of funding that is determined by an entitlement formula. Also eligible for entitlement funding are urban counties, which are counties of 200,000 or more population, excluding those living in metropolitan cities.

Funding for entitlement jurisdictions is determined using a dual-formula system, under which grantees receive the larger of two funding allocations as determined under the two formulas. The dual-formula was a compromise enacted in 1977 by Congress to deal with criticism that the first formula favored communities in the South and the West. The first formula (Formula A) uses shares of population, extent of poverty, and extent of housing overcrowding, with poverty given a 50% weight and the other two indicators weighted 25% each. An alternative formula was proposed (Formula B) that uses population growth lag, extent of poverty, and age of housing. Growth lag is weighted 20%, poverty 30%, and age of housing 50%. Growth lag is the shortfall between population in a city and the population it would have had if it had grown at the average rate of all metropolitan cities since 1960. Using the age of housing indictor clearly benefited jurisdictions in the East and Midwest. Studies performed at the time of the 1977 change, however, showed that age of housing was a better indicator of housing deterioration and neighborhood problems than the overcrowding measure used in the first formula.

When originally adopted in 1974, CDBG required entitlement jurisdictions to prepare plans that HUD had 75 days to review. HUD had limited authority to reject these plans, but plans were automatically approved if HUD did not act within the 75 days. In the 1981 Omnibus Reconciliation Act, pushed by the Reagan administration, legislation was adopted that ended HUD's front-end power to reject an entitlement jurisdiction's CDBG expenditure plan. Instead, jurisdictions were required only to submit a "submission package" consisting of goals, objectives, and projected uses of funds. HUD still retained the right to determine whether actual CDBG performance was in accordance with the law and with regulations.

In most states, Small Cities CDBG jurisdictions continue to compete statewide for CDBG funds in the same manner as in categorical programs. Moreover, in many states, subcompetitions are conducted for specific categories of CDBG expenditure, such as housing rehabilitation, public works, or economic development. Consequently, among small cities there is a strong resemblance between the old categorical grants that CDBG was designed to replace and current grant selection systems. According to a 1993 study of CDBG, over time many states have revised their selection criteria to place less emphasis on targeting CDBG Small Cities funds to communities with high levels of need and to activities that primarily benefit low- and moderate-income households. The result has been a spreading of CDBG funds to a higher number of jurisdictions than had been the case when the program was administered by HUD.

In terms of citizen participation, communities are required to have only one public hearing, which must occur after the city has provided information concerning availability of funds, objectives, and how the money is proposed to be spent, including benefits to low- and moderate-income households and details on displacement. Community comments must be considered by the jurisdiction in preparing a final statement of objectives.

Since passage of the 1990 National Housing Affordable Act, each CDBG entitlement recipient must also complete a Comprehensive Housing Affordability Strategy (CHAS). In addition, each CDBG recipient jurisdiction must prepare a grantee performance report to review its record in spending CDBG funds.

Overall, what has been learned from the CDBG program? A 1993 review of the CDBG program offered several significant insights. First, the targeting of CDBG funds to needy places and people has declined over the first 20 years of the program. The spreading of benefits occurred as more jurisdictions were able to participate in the CDBG program, thereby decreasing the funds going to the most distressed communities. This occurred at both the federal level, with the entitlement program, and at the state level, with the Small Cities program. Case study evidence also has shown that benefits were substantially spread within CDBG entitlement communities over the course of the program.

However, the amount of targeting does vary across jurisdictions and is influenced by the geographic distribution of poverty in a community, the structure of local decision making in terms of the relative strength of CDBG planners and elected officials in determining program expenditures, the influence of higher levels of government, and the organization of local benefits coalitions. Local benefits coalitions are more successful in lobbying for the targeting of CDBG funds to poor people and neighborhoods when they are able to ally with other important actors, such as the local chief executive, the federal courts, or HUD.

In the end, CDBG continues to confront a tension between state and local government flexibility and the national policy objective of using CDBG to principally benefit low- and moderate-income households. The flexibility that has been built into CDBG reinforces political tendencies to broadly distribute benefits, thereby reducing the benefits to low- and moderate-income households. The tension between flexibility and CDBG's principal objective is heightened by the fact that the spreading of CDBG benefits to less distressed areas and people has occurred at the same time that the level of distress in our nation's cities has continued to worsen. (SEE ALSO: *Housing and Community Development Act of 1992; Section 108*)

—*Charles E. Connerly and Y. Thomas Liou*

Further Reading

Hays, R. Allen. 1995. *The Federal Government and Urban Housing: Ideology and Change in Public Policy.* Albany: State University of New York Press.

Rich, Michael J. 1993. *Federal Policy Making and the Poor: National Goals, Local Choices, and Distributional Outcomes.* Princeton, NJ: Princeton University Press.

Teaford, Jon. 1990. *The Rough Road to Renaissance; Urban Revitalization in America, 1940-1985.* Baltimore: John Hopkins University Press.

U.S. Department of Housing and Urban Development. *Annual Report to Congress on the Community Development Block Grant Program.* Washington, DC: Author.

Wallace, J. E. 1995. "Financing Affordable Housing in the United States." *Housing Debate Policy* 6:785-814. ◄

▶ Community Development Corporations

A community development corporation (CDC) is a nonprofit community-based organization. Although they often address issues such as job training and development and commercial revitalization, CDCs are best known for dealing with housing problems, especially the replacement of substandard and abandoned housing and providing more affordable housing for low- and moderate-income residents. They are tax exempt under Section 501(c)(3) of the Internal Revenue Code.

Organization and Profile

CDCs have a staff, usually a combination of paid staff and community volunteers, and a board of directors. The only comprehensive, empirical national study of CDCs (130, in 29 cities surveyed in 1988) found that these CDCs, with an average life of 12 years, had an average staff size of 19, with the median staff size being 7.

Typically, staff members, especially the executive directors, play the key role in the development of policy and the implementation of programs. CDCs have boards of directors, which also play a role in the determining the CDC's direction. CDCs generally encourage community participation, and their boards have strong community representation. CDCs usually operate in one or more neighborhoods, with their service boundaries generally identified with constituent neighborhoods. In the national study, community residents and clients were a board majority in 41% of CDC boards studied and a plurality in 72% of those surveyed. Other local institutions (e.g., churches, civic organizations, financial lenders, and business groups) are usually represented. Outside representatives typically include city government, financial lenders, and religious groups.

CDCs rely on a wide variety of funding sources, both for operating support and for project development and management. The primary sources are government (federal, state, and local) and the private sector—philanthropic foundations, corporations, and banks.

In two-thirds of cities with populations more than 100,000, networks of CDCs have formed, joining together individual CDCs in citywide organizations. Examples include the Boston Housing Partnership, the Cleveland Housing Network, and the Pittsburgh Partnership for Neighborhood Development. National and local intermediaries, usually foundations, banks, corporations, and government, have provided key support. These include the Enterprise Foundation; the Local Initiatives Support Corporation (LISC), associated with the Ford Foundation; and the Council for Community-Based Development.

A special example of CDCs linked locally and a national network is the Neighborhood Housing Service (NHS). This form of CDC originated in Pittsburgh in the 1970s to provide market rate financing, primarily for housing rehabilitation and home purchase by moderate-income, first-time homebuyers in minority neighborhoods where "conventional lending" was unavailable because of mortgage "redlining" by financial institutions. NHS grew into a national network, supported by the federally organized National Reinvestment Corporation.

Statewide networks of CDCs have emerged in 17 states and the District of Columbia. They have lobbied for and received support from state-funded housing programs. A leading example of this is Massachusetts.

Origins

CDCs trace their origins to the late 19th-century urban and housing reform movements. Reformers seeking to improve conditions in urban slums, especially those inhabited by immigrants, in lieu of any governmental programs for housing the poor, sought to convince enlightened philanthropists to invest in limited-dividend housing. However, this effort failed to attract many investors. It was not until the advent of federally financed public housing in 1937 that the U.S. government finally became involved in addressing low-income housing problems.

The modern CDC movement traces its origin to 1966, when Senator Robert F. Kennedy successfully sponsored "special impact" legislation to authorize the federal antipoverty program to fund large-scale CDCs in selected neighborhoods. The first of these was the Bedford-Stuyvesant Restoration Corporation in Brooklyn, which is still in existence. The anti-redlining movement of the 1970s spawned more CDCs, seeking to promote reinvestment in poor, mostly urban neighborhoods, where there was much poverty and poor housing and little market activity. Under the Carter administration, the Neighborhood Self-Help Development (NSHD) program was created in the Department of Housing and Urban Development (HUD) to provide direct federal assistance to CDCs. Although this program was judged to be a success, it was terminated by the Reagan administration in 1980.

Despite the elimination of federal programs like this, the antipoverty effort, the VISTA (Volunteers in Service to America) program and others, CDCs still continued to grow in the 1980s, as did the problems of inner-city neighborhoods. Although HUD's budget drastically declined under the conservative Reagan administration, CDCs were able to obtain federal funding available through cities (primarily the Community Development Block Grant program) and entice private investment, primarily through a low-income housing tax credit attractive to corporate investors. CDCs gained experience and credibility and were able to make some progress in improving conditions within their territory.

Accomplishments

According to a survey of 834 CDCs, these groups had developed approximately 125,000 new and rehabilitated housing units through late 1988, when it was estimated that there were between 1,500 and 2,000 CDCs in the United States, compared with only 200 in the mid-1970s. Several other studies have provided considerable data on CDC achievements and the status of the CDC movement. Many CDCs have expanded their activities, and their development projects have grown larger in scale. Profiles of individual CDCs and their leaders are often inspirational.

Prospects and Problems

CDCs have compiled an impressive record with limited resources over a quarter century. Yet the social and economic problems confronting them have grown and often threaten to surpass their ability to respond effectively. In largely abandoned areas such as the South Bronx in New York City and riot-torn neighborhoods such Liberty City in Miami, CDCs have taken on daunting problems. CDCs alone, even if well supported, are not a substitute for government and cannot be expected to solve all of these problems.

CDCs must develop a political constituency to obtain the funding necessary for their survival. In cities with strong and long-standing CDC networks, this has occurred locally. Nationally, this is more difficult. However, CDC advocates did succeed in persuading Congress to provide special "set-asides" for CDC-sponsored housing when it authorized federally subsidized newly constructed housing in 1990. In 1993, Congress made the low-income housing tax credit permanent.

As long as CDCs continue to receive governmental, philanthropic, and corporate financial support, they will be able rehabilitate and build housing and revitalize their neighborhoods. However, physical development is not likely to resolve urgent other problems in poor neighborhoods (e.g., poverty, unemployment, crime and drug abuse, declining public services, poor schools, racial conflicts, etc.). Many modern CDCs were created by social action organizations. These parent organizations have almost all disappeared, and CDCs rarely engage in community organizing leaving a vacuum. Without organized neighborhoods, the complexity of problems may overwhelm CDCs and more traditional social service delivery agencies. (SEE ALSO: **Community-Based Housing; Urban Redevelopment**)

—*W. Dennis Keating*

Further Reading

Goetz, Edward G. 1993. *Shelter Burden: Local Politics and Progressive Housing Policy.* Philadelphia: Temple University Press.

Keating, W. Dennis, Keith P. Rasey, and Norman Krumholz. 1990. "Community Development Corporations in the United States: Their Role in Housing and Urban Redevelopment." Pp. 206-218 in *Government and Housing: Developments in Seven Countries,* edited by Willem van Vliet-- and Jan van Weesep. Newbury Park, CA: Sage.

Mayer, Neil S. 1990. "The Role of Nonprofits in Renewed Federal Housing Efforts." In *Building Foundations: Housing and Federal Policy,* edited by Denise DiPasquale and Langley C. Keyes. Philadelphia: University of Pennsylvania Press.

National Congress for Community Economic Development. 1989. *Against All Odds: The Achievements of Community-Based Development Organizations.* Washington, DC: Author.

Pierce, Neil R. and Carole F. Steinbach. 1987. *Corrective Capitalism: The Rise of America's Community Development Corporations.* New York: Ford Foundation.

Rasey, Keith P. 1993. "The Role of Neighborhood-based Housing Nonprofits in the Ownership and Control of Housing in U.S. Cities." In *Ownership, Control, and the Future of Housing Policy,* edited by R. Allen Hays. Westport, CT: Greenwood.

Rasey, Keith P., W. Dennis Keating, Norman Krumholz, and Philip D. Star. 1991. "Management of Neighborhood Development: Community Development Corporations." Pp. 214-36 in *Managing Local Government: Public Administration in Practice,* edited by Richard D. Bingham et al. Newbury Park, CA: Sage.

Vidal, Avis C. 1992. *Rebuilding Communities: A National Study of Urban Community Development Corporations.* New York: New School for Social Research, Graduate School of Management and Urban Policy, Community Development Research Center.

Walker, Christopher. 1993. *Nonprofit Housing Development: Status, Trends, and Prospects.* Washington, DC: Federal National Mortgage Association. ◄

► Community for Creative Non-Violence

The Community for Creative Non-Violence (CCNV) began in Washington D.C. in 1970 as an opposition to the war in Vietnam. In 1972, responding to the needs of people who were homeless and hungry in the nation's capital, CCNV members opened a soup kitchen. Shelters, free clinics, legal services, and other programs to serve Washington's poor followed. They also took the lessons of social change learned through the antiwar and civil rights movements and applied them to issues of poverty.

In 1997, CCNV operated the largest shelter in the United States with 1,400 beds and a full array of services, including a medical clinic, a 32-bed infirmary, a dental clinic, social services, job programs, drug and alcohol treatment, legal services, and art and education programs.

CCNV has continued to be a community of resistance, challenging the forces that create and sustain poverty and homelessness. In the 1980s, when there was a dearth of leadership on issues of poverty, CCNV emerged as a powerful voice for the poor. In 1982 and 1984, it helped the House Subcommittee on Housing and Community Development organize the first national hearings on homelessness. Through fasts, lawsuits, voter initiatives, acts of civil disobedience, national marches, and legislative advocacy, it brought homelessness and the lack of affordable housing into national focus.

In 1984, it engaged the Reagan administration in a two-year battle to retain and renovate the Federal City Shelter. This campaign included multiple arrests, three fasts, congressional hearings, and finally, intervention by Congress before $14 million in renovation was allocated to create a model facility for homeless people. In 1986, CBS television produced a film, *Samaritan: The Mitch Snyder Story,* about this struggle to build the CCNV shelter, starring Martin Sheen and Cicely Tyson. In 1988, an independent film producer released *Promises to Keep,* a documentary about the same issue that was nominated for an academy award. In 1987, CCNV played a key part in ensuring the passage of the McKinney Homeless Assistance Act, which continues to provide important programs nationally to homeless people.

CCNV also had a central role in the organization of the national Housing Now march in 1988, which brought 250,000 people to Washington, D.C., to demand affordable housing. In 1991, it created the Trust for Affordable Housing with $4.5 million obtained as a result of a class action suit against the District of Columbia. The trust provides funds for the creation of permanent housing for single homeless people in Washington, D.C.

Although CCNV lost its most powerful advocate and spokesperson, Mitch Snyder, in 1990, it has continued its work of more than two decades, combining its historical traditions of service, resistance, and spirituality. (SEE ALSO: **Homelessness**)

—*Carol Fennelly* ◀

▶ Community Housing Partnership

The Community Housing Partnership (Subtitle B of the HOME Investment Partnership Act, a component of the 1990 Cranston-Gonzalez National Affordable Housing Act) was federal legislation to direct federal housing funds to community-based nonprofit development organizations for the purpose of building and rehabilitating affordable housing.

The legislation emerged from discussions between Boston Mayor Raymond L. Flynn and Congressman Joseph P. Kennedy of Massachusetts. Boston, like many other cities, had seen a significant growth in the number and capacity of community development corporations and other nonprofit housing development groups during the 1980s, but the cutbacks in federal housing funds had hampered their ability to expand the housing supply.

The legislation was originally filed in Congress in 1987. Congressman Kennedy filed the bill in the House of Representatives. Senator Frank Lautenberg of New Jersey sponsored the bill in the Senate. Congress never acted on the bill as separate legislation, but it was folded into the Cranston-Gonzalez National Affordable Housing Act (NAHA) of 1990 as a separate program. The comprehensive housing bill was sponsored by Congressman Henry Gonzalez of Texas and Senator Alan Cranston of California, chairs of the House and Senate committees in charge of housing policy; it was passed by Congress and signed by President Bush in 1990.

One of the key provisions of the NAHA was the creation of the HOME program. This program targets federal funds to cities and states, exclusively for housing production and rehabilitation, through an allocation formula based on population size, poverty, and other factors. The Commu-

nity Housing Partnership (CHP) program was incorporated into the HOME program. The CHP part of this legislation required that each city use at least 15% of all HOME funds for community-based nonprofit housing developers, called community housing development organizations (CHDOs). A small portion of these funds are set aside for groups that provide technical assistance to nonprofit development organizations. Funds can be used only to provide housing for low-income and moderate-income households. The CHP program represents the first federal housing program designed exclusively to support nonprofit sponsors of family housing. (The Section 202 program supports nonprofit sponsors of elderly housing.)

Funding for HOME and its CHP component began in FY 1992. In that year, Congress authorized $1.46 billion in HOME funds. In its first year, cities exceeded the 15% threshold target for CHDOs, allocating 20.9% ($305.5 million) of all HOME funds to these nonprofit groups. A number of major cities allocated half or more of their HOME funds to CHDOs.

CHDOs used about three-quarters of the funding commitments to rehabilitate substandard housing, using the rest for new construction or for subsidies to tenants. About 30% of the CHDO funding commitments targeted rental housing. The remaining CHDO funds helped provide homeownership, primarily by helping existing low-income owners (including small landlords) repair their deteriorating properties. (SEE ALSO: *Cranston-Gonzalez National Affordable Housing Act of 1990*)

—*Peter Dreier* ◀

▶ Community Land Trust

A community land trust (CLT) is a nonprofit organization created to purchase and hold land on behalf of a community. CLTs are formed to keep land in a trust for the benefit of local residents, to ensure continued access to the land, and to provide for affordable housing. The CLT keeps housing affordable by removing the land from the speculative market so that the increased value of land never adds to the cost of the housing. The land trust is an old idea that has recently been tried on a limited, experimental basis in parts of the United States. This technique for maintaining housing affordability became popular among community-based affordable housing advocates during the 1980s. In 1996, there were more than 100 CLTs in development or operation across the United States, accounting for more than 4,000 units of housing.

Objectives

The objectives of a CLT are multiple. In a general sense, CLTs allow communities to maintain ownership and control over their land assets, and they allow more democratic control over the land development process. More specifically, land trusts have been formed to provide affordable housing for low- and moderate-income households. Land trusts can also be used to provide affordable leases to small, locally owned businesses or to ensure continued agricultural production on land threatened by other uses.

The ownership of land and control over the use of that land allows CLTs to pursue additional objectives as well. The secondary objectives of a land trust generally relate to the achievement of social goals such as encouraging housing integration, land conservation, and small business development programs.

How CLTs work

The land trust model separates ownership of land from ownership of the improvements made on the land. CLTs typically receive private financial donations or government subsidies to purchase property. Income from long-term leases allows further purchases and expansion of the land trust. CLTs add to the stock of affordable housing in two ways; they purchase undeveloped land and then arrange for the development of housing on their parcels, or they purchase land with housing already on it. In either case, CLTs remove the cost of the land from the purchase price of the housing and restrict resale prices to limit profit taking from housing transactions.

CLTs are democratically structured nonprofit corporations, with membership open to interested local community members. The members elect a board of directors, typically community residents who are not leaseholders on land trust property, community organization representatives, and other public interest representatives. The board sets policies for the CLT and directs the development or expansion strategy for the organization. Annual membership meetings are held in which goals, directions, and accomplishments of the board are reviewed. This broader community control and representation distinguishes land trusts from co-op models of housing.

The CLT attempts to serve the interests of both homeowners and the larger community. The needs of homeowners, for tenure security and equity, are guaranteed through a lease. CLTs grant long-term land leases to homeowners or business owners whose buildings occupy CLT land. The leases are set at a level affordable to low- and moderate-income households. Restrictions on resale of the improvements are placed on the lease that allow leaseholders a fair compensation for their investment but do not allow the taking of profits from increases in the market value of the land. Typically, the costs of the improvements can be recovered during resale, as can normal inflationary costs. Beyond that, however, the price is limited.

The homeowner or business owner receives security through the long term of the lease (often 99 years) and, typically, can pass the housing on to future generations. The homeowner also builds up equity in the house based on mortgage repayments and improvements made.

The resale of the home allows the seller to recover costs and realize equity while keeping the price below what it would have been with the value of the land included. Thus, although homeowners are guaranteed tenure security and equity, the community retains control over the land and is able to ensure the continued availability of affordable housing to low-income households. For example, CLTs generally retain the right of first refusal on sales that take place,

allowing them to find income-eligible families to purchase the homes. The CLT model can be combined with a number of housing types, including cooperative or mutual housing, private ownership of single-family homes, or rental housing.

History

The history of land trusts is a long one. Religious and spiritual beliefs among Native Americans, for example, regard the land as having a special status in that it cannot be owned by private individuals but, rather, exists for the use of all. This orientation toward land and the communal properties and benefits of land is captured in the land trust concept. In the late 19th century, Henry George, in *Progress and Poverty,* suggested that private ownership of land and the unequal distribution of land were the core reasons for persistent poverty in a country with abundant resources. His solution was a single tax of such scale that privately owned real estate would no longer provide income to its owners, with the revenue diverting, instead, to the state. It was a short step from this idea to the concept of land trusts, introduced by Ralph Borsodi during the 1930s as an alternative to private property.

The concept has been applied on a broader scale in countries other than the United States. In India, the *Gramdan* movement, an extension of Ghandian political and social ideas, is a grassroots, primarily rural, political movement that has pushed for community ownership of land and other economic assets. The *Gramdan* movement focuses on redistribution of wealth through land and advocates decentralized, communalized, decision making and governance in *Gramdan* villages. In Israel, the Jewish National Fund is a nongovernmental institution with enormous land holdings throughout the country. The fund leases land to those who use it in the long-term public interest.

In the United States, the first CLT was established in Georgia, in 1968. However, during the 1980s, CLTs emerged as a local response to housing affordability problems. In 1983, there were fewer than 15 CLTs nationwide. By the mid-1990s, there were more than 80 operating CLTs and more than 20 more in development.

The Institute for Community Economics (ICE) currently acts as the leading advocate of the CLT model. Initially formed in 1967 as the International Independence Institute, ICE provides technical assistance to groups wishing to form CLTs. Their services include board and staff training, project development assistance, finance packaging and ongoing consultations.

The 84 CLTs operating in the United States in 1996 had brought more than 4,000 units into the stock of permanently affordable housing. According to figures from ICE, more than half of the communities with CLTs are in rural areas or small cities. This is due, to some extent, to the large number of CLTs throughout Vermont, where the state supports CLT formation, and New Hampshire. CLTs are supporting a variety of housing types across the United States; 27% of CLT housing units are single-family homeowner units, 43% are rental, 15% are cooperative units, and the rest are either condominium, transitional housing, or vacant/under construction. CLTs also support emergency homeless shelters. Most (89%) of the residents of CLT housing are low-income or very low-income households (less than 50% and 80% of the area median income, respectively).

One obstacle facing the CLT movement is the reluctance of private lenders to finance CLTs because of the unorthodox nature of the ownership relations. Lenders are made uneasy by the fact that, unlike conventional models of property ownership, CLT homeowners own the physical structure but not the land under it. The more widespread use of CLTs as affordable housing strategies will depend on greater availability of private financing.

Government support for CLTs has been minimal, although it has grown in recent years. At the federal level, CLTs are now recognized as eligible community housing development organization recipients for HOME program funds. The Farmer's Home Administration (FmHA) has also committed funds to projects in Washington, Maine, and Connecticut. State and local governments have expanded their support of CLTs as well. The state of Connecticut appropriated $4 million to a fund to support CLTs as part of their Forever Housing initiative, aimed at creating a permanently affordable housing stock. Minnesota provides start-up assistance to groups developing CLTs. At the local level, the city of Burlington, Vermont, provided $200,000 to help initiate a land trust in that city, and Durham, North Carolina, committed close to a half a million dollars for no-interest second mortgages to leaseholders of North Carolina CLTs.

CLTs represent a community-based solution to affordable housing needs. During the 1980s, when CLTs began to emerge in significant numbers, the federal government was drastically cutting its housing assistance budget. Local governments were forced to fill in the gap and were prodded in that direction by a decentralized, community-based movement of affordable housing activists. This movement advocated various forms of housing assistance programs. CLTs embody several principles attractive to local activists, including continuous affordability of housing, the establishment of a strong nonprofit sector in housing, community-based control over land development, and alternatives to private market ownership models. (SEE ALSO: **Community-Based Housing; Tenure Sectors**)

—*Edward G. Goetz*

Further Reading

Davis, J. E. 1984. "Reallocating Equity: A Land Trust Model of Land Reform." In *Land Reform, American Style*, edited by C. C. Geisler and F. J. Popper. Totowa, NJ: Rowman & Allanheld.

Gunn, Christopher and Hazel Dayton Gunn. 1991. *Reclaiming Capital: Democratic Initiatives and Community Development*. Ithaca, NY: Cornell University Press.

Institute for Community Economics. 1982. *The Community Land Trust Handbook*. Emmaus, PA: Rodale.

International Independence Institution. 1972. *The Community Land Trust: A Guide to a New Model for Land Tenure in America*. Cambridge, MA: Center for Community Economic Development.

Soifer, Steven D. 1990. "The Burlington Community Land Trust: A Socialist Approach to Affordable Housing?" *Journal of Urban Affairs* 12:237-52.

White, Kirby and Charles Matthei. 1987 "Community Land Trusts." In *Beyond the Market and the State*, edited by Severyn T. Bruyn and James Meehan. Philadelphia: Temple University Press. ◄

► Community Reinvestment Act

A community reinvestment movement has emerged in urban areas of the United States. During the past 20 years, lenders in more than 100 cities have committed more than $210 billion in urban reinvestment initiatives to counter the effects of redlining in these communities. These commitments followed documentation by several academic, community organization, and government publications of credit availability problems in older urban communities and racial bias in mortgage-lending practices, facts of life that many residents had long experienced firsthand. One result of this movement, and a key factor accounting for these recent commitments, is the federal Community Reinvestment Act (CRA), passed in 1977. Under the CRA, federally regulated depository institutions that offer mortgage loans have an affirmative and continuing obligation to assess and respond to the credit needs of their entire service area, including low- and moderate-income neighborhoods, consistent with safe and sound lending practices.

Under the statute, federal financial regulatory agencies conduct periodic evaluations of the CRA performance of those institutions they regulate. Exams have focused on lender ascertainment of credit needs, extent to which they work with community organizations, and other procedural issues. Future exams will focus more on lender performance in terms of loans made, banking services provided, and investment activity. After the examinations are completed, lenders are given one of the following four CRA ratings: (a) outstanding, (b) satisfactory, (c) needs improvement, or (d) substantial noncompliance. The ratings and supporting rationale must be made publicly available.

In addition to the periodic CRA evaluation, this law requires regulators to consider the community reinvestment performance of the lenders they oversee whenever the lenders file applications for changes in their business operations, such as new charters, new levels of deposit insurance, new branch offices, relocations of offices, mergers or consolidations with other institutions, and acquisitions. The "teeth" in the law, however, come from challenges to these applications that third parties, including community organizations, can file with the regulatory agency. If there is evidence that a financial institution has not met its obligations under the CRA (e.g., analysis of Home Mortgage Disclosure Act data revealing that certain markets have not been served, evidence of gerrymandered service areas, failure to prepare a complete CRA statement), the regulator can approve the application with some conditions, delay consideration of the application, deny the application, or approve it. Although regulators have rarely, on their own initiative, rejected an application for CRA-related reasons, frequently they have delayed an application or attached conditions because of a third-party challenge. Often, the regulator will delay consideration while the financial institution and challenging party negotiate an agreement that will satisfy those filing the challenge and enable the lender to proceed with the application. Delays can be costly, and generally, the lender endeavors to minimize those costs. This is the basic process that has resulted in the reinvestment agreements that lenders have negotiated with community groups in cities around the United States.

A CRA challenge is not, however, a straightforward process. Invariably, it involves substantial work and support by various groups in a community. A $65 million commitment by Atlanta area banks followed a Pulitzer Prize winning series "The Color of Money" written by Bill Dedman for the *Atlanta Journal/Constitution*. A sympathetic state banking commissioner, a coalition of community groups (including civil rights, labor, church, and business organizations), and a series of reports by David Everett in the *Detroit Free Press* led to a $2.9 billion commitment by a consortium of banks in Detroit. A coalition of housing, economic development, and other community organizations in Chicago negotiated a $174 million, five-year commitment with three lenders that, when it expired, was extended for an additional five years, calling for $200 million more in housing and business loans. In Boston, research by the Federal Reserve Bank, aggressive organizing by a coalition of community groups, and active leadership by Mayor Raymond Flynn led to a $400 million commitment by that city's major lenders.

As significant as these financial commitments are, more than money has been negotiated in these agreements. In some cases, lenders have agreed to open a new branch bank in a previously underserved area. Affirmative action commitments to increase minority employment have been part of some agreements. Additional components of various agreements include more flexible underwriting standards, innovative marketing efforts to reach minority communities, creation of housing counseling programs for first time homebuyers, grants to nonprofit community organizations, reduced down payment requirements and closing costs, commitments to contract with minority vendors, and many others. Several lenders have applauded these community reinvestment initiatives, frequently observing that community reinvestment is consistent with good business. But the basic response of the industry has been to lobby for the dilution of CRA. Their "reform" proposals have generally contained two provisions. One is a "safe harbor" that would exempt any lender with an outstanding or satisfactory rating from most CRA requirements. Given that approximately 90% of institutions have received one of these two ratings, consumer groups object to excluding such a large segment of the lending community from the CRA.

A second provision is an exemption for small lenders from many of the reporting requirements because, it is argued, they cannot afford to hire the staff that larger institutions can employ to process these government mandates. Consumer groups respond that small lenders have been no more effective in complying with CRA than large ones and point to a study by the Office of Management and Budget concluding that the time required for a lender to meet federal paperwork requirements associated with CRA averages six hours per year, as reasons for rejecting this proposal.

In light of the Clinton administration's proposals for community development banking, some lenders have taken a third approach, suggesting that they be able to meet their CRA obligations by making a deposit or investment in a "community development finance institutions." Consumers have responded that even if a few million dollars become available through a small number of development banks or other types of community development finance institutions, lenders who control the vast majority of the billions of dollars invested by financial institutions should remain subject to the CRA.

The CRA has proven to be an effective vehicle for encouraging lending in redlined areas and responding to the racial gaps in lending patterns. Yet lenders have continuously pushed for regulatory "reforms" that would reduce the requirements of the law, and community groups have stressed the need for more aggressive enforcement. Redlining and community reinvestment remain highly contentious issues. There is little doubt that they will be part of urban policy debates for years to come. (SEE ALSO: *Blockbusting; Home Mortgage Disclosure Act; Redlining; Urban Redevelopment*)

—*Gregory D. Squires*

Further Reading

Bostic, Raphael W. and Glenn B. Canner. 1997. "New Information on Lending to Small Businesses and Small Farms: The 1996 CRA Data." *Federal Reserve Bulletin* 84(1):1-21.

Evanoff, Douglas D. and Lewis M. Segal. 1996. "CRA and Fair Lending Regulations: Resulting Trends in Mortgage Lending." *Economic Perspectives* 20 (6):19-46.

Marsico, Richard D. 1996. "The New Community Reinvestment Act Regulations: An Attempt to Implement Performance-Based Standards." *Clearinghouse Review* (March):1021-33.

Reinvestment Works. Periodic newsletter of the National Community Reinvestment Coalition, Washington, DC.

Squires, Gregory D., ed. 1992. *From Redlining to Reinvestment: Community Responses to Urban Disinvestment*. Philadelphia: Temple University Press.

Tholin, Kathy. 1996. *Tools for Promoting Community Reinvestment*. Chicago: Woodstock Institute. ◄

► Commuting

The separation of work and home is as old as work itself. Yet we tend to think of commuting to work, at least on a massive scale and over long distances, as primarily an outcome of 20th-century industrialization and urbanization. Shorter working hours and rising real incomes provided both the time and the means to commute over longer distances; improved transportation technologies (trams, railroads, and then the automobile) and new suburban housing construction provided the opportunities and the incentives to commute; the increasing size of firms and the widespread dispersion of jobs through suburbanization provided the necessity to commute. Postwar planning ideologies, especially the use of single-purpose zoning, also contributed by encouraging the rigid separation of land uses and thus increasing the isolation of jobs from housing.

Commuting, or, more precisely, the journey to and from work, is important here for several reasons. It is the largest single source of travel and one of the major consumers of energy and urban land. It is also, when dominated by the automobile, the principal contributor to atmospheric pollution, to accidental death and injuries, and to congestion within cities. Attempts to alleviate traffic congestion, in turn, consume even more urban land and public capital through expansion of highway and transit systems. The residential form of cities mirrors their journey-to-work characteristics.

Commuting is directly relevant to the study of housing because it is the physical and spatial manifestation of the complex links between the production and consumption spheres in cities. Commuting brings labor to the place of production and in effect returns income and wages to the location of the worker's household and local neighborhood. It also serves various other social functions, such as intermediate stops for shopping or day care. The time and distance that people are prepared to commute, or are capable of commuting, also shape the organization of local labor and housing markets. As such, decisions about housing provision (investment and construction) and consumption (purchase or rental) are intimately tied to the demands of firms for labor and to the changing spatial organization of production. Housing and employment locations are therefore simultaneously determined but with interesting and complex variations.

Classical theories of urban form offer one view of this relationship. Typically, they assume that all employment is located at the city center and all residential location decisions are made with respect to this center. Households then sort themselves out geographically according to the trade-offs they are willing (or are forced) to make between the advantages of living close to their work (thus living at higher densities near the center) and living farther out at lower densities, where land and housing are less expensive but commuting is more costly (in terms of time, effort, and money). In theory, the price of urban housing (especially residential land) is the inverse of transportation costs.

These models, however, have been widely criticized on two principal grounds: first, the assumption that all jobs are in the city center, and second, the implication that households select a residential location based on a fixed employment location. In contemporary metropolitan America, less than 10% of all employment on average is concentrated in the downtown area. In Western Europe and Japan, city centers have retained much higher proportions of total metropolitan employment, but the trend is in the same direction. The overwhelming majority of jobs are now in the suburbs, either widely dispersed or concentrated in a few new industrial districts and business nodes. The monocentric city has been replaced by a decentralized but polycentric urban form.

There is now considerable evidence that individuals often select job locations after or in response to a given choice of housing (or choice of living arrangement) rather than the reverse. This trend is perhaps most evident for two-earner households, single-parents, and others with particular housing needs, preferences, or constraints and is

reinforced by the fact that the rate of job turnover now often exceeds housing turnover. Despite the volume and complexity of commuting, the configuration of the journey to work has remained relatively constant over the last few decades, at least in North America. The frequency distribution of commuting distances follows an easily recognizable shape: Few people work at home (perhaps 5%) or close to home, and proportionally few commute very long distances of more than an hour. The majority, roughly 50%, commute between 20 and 35 minutes, with the actual distances varying by mode of transport (auto or transit) and region, and on average, these distances tend to increase the larger the urban area. Generally, travel times by automobile are shorter than by transit in North America. In Europe, average commuting distances are much shorter, but times are often longer, and relatively more workers commute by transit, by bicycle, and on foot.

Although average commuting distances have increased almost everywhere since the 1970s, and in some areas substantially, actual travel times have not, principally because of improved transportation facilities. Whereas metropolitan areas have grown dramatically in terms of their physical size, commuting times have not grown in proportion, in part because of the suburbanization of employment. As a consequence, origin-destination patterns have shifted, with suburb-to-suburb commuting now representing the largest single component (36%) of all work trips. On the other hand, in large and rapidly growing metropolitan areas, long-distance commuting has increased as households search for more affordable housing in outer suburbia or adjacent small towns and exurban areas. The social and environmental costs of these extended commuting patterns are high, notably for households with low or unstable incomes and for those with two or more workers.

Projecting current trends ahead, it is possible to argue that despite the communications revolution, and the resulting increased capability to work at home or away from the office, linked by computer networks, commuting could become even more important in influencing residential location decisions and the operation of local housing markets in the future. Most employment growth will be in the new suburbs but will be serviced only partly by new and expanded highways and even less so by new public transit. Cross-commuting between suburbs will continue to grow, and thus congestion will likely increase further. Large employers will tend to attract pools of labor to adjacent residential communities, in theory reducing longer commuting, but exclusionary zoning and high house prices will ensure that a labor demand-supply mismatch continues. Although some employers will read the changing suburban social landscape and locate their firms so as to access the required kinds of labor, often female, this will not be widespread. The result of these trends, on balance, is difficult to predict, but they are likely to lead to an increase in the volume of commuting, a gradual increase in commuting times, and longer commuting distances.

At the same time, commuting will continue to be a severe constraint for many households and workers, particularly given the increasing diversity of household types and continued residential segregation. In the United States, low-skilled inner-city residents, notably minorities, may find themselves trapped, increasingly cut off from expanding employment in the outer suburbs by distance or travel cost, by limited access to an automobile and poor public transit, or by physical or time constraints. Those residents may also be trapped by being excluded or discouraged from moving closer to those suburban jobs through restrictive zoning, the lack of affordable housing, racial discrimination, or a simple lack of information. In Europe, where many of the low-income housing estates are located in the suburbs, similar constraints may apply but often in the reverse direction.

Women show consistently shorter commuting distances than men, and for many, this is a matter of choice. For most others, however, particularly working women with children and low incomes, and single parents, the task of combining home and work responsibilities in the absence of universal day care, flexi-time, and reliable public transit, poses severe constraints on commuting and thus on their daily routine. These constraints, in turn, limit their access to housing and employment opportunities and thus reduce both long-term income levels and career choices. Given the increase in female labor force participation rates and single parenting, such constraints are likely to grow in importance.

Finally, it is worth reiterating that commuting is not just a transportation problem, however substantial that problem is. It is at the same time a social and environmental issue, as well as a component of urban labor markets and a significant factor in both the production of housing and the location decisions made by housing consumers. (SEE ALSO: *New Urbanism*)

—*Larry S. Bourne*

Further Reading

Aitken, Stuart C. and Timothy J. Fik. 1988. "The Daily Journey to Work and Choice of Residence." *Social Science Journal* 58(4): 463-75.

Clark, William A. V. and Marianne Kuijpers-Linde. 1994. "Commuting in Restructuring Urban Regions." *Urban Studies* 31(3): 465-83.

Giuliano, G. and A. Small. 1993. "Is the Journey to Work Explained by Urban Structure?" *Urban Studies* 30:1485-1500.

Gordon, P., H. Richardson, and Myung-Jin Jun. 1991. "The Commuting Paradox." *Journal of the American Planning Association* 57:416-21.

Kenworthy, Jeff, Felix Laube, Paul Bartner, Peter Newman, et al. 1998. *Cities and Automobile Dependence, 1960–1990: An International Urban Data Compendium*. Boulder: University Press of Colorado.

Madden, J. and L. Chen Chiu. 1990. "The Wage Effects of Residential Location and Commuting Constraints on Employed Married Women." *Urban Studies* 27:353-69.

Pisarski, A. 1987. *Commuting in America: A National Report on Commuting Patterns and Trends*. Westport CT: Eno Foundation.

Rutherford, B. and G. Wekerle. 1988. "Captive Rider, Captive Labor: Spatial Constraints and Women's Environments." *Urban Geography* 9:126-37.

Singell, Larry D. and Jane H. Lillydahl. 1986. "An Empirical Analysis of the Commute to Work Patterns of Males and Females in Two-Earner Households." *Urban Studies* 23:119-29. ◀

▶ Company Housing

In the United States, employers have provided housing to workers in mining, lumbering, textiles, and a handful of other industries. Company towns developed largely because the worksites were located in isolated, largely unsettled regions; some manufacturers in suburbs housed skilled workers to cut turnover. Most employers did not charge monopoly rents, although at times they used restrictive housing leases to limit collective action.

Company housing was important in several industries. New England textile mills housed young women in the 1800s to monitor the women's work in the mills and their social activities. Some manufacturers on the outskirts of cities in the early 1900s housed their most skilled workers. Mining towns and timber camps typically housed the majority of the workforce in isolated mountainous areas with little prior settlement. Most company towns met their demise as the surrounding areas became more densely settled and transportation improved.

The quality of company housing depended on the permanence of the worksite and the incomes of workers. In the timber industry, where companies exhausted the resource and moved on within a year or two, the housing was often rough-hewn, temporary bunkhouses that housed several men. In the coal industry where it might have taken 10 to 30 years to mine a seam, the companies built towns with a range of housing. Bachelors and young men seeking to save money often lived in cheap shanties, poorer families lived in smaller houses or boarded workers in larger houses, whereas families of skilled workers tended to live in more spacious houses. The newly built housing was often an improvement over the existing rural housing in areas such as the Appalachian mountains. Typically, the quality of sanitation and other services was about the same as in noncompany towns of similar size. On the other hand, the impermanence of the resource base often meant that the houses were not as longlasting as houses in towns with a broader industry base.

The quality of life also varied substantially across company towns. Towns with larger populations offered more amenities because economies of scale lowered their costs. Newer towns offered better and more modern facilities, whereas older towns relied on the technologies extant when they were built. Towns blessed with more spacious and less mountainous locations could usually offer better housing at lower cost. Some of the larger mining employers also experimented with welfare capitalism, providing better housing and services to attract and keep more productive workers.

Company towns were controversial institutions. Economics textbooks cite them as classic examples of monopoly, and many cite them as a tool for union busting. This emphasis on monopoly is incorrect. The rents for company housing were often lower than the rents workers paid in independent cities and towns. The rates of return on investment in company housing were similar to normal rates of return in competitive industries. The companies were unable to exploit a local monopoly because they had to attract workers from other areas in a regional labor market where they would compete with several hundred other employers. Typically, companies had to match higher monthly rents with similar increases in wages.

The union-busting image of company towns stems in part from the nature of the housing leases. The leases made the housing contingent on employment, allowed less notice for eviction, and allowed the company control over visitors. Many companies claimed that they rarely enforced the leases, and they often deferred rental payments when work was slow or the tenant was sick or injured. On the other hand, some companies used the clauses to keep union men out and to evict workers when they struck. The evictions and the presence of company guards at times contributed to the greater incidence of violent strikes in company towns.

Union busting, however, was not the primary reason for company towns. Fear of unionization was common to all employers, but company towns existed largely in areas isolated from other industrial and agricultural activity. The isolation gave both workers and employers reasons for seeking company housing. By renting, workers could avoid being tied to a single mine and could eliminate the risk of capital loss on a house in a town dependent on a cyclic industry. Employers had incentives to own the housing to avoid giving independent contractors a local monopoly position. An independent could freely exploit the monopoly at the expense of the employer, who was forced to pay higher wages to compete for workers in the regional labor market. Employers eliminated this wealth transfer by owning the housing themselves.

Manufacturers in the suburbs of cities also cited isolation as their reason for housing workers. They seemed to focus on cutting turnover of skilled workers, however, because they typically housed less than 30% of their workers and the housing was usually of higher quality than in other towns. In the late 1990s, employers still experimented with company housing in settings to help them attract skilled workers, particularly in cities with relatively high housing costs. (SEE ALSO: *Employer-Assisted Housing*)

—*Price V. Fishback*

Further Reading

Allen, James B. 1966. *The Company Town in the American West.* Norman: University of Oklahoma Press.

Brandes, Stuart D. 1976. *American Welfare Capitalism, 1880-1940.* Chicago: University of Chicago Press.

Buder, Stanley. 1995. *Pullman: An Experiment in Industrial Order and Community Planning, 1880-1930.* New York: Oxford University Press.

Crawford, M. 1995. *Building the Workingman's Paradise: The Design of American Company Towns.* New York: Verso.

Fishback, Price. 1992. *Soft Coal, Hard Choices: The Economic Welfare of Bituminous Coal Miners, 1890-1930.* New York: Oxford University Press.

Hall, Jacquelyn Dowd, James Leloudis, Robert Korstad, Mary Murphy, Lu Ann Jones, and Christopher B. Daly. 1987. *Like a Family: The Making of a Southern Cotton Mill World.* Chapel Hill: University of North Carolina Press.

Magnusson, Leifur. 1920. *Housing by Employers in the United States.* Bureau of Labor Statistics Bulletin No. 263. Washington, DC: Government Printing Office.

Reps, John. 1991. *The Making of Urban America: A History of City Planning in the U.S.* Princeton, NJ: Princeton University Press.

Shifflett, Crandall. 1991. *Coal Towns: Life, Work, and Culture in Company Towns of Southern Appalachia, 1880-1960.* Knoxville: University of Tennessee Press. ◄

► Comprehensive Housing Affordability Strategy

The National Affordable Housing Act of 1990, also known as the Cranston-Gonzalez Act, gave new resources to states and local governments to address affordable housing needs. The act also required that jurisdictions receiving funds under a variety of federal housing programs prepare a comprehensive housing affordability strategy (CHAS). This five-year planning tool involved identifying critical housing needs and setting priorities for addressing these needs. It must be approved by the U.S. Department of Housing and Urban Development (HUD) and is updated every year.

The CHAS replaces the housing assistance plan (HAP), a requirement for Community Development Block Grant recipients, and the comprehensive homeless assistance plan (CHAP), which had to be prepared for participation in McKinney Act homeless programs.

When the HAP was first mandated by the Community Development Act of 1974, it represented a significant innovation. The act itself consolidated a variety of categorical programs, distributed housing assistance consistent with local plans, and required the dispersion of housing assistance. The HAP made local governments identify housing needs and set goals for addressing these needs. In so doing, they were to ensure that housing assistance did not result in concentration and segregation of poor households. Applications for federal housing assistance were turned down if the proposed projects were not included in the local HAP. This gave local governments a measure of control that had not been available to them prior to the enactment of the Community Development Act.

The usefulness of the HAP was diminished by its emphasis on numerical definition of needs and goals with little consideration given to how these goals were to be met. The absence of federal resources meant that most of the needs could not be addressed. As a result, the HAP came to be widely viewed as little more than a bureaucratic hurdle.

The CHAS, in contrast, placed emphasis on strategic planning, including identifying where intervention would be most effective, coordinating public and private housing activities, and providing means for public participation and review. Coupled with an increase in federal housing resources made available through the Cranston-Gonzalez National Affordable Housing Act, the CHAS was expected to be a more effective planning tool than the HAP. It was also expected to serve as a means of monitoring affordable housing goals accomplishment.

Preparation of a CHAS involved completing a detailed set of tables summarizing housing assistance needs, local market conditions, and the housing inventory. A five-year strategy had to identify general priorities. This was followed by a one-year plan identifying available programs and resources and setting goals for the numbers and types of households to be assisted. The census provided special data tabulations to assist in this work.

It was possible for local governments to take the CHAS a step beyond federal requirements. For example, in Oregon, the cities of Gresham and Portland and Multnomah County decided to develop a countywide CHAS. This led to the development of a Housing and Community Development Commission to address housing problems on an interjurisdictional basis, at the same time eliminating and consolidating local advisory boards. A countywide housing authority was also created by expanding the authority of the Housing Authority of Portland. Consideration was given to addressing the CHAS at the regional level to ensure compatibility and coordination among the five counties that make up the Portland metropolitan area.

Although the CHAS did raise the visibility of housing as an issue, it did not fulfill HUD's expectations. The agency wanted a more holistic approach; the structure of the CHAS was such that it was prepared independently from nonhousing issues and strategies. The lack of resources continued to be a constraint. In response, HUD Secretary Henry Cisneros called for CHAS to be replaced by a consolidated plan to address housing *and* community development needs. Citizen participation, once again, is strongly emphasized. The document must contain a housing and community development profile that includes a description of needs; an analysis of market conditions, issues, and trends; an assessment of barriers to housing assistance and community development; and identification of resources, assets, and opportunities. A vision plan with a 10- to 20-year time frame along with a 5-year strategy and a 1-year projected use of funds also must be an integral part of the consolidated plan. These new requirements took effect on July 1, 1995.

The HAP, the CHAS, and the consolidated plan all represent efforts intended to ensure that planning serves as the basis of decisions regarding the public use of funds for housing and related needs. That the ground rules keep changing stems from the very real constraints associated with the absence of resources and the recognition that what resources are available must be used effectively and efficiently. In developing a standardized process and product that is supposed to be applied uniformly throughout the United States, it is easy to see how the end result can become bureaucratic and cumbersome. To the extent that local governments can see beyond such constraints and take ownership of the planning process, they will be able to produce a decision-making tool that plays an important role in addressing local needs. (SEE ALSO: *Community Development Act of 1974; Cranston-Gonzalez National Affordable Housing Act of 1990; Stewart B. McKinney Homeless Assistance Act*)

—*Deborah A. Howe* ◄

► Computer-Aided Design

For the most part, computer-aided design (CAD) technologies have been developed with little regard to particular building types or uses. However, housing was at the heart of at least two early CAD research and development efforts. In the late 1960s and early 1970s, Aart Bijl and the Edinburgh Computer Aided Architectural Design group (EdCAAD) developed software to help the Scottish Special Housing Association to design their wood frame buildings. This program enabled designers to select from materials and specifications given by the Housing Association and to lay out and analyze housing designs to a quite detailed level. It provided in an integrated form early versions of features that are still lacking in present-day CAD programs: analysis of daylighting and heat loss, testing of furniture layouts, generation of plumbing and electrical layouts, and production of bills of materials.

A second early CAD effort devoted to housing was developed in the Netherlands. Researchers at the Foundation for Architectural Research (SAR) in Eindhoven developed software to automate analysis in the SAR's "support-infill" method of mass-housing design. By distinguishing between shared and individual control of building elements and spaces, the SAR method enabled designers to analyze the variability of proposed housing designs. It provided a means to count the ways in which different dwelling unit types, designed according to certain dimensional norms, could be realized within a given design. The SMOOC (Sar Methode Onderzoek Ontwerpen met de Computer) program automated this analysis, which was valuable but tedious to carry out by hand. The computer, so often employed to stamp out identical designs, in this instance automated a method that promoted variation.

With the advent of desktop computing in the 1980s, software for computer-aided drafting, three-dimensional modeling, and architectural rendering has increasingly become a part of design offices. Commercial CAD packages enable designers to quickly construct and edit drawings, to separate components onto different drawing layers, to view three-dimensional models from viewpoints inside or outside the building, and even to produce animated simulations of movement along a path through or around a design. Rendering programs simulate the visual appearance of a building, showing lighting, shadows, colors, and textures. Renderings can then be placed seamlessly into photographs and video sequences of a site. Using head-mounted video displays and tracking users' eye, hand, or body movements, "virtual reality" software can simulate for the designer the experience of walking around in a proposed design. The question remains, however: Will this result in better buildings?

Conventional CAD programs ease the production and editing of drawings, but only in a limited sense can they be said to support designing. Even virtual reality extensions to rendering and animation programs can help designers understand only the visual appearance of designs. Many CAD researchers believe that future programs can do more to assist designers. For example, although a great deal of work has been done on simulating the energy, acoustical, lighting, and structural behavior of buildings, programs to analyze and evaluate design performance are still poorly integrated with CAD systems and they require significant effort to use. Another promising area in CAD is electronic libraries of building components, including geometric, functional, and cost and availability data. However, although electronic component libraries are technically feasible today, their widespread use awaits development of and agreement on a detailed standard scheme for describing component characteristics. Although such efforts have been under way for more than 10 years, acceptance and agreement has been disappointingly slow. These examples illustrate that many technically possible and conceptually simple extensions to present-day computer aided design have not yet been put into practice. Architectural CAD in 1998 remains overwhelmingly devoted to the management of building geometry.

The application of artificial intelligence technologies combined with continuing improvements in digital imaging and low-cost storage promises to extend present-day CAD in several interesting directions that may help designers of housing. For example, using a retrieval scheme known as case-based reasoning, a designer can find designs stored in a library that share features with the task at hand. A prototype "case-based design aid" has been developed that retrieves stories about problems and responses in past design cases from a library of postoccupancy evaluations. By searching libraries of annotated previous designs that store plans, photographs, videos, and interviews with users and clients, a housing designer may be able to anticipate problems and take advantage of opportunities. Another artificial intelligence technology, "rule-based systems," has been employed to automate logical reasoning in highly specific domains such as medical diagnosis and legal reasoning. Rule-based systems have been used to build programs for critiquing designs, for example, to automate checking of fire and building construction codes. Although these programs are still quite limited, this approach can be extended to check designs against databases of housing standards and norms.

—*Mark D. Gross*

Further Reading

Dinjens, P. J. M. and W. Hermens. 1978. "The Use of SMOOC: A Design Aid." *Open House International* 3(4):71-79.

Domeshek, Eric and Janet Kolodner. 1992. "A case-based design aid for architecture." In *Artificial Intelligence in Design,* edited by John Gero. Netherlands: Kluwer.

Flemming, Ulrich and Skip van Wyk, eds. 1993. *Computer Aided Architectural Design (CAAD) Futures '93.* Amsterdam: North Holland.

Kalay, Yehuda, ed. 1992. *Evaluating and Predicting Design Performance.* New York: John Wiley.

Mitchell, William J. 1977. *Computer-Aided Architectural Design.* New York: Petrocelli/Charter. ◀

▶ Condominium

Condominium as a Form of Tenure

The condominium is a form of real property in which the owners hold a simple fee title to a unit and share the ownership of the structure and any facilities intended for common use. A condominium association administers the common elements, and the unit owners share the costs. Most condominiums are housing complexes, but other frequently encountered forms include professional offices, manufacturing spaces, and even marinas.

Residential condominiums offer outright ownership in multifamily housing, with its inherent advantages: the buildup of equity, tax concessions, and management control. Condominiums are also set up to provide common facilities in complexes of single-family housing. Unit owners are financially responsible for their own property. They have to pay the taxes and other charges, and they can use the dwelling as collateral for a mortgage loan. The association can put a lien on this property if the owner fails to pay the dues.

The advantage over other forms of common property is that condominium laws have been enacted in many countries, spelling out the rights and obligations of the owners. Uniform deeds facilitate the legal and financial aspects of ownership, and because the title is clear, the value is not lowered by legal uncertainty.

Early History

Condominiums have ancient origins. They became a conventional form of property during the Middle Ages in areas with a scarcity of building sites, such as walled cities and mountainous regions. Later, the Napoleonic *Code Civil* provided an explicit legal basis. Its Article 664 recognizes the right of a person to own a building or part of a building on land that belongs to someone else. Many European countries adopted this civil code as the basis for their own judicial system.

As condominium units became more numerous and the demand for standard deeds became stronger, condominium statutes were widely adopted. Belgium passed its first Condominium Act as early as 1924; Italy adopted a similar law in 1934, France in 1938, Germany and the Netherlands in 1951. These acts clarify the rights and obligations of unit owners and set standards for the bylaws of the condominium associations.

With the civil code, the condominium concept spread from Europe to Latin America. Already in 1928, its use was regulated in Brazil. The Cuban condominium act helped to pioneer condominiums in Puerto Rico, where U.S. developers became acquainted with it. Puerto Rican developers and financial institutions petitioned the Federal Housing Administration (FHA) to make mortgage insurance available to secure loans on condominiums. Typically, such insurance is required to obtain favorable terms on long-term financing. In 1961, President Kennedy signed a Housing Act in which Article 234 authorized the FHA "to insure first mortgages on one-family units in multi-family structures and an undivided interest in the common areas and facilities that serve the structure."

The Introduction and Growth of Condominiums in the United States

It is a general principle of common law that a landowner owns not only the surface but everything above and below the property. It is equally well established that the owner may subdivide the property, also above or below the surface. New York's Park Avenue was built on the air rights over the tracks of the New York Central Railroad; Chicago's Merchandise Mart—built in the 1920s over the tracks of the Chicago and Northwestern Railroad—followed suit, as did buildings in many other cities. In housing, specially drafted deeds made individual ownership of apartments possible. The impulse for this early post-World War II legal innovation was the prohibition in the Veterans Administration Bill from making mortgage loans on anything less than fee-simple ownership.

The condominium created a homeowner market in multifamily structures. Developers saw its potential. But first the concept had to become accepted by builders, insurers, lenders, and realtors. The concerted efforts of some developers led to the adoption of condominium statutes by most states in the 1960s. By acknowledging the condominium concept, state legislatures stimulated serious activity in the field. Such an act made it legal for institutional lenders and government bodies to offer mortgage loans.

At first, the condominium was largely restricted to recreational and retirement complexes in Florida and California. Developer abuses—illegally retaining control over the association and overcharging the unit owners for recreational facilities—tainted the sector. But after additional legislation was adopted to clarify the rights and obligations of the various parties, condominiums became popular. The 1970 census recorded only 60,000 units throughout the United States, but within a decade they numbered more than 2 million units.

Since then, the condominium has taken its place as a standard form of housing tenure in the United States, as it has elsewhere. But to most people, it is still seen as housing for specific population groups (e.g., young singles, empty-nesters) for whom single-family homes are less suitable or too expensive. (SEE ALSO: *Common Interest Development; Condominium; Condominium Conversion; Cooperative Housing; Market Equity Cooperatives; Multifamily Housing*)

—*Jan van Weesep*

Further Reading

Eilbott, Peter. 1985. "Condominium Rentals and the Supply of Rental Housing." *Urban Affairs Quarterly* 20(3):389-99.

Miller, Joel E. 1997. "Condominiums and Cooperatives." *Journal of Real Estate Taxation* 24(Winter):152-55.

U.S. Department of Housing and Urban Development, Office for Policy Development and Research. 1975. *HUD Condominium/Cooperative Study.* Washington, DC: Author.

Van Weesep, J. 1988. "The Creation of a New Housing Sector: Condominiums in the United States." *Housing Studies* 2:122-33. ◄

► Condominium Conversion

The sale of private rental housing for owner occupation is a well-established phenomenon. Until the end of the 1960s, this process of tenure transformation was almost entirely restricted to single-family housing. The 1970s, however, witnessed the rapid growth of a new phenomenon in the housing markets of Western countries. This is the conversion of rental apartment blocks—or single-family housing complexes—to condominiums. The precise arrangements of the conversion vary from one country to the next. But everywhere the conversion results in fragmentation of ownership, as a single title is subdivided into separate units. Usually, the tenure changes as well. Most units convert from rental to owner-occupier tenure as the buyers of the individual units take up residence. But invariably, some units are purchased by speculative investors for short-term rental and eventual resale.

In the United States, the conversion of rental apartments to condominiums increased rapidly in the 1970s; by the end of the decade, some 150,000 units were converted each year. Until that time, most condominiums had been found in the Sunbelt states. Older industrial metropolises also witnessed increasing activity in the condominium sector. In European countries, condominium conversion, by and large, is restricted to the large cities. In smaller settlements, multifamily structures are scarce. Housing complexes with shared (recreational) facilities are also less common in Europe than in the United States.

In many countries, condominium conversion has contributed significantly to the decline of the private rental sector. Problems such as evictions and displacement of the poor, even homelessness, have been linked to the conversions.

Explanations

In its major 1980 study of condominium conversion, the U.S. Department of Housing and Urban Development emphasized the role of consumer demand in overheated owner-occupier markets as its major cause. The changing nature of housing demand has certainly played a part in the conversion process. For many buyers—be they sitting tenants or purchasers in the open market—buying a condominium can offer advantages above renting. But some analysts have argued that the underlying rationale lies on the supply side of the housing market. Prospective residents have virtually no alternative but to buy when many landlords decide to sell rather than to rent their properties.

Savvy owners of residential properties view their housing investments as commodities that can turn a profit. This profit can be realized through renting or selling. When the value of a vacant house that can be sold to a homeowner rises above the current value of future rental income, landlords will be induced to convert. In recent decades, the gap between the two values has increased as an effect of house price inflation and the long-term rise in interest rates.

The introduction of rent control—or the threat thereof—may also stimulate conversion rates. The general view has been that investment in rental property had generally become a poor alternative relative to most other investments. A combination of greatly increased values of owned units and real or perceived constraints on operating margins (in rental housing) provided a strong inducement for property owners to sell and reinvest elsewhere.

Timing

Although the underlying economic incentives for the sale of private rental property are broadly similar in Western countries, they do not explain why the sale of apartments in rental buildings began only around 1970. The timing is related to specific conditions in each country. Different causes triggering the conversion emerge from a comparison of three countries: the United States, England, and the Netherlands.

In the United States, condominiums were a relatively new phenomenon in the 1970s. Conversion of existing rental complexes could not take place until the newly built condominiums had become accepted by the real estate industry and the public at large. The ensuing wave of conversions is frequently explained by the growth of demand during the rapid house price inflation of that period. Another factor, supposedly, is the lack of inflation correction in the income tax scales, making deductions for homeowner costs more attractive.

Looking at the supply side, a strong inducement for conversion was the change in taxation on investment in rental property. Depreciation rules changed, and some of the residential tax shelters were eliminated in the mid-1970s. The Internal Revenue Service thereby reduced the profitability of rental investment and encouraged investors to turn elsewhere.

In Britain, a number of more or less simultaneous influences changed the way that landlords perceived their investment portfolios. Rapid house price inflation and steep rises in interest rates widened the value gap between rental and homeowner properties. Meanwhile, political uncertainty about rental property coincided with changes in the tax code to erode the will of investors to keep their property. Then, a strong political support for homeownership turned the tide definitively.

In the Netherlands, a number of other factors caused the number of conversions to grow since the late 1960s. In older housing, major investments were needed to rectify code violations. Rent control prevented the recapture of the investment, and it also kept the value of rental properties low. Investors turned increasingly to commercial property. But the main trigger for the conversions was a small but essential change in housing allocation rules. Because of a compelling housing shortage, all vacant rental units had been allocated by the housing authorities to people on waiting lists. In 1969, the law was changed to grant owners of vacant dwellings permission to occupy the property irrespective of their position on the waiting list. Thus, many households with a low priority were induced to buy a unit to jump the queue. Ease of access has continued to be the primary motive for the purchase of a condominium.

Political Commotion over Conversion

The conversion of much rental property caused many problems. This sparked political action. The nature of the problems, however, varied.

In the United States, professional converters frequently used hard-sell tactics. High financing and holding costs were a powerful incentive to sell out as quickly as possible. Some tenants welcomed the opportunity to buy their apartments. But many either did not want to buy or could not afford the cost of homeownership. In the virtual absence of tenure protection laws, many tenants were evicted. In some cities, opposition to conversion led local governments to pass tough anticonversion regulations and tenant protection laws. But many local officials were torn between the need to protect the rental sector and the desire to promote homeownership with its concomitant increase in tax revenues.

By the early 1980s, many states had enacted conversion controls. Tenants had to be informed of a pending conversion. They were granted the right of first refusal or a grace period after the expiration of their lease. In some places, conversion could take place only if a minimum number of the tenants agreed. But most housing officials have been reluctant to pass strict controls, because they remain convinced that these are disruptive in the long run. Controls are believed to make investors shy away from new rental investment.

In England, few tenants have been displaced by conversion. Because of the peculiarities of British law, the buyers of the units ran into problems. Until the mid-1980s, condominium ownership was not legally possible in England. Instead, long-term leases were sold. The owner-converter retained the title to the building even when all the units had been sold on long leases. Condominium associations did not exist, and the owner remained responsible for operation and maintenance. Where the sale of the units was slow or operating costs high, the owner often sought to improve the cash flow by increasing the service charges or decreasing the level of services and maintenance or both. Residents often found it difficult to enforce basic maintenance commitments. Worse yet, they had little control over expenditures. Moreover, the long leases decreased in value as time went by.

Eventually, legislation was passed to strengthen the position of long-lease holders. But British local authorities have remained virtually powerless to enact legislation. The central government has shown little interest in the problems experienced in a relatively small number of local authorities.

In the Netherlands, the existence of relatively strong tenant protection legislation has meant that tenants did not face eviction. Instead, it was the housing authorities that became alarmed over the decrease of the low- and moderately priced rental stock. They found it more and more difficult to guarantee decent, affordable housing to low-income groups. City governments acted to control unit sales and occupancy by the owners and tried to curb conversions.

By the end of the 1980s, conversion had caused fewer problems than anticipated because of the depressed housing market. For various reasons, the strong emphasis on rental housing was also abandoned at the time in favor of the promotion of homeownership. Rather than to try to stop all conversions, the new regulations were geared toward preventing specific conversion problems and limiting the conversion of low-cost housing. (SEE ALSO: *Condominium; Displacement; Gentrification*)

—*J. van Weesep*

Further Reading

Hamnett, C. and W. Randolph. 1986. "Tenurial Transformation and the Flat Break-Up Market in London: The British Condo Experience." Pp. 121-52 in *Gentrification of the City*, edited by N. Smith and P. Williams. London: Allen & Unwin.

Harloe, M. 1984. *Private Rented Housing in the United States and Europe*. London: Croom Helm.

U.S. Department of Housing and Urban Development, Office of Policy Development and Research. 1980. *The Conversion of Rental Housing to Condominiums and Cooperatives. A National Study of the Scope, Causes, and Impact.* Washington, DC: Author.

U.S. Department of Housing and Urban Development, Office of Policy Development and Research. 1981. *The Conversion of Rental Housing to Condominiums and Cooperatives. Impacts on Housing Costs.* Washington, DC: Author.

Van Weesep, J. and M. W. A. Maas. 1986. "Housing Policy and Conversion to Condominiums in the Netherlands." *Environment and Planning A* 16:1149-61. ◀

▶ Congregate Housing

The term *congregate housing* describes a broad variety of community residential settings with communal features that integrate housing and supportive services targeted to frail elders who do not require 24-hour care. In some states, congregate housing initiatives have included group homes or shared housing for two or more unrelated elders. Typically, however, as exemplified by the federal Congregate Housing Services Program initiatives, congregate housing refers to multi-unit, age-segregated buildings in which supportive services are available for elders who, on a rental basis, live either alone or with their spouse in their own apartments (including kitchen, bathroom, and living space).

Apartments generally have special design features (e.g., no thresholds between rooms, emergency pull cords, grab-bars in bathrooms). Included on the premises are areas for social functions and for congregate supportive services, with a congregate meals program available at a minimum. More enriched congregate housing programs offer additional services (e.g., transportation, light housework, help with chores, personal care). There is often office space for social services and case managers and for resident councils and the like, and the place can include a convenience store, a library, space for health promotion activities (e.g., exercise facilities), and office space for other community service providers (e.g., visiting nurses).

It has been estimated that, in 1990, between 400,000 and 500,00 elders in the United States lived in congregate housing. However, in many congregate housing complexes, a sizable number of tenants are not frail and may not be eligible for some services (e.g., personal care, light housework). Generally, a case management system exists to determine individualized needs for specific services available under building auspices.

Although there are unsubsidized for-profit congregate housing complexes, governmentally supported congregate housing predominates. Since the mid-1970s, some states have supported congregate housing by helping to subsidize services in new or existing buildings for elders. In a number of states, the housing finance agencies (HFAs) have provided tax-exempt bond financing for congregate housing in which developers are responsible for providing supportive services; states also use HUD/FHA [221(d)(4)] mortgage insurance. In many respects, federal funding has been pivotal, not only by providing various funding sources but by directly financing or providing loans for constructing congregate housing. These include federally supported public housing authority (PHA) buildings as well as nonprofit and limited-profit housing for lower-income elders, although a number of apartments in the latter two groups may represent market rentals as well.

Housing for elderly and physically impaired adults with integrated services existed prior to 1970. However, prior to that time, federal policy, as administered by the Department of Housing and Urban Development (HUD), did not support the financing of common spaces for supportive services, such as kitchen and dining space for congregate meals. Funding for the construction of housing with space for congregate meals was provided for the first time by the Housing Act of 1970 and was reinforced by Section 7 of the Housing and Community Development Act of 1974, which encouraged housing to meet occupants' special needs. No funding for services was included in either of these acts.

Under Title IV of the Housing and Community Development Amendments Act of 1978, however, the HUD-sponsored Congregate Housing Services Program (CHSP) was authorized and implemented on a demonstration basis. It was intended to complement existing community service programs and to ensure adequate funding for the delivery of meals and other nonmedical supportive services needed to maintain independent living.

Awardees were PHAs and Section 202 housing sponsors successfully demonstrating that community services were insufficient or not adequately available to meet the needs of their frail residents. (Section 202 sponsors are private, nonprofit groups receiving long-term federal loans to finance rental housing and related facilities for the elderly and/or handicapped.) To maintain an atmosphere of independent living, the CHSP was to serve only about 20% of the residents in the awardee apartment buildings (although exceptions were permitted). On a voluntary basis or funded from other than CHSP sources, a professional assessment committee (PAC) in each project was to be appointed to screen tenants for eligibility, admission to the program, and termination of services. Although awardee implementation approaches could vary, program participants were to receive two on-site meals seven days a week along with nonmedical supportive services as needed. Residents were to pay for services on a sliding scale basis.

With some revised guidelines, the CHSP became a permanent HUD-sponsored program in 1987. CHSP eligibility requirements have become increasingly "tighter," the mandatory two-meals a day requirement reduced to seven per week, and the copayment mechanism standardized. CHSP participants were required to pay a minimum fee of 40% of their adjusted monthly income for supportive services *and* housing.

In 1990, Section 802 of the Cranston-Gonzalez National Affordable Housing Act established a new Congregate Housing Services Program. The goals of this new program included the maintenance of independent living and improvement of resident assessment and coordination of services for residents with at least three limitations in activities of daily living whose needs are not being met through their informal support networks. Under this program, HUD and the Farmers Home Administration (FmHA) in 1993 awarded five-year renewable grants to federally supported housing developments, including FmHA 515 projects as well as PHA and federally supported private housing (e.g., Section 202 developments) to provide supportive services to eligible elderly and nonelderly residents. With some exceptions, the awardee projects cover at least 50% of the costs, HUD covers a maximum of 40%, and participants (unless impoverished) cover 10% of the service costs.

A congregate housing program not restricted to frail elders, the Supportive Services Program in Senior Housing (SSPSH), was also initiated in 1990 under a national demonstration supported by the Robert Wood Johnson Foundation (RWJF). Using RWJF funds as a base, 10 state HFAs worked with housing developments and service providers to finance and provide services in more than 400 HFA-administered age-segregated buildings for low- and moderate-income persons. Most of these housing projects used the services of a service coordinator (with lower qualifications than that of a case manager). The services for a specific housing project are dictated by resident preferences. In some instances, services requested are paid for entirely by participants on a fee-for-service basis.

The success of SSPSH in involving participation of developers and its low cost has generated enthusiasm and led to its replication in a program titled No Place Like Home (NPLH). Under NPLH, 11 state and local HFAs brought services to 50 additional housing projects.

Evaluations of the two CHSP initiatives as well as the SSPSH and NPLH consistently revealed the feasibility of congregate housing programs. An evaluation of the impact of the initial CHSP efforts completed in the mid-1980s, led to some changes in the program (e.g., elimination of the two meals a day requirement).

The implementation of CHSP was highly successful. The projects were effective in enlisting volunteer assistance of professionals in assessing a tenant's needs. The CHSP services complemented rather than substituted for other programs, and the cost was comparable to similar community social service programs. CHSP projects were generally successful in reaching the most vulnerable residents. Positive quality-of-life effects were found on tenant and self-satisfaction. Some impact was found on reducing institutional days (although not necessarily permanent placement), but not until the second year after program implementation. However, possible effects on admission practices—most particularly, the practice of accepting increasing numbers of at-risk residents—were not reflected. A separate analysis revealed that the proportion of deinstitutionalized elders among those moving into the CHSP sites after program implementation was more than five times that of the nonawardee control sites.

The potential of congregate housing as an alternative living arrangement for at-risk older adults was demonstrated as well in a study of a federally sponsored enriched public housing project for low-income frail elderly and handicapped adults conducted prior to the 1978 enactment of CHSP. Congregate services were possible through a special arrangement whereby land was given to the local PHA by the adjacent hospital for the chronically ill in return for

renting space for $1 a year for the life of the project. In addition to congregate meals and community supportive services (e.g., housekeeping and personal care), on-site health services were available, including outpatient clinic treatment rooms and physical and occupational therapy.

The study compared outcomes over a five-year period of entrants and comparable applicants waiting for residency. In addition to quality-of-life benefits, institutional days were reduced throughout the five-year study period. A three-year cost-savings analysis revealed a cumulative benefit-to-cost ratio of 2.21—that is, for every dollar of actual costs incurred for a resident, $2.21 was saved.

Also, using a database derived from several diverse longitudinal studies, outcomes of elderly persons with different levels of institutional risk were compared across elderly and conventional housing sites, classified by whether or not these persons received case management services. Although no differences were found for low-risk elderly, significantly fewer institutional days were observed for high-risk persons in elderly housing with case management.

The cumulative findings from these studies strongly indicated that congregate housing, whether it includes case management or only service coordination, can have positive benefits, at least with respect to increasing the quality of life of its tenants. Congregate housing with case management can have a positive effect on reducing institutional stays for high-risk elders who do not need round-the-clock care provided by nursing homes. (SEE ALSO: *Assisted Living; Continuing Care Retirement Communities;* **Elderly**)

—*Sylvia Sherwood and Shirley A. Morris*

Further Reading

Morris, John N., Caire E. Gutkin, Hirsch S. Ruchlin, and Sylvia Sherwood. 1987. "Housing and Case-Managed Home Care Programs and Subsequent Institutional Utilization." *The Gerontologist* 27:788-96.

Newman, Sandra. 1990. "The Frail Elderly in the Community: An Overview of Characteristics." Pp. 3-24 in *Aging in Place: Supporting the Frail Elderly in Residential Environments,* edited by David Tilson. Glenview, IL: Scott, Foresman.

Sherwood, Sylvia, David S. Greer, John N. Morris, Vincent Mor, and Associates. 1981. *An Alternative to Institutionalization: The Highland Heights Experiment.* Cambridge, MA: Ballinger.

Struyk, Raymond J., Douglas B. Page, Sandra Newman, Marcia Carroll, Makiko Ueno, Barbara Cohen, and Paul Wright. 1989. *Providing Supportive Services to the Frail Elderly in Federally Assisted Housing.* Urban Institute Report 89-2. Washington DC: Urban Institute Press. ◄

► Consortium for Housing and Asset Management

The Consortium for Housing and Asset Management (CHAM) works to expand the capacity of community-based organizations and others working in the nonprofit housing industry to responsibly own and professionally manage affordable housing. CHAM was founded in 1993 by the Enterprise Foundation, the Local Initiatives Support Corporation, and the Neighborhood Reinvestment Corporation. In 1995, it began an initiative to develop training and technical assistance programs at the local level. It provides asset and property management training for low-income, nonprofit housing providers. Regional branches across the country offer housing management training programs and provide advice on-site about specific property and asset management issues. In its national office, CHAM has developed a clearinghouse of information on property and asset management available to nonprofit housing providers. Its budget is supported by contributions from several foundations. Managing Coordinator: Roland C. Diggs. Address: American City Building, 10227 Wincopin Circle, Suite 500, Columbia, MD 21044-3400. Phone: (410) 715-3624. Fax: (410) 964-1918.

—*Mara Sidney* ◄

► Construction Industry

The U.S. construction industry is a vast complex of construction firms, product manufacturing companies, wholesale and retail outlets for building materials, transportation organizations, design service providers, codes and standards writing bodies, union and nonunion workers, membership associations advocating various interests, research groups in the federal government and in private companies, federal and local government regulatory agencies, and financing institutions. One of the most complex and dispersed components of the U.S. economy, the industry is difficult to analyze and evaluate. The very large "informal" construction and renovation activity by homeowners further complicates the picture.

The construction industry affects virtually every institution, company, and household by virtue of the need for shelter and infrastructure, both public and private. The construction industry provides houses, offices, shopping centers, factories, museums, airports, schools, public buildings, dams, bridges, roads, sewer systems, and the materials and processes to upgrade, maintain, and rebuild all of these facilities.

A key characteristic of the construction industry is that it is project based. Unlike manufacturing, which is organized to make many of one thing, construction is characterized by one-of-a-kind production of artifacts designed for a specific place and client. Thus, construction begins where manufacturing stops.

Another distinguishing feature of construction is that the object being made is stationary, with crews of workers moving around, through, and past it. This contrasts with assembly line production, in which it is usual for the object being made to move past workers or work groups. A further distinction is that in each construction project, a unique team of contractors, subcontractors, designers, and other specialists is organized and then disbands after its completion.

Construction constitutes a large sector of the U.S. economy. As a whole, the U.S. economy produced a gross domestic product (GDP) of nearly $7.5 billion in 1996. In that year, $570 billion of new construction was put in place in all classes of private and public residential and nonresidential work, including additions and alterations, public utilities, dams, and highways, making the construction industry itself approximately 13% of total GDP. It has consistently represented 5% to 12% of the GDP. The industry employs 5.3 million employees and 1.5 million self-employed proprietors and working partners.

Despite the magnitude of the construction sector, it represents a declining percentage of total economic activity, representing 8.5% in 1955 and 5.7% in 1982. Of the total investment in construction, the value of residential construction in 1996 was $247 billion, 43% of total building construction, and approximately 3.3% of GDP. From another point of view, roughly one-fifth of the U.S. economy is related to housing. Furthermore, for every dollar of new residential construction, $6 of other business is generated. Existing home sales in 1995 created $3.50 in other business for every dollar invested.

Housing starts in 1996 were 1.4 million, of which 1.16 million were single-family detached units (compared with apartments and condominium units). Residential renovation and maintenance accounted for additional investments of more than $110 billion in 1995 alone.

Construction activity is subject to cycles coinciding with national and international economic swings. This has always affected the labor and materials-producing components of the industry. Economic cycles also affect design service and construction organizations, which are the first to show the effects of an economic decline and one of the first to indicate an upturn, because owners must begin planning in advance of actual investment in construction. Construction activity is also sensitive to local weather conditions, because on-site construction is difficult to shield from inclement weather.

The Chain of Production
There are many closely linked parties both upstream and downstream of the actual construction site activity. Upstream are the many agents who acquire sites, prepare design documents and financing, and organize all preconstruction activities. This also includes the production for the market of a huge array of products.

The basic materials and commodity industries exist to provide products for construction. These include the manufacturers of commodity products such as plywood and dimension lumber and new composite wood-based products, steel and other metals, gypsum products, glass, stone, clay products, chemicals, plastics, adhesives, and other relatively low-value-added parts. Also included on the upstream side are the producers of more complex value-added elements, such as packaged heating and ventilation equipment, electrical subsystems, light fixtures, compressors, fixtures, and the like. This class of upstream parts is often known as industrially produced parts, their characteristic being that they are made in whole or in most respects on the initiative of the producer, for stock and sale.

Also upstream of actual on-site activities are the producers of elements such as windows, trusses, kitchen cabinets and counters, and other items that are made off-site for a specific project but that are not fully specified and produced until orders from the party controlling the site are placed.

These are often called prefabricated parts, produced in large or small quantities in factories or assembly centers. Most of these producers use industrywide or corporate standard parts but add to or modify them to suit a customer's order.

Supporting this on-site activity are the construction equipment manufacturers, who contribute cranes, earthmoving equipment, scaffolding, a vast array of power tools, and other equipment used in handling, shaping, and assembling parts.

Downstream of the on-site activities are other parties that acquire, use, sell, and manage constructed projects as real estate investments. These include leasing agents, brokers, marketing consultants, financial institutions, and users of all kinds. We also count as downstream agents those involved in rehabilitating and maintaining existing buildings and facilities and adapting buildings for new uses, or in demolition, disposal, and recycling of obsolete facilities.

Regulation

The construction industry has been socially mediated since the earliest times. Public health, safety, and welfare issues are intricately bound up in each stage of product manufacturing, on-site construction, building use, and adaptation.

Two forms of social mediation are prominent. On the one hand, building standards are agreements reached by volunteer bodies of experts such as the American Society of Testing and Materials (ASTM). Based on technical data, their purpose is to define and coordinate the work of both the design and construction professions and manufacturing industries. Examples include standards of terminology, dimensions, testing methods, and performance measures.

Building regulations, on the other hand, are rules based on these standards. These rules are adopted and enforced by state and municipal government officials. All professionals, industries, and the public must conform to them. Model building codes have been developed to harmonize regulations across wide areas of the United States. They are modified and adopted into law by local jurisdictions.

The earliest regulations in the United States concerned water supply and sanitary conditions, leading to the requirement in 1890 that houses have indoor plumbing. The first fire protection regulations came shortly after that, following disastrous fires in large cities. Energy conservation requirements were adopted in the 1970s following the oil shortages, and indoor air quality regulations and accessibility requirements came into force in the 1990s. Zoning, land use, and environmental protection regulations have been in use in most jurisdictions since the 1920s. Although building codes are locally enforced, some environmental and accessibility rules are federally mandated.

Advancements

Construction technology is usually developed outside the industry's core activities of design and construction, in companies that manufacture equipment and products, as well as in university and government laboratories. Construction companies in the United States for example, spend only about 0.02% of sales on research and development, and when materials and equipment suppliers are included, the industry expends about 0.4% of its sales on research and development investments.

Federal expenditures on building research occur through federal labs, such as the National Institutes of Standards and Technology in the U.S. Department of Commerce (with an annual budget in the 1990s of approximately $10 million), and also in various government agencies and departments, such as the U.S. Army Corps of Engineers Construction Engineering Research Laboratory, the Departments of Energy and Housing and Urban Development, and the Environmental Protection Agency. The Building Research Board of the National Academy of Sciences advises government agencies on building research, and the National Science Foundation supports academic research leading to new knowledge in engineering and materials sciences.

The federal role in housing research has never been substantial or sustained and is usually only indirect, following the view that the private sector is best suited to pay for and conduct research (notable exceptions have been the Forest Products Research Laboratory and the Department of Energy Laboratories). This pattern of reliance on the private sector for developing new processes and products changed briefly in the late 1960s and early 1970s, when Operation Breakthrough sought new technology and streamlined regulations, in large part owing to the urban crises that many major U.S. cities experienced in the late 1960s. After its termination in 1974, the federal government has mounted few similar comprehensive efforts to stimulate or direct housing research. Most of the federal efforts in housing research have focused on reducing energy consumption in housing production and use.

Many leading product manufacturing companies formed major research components in the 1970s, but during the 1980s, most were drastically cut back or eliminated. Very little remains of a robust private sector research capability in the construction and housing industries.

Internationalization

Increasingly, the domestic construction industry is affected by the global economy. Trade in materials and services between the United States and foreign concerns has grown dramatically in the past decade and, according to most experts, will continue to grow. The demand for an increased variety of products and services in all countries is stimulating the harmonization of international standards and other agreements affecting trade, including the introduction of metric measurements in products.

The U.S. construction industry is affected by these developments in several ways. First, foreign demand for U.S. products and services has increased. The larger product manufacturers, contractors, builders, and architectural/engineering firms have increased the percentage of their gross revenues gained from foreign contracts. Second, foreign products and services are increasingly marketed in the United States, particularly those related to interior systems, such as kitchens and kitchen appliances, heating and ventilating equipment, and specialty floor and wall finishes. Third, foreign companies have acquired significant shares in U.S. product manufacturing companies and architectural/engineering firms. Finally, with the increasing ease of travel, communication, data transmission by computers, and electronic mail, the importance of the international exchange of knowledge and information is increasing.

Housing

Up until the 1930s, a "housing industry" per se did not exist. There was little mass building or marketing of for-sale houses, although in the early part of the 20th century many large rental housing projects were built. There was no financing geared specifically to housing and little effort to spur technical advancements. Most single-family houses were framed with light wood frame technology, by small homebuilding companies operating locally but depending on a widening array of standard products. Most multifamily apartment buildings and urban row-type housing were built with masonry walls and wooden floor construction much like modest, low-rise office and other buildings of the time.

With the introduction of the 30-year fixed-rate mortgage and the Federal Housing Administration in 1933, a financing industry emerged that helped launch "the housing industry" as we know it. Before the mortgage reforms of the New Deal in the 1930s, most families lived in rental dwellings, and it was only in the late 1940s that a majority of U.S. households were homeowners. By 1950, 55% were homeowners, and by the 1980s, more than two-thirds owned their homes. Although no real construction of housing occurred at any scale immediately before and during WWII, pent-up demand produced a huge increase in housing production following the war, both public and private, and a parallel boom in the introduction of new products, building methods, and tools of construction, supported by large expenditures in research.

In 1945, total public and private housing starts numbered 326,000. In 1946, the number jumped to 1,023,000 and to 1,952,000 units in 1950. This activity was fueled by increased suburban development, spurred in kind by the extension of both public transportation and highways and the federal mortgage and tax policies. Advances in heating and air-conditioning, sanitary, and kitchen equipment were rapidly accepted as standards in housing construction, adding to a rapid increase in the quality of housing for many U.S. families.

Whereas mortgage financing set the stage, new organizations of builders emerged to deliver houses. Homebuilders, most of them in small firms, produced more than 45 million new houses between 1946 and the early 1990s. During the 1960s, many building and land development companies expanded and merged, but during the 1973-75 recession, many went bankrupt. In the inflation-fed late 1970s and early 1980s, large firms reemerged. But by the 1990s, the housing industry was characterized by a preponderance of small- to medium-size "merchant builder" companies, with only a few really large companies operating nationally.

All of these "merchant builders" have the same functions: securing land, arranging financing, putting in streets and services, managing the construction of the houses, and marketing them. During boom years, developers can build in anticipation of demand on a "speculative basis," but during lean times, builders will commence building only after units are sold, while still having to carry the land development costs. Unlike most producers, many if not most builders take their "product," a house on a lot, directly to the consumer with no middleman or dealer as in the automobile or appliance industries.

The housing industry is characterized by gradual change. It moves in very small and incremental steps. In part, this is because of its dispersion, its rootedness in local traditions and regulation and, in part, because it is so closely entwined with slowly changing cultural and social institutions and traditions. Despite the slow overall changes in construction processes, small-scale innovation in the construction industry is pervasive, particularly focused on the introduction of substitutes for existing products and methods. These gradual improvements are difficult to notice because of the magnitude and complexity of the industry and because it takes so long to move from concept into general practice—often, 25 years or more. (SEE ALSO: *Construction Technology; Residential Development*)

—*Stephen Kendall*

Further Reading

Building for Tomorrow: Global Enterprise and the U.S. Construction Industry. 1988. Washington, DC: National Academy Press.

Eichler, Edward P. and Marshall Kaplan. 1966. *The Community Builders.* Berkeley: University of California Press.

———. 1982. *The Merchant Builders.* Cambridge: MIT Press.

National Association of Homebuilders. 1993. *Housing Market Statistics.* Washington, DC: Author.

National Institute of Building Sciences. 1984. *Proceedings of the Next Generation of Housing Technology.* Washington, DC: U.S. Department of Housing and Urban Development, Office of Policy Development and Research.

U.S. Congress, Office of Technology Assessment. 1986. *Technology, Trade and the U.S. Residential Construction Industry—Special Report.* OTA-TET-315. Washington, DC: Government Printing Office.

U.S. Department of Commerce, Bureau of the Census. 1997. *Statistical Abstracts of the U.S.* Washington, DC: Author. ◄

► Construction Technology

The U.S. construction industry and the housing industry that is part of it depend on a stable but evolving base of methods, materials, and products, broadly known as construction technology. Because construction technology concerns the making of artifacts with long lives, representing deeply held cultural values, institutions, and social relations, the relatively slow rate of change of construction technology has no parallel in the production of other artifacts.

Construction technology, methods, and products are interdependent, having reached the present state of evolution over centuries, for the most part from trial and error and only a modest amount of organized research. In this evolutionary process, changes in products or introduction of new ones have often brought about developments in methods and practices, which are then put to use, leading to further evolution to correct problems. For instance, improvements in manufacturing of metals of increased strength and ductility, coupled with improvements in mass production methods, led to the invention of mass-produced nails in the early 19th century. This, in conjunction with the development of efficient and even portable saw mills, helped bring to use the wooden technology known originally as the "platform frame," first adopted in practice in Chicago in the 1830s, and now known as Western framing or light wood frame technology. This method is, in modified form, the most prevalent way of building house frames in the United States today.

Two basic ways of building predominate: (a) monolithic construction of bricks, stone, and concrete and (b) the method of erecting light frames of wood or steel posts or light studs and beams covered by various kinds of cladding.

The United States has been a pioneer in frame construction in both wood and steel. The availability of inexpensive labor, the large forests, and the resulting wood culture, historically, have been important factors in housing construction technology in the United States and Canada. Domestic construction in the United States has been and continues to be predominantly wooden, in structure and enclosure materials.

Another result of the convergence of mass production and the development of improved steel was the introduction of standard steel shapes, which in the last half of the 19th century, enabled the evolution of high-rise, steel-framed buildings, supported in kind by the development of the elevator and indoor plumbing.

The development of construction technology shows certain thresholds. Since the period of intensive innovation between 1830 and 1880 in the framing of building structures, little development of equal significance has occurred, although advances have been made in individual products and methods of installation and in specialized ways of building, such as mast and fabric structures, space frames, and subgrade construction, as well as in contracting practices.

Beginning in the 1950s, significant advances occurred in resource distribution equipment, particularly in heating, ventilating and air-conditioning, electrical power, and communications systems. In the same period, advances also were taking place in the materials used for the building envelope, with emphasis on decreasing weight, resistance to decay and deterioration, and improved weather resistance and energy conservation performance.

Kinds of Products

The magnitude, geographic distribution, and disaggregation of the market for products and services strongly influences the development of construction technology. To satisfy the wide variety of user needs, products must be low cost and of widespread utility. These "simple" products are then available for assembly into many specialized products aimed at more limited markets or specific projects.

A basic classification of products to make buildings and houses uses three groupings:

► *Commodity products* are basic materials relatively low in value added (low labor content and few technical operations needed in production) and produced in mass quantities. This category includes *raw materials,*

such as gravel, lime, cement, and *formed or processed products,* such as dimension lumber; exterior siding; roofing shingles; tubes and sections of plastic and steel; fasteners; most masonry units of fired clay and concrete; floor and wall coverings, such as tile and vinyl sheets; sealants and adhesives; and sheet material, such as glass, gypsum panels, and plywood.

► *Ready-made products* include higher value-added products, such as plumbing fixtures; doors and prehung doors with frames; stock windows; many electrical items, such as light fixtures; and heating, ventilating, and air-conditioning products. This category is more susceptible to consumer-style trends than most commodity products.

► *Made-to-order products* include elements such as most windows, skylights, sunrooms, roof trusses, facade panels for large buildings, staircases, precast concrete elements, kitchen and bath cabinets and countertops, and other objects following established designs and production operations but made only when an order is received.

Construction products can also be classified according to which party takes the initiative to produce them. On the one hand, both commodity products and ready-made products are characterized by the fact that they are produced for sale. The producer takes the initiative and takes the risk that the product will be used. Because of the high risk and large investments in research, product development, and production facilities, such "hardware" tends to be "general" to meet the largest market of users.

On the other hand, made-to-order products are produced by one party on the initiative of the products user. The risk usually falls on the party ordering the product, with the producer providing a service on a contractual arrangement. Producers of this kind are organized to respond to orders in the most efficient way, with standard production methods and tools and material supplies on hand.

An increasing number of products now fall into the made-to-order category, as manufacturing organizations learn how to lower production costs through flexible manufacturing, small-batch production, and mass customization methods to meet increasing demands for products matching user preferences. Even these, however, depend on a vast array of standard products in the market to reduce their costs. Because of these new production efficiencies, an increasing proportion of the total value of residential and other construction is in off-site facilities where parts production can take place in safe, environmentally controlled facilities. From off-site manufacturing, on-site construction can commence.

Methods

Methods include the tools used to move, cut, bend, assemble, and install products. The array of tools used in construction, both on- and off-site, is large and growing and includes familiar items such as hammers, saws, chisels, wrenches, drills, staplers, and tools for aligning and squaring, clamping, and the like, each of which has dozens of special types. In addition, an array of portable and fixed-in-place power tools have been developed since the 1940s, including tools for cutting and sawing, boring, grinding and shaping, fastening (including glue, nail, and staple guns), transporting, lifting and hoisting, aligning (laser technology for leveling suspended ceilings), and many more.

In addition, there is a vast array of "software" associated with construction technology, concerned with organizing, delivering, shaping, and installing parts. These now include computer software for construction phase management, estimating, and manufacturing. Training and apprenticeship activities are needed to keep those doing the work abreast of the most current methods and hardware.

Efficiency

Construction technology has long been associated with a kind of work that, although supported by tools of increasing sophistication, is basically handwork. Manufacturing and other off-site organizations turned to mechanical means to make products, employing robotic assembly techniques in advanced manufacturing and prefabrication facilities that reduced the labor content of production. However, construction has not similarly developed. In part, this is explained by the fact that each construction project is a one-of-a-kind product, a reality that has made factory-like manufacturing approaches, with few exceptions, largely ineffective, even on the most repetitive construction work.

The efficiency of on-site construction, organized by general contractors using current construction technology, has been among the most difficult challenges, which may be related to a decline in the value added by general contractors as a percentage of total construction project value. In contrast, subcontractors using specialized, high-value-added products and systems produced by manufacturers using advanced production techniques and incorporating many design and assembly decisions are gaining a larger share of value-added on-site construction activity. U.S. general contractors are known to be among the most efficient construction organizations in the world.

For its part, efficiency in the product manufacturing industries in general had, until the 1980s, been associated with variety reduction, top-down decision making, mass production, and both technical and organizational "integration." These efforts corresponded with a view that "market aggregation" was the key to higher quality and cost reduction. It was thought that the costs of construction could also be reduced by similar standardization and uniformity. However, this concept conflicted with the inherent variety of the market and the uniqueness of each project. As a result, in many instances, rework and handcrafted changes were made on-site to meet the building specifications.

Two things came to change this view of standardization. One was the availability, at low cost and in user-friendly ways, of computational support for information handling. Variety was no longer impossible or too expensive to organize. Second, there was a new recognition that the U.S. system was highly disaggregated and characterized by demands for variety along with efficiency. These two forces

led to new approaches to parts production in the manufacturing sector. Flexible design and production became a goal.

In contrast to the manufacturing industries for products such as appliances, automobiles, tools, and a broad array of consumer items, the construction industry has had a more difficult time making this transition. Construction technology has been slower to adjust, partly because it operates in another, more complex set of circumstances than manufacturing. Construction technology has a unique relation to local as well as regional, national, and international economic and organizational forces.

Evolution of the Technical Repertoire

Most of the evolution in construction technology occurs on a continuing basis by *substitution* of one material or element for another. For instance, nonorganic vinyl and aluminum exterior siding is now available as a substitute for wood siding. Light steel framing is being explored as a substitute for wood framing in residential construction as the increasing cost of wood makes substitutes attractive. Vinyl window frames are now available as substitutes for wood; plastic pipes are a substitute for copper or cast iron. New synthetics and recycled materials are now used to make certain interior finishes as a substitute for unprocessed natural materials.

At a certain point, a long process of substitutions may give way to a basic paradigm shift when a new material or method becomes available and coincides with new demands. The introduction of concrete in the Roman times was such a shift, making large nonrectangular forms possible for the first time, a development that corresponded with the demand for large gathering spaces. Another was the introduction of aluminum, which made light space frames possible, spanning vast areas unheard of with heavier materials.

The value of substitution as a form of technical evolution is that it does not disturb the culturally embedded image of the object or the social structure of producers and users controlling it. If a new product disturbs the existing socioeconomic fabric or the image too much, its acceptance will be more difficult.

Only rarely is a truly new system added to the technical repertoire. One such addition was environmental control systems—heating, ventilation, and air-conditioning. The principles of these systems have been known for centuries and were in fact in use as early as the Roman times. However, their reemergence in the 1950s, supported by inexpensive fossil fuels and advances in electric motors, compressors, and fans, has had a vast impact on buildings, cities, and human comfort. Other thresholds include the introduction of electricity and communication systems, plumbing, pumps and motors of all kinds, the invention of the silicone wafer for photovoltaic cells, and the availability of computers.

One final development of significance in construction technology concerns the depletion of relatively inexpensive fossil fuels. From 1860 to the present, the supply of fossil fuels has allowed building technology to move along a path of excess in energy embodied in building materials (the total energy required to make and put into place a material or component) and in low thermal performance. The assumption of an abundance of low-cost energy supplies is now in question. Energy efficiency and conservation through improved building technology is slowly becoming recognized as inevitable. This will stimulate a further evolution in construction technology and buildings. (SEE ALSO: *Construction Industry;* **Industrialization in Housing Construction;** *Manufactured Housing*)

—*Stephen Kendall*

Further Reading

Dertzouzos, M., R. Lester, and R. Solow. 1989. *Made in America: Regaining the Competitive Edge.* Cambridge: MIT Press.

Dibner, David and Andrew Lemer, eds. 1992. *The Role of Public Agencies in Fostering New Technology and Innovation in Building.* Washington, DC: National Academy Press.

Elliott, Cecil D. 1992. *Techniques and Architecture.* Cambridge: MIT Press.

Fitch, James Marsten. 1947. *American Building: The Environmental Forces That Shape It.* New York: Shocken.

Fitchen, John. 1986. *Building Construction Before Mechanization.* Cambridge: MIT Press.

Hounshell, David A. 1984. *From the American System to Mass Production, 1800-1932.* Baltimore: Johns Hopkins Press.

Iselin, Donald and Andrew Lemer, eds. 1993. *The Fourth Dimension in Building: Strategies for Minimizing Obsolescence.* Washington, DC: National Academy Press.

McKellar, J. 1993. "Building Technology and the Production Process." Pp. 136-54 in *House, Home, and Community: Progress in Housing Canadians, 1945-1986,* edited by J. R. Miron. Montreal: McGill-Queen's University Press.

Russell, Barry. 1981. *Building Systems, Industrialization, and Architecture.* New York: John Wiley. ◀

▶ Continuing Care Retirement Communities

Continuing care retirement communities (CCRCs), sometimes called life care communities, offer planned shelter, residential services, personal assistance, and nursing care to persons who are generally in their late 70s or older. Among the alternative housing options specifically marketed to elderly persons, the CCRC is the most likely to offer a full continuum of shelter and care facilities that can accommodate elderly persons who are both relatively active and those who suffer from serious age-related physical and mental disabilities. Most CCRCs require their residents to sign a long-term contract specifying the rules under which occupancy is allowed, the shelter and care obligations of the CCRC, and the costs of these benefits. In return for some combination of a lump sum entry fee and adjustable monthly fees, many CCRCs guarantee their elderly residents lifetime shelter and care. Consequently, some observers consider CCRCs not only a housing alternative but also a strategy for financing the elderly population's long-term care.

CCRCs appeared in the United States as early as the 1890s, but the modern continuing care industry had its roots in the post-World War II era of the late 1940s. These facilities were traditionally owned or sponsored by not-for-profit, charitably oriented groups with religious, fraternal, or sororal affiliations. In the earliest facilities, participants

would bequeath all their personal assets in exchange for shelter and nursing care to be provided for the balance of their lives (hence, the terminology, "life care"). Although the lifetime commitment still usually exists, this asset transfer arrangement is now uncommon. Most CCRCs now charge a one-time, up-front entrance fee and regular monthly maintenance fees. These monthly fees may increase over time, reflecting the usual rise in living costs. Alternatively, they may increase when the physical or mental disabilities of older residents demand that they occupy more supportive accommodations for which the CCRC has a higher rate structure. A smaller number of CCRCs offer a full continuum of care but charge only a monthly fee and not an entry fee. A very few CCRCs allow their residents to purchase their independent dwelling units under a condominium or cooperative (equity payment) arrangement.

The greatest growth of this retirement housing option started in the 1960s. By 1995, there were more than 1,000 CCRCs (about 800 of them having entrance fees) in the United States, housing approximately 350,000 elderly persons. The number of facilities is expected to double in the next 10 years. Although several large for-profit corporations entered the CCRC market in the 1980s, about 98% of CCRCs are still sponsored and managed by not-for-profit organizations. They are found throughout the United States, but especially in the states of California, Florida, Virginia, and Pennsylvania and to a lesser extent in North Carolina, Illinois, Indiana, Ohio, Kansas, and Texas. The principal source of data about CCRCs is the American Association of Homes and Services for the Aging (AAHSA).

Cost of Living in CCRCs

In 1990, entry fees ranged from an average of just more than $22,000 to almost $158,000. The lowest entry fee reported was just more than $1,000, the highest $625,000. Monthly fees ranged from an average low of $614 to an average high of $1,692. These fees reflect common factors such as market location and the size and luxuriousness of facilities. They also are higher when the CCRC guarantees the future availability of personal care, health services, and nursing home care for fees that will be below market rates. It is estimated that if elderly incomes keep up with inflation, 15% of those over the age of 75 could afford CCRCs in the year 2000 and 25% in 2020.

Types of Shelter and Service Arrangements

The *independent living units* (ILUs) or congregate care facilities of the CCRC are occupied by healthy older persons who can perform their usual activities of living without supervision or assistance. Their accommodations may consist of efficiency, one-bedroom, and two-or-more bedroom unit apartments, cottages, cluster homes, or single-family dwellings. These dwelling units are conventional in every way except that they may contain special design features such as grab-bars in the bathrooms and a continually monitored emergency call alarm system. Female residents in ILUs have an average age of 81.2, and males, 80.8.

Residents are usually offered housekeeping service, scheduled transportation, heavy housecleaning, prescribed diets, flat-linen laundering, personal laundry facilities, and special care during illness (e.g., tray service and health supervision). Most CCRCs have at least one activity director who organizes an extensive array of leisure and recreational activities. Residents usually eat in a central dining room and have access to common areas that accommodate various social and recreational activities (game room, beauty or barber shop, craft room, place for religious services, exercise room). Most CCRCs give their ILU residents on-site access to a variety of outpatient organized health-related services, such as annual physical examinations, podiatry care, physical and speech therapy, and occupational care. Medical and nursing staff will often hold regular office hours at certain times of the week.

The *assisted-living units* (ALUs, also called personal care or residential care units) of the CCRC are designed for less independent older persons with physical or mental impairments who require some supervision and assistance with bathing, eating, toileting, grooming, walking, and taking medications but not to the extent of needing the more intensive and round-the-clock nursing or rehabilitative care of the nursing home. Female ALU residents have an average age of 85.6, and males, 83.2. ALUs usually consist of single rooms, often furnished like conventional bedrooms and usually containing toilet and bathing facilities. They consist less commonly of small studio or one-bedroom units that contain a scaled-down kitchen. Residents are offered a similar array of residential and health services as those in ILUs. ALU facilities may contain their own common areas for dining, social, and recreational activities, but these are usually less extensive than found in the ILU accommodations. Although most CCRCs provide their residents with assistance with their activities of daily living, not all provide this care in designated ALUs. About 13% of CCRCs offer this assisted care only to residents in their ILU apartments, and 25% offer personal assistance only in their nursing home facilities.

Nursing home accommodations consist of one-room, furnished units with a bathroom. These units may be occupied by two or more persons. Skilled nursing home care is provided to residents whose chronic health conditions, personal disabilities, or mental impairments (such as Alzheimer's disease) are of sufficient severity that they require constant supervision, assistance, and care or those who benefit from rehabilitative therapy to improve or maintain their abilities. Female nursing home residents have an average age of 87.1, and males, 84.9. With their nursing stations and rehabilitative equipment, these facilities often have the look and feel of a wing in a hospital.

The number of ILUs, ALUs, and nursing home beds found in a CCRC can vary greatly. On average, they contain about 200 ILUs, 44 ALUs, and 91 nursing home beds. The total number of CCRC units ranges from under 70 to more than 2,000, with the average size being more than 300. These various shelter types will be housed on different floors or wings of a single high-rise building or in physically adjacent buildings (garden apartments, cottages, duplexes, and midrise and low-rise buildings) in a campus setting.

Various combinations exist: ALUs may be located in the same building as the congregate care or independent units but on a different floor or wing; they may be located in the same building as the nursing home but again on a different floor or wing; alternatively, they may occupy a freestanding building. The nursing home may be a freestanding building on the campus grounds but at some distance from the congregate housing.

Contracts and Guarantees

The contracts that CCRCs offer their residents specify the shelter arrangements, residential services, personal and health care, and nursing care that residents are guaranteed over their lifetime in the facility and what all this will cost. They also address the conditions under which the CCRC may cancel its agreement, under what circumstances the up-front or entrance fee will be refunded, and the conditions under which residents must transfer from one type of accommodations to another. Most CCRC contracts provide entitlement to occupancy in return for an entry fee but do not give ownership status of the living unit or undivided interest in the real property. Entry fees may be nonrefundable or partially refundable, whereupon some part of the fee may be returned, depending on the length of the resident's period of stay.

The AAHSA distinguishes three principal contract types offered by CCRCS: extensive, modified, and fee-for-service. These reflect the differences in the way CCRCs charge for their personal assistance and nursing care and the extent to which they guarantee this care without additional costs to residents who become physically or mentally frail. Both extensive (earlier called all-inclusive) and modified contracts guarantee residents shelter, residential services, and amenities along with personal assistance and nursing care in return for an initial entrance fee and a monthly payment schedule. Extensive contracts offer unlimited personal assistance and nursing care services with the promise that monthly fees will not be increased over the resident's occupancy duration except to compensate the CCRC for normal operating cost increases and cost-of-living adjustments. The ability of the CCRC to guarantee the availability of care to address future mental or physical disabilities is considered a form of long-term care insurance, by which the resident is prepaying for rights to health and nursing care. From the CCRC's perspective, a portion of the charged fees represents an insurance premium paid by all the residents for the health and long-term care that will be used (at any time) by a relatively small group of residents. To guarantee that the CCRC ownership can meet its long-term care commitments, an increasing percentage (over 15%) of CCRC operators are themselves purchasing commercial insurance to underwrite their future costs.

Modified contracts, on the other hand, offer only a specified amount (days per year) of assisted and nursing home care at a reduced per diem rate, after which residents must pay either a full or a discounted per diem rate for these health-related services. Fee-for-service contracts guarantee only specified residential services to its occupants, and others can be purchased on an à la carte basis. Although residents are offered priority admission into a CCRC's assisted care or nursing care facilities, they must pay full per diem rates for this care (that is, assisted care or nursing home daily rates at the time they are admitted). Irrespective of contract type, CCRCs differ as to what residential, assisted, and nursing care services they offer in return for residents' regular fees and for which ones, although available in the facility, they charge extra. In practice, residents are often offered contract arrangements combining one or more elements of these three prototypic types.

Resident Motives and Characteristics

The average CCRC houses 340 older residents. Persons who move into the ILUs of a CCRC have an average age of 79; are disproportionately female, white, and single (i.e., unmarried, divorced, or widowed); and have middle to upper-middle incomes. When they enter the ALUs, their average age is 83.7 and 84.2 when they occupy the nursing home facility. Older persons attracted to the CCRC alternative are characterized as need driven. That is, their moves are often precipitated by their insecurity about living alone and by disabilities that jeopardize their ability to live on their own. The CCRC holds in particular four key attractions to the older person: a sense of security in knowing that help is available on an emergency basis; the prospects of living independently in a setting requiring minimum upkeep; the contractual guarantee that health services, personal care, and nursing home care will be available when needed; and a desire not to be a burden on one's children or friends. Most CCRC residents express considerable satisfaction with their accommodations.

Entry Requirements

Older residents who seek admission to a CCRC, usually to its ILUs, are required to meet various requirements. They must have sufficient assets and income to cover the entry and monthly fees. Depending on the orientation of the CCRC, ethnic, religious, or fraternal order affiliations may also be important. The vast majority of CCRCs (96%) require older people to undergo some form of medical examination to assess their physical and mental status. Some CCRCs require prospective residents to have both Medicare Part A and Part B coverage. Selected preexisting diagnostic conditions will cause a CCRC to reject older persons as occupants of their ILUs. These include Alzheimer's disease, chronic obstructive pulmonary disease (e.g., emphysema), stroke, general organ failure (e.g., kidney), and loss of sight or hearing. Functionally, older people are generally rejected if they are incontinent, have severe cognitive impairment, require assistance to perform activities of daily living, and are unable to self-medicate without supervision. About one-third of the CCRCs admit less able older people directly into their ALUs and more than 60% admit them directly into their nursing home facilities. In both instances, residents then pay on a monthly fee-for-service-basis.

Problems and Regulatory Responses

The most serious problems confronted by the CCRC industry result from projects that have become financially insolvent and cannot fulfill their contractual obligations to provide health and nursing home care. CCRC financial failures have primarily occurred for the following reasons: occupancy rate expectations were unrealistic and not reached; the resident turnover rates in their ILUs (and the resulting generation of new entry fees) were mistakenly assumed to be too large; the monthly fees could not be contractually increased to meet higher operating costs; and the financial reserves were insufficient. The CCRC management should be able to cover five categories of costs: debt service (paying the mortgage on the property), equipment replacement (to replace aging facilities), health care (to fulfill the health care coverage commitments), financial aid (to assist older occupants who are unable to pay some or all of the fees required for their residence or health), and unexpected contingency funds (to cover unplanned operating costs). A very small number of CCRC owners have also deliberately victimized residents, have been involved in unfair business practices (e.g., misleading advertising), or have offered unfair contract terms (e.g., denial of resident rights with respect to transfer policies). Given that residents have invested all or a major part of their assets (often through the sale of their home) as part of their entry fee, the prospect of provider default, fraud, or negligence has potentially disastrous human consequences.

In an attempt at voluntary regulation of the CCRC industry, accreditation status is given by the Continuing Care Accreditation Commission (CCAC). CCAC is an independent commission sponsored by the AASHA. CCRCs are accredited that effectively meet their own mission and meet the commission's standards of governance and administration, resident life, finance, and health care. Only a small percentage of CCRCs have initiated the application process and are currently accredited.

The federal government's role in regulating the CCRC industry has been limited to the imposition of nursing home standards under the Medicare and Medicaid programs. Most oversight is provided by state governments. Regulatory standards are imposed in some 37 states, a number that has increased greatly over recent years. Much variation exists, however, in their certification, reporting, and enforcement practices. One major criticism is that state authority is diffused among too many different agencies: the health department (nursing home licensure and certificate of need), insurance department (financial requirements), the welfare department (personal care and assisted living), and the office on aging (patient rights and ombudsman). (SEE ALSO: *Assisted Living; Congregate Housing; Elderly*)

—*Stephen M. Golant*

Further Reading

American Association of Homes for the Aging and Ernst & Young. 1993, *Continuing Care Retirement Communities: An Industry in Action.* Vol. 1. Washington, DC: American Association of Homes for the Aging.

Cassel, Edythe J. 1993. *The Consumers' Directory of Continuing Care Retirement Communities.* Washington, DC: American Association of Homes for the Aging.

Gordon, Paul A. 1993. *Developing Retirement Communities.* 2d ed. Vol. 1. New York: John Wiley.

Somers, Anne R. and Nancy L. Spears. 1992. *The Continuing Care Retirement Community*. New York: Springer.
Winklevoss, Howard E. and Alwyn V. Powell. 1984. *Continuing Care Retirement Communities: An Empirical, Financial, and Legal Analysis*. Homewood, IL: Irwin. ◄

► Contract Rent

Housing market participants, tenants, landlords, and property managers need to understand the nature of the rent payment that the tenant makes to the landlord for the use of the dwelling unit. Several important aspects of contract rent need to be recognized, and it needs to be distinguished from other forms of rent.

Contract rent is the amount of dollars per time period that the tenant pays the landlord by mutual agreement in the lease for the use and possession of real property. In the residential context, contract rent is the monthly payment that the tenant pays to the landlord for the use and possession of an apartment unit or a single-family dwelling unit. The contract rent is determined by the agreement reached between the tenant and the landlord or the landlord's representative, the property manager, during the negotiation of the residential lease.

In a multifamily dwelling unit, contract rent is a gross payment from the tenant to the landlord. The landlord uses the funds to cover the operating expenses of the property, such as property taxes, casualty and liability insurance, maintenance, and repairs. The gross rent payment is also used to cover the mortgage loan payments the property owner must make. In a single-family dwelling unit, contract rent can be a gross payment as in the case of the apartment, or it can be a form of net payment. In a net payment arrangement, the tenant can agree in the lease to pay a fixed amount to the landlord on a monthly basis and also to pay the property taxes on the dwelling. In a net-net, or double net, agreement the tenant agrees to pay a fixed amount to the landlord and also to pay the property tax bill and the insurance on the property. In a net-net-net (triple net) agreement, the tenant also agrees to maintain and repair the structure.

Conceptually, contract rent is not market rent. Market rent is the rent level that is currently being received by properties that are very similar or comparable to the dwelling units being supplied to the market by the property in question. Contract rent can be equal to market rent when the lease is negotiated and signed. However, as time passes and the demographic and economic circumstances in the market change, contract rent can be less than or greater than market rent. In addition, contract rent may be different from market rent because the tenant and/or the landlord are not aware of the current market rent and they negotiate a rent that does not reflect current market circumstances.

Contract rent is an effective price, not an asking price. Landlords or property managers can ask or quote a rent that they would like to receive. However, the tenant may actually pay less than the asking rate because the tenant negotiates and receives additional benefits beyond the use and possession of the dwelling unit. For example, the tenant

might agree to pay $500 per month, but the lease agreement provides for a free month. The asking rent for the 12-month period is $500 per month or $6,000 for the year, but the tenant only pays $5,500 for the year or $458.33 per month. This same principle applies if the tenant receives a gift for signing the lease or receives any other benefit. (SEE ALSO: *Private Rental Sector*)

—*Joseph S. Rabianski*

Further Reading

Blankenship, Frank J. 1989. *The Prentice Hall Real Estate Investor's Encyclopedia*. Englewood Cliffs, NJ: Prentice Hall.
Jacobus, Charles J. 1996. *Real Estate Principles*. Upper Saddle River, NJ: Prentice Hall.
McKenzie, Dennis J. 1996. *Essentials of Estate Economics*. Upper Saddle River, NJ: Prentice Hall.
Peiser, Richard B. 1992. *Professional Real Estate Development: The ULI Guide to the Business*. Washington, DC: Urban Land Institute.
Shim, Jae K. 1996. *Dictionary of Real Estate*. New York: John Wiley.
Tosh, Dennis S. 1990. *Handbook of Real Estate Terms*. Englewood Cliffs, NJ: Prentice Hall. ◄

► Cooperative Housing

Housing cooperatives are democratically governed nonprofit corporations whose resident shareholders jointly own multiple-unit properties. Cooperatives resemble rental housing in that residents pay a monthly fee in return for the occupancy of a unit that they do not own. These fees buy down the collectively held mortgage, pay for expenses, and capitalize reserve funds. Cooperatives also resemble homeownership in that residents own shares in the total property; enjoy the security, control, and tax advantages of homeownership; and are responsible, through an elected board, for the maintenance and management of the property. Cooperatives are unique, however, in being collectively owned and governed.

Cooperatives differ by the type of shares issued. Market equity cooperatives allow shares to be traded at market value as if units were individually owned. These cooperatives, composed predominantly of middle-class households, sometimes allocate shares on the basis of unit size or desirability. Limited-equity cooperatives restrict the amount of return on the sale of a share to the amount of equity accrued during occupancy, sometimes adjusting for inflation, interest, or improvements. Leased cooperatives do not own their properties but, rather, hold long-term leases from a community land trust, mutual housing association, or limited partnership that grants them many but not all of the rights as other cooperatives. These may include an option to buy.

Cooperatives account for less than 1% of the U.S. housing stock, compared with as much as 25% in some Scandinavian countries. In 1993, the American Housing Survey counted 729,000 cooperative housing units, whereas the National Cooperative Bank estimates the number to be about 1 million. Although disproportionately located in New York, cooperatives exist in other large cities and in more than 30 states, especially California, Florida, Illinois, Michigan, New Jersey, and Pennsylvania. During the 1970s, few conversions of rental housing involved cooperatives, whereas in the 1980s, about half did. Recent changes in the housing market and federal housing legislation suggest that the number of cooperatives has been increasing.

Early History

The first known housing cooperative existed in Rennes, France in 1720. But the first major cooperative movement developed in Britain during the early 19th century. The "pioneers" of Rochdale, England, laid down what has become the manifesto of the international cooperative movement. The Rochdale Principles, formulated during the 1840s, include the following:

- ► Democratic control by residents (one-share, one-vote)
- ► Open, voluntary membership
- ► Limited returns on investment and return of surplus to the members
- ► Open disclosure, active participation, and continued education
- ► Expansion of services to members and to the community
- ► Cooperation between cooperatives

In the late 19th and early 20th centuries, the Scandinavian countries developed the European model of housing cooperatives. "Mother" cooperatives are building societies that collect capital and construct individual "daughter" cooperatives. Although the residents collectively own and govern daughter cooperatives, they receive ongoing assistance from the mother. Almost all European cooperatives are limited-equity in form, but in recent years, some cooperative units have been traded at market rates. The arrangement in which housing associations develop and oversee numerous smaller cooperatives has been institutionalized throughout Europe, particularly in Scandinavia and Germany. The German model, in which residents have control but not full ownership, is the prototype for the U.S. mutual housing associations (MHAs) that have proliferated during the last decade.

The first U.S. housing cooperatives—called "home clubs"—were formed in New York City in the 1870s and 1880s. A minority of the residents, who tended to be affluent, actually owned the property. Shares were distributed by the size and location of the apartment and could be sold at a profit. Market rate cooperatives also flourished in New York during the 1920s where they were built on speculation. Cooperative screening of prospective members appealed to the affluent's desire for exclusivity. In the 1940s, the institution of rent control fueled conversions of rental housing to market rate cooperatives. Another wave took place in the 1980s when inflated property values, tax changes, and relaxed rent control made it attractive to convert luxury rental properties to cooperatives. However, where property values declined and the sponsor or former landlord still held units, cooperative defaults increased.

In contrast, the first U.S. cooperative built in conformity with the nonprofit Rochdale Principles was organized in 1916 by Finnish groups in Sunset Park, Brooklyn. It exists to this day. Growing out of the much larger consumer cooperative movement of the period and encouraged by the New York State Limited Dividend Housing Companies Act of 1926, union, ethnic, and political groups constructed more than a dozen working-class cooperatives, particularly in the Bronx. In 1929, there were 45 U.S. housing cooperatives, accounting for only a few thousand units. Most were in New York, but Chicago, Detroit, San Francisco, and Philadelphia also had them. By 1934, however, more than 75% of Chicago and New York cooperatives collapsed. Few survived the Depression.

Postwar Developments

The fortunes of cooperatives improved with postwar federal and state legislation. Over time, government subsidies shifted from an initial focus on the middle-class toward moderate- and low-income cooperative housing.

In 1942, an amendment to the Internal Revenue Code allowed cooperative shareholders to deduct real estate taxes and mortgage interest from their personal income taxes. Section 216, as this provision is known, serves as a potent incentive for more affluent taxpayers to join cooperatives. In 1952, Section 213 made Federal Housing Administration (FHA) insurance available for cooperatives' blanket mortgages. Most Section 213 cooperatives have been middle income, and few defaulted on their market rate mortgages. The provision produced few low-income or limited-equity cooperatives and did not originally mandate membership participation in development.

By the end of the 1950s, almost a third of the nation's cooperative units were located in New York State. Private developers constructed most of the state's cooperative housing. After the 1980s wave of market rate cooperative conversions, the Federation of New York Housing Cooperatives attained a membership of 100,000 cooperative residents.

Other postwar developers of cooperative housing joined together in 1960 to form the National Association of Housing Cooperatives (NAHC), the primary nationwide technical service provider and lobby for cooperative housing. NAHC publishes the *Cooperative Housing Journal* as an educational and networking tool for its membership of affiliated cooperatives.

Cooperative Services, Inc. (CSI), which originated in a postwar dairy cooperative, concentrated on organizing cooperatives for the elderly in the 1960s and 1970s with the assistance of Section 202 subsidies. Recently linked with Section 8 rent subsidies, Section 202 offers 40-year below-market interest rates, loans for rental housing for the elderly and handicapped. CSI is not a nonprofit but a consumer cooperative in which residents of all its projects are shareholders. Its individual cooperatives are self-managed but not financially independent. Thus, CSI is officially a mutual housing association with rental buildings. Originally, CSI focused on Michigan but has expanded across the nation. By winning a judgment in 1979 after a long dispute with the Department of Housing and Urban Development (HUD), CSI forced the release of 2,000 elderly housing units. Although the number was cut back to 1,300, CSI now has 3,485 units in 21 co-ops for seniors.

The Foundation for Cooperative Housing was established in 1952 as a national research, training, and technical assistance provider. Under the leadership of Roger Willcox, the foundation first converted several large housing projects to cooperatives, including the federally built, Depression Era new town of Greenbelt, Maryland, and several public housing projects. During the 1960s, with the assistance of new federal programs, the foundation expanded into new construction in 30 states. Called the Cooperative Housing Foundation since 1980, the organization has sponsored 60,000 homes for low- and moderate-income people in 400 co-ops. It is probably the largest cooperative sponsor in the nation. Today, the foundation has given priority to developing cooperatives in the South and Southwest.

The foundation, like other developers, took advantage of federal policies of the 1960s targeted at the lower end of the housing market. The Section 221(d)(3) program, introduced in 1961 and responsible for 23,000 units in 313 cooperatives, offered below-market interest rate loans for moderate-income, nonprofit cooperatives. In 1965, a rent supplement was made available to low-income cooperative members. Although successful, the program was expensive and was superseded by Section 236 in 1968. Section 236, which subsidized interest rates and rents on new or rehabilitated multifamily housing, produced 16,000 cooperative units by 1976. Most 236 nonprofit cooperatives were moderate income and well managed, but 57% failed when fixed subsidies could not cover escalating costs. In contrast, only 33% of limited partnership cooperatives went under. Some defaulting cooperatives were sold by HUD, whereas others received Section 8 subsidies as a hedge against inflation. Increasingly, new low-income cooperatives use Section 8 subsidies to finance cooperative shares and rents. As the 20-year use restrictions on the 60,000 Section 221(d)(3) and 236 cooperative units began expiring, the 1990 Housing Act created a mechanism to maintain them as affordable housing.

During the 1970s, attempts to overcome the weaknesses of earlier programs and to reach a lower-income population led to a series of pilot programs incorporating community nonprofits in partnerships with the public and private sector. One example is the Section 510 Multi-Family Homesteading Cooperative Demonstration, which made use of Section 312 rehabilitation loans and other subsidies and private contributions. But only half of the 510 projects survived. New York's Tenant-Interim Lease program improved on the idea of converting abandoned buildings to cooperatives. Tenants have formed more than 366 low-income, limited-equity cooperatives—over 12,000 units—in rehabilitated, landlord-abandoned, tax-delinquent, city-owned housing. Despite some financial problems, resident satisfaction is high. Some of these cooperatives also formed an MHA, the Self Help Works Consumer Cooperative, to share the costs of professional assistance, credit, supplies, and insurance.

In 1980, the National (Consumer) Cooperative Bank was founded to make loans and provide technical and financial services to cooperatives throughout the United States. However, the bank has not helped many low-income cooperatives because it lends, like a private institution, at market interest rates and requires large down payments. In the early 1980s, a secondary market in cooperative obligations developed, encouraging private sector loans to cooperatives. Fannie Mae (formerly known as the Federal National Mortgage Association) trades in cooperative share loans and, like Freddie Mac (Federal Home Loan Mortgage Corporation), in cooperative blanket mortgages, including Section 213-insured ones. Ginnie Mae's tandem programs can also assist some federally subsidized cooperatives.

Cooperatives Today

The Reagan and Bush administrations severely cut the federal housing budget. However, several policies still provide federal assistance to cooperatives. Among the best-known programs is the sale of public housing to tenants. As was done in earlier decades, the Public Housing Home-ownership Demonstration sold several multifamily projects to cooperatives with Section 5(h) authority. Although the cooperatives were more likely to raise private capital than the fee-simple sales programs, they were a mixed success. For example, the Denver "cooperative" did not actually sell shares, relied on Section 8 rent subsidies, and experienced considerable resident turnover. Nevertheless, Section 123 of the 1987 Housing Act extended the right to buy public housing to cooperatives formed by eligible resident management corporations. This policy was further institutionalized in Title IV of the Cranston-Gonzalez National Affordable Housing Act of 1990. The HOPE (Homeownership and Opportunity for People Everywhere) program provides planning and implementation grants for the creation of limited-equity cooperatives in public housing. Nonprofit cooperatives may also receive grants to sponsor such conversions. Finally, a pilot demonstration to develop limited-equity cooperatives on government land in Army housing has been launched under the Soldier Housing and Retirement Equity (SHARE) program.

The 1990 act created the HOME (Homeownership Made Easy) program. With state and local matching grants, funds are provided specifically for nonprofit community development corporations to construct or rehabilitate affordable housing, including cooperatives. HOME reflects rising national interest in the mutual and community-based housing approach. Although most MHAs do not sell shares to residents and treat their properties as rentals, the model, as CSI has shown, can be extended to cooperatives. Other national and regional nonprofits assisting low-income-housing cooperatives include the Consumer-Farmer Foundation, Institute for Community Economics, BRIDGE, LISC, and the Enterprise Foundation. Unions, through pension fund investments or construction, have also contributed to limited-equity cooperatives. These associations have become repositories of technical information accumulated from the experiences of many local cooperatives, community development corporations, and community land trusts, encouraging replication.

Federal cutbacks during the 1980s also increased the role of states and municipalities in low-income housing development. For example, some developed public-private housing partnerships that assisted cooperatives by waiving property taxes, writing down loans, clearing or even donating city-owned land, and redistributing scarce Community Development Block Grant (CDBG) funds toward housing.

Tax changes in the 1980s both hurt and helped cooperatives. On the one hand, the 1986 Tax Reform Act eliminated the tax shelter incentives for syndication of cooperative property. In addition, the act reduced the value of Section 216 deductions. The Internal Revenue Service also sought to tax interest on cooperative reserve funds at a corporate (membership organization) rate. Cooperatives maintain that earnings on reserves represent patronage refunds or price adjustments to the shareholders.

On the other hand, the Tax Reform Act instituted the low-income housing tax credit. For 10 years, private investors in new or rehabilitated cooperatives gain an annual 9% credit (4% in subsidized or acquired projects), provided that fixed percentages of the units are occupied by low-income or very low-income households. The tax credit reinstates a mechanism to raise capital through syndications. Often, this capital must be further supplemented with public subsidies. Moreover, cooperatives lose Section 216 deductions with syndication. But they retain considerable autonomy as specified in a master lease with the limited partnership. The Tax Reform Act also made limited-equity cooperatives eligible for tax-exempt "private activity bond" financing from state housing finance agencies. Accepting such loans also makes the cooperative ineligible for Section 216 deductions.

Organization

Private developers, government, community nonprofits, or the prospective members themselves may initiate the organization of cooperatives. In recent decades, low-income tenants have launched cooperatives in the wake of rent strikes or as a form of protest over abandonment or disrepair.

Typically, the organization of cooperatives moves through a number of steps. The property must first be acquired. When government has obtained the property through default, it often assists in facilitating the sale. In other cases, raising seed money is necessary to pay for legal and financial professionals or MHA assistance. If rehabilitation precedes acquisition, there are further delays for fund-raising and selecting a contractor. As interest rates rose in 1970s, Section 312 loans for home improvement were discontinued; low-income cooperatives must now rely on CDBGs, nonprofit housing trusts, and local sources to finance construction. Some nascent cooperatives and MHAs, such as Habitat for Humanity, may contribute sweat equity.

The cooperative must also legally incorporate to receive the property and any subsidies or loans and to protect shareholders from personal liability. The cooperative bylaws spell out the rules under which the corporation is governed and codify the rights and responsibilities of shareholders and the board of directors. Cooperatives must also establish procedures to select members. Background and credit checks, interviews, and even home visits are frequently conducted. Confidentiality must be preserved, discrimination avoided, and regulations met, particularly if subsidies are involved.

Once members are selected and collectively approved, they undergo training in cooperative living. The importance of this stage cannot be underestimated. Many cooperatives fail because the members do not fully understand this relatively rare form of tenure. Explaining the technical bylaws and financial structure of the cooperative is particularly important. Some cooperatives also offer household budgeting, home repair, and leadership training. Training enables members to participate in the full range of cooperative activities. Self-governing cooperatives sometimes make attendance at meetings or committee work mandatory. Board membership may also rotate. Some cooperatives manage their buildings and must coordinate tasks; others supervise hired managers.

As cooperatives mature, reserve funds are capitalized and blanket mortgages amortize. Once the collective mortgage is paid, monthly costs may actually decline. Accumulated equity or a decision to raise monthly payments enables expansion, from capital improvements to new ventures such as cooperative nurseries and grocery buying.

Advantages and Weaknesses of Cooperative Housing

Cooperatives offer many advantages over rental housing. They cost less than virtually any other kind of subsidized housing. A 1972 Urban Institute study found that federally assisted cooperatives reduced litter, vandalism, turnover, and administrative and operating costs below the levels of comparable limited partnerships and noncooperative Section 221(d)(3) and Section 236 projects. Section 8 cooperatives outperformed rental Section 8 limited partnerships on total operating expenses by lowering repair, maintenance, and especially administrative costs. It can cost less to convert a rental unit to a limited-equity cooperative than it does to support a Section 8 unit for one year.

Members also enjoy financial advantages over renters. Provided that the cooperative owns the property, individuals can deduct their share of real estate and mortgage interest payments from income taxes. They also accrue equity, even when it is limited. They can participate in decisions regarding capital improvements to the property.

Although decision making may be less efficient, cooperatives operate more efficiently than rental housing. Because residents live in the building, they have lower information costs than absentee managers, easily collect delinquent rents, quickly fill vacancies, reduce vandalism, and handle many repairs and operations themselves. Thus, residents report that cooperatives offer better services at a lower cost than their previous landlords.

Participatory cooperative management has psychological as well as economic benefits. More involved residents— especially women—are more satisfied with their housing and their lives. Virtually everyone rates participatory management highly, believing that cooperative members best understand their own needs.

Members are also more secure than tenants. Limited-equity cooperatives have additional advantages of reducing the risk of displacement from escalating rents and maintaining the long-term affordability of housing in booming markets.

Therefore, low-income households overwhelmingly prefer cooperatives to private renting and tend to be very satisfied with cooperative living. They most appreciate the affordability, control, security, and sociability of cooperatives compared with rental housing. Cooperatives offer lessons in home maintenance, housing law, and getting along with people.

Organizing cooperatives is a slow and complex process, even for experienced professionals. Without technical assistance and federal subsidies, it is particularly risky. The survival rate of cooperatives is not high. When costs skyrocket or receipts fall, many cooperatives, even subsidized ones, default. Despite reserve funds and fixed mortgages, cooperatives have failed when unemployment rates or operating costs rose, balloon payments were not met, turnover produced excessive vacancies, or local real estate markets collapsed. Without information and technical skills to cope with these problems, members can be overwhelmed.

Cooperatives are also socially precarious. Overall participation may wane if the board professionalizes or management bureaucratizes or if rental subsidies make members feel like tenants. Waning commitment, high turnover, factionalism, linguistic or racial conflicts, distrust and poor communication between members and the board or professionals all threaten solidarity. Particularly divisive issues are evictions, favoritism, abuse of authority, and whether to convert from limited-equity to market equity status. (SEE ALSO: *Affordability; Common Interest Development; Community-Based Housing; Cooperative Housing Foundation; Home Ownership Made Easy [HOME] Investment Partnerships Act; Housing Abroad: Canada; Limited-Equity Cooperatives; Market Equity Cooperatives; National Association of Housing Cooperatives [NAHC]; Section 213;* **Tenure Sectors**)

—*Hilary Silver*

Further Reading

Dolkart, Andrew. 1993. "Homes for People: Non-Profit Cooperatives in New York City—1916-1929." *Cooperative Housing Journal* (annual):13-22.

Heskin, Allan and Jacqueline Leavitt, eds. 1995. *The Hidden History of Housing Cooperatives.* Davis: University of California, Center for Cooperatives.

Leavitt, Jacqueline and Susan Saegert. 1990. *From Abandonment to Hope: Community-Households in Harlem.* New York: Columbia University Press.

Miller, Joel E. 1997. "Condominiums and Cooperatives." *Journal of Real Estate Taxation* 24(Winter):152-55.

Parliament, Claudia, Stephen Parliament, and Anita Regmi. 1988. "The Effect of Ownership on Housing Operating Costs: Cooperative versus Rental." *Cooperative Housing Journal* (annual): 15-20.

Rohe, William M. 1995. "Converting Public Housing to Cooperatives: The Experience of Three Developments." *Housing Policy Debate* 6(2):439-79.

Siegler, Richard and Herbert Levy. 1986. "Brief History of Cooperative Housing." *Cooperative Housing Journal* (annual): 12-19.

Silver, Hilary. 1991. "State, Market, and Community: Housing Cooperatives in Theoretical Perspective." *Netherlands Journal of Housing and the Built Environment* 6:185-203.

Van Ryzin, Gregg. 1992. "How Participatory Management Can Benefit Residents and Your Co-op: An Overview of Findings from a Recent Study of Cooperative Housing for the Elderly." *Cooperative Housing Journal* (annual): 11-14.

Zimmer, Jonathan. 1977. *From Rental to Cooperative: Improving Low and Moderate Income Housing.* Beverly Hills: Sage. ◄

► Cooperative Housing Foundation

The Cooperative Housing Foundation (CHF) was established in 1952 as a private nonprofit organization to help low-income families overseas build better housing and communities. Since that time, the CHF has been working at the grassroots and at the municipal and national government levels in activities aimed at enabling people to live in better, healthier environments. The CHF provides pragmatic, innovative, private sector assistance in economic development, women's economic empowerment, settlements, and planning. Through approaches such as credit and finance, job creation, institution building, and policy formulation, the CHF enables families to invest their own resources to improve their economic situation and their living conditions.

The CHF helps to strengthen the capabilities of host governments and communities, donor agencies, small- and medium-size private businesses, and nongovernmental organizations (NGOs) in 80 developing countries. In addition, it engages in research and training to promote the transfer of knowledge and skills in the development of supportive public policies for the housing sector. The CHF has a multinational, multilingual, and interdisciplinary staff consisting of architects, economists, financial managers, lawyers, urban planners, and others. CHF projects are funded through grants, contracts, and contributions. President and CEO: Michael E. Doyle. Cooperative Housing Foundation, 8300 Colesville Road, Suite 420, Silver Spring, MD 20910. Phone: (301) 587-4700. Fax: (301) 587-2626. (SEE ALSO: *Cooperative Housing*)

—*Caroline Nagel* ◄

► Council for Affordable and Rural Housing

Founded in 1980, the Council for Affordable and Rural Housing CARH (formerly the Council for Rural Housing and Development) seeks to secure adequate funding for affordable housing programs and works to maintain a tax environment conducive to the continued construction of affordable housing. CARH has two annual conferences, three 1-day educational seminars, and five standing committees. CARH also offers telephone access to its national office staff and serves as an information clearinghouse for members. Its membership includes managers, developers, owners, architects, market analysts, accountants, and all other participants in affordable housing. CARH has three publications: *CARH News,* a monthly newsletter to members only; *CARH On-Site,* a quarterly newsletter for on-site managers for $20 per year; and *Housing Action Letter,* which is published as needed to cover legislative developments. CARH operates on an annual budget of $500,000, funded by membership dues, seminars, and publications.

Current membership: 330 members nationwide, including 21 state-affiliated organizations. Staff Size: 3. Executive Director: Anna M. Moser. Address: 1300 19th St. NW, Suite 410, Washington, DC 20036. Phone: (202) 296-5159. Fax: (202) 785-2008. (SEE ALSO: **Affordability;** *Rural Housing*)

—Laurel Phoenix ◀

▶ Council of Large Public Housing Authorities

Founded in 1981, Council of Large Public Housing Authorities (CLPHA) is composed of more than 60 of the largest housing authorities that own and manage more than 40% of the public housing stock in the United States. CLPHA's mission is the preservation and improvement of public housing programs addressing physical conditions, management and operations issues, legislation, and funding. CLPHA organizes quarterly meetings, engages in research and advocates on behalf of public housing before Congress and the Department of Housing and Urban Development. Publications include a monthly newsletter, a monthly mailer, training materials, research reports, and policy analysis reports. Current membership: 63 public housing authorities. Staff size: 6. Executive Director: Sunia Zaterman. Address: 601 Pennsylvania Ave. NW, Suite 825, Washington, DC 20004-2612. Phone: (202) 638-1300. Fax: (202) 638-2364. (SEE ALSO: **Public Housing**)

—Laurel Phoenix ◀

▶ Council on Tall Buildings and Urban Habitat

The Council on Tall Buildings and Urban Habitat is an international professional society founded in 1969 that holds world congresses every five years, organizes regional conferences, and acts as an information clearinghouse for its members. Formerly named the Joint Committee on Tall Buildings, the council publishes *The Times,* a newsletter that comes out three or four times a year (included in membership dues of $75/yr.); monographs and other tall-building volumes and proceedings; and videos. The council operates on a budget of $200,000 and is funded through private donations and grants. Current membership: 1,500 worldwide. Staff size: 8. Contact person: Dolores Rice. Address: Lehigh University, 13 E. Packer Ave., Bethlehem, PA 18015. Phone: (610) 758-3515. Fax: (610) 758-4522. E-mail: inctbuh@lehigh.edu. (SEE ALSO: *High-Rise Housing*)

—Laurel Phoenix ◀

▶ Cranston-Gonzalez National Affordable Housing Act (P.L. 101-625)

This legislation, enacted in 1990, constituted the largest set of changes in U.S. housing law since 1974, including the

reiteration of the 1949 goal that "every American family be able to afford a decent home in a suitable environment." The act included (a) new programs to encourage home-ownership for low-income families, (b) provisions to offer social services for certain groups in subsidized housing, (c) provisions to keep privately owned low-income housing available for low-income families, (d) attempts to improve the management of public housing, and (e) measures to keep the Federal Housing Association (FHA) solvent. The legislation also required state and local authorities to formulate a strategy to improve housing affordability, known as a CHAS (comprehensive housing affordability strategy), and to get Department of Housing and Urban Development (HUD) approval of those strategies to receive federal housing assistance.

One major new program created by the act was HOME Investment Partnerships (Title II of the act): Through a new HOME Investment Trust Fund, HOME was to provide block grants to state and local governments to help meet local housing needs according the strategies outlined by the implementing unit's CHAS. The grants were to be allocated on the basis of housing need and required local and state governments to provide matching funds of 25% to 50% of project costs. HOME funds cannot be used to subsidize public housing, rental assistance, the preservation of existing public housing, or administrative expenses.

HOME funds are not just for the use of agencies of state and local government but are also to be used to encourage the creation of partnerships between the public sector and private sector as well as for the expansion of nonprofit housing associations. Against the wishes of the Bush administration, Congress provided some HOME funding for new housing. The creation of HOME was accompanied by the termination of earlier programs, including Housing Development Action Grants, Nehemiah Grants, urban homesteading, and some other rehabilitation programs.

A second major new program, HOPE (Homeownership and Opportunity for People Everywhere), was established to encourage the sale of public housing units to their tenants (Title IV of P.L. 101-625; also became a new Title III for the U.S. Housing Act of 1937). To help replace the loss of low-income rental units that would result from the HOPE sales, the legislation authorized the use of Section 8 five-year certificates or vouchers to replace those units in the stock of assisted rental housing. The legislation also authorized grants to promote homeownership programs for properties owned or insured by HUD, the U.S. Department of Agriculture, the Veterans Administration, the Resolution Trust Corporation, and state and local governments.

A third initiative was the creation of the National Housing Trust, to help qualifying first-time homebuyers with subsidies for their down payment costs and to hold their mortgage interest rates at 6% or less.

The act also attempts to deal with the problems of specific groups. Title VIII, Subtitle D of the legislation, "The AIDS Housing Opportunity Act," provides grants to help states and local governments provide for the housing needs of those with AIDS. The Shelter Plus Care program (Title VIII of the act) calls for combining housing assistance to the homeless with social services. Another provision of the legislation allowed public housing agencies to house families whose housing problems would result in their children being placed in foster care. The legislation also provided for the offering of supportive housing services for the elderly and disabled. Furthermore, it gave public housing authorities the flexibility to allow rent payments that were larger than previously allowed by law.

The legislation also targeted recipients of subsidized housing for special efforts to promote their economic well-being. The Family Self-Sufficiency Program encourages public housing authorities to provide training and support services to both public housing and Section 8 recipients to become economically independent. The legislation contains economic incentives for recipients to move out of subsidized housing when they are able to afford unsubsidized housing.

The act attempted to preserve some of supply of low-income rental housing that was likely to be converted into normal (and higher-priced) housing units. Title VI of P.L. 101-625, also known as the Low-Income Housing Preservation and Resident Homeownership Act, dealt with the potential loss of housing for low-income families when owners of low-income rental units financed with federal mortgage subsidies tried to convert the units to nonsubsidized, higher-rent units. The original legislation had allowed for the owners to "prepay" the loans after 20 years and do what they pleased with the units. However, since 1987, Congress had maintained a moratorium on such prepayments to avoid reducing the supply of low-income housing units. In Title VI, HUD was required to provide incentives to owners to continue offering their units to rent to low-income families, based on owners being allowed to receive fair market value for their property.

The legislation introduced provisions to improve the management of public housing. It called for the measurement of local public housing authority (PHA) performance and the identification of badly performing PHAs by HUD. PHAs identified as being in trouble were then to come to agreement with HUD on how to proceed to improve their performance.

Finally, the legislation also attempted to deal with the financial problems of the FHA. The legislation altered FHA loan requirements to staunch losses in program but with the effect of making FHA loans more expensive. (SEE ALSO: Affordability)

—*Nathan H. Schwartz*

Further Reading

Congressional Information Service. 1991. *CIS Annual 1990: Legislative Histories.* Washington, DC: Author.

Congressional Quarterly, Inc. 1991. *Almanac.* 101st Congress, 2d sess., 1990. Vol. 46. Washington, DC: Author.

Milgram, Grace. 1994. *A Chronology of Housing Legislation and Selected Executive Actions, 1892-1992.* Congressional Research Service Report prepared for the Committee on Banking, Finance, and Urban Affairs and the Subcommittee on Housing and Community Development, House of Representatives, 103d Congress, 1st sess. Washington, DC: Government Printing Office. ◀

▶ Crime Prevention

Difficult crime problems have led housing and planning professionals to become more directly involved in supporting a wide variety of crime prevention activities. They have, for example, begun to work with other organizations in sponsoring preventive social services, such as job training and employment programs, in high-crime communities. They have also sought to design, or redesign, housing developments and neighborhoods in ways that reduce opportunities for criminal acts. Programs to teach residents how to protect themselves and their property are also being offered, and community groups are being organized to undertake surveillance activities, such as block watches, mobile patrols, and escort programs. New approaches to policing, including problem-oriented policing and community policing, are also being used to address crime-related problems and to involve community residents in crime prevention activities. Finally, public housing authorities are implementing special programs to evict tenants involved in criminal activities and to better control access to the properties they manage.

Crime has become a serious problem in U.S. society. Between 1987 and 1991, the overall rate of reported crime per 100,00 inhabitants rose by 6%, and the rate of violent crime rose 24%. Although crime rates have fallen slightly in more recent years, crime is still a major concern for most Americans. Crime and fear of crime is of concern to housing and planning professionals because it has been linked to neighborhood deterioration. Those who are able, often abandon high-crime areas for other locations. Moreover, vandalism, arson, and other property crimes directly undermine the livability of housing developments and neighborhoods.

Research on conventional policing has shown that it is only marginally effective in deterring crime. Many have come to believe that the key to deterring crime is involving community residents in crime prevention activities. Moreover, housing and planning professionals have realized that there is much they can do both to address the underlying causes of crime and to assist the police in maintaining safe communities.

Crime reduction strategies can be placed along a continuum that goes from an emphasis on prevention to an emphasis on control. Preventive approaches seek to address the underlying causes of crime, such as poverty, drug abuse, and unemployment, whereas control approaches seek to either deter or apprehend and punish the perpetrators of crime. Many believe that effective crime prevention programs must include both type of strategies.

Preventive Services Approach

A wide variety of social and educational services are offered as a means of addressing the major causes of crime, including poverty, unemployment, drug addiction, and social alienation. These programs include employment programs for both youth and adults, youth recreation programs, drug education and rehabilitation services, high school completion and other educational programs, and parenting programs. Many public housing authorities in cooperation with local social service providers have begun to offer one or more of these services on the sites of their larger, higher crime developments.

Environmental Approach

Research suggests that there is a relationship between the design and maintenance of buildings and grounds, and both crime and fear of crime. Research has found that shorter buildings with a smaller number of units per entryway, buildings with better opportunities for surveillance of entryways and stairwells, and buildings with real or symbolic barriers separating public and private spaces had lower crime rates. At the neighborhood scale, areas with more homogeneous residential land uses, fewer major roads, and less on-street parking have lower rates of crime. Housing and city planning professionals have begun to incorporate these concepts into their designs and plans.

Other research has shown that fear of crime in an area is largely determined by the presence of "incivilities," or signs of disorder, such as litter, graffiti, abandoned cars, vacant lots, and the like. Neighborhood cleanup programs have been organized to remove these signs of disorder in the hope of reducing fear.

The presence of certain types of land uses has also been associated with crime. Abandoned buildings in some neighborhoods, for example, are used as "crack houses" or for other illegal activities. Research has also shown that crime in the vicinity of bars, pawn shops, convenience stores, and certain other commercial establishments is higher than in other areas. This has led to efforts to close those businesses and to either raze or rehabilitate abandoned dwelling units.

Individual Protection Approach

This approach includes a variety of activities designed to assist individuals in protecting their persons and their properties. Civic groups, for example, encourage people to carry whistles or air horns to alert fellow residents of the need for assistance. Other programs encourage residents to upgrade their locks and other security equipment and to engrave their valuables with an identification number. Still other groups sponsor escort services during evening hours and provide educational activities on effective precautions against personal victimization.

Collective Surveillance Approach

Surveillance approaches are designed to actively involve community residents in watching out for suspicious or criminal activity and reporting it to the police. Possibly the most popular form of this approach is the community or block watch program. They have been called "the backbone of the nation's community crime prevention efforts." Typically, these programs are sponsored either by local police departments or local civic organizations. Participants are trained to recognize suspicious activities and report them to the police. Signs announcing the area as a community watch area are often displayed. A recent study of the effectiveness and use of neighborhood watch programs concluded that, at least in the short run, these programs can be effective in reducing certain types of crime, particularly residential burglary. They also found, however, that

participation typically dwindles over time and that these programs are less likely to be found in the higher-crime areas.

A more active form of surveillance involves citizens riding or walking around their communities and reporting suspicious or criminal behavior to the police, using radios or mobile telephones. Usually, those involved in these activities are unpaid volunteers, but typically, they receive some training from the local police department with whom they work closely. Little is known about their effectiveness at the present time.

Criminal Justice Approach

The traditional approach to policing has been called "incident-driven policing." This involves patrolling by car and responding to calls for service. It also involves response to individual incidents through investigations and arrests. Although still the dominant approach of many police departments, a variety of new approaches have been gaining favor in recent years. Two of the most popular new approaches are "problem-oriented policing" and "community policing."

Problem-oriented policing has been defined as "a departmentwide strategy aimed at solving persistent community problems. Police identify, analyze, and respond to the underlying circumstances that create incidents." The role of the police under this model is greatly expanded. They become problem analysts and general problem solvers rather than law enforcers. If, for example, abandoned houses in a particular area are being used for the sale and use of drugs, the police work with other city agencies to have the units rehabilitated or demolished. Or if a lack of constructive activities for youths is found to be related to vandalism, the police may work with other agencies to sponsor recreational or other constructive activities for young people. Police personnel are taught how to identify problems, analyze the causes of problems, design and implement solutions, and assess the effectiveness of the solutions.

Community policing is related to problem-oriented policing, but it puts greater emphasis on involving the community. The major goal of community policing is to engage the community in crime prevention activities and to establish communication and trust between police personnel and residents. A wide variety of activities may be part of a community policing program, including foot patrols, neighborhood or mini police precincts, neighborhood watch or patrol programs, community organization and problem solving, conflict mediation, and the organization of social and recreational activities. The central focus of all these activities, however, is to get the community to accept some of the responsibility for the safety of their neighborhoods.

Crime Prevention in Public Housing

Many public housing developments have serious problems with crime and illegal drugs. The characteristics of the residents, the design and location of the developments, and the management of the developments have all been offered as explanations for these problems. The residents of public housing are mostly very poor, female, unemployed, single parents. Many of the developments are large, dense, and located in inner-city, high-crime areas. Finally, the management of some public housing has been lax, resulting in situations in which adequate background checks are not conducted on applicants and those responsible for crime are not evicted.

Recently, public housing authorities have become more directly involved in crime prevention activities. They have been assisted by the Public Housing Drug Elimination Program funded under the Anti-Drug Abuse Act of 1988. This program provides local housing authorities with grants to sponsor antidrug and crime prevention efforts in public housing developments. Moreover, HUD has been urging local housing authorities to include specific language in their leases that allows eviction if anyone in the dwelling unit is using or selling illegal drugs.

Many believe that much of the crime in public housing developments is being committed by nonresidents. This has led some housing authorities to organize "sweeps" in which trespassers are removed from the grounds and inspections of each unit are conducted. Residents are then given identification cards, and security guards are posted at the entrances of buildings to control access. Improvements are also made to the units and grounds and residents are involved in helping to maintain security.

Many public housing authorities have also adopted more aggressive tenant screening and eviction policies. Extensive background checks are done to determine if applicants have criminal records. If they have, this may disqualify them for public housing. Moreover, if residents of public housing are convicted of a crime, they may be promptly evicted. In some instances, housing authorities have begun to evict people who have simply been charged with a crime under a "preponderance of evidence" standard.

Finally, some housing authorities have hired their own security forces to police their developments. Normally, these are fully deputized law enforcement officers who work closely with the local police forces. Many are also involved in community policing activities. (SEE ALSO: *Drugs and Public Housing*)

—*William M. Rohe*

Further Reading

Feins, J., J. Epstein, and R. Windom. 1997. *Solving Crime Problems in Residential Neighborhoods: Comprehensive Changes in Design, Management, and Use.* Washington, DC: U.S. Department of Justice.

Greenberg, S. W. Rohe and J. Williams. 1984. "Neighborhood Design and Crime: A Test of Two Perspectives." *Journal of the American Planning Association* 50(1):48-61.

Holzman, H. R., T. R. Kudrick, and K. P. Voytek. 1996. "Revisiting the Relationship between Crime and Architectural Design." *CITYSCAPE* 2(1):107-26.

Rainwater, Lee. 1966. "Fear and the House-as-Haven in the Lower Class." *Journal of the American Institute and Planners* 21:773-85.

Reppetto, T. 1974. *Residential Crime.* Cambridge, MA: Ballinger.

Rosenbaum, D. 1986. *Community Crime Prevention: Does It Work?* Beverly Hills, CA: Sage.

Rosenbaum, D. 1994. *The Challenge of Community Policing: Testing the Promises.* Thousand Oaks, CA: Sage.

Sherman, L. P. Gartin and M. Buerger. 1989. "Hot Spots of Predatory Crime: Routine Activities and the Criminology of Place." *Criminology* 27(1):27-55.

Skogan W. 1989. *Disorder and Decline: Crime and the Spiral of Decay in American Neighborhoods.* Berkeley: University of California Press.

Wekerle, G. R. and G. Whitzman. 1995. *Safe Cities: Guidelines for Planning, Design, and Management.* New York: Van Nostrand Reinhold. ◄

► Cross-National Housing Research

Housing research undertaken in two or more countries is usually labeled *cross-national.* The aspects of housing studied are generally the same as in single-nation housing research. What is distinctive about cross-national housing research is the scope of the data collected and the attempt to explain the similarities and differences observed.

A distinction needs to be made between two types of cross-national research: descriptive and explanatory. Descriptive studies are those that provide facts about the workings of housing systems. Essentially, they juxtapose descriptions of two systems but do no more than note the differences and similarities. Such studies have focused on housing policy, housing tenure (levels of owner occupation and the size of the nonmarket sector), housing production, financing systems, house prices, housing cost-income ratios, and housing space levels. Early examples include the work by Edith Elmer Wood and Catherine Bauer. The great virtue of such research is that it provides a corrective to the "ethnocentric" approach one gains from being familiar with only one society. The culture shock of discovering how different housing systems are can be a stimulus to understanding. For example, for people in the United States to discover that in Western Europe housing policy is a major element of the welfare state rather than simply a matter of helping the private building industry maximize its output of dwellings can be an enlightening insight. To learn that in 1980 the level of owner-occupation was 90% in Bangladesh but only 30% in Switzerland also conflicts with common assumptions—as well as suggesting that the quality of what is owned needs to be taken into account in comparisons. The Swedish practice of creating public land banks to cut down the cost of house production may sound like a dream or a nightmare depending on one's political standpoint, but it is certainly a novel idea. Finally, for Westerners to discover that under state socialism the highest-status groups obtained the best housing likewise jolts preconceptions about intersystem differences. Descriptive research can also lead to the introduction of housing innovations, through the transfer of policies from one society to another. However, whether such transfers are successful or not depends on the similarity of the wider social economic and political context of the two societies and on how strongly rooted in this context the original policy is.

Explanatory cross-national housing research on the other hand goes further and asks why the differences and similarities between housing systems, policies, and so on have come about. Historians would say that the answer lies in the past with every society following a unique path. Social scientists however respond in two ways. Some believe that these historical sequences reveal broad regularities, and they are interested in what is common among them. Others ignore historical processes and seek explanations in features of the present-day context of societies. They regard explanatory cross-national research as a synonym for cross-national comparative analysis.

To bring out the advantages of explanatory cross-national housing research, let us consider the case of single-nation research. The obvious defect of the latter is that any finding, such as a relationship between household income level and housing demand, may be specific to that country. This may not matter from the point of view of the policymaker. But social scientists are interested in making generalizations about relationships that hold across societies. It is rare to find a relationship that is not affected by the institutional context of the country concerned. For example, the relation between household income and what housing households get access to depends greatly on the lending criteria of financial institutions and on the extent of the nonmarket housing sector, both of which vary greatly between countries. Attempts to generalize from the United States to societies with a strong welfare housing sector are likely to be risky.

Explanatory cross-national housing research can thus be seen as a response to the context-dependent character of single-nation findings. Its attraction is that it allows the study of the effect of variations in contextual features that are fixed when one society is studied at one point in time. Something that is a parameter for one society—for example, in the case of the United States, a capitalist socioeconomic system and weak political influence of labor—can be treated as a variable if comparisons are made with other societies that lack these features—for example, state socialist societies and welfare capitalist societies.

The logic of explanatory cross-national housing research can be seen by comparing Sweden and the United States, which differ in the level of political influence of labor. If the only difference between the United States and Sweden were that Sweden showed a much stronger labor influence on government and it was considered likely on theoretical grounds that this influenced housing policy, then differences between United States and Swedish housing policy, such as their level of welfare orientation, could be attributed to their different level of labor influence. In this way, a parameter of the U.S. context can be studied as a variable. The conclusion would be that the weak political influence of labor in the United States was likely to reduce the likelihood of U.S. housing policy having a welfare orientation. That conclusion may not matter to policymakers but could be of interest to pressure groups trying to mobilize for change in housing policy as well as to social scientists.

If cross-national research includes a more diverse range of societies—for example, developed capitalist and developed state socialist—the same logic applies. Two types of result are possible. If differences between housing features (e.g., the level of state-provided housing) are found in, say, the United Kingdom and Hungary (pre-1989), these would typically be seen as effects of the difference in socio-economic system. On the other hand, if similarities are found (e.g., both countries have a policy of privatizing state housing), these would be explained as due to either common systemic features or common pressures. In the case of the United Kingdom and pre-1989 Hungary, a common socioeconomic system is excluded, but other systemic features may be similar—for example, an "overburdened" public sector. On the other hand, if a similar pressure acts on both state socialist and capitalist societies and produces a similar outcome, this would imply that the type of socioeconomic system was not all-important. For example, it can be argued that state socialist societies are just as affected by the international economic system as capitalist societies. In all these cases, cross-national research allows the effect of system-level or contextual features to be identified.

A brief reflection on the argument so far will reveal some of the limitations of explanatory cross-national housing research.

First, the conclusions drawn above all rest on *assumptions* about the generality of the cause-and-effect links that exist in reality. For example, the comparison of the United States and Sweden *assumes* that if the labor movement was strong politically in the United States it would have the same effect as in Sweden—namely, to shift housing policy in the welfare direction. This can only be an assumption. Likewise it is only an assumption that if those with the highest-status jobs get the best housing in both capitalist and state socialist societies, this shows that there is a common logic of industrialism. These cases illustrate how the conclusions of cross-national research rest on assumptions about causal processes that are debatable. An alternative model would be that cause-and-effect relations are not as universal as the law of gravity but vary between groups of societies—although probably not between individual societies.

A second limitation of explanatory cross-national research is a more practical one—namely, that it is exceptional to find only one contextual feature differing between countries as in the above examples. The philosopher J. S. Mill who first codified the principles of research design imagined that by choosing cases appropriately one could compare pairs of cases that differed in only one respect. This may be possible in experiments on crop yields where more variables are under the researcher's control. In cross-national social research, however, where there is a limited number of societies each with characteristics beyond the researcher's influence, it is much more likely that there will be numerous differences between societies. Sweden and the United States differ in population size, ethnic diversity, and industrial structure as well as in the political variable referred to; Hungary and the United Kingdom differ in population size, level of urbanization, industrial structure, and income level as well as in type of socioeconomic system. Today, Mill's view is rejected, and it is acknowledged that differences between societies that are irrelevant to the topic in question can be ignored. For example, differences in climate and position on the globe may affect housing design but probably have no other influence on housing policy. But such judgments, which reflect prevailing theories about the determinants of housing policy, always remain open to debate. The fact that two societies will display more than one difference theoretically relevant to the topic in question remains a real obstacle to cross-national research.

This discussion of the advantages and disadvantages of explanatory cross-national research leads directly to the question of the choice of countries in such research. The two main approaches are known as "most similar systems" and "most different systems," following Przeworski and Teune. In the former, which is usual in housing research, countries are chosen that have a broad resemblance. For example, developed capitalist countries or developed state socialist countries are standard choices. The justification for this is that the influence of societal features, such as type of socioeconomic system or level of economic development is considered so important that if it is not "held constant," it would interfere with any results obtained. Research by Headey following this design has, for example, examined differences in housing policy among developed capitalist societies and has attributed them to differences in the strength and patterns of interaction between labor, employers, government, housing producers, and housing consumers and to party political strategies. Other research, by Harloe on housing tenure, has revealed the importance of industrialization in leading to an increase in rental housing in cities in Europe and the United States in the late 19th and early 20th centuries.

The drawback of the most similar systems approach—which amounts to dividing societies into different families—is that common causes operating on all families or processes of convergence between families are not studied. For example, a study by Schnore started from the observation that in Latin America the elite live in central locations and the poor in the outskirts, whereas in the United States, the reverse spatial pattern is found. Rather than treat Latin America and the United States as belonging to different families, Schnore insists that their different spatial patterns are due to a simple general pattern that reflects the impact of levels of technological development. Schnore argues that as transportation and communications technology improves, the elite no longer need to live in central locations but can exercise control from peripheral locations. Whether he is right or wrong, the point is that if societies are categorized into most similar groups, any such general explanations that apply across highly diverse societies will be missed.

The most different systems approach works on the reverse assumption to most housing research. It argues that system characteristics are irrelevant—and therefore considers all countries together. It seeks relationships that hold across diverse societies—that are invariant to context. This approach is used in psychology but is little used in research on housing or in the other social sciences. Schnore's study, an example of the most different systems approach, is rare in regarding technology as the determinant force in housing systems.

A key issue in explanatory cross-national housing research is what types of explanatory variable should be included. Aspects of the societal context such as the strength of labor or of producer interests are often used. But these are essentially cross-sectional and do not capture trends over time. Research that does try to do this takes two directions. The first is to periodize history and say that as societies move between one stage and another, so housing systems will change. Marxist analyses in terms of modes of production and/or phases of capitalism (commercial, industrial, corporate) are examples of this. The second is to identify a continuous historical process, such as industrialization, urbanization, or the growth of gross national product per capita and argue that it explains changes in housing system. The problem is to theorize historical processes and avoid saying that every society has a unique history.

In sum, a majority of cross-national housing research has been descriptive rather than explanatory. It has drawn attention to the substantial differences between housing systems in different countries. Explanatory cross-national housing research has been less developed partly because of the logical and practical problems outlined above. But both have great potential for helping our understanding of housing processes. (SEE ALSO: *Centrally Planned Housing Systems; Housing Abroad: Canada; Housing Abroad: Western and Northern Europe; Housing Distribution Mechanisms;* **Third World Housing;** *Welfare State Housing*)

—*Christopher G. Pickvance*

Further Reading

Ball, M., M. Harloe, and M. Martens. 1988. *Housing and Social Change in Europe and the USA.* London: Routledge.

Clapham, David. 1996. "Housing and the Economy: Broadening Comparative Housing Research." *Urban Studies* 33(4-5):631-47.

Harloe, M. 1985. *Private Rented Housing in the United States and Europe.* London: Croom Helm.

Headey, B. 1978. *Housing Policy in the Developed Economy.* London: Croom Helm.

Karn, Valerie and Wolman, Harold. 1992. *Comparing Housing Systems: Housing Performance and Housing Policy in the United States and Britain.* Oxford, UK: Clarendon.

Kemeny, Jim. 1995. *From Public Housing to the Social Market: Rental Policy Strategies in Comparative Perspective.* London: Routledge.

Papa, Oscar. 1992. *Housing Systems in Europe: Part II. A Comparative Study of Housing Finance.* Delft, The Netherlands: Delft University Press.

Pickvance, C. G. 1986. "Comparative Urban Analysis and Assumptions about Causality." *International Journal of Urban and Regional Research* 10:162-84.

Przeworski, A. and H. Teune. 1970. *The Logic of Comparative Social Inquiry.* London: Wiley.

Schnore, L. F. 1965. "On the Spatial Structure of Cities in the Two Americas. In *The Study of Urbanization,* edited by P. M. Hauser and L. F. Schnore. New York: John Wiley.

Turner, B., J. Hegedus, and I. Tosics. 1992. *The Reform of Housing in Eastern Europe and the Soviet Union.* London: Routledge.

Van Vliet--, Willem, ed. 1990. *International Handbook of Housing Policies and Practices.* Westport, CT: Greenwood. ◀

▶ Crowding

Household crowding is the stress associated with living in an overpopulated household. When too many people occupy a household, residents may experience unwanted social interactions, a lack of privacy, and an inability to move about freely in their home. Prolonged exposure to crowding can have adverse effects on interpersonal relations, psychological well-being, and physical functioning.

Distinction between Household Density and Crowding

Density is an objective property of the physical environment, whereas crowding refers to residents' subjective psychological experience. The negative effects of density on human health and behavior appear to be strongest when individuals feel crowded by the density. Density is typically measured by calculating a ratio of the number of people to a given amount of space. Household density can be assessed in several ways. The most common measure is the ratio of persons per room in a household. The more people there are per available space in a house, the greater the density. Crowding is more likely to be experienced in higher-density than in lower-density homes. However, for reasons discussed below, high density is not always sufficient to cause people to feel crowded.

Levels of Household Density in the United States

Households with greater than 1.0 person per room are considered to be overly populated. During the early decades of this century, there was great concern among urban policymakers and housing officials about the effects of crowding on health, safety, and general well-being. In the 1920s and 1930s, research at the University of Chicago focused on crime rates and other correlates of high densities. From 1940 to 1980, according to U.S. census data, the proportion of households with more than 1.0 person per room declined steadily from 1940 to 1980 and then increased slightly in the latest decade (see Figure 8). Although the proportion of households with greater than 1.0 person per room appears rather small in 1990, it represents more than 4.5 million households. The majority of crowded households are headed by whites (47%), followed by blacks

(22%), individuals of Hispanic origin or other races with Spanish surnames (21%), and Asians and Pacific Islanders (10%). However, an examination of the proportion of crowded households within each racial group reveals that crowding is disproportionately high among certain racial subgroups. Approximately one-quarter of households headed by individuals of Hispanic or Asian origin have greater than 1.0 person per room. In contrast, only 10% of households headed by blacks and 3% of households headed by whites have greater than 1.0 person per room.

The census provides a reasonable estimate of the number of high-density households. However, the exact number of people living in high-density or crowded conditions is unknown. It is possible to estimate the number of people living in crowded conditions by multiplying the number of households with greater than 1.0 person per room by the median number of persons per household, which was approximately 2.3 in 1990. Using this procedure, there were approximately 10 million people living in overpopulated households in the United States in 1990. This figure is, however, likely to be an underestimate because households with more than 1.0 person per room probably have more than an average of 2.3 residents.

Effects of Crowding on Health and Behavior

Research on household crowding suggests that it contributes to a wide range of health and social problems. High household density is associated with mortality, fertility, public assistance, and juvenile delinquency rates. In comparison with people from relatively low-density households, people from high-density households also tend to have fewer friends, more difficulties with neighbors, and less emotional support from their social networks. High household density is also related to negative moods and symptoms of depression and anxiety in adults.

Children appear to be more negatively affected by high density than are adults, partly because children have less control over their environment than do adults. In comparison with children from low-density homes, those from high-density homes tend to have more behavioral problems in school, more anxiety, greater distractibility, more conflicts, lower achievement motivation, and poorer verbal abilities. In addition, parents from high-density homes are less likely to interact with or monitor the behaviors of their young children than are parents from relatively low-density homes.

Household crowding also is a frequent problem among college students, who often double up in apartments to save money. The effects of crowding on students are similar to those observed in community samples. In comparison with students in low-density housing, crowded students have more unwanted social interactions, more frequent negative moods, and exhibit greater social withdrawal and insensitivity to others' needs.

Caution must be applied when interpreting associations between household density and human health and behavior. Household density often accompanies other conditions that could have negative effects on humans. Noise, for example, is likely to be greater in high-density households than in low-density households, and noise can interfere

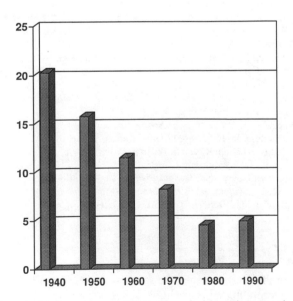

Figure 8. Percentage of U.S. Households with More Than One Person per Room (1940–1990)
SOURCE: U.S. Bureau of the Census.

with children's attention, hearing, and learning abilities. People in high-density households also are likely to be poorer than those in relatively low-density households. Poverty could be the cause of both crowding and the various health and behavior problems that often are attributed to crowding. Thus, it is difficult to establish what role, if any, household crowding has in the etiology of social and health problems in humans. Some researchers use statistical techniques to attempt to examine the effects of density independent of other social factors. The problem with such techniques is that researchers are never certain that they have identified all relevant factors that influence both density and the outcome of interest.

Cultural Adaptations

In a cross-cultural study on the effects of household density on psychological distress symptoms, researchers found U.S. college students were more adversely affected by high density than were adult males living in India. Other researchers have found that Japanese and Chinese families appear to be relatively unaffected by living in high-density homes. Finally, among different U.S. ethnic groups, high household density appears to have a stronger negative effect on the mental health and social relations of black Americans than of white Americans, and only a weak effect on Americans of Hispanic origin.

Why do different cultural groups have unique reactions to household density? One possibility is that groups of people who have been exposed to high-density living conditions develop methods of coping with the crowding. For example, in crowded Chinese households, family members will eat at different times to reduce the amount of crowding during meals. Cultures with a high population density also appear to have developed customs and rules of behavior that reduce the stress of crowding by making the social environment more predictable and controllable. Customs related to the amount of distance between people when interacting socially could also influence preferences and expectations for personal space. If high population density causes people to interact at distances that violate their space preferences, then the high density is likely to be perceived as crowded and undesirable.

Social Influences

The quality of the social environment can have a strong influence on the relationship between crowding and psychological distress symptoms. For example, people living in high-density households who have frequent hassles with their housemates are more likely to be psychologically distressed than are people living in high-density households with relatively few hassles or people living in low-density households. People living in high-density households may be particularly distressed by social hassles because it is more difficult to avoid or escape the hassles in a high-density home than in a low-density home. In contrast, people living in high-density households who have supportive housemates are less distressed than are people living in high-density households with relatively unsupportive housemates or people living in low-density households.

Theories of Crowding

There are three prominent theories to explain the effects of household crowding on human behavior and psychological functioning: behavioral constraint, control, and overload/arousal theories.

According to the behavioral constraint theory, high density interferes with individuals' goal obtainment by restricting their movements and behaviors. The diminished freedom makes the high density noxious and undesirable. In high-density settings that do not thwart their goal-directed behaviors, people tend to be less negatively affected by the high density than when their goals are thwarted. Imagine, for instance, that before starting their day, two men have the same daily routine, which includes exercising, showering, shaving, eating breakfast, and getting dressed. Now imagine that one man does his routine while dodging other people in the house and competing to use the bathroom and the toaster, whereas the other man has risen early in the morning and has no obstacles to interfere with his morning routine. Although both men may live in similarly crowded households, the crowding does not interfere with or bother the man who rose from bed before the rest of his household. The other man, however, is likely to be bothered by crowding because it constrains his behaviors and interferes with his routine.

Control theories of crowding maintain that high density is undesirable and harmful because it renders the environment more unpredictable and exposes individuals to uncontrollable situations. A lack of control in high-density settings has been shown to exacerbate the negative effects of density on humans, whereas the availability of control has been shown to reduce the negative effects of density. Sometimes perceived control rather than actual control in a situation is sufficient to reduce the negative effects of high density on performance or mood. However, if one's expectations for control in a high-density situation do not match the actual availability of control, then the high density can be more disturbing than if one expected little control in the situation. Thus, it appears that control or beliefs and expectations about control in high-density environments influence how strongly humans are affected by crowding.

A final theory posits that high density increases pathology because of sensory overload from excessive stimulation. Humans have a limited capacity to process information; in high-density settings, the information available in the environment exceeds that capacity. Sensory overload often leads to overarousal, which can diminish complex task performance and contribute to health problems. Several researchers have shown that high density is associated with elevated blood pressure and heart rate, which supports the overload/arousal theory.

Conclusions

Household crowding seldom results in extreme social pathologies. Unlike lower animals, humans appear to be able to adapt to and cope with high-density situations with a good deal of tolerance. However, adverse psychological, somatic, and social reactions to living in crowded households do occur, particularly when people are chronically crowded. Community services and facilities could be developed in response to these problems. In areas where residential densities are quite high, parks, recreational, and athletic facilities could provide a respite from crowding. To the extent that household densities are a function of poverty and lack of affordable housing, appropriate policies might help alleviate the problems. An initial step is for communities to plan their growth more carefully and make housing development responsive to the needs of the population. Finally, individuals may find their own ways to adapt to crowded living conditions by limiting their exposure to high density through architectural interventions and careful planning of space usage. (SEE ALSO: *Doubling Up; Size of Unit*)

—*Stephen J. Lepore*

Further Reading

Baum, Andrew and Paul Paulus. 1989. "Crowding." Pp. 533-70 in *Handbook of Environmental Psychology,* edited by Daniel Stokols and Irwin Altman. New York: John Wiley.

Edwards, John N. et al. 1994. "Why People Feel Crowded: An Examination of Objective and Subjective Crowding." *Population and Environment: A Journal of Interdisciplinary Studies* 16(2):149-73.

Evans, Gary, and Stephen Lepore. 1992. "Conceptual and analytic issues in crowding research." *Journal of Environmental Psychology* 12:163-73.

Gove, Walter and Michael Hughes. 1983. *Overcrowding in the Household.* New York: Academic Press.

Myers, D., W. C. Baer, and S-Y. Choi. 1996. "The Changing Problem of Overcrowded Housing." *Journal of the American Planning Association* 62(1):66-84.

Wirth, L. 1988. "Urbanism as a Way of Life." *American Journal of Sociology* 44:1-24. ◀

▶ Cultural Aspects

The forms and meanings of housing vary across cultures and over time. Considering the role that culture plays in the area of housing helps explain these variations. Both *housing* and *culture,* however, need to be defined. Housing is not a self-evident concept, and culture is too general and too abstract to be useful. Using specific and concrete expressions of culture and relating them to a definition of housing has major implications for understanding housing (and hence for housing research) and for housing policy and design.

Variability of Housing

Housing is the most ubiquitous form of built environment. Because all human groups construct housing, it is found wherever and whenever humans have lived. Fairly elaborate and complex houses and groups of houses have been built for at least 300,000 years, possibly even by precursors of Homo sapiens. There is, therefore, a large body of evidence regarding the ways in which humans have dwelled. When considered cross-culturally and over time, the extraordinary variety of housing is striking, with hundreds of distinct kinds of houses. Houses vary in (a) form, shape, construction, materials, size, spatial organization, and siting; (b) the meanings they communicate and how they do so; (c) their relation to one another (scattered, in settle-

ments of different kinds, communal dwellings where settlement and house are one); and (d) their use, both individually and as parts of a larger system.

This variety of form and meaning needs to be explained, particularly given the much smaller range of functions and activities that occur in housing. The most general explanation is that this variety is due to the role of culture.

Houses have many purposes: to provide shelter from climate, human and animal enemies, and supernatural powers; to help achieve privacy; to mark location; to express or emphasize social identity and status; and to communicate other meanings. Each purpose can be interpreted and addressed in many ways. For example, there is great variability in how "shelter" and "protection" are defined, even where climate imposes stringent demands. How much shelter people require, how they provide it, for what purposes they use it, and other similar decisions often seem to be made on the basis of nonclimatic and nonphysiological criteria, leading to a range of heating, acoustic, and other standards even among technologically advanced modern societies. Another source of housing variation is the sacred character of many traditional houses and settlements. By imposing a sacred order, using rituals and replicating sacred schemata, housing forms a humanized, safe space in a profane and potentially dangerous world. Its form carries symbolic meaning, meaning that varies according to culture.

The rich variety of house forms can thus be understood if cultural factors are considered to be primary; climate, technology, materials, and economy are secondary or modifying factors and may impose constraints. Although the interplay of all factors best explains the form of dwellings, cultural aspects can be seen as primary because a house is more than a material object or structure that provides shelter; it is a cultural and social unit of space created to support a way of life.

To understand the role culture plays in housing, however, it is necessary to clarify the meaning of the terms housing and culture and to suggest ways in which the relationships between them can be studied.

What Is "Housing"?

If housing's variability is central for understanding the relation between housing and culture, then research must be comparative: cross-cultural and historical. That, in turn, requires comparable units—that is, units that are culturally neutral. It can be shown that housing is best conceptualized as that part of the built environment in which particular activities take place; in different cultures, the same activities occur in different settings, including many settings beyond the confines of the house. These other settings include open spaces, barns, stables, streets, coffee- and teahouses, shops, markets, taverns, clubs, men's houses, and so on. Asking who does what, where, when, including or excluding whom, and why shows that housing is best conceptualized as that system of settings in which particular (specified) systems of activities occur. These systems of settings need to be studied and compared.

Thus, what constitutes a dwelling or housing in any given case needs to be discovered rather than assumed. Because the definition depends on the distribution among settings of particular activities, what is included or excluded must be defined carefully in terms of cultural specifics. In some cases, it may include unexpected parts of the settlement or even areas beyond; there may be multiple dwellings and settlements; in some cases, such as communal dwelling found in various traditional societies, the dwelling and settlement are one. In all cases, how the included settings are used, combined, linked or separated, and closed off or opened up will vary. One result is that housing that is judged as overcrowded or too dense by one culture's standards is perfectly adequate when these additional settings are taken into consideration.

What Is Meant by *Culture*

Culture is the property of all human beings, the attribute that makes us human and thus defines the (single) species. At the same time, however, culture also creates a large number of diverse groups.

It is important to realize that culture is not a "thing"; rather, it is a theoretical construct. No one ever has, or ever will, see or observe culture—only its effects and products. On the basis of these, inferences are made about an unobservable entity; they become indicators and referents of the "culture."

First used in anthropology in the late 19th century, the concept of culture was then adopted by other social sciences. It is used differently in various disciplines and countries, and there is continuing and heated discussion about the concept. Definitions of culture seem to fall into six broad classes, which are complementary rather than in conflict. Briefly, (a) one defines culture as a way of life typical of a group, (b) another as a system of symbols, meanings, and schemata transmitted through symbolic codes, and (c) the third as a set of adaptive strategies for survival, related to resources and ecology. One might also define culture in terms of what it does. Again, three general answers have been given: (d) Culture is the distinctive means whereby populations maintain their identity—that is, culture distinguishes among groups; (e) cultures are control mechanisms and carry information about how behavior and artifacts are to be (and can be thought of as an analog to DNA); (f) culture provides a structure or framework that gives meaning to particulars. For any given question, one definition of culture may prove most useful and be emphasized.

Culture is thus a complex concept, too complex to use in analyzing or designing housing because as a theoretical construct it is both too broad (or general) and too abstract to be related to housing (or any built environment). One needs to identify components or expressions of culture, to "dismantle" culture and study the interrelationships of the components with one another and with housing.

Two types of dismantling seem to be useful. The first, addressing the view that "culture" is too abstract, makes it more concrete by identifying its observable social manifestations. Thus, groups, family and kinship structures, institutions, roles, social networks, status relations, and the like often have settings associated with them or otherwise in-

fluence various aspects of housing. It then becomes feasible to relate these two sets of variables. The problem of generality, which also makes it virtually impossible to relate culture to housing or to use in design, can be overcome by moving to ever more specific aspects of culture, starting with worldviews, through values, to lifestyle (with its associated activity systems). Particular components of housing can be related to values and especially to lifestyles (and activity systems).

Using lifestyle and social variables, and conceptualizing housing as that system of settings within which certain systems of activities take place, helps explain the role of culture in the variability of housing (seen cross-culturally and historically) and also helps in responding to it.

The Relationship between Culture and Housing

All possible explanations of the role of culture in housing diversity are variations on a single theme: people with different values, ideals, norms, and schemata respond differently to various imperatives—physical, economic, social, and ritual. Because housing was always more than shelter, "function" was always much more than a physical, utilitarian, or instrumental concept, and meaning is frequently a most important aspect of function or activities.

Any activity can be conceptualized as having four components: (a) the activity itself (its most instrumental or manifest aspect), (b) how it is carried out, (c) how it is associated with other activities to form activity systems, and (d) the meaning of the activity (its most latent aspect). Cultural variability increases as one moves from manifest to latent aspects.

For example, preparing food is a cultural universal because all human groups transform raw food. However, how they do so varies considerably. The meaning of food preparation varies most: It may be a core ritual, central in establishing status hierarchies or in maintaining group identity; it may involve notions of purity or pollution or need to be hidden. Therefore, the nature of the settings for preparing food, their location, relation to other settings, size, organization, equipment, and so on will vary greatly.

To understand cultural aspects of housing, one must consider wants more than needs, because there are always many ways to satisfy needs. As a result, choice is important and central in understanding housing. There are, of course, also constraints. The nature and relative role of choices and constraints, including prejudice, sumptuary laws, resource availability, and political relationships, are an important aspect of the study of cultural aspects of housing.

The choices made among available alternatives on the basis of wants vary according to culture. The specific ways of living with, interacting with, or avoiding others; the specific nature of activity systems and so on; and the nature of the related systems of settings are all highly variable and have particular meanings for members of particular groups. Design of housing, as of all built environments, can be understood in terms of a series of choices among the alternatives available within more or less severe constraints. These choices express ideals, values, norms, and worldviews and may communicate identity and express status.

Implications

One immediate consequence is that many standards are not universal; what constitutes "good housing" varies with culture. The standard of environmental quality is an example. It can be described by a set of attributes that are evaluated as desirable or undesirable and either chosen, rejected, or modified. Four things can vary: the attributes, their ranking, their importance vis-à-vis other things, and whether they are positive or negative. Like housing, environmental quality needs to be discovered not assumed, because it also is culture specific. All attributes of housing can be evaluated differently: climatic response, settlement form, location, size, internal and external spaces and their organization, vegetation, density, privacy, what is communicated and how, and so on.

The choices among alternatives that lead to the specific system of settings and its environmental quality profile are made to achieve congruence with lifestyle and to be supportive of it. Lifestyle can be conceptualized as the result of choices about how to allocate resources—time, money, and effort. This is why lifestyle is generally a particularly useful expression of culture. It becomes even more useful in large, complex societies such as the United States where there are many cultural groups and where ethnicity, race, religion, and class cross-cut in complex ways and rarely have direct influence on housing. Income is an enabling rather than determining variable; people with identical incomes can make very different choices. The social expressions of culture such as family structure, roles, and life cycle stages remain useful, but they also typically have lifestyle implications, as may ethnicity, race, class, religion, and income. Lifestyle (and the resultant activity systems) also encompass what, in housing research and studies of environment and behavior, have been called "special-user groups"—single parents, singles, people with disabilities, immigrants, and other groups. Lifestyle also helps to differentiate excessively broad special-user groups, such as the "elderly," the "urban poor," the "homeless," "squatters," and the like. Lifestyle applies to all groups, which makes it the most useful concept in considering the relationship of housing to culture.

Lifestyle is also applicable to culture change, whether in developed or developing countries. Thus, changing sex roles influence housing through changing lifestyles and associated activity systems. A typical change in traditional societies is a great increase in possessions. This apparently trivial change in lifestyle can be shown to have a whole set of consequences for housing. These include an increased need to communicate resulting status variations, increased amount of space needed, more numerous and specialized settings, and increased physical enclosure to protect possessions. These, in turn, greatly influence social interaction, privacy, and so on.

In the past, diversity occurred mainly cross-culturally. Any one traditional society tended to be remarkably homogeneous, as were its dwellings and settlements, which frequently symbolized the group. Moreover, change tended to be slow. Today, in countries such as the United States, and increasingly in others, extraordinary cultural diversity is found within the society in lifestyles and hence in allocation of resources, ideals and values, forms of privacy, and unspoken rules about age and sex division, space use, communication of status or identity, family arrangements, trajectories of life cycle stages, and so on. But although lifestyle diversity increases greatly (some house-marketing analyses use approximately 50 lifestyle groups), housing diversity seems to decrease. There is thus a potential and possibly an actual mismatch: A policy and design goal should be increased housing diversity to reflect the diversity of users. Because housing is usefully conceptualized as a system of settings, the larger milieu (for example, neighborhoods) may also play a role in responding to specific lifestyles in their forms of social homogeneity, specialized institutions and settings, spatial organization, unwritten rules, and cues communicating various meanings (such as status and identity).

Rates of change in all these variables also increase, as they do in forms of sharing of dwellings, recreation, technology and work patterns, roles, organization of time, and so on. The greater diversity of housing thus requires open-endedness, which is also needed to allow rapid change and to allow users to participate, to express their individuality, and to complete, add, and modify. Also, meanings are increasingly communicated through objects, furnishings, landscaping, and the like—through personalization. Open-endedness thus becomes an important latent function of housing, particularly because the need to communicate identity also increases with rapid change and the proliferation of groups.

Whereas all built environments need to be culture specific, this applies particularly to housing. As the primary setting for life, it needs to be highly supportive of the various aspects of culture; there are even cases in which the very survival of certain cultures may depend on such supportive environments. Thus, the imposition of ever-more uniform standards, both in developed and developing countries, may be misguided. To understand housing properly, and to make informed policy and design decisions, the study of the relationship of culture to housing is an essential first step. (SEE ALSO: **Behavioral Aspects;** *Earth-Sheltered Housing; Energy Conservation; Home; Religion; Vernacular Housing; Yanomami Shapono*)

—*Amos Rapoport*

Further Reading

Altman, I., A. Rapoport, and J. F. Wohlwill, eds. 1980. *Culture and Environment.* Vol. 4 of Human Behavior and Environment. New York: Plenum.

Kent, S., ed. 1990. *Domestic Architecture and the Use of Space: An Interdisciplinary Cross-Cultural Study.* Cambridge, UK: Cambridge University Press.

Low, S. M. and E. Chambers, eds. 1989. *Housing, Culture and Design: A Comparative Perspective.* Philadelphia: University of Pennsylvania Press.

Rapoport, A. 1969. *House Form and Culture.* Englewood Cliffs, NJ: Prentice Hall.

Rapoport, A. 1985. "Thinking about Home Environments: A Conceptual Framework." Pp. 255-86 in *Home Environments.* Vol. 8 of Human Behavior and Environment, edited by I. Altman and C. M. Werner. New York: Plenum.

Rapoport, A. 1986. "Culture and Built Form: A Reconsideration." Pp. 157-75 in *Architecture in Culture Change: Essays in Built Form and Culture Research,* edited by D. G. Saile. Lawrence: University of Kansas.

Rapoport, A. 1990. "Systems of Activities and Systems of Settings." Pp. 9-20 in *Domestic Architecture and Use of Space,* edited by S. Kent. Cambridge, UK: Cambridge University Press.

Rapoport, A. and N. Watson. 1972. "Cultural Variability in Physical Standards." Pp. 33-53 in *People and Buildings,* edited by R. Gutman. New York: Basic Books. ◄

► Custom-Built Homes

Throughout this century, the American dream of home-ownership fueled the housing market in the United States. Coupled with a seemingly endless supply of land, it created a strong demand for the single-family home located on its own site. Households with sufficient financial resources and aspiring to an individual home, different from a generic house developed for a mass market, typically seek to build a uniquely designed custom home.

The custom home market caters to middle- and upper-middle class households. Customized homes are more expensive to plan and build than tract homes. Successful custom homes require that the client have an adequate budget and the desire to participate in the design process.

Nationally, in 1994, custom homes accounted for 34% of all housing starts. Custom homes cost up to twice as much per square foot as a tract house. In addition, custom building takes time, demanding a commitment to a year or more by the client or a paid representative.

The design professional, developer, or contractor must ensure that the custom home meets the client's needs, desires, and resources. Prior to proceeding with any building design, an architectural program must be established. The program, developed jointly by the design professional and client, outlines the requirements of the building. The former starts the design process, eventually specifying the site, size, layout, and aesthetics, within an overall budget. The first step of a program focuses on the client's wish list, represented by pictures, scrapbooks, sketches, or questionnaires. These tools help the client communicate and analyze the relationships between design and aspects of quantity, quality, and cost.

A team approach offers the client the best results. The client should select an architect, engineer, and contractor prior to proceeding with the design process, providing each team member the opportunity to contribute ideas to and evaluations of design. Architectural fees may range from 5% to 10% of construction. Alternatively, a client may hire an architect to prepare the architectural design, selecting the contractor afterward. This more common process may not produce the best results, because the architectural design is developed without the benefit of the contractor's expertise in construction methods, systems, and costs. (SEE ALSO: *Architectural Review Boards*)

—*Fred Andreas*

Further Reading

Alexander, Christopher. 1985. *The Production of Houses.* New York: Oxford University Press.

Frampton, Kenneth. 1985. *Modern Architecture.* London: Thames & Hudson.

Lennark, Suzanne. 1979. *Exploration in the Meaning of Architecture.* Woodstock, NY: Gondolier.

Moore, Charles. 1980. *The Place of Houses.* New York: Holt, Rinehart & Winston.

D

▶ Deed

A deed is a legal instrument that serves as evidence of title to real estate. Four basic characteristics underlie a valid deed. First, it must be in writing. Second, it must identify the individual or individuals to whom the title is conveyed as well as the nature of his/her/their ownership. For example, real estate could be owned by a single woman as her sole and separate property or by a couple as tenants in common or joint tenants with right of survivorship. The parties (or party) who will hold title must have the legal capacity for doing so, and the parties (or party) who grant the title must also enjoy such capacity. If a legal entity such as a corporation or partnership is involved, all relevant parties must have the authority to act on its behalf. Third, the deed must adequately identify the property in question and must further make clear the intention to transfer title. Finally, the deed must be signed by the person(s) making the conveyance.

There are different categories of deeds, but the one that includes the most complete set of warranties by the grantor is the general warranty deed. In it, the grantor provides assurances that he or she owns the property. Furthermore, no obstacle exists that would harm the transfer of the property. The deed must also represent that no unstated claims exist against the property (e.g., prior liens or unpaid property taxes). Last, the deed must assure that no one can interfere with the new owner's use or possession of the property, a condition commonly termed "quiet enjoyment."

A deed is also a matter of public record but not before it is "recorded" in the appropriate office of local land records. The recording system establishes that the property as described is now owned by the individual(s) who have taken title.

—*Daniel J. Garr*

Further Reading

ABA Standing Committee on Lawyers' Title Guaranty Funds. 1991. *Buying or Selling Your Home: Your Guide to: Contracts, Titles, Brokers, Financing, Closings.* Chicago: American Bar Association.
Blankenship, Frank J. 1989. *The Prentice Hall Real Estate Investor's Encyclopedia.* Englewood Cliffs, NJ: Prentice Hall.
Jacobus, Charles J. 1996. *Real Estate Principles.* Upper Saddle River, NJ: Prentice Hall.
McKenzie, Dennis J. 1996. *Essentials of Estate Economics.* Upper Saddle River, NJ: Prentice Hall.
Peiser, Richard B. 1992. *Professional Real Estate Development: The ULI Guide to the Business.* Washington, DC: Urban Land Institute.
Stach, Patricia Burgess. 1988. "Deed Restrictions and Subdivision Development in Columbus, Ohio, 1900-1970." *Journal of Urban History* 15(November):42-68.
Tosh, Dennis S. 1990. *Handbook of Real Estate Terms.* Englewood Cliffs, NJ: Prentice Hall. ◄

▶ Delinquency on Loan

About 85% of homeowners in the United States purchase a home with a down payment and a mortgage loan. The mortgage, which pledges buildings or property as collateral, is repaid on a regular schedule over a given period of time. The failure to meet a scheduled loan payment poses a threat to keeping the home. Technically, the borrower is in default when a scheduled mortgage payment is missed. However, in the United States, borrowers are usually considered to be in default only after they have missed three consecutive mortgage payments. It is traditional to give the borrower the benefit of the doubt and to refer to missed loan payments of less than three months as "delinquent." The role of lender discretion is reflected in statistical categories: Loans are said to be 30, 60, and 90+ days delinquent and "in foreclosure." Furthermore, the lender can choose to renegotiate a delinquent loan at any point in the process.

Although less serious than default, the costs to the delinquent borrower can include monetary penalties, lower credit ratings, and emotional stress. The majority of delinquencies are cured within 30 days and do not result in foreclosures. The 30-day delinquency rate consistently runs more than four times the 60- and 90-day rates.

The study of loan delinquency has not received as much attention as default-foreclosure, in part because it is less serious and costly. One issue is the extent to which delinquency is distinct from or related to default-foreclosure. Although the popular view is that delinquency is a step on the way to default, the research findings present a somewhat different picture.

The dominant model, as with default, has been the borrower payment decision. What factors account for the decision to delay as opposed to pay, prepay, or stop payment? Are they primarily related to the loan and lender or to the borrower? Compared with data from studies of default, borrower-related characteristics appear to be more important to the delinquency decision than loan characteristics.

Studies in the 1970s, which took the lender's perspective, identified borrower-related characteristics known at the time of mortgage origination (occupation—an indicator of income stability—and household income) in addition to the loan factors more consistent with the default literature (loan-to-value ratio, secondary financing).

Subsequent research, taking the borrower's perspective, has, for example, simulated borrower behavior when income declined. Defining potential delinquency as changes in the mortgage payment to income ratio, a 1982 study examined borrower risk under alternative (variable-rate) mortgage instruments and found that occupational differences related to income variability and sources of income were important in determining potential delinquency—in particular, severity and duration—regardless of the mortgage instrument. There was also evidence that alternative mortgage instruments with the widest variability produced potential delinquency regardless of borrower characteristics.

Some argue that the distinction between mortgage default and delinquency is somewhat artificial. Borrower-related characteristics such as crisis events or job mobility are important to default decisions as well as to delinquency and are not well understood. However, the borrower payment model has its weaknesses. It assumes clear payment choices, whereas in reality, we know only retrospectively whether the borrower has chosen to delay or stop payment. Hence, some call for a model that incorporates delinquency along with default to explore whether these decisions are related and/or sequential.

Trends

Historically, delinquencies (and foreclosures) have varied with the economy at local, regional, and national levels. In the post-World War II period, they fluctuated within a narrow range. Beginning in the 1970s, they fluctuated more widely and generally rose. In the 1980s, they rose steeply and peaked in 1986. Between 1965 and 1989, the total number of delinquent loans rose 265%, from 465,000 to 1.7 million loans. A study of all problem residential loans found that the unemployment rate, with a time adjustment for foreclosures, explained delinquency rates.

Also contributing to rising rates of delinquency were institutional changes in the home finance system in the 1980s. The newly deregulated system is marked by a competitive market of lenders who "sell" loans; by new instruments such as the adjustable-rate mortgage, which shifts the risk from lender to borrower; and by the increased use of mortgages as financial instruments. Delinquent borrowers with mortgage repayment difficulties in the 1980s had more difficulty negotiating forbearance agreements because most loans had become nonlocal and subject to the centralized decision making of large financial institutions who were less responsive to local or regional economic conditions and more difficult to contact.

It seems likely that delinquency and default rates will remain relatively high in the near future because of the changing economy and accompanying unemployment, the fluctuating real estate market, and continuing changes in the home finance system. Given the continuing importance of homeownership, this may direct more attention to understanding delinquency and its relation to default. (SEE ALSO: **Affordability;** *Foreclosures*)

—*Lily M. Hoffman*

Further Reading

Heisler, Barbara S. and Lily M. Hoffman. 1987. "Keeping a Home: Changing Mortgage Markets and Regional Economic Distress." *Sociological Focus* 20(3):227-41.

Holloway, Thomas M. and Robert M. Rosenblatt. 1990. "The Trends and Outlook for Foreclosure and Delinquencies." *Mortgage Banking* 51(1):45-59.

Mortgage Bankers Association of America. 1996. "National Delinquency Survey, Historical Series" and Quarterly Data Series for specific years. Available from the Mortgage Bankers Association of American, Washington, DC 1-800-793-MBAA.

Quercia, Roberto G. and Michael A. Stegman. 1992. "Residential Mortgage Default: A Review of the Literature." *Journal of Housing Research* 3(2):341-79.

Rosenblatt, Robert and Thomas Holloway. 1990. *Problem Loans: Trends, Causes, and Outlooks for the Future.* Washington, DC: Mortgage Bankers Association of America, Economic Department.

Webb, Bruce G. 1982. "Borrower Risk under Alternative Mortgage Instruments." *Journal of Finance* 37(1):169-83. ◄

► Demand-Side Subsidies

Demand-side subsidies go to consumers and reduce the costs of consumption. Such housing subsidies make it cheaper for households to rent or buy housing. They can take the form of a direct cash payment by the government, or they can involve preferential tax treatment that reduces the payments a household would otherwise have to make. The principal forms that demand-side subsidies take are housing allowances or housing vouchers and homeownership tax concessions or tax expenditures. Housing allowances are sometimes called *subject subsidies*.

There are many systems of housing allowances throughout the world. In most countries, allowances are payable only to tenants, but in some countries, Germany and France for example, a small proportion of low-income homeowners are eligible. Housing allowances typically involve a cash payment to households although, for administrative purposes, the money is sometimes paid to landlords who then reduce the rents they require from tenants. The value of housing allowances is, under most systems, an increasing function of rent and household size and a decreasing function of income. The payment to any household is usually limited by various rules that include, for example, a restriction on the eligible rent and a maximum income above which no allowance will be paid.

Housing allowances have different names in different countries. In the United States, they are "housing vouchers"; in the United Kingdom, "housing benefits"; in Germany, *Wohngeld;* and in France, *aide à la personalisée au logement.*

The payment of housing allowances is usually tied to living in housing of an officially determined minimum standard or above. This has, for example, been the case with the Section 8 housing voucher program in the United States in which low-income families who can find acceptable accommodation in the private sector pay a rent that they can afford related to their income level and the landlord receives from the authorities the difference between this and a contract rent that is based on an administratively determined so-called fair market rent.

As instruments of housing policy, housing allowances have gained in relative significance in Europe and North America in recent decades as supply-side subsidies have declined. Their use reflects a perception that housing problems are principally a question of affordability and that by giving low-income households more buying power, they are better able to meet housing costs and better able to compete in housing markets. Politicians who favor such allowances often support a market-oriented approach to housing moderated only by this form of assistance. They advocate allowances on the grounds of targeting assistance to where it is most needed and giving households choice over how they allocate resources. On distributional grounds, allowances are thus argued to be superior to production subsidies.

Housing allowances are intended to increase the quality of housing consumed by low-income households, and they are sometimes expected, also, to reduce the proportion of household income devoted to housing costs.

Critics of this approach to housing subsidies claim that many households, although eligible, fail to claim allowances out of ignorance or administrative barriers and thus the "take-up" of allowances is often low. It is also claimed that because allowances are related to income levels, as incomes increase, the level of allowance falls, and this can be a disincentive to work. In practice, however, many recipients are not participants in the labor market.

The demand-increasing effects of allowances may, if housing supply is restricted, result mainly in higher rents, and thus the benefit of the subsidy is passed on to housing suppliers. The more radical detractors of this demand-side approach suggest that it will not work, because it fails to reduce fundamental shortages in supply.

Preferential tax treatment of homeowners is another form of demand-side subsidy that is common throughout the world. This does not involve any direct payments by government; rather, it involves homeowners paying less tax. This type of subsidy is often termed *tax expenditures.* The principal type of tax expenditure involves interest payments on loans for house purchase. Typically, these payments are deductible from the taxpayer's gross income, thus reducing the volume of income on which tax is charged and reducing the tax bill. In some countries—for example, the United Kingdom and France—there are rigid limits on this tax deductibility. In others—for example, the Netherlands and Denmark—the system is more open-ended and the cost of the concession is very large.

In the United States, as in many other countries, housing-related tax expenditures involve substantial losses to the Treasury that make them easily the most costly form of housing subsidy. The deductions from federal income tax returns include not only mortgage interest payments but also local real estate taxes. There are, furthermore, capital gains tax concessions for homeowners. They do not, in addition, have to pay tax on the imputed rental income from the equity in their home, as some economists argue they should.

These indirect demand-side subsidies to homeowners have been the subject of much criticism from economists in the many countries in which they apply. They are said to be inequitable in that they give the biggest benefits to those with high incomes and expensive houses and inefficient in that there are potentially much cheaper ways to promote homeownership. Housing tax expenditures may also cause real estate values to be higher than would otherwise be the case and thus, arguably, distort choice and resource allocation in favor of homeownership compared with other investments and other forms of housing. This form of housing subsidy has typically arisen as a result of fiscal policy rather than of housing policy.

Demand-side housing subsidies in the form of housing allowances and tax expenditures may be appropriate if lack of demand is the problem and increasing demand brings about significant benefits without creating inefficiencies and injustices. Their critics argue that they do both of these. (SEE ALSO: **Subsidy Approaches and Programs;** *Supply-Side Subsidies*)

—*Michael Oxley*

Further Reading

Dolbeare, Cushing. 1986. "How the Income Tax System Subsidizes Housing for the Affluent." Pp. 264-71 in *Critical Perspectives on Housing,* edited by Rachel G. Bratt, Chester Hartman, and Ann Meyerson. Philadelphia: Temple University Press.

Grigsby, William. 1990. "Housing Finance and Subsidies in the United States." Pp. 25-45 in *Affordable Housing in Britain and America,* edited by Duncan Maclennan and Ruth Williams. York, England: Joseph Rowntree Foundation.

Hallett, Graham, ed. 1993. *The New Housing Shortage: Housing Affordability in Europe and the USA.* London and New York: Routledge.

Hartman, Chester. 1986. "Housing Policies Under the Reagan Administration." Pp. 362-76 in *Critical Perspectives on Housing,* edited by Rachel G. Bratt, Chester Hartman, and Ann Meyerson. Philadelphia: Temple University Press.

Howenstine, E. Jay. 1986. *Housing Vouchers: A Comparative International Analysis.* New Brunswick: Rutgers, State University of New Jersey, Center for Urban Policy Research.

Oxley, Michael. 1987. "The Aims and Effects of Housing Allowances in Western Europe." Pp. 165-78 in *Housing Markets and Policies under Fiscal Austerity,* edited by Willem van Vliet--. Westport, CT: Greenwood.

Oxley, Michael and Jacqueline Smith. 1996. *Housing Policy and Rented Housing in Europe.* London: E. and F. N. Spon.

Wood, Gavin. 1990. "The Tax Treatment of Housing: Economic Issues and Reform Measures." Pp. 43-75 in *Housing Subsidies and the Market: An International Perspective,* edited by Duncan Maclennan and Ruth Williams. York, England: Joseph Rowntree Foundation. ◄

▶ Dementia

Alzheimer's disease and related dementias are becoming increasingly common, affecting more than 50% of people over the age of 80 in the United States. Given the lack of medical treatment or cure, attention is increasingly focused on care and management strategies, including the potentially supportive role of the physical environment. Most people with dementia eventually live in a supportive residential setting such as a group home, assisted-living center, or nursing home. Design of appropriate environments must be based on an understanding of the unique needs of this population. The following will briefly examine these needs, explore their environmental implications, highlight the application of these design principles in long-term care settings, and review the state of research on settings for people with dementia.

Understanding the Population

Dementia is characterized by profound memory loss, especially for recent events, disorientation, behavior changes, loss of language functions, inability to think abstractly, inability to care for oneself, emotional instability, and loss of sense of time or place. Alzheimer's disease, the most well-known form of dementia, is progressive and irreversible, accounting for as much as 66% of all cases of dementia, but many other conditions can also cause dementia. Understanding the cognitive and behavioral manifestations of dementia is a critical precursor to designing supportive residential settings.

At first, the person with dementia may have only mild symptoms, such as forgetting where familiar objects (e.g., keys) have been placed or forgetting to do things. Lists and reminders are often effective at minimizing the consequences of these problems. This is typically followed by problems experienced when traveling to unfamiliar locations (inability to figure out how to get from here to there) and difficulties completing complex tasks such as planning a meeting or party or balancing a checkbook. The person may not be able to remember names of people recently introduced or the content of material just read. Soon, the individual cannot recall the names of family members, frequently misplaces objects, and cannot handle money in simple transactions. Eventually, the person becomes incapable of finding his or her way around the house (cannot find the bathroom or bedroom), cannot get dressed or prepare a meal, has problems with judgment, often has some language impairment, and has no self-direction, requiring someone else to set up every activity and provide step-by-step directions. As the disease progresses, the person with dementia loses ability to speak coherently or understand and respond to simple directions or questions; requires complete assistance in bathing, grooming, and eating; and may lose the ability to walk. Essentially, the brain is no longer capable of telling the body what to do.

Environmental Implications

Recognizing the special needs of this cognitively impaired population, several environment-gerontology researchers have developed an array of therapeutic goals for long-term care environments for people with dementia. The goals vary in terms of specificity of the problem being addressed: Some target specific behavioral manifestations of dementia (such as disorientation to place); others are directed toward more generalized needs of individuals living in long term care settings (e.g., general comfort). The following therapeutic goals provide a framework for understanding how specific facilities have addressed the design of settings for people with dementia.

Awareness and Orientation: People with dementia become increasingly less capable of maintaining awareness and orientation within a space, even one that is familiar, such as a home of many years. Thus, the move to a new living arrangement can make it even more difficult to understand or remember where certain rooms are located. This goal suggests that physical environments should minimize elements of the environment that can be confusing to the person with dementia (e.g., long hallways with many identical doorways). Orientation can be supported through appropriate use of signage, increased visibility of important locations (such as bedrooms and bathrooms), and other strategies to maximize awareness and understanding of one's presence and location within space.

Application of Design Principle: The Corinne Dolan Alzheimer Center at Heather Hill in Chardon, Ohio, created a display case (5 in. × 16 in. × 48 in.) secured by a lockable glass door located in the wall adjacent to each bedroom entrance, where personal mementos act as orientation cues. This facility also has an open design, like the Weiss Pavilion at the Philadelphia Geriatric Center, so that the country kitchen and dining room are easily visible from every bedroom entrance, eliminating the need to traverse hallways looking for public areas.

Environmental Stimulation and Challenge: Recognizing that people with dementia have decreased ability to interpret information or environmental stimulation has led to the suggestion that all sources of stimulation should be eliminated to the greatest extent possible. Alternatively, it is argued that institutional environments, which are often impoverished and sensorially deprived, can cause a variety of negative behavioral and cognitive consequences. Thus, architectural design should provide stimulation that is appropriate to the needs of the residents. This means regulating or eliminating noxious and distracting stimuli (noisy alarms and call bells, noise from cart traffic, glare from highly polished floors or from glass on pictures, etc.) and providing appropriate stimulation that is meaningful to the residents (tactile wall hangings that encourage exploration, acoustic treatment to absorb noise, personalized orientation cues at bedroom entrances, etc.) Stimulation may be visual, auditory, tactile, olfactory, mental, or physical.

Application of Design Principle: Woodside Place, in Oakmont, Pennsylvania, is a 36-bed facility, with three "household" units of 12 beds each. The small size of the unit limits the number of people involved in group activities to a number that is manageable by most residents. Each house-

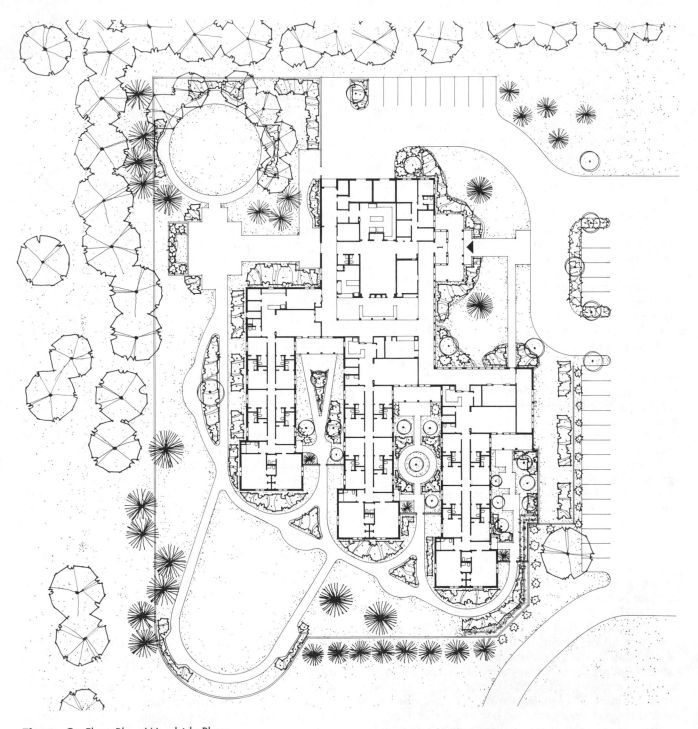

Figure 9. Floor Plan, Woodside Place

hold has its own kitchen, dining room, and living room, encouraging participation in familiar and meaningful activities, such as cooking. Each household includes several thematic quilt wall hangings, providing tactile and visually meaningful stimulation (see Figure 9).

Support Functional Abilities: The progressive nature of dementia means that residents are increasingly less capable of

independently completing activities of daily living (i.e., dressing, bathing, grooming, eating, walking). Loss of these skills, once taken for granted, can be a continually traumatic and negative experience. The environment, therefore, should compensate for these losses and support remaining skills as much as possible through the provision of appropriate prosthetics, such as grab-bars and railings. Tubs that are easy to get in and out of, closets that provide easy accessibility to a moderate (i.e., not overwhelming) selection of clothing, and places designed for grooming can serve to maintain the residents at their highest functional levels.

Application of Design Principle: Incontinence is a common problem, which often stems, in the midstages of the disease, from inability to locate the bathroom in time to use it. The Corinne Dolan Alzheimer Center designed the bathrooms within each bedroom to be highly visible, making it easier for residents to find and use the bathroom independently.

Safety and Security: Wandering is one of the hallmark symptoms of dementia. Combined with problems in spatial orientation, individuals who walk away can easily become lost and are at risk for injury or exposure to the elements. In addition, dementia impairs a person's ability to judge the safety of one's actions. Thus, the environment needs to provide special protection for the residents, minimizing access to unsafe areas without being overly restrictive (e.g., locking every door) (see Figure 9).

Application of Design Principle: Several facilities, including the Alzheimer's Care Center of Kennebunk Valley (Gardner, Maine), the Corinne Dolan Alzheimer Center, and Woodside Place, limit residents' egress through the front entrance but provide unrestricted access (except in inclement weather) to a spacious but secured courtyard in back.

Positive Social Milieu: Verbal communication becomes increasingly difficult, particularly in large-group situations where several things may be going on at once. The environment should be designed to encourage interaction in and between small groups of two to six people.

Application of Design Principle: Several facilities (Woodside Place, Corinne Dolan Center, Washington Home in D.C., and Alexian Village in Milwaukee) provide a variety of different places to interact with others: kitchen/nourishment centers, dining rooms, living rooms, activity rooms, music rooms, visiting nooks, and patios or courtyards. The variety of spaces increases opportunities for social interaction of different sized groups.

Privacy: The need to be alone is a fundamental human need that does not disappear with age or dementia. Most long-term care settings provide little or no opportunity for residents to be private, either in common rooms (private visiting rooms) or in the bedroom. Privacy is not defined solely in visual terms, so a "privacy curtain" between beds in a shared bedroom does not constitute privacy. The environment should provide opportunities for residents to be alone.

Application of Design Principle: Although private bedrooms can provide opportunities for privacy, Alexian Village has semiprivate bedrooms that are designed in an L shape, with beds placed on each arm of the L and the entrance and bathroom at the joint of the two arms. Other facilities (Corinne Dolan Center, New Perspective Group Home Number 4, Mequon, Wisconsin, and the Washington Home)

include a variety of spaces that provide several opportunities to be alone or with one other person.

Healthy and Familiar Environment: The move to a long-term care setting means a loss of much that is familiar—both in terms of home and possessions and in terms of the general daily routine. The physical environment should therefore help residents maintain ties with what is familiar, typically, a home. Arrangement of spaces within the unit should provide for transitions between more private areas (bedrooms and bathrooms) and public areas (dining and activity rooms). Bedrooms should be furnished with residents' possessions, such as furniture or memorabilia. Encouraging personalization through the provision of shelves and display areas can also help to enhance sense of self and identity. Furthermore, dining rooms and meal service should also be designed to re-create familiar experiences (i.e., not tray service).

Application of Design Principle: Several facilities (Woodside Place, Corinne Dolan Alzheimer Center, Franciscan Woods in Milwaukee) have been able to eliminate the traditional large, centrally located, institutional nurses station. It is typically replaced with staff working space located in the kitchen or at a desk in a small lounge associated with a cluster of bedrooms. Woodside Place encourages personalization of each bedroom with a plate rail 18 inches from the ceiling. Wealshire, in Lincolnshire, Illinois, reflects a familiar and homelike setting by creating households based on traditional residential room arrangements. The symbolic main entrance to the unit is from the courtyard through the front door, which is located adjacent to the living and dining rooms. The back door enters into the kitchen area, just as it typically does in a house. A transitional hallway separates these public areas of the household from the more private bedroom and bathing areas (see Figure 10).

Control: Moving to a long-term care setting involves loss of control over many aspects of one's life. The cognitive deterioration caused by dementia also results in loss of control. Control over one's life (or parts of one's life) is often considered to be of paramount importance: Therefore, the environment should support opportunities for individual control. For example, access to a kitchen allows residents to have a snack when they feel hungry, instead of being given a snack at a scheduled time.

Application of Design Principle: Unrestricted access to secure outdoor courtyards (or access that is restricted only at night and during inclement weather) provides residents a measure of control over their lives. Individualized heating and air-conditioning units for each room, easily operable windows, and color-contrasted light switches (Corinne Dolan Alzheimer Center) also support individual control over personal living spaces.

Research on Environments for People with Dementia
Research on environments for people with dementia has been slow to emerge but is increasing. The 1992 Office of Technology Assessment report on special care units for

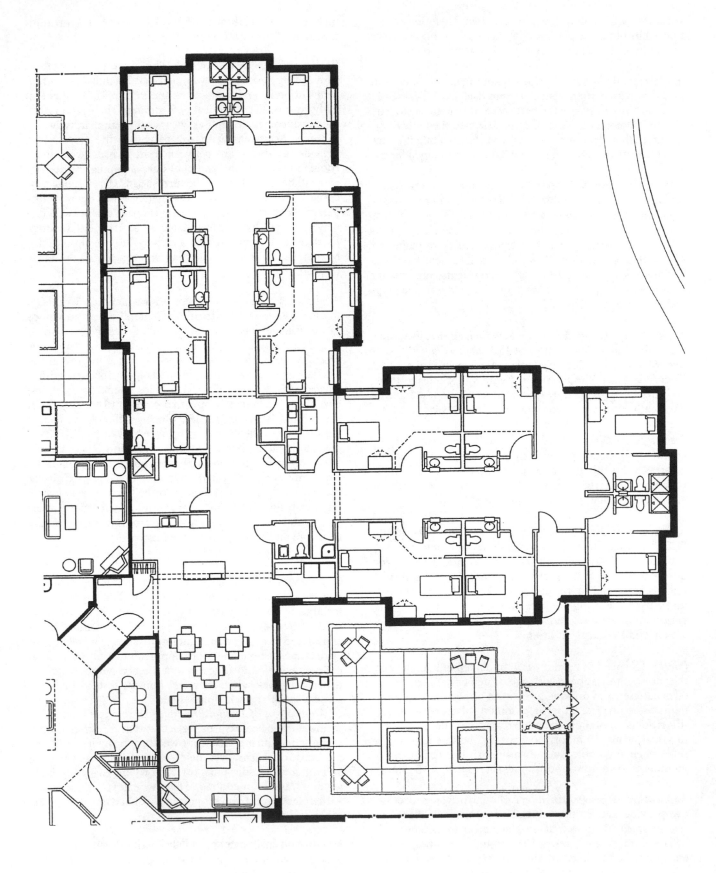

Figure 10. Floor Plan, Wealshire

people with dementia documents 15 evaluative studies and 16 descriptive studies of special care units for people with dementia. A few facilities are actively engaged in environmental research to test the efficacy of specific environmental features, such as different orientation cues (Corinne Dolan Center, Alzheimer's Care Center in Kennebunk Valley, and Woodside Place). Other research has used a more global conceptualization of the environment, treating it in an undifferentiated manner (Weiss Institute, Wesley Hall in Chelsea, Michigan, and Grace Presbyterian Village, Dallas). Results of this research generally find positive benefits of a therapeutic environment, although it is often difficult to determine the exact benefits of specific environmental characteristics. Recognizing the need for more research, the National Institute on Aging has recently funded 10 separate studies to evaluate the effectiveness of special care units for people with dementia. Preliminary results suggest that although SCUs do not necessarily have more positive clinical or biomedical outcomes, they are often associated with higher staff and family sataisfaction scores. (SEE ALSO: **Elderly;** *History of Housing*)

—*Margaret P. Calkins*

Further Reading

Calkins, M. P. 1988. *Design for Dementia: Planning Environments for the Elderly and the Confused.* Owings Mills, MD: National Health Publishing.

Cohen, U. and J. Weisman. 1991. *Holding on to Home.* Baltimore, MD: Johns Hopkins University Press.

Lawton, M. P. 1981. "Sensory Deprivation and the Effect of the Environment on Management of the Patient with Senile Dementia." In *Clinical Aspects of Alzheimer's Disease and Senile Dementia,* edited by N. Miller and G. Cohen. New York: Raven.

Sloane, P. D. & L. J. Mathew. *Units in Long-Term Care.* Baltimore, MD: Johns Hopkins University Press.

Teresi, J., M. P. Lawton, M. Ory, and D. Holmes. 1997. *Measurement in Elderly Chronic Care Populations.* New York: Springer.

U.S. Congress Office of Technology Assessment. 1992. *Special Care Units for People with Alzheimer's and Other Dementias: Consumer Education, Research, Regulatory and Reimbursement Issues.* Vol. No. OTA-H-543. Washington DC: Government Printing Office.

Weisman, G., M. Calkins, and P. Sloane. 1994. "The Environmental Context of Special Care." *Alzheimer's Disease International* (Special Issue, Spring):S308-20.

Zeisel, J., J. Hyde, and S. Levkof. 1994. "Best Practices: An Environment Behavior (E-B) Model for Alzheimer Special Care Units." *American Journal of Alzheimer Care and Related Disorders and Research* (March/April):4-21. ◀

▶ Demonstration Cities and Metropolitan Development Act of 1966

On January 26, 1966, in a message to Congress, President Lyndon Johnson proclaimed that "Nineteen-sixty-six can be a year of rebirth for American cities." Congress would later propose broadscale programs to renew cities. The Demonstration Cities and Metropolitan Development Act of 1966 (PL 89-754) represented the primary vehicle for these programs and served as the capstone for President Johnson's "Great Society." It was a bold, experimental,

and innovative program. President Johnson, on signing the act on November 3, 1966, commented that it was "a measure for all cities and that the success of the Act depended upon the partnership of the States and local governments, the private sector as well as the participating communities." Johnson also referred to the act as "Model Cities"—a term that was later adopted by U.S. Department of Housing and Urban Development officials and others.

Looking at the "total environment" of neighborhoods, Title I, Section 1, of the act stated that a major purpose of the program was to provide "additional financial and technical assistance to enable cities of all sizes (with equal regard to the problems of small as well as large cities) to plan, develop, and carry out locally prepared and scheduled comprehensive city demonstration programs containing new and imaginative proposals to rebuild or revitalize large slum and blighted areas." The financial and technical assistance was to be used

> to enable cities of all sizes (with equal regard to the problems of small as well as large cities) to plan, develop, and carry out locally prepared and scheduled comprehensive city demonstration programs containing new and imaginative proposal to rebuild or revitalize large slum and blighted areas; to expand housing, job, and income opportunities; to reduce dependence on welfare payments; to improve educational facilities and programs; to combat disease and ill health; to reduce the incidence of crime and delinquency; to enhance recreational and cultural opportunities; to establish better access between homes and jobs; and generally to improve living conditions for the people who live in such areas, and to accomplish these objectives through the most effective and economical concentration and coordination of Federal, State, and local public and private efforts to improve the quality of urban life.

The act focused on entire neighborhoods, not simply parts of neighborhoods. It was hoped that the selected communities would be showcased to illustrate to other cities what could be done to rebuild communities by improving conditions in substandard neighborhoods.

Cities competed with other cities for federal funding under the Model Cities program. Criteria were used to review the various applications. For example, applications were reviewed to determine if the local plan provided for relocating individuals and businesses displaced by urban renewal activities. Second, the degree to which the proposed project contributed to the development of the entire city was determined. Third, the likelihood that the proposed project could be started quickly was also assessed.

The Demonstration Cities and Metropolitan Development Act of 1966 authorized or expanded a number of individual programs. It authorized an insurance program for single-family homes and low- and moderate-income families and for people displaced by the renewal activities—Section 221(h). Additional titles in the act authorized (a) loans for newly constructed homes; (b) rural housing from the Farmers Home Administration; (c) mortgage insurance for group practice facilities of medicine, dentistry, or optometry; (d) funding encouraging comprehensive and areawide planning; (e) localities to credit expenses (either $3.5 million or 25%, whichever is less) for urban renewal costs associated with facilities used for civic or cultural functions or certain specified medical facilities at public universities; (f) funding for the establishment of urban information and technical assistance centers; and (g) recognition that planning for historic and architectural preservation activity be considered legitimate urban renewal costs.

The establishment of new towns or communities within selected cities was also authorized. These new towns were to be better designed (economically, socially, and environmentally) than the existing cities. To assist in the establishment of these new towns, developers were eligible to receive up to $25 million in Federal Housing Administration mortgage insurance to help finance land acquisition and improvements. (SEE ALSO: **Urban Redevelopment**)

—*Roger W. Caves*

Further Reading

Frieden, Bernard J. and Marshall Kaplan. 1977. *The Politics of Neglect: Urban Aid from Model Cities to Revenue Sharing.* Cambridge: MIT Press.

Gelfand, Mark I. 1975. *A Nation of Cities: The Federal Government and Urban America, 1933-1965.* New York: Oxford University Press.

Gorham, William and Nathan Glazer, eds. 1976. *The Urban Predicament.* Washington, DC: Urban Institute Press.

Lord, Tom Forrester. 1977. *Decent Housing: A Promise to Keep.* Cambridge, MA: Schenkman.

McFarland, M. Carter. 1978. *Federal Government and Urban Problems: HUD: Successes, Failures, and the Fate of Our Cities.* Boulder, CO: Westview. ◀

▶ Density Bonus

Efforts to expand the supply of affordable housing often pit public officials against the economic realities of high land and building costs. Consequently, many low- and moderate-income families find it very difficult or impossible to purchase or rent housing in some communities. Local zoning and planning authorities can contribute to the problem by imposing substantial minimum lot size requirements on developers. Providing a density bonus to developers forms a means to increase the supply of affordable units without significant public outlays.

A density bonus enables a builder to develop a parcel of land more intensively than existing zoning rules would allow. In principle, the increase in the number of units will enhance the project's profitability and enable a builder to offer some units at a below-market price. Many of the municipalities in New Jersey, which has been at the center of affordable housing litigation and legislation, have relied on density bonuses to meet state-mandated low- and moderate-income housing requirements.

Typically, developers receive a density bonus in either of the following two types of situations. One possibility is to provide a density bonus when requiring developers of new

housing to set aside some units at a price sufficiently low to make the housing affordable for persons with limited incomes. A second approach is to use a density bonus to establish incentives for builders to create inclusionary developments containing both market- and below-market-priced units. A common pattern in New Jersey has been for a municipality to change zoning from one unit per acre to six units per acre and require the developer to provide one affordable unit for every four market rate units.

The success of a density bonus depends on numerous factors. The overall demand for housing must be sufficiently strong to allow builders to set a price high enough to make a satisfactory return on their investment while at the same time providing enough additional revenue to subsidize units sold at prices below costs. The size of a density bonus and the extent of the affordable housing exaction also influence a developer's willingness to participate in this approach to expanding the supply of affordable housing.

During the 1980s, density bonuses proved to be a useful mechanism to encourage the construction and sale of affordable housing. At the same time, more intense land development carries with it other consequences for municipalities. Greater demands on public services and less open space are two important correlates of an affordable housing policy based on the use of density bonuses. Furthermore, a density bonus may result in increases in the supply of housing in one community that will indirectly affect the prices of other homes in the community and surrounding towns. At this time, there is a need for additional data on the possible negative consequences associated with the use of a density bonus. This knowledge is needed to inform public policy decisions regarding the appropriate use of a density bonus in the overall effort to increase the supply of affordable housing. (SEE ALSO: *Impact Fees; Linkage*)

—*Jeffrey Rubin and Joseph J. Seneca*

Further Reading

Franzese, Paula A. 1988. "Mount Laurel III: The New Jersey Supreme Court's Judicious Retreat." *Seton Hall Law Review* 18:30-54.

Lamar, Martha, Alan Mallach, and John M. Payne. 1989. "*Mount Laurel* at Work: Affordable Housing in New Jersey, 1983-1988." *Rutgers Law Review* 41:1197-1277.

Mallach, Alan. 1984. *Inclusionary Housing Program: Policies and Practices.* New Brunswick: Rutgers, State University of New Jersey, Center for Urban Policy Research.

Rubin, Jeffrey I. and Joseph J. Seneca. 1991. "Density Bonuses, Exactions, and the Supply of Affordable Housing." *Journal of Urban Economics* 30:208-23.

Rubin, Jeffrey I., Joseph J. Seneca, and Janet G. Stotsky. 1990. "Affordable Housing and Municipal Choice." *Land Economics* 66:325-40. ◀

▶ Department of Housing and Urban Development Act of 1965

On several occasions in 1961, President John F. Kennedy called for the creation of a cabinet-level department, the Department of Urban Affairs and Housing, which would assume responsibilities for the increasing number of federal activities in housing and in rebuilding the nation's urban areas. It encountered a great deal of opposition from the public and private sectors. Some individuals and groups argued that the proposed department would create more federal bureaucracy, create confusion among existing agencies, pirate away functions from existing agencies, and expand the federal role in an area that should be the province of the states and local governments. The proposed department was rejected during the Kennedy administration.

President Lyndon Johnson continued the fight for the creation of the cabinet-level department that would provide vision, commitment, and leadership in minimizing city problems. Johnson reworked and revised Kennedy's ideas and renamed the proposed department the Department of Housing and Urban Development. In his January 4, 1965, State of the Union Message, Johnson called for its creation to spearhead efforts to improve U.S. urban areas.

During a March 2, 1965, message on cities, President Johnson acknowledged that the proposed department would "be primarily responsible for federal participation in metropolitan area thinking and planning" and would "provide a focal point for thought and innovation and imagination about the problems of our cities."

The U.S. Department of Housing and Urban Development was created when President Johnson signed the Department of Housing and Urban Development Act of 1965 (PL 89-174) on September 9, 1965. Section 2 of the act indicates that Congress "declares that the general welfare and security of the Nation and the health and living standards of our people require, as a matter of national purpose, sound development of the Nation's communities and metropolitan areas in which the vast majority of its people live and work." Moreover, a new executive department is needed because it is

> desirable to achieve the best administration of the principal programs of the Federal Government which provide assistance for housing and for the development of the Nation's communities; to assist the President in achieving maximum coordination of the various Federal activities which have a major effect upon community, suburban, or metropolitan development; to encourage the solution of problems of housing, urban development, and mass transportation through State, county, town, village, or other local and private action, including promotion of interstate, regional, and metropolitan cooperation; to encourage the maximum contributions that may be made by vigorous private homebuilding and mortgage-lending industries to housing, urban development, and the national economy; and to provide for full and appropriate consideration, at the national level, of the needs and interests of the Nation's communities and of the people who live and work in them.

The new department would take over the powers and duties of the Housing and Home Finance Agency, Federal Housing Administration, Public Housing Administration, the Federal National Mortgage Association (now known as Fannie Mae), and would be directed by a secretary appointed by the president and confirmed by the Senate. The secretary would advise the president on housing and urban development matters, carry out federal programs affecting cities, provide technical assistance to the lower levels of government on housing and urban development matters, and consult and cooperate with other federal agencies on housing and community development matters. Robert C. Weaver, administrator of the Housing and Home Finance Agency, became the first secretary of the U.S. Department of Housing and Urban Development in 1966. (SEE ALSO: *U.S. Department of Housing and Urban Development*)

—*Roger W. Caves*

Further Reading

Cleaveland, Frederick N. 1969. *Congress and Urban Problems*. Washington, DC: Brookings Institution.

Gelfand, Mark I. 1975. *A Nation of Cities: The Federal Government and Urban America 1933-1945*. New York: Oxford University Press.

McFarland, M. Carter. 1978. *Federal Government and Urban Problems: HUD: Successes, Failures, and the Fate of Our Cities*. Boulder, CO: Westview.

Willmann, John B. 1967. *The Department of Housing and Urban Development*. New York: Praeger. ◄

► Department of Veterans Affairs

The Veterans Administration (VA) was created in 1930 by President Herbert Hoover to consolidate three federal agencies that provided benefits to veterans. In 1989, Congress passed legislation elevating the agency to cabinet status and renaming it the Department of Veterans Affairs. Initially, the VA provided medical services and pensions to veterans. In 1944, passage of the Servicemen's Readjustment Act added home loan and education benefits to its array of services.

The VA is composed of three organizations that administer veterans programs: the Veterans Benefits Administration, the Veterans Health Administration, and the National Cemetery System. The first organization is responsible for the home loan guaranty program, which includes appraising properties, supervising construction of new properties, establishing the eligibility of veterans for the program, passing on their ability to repay a loan and the credit risk, servicing and liquidating defaulted loans, and disposing of property acquired as a consequence of defaulted loans. Often, lenders will make VA-guaranteed loans without down payments and with interest rates lower than the market rate.

The VA's greatest impact on U.S. housing occurred in the years following World War II. The federal government had become increasingly active in the housing market since the Great Depression, through Federal Housing Administration (FHA) mortgage insurance programs, highway construction, public housing, and urban renewal. Congress intended the Servicemen's Readjustment Act of 1944, also called the GI Bill of Rights, to express gratitude to World War II soldiers and to facilitate their transition to civilian life. The act included a loan guaranty program that was a direct government subsidy, unlike the existing FHA mortgage insurance, which was financed by an insurance premium. Although other subsidy programs, such as public housing, drew opposition in Congress, support for veterans' benefits was nearly unanimous. In 1945, the program became permanent, rather than one oriented to temporary "readjustment" needs.

From 1945 through the 1950s, the Cold War and the Korean conflict produced a continuous stream of veterans who wanted to take advantage of this and other VA programs. At times, Congress gave veterans priority status to take advantage of other programs; the Veterans' Emergency Housing Act of 1946, for example, stated that housing built with federal assistance had to be offered for sale exclusively to veterans for 60 days. From about 1946 to 1956, VA financing contributed significantly to housing industry activities. VA-guaranteed loans represented between 12% and 28% of all loans made during these years, with a high of 28% in 1947 and another peak of 25% in 1955. VA funds guaranteed loans for 25% of housing starts in 1947, and 30% in 1955. In 1995, by contrast, VA-guaranteed loans consisted of about 5% of mortgage originations.

The decline in the share of VA loans relates to the changing economy, the changing veteran population, and changes in lenders' perceptions of the VA loan program. A 1991 evaluation notes that one of the program's initial goals was to stimulate spending of savings accrued during World War II. Poor economic conditions in the early 1980s resulted in high VA loan foreclosure rates, which contributed to lenders' declining willingness to participate in the VA loan program. In addition, lenders disliked the program's increasing complexity and distrusted VA appraisers. Finally, the program's primary benefit to veterans, relative to conventional and FHA loans, is that it reduces or eliminates the down payment. This is most advantageous to first-time homebuyers; as the average age of veterans increases and as the number of veterans declines, fewer veterans are first-time buyers.

During the post-World War II era, the FHA and VA worked closely with large development companies, who used their loan programs to attract homebuyers and facilitate the sales process. Levitt and Sons, for example, could obtain FHA commitments to finance thousands of homes before clearing any land. Salespeople on-site in a model home could close 350 sales per day, arranging for hundreds of FHA and VA mortgages at once. In the late 1940s, veterans could buy in Levittown with no down payment and installments of $56 a month. Indeed, by 1950, 60% of VA loans were for newly built homes, primarily in the growing suburbs. VA financing programs, along with other government housing programs during this period, contributed to the perpetuation of residential segregation by adhering to contemporary real estate industry practices that devalued property or refused to insure mortgages in minority or integrated neighborhoods. Until 1948, racial covenants were common elements of deeds, meant to restrict minority access to some neighborhoods.

In 1993, the population of veterans was about 27 million. Various active duty requirements determine eligibility for VA-guaranteed loans; about 1% of veterans hold VA

loans. Loans may be used for home purchase, homebuilding, home improvements, and refinancing. The Department of Veterans Affairs is located at 810 Vermont Avenue, NW, Washington, D.C. 20420; (202) 273-5700. The toll-free number (800) 827-1000 connects to the nearest regional VA office. (SEE ALSO: **Federal Government**; *Levittown*)

—*Mara Sidney*

Further Reading

Congress and the Nation. 1965-1993. Vols. I-VII. Washington, DC: Congressional Quarterly Service.

Veterans Housing Loan Program Evaluation Final Report. 1991. Washington: DC: Office of Program Coordination and Evaluation. ◄

► Depreciation Allowance for Landlords

The depreciation allowance for landlords is a special type of income tax deduction associated with investment property. When combined with normal operating expenses and mortgage interest, depreciation can be subtracted from a property's gross income to arrive at its taxable income. However, in practice, depreciation allowance for landlords is a significant component of real estate strategy that has been used to generate a loss for tax purposes even though the property may be appreciating in value while generating a positive cash flow before taxes.

There are two conditions under which the depreciation allowance for landlords works to generate a loss for tax purposes. First, although real property suffers physical deterioration over time, if it is well located and well maintained, its useful economic life will be far longer than that suggested by the depreciation guidelines of the Internal Revenue Service (IRS). In 1986, the IRS lengthened the depreciation periods for nonresidential properties from 18 to 30.5 years and for residential properties from 18 to 27.5 years. Clearly, many properties will continue in profitable use far beyond these parameters. By lengthening the depreciation period for these properties, the value of their depreciation allowance decreased.

Second, depreciation deductions are hardly ever used to create a cash reserve to purchase replacement property and/or equipment at the end of the depreciation period. When depreciation deductions are taken that are greater than the actual economic erosion of the property's value, a tax shelter results. If the property is not declining in value and if no fund is created to purchase a replacement, a tax deduction is realized because no cash has been expended for replacements and the property has not declined in value.

Such a real estate investment strategy, of which the depreciation allowance for landlords is a prime component, would have great appeal to those who do not wish to own property only for a flow of income and future capital gains but who wish to reduce their taxable income by generating a loss for tax purposes, an outcome greatly enhanced by the depreciation allowance for landlords.

Not surprisingly, the burgeoning federal budget deficit of the 1980s compelled Congress to scrutinize the impact of tax-sheltered investments such as real estate. The depreciation allowance was viewed as the most significant of these real estate tax shelters because, as seen above, it is a noncash deduction and because real estate is a very high value investment; it generates very large deductions for income tax purposes.

The 1986 Tax Reform Act was nothing less than a frontal assault on real estate tax shelters. The act created three types of income. *Portfolio income* refers to returns from investments made in financial instruments such as stocks, bonds, and certificates of deposit. *Active income* results from the conduct of a trade or business in which the taxpayer materially participates. The income from salaries, wages, small businesses, and the professions fits into this category.

It is the third category of income, passive income, that has had a major impact on real estate. Passive income refers to funds generated by the operation of a business in which the taxpayer does *not* "materially participate." A good illustration of such income is that created in the operation of a limited partnership, a popular form of real estate ownership for a small investor but one certainly not limited to real estate. A limited partner receives a specified return on the proceeds of the investment activity; in real estate limited partnerships, those proceeds were passed on in the form of income and tax shelters. The limited partner plays no role in the day-to-day operations of the business, which is the function of the general partner(s). As a result, a limited partnership is automatically a passive activity.

However, if an individual actually owns an apartment house or an office building, one would expect this to be considered an investment that generates active income because owning real estate is a very "active" and "hands-on" business. Paradoxically, however, *all* real estate activities are treated as passive. Thus, the losses generated by real estate tax shelters are passive losses and are limited by 1986 Tax Reform to a maximum of $25,000 per year. However, even that sum is phased out at 50 cents per dollar for taxpayers with adjusted gross incomes of more than $100,000.

Consequently, individuals who invested in real estate prior to 1986 witnessed the nullification of their tax avoidance strategies, especially those with incomes of more than $100,000 or with passive losses exceeding $25,000, or both. The implications of this are twofold. First, passive income must be created so that passive losses can be offset. The most popular way this has been achieved is through partnerships that purchase properties solely for their ability to generate income. As a result, little or no debt is used to purchase them because interest expenses no longer have a tax shelter value. Second, properties that were brought into existence largely for their tax shelter characteristics, such the landlord's depreciation allowance, have seen a drastic decline in their economic value, and many have gone into mortgage default; the recession of the early 1990s exacerbated that situation further. Consequently, many savings and loan and banking institutions can in part attribute their difficulties to holding mortgages on such properties.

Tax shelters, such as the depreciation allowance for landlords, have been criticized not only for their impact on the federal deficit but because they have enjoyed a favored status compared with other investments that, except for municipal bonds, enjoy no tax-related advantages. Real estate will become an attractive investment once again when its return on dollars invested competes favorably with other investment alternatives. Without tax shelters such as the depreciation allowance for landlords, real estate will have to produce a significantly higher return than stocks and bonds, both of which can be easily sold at any time; in contrast, real estate is illiquid and requires months or even years to be marketed to a buyer.

Beyond the 1990s, investment real estate will become a very difficult sector of the economy in which to invest successfully unless acquisition cost reflects the elimination of tax shelters such as the depreciation allowance for landlords. (SEE ALSO: *Depreciation of Property; Tax Incentives*)

—*Daniel J. Garr*

Further Reading

Real Estate Desk Book. 1978. 5th ed. Englewood Cliffs, NJ: Institute for Business Planning (see pp. 181-209).

Wurtzebach, Charles H. and Mike E. Miles. 1995. *Modern Real Estate.* 5th ed. New York: John Wiley (see pp. 203-211). ◄

► Depreciation of Property

Depreciation is a special type of income tax deduction associated with investment property. Along with operating expenses and mortgage interest, depreciation can be subtracted from a property's gross income to arrive at its taxable income. In the United States, income taxes have always been levied on net income, not on gross revenue. This implies that a set of rules exists to determine the exact amount of taxable income. These rules, contained in the Internal Revenue Code, have always recognized that a portion of gross investment income must be used to replace assets that gradually lose their value over time. Such losses occur as a result of use, wear, and tear or the actions of natural elements. All physical assets–the structure itself, machinery, appliances, and the like—are considered depreciable assets. However, land is deemed to have a permanent life of its own and therefore is not depreciable. Although land that produces minerals and ores may be depreciable, this is a special and limited case with unique applications and special legislation and it will therefore be excluded from this discussion.

A property must meet certain criteria if depreciation for tax purposes may be applied:

1. The property must have a useful life span of a finite nature.
2. That useful life span must be determinable.
3. The property must be used in the course of earning a taxpayer's income (e.g., capital equipment or property used in the generation of income).
4. Depreciable property may not be an item of business inventory. A widget dealer cannot depreciate widgets, nor can a homebuilder depreciate an inventory of unsold units in a subdivision.

5. The property must not be the taxpayer's residence because it is not considered to be income producing.

Because land is a nondepreciable asset, total purchase price must reflect an allocation between land and improvements because the latter must be assigned a separate value. If an appraisal does not assign a percentage of total purchase price to improvements, the IRS will accept a prior year's property tax determination. Although that particular value may be obsolete and thus not representative of actual purchase price, the percentage of total value allocated to improvements may be used. The higher the percentage of total value assigned to improvements, the greater the annual depreciation deduction and, as a result, the lower the tax obligation will be.

Government has modified depreciation schedules over the years. For example, if a piece of equipment has a useful life of 15 years and an original cost of $15,000, for tax purposes, it could be depreciated at $1,000 per year. But, when that equipment needs to be replaced, it may very well cost $30,000. Between 1981 and 1986, the IRS had been taking notice of the relatively high rate of inflation by shortening depreciation periods and accelerating recovery of capital with greater depreciation write-offs. This system allowed faster recovery of capital in anticipation of higher replacement costs. As a result, the term *depreciation* tends to be synonymous with *cost recovery*. (SEE ALSO: *Depreciation Allowance for Landlords; Limited-Dividend Development; Obsolescence*)

—*Daniel J. Garr*

Further Reading

Follain, James R. 1986. "The Impact of the President's Proposals and H.R. 3838 on the Housing Market." Pp. 61-85 in *Tax Reform and Real Estate,* edited by James R. Follain. Washington, DC: Urban Institute Press.

Hulten, C. R., ed. 1981. *Depreciation, Inflation and the Taxation of Income from Capital.* Washington, DC: Urban Institute Press. ◀

▶ Development Permit

Development permit is a term with two distinctly different uses. In one use, it refers to a streamlined approach to the administration of suburban development and urban redevelopment schemes. Traditionally, land use development has required public approval or amendments separately under each of several distinct ordinances: for example, zoning and land subdivision. Some municipalities have adopted a unified development ordinance that integrates such provisions into a single code and in which approval takes the form of a development permit. A development permit can be thought of as regulatory reform that enables one-stop application filing for developers.

In its other use, development permit refers to a system of discretionary planning wherein a land owner's ability to develop is limited not by the stated requirements of a zoning bylaw but by the requirement that permission of the planner be obtained. Proponents argue that development permits give planners the flexibility they need to negotiate with developers in face of the increasing complexity of land use planning. Critics argue that development permits put too much discretionary power in the hands of planners. However, most planning ordinances do afford planners some degree of discretion; the debate is essentially about how much discretionary power should be given and how it should be exercised. (SEE ALSO: *Zoning*)

—*John R. Miron*

Further Reading

Blankenship, Frank J. 1989. *The Prentice Hall Real Estate Investor's Encyclopedia.* Englewood Cliffs, NJ: Prentice Hall.

Chapin, F. S. and E. J. Kaiser. 1994. *Urban Land Use Planning.* 4th ed. Urbana: University of Illinois Press.

Dowall, David E. 1990. The public real estate development process. *Journal of the American Planning Association* 56:504-12.

Jacobus, Charles J. 1996. *Real Estate Principles.* Upper Saddle River, NJ: Prentice Hall.

McKenzie, Dennis J. 1996. *Essentials of Estate Economics.* Upper Saddle River, NJ: Prentice Hall.

Peiser, Richard B. 1992. *Professional Real Estate Development: The ULI Guide to the Business.* Washington, DC: Urban Land Institute.

Schultz, M. S. and V. L. Kasen. 1984. *Encyclopedia of Community Planning and Environmental Management.* New York: Facts on File.

Tosh, Dennis S. 1990. *Handbook of Real Estate Terms.* Englewood Cliffs, NJ: Prentice Hall. ◀

▶ Discrimination

Housing market discrimination is the wide variety of illegal acts undertaken by property owners, real estate and home insurance agents, and households who disfavor home seekers on the basis of one or more of their characteristics that are protected by law. Race, nationality, gender, religion, color, familial status, and handicap are classes currently protected by federal law. Housing market discrimination tries to make housing less desirable, affordable, and/or available to individual households who are victimized. At the societal level, housing market discrimination on the basis of race perpetuates segregation of neighborhoods. Discriminatory acts continue to be perpetrated frequently in U.S. housing markets—federal, state, and local laws to the contrary notwithstanding. This entry explains what forms housing discrimination takes, why it occurs, how often it occurs, what harms it has, and legal issues surrounding efforts to eliminate it.

The primary focus is on discriminatory acts, especially promulgated on the basis of race, that occur during housing search, sale, or rental transactions. Discrimination involving the marketing or insurance of homes will also be considered. The topic of discriminatory home mortgage lending is, however, beyond the scope of this entry and is covered elsewhere.

Racial discrimination in other aspects of the housing market beyond the transaction itself is less well identified. Several studies have found a distinctly lower rate of homes advertised for sale in newspapers or having open houses when those houses are in predominantly minority communities. Advertising has also been identified that, through selective use of complimentary adjectives, "code words," or human models, suggests the racial composition and desirability of the area in which the advertised property is located. Thus, advertising discrimination not only represents a form of inferior service to homeowners in minority-occupied areas but a form of racial steering as well.

There is mounting evidence that racial discrimination occurs in the area of home insurance. The Association of Community Organizations for Reform Now (ACORN) in 1992 conducted statistical analyses of home insurance patterns in five metropolitan areas, and paired auditors posing as insurance seekers made phone calls in 13 others. The results showed the following:

1. Predominantly minority-occupied city areas had proportionately fewer single-family homes covered by insurance, and the type of coverage was inferior but was supplied at a higher premium than in predominantly white-occupied suburban areas.
2. In predominantly minority-occupied areas of Chicago and Milwaukee, the rates of cancellation and nonrenewal were much higher than in white-occupied areas.
3. Agents erected an array of obstacles to residents seeking insurance for properties in predominantly minority-occupied areas, such as refusal to quote premiums, specification of low minimum policy amounts, and requiring property inspections.

There have been a few cases alleging insurance discrimination, a federal court held in 1992 that such acts were prohibited by the 1968 Fair Housing Act; this ruling subsequently was let stand by the Supreme Court in 1993.

Discriminatory acts related to the housing transaction can take many forms but can be categorized into six general types:

1. *Exclusion:* The discriminator tries to limit or distort information about housing vacancies or otherwise tries to discourage certain home seekers from moving in. This can involve, for example, failure to communicate with prospects or lying about dwelling availability or terms and conditions of the transaction.
2. *Steering:* The discriminator tries to convey a selected body of information about the geographic distribution of housing vacancies based on the race of the home seeker. A real estate agent, for example, may show minority home seekers vacancies only in predominantly minority-occupied neighborhoods.
3. *Harassment:* The discriminator tries to intimidate or physically or psychologically abuse particular households already living nearby in an attempt to convince them to move out. Neighbors might, for instance, spray paint racist graffiti or burn crosses on the property of a minority household who has moved into the area.

4. *Blockbusting:* The discriminator tries to encourage panic selling by homeowners by promulgating fears that many members of a minority group will be moving into the area. As illustration, a blockbuster may send out minority agents into an Anglo-occupied neighborhood to solicit homes for sale and spread rumors of impending "invasion" by minorities.
5. *Exploitation:* The discriminator tries to extract special financial terms or conditions from protected home seekers. Exploitation may take the form of quoting higher prices or rent and requiring higher security deposits.
6. *Service quality:* The discriminator fails to provide equally helpful, qualitative, and respectful services to all home seekers. As illustration, the victim of discrimination may be forced to wait longer for service, provided inferior quality of services, offered less complete information or assistance, and afforded fewer professional courtesies.

Typically, all these forms of discrimination, with the exception of harassment, are difficult for victims to detect because they have no standard against which to judge the agent's behavior. Many sorts of discriminatory acts can be camouflaged by an aura of courtesy and the appearance of honesty.

Why housing market discrimination occurs is the subject of much professional disagreement. Theories can be divided into six variants:

1. *Agent prejudice:* Brokers, agents, or landlords have a personal animus and treat members of a protected class poorly, even at economic cost to themselves.
2. *Customer prejudice:* Brokers steer home seekers to areas where their group predominates on the presumption that they would feel "more comfortable" there and the agent would be wasting time showing them elsewhere. Landlords exclude prospective tenants of a different group if their current tenants are thought to be prejudiced against that group.
3. *Potential customer prejudice:* Brokers or landlords exclude members of a group from a particular neighborhood or apartment complex if they fear that the introduction of said group would enrage current residents, who would retaliate by not listing homes with the offending broker or would move out of the complex en masse.
4. *Expected discrimination:* Brokers or landlords exclude applicants who are members of a protected class from a neighborhood or building or steer them away from particular lenders because they believe that these applicants will be discriminated against by other households or institutions.
5. *Inferior tenant:* Landlords exclude tenants from certain groups because of statistical discrimination; their experience indicates that group identity correlates with unstable rent payments, poor maintenance, severe damage, disruptive behaviors, or other undesirable attributes. Advertisers and home insurers may

similarly discriminate based on stereotypes derived from their previous experiences with properties in minority-occupied neighborhoods.

6. *Pure-profit maximizing:* If the protected group's elasticity of demand is less than that of another, agents attempt price discrimination (higher sales prices, rents, or insurance premiums). An alternative version of this motive is that by promulgating racial segregation with transitional buffer neighborhoods, brokers can intensify racially motivated residential turnover and thereby reap superior commission revenue. Another version is that certain forms of property advertising have differential payoffs in different neighborhoods.

Unfortunately, extant empirical work is ambiguous, yielding some support for all of the above motivations but no unanimous support for one. These findings could be interpreted at face value: Different discriminators in different contexts have different motivations. They may also indicate that most discriminatory acts are undertaken without clear motivations other than generalized habit; those engaging in such habitual responses may simply have few personal or contextual characteristics in common. What does seem clear, however, is that agents' stereotyped beliefs about the characteristics, beliefs, or preferences of others is the dominant cause of discrimination in housing transactions.

For decades, what we now call housing discrimination was not only legal but required in various ways. Sellers could affix restrictive covenants to the deeds of their properties that could preclude their subsequent sale to members of those groups so restricted; the enforceability of these covenants was overturned by the Supreme Court in the 1948 *Shelley v. Kramer* case (334 U.S. 1). For decades, real estate agents' code of professional ethics forbade them from introducing minority groups into neighborhoods, out of the belief that such integration would reduce property values; only in the 1960s were these codes revised. The Federal Housing Administration used similar logic from 1935 to 1950 to justify segregation as an eligibility criterion for receiving low-interest mortgage loans for new suburban housing developments. Thus, much of the contemporary momentum of housing discrimination can be traced to explicit private and public policies of the not-so-distant past.

The predominant technique for investigating how often racial discrimination occurs in the context of seeking a home or apartment is known as the "fair housing audit." Auditing is an investigative technique that consists of a two-person team of different races or ethnicities who are matched on intrinsic characteristics such as age and are supplied with fictitious but essentially similar identities. Teammates both approach (typically within one or two hours of separation) a sampled apartment or agency, pose as bona fide home seekers and independently record the characteristics of their experience. Systematic differences in treatment thus revealed would indicate discrimination. Studies sponsored by the U.S. Department of Housing and Urban Development (HUD) reached three important conclusions:

1. Housing discrimination against black and Hispanic home seekers and apartment seekers occurs in roughly half of the instances when these persons interact with an agent.
2. Typically, this discrimination is subtle in nature and therefore difficult for the individual to detect.
3. The frequency of discrimination against African Americans has not changed noticeably since 1977.

By contrast, comparatively little is known about discrimination during housing transactions that is directed at racial groups other than African Americans and Hispanic Americans, or at other protected classes. A pilot study in a small Ohio town using trios of auditor teammates found extremely high frequencies of discrimination in the apartment market against single women both with and without children, compared with single, childless men. Boston audit data have shown that testers posing as a couple with a child under six years old inspected fewer homes for sale than testers posing as single adults. The large proportion of complaints to HUD alleging gender and familial status discrimination juxtaposed against this evidence from pilot-testing studies makes it imperative that comprehensive, rigorous testing studies be conducted to ascertain the nature and extent of such acts.

Racial discrimination in the housing market has several harms. From a societal perspective, discrimination encourages segregation by preserving all-white enclaves and channeling minority housing demands in ways that promote resegregation. Discrimination may also indirectly affect segregation by altering minorities' perceptions of the housing market. These perceptions guide how, where, and from whom individual minority home seekers gather information and thus influence the spatial search patterns these home seekers employ. By foreclosing desirable alternatives, discrimination helps consign minorities to inferior dwellings and neighborhoods; by limiting opportunities for minority homeownership and the appreciation of minority-owned property, it retards their wealth accumulation. The minority home seeker also suffers psychological damage and wasted time, effort, and expenses. One study estimates that discrimination costs the average minority family $1,500 each time it searches for housing. Racial residential segregation caused by discrimination translates into segregated, underfunded, ineffective school systems for minority youth, who, not surprisingly, tend to underachieve and drop out as a result. Lower-paying, less stable occupational attainment results. Segregation also limits access to and information about burgeoning employment opportunities in the suburbs. It permits the formation of distinctive minority subcultural traits, behaviors, and norms that abet discrimination against minorities in labor markets. Collectively, these effects intensify interracial economic disparities.

These inequalities, of course, merely serve to legitimate the original racial stereotypes, and with little residential contact, there are few contexts in which these stereotypes can be challenged. In addition, economic inequalities feed back to segregation by rendering fewer and fewer minorities capable of affording the sort of housing found in white-occupied neighborhoods; increasingly, they are confined to inferior areas. Thus, housing market discrimination forms a key link in a self-perpetuating cycle of segregation, inequality, prejudice, and discrimination in many markets.

The aforementioned acts of housing market discrimination have been forbidden by federal law since Title VIII of the Civil Rights Act was passed in 1968. Handicap and familial status were added to the list of protected status by the Fair Housing Amendments Act of 1988. The 1988 act also created a system of HUD administrative law judges designed to speed resolution of complaints and established stiffer fines (up to $100,000 in punitive damages for repeat offenders). Most states and many localities have their own fair housing statutes that are substantially equivalent to the federal laws; some even prohibit additional bases of discrimination, such as age, health, and sexual preference.

The juxtaposition of a continuing high incidence of housing market racial discrimination nationwide and long-standing laws to the contrary raises a troubling issue about the adequacy of enforcement. The fair housing laws are seen by some as failing to generate an adequate deterrent. Because discriminatory acts are camouflaged, victims seldom recognize when their rights have been violated. Yet the enforcement mechanism is triggered by victims complaining to the appropriate authorities. As a result, many discriminators probably believe that they can violate the law with impunity. HUD itself has estimated that less than 1% of the incidents of housing market discrimination generate complaints from victims.

The foregoing suggests that changes are needed if existing fair housing laws are to be more successful in deterring discrimination. Several initiatives have been tried recently.

Campaigns to educate the public about their rights, how to spot discrimination, and how to seek redress have recently been promulgated by the National Fair Housing Alliance, the umbrella organization for the approximately 60 private, nonprofit fair housing organizations nationwide. The 1988 Fair Housing Amendments Act gave HUD the authority to initiate pattern-and-practice suits, presumably based on auditing evidence, although thus far HUD has been reluctant to pursue such an avenue directly. However, in the 1987 Housing and Community Development Act, Congress authorized a Fair Housing Initiatives Program (FHIP) that provided financial resources via HUD to private fair housing groups to conduct housing discrimination auditing. In 1992, the Department of Justice began conducting enforcement audits in a number of metropolitan areas; at this writing, dozens of suits have been brought on the basis of the evidence gained thereby, but initial results have been encouraging.

These initiatives show promise because they remove the burden of detecting and litigating cases of suspected discrimination from the victim and place it on a private, nonprofit, or public organizations. These organizations have additional tools (e.g., audits) and human and financial resources available, thus offering comparative advantages in the fight against housing market discrimination. (SEE ALSO: *African Americans; Asian Americans; Blockbusting; Exclusionary Zoning;* **Fair Housing Amendments Act of 1988;** *Fair Housing Audits; Fair-Share Housing; Hispanic*

Americans; Home Mortgage Disclosure Act; Inclusionary Zoning; Restrictive Covenant; Segregation; Steering; Yonkers)

—*George C. Galster*

Further Reading

Becker, Gary. 1957. *The Economics of Discrimination.* Chicago: University of Chicago Press.

Galster, George. 1990. "The Great Misapprehension: Federal Fair Housing Policy in the Eighties." Pp. 137-57 in *Building Foundations: Housing and Federal Policy,* edited by Denise DiPasquale and Langley Keyes. Philadelphia: University of Pennsylvania Press.

———. 1992. "Research on Discrimination in Housing and Mortgage Markets." *Housing Policy Debate* 3:639-84.

Goering, John, ed. 1986. *Housing Desegregation and Federal Policy.* Chapel Hill: University of North Carolina Press.

Helper, Rose. 1969. *Racial Policies and Practices of Real Estate Brokers.* Minneapolis: University of Minnesota Press.

Ross, Stephen L. 1996. "Mortgage Lending Discrimination and Racial Differences in Loan Default." *Journal of Housing Research* 7(1):117-26.

Turner, Margery. 1992. "Discrimination in Urban Housing Markets: Lessons from Fair Housing Audits." *Housing Policy Debate* 3:185-216.

Yinger, John. 1995. *Closed Doors, Opportunities Lost: The Continuing Cost of Housing Discrimination.* New York: Russell Sage. ◄

► Displacement

Urban Renewal and Freeways

Displacement of urban households first became an issue in the 1950s when the federally assisted urban renewal program displaced large numbers of residents in inner-city neighborhoods. Slum clearance typically targeted poor neighborhoods with substandard housing labeled "blighted." These were often inhabited by racial minorities.

Urban freeways, which were built in conjunction with the interstate highway system authorized in 1956, also were built by clearing inner-city neighborhoods. The same type of poor neighborhoods (with low land values and little political influence) were often chosen for these routes, dissecting or eliminating many urban neighborhoods.

Both the urban renewal and freeway programs provided relocation assistance and guaranteed replacement housing. However, as critics argued and, later, congressional investigations and litigation showed, hundreds of thousands of displacees did not receive relocation assistance or decent replacement housing. Very little replacement housing was actually built on urban renewal sites.

The protests of neighborhood residents against displacement finally led to the requirement in the late 1960s that residents be involved in redevelopment planning and that demolished low-income housing be replaced on a one-for-one basis. These protests also led to the 1970 uniform federal relocation reforms that significantly improved relocation benefits for displacees affected by federally financed programs.

Displacement Caused by Gentrification

In the late 1970s, displacement again became an issue. However, it was not government programs that were the primary cause of displacement but, rather, private actors in the housing market, although these private actors often participated in government housing subsidy and insurance programs. The growing inflation greatly increased housing prices, reducing the production of new housing because it became too expensive for many homebuyers. This and the increasing attraction of some historic urban neighborhoods to many younger people in search of downtown housing combined to promote "gentrification." Higher-income households competing for existing housing, which they could afford to renovate, led to the displacement of lower-income occupants of this housing in many cities. Renters were especially vulnerable. Speculative realtors and developers could buy rental housing, evict tenants on short-term leases, and renovate it for conversion to condominiums.

Many observers believed that gentrification reflected a "back-to-the-city" movement from the suburbs. However, empirical studies do not support this concept. Instead, most urban gentrifiers, typically professionals employed in downtown businesses, already lived in the central city prior to moving into gentrifying neighborhoods. Whereas most gentrifiers were white, they displaced not only lower-income minorities but also poor white residents in gentrifying neighborhoods. The gentrification process was usually gradual, and the displacement impact was difficult to measure.

Other Forms of Displacement

Gentrification has been only one form of private displacement. An unknown but large number of poor tenants are routinely evicted because of their inability to pay market rents. This has been exacerbated by the loss of much lower-rent housing in central cities without replacement because of cutbacks, beginning in the early 1980s, in the production of subsidized housing. The most dramatic example of this pattern has been the demolition or conversion of much of the single-room occupancy (SRO) housing stock. Usually located in downtown "skid row" sections, SRO housing has historically provided cheap housing for retirees and transients, including new immigrants.

Much housing in older, inner-city neighborhoods has been lost to disinvestment. Absentee landlords "milked" buildings, often occupied by poor tenants dependent on public assistance for income, reducing and eventually eliminating basic services before abandoning these buildings. They then sit empty, at least until they are demolished or are destroyed by arson.

In some parts of the United States, displacement has been triggered by economic decline. Workers in obsolete or declining industries have been fired or laid off. Without replacement employment at comparable wages, these workers have often lost their homes because of their inability to pay their mortgages or rent. The ultimate version of this form of displacement is homelessness.

Antidisplacement Reforms

As displacement became controversial and more widespread in the 1970s, attempts were made to document its magnitude. Although antidisplacement organizers estimated annual displacement in the United States as high as 2 million persons, a 1979 HUD study concluded that only a few hundred thousand Americans were displaced annually and that it was a local rather than a national issue.

Antidisplacement efforts focused on the federal, state, and local governments and the courts. In addition to enforcement of existing federal relocation legislation, coverage was broadened to provide relocation benefits to those displaced privately but in projects receiving some form of federal assistance. A major threat to residents of below-market, federally subsidized housing, privately owned and operated, came in the 1980s when many developers sought to prepay or terminate their subsidy contracts as they expired. Their motivation was to convert this rental housing to more profitable uses, such as market-rate condominiums. In 1987, Congress declared a temporary moratorium and in 1990 regulated such conversions to minimize the loss of this housing stock. Without this legislation, several hundred thousand housing units and their low- and moderate-income occupants would have been at risk of displacement.

At the state and local level, a variety of antidisplacement legislation was proposed and, in many instances, adopted. These legislative reforms included (a) rent and eviction controls; (b) condominium conversion controls, including temporary moratoria in "tight" housing markets, lifetime leases for certain categories of tenants such as the low-income elderly, one-for-one replacement of converted units, and relocation assistance; (c) demolition controls; and (d) antispeculation taxes. These types of regulatory measures were strongly opposed politically by the real estate industry and regularly challenged in the courts. Nevertheless, many of these regulatory reforms have survived and remain in place, even with changing housing conditions.

Displacement due to neighborhood disinvestment and housing abandonment has been the subject of efforts, beginning in the mid-1970s, to force federally regulated lenders to invest in inner-city neighborhoods. Through the leverage of the Community Reinvestment Act, many institutional lenders have agreed to invest in housing. This funding has most often been used by nonprofit community development corporations, which have typically been involved in the rehabilitation of substandard or vacant housing. Their activities, where successful, have promoted neighborhood revitalization and stemmed the tide of housing abandonment and further displacement of poor, inner-city residents.

Displacement is a continuing phenomenon. Its magnitude and locus depend on differing conditions in the housing market, shifts in the economy, demographic trends, and public development policies and programs. The critical issue remains as to how comprehensive and effective governmental guarantees are to prevent or minimize displacement and to provide adequate relocation housing for those who are displaced. (SEE ALSO: *Blight; Eviction; Gentrification; Uniform Relocation Assistance and Real Property Acquisition Policies Act of 1970; Uniform Residential Landlord and Tenant Act;* **Urban Redevelopment**)

—*W. Dennis Keating*

Further Reading

Bryant, Don C. and Henry W. McGee. 1983. "Gentrification and the Law: Combatting Urban Displacement." *Journal of Urban and Contemporary Law* 25:43-144.

Gale, Dennis E. 1984. *Neighborhood Revitalization and the Postindustrial City: A Multinational Perspective.* Lexington, MA: Lexington Books.

Hartman, Chester, Dennis Keating, and Richard LeGates. 1982. *Displacement: How to Fight It.* Berkeley, CA: National Housing Law Project.

Keating, W. Dennis. 1985. "Urban Displacement Research: Local, National, International." *Urban Affairs Quarterly* 21:132-36.

National Urban Coalition. 1979. *Neighborhood Transition without Displacement: A Citizens' Handbook.* Washington, DC: National Urban Coalition.

Nelson, Kathryn P. 1988. *Gentrification and Distressed Cities.* Madison: University of Wisconsin Press.

Palen, John J. and Bruce London, eds. 1984. *Gentrification, Displacement, and Neighborhood Revitalization.* Albany: State University of New York Press.

Schill, Michael H. and Richard P. Nathan. 1983. *Revitalizing America's Cities: Neighborhood Reinvestment and Displacement.* Albany: State University of New York Press.

U.S. Department of Housing and Urban Development. 1979. *Displacement Report.* Washington, DC: Author.

Vaughn, Susan J. 1980. *Private Reinvestment, Gentrification, and Displacement: Selected References with Annotations.* Chicago: Council of Planning Librarians. ◄

► Do-It-Yourself

Do-it-yourself is a term used generally to describe activities undertaken by consumers, individually or cooperatively, in lieu of paying someone else to do them and, specifically, to describe renovation and repair activity undertaken by homeowners. The growing importance of do-it-yourself renovation and repair is indicated both by the proliferation of retailers of building materials and home improvements and by the increasing availability of factory-assembled components and component systems that require little or no skill or training to install.

In the 1993 American Housing Survey (AHS), owner-occupier households (61.25 million in total) are categorized on the basis of whether specified renovation and repair activity in the past two years was done mostly by household members or by others (see Table 5). Activities that require less skilled labor—such as remodeling or addition of a kitchen or bathroom, the addition of insulation, addition of a new room, and the installation of storm doors and windows—are popular among do-it-yourselfers. Because of the need for physical dexterity, do-it-yourself is less common among elderly households. Do-it-yourself activity is also more commonplace in rural areas and small towns, in part because of the absence of skilled trades.

More broadly, do-it-yourself decisions can be viewed within a household production framework. In this framework, consumers are seen to purchase various goods and services, then use these within the household to produce "commodities" that the household desires. At one level, do-it-yourself can be seen, for example, simply as the trade-off between bread bought in a store and bread made in the

TABLE 5 Owner-Occupier Households Reporting Repairs, Improvements, or Alterations in Last Two Years, United States, 1993 (in millions)

	Total	Mostly Done by Household	Mostly Done by Others	Worker Not Stated
Bathroom remodeled or added	6.126	3.420	2.601	0.105
Insulation added	3.827	2.003	1.670	0.154
Kitchen remodeled or added	5.033	2.568	2.399	0.067
Additions built	2.469	1.092	1.342	0.035
Storm doors/windows bought and installed	7.180	3.221	3.817	0.142
Siding replaced or added	3.212	0.949	2.192	0.071
Roof replaced	9.722	2.321	7.238	0.163
Major equipment replaced or added	6.053	1.093	4.873	0.087
Other major work	12.096	4.073	7.655	0.369

SOURCE: U.S. Department of Commerce, Economics and Statistics Administration, Bureau of the Census. American Housing Survey for the United States in 1993, p. 137.

home. At another level, commodities can be exemplified by "good health": something that the household produces for itself by combining a diet of foodstuffs, exercise, medical care, and lifestyle. At either level, the extent of do-it-yourself activity is determined by the opportunity cost of the household's time (access to paid work and wage rate for each household member) and the availability and price of substitutes. Substitutes for do-it-yourself activity include both store-bought goods and purchased services. Cast more generally than just renovation and repair, this would include, for example, the purchase of property management services by a condominium owner in lieu of doing one's own property management in other forms of home-ownership.

In 1993, U.S. homeowners living in single-dwelling properties spent $13 billion on construction materials for repair and improvement work done by the household itself. They also purchased construction materials in the amount of $3 billion for work that was subsequently completed by contractors. These amounts are part of the $71 billion in total household expenditures in 1993 on residential improvements and repairs, including labor and materials supplied by contractors. (SEE ALSO: *Owner Building*)

—*John R. Miron*

Further Reading

Fallis, G. 1993. "The Suppliers of Housing." Pp. 76-93 in *House, Home, and Community: Progress in Housing Canadians, 1945-1986,* edited by J. R. Miron. Montreal: McGill-Queen's University Press.

Hill, T. P. 1979. "Do-it-yourself and GDP." *Review of Income and Wealth* 25(1):31-39.

U.S. Department of Commerce. Quarterly. *Expenditures for Residential Improvement and Repair.* Quarterly report of the Construction

Repair Branch, Bureau of the Census, Economics and Statistics Administration. Washington, DC: Author. ◄

► Doubling Up

Doubling up occurs when two or more families or unrelated individuals share a dwelling unit designed for occupancy by a single party. This arrangement is considered problematic in the United States and other Western nations because it violates an important housing norm: that all families should have a home of their own, for their exclusive use. The strength of the one-family-per-unit norm means that few people in the United States see doubling up as a desirable strategy. In most cases, economic deprivation or an equally compelling reason must exist before members of separate households will resort to doubling up. Their reluctance is easy to understand. A variety of potential costs are thought to be associated with sharing a residence, including reduced space, privacy, and autonomy.

Despite its negative aspects, doubling up happens relatively often and takes several forms. On the basis of 1996 estimates for the United States, approximately 153 million people 18 years of age or older belong to family groups. Of these people, nearly one-fifth—or 30 million—are doubled up in housing units that they do not own or rent. The majority are adult children who have never left their parents' home or who have "returned to the nest," either alone or accompanied by their own families. Elderly parents, other relatives, and members of unrelated subfamilies are also represented among the 30 million. This figure, although impressive, understates the magnitude of the doubling-up phenomenon. Children under 18 are excluded, as are unrelated persons (outside of families) who live in the same unit.

Contemporary interest in doubling up can be traced to the prominence of the homelessness issue since the early 1980s. Some researchers have treated the doubled up as part of the "hidden homeless" population, whereas others have put them in the "marginally housed" category. Regardless of how they are labelled, their vulnerability is the common theme. Yet certain historical instances of doubling up suggest a more routine—and secure—situation. In preindustrial Europe, for example, a high proportion of young adults worked as servants prior to marriage, appending themselves to a succession of families in response to shifting demands for labor. And in the United States during the early 20th century, unmarried individuals rarely established independent households. Most of those under the age of 30 continued to reside with their parents. For the unmarried 30 and older, boarding in a nonrelative's home supplemented the stay-with-parents option.

The Great Depression and World War II presented a different set of historical circumstances. Because of the housing shortages and extreme financial hardship of the period, doubling up reached record levels. Sharing a dwelling was often the only way that a new family could find housing or that an existing family—by taking in additional members—could afford to keep its unit. One occasionally hears parallels drawn between the Great Depression era

and the housing market faced by Americans today. Given similar concerns voiced then and now over shelter security, there may be some validity to the comparison. With respect to doubling up, however, the current roots of the problem appear more complex.

A key underlying factor during the 1980s and 1990s is what has been called the "housing squeeze." Although the U.S. housing stock has grown substantially in recent decades, many consumers find their choices constrained. This is because housing prices and rents have increased more rapidly than income. Steep land and construction costs have also played a role, steering builders toward the high end of the market and greater profit potential. Trends in household formation and dissolution further exacerbate the squeeze. The number of smaller households has risen with increases in (a) divorce and separation, (b) young adults "on their own" (out of their parents' homes) before marriage, (c) persons never marrying, and (d) residential independence among the elderly, particularly after the death of a spouse. The consequence of these forces is a mismatch between supply and demand: There are not enough affordable units to meet the needs of less affluent families and individuals.

In the face of limited housing availability, doubling up becomes a reasonable strategy for keeping a roof over one's head. People pursue this strategy at predictable points in the life cycle. It appears most closely linked with status transitions, such as beginning or ending a marriage, having a child, leaving the workforce (via unemployment or retirement), finishing school, or being discharged from an institution or the military. The financial pressures that accompany such transitions can place housing out of reach, at least temporarily. One way to adapt is by moving in with relatives or friends or by finding another party willing to share housing space and costs.

Members of minority groups, especially African Americans and Latinos, are at greater shelter risk throughout their lives and therefore rely more heavily on doubling up than do whites. The prevalence of doubling up varies across communities as well as across groups. In metropolitan areas with large percentages of older residents, single-person households, and foreign-born residents, the doubling-up rate tends to be high. Another metro characteristic that boosts the rate is a tight, expensive, owner-dominated housing market. Based on these patterns, it seems safe to conclude that both personal disadvantage and structural context influence the likelihood of having to share a dwelling unit.

Much less is known about the consequences of doubling up. Obviously, when two families occupy a single unit, space may be at a premium. Crowded conditions can lead to feelings of diminished privacy and control, which in turn could heighten conflict over household responsibilities, child care, and the like. But positive outcomes are also possible. In the case of parent-adult child coresidence, for instance, companionship, resource pooling, and help with daily tasks are a few of the benefits available to household members. The more equitable the distribution of these benefits—and of any costs—the less negative we might anticipate members' evaluations of their living arrangements to be. Even in a balanced situation, however, doubling up contradicts the significant normative expectation that adults should have their own homes. Because not having a home of one's own is viewed as a deficiency, most people are hesitant to turn to others for shelter assistance. Doubling up thus constitutes a type of housing adaptation used only when absolutely necessary. (SEE ALSO: *Crowding; Housing Occupancy Codes; Lodging Accommodation*)

—*Barrett A. Lee*

Further Reading

Adams, John S. 1987. *Housing America in the 1980s.* New York: Russell Sage.

Goldscheider, Frances K. and Linda J. Waite. 1991. *New Families, No Families? The Transformation of the American Home.* Berkeley: University of California Press.

Hajnal, J. 1983. "Two Kinds of Pre-Industrial Household Formation System." Pp. 65-104 in *Family Forms in Historic Europe,* edited by Richard Wall. Cambridge, UK: Cambridge University Press.

Morris, Earl W. and Mary Winter. 1978. *Housing, Family, and Society.* New York: John Wiley.

Mutchler, Jan E. and Lauren J. Krivo. 1989. "Availability and Affordability: Household Adaptation to a Housing Squeeze." *Social Forces* 68:241-61.

Myers, Dowell and Jennifer R. Wolch. 1995. "The Polarization of Housing Status." Pp. 269-334 in *State of the Union: America in the 1990s,* edited by Reynolds Farley. New York: Russell Sage.

Saluter, Arlene F. 1994. "Marital Status and Living Arrangements: March 1995 Update." *Current Population Reports.* Series P20-491. Washington, DC: U.S. Bureau of the Census.

Sweet, James A. and Larry L. Bumpass. 1987. *American Families and Households.* New York: Russell Sage.

Ward, Russell A. and Glenna Spitze. 1992. "Consequences of Parent-Adult Child Coresidence: A Review and Research Agenda." *Journal of Family Issues* 13:553-72. ◀

▶ Down Payment

In a traditional real estate transaction, a long-term mortgage is secured for part of the cost of the purchase. The difference between the purchase price and the mortgage is the required equity contribution. This equity contribution is often referred to as the *down payment.*

The down payment is an important element in the residential financing system. For example, research has shown that default rates (i.e., the proportion of loans in which the borrower no longer makes periodic payments as a percentage of total loans) vary inversely with the size of the down payment. In other words, where there is a sizable ownership interest in the property, there is a greater likelihood that the borrower will not allow the loan to become default. It is not surprising that lenders regard the size of the down payment as an important parameter in making the decision of whether or not to approve a mortgage loan.

Suppose that property values fall unexpectedly. The size of the down payment is important in determining whether or not the owner continues to enjoy an economic interest in the property. If the down payment is small, it is quite possible that the owner will be "out of the money" (i.e., the value of the property becomes less than the outstanding balance of the mortgage). In this case, the down payment

is a critical factor for what will happen to the property in the future.

On the other hand, increasing the down payment is equivalent to reducing the degree of financial leverage. By contributing more equity, the investor uses less debt financing. Because debt financing is generally cheaper than equity financing, the rate of return to the investor resulting from financial leverage will be reduced with higher down payments. It is true that the debt service payments will also be lower, but so will the rate of return on equity.

Finally, one of the obstacles to homeownership is the requirement of a substantial down payment. In some markets, the most difficult requirement for first-time buyers is the amount required for the down payment. Although a substantial down payment may be necessary for the financial institution to be willing to make the loan, if the property under consideration is expensive, the amount required as a down payment can be a very difficult test to meet. The amount of the down payment also differs cross-nationally. In some countries, the down payment required may be as high as 50% or more of the purchase price for the typical transaction.

—*Austin J. Jaffe*

Further Reading

ABA Standing Committee on Lawyers' Title Guaranty Funds. 1991. *Buying or Selling Your Home: Your Guide to: Contracts, Titles, Brokers, Financing, Closings.* Chicago: American Bar Association.

Blankenship, Frank J. 1989. *The Prentice Hall Real Estate Investor's Encyclopedia.* Englewood Cliffs, NJ: Prentice Hall.

Brueckner, Jan K. 1986. "The Downpayment Constraint and Housing Tenure Choice." *Regional Science and Urban Economics* 16:519-25.

Jacobus, Charles J. 1996. *Real Estate Principles.* Upper Saddle River, NJ: Prentice Hall.

Jaffe, Austin J. and C. F. Sirmans. 1995. *Fundamentals of Real Estate Investment.* 3d ed. Englewood Cliffs, NJ: Prentice Hall.

McKenzie, Dennis J. 1996. *Essentials of Estate Economics.* Upper Saddle River, NJ: Prentice Hall.

Peiser, Richard B. 1992. *Professional Real Estate Development: The ULI Guide to the Business.* Washington, DC: Urban Land Institute.

Tosh, Dennis S. 1990. *Handbook of Real Estate Terms.* Englewood Cliffs, NJ: Prentice Hall. ◀

▶ Drugs and Public Housing

Drugs have been an increasing problem in public housing since the mid-1980s, with the appearance of crack cocaine—cheap, easily packaged and consumed—a contributing factor. Although studies have shown that, on a per capita basis, both usage and sales are higher in poor neighborhoods outside of public housing than within it, media attention has largely focused on the problem within public housing. Part of the explanation lies in the very visible efforts of many housing authorities to combat the problem and in the process, of course, calling attention to it.

Some authorities' efforts to combat drug use and sales and the violence that often accompanies them, particularly the highly promoted actions of the Chicago Housing Authority in conducting police sweeps of apartments and instituting strict eviction policies in which whole families are held responsible for the actions of a single member, have put such authorities in conflict with civil liberties concerns. Legal services offices representing individual tenants threatened with eviction often end up taking positions at variance with those of many tenants' groups, for whom drug use is a plague. Tenant patrols, education, and treatment services, often well organized in public housing, probably present a more effective solution in the long run; tenant management in general discourages drug use effectively, in part through exceptionally strict enforcement of rules of occupancy.

The architecture of public housing is sometimes held responsible for drugs in public housing, but there is no evidence suggesting that similar groups in other housing witness less of a problem.

The "one strike and you're out" executive order of President Clinton, issued in the spring of 1996 as a guideline to local housing authorities, is only the most recent instance of the difficulties of dealing with social problems through housing management. The instruction to housing authorities to act swiftly when a tenant in public housing is convicted of a serious crime (drug dealing being a clear example) leaves the nature of the appropriate action still in the discretion of the local authority. Whether, for instance, a whole family can be evicted because of the transgression of one of its members remains a matter to be decided locally, depending on the circumstances of each case. The issuance of the executive order shows the high visibility of the problem of drugs in public housing and its potent perceived political role. (SEE ALSO: *Crime Prevention*)

—*Peter Marcuse*

Further Reading

Cisneros, Henry G. 1995. *Defensible Space: Deterring Crime and Building Community*. Washington, DC: U.S. Department of Housing and Urban Development.

Dunworth, Terrence. 1994. *Drugs and Crime in Public Housing: A Three-City Analysis*. Washington DC: U.S. Department of Justice Office of Justice.

Keyes, Langley C. 1992. *Strategies and Saints: Fighting Drugs in Subsidized Housing*. Washington, DC: Urban Institute Press.

United States. Congress. House. Committee on Government Operations. 1988. *Just Saying No Is Not Enough: HUD's Inadequate Response to the Drug Crisis in Public Housing*. 48th Report. Washington, DC: Government Printing Office.

United States. Congress. House. Select Committee on Narcotics Abuse and Control. 1989. *Drug Problem and Public Housing*. Hearing. 101st Cong., 1st sess. June 15, 1989. Washington, DC: Government Printing Office.

United States. Congress. Senate. Committee on Governmental Affairs. Permanent Subcommittee on Investigations. 1989. *Drugs and Public Housing*. Hearing. 101st Cong., 1st sess. May 10, 1989. Washington, DC: Government Printing Office.

U.S. Department of Housing and Urban Development. 1991. *Together We Can—Meet the Challenge: Winning the Fight against Drugs*. Washington DC: Office of Policy Development and Research, Office of Public and Indian Housing.

E

▶ Earth-Sheltered Housing

Earth-sheltered housing is a term commonly used to characterize residential structures enveloped by earth. This can take place in supraspace structures (above ground) covered by at least half a meter of earth mass or in below-ground locations. Similar terms indicate the degree to which the structure is beneath the earth.

Historically, earth-sheltered space is the most ancient type of habitat used by humankind. Almost every civilization has started with this form of living. Natural caves or human-made earth-sheltered structures continue to have varied uses, as they have for millennia. The most common use for earth-sheltered structures has been for their diurnal and seasonal sheltering from environmental extremes. They have been used for protection against extreme climates such as hot-dry (in the Sahara) or cold-dry (as in central Canada or central Siberia); for special storage, such as for grains or large quantities of water to minimize loss by evaporation; for irrigation canals (in Iran); for religious space to achieve tranquillity and ideal circumstances for contemplation; and for defense.

With the energy crisis of the 1970s, a rising interest in modern earth-sheltered space has developed. The intentions are to use such space for all types of land uses, including residential, commercial, industrial, educational, and cultural centers.

Earth-sheltered habitats have been in use primarily, but not exclusively, in three large concentrations in the world. The first and largest are built into the loess soil of northern China, where an estimated 30 to 40 million people are still living in so-called human-made cave dwellings. Historically, this way of living has taken place for more than four millennia in China. The prime historical reasons for using this style of dwelling are to conserve the land for agricultural uses and to eliminate the need for building materials. The loess soil maintains itself firmly when it is free from water contact and high humidity.

The second major concentration of earth-sheltered dwellings is located in the northern Sahara desert bordering southern Tunisia, on the Matmata Plateau. Twenty-two communities still exist along this plateau. The most com-

monly known among them is the village of Matmata. The prime motivation for such usage is to protect the residents from the intense outdoor heat and to defend against invaders. This way of life has been in use by the Berber in North Africa for more than two millennia. This region is also characterized by semi-arid to extreme arid conditions. The Romans, during their invasion of North Africa, used subsurface summer villas in the city of Bulla Regia located in northern Tunisia. This city region was the breadbasket of the Roman Empire. Both the Tunisians and the Chinese have used two types of earth-sheltered habitat (the pit and the cliff).

The third concentration of earth-sheltered housing is in Cappadocia at the center of the Turkish Plateau, some 400 kilometers southeast of Ankara, the modern capital of Turkey. Here, too, the climate is hot-dry in the summer and cold and partly snowy in the winter. As in the Chinese and the Tunisian cases, the dwellers practice the method of "cut and use" without the necessity of building materials. Earth-sheltered habitats in Cappadocia may have started in the second millennium B.C. There were two types of earth-sheltered construction: cliff dwellings and underground cities. The cliff dwellings were used almost continuously until the middle of the 20th century. The underground cities were probably used until around the 10th century A.D. The cliff dwellings—and probably the underground cities also—were used by the Hittites, Greeks, Romans, and Byzantines. Geologically, the area is covered with volcanic tufa, which has been shaped by intense wind deflation and water erosion, creating pinnacles of relatively soft, stone cones. Settlers cut into the cliffs of these cones and created an integrated nest of 8- to 10-story dwellings.

A distinction of the Cappadocian underground cities is the complicated network connecting the underground dwellings and establishing a city pattern for dwellings, defense purposes, religious centers, wine industry, storage, and stables for raising horses. It is estimated that there are hundreds of underground cities located in seven Turkish provinces at the heart of Cappadocia. In 1997, only a few dozen cities had been uncovered, and even fewer were open to the public.

Other noticeable places in the world where earth-sheltered habitats have been used throughout history are the south-

Figure 11. Chinese Cliff Dwelling. This home, built in loess soil, is located in the village of Gao Me Wan in Shaanxi Province of Northern China.
Photograph by Gideon S. Golany.

western United States, central Spain, southern France, the Trulli region and Matera of southern Italy, southern Israel and Jordan, and India. Almost all these regions have semi-hot-dry climates.

Indigenous usage of below-ground habitats has been historically associated with negative images such as darkness; poor ventilation; limited sunshine or size; structural instability; conditions lending themselves to the support of mold, snakes, and rats; and many other negative connotations associated with poverty. Modern technology can overcome all these problems and can produce spaces larger in size, with good ventilation and full of light and sunshine. The most difficult problem to overcome remains the psychological constraints of stigma and claustrophobia.

Some technologically advanced countries have developed sophisticated below-ground spaces for diversified land uses. Japan has become a leading country in the development of below-ground space with the design, construction, and maintenance of 78 below-ground shopping centers. These shopping centers include department stores, restaurants, a variety of shops, entertainment facilities, food markets, and offices. Canada has also developed a large modern shopping center below ground in Montreal. The United States has constructed a large number of earth-

sheltered habitats, scattered throughout California and other regions of the country.

Improved thermal performance is a major benefit when constructing spaces below ground. The mass of the earth has its own pattern of thermal behavior entirely different from that of structures above ground. This mass, especially between 0 and 10 meters in depth, functions as a temperature insulator and heat retainer. The basic rule is the more shallow the depth, the more the temperature fluctuation and vice versa. At depth of around 10 meters, the temperature will become stable diurnally and seasonally.

On the surface of the ground, outdoor temperatures penetrate lightly into the soil throughout the day, to a depth of around 7 centimeters. This shallow penetration forms a thermal wave moving toward the deeper part of the earth and reaches a depth of 10 meters after one thermal season. Thus, the summer outdoor temperature will reach a given below-ground space at a depth of 10 meters in the wintertime (around 20 °C). Similarly, outdoor winter temperatures will reach the same given space in the summertime (around 10 °C). Thus, the underground space will be cool in the summer and warm in the winter, an ideal condition that reduces energy consumption. Moreover, compared with above-ground structures, the stability of the temperature and relative humidity at the depth of 10 meters provides comfortable ambient conditions, as well as a healthful and tranquil environment. It also shortens recuperation time after surgery.

To achieve optimal conditions for below-ground habitats, it is necessary to avoid construction of earth-sheltered habitats in flat areas, as is the case with the pit type of shelter. A better alternative is to locate the habitat on a slope with a medium or high gradient. Sloped underground habitats can provide plenty of natural light and sunshine, ventilation, and good air quality. They can also minimize the hazards of earthquakes, floods, and storms. Additional advantages are associated with the natural environment, a pleasing view of the lowlands, and the creation of private space.

To ease or eliminate the psychological problems associated with living in an earth-sheltered habitat, some design and site selection measures need to be considered. These measures include the introduction of large space, high ceilings, light colors, integration of nature into the house (plants, flowers), good ventilation, indoor-outdoor visual contact, and the selection of a sloped site. (SEE ALSO: *Cultural Aspects; Energy Conservation; Vernacular Housing*)

—Gideon S. Golany

Further Reading

Golany, Gideon S. 1983. *Earth-Sheltered Habitat: History, Architecture, and Urban Design.* New York: Van Nostrand Reinhold.

Golany, Gideon S. 1988. *Earth-Sheltered Dwellings in Tunisia: Ancient Lessons for Modern Design.* Newark: University of Delaware Press.

Golany, Gideon S. 1989. *Urban Underground Space Design in China: Vernacular and Modern Practice.* Newark: University of Delaware Press.

Golany, Gideon S. 1990. *Design and Thermal Performance: Below-Ground Dwellings in China.* Newark: University of Delaware Press.

Golany, Gideon S. 1992. *Chinese Earth-Sheltered Dwellings: Indigenous Lessons for Modern Urban Design.* Honolulu: University of Hawaii Press.

Golany, Gideon S. 1992-1993. "Soil Thermal Performance and Geo-Space Design." *Proceedings of JSCE, Geotechnical Engineering, Japan Society of Civil Engineering* 445/III-18:1-8.

Golany, Gideon S. and Toshio Ojima. 1996. *Geo-Space Urban Design.* New York: John Wiley.

Moreland, Frank L., Forrest Higgs, and Jason Shih, eds. 1978. *Earth Covered Building: Technical Notes.* Proceedings of a conference on the Use of Earth Covered Settlements, Ft. Worth, TX. Vols. 1 and 2. Funded by and prepared for the U.S. Department of Energy, sponsored by the University of Texas at Arlington.

Sterling, Raymond L. and John Carmody. 1993. *Underground Space Design.* New York: Van Nostrand Reinhold. ◀

▶ ECHO Housing

ECHO (Elder Cottage Housing Opportunity) houses are the U.S. version of an Australian concept, "granny flats." They are small, movable homes designed with features that accommodate frail and disabled people and intended to be placed temporarily in the side or rear yard of a family member's home.

ECHO homes provide "privacy with proximity." They make it possible for adult children to support aging parents.

The same concept can be used to allow parents to support a disabled child or a single parent. ECHO homes can also be placed next to nursing homes so that a spouse who does not need such intensive care as the partner in the nursing home can live nearby.

The acronym *ECHO* was coined by Leo Baldwin, housing coordinator for American Association of Retired Persons in the early 1980s. His purpose was to avoid the negative reaction many older men had to living in something called a granny flat. The acronym was also chosen to indicate that the ECHO home is an echo of the larger home it sits next to.

Benefits

The financial benefits of ECHO housing are as great as the more obvious social benefits. Personal care for sick older people is very expensive. Affordable institutional facilities are often accused of providing poor services. ECHO housing makes it possible for families to care for their own at a lower cost to the family's collective resources.

In addition, because about 80% of older people in the United States own their own homes, 62% of them without any mortgage, the move into an ECHO home can improve their finances substantially. A typical move would involve the sale of a $100,000 home to move into a $30,000 ECHO home, resulting in $70,000 that can go into savings and earn at, 5% interest, $3,500 a year in added income. The $30,000 is a rough estimate for the cost of purchasing and siting the ECHO home.

Zoning

Efforts to amend zoning to permit ECHO housing often run into resistance. The typical fears are that the units will be unattractive, that they will not be removed once they are no longer needed, and that they will be used for other purposes. Zoning variances make ECHO housing a special exception use, require annual or biannual permit renewals, require the ECHO home to be compatible in color and design with the main home, and require that the site be relandscaped when the unit is removed. Another common provision prohibits use of mobile homes as ECHO homes. It is also possible to require, as in Victoria, Australia, that the unit be owned or controlled by a third party, such as a housing agency or hospital. Doing so increases community confidence that the unit will be removed when no longer needed.

Conditional use permits can be tailored to an individual occupant's needs, because the unit will be removed when no longer needed. For example, it may not be necessary to require a parking space for someone who no longer drives. Another unique zoning issue is the need for speed in zoning and building permit approval. Often, decisions about where to live in old age are triggered by a health crisis. The occupant of an ECHO unit may be someone coming out of a hospital. Provision should be made to allow for rapid permit approval in such circumstances.

Lack of Market Response

ECHO housing, or granny flats, has been successful in Australia. Mobile homes are the only housing type used in great numbers for ECHO purposes in the United States. Frederick County in Maryland has allowed ECHO housing in its agricultural zone. The county permitted temporary placement of mobile homes when used to provide care for an aging or disabled relative. As of 1991, 38 permits were held for such mobile homes.

No other U.S. community has had similar success with purpose-built ECHO housing. As of early 1993, the largest numerical success of any U.S. local government is in Warren County, New Jersey, where four such units are in place. Reasonable zoning for ECHO units has existed in 20 or 30 communities in New York, New Jersey, Pennsylvania, and Connecticut since at least 1985.

Several marketing problems are among the obstacles to more widespread use of ECHO housing. Potential residents may be too frail to shop for ECHO units. In addition, families often feel that the $25,000 to $30,000 cost for an ECHO unit is money not well spent, because they do not expect their elderly parent to live much longer. In fact, however, many elderly parents live longer than their families expect; although average life expectancy is now about 75 years, few people realize that a woman who reaches the age of 75 still has a life expectancy of about 12 years.

In Australia, where the concept has been most successful, granny flats are the property of the Ministry of Housing of the State of Victoria and are rented. This has made acceptance easier because families do not have to commit as much money. In the United States, the only rental program has been in Warren, New Jersey.

The Future of ECHO Housing

The Department of Housing and Urban Development was required by Congress in 1992 to fund an ECHO housing demonstration project as part of Section 202 Elderly Housing Program. Although that project may bring results, the viability of ECHO housing in the United States is an open question at present. There is no question that a good unit can be produced and sold. The fact that units have been placed in a variety of communities also indicates that amending zoning is not that difficult.

The real issue in ECHO housing is marketing. No one has been able to market the concept in significant numbers to people in a single jurisdiction, with the exception of Frederick County, Maryland, where the homes are not in fact ECHO units but mobile homes.

Two other points are worth noting based on the Australian experience. First, in Victoria, unsubsidized use of ECHO units dropped dramatically when zoning for accessory apartments was introduced. Second, costs for removing, storing, repairing, and resiting used ECHO units have been much higher than initially expected. (SEE ALSO: *Accessory Dwelling Units;* **Elderly;** *Manufactured Housing*)

—*Patrick H. Hare*

Further Reading

Hare, Patrick H. and Linda E. Hollis. 1983. *E.C.H.O. Housing: A Review of Zoning and Other Issues.* Washington, DC: American Association of Retired Persons.

Lazarowich, N. Michael. 1991. "Granny Flats as Housing for the Elderly: International Perspectives." *Journal of Elderly Housing* 7(2).
Mace, Ronald L. and Ruth Hall Phillips. 1984. *E.C.H.O. Housing: Recommended Construction and Installation Standards.* Washington, DC: American Association of Retired Persons. ◀

▶ Efficiency Apartment

In the United States, dwelling units intended for long-term occupancy and built on top of each other with their own kitchens and baths are called "apartments." An efficiency apartment consists of a single habitable room plus a kitchen and a bathroom. Such apartments are intended primarily for single people and are therefore often found in elderly and other nonfamily housing. The kitchen may be very small and located within the single room. The apartment is nonetheless self-sufficient and independent of other apartments. Efficiency apartments may also be called "studios," limited living units, or minimum dwelling units.

The first efficiency units were probably created in hotels in the 1920s and provided long- or short-term furnished accommodations. Such units are again appearing in hotels. Single rooms of smaller size and with smaller kitchens than efficiency apartments are also being built in single-room occupancy (SRO) housing. However, unlike most apartment buildings, SRO housing almost always incorporates some spaces to be shared by residents of different units. Some cities, such as San Francisco, have recently amended their building codes to permit the construction of complete apartments that are smaller than the minimum required size of efficiency apartments. (SEE ALSO: *Accessory Dwelling Units; Single-Room Occupancy Housing*)

—*Karen A. Franck*

Further Reading

Cromley, Elizabeth C. 1991. *Alone Together: A History of New York's Early Apartments.* Ithaca, NY: Cornell University Press.
Groth, Paul. 1994. *Living Downtown: The History of Residential Hotels in the United States.* Berkeley: University of California Press. ◀

▶ *EKISTICS: The Problems and Science of Human Settlements*

First published in 1955, *EKISTICS* is a bimonthly, English-language publication of the Athens Center of Ekistics. It is concerned with the interests of architects, urban and regional planners, scholars, and policymakers in related fields. Issues contain original articles on themes regarding human settlements, including housing, civic design, urban economics, and planning practice and education. Illustrations are included. Circulation: 3,000 worldwide. Price: $100.00 per year. Editor: P. Psomopoulos, President, Athens Center of Ekistics, 24 Strat. Syndesmou Street, 106 73 Athens Greece. Phone: 3623-216. Fax: 3629-337.

—*Caroline Nagel* ◀

TABLE 6 Continuum of Housing

Housing Options for the Elderly	Independent[a]		Semi-Independent[b]				Dependent[c]		
Single-family housing	•	•							
Conventional apartments	•	•							
Accessory units	•	•							
ECHO housing		•	•						
Shared housing		•	•						
Retirement communities		•	•						
Age-specific communities		•	•	•					
Single-room occupancy (SRO) hotels			•	•	•				
Congregate housing			•	•	•				
Assisted living						•	•	•	
Skilled nursing care									•
Continuing care retirement community			•	•	•	•	•	•	•

Notes:

(a) *Independent:* Living arrangements designed for individuals and couples capable of handling their own housekeeping, cooking, and personal care needs.

(b) *Semi-independent:* Living arrangements designed for those with some chronic limitations. Residents are self-sufficient and capable of self-care but may rely on facility for meals, housekeeping, and transportation.

(c) *Dependent:* Living arrangements that provide 24-hour nursing care for severely impaired individuals.

► Elderly

Over the last several decades, older persons in the United States, who accounted for 32 million persons in 1990 (12% of the total population), have benefited from general policies that have encouraged homeownership and attempted to reduce housing costs as well as particular programs designed for the elderly, such as age-specific housing. These policies, along with income-related programs, such as Social Security and Supplemental Security Income (SSI), have assisted the elderly in improving their housing situation. Nevertheless, a substantial number of older persons still pay an excessive amount of their income for housing or live in physically deficient housing. Moreover, during the late 1980s, it became apparent that policies failed to address the needs of the growing population of persons more than 85 years of age. This group numbered 2.8 million in 1990 and by the beginning of the 21st century will exceed 4.9 million persons. To stay out of more institutional settings such as nursing homes, these older persons need physically supportive housing linked with services and health care. The resolution of these problems is constrained, however, by continuing reductions in the Department of Housing and Urban Development's (HUD's) overall budget, the difficulties in bridging housing and services, and increasing competition from other needy groups (e.g., people with AIDS, homeless and very low-income families, people with disabilities, and Native Americans in need of adequate housing) for scarce public funds.

The Continuum of Housing

The vast majority of persons enter old age living in homes or apartments chosen for their appropriateness in midlife when they were healthy and independent. However, as persons age, they may find that these settings no longer fit their needs or preferences. Consequently, a variety of housing types have arisen over the last several decades that provide a range of lifestyle, environment, and service options. The concept of a continuum in housing recognizes that as persons age and become less independent, they may require increased services and more physically supportive environments in terms of accessibility, features, and design. Tables 6 and 7 illustrate how different housing types, such as single-family homes, conventional apartments, ECHO housing (small detached units adjacent to a home), congregate housing (multi-unit living arrangements that generally provide group meals), assisted living (residential settings that provide individualized health and personal care services), and nursing homes generally meet the needs of older persons who are categorized as independent, semidependent, and dependent. Although semidependent and even dependent older persons can be found throughout the housing continuum, supported by caregivers and the formal service system, independent older persons are unlikely to reside in housing types such as assisted living. A major exception are the approximately 900 continuing care retirement communities (CCRCs), which house 400 to 600 older persons each in settings that include independent living, assisted living, and nursing care.

TABLE 7 Special Housing Options for the Elderly

1. *ECHO Housing*—A temporary movable unit designed for use by the elderly and handicapped. It is small enough to be installed in a side or rear yard and made well enough to withstand repeated moves.
2. *Accessory Units*—Separate units typically created in the surplus space within single-family homes.
3. *Shared Housing*—Arrangements in which two or more unrelated people share a house or an apartment. Usually, private sleeping quarters are available; the rest of the house is shared.
4. *Retirement Communities*—Settings that cater to persons aged 55 and older; they often include community facilities and amenities related to recreation.
5. *Single-Room Occupancy (SRO)*—Renter-occupied one-room housing units in an apartment building or a residential hotel available to low-income older adults.
6. *Congregate Housing*—Housing multi-unit living arrangements that generally provide meals and some other limited services.
7. *Assisted Living*—A residential setting in which residents have their own private living units with individual health and personal care provided by the housing sponsor.
8. *Continuing Care Retirement Communities*—Complexes for 400 to 600 older persons that, in addition to living arrangements for independent older people, include assisted living and personal care services, nursing home accommodations, and a range of community facilities paid for through a combination of individual entrance fees and monthly charges.

Ideally, the needs and preferences of older persons would be matched to appropriate housing types. Unfortunately, the reality is that many housing options are limited by political opposition, unaffordability, and the bias of long-term care expenditures toward institutional care. For example, ECHO housing, accessory units, and shared housing often face local zoning restrictions. Congregate housing, assisted living, CCRCs, and retirement communities are often affordable only by middle- and upper-middle-income older persons. Policies that encourage "aging in place" of residents through retrofitting housing with special features have been very uneven and underfunded. Finally, Medicaid reimbursement policy has provided incentives for housing very frail older persons in settings such as nursing homes rather than providing funds for community-based services. Over the last several decades, efforts have occurred to expand housing options for older persons through a number of housing approaches that encourage aging in place and the development of new supportive housing arrangements.

Tax Policy and Supply-Side Programs

Although not a direct component of housing legislation, tax policy allowing homeowners to deduct mortgage interest from income taxes is the largest housing subsidy program. In the mid-1990s, such deductions amounted to about 60 billion dollars, compared with a much smaller amount of housing assistance payments. Although most older homeowners have paid off their mortgages and are not current beneficiaries, these policies have contributed to more than three-quarters of elderly householders becoming homeowners, the highest percentage of all age groups. Older homeowners have also been assisted at the local level by property tax relief programs that reduce the financial burden of low-income homeowners.

Although supply-side programs that seek to increase the supply of affordable housing through construction or support for housing providers have assisted fewer persons than overall tax policy, they have been targeted to low- and moderate-income persons, many of whom are the elderly. The largest supply-side program has been public housing. Emanating out of the Great Depression, public housing had among its original goals the elimination of slums and the housing of temporarily poor families. The 1937 Housing Act did not include elderly persons in the eligibility criteria for public housing. After a congressional report revealed that the elderly constituted a disproportionate number of persons living in substandard housing, in the 1956 Housing Act, Congress redefined "low-income family" to include single elderly persons. Consequently, the elderly, viewed as the "deserving poor," a group in need of better housing and tenants who caused few management problems, became a major constituency of the program. By the mid 1990s, elderly families occupied approximately 37% of the public housing stock of 1.4 million units, with about 300,000 older persons living in public housing units designated for the elderly and handicapped.

The Section 202 program was created by the Housing Act of 1959 as a new construction program to house moderate-income independent older persons by providing age-specific housing, often with services such as meals provided in central dining rooms, design accessibility for wheelchairs, and special features such as emergency call buttons in apartments. Sponsorship was restricted to nonprofit organizations. In 1978, HUD established set-asides for handicapped individuals in Section 202 housing. In the 1980s, new Section 202 housing was targeted to lower-income persons eligible for Section 8 rental certificates, and in 1991, Section 202 was amended to limit participation to the elderly. By 1998, the Section 202 program consisted of approximately 325,000 units.

Although not intended solely for the elderly, federal housing programs such as Section 221(d)(3) and Section 8 new construction have, over time, housed substantial numbers of older persons. Section 221(d)(3), similar to the 202 program, targeted moderate-income persons who were not eligible for public housing. Many of the original tenants "aged in place" and by the 1990s were in their 70s and 80s. Approximately 80% of Section 8 new and rehabilitated units have been for the elderly.

Housing providers and advocates have judged supply-side programs such as public housing for the elderly and Section 202 to be relatively successful in terms of high levels of tenant satisfaction, the formation of new friendships, an absence of serious management problems, and the provision of supportive environments, but these programs have never been very large in magnitude. Public housing, even for the elderly, has suffered from the overall negative image of "bare-bones" subsidized housing, a lack of funds for routine maintenance and security, and many undesirable locations. Despite intense lobbying by organizations such as the American Association of Retired Persons (AARP) and the American Association of Homes and Services for the Aged (AAHSA), conservative critics have long considered Section 202 an expensive program and have constantly threatened to eliminate it. By the 1990s, production of new Section 202 projects had dwindled to approximately 6,000 to 8,000 new units per year, compared with almost 20,000 units in the late 1970s.

Demand-Side Programs

In the late 1970s, policy analysts concluded that excessive housing costs had replaced poor housing conditions as the major housing problem and that vouchers or certificates for use in existing housing were two to three times less expensive than building new units. During the 1980s and early 1990s, these arguments were used by the Reagan and Bush administrations, which did not want the federal government involved in housing production, to virtually halt supply-side programs. Demand-side programs have been considered especially advantageous for the elderly because they offer participants the choice of "aging in place" in their own dwelling units and neighborhoods rather than moving to a housing complex in a particular location. In 1990, approximately 26% of the 813,000 Section 8 existing units, the government's major demand-side program, were occupied by the elderly.

The Persistence of Housing Expenditure and Condition Problems

Despite the demand and supply-side interventions, many elderly still experience serious housing expenditure problems and live in physically deficient dwelling units. For example, in 1991, 48% of elderly owners with a mortgage, 20% of elderly owners without a mortgage, and 66% of elderly renters were paying more than 30% of their income for housing. In each of these categories, the rates of excessive housing costs for the elderly are from one and a half to two times greater than those experienced by the nonelderly. Approximately 9% of the elderly live in units judged to be moderately or severely deficient, a rate similar to that of the nonelderly. The most severely affected older persons in terms of excessive housing costs and physical deficiencies are renters, members of minority groups, and women living alone.

Developing Supportive Housing

During the 1980s, evidence mounted concerning a new housing-related problem: An increasing number of frail older persons needed physically supportive housing linked with services. Research has found, for example, that approximately 7% of the 1.6 million elderly in federally assisted housing needed help with at least one activity of daily living (ADL), such as bathing and ambulating. Of older tenants, 12% to 17% needed assistance with at least one instrumental activity of daily living (IADL), such as cooking, shopping, and cleaning. Altogether, about 350,000 tenants needed assistance to remain in government-assisted housing.

Almost all housing, however, has been developed for persons who can function independently. For example, the great majority of the homes owned by older persons lack basic supportive features such as grab-bars or ramps. Older persons living in particularly small units may not have adequate space for caregivers to easily provide help. Even federally assisted housing has very few available common spaces for or connections with services to support the increasing number of frail residents who have aged in place.

Although many federally assisted housing complexes, especially those under the Section 202 program, attempted to adapt their facilities by adding services, overall there has been no coherent policy or system to meet the needs of frail older persons. In fact, Medicare and Medicaid reimbursement policies have provided incentives to house frail older persons in nursing homes. To address this problem, elderly interest groups, after a decade of lobbying, convinced Congress in 1978 to fund a 63-site HUD congregate housing services program (CHSP) demonstration to test the viability of providing service coordination and services such as meals and homemaking for approximately 3,000 frail older tenants in public and Section 202 housing. HUD concluded that the CHSP did not demonstrate that it saved money that otherwise would have been spent on nursing home care and therefore was not cost-effective. The CHSP evaluators, however, contended that their study was never intended to directly answer questions of cost-effectiveness. As evidence of the program's potential, they pointed out that the CHSP served a number of older persons who had previously lived in nursing homes. In any case, HUD had little incentive to fund services because any savings in nursing home admissions would accrue to the Department of Health and Human Services. Nevertheless, at the insistence of Congress, the CHSP program was continued and expanded; however, because of budget constraints and HUD's reluctance, it has remained small.

In the early 1990s, advocates of policies to better meet the needs of frail elderly took advantage of a major legislative effort to pass the first major housing bill in over a decade by including several provisions related to supportive housing in the National Affordable Housing Act (NAHA) of 1990. First, the Section 202 program was revised to allow sponsors to build supportive housing for elderly with up to 15% of the funds usable for services. Second, service coordinators became an allowable cost in Section 202 housing that included a high percentage of frail elderly. Third, up to $51 million was authorized to expand the CHSP and retrofit individual units and common spaces for frail older and younger disabled persons. Fourth, a demonstration (HOPE for Elderly Independence) involving 1,500 frail older participants was created to test the effectiveness of a combination of housing vouchers and services on their independence. Fifth, the act supported the concept of older persons using the equity in their homes to generate income by providing federal insurance for up to 25,000 home equity conversion mortgages.

Future Directions

In the next decade, elderly interest groups are likely to continue lobbying for general housing policies to address problems of housing affordability and physically inadequate housing while focusing attention on meeting the needs of frail older persons who will need supportive residential settings with services. The projected increase of the population most likely to require assistance—those more than age 85—will number approximately 8 million persons by the year 2030. Developments in this area will be driven by three forces: the high costs of nursing home care, the realization that many frail older persons can be cared for more humanely in residential settings, and the demand for alternatives to institutionalization by older persons and their families. A number of policy options are under consideration, including expanding supportive housing, developing assisted living, and unbundling services from housing. The prospects for such initiatives, however, must be evaluated in the context of constrained federal housing budgets, continued disagreement about who will pay for the housing and services, and increasing competition from other groups for scarce housing resources.

Expanding Supportive Housing

As support for federally assisted housing for independent elderly has waned, advocates for the elderly have viewed the existing 20,000 federally assisted housing complexes for the elderly as an invaluable and irreplaceable resource, especially for frail, low-income older persons who have few housing alternatives other than board-and-care or nursing homes. There is the continual danger that some of this housing will be judged marginal because of its age, location, or condition. Important steps to ensure the use of this stock for frail older persons include adding service coordinators to all projects, retrofitting projects, and adding services to these complexes. These efforts, however, may be hindered by competition with other groups for housing, the lack of funds to renovate older projects, difficulties in obtaining service commitments (both the CHSP and service coordinator program are in jeopardy), and proposals that would replace project-based operating subsidies with vouchers. In addition, planning conversions from independent to supportive housing face problems of ensuring a high quality of life and cost-effectiveness.

Developing Assisted Living

Assisted living, a form of very supportive housing, is also likely to be high on the public agenda. This type of housing is intended for persons who require on-site supervision, services, and supportive environments. Assisted living tends to serve very frail persons, including those with early to middle stages of Alzheimer's disease. Typically, residents have their own individual small apartment units made up of a sleeping and living area, kitchenette, and bathroom. Shared spaces include a dining room, recreation, and social areas. Services include meals, housekeeping, personal care, and monitoring of medications. In some cases, assisted living has been developed on a separate floor or wing of an existing housing complex, thereby allowing seniors with deteriorating physical and mental capacities the opportunity to continue residing in their same building but receive more intensive assistance.

Assisted living combines elements of housing and personal care services within a homelike or noninstitutional environment. By 1997, the top 30 private providers in this field reported housing approximately 75,000 persons. Costs ranged between $1,500 and $3,000 per month. The challenge for public policy is how to ensure that assisted living remains residential and becomes affordable for low- and moderate-income persons. The achievement of such objectives will depend on developing appropriate state regulations that stress autonomy and privacy and the melding of income streams from diverse programs, such as Section 8 housing certificates, SSI, and Medicaid funds.

Untying Housing and Services

NAHA reforms to create more supportive housing represent a housing-based approach. However, it has been difficult to obtain funds for services from HUD. Moreover, many older persons with severe problems live in other parts of the housing stock. A progressive policy would untie or unbundle housing and services for nursing-home-eligible persons. It would provide a range of physically supportive environments and portable services so that frail older persons would have a wide array of residential options from which to choose, including multi-unit apartments, assisted-living complexes, congregate housing, and individual homes or apartments to support aging in place. Such a policy, which has evolved in Scandinavia, would provide frail older persons who would otherwise be forced into nursing homes with greater choice in selecting residential settings based on their preferences for privacy or sociability. For reasons of efficiency, specialized facilities could be restricted to those with high levels of disability who would benefit from companionship, close support of resident staff, and proximity to communal facilities.

Creating a policy that provides an adequate range of housing options and supportive services would move the United States closer to Northern European countries, where the trend has been to halt the construction of nursing homes, convert them into a form of sheltered housing, create new types of supportive housing, and provide services in a variety of residential settings.

Conclusions

Over the last 10 years, there has been a reliance on general policies to address problems such as physically deficient housing and excessive housing costs with a new emphasis on creating special programs to meet the housing and service needs of a growing number of frail older persons. The latter approach has included strategies to convert government-assisted housing for independent elderly into supportive housing, modify homes to add supportive features such as grab-bars and ramps, develop assisted living, and provide services to individuals living in a variety of housing types. These types of solutions call for a reconceptualization of housing beyond bricks and mortar and require the close coordination of housing, social services, long-term care, and health services.

The extent to which a combination of general and age-specific policies can meet the needs of the elderly must be viewed not only in the context of limited federal funds but also of competition from other groups. For example, over the past decade, increasing numbers of younger persons with disabilities, faced with few other alternatives, moved into public housing primarily inhabited by the elderly. Many concerns were raised about mixing these two population groups, and advocates for the elderly, meeting informally through the Elderly Housing Coalition, concluded that a valuable resource for the elderly was fast disappearing. Organizations such as AARP, AAHSA, and the National Association of Housing and Redevelopment Officials convinced Congress to pass legislation allowing public housing projects, under certain specific circumstances, to be designated as elderly only. On the other hand, some analysts have concluded that because they have been viewed as the deserving poor and represented by vocal and powerful interests, the elderly have benefited disproportionately from government assistance. For example, some researchers, noting that 35% of eligible elderly renters compared with 28% of eligible younger families received rental assistance, argue that on the basis of worst-case needs, assistance should be targeted more heavily to families and less to the elderly. Analysts on the other side point out that poverty is more persistent among the elderly than other age groups, and the cohort of elderly renters who are more than 75 years of age—a group with fewer assets than younger renters—will grow dramatically by the year 2000. In the face of limited and dwindling government resources for housing, such arguments about the relative needs of different groups, targeting strategies, how best to meet the needs of the growing population of frail elderly, and the role of age-specific housing are likely to persist. (SEE ALSO: *Accessory Dwelling Units; American Association of Homes and Services for the Aging; American Seniors Housing Association; Assisted Living; Assisted Living Facilities Association of America; Board-and-Care Homes; Congregate Housing; Continuing Care Retirement Communities; Dementia; ECHO Housing; Elderly Housing Coalition; Foundation for Hospice and Homecare; Home Care; Home Matching; Hospice Care; Hospice Foundation of America; Journal of Housing for the Elderly; National Institute of Senior Housing; National Resource and Policy Center on Housing and Long Term Care; Nursing Homes; Retirement Communities; Section 202; Section 231; Section 255; Senior Citizens Housing Act of 1962; Shared Group Housing*)

—*Jon Pynoos*

Further Reading

Callahan, James J., ed. 1993. *Aging in Place*. Amityville, NY: Baywood.
Golant, S. 1992. *Housing America's Elderly: Many Possibilities/Few Choices*. Newbury Park, CA: Sage.
Khadduri, J. and K. Nelson. 1992. "Targeting Housing Assistance." *Journal of Policy Analysis and Management* 11:21-41.
Lawton, M. P., M. Moss, and M. Grimes. 1985. "The Changing Service Needs of Older Tenants in Planned Housing." *The Gerontologist* 25:258-64.
Pynoos, J. 1990. "Public Policy and Aging in Place: Identifying the Problems and Possible Solutions." Pp. 167-208 in *Aging in Place: Supporting the Frail Elderly in Residential Environments*, edited by D. Tilson. Glenview, IL: Scott Foresman.
Pynoos, J. 1993. "Linking Federally Assisted Housing with Services for Frail Older People." *Journal of Aging and Social Policy* 4(3&4):157-77.
Pynoos, J. and P. Liebig, eds. 1995. *Housing for Frail Older Persons: International Policies, Prospects, and Perspectives*. Baltimore: Johns Hopkins University Press.
Regnier, V. 1994. *Assisted Living Housing for the Elderly*. New York: Van Nostrand Reinhold.
Struyk, R. 1987. "Home Adaptations: Needs and Practices." Pp. 259-76 in *Housing the Aged: Design Directives and Policy Considerations*, edited by V. Regnier and J. Pynoos. New York: Elsevier.
Struyk, R., D. Page, S. Newman, et al. 1989. *Providing Supportive Services to the Frail Elderly in Federally Assisted Housing*. Washington, DC: Urban Institute Press. ◀

▶ Elderly Housing Coalition

The Elderly Housing Coalition is composed of organizations concerned with the maintenance and growth of affordable and suitable housing for older Americans. Representatives from groups such as the American Association of Homes and Services for the Aging, the American Association of Retired Persons, B'nai B'rith, the National Council of Senior Citizens, and the National Association of Housing and Redevelopment Officials meet monthly in Washington D.C. to discuss pending legislation, regulations, research findings, and strategies for coordinating housing and aging. On occasion, the coalition prepares position papers and meets with members of Congress and the Department of Housing and Urban Development. Contact person: Larry McNickle, American Association of Homes and Services for the Aging, 901 E Street, NW, Suite 500, Washington, D.C. 20004-2037. Phone: (202) 508-9428. Fax: (202) 783-2255. (SEE ALSO: **Elderly**)

—*Jon Pynoos* ◀

▶ Emergency Low-Income Housing Preservation Act of 1987

The Emergency Low-Income Housing Preservation Act (ELIHPA) was adopted by Congress in Title II of the Housing and Community Development Act of 1987. It responded to the emerging crisis presented by the potential loss of nearly 1 million affordable apartments through prepayments of federally insured and financed mortgages and terminations of Section 8 rental assistance contracts.

During the 1960s and 1970s, hundreds of thousands of rental units were developed nationwide by for-profit groups who received low-interest federal loans and agreed to keep rents affordable to low- and moderate-income families in exchange for tax-sheltered depreciation benefits. The loans were provided under the Rural Housing Service (RHS) (formerly the Farmers Home Administration [FmHA]) Section 515 program and Department of Housing and Urban Development (HUD) Section 221(d)(3) and Section 236 programs. The use of the projects was controlled through regulatory agreements that were coterminous with the terms of the loans. However, although loan maturities ran from 40 to 50 years, administrative decisions by the FmHA (RHS) and HUD allowed owners to prepay their loans and, thereby, remove all regulatory restrictions on admissions and rents well before loan maturity.

In some cases, very-low-income tenants who could not even afford the mortgage-subsidized rents under these programs received supplemental project-based subsidies through the Section 8 Program that covered the difference between 25%, later increased to 30%, of adjusted gross income and the unit rent. Other projects were developed under the Section 8 New Construction and Substantial Rehabilitation Programs with nonsubsidized mortgages but long-term commitments of Section 8 by HUD.

ELIHPA was Congress's attempt to stanch the potential for massive tenant hardship and displacement when these subsidies were terminated. First, Congress took a page from a 1986 moratorium on FmHA (RHS) prepayments and passed a 2-year moratorium, in lieu of a permanent solution, to restrict prepayments on the estimated 400,000 HUD units with 20-year prepayment restrictions. Second, Congress imposed permanent prepayment restrictions on prepayments of the estimated 110,000 pre-1979 FmHA (RHS) units, which had no prepayment restrictions. This replaced the emergency legislation adopted in 1986. Finally, Congress authorized HUD to increase rents in the estimated 465,000 units with expiring Section 8 contracts, which ran 5 years or more, and imposed a longer notification period for tenants prior to termination.

The HUD prepayment restrictions were contained in Subtitle B of Title II. The unilateral ability of owners to prepay was greatly restricted and, instead, owners were provided financial incentives to extend project affordability or transfer the property. After filing an initial notice of intent to prepay, owners were required to submit a "plan of action" to HUD. The plan was to include information describing any proposed changes in the status of the mortgage and regulatory agreement, the low-income affordability restrictions and current ownership, and the impacts of any such changes on the existing tenants and supply of affordable housing in the community.

An owner could pursue any of three options: (a) request incentives from HUD to continue the low-income use of the housing; (b) sell the housing to a qualified nonprofit organization, limited-equity tenant cooperative, public agency, or other acceptable entity that agreed to maintain affordability; and (c) try to prepay. The incentives included an equity loan for 90% of the appraised value of the property, increased rate of return on investment, a capital improvements loan, and in the case of a transfer, an acquisition loan. Additional Section 8 funds were authorized to support the higher rents after provision of incentives and still keep units affordable. To approve a "stay-in" or sale, HUD must find that there will be no tenant displacement, low-income tenants will pay no more than 30% of income, and affordability will be maintained for the remaining term of the original mortgage. To approve a prepayment, HUD must find that termination will not cause economic hardship or displacement of current tenants or adversely affect the supply of affordable housing or housing opportunities for minorities in the community.

Subtitle B was replaced in 1990 by Title VI of the National Housing Act, the Low-Income Housing Preservation and Resident Homeownership Act (LIHPRHA), which provided a different and much more detailed procedural scheme for processing prepayment-eligible projects. Certain owners were permitted to reserve the option to proceed under the "old" law or the "new" law.

The FmHA (RHS) prepayment provisions, contained in Subtitle C of Title II and titled Rural Rental Housing Displacement Prevention, created a three-part prepayment process. First, after filing a prepayment notice, the FmHA (RHS) was obligated to offer financial incentives to owners to induce them to extend project affordability or transfer the property to a qualified purchaser—nonprofit or public agency—who agreed to extend affordability for a minimum of 20 years. The main incentives were an equity loan for up to 90% of the difference between the appraised value of the property and outstanding loan balance and increased rate of return on investment. In the case of a transfer, the main incentives were an acquisition loan for up to 102% of the appraised value of the property and $10,000 advance to cover some direct preacquisition costs. In both cases, the FmHA (RHS) could authorize additional rental assistance to ensure continuing affordability.

Second, owners who rejected the incentives and still sought prepayment were subject to three prepayment requirements or exceptions that had to be satisfied before the FmHA (RHS) could grant prepayment approval. Under the first exception, an owner could prepay by agreeing to continue operation of the housing in low-income use for 20 years from the date the loan was made and, thereafter, sell the property to a qualified nonprofit or public agency. The second and third exceptions required owners to demonstrate that the prepayment would not materially affect the housing opportunities of minorities and not cause displacement or adversely affect the supply of safe, decent, and affordable housing.

If owners rejected all incentives and could not meet any of the prepayment exceptions, the FmHA (RHS) was required to force a sale to a nonprofit organization or public agency at the property's appraised market value. Local purchasers had the first right of purchase. If, however, after a 180-day marketing period, a bona fide purchase offer was not executed, owners had an unrestricted right to prepay and convert their properties to market-rate housing.

Finally, Subtitle D required owners to give tenants and HUD one year's notice of termination of a project-based Section 8 contract and no less than 90 days for termination of a housing certificate or voucher. Moreover, HUD was directed to adjust contract rents to reflect true market comparables, up to 120% of the fair market rents in HUD's Section 8 existing program. (SEE ALSO: **Affordability**)

—*Robert Wiener*

Further Reading

Collings, Art. 1987. *Displacement of Tenants Through Prepayment of FmHA Section 515 Loans*. Washington, DC: Housing Assistance Council.

National Task Force on Rural Housing Preservation. 1992. *Preserving Rural Housing*. Sponsored by California Coalition for Rural Housing Project, Housing Assistance Council, and National Housing Law Project, Washington, DC.

U.S. General Accounting Office. *Rural Rental Housing: Impact of Section 515 Loan Prepayments on Tenants and Housing Availability.* 1988. Washington, DC: Author.

Wiener, Robert. 1987. "Rural Housing: The Prepayment Debate." *Economic Development and Law Center Report* 17(3-4):1-11.

Wiener, Robert. 1990. "Prepayment Hits Countryside Hard." *Shelterforce* (March/April):18-21. ◀

▶ Employer-Assisted Housing

The term *employer-assisted housing* collectively stands for the various techniques used by employers to provide housing assistance or housing benefits to employees.

Unlike employer housing programs from an earlier era, today all employer-assisted housing programs in the United States are organized on an "arms-length" basis between employer and employee. Frequently, a governmental or private sector intermediary will stand between the employer and employee, playing the role of benefit organizer, provider, manager, or implementer. Modern employer-assisted housing programs are also organized differently than are programs in Western Europe (principally in the Scandinavian countries and Germany) and in Japan. In these nations, employer-assisted housing was an important method of housing finance during the two decades following the end of World War II. In Europe, employer-assisted housing grew as a result of tripartite agreements between the government, employers, and large labor unions, in which employers invested and contributed to savings-and-loan-like institutions owned by the unions and insured by the government. These institutions, in turn, financed and frequently built the housing for workers. In Japan, where housing remains in short supply and very expensive, large corporations continue to provide housing benefits for workers, particularly single employees, for whom dormitory-like facilities are provided, amounting to about 4.5% of that nation's housing stock in 1995.

Employer-assisted housing programs can be generally categorized as either demand or supply programs. Demand programs enhance the affordability of existing housing and generally facilitate homeownership, whereas supply programs assist in the construction or substantial rehabilitation of specific housing units and can be used to create both multifamily and single-family housing. The program selected is often a function of the type of problem the employer is seeking to address and the cash, risk, and debt capacity of the employer.

Types of Programs

DEMAND PROGRAMS

Group mortgage origination: Employers (and labor unions) can negotiate with lenders for discounted mortgages (reduced fees, points, interest rates, and underwriting concessions) on the basis of the employer's providing exclusive marketing access to employees (including providing and endorsing informational materials provided by a specific lender or even providing applicant presceening for the lender). The discounts obtainable vary with market conditions, the demographics of the employee population, and likely volume.

Closing cost assistance: Employers can pay various closing costs, especially points. Providing this assistance is taxable income for the employee, but points are generally tax deductible, creating an offset that is, in effect, a tax-exempt benefit.

Mortgage guarantees: Employers can guarantee all or a portion of a mortgage (typically the down payment portion), enabling homebuyers to purchase housing with little or no down payment. Guarantees have the advantage for employers of requiring no "up-front" cash expenditure, although a contingent liability is created.

Group mortgage insurance: Lenders and public agencies can establish insurance pools into which employers pay one-time premiums on behalf of their employees. Besides saving mortgage insurance costs for their employees, group mortgage programs avoid the underwriting strictures of federal and private insurance programs, enabling innovative underwriting and down payment requirements. Unlike mortgage guarantees, insurance programs require cash expenditures, but in return, no contingent liability accrues to the employer.

Down payment assistance: Employers are providing down payment assistance directly by making loans and grants and indirectly by guaranteeing lender loans. Some employers offer forgivable loans tied to an employee turnover prevention strategy. In these programs if the employee stays with the employer for a period of years, the loan is forgiven, but if the employee leaves before that time, the loan is immediately due. The rate of forgiveness can be set at a rate that is less than the cost of turnover, ensuring that whether the employee stays or leaves the employer has at worst not lost any money and at best saved money.

SUPPLY PROGRAMS

Direct cash assistance: This can be provided at many different levels. Smaller investments might yield an employee priority in obtaining a lease, and larger cash contributions would result in rent rate discounts.

Construction loan guarantees: Providing a developer with a loan guarantee for all or a portion of a construction loan not only can save a developer substantial insurance and interest fees, it can also change the underwriting on a loan, because an additional balance sheet is providing security. In return for these savings, employers can seek rent rate and sales price concessions for employees.

Purchase guarantees: These guarantees enable a developer to know in advance the maximum cost of construction interest, ensuring that if units are not purchased in the marketplace by an agreed-on date, the employer agrees to purchase the units. The cost of this purchase to the employer is not the cost of purchasing but the cost of holding the units until they can be sold. In return for providing the guarantee, the developer would be expected to market units at a discount to employees. If all of the units are sold before the agreed-on "take-out" date, this program has essentially no cost to the employer.

Land: Many firms have excess lands that can be sold or leased on a long-term basis to developers. The value of the land subsidy is then calculated in a sales or rent rate discount.

Master leases: Employers who agree to long-term leases can obtain rent rate concessions from property managers. Although the employer is guaranteeing the reduced rent, the employer pays the reduced rent only in the event of a unit vacancy.

Housing trust funds: These are a specialized form of housing benefit. Trust funds are typically used by unionized firms in which a sum per hour per worker is placed in a fund controlled by management and labor. The fund can be used to provide a variety of supply and demand programs.

Employer Motivations

Employers have a variety of motivations for offering housing benefits. One reason is as an employee recruitment and retention tool in which housing benefits are used to directly overcome high housing costs or housing scarcity. Housing benefits are more efficient solutions to the problems of housing affordability and scarcity than higher wages, even if higher wages are economically possible. This is because higher wages are subject to taxation (some housing benefits can be structured to avoid taxes), because there is no ensurance that housing prices will not rise faster than wages (as took place on both coasts during much of the 1980s), and because raising wages for a relatively few employees within a labor market will not spur additional construction.

Other employers will be motivated by neighborhood revitalization or safety needs. Employers operating in undesirable areas have increased security and insurance costs and trouble attracting customers. By targeting a benefit to neighborhoods surrounding the employer, the employer can create a buffer zone of responsible employees with a strong interest in neighborhood security and revitalization. Targeting a benefit may also be a useful political strategy for firms seeking to expand their facilities into a potentially hostile neighborhood. By having employees as neighborhood residents, employers can combat NIMBYism ("not in my back yard").

The federal Clean Air Act is also providing an incentive for locationally targeted housing benefit programs. The act requires large employers (those having more than 100 employees) and operating in areas that do not meet federal air quality standards to take actions to reduce the air pollution resulting from their employees' commute to work. In California air quality management districts, employer housing programs that reduce commuting distances or enable employees to obtain housing proximate to mass transit that serves the employer are being accepted as appropriate air pollution abatement strategies.

Employer Partnerships

Housing benefits, unlike other employee benefits, attract public and private partners. For example, lenders wanting to achieve Community Reinvestment Act goals are frequently willing to provide various mortgage discounts, and real estate brokers and appraisers can provide volume-driven discounts.

Similarly, both the public and nonprofit sectors are in the business of providing assistance to low- and moderate-income families. However, these programs are typically very underfunded in contrast to the need. Employer programs that assist these income groups can be linked to public and nonprofit programs, providing both the employer and the public or nonprofit provider with the opportunity to leverage resources with the other's program.

Partnerships of these kinds are often important to the employer, not simply as a way of enhancing a benefit program but as a way of working with a partner who is familiar with housing, because the employer frequently is not. Housing benefit programs are not standardized, as the list of program varieties indicates. However, Fannie Mae's (formerly, the Federal National Mortgage Association) MAG-NET program attempts to standardize housing assistance by (a) seeking specific (and fewer) types of employer interactions, principally down payment loans, forgivable loans, grants and closing cost assistance, and (b) establishing minimal levels of employer assistance (2% of the mortgage amount). Using its market power, advertising and lender relationships, Fannie Mae is able to direct employer interest toward those interventions specified in the MAGNET program. Although this has the effect of truncating choice for employers (and employees), it also enables employers to provide some assistance, in a safe manner, without having to do much research or program customization.

To overcome employer unfamiliarity with housing and to advance local housing agendas, a number of local governments have taken the lead in forming civic housing partnerships in which the local government organizes a program and provides some resources in return for employer participation. (SEE ALSO: **Affordability;** *Company Housing*)

—Daniel Hoffman

Further Reading

Daunton, M. J. 1990. *Housing the Workers: A Comparative History, 1850-1914.* London: Leicester University Press.

Ferlauto, Richard C., Daniel N. Hoffman, and David C. Schwartz. 1991. "Housing Benefits: New Resources for Managers." *Public Management* 73(1):13-17.

Hoffman, Daniel and David Schwartz. 1992. "Low-Cost Housing: A Benefit for Employers." *Journal of Housing* 49(5):233-38.

Schwartz, David C., ed. 1990. "Coming of Age: Employer-Assisted Housing." *Shelterforce* 8(3 [Special issue]).

Schwartz, David C. and Daniel Hoffman. 1989. "Employers Help with Housing." *Journal of Real Estate Development* 5(1):18-22.

Schwartz, David C. and Daniel N. Hoffman. 1989-90. "Employer-Assisted Housing: A Benefit for the '90s." *Employment Relations Today* 17(1):21-29.

Schwartz, David C., Daniel N. Hoffman, and Richard C. Ferlauto. 1992. *Employer-Assisted Housing: A Benefit for the 1990s.* Washington, DC: Bureau of National Affairs Books. ◄

► Energy Conservation

Since the Arab oil embargo of 1973, which heralded the end of an era of inexpensive fossil fuels, energy use in the residential sector of the United States has been the focus of considerable research; demonstration programs; and federal, state, and local policies. Although the nation has become more energy efficient since the energy crisis that followed the embargo, energy conservation measures in the residential sector are not implemented at levels consistent with available technology.

Housing in the United States accounts for about one-third of the nation's total energy consumption. Although the actual numbers vary slightly from year to year and by geographic region, in the annual energy budget of the average U.S. household, heating, the largest single category, represents more than half or about 54%; water heating accounts for 18%; appliances use about 23%; and air-conditioning uses about 5%.

Although the use of energy in housing cannot be totally eliminated, it can be reduced. In existing homes, energy conservation features, such as insulation, caulking, and weatherstripping can be added. Households can also take actions to reduce energy use, such as lowering thermostat settings for heating systems, substituting fluorescent tubes for incandescent lightbulbs, and purchasing energy-efficient appliances.

Although new homes can be constructed with optimal levels of energy efficiency built into them, a majority of new homes are built to the minimum levels specified by building codes. Whether optimal levels are designed into newly constructed homes depends on a number of inter-related factors, including the nature of the U.S. housing industry. The provision of residential shelter in the United States is an activity with a vast number of participants, including landowners, developers, architects, government officials, laborers, lenders, real estate brokers, and owners. The housing industry can be characterized as a highly fractionalized activity, involving many small producers and consumers.

The localized nature of housing markets and the increasing complexity of the technology of energy efficiency inhibit the diffusion of innovations. In addition, the high initial costs characteristic of energy-conserving homes deter buyers, who typically are cost sensitive, and lenders, who have been slow to recognize the value of these homes. Progress has been made in the development of home energy rating systems and energy-efficient mortgages, which may eventually help to overcome these barriers.

The evolution of energy-efficient design in the residential sector of the United States can be considered to be in the third phase of its development. The first phase, begun shortly after the 1973 embargo, was dominated by an interest in active solar systems. Initial costs of the mechanical equipment used in these systems and the large spaces necessary for energy collection and storage were drawbacks, especially in northern regions of the United States. The second phase was marked by a resurgence of the centuries-old technique of passive solar design, in which the solar energy collection, storage, and distribution systems were incorporated into the components of a structure. Initial costs, again, and uncertainty about energy savings hindered the widespread use of passive solar architecture.

The third phase of energy-efficient design reduces the need for any type of energy source, whether it is oil, gas, electricity, or sunlight, through superinsulation. The term *superinsulation* refers to a method of house construction that is unique—not because of the materials that are used, but because of the *way* they are used. Although there is no agreement regarding a precise definition for superinsulated housing, the term generally refers to extremely high levels of insulation materials in the building envelope, a continuous air/vapor retarder, and an effective ventilation system. Although a superinsulated house may cost as much as 10% more to build than a house of conventional construction, energy savings are very high. A superinsulated house can be designed to require as little as 5% of the energy required by its conventional counterpart. Furthermore, less uncertainty is associated with the savings achievable from superinsulation than with those from solar designs, because solar gain is less crucial to a house that requires little energy for heating or cooling.

The barrier between a house's interior and exterior environments is formed by the building shell, or *thermal envelope*. The thermal envelope encloses living space where the thermal conditions, or temperature, need to be maintained. In a home, the thermal envelope is the layer or layers of materials that stop or slow down heat transfer between the living space and the outdoors. Figure 12 shows a section, or vertical cut, through a two-story house with a basement and attic. The heavy line represents the thermal envelope.

The walls, ceilings, and floors that define the thermal envelope can be filled and covered with various types of insulation, a material that slows or prevents heat loss (during the heating season) or heat gain (during the cooling season). The insulating value of a material, such as fiberglass, cellulose, or polystyrene, is indicated by the material's resistance to heat flow. This resistance is expressed in a number, called an R-value. A material with a high R-value is a good insulator. A material with a low R-value is a poor insulator.

Wall insulation is important because it is an integral part of the thermal envelope and because such a large area of the house is exposed to the outdoors. Of all the elements of the thermal envelope, windows typically are the least energy efficient, with even double-pane windows losing heat up to 20 times faster than the average insulated wall.

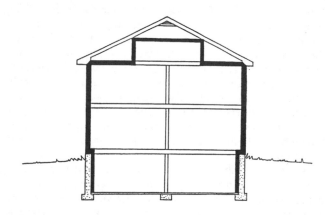

Figure 12. Thermal Envelope of a House

The energy efficiency of existing windows in a home can be improved with additional panes of glass or plastic, caulking, and weatherstripping; or units can be replaced with energy-efficient windows that take advantage of technological developments. Low-emissivity (or low-E) windows consist of two or more panes of glass, with a microscopically thin metallic coating applied to one of the panes. This low-E film reflects radiated heat back to its source, thereby keeping a home warmer in the winter and cooler in the summer. Other windows use low-E films in conjunction with special gas, such as argon, between the glass panes. The gas filling is heavier than ordinary air and acts as an additional insulator. Other windows use electrochromic coatings—thin, solid-state materials that change optical properties when exposed to electric current, allowing home occupants to exercise greater control over the amount of sunlight and heat that enter a structure.

Paralleling developments in energy conservation technology is an interest in the effects of these measures on indoor air quality. An avoidable consequence of reducing a home's rate of exchange between indoor and outdoor air is a buildup of pollutants that can include formaldehyde from particle board and plywood, nitrogen oxides from unvented gas-burning appliances and tobacco smoke, organic compounds released during the aging of synthetic materials, and radon gas from certain soils that come into contact with the house or water supply. Moisture levels in houses with inadequate ventilation can be high enough to cause structural damage. Design solutions to these problems include ventilation systems called *air-to-air heat exchangers* that, during the heating season, capture the heat from outgoing stale air and transfer it to incoming fresh (but cold) air. This flow is reversed during the cooling season. These systems prevent improvements in a home's level of energy efficiency from creating an unhealthy indoor environment.

Energy conservation in a home can also be affected by choices made in heating and cooling equipment. Furnaces and boilers are now manufactured to very high levels of annual fuel utilization efficiency (AFUE) ratings. AFUE measures a heating system's efficiency over an entire heating season and refers to what percentage of fuel used by a system is converted to useful heat. AFUE ratings of more than 95% are possible with new equipment. Much of the savings in these systems results from the way in which products of combustion are expelled from the furnace or boiler. Older systems used chimneys for this purpose, which relied on the principle that warm air rises. Some heat produced by the furnace or boiler was deliberately directed up the chimney to keep it warm. The rising warm air would then carry away combustion products, including carbon monoxide and nitrogen dioxide. Instead of heating a chimney for this purpose, an energy-efficient furnace or boiler uses direct venting of combustion products through a pipe that passes through a wall, in much the same way that a clothes dryer is vented.

Another energy-conserving heating system that can be used to heat an entire house or be used as a supplementary system is electric thermal storage (ETS). ETS systems are popular in areas served by electric utility companies that offer lower rates at night, which are called "off-peak" rates. These rates are usually half the daytime rates. An ETS system consists of a cabinet that contains ceramic bricks heated with electric current. Once charged, the bricks radiate heat without the use of current during the times of higher rates. ETS represents a practical alternative to conventional electric resistance heating systems in areas where discounted fuel rates are available to consumers.

In 1852, the heat pump was first proposed as a heating device by William Thompson (later known as Lord Kelvin), who referred to the device as a "heat multiplier." Although it is not a practical alternative in all areas of the United States because of decreases in efficiency at colder outdoor temperatures, a heat pump can be used to both heat and cool a dwelling. It generates more energy than it uses by removing heat from air, water, or the earth and making it available for water or space heating needs. The system operates on the principle that heat is available for such a transfer until temperatures drop to absolute zero (−450 °F). The operation of the heat pump can be reversed for cooling in the summer.

If air-conditioning is a requirement in a home and a heat pump is not a practical alternative, other choices are window- or wall-mounted units to cool individual rooms or a central cooling system for an entire house. Efficiency is rated differently for each type. Central systems use a seasonal energy efficiency ratio (SEER) that takes climate into account for its calculation. For room air conditioners, the energy efficiency ratio (EER) is used. In both cases, the higher the number, the more efficient the unit.

Avoiding energy losses in a home is not limited only to things that can be done to the dwelling. Outdoors, some techniques can help minimize heating and cooling bills. Windbreaks can be placed between a home and the direction of prevailing winds to block and divert them from or up and over the home. Evergreen trees—those that keep their foliage during cold weather—serve well as windbreaks. When properly placed, they can reduce winter wind speeds by 50% or more, which will significantly reduce heating costs.

Leaf-bearing trees should be placed on the south, east, and west sides of a home for maximum summer shade. This strategy will make it easier to keep a home cool during warm weather and reduce the cost of air-conditioning.

Incorporating principles of energy efficiency in a home as it is being built or renovated is not a difficult task. Doing so benefits individual households by decreasing energy expenses. It also benefits society by reducing environmental damage caused by burning fossil fuels to produce heat and electricity. (SEE ALSO: *Cultural Aspects; Earth-Sheltered Housing; Environment and Housing; Home Energy Rating Systems; Solar Housing*)

—*Joseph Laquatra*

Further Reading

Bell, M., R. Lowe, and P. Roberts. 1996. *Energy Efficiency in Housing.* Aldershot, UK: Avebury.

Hop, Frederick Uhlen. 1989. *The Energy-Saving House Design Handbook.* Englewood Cliffs, NJ: Prentice Hall.

Laquatra, Joseph. 1987. "Energy Efficiency in Rental Housing." *Energy Policy* 15(6).

Nisson, J. D. Ned and Gautam Dutt. 1985. *The Superinsulated Home Book.* New York: John Wiley. ◄

► Enterprise Foundation

Founded in 1981, the Enterprise Foundation's mission is to provide low-income households with opportunities to access fit and affordable housing and to move out of poverty. Working at more than 153 locations, it has helped to produce more than 36,000 housing units by providing training and technical assistance to more than 350 neighborhood-based nonprofit organizations. The foundation produces several publications, including *Cost Cuts,* a quarterly newsletter; *Network News,* a triennial newsletter; *Hope from Homes,* a triennial newsletter; a publications catalog; and an annual report. Annual budget: $13 million. Staff size: 130. Director of Communications: Sandra Gregg. Address: 10227 Wincopin Cr., Suite 500, Columbia, MD 21044. Phone: (410) 964-1230. Fax: (410) 772-2701. (SEE ALSO: *Local Initiatives Support Corporation; Neighborhood Reinvestment Corporation*)

—*Laurel Phoenix* ◄

► *Environment and Behavior*

Environment and Behavior is a bimonthly journal concerned with research in environmental design. It contains primarily empirical articles that focus on the environment-behavior interface, including housing. The readership is multidisciplinary, including architects, environmental psychologists, geographers, sociologists, landscape architects, and environmental designers. It was first published in 1969 and has an international circulation of 1,232. Price: $79 for individuals; $280 for institutions. Editor: Robert B. Bechtel, Department of Psychology, University of Arizona, Tucson, AZ 85721. Phone: (520) 621-7430. Fax: (520) 621-9306. (SEE ALSO: **Behavioral Aspects**)

—*Caroline Nagel* ◄

► Environment and Housing

Environment-Housing Links and Trends

The amount, types, and ratios of labor, energy, and raw materials used during the production and consumption of housing and the treatment of waste that results from these processes all have implications for the environment. Housing production has become less labor intensive and more dependent on manufacturing processes that require more energy. Rather than using local materials, housing construction increasingly uses imported products, either from other regions or from other parts of the world. Indiscriminate use of scarce or nonrenewable resources such as hardwoods from tropical forests or Wyoming marble has detrimental effects on the global environment. Transportation of such products over long distances also consumes more energy.

The design of the unit, its size and configuration, reflects and influences lifestyle patterns that determine energy and resource use in heating and cooling, refrigeration, water consumption, and, in general, use of small and large appliances. Although households are using more energy-saving products, there is a counter-trend of growing demand for durable consumer products, such as refrigerators and air-conditioning and microwave units. The design of a house also affects the treatment of waste products and recycling. There are various model designs aimed at resource recycling and waste treatment. However, these designs have not yet become widely accepted and used in the market. Finally, the adaptability and flexibility of the house for reuse and the reusability or biodegradability of building materials at the end of the life of a house affect resource conservation and environmental pollution. Although there is an ongoing effort for the reuse of old buildings or old components, it is offset by the greater volume of new development.

In addition to the design and production processes, residential land use characteristics, such as the location, size, type, density, and pattern of housing, also have environmental consequences. Whether or not residential development occurs on floodplains, wetlands, ecologically unique or sensitive areas, or lands that pose natural hazards, such as earthquakes, fires, or avalanches, has immediate and long-term environmental, social, and economic impacts. Low-density developments often threaten wetlands, forests, riparian habitats, and endangered species. Conversion of these lands to urban use can be prohibited through regulatory procedures or by public purchase of them.

The location of housing also affects the commuting distance to employment and services and, consequently, dependence on transportation. The national trend in this respect has been toward location of new developments away from urban concentrations; the United States experienced massive suburbanization after World War II. This trend continues. Between 1950 and 1990, the total suburban population increased by 181%, compared with a 65% increase in total population. During that same time, the proportion of Americans living in suburbs rose from 27% to 46%. At the same time, a large number of deteriorated units are being abandoned, mostly in inner cities. An indirect effect of location preferences for new development has been reduction of public resources allocated to older urban areas as well as infill opportunities.

Traditional zoning created a separation of land uses so that in traditional suburban developments, services and businesses essential for daily living are far from residential environments. Not only are they not within walking distance, these places are becoming increasingly far even in terms of driving distances. As a result of both of these factors, the choice to live in the country and the pervasiveness of separation of uses, average vehicle miles traveled per year per household increased 29% between 1983 and 1990. Between 1975 and 1990, the vehicle miles traveled increased three times faster than population growth. This sprawl-generated transportation system burns up 69% of the nation's oil (half of which is imported), with concomitant increases in carbon dioxide emissions, as well as fragmentation and disruption of ecosystems and wildlife habitat because of highway construction and land consumption.

Consistent with increased ex-urban residential development has been the continuing desire (of 73% of the population) for ownership of single-family dwellings of increasing lot and house size. Of 109.5 million occupied housing units counted in the 1996 census, nearly two-thirds, or 66.2 million housing units, were single-family detached homes. In the 46 years between 1950 and 1996, the average size of a unit more than doubled from 800 square feet to 1,686, whereas average household size declined from 3.37 to 2.62 persons. These trends have direct implications for more energy and resource consumption within each household. They lead to loss of open space and creation of lawns that require water and pesticides (such as Kentucky bluegrass lawns), and they introduce exotic species that often compete with and degrade native species in adjacent natural areas. More roads and driveways mean increased impervious surfaces and more runoff resulting in increased soil erosion and groundwater pollution because of nonpoint sources, such as the salt used in the winter to melt ice on roads.

In addition to effects of overall lower densities, internal geometry of residential developments and siting of individual units affect the amount of energy used for heating/cooling and transportation. For example, modifying the uniform lot subdivisions that are common in traditional subdivision ordinances can encourage more flexible plans that allow for more sensitive responses to site conditions and preserve more natural and continuous open space. Attached housing with shared walls, aligning streets and siting buildings with regard to the sun, planting trees along streets and with respect to the sun, all affect solar access and reduce energy use. Cluster and solar access zoning are among regulations designed for these purposes.

Regulatory Responses: Issues and Innovations

The federal legislation aimed at protecting sensitive and critical environments and controlling pollution has resulted in a significant amount of local activity. After the National Environmental Policy Act (NEPA) of 1969, many states enacted their own versions, known as "little NEPA" acts. The Clean Air Act of 1970; the Clean Water Act of 1972; Marine Protection, Control, and Sanctuaries Act of 1972; the Coastal Zone Management Act of 1972; the Safe Drinking Water Act of 1974; the Resource Conservation and Recovery Act of 1976; the Superfund Act of 1980; and their amendments in the 1990s all set the context for efforts to mitigate the effects of residential development on the natural environment. The federal government has influenced activities at the local level by providing incentives, by sanctions for obtaining federal funding, or by example. However, the jury is still out on how significant the effects have been. Urban homesteading or other sweat equity approaches, combined with solar rehabilitation and other energy-saving measures, relandscaping of private urban spaces using ecological principles, linking and restoring derelict urban lands and brownfields for public use, and reclaiming patches of urban wilderness to improve ecological conditions of residential environments are examples of strategies that need to be applied more widely.

As a response to the environmental concerns, planners developed a comprehensive approach to land planning based on land capability analysis. This approach, initially developed by Ian McHarg, involves detailed study of natural conditions of an area to decide on the land uses that are most suitable and not exploitative or destructive of natural processes. It is widely used by land planners.

Concern over the declining quality of the environment in general, and of residential areas in particular, has led to a "quiet revolution" in land use regulation, resulting in various forms of growth management at the state and county levels, the most notable among these being in New Jersey and Oregon. Many local governments also control growth by capping residential and commercial development. These regulations affect the intensity, type, and location of residential development. The typical growth management strategy also includes preservation of open space at the edge of a community or a region. However, in places where such greenbelt strategies have been applied (e.g., Boulder, Colorado), selection of lands for public purchase may be based less on ecological than on recreational criteria.

"New urbanism" is another approach to residential land use planning that has found support among planners and the real estate market. Its principles require denser, mixed-use developments, served by public transportation. This approach holds promise for reducing some of the potential negative environmental impacts of new residential developments. However, most examples developed so far are on the urban fringe and have not yet appreciably changed commuting patterns and automobile usage. Also, new urbanism has been slow to implemente more aggressive design principles for ensuring sustainable environments, such as native landscaping, communitywide sustainable energy and waste treatment systems, and green architecture for the individual units.

Environmental regulations and growth management planning may have unequal social and economic consequences. Higher-income groups are in a better position to buy their way out of polluted areas and resource-inefficient homes, leaving the poor to occupy the most environmentally degraded environments. Growth management strategies and the new urbanism approaches have been implemented in places where the market forces have been favorable and benefited those who are able to afford the improved environments. In a market-led economy where economic health is measured by "new housing starts," the environmental gains will inevitably be selective, will be very limited in scope, and will not favor infill, reuse, or labor-intensive building methods.

An effective strategy to address housing-environment issues will have to consider socioeconomic distributional consequences of proposed solutions rather than framing the challenge simply as a balancing act between goals of sustained economic growth and preservation of the environment. Approaches should emphasize user and community control over the supply of housing (its production, delivery, and management), consider the housing provision as a source of jobs for potential residents, and bring environmentally desirable building, landscaping, and planning practices to a wider segment of the population. (SEE ALSO: *Brownfields; Energy Conservation; Growth Management*)

—*Fahriye Hazer Sancar*

Further Reading

Beatley, T. and K. Manning. 1997. *The Ecology of Place: Planning for Environment, Economy and Community.* Washington, DC: Island Press.

Bhatti, Mark. 1996. "Housing and Environmental Policy in the UK." *Policy and Politics* 24(2):159-70.

Calthorpe, P. 1993. *The Next American Metropolis: Ecology, Community and the American Dream.* Princeton, NJ: Princeton Architectural Press.

Coates, G., ed. 1981. *Resettling America: Energy, Ecology, and Community.* Andover, MA: Brick House Publishing.

Downs, A. 1994. *New Visions for Metropolitan America.* Washington, DC: Brookings Institution.

Hayden, D. 1984. *Redesigning the American Dream.* New York: Norton.

McHarg, I. 1969. *Design with Nature.* New York: Natural History Press, Doubleday. ◀

▶ Environment and Planning A: Urban and Regional Research

First published in 1969, this monthly journal covers issues in urban planning, including housing and topics such as environmental values, long-term urban growth, and locational analyses of industries. Its circulation is international. Price: $805 per year (12 issues); individual subscriptions available on request. Submit papers in triplicate to R. J. P. Whitehead, Editorial Assistant, 19 Dunstarn Lane, Leeds LS16 8EN, United Kingdom. Address: Pion Limited, 207 Brondesbury Park, London NW2 5JN, United Kingdom.

—*Laurel Phoenix* ◀

▶ Environment and Planning B: Planning and Design

This is bimonthly journal concerned with the interests of urban and regional planners, architects, and design theorists. It focuses on mathematical, computer, and systems approaches to cities, regions, buildings, and urban morphologies. Articles cover new computer methods in planning and design, new developments in planning theory, design simulation models, analyses of urban and architectural structure, land development processes, and planning innovations. It was first published in 1974 and is circulated worldwide. Price: $300 per year (six issues); individual subscriptions available on request. Submission editor: Professor Michael Batty. Address: Centre for Advanced Spatial Analysis, University College London, 1-19 Torrington Place, Gower Street, London WC1E 6BT, United Kingdom.

—*Caroline Nagel* ◀

▶ Environment and Planning C: Government and Policy

Environment and Planning C: Government and Policy is an interdisciplinary and international quarterly journal reporting original research in economics, political science, public administration, geography, urban and regional studies, and related disciplines. Articles focus on housing, planning, and land use; urban, environmental, and regional policy; transportation issues; tax and fiscal policy; and urban economic development. First published in 1983, it has an international circulation. Price: $205 per year (four issues); individual subscriptions available on request. Submissions editors: Professor H. Wolman, Department of Political Science, Wayne State University, Detroit, MI 48202. R. J. Bennett, Department of Geography, University of Cambridge, Downing Place, Cambridge CBZ 3BN, UK. Susan E. Clarke, Political Science, University of Colorado, Campus Box 333, Boulder, CO 80309. Phone: (303) 492-2953. Fax: (303) 492-0978.

—*Caroline Nagel* ◀

▶ Environment and Planning D: Society and Space

Environment and Planning D: Society and Space is an international multidisciplinary bimonthly journal focusing on theoretical and empirical research that addresses the relations between society and space. First published in 1983, it features articles on social theory and human geography; constructions of nature and culture; international environmental politics; industrial restructuring, underdevelopment, and uneven development; the built environment and urban and regional planning; and social movements, social struggles, and the geography of social change. Price: $285 per year (six issues), per institution. Individual subscription rates available on request. Editor: Professor Geraldine Pratt, Department of Geography, University of British Columbia, Vancouver, B.C. Canada V6T 1Z2. Phone: (604) 822-2663. Fax: (604) 822-6150.

—*Caroline Nagel* ◀

▶ Environment and Urbanization

Environment and Urbanization is a biannual publication serving to encourage Third World researchers, teachers, nongovernmental organization (NGO) staff, and professionals to debate on issues and exchange information on their activities and publications. Each issue contains 7 to 10 papers on a particular theme. In addition, each issue includes a guide to the literature, profiles of innovative Third World NGOs, book reviews, and research reports in English, Spanish, French, and Portuguese. The audience consists of academics and professionals working on housing and urban issues in the Third World. First published in 1989, its circulation is 2,400, internationally. Submissions editor: David Satterthwaite. Address: IIED, 3 Endsleigh St., London,

WC1H 0DD, United Kingdom. Phone: 44 171 388 2117. Fax: 44 171 388 2826. E-mail: humansiied@gn.apc.org. (SEE ALSO: **Third World Housing**)

—Laurel Phoenix ◀

▶ Environmental Contamination: Asbestos

Asbestos is a term that refers to a group of naturally occurring mineral fibers that have been used by mankind since the Stone Age, when they were mixed with clay to reinforce pottery. When mining and milling techniques were refined during the Industrial Revolution, asbestos began to be used more extensively. From that time, asbestos was added to thousands of products until the 1970s, when dangers of the material began to be widely recognized.

Studies of laboratory animals and asbestos workers and their families have shown that exposure to airborne asbestos is extremely hazardous. Several potentially fatal diseases can be caused by this exposure: lung cancer; mesothelioma, a cancer of the membrane that lines the chest and abdominal cavities; and asbestosis, an irreversible scarring of the lungs. Symptoms of these conditions are not usually noticeable until 20 to 30 years after the initial exposure to asbestos.

Most consumer goods produced today do not contain asbestos. Those that do contain asbestos fibers that could be inhaled are required to be labeled as hazardous. The potential for asbestos exposure exists in homes built before the 1980s as well as in older household products, such as appliances; walls, pipes, and ducts; floor tiles; and exterior roofing and siding materials. Asbestos was added to these products to strengthen them, to add thermal or acoustical insulation properties, or for fire protection. Typical building components and products that may contain asbestos in homes include the following:

- ▶ Steam pipes, boilers, and furnace ducts insulated with asbestos-containing materials
- ▶ Resilient floor tiles, backings on vinyl sheet flooring, and adhesives used for installing floor tiles
- ▶ Materials used as insulation around furnaces, boilers, and wood stoves, such as cement sheet, millboard, paper, or plaster mixtures
- ▶ Gaskets used around doors in furnaces and wood or coal stoves
- ▶ Soundproofing or decorative sprayed-on materials on walls or ceilings
- ▶ Spackling compounds used in finishing gypsum board and textured paints
- ▶ Asbestos cement roofing or siding materials
- ▶ Artificial ashes and embers used in gas fireplaces
- ▶ Older consumer products, including fireproof gloves, stove top pads, and covers for ironing boards, hair dryers, and toasters

Although asbestos cannot be positively identified through a visual inspection, building, plumbing, and heating contractors who have frequently worked with asbestos can usually make a reasonable judgment about whether asbestos is present in a material. In some situations, the best recourse is to contract for the services of professional environmental technicians who are trained in safely removing suspected asbestos-containing materials and analyzing them. Once asbestos is identified in a home, an analysis for potential safety hazards is conducted. In many cases, if asbestos is in good condition and not friable or releasing fibers into the air, it is best left alone. Disturbing asbestos that is not friable may create a hazard where none existed.

If asbestos that is not creating a hazard is identified in a home, it should be inspected regularly for signs of weakening or damage, such as ripping and water damage. This should be a visual inspection that does not include touching the material. If damage is noticed, corrective steps should be taken. These actions—repair or removal—should always be done by a professional.

Typically, repair involves sealing or covering asbestos-containing material. Sealing, also called encapsulation, refers to treating the material with a sealant that binds asbestos fibers together or coating the material to prevent the release of fibers. Covering, also called enclosure, is done by placing a protective wrap around asbestos-containing material. Whether either type of repair is undertaken, asbestos remains in the home, which can make later removal more difficult. However, repairing asbestos-containing materials is much less costly than removal.

Removal is the most expensive method for dealing with asbestos contamination. It is also the most hazardous because of the potential for release of fibers. Because many states and localities regulate the removal and disposal of asbestos, this work should be undertaken only by trained professionals, who are referred to as *asbestos abatement contractors*. Usually, these contractors must undergo special training and licensing procedures.

Regulations regarding asbestos repair and removal services vary across the United States, and in many areas, they do not apply to private homes but only to schools and other public buildings. Nevertheless, homeowners who decide to take corrective measures for an asbestos problem or who are planning remodeling projects that may disturb asbestos, should still rely on trained professionals to guard against risking the health of family members. (SEE ALSO: *Environmental Contamination: Lead-Based Paint Hazards; Environmental Contamination: Radon; Environmental Contamination: Toxic Waste*)

—Joseph Laquatra

Further Reading

American Lung Association; United States Consumer Product Safety Commission; United States Environmental Protection Agency. 1990. *Asbestos in Your Home.* U.S. GPO: 1990—276-202. Washington, DC: Government Printing Office.

Bower, John. 1989. *The Healthy House: How to Buy One, How to Build One, How to Cure a "Sick" One.* New York: Carol Communications.

Croke, Kevin et al. 1989. "Asbestos Removal and Treatment Impacts on Housing and Urban Neighborhoods." *Journal of Environmental Systems* 18(2):123-31.

National Association of Home Builders. 1989. *Asbestos Handbook for Remodelers: How to Protect Your Business and Your Health.* Washington, DC: Author.

U.S. Environmental Protection Agency; United States Consumer Product Safety Commission. 1988. *The Inside Story: A Guide to Indoor Air Quality.* EPA/400/1-88/004. Washington, DC: Author. ◄

► Environmental Contamination: Lead-Based Paint Hazards

Extent of the problem

Of all occupied housing units built before 1980, approximately 74%, or about 57.4 million homes, are estimated to have lead-based paint. Lead does not dissipate, biodegrade, or decay, so lead deposited into dust and soil becomes a long-term source of exposure. Anyone is susceptible to its pernicious effects, but young children are especially vulnerable.

An estimated 88% of homes built before 1940 have lead-based paint. By contrast, it was used in only about 76% of homes built between 1960 and 1979, the year that lead in house paint was prohibited. Of particular concern are the 14 million housing units that contain deteriorated lead-based paint and the 3.8 million deteriorated units that are occupied by young children.

During the past 10 to 15 years, average blood lead levels have decreased in the United States and in several European countries at about 5% to 15% per year. This decrease is because of the reduction of lead in gasoline, canned food, and drinking water. The most notable improvement in the United States has been that of ambient air. Lead in house dust and urban soil, which mostly originates in lead paint, has grown in relative importance in recent years as other sources of childhood lead exposure have declined.

Measurement of blood lead levels is the standard means of establishing lead exposure. According to the Centers for Disease Control and Prevention (CDC), the threshold level of lead associated with deficits in neurological development in children is 10 to 14 ug/dl.

Results from Phase 2 of the third National Health and Nutrition Examination Survey (NHANES), conducted during the period of October 1991 to September 1994, indicate that 4.4% of children in the United States aged one to five years, representing an estimated 930,000 children, have blood lead levels at or above 10 ug/dl. This represents a 51% reduction in the prevalence of elevated blood lead levels from Phase 1 of NHANES III, conducted during the period of October 1988 to September 1991. Independent predictors of elevated blood lead levels found in Phase 2 are non-Hispanic black race/ethnicity, low income, and living in housing built before 1946. For children living in housing built before 1946, 21.9% of non-Hispanic black children, 13.0% of Mexican American children, and 5.6% of non-Hispanic white children had elevated blood levels. For children who lived in housing built before 1946, about one in six (16.4%) living in low-income households and 11.5% in cities with populations more than 1 million had elevated blood lead levels.

Consequences of Exposure

Lead is a powerful toxicant with no known beneficial purpose in the human body. High levels of internal exposure that are left untreated can cause convulsions, coma, and death. The primary target organ is the central nervous system. Because children are still developing neurologically, they are more at risk from exposure to lead than are adults. Their bodies absorb and retain a larger percentage of ingested lead per unit of body weight than do adults, and their frequent hand-to-mouth activity brings them in greater contact with the environment. In homes that are poorly maintained, deteriorated lead-based paint and lead-contaminated soil are likely to be the most important sources of exposure to lead. However, in well-maintained homes in areas with little soil contamination, children are more often exposed to lead as a result of renovation methods that generate lead dust, such as the use of power sanders to remove old paint.

Studies show that low-level lead exposure poses risk of developmental neurotoxicity in children, causing abnormalities in their neurobehavioral and neuropsychological development. Lead exposure during early childhood produces deficits in learning and short-term memory.

Lead also causes nonspecific, decremental loss of tissue and organ function. Most research has been on the effects of lead on IQ, but lead also affects attention, language function, and academic performance. It can also affect human reproduction. There can be a long delay between the onset of low-level lead exposure and the development of toxic manifestations. Evidence is accumulating to suggest that lead stored in bone is not "inert" and that there may be a significant mobilization of lead during pregnancy, possibly causing adverse effects on fetal development and growth of infants. Recent analyses suggest that the threshold for lead toxicity is at nanogram rather than microgram range of blood lead concentrations and hence lower than current standards.

National, State, and Local Efforts

The Residential Lead-Based Paint Hazard Reduction Act of 1992 contains federal provisions pertaining to the control of lead-based paint hazards and the reduction of lead exposure.

This legislation also authorizes a grant program for state and local governments for the evaluation and reduction of lead-based paint hazards in privately owned housing built before 1978 and occupied by families of low and moderate income. Housing units are treated for lead paint hazards with methods ranging from specialized cleaning and maintenance efforts to full-scale removal of all lead paint.

The recipients of the grants provide matching funds for additional rehabilitation and lead hazard reduction. This includes in-kind services such as blood testing and screening of children under the age of six, public education and outreach programs directed to low-income families in priority housing in target neighborhoods, and rehabilitation in conjunction with lead hazard reduction efforts. The act also prescribes hazard evaluations and reductions for federally assisted housing built before 1978. National surveillance of blood lead levels helps in targeting interventions, tracking programs in eliminating childhood lead poisoning, and evaluating lead exposure in workers.

CDC funds the initiation and expansion of state community-based programs for the following:

► Screening large numbers of young children for lead poisoning
► Identifying possible sources of lead exposure
► Monitoring medical and environmental management of identified children
► Providing information to the public, health professionals, and policymakers
► Encouraging community action programs to eliminate childhood lead poisoning

Funds may also be used to develop the infrastructure needed to ensure timely and effective screening of children and identification and remediation of environmental lead hazards. HUD estimates that during fiscal years 1994, 1995, and 1996, the average annual expenditure related to the reduction of lead hazards by HUD grantees and public and Indian housing authorities was approximately $107 million.

Sources for More Information
The Residential Lead Hazard Control Reference Library brings together a wealth of information about lead-based paint dangers and their remediation on a CD-ROM. Its contents include HUD's Guidelines for the Evaluation and Control of Lead-Based Paint Hazards in Housing, key Federal rules and regulations on lead-paint abatement and disclosure, important public reports and scientific studies on residential lead hazards, and a widely used pamphlet that promotes lead-paint awareness. The Reference Library (AVI77) is available from HUD USER for $25.

Publications from HUD USER are available from: HUD USER, P.O. Box 6091, Rockville, MD 20850. Phone: 1-800-245-2691 and 1-800-483-2209 (TDD). FAX: (301) 519-5767.

Recent updates on laws, regulations, guidelines, and other activities at HUD's Office of Lead Hazard Control are available at: http://www.hud.gov/lea/leahome.html.

The National Lead Information Center is another important source of information on the hazards of lead and how to comply with the recent law regarding disclosure of lead-based paint to buyers and renters. It maintains a hotline phone number at 1-800-LEAD-FYI, which distributes information packets. Its clearinghouse phone number is 1-800-424-LEAD, which is a source of more technical information. Requests and comments may be sent via the Internet to: leadctr@nsc.org. It also contains a website at: http://www.nsc.org/ehc/lead.htm. (SEE ALSO: *Environmental Contamination: Asbestos; Environmental Contamination: Radon; Environmental Contamination: Toxic Waste; National Center for Lead-Safe Housing; Residential Lead-Based Paint Hazard Reduction Act of 1992*)

—*Barbara A. Haley*

Further Reading
Brody, D. J., J. L. Pirkle, R. A. Kramer, K. M. Flegal, T. D. Matte, E. W. Gunter, and D. C. Pashcal. 1994. "Blood Lead Levels in the U.S. Population." *Journal of the American Medical Society* 272(4):277-316.

Centers for Disease Control and Prevention. 1997. "Update: Blood Lead Levels—United States, 1991-1994." *Morbidity and Mortality Weekly Report* 46(7):141-46.

Davis, M. J., R. W. Elias, and L. D. Grant. 1993. "Current Issues in Human Lead Exposure and Regulation of Lead." *NeuroToxicology* 14(2-3):17-27.

Fraas, Arthur and Lutter, Randal. 1996. "Abandonment of Residential Housing and the Abatement of Lead-Based Paint Hazards." *Journal of Policy Analysis and Management* 15(3):424-29.

Grandjean, P. 1993. "International Perspectives of Lead Exposure and Lead Toxicity." *NeuroToxicology* 14(2-3):9-12.

U.S. Department of Health and Human Services. 1991. *Preventing Lead Poisoning in Young Children: A Statement by the Centers for Disease Control.* Atlanta, GA: Public Health Service.

U.S. Department of Housing and Urban Development. 1997. *Moving toward a Lead-Safe America: A Report to the Congress of the United States.* Available at http://www.hud.gov/lca/lcahome.html and from HUD's Office of Lead Hazard Control at (202) 755-1785, Ext. 114.

U.S. Department of Housing and Urban Development. 1991. *Comprehensive and Workable Plan for the Abatement of Lead-Based Paint in Privately Owned Housing: Report to Congress.* HUD-PDR-1295. Available from HUD User at 1-800-245-2691.

U.S. Environmental Protection Agency. 1995. *Report on the National Survey of Lead Based Paint in Housing.* EPA 747-R95-005. Available from 1-800-424-LEAD. ◄

► Environmental Contamination: Radon

Radon is a colorless, odorless, and tasteless radioactive gas produced from the decay of the element radium, itself a decay product of uranium. Because trace amounts of radium are present in most rocks and soils, radon exists nearly everywhere in the world as a soil gas and groundwater contaminant. Because it is a gas, radon can travel easily in the earth, between soil and rock particles. It enters buildings through cracks and openings in foundation walls and floors. Radon can also enter a house through the water supply, from which it can be released into house air through aeration at faucets, shower heads, and water-using appliances.

Radon is a health concern because it continues to decay into solid radioactive elements, referred to as radon decay products or progeny, which can become attached to dust and other particles in the air and then be inhaled by people. Radiation released by radon decay products that become lodged in lung tissue is considered to be the cause of between 7,000 and 30,000 lung cancer deaths annually in the United States. Some controversy exists regarding the accuracy of this estimate, but its supporters defend it by citing rates of lung cancer that have been observed in epidemiological studies of uranium miners. In addition, all major health organizations agree that radon causes thousands of deaths every year from lung cancer, which could be prevented. This is especially true among smokers, because risks of developing lung cancer from exposure to radon gas are much higher for smokers than for nonsmokers.

Radon concentrations are usually expressed in radiation units known as *picocuries per liter of air* (pCi/L). One pCi/L corresponds to 0.037 radioactive disintegrations per second in every liter of air. The U.S. Environmental Protection Agency (EPA) considers concentrations above 4 pCi/L as potentially dangerous.

The presence of radon can be detected under different testing conditions and with a number of special devices. Activated charcoal canisters are typically used for short-term tests. Long-term tests can be conducted using electret or alpha track detectors. Radon levels vary over time, and for this reason, long-term tests provide more reliable estimates of average concentrations in a home. Because radon levels vary within communities and even neighborhoods, testing is the only way to determine a particular home's radon level.

Steps taken to reduce indoor radon levels are referred to as *mitigation measures,* which in some cases consist of procedures as simple as sealing cracks in floors and walls. In other cases, special ventilation systems that draw radon gas from beneath concrete slabs and then expel the gas to the outdoors are used. This approach is referred to as *sub-slab depressurization* and consists of plastic pipe and exhaust fans. Costs for these systems can be minimal if they are installed at the time homes are constructed. For existing homes, installation costs vary, but they are similar to other home repairs.

No reliable method exists for testing a site for radon gas and using the test results to predict radon levels in a home to be built there. Although the site could be characterized with tests, modifications made to the site during construction are likely to change its characteristics. In addition, the depth at which tests are conducted may prevent the identification of problem spots. However, the increasing awareness of radon gas is causing both builders and buyers alike to seek assurances that new homes will be free of this environmental hazard. To address this concern, radon-resistant construction features can be included in the design and construction of new houses. If implemented during house construction, these measures may add little to the cost of the house and may be low-cost insurance against radon contamination. Radon-resistant construction features rely on familiar building materials and existing technologies. Many of these features rely on the same principles and techniques used to make new homes watertight. Some states have even mandated the use of radon-resistant construction features through provisions in building codes.

Buyers of existing homes, as well as lenders who finance them, who are concerned about radon levels in homes they may purchase, may request information about radon levels before a purchase is completed. This procedure is similar to information that is routinely gathered about other aspects of a home before a sale is completed. For this reason, homeowners are encouraged to test for radon and take any necessary mitigation measures before a home is placed on the market for sale. (SEE ALSO: *Environmental Contamination: Asbestos; Environmental Contamination: Lead-Based Paint Hazards; Environmental Contamination: Toxic Waste*)

—Joseph Laquatra

Further Reading

Bower, John. 1989. *The Healthy House: How to Buy One, How to Build One, How to Cure a "Sick" One.* New York: Carol Communications.

Clarkin, M. and T. Brennan. 1991. *Radon Resistant Construction Techniques for New Residential Construction.* EPA/625-2-91/032. Research Triangle Park, NC: U.S. Environmental Protection Agency.

Edelstein, Michael R. and William J. Makofske. 1996. *Radon's Deadly Daughters.* Lanham, MD: Rowman & Littlefield.

Soderqvist, Tore. 1995. "Property Values and Health Risks: The Willingness to Pay for Reducing Residential Radon Radiation." *Scandinavian Housing & Planning Research* 12:141-53.

U.S. Environmental Protection Agency; U.S. Department of Health and Human Services; U.S. Public Health Service. 1992. *A Citizen's Guide to Radon.* 2d ed. Washington, DC: Government Printing Office.

Weinstein, Neil D. and Peter M. Sandman. 1992. "Predicting Homeowners' Mitigation Responses to Radon Test Data." *Journal of Social Issues* 48(4):63-83. ◄

► Environmental Contamination: Toxic Waste

In the years since 1978, when the contamination of the Love Canal neighborhood was first publicly recognized, contamination of the residential environment by toxic chemicals or radioactive materials has garnered increasing attention and study, joining the continuing concern for contamination owing to biological pollutants. The residential environment is affected by global, regional, local, or site-specific contaminants. Multiple media may be implicated, including the outdoor air, water, soil, indoor air, building materials, or some combination thereof. Residential contamination is often "bounded" within a given neighborhood or region. Overall, many thousands of contaminated communities have been identified worldwide, bridging urban, suburban, rural, and tribal settings.

In the United States alone, some 4,500 sites were identified as "Superfund sites" (sites designated for remediation or cleanup) in 1994, with some 1,320 of these listed on the National Priority List. That this was only a partial representation of contamination is indicated by the fact that only 29 of 130 contaminated Department of Energy (DOE) installations are on the National Priorities List. Against a projected cost of $75 billion to address Superfund sites, these DOE sites are anticipated to cost $360 billion to remediate. And an accounting of the potentially contaminated sites associated with DOE activities on military installations approaches 30,000. In addition, one must consider industrial and other sites being addressed through state Superfund programs, contamination sites administered under the Resource Conservation and Recovery Act, and other radionuclide sites. Even this list must be considered to be only a partial accounting of the known extent of the contamination of the environment in the United

States. Other parts of the world, such as the former communist bloc countries, are thought to be even more seriously contaminated.

Contamination has also resulted from actions taken by residents themselves—for example, in applying pesticides and fumigants inside the home or to their lawns, in the choice of insulating materials, or by improper care and disposal of fuels, lubricants, or other hazardous materials. In other situations, residents are purely the victims of others' actions or of natural events, the outcomes of which are the release of hazardous materials to the biosphere so as to affect the home environment and the water sources, food sources, or air that supports the residents.

The "Built" Environment Versus Nature

Life in a contaminated community has become the normative experience for an increasingly growing percentage of the world's population. Despite the frequency with which contamination is encountered, problems of pollution continue to be treated as though they were separate and separable from problems of housing. There has been surprisingly little attention from designers and planners to the challenge of creating habitable housing in the midst of an increasingly uninhabitable environment. Such problems fall through the disciplinary cracks separating architecture, planning, engineering, and environmental studies. Society selectively attends to the reality of the "built" environment, ignoring the contextual suprareality of the "natural" environment. Contamination speaks to the increasing lack of separability of either realm.

Contamination and the Safety of Housing

Contamination of the ambient environment within which people live has created a background of hazards. It is rare to have clear causal evidence relating a particular environmental contaminant to specific health outcomes.

The link between lead and neurological symptoms, such as learning deficits in children, illustrates an area of relative epidemiological certainty. The home environment may contain lead in exterior soils, resulting from air deposition from ambient environmental sources (such as leaded gasoline, lead-emitting factories, and exterior paints) or from improper disposal of hazardous wastes (such as lead-contaminated fill material or improperly disposed batteries). In turn, such soils can become a source of lead exposure through gardens, play in the backyard, or dust entering the home. The interior environment, in addition, may evidence lead from sources such as peeling paint. Lead may also be present in drinking water, often the result of the deterioration of lead solders, pipes, or other components of house plumbing.

There is much yet to be learned about lead exposures and infinitely more to be learned about other single-contaminant exposures, multiple-contaminant interactions, and cumulative impacts. Effects from such contaminants may include cancers, reproductive problems, and somatic problems, ranging from allergies and asthma to skin or kidney problems. Neuropsychological and behavioral problems may result. Emotional stress and even psychopathology may be linked to pollution. Many of these symptoms and consequences may have multiple causes, may never be clearly linked with contamination, may occur at low chronic "background" levels as opposed to acute episodes, or may have latent outcomes that place them temporally distant from exposures. As with individual contaminants, the symptoms may also interact mutually in complex and unknown ways.

Uncertainty and Residential Contamination

Environmental contamination is often invisible. There may be a considerable delay between the release of the contaminant, the onset of exposure, and the point of discovery. Spatial connectors linking contamination and its source may be obscure or absent. Potentially complex interactions between multiple factors are likely to occur. When these factors are combined with ambiguity over the degree and consequences of exposure, it can be seen that uncertainty is a major factor in contamination of housing. The greater the uncertainty, the easier it is to deny that the contamination has occurred or to play down its consequences. Interestingly, many parties have a stake in such denial—including the owners, investors, developers, realtors, neighbors, and even residents of the contaminated property. Uncertainty makes it likely that different interpretations of the contamination will occur. Divergent beliefs about contamination are a key basis for community conflict and for disagreements between government officials and residents. Finally, uncertainty adds to the stress experienced by "victims" of contamination, principally those who believe that they have suffered hazardous exposures.

Contamination and Residential Life

Because contamination itself is often invisible and silent, people are more aware of the events surrounding the contamination than of the contamination itself. Contamination events may occur with a bang or a whimper. The precipitating cause may be part of the event—as with industrial accidents spewing immediate hazards—or a period of incubation may occur before the event begins, as with an unseen leaking gas tank buried beneath ground and discovered only after considerable delay. For unseen pollutants, the public announcement of the contamination is often the beginning of the event. Media publicity, government communications, various meetings, community activism, health studies, cleanup activities, hearings, lawsuits—all are included in the events surrounding the contamination. A chronology of these events represents a continuous intrusion into affected residents' previously normal lives.

Personal lifestyle changes often result from the unfolding events associated with environmental contamination. Areas of the home or yard may be defined as out-of-bounds. Warnings against the use of tap water may result in the importation of water for drinking, cooking, and personal hygiene. Showering, dish washing, clothes washing, and watering of gardens may become marginal areas of safe activity. When young children are present in the family, particular diligence may be required to ensure protection. Residents may attend meetings with government officials, with their communities, and with other citizens. Such forced behavioral changes are not only nuisances, but they serve as reminders of the underlying contamination, with its multiple meanings of threat and victimization. Related stresses complicate both intra-and extrafamily relationships.

Forced or voluntary residential relocation—sometimes temporary, sometimes of indeterminant length, and sometimes permanent—can be a consequence of contamination. At times, the contents of the residence and even the residence itself must be abandoned. Although forced relocation is not common, it is a major source of life stress, changing the basic shape of the person's life, demanding key decisions, and entailing perhaps major and irreplaceable losses. More frequently, the resident remains living at home during the protracted process for determining the type and degree of environmental cleanup that will occur. Remediation activities may be conducted around the house—for example, excavation, removal, and replacement of the soil—with the cleanup workers wearing protective clothing while residents watch unprotected. Mitigations, such as on-site incinerators, in-place storage, or treatment plants, may be constructed, augmenting worries from the original contamination with what residents see as new hazards. For invisible contaminants, it is difficult to know when or to trust that an environment formerly labeled as contaminated is now clean.

Contamination and (or of) the Meaning of Housing

Contamination events go beyond lifestyle impacts to affect the lifescape, the set of basic assumptions made by people in the course of everyday life. Thus, although it is common for people to assume that they will be healthy, contamination results in a contrary assumption of impending illness. People further frequently assume that they exert extensive control over the course of their own lives; this assumption is directly refuted by a contamination event. It is common to assume that bad things (such as contamination) will not occur to good people. When they do, the victims are angry and hurt. Furthermore, it is common to trust that others— and particularly the government—will come to the aid of contamination victims. When such aid is absent, insufficient, or inappropriate, there is a further erosion of trust.

Part of the lifescape involves people's understanding of their place in the environment. Although it is commonly assumed that the environment is a benign force present to serve human needs, after contamination, people may realize their immersion in and dependence on the environment. At the same time as they now see themselves as part of a defiled "nature," their assessment of nature shifts from a benign view to one recognizing the environment as a source of potential danger against which there is relatively little they can do to protect themselves and their families.

Finally, home is an essential part of the lifescape. For those achieving the ideal, home is a place offering a sense of security, personal control, and identity. Homeownership further offers the opportunity to invest in property and to garner status. Contamination frequently results in an inversion of this concept of home. What was a place of security is now a place to fear. What was a place to exert personal control is now controlled by others and by contamination events. What once was a place imbued with personal identity related to family life, taste, and achievement is transformed into a place of stigmatized identity.

If these characteristics of the psychological value of home were all that was affected by contamination, people could sell their homes and escape. It is a particularly cruel aspect of the inversion of home that residents are frequently trapped in the home at the very point they wish to escape. Environmental stigma causes this trap—namely, because contamination diminishes the value and desirability of the home to others. Homeowners holding mortgages and having invested their equity at home most often lack the resources to walk away from their property. Thus, fearing the consequences, most remain, trapped in a home where dreams have been turned into nightmares.

Although contamination causes environmental stigma, it is often the case that contamination reflects social stigma. Thus, a growing literature converges on the conclusion that poorer people are most likely to be affected by contamination and, even more conclusively, that racial minorities are its most likely victims. Although victimization through contamination always represents an environmental injustice, regardless of whom is affected, such demographic patterns suggest an underlying "environmental racism." Not only does the cause and consequence of contamination reflect complex social as well as environmental dynamics, these issues also complicate questions of remediation, as well, as discussed below.

Liability and Regulation

Residential contamination invites complex questions of legal liability and efforts to provide protection through government regulation. These issues include the growing stringency to which sellers of property (and the realtors who assist them) are held responsible for contamination found by the buyer. In the United States, the federal Superfund law and its state counterparts set forth procedures whereby government can implement actions to protect people and the environment from contamination, seeking to identify and hold liable "potentially responsible parties" at a later point. Federal and state legislation governing the handling and disposing of hazardous materials has also created numerous regulatory frameworks, allowing for government and citizens to take protective actions to avoid or address contamination. Finally, civil litigation provides an avenue for residents to seek redress for environmental contamination. Although none of these remedies comes close to resolving the issues posed by contamination for housing, they have become part of the overall social environment surrounding the built environment.

Such governmental remedies for contamination have confronted various barriers to success, stemming from cost factors, rigidity of bureaucratic structures, the sheer numbers of sites requiring action, opposition from business and local government, and questionable remedies. Often, the affected communities have not been adequately involved in devising solutions, adding to the stress and conflict associated with the social response to contamination. A nagging issue has involved how "clean" the cleanup of the site must be, with the community generally seeking the highest level of cleanup, the responsible parties the lowest level, and government falling in between. Removal of all pollutants in an effort to return sites to a condition comparable to the precontamination status has given way to decreasing levels of cleanup. There has been a further shift from remediation requiring removal of contaminants to on-site treatment. This shift avoids the difficult moral, economic, and political questions entailed by moving problems from one "backyard" to another and of creating new hazardous waste disposal sites in areas previously uncontaminated. However, it raises issues of "perpetual jeopardy" because on-site treatment techniques, such as incineration and containment, entail their own risks and may be viewed as a lower level of cleanup. On-site approaches also perpetuate environmental stigma and the potential disruption of life in the affected area. Such mitigations are likely to be viewed as unfair because the injustice involved in the original contamination is never offset by restoration to the original state. Furthermore, it may difficult to identify a point when the environment has been cleansed of invisible contaminants. Also, it may be hard to determine the required duration for remedial strategies, such as pumping and treatment of groundwater, and to identify the need for ongoing environmental monitoring. Environmental trust cannot, therefore, be restored.

Remediation is not free of questions relating to environmental justice and racism. For example, does the blocking of sites for new hazardous waste facilities in politically powerful communities mean that new sites will be located only in less powerful areas, perpetuating patterns that reflect bias? The move to on-site disposal of contaminants similarly perpetuates bias, given that these contaminants have already been disproportionately found in minority areas. Is on-site disposal, therefore, a means of keeping the contamination in minority neighborhoods? Finally, is some "affirmative action" now demanded to create equity of risk?

Both removal and on-site approaches frequently invite questions about whether today's state-of-the-art solutions represent tomorrow's hazardous waste problems. Bioremediation techniques available for some types of contamination offer the potential for using living organisms to restore ecosystems. Such approaches may prove to be the only real hopes for true restoration.

The overall challenge of remediating contaminated environments has invited political and scientific efforts to contain the need for such costly interventions. These shifts away from precautionary protection involve redefining the levels of risk that are considered unsafe and thus trigger cleanup, the balancing of known economic impacts of cleanup against the "benefits," and the differentiating of levels of cleanup by the intended use of a property—with industrial sites having higher tolerated levels of contamination than residential sites. Rather than a general approach that seeks environmental decontamination, these shifts promote social adjustment to a contaminated environment. Geographic areas of permanent contamination would be accepted and the use of the term *contamination* more selective.

Conclusion

As the consequences of environmental pollution have been increasingly "brought home" to the residential environ-

ment, people have been confronted with true dilemmas. If the home does not protect against the dangers faced in the larger environment, where one exercises considerably less control than one does at home, then is there no refuge from unwanted ecological change and contamination? Do we then learn to accept and live with (or die from) environmental pollutants that are now seemingly ubiquitous, as though these pollutants are part of our definition of the "normal" environment? Or do we reject this alteration of the ambient environment, striving to reattain conditions that we consider to be more "natural" and, therefore, safe?

These questions are not abstractions. Rather, they undergird key policy questions of the late 20th century regarding acceptable risk, allowed use of known "cancer-causing" substances in foods and drugs, and responsibility for causing environmental change. They also undergird the more personal decisions that people make within the choices offered in the marketplace—over which products to buy and consume or use in their homes, or, for that matter, which home is safe enough to buy or rent or occupy. Although there may be changes in the social definition of contamination and in the technological and social control of pollution, the issue of environmental safety is likely to be a persistent one for the built and residential environment. (SEE ALSO: *Brownfields; Environmental Contamination: Asbestos; Environmental Contamination: Lead-Based Paint Hazards; Environmental Contamination: Radon; Locally Unwanted Land Use; Not in My Back Yard; Temporarily Obsolete Abandoned Derelict Sites*)

—*Michael R. Edelstein*

Further Reading

Brown, Phil and Edwin J. Mikkelsen. 1990. *No Safe Place: Toxic Waste, Leukemia, and Community Action*. Berkeley: University of California Press.

Bullard, Robert. 1990. *Dumping in Dixie: Race, Class, and Environmental Quality*. Boulder, CO: Westview.

Edelstein, Michael R. 1988. *Contaminated Communities: The Social and Psychological Impacts of Residential Toxic Exposure*. Boulder, CO: Westview.

Fitchen, Janet M. 1989. "When Toxic Chemicals Pollute Residential Environments: The Cultural Meanings of Home and Homeownership." *Human Organizations* 48(4):313-24.

Kiel, Katherine A. 1995. "Measuring the Impact of the Discovery and Cleaning of Identified Hazardous Waste Sites on House Values." *Land Economics* 71(4):428-35.

Kohlhase, Janet E. 1991. "The Impact of Toxic Waste Sites on Housing Value." *Journal of Urban Economics* 30:1-26.

Kroll-Smith, J. Stephen and Stephen R. Couch. 1990. *The Real Disaster Is Above Ground: A Mine Fire and Social Conflict*. Lexington: University of Kentucky Press.

Levine, Adeline G. 1982. *Love Canal: Science, Politics and People*. Boston: Lexington Press.

Mohai, Paul and Bunyon Bryant, eds. 1992. *Race and the Incidence of Environmental Hazards*. Boulder, CO: Westview.

Oakes, John Michael, Douglas L. Anderton, and Andy B. Anderson. 1996. "Longitudinal Analysis of Environmental Equity in Communities with Hazardous Waste Facilities." *Social Science Research* 25:125-48. ◀

▶ Environmental Design Research Association

The Environmental Design Research Association (EDRA) is the longest established organization dedicated to improving the quality of human environments through research-based design. Founded in 1968 by design and social science professionals, academicians, and students, EDRA seeks to advance the art and science of environmental design research, improve understanding of the interrelationships between people and their built and natural surroundings, and help create environments responsive to human needs.

EDRA has an international membership of about 800, whose interests and expertise represent the broad domain of this multidisciplinary field. They are concerned with applied and theoretical issues such as (a) the use and meaning of the designed environment (including housing) for diverse and special-user groups; (b) programming, design, construction, and management processes that incorporate information about user requirements, facilitate user participation, and contribute to equitable and sustainable environments; and (c) more generally, design and research methods and the ongoing generation and refinement of theories within the field. Among the multiple generations that encompass its membership are some of the most important contributors to environmental design research and practice. Such excellence in the field is recognized by yearly awards, including the EDRA Career, Achievement, Service, and Student Awards.

EDRA members and other interested registrants meet at an annual conference to share ideas and practices in research papers and design presentations; integrative symposia; exploratory workshops; interactive poster sessions; open-ended working groups; film, video, and slide-tape presentations; and other special events related to a conference theme. Conferences have been held in Canada, Mexico, and the United States, most recently to address the themes of healthy environments, sustainable and equitable habitats, power by design, public and private places, space design and management, and other issues of general interest to the EDRA membership.

Members are kept informed during the rest of the year through the *Design Research News,* a quarterly publication reporting on current developments in the field, updates on research in progress, reviews of recent publications, announcements of upcoming conferences and events, and contributions by members. EDRA networks, formed ad hoc by members with similar areas of special interest, communicate during the year by newsletters, correspondence, and informal communication. A directory of "Resources in Environmental Design Research" lists the membership and their areas of expertise as well as EDRA programs and publications.

EDRA is committed to dissemination of environmental design research and practice. EDRA conferences are chronicled in annual proceedings. EDRA has cooperated with publishers to support two book series, *Environmental Design Research: Direction, Process and Prospects* and *Advances in Environment, Behavior and Design,* and two journals, *Environment and Behavior* and the *Journal of Architectural and Planning Research.*

EDRA is administered by a board of directors elected by the membership and administered by an association management firm. Information may be obtained by contacting the EDRA Business Office, P.O. Box 7146, Edmond, Oklahoma 73083-7146. Phone: (405) 330-4863. Fax: (405) 330-4150. E-mail: edra@telepath.com. Information is also available on the EDRA web site: http://www.aecnet.com/EDRA.

—*Roberta M. Feldman* ◀

▶ Environmental Hazards: Earthquakes

Earthquakes and faulting affect housing in vast portions of the world, placing hundreds of millions of people at risk. Some of the largest earthquakes recorded in history have taken place in China, where an earthquake in 1556 in Shensi claimed 830,000 lives and another earthquake in 1976 claimed about 650,000 lives. Nations that have suffered particularly destructive earthquakes in the past 50 years include Egypt, Morocco, Chile, Peru, Nicaragua, Guatemala, Indonesia, the Philippines, Turkey, Iran, Algeria, and Italy. In each of these events, there were many thousands of fatalities and numerous injuries. There was also serious disruption of communities and massive property loss. In the United States, earthquakes that have been accompanied with losses of life have occurred in South Carolina, Alaska, Montana, Washington, and California. Deaths associated with earthquakes are usually caused by structural failures—the collapse of houses, bridges, and other built structures.

Earthquake risk is calculated on the basis of known geologic conditions and previous earthquakes in the historical record. To identify the areas where future earthquakes are likely to take place, it is necessary to identify the plate boundaries or fault zones and the historical record of earthquakes. In the United States, plate boundaries exist along the western portion of California and southern Alaska. Other major faults are located along the Wasatch mountains in Utah and the St. Lawrence River. However, large earthquakes, including several within the United States, have taken place far from the plate boundaries. For example, major earthquakes in excess of M (magnitude) 8.0 have occurred near New Madrid, Missouri (1811 to 1812), and one in excess of M 7.0 near Charleston, South Carolina (1886).

In the United States alone, an estimated 70 million people in 39 states are at high risk from earthquake hazards, and some 120 million people are exposed to at least moderate risk. The likelihood of experiencing a major destructive earthquake varies from "high" in parts of California; Alaska; the Puget Sound Region; the Wasatch front in Utah, Idaho, and Montana; the central Mississippi Valley; the Charleston region; the Boston region; and upstate New York to virtually "none" in south Florida, central and south Texas, southern Mississippi, Alabama, and north Florida. Similarly, most of the central portions of the United States, including Colorado, North and South Dakota, Minnesota,

Iowa, Wisconsin, and Michigan, have relatively little earthquake risk.

Although they occur less frequently, earthquakes in the eastern part of the United States can be devastating in the areal extent of damage. The New Madrid, Missouri, earthquake was felt over an area of 2 million square miles—from Canada to the Gulf of Mexico and from the Rocky Mountains to the Atlantic Ocean. Luckily, the region was sparsely populated at that time. However, a recurrence of the New Madrid earthquake would cause widespread damage to Missouri, Tennessee, Kentucky, Illinois, and neighboring states. The impact zone for eastern U.S. earthquakes is as much as 20 times greater than for western U.S. earthquakes of the same magnitude.

There are at least four sources of potential damage associated with the earthquake. The first is termed *ground rupture* and refers to the sudden fracturing of the earth's surface. Although photographs of cracks in the earth following a major earthquake are dramatic, this type of damage to the earth is not the only by-product of seismic activity. A second, and more pernicious, form of damage can be attributed to *fault creep,* the slow and gradual fracturing of the earth that characterizes regions underlain with surface faulting. This source of damage is continually at work in some areas, causing continuing damage to streets or buildings along the fault races. A third type of damage is *ground failure* accompanying an earthquake. Depending on the soil structure on which construction has taken place, ground failure can involve landslides or possibly even liquefaction, the transformation of loose materials such as sand and silt into a fluidlike state, resulting in ground materials becoming converted into something resembling quicksand. Liquefaction is a major hazard in areas where land has been "reclaimed" from bays or estuaries and major construction has been permitted on loosely packed, relatively fine sands or silts where the water table is close to the ground surface. Examples of areas subject to liquefaction are numerous and include much of the shore surrounding the San Francisco Bay. Another source of damage related to earthquakes is that associated with *tsunamis* or massive sea waves. These waves (sometimes popularly mislabeled as "tidal waves") have caused serious damage in Japan, Hawaii, and Alaska; they may reach heights of 130 feet.

A major earthquake in an area with a high population concentration is a recipe for a major catastrophe. Potential losses are a function of two factors: seismic hazard (risk from the physical properties of the environment) and vulnerability of the structures. Seismic hazard includes the intensity and nature of shaking, liquefaction, surface fault rupture, and other damage and disruption to the land surface caused by the seismic event itself. Vulnerability of the structures is a function of construction practices. Shaking from an earthquake may directly damage the building through oscillations of the structure or ground failure. In California, for example, structures built before 1934 of unreinforced masonry structures are at greatest risk. Some 8,000 of these structures are located in Los Angeles. Wood frame homes perform extremely well in earthquakes. Serious damage to wood frame structures usually results when they are poorly constructed, have unbraced cripple walls, are not

bolted to their foundations, or experience ground failure. Post-1940 construction codes require that buildings be astened to reinforced concrete perimeter foundations. Wood frame homes constructed on alluvial deposits, particularly if the water table is high, are still at serious risk of damage.

The actual loss of life from an earthquake varies depending on the time of day; the largest number of casualties is potentially associated with an earthquake that occurs during rush hour traffic. The fires following an earthquake may cause the greatest damage to life and property (as in the 1906 San Francisco earthquake).

Damage to housing can be reduced through the development and enforcement of strict construction codes as well as appropriate land use planning controls. For example, states or local governments can reduce the vulnerability of structures by adopting mitigation policies to guide development in hazardous areas, adopting ordinances requiring setbacks from surface fault traces, and controlling the location and density of development in hazardous areas. They can also adopt building codes, establish seismic-resistance building standards, retrofit standards for existing buildings, remove existing development, or prevent future development in particularly hazardous areas. Households may also reduce potential damage to their homes through adopting several rather simple and inexpensive mitigation measures. These include minimizing the probability of a fire following the earthquake by learning how to shut off gas, water, and electricity and reducing potential auxiliary damage by securing the water heater, heavy furniture, appliances, and air conditioners, as well as maintaining the stability of chimneys and the roof. (SEE ALSO: *Environmental Hazards: Flooding; Environmental Hazards: Hurricanes; Housing after Disasters*)

—*Risa Palm*

Further Reading

Berke, Philip R. and Timothy Beadey. 1992. *Planning for Earthquakes: Risk, Politics and Policy.* Baltimore: Johns Hopkins University Press.

Beron, K. J. et al. 1997. "An Analysis of the Housing Market before and after the 1989 Loma Prieta Earthquake." *Land Economics* 73(1):101-13.

National Committee on Property Insurance. 1989. *Catastrophic Earthquakes: The Need to Insure against Economic Disaster.* Boston: National Committee on Property Insurance.

Palm, Risa I. 1990. *Natural Hazards: An Integrative Framework for Research and Planning.* Baltimore: Johns Hopkins University Press.

Willis, K. G. and A. Asgar. 1997. "The Impact of Earthquake Risk on Housing Markets." *Journal of Housing Research* 8(1):125-36. ◄

▶ Environmental Hazards: Flooding

Floodplains—flat, natural areas paralleling rivers—are created by rivers to carry streamflow during times of high water and flooding. These lands have physical, ecological, and economic values. In particular, they offer a number of advantages for human use, including extensive flat areas, usually with well-drained soils; proximity to sources of water for consumption and waste disposal; and access to water-based transportation. For these reasons, floodplains have historically been sought as desirable sites for both agriculture and urban development, with housing being a primary use. In addition, in some places, floodplains offer prestigious locations for private homes. Thus, in the United States, for example, approximately 7% of all land is located in the 100-year floodplain, the area that has a 1% chance of being flooded in any given year. The percentage is much higher in coastal areas and along major rivers, where many large cities are situated. This translates into an estimated 9,600,000 households at risk from flooding in the more than 21,000 communities designated as flood prone. The number of households, and therefore people, at risk has been increasing over time, and not just in the United States. Most countries are experiencing greater vulnerability to flood losses because of increased occupancy of floodplains. In Bangladesh, for example, landless families, which constitute 30% to 50% of the households, are being forced to occupy the most hazardous areas within the floodplain.

The 100-year floodplain (or 1% flood) represents the recurrence interval that is frequently used in flood hazard management, flood control schemes, and flood insurance programs. However, this is an outer limit, because land (and therefore housing) closer to the river and lower in elevation are at greater risk. Whatever the probability, when floods do occur, they cause loss of life, property damage, and disruption of day-to-day patterns of living. In the United States and other developed countries, property damage is the greatest concern, because loss of life has generally been minimized through sophisticated warning systems. There are some distinct exceptions to this, where loss of life has been high because of a lack of warning time and because of the number of people using flood hazard areas. This is especially the case with flash floods, such as the 1976 Big Thompson Canyon flood in Colorado, in which at least 139 people died.

Flooding can cause damage to both the structure and contents of houses. It is difficult to separate residential losses from other losses to flooding because of data collection procedures, so specific figures on residential losses are not readily available. However, direct flood losses average more than $1 billion annually in the United States, and much of this is felt among those living in flooded areas. This average has been exceeded a number of times even with single events, such as flooding caused by Tropical Storm Agnes in 1972, which exceeded $2 billion and the 1993 midwestern flooding for which property damage has been estimated at $12 billion.

The extent of damages to housing depends on the nature of flooding. Floods of rapid velocities may cause structural damage, sometimes by knocking a house off its foundation. Floods of long duration may also cause structural damage because water can infiltrate the walls and floors, ultimately requiring replacement. Damage to contents occurs when water enters houses. The extent of damage depends on the velocity, depth, and duration of the flood. However, even a flood of negligible depth, perhaps entering only basements, can be damaging if furnaces, water heaters, and electrical supplies are located there. Thus, not only is the location of housing in floodplains a problem but so is the design of that housing.

In many countries even today, and until the early 20th century in the United States, living in a flood-prone area was seen as a personal or local responsibility. Following a series of disastrous floods in the early 20th century, U.S. government attempts to protect housing and other development from flooding centered on flood control structures, such as dams, levees, and floodwalls. However, such an approach, in fact, encourages development of floodplains, because the structures provide a false sense of security. Hence, more recent government policy in the United States has focused on either avoiding residential development in floodplains or flood proofing any structures that are built. The National Flood Insurance Program (NFIP), initiated in 1968 following a federal task force report on increasing flood losses, requires that communities avoid building in floodplains, unless such building is flood proofed. Flood proofing generally refers to emergency or permanent measures designed to minimize vulnerability of buildings and their contents. Examples of flood proofing include elevating houses above the expected flood level on stilts, pedestals, walls, or on fill and putting furnaces and water heaters on a floor that is above expected flood levels. This is appropriate for existing houses in the floodplain, as is installing a wall around a house to keep floodwaters out. The design heights are based on probabilities of expected flood frequency and depth. None of these actions prevents flooding from occurring, as flood control structures such as dams, levees, and floodwalls are designed to do. However, these structures are expensive to build and maintain, and they have become increasingly controversial because they cannot prevent all flooding and because of the environmental damage they cause. In contrast, land use and flood-proofing regulations are designed to minimize the damage from flooding. In addition, flood insurance is available through the federal government to insure against losses, although this may be conditional, based on flood-proofing capabilities and extent of exposure to flooding. Unfortunately, the NFIP and other policies relating to the flood hazard have not been entirely successful in reducing losses, partly because only 15% to 30% of flood-prone structures in the United States are insured.

Most floodplains in the world, and certainly in the United States, are densely developed because of the amenities such locations offer. Restrictions on such development exist in some developed countries, perhaps most notably in the United States. However, the experience in developing countries is somewhat different, with increasing vulnerability to flooding as more and more people live in flood hazard areas, either by choice or because of a lack of suitable alternative locations. The emphasis on recognizing and planning for flooding in residential development, both new and existing, represents a sound approach to dealing with the hazard but is useful only where such planning can be enforced. A similar approach has been suggested for the possible impacts of sea level rise. In New Zealand, for instance, some planning councils are considering adopting more stringent restrictions on development in coastal areas so that houses are set back farther from mean high tide to accommodate expected flood levels. (SEE ALSO: *Environmental Hazards: Earthquakes; Environmental Hazards: Hurricanes; Housing after Disasters*)

—*Burrell E. Montz*

Further Reading

Burby, Raymond J. and Steven P. French. 1995. *Floodplain Land Use Management: A National Assessment.* Boulder, CO: Westview.

Burton, Ian, Robert W. Kates, and Gilbert F. White. 1993. *The Environment as Hazard.* 2d ed. New York: Guilford.

Federal Interagency Floodplain Management Task Force. 1992. *Floodplain Management in the United States: An Assessment Report.* Document FIA-17. Washington, DC.: Federal Emergency Management Agency. ◀

▶ Environmental Hazards: Hurricanes

Throughout the United States, coastal areas that are dangerous to life and property are becoming more attractive to housing development. People are drawn to economic opportunities offered by tourism and ocean commerce and to the natural beauty of living on shorefronts, despite the risks posed by hurricanes. This entry reviews the potential for creating hurricane mitigation strategies for housing. An underlying argument is the need to integrate hazard mitigation with development management programs to achieve sustainable housing settlement patterns.

Basic Concepts of Hazard Management

Hazard management consists of four stages:

1. Planning for mitigation of long-term impacts
2. Emergency preparedness planning for quick response once a disaster strikes
3. Emergency response immediately after a disaster
4. Long-term recovery

A key distinction can be drawn between these stages based on the two key functions of hazard management. Emergency management deals with Stages 2 and 3 that occur before and immediately after an event. Preparedness plans are aimed at coordinating immediate disaster response activities of public and private organizations and individuals. Hazard mitigation involves long-term hazard reduction in Stages 1 and 4. Mitigation plans guide (re)development of housing and public infrastructure to avoid the catastrophic impact of future disasters.

Community hazard mitigation plans have three major goals. The first involves containing and modifying the hazard through engineered structures such as seawalls and bulkheads or through maintaining the mitigative functions of natural features such as dunes and beaches that act to reduce the effects of battering wave action. The second goal deals with reducing the vulnerability to damage by protecting housing and public facilities in hazard areas, as with building elevation and wind-resistant design. The third goal involves minimizing exposure to hazardous forces through limiting land uses in hazard areas, as with housing location and density regulations.

The type of hazard affects the choice of mitigation goals. In the case of hurricane force winds, mitigation efforts are designed to reduce the vulnerability to potential damage. After the collapse of many poorly constructed homes from Hurricane Andrew of 1994 in South Florida, the primary focus of mitigation efforts was on enforcing roof cladding and framing standards of building codes. In the case of coastal storm surge, mitigation seeks to contain or modify the hazard. There is considerable consensus about the mitigative benefits from surge of beaches and sand dunes. One dramatic case that may prove useful for accreting U.S. shorelines is Bangladesh's decision to institute a program to seed off-shore sand bars created by silt buildup from riverene flooding to form barrier islands that serve as barriers against storm surge. Mitigation can also seek to minimize exposure of structures to storm surge through relocation of structures in exposed areas, as was the case in relocating the Brownwood subdivision in Baytown, Texas, after Hurricane Alicia in 1983.

Development Management and Hurricane Mitigation

Development management programs focus on guiding various characteristics of development to achieve public interest objectives, such as hazard mitigation, provision of affordable housing, and environmental protection. These characteristics include the location, type, amount, density, quality, and rate of private and public development and redevelopment. Each of these characteristics can be closely linked to hurricane hazard mitigation.

Location. Mitigation deals with two types of hazard locations. In high-hazard areas, housing and other forms of development should be discouraged by limiting the density and intensity of use, redirecting reconstruction, preventing extension of public facilities (roads, bridges, and water and sewer), and requiring structural strengthening and land acquisition for open space. In low-hazard areas, incentives should be provided through transfer development rights from high-hazard areas or through increased density allowances.

Rate. The timing of development plays a key role in mitigation. It defines the carrying capacity of evacuation routes and emergency shelters and the capability of the natural environment and protective structures, and it determines the demands placed on these facilities and the environment under different levels of housing development. Setting limits on the development rate can lead to avoidance of overloads of housing development and ensure that capacities are not exceeded by demands under crisis conditions.

Type. The type of development specifies various classes of development permitted in high-hazard areas. For example, high-occupancy housing structures must be engineered for safety and sheltering in high-hazard areas. Public infrastructure that encourages high-intensity types of land development, such as multifamily housing units, should be extended away from high-hazard areas. Low-intensity types of land use, such as open space, parks, or low-density single-family homes, should be encouraged in high-hazard areas.

Amount. Population size, like the rate of development, must be balanced with capacity of evacuation, shelter, and protective systems. Actions that can be taken to ensure that emergency demands do not exceed the capacity of these systems include limiting the amount of population at risk and increasing system capacity through infrastructure investment and regulating development to protect the mitigation functions of the natural environment.

Public Cost. Public investments in infrastructure in high-hazard areas can be viewed as a subsidy to housing development, which, in turn, leads to increased exposure of people and property at risk. Mitigation efforts should identify and severely limit or prevent subsidies that increase risk. This action transfers the cost of development and hurricane protection from the public sector to the private sector.

Quality. Mitigation initiatives must stipulate high-quality design, construction, and conservation practices for all development located in hazardous areas. Construction practices should ensure protection of occupants and minimize damage.

Implementation Tools. Implementation tools available for guiding urban growth and housing development include land use plans, building codes, zoning, taxation schemes, capital facilities policy, and information dissemination. Implementation tools for guiding housing development rely on formal land use regulations such as zoning and subdivision design regulations, land taxation schemes, and property acquisition as well as on capital facilities policies that specify timing and location of public investments. Effective use of these tools requires considerable resources. Accurate maps are needed to identify permitted uses, set structural strengthening requirements, and determine infrastructure investments in different hazard zones. Sufficient numbers of appropriately trained building inspection staff are needed to monitor and enforce development practices.

Five types of local government powers can be used to achieve mitigation:

1. *Planning power* is used to achieve community agreement on a future course of action as indicated by a land use plan, using tools for education, participation, consensus building, visioning, persuasion, and coordination.
2. *Regulatory power* is used to direct and manage development to achieve land use patterns that mitigate hurricane hazards, using tools of zoning, subdivision regulations, building codes, and floodplain regulation.
3. *Spending power* attempts to coordinate public expenditures to achieve concurrence of infrastructure supply with community demand and hazard area avoidance by placing infrastructure outside hazard

areas, using tools for capital improvement programs and budgets.
4. *Taxing power* seeks objectives such as ensuring adequate infrastructure and enhancing mitigation, using tools such as hazard tax assessment, improvement districts, and land tax abatement for preserving open-space uses.
5. *Acquisition power* is used for purchase of lands in hazard areas, using tools such as eminent domain, purchase of development rights, and dedication of conservation easements.

In sum, local governments have substantial power to plan and regulate land use and development. This power can be used to effectively reduce the economic, social, and environmental impacts of hurricanes on housing.

Integrated Hazard Mitigation

Typically, mitigation efforts are directed toward single types of hazards. However, there are significant advantages to taking an integrated rather than a hurricane hazard-specific approach to mitigation. Increasingly, studies reveal that many types of hazard mitigation actions are generic across hazards. For many of the problems that arise in mitigating potential loss, it does not matter what specific type of hazard is involved. Generic actions can be taken for many hazards, whether the task be mapping, assessing community vulnerability, coordinating interorganizational actions, or initiating public awareness programs.

There are several practical reasons for the integrated approach:

1. It is cost-efficient in terms of time, money, effort, and other resources.
2. It is politically better because it mobilizes a wider range of groups affected by different hazards, thereby creating a more powerful constituency for mitigation.
3. It minimizes duplication, conflict, overlaps, and gaps in mitigation initiatives.
4. It increases the efficiency and effectiveness of organized efforts to cope with disaster.

Sustainable Development as a Long-Term Strategy for Mitigation

Sustainable development represents an overarching framework for hurricane hazard mitigation. Land use patterns that fail to account for the location of high-hazard areas are not sustainable. Furthermore, housing not built to withstand predictable hurricane forces is not sustainable.

The sustainable development concept also offers a useful framework for integrating hazard mitigation with other social, economic, and environmental goals. The location of new housing development should be evaluated by multiple sustainability criteria. Does it ensure sustainable use of, and impacts on, natural resources? Does it provide for social needs of different income and ethnic groups? Does it represent an efficient investment of financial resources that might otherwise have been available for other economic development investments? Sustainable development thus offers an important framework for integrating diverse environmental, social, and economic concerns.

A sustainable housing development strategy should be pursued at both local and national levels. It should integrate strategies for affordable housing, location, and timing of public investments, taxation policy, and natural resource conservation. A sustainable development strategy for hazard areas would take a long time. Actions that satisfy short-term needs are often given priority over hazard mitigation, especially in high-growth communities where short-term public facility investment demands are almost always of the highest priority. Actions that satisfy a short-term goal (locating second homes in dangerous beachfront locations) make little sense when long-term goals (the costs of shoreline erosion and increased coastal storm damage) are considered.

As an example, following Hurricane Andrew, there was an innovative effort to rebuild according to sustainability principles. An affordable housing project, called Jordan Commons, used community-building principles to achieve physical design, services for families, environmental sustainability, and community self-sufficiency. The project incorporates some impressive environmental features, including reflective metal roofs that reduce air-conditioning costs, hurricane-resistant steel frames, low-volume toilets, low-flow showerheads, and use of landscaping to shade buildings as part of a "cool communities" demonstration project. The Jordan Commons initiative also addressed social needs to foster self-governance.

In another impressive example of sustainable development, Soldiers Grove, Wisconsin, relocated its downtown from the Kickapoo River floodplain, while simultaneously achieving other sustainability objectives, including use of solar energy, life cycle analysis of building materials, and mixing housing into downtown commercial redevelopment.

Conclusion

Government leaders typically downplay the importance of hazard mitigation, especially as the memory of the last disaster fades from the public consciousness. Despite these problems, national governments throughout the world are increasingly recognizing natural hazards, including hurricanes. Countries such as Jamaica, New Zealand, and a consortium of Pacific Island nations now have at least a minimum framework that includes national physical plans that stipulate coastal shoreline buffers, and propose to provide infrastructure and increase allowable development densities in suitable sites away form high-hazard areas. The challenge for the United States is to pursue a sustainable development strategy that parallels the efforts of other countries that link hazard mitigation with other social, economic, and environmental goals. (SEE ALSO: *Environmental Hazards: Earthquakes; Environmental Hazards: Flooding; Housing after Disasters*)

—*Philip R. Berke*

Further Reading

Berke, Philip and Timothy Beatley. 1997. *After the Hurricane: Linking Recovery to Sustainable Development in the Caribbean.* Baltimore, MD: Johns Hopkins University Press.

Burby, Raymond, ed. 1997. *Overwhelming Hazards: Land Use Planning for Safer Communities.* Washington, DC: National Academy Press.

Godschalk, David, David Brower, and Timothy Beatley. 1989. *Catastrophic Coastal Storms: Hazard Mitigation and Development Management.* Durham, NC: Duke University Press.

Kriemer, Alcira and Mohan Munasinghe, eds. 1991. *Managing Natural Disasters and the Environment.* Washington, DC: World Bank. ◀

▶ Equity

In the context of housing and the built environment, the term *equity* has at least four distinct meanings. *Home equity* is the net amount of money that homeowners would receive if their house was sold at any given point in time. It is the value of homeownership. When buyers purchase a home, the amount of equity they must invest is the total cash down payment. For example, assume that a house costs $100,000 and closing costs, including financing costs, are $5,000. If the mortgage is $80,000, the buyer's equity will be $25,000 ($100,000 + $5,000 − $80,000). As the homeowner amortizes the mortgage over the years, the equity increases by an amount equal to the pay down on the mortgage. The equity also increases as the house appreciates in value. In the above example, suppose the house is worth $140,000 after five years and the mortgage balance has been reduced to $75,000. The homeowner's equity would now be $65,000. Although common use of the term home equity does not take into account selling costs, the true home equity should be reduced by selling costs. For example, if selling costs for sales commission and title policy in this case are $10,000, the net home equity would be $55,000.

Home equity mortgages are second (or third) mortgages backed (collateralized) by a homeowner's equity. If a homeowner's equity is, say, $50,000, he or she can obtain a home equity mortgage that is 50% to 80% of the value of the equity. Home equity mortgages are subordinated to first mortgages (have lower priority) and typically have higher interest rates and shorter terms. They became more popular after the Tax Reform Act of 1986, which eliminated tax deductions for interest on consumer loans while preserving the deductibility of interest on real estate loans. Borrowers switched to home equity loans to cover consumer needs because the interest is tax deductible.

In real estate development projects, equity is often used to mean one's ownership in a project—or *project equity*. Development projects often have multiple buyers of financing so that equity is different depending on one's position in the deal. Because the term equity has a variety of meanings to different participants in a real estate deal, one must be extremely careful to define how the term is being used. The value of one's equity depends on how the deal is structured.

The following example illustrates various common usages of the term *equity* by different real estate professionals. Suppose that an apartment project costs $1 million to build and is financed by a first mortgage from a bank for $700,000, a second mortgage (perhaps from the land seller) for $200,000, and cash from the developer of $100,000. In underwriting its first mortgage, the bank requires, say, 30% equity before it will make a $700,000 mortgage. Because the second mortgage and developer's cash are subordinated to the first mortgage, the bank will regard both sources as equity, for a total of $300,000. The bank's first mortgage will be paid off in full before anyone else receives anything from the sale or foreclosure of the property. The developer's equity in the example is $100,000. If he or she owns 100% of the project, all of the benefits from appreciation and amortization of both mortgages will increase the value of the equity.

Many new types of mortgages, such as convertible and participating mortgages, give the lender an equity interest in the real estate deal. In the preceding example, suppose that the first mortgage is a participating mortgage that gives the bank a 50% participation in the real estate. This means that if the apartment project was sold for $1.5 million, the bank would receive $250,000 (half of the appreciation of $500,000) in addition to repayment of its mortgage. Typically, developers will accept participating mortgages in exchange for below-market interest rates and higher loan-to-value ratios. In this example, the developer still owns 100% of the project but has a 50% equity interest in the "upside" of the project; that is, the developer receives 50% of the increase in value created by the apartment project.

The developer's equity is only $100,000—the amount of cash invested. Suppose that the developer has raised the cash from two investors who each put up the $50,000 in exchange for 25% of the deal. Their equity is then said to be 25%—their percentage of ownership in the deal.

Many aspects of real estate raise concerns about fairness. In this connection, *equity* refers to how the burden of paying for various costs associated with real estate development and ownership is divided. For example, equity in infrastructure financing refers to the degree to which all residents of a subdivision or city pay their fair share of the costs for water, sewers, and roads. New buyers are justifiably concerned about the "equity" of taxes and impact fees they pay for their homes. In recent years, many high-growth communities have passed impact fees that place a disproportionate share of the cost of public facilities on new homebuyers. Existing residents of the community are able to shift the burden of paying for public facilities, such as road and school improvements, to the newcomer—who is not able to vote on the tax or impact fee in the first place. This shift raises serious questions of equity in the distribution of tax burden—issues that have yet to be resolved. (SEE ALSO: *Down Payment*)

—*Richard B. Peiser*

Further Reading

ABA Standing Committee on Lawyers' Title Guaranty Funds. 1991. *Buying or Selling Your Home: Your Guide to: Contracts, Titles, Brokers, Financing, Closings.* Chicago: American Bar Association.

Blankenship, Frank J. 1989. *The Prentice Hall Real Estate Investor's Encyclopedia.* Englewood Cliffs, NJ: Prentice Hall.

Jacobus, Charles J. 1996. *Real Estate Principles.* Upper Saddle River, NJ: Prentice Hall.

McKenzie, Dennis J. 1996. *Essentials of Estate Economics.* Upper Saddle River, NJ: Prentice Hall.

Peiser, Richard B. 1988. "Calculating Equity-Neutral Water and Sewer Impact Fees." *Journal of the American Planning Association Journal* 54:38-48.

Peiser, Richard B. 1992. *Professional Real Estate Development: The ULI Guide to the Business.* Washington, DC: Urban Land Institute.

Shim, Jae K. 1996. *Dictionary of Real Estate.* New York: John Wiley.

Tosh, Dennis S. 1990. *Handbook of Real Estate Terms.* Englewood Cliffs, NJ: Prentice Hall. ◄

► Escrow Account

The concept of escrow involves a disinterested third person holding money or documents pending the satisfaction of the terms and conditions of a written agreement. Although escrow agreements represent a common vehicle for the closing of a real estate sale, the establishment of an escrow account is more commonly associated with the systematic accumulation of funds to pay taxes, hazard insurance, and other assessments on an annual or semiannual basis. These accounts are often labeled *impound accounts, reserve accounts,* and *trust accounts.* Depending on the state in which the borrower is located, the account either will be treated as a trust account administered by the lender or it will be commingled with the assets of the lender. In the latter case, the lender remains responsible for a proper accounting of the funds in the same manner as a trustee.

The mechanics of an escrow account for the purpose of accumulating funds involve a number of steps. First, the secured lender must estimate the annual charges for the payment of taxes, hazard insurance, and other assessments. These sums are based on the amount of the previous charge and any notices of change received by the secured lender. Second, the lender divides the total projected annual payments by 12. This sum is added to the borrower's monthly principal and interest payment. The portion of the monthly payment in excess of the principle and interest due are deposited into the escrow account. The borrower's failure to pay the amount needed to fund the escrow account is an event of default. The lender will make payments from the escrow account as the taxes, hazard insurance, and other assessments come due. This practice will continue so long as the debt remains unpaid.

In the past, some lenders required borrowers to keep excessive funds in escrow accounts. With lenders often serving as the administrators of these accounts, overfunding provided lenders a source of interest-free funds. In response, the law now limits the size of escrow impound accounts. The Real Estate Settlement Procedures Act (RESPA) of 1974 limits the size of the excess to one-sixth of the estimated annual charges. In effect, the most a lender may require in the account is the projected necessary monthly accumulation plus an additional two month's accumulation. The two-month cushion is intended to prevent timing problems between the collection and disbursement of the funds. Here is an illustration of the limit: If the borrower's annual tax liability is $1,200 and the annual hazard insurance premium is $600, the lender would need to accumulate $150 per month to cover these charges. Assume that both of these charges are due and payable July 1. Therefore, the maximum amount the lender could require the borrower to have on deposit in the escrow account on February 1 of the following year would be $1,350 ([July through January = 7 months × $150 = $1,050] plus [two months' cushion = $300] = $1,350). Although not mandated by RESPA, some states require lenders to pay interest to the borrowers on the balance in the impound account. (SEE ALSO: *Real Estate Settlement Procedures Act*)

—*Jeffery M. Sharp*

Further Reading

Hinkel, Daniel F. 1995. *Practical Real Estate Law*. 2d ed. Minneapolis/St. Paul: West.

Floyd, Charles F. and Marcus T. Allen. 1994. *Real Estate Principles*. 4th ed. Chicago: Real Estate Education Co.

French, William B. and Harold F. Lusk. 1979. *The Law of the Real Estate Business*. 4th ed. Homewood, IL: Irwin.

Mills, Edwin S. 1994. "The Functioning and Regulation of Escrow Accounts." *Housing Policy Debate* 5(2):203-18.

Seidel, George J. 1993. *Real Estate Law*. 3d ed. Minneapolis/St. Paul: West. ◄

► European Network for Housing Research

European Network for Housing Research (ENHR) is an organization of research institutes and individual scholars in the housing field. The purpose of the network is to direct research attention toward discovering the links between political, economic, and social factors in the international housing market; to develop methods for analyzing and measuring the housing situation; and to foster communication between social scientists working in this field. To encourage cross-disciplinary, international research projects, the network sponsors conferences and working groups that focus on issues such as national housing policies, housing models, housing education, real estate management, and residential migration patterns. In addition, it provides its members with the quarterly *ENHR Newsletter*, which provides a forum for debate and exchange of research ideas and results. Founded in 1989, ENHR maintains a membership of 917 individuals and 91 institutions throughout Europe. Contact person: Dr. Bengt Turner. Address: c/o Institute for Housing Research, Uppsala University, P.O. Box 785, S-80129 Gävle, Sweden. Phone: 46 26 147700. Fax: 46 26 147802. (SEE ALSO: *European Social Housing Observation Unit; Housing Abroad: Western and Northern Europe*)

—*Caroline Nagel* ◄

► European Real Estate Society

The European Real Estate Society (ERES) is an international association founded in 1994 providing a network between real estate centers and practicing professionals. Primarily focusing on commercial real estate, ERES seeks to address the needs of the European commercial property industry and academics who deal with the rapidly internationalizing property industry. Membership: Professionals and academics interested in commercial property research. Secretary/treasurer: Barry Wood. Address: Department of Town and Country Planning, University of Newcastle upon Tyne, Newcastle upon Tyne, England, NE1 7RU. Phone: 44 (0) 191 222 6389. Fax: 44 (0) 191 222 8811. E-mail: B.D.Wood@ncl.ac.uk (SEE ALSO: *Real Estate Developers and Housing*)

—*Laurel Phoenix* ◄

► European Social Housing Observation Unit

This organization publishes a journal, *Euro-Habitat,* in English and French that serves members of the construction industry in Europe as well as a broader international audience interested in the European construction industry. Each issue focuses on a different country, covering topics such as "Reform of Social Housing in Germany," "The Social Housing Guarantee Fund in the Netherlands," "Private Financing for Social Housing: The Role of the Housing Finance Corporation," and "The Sale of Publicly Owned Rented Housing in Italy." Price: 3,500 French francs per year. Secretariat: Sandrine Languerre. Address: Mission Europe—Union Nationale des Federations d'Organismes HLM, 14, rue Lord Byron, 75384 PARIS CEDEX 08 FRANCE. Telephone: 33 (1) 40 75 79 89. Fax: 33 (1) 40 75 79 89. WWW: http://ourworld.compuserve.com/homepages//observatoire. (SEE ALSO: *European Network for Housing Research; Housing Abroad: Western and Northern Europe; Social Housing*)

—*Laurel Phoenix* ◄

► Eviction

Because housing is a fundamental necessity and the point of access for political, social, cultural, and economic connections to the broader community, displacement through eviction often has devastating economic, psychological, and physical consequences. Eviction—the forced removal of people from their homes—is commonplace throughout the world, yet has been increasingly recognized by the international community as a violation of basic human rights. The Centre on Housing Rights and Evictions, an international nongovernmental organization, has documented thousands upon thousands of incidents of forced eviction on each continent. Concern about forced evictions led the United Nations to appoint a special rapporteur on Housing Rights in 1993, which has sparked calls by NGOs and others for enforceable international standards for security of tenure.

Although there is global interest in the issue of evictions, the legal norms governing evictions and social forces giving rise to evictions tend to be local. In the United States, the contemporary landlord-tenant relationship derives—as the nomenclature for the parties indicates—from the legal norms of feudal English land tenure. Like its feudal counterpart, the core of the modern relationship is a legal transaction for dominion over territory. That territory is, in modern day, more likely to be an urban apartment than acreage of land, and legal reforms such as the implied warranty of habitability and rent regulation have given present-day tenants greater rights than their feudal counterparts. But ultimately, the tenant's hold on that territory—his or her home—remains temporary, conditional, and subordinate to the rights of the landlord. When the temporary term has ended, when the condition for use has not been met, or when permission to use has been revoked, a landlord can obtain physical and legal possession from a tenant through eviction. The root of the English word *evict* is the Latin word, *vincere,* which means to conquer or overcome completely.

Under English common law (legal precedents established by judges), a landlord did not need a court order to evict and was permitted to use self-help, as long as the landlord used no more force than necessary to regain possession. Although this centuries-old principle remains in effect in some states of the United States, most jurisdictions now require a landlord to commence a court proceeding and obtain a judge's order before he or she may evict. In these jurisdictions, a self-help eviction is a tort, or civil wrong, and a tenant who has been put out of his or her home without court process can go to court and obtain an order to be restored to possession; the wrongfully evicted tenant can also seek monetary damages. Under a municipal law in New York City, a self-help eviction is a crime as well as a tort, and a landlord who removes a tenant without a court order is subject to criminal as well as civil penalties.

For the most part, the court procedures used for evicting tenants are summary proceedings—that is, streamlined legal procedures that move at a far more accelerated pace than ordinary civil litigation. These expedited and simplified summary eviction proceedings were instituted legislatively throughout most of the United States in the early 19th century as an alternative to the more drawn-out court proceedings that are applicable to most civil disputes. By providing landlords with a relatively quick method of obtaining court-ordered evictions, legislatures hoped to encourage landlords to seek court orders rather than use self-help to evict.

Summary eviction proceedings statutes always require a landlord to provide advance notice to a tenant that eviction is sought and that the tenant has an opportunity to be heard in court before a judgment authorizing eviction is rendered. Although nonpayment of rent or some other violation of the tenant's obligation under the lease is always a sufficient basis for an eviction, most jurisdictions do not require a landlord to provide any reason at all for an eviction once a lease has expired. Some jurisdictions, however, require landlords of certain categories of housing to prove that they have "just cause" to evict. New Jersey, for example, has a just-cause eviction statute that applies to landlords of certain multiple dwellings. And in housing that is government owned or subsidized, landlords must prove just cause to evict.

Poor people and members of minority groups are evicted in highly disproportionate numbers. In New York City in the 1980s and early 1990s, for example, more than 300,000 eviction proceedings were brought each year and almost 30,000 evictions a year were performed. The population facing eviction during that period was almost 90% African American and Latino, whereas the general population was only 50% African American and Latino. Similarly, a disproportionate 90% of the 195,000 tenants facing eviction in Baltimore's "rent court" in 1991 were African American, and the vast majority were in households with below-poverty-level incomes. And although eviction is often a complex legal proceeding, tenants are rarely able to obtain representation by counsel to defend themselves, whereas landlords are usually represented by counsel. Eviction proceedings thus tend to be very one-sided.

Eviction is generally played out as a conflict between an individual tenant and an individual landlord. In the most typical instance, a landlord seeks to evict a tenant for nonpayment of rent. But individual evictions generally reflect broad economic, social, and political factors. Evictions for nonpayment of rent rise in a region when the economy deteriorates or when public assistance benefits are reduced, and evictions rise in a neighborhood when that neighborhood is transformed into a more marketable location through the process known as "gentrification."

Organized resistance to eviction has frequently been used by tenants as a tool to wrest a concession from a particular landlord. A rent strike, or collective rent withholding by tenants in a single building, is a tactic often used in urban areas to make eviction proceedings prohibitively expensive and thus to pressure a landlord to remedy building conditions in need of repair. Organized resistance to eviction has also periodically been used to effect broad change in housing law or policy. In the mid-19th century, tenant farmers in New York State withheld rent for years, armed themselves, and fought off evictions farm by farm until they succeeded in convincing the state legislature to break up huge precolonial landholdings and grant them title to their farms. During World War I, the Socialist Party organized rent strikes and fought off evictions of workers who had flocked to Washington, D.C., New York City, and other urban areas in search of jobs in war-related industry, only to be squeezed out of their homes by the astronomical rents landlords could charge because of an acute housing shortage. This resistance led to the first rent control statutes. From members of the large, integrated Delta Cooperative Farm in Mississippi in the 1950s to the elderly, mostly immigrant residents of downtown San Francisco's International Hotel in the 1970s, people struggling against eviction have shaped, and have been shaped by, the politics of the times. (SEE ALSO: *Displacement; Housing Courts; Lease; Rent Strikes; Right to Housing; Tenant Organizing in the United States, History of*)

—*Andrew Scherer*

Further Reading

Bratt, R., C. Hartman, and A. Meyerson, eds. 1986. *Critical Perspectives on Housing.* Philadelphia: Temple University Press.

Christman, Henry. 1945. *Tin Horns and Calico.* New York: Henry Holt.

Kalodny, Lawrence. 1991. "Eviction Free Zones: The Economics of Legal Bricolage in the Fight against Displacement." *Fordham Law Journal* 18(Spring):507.

Lawson, Ron, ed. 1986. *The Tenant Movement in New York City, 1904-1984.* New Brunswick, NJ: Rutgers University Press.

Lempert, Richard. 1989. "The Dynamics of Informal Procedure: The Case of a Public Housing Eviction Board." *Law & Society Review* 23(3):347-98.

Monsma, Karl and Richard Lempert. 1992. "The Value of Counsel: 20 Years of Representation before a Public Housing Eviction Board." *Law & Society Review* 26(3):627-67.

Werner, Frances. 1990. "Requiring Just Cause for Eviction from Private Housing: A Survey of Litigation Strategies and Legislative Approaches." *Clearinghouse Review* (February):1088. ◄

► Exclusionary Zoning

An exclusionary zoning ordinance is an ordinance containing restrictive regulations of land use that have the effect of excluding lower-income groups from a community by driving up the cost of housing. Because racial and other minorities are overrepresented in lower-income groups, exclusionary zoning also bars minorities from suburban housing markets. Exclusionary zoning is most commonly adopted by suburban communities.

Exclusionary zoning is the subject of controversy, litigation, and legislation. Advocates of affordable housing blame exclusionary zoning for the rising cost of new housing. Planners blame exclusionary zoning for the concentration of unemployed or underemployed minorities in central cities, far from suburban job opportunities. In a growing number of states, courts and legislatures have invalidated exclusionary zoning ordinances and have mandated the adoption of inclusionary zoning techniques.

What Makes Zoning Exclusionary

It is important to distinguish exclusionary zoning from the nonexclusionary zoning ordinance that typically divides all land within a community into districts in which only certain uses are allowed. All zoning bars certain land uses or building types from certain districts. For example, U.S. zoning ordinances typically exclude commercial and industrial land uses from residential zoning districts. In addition, multifamily uses are excluded from single-family zoning districts. The exclusion of multifamily uses from single-family zoning districts was upheld by the United States Supreme Court in *Village of Euclid v. Ambler Realty Company* (272 U.S. 365) (1926).

The courts, therefore, do not invalidate zoning ordinances simply because they exclude certain uses from certain zoning districts. In fact, in most states, the courts will not invalidate zoning ordinances even if they exclude commercial and industrial uses from an entire community. But the courts will find zoning ordinances exclusionary if they improperly exclude from a community the affordable housing that would otherwise be available to lower-income groups and racial minorities that depend on such housing.

Exclusionary Zoning Techniques

A community's zoning becomes exclusionary when it totally excludes certain residential land uses or when it includes other restrictive provisions that limit the availability of land for housing that would be available to lower-income groups and racial minorities. In 1978, an American Bar Association Commission identified six zoning techniques as the principal devices used by suburban communities to exclude low-income households:

1. *Large-lot zoning.* By restricting development in a jurisdiction substantially to single-family dwellings on large lots (one-half acre or more), a community can limit the supply of housing sites and thereby increase residential land costs generally, particularly where the demand for new housing is strong. Large lot sizes can also add costs to land improvements by increasing the required linear feet of streets, sidewalks, gutters, sewers, and water lines.

2. *Minimum house size requirements.* By raising the lower limit of construction costs, such requirements can be the most direct and effective exclusionary tool.

3. *Prohibition of multifamily housing.* By limiting development to single-family dwellings, a community can effectively eliminate the most realistic opportunity for housing persons of low and moderate incomes. Multifamily housing generally represents such an opportunity because their higher densities usually mean lower land costs per unit, and federal subsidies are most frequently geared to this type of housing.

4. *Prohibition of mobile homes.* As the average price for a single-family conventional home continues to climb beyond the reach of a large proportion of the population, the only available nonsubsidized form of housing for those persons who wish to own rather than rent is a mobile home.

5. *Unnecessarily high subdivision requirements.* When a community requires a developer to provide land improvements far above the necessary minimum or to dedicate substantial amounts of land (or pay fees in lieu of such dedication) for open space, schools, recreation facilities, and so forth, these costs are passed on to the consumer and the cost of housing is effectively increased.

6. *Administrative practices.* Improper notice, ex parte contacts, administrative delays, the imposition of arbitrary development demands in exchange for local permits, and other local practices can also be used for exclusionary ends.

Not all commentators agree that all of these techniques are exclusionary. Some commentators, for example, question whether large lot requirements have as significant an exclusionary effect as some of the other devices just listed. Other studies have brought to light other exclusionary zoning techniques. For example, local regulations may limit the building of large rental units and thereby limit rental housing opportunities on which low- and moderate-income families must rely.

Some social scientists question whether any of the zoning provisions just described have a significant exclusionary effect. Particularly where suburban municipalities are small and numerous, it is argued, the pattern of development and land use regulation is fragmented and localized. As a result, one municipality's exclusionary zoning regulations will not necessarily bar low- and moderate-income households from suburban housing near suburban jobs. It is also argued that municipalities in balkanized suburbs lack the monopoly on zoning power that would allow them to use exclusionary zoning in their quest to attract only the kind of development that yields high tax revenues. Yet other commentators defend exclusionary zoning on principle, as a reasonable decision to avoid the costs and tax losses associated with low- and moderate-income populations.

The Attack on Exclusionary Zoning

Exclusionary zoning has come under attack in both state and federal courts. In state courts, the attack has generally centered on the economic effect of exclusionary land use regulations: lifting the cost of housing above the reach of low- and moderate-income households. In federal courts, the attack on exclusionary zoning has focused on the effect of exclusionary land use regulations on racial and ethnic minorities in violation of their federal constitutional or statutory rights.

New Jersey courts have led state courts in striking down suburban land use controls because they are economically exclusionary. In *Southern Burlington County NAACP v. Township of Mt. Laurel (Mt. Laurel I)* (N.J. 336 H. 2d 713, 1975), the New Jersey Supreme Court struck down Mt. Laurel's zoning ordinance. The court held that all "developing" communities in New Jersey must exercise their zoning power to ensure a "realistic opportunity for the construction of its fair share of the present and prospective regional need for low and moderate income housing." Applying that new "fair share" doctrine to Mt. Laurel's zoning ordinance, the court struck it down because the only residential development it allowed was for middle-income housing. In later decisions, New Jersey courts have revisited and clarified the fair share doctrine of *Mt. Laurel I.* The Supreme Court itself, in *Southern Burlington County NAACP v. Township of Mt. Laurel (Mt. Laurel II)* (N.J. 456 F. 2d 390, 1983), subsequently decided that the fair share obligation is incumbent on *all* communities in New Jersey, not just developing communities. In Massachusetts, New Hampshire, New York, and Pennsylvania, the courts have found and applied similar doctrines under their own constitutions and jurisprudence. Not all of these states adopted New Jersey's fair share doctrine, but they did adopt stringent rules that invalidate zoning that has an exclusionary effect.

In federal court, exclusionary zoning may be attacked only if it is an instance of racial or ethnic discrimination that violates the Fair Housing Act or the equal protection clause of the U.S. Constitution's Fourteenth Amendment. The U.S. Supreme Court imposed this restriction on federal courts in *James v. Valtierra* (402 U.S. 137, 1971) and in *Village of Arlington Heights v. Metropolitan Housing Development Corporation* (429 U.S. 252, 1977). In *James,* the Court held that a California community did not violate the equal protection clause when it used zoning (by referendum) to exclude low-income housing. Thus, the poor, as

such, are unprotected by the equal protection clause. Later, in *Arlington Heights,* the Court held that an exclusionary land use regulation will violate the equal protection clause only if "discriminatory purpose was a motivating factor" in the regulation's adoption. Thus, even racial minorities are unprotected by the equal protection clause, if they cannot prove discriminatory purpose.

The Supreme Court has not yet decided an exclusionary zoning case brought under the Fair Housing Act. The lower federal courts, however, have held that plaintiffs need not prove discriminatory purpose in cases brought under the Fair Housing Act. A discriminatory effect is enough, although the federal courts do not agree on what must be shown to prove a discriminatory effect. *Huntington Branch, NAACP v. Town of Huntington* (844 F. 2d 926, 2d Cir. 1988, *aff'd.* 488 U.S. 15), decided in 1988 by the U.S. Court of Appeals for the Second Circuit, is one of the more important of these cases. The court held that a refusal to rezone land in a white neighborhood for federally subsidized multifamily housing was a violation of the act. But like the equal protection clause in the U.S. Constitution's Fourteenth Amendment, the Fair Housing Act protects only racial minorities. The act extends no protection to the economically disadvantaged.

Exclusionary Zoning: An Assessment

Because federal courts afford little protection to low- and moderate-income groups, state courts and legislatures are the principal venues for successful attacks on exclusionary zoning. The number of states that enjoin exclusionary zoning continues to grow. For example, the New Hampshire Supreme Court embraced the Mt. Laurel fair share doctrine in 1991. In those states, courts and legislatures have fashioned a range of remedies for exclusionary zoning. These range from "builders' remedies" (e.g., orders to issue building permits for affordable housing) to mandated inclusionary zoning techniques. Some social scientists continue to question the exclusionary effect of exclusionary zoning, whereas others continue to defend its reasonableness. (SEE ALSO: *Civil Rights Act of 1968, Title VIII;* **Discrimination;** *Inclusionary Zoning; Mt. Laurel; Segregation; Zoning*)

—*Daniel R. Mandelker and Harold A. Ellis*

Further Reading

Babcock, R. F. and F. P. Bosselman. 1973. *Exclusionary Zoning: Land Use Regulation and Housing in the 1970s.* New York: Praeger.

Bergman, E. M. 1974. *Eliminating Exclusionary Zoning: Reconciling Workplace and Residence in Suburban Areas.* Cambridge, MA: Ballinger.

Fishman, R., ed. 1978. *Housing for All Under Law.* Report of the American Bar Association Commission on Housing and Urban Growth. Chicago: American Bar Association.

Haar, C. M. 1996. *Suburbs Under Siege: Race, Space, and Audacious Judges.* Princeton, NJ: Princeton University Press.

Johnston, R. 1984. *Residential Segregation, the State, and Constitutional Conflict in American Urban Areas.* Orlando, FL: Academic Press.

Kirp, David L., John Dwyer, and Larry A. Rosenthal. 1995. *Our Town: Race, Housing & Soul of Suburbia*. New Brunswick, NJ: Rutgers University Press.

Mandelker, D. R. 1997. *Land Use Law*. 4th ed. Charlottesville, VA: Michie.

Mandelker, D. R., R. A. Cunningham, and J. M. Payne. 1995. *Planning and Control of Land Development: Cases and Materials*. 4th ed. Charlottesville, VA: Michie.

Salsich, P. 1991. *Planning, Zoning, Subdivision Regulation, and Environmental Control*. Colorado Springs, CO: Shepard's/McGraw-Hill.

Tucker, W. 1991. *Zoning, Rent Control, and Affordable Housing*. Washington, DC: Cato Institute. ◄

► Experimental Housing Allowance Program

From its inception in the 1930s to the beginning of the 1970s, the primary thrust of U.S. federal housing policy was to provide tax and direct payment incentives to encourage the real estate industry to provide decent, safe, and sanitary housing for persons of low or moderate income. The main programs—public housing, mortgage insurance, and direct subsidies to lenders and developers of single-family and multifamily housing—were designed to increase the supply of decent housing. This emphasis on supply was accelerated as a result of the urban riots of the 1960s and the publication of a series of reports in 1968 calling for major increases in federal subsidies for housing development.

As federal housing programs expanded in the 1960s, a number of academics, housing advocates, and political leaders questioned the emphasis on supply and argued that policies designed to increase the effective demand of low-income persons by making available cash payments to supplement rental housing costs would be a more efficient and more equitable approach. The Rent Supplement Program, authorized by Congress in 1965, was a limited step in this direction. The Section 8 certificate program that began in 1974 included a component that could be applied toward the rent for existing housing units.

By the early 1970s, a sharp debate was taking place between the advocates of "supply-side" and "demand-side" government involvement. Both sides argued that its approach would result in a more effective use of available government resources for decent housing. In 1972, the Department of Housing and Urban Development (HUD) decided to conduct a comprehensive study of the demand-side approach by committing $200 million to an Experimental Housing Allowance Program (EHAP). HUD described EHAP as the "first major attempt to subject a housing program concept to systematic testing." EHAP operated for five years, beginning in 1973. Three separate experiments were conducted:

1. The *demand experiment* focused on how people respond to housing subsidies. Questions studied included these: How many eligible renters can actually be expected to take part in such a program? How do they actually use the money they receive? What factors affected their choice of whether or not to participate?

2. The *supply experiment* examined housing markets to determine what impact an infusion of cash subsidies to potential renters would have on the supply of housing. Question asked included these: What happens to the price and the quality of housing when eligible renters and homeowners are offered housing subsidies through an allowance program? To what extent are repairs made and the quality of housing improved as a result of the availability of a housing allowance? What effect does the allowance have on the quality of neighborhoods as a result of actions taken by people who participate in the program?

3. The *administrative agency experiment* looked at the administration of a housing allowance program, particularly the cost associated with implementation of the program. Eight public agencies were selected to administer programs using the same allowance formula that was used in the supply experiment.

EHAP followed families for five years. The program generated controversy even before the end of the experiment. For example, a 1978 report of the General Accounting Office (GAO) contended that because of "basic design flaws," the results of the experiment would be inconclusive. Although there were a total of 12 sites where the experiments were conducted, GAO concluded that they "were too few and lacked the characteristics typical of the major urban areas they were intended to represent, to permit a reasonable projection to a national program." For example, according to the GAO, neither South Bend, Indiana (chosen to represent low-growth, high-black areas, such as New York, Chicago, and Washington), nor Green Bay, Wisconsin (representing high-growth, low-black areas), had the characteristics of large metropolitan areas, such as racial ghettos and a high percentage of multifamily dwellings. In a report issued in 1980 at the conclusion of the study, HUD presented a number of findings:

1. More stringent housing quality standards reduced the number of participants. Although the incomes of approximately 20% of U.S. households would have made them eligible for a housing allowance at the beginning of the study, the majority of those households lived in housing that HUD considered substandard. To participate, eligible households had to live in, or move to, housing that met the program's standards. HUD found that eligible households who lived in lower-quality housing did not participate as readily as did their counterparts in higher-quality housing.

2. As the amount of housing allowances was increased, the number of households who participated increased, but not necessarily in the same ratio. At least part of the allowance had to be used to meet the costs associated with obtaining housing meeting the quality standards of the program. This reduced the number of eligible households who participated.

3. The selection of housing quality standards and payment levels had important implications for meeting program goals but had to be weighed against other goals, such as minimally acceptable housing quality and budgetary constraints.

4. Although a "complete understanding of an accurate estimate" of the effect of housing allowances on housing prices was not available at the end of the study, HUD stated that it was "confident that it is small." In HUD's view, the supply experiment "laid to rest the fear that a full-scale housing allowance program would drive up housing prices substantially . . . or do almost nothing to increase the supply of decent, safe, and sanitary housing."

5. A full-scale housing allowance program in 1980 would have reached about 7.2 million households of 17.5 million income-eligible households. The average monthly allowance would have been about $65. Total program costs would have been $7.4 billion per year, of which $1.7 billion would have gone to administration.

Although the results of EHAP could not answer all the questions that HUD set out to answer, HUD and other observers were persuaded that housing allowances could make a difference and could offer greater choices to tenants who are willing to take some of the risks associated with entering the private market. The results of this study, and the change from a Democratic to a Republican administration in 1980, contributed to a shift in federal housing policy from one of primary emphasis on supply-side investment to one of primary emphasis on demand-side investment.

This shift was accompanied by a sharp reduction in new fiscal commitments for housing, so the concept of a full-scale housing allowance was not implemented. For example, in 1979, subsidized housing starts totaled 175,000 units. By 1989, the number of new or substantially rehabilitated subsidized housing starts had dropped to less than 21,000, while more than 1 million households, composing 26% of all HUD-managed assisted units, were receiving HUD certificates or vouchers for rental assistance rather than subsidized housing units. In the 1990s, the Section 8 certificate and voucher programs became mired in controversy over management problems and neighborhood opposition. (SEE ALSO: *Housing Allowances*)

—*Peter W. Salsich, Jr.*

Further Reading

Bendick, M. and J. Zais. 1978. *Incomes and Housing: Some Insights from the Experimental Housing Allowance Program.* Washington, DC: Urban Institute.

Casey, C. 1992. *Characteristics of HUD-Assisted Renters and Their Units in 1989.* Washington, DC: U.S. Department of Housing and Urban Development, Office of Policy, Development and Research.

Controller General of the United States. 1978. *An Assessment of the Department of Housing and Urban Development's Experimental Housing Allowance Program.* Washington, DC: Government Printing Office.

Frieden, B. 1980. "Housing Allowances: An Experiment That Worked." *Public Interest* 15(59):15-35.

Hetzel, O., D. Yates, and J. Trunkto. 1978. "Making Allowances of Housing Costs: A Comparison of British and U.S. Experiences." *Urban Law and Policy* 1(229):229-73.

Montgomery, R. and D. Mandelker, eds. 1979. *Housing in America: Problems and Perspectives.* 2d ed. Indianapolis, IN: Bobbs-Merrill (see pp. 312-21).

Stegman, M. 1991. *More Housing, More Fairly: The Report of the 20th Century Fund Task Force on Affordable Housing.* New York: Twentieth Century Fund Press.

U.S. Department of Housing and Urban Development. 1978. *A Summary Report of Current Findings from the Experimental Housing Allowance Program.* Washington, DC: Government Printing Office. ◄

F

▶ Fair Housing Amendments Act of 1988

The Fair Housing Amendments Act of 1988 (FHAA) amended the federal Fair Housing Act (Title VIII of the Civil Rights Act of 1968) in a number of significant ways, most notably by adding handicap and familial status to the types of discrimination outlawed by the statute and by creating a new enforcement mechanism for handling administrative complaints to the Department of Housing and Urban Development (HUD). Other important changes included the liberalization of the punitive damages and attorney's fees provisions applicable to private lawsuits and the authorization for civil penalties and damage awards to persons aggrieved in Justice Department suits.

The origins of the FHAA can be traced to the early 1970s, when congressional hearings began to call attention to the inadequacies of Title VIII's enforcement mechanisms. Numerous studies during the next decade showed that racial discrimination in housing was continuing at alarming rates despite Title VIII's prohibitions. One HUD study concluded that some 2 million instances of housing discrimination against African Americans occur every year, a fact mentioned repeatedly in the legislative history of the FHAA to demonstrate the need to strengthen Title VIII.

The principal goal of the FHAA was to put "teeth" in Title VIII's enforcement mechanisms, to make it a more effective tool for combating housing discrimination in the United States. In particular, the HUD administrative procedure, which had been limited to seeking voluntary compliance among the parties, was changed to provide for serious sanctions (including damage awards and civil penalties) against those found to have committed discriminatory housing practices.

As with the original Fair Housing Act, FHAA complaints to HUD that come from states or localities with fair housing laws that are "substantially equivalent" to the FHAA must be referred to the appropriate state or local agency for handling. By 1995, more than half of the administrative complaints under the FHAA were being referred to the 32 states and 40 localities that had enacted substantially equivalent fair housing laws.

Most of the FHAA complaints that HUD retains are resolved by conciliation or some other means that does not involve a formal determination concerning the discrimination alleged. About 5% of the cases result in formal charges of discrimination. In about 60% of these charged cases, one or more of the parties "elects" to have the case decided in federal district court, where it is prosecuted by the Department of Justice and where the court may award equitable relief, actual and punitive damages to aggrieved persons, and attorney's fees. If there is no election to court, the case is prosecuted by a HUD lawyer and is tried before a HUD-appointed administrative law judge, who may award actual damages to the aggrieved person, civil penalties of up to $50,000 to the government, injunctive relief, and attorney's fees.

The FHAA, like the original Fair Housing Act, gives complainants the option of bypassing this entire administrative procedure and going directly to court, where they may be awarded equitable relief, actual and punitive damages, and attorney's fees. Indeed, the FHAA makes this option both easier to use and more attractive by extending the statute of limitations for private litigants from 180 days to two years and by eliminating Title VIII's $1,000 cap on punitive damages and its "financial inability" limitation on attorney's fees awards.

The FHAA also authorizes the Justice Department to intervene in private cases if the attorney general certifies that the case is "of general public importance." Conversely, an aggrieved person may intervene in a "pattern or practice" suit brought by the Justice Department and may obtain any relief in that suit that would be available in a private case. Even without such intervention, the Justice Department is authorized by the FHAA to seek monetary damages for aggrieved persons and civil penalties of up to $100,000 for the government in pattern or practice cases, in addition to the equitable relief that Title VIII has always authorized in such cases.

In the years following enactment of the FHAA, both private lawsuits and Justice Department pattern or practice cases produced dramatically higher verdicts and settlements, many of which exceeded $100,000 and a few of which topped $1 million. Some of the more noteworthy results occurred in anti-harassment cases and in suits challenging racial and national origin discrimination in advertising, mortgage lending, and insurance, as well as in the more typical cases involving steering and refusals to rent.

Passage of the FHAA also produced a substantial increase in the number of fair housing complaints filed with HUD, which rose from 4,000 to 5,000 annually in the years preceding the FHAA to more than twice this number by the mid-1990s. A large portion of these complaints were based on familial status and handicap, which became, after race, the two most common bases of complaints under the FHAA.

The FHAA's ban on familial status discrimination does not apply to "housing for older persons," but otherwise, it extends to all of the housing transactions covered by Title VIII. The law defines "familial status" as one or more individuals under the age of 18 living with a parent, a person having legal custody of such individual(s), or the designee of such parent or legal custodian; also included is any person who is pregnant or who is about to secure legal custody of someone under the age of 18.

The FHAA's prohibition of familial status discrimination means that housing providers may no longer refuse to deal with people because their households include children. Nor may tenants be evicted just because they have a baby or bring a child into the unit. In addition, limitations based on the number of children in a family are illegal, and families with children cannot be restricted to certain parts of an apartment building or housing complex. Landlords may not impose higher security deposits or rental charges on families with children, and such families cannot be denied access to recreational or other facilities and services that are made available to other tenants, although reasonable health and safety rules relating to the use of these facilities and services may be imposed. Advertising that indicates any preference, limitation, or discrimination based on familial status is also illegal.

The FHAA's familial status prohibitions also outlaw housing practices that have a disproportionate impact on families with children. Reasonable occupancy standards are allowed, but only if they are applied to all occupants and do not operate to discriminate against families with children.

The FHAA exempts housing for older persons from its prohibitions against familial status discrimination. This exemption extends to three separate categories of housing: (a) government-subsidized housing specifically designed and operated to assist elderly persons; (b) housing occupied entirely by persons 62 years of age or older; and (c) housing "intended and operated for occupancy by persons 55 years of age or older," which means that the housing must have at least 80% of its units occupied by at least one person aged 55 or older and must have taken other steps to indicate that it is reserved for older persons. The housing for older persons exemption applies only to the familial status prohibitions of the FHAA and does not allow such housing to discriminate on the basis of race, color, religion, sex, handicap, or national origin.

The FHAA's addition of handicap discrimination to Title VIII's prohibited bases was seen by Congress as "a clear pronouncement of a national commitment to end the un-

necessary exclusion of persons with handicaps from the American mainstream." Like all other modern federal laws dealing with disability rights, the FHAA broadly defines the protected class of handicapped persons to include the following:

► Everyone with a physical or mental impairment that substantially limits one or more of such person's major life activities
► Everyone with a record of having such an impairment
► Everyone regarded as having such an impairment

Also protected are nonhandicapped buyers and renters who reside or are associated with handicapped people (e.g., parents or roommates of handicapped persons).

Thus, coverage extends far beyond such obvious examples as wheelchair-bound and visually impaired people to include those who are limited by alcoholism, high blood pressure, emotional problems, mental illness or retardation, learning disabilities, and many of the difficulties associated with old age. The phrase "physical or mental impairment" includes various diseases, such as cerebral palsy, epilepsy, muscular dystrophy, multiple sclerosis, cancer, heart disease, and diabetes. Coverage also extends to persons suffering from communicable diseases, including AIDS and HIV infection. Drug addiction (other than addiction caused by current, illegal use of a controlled substance) is also included. The breadth of the FHAA's "handicap" definition means that it covers a large portion of the entire U.S. population, estimated at the time of the statute's passage to be about one of every six persons in the country.

In addition to banning handicap discrimination to the same extent that racial and other illegal forms of housing discrimination are outlawed, the FHAA added three special provisions that apply only in handicap situations. First, the law requires that handicapped persons be allowed, at their own expense, to make any reasonable modifications necessary for their full enjoyment of the premises. Examples of the modifications covered by this provision include widening doorways to make rooms more accessible, installing grab-bars in bathrooms, and lowering kitchen cabinets to a height suitable for persons in a wheelchair.

Second, the FHAA requires housing providers to "make reasonable accommodations in rules, policies, practices, or services" necessary to afford handicapped persons "equal opportunity to use and enjoy a dwelling." The concept of "reasonable accommodations" is derived from § 504 of the 1973 Rehabilitation Act and, in the housing area, includes examples such as allowing a blind tenant to have a Seeing Eye dog even if the building has a "no pet" policy and reserving a parking place for a mobility-impaired tenant closer to his or her unit than other tenants would be entitled to have.

Third, the FHAA requires that six specific accessibility-enhancing features be included in the design and construction of all "covered multifamily dwellings" constructed for first occupancy after March 13, 1991. Covered multifamily dwellings are buildings with at least one elevator that have four or more units and ground floor units in nonelevator buildings with four or more units. Examples of the features that must be included in these dwellings are accessible routes into and through the building and doors that are wide enough to accommodate wheelchairs.

The FHAA's prohibitions against handicap discrimination also have been used to strike down zoning and other governmental restrictions on communal housing opportunities for the handicapped (sometimes called "group homes"). The legislative history of the FHAA clearly shows that the statute was "intended to prohibit the application of special requirements through land-use regulations, restrictive covenants, and conditional or special use permits that have the effect of limiting the ability of [handicapped] individuals to live in the residence of their choice in the community." Thus far, a substantial amount of the litigation based on the handicap provisions of the FHAA has been devoted to group home cases, including the first FHAA case to reach the Supreme Court, *City of Edmonds v. Oxford House, Inc.* (1995).

Enactment of the FHAA was the most important development in fair housing law in 20 years. The FHAA helped to reinvigorate enforcement of the basic prohibitions of the 1968 Fair Housing Act and launched new efforts designed to challenge housing discrimination based on handicap and familial status. Implementation of the changes wrought by the FHAA has occupied a major part of the fair housing agenda during the 1990s and is likely to continue to do so well into the 21st century. (SEE ALSO: *Blockbusting; Civil Rights Act of 1866; Civil Rights Act of 1968, Title VIII;* **Discrimination;** *Fair Housing Audits; Fair Share Housing; Gautreaux Program; National Fair Housing Alliance*)
—*Robert G. Schwemm*

Further Reading

Franklin, D. 1996. "Civil Rights vs. Civil Liberties? The Legality of State Court Lawsuits under the Fair Housing Act." *University of Chicago Law Review* 63(4):1607-38.

Massey, Douglas S. and Nancy A. Denton. 1993. *American Apartheid: Segregation and the Making of the Underclass.* Cambridge, MA: Harvard University Press.

Schwemm, Robert G. 1990. *Housing Discrimination: Law and Litigation.* New York: Clark Boardman Callaghan. Annual supplements.

Schwemm, Robert G. 1993. "The Future of Fair Housing Litigation." *John Marshall Law Review* 26:745-73.

Turner, Margery A., Raymond J. Struyk, and John Yinger. 1991. *Housing Discrimination Study: Synthesis.* Washington, DC: U.S. Department of Housing and Urban Development.

U.S. Commission on Civil Rights. 1994. *The Fair Housing Amendments Act of 1988: The Enforcement Report.* Washington, DC: Author. ◀

▶ Fair Housing Audits

A fair housing audit is a survey technique that isolates the impact of a person's minority status on the way he or she is treated in the housing market. An audit consists of successive visits to the same landlord or real estate agent by two auditors who are equally qualified for housing but who differ in minority status. After the interview with the housing agent, each audit teammate independently completes a survey form to describe how he or she was treated. In a sample of audits, discrimination is defined as systematically less favorable treatment of minority auditors.

Before fair housing audits were developed, researchers studied housing discrimination indirectly by looking for its impacts on housing prices, housing quality, homeownership, and the pattern of racial residential segregation. Audits represent a major advance because they make it possible to measure discrimination directly; in fact, they literally catch housing agents in the act of discriminating. Moreover, audits make it possible to observe the circumstances under which discrimination occurs.

A Brief History of Fair Housing Audits

Fair housing audits, also called "tests," were first used by fair housing organizations to determine the validity of minority home seekers' complaints. The U.S. Department of Housing and Urban Development (HUD) sponsored the first nationwide audit study in 1977. This study conducted 3,264 audits in 40 metropolitan areas and found significant discrimination against blacks in both the sales and rental of housing. In addition, at least 72 audit studies were conducted in individual cities between 1974 and 1990, and most of them found strong evidence of discrimination against blacks or Hispanics. HUD sponsored a second national audit study in 1989, the Housing Discrimination Study (HDS), which conducted 3,745 audits in 25 metropolitan areas. HDS found more evidence of discrimination against blacks and provided the first national estimate of discrimination against Hispanics.

The audit technique imposes a time cost on audited housing agents, including those who do not discriminate. However, most observers agree that, given the long history of housing discrimination, fair housing audits, like income tax audits, are a legitimate tool for identifying people who break the law even though they impose a cost on some law-abiding citizens. Moreover, the U.S. Supreme Court ruled, in the 1982 *Havens* case, that fair housing audits are a legitimate investigative tool.

Audit Methodology

Audit Design

To isolate the impact of race or ethnicity on the treatment that a home seeker receives, an audit attempts to make the two members of an audit team as comparable as possible on all characteristics, except minority status, that might influence their ability to rent an apartment or buy a house. Researchers can enhance comparability through matching, assignment, training, and timing.

Matching involves the selection of audit teammates with the same fixed characteristics—namely, sex, age, and general appearance. Assignment gives each auditor an income, down payment (or deposit) capability, marital status, and number of children for the purposes of each audit. Teammates are assigned characteristics that are virtually identical and that qualify them for the housing unit on which the audit is based.

Training is intended to minimize differences in the way teammates behave. All HDS auditors were trained, for example, to express interest in all housing that met their price and size requirements and not to indicate any community or neighborhood preferences. Finally, the timing of visits to the agent is carefully controlled. Teammates initiate their visits to the housing agent within a relatively short time of each other so that the circumstances they encounter are virtually the same, and most recent audit studies randomize the order in which teammates contact the housing agent.

These steps are designed to make audit teammates equally qualified for the advertised housing unit. In a careful audit, teammates have only minor differences on any characteristic—except for minority status—that is known to affect their treatment in the housing market, so any systematic unfavorable treatment of minority auditors can be interpreted as discrimination.

Sampling

The central sampling issue in an audit study is the selection of housing units or agents to audit. Most studies draw a random sample of advertisements from the major newspaper in the area under investigation. Each advertisement becomes the focus of one audit, and the two audit teammates visit the agency that placed the advertisement to inquire about the availability of the advertised unit.

The focus on newspaper advertisements implies that audit studies do not necessarily measure discrimination against the average minority household but instead measure the discrimination that occurs when minority households inquire about housing that is advertised in the newspaper and for which they are qualified.

Measurement and Inference

The incidence of discrimination, which refers to the probability that a minority customer will encounter discrimination, cannot be measured without considering two distinctions.

The first distinction is between the concepts of gross and net discrimination. The gross concept, which is related to the logic of fair housing enforcement, is to say that the incidence of discrimination is the probability that a minority home seeker will encounter less favorable treatment than his or her teammate. The net concept, which is related to studies of wage and housing price discrimination, is to say that the incidence of discrimination is the difference in the probability that majority and minority home seekers will encounter unfavorable treatment.

The second distinction is between total and systematic unfavorable treatment. Total unfavorable treatment equals systematic unfavorable treatment—that is, unfavorable treatment caused by some systematic element of housing agent behavior, plus random unfavorable treatment. Conscious discrimination by an agent is an example of systematic unfavorable treatment, whereas not showing the advertised unit to the minority auditor because it was rented right after the white auditor's visit is an example of random unfavorable treatment. A researcher must decide whether to try to isolate systematic unfavorable treatment.

Two simple measures of the incidence of discrimination are readily defined. The simple gross measure is the share of audits in which the minority auditor encountered unfavorable treatment, whereas the simple net measure equals the simple gross measure minus the share of audits in which the majority auditor encountered unfavorable treatment. These two measures are appropriate for investigating total unfavorable treatment using the gross and net concepts, respectively. Moreover, if one assumes that all agent behavior is systematic, these two simple measures also are appropriate for a focus on the concepts of gross and net systematic unfavorable treatment. If one assumes instead that all unfavorable treatment of white auditors is random and that random factors are equally likely to favor minority and white auditors, the simple net measure indicates systematic unfavorable treatment according to the gross concept.

In short, the simple net and gross measures are appropriate for studying total unfavorable treatment but cannot be applied to systematic unfavorable treatment without strong assumptions about the role of random events. Measures of discrimination derived from more formal models that account for random factors, including unobserved differences between teammates, are now available, but they require advanced statistical procedures.

The simple net measure is associated with a simple test of the null hypothesis that no discrimination exists—namely the well-known paired difference-of-means test. More satisfactory tests of this null hypothesis can be obtained by applying advanced statistical techniques to formal models that contain random errors.

In some cases, it is useful to examine the severity of discrimination. One might want to know, for example, the average difference in the number of houses shown to minority and white auditors. For agent behavior that is binary, such as whether the advertised unit is shown, there is no difference between the incidence and severity of discrimination, but these concepts are quite different for behavior that can take on many values. The severity of discrimination can be measured by the difference in mean treatment of minority and white auditors, and the null hypothesis of no discrimination can be evaluated with a paired difference-of-means test.

Selected Audit Results from HDS

The HDS audits uncovered discrimination in many types of housing agent behavior, beginning with responses to the auditors' inquiries about the advertised unit. Using the simple net measure, the incidence of discrimination in the availability of this unit was 5.5% for blacks in both the sales and rental markets, 4.2% for Hispanic homebuyers, and 8.4% for Hispanic renters. Agents have more leeway to discriminate in showing units they have not advertised. Thus, the simple net measure indicates that blacks were recommended or shown fewer units than their white teammates 19.4% of the time in the sales market and 23.4% of the time in the rental market. The comparable figures in the Hispanic-white audits are 16.5% (sales) and 9.8% (rental). These net incidence measures are all statistically significant.

The severity of discrimination in housing availability also was quite high; that is, minority auditors were recommended or shown significantly fewer housing units than were their white teammates. Blacks learned about 23.7% fewer units in the sales market and 24.5% fewer units in the rental market. For Hispanics, the differences in total units made available were 23.6% (sales) and 10.9% (rental). These differences are all statistically significant.

HDS also uncovered statistically significant discrimination against minorities in information about available mortgages, offers of assistance in finding financing, requests to call back or make follow-up calls, and offers of special rental incentives, such as a month's free rent or a waived security deposit. Moreover, real estate brokers sometimes steered minority customers toward areas with a high minority concentration.

Implications of Audit Studies

Title VIII of the 1968 Civil Rights Act, also known as the Fair Housing Act, outlawed discrimination in housing but had weak enforcement provisions. Fair housing audits have proven to be an effective tool for an individual to use in a civil suit against a suspected discriminator. Moreover, thanks in part to evidence from the 1977 national audit study, Congress amended the Fair Housing Act in 1988 to give stronger enforcement powers to both HUD and the Department of Justice; and thanks in part to the HDS evidence of continuing discrimination, HUD and Justice have made extensive use of their new powers. Further audit studies will no doubt play a role in evaluating the success of these efforts. (SEE ALSO: **Discrimination; Fair Housing Amendments Act of 1988**)

—*John Yinger*

Further Reading

Galster, George C. 1990a. "Racial Discrimination in Housing Markets During the 1980s: A Review of the Audit Evidence." *Journal of Planning Education and Research* 9:165-75.

———. 1990b. "Racial Steering in Urban Housing Markets: A Review of the Audit Evidence." *Review of Black Political Economy* 18(3):105-29.

Reed, Veronica M. 1991. "Civil Rights Legislation and the Housing Status of Black Americans: Evidence from Fair Housing Audits and Segregation Indices." *Review of Black Political Economy* 19(3-4): 1-42.

Struyk, Raymond, Margery A. Turner, and John Yinger. 1991. *Housing Discrimination Study: Synthesis*. Washington, DC: U.S. Department of Housing and Urban Development.

Turner, Margery A., Maris Micklensons, and John G. Edwards. 1991. *Housing Discrimination Study: Analyzing Racial and Ethnic Steering*. Washington, DC: U.S. Department of Housing and Urban Development.

Wienk, Ronald E., Clifford E. Reid, John C. Simonson, and Frederick J. Eggers. 1979. *Measuring Discrimination in American Housing Markets: The Housing Market Practices Survey*. Washington, DC: U.S. Department of Housing and Urban Development.

Yinger, John. 1986. "Measuring Discrimination with Fair Housing Audits: Caught in the Act." *American Economic Review* 76:881-93.

Yinger, John. 1993. "Access Denied, Access Constrained: Results and Implications of the 1989 Housing Discrimination Study." Pp. 69-112 in *Clear and Convincing Evidence: Testing for Discrimination in America*, edited by M. Fix and R. Struyk. Washington, DC: Urban Institute. ◀

▶ Fair Market Rent

The fair market rent is the maximum allowable rent that can be charged under the auspices of the Section 8 Existing Housing Assistance Payments Program established by the Housing and Community Development Act of 1974. Fair market rents (FMRs) are the Department of Housing and Urban Development's (HUD's) determinations of the rents, including utilities (except telephone), ranges and refrigerators, parking, and all maintenance, management, and other essential housing services, that would be required to obtain, in a particular market area, privately developed and owned rental housing of modest design with suitable amenities.

Separate fair market rents are established by unit size (number of bedrooms), basic structure type (detached, semi-detached/rowhouse, walk-up, and elevator apartments) and occupant group (nonelderly family and elderly family, including handicapped) for individual market areas.

The concept of fair market rent is predicated on the need to control the cost of the Section 8 program. Without a maximum allowable rent, the difference in cost between an unpretentious unit of average or slightly below-average quality and a luxury unit would be borne by the taxpayers. Fair market rents are ceilings, and HUD will not automatically approve proposals calling for maximum rents. Actual rents must also meet a criterion of reasonableness when compared with similar unsubsidized units in the community. Rents for Section 8 units cannot be higher than rents charged for comparable unsubsidized units. And in communities with rent control, rents for Section 8 units subject to control must be comparable to other rent-controlled units.

Fair market rents are determined annually by HUD. Initially, the Economic and Market Analysis Division of HUD had wanted to set FMRs at the median rents of recently occupied units meeting all Section 8 standards, but data limitations resulted in the use of median rents of recently occupied units meeting only the plumbing standards. As a result, it concluded that initial FMRs were too low. This led to upward adjustments in response to requests by public housing authorities (PHAs). Not only did PHAs realize higher administrative fees that were tied to higher FMRs, but it became significantly easier for the certificate holder to find a suitable unit. Lower FMRs might result in lower program costs, but at the same time, they would limit the extent to which Section 8 certificate holders would have housing opportunities in otherwise inaccessible communities.

Fair market rents are a good illustration of policy trade-offs between per-unit cost and housing quality as well as integration of HUD-assisted populations into better quality communities. (SEE ALSO: **Affordability**; *Housing-Income Ratios; Section 8*)

—*Daniel J. Garr*

Further Reading

General Accounting Office. 1996. *Rental Housing: Use of Smaller Market Areas to Set Rent Subsidy Levels Has Drawbacks*. Washington, DC: Government Printing Office.

"Generally Applicable Definitions." 1996. P. 46 in *Code of Federal Regulations*. Vol. 24. Part 5m, Subpart A. Washington, DC: Government Printing Office.

"Section 8 Housing Assistance Payment Program: Fair Market Rents and Contract Rent Annual Adjustment Factors." 1996. Pp. 239-41 in *Code of Federal Regulations*. Vol. 24. Part 888. Washington, DC: Government Printing Office.

U.S. Department of Housing and Urban Development. 1995. "Fair Market Rents for the Section 8 Housing Assistance Payments Program: Amendments to Method of Calculating." *Federal Register* 60:11870-873. ◀

▶ Fair Share Housing

In 1970, the Miami Valley Regional Planning Commission in Ohio developed the "Dayton Plan," which first introduced the concept of fair share housing. Recognizing inequities in its housing system, the Miami Valley region developed a plan to allocate the total number of needed low- and moderate-income housing units among all localities in the larger metropolitan area. Fair share plans, also known as regional housing allocation systems, are attempts to improve the distribution of low-income housing units across a region in a way that will expand housing opportunities, and the benefits of good neighborhoods, for low- and moderate-income families. Many of the original plans emphasized the need to reduce the impact of exclusionary suburban zoning by making areas accessible to households historically confined to cities, including the poor and racial and ethnic minority groups. Within five years of the development of the Dayton Plan, approximately 40 regional agencies adopted a fair share plan.

The 1975 *Mt. Laurel I* decision by the New Jersey Supreme Court provided the legal basis for the fair share concept. In 1971, the National Association for the Advancement of Colored People (NAACP) sued the Camden, New Jersey, suburb of Mt. Laurel for using zoning to limit its housing to middle- and upper-income families. The court ruled unanimously that areas had an obligation to consider housing needs of the entire region in zoning decisions. The decision, influencing courts nationwide, also held that municipalities must allow all economic groups access to housing throughout the region. As a result, New Jersey municipalities would have to create opportunities for the construction of their "fair share" by zoning some land for lower-income households.

Despite the legal backing for the fair share concept, most areas were slow to adopt any type of allocation plan. Even in New Jersey, areas resisted opening what were often rather exclusive areas to the poor and even more so to racial and ethnic minorities. As a consequence of some areas' diffidence and some areas' refusal to provide more housing opportunities, New Jersey's 1983 *Mt. Laurel II* decision decried the continued exclusionary zoning. This decision required municipalities to adopt specific strategies, including using federal and state money for low-income housing and assisting developers in setting aside a portion of their development for moderate- and low-income people, to achieve fair share goals.

Many localities, in New Jersey and elsewhere, were troubled that fair share was not just a strategy that would open the suburbs to the poor but that such a strategy would also open areas to racial and ethnic minorities. Local officials had long been aware of the considerable overlap between the economically disadvantaged and racial and ethnic minorities; racial minorities, a disproportionate part of the poor, could benefit from fair share strategies. Exclusionary zoning, which fair share was meant to compensate for, often restricted or altogether barred access by racial and ethnic minorities by raising the price of land and thereby the cost of housing. Although outright housing discrimination against racial and ethnic groups had been made illegal by prior legislation and judicial decisions, it still occurred on a de facto basis, disguised by the wording of exclusionary zoning regulations. At least implicitly, exclusionary zoning created and re-created economically and racially homogeneous neighborhoods; fair share, on the other hand, had the potential to increase economic and racial diversity. The main issue of the initial Mt. Laurel suit was economic discrimination, but it was filed by the NAACP in light of the obvious negative impact of exclusionary zoning on racial minorities.

Since the 1970s, the most active time for fair share implementation, there have been federal and state legislation and other court decisions incorporating at least the spirit of fair share housing but rarely its actual intent. For example, in the 1970s, HUD developed the idea of housing assistance plans (HAPs) through which localities determined their own housing needs and identified those groups most in need of housing. HAPs were submitted by each locality to an area office that determined housing allocations, including the dispersion of Section 8 rent certificates. However, the federal government never officially adopted the fair share concept and did not mandate it as a condition for receiving federal money.

It is difficult to assess the effectiveness of fair share, per se, because it occurred within the context of other changes in the housing delivery system. Furthermore, it followed important civil rights legislation, including Title VIII of the Civil Rights Act of 1968, also known as the Fair Housing Act. All of these decisions had as one of their goals a reduction in housing discrimination. Yet an examination of the current low-income housing situation, the continuation of exclusionary zoning, the persistence of racial segregation, and the closed suburbs of many metropolitan areas show that these areas are not building for and renting to their fair share of lower-income people. Even in New Jersey, where one might expect to see the greatest impact, progress has been slow. Its fair share strategy mostly benefited white suburbanites and moderate-income homebuyers, not the poor and central-city residents and not racial and ethnic minorities. Fair share can be described as having had only limited effectiveness. (SEE ALSO: **Discrimination**; *Fair Housing Amendments Act of 1988; Mt. Laurel*)

—Judith McDonnell ◀

Further Reading

Baer, W. C. 1986. "Housing in an Internationalizing Region: Housing Stock Dynamics in Southern California and the Dilemmas of Fair Share." *Environment and Planning D: Society and Space* 4:337-49.

Hughes, Mark Alan and Therese J. McGuire. 1991. "A Market for Exclusion: Trading Low-Income Housing Obligations under Mt. Laurel III." *Journal of Urban Economics* 29:207-17.

Keating, W. Dennis. 1994. *The Suburban Racial Dilemma: Housing and Neighborhoods.* Philadelphia: Temple University Press.

Listokin, David. 1976. *Fair Share Housing Allocation.* New Brunswick: Rutgers, State University of New Jersey, Center for Urban Policy Research.

Tucker, William. 1990. *The Excluded Americans: Homelessness and Housing Policy.* Washington, DC: Regnery Gateway.

Yinger, John. 1995. *Closed Doors, Opportunities Lost: The Continuing Costs of Housing Discrimination.* New York: Russell Sage.

Zarembka, Arlene. 1990. *The Urban Housing Crisis: Social, Economic, and Legal Issues and Prospects.* Westport, CT: Greenwood. ◀

▶ Family Self-Sufficiency

Family self-sufficiency programs are designed to promote families' becoming financially independent from government housing, welfare subsidies, or both. Strategies for promoting family self-sufficiency include (a) providing services believed to promote self-sufficiency, such as education, job training, child care, and transportation; (b) promoting homeownership; or (c) moving families to new, often suburban, neighborhoods where greater opportunities for economic mobility and independence are thought to exist. The development of self-sufficiency programs indicates a shift in housing policy away from viewing housing assistance primarily as a means of providing decent housing toward viewing it as a means of promoting economic independence.

Strategies for Promoting Family Self-Sufficiency

Coordinating Services and Housing

Believing that many families receiving public assistance face multiple barriers to becoming economically independent, a number of programs promote self-sufficiency by coordinating housing with a range of social services. At the core of these service packages are usually adult education and employment programs. Additional support services, such as child care, medical assistance, transportation, life skills training, counseling, and case management are also often included to address other obstacles that households face in becoming self-sufficient.

A number of coordinated housing and service delivery programs have been developed in recent years. Early federal initiatives include two Department of Housing and Urban Development (HUD) demonstration programs: Project Self-Sufficiency and Operation Bootstrap. Various local initiatives were also developed in the 1980s, notable examples include Baltimore's Lafayette Courts Family Development Center and Charlotte's Stepping Stone and Gateway Programs.

The results of these programs have produced cautious optimism. Small but noticeable improvements in education,

employment, or both have normally been achieved by participating households, although complete self-sufficiency has been rare. Indeed, a few evaluations have noted an increase in government assistance among participating households during the early months of program participation, although assistance rates then drop over time.

Questions still remain about the appropriate mix of services required to promote self-sufficiency. Evidence suggests that child care increases employment opportunities, but the effect of other support services, such as transportation or counseling, on promoting self-sufficiency has not yet been thoroughly examined.

Although much remains to be learned about these programs and their effectiveness, several recent federal programs have incorporated many of the features of these early efforts. For example, the Family Self-Sufficiency Program (FSS), enacted as part of the Cranston-Gonzalez National Affordable Housing Act of 1990, mandates that all public housing authorities provide families with coordinated employability programs, with the number of families that must be served tied to the number of new units provided to the housing authority. Public housing authorities were also encouraged in this legislation to develop Family Investment Centers, facilities located on or near public housing developments where residents could access a variety of social services.

Promoting Homeownership

Although homeownership has traditionally been a measure of economic independence, several programs have used homeownership as a means of encouraging self-sufficiency. These programs allow low-income households to become homeowners with the intent that this will result in positive benefits to both the households and the community. Recent efforts at promoting homeownership have shown mixed success. Although some low-income households have been able to purchase homes, qualifying households are often already more self-sufficient. Nor has it been shown that becoming a homeowner leads to other changes that make families more self-sufficient. Nevertheless, several homeownership demonstration programs—HOPE grants, for Homeownership and Opportunity for People Everywhere—were included in the Cranston-Gonzalez National Affordable Housing Act of 1990.

Helping families save for homeownership has also been used as an incentive in several self-sufficiency programs. In Charlotte's Gateway Housing Program, an escrow account was set up so that families could eventually have money to purchase a home. As in a regular public housing program, the rent of participating households is tied to annual income. However, any difference between the actual operating cost of the unit and the rent paid is put into the escrow account. On the basis of the experiences of the Charlotte program, the recently enacted FSS also features an escrow account.

Changing Neighborhoods

A few programs have tried to promote self-sufficiency by moving families to neighborhoods where greater economic and social opportunities are thought to exist, often in the suburbs. The Gautreaux program in Chicago is the major example of this approach. In the Gautreaux program, Section 8 certificate recipients are allowed to move throughout the Chicago metropolitan area, and, with the assistance of housing locators and counselors, a number have moved to suburban locations. Research has shown that both parents and children obtain significant employment and education benefits by moving to these neighborhoods.

Based on the success of the Gautreaux program, a five-city demonstration program called Moving to Opportunity was included in the Housing and Community Development Act of 1992. Opposition to the program led Congress to rescind funding for all but the first phase of the program, however.

Measuring Self-Sufficiency

The ultimate measure of success for a self-sufficiency program is whether a family needs public assistance. However, most programs have not used such a strict criterion. Instead, more flexible definitions have been used, emphasizing the process of becoming self-sufficient, noting if families are taking steps that can lead toward more independence (e.g., obtaining more education). These flexible definitions recognize both the structural situations that families are in, such as a weak job market, low minimum wages, and tight or expensive housing markets, as well as the individual problems that impede movement toward self-sufficiency.

Role of Housing

Traditionally, the goal of housing assistance has been to provide decent housing. Although additional services have occasionally been provided, the focus has normally been on the "bricks and mortar." With a growing concern about promoting self-sufficiency, however, factors beyond bricks and mortar have had to be taken into account. The result has been the development of various self-sufficiency programs that try to use housing as a tool or an inducement for becoming economically independent. (SEE ALSO: *Gautreaux Program; Home Ownership and Opportunity for People Everywhere; Investment Partnerships Act*)

—*C. Scott Holupka*

Further Reading

Nenno, Mary. 1991. "Housing, Public Welfare and Human Services: New Progress Toward Integrated Systems." In *Family Self-Sufficiency: Linking Housing, Public Welfare and Human Services,* edited by the National Association of Housing and Redevelopment Officials and the American Public Welfare Association. Washington, DC: National Association of Housing and Redevelopment Officials.

Newman, Sandra and Ann Schnare. 1992. *Beyond Bricks and Mortar: Reexamining the Purpose and Effects of Housing Assistance* (Report 92-3). Washington, DC: Urban Institute.

Polit, Denise and Joseph O'Hara. 1989. "Support Services." In *Welfare Policy for the 1990's,* edited by Pheobe Cottingham and David Ellwood. Cambridge, MA: Harvard University Press.

Rohe, W. and R. Garshick Kleit. 1997. "From Dependency to Self-Sufficiency: An Appraisal of the Gateway Transitional Families Program." *Housing Policy Debate* 8(1): 75-108.

Shlay, Anne. 1993. "Family Self-Sufficiency and Housing." *Housing Policy Debate* 4:457-95.

Shlay, Anne B. and C. Scott Holupka. 1992. "Steps toward Independence: Evaluating an Integrated Service Program for Public Housing Residents." *Evaluation Review* 16:508-33. ◀

▶ Fannie Mae

Fannie Mae (formerly known as the Federal National Mortgage Association) is a private corporation with a public mission to facilitate homeownership. For more than half a century, Fannie Mae has pursued this mission by providing financial products and services that increase the availability and the affordability of mortgage credit for low-, moderate-, and middle-income households.

Operating centrally in the housing finance market—a $4 trillion industry—Fannie Mae provides a secondary market for mortgage loans. It purchases residential home loans from mortgage-lending institutions, thereby replenishing their supply of mortgage funds available for lending.

Fannie Mae converts packages of loans into mortgage-backed securities (MBSs) that it then guarantees for sale to investors, or it purchases and retains these mortgages in its portfolio. To finance these mortgage purchases, Fannie Mae raises funds in capital markets in the United States and overseas. By so doing, Fannie Mae channels billions of dollars each year from the world's capital markets into the U.S. housing finance market.

Fannie Mae purchases loans from a nationwide network of approved mortgage lenders through its five regional offices in Atlanta, Chicago, Dallas, Pasadena, and Philadelphia. The corporation is headquartered in Washington, D.C.

The largest corporation in the United States, Fannie Mae had $392 billion in assets and an additional $710 billion in MBSs outstanding in 1997. Next to the U.S. Treasury, it is the second largest borrower in the capital markets. Its stock is traded on the New York Stock Exchange (FNM) and other major exchanges. Fannie Mae has grown to be the nation's largest investor in home mortgages. The corporation expects to help well over 20 million households buy homes in the 1990s.

Fannie Mae's History

In response to the massive upheavals in the housing finance system experienced during the Great Depression, President Franklin D. Roosevelt requested the establishment of a national mortgage association. On February 10, 1938, the Federal National Mortgage Association (now Fannie Mae) was chartered as a wholly owned subsidiary of the Reconstruction Finance Corporation, a federal agency. Just four years earlier, in 1934, the Federal Housing Administration (FHA) had been established to exert a stabilizing influence on the mortgage and residential real estate markets. The FHA mortgage, a long-term, self-amortizing loan on which principal was repaid gradually over the life of the loan, was vital to the development of a national mortgage market. Fannie Mae had been created to buy these FHA-insured loans from mortgage lenders so that they, in turn, could make more loans to consumers. It was the start of the modern secondary mortgage market. In 1948, Fannie Mae was given authority to purchase loans guaranteed by the Veterans Administration (VA) as well.

In 1954, the ownership of Fannie Mae changed, as the corporation became partly owned by private stockholders. Fannie Mae issued nonvoting preferred stock to the Secretary of the Treasury and nonvoting common stock. The latter was purchased by mortgage lenders who were required to own certain amounts of stock before they could sell mortgages to Fannie Mae.

The most significant step in Fannie Mae's evolution, however, occurred in 1968 when Congress divided the original Fannie Mae into two organizations: the current Fannie Mae and the Government National Mortgage Association (GNMA, or "Ginnie Mae"). Ginnie Mae remains a government agency within the Department of Housing and Urban Development, helping to finance government-assisted housing programs. Ginnie Mae provides liquidity to the FHA/VA market primarily through its MBS guaranty activities. Under this program, lenders originate loans and then package them into pools of FHA-insured and VA-guaranteed mortgages. Ginnie Mae guarantees timely payment of both principal and interest to the investor. The lenders issue securities backed by the mortgages and then sell the securities to institutional and other investors.

The legislation splitting Fannie Mae from Ginnie Mae in 1968 marked the beginning of Fannie Mae's conversion into a privately owned and managed corporation. Congress rechartered the newly private Fannie Mae with a mandate to enhance the efficient flow of funds through the secondary market to U.S. mortgage lenders. Congress also mandated that the corporation operate with private capital on a self-sustaining basis. Its transition to private status was completed in 1970.

Two other major steps in Fannie Mae's growth came in 1972 and 1981. In 1972, Fannie Mae began purchasing conventional loans—mortgages that are not insured or guaranteed by a government agency—and in 1981 the company began issuing MBSs.

In 1970, Congress also chartered another principal participant in the secondary mortgage market, the Federal Home Loan Mortgage Corporation (Freddie Mac). Originally, Freddie Mac was chartered to serve the member institutions of the Federal Home Loan Bank System, primarily federally insured savings and loan associations. Today, Freddie Mac's federal charter calls for a stockholder-owned corporate structure essentially the same as Fannie Mae's.

In late 1992, after three years of intense governmental study and scrutiny of the secondary mortgage market and the operations of Fannie Mae and Freddie Mac, the Federal Housing Enterprises Financial Safety and Soundness Act was signed into law. This landmark legislation modernized Fannie Mae's and Freddie Mac's capital standards and regulatory oversight. The act also affirmed that the housing finance system now in place, with Fannie Mae as its leading institution, is the system of the future.

In the 1980s and the 1990s, Fannie Mae's management had accelerated changes to be responsive to the needs of the market and the challenges of affordability. Operating within its charter, Fannie Mae serves as the nation's housing partner, providing financial products and services that increase the availability and the affordability of housing for low-, moderate-, and middle-income households. Contact: Fannie Mae, 3900 Wisconsin Ave., NW, Washington, DC 20016-2892. (SEE ALSO: **Federal Government**; *FHA Title I Home Improvement Loan Program*; *Government National Mortgage Association*; *Secondary Mortgage Market*)

—*Lawrence Q. Newton*

Further Reading

Hays, R. Allen. 1995. *The Federal Government and Urban Housing: Ideology and Change in Public Policy.* Albany: State University of New York Press.

MacDonald, Heather. 1995. "Secondary Mortgage Markets and Federal Housing Policy." *Journal of Urban Affairs* 17(1):53-79.

Moran, Michael J. 1985. "The Federally Sponsored Credit Agencies: An Overview." *Federal Reserve Bulletin* 71:373-88. ◄

► Fannie Mae Foundation

Fannie Mae Foundation, a private foundation, supports national and local nonprofit organizations working to provide decent and affordable housing and otherwise to improve the quality of life in communities throughout the United States. To promote homeownership, the foundation sponsors public service outreach efforts, including consumer education and homebuying fairs. The foundation has a significant research and policy analysis focus. It conducts, sponsors, and publishes research on a variety of housing and urban issues. The foundation's sole source of support is Fannie Mae (formerly known as the Federal National Mortgage Association). It is headquartered in Washington, D.C. and has regional offices in Atlanta, Chicago, Dallas, Pasadena, and Philadelphia.

The foundation's research and policy analysis functions include those previously performed by the Fannie Mae Office of Housing Research, which was created in 1989 to pursue research that would lead to innovative solutions to the nation's housing and urban problems. To achieve its goal, the foundation uses a multidisciplinary research approach, drawing on the skills of staff experts and a broad network of outside scholars and professionals. Foundation research spans the fields of finance, economics, urban planning, demographics, geography, technology, ethnography, sociology, and more. Using state-of-the-art statistical and econometric models, advanced technology, and policy analysis, the foundation covers most aspects of housing and urban issues, including the housing finance industry, mortgage markets, and government policies.

Through its publications and activities, the foundation offers a forum for meaningful debate and provides insight for public, private, and nonprofit organizations seeking to ensure affordable homeownership and rental opportunities in the United States. The foundation publishes two scholarly journals, *Housing Policy Debate* and *Journal of Housing Research,* and a wide variety of special publications. Foundation activities include the Annual Housing Conference and Research Roundtable Series as well as sponsorship and participation in a variety of industry and research initiatives. For more information, contact the Fannie Mae Foundation, 4000 Wisconsin Avenue, NW, North Tower, Suite One, Washington, DC 20016-2804. Phone: (202) 274-8000. Fax: (202) 274-8100. E-mail: fmfpubs@fanniemaefoundation.org. World Wide Web: http://www.fanniemaefoundation.org.

James H. Carr ◄

► Farmers Home Administration (Rural Housing Service)

The Farmers Home Administration (FmHA) is an agency of the U.S. Department of Agriculture. It operates a variety of farm and nonfarm programs in rural areas. These programs offer financial assistance for the acquisition, rehabilitation, construction, and operation of new and existing single-family and multifamily housing; acquisition of farmland; improvement of farm buildings; operation of farms and conservation of water and soil; installation, improvement, operation, and maintenance of community water, sewer, and solid waste disposal systems; and development of small and emerging private business enterprises.

Federal rural housing assistance had its beginnings in the New Deal programs of the 1930s. The Resettlement Administration and its successor, the Farm Security Administration, provided loans and grants to depression-stricken families for rural rehabilitation and resettlement to areas where they could become self-sufficient on family farms. In 1946, the Farm Security Administration was terminated. Its successor, the FmHA, inherited the extensive network of county offices developed to assist farm households.

The current housing programs of the FmHA are authorized by the landmark Housing Act of 1949, as amended, which laid the foundation for the federal effort to provide safe, decent, and sanitary housing for all U.S. citizens, urban and rural. The 1949 act is the edifice on which all FmHA programs are built. The first of these programs introduced in 1949 was the Section 502 homeownership loan program, providing assistance for home purchase and improvement. This program is still the mainstay of the FmHA's housing component, resulting in more than 1.5 million loans totaling nearly $25 million.

Initially, the FmHA provided mortgage credit only to qualified farmers rather than to the general rural low-income population. Not until the Housing Act of 1961 was the FmHA directed to make loans to nonfarm rural families. The Senior Citizens Housing Act of 1962, for the first time, authorized the FmHA to make loans for rental housing for elderly households who could not afford homeownership or could no longer maintain their housing and required alternative accommodations. Subsequent acts in the 1960s and 1970s extended housing assistance to other rural renter households and farmworkers. Today, the FmHA provides subsidies for homeownership, rental housing, home repair, and rehabilitation for both farm and nonfarm families.

The FmHA housing programs were set up to meet the unique service and credit needs in rural parts of the country that the U.S. Department of Housing and Urban Development (HUD) and its predecessors were ill suited to meet. First, unlike HUD, the FmHA delivery system is highly decentralized, to best reach the vast number of rural communities. Services are delivered through a system of 1,600 county, 250 district, and 47 state offices. Applications for single-family housing are submitted to FmHA county offices, who also perform counseling, technical assistance, and loan servicing. Applications for multifamily housing, community facilities, and business development are serviced at the district level.

Second, a system of direct lending was implemented, recognizing the scarcity of private credit resources in rural areas. Unlike HUD's major lending programs for homeownership, rental housing construction and housing rehabilitation, the FmHA has provided what in effect are direct government loans rather than mortgage insurance or interest reduction payments on loans originated by private lenders and other nonfederal sources. The agency also makes limited grants. Nearly all money loaned by the FmHA comes from loan repayments and from the sale of government securities on the private commercial market. This has enabled the FmHA over the years to reduce its dependence, in part, on direct appropriations and insulate it from the budget process. The FmHA is the federal government's largest direct lender, with a portfolio of loans amounting to more than $63 billion. Increasingly, in more recent years, Congress has sought to limit the FmHA's role as direct lender and shift to guaranteed loans.

Although committed to the extension of federal credit for housing in rural areas, the FmHA is also mandated to act as a lender of last resort. This means that applicants for most FmHA loans must first satisfy a "credit elsewhere" requirement by demonstrating that they cannot secure the credit necessary for such housing from other sources on reasonable terms and conditions. In other words, the borrower must not have sufficient personal resources and must be unable to obtain financing at loan amounts, interest rates, maturities, and other terms and conditions that would make the housing affordable. Furthermore, the borrower must agree to graduate and prepay or refinance the remaining balance of the loan at the earliest opportunity, whenever alternative credit can be secured at reasonable rates and terms. This "graduation" requirement contributed to the problem that emerged in the late 1970s and 1980s when owners of FmHA rental housing prepaid their loans, removed FmHA rental controls, and converted the housing to market rate operation.

With the exception of the FmHA's farm labor housing program, which can be used in urban areas, all funds are restricted to rural areas. The definition of rural has changed many times since 1949, primarily through expansions in maximum population sizes. Generally, the FmHA defines rural as open country, or any town, village, city, or place, including the immediately adjacent densely settled area, that is not part of or associated with an urban area, has a population not in excess of 10,000, and is rural in character.

Towns, villages, cities, and places with populations up to 20,000 may be considered rural if they are not contained within a metropolitan statistical area and have a serious lack of mortgage credit for low- and moderate-income households. Legislation in 1990 "grandfathered" in areas with populations up to 25,000 that were classified as rural prior to the 1990 decennial census. Five percent of appropriations are set aside in rural areas defined as "underserved" based on high rates of poverty and substandard housing and low per capita rates of FmHA assistance.

Eligible families and individuals must have very low to low incomes to qualify for most FmHA housing programs. The exceptions are the Section 502 guaranteed homeownership and Section 515 rental housing loan programs, which permit occupants with moderate incomes to participate. Although the ceilings for "very low" and "low" mirror most federal programs—50% and 80% of area median income, respectively—the FmHA is unique in setting the moderate-income ceiling at $5,500 above the low-income ceiling.

The major FmHA housing loan and grant programs are listed below:

Section 502 Homeownership. This program provides unsubsidized, guaranteed loans to families with incomes less than 115% of area median income and subsidized, direct loans with interest rates down to 1% and terms up to 38 years to low- and very low-income families to enable them to purchase, build, repair, rehabilitate, relocate, and refinance houses; purchase and prepare sites; and install water and sewer facilities for homeownership.

Section 504 Housing Repair. This program provides 20-year, 1% loans up to $15,000 to very low-income homeowners and grants up to $5,000 for elderly homeowners for repairs, improvements, and removal of health and safety hazards.

Section 515 Rural Rental and Cooperative Housing. This program provides loans with interest rates down to 1% and terms up to 50 years (changed to 30 years in fiscal year 1998) to nonprofit corporations, limited-equity cooperatives, public agencies, individuals, partnerships, limited partnerships, and for-profit corporations for the construction or acquisition and rehabilitation of rental housing for very low, low-, and moderate-income households.

Section 514/516 Farm Labor Housing. This program provides 33-year loans at 1% interest to farmers, associations of farmers, nonprofit organizations, and public agencies and grants up to 90% of development cost to nonprofit organizations, Indian tribes, public agencies, and farmworker associations for the construction, acquisition, improvement, and repair of housing, sites, and related facilities for farm laborers living in rural and nonrural areas.

Section 521 Rural Rental Assistance. This program provides 5-year, project-based rental assistance for very low and low-income families living in FmHA rental housing to cover the difference between 30% of adjusted monthly income and monthly unit rent, including the cost of utilities.

Section 523 Self-Help Housing. This program provides two-year, administrative funding contracts to nonprofit organizations and public agencies sponsoring self-help housing developments and two-year loans at 3% interest to

purchase and develop building sites and construct streets and utilities for low- and moderate-income families.

Section 524 Rural Housing Sites. This program provides two-year, market rate loans to nonprofit organizations and public agencies to purchase and develop building sites and construct streets and utilities for low- and moderate-income families in non-self-help developments.

After nearly 50 years of operation, legislation enacted by Congress in 1994, the Department of Agriculture Reorganization Act, radically restructured administration of the housing programs of the U.S. Department of Agriculture. In FY 1995, FmHA was reorganized and most of its programs were located within a new supra-agency called Rural Economic and Community Development (RECD). Its housing and community facilities programs were consolidated in a new Rural Housing and Community Development Service (RHCDS) within RECD. In FY 1996, both names were shortened—RECD to Rural Development (RD) and RHCDS to Rural Housing Service (RHS). To achieve budget savings, the legislation also authorized the secretary to proceed with the closure or consolidation of approximately 1,100 Department of Agriculture field offices over a five-year period. (SEE ALSO: *Rural Housing; Section 502; Section 515*)

—*Robert Wiener*

Further Reading

Collings, Art and Linda Kravitz. 1986. "Rural Housing Policy in America: Problems and Solutions." In *Critical Perspectives on Housing*, edited by Rachel Bratt, Chester Hartman, and Ann Meyerson. Philadelphia: Temple University Press.

A Guide to Federal Housing and Community Development Programs for Small Towns and Rural Areas. 1994. Washington, DC: Housing Assistance Council.

Rural Housing Programs: Long-Term Costs and Their Treatment in the Federal Budget. 1982. Washington, DC: U.S. Congress, Congressional Budget Office. ◀

▶ Farmworker Housing

The farmworker population in the United States is mostly poor, minority, foreign-born, and undereducated. It is also mobile. Few agricultural areas provide year-round employment for agricultural workers. Most of the work is seasonal, with some crops requiring field workers for as little as several weeks or a month, others for much longer periods, and still others requiring temporary workers at different times of the year. These seasonal workloads contribute to a migratory labor force that follows the crops throughout states, across state borders, and across national borders—sometimes alone, sometimes with spouses and children.

This flow of workers has few parallels in other industries, providing farmworkers with both a unique identity and a unique burden. Laden with household belongings and, often, family members, the farmworker is a modern-day nomad, in constant search of work and shelter. The same communities and employers that beckon farmworkers for their seemingly tireless energy at planting or harvest time, often deny these same workers the rewards of hard work: decent wages, good living conditions, participation in community life. Instead, too often, farmworkers and their families live in the lowliest of conditions: in shacks, barns, and chicken coops; along riverbeds; in hand-dug caves, and with unsafe electrical wiring, raw sewage, and polluted water.

Farmworkers in the United States

The National Agricultural Worker Survey (NAWS), required by the Immigration Reform and Control Act (IRCA) of 1986, provides the most comprehensive data on the nation's agricultural labor force. According to NAWS, farmworkers in the United States are increasingly of Latin American origin, primarily Mexican Americans, Mexicans, and Central Americans; poor; reliant on agriculture for most, if not all, of their incomes; mostly unskilled; and generally possessing limited education and English language ability. An indeterminate but presumably large number of farmworkers are undocumented.

On what is referred to as the "Delmarva Peninsula," encompassing the Atlantic regions of Delaware, Maryland, and Virginia, the farmworker population in the mid-1980s was mostly African American and Afro-Caribbean. By 1992, this population had shifted almost entirely to Hispanics (84%), with a growing number of Central American workers, often undocumented. Most of the Mexican or Mexican American workers tend to travel with their nuclear and extended families, whereas the remaining African American workers are mostly single, young males. The Haitian-born population, once prominent in the migrant stream, is tending to "settle out" in local communities.

In the agricultural areas along the Mexican border in Texas and New Mexico, workers stream across the border on a daily basis to work in the fields. Families with young children leave their home base just north of the border to travel the migrant stream that runs to the Northwest or to the upper Midwest, through the Mid-Atlantic states to the Northeast. More and more, these workers are non-English speaking. Many cannot read in any language.

In California, the nation's leading farm state, an estimated 700,000 individuals do farm work during the peak harvest season. In 1997, about 92% of field workers were foreign-born, a sharp increase from the 50% figure found in 1977. Most were young Mexican men who first came to the state sometime after 1978, nearly half of whom migrate each year from their usual place of residence to secure farm employment. In the late 1990s, the migrant proportion of the state's total farmworker population (47%) is higher than was found in early 1983 (39%). Increasingly, newly arriving migrants are indigenous peoples. One study found 50,000 Mixtec Indians, entire families from the Mexican state of Oaxaca, living in mountain encampments in San Diego County, along the border with Mexico. Many spoke only the Mixtec dialect, were undocumented, and had limited job skills.

Finally, farmworkers are extremely poor. NAWS data indicate, for example, that the annual median personal income of California farmworkers is between $5,000 and $7,500. No fewer than 48% live in poverty. The main factors contributing to their poverty are low wages and the temporary nature of most farm jobs. Despite their low incomes, only 13% of farmworkers receive any kind of needs-based services from government social service programs, and only 3% report public (low-income) housing assistance.

Housing Development Issues

The type of housing needed by farmworkers (single family, multifamily, barrack style, rental, homeownership), ownership entity (grower, farmworker association, etc.), permanence of the housing (seasonal or year-round), affordability levels (level of subsidy required to ensure affordability), and social service needs (health, education, translation, job training, among others) are all critical issues in housing development that flow from the makeup and lifestyle of this population. Developers must wrestle with whether or not to build in the farmworker's "home base"—the place to which the migrant returns or where the farmworker lives year-round, even if the work is only seasonal. If the farmworker population is mostly migratory, developers must know whether farmworkers are traveling alone or accompanied by family members, the period of time the housing will be occupied, and whether it should be winterized or can meet some lesser occupancy standard. The peculiar circumstances and increasing diversity of the farmworker population present difficult challenges to affordable housing developers and service providers.

Land and Infrastructure

Farmworker housing development is often tightly woven into the debate over replacing agricultural land with housing of any kind and considerations about building in areas of possible or actual environmental hazard. Many sites are eliminated because of environmental hazards such as lead arsenic and DDT used on fruit trees. Prior agricultural land may have contained leaking drums of other pesticides, fuel for farm equipment, or both. Should this land offer the only available or affordable building sites, the developer must factor in the potentially substantial costs of mitigating the hazard. Housing lending agencies, generally, do not want to pay for mitigation, and the costs, unless grants funds can be obtained, would render many developments financially infeasible for the eventual occupants.

Moreover, the difficulty of developing farmworker housing is often compounded by inadequate infrastructure. Although funding agencies, such as the U.S. Department of Agriculture (USDA) Rural Housing Service (RHS), prefer that housing for farmworkers be situated within communities with essential services, many rural communities lack water and sewer systems to enable any development. For example, a common part of the rural landscape along the Mexican border is the "colonia"—a subdivision containing minimal shelters and inhabited by mostly low-income, Hispanic households. The remote locations of many colonias make the provision of water and sewers infeasible, which impedes the development of truly decent housing for the farmworker families who hold title to such property. The alternative is septic tanks, but finding appropriate soils on adequately sized drain fields is problematic at best.

Finally, even where land and infrastructure are available, such as in small towns and urbanizing areas close to where farmworkers work, exclusionary zoning and land use restrictions may preclude farmworker housing development. For example, some rural and suburban communities do not zone for multifamily housing, and some have strict development standards that prohibit housing, such as barrack- or bunk-style units, that would be acceptable in a more rural setting.

Community Response

Although local community response to the development of housing for agricultural workers varies, NIMBY (not in my back yard) opposition is relatively common. Many communities that encourage the influx of agricultural workers during the harvest season will oppose efforts to improve the workers' housing conditions, particularly if those improvements will encourage year-round residency. Red flags are raised: overcrowded schools (even though children of agricultural workers may already attend the local schools), criminal activity, and undocumented workers. In some cases, developers are forced to conduct very expensive environmental studies that might otherwise not be required. Some developers attempt to counter opposition by carefully selecting sites that do not require any public hearings on easements, conditional use permits, or variances. This is difficult to achieve, however, in those rural communities attempting to raise revenues through the land use and permit process to pay for essential services.

Opposition to farmworker housing improvement or development is widespread but not universal. Colorado growers, for example, have formed partnerships with local and state government to finance and manage housing that will encourage reliable workers to return to their areas each year. In New Mexico, community education emphasizing the welfare of agricultural workers' children has created a positive attitude toward farmworker housing in some communities. Nonprofits are taking over management of poorly run, private farm labor camps, and local leaders are serving on the boards of directors of single-purpose housing sponsors created for each development.

Developer Capacity

Farmers once provided much of the housing occupied by farmworkers. With increased enforcement of federal and state health and safety or employee housing laws, however, the absolute number of grower-operated units has dramatically declined. In California alone, the number of state-licensed labor camps went from 5,000 in 1968 to 1,000 in 1994, coinciding with increased penalties for violations of that state's Employee Housing Act.

Although individual growers or grower associations are eligible to obtain loans from the USDA's RHS to build or improve housing for their workers, the trend has been toward non-employer-provided housing. Public housing authorities, nonprofit housing development corporations, and even some farmworker associations and migrant health clinics have stepped forward as sponsors and/or managers of housing for agricultural workers. Finding willing sponsors, however, can be difficult. The process of locating suitable land, preparing one or more funding applications,

winning local approvals for improvements and services, and gaining federal financing is time-consuming. Very few local organizations have the funds to support staff during the several years it might take to get a development approved, under construction, and occupied. And development fees are insufficient to compensate sponsors for the up-front cost of application.

Development Costs versus Resources

Finally, development needs far surpass the available funding resources. A federally funded migrant housing needs assessment performed in 1980 concluded that more than 750,000 units of housing were required to meet needs, nationally. Yet federal resources for farmworker housing are extremely limited. They include (a) a technical assistance funding program of the Department of Labor, which enables long-time grantees of the department to assist others in developing farmworker housing; (b) discretionary funding through the Department of Health and Human Services Community Block Grant, available on a competitive basis for a small number of farmworker housing developments; and (c) loan and grant funds under the Department of Agriculture's Rural Housing Service Section 514/516 Farm Labor Housing Program. Given inflation in housing and land development costs and greatly reduced levels of federal program funding, it would take many years just to meet existing needs.

The Section 514/516 loan and grant program for the development of year-round or migrant housing is the principal response to the housing needs of farmworkers in the United States. The program provides grants to nonprofit and public agency sponsors for up to 90% of development cost; low-interest loans to any eligible sponsor, including individual growers and grower associations; and rental assistance covering the difference between rents and 30% of adjusted family income. Many farmworker families have been provided homeownership assistance through RHS's Section 502 single-family program, but funds are not earmarked specifically for farmworkers.

Only a small number of communities receive farmworker-housing assistance. Federal funding has greatly declined since the late 1970s and has remained relatively static in the 1990s. For example, funding in FY 1995, $26,815,00 for both 514/516, was expected to assist the development of only 447 units nationwide. This compares with 2,500 units assisted in FY 1979 when farmworker housing was given higher priority in federal funding—$68,500,000 for both programs. In the absence of improved funding at the federal level, several states have created programs specifically to aid in the development or rehabilitation of farmworker housing or have made general housing programs available for this purpose.

Conclusion

The seasonal and migrant farmworker population is one of the most economically disadvantaged groups in our society. Because of low wages and the seasonal nature of their work, farmworkers have few resources to obtain decent housing. In many areas, growers are moving out of the housing business, finding it too costly and too regulated. Rural communities often ignore the needs of these temporary workers.

This leaves to nonprofit organizations, and some government agencies, the task of obtaining financial resources, overcoming local opposition, and developing adequate housing. These groups rely primarily on the federal government for help, and some may benefit from a handful of state governmental programs, but the problem they are attempting to address is large, and the resources are scant. No less than a major public commitment of funds and improvement in the housing delivery system will resolve the long-standing shelter problems of the nation's farmworkers. (SEE ALSO: *Immigration; Rural Housing*)

—*Susan Peck*

Further Reading

Housing Assistance Council. 1986. *Who Will House Farmworkers? An Examination of State Programs.* Washington, DC: Housing Assistance Council.

Housing Assistance Council. 1992. *Who Will House Farmworkers? An Update on State and Federal Programs.* Washington, DC: Housing Assistance Council.

InterAmerica Research Associates. 1980. *National Farmworker Housing: Final Report.* Rossly, VA: Author.

U.S. Farmworkers in the Post-IRCA Period. 1993. Based on data from the National Agricultural Workers Survey (NAWS). Washington, DC: U.S. Department of Labor. ◄

► Federal Government

Since 1937, the federal government in the United States has taken the lead in assisting in the provision of housing for lower-income households who could not afford standard housing available on the private market and in eliminating slums and blighted areas. What began as limited, single-focused initiatives evolved over a period of 40 years, from the 1930s through the 1970s, into a significant array of federal assistance programs, but not without ups and downs in activity and funding. The cabinet-level Department of Housing and Urban Development was created in 1965. The 1980s witnessed a sharp decline in federal government support, and not until the passage of the Cranston-Gonzalez National Affordable Housing Act of 1990 was new momentum established with the authorization of the HOME housing assistance program. The initiatives of the Clinton administration are relatively modest in funding, owing to the continuing tightness of the federal budget.

A major reason for the instability of federal housing and urban development efforts has been the perception that they require only short-term attention in response to special crisis situations and the fact that they have been isolated from the mainstream forces shaping life in the United States, such as the move of city populations to suburban areas following World War II. The decade of the 1990s provides an opportunity to forge new relationships between central cities and outlying areas.

The Evolutionary Framework for Federal Action

As early as 1892, Congress expressed concern about the slums in larger U.S. cities by appropriating $20,000 to investigate these conditions. Federally funded emergency housing for World War I shipyard workers set an early though limited precedent for direct housing activity by the federal government. The Great Depression, beginning in 1929, brought together the economic circumstances—unemployment, decline of the housing industry, and loss of family homes—to generate support for more direct public action. Early federal government efforts to cope with depression-related housing conditions were directed at assisting individual homeowners to avoid foreclosures and stimulating activity for the hard-hit homebuilding industry. This included the Federal Home Loan Bank System (1932), the Home Owners' Loan Corporation (1933), and the chartering of the Federal Savings and Loan Association and the Federal National Mortgage Association (1934). All of these institutions were focused on private industry and individual homeowners. Of particular importance was the establishment of the Federal Housing Administration (FHA) (1934), which became a major force in stimulating homeownership for middle-income families by insuring home mortgages. National housing policy has consistently supported the preference of Americans for single-family houses. In addition to insured home mortgages under the FHA, the Veterans Administration insured home loans after World War II, and beginning in 1946, the Farmers Home Administration provided housing assistance for farm owners and farmworkers in rural areas. A major element supporting homeownership in national housing policy is the tax deduction authorized for interest on mortgages and local real estate taxes under the federal income tax; higher-income families who can afford the largest mortgages get the biggest tax breaks. In 1996, the combined total of these deductions was estimated at about $68 billion. This amount was a multiple of the payments for HUD's low-income housing assistance programs, which in FY 1995 were estimated at $19.3 billion.

The most significant and long-lasting federal effort to provide housing for low-income families began in 1933 under the National Industrial Recovery Act, which under the Public Works Administration (PWA) authorized federal funds to finance low-cost housing, slum clearance, and subsistence homesteads. This early initiative became a permanent program called "public housing" under the U.S. Housing Act of 1937 and provided a framework that influenced the course of federal action over the succeeding decades. This framework involved direct assistance by the federal government to local public housing authorities for the construction of housing for low-income families and for slum clearance, which ultimately became "urban redevelopment" (urban renewal) under the Housing Acts of 1949 and 1954 and Community Development Block Grant (CDBG) assistance under the Housing and Community Development Act of 1974. It provided the legal basis for federal assistance in housing and urban development for the programs to follow.

The Shifts in Assistance Programs

The story of federally assisted housing and urban development is one of changes in program mechanisms. From 1933 through 1980, 16 pieces of congressional legislation initiated or amended housing and urban development programs. The most significant of these were (a) the U.S. Housing Act of 1937, which created the public housing program;

(b) the Housing Act of 1949, which created the urban redevelopment (urban renewal) program; (c) the Demonstration Cities (renamed Model Cities) and Metropolitan Development Act of 1966, which set the first directions for the cabinet-level Department of Housing and Urban Development established in 1965; (d) the Housing and Urban Development Act of 1968, which set 10-year housing goals, created new moderate-income rental and homeownership programs, and expanded the scope of federal support for housing and urban development; and (e) the Housing and Community Development Act of 1974, which consolidated the previous categorical programs under urban renewal into a consolidated block grant of assistance (CDBG) to states and localities and enacted a new federal assistance program called "Section 8," which provided family-based support both for newly constructed housing for low-income families and for such families living in private housing in the community.

The Ups and Downs of Federal Assistance

Over the 60-year history of federal involvement in housing and urban development programs, a consistent pattern of "ups and downs" in the national commitment emerged in response to the changing political, economic, and social climate of life in the United States. Federal assistance has generally come as a political response to national conditions: the economic Great Depression of the 1930s, the turmoil in U.S. cities in the 1960s, and the conservative climate of the 1980s. This political response can be observed in the activity of housing interest groups ranging from the conservative real estate community to cities and public agencies to neighborhood community development corporations to low-income advocacy organizations. This influence is reflected in a series of changes in programs and priorities. Beginning in 1937, there have been 10 different housing assistance mechanisms:

1. Public housing (1937)
2. Section 202 elderly/handicapped housing (1959)
3. Section 221(d)(3) below-market interest rate rental housing (1961)
4. Rent supplements (1965)
5. Section 235 below-market interest rates for homeownership (1968)
6. Section 236 below-market interest rates for rental housing
7. Section 8 family-based support for new construction (1974)
8. Section 8 family-based support (certificates) for occupancy of private housing
9. Housing Development Action Grants (HODAG) (1983)
10. The HOME matching grant affordable housing program (1990)

Of these programs, only public housing, Section 202, Section 8 certificates in existing private housing, and the HOME program survived in 1996. The average life of a federal housing assistance program is seven years, excluding public housing (58 years), the Section 202 program (36 years), the Section 235 homeownership program (21 years, terminated in 1989), and the Section 8 certificate/voucher program (21 years). Even the long-lived programs of public housing and Section 202 have seen constant shifts in policies and funding. In 1998, a standing inventory of 4.5 million HUD-assisted rental housing units existed, of which the largest numbers were Section 8 certificates/vouchers (1.4 million) and public housing units (1.2 million). Although 4.5 million units seems like a large number, they make up less than 5% of the total of more than 100 million housing units in the United States.

Beginning in the late 1970s, there was a significant shift in federal assistance away from the construction of new housing developments to the use of family-based certificates/vouchers, which rose from 40% in 1979 to a high of 87% in 1987. Due to the tight federal budget, there were no new (incremental) Section 8 certificates/vouchers or new public housing units authorized in the fiscal year 1998 Appropriations Act; there was support only for existing units. The HUD budget proposals for FY 1996 advocated the complete replacement of low-income housing construction as well as existing developments with certificates/vouchers. Two major reasons were behind the shift from new construction to family-based certificates/vouchers. The first was conservative opposition to the role of government in providing housing assistance; there has been a clear trend for providing assistance directly to individual families. The second ws the perception that certificates/vouchers are less costly to the federal government than new low-income housing construction. However, this advantage depends on the factors used, in particular in calculating the percentage of the cost of locally available housing in establishing fair market rents. In 1995, HUD proposed, and Congress approved, a drop from 45% to 40% in this percentage, thus reducing the cost to the federal government. This change also reduced the range of housing units available for assistance, particularly for lower-income worker families. In addition, these calculations do not consider the long-range cost benefits that result from the construction of a publicly owned physical asset (a housing development) that can be used for a period up to 50 years to house successive numbers of low-income families.

Another factor influencing the direction of affordable housing construction has been the change in housing finance that occurred in the 1980s. This included the growth of public-private housing partnerships (covered in more detail in the entry on Local Government). Under these partnerships, private investors acquire equity in affordable housing developments and receive low-income housing tax credits authorized under the Tax Reform Act of 1986; this provides an important incentive for private investment.

There have also been shifts in federally assisted urban development programs, similar to those in assisted housing. Beginning in 1949, there have been six different community and economic development and redevelopment programs:

1. Urban Redevelopment (1949), a long-term comprehensive land reuse and redevelopment program re-

named urban renewal in 1954 with a new focus on neighborhood improvement (terminated in 1974)

2. The Economic Opportunity Program (1964-1971), focused on improving the economic status of low-income families
3. The Model Cities Program (1966-1974), a comprehensive effort to improve the condition of distressed central-city areas and their residents
4. The Community Development Block Grant (CDBG) program (1974 to date), a comprehensive set of eligible activities funded as a block of federal money based on a formula calculation of need in states and localities
5. Urban Development Action Grants (UDAG) (1977-1989), a project-based assistance program designed to retain and expand economic activity and jobs in severely distressed communities
6. Empowerment zones and enterprise communities (1993 to date)

The longest-lived program was urban renewal (26 years); at the time of termination in 1974, only 50% of urban renewal projects had been completed but had resulted in more than 98,000 acres of land turned over for private development, 410,000 new or rehabilitated housing units, and 10,000 new or rehabilitated nonresidential buildings. The emphasis in the urban renewal program on large-scale clearance of blighted residential areas resulted in some cities in massive displacement of low-income residents and brought about significant changes in the program structure, principally a shift to central business district economic development and neighborhood rehabilitation rather than clearance. In 1998, the extant programs were CDBG (24 years), with an appropriation of $4.6 billion in FY 1995, and Empowerment Zones and Enterprise Communities, adopted in 1993 and providing $1 billion in federal grants and $2.5 billion in tax incentives over five years to stimulate the economic revitalization of distressed neighborhoods.

Important changes taking place in U.S. urban areas during the 1990s could alter the direction of federal housing and urban development programs and policies. Among the more significant changes is the restructuring of the U.S. economy in response to worldwide economic influences. This is not a temporary situation that will revert to earlier patterns but a basic shift in economic and political structure. The "urban region" is increasingly becoming the focus of economic activity, with central cities assuming a place different from before in the total region. Most of the nation's economic strength is centered in these regions. In addition, the 1980s saw a significant expansion in housing and urban development action and capability at state and local levels, outside of federally assisted programs. Both of these factors have regenerated interest in creating a national urban policy. (SEE ALSO: *Department of Veterans Affairs; Fannie Mae; Federal Home Loan Bank System; Federal Housing Administration; Federal Housing Finance Board; Federal New Communities [Title IV] 1968 and [Title VII] 1970; FHA Title I Home Improvement Loan Program; Government National Mortgage Association [Ginnie Mae]; History of Housing; House of Representatives, Committee on Banking* *and Financial Services; Housing and Home Finance Agency; Local Government; National Commission on Urban Problems (Douglas Commission); Moratorium on Federally Assisted Housing Programs; Senate Committee on Banking, Housing and Urban Affairs; State Government; U.S. Bureau of the Census; U.S. Department of Housing and Urban Development*)

—Mary K. Nenno

Further Reading

Garvin, Alexander. 1996. *The American City: What Works, What Doesn't*. New York: McGraw-Hill.

Hanson, Royce. 1986. *Urbanization and Development in the United States: The Policy Issues*. Washington, DC: National Association of Housing and Redevelopment Officials.

Hayes, R. Allen. 1995. *The Federal Government and Urban Housing: Ideology and Change in Public Policy*. 2d ed. Albany: State University of New York Press.

Library of Congress, Congressional Research Service. 1993. *A Chronology of Housing Legislation and Selected Executive Actions: 1892-1992*. Prepared for the Subcommittee on Housing and Community Affairs, Committee on Banking, Housing and Urban Affairs, U.S. House of Representatives, Committee Print 103-2, Washington, DC.

National Association of Housing and Redevelopment Officials. 1986. *Housing & Community Development: A 50-Year Perspective*. Washington, DC: Author.

Nenno, Mary K. 1996. *Ending the Stalemate: Moving Housing and Urban Development into the Mainstream of America's Future*. Lanham, MD: University Press of America.

Real Estate Research Corporation. 1974. *Future National Issues Concerning Urban Development*. Washington, DC: U.S. Department of Housing and Urban Development.

Teaford, Jon C. 1990. *The Rough Road to Renaissance: Urban Revitalization in America, 1940-1985*. Baltimore, MD: Johns Hopkins University Press. ◄

► Federal Home Loan Bank System

In 1928, there were approximately 12,000 savings and loan (S&Ls) institutions in the United States. They provided most of the financing for homebuyers through home mortgages. They were not subject to any federal regulation, although most were state chartered and subject to state regulation. In the wake of the stock market crash of 1929 and the ensuing Great Depression, there were massive homeowner mortgage defaults and foreclosures, leading to the bankruptcy and threatened bankruptcy of thousands of S&Ls.

This led to the creation of the Federal Home Loan Bank (FHLB) system. The Federal Home Loan Bank Act, signed by President Herbert Hoover on July 22, 1932, created an FHLB system modeled after the Federal Reserve System. The FHLB, governed by five members, had three objectives: (a) to provide secondary liquidity to mortgage-lending institutions that had temporary cash flow problems, (b) to transfer loanable funds from surplus to saving deficit areas, and (c) to attempt to stabilize the residential construction and financing industries. The FHLB would sell stock to its members and then use this as a source of loans. Twelve regional FHLBs would supervise member institutions.

To ensure that homeowners would enjoy the benefits of this reform, Congress passed the Home Owners Loan Act in 1933. This created the Home Owners' Loan Corporation (HOLC), which over its life refinanced 1.8 million delinquent mortgages before it was liquidated in 1954. In 1934, Congress created the Federal Housing Administration (FHA) to reform home mortgages and to provide federal mortgage insurance to homebuyers. Congress also created the Federal Savings and Loan Insurance Corporation (FSLIC) to insure savings accounts in the thrift industry.

The combination of these reforms and the creation of these new federal agencies changed the nature of home purchasing in the United States and bailed out the thrift industry. With the creation of the Federal National Mortgage Association (now Fannie Mae) in 1938, the federal government provided a mechanism for S&Ls to sell federally insured mortgages in a secondary market.

For three decades (1935-1965), the FHLB was a quiet success, and the thrift industry prospered during the post-World War II suburban homebuilding boom. Then in the latter half of the 1960s, thrift institutions began to experience "disintermediation" (the withdrawal of funds from S&Ls because of higher rates of returns available from other investments). In an era of rising interest rates and inflation, S&Ls were at a competitive disadvantage with the commercial banks. In the 1970s, the FHLB proposed that the scope of authorized S&L activities be broadened and that the dependence of the thrift industry on federal interest rate controls be ended. This came in an era of expanded federal regulation of thrift institutions (e.g., the Home Mortgage Disclosure Act and Community Reinvestment Act of 1975 and 1977).

March 31, 1980, marked the start of a new era for the FHLB system and S&Ls. Congress passed the Depository Institutions Deregulation and Monetary Control Act (DIDMCA). This legislation was designed to gradually (over six years) allow thrift institutions to expand their activities, making them more similar to commercial banks. A key reform was to allow S&Ls to invest in commercial real estate, a speculative area previously beyond their focus, which was almost exclusively home mortgages and usually within the immediate service area of the locally based thrift institutions. The Depository Institutions Act of 1982 continued this trend.

The consequences of these reforms proved to be disastrous, resulting in one of the most expensive governmental scandals in U.S. history. S&Ls began to engage in speculation that led to the bankruptcy of many institutions. The FDIC (Federal Deposit Insurance Corporation) and FSLIC were responsible for liquidating failed thrifts and ensuring that depositors were protected. By 1986, FSLIC's insurance fund was broke, despite an increase in its membership fee, after it had supervised hundreds of liquidations and mergers. Congress had to come to the rescue in 1987, authorizing the U.S. Treasury to cover FSLIC's activities through the FHLB.

The S&L scandals led to the demise of the FHLB system. In 1989, Congress passed the Financial Institutions Reform, Recovery, and Enforcement Act (FIRREA). It created a Resolution Trust Corporation (RTC) to oversee the sale of the assets of failed thrifts, with funding from the U.S. Treasury. The FHLB was dissolved and replaced by an Office of Thrift Supervision under the guidance of the Treasury. The FDIC was to continue and oversee deposit insurance. (SEE ALSO: **Federal Government;** *Federal Home Loan Mortgage Corporation;* **Mortgage Finance**)

—*W. Dennis Keating*

Further Reading

Congressional Budget Office. 1993. *Resolving the Thrift Crisis.* Washington, DC: U.S. Congress.

Eichler, Ned. 1989. *The Thrift Debacle.* Berkeley: University of California Press.

Federal Home Loan Bank Board. 1982. "Fifty Years of Service: Federal Loan Bank Board." *Federal Home Loan Bank Board Journal* 15:1-108.

Mayer, Martin. 1990. *The Greatest-Ever Bank Robbery: The Collapse of the Savings and Loan Industry.* New York: Scribner's.

White, Lawrence J. 1991. *The S&L Debacle: Public Policy Lessons for Bank and Thrift Regulation.* New York: Oxford University Press.

Woerheide, Walter J. 1984. *The Savings and Loan Industry: Current Problems and Possible Solutions.* Westport, CT: Quorum. ◀

▶ Federal Home Loan Mortgage Corporation

The Emergency Home Finance Act of 1970 established the Federal Home Loan Mortgage Corporation (FHLMC, more commonly known as "Freddie Mac"). Its purpose was to provide a secondary mortgage market for conventional mortgages. Although Freddie Mac is able to purchase Federal Housing Administration (FHA) and Department of Veterans Affairs (VA) loans, the main thrust of its activities is to make a market for conventional mortgages, including, more recently, adjustable-rate mortgages (ARMs).

The development of support operations for the secondary mortgage market for government-insured and government-guaranteed loans was well established by the end of the 1960s through the Federal National Mortgage Association (now known as "Fannie Mae") and the Government National Mortgage Association (GNMA, more commonly known as "Ginnie Mae"). However, it is misleading to think that a majority of the residential mortgage loans in the United States were government sponsored. By 1970, 65% to 75% of the total mortgage loans in the United States originated in private institutions. These "conventional" mortgages needed to be packaged and sold for the same reasons that government mortgages did: to provide liquidity to the U.S. mortgage market and for the financial system as a whole.

Freddie Mac accomplishes this goal by issuing a variety of institutional investments to finance its activities. These include Freddie Mac pass-through securities (called "mortgage participation certificates" or "PCs") and collateralized mortgage obligations (CMOs). By 1990, Freddie Mac had issued more than $300 million in pass-throughs and $4.5 million in CMOs.

Freddie Mac's influence in assisting the mortgage finance system is likely to grow in the years ahead. As the secondary mortgage market continues to expand, Freddie Mac is likely to be a dominant force for conventional loans from private banks, thrift institutions, and other sources. In future years, Freddie Mac may be viewed as one of the critical players in the reformation of the financial system. (SEE ALSO: *Federal Home Loan Bank System*)

—*Austin J. Jaffe*

Further Reading

Clauretie, Terrence M. and James R. Webb. 1993. *The Theory and Practice of Real Estate Finance*. Fort Worth, TX: Dryden.

Lore, Kenneth G. 1987-1988. *Mortgage-Backed Securities: Developments and Trends in the Secondary Mortgage Market*. New York: Clark Boardman.

Sellon, Gordon H., Jr. and Deana VanHahman. 1988. "The Securitization of Housing Finance." Federal Reserve Bank of Kansas City. *Economic Review* (July/August):3-20 ◄

► Federal Housing Administration

The Federal Housing Administration (FHA), created by the Housing Act of 1934, provides mortgage insurance, primarily for homebuyers with small down payments. The homebuyer pays an insurance premium to protect the mortgage lender. If the borrower defaults on the loan, the FHA pays the lender and takes possession of the house. The borrower is not protected in any way. The program has been very active and has successfully helped many people in the United Sates to become homeowners. In addition, it played a major role in reforming the mortgage-lending industry and helping to create the secondary mortgage market, as well as affecting the spatial organization of modern U.S. cities.

The FHA is one of the longest lived and most successful of U.S. federal housing programs. It was begun by the Housing Act of 1934 in part to help remedy the problem of home foreclosures caused by the Great Depression and to stimulate the construction industry and increase employment. The act has been amended frequently since 1934, but its most important part remains Section 203(b), the mortgage insurance program. Under this program, the FHA insures mortgage loans made by lending institutions for the purchase of one- to four-family owner-occupied housing. Should the borrower default, the lender is protected from loss. The homebuyer pays for this insurance, but the lender, not the borrower, is protected. The advantage for the lending institution is insulation from the majority of loss associated with defaults and foreclosures. If the homebuyer defaults, the FHA pays the lender the amount owed and receives the property. The advantages for the homebuyer include a much lower down payment, standardized terms, and in many cases, the ability to buy a home at all.

Other FHA programs included in the Housing Act of 1934 and its later amendments covered a range of housing types and needs. These programs waxed and waned with different administrations and different problems. None has been as active or as long-lived as the 203(b) program, and coverage of the entire set is beyond the scope of this entry. However, examples include Section 203(k) (mortgages for home repair), Sections 203(n) and 213 (for units in cooperative housing developments), Section 221(d)(2) (fo-

cused on households displaced by urban renewal), Section 221(d)(3) (mortgages for low- and moderate-income rental properties), Section 223(f) (mortgages for existing rental complexes), Section 251 (adjustable-rate mortgages), Section 243(c) (condominium mortgage insurance), and Section 245 (graduated-payment mortgages). The remainder of this discussion focuses on the 203(b) program.

FHA insurance completely altered mortgage lending in the United States in the 1930s and has had far-reaching consequences for all aspects of residential financing. Prior to its existence, the ordinary mortgage loan had a large down payment (50% or more), was for a relatively short term (10 years or less), and ended its term with a large lump sum payment of principal due to the lender (known as a balloon payment). Most homebuyers planned to refinance this balloon and go on making payments. The Great Depression made refinancing difficult if not impossible. Individual homebuyers made payments on a loan for 10 years and, in fact, might have paid most of the cost of the property, only to lose it to foreclosure because they could neither refinance the balloon payment nor afford to pay it completely. This contributed to the large increase in the numbers of defaults and foreclosures on mortgage loans during the Great Depression; at times in the early 1930s, half of the residential mortgages in the United States were in default. Because few people could buy homes without mortgages and mortgages were not readily available, there was a substantial impact on the home construction industry as well. Housing starts averaged about 700,000 new units per year in the 1920s but had dropped to 93,000 in 1933.

The federal government attempted to ameliorate some of the housing market difficulties caused by the Great Depression in several ways. In 1932, it established the Federal Home Loan Bank System to act as the governing body for savings and loan associations. In 1933, the Home Owners' Loan Corporation came into being to help refinance existing mortgages. The creation of the FHA in 1934 was specifically aimed at the goals of making money available for mortgages, increasing new construction, reforming mortgage lending, raising housing standards, and broadening the opportunities for homeownership.

The FHA created the mortgage loan that Americans think of as normal today. It pioneered the long-term loan (20 to 30 years) with a low down payment (roughly 3% to 5%, although there is a formula to determine the specific amount in each case). In addition, FHA-insured loans were fully amortized; that is, if one paid the same amount every month according to the payment schedule for the full term of the loan, the entire principal and all interest would be paid; there would be no balloon payment due. Until 1983, the interest rates allowed on FHA-insured loans were slightly lower than market rates, with sellers making up the difference by paying fees called *points* to the lenders. The FHA interest rate today is more market determined.

The FHA insurance program is not an explicit subsidy. Nonetheless, the simple fact of being a federal program provides support. Because every borrower pays the same premium regardless of risk, there is also some subsidy from lower-risk to higher-risk clients. In addition, the combination of FHA insurance with the secondary mortgage market (see below), also created by the federal government, constitutes at least an implicit subsidy.

FHA's mortgage terms were revolutionary at the time, and the insurance aspect was necessary to convince lenders to try them. The concept proved so successful, however, that the terms are now commonplace (except for the very low down payment) and a private mortgage insurance industry has sprung up to handle lower-risk cases. The FHA remains the insurance organization for very low down payment loans and low- to moderate-income homebuyers.

One of the important consequences of FHA insurance was to greatly broaden the potential market for homeownership. Notwithstanding a number of fluctuations and important differences among population groups, in the late 1990s, about 65% of U.S. households owned or were buying a home (this dropped slightly in the 1980s but rose to earlier levels in the 1990s). Another important effect was to standardize many development and planning practices, leading to improved design in new neighborhoods. Standardized mortgage terms and standards of actuarial soundness allowed the development of a secondary mortgage market on which mortgage loans could be bought and sold, thus making capital far more mobile nationwide. One no longer needed to know anything about a specific property in another place; if it was an FHA-insured loan, it had to meet certain standards and the insurance made it a safe investment, even if the borrower defaulted. This allowed investors of all kinds from many different regions to put money into real estate all over the United States.

After World War II, the FHA insured a phenomenal number of mortgage loans and had a significant role in financing growth in U.S. suburbs (along with the Veterans Administration mortgage guarantee program that joined it). There was an emphasis on new homes, on homogeneous (i.e., white) communities, and on actuarial soundness. In fact, the FHA's primary insurance program made money. Community homogeneity was considered necessary to the preservation of property values in neighborhoods. The agency did not originate this concept but, rather, reflected U.S. society (and particularly the various aspects of the housing industry) at the time. The FHA's appraisal manuals and rules explicitly required racial segregation in areas where it insured homes. Discrimination remained acknowledged practice until the 1960s. With the creation of the Department of Housing and Urban Development (HUD) in the 1960s, the FHA became a unit within HUD.

In the late 1950s and the 1960s the FHA became victimized by its own success. First, having shown that mortgage insurance could be profitable, it encouraged the growth of the private mortgage insurance industry. These companies take on the lower-risk insurance cases (usually a higher down payment—up to 20% down) and charge a lower premium, so homebuyers prefer to use them when possible. They do not, however, insure the entire loan amount. Second, the FHA had a reputation for being cumbersome and slow, so many institutions and borrowers preferred to work with private companies. Third, by emphasizing actuarial soundness, white homebuyers, and sub-

urban neighborhoods, the FHA helped to create many of the problems of segregation and inner-city blight that began to surface in U.S. central cities.

The federal government undertook several changes in the FHA program to attempt to end its role in exacerbating urban problems. Racial discrimination was officially outlawed by executive order in the early 1960s and by legislation in the late 1960s. Also in the late 1960s, legislation required some programs within the FHA to use criteria of "acceptable risk" rather than "actuarial soundness" in areas where riots had occurred or where they might occur. Programs other than the 203(b) program were especially active in these efforts. The neighborhoods affected were often "redlined" by conventional lenders (i.e., those not using FHA insurance or VA guarantees). Lenders refused to make mortgage loans on properties in those neighborhoods no matter what the buyer's qualifications. This created a self-fulfilling prophecy whereby a lender would refuse to loan in an area on the grounds that the neighborhood was declining, but because no one could get a mortgage or home improvement loan, the area did decline.

When FHA-insured loans became available in these central-city neighborhoods, lenders turned to them almost exclusively. Unfortunately, the results were often negative. Large numbers of homes in specific neighborhoods would be sold with FHA insurance, sometimes with inadequate inspections or care as to the qualifications of the buyers. Nonwhite homebuyers were disproportionately represented in the FHA clientele, so they often suffered from the inadequate inspections. In at least some cases, the result was houses in poor condition sold to buyers who did not have the income to handle the loan or the necessary maintenance and repairs. This led to defaults, foreclosures, and boarded-up units that belonged to the FHA. Some of these FHA units were not cared for as well or sold as quickly as they should have been and were often vandalized. A few such units in a neighborhood could certainly bring about rapid decline. In the early 1970s, President Nixon declared a moratorium on all federal housing programs.

The FHA program soon insured loans again, but there were now more attempts to monitor the activities of the lenders in the neighborhoods (starting with the Home Mortgage Disclosure Act, and continuing most recently with Financial Institutions Reform, Recovery, and Enforcement Act). The FHA continues to operate, and although it does not do the volume of business that it did in the 1950s, it is still very important for certain kinds of borrowers. Usually, these are first-time homebuyers with limited ability to make down payments. There is an upper limit on the cost of the house one can purchase with the insurance, and the limit varies by region, so more expensive home sales do not involve the FHA. In addition, the FHA continues to play a demonstration role. Just as it originally proved to lenders that a low down payment with a long-term, fully amortized loan would work, in the 1990s, provisions of the FHA program are used to try out variable-rate loans, reverse-equity mortgages, and other new mortgage loan types.

The FHA played a major role in shaping U.S. suburbs and contributing to problems of racial segregation and jurisdictional fragmentation. These problems arose in part from explicit factors such as the FHA's racial discrimination and insistence on all-white communities and in part from implicit factors, such as the suburban expansion required by the sheer volume of loans insured. On the other hand, it succeeded in helping to create jobs and has been very popular with private business. The insurance has undoubtedly helped a great many people become homeowners— something that Americans value highly. FHA mortgage insurance has existed for 60 years. It is one of the few U.S. federal programs that has not been shelved or changed beyond all recognition during its lifetime. (SEE ALSO: *Balloon Mortgage;* **Federal Government;** *FHA Title I Home Improvement Loan Program*)

—*Hazel A. Morrow-Jones*

Further Reading

Boyer, B. D. 1973. *Cities Destroyed for Cash: The FHA Scandal at HUD.* Chicago: Follett.

Brueggeman, William B. and Jeffrey D. Fisher. 1993. *Real Estate Finance and Investments.* 9th ed. Homewood, IL: Irwin.

Daye, Charles E., Daniel R. Mandelker, Otto J. Hetzel, James A. Kushner, Henry W. McGee, Jr., Robert M. Washburn, Peter W. Salsich, Jr., and W. Dennis Keating, eds. 1989. *Housing and Community Development: Cases and Materials.* 2d ed. Durham, NC: Carolina Academic Press.

Haar, C. M. 1960. *Federal Credit and Private Housing: The Mass Financing Dilemma.* New York: McGraw-Hill.

Jacobs, Barry G., Kenneth R. Harney, Charles L. Edson, and Bruce S. Lane. 1986. *Guide to Federal Housing Programs.* 2d ed. Washington, DC: Bureau of National Affairs.

Miles, Mike E., Emil E. Malizia, Marc A. Weiss, Gayle L. Berens, and Ginger Travis. 1991. *Real Estate Development: Principles and Process.* Washington, DC: Urban Land Institute.

Vandell, Kerry D. 1995. "FHA Restructuring Proposals: Alternatives and Implications." *Housing Policy Debate* 6(2):299-393. ◀

▶ Federal Housing Finance Board

The Finance Board was created by the Financial Institutions Reform, Recovery, and Enforcement Act (FIRREA) of 1989 to supervise the operations of 12 federal home loan (FHL) banks and the Office of Finance. The five-director board oversees the FHL banks' financial performance and operations and administers the Affordable Housing, Community Investment, and Community Support programs. Member institutions borrow funds from FHL banks at interest rates lower than the commercial market to give them the liquidity to offer loans to homebuyers. In addition, FHL banks also loan money to members to support the Affordable Housing and the Community Investment Programs. The FHL bank system was created in 1932 to provide a credit system to strengthen the housing market weakened by the Great Depression. Assistant Secretary for Housing, Federal Housing Commissioner: Nicolas Retsinas. Address: 1777 "F" Street,

NW, Washington, DC 20006. Phone: (202) 408-2500. Fax: (202) 408-1435. (SEE ALSO: **Federal Government**)

—*Laurel Phoenix* ◀

▶ Federal New Communities (Title IV) 1968 and (Title VII) 1970

The federal government's renewed support of new communities since the New Deal's greenbelt towns came about with the passing of (Title IV of) the Housing and Urban Development Act of 1968 and (Title VII of) the Housing and Urban Development Act of 1970. Mortgage and loan insurance were the main forms of aid to developers. By 1975, 13 new communities were enrolled in Title VII, but the program was terminated in 1983 because of the insolvency of all developers except one.

During the 1970s, many believed that the publicly built British and European new towns such as the privately sponsored Columbia, Maryland; Reston, Virginia; Irvine, California, and others then under development were the type of large-scale new communities that could control urban sprawl and shape national growth.

Loan guarantees up to $50 million and supplementary grants for the developers of a single community under Title IV enabled St. Charles Communities, Maryland; Park Forest South, Illinois; Flower Mound, Texas; and Maumelle, Arkansas, to be insured.

Title VII improved on Title IV, established a revolving fund, and provided technical assistance. Developers and local public agencies where these new communities were located were encouraged to combine Title VII benefits with other federal programs. Thirteen new communities, with a combined projected population of 785,000 over 20 years, were insured (or reinsured from Title IV): Cedar-Riverside, Minnesota (the only new-town-in-town); Flower Mound, Texas; Ganada, New York; Harbison, South Carolina; Jonathan, Minnesota; Maumelle, Arkansas; Newfields, Ohio; Park Forest South, Illinois; Riverton, New York; St. Charles Communities, Maryland; Shenandoah, Georgia; Soul City, North Carolina; and The Woodlands, Texas.

In 1975, HUD, which administered the program, instituted a moratorium on new commitments after finding that within three years after the new communities were started, most had run out of funds. Many developers had little or nothing to show after spending millions. Only The Woodlands did not default in its loan payments and is still under development. In 1981, HUD foreclosed on nine of the new towns. It terminated the program in 1983.

A 1983 HUD study concluded that "while Title VII accomplished noteworthy objectives, it represented a costly loss" in excess of $561 million to the taxpayer for defaulted loans, interest, and administrative costs. Compared with conventional developments, it was found to be "an inefficient way of achieving high quality, a balance of land uses, and communities with diverse population characteristics."

The enterprise was high risk. Timing was poor: Two national recessions, one in the early 1970s and the other in the early 1980s, slowed sales and exacerbated the developers' cash flow difficulties. National leadership and priorities changed with political party in a period dominated by the Vietnam War, social unrest, and oil embargoes.

Key Findings

▶ No community achieved its anticipated growth rate owing, in some cases, to poor site selection and overly optimistic projections.

▶ All communities were undercapitalized, except The Woodlands.

▶ HUD was inadequately led, staffed, and supported under changing national governments.

▶ HUD's ultraconservative valuation of the land for loan guarantees and excessive delays in approvals cost some developers dearly.

▶ Only three communities were successful in achieving a mix of rental and ownership, affordability, and racial mix in housing.

▶ All communities had a positive effect on adjoining communities by more efficient land use and conserving the environment.

▶ Few notable physical design differences between these new communities and conventional development were apparent.

▶ Only four had some success balancing housing and jobs.

▶ Because of their mostly suburban locations, residents took as much time as the national average to get to work.

▶ Technological, design, and organizational innovations in the process of community building and construction were limited and not unique to these federal programs.

(SEE ALSO: **Federal Government**; *New Towns*)
—*Michael McDougall*

Further Reading

Bailey, James, ed. 1973. *New Towns in America: The Design & Development Process.* New York: John Wiley.

Burby, Raymond J. and Shirley Weiss. 1976. *New Communities, U.S.A.* Lexington, MA: Lexington Books.

Division of Policy Studies, Office of Policy Development, U.S. Department of Housing and Urban Development. 1985. *An Evaluation of the Federal New Communities Program.* Washington DC: U.S. Department of Housing and Urban Development.

Morgan, George T., Jr. and John O. King. 1987. *The Woodlands: New Community Development, 1964-1983.* College Station: Texas A&M University Press. ◀

▶ Feminist Housing Design

Feminist housing design holistically addresses the political, social, and economic dimensions of dwelling in ways that seek to better women's lives. The word *feminism* does not have a common, universal definition. Therefore, this entry begins with a necessary explanation of the various meanings of the term. An analysis of the relationship between gender and housing follows. The third section describes a number of historic and contemporary housing schemes that embody feminist goals and values. The conclusion suggests a feminist agenda for housing and neighborhood development beyond the 1990s.

Simply put, feminism means supporting women's rights and choices. Feminist theory, of which there are many variations, constructs an analytic framework for understanding the root causes of women's inferior position in male-dominated (patriarchal) cultures. Common to all definitions is a political perspective that recognizes the oppression of women as a class (while acknowledging the differing conditions of women's oppression based on socioeconomic class, race, and sexual orientation) and a commitment to eliminating the oppression of all women.

How does the socially constructed category of gender, which assigns to men relative power and status and to women dependence and marginality, affect women's housing rights and choices in North America? An examination of five interrelated aspects of women's lives that have traditionally defined women's "social place" in patriarchy is revealing: economic discrimination, aging, marital status, domestic violence, and responsibility for children.

Despite the indefatigable efforts of the women's movement throughout the last three decades to establish equal employment opportunities for women, the median earnings of women working full-time are still only 64% of the median earnings of a comparable group of men, regardless of occupation. The economic disparity between women and men means that fewer housing units are affordable for women and that women must spend a greater relative portion of their incomes on housing than do men for the same housing, often up to an exorbitant 50% or more, leaving fewer financial resources for food, clothing, health care, and education. Poverty and the related inability to pay for adequate shelter is overwhelmingly a "woman's problem." In 1996, more than 4 million female-headed families and 21 million women were living below the poverty level.

Along with economic discrimination, aging has a dramatic effect on how and where women live. Widowed and single elderly women are among the most poverty stricken of all people in the United States, especially if they are Hispanic, African American, Asian American, or rural women. They have the smallest incomes, the smallest budget for housing and related services, live in the worst housing conditions, and have the most health problems.

Third, marital status has consistently determined how women are housed. Historically, for most women, owning a house was realized through marriage, divorce, widowhood, or inheritance. Without a husband's income, women could not afford to buy or maintain a dwelling. Without his credit and signature, they could not legally own property. Until the Equal Credit Opportunity Act was passed in 1977, mortgage lenders considered it sound business practice to discount the income of any woman of childbearing age by half when determining loan eligibility because they assumed she would get pregnant and drop out of the workforce. Today, the majority of unmarried women are still renters.

The fourth condition of womanhood that adversely affects women's housing tenure is domestic violence. Every

12 seconds, a woman is beaten in the United States. More than half of all married women are physically abused, and no age group, race, or class is immune. Countless women remain in violent relationships because they have no place to go. One third of the 1 million battered women who seek emergency shelter each year in the United States can find none. In the 1990s, almost half of all homeless women (the fastest growing segment of the homeless) were refugees of domestic violence.

Fifth and finally, responsibility for childbearing and child rearing affect women's housing options. Even though recent amendments to the Fair Housing Act of 1968 made discrimination against children in rental housing illegal, many families with children, particularly single-parent female-headed families, still face difficulties in finding a landlord who will agree to rent to them. Furthermore, women with children seeking housing have the added locational considerations of schools and child care. In 1985, 70% of U.S. women with children between 6 and 17 years old were working for wages or looking for work. Yet affordable child care is still unavailable to the majority of working mothers. There is also the overwhelming problem of teenage pregnancy, which perpetuates the cycle of poverty and illiteracy among women.

Solving these and other problems that women face through developing new housing arrangements is not a recent idea. Throughout the 19th and 20th centuries, women promulgated proposals for liberating themselves from restrictive gender roles through redesigning the domestic environment and their relationship to it. For example, in 1868 a Cambridge, Massachusetts, housewife, Melusina Fay Pierce, organized women into producers' and consumers' cooperatives to perform domestic work collectively and charge their husbands retail prices equivalent to men's wages. In 1898, Charlotte Perkins Gilman, grandniece of abolitionist Harriet Beecher Stowe, published *Women and Economics,* in which she argued that human evolution was being retarded by women's confinement to housework. She supported a residence hotel of kitchenless apartments for working women and their families, with linen service, child care, and public dining. In 1916, Alice Constance Austin, a self-taught architect, presented her visionary model of a feminist socialist city for 10,000 people to be built in California. Houses were kitchenless, furniture was built in, beds rolled away, heated tile floors eliminated carpeting, and hot meals were delivered through underground tunnels from a central kitchen.

All of the architects of these schemes demanded the elimination of unpaid domestic work and women's confinement to it. They argued that the physical separation of domestic space from public space and the economic separation of the domestic economy from the political economy—both the results of industrial capitalism—had to be overcome if women were to be fully emancipated.

The feminist critique of domestic architecture and the nuclear family reemerged in the 1960s and 1970s with the second wave of feminism. In 1973, anthropologist Margaret Mead called the suburban home a "horrible little misery"; feminist Betty Friedan blamed it for women's loneliness and depression. Shulamith Firestone, in *The Dialectic of Sex: The Case for Feminist Revolution,* proposed replacing the biological/legal family with households of choice that would live in campuslike communities in prefabricated housing that could be easily set up and dismantled near permanent buildings that would fill community needs for socializing, dining, working, and studying.

In 1980, architect Dolores Hayden developed a scheme for spatially and economically reorganizing the typical suburban neighborhood by incorporating local employment, child care, transportation, and communal living. She suggested that individual backyards be joined together to form public commons and gardens; that side yards and front lawns become fenced, private spaces; and that private garages be converted to public garages, laundries, day care facilities, community centers, kitchens, and rental apartments. In 1984, Hayden's groundbreaking book, *Redesigning the American Dream,* was published, a visionary feminist blueprint for reshaping the future of housing, neighborhoods, work, and family life.

Throughout the 1980s, women entered the labor force in unprecedented numbers, and the percentage of households of singles, single parents, the elderly, two-paycheck couples, and unrelated adults burgeoned. In 1984, a national competition was held to design the "new American house," suitable for the "contemporary American family." The winning team—Jacqueline Leavitt, a planner, and Troy West, an architect—designed affordable town houses for active households with little time for housework that included office space for work at home and on-site child care. A modified version, Dayton Court, was built in St. Paul, Minnesota.

During the mid-1980s, architects Kathryn McCamant and Charles Durrett began promoting and designing cohousing communities in the United States after studying and living in them for more than a year in Sweden, the Netherlands, and Denmark. Cohousing is a cooperative community where individual households are clustered around a common house with shared dining, child care, workshops, and laundry. Each home is self-sufficient with a complete kitchen, but community dinners are available and encouraged in the common house. These developments are also unique in that they are organized, planned, and managed by the residents themselves.

Although relatively few of these provocative proposals that free women from the sole responsibility for domestic chores and child rearing have been built in the United States, in Canada, cooperative housing projects developed by and for women have proliferated since the early 1970s. Gerda R. Wekerle has documented these projects. User participation and control over the development process, residents' management, and the experience of living in a supportive community of women are critical aspects of all Canadian women's housing cooperatives.

Similar goals are central in the architectural developments built in the United States in the 1970s and 1980s to promote self-sufficiency and security for women who are at risk, such as pregnant teens, battered women, homeless and low-income elderly women, and single-parent women. Supportive housing combines residential accommodations with services, such as health care, psychological and financial counseling, child care, legal assistance, and job training. It can be emergency, transitional, or permanent housing. Supportive housing for women has been sponsored, designed, financed, and built by organizations, women, and men who are dedicated to creating places of safety, refuge, and empowerment for disadvantaged women where they can reshape their lives, futures, and the welfare of their children.

Fifty such projects have been documented and illustrated by architect Joan Forrester Sprague. Included in her case studies is a history of Women's Advocates in St. Paul, Minnesota, the first battered-women shelter in the United States, begun in 1971 by a women's collective. One of the founders, Mary Vogel, was entering architecture school when the shelter was being designed. In the early 1990s, she became the project programmer and designer for Garfield Commons in Cincinnati, a 43-unit rental apartment development that was created in an abandoned elementary school to serve low- to moderate-income single mothers and elderly women. This project's developer, the Women's Research and Development Corporation (WRDC), was founded in Cincinnati in 1988 to create affordable housing with economic opportunities for women The founding members of the WRDC, all women, were diverse in education, income, and race.

The first organization of this sort in the United States, the Women's Development Corporation, was founded in the mid-1970s by three architects—Katrin Adam, Susan Aitcheson, and Joan Forrester Sprague. Realizing that housing security for poor women and their children is inextricably related to jobs and marketable job skills, in their first model housing development (Villa Excelsior in Providence, Rhode Island, consisting of 76 rehabilitated, scattered-site units), their program called for training future residents in construction, maintenance, and management of the property and the creation of several resident-operated businesses that would also provide services essential to working mothers, such as a food cooperative, a cafeteria, and a child care center.

Other successful housing corporations were founded and operated in the 1970s and 1980s by women residents of some of the most deplorable public housing in the United States. In 1972, at the age of 25, Kimi Gray was a divorced mother of five living on welfare in Kennilworth Parkside, a Washington, D.C., development housing 3,000 people. Today, she chairs the Kennilworth Parkside Resident Management Corporation, a multi-million-dollar corporate enterprise that has created jobs for residents in new businesses, including a laundry; catering, roofing, moving, and construction companies; a health center; a child care center; a food co-op store; and a beauty shop. Similar successes have been achieved at Cochran Gardens in St. Louis, under the charismatic leadership of Bertha Gilkey, founder of the Cochran Tenant Management Corporation, which has built 700 new town houses for low- and moderate-income households since 1975. The self-help efforts of working-class and low-income women—the majority of them women of color—have produced organizations all across the United States, such as the National Congress of Neighborhood Women in Brooklyn and the Women's Community Revi-

talization Project in Philadelphia, that are bringing a women's perspective to housing and neighborhood development, providing affordable, safe homes with supportive services for women and families, and developing impressive leadership skills among their constituents. These organizations are based on a recognition that shaping shelter through participation in all aspects of the housing production process—from planning neighborhoods and housing developments to designing, financing, building, rehabilitating, and maintaining them—is as important to the quality of women's lives as is the actual product.

Whether they are built projects or theoretical proposals, historic or contemporary examples, the central tenet of feminist housing design is empowerment for women. The strategies for accomplishing this differ among feminists. Some argue that women should not be defined as a special-needs group but should be provided with enough income and financing to compete in the existing housing market. Others argue that women are paying an intolerable and unnecessary price in trying to function as wage earners, wives, and mothers in conventional housing and neighborhoods designed around the specious assumption that all women get married, stay married, have children, stay at home to raise them, and are financially supported by a husband. Although this assumption is true for some women, it is no longer true for most and has never been true for all. Women have always worked outside the home, for the same reasons men do—to support themselves and their families and to contribute to society.

To alter women's traditional relationship to domesticity and move beyond the conventions of gender embedded in traditional housing and neighborhood design, the gender-based segregation of workplace and dwelling, cities and suburbs, and private life from public existence must be spatially, economically, and socially reintegrated to support women's and men's equal participation in both the paid labor force and the unpaid labor associated with housekeeping and child rearing. Networks of community-based social and domestic services must be established, including quality, public child care paid for through taxes, just like public schools. Existing dwellings must be adapted to make them more affordable and appropriate for diverse household sizes, types, and incomes (for example, through the legalization of accessory apartments in suburban areas). New forms of congregate and cohousing must be developed to maximize recreation and neighborly support; reduce residential segregation by marital status, age, race, and income; minimize wasteful energy consumption; increase opportunities for homeownership and security of tenure in rental housing; and support the economic independence and personal choices about social life and child rearing that most U.S. households value.

Understandably, housing and planning innovations are difficult to accomplish. They require considerable money, changes in zoning regulations and building codes, and the imagination and commitment of architects, planners, builders, developers, and communities. The building professions and trades have long traditions of male domination. As recently as 1997, less than 10% of the 45,855 professionally licensed members of the American Institute of

Architects were women, the group most likely to bring an awareness of women's housing needs to design practice. Among this small percentage, few have identified themselves as feminists and acted accordingly as practitioners or educators.

Furthermore, architects—be they women or men—have had relatively little impact on what actually gets built, where, how, and for whom. Investment builders, engineers, developers, governmental agencies, city managers, the real estate industry, corporations, and financial institutions make most of these decisions. Few women are in important decision-making positions in these occupations and businesses either.

The built environment exists fundamentally as the expression of an established social order. From penthousing to public housing, the spatial arrangements of our communities and dwellings reflect and reinforce the nature of traditional gender, race, and class inequalities. These patterns will not be easily changed until the society that produced them is changed. (SEE ALSO: *Women as Housing Producers; Women as Users of Housing*)

—*Leslie Kanes Weisman*

Further Reading

Allen, Polly. 1988. *Building Domestic Liberty: Charlotte Perkins Gilman's Architectural Feminism.* Amherst: University of Massachusetts Press.

Franck, Karen A. and Sherry Ahrentzen, eds. 1989. *New Households, New Housing.* New York: Van Nostrand Reinhold.

Hayden, Dolores. 1981. *The Grand Domestic Revolution: A History of Feminist Designs for American Homes, Neighborhoods, and Cities.* Cambridge: MIT Press.

———. 1984. *Redesigning the American Dream: The Future of Housing, Work and Family Life.* New York: Norton.

McClain, J. with C. Doyle. 1994. *Women and Housing: Changing Needs and the Failure of Policy.* Toronto: James Lorimer.

Sprague, Joan Forrester. 1991. *More Than Housing: Lifeboats for Women and Children.* Boston: Butterworth Architecture.

Weisman, Leslie Kanes. 1992. *Discrimination by Design: A Feminist Critique of the Man-Made Environment.* Urbana: University of Illinois Press.

Wekerle, Gerda R. 1988. *Women's Housing Projects in Eight Canadian Cities.* Ottawa: Canada Mortgage and Housing Corporation. ◀

▶ Feng-Shui

Feng-shui is a traditional practice used to harmonize people with their environment. Originating from ancient China, *feng-shui* literally means "Wind and Water." To avoid cold wind and to have water are the most fundamental feng-shui principles for house site selection. However, feng-shui has provided guidance for harmonious relationships more broadly.

Feng-shui has developed into two main fields: one is yang-house feng-shui, applied to cities, villages, palaces, and housing; the other is yin-house feng-shui, applied to landscape analysis and graves. There are two major schools: the form school and the compass school. The first focuses on the reaction between Qi (vital energy flow) and the forms of land, whereas the second pays more attention to astro-nomical factors and positionss with the feng-shui compass. The two are often combined in practice. Yin-house feng-shui, which refers to landscape planning, is more involved with the form school, whereas yang-house feng-shui concerns itself more with the compass school.

Feng-shui analyses of house design take into account the environment as a whole, including time, space, and people. The design considers forms in relation to arranging Qi. It emphasizes orientation and position at various scales, from bed, door, courtyard, and neighborhood to the surrounding landscape, including mountains and water. According to feng-shui, different years have different favorable positions and orientations and people's birthdays indicate different preferences, which influences the architectural and site design.

Origin

Feng-shui is deeply rooted in ancient Chinese cosmological beliefs, such as Qi, yin, and yang, the five-element theory, and the divination of "I Ching." It evolved further with Taoist, Confucian, and Buddhist philosophy. Its origins can be traced to the "I Ching," or the Book of Change, 3,000 years ago. It reflects the organic and dialectical worldview of traditional China: All things in the universe are interrelated and integrated in a unified whole. Feng-shui holds that what works in a microcosm also works in a macrocosm. Therefore, the criteria for selecting the sites of cities, towns, houses, and even graves are almost the same. In the Western Zhou dynasty (1100–771 B.C.), a primitive form of geomancy was used to determine locations of houses or graves. Until the 19th century, feng-shui was firmly connected to events of every day life, from site selections for cities to interior design, from selecting a day for marriage to naming a child. When Western science took hold, feng-shui became labeled as backward superstition. After 1949, when communism took over, feng-shui practice was forbidden until the 1990s. However, in the countryside and among the expatriate Chinese, the adherence to feng-shui continued.

Feng-shui has shaped the built environment of China throughout history. For example, research has shown how feng-shui models structured the traditional Beijing courtyard houses. In fact, the courtyard, a central opening enclosed by buildings, has been a basic model for traditional Chinese built environments, including cities, houses, and gardens. The Beijing courtyard house was the basic unit of the city. Formal city planning arranged courtyard dwellings on a grid system.

According to feng-shui, a favorable site for a dwelling is enclosed by surrounding hills, called "tiger and dragon hills," and faces south, with a view of a mountain peak. In front of the site is an open space containing either a lake or a meandering river. In this space, living Qi abounds. The ancient Chinese brought this model of their ideal habitat to their designs when creating a human-made space, such as a house. Buildings corresponded to mountains, roads to rivers, and walls to hills. The spatial form of the Beijing courtyard dwelling is a physical embodiment of the ideal feng-shui model of landforms. Emphasizing favorable orientations and positions, the layout of the Beijing courtyard dwelling arranged Qi, as derived from the I Ching diagrams expressing Chinese cosmological beliefs, such as the Luo Book, the Nine Chamber Diagram, and the Later Heaven Sequence. These feng-shui principles were used in a way that reflected central control within a hierarchical family system and were applied to reinforce the strict class system of the traditional Chinese society.

Feng-shui practice has begun to influence housing in the United States. In contrast to the Western-influenced skepticism of 19th-century China, scholarly interest has resurfaced in the potential contributions of feng-shui to contemporary environmental science. Furthermore, the general public's attention to feng-shui has been growing fast and has taken on the characteristics of a popular vogue and a profitable business.

For example, the number of feng-shui consultants in the United States has been increasing rapidly, and a variety of workshops and seminars offer commercially marketed feng-shui versions. Feng-shui consultants will design or remodel interiors, including orientations of the doors, rearrange rooms and furniture, and add decorations, sometimes in consideration of the resident's birthday. They also suggest ways of arranging plants, pools, and rocks in gardens and recommend sites for house building.

In contrast to the lack of written documents on many prehistoric landscapes and house designs, feng-shui has been written about for 3,000 years. It provides a rich anthropological source on human settlements. Most of it was written in ancient Chinese poems, their meaning ambiguous, partly to keep its myth. Moreover, there are many different versions and schools, which often conflict with each other. The mainstream of classic Chinese literature often excluded feng-shui. However, it was spread widely among the general public. The English versions mirror these characteristics. Despite the growing commercial interest in the United States, there has been no complete and scholarly authoritative translation of feng-shui text. (SEE ALSO: *Cultural Aspects; Home*)

—*Ping Xu*

Further Reading

Chiou, S. & R. Krishnamurti. 1997. "Unraveling Feng-Shui." *Environment and Planning B* 24(4):549-72.

Eitel, E. J. [1873]1988. *Feng-Shui: The Science of Sacred Landscape in Old China.* With commentary by John Michell. London: Synergetic Press.

Feuchtwang, S. D. R. 1974. *An Anthropological Analysis of Chinese Geomancy.* Laos: Editions Vithagna.

Rossbach, S. 1983. *Feng-Shui: The Chinese Art of Placement.* New York: E. P. Dutton.

Skinner, S. 1982. *The Living Earth Manual of Feng-Shui: Chinese Geomancy.* London: Routledge & Kegan Paul.

Xu, Ping. 1998. "*Feng-shui* Models Structured Traditional Beijing Courtyard Houses." *Journal of Architectural and Planning Research* 15(4). ◄

► FHA Title I Home Improvement Loan Program

The Title I Home Improvement Loan Program was established by Congress in 1934. It was designed to provide homeowners who have little equity in their homes the ability to finance home improvements. Under the program, a bank or qualified lender makes loans from its own funds to eligible borrowers. The Department of Housing and Urban Development (HUD) insures the lending institution against a possible loss on the loan, if the loan proceeds are used for qualified improvements, proper underwriting procedures were used, and regulations were followed. The 1994 claims rate for HUD-insured loans in default was 1.34%. HUD insures up to 90% of any individual loan and insures 10% of the lender's entire portfolio.

Underwriting flexibility allows Title I loans to be made with higher debt-to-income ratios than most other loans. The maximum debt ratio allowed under HUD regulations is 45%. Higher debt ratios may be acceptable when the lender applies disposable or discretionary criteria to the decision process. Thus, Title I offers an attractive financing alternative to property owners who have good credit but little equity in their homes. It is particularly useful for recent purchasers.

The Title I loans cover a wide variety of alterations and repairs to improve or protect the basic livability and utility of the property: structural additions and alterations, siding, roofing, insulation, plumbing, heating and cooling systems, solar energy systems, interior finishing, and landscaping. Most of Title I loans are for improvements to single-family homes. However, Title I loans can also be obtained for multifamily and manufactured homes and for nonresidential structures.

Title I property improvement loans are originated in one of two ways: *Direct loans,* for which the borrower makes application directly to a lender without assistance from a dealer or contractor. The loan proceeds are given to the borrower, who is responsible for hiring the contractor and controlling the funds. *Dealer loans,* with which a home improvement contractor assists the borrower in obtaining the loan. The loan proceeds are given to the dealer but only when the work is completed to the borrower's satisfaction.

Of the 83,034 loans originated in 1997, for example, 60% were direct loans for a total of $881 million and 40% were indirect (dealer) loans amounting to $398.1 million. The percentages of dealer and indirect loans have been steady during the 1990s. The average size of a Title I home improvement loan increased between 1996 and 1997 from $14,386 to $14,593.

Lending institutions that participate in the program include national and state banks, savings banks, savings and loan associations, federal and state credit unions, mortgage companies, and finance companies. The program is also widely used with state and local housing assistance and neighborhood revitalization programs.

The following maximum loan limits and terms apply:

- ► *Single family*—Maximum loan amount is $25,000. Maximum term is 20 years, 2 months.
- ► *Multifamily*—Maximum loan amount is $60,000 or an average of $12,000 per dwelling unit, whichever is less. Maximum term is 20 years.
- ► *Nonresidential*—Maximum loan amount is $25,000. Maximum term is 20 years.

▶ *Manufactured home*—Maximum loan amount on a manufactured home that qualifies as real property is $17,500; if personal property, it is limited to $5,000. Maximum term is 15 years for real property and 12 years for personal property.

(SEE ALSO: **Federal Government;** *Federal Housing Administration; Fannie Mae; Title One Home Improvement Lenders Association*)

—*Peter Bell*

Further Reading

Bell, Peter. 1997. *The Title I Deskbook*. Washington, DC: Title One Home Improvement Lenders Association.

Emerling, Robert. 1997. *The Complete Guide to Getting Your Title I Claims Paid*. Washington, DC: Title One Home Improvement Lenders Association and Old Newport Funding. ◀

▶ Filtering

Filtering is a concept that describes changes in the housing market and in neighborhoods. The general notion is that of a trickle-down process whereby older houses, as they decline in value and rent, are transferred from richer to poorer families. It is assumed that within a metropolitan housing market, units are implicitly ranked hierarchically by their size and quality, location, and neighborhood (the last by their physical and social characteristics, as well as the level of city services delivered—e.g., utilities, parks, protection against crime). Price and rent roughly approximate this hierarchy. The housing hierarchy declines with time because of age and obsolescence, whereas the upper levels are replenished by new construction.

At root, filtering is a longitudinal process that matches households with different income levels, to housing with different price and quality levels. Thus, the highest-income households tend to have the best and most expensive housing, whereas the poorest have the least expensive and lowest quality. Filtering involves *change over time* in this matching.

Controversy surrounds the assumption that filtering can work to improve the housing of the poor. In principle, construction of new housing for middle-income people frees up their old housing for lower-income people to occupy. A number of obstacles may reduce this beneficial effect. Analysis of vacancy chains has shown that households replacing movers to new units are not much lower on the income scale. Formation of new households or migration into the market causes new middle-income households to intercept the vacancy before it trickles down to the poor. More generally, the downward filtration depends on the ratio of supply and demand in the market area: If an excess of housing exists, more of the middle-income housing may fall into the possession of lower-income households. However, declines in price are usually matched by declines in quality because landlords decrease maintenance. For filtering to provide a net gain to the poor, prices would need to drop more quickly than buildings can deteriorate. Whenever such a free fall in housing prices has occurred, the housing stock and neighborhoods typically have also declined in quality.

An alternative view of filtering emphasizes both the movements of households and the changes in the available housing. Typically, households experience rising incomes over their life cycle and move upward in the housing market through a succession of housing units. This upward mobility takes them through a series of units, each of which may be on a path of downward filtration. Thus, filtering may be seen as a down escalator, with people moving up the housing escalator at the same time as the housing units decline in value and quality. Whether people are made better off by this process depends on both how fast the units are declining and how fast the households are moving upward in their housing careers.

For reasons of diversity of life experiences and market conditions, the outcomes of filtering are extraordinarily complex. Low-income households sometimes live in expensive units. These may be elderly with low retirement income who bought their house decades earlier, when the price was lower and their income was higher. These and other exceptions have defied efforts to precisely define filtering. The professional disarray is further emphasized by the anomaly of gentrification in the 1970s and the surge in house prices and rents in the 1980s—both contrary to the prior emphasis on their decline. Neither should be considered as an exception to the filtering process—only a particular outcome of upward and downward forces played out on the housing escalator.

Understanding of filtering must emphasize time, for filtering is a longitudinal process. The great durability of the housing stock, with a life of say at least 40 years for apartments and 80 to 100 for single-family houses, means that they last for long periods of time, serving multiple households with different characteristics and attitudes about housing as they successively occupy the housing unit. This durability means that construction at any one point in time—a particular vintage—reflects the technology and fashion preferences of consumers in that era, but that unit continues on for multiple eras, perhaps cycling in and out of fashion stylistically, although accruing increasing obsolescence. These changes bring forth remodeling efforts to remedy this obsolescence, which in turn alters the unit so that its characteristics at the end of its life (just prior to demolition) may be rather different from when it was first constructed.

Housing units are not the only things that change. Households also go through life cycles (young, middle-aged, and elderly, perhaps at times with children). Thus, these households must periodically match their housing with their needs of a particular cycle. Making things more complex in this match, however, is the reality that the young or elderly households of a particular cohort in time, may display different earnings and preferences than other cohorts at that same position in their life cycle. In short, each cohort of households and each vintage of housing unit may follow a life cycle different from their predecessor's. So also for neighborhoods of households and housing units.

Thus, there is more of a roller-coaster effect from changing historical conditions, including variations in population growth rates, economic cycles, technological change, and modifications in governmental programs. For example, just because in the 1930s to the 1950s, many elderly moved out of large Victorian homes to more efficient apartments, the elderly at present occupying homes built during the 1950s to the 1970s will not now also leave their units. Their single-level, easier to maintain, tract homes may allow them to age in place. Nor do these newer units lend themselves to (nor do modern zoning ordinances permit) being carved up into smaller units in the way of the earlier Victorian units. So also suburban flight today is more pronounced than in earlier eras, which further affects housing preferences and results in patterns of different age cohorts.

Even in the same time period, filtering may proceed at different rates for different segments of the market. Housing is a heterogeneous product that precludes an even flow. Through market segmentation, various submarkets of substitutable units (e.g., two- and three-bedroom units) appeal in varying degrees to different submarkets of households (e.g., different ethnicities, size, and income). Racial-ethnic discrimination in mortgage lending, selling, and renting further segments the market, impeding filtering processes. Accordingly, filtering works by fits and starts so that what happens in the filtering flow of one submarket only partially affects another. Some submarkets will filter down more rapidly when demand slackens for that type of unit in that location. At the same time, increasing demand in other submarkets may reverse deterioration by pushing prices and quality higher, causing upward filtering.

Finally, there is the normative issue of whether filtering should be relied on as a beneficial process or whether filtering has the unacceptable outcome of crowding the poor into low-quality housing. Both scenarios have been assumed and experimented with in public policy over the decades, with mixed results. Filtering as a process helps different income groups with different needs adjust to the ever-changing housing stock. To this extent, it is beneficial. But it does not solve the housing problem by itself. Poor households get poor-quality housing in the filtering process. Nevertheless, it is wrong to assume that there is no merit to this process simply because it does not redistribute good housing from rich to poor. No market process can be expected to do that. Government assistance is required. To date, U.S. society has been unwilling to vote the amount of assistance necessary to allow all households to live in good housing. (SEE ALSO: *Housing Market; Vacancy Chains*)

— *William C. Baer and Dowell Myers*

Further Reading

Baer, William C. and Christopher B. Williamson. 1988. "The Filtering of Households and Housing Units." *Journal of Planning Literature* 3(2):128-52.

Downs, Anthony. 1969. "Housing the Urban Poor: The Economics of Various Strategies." *American Economic Review* 59(4):646-51.

Lowry, Ira S. 1960. "Filtering and Housing Standards." *Land Economics* 35:362-70.

Myers, Dowell. 1990. "Filtering in Time: Rethinking the Longitudinal Behavior of Neighborhood Housing Markets." Pp. 274-96 in *Housing Demography,* edited by Dowell Myers. Madison: University of Wisconsin Press.

Rothenberg, Jerome, George C. Galster, Richard V. Butler, and John R. Pitkin. 1991. *The Maze of Urban Housing Markets.* Chicago: University of Chicago Press.

Weicher, John and Thomas Thibodeau. 1988. "Filtering and Housing Markets." *Journal of Urban Economics* 23:21-40. ◀

▶ Fixed-Rate Mortgage Loan

In the financing of U.S. real estate, the fixed-rate mortgage loan has had a dominating role. For most kinds of working property, including farms, industrial property, offices, commercial buildings, apartments, and single-family homes, the long-term, fixed-rate mortgage loan has been a central feature of ownership. Only in the development and construction phases of real estate is this loan eclipsed by other types of debt.

The Fixed-Rate, Level-Payment Loan

Especially significant in U.S. real estate has been the level-payment form of fixed-rate debt. It has been critical in permitting the rate of homeownership in the United States to rise from around 45% during the 1930s to near 65% by 1980. During that same period, it also enabled home mortgage debt to rise from around 7% of all debt owed by governments, households, and nonfinancial enterprises to around 25%. It is by far the largest type of private debt in the U.S. economy, sometimes approaching the magnitude of all government debt combined. The fixed-rate, level-payment loan has been a major enabling factor to propel the suburbanization of the United States in the decades since World War II.

Variation in Fixed-Rate Loans

Fixed-rate mortgage loans have a wide range of forms, differing by the manner in which principal is repaid. The interest-only form and the level-amortization form are especially important as benchmarks. The first, often called a *straight term loan,* is similar to a standard corporate or government bond in that payments include no principal. Rather, the entire loan is paid back at the end of term. By contrast, a *level-amortization loan* schedules principal repayments just sufficient to pay off the loan at the end of the term while maintaining level payments throughout. This pattern necessitates that the amount of principal repayment be small in the first payment, gradually increasing until it constitutes most of the payment at the end. A *partially amortized loan* has level payments that include some principal amount, but not enough to pay off the loan by the end of the term. Instead, a large final payment of principal is required, known as the *balloon payment.*

Some types of fixed-rate loans involve *negative amortization.* In other words, the payment on the loan is less than the amount of interest being charged, and the unpaid interest is added each period to the principal balance. The payments usually increase over time. For example, in a graduated-payment mortgage, the payments are scheduled to increase by a predetermined percentage each year until they are sufficiently large to stop the negative amortization and to begin amortizing the loan over the remaining term of the loan.

There can be significant differences between fixed-rate mortgages in the rights of the borrower and lender. For example most "conventional" home mortgages (those not insured or guaranteed by an agency of the U.S. government) contain a "due-on-sale" clause allowing the lender to declare the outstanding balance immediately due if the property is sold. On the other hand, most government-insured or -guaranteed home loans do not contain a due-on-sale clause. As another example, most standard home mortgage loans give the borrower the right to prepay the loan at any time, without penalty. But without a clause explicitly giving the borrower this right, prepayment traditionally is prohibited or may be allowed only with a substantial fee. Most mortgages other than home loans do not allow free prepayment.

There is little limit to the creative use of fixed-rate mortgage loans in combination. A common example is a fixed-rate purchase money mortgage. Suppose, for example, a homebuyer would like to preserve an attractive mortgage that already exists on the home to be purchased, but the outstanding balance is only half the value of the property. Typically, the buyer would find it hard to assemble sufficient cash to cover the difference. But if the seller is willing to accept installment payments for, say, half of the difference between the price and loan amount, the buyer has preserved the old mortgage and reduced the down payment to 25%, achieving a much more manageable transaction.

As another example of combining fixed-rate loans, the purchase money mortgage above might be a *wrap-around mortgage*. That is, instead of the seller's accepting a standard second mortgage note for 25% of the purchase price, the seller accepts a note for 75% of the purchase price (equal to the old loan plus new), with a correspondingly larger payment. The seller, rather than paying off the old loan, continues making its payments out of the payments received on the wrap-around loan. The buyer, instead of making payments on two loans, makes a single payment to the seller on the 75% wrap-around.

The advantage of the wrap-around is that it gives the seller greater security against default. Because the seller is making the payments on the old (senior) loan, the seller can have more confidence that the old loan will not threaten the junior loan through default and therefore can offer better terms to the buyer.

The Evolution of the Fixed-Rate Home Loan
The evolution of the home loan is the central story of mortgage finance. Before the 1930s, home mortgage lending was very local in nature and terms were very restrictive. The ratio of loan to value seldom exceeded 50%, and the loan term seldom exceeding five to seven years.

The form of home mortgage loans began to change drastically with the creation of the Federal Housing Administration (FHA) in 1934. At the heart of FHA was the mortgage loan insurance program now embedded in the fabric of U.S. housing finance. With FHA insurance available to protect lenders from the uncertainties of a long-term commitment, they were, for the first time, willing to make long-term, high loan-to-value home mortgage loans. These loans started out with terms of little more than 20 years and with loan-to-value ratios of little more that 75%. Their success paved the way for the modern home mortgage loan, which can exceed 90% of value (97% for small FHA-insured loans) and can now exceed 30 years in term.

Although the fixed-rate, level-payment home loan remains dominant in the United States, its risks have increased since the 1970s. This type of loan works best in an economic environment with low inflation and low interest rate uncertainty. But both inflation and interest rate changes have become greater risks as the U.S. economy has become more internationalized. Two responses to this have occurred. First, the use of alternative mortgage instruments, especially adjustable-rate mortgages (ARMs), has increased significantly. Second, traditional sources of home loan funds, such as the thrift institutions (savings and loan associations and savings banks), have been supplanted by more sophisticated investors through secondary mortgage markets and mortgage securitization.

Because of their low default risk, their long life, and their sheer volume, fixed-rate home loans have now become the basis for creating securities. During the 1980s, mortgage-backed securities (MBSs), and extremely complex "derivative" securities based on the MBSs, became major investment vehicles on "Wall Street." Thus, the long-term, level-payment, fixed-rate mortgage loan that has been central to the transformation of "Main Street" and the suburbs of the United States since the 1930s, is now transforming "Wall Street" as well. (SEE ALSO: **Mortgage Finance**)

—*Wayne R. Archer*

Further Reading
Brueggeman, W. B. and J. D. Fisher. 1993. *Real Estate Finance and Investments*. 9th ed. Homewood, IL: Irwin.
Clauretie, T. M. and J. R. Webb. 1993. *The Theory and Practice of Real Estate Finance*. Fort Worth, TX: Dryden.
Dennis, M. W. and M. J. Robertson. 1995. *Residential Mortgage Lending*. Englewood Cliffs, NJ: Prentice Hall.
Herzog, J. P. and J. S. Earley. 1970. *Home Mortgage Delinquency and Foreclosure*. National Bureau of Economic Research, General Series 91. New York and London: Columbia University Press. ◀

▶ Flat-Rent Concept

Traditionally in public housing programs, residents pay 30% of their income for housing. As income rises, so does the rent, and as income declines, the monthly rent payment declines as well. Residents are required to report changes in income as they occur. Often, increases in rent act as a disincentive for residents of public housing to accept employment that offers nominal increases in income because there is an offsetting loss in public benefits. Too many times, the value of the loss in benefits is greater than the increase in income. Furthermore, the intervention by public authorities in the lives of residents is a constant reminder that one is living "on the public dole." The flat-rent concept is designed to mimic the private market where a fixed rent is established for a one-year period and must be paid, regardless of changes in income.

In Boulder, Colorado, a flat rent is used for all properties that have equity participation through the Community Housing Assistance Program (CHAP). The CHAP is a local housing trust fund financed from revenues derived from a housing excise tax paid on residential and nonresidential development and an .8 mill property tax. These dollars are allocated to nonprofit and private sector developers in the form of equity grants in return for retaining permanent affordability of the units. Through a combination of CHAP funds, federal grants, low-income housing tax credits, and below-market-rate financing, units are acquired or built that can be leased at rents 65% to 80% of the market rent for a comparable unit. Residents of these housing units pay a rent based on 30% to 35% of the monthly income of a household earning 50% of the local area median income. Generally, it is assumed that 1.5 persons per bedroom will be living in a unit as the baseline for establishing the maximum rent. Some entities charge less rent to house persons at the lower end of the income spectrum.

Once a rent is established for a particular type and bedroom size, it is advertised as available to the general public. Screening by the property managers ensures that residents earn incomes at or below those established for the program. A one-year lease is negotiated with the tenant. The lease stipulates the monthly rent that must be paid for the entire year. This rent is paid by the tenant, regardless of any changes in income that may occur during the year. In this respect, the rent amount is exactly what a resident would encounter in the private market where the lease is treated as a contract and the agreed-on rent is paid monthly. At the end of the lease term, the tenant must be recertified as to income. Generally, rents are then adjusted to compensate the property owner for increases in operating costs or by the consumer price index. Residents sign a new lease for the year with the new rent reflecting any change in income. Once residents earn 120% of the maximum income allowed in the program, they have one year of reduced rent payments before paying the fair market rent.

The advantage of a flat-rent program is that residents have an incentive to earn additional income, and they have choices about how to use it. Many residents in these programs have saved enough funds at the end of a two- to three-year period to allow them to purchase a small multifamily unit. They report that if their rents had increased as income increased, the opportunity to save funds would have been lost. Also, knowing that rents are a fixed cost for a one-year period, with reasonable adjustments to occur on an annual basis, has provided an incentive to increase earnings over time. (SEE ALSO: **Affordability; Private Rental Sector**)

—*Kathy McCormick* ◄

► Flexible Housing

The concept of *flexible housing* is used by housing experts who address problems of rapid social change, while at the same time trying to harness new industrial processes,

materials, and scientific management practices in housing. Architects in many countries, observing the rigidity of multifamily housing and urban environments and sharing the idea that large-scale housing should not be simplistically repetitive, have sought flexibility by designs in which parts could move and large superstructures could be filled in later. These designers included, in the mid-1950s to 1960s, Yona Friedman in France, the Metabolist group in Japan, and Archigram in England. John Habraken, working in the Netherlands at the same time, explored the concept that each dwelling should be changeable according to changes in the occupants' lives.

Flexible housing in general involves two seemingly conflicting goals. The first goal is a variety rather than uniformity of dwellings. This goal recognizes the heterogeneity and demographic dynamics in most housing markets; decentralized regulatory structures; wide regional differences; and the active roles taken by households in housing decisions concerning location, price, and quality.

The second goal is efficiency. This is principally concerned with price reduction, speed, reduction in conflict among the many parties involved, and quality of production. To date, much work to bring efficiency to housing processes has focused on adopting measures that proved successful in mass production, standardization, and scientific management occurring in industry in general. In housing, this has led many to accept the concept that maximum efficiency can be obtained by standardizing dwelling plans or major elements of houses, such as bathrooms and kitchens, and mass producing them.

At the same time that some experts advocate the production of standard dwellings as a means to achieve efficiency, the market of households shows evidence of a need for increased variability as a result of social and economic changes and the natural variety of a heterogeneous society. Two changes are at the forefront. One involves the changing composition of households. The five-person family is no longer the norm. Instead, we have smaller households, more elderly households, and more single people. The second change concerns the need to rehabilitate older buildings, designed for one kind of household and technical standard but now needing to be different.

The tension between the need for flexibility and efficiency was solved on a short-term basis in public and rental housing by the elimination of households as active agents in the housing process: Households would presumably be satisfied to have a dwelling designed by experts using demographic surveys as a basis of determining standard unit types and mix. The private for-sale market sought to aggregate and align housing demand to suit the scientific management principles adopted from industrial mass production. But because nonpublic housing demand has never responded to market aggregation strategies, the conflict between variety and efficiency remained a central problem in flexible housing design theory and practice.

Another use of the term *flexible housing* refers to houses in which walls are movable, such as traditional Japanese houses using sliding screen walls to open or separate adjoining spaces at will. The term has also been used to refer to expandable or demountable houses. Distinctions are usually made between flexibility and changeability in the short and long term.

There is widespread recognition that single dwellings, and buildings containing many dwellings, are successful in the long term to the degree to which they can transform, or be "flexible," in relation to the changing needs and preferences of the occupants. Although some elements of detached houses or multifamily buildings remain stable over long periods, such as the basic structure and property lines, other aspects change, such as exterior and interior finishes, equipment, fixtures, and in some cases, room arrangements.

In summary, flexible housing, considered in terms of both technology and decision control, is housing in which changes can be made without difficulty. As both technology and decision-making patterns have become more complex and intertwined, flexibility in completed buildings is increasingly difficult. Walls and floors are no longer simple structural forms of wood or concrete but are filled with pipes, cables, ducts, and other elements serving equipment distributed throughout a single house. In multifamily buildings, walls and floors are frequently filled with pipes and cabling serving several dwellings, making change of one unit impossible without disturbing the others. Walls and floors are also associated with acoustical and visual privacy, interior decor, weather resistance, and other important attributes. This physical and organizational interplay highly constrains flexibility.

This "entanglement" of building parts and the parties designing, approving, and installing—and later changing—each part makes flexible housing very difficult to achieve. Certain construction framing technologies, such as the conventional wood frame system and wood post-and-beam construction lend themselves to adaptation or flexibility with more ease than do heavy concrete panel building systems, in part because piping and wiring in these types of construction can be reached rather simply by removing wall and ceiling surfaces. But because of the limitation of these methods of housebuilding, more advanced methods of disentangling the resource lines from walls and floors are being introduced. (SEE ALSO: **Industrialization in Housing Construction;** *Modular Construction; Support-Infill Housing*)
—*Stephen Kendall*

Further Reading

Bender, Richard, and Forrest Wilson, ed. 1973. *A Crack in the Rear View Mirror: A View of Industrialized Building.* New York: Van Nostrand.

Habraken, N. J. 1970. *Three R's for Housing.* Amsterdam: Scheltema & Holkema.

———. 1972. *Supports: An Alternative to Mass Housing.* English language edition. London: Architectural Press.

Hamdi, Nabeel. 1991. *Housing without Houses: Participation, Flexibility, Enablement.* New York: Van Nostrand.

Howe, Deborah A. 1990. "The Flexible House: Designing for Changing Needs." *Journal of the American Planning Association* (Winter):69-78.

Kelly, Burnham et al. 1959. *Design and the Production of Houses.* New York: McGraw-Hill.

Russell, Barry. 1981. *Building Systems, Industrialization and Architecture.* New York: John Wiley.

"The Twentieth Century Fund." 1944. In *American Housing: Problems and Prospects,* Miles L. Colean. New York: Twentieth Century Fund. ◄

► Floor-Area Ratio

Floor-area ratio (FAR) is a measure of a building's size in relation to the amount of land it occupies. FAR is the total floor area of a building on all habitable stories divided by the total lot area. FAR regulations are enacted by elected local officials and enforced by planning boards and building review staff to govern the bulk of new construction. FAR values are distinct from ground coverage or "footprint" values, which do not take into account multiple levels. FARs indicate density more accurately than simple ground coverage. For example, a typical suburban house with 800 square feet on the main floor and 600 square feet on the upper floor, built on a 7,000-square-foot lot, has a footprint of only 800 square feet, a total floor area of 1,400 square feet, and a FAR of 1,400 square feet divided by 7,000 or .20 FAR.

The FAR method of regulation can help maintain the scale of historic low-density districts, particularly where traditional older height and setback regulations never anticipated the full build-out of the entire permitted building envelop. In such districts, FAR limitations overlay the standard setback regulations and protect the existing or historic scale of the neighborhood. Where infill and redevelopment are encouraged by governing authorities, particularly where additional units take the form of alley houses or "granny flats," FAR regulations can define the scale of new construction. For example, in a district of freestanding houses, newly constructed units along the rear alley might be limited to less than half the FAR of the front unit, to maintain the primacy of the street frontage and simulate the appearance of mews or converted garages at the rear.

However, not all FAR regulations are primarily intended to protect the scale of an urban district. In some cities, FAR regulations are implemented for greater design freedom. For example, the New York City 1916 zoning ordinance specified the uniform and familiar "ziggurat" or stepped setback building masses that characterized much of midtown. Compared with these original and more prescriptive regulations, a variation of FAR regulation was adopted in 1960 that allows greater floor area and height in return for public plazas at the ground level. The character and scale of previously uniform districts has become disrupted by towers that are out of context with the street and by a plethora of wind-swept open spaces. Thus, FAR regulations enacted in the name of design flexibility can result in the erosion of consistent urban fabric.

On a regional basis, average FARs can be used to indicate the intensity of land development. FAR as an indicator of the intensity of land use is related to, and limited by, the predominant modes of transportation. A developer considering a tract of land might apply rule-of-thumb FAR figures to determine very early in the planning process what a particular large parcel might realistically support. If easy circulation and parking for automobiles is the chief consideration, retail developers in the United States have a strict maximum FAR guideline of .25: an industry-standard figure that most national chains will not exceed. When housing and office spaces are included, the FAR might be pushed to .40, but more intense development requires the expense of parking garages or convenient mass transportation.

Where land is relatively inexpensive and automobile usage is subsidized through highway construction programs, development will tend to sprawl rather than intensify through higher FARs. Traffic congestion and commute times become unacceptable as land use intensities fall below the 2.0 FAR generally considered the minimal regional density that will support mass transit. Thus, through dissipation and sprawl, the high ratios (5.0 FAR and above) associated with preautomotive, pedestrian-oriented districts and efficient metropolitan transit often will not evolve. (SEE ALSO: *Infill Housing; New Urbanism; Setback Requirement; Space Standards*)

—*Paul Anthony Saporito*

Further Reading

De Chiara, J., J. Panero, and M. Zelnik. 1995. *Time Saver Standards for Housing and Residential Development* (2nd ed.). New York: McGraw Hill.

Eisner, S., A. Gallion, and S. Eisner. 1993. *The Urban Pattern*. 5th rev. ed. New York: Van Nostrand.

Garreau, Joel. 1991. *Edge City*. New York: Doubleday.

Krieger, Alex. 1991. *Towns and Town Making Principles: Andres Duany and Elizabeth Plater-Zyberk, Architects*. New York: Rizzoli International. ◄

▶ Foreclosures

Residential or home mortgage foreclosure must be understood in the context of a home finance system that allows buyers to finance their homes over a long period of time by means of scheduled payments. At the heart of this system are the institutions that make money available—mortgage-lending institutions—and the instruments that define the manner of repayment—the mortgage. Typically, the buyer pays a proportion of the purchase price as a down payment and takes out a mortgage loan for the remainder. The mortgage pledges property or buildings as security. If the borrower defaults on his or obligation, the lender can start foreclosure proceedings and claim title to the property. In the United States, approximately 90% of homes are purchased with mortgages. In 1993, only 37% of all homeowners did not have home mortgages.

Because few homebuyers purchase their homes outright, homeownership, in reality, is a process that consists of two analytically separate components: buying (and selling) and keeping a home. Buying requires making houses and money available, often in conjunction with financial incentives for homeowners such as tax relief. Keeping a home requires meeting a schedule of payments and involves policies and programs that permit households to continue to make monthly mortgage payments even through their income and general revenues may suffer temporary setbacks. Although the viability of homeownership depends on keeping as well as buying, there has been relatively little attention in the home finance literature or in the public policy arena to keeping homes, until the last decade when the rate of delinquencies and foreclosures started to rise.

Delinquency, Default, and Foreclosure

The borrower who does not make a scheduled mortgage payment (or fulfill any covenant of the mortgage agreement) is technically in default. The U.S. Department of Housing and Urban Development (HUD) limits legal default to beginning thirty 30 after failure to live up to the mortgage agreement. However, it is traditional to consider one or two missed payments as delinquent (late or delayed payment), extending a "grace period" and thereby giving the borrower the benefit of the doubt, and to declare default (the borrower stops payment) only with the third missed payment. The importance of lender discretion in this process is reflected in the manner in which the Mortgage Bankers Association reports statistics. It characterizes loans as past due 30, 60, and 90+ days and foreclosures started.

Default-foreclosure is a legal process that transfers ownership of the property from the borrower to the lender. Depending on the state, this can be accomplished nonjudicially as well as judicially. In a judicial foreclosure, the court orders the foreclosure and conducts a public sale with the proceeds going to fulfill the debt. This process is costly and time-consuming. There is a specific period of time before foreclosure can take place and an equitable redemption period during which the borrower may pay off the entire debt and avoid the foreclosure.

In a nonjudicial foreclosure, an independent party (attorney, foreclosure service, etc.) conducts the sale after the equitable period of redemption. These foreclosures, also referred to as power-of-sale foreclosures, foreclosure by advertisement, or trustee's sale are available in approximately one-half of the states, take less time, and are much less costly than the court-supervised judicial process.

Default-foreclosure, however, is costly to all the involved parties. Borrowers, for example, face lower credit ratings, loss of fees, and equity as well as emotional distress. Collectively higher rates of default and/or mortgage loss lead to higher interest rates for borrowers. Lenders and insurers may suffer financial losses and face regulatory restrictions.

Lenders have options other than foreclosure. They can forebear on a loan in default (i.e., give the borrower extra time to make the loan current), renegotiate the loan with lower payments, or take a deed-in-lieu-of-foreclosure. Federal Housing Administration (FHA) and Department of Veterans Affairs (VA) rules, for example, specifically encourage some of these alternate procedures and set specific time periods for forbearance.

The Foreclosure Literature

Studies of residential mortgage default are a fairly recent phenomenon. They began in the late 1960s and early 1970s to help lenders and loan-insuring institutions assess the risk of various loan and borrower characteristics. The key explanatory variable for foreclosures is the initial loan-to-value ratio.

Early studies focused on the individual borrower and examined the characteristics of loans and borrowers at the time of origination. The findings emphasized loan characteristics, such as equity, in predicting the default decision.

Second-generation studies created models of consumer behavior to explain decision making. The most influential of these is an option-based model in which the borrower weighs the market value of the mortgage and the equity in the home. From this perspective, default is viewed as purely financial, and borrower characteristics such as employment status and income do not matter. However, some studies have shown that nonequity factors play an important role in the default decision.

Third-generation studies were spurred by high rates of default—particularly for the FHA and the VA in the late 1970s and 1980s—and take an institutional perspective. Their main contribution is methodological, using measures of risk, such as mortgage loss, that better reflect institutional concerns. For example, although two defaults will contribute equally to a default rate, their contributions to expected loss can vary significantly depending on the loan amount.

The dominant view has been to treat the default decision as an option driven primarily by loan characteristics, mainly home equity. Lender decisions on whether to foreclose or work with delinquent loans are affected by the legal costs of foreclosure. The role of other borrower-related factors, transaction costs (reputation), or crisis events (divorce or economic events, such as job change or transfer) is less well understood mainly because contemporaneous data are not available.

Driven by the needs of lenders and institutions, default-foreclosure studies have also neglected the impact of widespread foreclosures on communities and regions.

The Old and the New Home Finance System

Historically, foreclosure has been important to the history of housing and home finance in the United States. During the Great Depression, a significant proportion of all residential mortgages were in default, and accounts cite as many as 1,000 foreclosures daily between 1931 and 1933. The effect of foreclosures on homeownership and on financial institutions was to spur legislation that created a sheltered and government-protected capital market for residential homes.

This "old" system, in place until recently, consisted of functionally discrete and heavily regulated mortgage lenders—the savings and loans (S&Ls) or "thrifts"; a standard fixed-rate, long-term, self-amortizing mortgage instrument (FRM); and a federally sponsored secondary mortgage market that enabled low-equity home purchase and redistributed funds regionally.

In the recent period, there have been major changes in the home finance system. Beginning in the mid-1970s, with the crisis of the thrift industry and the demands of the large banks, financial deregulation began to restructure the system into a competitive market of lenders, new adjustable-rate mortgages (ARMS), increased securitization (the movement of mortgages into the secondary market), and trading of mortgage-backed securities. What impact have these changes in the home finance system had on keeping a home?

Foreclosure in the New Financial Environment

Home mortgage foreclosures have become an increasing problem in Canada and Europe as well as in the United

Table 8 Delinquency and Foreclosure Rates in the United States, 1955 to 1997

Year	Percentage of Loans in Foreclosure (all loans)	Percentage of Loans 90+ Days Delinquent (all loans)
1955	a	.16
1960		.30
1965	.38	.33
1970	.32	.32
1975	.44	.49
1980	.34	.57
1981	.43	.64
1982	.59	.77
1983	.68	.84
1984	.68	.89
1985	.78	.94
1986	.92	1.01
1987	1.08	.93
1988	1.01	.85
1989[b]	1.00	.77
1990	.94	.71
1991	.99	.80
1992	1.04	.80
1993	1.00	.77
1994	.94	.76
1995	.88	.74
1996	.99	.63
1997[c]	(1.07)	(.57)

SOURCE: Mortgage Bankers Association of America National Delinquency Survey, Quarterly Data Series, National Delinquency Historical Series.
NOTE: Figures represent the annual average of percentage of loans for four quarters.
a. Data were not collected prior to 1962.
b. Expanded sample.
c. Average of first 3 quarters of 1997.

States. In the post-World War II period, the general pattern in the United States has been for foreclosures (and delinquencies) to vary along with general (as well as local and regional) economic conditions, fluctuating within a narrow range (see Table 8). Between 1969 and 1979, foreclosures fluctuated more widely and generally rose. Beginning in the 1980s, the foreclosure rate climbed steeply, peaked in 1987, and has not dropped significantly since. Total loans in foreclosure have dramatically risen along with the number of loans serviced. In 1965, approximately 45,000 loans were in foreclosure, and in 1989, the number was 300,000, an increase of 590%. FHA and VA rates have been particularly high. The percentage of loans in foreclosure in 1996 for all loans is .99, VA loans 1.56, and FHA loans 1.64.

Although the recent sharp rise in foreclosure rates can be traced to numerous social and economic factors not directly associated with changes in the home finance system—among them, unemployment, divorce, and inflation accompanied by high real estate prices until the 1990s—the new institutional framework has exacerbated existing social and economic trends.

On the whole, the deregulated home finance system favors buying over keeping. Borrowers are encouraged to "shop around" among a wide choice of lenders, rates, and instruments. Competition has encouraged lenders to be salesmen of potentially risky loans rather than careful assessors of risk. Borrowers who confronted mortgage repayment difficulties in the 1980s had more difficulty negotiating forbearance agreements because the loans had "left home" and were increasingly held by nonlocal institutions with centralized decision making, less responsive to local or regional economic conditions. In addition, the risk associated with individual loans had been diffused by secondary market activities.

New variable-rate instruments (e.g., ARMS) may have introduced new liabilities for borrowers, although the studies to date are few. Introduced at a time when mortgage interest rates were high and expected to rise, ARMS have been associated with low equity or inflated property assessments and credit evaluations, suggesting that ARMS may encourage economically rational default-foreclosure when interest rates rise or housing prices fall.

Finally, congressional response to the S&L crisis has also contributed to high default rates. The Resolution Trust Corporation's large-scale sale of nonperforming loans in the late 1980s led to an influx of below-market-value real estate and to defaults by borrowers whose home equity and loan-to-value ratio were depressed.

In the past, market rationality dictated responsiveness to changing local and individual needs, and the interests of mortgage borrowers and lenders tended to coincide. In the new system, home mortgages have lost their once-sheltered position in the credit markets and become more of a commodity. This has made for a greater separation between mortgage originators and homeowners, both in terms of number of intermediaries, geographic distance, and interests. Lenders who can sell their mortgages on the secondary market do not have the same incentive to avoid delinquencies and foreclosures. The net effect of the new home finance system has been to dissociate the two components of homeownership—buying and keeping.

Responses and solutions to the surge of delinquencies and foreclosures range from proposals for formal legislation to informal agreements and arrangements. The first to act on these issues were threatened homeowners, particularly the unemployed in the industrial belt in the early 1980s. Along with their advocates, they negotiated forbearance agreements and mortgage moratoria and introduced bills in state and federal legislatures. As foreclosures rose in response to the insolvency of the thrifts and spread beyond regional pockets, institutional actors—both federal and private, both lenders and insurers—have been motivated to fashion solutions, ranging from identifying and contacting problem loans, to workout strategies to allow borrowers to remain in their homes, to applying more stringent lending criteria to reregulation.

Given the likelihood of continuing instability in the general economy, in real estate, and in home mortgage finance, it seems likely that foreclosure will remain a significant policy issue in the future. (SEE ALSO: **Affordability;** *Delinquency on Loan*)

—*Lily M. Hoffman*

Further Reading

Carroll, Thomas M. et al. 1995. "HUD versus Private Bank Foreclosures: A Spatial and Temporal Analysis." *Journal of Housing Economics* 4:183-94.

Clauretie, Terrence M. 1987. "The Impact of Interstate Foreclosure Cost Differences and the Value of Mortgages on Default Rates." *Journal of the American Real Estate and Urban Economics Association [AREUEA Journal]* 15(3):152-67.

Dunaway, B. 1985. *The Law of Distressed Real Estate.* New York: Clark Boardman.

Heisler, Barbara S. and Lily M. Hoffman. 1987. "Keeping a Home: Changing Mortgage Markets and Regional Economic Distress." *Sociological Focus* 20(3):227-41.

Hoffman, Lily M. and Barbara Schmitter Heisler. 1988. "Home Finance: Buying and Keeping a House in a Changing Financial Environment." Pp. 149-65 in *Handbook of Housing and the Built Environment in the United States,* edited by Elizabeth Huttman and Willem van Vliet--. New York: Greenwood.

Holloway, Thomas M. and Robert M. Rosenblatt. 1990. "The Trends and Outlook for Foreclosure and Delinquencies." *Mortgage Banking* 51(1):45-59.

Mortgage Bankers Association of America. "National Delinquency Survey, Historical Series" and Quarterly Data Series for specific years. Available from the Mortgage Bankers Association of America, Washington, DC 1-800-793-MBAA.

Quercia, Roberto G. and Michael A. Stegman. 1992. "Residential Mortgage Default: A Review of the Literature." *Journal of Housing Research* 3(2):341-79. ◀

▶ Foundation for Hospice and Homecare

The Foundation for Hospice and Homecare (FHH), now defunct, was a national organization dedicated to establishing and maintaining quality standards for home health care services. Formerly the National Home Caring Council, the FHH was credited with creating a national accreditation program for home health aide services and a curriculum for the training of home care paraprofessionals in 1978. The foundation also developed the National Certification Program, in conjunction with the Administration on Aging, to address the absence of national standards for paraprofessionals in home care. Implemented since 1990, the certification program protects the consumer of home health care services by setting minimum standards for the preparation of paraprofessionals, a common curriculum, and competency testing mechanisms.

Besides setting industry standards, the FHH conducted a series of conferences and seminars each year to educate home health caregivers and agencies. The foundation also published a variety of educational materials for consumers and caregivers on health care training, research, management, and public relations and distributed free consumer guides on hospice and home care.

The FHH was committed to researching the adequacy of health and social policies in the United States and to developing alternatives to institutionalization. In its final years of existence, research efforts focused on pediatric home care and hospice care for AIDS patients. (SEE ALSO: **Elderly;** *Home Care*)

—*Caroline Nagel*

G

► Gautreaux Program

The Gautreaux program is the result of a 1976 Supreme Court consent decree in a lawsuit on behalf of black public housing residents, charging intentional segregation against the Chicago Housing Authority and the U.S. Department of Housing and Urban Development (HUD) (*Hills v. Gautreaux,* 425 U.S. 284, 1976). The program allows public housing residents and those who had been on the waiting list for public housing to receive Section 8 housing certificates and move to private apartments either in mostly white suburbs or in the city of Chicago. The program provides extensive housing services to find landlords willing to participate in the program and to counsel families about these moves. Since 1976, over 6,000 families have participated, and over half moved to middle-income white suburbs. The program gives participants rent subsidies that permit them to live in private apartments for the same cost as public housing. Participants move to a wide variety of over 115 suburbs throughout the six counties surrounding Chicago. Suburbs with less than 70% whites were excluded by the consent decree, and very high rent suburbs were excluded by funding limitations of Section 8 certificates. Yet these constraints eliminate very few suburbs. The receiving suburbs ranged from working class to upper-middle class and were located from 30 to 90 minutes driving time from the participants' former addresses.

To provide mutual support, the program tries to move two or three families to each neighborhood, but it also avoids moving many families to any one neighborhood or community, and in fact, it succeeded in this goal. As a result, the program had low visibility and low impact on receiving communities.

The program has three selection criteria: Families must have no more than four children, no large debts, and acceptable housekeeping standards and behaviors. With these criteria, program designers sought to avoid overcrowding, late rent payments, and building damage. But none of these criteria was extremely selective, and all three reduced the eligible pool by less than 30%. Although these selection criteria make this an above-average group compared with housing project residents, they are not a "highly creamed" group. The program's procedures create a quasi-experimental design. Although all participants come from the same low-income black city neighborhoods (usually public housing projects), some move to middle-income white suburbs, whereas others move to low-income black urban neighborhoods. In principle, participants have choices about where they move, but in actual practice, participants are assigned to city or suburb locations in a quasi-random manner. Apartment availability is unrelated to client interest; it is determined by housing agents who do not deal with clients. As units become available, counselors offer them to clients, according to their position on the waiting list, regardless of clients' locational preference. Although clients can refuse an offer, very few do so because they are unlikely to get another. As a result, participants' preferences for city or suburbs have little to do with where finally move. Because this is virtually the only program that has moved thousands of low-income black families in quasi-random fashion to a wide diversity of suburban and city neighborhoods, it provides a unique opportunity to understand the effects of neighborhoods on residents.

Studies of Gautreaux participants found impressive effects. Compared with youths who moved to city neighborhoods, suburban movers were much more likely to graduate from high school, to attend college, to attend four-year colleges (rather than two-year schools), and if they did not attend college, they were more likely to have jobs and to have better jobs (jobs with benefits and higher wages). The differences were large and statistically significant. Because of the quasi-random design, these results are powerful evidence of neighborhood effects.

The results prompted the Bush administration to initiate a national demonstration program, Moving to Opportunity (MTO), explicitly based on these findings. The residential mobility program has also been a central element in Secretary Cisneros's agenda for HUD, and several commentators have called the Gautreaux program one of the most promising approaches to race relations in the United States today. (SEE ALSO: **Discrimination; Fair Housing Amendments Act of 1988**)

—*James E. Rosenbaum*

Further Reading

Husock, Howard. 1995. "A Critique of Mixed Income Housing: The Problems with Gautreaux." *The Responsive Community* 5(2):34-44.

Peroff, K. A., C. L. Davis, and R. Jones. 1979. *Gautreaux Housing Demonstration: An Evaluation of Its Impact on Participating Households.* Washington, DC: U.S. Department of Housing and Urban Development, Office of Policy Development & Research, Division of Policy Studies.

Popkin, S., J. E. Rosenbaum, and P. Meaden. 1993. "Labor Market Experience of Low-Income Black Women in Middle-Class Suburbs." *Journal of Policy Analysis and Management* 12(3):556-73.

Rosenbaum, J. E., N. Fishman, A. Brett, and P. Meaden. 1993. "Can the Kerner Commission's Housing Strategy Improve Employment, Education, and Social Integration for Low-Income Blacks?" *North Carolina Law Review* 73(5):1519-56.

———. 1995. "Changing the Geography of Opportunity by Expanding Residential Choice: Lessons from the Gautreaux Program." *Housing Policy Debate* 6(1):231-69. ◀

▶ Gentrification

"Gentrification" is the process by which central urban neighborhoods that have undergone disinvestment and economic decline experience a reversal, reinvestment, and the in-migration of a relatively well-off, middle- and upper-middle-class population. As a process, gentrification is highly significant because it contradicts most of the traditional 20th-century North American and European models of urban change, which assumed that urban growth takes place in a geographically outward direction through suburbanization. In contrast to the predictions of these models, gentrification represents urban growth—remaking the urban landscape—at the center. The process is both heralded by urban politicians and planners for its economic effects on city tax bases and criticized by scholars and activists for the displacement of working-class and poor residents that it causes.

The word *gentrification* was coined by Ruth Glass to refer to the residential rehabilitation of housing in London's Islington:

> One by one, many of the working-class quarters of London have been invaded by the middle classes—upper and lower. Shabby, modest mews and cottages—two rooms up and two rooms down—have been taken over, when their leases have expired, and have become elegant and expensive residences. Larger Victorian houses, downgraded in an earlier or recent period—which were used as lodging houses or were otherwise in multiple occupation—have been upgraded once again. . . . Once this process of "gentrification" starts in a district it goes on rapidly until all or most of the original working-class occupiers are displaced and the whole social character of the district is changed. (p. xviii)

Explanations for gentrification have been a source of considerable controversy in the housing literature. The traditional position, widely represented in the public media as well as in the urban social science literature, suggests that gentrification results from altered patterns of urban lifestyle and consumption. Authors generally cite the maturation of the baby boom generation following World War II, the expansion of dual-earner families, the proliferation of single-member and nonfamily households, and a decisive cultural rejection of suburban patterns of consumption as reasons for an altered profile of consumer demand. Less individualistic and more societal versions of this argument ask what social forces produced gentrifiers as a class and make the connection to shifting class structures and employment structures, the emergence of a "new middle class" or a class of "yuppies"—young upwardly mobile professionals.

Alternative theories of gentrification disavow such an exclusive focus on patterns of consumption and stress instead the primary role of capital investment and disinvestment in the land market. Gentrification, according to these theories, is to be explained less by changing patterns of consumption than by a consistent pattern of disinvestment in many older inner-city neighborhoods. This disinvestment may take various forms—delayed maintenance and repairs, redlining of neighborhoods, property tax arrears, denial of needed public and private services, physical and economic abandonment—and it results not only in the devaluation of affected properties but also in the commensurate devaluation of neighborhood land. This devaluation, in turn, leads to a rent gap between the actual ground rent for a neighborhood or parcel of land and the potential ground rent that could be demanded under a higher and better use in a given location. Thus, disinvestment creates the possibility of profitable reinvestment in the form of gentrification.

Gentrification has been the touchstone for a broad range of debates in urbanization and housing, and there have been various attempts to reconcile these production-side and consumption-side explanations. Equally important, however, is the history of the gentrification process itself, which has changed significantly from the narrow residential process observed so vividly by Ruth Glass more than three decades ago. Indeed, many neighborhoods affected by gentrification have historically been working-class residential areas—New York's Lower East Side, London's inner boroughs, Baltimore's Fells Point, Glebe in Sydney, inner Amsterdam, Society Hill in Philadelphia, much of the Left Bank in Paris. But beyond these classical and early examples of gentrification, the process has developed into a much broader and deeper urban phenomenon. If gentrification was indeed a product of the housing market in the early 1960s, by the 1990s, with the intervening economic recession from 1973, the expansion of the 1980s, and the depression after 1989, gentrification has become part of a much larger pattern of urban and economic restructuring. Obsolete industrial and commercial areas at the urban core, long abandoned following disinvestment, have also been subject to systematic redevelopment for luxury residential and recreational purposes in and around the inner city. This trend has been stimulated by the redevelopment of global, national, and regional city centers built on their financial and administrative functions. Far from being an isolated transformation of the housing market, gentrification has become increasingly integrated within wider circuits of global and urban restructuring.

Gentrification is a highly contested process. In the first place, insofar as it leads to higher rents and property taxes, and more generally presents landlords with an incentive to evict tenants, gentrification leads to displacement. Data on the number of people evicted and displaced by private market gentrification are very limited, but by the early 1980s, government surveys estimated that of approximately 3 million people displaced each year in the United States, a substantial minority of these were gentrification related. Those figures rose in the late 1980s, diminished again in the early 1990s with the selective decline in gentrification activity, and rose again with the economic boom of the mid 1990s. Of the people displaced, many became homeless, but again, data on this issue are not readily available. Many gentrifying neighborhoods also experience an increase in squatting as an alternative to gentrification, either as a result of displacement or because of the availability of vacant housing that is not yet gentrified. This may be the case in London and Amsterdam, Christianhaven in Copenhagen, and New York's Lower East Side, and elsewhere, but the consumption of vacant buildings was largely complete by the mid 1990s, and displacement from occupied buildings ensued.

Gentrification, then, is part of a larger spatial restructuring of the city. This connection between gentrification and "deconcentration" has been noted by housing activists and scholars alike. According to Harold Rose, if "the evolving spatial pattern of black residential development is not significantly altered" (p. 139), the next generation of "ghetto centers will essentially be confined to a selected set of suburban ring communities located in metropolitan areas where the central city black population already numbers more than one-quarter million." These are precisely the same suburbs, on the cusp of disinvestment, where in the 1980s and 1990s, recent immigrants from Latin America, Asia, and Eastern Europe have begun to locate. This would seem to suggest an outward diffusion of economic disinvestment, leaving opportunities for reinvestment and gentrification in its wake.

Whether quite such a neat geographical scenario will happen is difficult to say. The economic depression of the late 1980s and early 1990s has seriously affected gentrification in many (but by no means all) cities in North America, Europe, and Australia. If one believes the consumption-side argument and if one believes, further, that the baby boom and other such demographic and cultural effects on demand were an oddity of the 1970s and 1980s, the tendency would be to assume that gentrification would not again be such a powerful force shaping the urban geography of advanced capitalist cities. If, however, one accepts the importance of cyclical patterns of capital investment and especially disinvestment, the depression of the late 1980s and early 1990s can be seen to have created a deeper and broader disinvestment and, hence, to have enhanced the conditions for a new and more intensive round of gentrification. (SEE ALSO: *Condominium Conversion; Displacement; Historic Preservation; Incumbent Upgrading; Speculation;* **Urban Redevelopment**)

—*Neil Smith*

Further Reading

Beauregard, Robert. 1984. "Structure, Agency and Urban Redevelopment." Pp. 51-72 in *Capital, Class and Urban Structure,* edited by M. Smith. Beverly Hills, CA: Sage.

Butler, T. 1997. *Gentrification and the Middle Classes.* Brookfield, MA: Ashgate.

Clark, Eric. 1987. *The Rent Gap and Urban Change: Case Studies in Malmo 1860–1985.* Lund, Sweden: Lund University Press.

Glass, Ruth. 1964. *Aspects of Change.* London: Centre for Urban Studies & MacGibbon & Key.

Hamnett, Chris. 1984. "Gentrification and Residential Location Theory: A Review and Assessment." Pp. 283-319 in *Geography and the Urban Environment, Progress in Research and Applications,* edited by D. Herbert and R. Johnston. Chichester, UK: Wiley.

Lees, Loreta. 1994. "Gentrification in London and New York: An Atlantic Gap?" *Housing Studies* 9(2):199-218.

Ley, David. 1997. *The New Middle Class and the Remaking of the Central City.* Oxford, UK: Oxford University Press.

Rose, Harold M. 1982. "The Future of Black Ghettos." P. 139 in *Cities in the 21st Century.* Urban Affairs Annual Review, Vol. 23, edited by G. Gappert and R. Knight. Beverly Hills, CA: Sage.

Smith, Neil. 1996. *The New Urban Frontier: Gentrification and the New Urban Frontier.* London: Routledge.

Smith, Neil. 1979. "Toward a Theory of Gentrification: A Back to the City Movement by Capital Not People." *Journal of the American Planning association* 45:538-48.

Smith, Neil and Peter Williams, eds. 1986. *Gentrification of the City.* Boston: Allen & Unwin.

Zukin, Sharon. 1987. "Gentrification: Culture and Capital in the Urban Core." *Annual Review of Sociology* 13:129-47. ◄

► Global Strategy for Shelter

In 1988, the U.N. Centre for Human Settlements published its *Global Strategy for Shelter to the Year 2000.* This publication marked a major shift in policy of the United Nations as well as that of other multilateral and bilateral development assistance organizations and housing policy experts. What became the new policy focus and why was there a change?

Before the 1980s, human settlements policies exclusively focused on providing housing for the poor. Most multinational and bilateral aid agencies and national governments concentrated on squatter settlement upgrading and sites and services programs to help the poor obtain housing. According to nearly all accounts, the "housing for the poor," strategy has not worked. Administrative and financial problems thwart the widespread replication of squatter settlement upgrading programs. Sites and services programs fared no better, and there are few examples for which the output of sites and services programs have come close to matching the housing needs of the poor. Problems with replicability and implementation as well as the sheer scale of housing needs of the poor quickly revealed the limitations of the housing for the poor strategy. As Michael Cohen pointed out, by the end of the 1970s, it was evident that the notion of replicability could no longer mean doing more of the same things, but rather, it had to involve seeking new ways to increase the scale in the provision of housing, whether through public or private sector efforts or some new combination of the two.

The Global Strategy for Shelter replaces sites and services and squatter upgrading policies with a totally new policy initiative referred to as the "enabling approach." This new policy directive, articulated by the United Nations, the World Bank, and other multilateral and bilateral agencies concerned with housing, focuses on implementing reforms to improve the overall efficiency and effectiveness of housing markets. Instead of direct interventions to improve squatter housing conditions or provide sites and services, the enabling approach works to revise or eliminate policies or regulations that impede the provision of housing, and it places greater reliance on private and individual initiatives to produce housing. International development assistance agencies have begun to concentrate on providing technical assistance and policy-based lending to developing countries aimed at enabling housing markets to work better. Most efforts try to alter counterproductive policy and institutional environments that constrain housing supply and drive up housing costs.

The provision of infrastructure, land, and housing development controls, building codes, land titling and registration systems, and rules regarding housing finance have significant adverse impacts on housing production and housing prices. In rapidly growing cities, infrastructure services are not expanding fast enough to accommodate housing demand. This is usually due to unrealistic and impractical infrastructure financing and cost recovery policies. By setting service charges too low, most infrastructure service providers do not have the resources to expand and maintain their networks. As a result, informal and unauthorized housing development predominates. If infrastructure services could be expanded in step with housing demand, housing markets would be more efficient and low- and moderate-income households would have better access to housing services.

If land and housing development regulations are overly restrictive, it is difficult for housing supply to respond to housing demand pressures. Unfortunately, most land use planners do not consider the cost and affordability implications of the master plans, zoning ordinances, and subdivision development controls they design. If standards are set too high, low- and moderate-income households are forced into the informal housing sector, where housing development is unregulated. In Thailand, for example, where land and housing development controls are limited, housing is very affordable and informal settlements are decreasing in relative terms. In other countries, such as Trinidad and Tobago, development controls are very strict and housing standards are very high. There, nearly 70% of all housing construction activity is unauthorized because most people simply cannot afford housing built to the high standards.

In sub-Saharan Africa and parts of Latin America, land titling and land registration systems do not function, making it extremely difficult and expensive to obtain land with secure title. Without secure titles, owners cannot obtain financing for housing construction. Courts in Accra, Ghana, are clogged with over 10,000 landownership disputes, and suburban development in city has come to a halt because of ownership uncertainty.

A final problem is the lack of housing finance. Without a well-developed housing finance system, it is difficult for housing markets to work efficiently. Although a few builders will build for the wealthy, low- and moderate-income households will gradually build their own homes using their own labor or small contractors. Without a large-scale housing development industry, the physical provision of housing is inefficient, and it is difficult for housing markets to keep pace with rapidly expanding housing demand.

In response to these policy and institutional impediments, the Global Strategy for Shelter encourages developing country governments to undertake a range of institutional reforms, including the following:

► Clarifying and strengthening property rights related to the ownership and transfer of land and real property
► Developing housing and construction finance institutions to mobilize and disperse capital for housing construction and purchase
► Rationalizing subsidies for housing and infrastructure to make them more transparent, better targeted, and affordable
► Improving the provision of infrastructure by better coordinating service delivery, rationalizing financing and cost recovery and target service delivery to underused and vacant parcels in existing urban areas
► Reforming land and housing development regulations to better balance costs and benefits
► Promoting competition in the housing construction and building materials industry
► Developing a market-oriented institutional framework for implementing enabling reforms that combine public, private, and nongovernmental actors

In conclusion, the enabling approach reflects an important shift in international housing policy. It emphasizes the role of the market, not the government, in housing delivery, and it focuses interventions on policy and institutional reform, not on direct housing production. Although it is too early for a comprehensive assessment of the enabling approach of the Global Strategy for Shelter, preliminary evidence indicates that housing market conditions are improving in cities following the enabling approach. (SEE ALSO: *Habitat: U.N. Conference on Human Settlements;* **Third World Housing;** *U.N. Centre for Human Settlements*)

—*David E. Dowall*

Further Reading

Acquaye, E. & Associates and Kasim Kasanga. 1990. *Institutional/ Legal Requirements for Land Development: Case Study.* Nairobi, Kenya: U.N. Centre for Human Settlements.

Cohen, Michael. 1983. "The Challenge of Replicability: Towards a New Paradigm for Urban Shelter in Developing Countries." *Regional Development Dialogue* 4(1):90-99.

Dowall, David E. 1989. "Bangkok: A Profile of an Efficiently Performing Housing Market." *Urban Studies* 26:327-39.

U.N. Centre for Human Settlements. 1988. *Global Strategy for Shelter to the Year 2000.* Nairobi, Kenya: Author.

World Bank. 1993. *Housing: Enabling Markets to Work.* Washington, DC: Author. ◄

▶ Government National Mortgage Association

With the passage of the Housing and Urban Development Act in 1968, which created the Department of Housing and Urban Development (HUD), Congress created the Government National Mortgage Association (GNMA, more commonly known as "Ginnie Mae"). Its activities were formerly part of the agenda of Fannie Mae (formerly known as Federal National Mortgage Association).

Ginnie Mae was formed to achieve three goals: (a) to manage and sell loans previously held by Fannie Mae, (b) to guarantee government-sponsored mortgage pools, and (c) to assist with the development of certain low-income mortgage assistance programs. A well-known example of the latter program is called the Tandem Plan, where Ginnie Mae makes loans to low-income borrowers and Fannie Mae purchases the loans from Ginnie Mae and resells them in the secondary mortgage market.

A second function for Ginnie Mae has proven to be the most important. Ginnie Mae assists in the packaging, standardizing, and guaranteeing of cash flows from government-sponsored mortgages for investors in the secondary mortgage market. As a result, Ginnie Mae has been essential in the development of the mortgage-backed securities (MBSs) market. The primary instrument is called a "GNMA pass-through" (or more generally, a mortgage pass-through). It enables investors to receive cash flows based on the repayment characteristics of a pool of mortgages. Investors find these instruments attractive for all of the reasons associated with mortgage securitization, but in addition, GNMA pass-throughs have lower default risk due to FHA insurance or VA guarantees. Thus, these securities resemble other government securities in terms of the absence of default risk, but prepayment risk remains.

GNMA pass-throughs account for more than 40% of all mortgage pass-throughs issued during the early 1990s. This amount is almost 15% of the total of all MBSs issued in the secondary mortgage market. On the basis of these rates alone, Ginnie Mae is properly regarded as a major and essential player in the MBS market and thus important in the modern mortgage finance system in the United States. (SEE ALSO: **Federal Government;** *Fannie Mae;* **Mortgage Finance**)

—*Austin J. Jaffe*

Further Reading

Clauretie, Terrence M. and James R. Webb. 1993. *The Theory and Practice of Real Estate Finance.* Fort Worth, TX: Dryden.

Fabozzi, Frank J. and Franco Modigliani. 1992. *Mortgage and Mortgage-Backed Securities Markets.* Boston: Harvard Business School Press.

Lore, Kenneth G. 1987-1988. *Mortgage-Backed Securities: Developments and Trends in the Secondary Mortgage Market.* New York: Clark Boardman.

Sellon, Gordon H., Jr. and Deana VanHahman. 1988. "The Securitization of Housing Finance." Federal Reserve Bank of Kansas City. *Economic Review* (July/August):3-20.

Winger, Alan R. 1987. "The Securitization of Real Estate Finance." Federal Home Loan Bank of Cincinnati. *Review* (August):2-5. ◄

▶ Graduated-Payment Mortgage

The graduated-payment mortgage (GPM) was one of the initial alternative mortgage instruments (AMIs). The GPM was invented to moderate the consequences of the tilt effect phenomenon, which shifts the real burden of debt repayment to the borrower during periods of anticipated high inflation.

First used by the Federal Housing Administration (FHA) in its Section 245 Program, the GPM offers a payment plan in which monthly payments during the first three to five years are lower than they would be under a fixed-rate mortgage (FRM). The specification of the schedule is given in the instrument, usually with a base amount increasing at a constant rate per year or as a percentage of the FRM payment. After the initial period, the outstanding balance is then amortized over the remaining life of the mortgage. The life of the loan is not extended, despite the change in debt service payments at the end of the initial period.

The intent is to alleviate the burden of making an FRM debt service payment when nominal interest rates are high due to anticipated future inflation. Over time, it is expected that income levels will rise with inflation so that after the initial three- to five-year period, the real cost of making a higher debt service payment will be reduced. The concept is to reduce the nominal payment in the beginning and increase the payment after that time.

Whenever the GPM debt service portion of the monthly payment is less than it would be under an FRM (assuming that the interest rate on both mortgages would be the same), negative amortization will occur. Negative amortization means that the borrower owes more interest than he or she has paid in a given month, and rather than principal being set aside to repay the original loan, an additional amount of interest accrues. The negative amortization amount typically becomes part of the outstanding balance of the loan. In these cases, the outstanding balance grows rather than falls during the first several years of the loan.

Lending institutions find negative amortization to be very unattractive because their risk exposure increases over time. Borrowers might not like this result either, but they benefit from having a reduced obligation during the early years of a GPM. In any event, the private mortgage market has not used the GPM very often, although other AMIs (e.g., adjustable-rate mortgage) have been used extensively. In addition, lenders believe that the GPM would be difficult to sell to the secondary mortgage market.

Most observers believe that due to additional default risk associated with the GPM, its interest rate should be higher than on the FRM. This would offset some of the benefit to the borrower who seeks relief during the early periods of the loan when inflation is expected. Although the mechanics of the GPM are not difficult to work out, it remains to be seen whether there will be renewed interest in this AMI in the future. (SEE ALSO: *Alternative Mortgage Instruments;* **Housing Finance; Mortgage Finance**)

—*Austin J. Jaffe*

Further Reading

Brueggeman, William B. and Jeffrey D. Fisher. 1993. *Real Estate Finance and Investments.* 9th ed. Homewood, IL: Irwin.

Fabozzi, Frank J. and Franco Modigliani. 1992. *Mortgage and Mortgage-Backed Securities Markets.* Boston: Harvard Business School Press.

Tucker, Donald. 1975. "The Variable-Rate Graduated-Payment Mortgage." *Real Estate Review* (Spring):5:71-82. ◀

▶ Growing-Equity Mortgage

A growing-equity mortgage (GEM) is a scheme for paying off a fixed-rate mortgage (FRM) more rapidly by having monthly payments increase over time. The initial monthly payment is the same as for an FRM. Increments in the monthly payment are prepayments—they directly amortize the principal—and thus abbreviate the term of the mortgage and the total interest ultimately payable under the mortgage. For consumers who expect their incomes to rise faster than inflation, a GEM offers an attractive scheme of enforced saving.

In its operation, a GEM is like a graduated-payment mortgage (GPM). However, a GPM generally initially involves negative amortization: principal owed actually increases early in the mortgage term because the first few payments are not sufficient to cover interest expense. In a GEM, by contrast, the principal always decreases. To illustrate, suppose a consumer takes out a 25-year term mortgage for $100,000 at an annual uncompounded interest rate of 8.5%. If taken as an FRM, the monthly payment would be $805.23 for each of the 300 months in the term. Over the 300 months, the consumer would incur interest expenses that total $141,566. Suppose instead, the consumer takes this as a GEM with payments that increase monthly at a compounded rate of 0.25% (that is, just over 3% per year). The Month 1 payment would still be $805.23, but by Month 191, when the loan is fully repaid, the monthly payment would have risen to $1,274.81. In all, the consumer would have paid only $88,613 in interest. If, alternatively, the consumer takes this as a GPM (again with a monthly increment of 0.25%), the Month 1 payment would be only $614.80, the Month 300 payment would be $1,297.07, and the total interest paid would be $174,208. With both a GEM and an FRM, principal owing falls steadily over time from its starting value of $100,000; in a GPM, principal owing rises initially, peaking at $103,809 in Month 72. (SEE ALSO: *Alternative Mortgage Instruments*)

—*John R. Miron*

Further Reading

Downs, A. 1985. *The Revolution in Real Estate Finance.* Washington, DC: Brookings Institution.

Fabozzi, F. J. and F. Modigliani. 1992. *Mortgage and Mortgage-backed Securities Markets.* Boston: Harvard Business School Press.

Pozdena, R. J. 1988. *The Modern Economics of Housing: A Guide to Theory and Policy for Finance and Real Estate Professionals.* New York: Quorum Books (see Chapter 8). ◀

▶ Growth Machines

A *growth machine* is a conceptual term for the view that cities are under the control of development interests who use local government as a tool for their growth goals. First coined by sociologist Harvey Molotch, cities as "growth machines" means that those who derive benefit from commercial land use markets—builders, realtors, local financial institutions, and their supporting professionals (real estate attorneys, accountants, civil engineers) are the most potent force at the urban level. Local newspapers, dependent on regional expansion for their own market growth, are also a key constituent of the "growth coalition." As repeated across the national landscape, growth machine dynamics become the important political and moral force shaping the national urban system—what goes on within cities as well as relations between them. Congruent ideas have been put forward by political scientists Clarence Stone (based on Atlanta), John Mollenkopf (based on San Francisco), Todd Swanstrom (based on Cleveland), planner Susan Fainstein (using many cities), and Molotch's collaborator, sociologist John Logan.

Contrasts with Other Schools of Thought

The growth machine concept differs from "power elite" theory, with which it is sometimes associated, because it stipulates that only a segment of the business elite, those with interest in local development per se (as opposed, for example, to corporate owners and managers) play the important local political roles. Although growth elites may act in concert with other business groups and, indeed, help "prepare the ground" for their land use needs, there is a distinction between those who do business in a particular place and those who make money from the manipulation of place itself. This sets up at least the possibility of conflict within the business community as to how local land use should be managed.

The growth machine concept differs from the views of most economists who tend to view any form of development as intrinsically good for the locality, at least those emerging from the private market. But using growth machine theory, the consequences of development are suspect because of the way market conditions are structured under elite domination. For growth machine theorists, unless localities capture the true costs of development, as carefully measured, any given project is a net loss and the locality would be better off without it. Development is not necessarily a good, even in the terms most often represented as its primary advantages—easing problems of unemployment, high housing costs, or fiscal crises. Researchers have found that cities that are larger, denser, or have higher rates of growth do not outperform other cities on standard indicators of public well-being.

Another distinguishing attribute of growth machine theory is its insistence on questioning how specific groups will be differently affected by a given project or by growth policies overall. Some analysts treat a city's rank in population, aggregate production, or volume of construction as proxy for benefit. But again, effects on diverse urban groups need to be traced; growth machine theory invites investigation of just how one sector of the population may be manipulating institutions and geographic outcomes in ways that harm others, sometimes even the majority.

Affordable Housing

The growth machine model holds that those who use the city to make money have little inducement to provide housing for low-income residents or service their other needs. Such housing is not lucrative for developers and, more profoundly, may encourage the residence of people whose lifestyles and low buying power may dampen other growth prospects. That is one reason why, through this view, U.S. cities have done a relatively poor job of providing low-cost housing. Indeed, evidence suggests that the localities less beholden to growth coalition domination, as evidenced by their tougher environmental regulations, do a better job providing affordable housing than other types of places.

There are two possible reasons for this. First, places that have tight controls tend to allow density bonuses for affordable housing—a provision that is effective precisely because restrictions are otherwise in place. The second reason is that, also common in growth control cities, permission for market rate projects is made contingent on contributions to the affordable housing stock—again, effective only because otherwise, developers would not get access to their most lucrative sites. In terms of market housing, studies of the effects of strict growth controls on prices indicate no tendency for cities with tighter regulations to experience higher housing price inflation as a result. Such findings strengthen the conviction that the concentration of urban power in the hands of development interests does not serve the poor. Rather than aligned with NIMBY-ism (a charge sometimes made), growth machine theorists see a climate of overall restriction as a route toward inclusion rather than exclusion.

Competition across Cities

One of the consequences of cities functioning as growth machines is that local policy becomes guided not by the search for ways to enhance the life of residents but, rather, for ways to increase development. This yields a competition between places for the limited amount of growth that can take place at any given time. As part of this competition, localities often bid against one another, offering special inducements such as tax breaks, fee waivers, or relaxation of environmental standards. Such inducements are a waste if the project would have come even without them. Local subsidies are often irrelevant because they form so small a portion of investors' operating costs or because they pale compared with other considerations, such as access to a preferred labor supply or customer base.

Sometimes the local problem is so great that no amount of subsidy will stimulate a successful outcome. Some very large undertakings, including those using "private-public partnerships" have proven wasteful failures. Flint, Michigan's downtown rejuvenation project, based on making a tourist attraction out of the declining city's role as a former auto capital, is one of the better known instances. Such projects result in terrible losses to the cities that subsidize

them as well as to the hapless nonprofit foundations that may hold a stake. Other ambitious projects, widely hailed as unquestionable successes, such Baltimore's Harbor Place or Milwaukee's Grand Avenue, do little for most of their cities' neighborhoods, which census data reveal, continue to decline at a faster rate than their centers manage to rejuvenate.

The biggest problem, at the national level, is that the whole system generates no net benefits across cities and counties. The contest is for the *distribution* of the putative benefits from growth rather than for an increase in those benefits for the United States as a whole. It is not an efficient method for generating investment in human skills or even economic infrastructure, much less enhancing the quality of life or of ecological sustainability. Cities compete with other cities by lowering rather than raising the standards for wages, the safety net under the unemployed, or levels of environmental protections.

Federal and most state governments have encouraged the growth machine system through redevelopment (by exempting redevelopment bonds from taxation, for example), fostering enterprise zones, and limiting alternative revenue sources that could help localities deal with their social and physical problems without chasing the putative solution of more development.

United States Compared with Other Countries

The U.S. urban system is especially conducive to development through growth machines. Local land use autonomy, strong in the United States compared with other countries, remains one of the few arenas of meaningful "home rule." This local autonomy has a double-edged sword. At the same time that localities have been free to control land use, they have also been significantly responsible for funding local services and coming up with the money to build and maintain physical infrastructure. This makes the provision of local services, at least plausibly, dependent on local growth. Whether or not growth pays for itself is another matter, but using development to enhance services at least *appears* plausible. In parts of Europe, by contrast, there has been no such link because services are supported by the central state, with less regard to where a citizen happens to live.

Another distinctive feature of the U.S. urban scene has been the relatively free-market status of land and buildings. Urban land is largely held privately. Real estate can be bought and sold, razed and manipulated, as entrepreneurial strategies dictate. Most industrial societies have tended to put greater amounts of city land under public ownership and required that various cultural, welfare, or technocratic goals be met as part of the use of urban space. The U.S. conditions foster an activist growth elite who can use local government as a primary tool for entrepreneurial effort. Indeed, fortunes are made or lost by the kind of zoning that can be gained and the way public investments (parks, roads, water) can be structured. Free to buy, sell, and build, members of the growth coalition invest heavily in political campaigns to elect politicians sympathetic to their goals—a process that increases their influence on all manner of local issues.

By the late 1980s, many societies were becoming more like the United States on these issues. The Thatcher government in Britain privatized land use in new ways and decentralized some aspects of government. Even in the welfare states of Scandinavia and The Netherlands, pressures mounted for greater reign for the free market in land and buildings. European integration spurs place competition as each country promotes its cities as development sites at the expense of other countries and as the former Soviet republics join the chase. Mexico and other less developed societies are retreating from the kinds of land reform that limited who could own property and how it could be bought and sold. The United States may be exporting its growth machine system to the world.

Criticisms of the Growth Machine Perspective

For followers of rival schools of urban political economy, the forces that determine which places grow are too powerful to be much influenced by parochial elites, no matter how dominant their local power. Changes in markets and productions systems, now global in scale, have their own logics that overpower local dynamics. It may even be an overstatement that growth elites inevitably gain control of local government. Although development groups may have influence, the success of environmental movements in various parts of the United States indicates a countervailing force. Environmentalists in cities such as Boulder, Colorado, for example, have found fertile cultural soil in which to virtually take over local government. The power base of U.S. cities and counties may be more open and participatory than that implied under growth machine thinking as evidenced by the give-and-take of the regulatory apparatus in many places.

Whatever failings the current U.S. system of urban development may possess, it has resulted in both robust economic growth (at least over the long haul) and specific instances of environmental innovations. Growth machine theory tends to be critique, providing little by way of substantive information on how growth can best be accommodated, either in terms of environmental protection or increasing economic productivity. On the other hand, it is a perspective that has explained a great deal of what goes on in an increasing number of the world's places and sets a challenge for competing explanations of the urban system. (SEE ALSO: *Growth Management; Speculation*)

—*Harvey Molotch*

Further Reading

Appelbaum, Richard P. 1978. *Size, Growth, and U.S. Cities*. New York: Praeger.

Broadbent, Jeffrey. 1990. "Strategies and Structural Contradictions: Growth Coalition Politics in Japan." *American Sociological Review* 54(5):707-21.

Fainstein, Susan. 1986. *Restructuring the City: The Political Economy of Urban Redevelopment*. New York: Longman.

Fasenfest, D. 1986. "Community Politics and Urban Redevelopment: Poletown, Detroit, and General Motors." *Urban Affairs Quarterly* 22(1):101-23.

Logan, John R. and Harvey Molotch. 1987. *Urban Fortunes: The Political Economy of Place*. Berkeley: University of California Press.

Mollenkopf, John. 1983. *The Contested City*. Princeton, NJ: Princeton University Press.

Molotch, Harvey. 1976. "The City as a Growth Machine." *American Journal of Sociology* 82(2):309-30.

Squires, George D., ed. 1989. *Unequal Partnerships: The Political Economy of Urban Redevelopment in Postwar America*. New Brunswick, NJ: Rutgers University Press.

Stone, Clarence N. 1989. *Regime Politics: Governing Atlanta, 1946-1988*. Lawrence: University Press of Kansas.

Swanstrom, Todd. 1985. *The Crisis of Growth Politics: Cleveland, Kucinich, and the Challenge of Urban Populism*. Philadelphia: Temple University Press. ◄

► Growth Management

Growth management programs help rural, suburban, and urban communities achieve their land use goals and objectives. In recent years, typical growth management objectives have been to slow development until the necessary public services become available and to avoid the unnecessary costs that often accompany new growth. In particular, communities adopt policies that they anticipate will minimize the negative environmental, social, and fiscal impacts of growth. Less common growth management systems are those designed to encourage beneficial industrial, commercial, and residential development.

Examples of specific growth management policies include conventional zoning and subdivision regulations, annexation controls, urban limit lines or greenbelts, and infrastructure service and timing requirements. Sometimes a distinction is made between growth management policies and growth controls, the latter being more narrowly defined as policies designed to limit population growth, housing construction, economic growth, or all three. Growth controls include measures such as population growth caps, residential building permit caps, and development moratoria. Growth controls are generally considered to be part of the broader class of growth management policies; however, because the vast majority of growth management policies include growth control measures, the two phrases are often used interchangeably. In the 1950s and 1960s, most communities in the United States vigorously promoted growth. This progrowth attitude changed in the 1970s as concerns about the environmental conditions grew and as baby boom children became young adults, dramatically increasing the demand for housing. Increasingly, small suburban "bedroom" communities in rapidly growing areas (typically, communities with relatively high-income, well-educated residents) began implementing policies to limit or control their residential growth. Growth controls were expected to reduce noise and air pollution, traffic congestion, and damage to ecologically sensitive areas and, generally, to preserve the quality of life of a community. Thus, environmental groups were key supporters of these measures. The trend was most evident in California where over 100 communities adopted some form of growth control and where some city governments (e.g., San Diego, San Francisco, and Los Angeles) passed legislation limiting the rate of development. However, growth management programs were not limited to California but were implemented in all regions of the United States. A recent study identified 567 local governments that had some form of growth control.

The growth management policies of the early 1970s usually included some type of explicit or implicit restriction on the rate or quality of residential housing construction. The implementation of these measures inevitably involved significant externalities or spillover effects; that is, the restrictions often resulted in development's being pushed into the surrounding unrestricted communities. Thus, it was argued that the problems were exported rather than solved. And because the restrictions often resulted in people moving farther away from employment centers, growth controls sometimes caused an exacerbation of problems such as traffic congestion and air pollution.

Not surprisingly, these restrictive growth measures were often opposed by builders, developers, and property owners who were used to and who personally benefited from a laissez-faire system. A frequent criticism of these early growth management policies was their exclusionary nature. The restrictions on housing supply generally increased housing prices and rents, disproportionately affecting minorities. In addition, growth controls were found to change housing characteristics (e.g., more larger houses built), making housing less affordable to low-income families. This exclusionary effect raised the issue of the legality of controls. However, the growth control ordinances stood up to the legal challenges put forth by property owners, builders, and developers.

In the 1980s and 1990s, communities implemented more comprehensive and sophisticated growth management policies. These new measures, which included development or impact fees and permit allocation systems, typically linked the rates of commercial, industrial, and residential development to the provision of key public facilities (e.g., schools, parks, and libraries) and regional infrastructure (e.g., sewage and waste disposal systems). Often accompanying these new measures was some type of provision to promote affordable housing. These provisions included giving preference in the building permit queue to low-income housing, exempting affordable housing from the restrictions, or specifying that a certain percentage of all new housing should be for low-income individuals. These more comprehensive growth ordinances are generally supported by a wider range of community groups than were the earlier policies. In Boulder, Colorado, for example, numerous groups, including the local Board of Realtors and the Chamber of Commerce support the current comprehensive growth management policies.

Growth management initiatives are constantly being modified and altered as the particular circumstances of the communities and states change. Recently, some states experiencing growth pressures have passed legislation requiring communities to provide plans showing where growth is anticipated and how the communities will provide the necessary services. Florida, New Jersey, and Oregon are examples of states that have experimented with state-level mandates.

There is no consensus within the planning profession as to which methods of growth management are the most effective. Numerous studies have been conducted on the effects of these policies, particularly the effects on housing quantity and quality and their impact on the poor. What researchers have found is that the price effects of growth control are quite varied. For example, in rapidly growing metropolitan areas where growth control measures are ubiquitous and where controls may grant monopoly power to builders, average housing prices may be as much as 35% to 40% higher than similar communities without controls. In communities located in metropolitan areas where growth controls are less prevalent, growth controls may have little or no effect on housing prices, but they increase building activity in surrounding communities. Most growth control communities lie between these two extremes, with typical housing price effects between 10% and 20%. It has been found that even if growth controls are not actually in place, if the city planners favor such controls or if residents anticipate the implementation of such controls, prices may rise in anticipation of controls.

In the future, researchers need to study not just the effects of growth management policies on housing but also their effects on the rate of job creation and on the negative externalities and congestion costs associated with growth (e.g., the levels of traffic congestion and air pollution). (SEE ALSO: *Environment and Housing; Growth Machines*)

—*Jane H. Lillydahl*

Further Reading

Brower, David, David R. Godschalk, and Douglas Porter. 1989. *Understanding Growth Management: Critical Issues and a Research Agenda*. Washington, DC: Urban Land Institute.

Dowall, D. 1982. "An Examination of Population-Growth-Managing Communities." In *Environmental Policy Implementation*, edited by D. Mann. Lexington, MA: Lexington Books.

Landis, John D. 1992. "Do Growth Controls Work? A New Assessment." *Journal of the American Planning Association* 58(4):489-508.

Lillydahl, Jane H. and Larry Singell. 1987. "The Effects of Growth Management on the Housing Market: A Review of the Theoretical and Empirical Evidence." *Journal of Urban Affairs* 9:63-77.

Navarro, Peter and Richard Carson. 1991. "Growth Controls: Policy Analysis for the Second Generation." *Policy Sciences* 24:127-52.

H

▶ Habitat for Humanity International

Habitat for Humanity International is an ecumenical, Christian organization that seeks to eliminate substandard, poverty housing through the construction of low-cost homes and the rehabilitation of existing ones. Habitat for Humanity programs are joint ventures that involve residents and community volunteers in the building and renovation process. Former President Jimmy Carter's participation helped provide publicity for these efforts. New houses are sold at no profit to partner families, and no-interest mortgages are issued over a fixed period. Small monthly mortgage payments, taxes, and insurance are repaid over a 7- to 20-year period and deposited into a revolving fund that supports the construction of new houses. Since its founding in 1976 by Millard and Linda Fuller, Habitat for Humanity has constructed more than 60,000 homes worldwide. Currently, there are more than 1,393 affiliated projects in the United States and 302 affiliates abroad.

In addition to its building projects, Habitat for Humanity also coordinates services for the homeless, sponsors educational programs for schools and the general public, and distributes *Habitat World,* a free bimonthly newsletter that reports project news and information on affordable housing.

An ecumenical committee of 27 people and a staff of 301 paid employees manage the organization. The organization's programs depend primarily on volunteer labor, and local volunteer boards oversee individual projects. Habitat for Humanity operates on an annual budget of more than $71 million and is funded through grants and private donations of money and materials; it accepts government funds only for the acquisition of streets, utilities, and land and for property rehabilitations. Contact person: William Mecke, Director of Marketing. Address: 121 Habitat Street, Americus, GA 31709. Phone: (912) 924-6935. Fax: (912) 924-6541. E-mail: Public_Info@Habitat.org. World Wide Web: http://www.Habitat.org. (SEE ALSO: *Habitat World*)

—*Caroline Nagel* ◀

▶ Habitat International

First published in 1976, *Habitat International* is a quarterly journal on research into urban issues in a development context. It addresses aspects of urban policy and implementation, the links between planning, building, and land finance and management; the provision of urban services; and other related problems. Its audience is made up of individuals concerned with the study, design, production, and management of human settlements. The annual subscription price for this internationally circulated journal is $745. The editor welcomes manuscripts that stimulate debate to discover solutions to these outstanding issues. Editor: Charles L. Choguill. Address: Department of Landscape, Environment and Planning, Royal Melbourne Institute of Technology, GPO Box 2476V, Melbourne, Victoria 3001, Australia. Fax: 613 9525 4908. (SEE ALSO: **Third World Housing**)

—*Laurel Phoenix* ◀

▶ Habitat International Coalition

Habitat International Coalition (HIC) is a broad-based independent alliance of more than 350 nongovernmental and community-based organizations working in the field of housing and human settlements in nearly 80 countries. HIC has its roots in nongovernmental organizations (NGOs) at the first U.N. Conference on Human Settlements, Habitat, held in 1976. The HIC international secretariat is based in Mexico City; focal point organizations in Asia, Latin America, West Africa, East Africa, Europe, and North America coordinate the HIC networks in their respective regions.

The primary focus of HIC is creating conditions that lead to the implementation of the right to a place to live in peace and dignity. HIC's most intense activities deal with the following:

1. Advancing housing rights
2. Opposing forced evictions

3. Promoting the gender perspective within shelter issues
4. Building alliances with grassroots and community-based organizations
5. Promoting community-based solutions to the global housing crisis
6. Linking housing and environmental issues and promoting sustainability in human settlements development
7. Advancing community-based housing finance solutions

Coalition members share the conviction that housing access problems are essentially structural in nature, linked to issues of domination by either the market or the state. Thus, the problem is not simply one of individuals' or communities' lacking adequate housing, which can be solved by traditional policy-oriented or project approaches. Rather, and in recognition of the fact that the majority of housing in the world is produced by the people themselves, HIC believes that solutions lie more in creating the necessary social, economic, political, and cultural conditions through which people can demand, gain, and retain adequate housing and living conditions.

As a coalition of existing local, national, regional, and international organizations, the bulk of HIC's work is geared toward building a unified global movement in support of people-based and people-directed measures devoted to ensuring for everyone in the shortest possible time the right to a safe and secure place to live. A discussion of some of the main global HIC programs follows.

Global Campaign for Housing Rights and Against Evictions
Since 1989, HIC has been very active in campaigning for the human right to adequate housing. The campaign has been carried out locally as well as internationally within institutions such as the U.N. Commission on Human Rights and other legal bodies. The HIC Housing Rights Subcommittee coordinates the campaign.

As part of the campaign, HIC carries out fact-finding missions to various countries to examine housing rights violations and other problems as well as to urge governments to fulfill their obligations with respect to housing. HIC has carried out missions to the Dominican Republic, Panama, Nicaragua, Hong Kong, Israel, and Palestine, and to the cities of Calcutta, Seoul, and Rio de Janeiro.

Some of HIC's most influential work has been accomplished within the U.N. human rights arena. Since the early stages of the campaign, HIC has been able to secure the adoption of numerous U.N. resolutions on housing rights issues and forced evictions, the appointment of a U.N. Special Rapporteur on Housing Rights, and the official condemnation of several governments because of the existence of housing rights violations. HIC has facilitated the direct participation of local NGOs and grassroots groups from Colombia, India, Mexico, Canada, Italy, Belgium, the Philippines, Panama, Brazil, and other countries within the United Nations, and it welcomes such involvement from even more countries.

Parallel to the housing rights campaign, HIC is also involved in various efforts against the widespread and often violent practice of forced evictions. In addition to monitoring the prevalence of evictions throughout the world, HIC has actively sought to pressure governments notorious for their lax attitude toward the involuntary removal of persons from their homes and to reconsider pending eviction plans. Within the United Nations, delegations representing HIC have been able to obtain strong denunciations against various governments propagating evictions, such as the Dominican Republic and Panama.

The Women and Shelter Network
Women play the singly most important functions within both the home and their communities, but they rarely have equal rights or access to land or dwellings. For that reason, the Women and Shelter Network was established to provide women with a forum for sharing information and for international campaigning for their equal right to housing. The network, which operates independently, carries out a variety of activities, including the publication of a newsletter and the mobilization of grassroots organizations.

Habitat and Environment
The HIC Habitat and Environment program was established during the preparations for the U.N. Conference on Environment and Development, the Earth Summit, held in 1992, to emphasize the critical relationships between sustainable and equitable human settlements development and growing world environmental concerns. The program has sponsored international research on, for example, the development and implementation of sustainable and community-controlled technologies for the provision of drinking water and sanitation services in urban settlements. The HIC Habitat and Environment Subcommittee maintains a presence within the U.N. Commission on Sustainable Development, works to further develop synergetic links with important environmental organizations, and publishes a newsletter for the dissemination of its platform and work.

Publications
HIC publishes a variety of materials, including *HIC News*, to inform its members of recent developments in the issues it works for and to generate increased attention to housing problems. HIC's affiliate, the Centre on Housing Rights and Evictions (COHRE), publishes some 10 documents annually, designed to empower local groups and communities in effectively invoking and using housing rights and other legal resources at the local and national levels.

Membership in HIC is open to all nonprofit community-based organizations, NGOs, voluntary agencies, and research, scientific, and educational institutions whose activities deal with housing, human settlements, or human rights. Individuals may also join HIC. To receive further information, contact: HIC Secretariat, Habitat International Coalition/ Cordobanes 24/ Col. San Jose Insurgentes/ Mexico D.F. 03900. Phone: (525) 651 6807. Fax: (525) 593 5194. (SEE ALSO: *Right to Housing*)

—Scott Leckie ◄

► Habitat: U.N. Conference on Human Settlements (Vancouver Conference)

In 1976, the United Nations convened its first major Conference on Human Settlements. The agenda's conference was twofold: (a) to discuss and adopt a plan of action for the international community and for member governments and (b) to decide on the appropriate institutional arrangements to make the U.N. system more responsive to the challenge posed by human settlements development, particularly in developing countries.

The plan of action adopted by the conference consisted of a Declaration of Principle and of 64 recommendations for national action. The recommendations covered six major areas:

1. Settlement policies and strategies
2. Settlement planning
3. Shelter, infrastructure, and services
4. Land
5. Public participation
6. Institutions and management

Under these headings, individual recommendations referred to a very broad range of issues that can be traced in the Human Settlements Programme of the United Nations and in specific programs launched and developed by the U.N. Centre for Human Settlements (Habitat) and by its international partners in human settlements development.

The institutional arrangements outlined by the conference resulted in the consolidation of various existing units and activities of the secretariat into the U.N. Centre for Human Settlements (Habitat), headquartered in Nairobi, Kenya, to act as the focal point for human settlements within the U.N. system, as the secretariat of the U.N. Commission on Human Settlements, and as the executor of an integrated program, including research, technical cooperation, and information dissemination. (SEE ALSO: *Global Strategy for Shelter;* **Third World Housing**)

—*Iouri Moisseev* ◄

► *Habitat World*

Habitat World is the educational, informational, and outreach publication of Habitat for Humanity International, a nonprofit, ecumenical Christian organization dedicated to eliminating poverty housing worldwide. Habitat works in partnership with people in need to build simple, decent shelter that is sold at no profit, through no-interest loans. Habitat has built more than 60,000 homes worldwide and is in all 50 states and 46 countries. First published in 1984, *Habitat World* has a circulation of 1.5 million worldwide. Articles are written by staff writers, freelancers, and affiliates of Habitat for Humanity. Editor: Milana McLead, 121 Habitat Street, Americus, GA 31709. Phone: (912) 928-6935. Fax: (912) 928-4157.

—*Caroline Nagel* ◄

► Halfway House

A halfway house is a residential facility programmatically designed to assist individuals to either live in society or re-enter the community. Typically, these individuals come from a more institutional living arrangement, such as a mental health hospital or prison. Occasionally, as emergency shelter providers become overloaded with homeless people, halfway houses may offer shelter to someone directly from the streets. Some halfway houses are operationally designed for long-term care of residents, whereas others may have the goal of assisting people re-entering the community. In the latter case, the halfway house will have time limits on the length of stay. Halfway houses are intended to be located in residential areas.

The main purpose of a halfway house is to create a family-style living arrangement to facilitate the process of "normalization" as part of deinstitutionalization. *Deinstitutionalization* is the policy of state mental health providers to release clients and not to accept individuals who do not pose a threat to themselves or others. With the introduction of Thorazine in the late 1950s, the courts' recognition of liberty interests of mentally ill and mentally retarded people in the early 1960s, and the consensus emerging in the 1960s in the psychiatric community that mental institutions perpetuated the illnesses they were designed to cure, many mentally ill people have since either been released from or not admitted to state mental health hospitals. This greatly increased the demand for halfway houses. Before reentry into the community, deinstitutionalized people need to learn "normalcy," and the best way to achieve this is through a supportive, family-style living arrangement in a residential setting. Normalcy generally means the ability to take care of oneself without posing a threat to oneself or others. Halfway houses are usually staffed 24 hours a day with at least one professional caretaker. Ideally, halfway houses shelter about six individuals. Housing and fire codes generally limit the number of unrelated individuals who can reside together in a residentially zoned property to less than 12, depending on the building structure and the local ordinance. The halfway house represents the point halfway between the institution and an independent living arrangement. Funding comes from a variety of sources—state reimbursement, donations, and income of the residents.

Many state and local government agencies and many nonprofit organizations now run successful halfway houses, in both residential and nonresidential neighborhoods. There is no national estimate of the total number of halfway houses. Although very few communities are without at least one, local resistance remains common.

Often, communities and neighborhoods resist the siting of a halfway house (the so called NIMBY syndrome—"Not in My Backyard"), generally citing concerns about property values. A nationwide study performed by the Washington, D.C., Department of Corrections indicated a decrease in the property values of nearby houses after the opening of a halfway house for parolees. Another study performed by the Green Bay Planning Commission in Green Bay, Wisconsin, found no diminution in value in the purchase price or assessed value of residential property within a two-block radius of a halfway house over a four-year period of operation. The actual evidence of increase or decrease of value of surrounding residential, commercial, or industrial is mixed because it is difficult to ascertain a baseline of property value. Also, the type of halfway house varies greatly. A question in this regard arises as to how much of a "family" the residents are. Courts usually give broad deference to legislative definitions of family. The main issue is the authority of a locally elected body to determine the living arrangements of its citizens in terms of public health, safety, and welfare (*Belle Terre v. Boraas*, 416 U.S. 1 [1974]). However, the 1988 Amendments to the Fair Housing Act prohibit discrimination against families with children and may open the way for more halfway houses. (SEE ALSO: *Psychiatric Disabilities, Housing of Persons With*)

—*Robert W. Collin*

Further Reading

Collin, Robert. 1992. *Homelessness in the United States: 1980-1990. Journal of Planning Literature* 22(1):22-37.

Hope, Marjorie and James Young. 1984. "From Back Wards to Back Alleys: Deinstitutionalization and the Homeless." *Urban and Social Change Review* 7(2):16-24.

Okolo, Cynthia and Samuel Guskin. 1984. *Community Attitudes toward Community Placement of Mentally Retarded Persons*. Bloomington, IN: Academic Press.

Seltzer, Marsha Mailick. 1984. "Correlates of Community Opposition to Community Residences for Mentally Retarded Persons." *American Journal of Mental Deficiency* 89:1-8.

Smith, Thomas P. 1989. "Saying Yes to Group Homes." *Planning* 55(12):24-25. ◀

▶ Health

Florence Nightingale believed that the association between the quality of people's housing and their health was "one of the most important that exists." Certainly, when we think about the major social reforms of the Victorian era, including the clearing of slum housing, the engineering of proper sewers, and the introduction of clean domestic water supplies, we have no difficulty in understanding the huge impact that these had on people's health. But at present, many of us find it harder to appreciate that a link still exists between housing and health. Perhaps we underestimate it because many of today's problems are not as gross as those of former years. Perhaps the linkage to housing seems small because we now put so much emphasis on other determinants of health—smoking, drinking, and even genetics. Perhaps it is because we have come to expect high standards of proof when looking at the causes of ill health, and firm scientific evidence is often scarce when it comes to housing. So just how strong is the evidence that our homes affect our health, and in what ways do they do so? The following discussion focuses on Western countries and does not cover the often more serious housing-related health problems found in the Third World.

Definitional and Conceptual Questions

It is surprisingly difficult to prove that there is a link between housing and health. For a start, what to study is hard to decide. *Housing* has no simple definition. To most of us, our house is the "bricks and mortar" of the dwelling in which we live. But depending on the health effect we want to study, we may need to use a much wider definition. If we are interested in the possible effects of housing on chest disease, we need to think of the structure of the building: Is it warm, dry, and well ventilated? We also need to look at the way in which the building is furnished: Is the heating system generating irritant waste gases? Is formaldehyde being released from foam-filled furniture? If we want to study the effects of housing on mental health, we may need to take an even wider view: Is the house close to shops and other amenities? How much social support is there in the area? Is there a safe play area for the children? By choosing an inappropriate definition of housing, we may miss very real effects.

Of course, the definition of *health* is also important. The World Health Organization defines health as "a state of total mental, physical, and social well-being," but there is no practical way in which this could be measured. Any scientific study must take a more restricted view, and this can cause problems when the results are interpreted. The numerous studies into possible effects of housing on children's respiratory health illustrate the difficulty. Some have used recorded consultations with doctors as the measure of ill health. But not all people consult a doctor every time they feel ill, not all medical records are reliable, and people who suspect that their poor-quality housing is affecting their child's health may consult doctors more in the hope of being helped to find different housing than because they have a clear medical need. But if scientists try to use a more objective measure—perhaps by testing children's respiratory function in a laboratory—they may miss important problems that are present only intermittently or only when the child is at home.

Another major barrier to the study of housing and health is the difficulty of teasing out the relative contributions of the many factors that operate together. For example, we know that children who live in cold, damp houses are more likely to have asthma than those from warm, dry homes. But they are also much more likely to come from low-income families. We also know that adults on low incomes are more likely to smoke and that childhood asthma is strongly linked to passive smoking. Proving that the cold damp home rather than the parental smoking is causing the child's problems is difficult.

Cause or Effect?

Another problem in studying housing and health is trying to separate cause from effect. Many studies have shown that people living in poor housing are more likely to suffer ill health than those in good homes. It is tempting to conclude that the housing problems have made the residents ill. But in many countries today, good housing is a fairly scarce commodity, and like anything in short supply, it is most likely to be taken by the fit, the articulate, the advantaged, or those with good advocates. Ill people may not be well placed to compete for the limited number of good homes. Their poor-quality homes may be the result of their ill health rather than its cause.

One way of trying to unravel these issues is to attempt formal studies of the effects of rehousing. Several U.S. studies have shown that people's reported health improves after they have been rehoused in slum clearance programs. This has led some workers to conclude that the slum housing was making people ill. But others have pointed out that people who want to be provided different housing may exaggerate their health problems in the hope of speeding up the process. Any subsequent improvement in their health may, therefore, be more apparent than real.

Strong Evidence

Despite the problems of making a scientific study of housing and health, some of the links are so strong or have been studied so extensively that most people accept them as proven. Perhaps the strongest scientific evidence concerns the effects of cold, damp conditions. It is well established that certain medical conditions are more common in people living in cold homes. The incidence of strokes and heart attacks increases in old people as their living room temperature drops below about 16 °C. At very low temperatures, they are also at risk of hypothermia.

Chest complaints are also strongly correlated with living in cold homes, but here, the main link seems to be with damp. A cold home is usually also a damp one, and damp houses encourage the growth of molds that release spores known to cause lung diseases, especially asthma. But surprisingly, asthma is also common in people living in very warm homes. In many Scandinavian countries, homes are built to be warm and energy efficient, with tightly fitting doors and windows and efficient recirculation of warmed air. People living in such homes have been found to be at increased risk of chest complaints. The warm, dry atmosphere and poor ventilation encourage the growth of various organisms, including house dust mites, a major cause of asthma. Indeed, the ways in which the housing environment can affect the lungs are numerous. Cold, damp, mold, mites, formaldehyde released from preformed foam furniture, carbon monoxide leaks from faulty heating systems, radon, asbestos released during renovation, and smoke and other pollutants can all harm the residents. Trying to untangle their relative contributions can be a nightmare, but the evidence is gradually accumulating, and in Britain, there have been several successful court cases in which residents have won compensation from their landlords after developing chest problems while living in badly maintained, cold, and damp homes.

Safe Houses

Perhaps the most common way in which housing affects health is through domestic accidents. We like to think of homes as safe places, but in Britain, nearly two-fifths of all fatal accidents occur in people's own homes, and domestic accidents cost the National Health Service more than £300 million a year. Many of these accidents could be prevented, and safety is increasingly being "built into" new homes. Domestic accidents are also strongly correlated with factors such as poverty, low social class, overcrowding, and the like. Substandard housing is at greatly increased risk from fire—usually secondary to electrical faults. Escape from grossly overcrowded buildings can be very difficult. Countless lives could be saved each year if every home was fitted with smoke alarms to give residents enough warning to be able to escape.

Other obvious hazards around the home include badly lit stairwells, nonsafety glass in doors and windows, easily opened window catches, and unprotected balconies. And although domestic accidents are more common in poor families, the rich are also at risk—one of the commonest places for children to drown is in uncovered domestic swimming pools. Fortunately, many of the accidents that occur in houses can be predicted, and realistic steps to prevent them are often possible. Domestic accidents are most likely to affect the very old and young children, and this too can help to predict where changes should be made. Many of the effects of housing on health are subtle and difficult to spot, so it is a particular tragedy that in this area, where the effects are obvious and well researched, so many people continue to be injured or killed each year.

Homelessness

No review of the health effects of housing could ignore the most serious issue of all—the effects of homelessness. Throughout the developed world, homelessness is becoming an increasingly obvious problem, and most major cities now have a very visible street population. For people sleeping on the streets, the health problems are obvious. Cold, hunger, and dirt increase the chances of serious illness. Illnesses that would be fairly trivial for other people can develop into major problems. Simple cuts and scratches rapidly become infected and are difficult to heal. Chest and gastrointestinal infections are common and serious. Malnutrition can develop. Some conditions that were familiar two generations ago but had almost disappeared are now reemerging among the homeless. Perhaps the most publicized is the rapid spread of tuberculosis. This is made worse by the fact that homeless people may find it impossible to follow a complicated treatment program and default on treatment, therefore contributing to the emergence of resistant strains of the disease.

If street people have preexisting medical problems, they face additional hardship. How can people with diabetes control the condition and avoid complications if they have no access to a proper diet, nowhere to store their insulin, and no adequate means of keeping clean? How likely is it that people with schizophrenia will comply with their medication when they also have to cope with the problems of sleeping on the streets? How can people be rehabilitated after a stroke or heart attack if they have no home?

But the problems of homelessness go beyond this obvious group of street people. Increasingly, researchers are recognizing a large group of "hidden homeless" people who do not appear on any official statistics and who may not be easy to identify but who have very specific health needs and problems. They often live in grossly overcrowded conditions—perhaps camping out with relatives—or in houses that are really unfit for human habitation. Infections can spread rapidly in such conditions. A proper diet may be impossible if there is inadequate access to cooking and food storage facilities. Accidents, including scaldings and burns, are more likely in overcrowded rooms. Surveys of homeless women placed in temporary accommodation in London have found that they are three times as likely to need hospital admission during pregnancy as other women, and a quarter of their babies are of low birth weight compared with 1 in 10 of the general population.

The health problems of homeless people are compounded by the difficulties of providing health services for them. Some homeless people have personal or psychological handicaps that complicate the issue. Motivation to seek help and to complete a course of treatment is often low. Many fail to keep follow-up appointments. It can be difficult to keep accurate records about someone who is constantly on the move, and written communication between health workers and patients is almost impossible. Such difficulties can antagonize the providers of mainstream services, and if they express this hostility, it becomes even less likely that homeless people will consult them. Homeless people using mainstream services may also have to cope with hostility from their fellow patients—few people are happy to share a waiting room with someone who has not been able to bathe, shower, or change clothes for weeks.

These difficulties have led some health workers to develop separate services for homeless people. In London, for example, there are several walk-in clinics where homeless people can obtain basic health care, including referral to the hospital if appropriate, without registering or making an appointment. Such schemes are popular, especially with the relatively small group of homeless people who have made a conscious decision to opt out of mainstream society. But they also increase the marginalization of this already disadvantaged group, and by hiding the scale of the problem from the main service providers, they may allow its size to be underestimated. (SEE ALSO: *Acquired Immune Deficiency Syndrome; Health Codes; Physical Disabilities, Housing of Persons with; Psychiatric Disabilities, Housing of Persons with; World Health Organization*)

—*Stella Lowry*

Further Reading

Conway, Jean. 1995. "Housing as an Instrument of Health Care." *Health & Social Care in the Community* 3:141-50.

Engels F. 1892. *Conditions of the Working Class in England in 1844.* London: George Allen & Unwin.

Essen, Juliet, Ken Fogelman, and Jenny Head. 1978. "Children's Housing and Their Health and Physical Development." *Child: Care, Health and Development* 4:357-69.

Hopton, Jane and Sonja Hunt. 1996. "The Health Effects of Improvements to Housing: A Longitudinal Study." *Housing Studies* 11(2):271-86.

Lowry, S. 1991. *Housing and health.* London: British Medical Journal.

Smith, Susan J. 1990. "Health Status and the Housing System." *Social Science Medicine* 31(7):753-62.

Warsco, K. 1992. "Explaining Housing-Related Illness." *Housing and Society* 19(3):49-62.

World Health Organization. 1988. *Guidelines for Healthy Housing.* Copenhagen: Author. ◄

► Health Codes

The justification for public involvement in setting construction and maintenance standards through building, housing, and other codes has its origins in 19th-century concerns about the spread of disease and pestilence. Outbreaks of tuberculosis, typhoid, cholera, and smallpox drew reformers' attention to deplorable housing conditions. These included extreme overcrowding; little, if any, ventilation; the absence of sanitation facilities (in many cases resulting in defecation and urination in public hallways); and dangerous accumulations of garbage and waste that attracted rats and other pests. Some cities tried to address the worst of these concerns by adopting building and health regulations.

The housing codes adopted by many cities in the 20th century had their derivation in these concerns with health. In contrast with building codes that address materials, products, and construction techniques, housing codes focus on the living environment, with emphasis on health-related standards, such as the elimination of lead-based paint and the provision of functional toilets. In communities with housing codes, health codes play less visible roles, concerned primarily with issues of sanitation, particularly the accumulation of garbage.

In rural areas, health codes play a more significant role in housing quality. Sanitarians (i.e., public health experts) are involved in approving septic system design and ensuring the availability of a safe water supply in addition to addressing sanitation concerns. Typically, perc tests are required to assess the rate at which water percolates into the ground. This establishes whether a given piece of property can accommodate a septic system. Other health code standards set, for example, minimum distances between a well and a septic system. (SEE ALSO: *Building Codes; Code Enforcement;* **Health;** *Housing Codes*)

—*Deborah A. Howe*

Further Reading

American Public Health Association. 1948. *Standards for Healthful Housing: Vol. 1. Planning the Neighborhood.* Chicago: Public Administration Service.

American Public Health Association. 1950. *Standards for Healthful Housing: Vol. 2. Planning the Home for Occupancy.* Chicago: Public Administration Service.

American Public Health Association 1951. *Standards for Healthful Housing: Vol. 3. Construction and Equipment of the Home.* Chicago: Public Administration Service.

Duffy, John. 1990. *The Sanitarians: A History of American Public Health.* Chicago: University of Illinois Press.

Rosen, George. 1993. *A History of Public Health.* Baltimore, MD: John Hopkins Press. ◄

► High-Rise Housing

Few architectural issues have generated as much debate as the use of high-rise buildings for public housing. In the first 20 years following World War II, large public housing authorities in the United States constructed thousands of units in elevator buildings. But by the early 1970s, the practice had come under tremendous criticism for allegedly promoting antisocial behavior in project residents. Today, the development of high-rise buildings for low-income families is almost universally deplored, as public agencies and nonprofit developers endeavor to find low-rise solutions for high-density projects. The focus of the following discussion is on high-rise, public housing for low-income families and does not include consideration of high-rise buildings for middle-class and affluent families or for the elderly.

The widespread construction of high-rise public housing was largely a phenomenon of the 1950s and 1960s. The early public housing programs of the New Deal era produced low-rise projects typically organized according to "superblock" principles. A superblock was an assemblage of at least three city blocks in which the existing street system was erased and low buildings were arranged in simple slab or L-shaped configurations in the resulting open space. This type of site planning was believed to have multiple advantages over traditional urban housing forms, such as tenements and row houses. It produced development and construction economies by reducing expensive street frontage and simplifying building configurations, while simultaneously introducing high standards of light, air, and open space.

Congress resurrected the public housing programs of the 1930s in the Housing and Redevelopment Act of 1949. Under this first postwar housing program and its various amendments, many cities began to take the superblock arrangement and extrude the buildings upward to anywhere from 6 to 30 stories. The resulting form was the now familiar high-rise, inner-city housing project consisting of a group of towers or slabs arranged on a large superblock and separated by wide plazas. Large cities such as New York, Chicago, and St. Louis constructed virtually all their postwar public housing in high-rise, high-density projects.

Although the alleged effects of high-rise living have received tremendous attention from environmental psychologists and architects, the question of why federal and local housing authorities chose to build high-rise buildings in the first place has gone largely unaddressed. The most generally accepted interpretation attributes the phenomenon to U.S. architects' infatuation with the principles espoused by the French architect LeCorbusier and the Congress Internationale d'Architecture Moderne (CIAM). In his many writings, LeCorbusier promoted the "tower in the park" as part of his vision of the architecture of the new machine age, which would liberate humanity from the dense, congested, and unhealthy industrial city. In the highly competitive climate of the immediate postwar period, new and ambitious U.S. architecture firms such as Skidmore, Owings, Merrill (SOM) saw public housing work as a way of gaining professional recognition by constructing their own versions of LeCorbusier's Unité d'Habitation. According to this view, local housing authorities followed the architects' lead in deciding to develop high-rise projects.

Whereas architectural historians have tended to ascribe to the proposition that the high-rise public housing episode was the result of modernist design ideology, urban historians have constructed an interpretation that emphasizes instead the politics of urban renewal and the economics of public housing development. According to this view, local housing authorities perceived high-rise buildings as a way to accommodate federally mandated restrictions on development costs, while at the same time locating projects on expensive inner-city slum sites. High-rise public housing was the result of racially discriminatory site selection practices at the local level coupled with intense pressure by federal officials for development economies.

Large housing authorities in the postwar era routinely selected public housing sites in the center of African American neighborhoods. This resulted in part from a 1954 Supreme Court ruling that prevented cities from designating projects as exclusively for people of one racial or ethnic group. In many cities, the largely Caucasian suburbs violently resisted any efforts to develop integrated public housing projects within their borders. Housing officials and city councils tended to respond to this political pressure by selecting sites almost exclusively in inner-city, African American neighborhoods.

These site selection practices were reinforced by many cities' postwar redevelopment objectives. The 1949 Housing and Redevelopment Act made funds available to localities to acquire and clear blighted property and then sell it to private developers. Typically, cities used the funds to clear deteriorated residential areas close to the central business district and turn the property over for revenue-generating commercial development. Public housing projects were constructed at some distance from the redevelopment areas so as to accommodate the low-income, usually African American, households displaced by the slum clearance. An explicit part of this overall strategy was to remove low-income families from the areas closest to the central business district, where the growth of slums was threatening to erode downtown property values. Historian Arnold Hirsch has described this process of confining African Americans households to dense inner-city housing projects as an overt strategy of "making the second ghetto."

The inner-city sites selected by local housing authorities were tremendously expensive. Not only were the property values inflated, but the city had to expend large sums to condemn and clear the sites before public housing construction could commence. These rising expenditures for site acquisition conflicted with a federally imposed cap on the amount local authorities could spend per unit of public housing.

Local housing authorities viewed high-rise buildings as a way to resolve the dilemma between high site acquisition costs and the need to keep overall development costs to a minimum. Using a small number of tall buildings instead of a larger number of walk-ups allowed the local authority to increase project densities and thereby average the acquisition and clearance costs over a larger number of units. It was also believed that the simple high-rise slabs were cheaper to construct than low-rise, high-density forms with complicated building configurations.

The work by urban historians on the history of postwar urban redevelopment and housing policy demonstrates that the high-rise phenomenon must be understood as more than simply a matter of architectural fashion. The decision to abandon the low-rise forms of the 1930s was also motivated by racially discriminatory site selection practices, the economics of project development, and the urgency of postwar urban redevelopment activities.

Although the question of why high-rise public housing became so popular in the postwar era has received little attention, there is a great deal of literature devoted to the alleged behavioral consequences of living in a high-rise environment. Long before World War II, many architects, policymakers, sociologists, and psychologists voiced concerns that high-rise living was incompatible with certain aspects of U.S. family life. As the first highly visible high-rise projects came to be constructed in the 1950s, researchers began to investigate that question.

One of the first studies on high-rise housing was commissioned by the Philadelphia Housing Authority, which hired anthropologist Anthony Wallace to compare tenant behaviors in a low-rise and a high-rise project. Writing in 1954, Wallace articulated what today remain many of the central critiques of high-rise housing for families with children. He concluded that the lack of privacy in high-rise buildings led to excessive and unwanted social contact, that there was not enough privately controlled outdoor space in which mothers could supervise their children, and that the projects required a high level of involvement by management.

The early 1970s saw the publication of a number of investigations of high-rise public housing, prompted in part by the growing notoriety of St. Louis's ill-fated Pruitt-Igoe project. This work argued that high-rise projects contained spaces the residents could not control and keep safe. Some designers were particularly critical of Pruitt-Igoe's large, unprotected plazas, which offered tenants little ability to monitor who entered or exited the buildings.

The literature arguing that there is a link between criminal behavior and high-rise architecture is not without its critics. Many have pointed out that simply correlating high crime rates to the number of stories in a building fails to demonstrate a causal link. Others have pointed out that environmental design arguments are a subtle form of blaming the victim, because they presuppose that people of a particular income level or race unavoidably bring with them behavior problems that have to be designed against. Such arguments naturalize the presence of crime rather than seeing it as a product of institutionalized racial and economic oppression. Sociologist Lee Rainwater, who conducted extensive field work at Pruitt-Igoe in the 1960s, interpreted resident behaviors as a response to the racial discrimination and economic injustice they faced.

It is unlikely that any definitive conclusions will ever be drawn about the behavioral consequences of high-rise architecture. However, the links made in the 1970s between crime and high-rise design certainly contributed to unfavorable public opinion about public housing and helped set the stage for the 1972 decision by the Nixon administration to discontinue the public housing program. Housing professionals at both federal and local levels now actively discourage high-rise housing for families with children. It is doubtful that we will ever see a revival of the type of high-rise housing design that characterized the immediate postwar era. (SEE ALSO: *Children; Council on Tall Buildings and Urban Habitat; Multifamily Housing; Postoccupancy Evaluations; Pruitt-Igoe;* **Urban Redevelopment**)

—*Kate Bristol*

Further Reading

Bauman, John. 1987. *Public Housing, Race and Renewal: Urban Planning in Philadelphia 1920-1974*. Philadelphia: Temple University Press.

Bristol, Katharine G. 1991. "The Pruitt-Igoe Myth." *Journal of Architectural Education* 44(3):163-71.

Hirsch, Arnold R. 1983. *Making the Second Ghetto: Race and Housing in Chicago, 1940-1960*. Cambridge, UK: Cambridge University Press.

Jeffery, C. R. 1971. *Crime Prevention through Environmental Design*. Beverly Hills, CA: Sage.

Marmot, Alexi Ferster. 1981. "The Legacy of LeCorbusier and High Rise Housing." *Built Environment* 7(2):82-95.

Rainwater, Lee. 1970. *Behind Ghetto Walls: Black Family Life in a Federal Slum*. Chicago: Aldine.

Wallace, Anthony. 1952. *Housing and Social Structure: A Preliminary Survey with Particular Reference to Multi-Storey, Low-Rent Public Housing Projects*. Philadelphia: Philadelphia Housing Authority. ◀

▶ Hispanic Americans

The Hispanic population represents one of the fastest growing and most diverse populations in the United States. This population and its housing are concentrated in the West, primarily in California (43%); the South, mostly in Texas and Florida (32%); and the Northeast (18%) regions of the country. Hispanics are the most urban (90%) ethnic/racial group and the second most likely to reside in an inner city (52%) after African Americans (59%). But the group is also highly heterogeneous. First recognized as comprising primarily those of Mexican, Puerto Rican, and Cuban ancestry, it is now recognized for its much larger diversity of individuals from various Central and South American countries. This diversity also includes an array of cultural, social, and political beliefs and values with origins in countries extending from Mexico to the tip of South America. However, one value held in common by all Hispanic families is the importance of the family unit and the goal of homeownership.

This strong desire to own a home is exemplified in the growing rates of homeownership among all Hispanic groups, despite low levels of income. For example, one of every five (21%) Hispanic families living in poverty owns its home. The high priority placed on ownership by Hispanic families means that they are willing to sacrifice other important or essential items, such as health insurance or a private means of transportation, to attain homeownership.

The housing owned by Hispanics is generally older, in low-income areas, and often inadequate in terms of housing condition and size. The stability of ownership often outweighs the issue of housing quality. In many cities, however, even this opportunity is being all but eliminated by rising

TABLE 9 Housing Characteristics in the United States, by Race and Ethnic Origin

Housing	White Non-Hispanic	Black	Hispanic
Owner occupied (%)	69.9	42.8	38.8
New construction (< 4 years, %)	5.9	3.1	3.5
Reside in mobile homes (%)	7.0	3.4	3.3
Severe/moderate physical problems (%)	6.8	20.9	13.3
Moved in past year (%)	16.0	21.2	27.6
Urban residence (%)	69.7	87.4	90.4
In (P)MSAs (%)			
In central city	25.8	59.1	51.6
In suburbs	49.8	26.8	38.0
Outside (P)MSAs (%)	24.3	14.2	10.4
Northeast (%)	21.1	17.8	18.3
Midwest (%)	26.7	19.8	6.6
South (%)	32.7	53.4	31.7
West (%)	19.5	8.9	43.4
Household income	$35,589	$18,535	$22,578
Median age of householder (in yrs.)	51.0	43.0	40.0
Persons per unit	2.0	2.4	3.1
Average number of rooms	5.4	5.0	4.6
Persons per room	0.6	1.1	4.3
Female households (%)	12.6	14.3	8.0
Structure (%)			
New	10.8	6.2	7.9
Recent	30.4	28.0	26.1
Postwar	37.3	40.7	41.7
Prewar	21.5	27.0	22.1
Median year built	1964	1958	1960
Cooperative/condominium (%)	4.3	2.3	4.0
Central-city housing (in thousands)	19,023	6,396	3,216
New structure (< 4 years, %)	3.6	1.4	2.0
Severe physical problems (%)	2.6	5.0	5.4
Moderate physical problems (%)	3.7	11.8	11.4
Moved in past year (%)	20.3	21.1	28.6
Poverty households (%)	10.4	29.9	24.1
Total occupied housing (in thousands)	73,625	10,832	6,239

SOURCE: U.S. Bureau of the Census, *American Housing Survey*, 1994.

housing costs, loss of job opportunities, and limited availability of affordable housing.

Unfortunately, the housing picture for low-income Hispanics remains bleak. As indicated in Table 9, these households must endure living in some of the country's poorest, smallest, and most overcrowded housing, using large proportions of family income for housing and generally being limited to renting instead of owning. Composed of a younger and less financially secure population, many Hispanic households must overcome severe economic obstacles and barriers in their attempt to acquire appropriate housing. This is particularly true for the Hispanic family seeking to purchase its first home.

Beyond economic obstacles, housing discrimination and segregation continue as pervasive and enduring problems affecting the housing choices of Hispanic populations. Despite federal legislation barring discrimination, Hispanics were more residentially segregated in 1990 than in 1980. Residential choice is due in large part to affordability but also to restrictive access to housing based on ethnic or national background, skin color, and proficiency of the English language. Even today, Hispanic housing remains concentrated in primarily Hispanic neighborhoods or barrios.

Of all social problems experienced by Hispanic populations, probably the most pressing problem—even beyond the issues of jobs, transportation, and schools—is the availability of housing that is affordable and in good condition. For the Hispanic family, a home is much more than shelter. It is the center for the development of family relations; it provides the basis for social relationships as well as economic and political participation.

Although the importance of the home is common to all Hispanic populations, differences exist in housing characteristics by subgroup origins. In general, Cubans continue to have better housing when compared with other subgroups, whereas Puerto Ricans have the worst housing. These differences are even greater among all subgroups within the Hispanic population. Those with a longer history in the United States have better housing than those whose residency is temporary or who are recent arrivals to the

United States. All these differences reflect not only economic status but also cultural factors, generational differences, regional differences, immigration status, and length of residency in the United States.

Of all Hispanic groups, recent immigrants have the most difficulty in securing decent housing. The distinction must be made between economic immigrants and political refugees. The housing choices of economic immigrants, who are not eligible for government subsidized housing, are limited to what is available in the private sector. For many of these families, this means a less than healthy housing situation, such as doubling up with other families or living in structures never intended to be used as a residence. Political refugees, on the other hand, fall into a category in which they are provided services to help establish themselves in local communities. Unfortunately, owing to severe cutbacks in all forms of government housing there is simply not enough housing to meet the need.

There are also differences among subgroups in housing choice, type, and form of tenancy. For example, the more cosmopolitan subgroups, Cubans and Puerto Ricans, are more willing to accept and participate in alternative living arrangements, such as cooperative or shared housing. Persons of Mexican origin, by contrast, have more rural origins and a relatively narrow view of private property ownership. They are reluctant to accept different forms of ownership such as limited-equity or communal living. These differences also affect the level of participation, leadership, and responsibility by Mexican origin and non-Mexican origin persons in tenant organizations found in alternative forms of housing.

As Hispanics continue to be overrepresented in the lower-income strata of the economy, they will become a greater part of government-assisted housing populations. Historically, many Hispanic populations were denied access to subsidized housing programs as a result of both acts of racism and discrimination occurring in the distribution and administration of these programs. However, as Hispanic populations grew and gained a greater presence in local political arena, blatant attempts to exclude them from publicly funded housing were no longer tolerated or attempted.

The exclusion of Hispanic populations from publicly assisted housing programs is a direct result of the political context in which these programs were enacted and developed. Policies for low-income housing were generally seen as an issue between blacks and whites. As a result, housing policies reflected the economic, social, and demographic characteristics of these two groups, including the characteristic racial segregation of housing.

Today, within many of the subsidized housing communities, the stability and unity of Hispanic families has had a positive affect on the community itself. The desire to improve the quality of their homes, particularly for their children, helped foster positive changes in many of these communities. This stability, despite poverty, reflects the desire of Hispanic families to provide a positive atmosphere for their children.

Whether referring to private sector or publicly assisted housing, the positive attributes associated with Hispanic homes are the direct result of Hispanic women, who continue to bear the primary responsibility for the condition and security of their families' homes. Their involvement in resident organizations and homeowner associations reflects a desire that their children be provided with safe, adequate, and decent housing. Potential threats to the community are seen as a direct threat to the stability of the Hispanic home and become, therefore, areas for involvement. At the same time, however, existing locational patterns and cultural roles of the Hispanic home continue to reinforce a division of labor based on gender and to perpetuate a male-dominated Hispanic household.

Overall, U.S. housing policy has provided few guidelines for Hispanic populations. Documentation for housing assistance programs continues to be incomprehensible to primarily Spanish language renters and home seekers. Mortgage loan policies still respond to the primary earner and have not adapted to families as economic units. Scarcities in affordable housing require Hispanic families to double up, challenging occupancy regulations. Those present without documentation are ineligible for public housing or rental vouchers, and the trend is to deny these benefits to legal arrivals as well. Finally, although mutual assistance is central to Hispanic culture, resulting in benefits when seeking employment, retaining connections with the homeland, and providing opportunities to pool funds for large purchases, these mutual efforts do not extend to homebuying or creating affordable housing.

Hispanics are most likely to reside in areas with a scarcity of quality, affordable housing. Affordability and subtle barriers that persist despite efforts to lessen discriminatory practices hamper accessibility to better housing. (SEE ALSO: **Discrimination**)

—Ronald Brooks and Leo F. Estrada

Further Reading

Betancur, J. J. 1996. "The Settlement Experience of Latinos in Chicago: Segregation, Speculation and the Ecology Model." *Social Forces* 74(4):1299-1324.

Frey, W. H. and R. Farley. 1996. "Latino, Asian, and Black Segregation in U.S. Metropolitan Areas: Are Multiethnic Metros Different?" *Demography* 33(1):35-50.

Gilderbloom, John and John P. Markham. 1993. "Hispanic Rental Housing Needs in the United States: Problems and Prospects." *Housing and Society* 20(3):9-25.

Goering, J. M., ed. 1986. *Housing Desegregation and Federal Policy.* Chapel Hill: University of North Carolina Press.

Massey, Douglas. 1992. "Racial Segregation in U.S. Metropolitan Areas." *Urban Geography* 32(6):127-39.

Massey, Douglas S. and Nancy A. Denton. 1989. "Residential Segregation of Mexicans, Puerto Ricans, and Cubans in Selected U.S. Metropolitan Areas." *Social Sciences Review* 73(2):73-83.

Orfield, Gary. 1991. "Hispanic Re-Segregation and Schools." *Education and Schools* 19(4):67-78.

Rodriguez, G. & C. Flournoy. 1985. "Illegal Segregation Pervades Nation's Subsidized Housing." *Shelterforce* 9(May/June):8-9.

U.S. House, Subcommittee on Housing and Community Development of the Committee on Banking, Finance and Urban Affairs. 1985. *Housing Needs of Hispanics.* 99th Congress, 1st sess. Washington, DC: Government Printing Office. ◄

► Historic Preservation

Over the past quarter century in the United States, there has been a substantial increase in the number of housing units created or retained through the renovation and restoration of older buildings. Both residential and nonresidential structures have been improved to provide living space while their essential architectural scale, materials, and features have been preserved. Combining the twin goals of housing supply and historic preservation has contributed significantly both to human welfare and to the historic, architectural, and cultural legacy of the nation's communities. A primary objective of historic preservationists is to protect and restore structures of architectural and historic merit. But in doing so, many have added to the stock of habitable dwelling units available to people of varying incomes. Although new construction is the chief method by which our national housing supply grows, historic preservation techniques have contributed substantially since the 1970s.

The principal purpose of those who attempt to conserve and restore older structures has always been to ensure that their aesthetic, cultural, and educational attributes will survive for present and future generations of people to study and appreciate. Until the 1960s, preservationists' efforts were largely limited to a few building types, such as churches, meeting houses, or town halls, or to various kinds of residential properties, such as plantation houses, mansions, and estates. Interest centered on properties built in the periods before the Civil War. For the most part, preservationists attempted to maintain these buildings in their original use. Historic houses were either occupied as dwellings—and occasionally opened to the public by their occupants—or converted to unoccupied house museums or historic society headquarters open on a regular basis for educational purposes. Particularly distinguished examples of house museums include Monticello, the home designed by Thomas Jefferson, near Charlottesville, Virginia, and Mount Vernon, the plantation home of George Washington.

Almost invariably, historic protection was accorded a building only if it was associated with important figures or noteworthy events in history or with prominent architects. In some cases, obsolete forms of architecture such as fortresses or stockades were deemed worthy of protection and restoration. If older buildings survived in the absence of any of these attributes, as housing sometimes did, it was usually because they continued to serve a locally worthy purpose and were not in a high-demand location. For example, although single-family houses in central cities were eliminated to provide space for larger, more profitable commercial buildings, in small towns and rural areas they often persisted, albeit in altered form, for decades.

The Great Depression of the 1930s, however, contributed to substantial disinvestment in older urban neighborhoods. After World War II, demand for suburban housing, the rise of suburbanization, the widespread use of automobiles to commute between home and workplace, and the federally subsidized construction of the interstate highway system all contributed to declining consumer demand for older neighborhoods in cities and towns. Vacancies, abandonment, arson, and neglect occurred in many neighborhoods, and in others, large numbers of poor and minority households supplanted working- and middle-class families. These influences contributed still further to the outmigration of middle-income families to the suburbs. The net effect of these forces was to massively undercut the demand for real estate in older sections of urban communities. Slums and blighted areas multiplied.

Meanwhile, as older U.S. cities and towns suffered, the historic preservation movement in the nation was slowly spreading. One of the most critical influences occurred in 1966 when Congress enacted the National Historic Preservation Act. The new legislation established the National Register of Historic Places, a listing of federally recognized historic resources across the nation. It also provided a program of grants, loans, and technical assistance through the National Park Service to aid state, local, and nonprofit organizations in historic preservation projects and programs. One impact of these measures was that the preservation movement enjoyed newfound respect and visibility and won many new adherents. But the act also encouraged preservationists to think beyond traditional norms in evaluating structures and spaces for historic preservation.

Closely linked to the 1966 act was the Congress's decision in 1968 to amend the federal Urban Renewal Program, renaming it the Neighborhood Development Program (NDP). NDP encouraged local authorities to attempt to rehabilitate or restore older buildings for reuse rather than demolishing them and replacing them with new structures. In part, the result of active lobbying by organizations such as the private nonprofit National Trust for Historic Preservation, the NDP became a tool both for rehabilitating housing and for maintaining the traditional form, scale, and materials of many 19th- and early 20th-century urban neighborhoods. Boston's South End and Baltimore's Federal Hill are but two of many neighborhoods improved through the NDP.

These and other influences meant that, as the U.S. historic preservation movement grew during the 1960s and 1970s, preservationists gradually embraced a much larger range of stylistic periods and types of buildings. Earlier, preservationists had restricted their attentions to Georgian, Colonial, Federal, and Greek Revival styles. As historic preservation took on broader values and new meanings, Queen Anne, French Second Empire, Stick, and Shingle-style architecture of the late 19th and early 20th centuries began to attract the attention of preservation advocates. Long-neglected structures such as old post offices, banks, railroad stations, warehouses, factories, hotels, schools, stores, and offices were examined with new interest. Although not necessarily associated with important events or figures in history, many of these properties were appreciated because they were of unique design, were characteristic of a particular style or structural method, or typified a generic building type of importance to the history of the region in which they were located. But it was apparent that only a few of these buildings could be converted to museums and that the original purposes for which many were designed were no longer economically viable. As a result, preservationists began to place more emphasis on "adaptive use," or the recycling of older properties for new uses.

Meanwhile, a growing population in the nation meant that the demand for additional housing was rising. Children born after World War II, when returning servicemen and women contributed to a rapid rise in birth rates, were growing to adulthood in the late 1960s and 1970s. Unlike many of their parents, the new generation had not grown up in city neighborhoods and few had any perception that success in life meant moving to the suburbs. In addition, their unprecedented numbers placed new demands on the nation's residential stock, and older buildings represented a relatively inexpensive source of housing. Many of these young middle-class adults were attracted to older neighborhoods with interesting architectural features and convenience to downtowns and employment centers. They purchased dwellings, as well as commercial and industrial properties, and renovated them for their own occupancy or for rental investment purposes. The trend came to be called *gentrification*. Borrowed from the British (where comparable events were occurring), the term implied the return of the "gentry" or middle-class to central-city living. Gentrification was also occurring in some Canadian, Australian, and Western European cities. Mostly singles and couples without children, the gentrifiers usually had college educations and worked in professional, technical, sales, managerial, and financial occupations. With higher incomes than those they replaced, their lifestyles sometimes brought them into conflict with incumbent residents. Although many incumbents profited handsomely from the sale of their properties to affluent newcomers, other residents were "displaced" or forced to move away because of evictions, higher rents, and rising property taxes.

During the 1970s and 1980s, this process helped to physically and economically revitalize hundreds of neighborhoods, such as New York's Park Slope, Washington's Dupont Circle and Capitol Hill, Cincinnati's Mount Adams, and St. Louis's Lafayette Square. Combining the need for housing, as well as office, retail, and other spaces, gentrification-induced private market reinvestment became perhaps the most prominent expression of efforts to wed the interests of historic preservationists and housing advocates.

However, it became apparent to many housing activists, developers, bankers, and historic preservationists during the 1970s that older buildings could be reused not only for middle-class households but with proper financing and public policies, for low- and moderate-income families as well. Inner-city revitalization, in other words, could address goals of both preservationists and housing advocates. Government officials, anxious to stimulate investment in communities and counteract the flight of families to the newer suburbs, were generally supportive, if not enthusiastic, about wedding these two goals. As it turned out, historic preservationists and housing advocates had found common ground. These twin interests were reflected in the federal Urban Homesteading and Neighborhood Housing Services programs enacted by Congress in the early 1970s.

Related to gentrification have been efforts by historic preservationists to encourage adaptive use of older structures that have outlived their original purposes. Railroad stations, mills, warehouses, churches, schools, offices, hotels, and other buildings with noteworthy architectural character have been converted to other uses, including apartments, condominiums, and houses. For example, the former Chickering Piano Factory in Boston was converted to housing, studio, and exhibition space for artists. The Cast-Iron Building in Manhattan, formerly an industrial loft, was transformed into apartments. An 1893 firehouse in San Francisco was renovated into a private home. Former streetcar barns in Washington, D.C., and in New Bedford, Massachussets, have been turned into apartment complexes. Thousands of such buildings have been adaptively used over the past 20 years. In some cases, entire industrial areas, such as Seattle's Pioneer Square, Boston's Atlantic Avenue, and New York's SoHo, have been transformed for mixed residential, office, and commercial uses. In nearly all cases, preservationists attempt to restore building exteriors to their original or near-original appearance; interiors may be altered to fit the new uses, although wherever possible, attempts are made to preserve original details and materials, incorporating them into the new spaces.

The full contribution of preservation forces such as gentrification, federal neighborhood and housing programs, and adaptive use can only be estimated. One federal study found that 4.9 million housing units were added to the national housing stock through building conversions between 1970 and 1980.

One of the obvious attractions of renovating or restoring older buildings in urban settings is that, generally, housing can be produced at lower costs than units provided through new construction. In new construction, one must acquire a property, demolish existing structures, clear the debris (and pay for its disposal in a landfill), and in some cases, reconfigure sewer, water, and other utility connections. Then, site engineering, excavation, and foundation work are necessary, followed by construction of new buildings. Landscaping, paving, and site finishing work are also included. All of this work occurs, of course, at current costs for labor and materials. In rehabilitation of older structures, however, existing foundations and exterior walls are used with little or no improvements beyond repairs, cleaning, and cosmetic work. Interior spaces are often reconfigured; new kitchens, bathrooms, plumbing, and electrical wiring are installed; and modern safety features, such as fire-retardant party walls, fire escapes (for multifamily buildings), and sprinkler systems, are added. Energy-efficient windows and doors and new roofing may be necessary. In any event, rather than drastically alter the stylistic character of the building, the preservationist approach is to attempt to restore or approximate the original appearance of the building. Thus, instead of installing expensive vinyl or aluminum siding, the original wood, shingle, or masonry exterior will be cleaned and reused. Period details such as intricate bargeboard (gingerbread), delicate spindlework, grandiose bracketry, walnut newel posts, ornate cast iron cresting and fencing, carved stone, and stained glass will be retained whenever possible. As a result, reuse of existing structures typically costs considerably less than new construction.

But these advantages may be offset in some circumstances. New construction usually means much more efficient use of a given building site, allowing space for parking, mechanical systems, and storage to be incorporated underground. Greater building heights permit stacking of more dwelling units. Although usually more costly, new construction may serve more households on a given site than rehabilitated residences. But these gains may be offset by other problems that arise from high-density residential construction. During the 1950s and 1960s, especially prior to enactment of the NDP, many high-rise apartment complexes were built for low- and moderate-income families through government programs such as public housing. Some of these became havens for crimes such as prostitution, drug dealing, rape, and assault. Vandals destroyed hallways, windows, and vacant apartments. Trash was strewn about the grounds and stripped cars were abandoned. Infamous examples include Chicago's Cabrini-Green and St. Louis's Pruitt-Igoe projects. Thus, although housing large numbers of people per unit of land, the efficiencies of many of these high-rise projects were offset by the costs of policing, repair, and maintenance that became necessary. In addition, renewal programs often replaced low-density older housing with commercial, office, and institutional projects, many of which were also incompatible with their surroundings. As a result, in the late 1960s, the federal government began to back away from sponsoring more such buildings for poor families and instead adopted policies for lower-density, "scattered-site" housing. Hence, from the early 1970s, federal policy shifted to lower-density new construction and renovation of one- through four-story older buildings, boosting preservationists' efforts to protect architectural and neighborhood character.

Without doubt, one of the most influential catalysts uniting historic preservation and housing was the historic preservation tax credit program maintained by the U.S. Internal Revenue Service. First enacted by Congress in 1976 and liberalized in the early 1980s, the tax credit program permitted those who renovated qualified older buildings for commercial reuses according to National Park Service preservation standards to offset their personal tax liabilities by receiving a credit for allowable renovation and restoration expenditures. In addition, these developers could sell tax credits to investors, using their proceeds to help finance the project. Investors, in turn, could use the credits to reduce their liability to the IRS. With passage of the 1986 Tax Reform Act, financial incentives in the tax credit program were substantially reduced and the rate of building renovations declined thereafter. Nonetheless, according to the National Trust for Historic Preservation, between fiscal years 1982 and 1985, more than 62,000 housing units were rehabilitated under tax credit subsidies; almost 30,000 of these were created by conversions of nonresidential buildings to residential use. Although most of these units were affordable only for middle- or upper-income people, several thousand were intended for low- and moderate-income occupancy. Other federal initiatives contributing to the reuse of older buildings for housing include the Community Development Block Grant and Federal Housing Administration mortgage guarantee programs, as well as the now-defunct Urban Development Action Grant and Section 8 Housing Rehabilitation programs. Among state and local governments, several, such as Washington State and North Carolina, maintain programs that reduce or limit property tax liabilities for homeowners who restore or renovate their homes according to specified historic preservation standards. Some of these units have also been successful in raising public and private funds from local organizations to create housing conservation and rehabilitation revolving funds.

In short, housing advocates and historic preservationists have been successful in many communities in fostering common methods to advance common goals. Completed projects around the nation have demonstrated that socio-economic and class divisions that once separated the activities of the two groups have diminished with time. As a result, housing and preservation activists have added many thousands of units to the U.S. housing supply while conserving neighborhoods and historic resources. The net effect has been to contribute significantly to housing the nation's population while protecting the historic legacy of its spaces and places. (SEE ALSO: *Gentrification; Incumbent Upgrading; Infill Housing;* **Urban Redevelopment**)

—*Dennis E. Gale*

Further Reading

Advisory Council on Historic Preservation. 1979. *The Contribution of Historic Preservation to Urban Revitalization.* Washington, DC: Government Printing Office.

Andrews, Gregory, ed. 1981. *Tax Incentives for Historic Preservation.* Washington, DC: Preservation Press.

Carlson, Daniel. 1991. *Reusing America's Schools: A Guide for Local Officials, Developers, Neighborhood Residents, Planners and Preservationists.* Washington, DC: Preservation Press.

Escherich, S. M. 1996. *Affordable Housing through Historic Preservation.* Upland, PA: Diane.

Listokin, David, ed. 1983. *Housing Rehabilitation: Economic, Social and Policy Perspectives.* New Brunswick: Rutgers, State University of New Jersey, Center for Urban Policy Research.

Maddex, Diane, ed. 1985. *All About Old Buildings: The Whole Preservation Catalog.* Washington, DC: Preservation Press. ◀

▶ History of Housing

Although shelter, like food, is often considered a basic human need, in the United States, housing is not recognized as a right to which all persons are entitled. This has resulted in inadequacies and inequities for some individuals and in private efforts and public policies to address those inadequacies and inequities.

This entry briefly examines from a historical perspective how housing has traditionally been provided or acquired in the United States and considers two major housing issues related to that provision. These issues concern the inability of some to obtain adequate housing that they can afford and the limitations on others' residential location, irrespective of quality or cost, because of their race or ethnicity. It surveys both private and public efforts to address these issues and places them in their larger economic, political, and social context.

Since its founding, the United States has been in principle a capitalist democracy. As such, it has relied on a market system to provide most goods and services, in the belief that producers will strive to meet the demand that consumers express by their purchases. Although there have been instances of owner building, housing has been largely a consumer good, regardless of its physical form or terms of occupancy. One "purchased" shelter by renting an apartment or buying a house (or the land and materials to build one). Moreover, for most of U.S. history, the private sector has provided the buildings. The public sector (i.e., some level of government) has often regulated the size, type, or location of housing or set standards for its construction or occupancy, but except in a very few instances, it did not build the structures.

In colonial times and for decades afterward when settling new areas, individuals who owned land either built their own dwellings or hired someone else to do so. Those who were not landowners lodged or roomed with another household or rented or leased a dwelling independently. As the population of the United States increased in size, urban (and then suburban) areas expanded, more or less standard dwelling types and tenure arrangements appeared. Dwellings could be single-family detached structures; duplexes, triplexes, or quadruplexes; row houses, where two or more dwellings were joined together linearly; or apartment buildings containing dozens of dwelling units. A household might own its own home, rent or lease it from a landlord, or own it cooperatively. Although an individual or household might occupy any type of structure under any tenure arrangement, the common perception has been that people rented an apartment until they could afford to own a single-family home.

Like other consumer goods, housing eventually wore out and had to be replaced or required maintenance to prolong its life. Also, different types of households—single individuals, families with children, extended families, elderly couples—had different housing needs, and the same individual or household could have different needs at different times. Unlike some other goods, however, housing was not generally convertible or portable. Once built, an apartment building remained in its original location, and although a single-family home could be subdivided to house more than one household or moved to another site, neither was easily or cheaply accomplished. These constraints mean that at times, the available housing supply was not appropriate to meet individual and collective housing needs. Moreover, for some population groups, characteristics of housing itself and U.S. reliance on the private sector to provide it have created housing problems such as high cost, low quality, and discriminatory practices.

Problems—and Efforts to Solve Them

Adequacy and Affordability

There have always been some households whose dwellings were inadequate in size, condition, or some other respect and some who paid more than they could afford for their housing. This circumstance came to be viewed as a problem, however, only in the late 19th century. At that time, cities in the United States were growing rapidly because of immigration from rural areas and foreign coun-

tries. Cities also felt the greatest impact of increasing industrialization. As the population grew faster than new dwellings could be constructed, property owners produced new housing wherever and however they could. They built small dwellings in the rear or side yards of existing structures, converted basements and attics into living space, and subdivided single-family homes, duplexes, or apartments to create still more dwellings of ever-decreasing size. At the same time, the wages paid to the factory workers who swelled the cities were so low that a small apartment might be all a large family could afford and many households took in lodgers to help pay the rent.

The resulting overcrowded dwellings were a problem not only for their residents but also for the city as a whole. Properties deteriorated because of overuse and undermaintenance, becoming hazardous to their tenants. At the same time, inadequate urban water and sewer systems—where they existed—could not serve the populations' needs, so disease flourished. In addition, some individuals involved in reform efforts at the turn of the century believed that poor living conditions fostered depraved behavior. When rates of disease, death, and crime reached such levels that the public at large felt threatened, reformers turned their attention to the overcrowded slums and tenements.

The first private attempts to address the problem involved model tenements and limited-dividend corporations. "Model" tenements would be structurally sound, have more light and air than existing slum dwellings, and be large enough to allow some privacy to their residents. The rents would be low enough for a factory worker's income but high enough to cover the cost of production with no operating profit (thus being philanthropic housing) or produce a "limited" dividend on the investment. The "dumbbell" tenement is perhaps the best known effort at improved tenement house design. Its name derives from the building's footprint, because courtyards on each side made the tenement narrower in the middle while allowing some interior rooms to have windows. Although such efforts improved the quality of housing for their residents, they were too few to have much impact on the overall problem.

The reformers then turned to regulation. They actively worked for adoption of tenement laws, housing codes, building codes, and eventually, zoning ordinances that would govern construction of new housing and limit population density. Such regulatory efforts did not prohibit occupancy of existing inadequate dwellings, however, and they also depended on municipal enforcement for their effectiveness. Moreover, raising standards could potentially raise the cost of construction; thus, regulations could aggravate the affordability problem some households faced. Despite the drawbacks of regulatory efforts and the fact that the profit-seeking private market could not produce safe, healthful dwellings that the lowest-income households could afford, few early 20th-century reformers suggested that government should actually provide housing. For example, New Yorker Lawrence Veiller, long a champion of regulation and important in the development and passage of New York City's landmark zoning code, was adamant that government should not build dwellings.

Not until the Great Depression of the 1930s did direct public provision of housing gain some acceptance. The federal government had always provided some housing for those serving in the military and had expanded this practice to include defense industry workers during World War I, so a precedent existed for public efforts to meet a particular need. With unemployment at 25% and many of the employed working fewer hours, families and individuals found themselves unable to meet their rent or mortgage payments. No longer was inadequate housing a problem only for the poor and working classes, for evictions and foreclosures forced many middle-class households to double up or find shelter wherever they could. On the basis of research by housing reformer Edith Elmer Wood, President Franklin Roosevelt announced that one-third of the nation was ill housed.

Legislation passed as part of Roosevelt's New Deal, particularly the Housing Acts of 1934 and 1937, made some element of housing provision a public responsibility at the federal level. Although there were and have been a variety of efforts since the 1930s, most federal housing policy has been directed at one of two goals that date from the New Deal. These are promoting homeownership and providing housing assistance for low-income persons.

The Housing Act of 1934 created the Federal Housing Administration (FHA) to establish and operate a mortgage guarantee program. Although the government would not directly provide mortgage money to homebuyers, it would guarantee private mortgage lenders that they would be paid if the purchaser defaulted. To minimize risk, lenders could offer FHA-backed mortgages only for those properties and prospective buyers that met criteria set by the FHA. The major impact of the FHA came from changes in mortgage practice, however, rather than in the criteria. FHA mortgages were for a longer time period, had a higher loan-to-value ratio (thus lowering the necessary down payment), and were self-amortizing. By broadening the pool of potential homebuyers, these changes supported the government's goal of promoting homeownership. The impact was increased as lenders applied similar criteria and practices to mortgages that were not guaranteed by the FHA. Moreover, since the end of World War II, the Department of Veterans Affairs has operated a similar program for those who served in the military.

The Housing Act of 1937 created the U.S. Housing Authority to fund the construction of housing for low-income households. Ideas of long-time housing advocate Catherine Bauer greatly affected the form of the low-income housing program. Using federal funds, local housing authorities acquired land and built dwellings or contracted for their construction, then acted as the landlord. With construction costs underwritten and project income needing to cover only operating expenses with no profit, rents could be low enough for those unable to afford private market rate housing. The early subsidized housing projects were small scattered developments of low-rise garden apartments and row houses intended for the working poor on their way up the economic ladder. By the 1960s, because of economic constraints and changes in tenant selection criteria, subsidized housing was more often in large, high-rise projects filled with the permanent poor, differing little

in some respects from the slum tenements that had concerned housing reformers 60 years earlier.

Specifics of the various programs aimed at promoting homeownership or assisting low-income families with housing have changed since the 1930s, but the two major policy goals have not. Moreover, although there has been wide support for homeownership, there has been strong and repeated opposition to any form of government-subsidized housing. Historically, too little money has been appropriated to meet stated legislative goals for dwelling unit construction, and residents have protested construction of "projects" in their neighborhoods. Meanwhile, the aging of the housing stock, continued growth in the population, and deindustrialization of the national economy mean that many individuals and families still find themselves inadequately or too expensively housed.

Discrimination

Unfortunately, dwelling cost, quantity, and quality have not been the only housing problems in the United States. Discrimination has limited some individuals' choice of residential location. Although people have experienced discrimination based on their age, gender, or parental status, the most pervasive form has been discrimination based on race or ethnicity. Moreover, if the United States was slow to tackle the housing problems of the poor, it was even slower to tackle those of minorities.

Historically, the terms of their employment as servants or slaves and their lack of income determined where minorities, particularly African Americans, lived in the United States. Also, individual property owners could and did refuse to sell or rent to blacks or other ethnic minorities. However, in the rapidly growing, industrializing, chaotic cities of the early 20th-century, housing discrimination against minorities became much more widespread.

Developers such as Kansas City's Jesse Clyde Nichols, who planned exclusive new subdivisions and suburbs for upper-income homebuyers, inserted restrictive covenants in the deeds governing various aspects of land use. They prohibited resubdivision of large lots, limited construction to one single-family home, set high minimum floor areas or building costs, and often specified building materials and design. They could then assure potential purchasers that their property values would not decrease because of future unpleasant intrusions into the neighborhood, because restrictions applied to subsequent owners as well the original purchaser.

Among the "unpleasant intrusions" developers guarded against were minority groups. In the 1910s, and even more so in the 1920s, it became standard practice for many developers to write deed restrictions for their subdivisions prohibiting ownership, rental, lease, or occupancy by specified minorities. The group specified varied somewhat by region of the United States; Asians were singled out in California, the Irish in Boston, Italians or East Europeans in midwestern industrial cities, and blacks everywhere. Race restrictions applied not only to exclusive areas, where few minorities would have had the financial resources to buy, but also to middle-income neighborhoods.

The use of race restrictions remained widespread until the 1948 U.S. Supreme Court decision in *Shelley v. Kramer* (334 U.S. 1) ruled that they were not legally enforceable. Even after that, however, they occasionally appeared. Because race restrictions were written at a time when cities were expanding and much new land was being developed and because they could be in effect for several decades, minorities' choice of residential location was often quite limited.

Discrimination in the mortgage banking industry also limited residential options. Even when buyers were willing to sell to minorities, lenders often refused to write mortgage loans for minority individuals or on properties in minority neighborhoods.

In addition, federal agencies practiced racial discrimination. In its early years, the FHA required that those developing new subdivisions for FHA-guaranteed mortgages insert racial covenants into the deeds. When this was no longer a legitimate practice, the FHA established underwriting criteria for their mortgage applicants—both people and properties—that greatly limited the pool of potential African American homeowners. At the same time, public housing authorities in many cities maintained segregated projects, building and renting apartments for blacks in minority neighborhoods and for whites in white neighborhoods. Moreover, the number of apartments available to each race was rarely proportional to the population in need, so that minorities waited much longer than whites for assisted housing and lived in inadequate or too expensive housing in the meantime. Chicago's *Gautreaux* case is legendary for documenting discrimination by a public housing authority.

Finally, local governments have played a part in housing discrimination. Early in this century, when local public land use control by zoning was in its infancy, some cities wrote ordinances designating occupancy of blocks in the city by race, prohibiting African Americans from living in white neighborhoods and vice versa. Legal challenges gradually ended that practice. Since then, cities have often administered their zoning or housing codes in a discriminatory manner, maintaining high standards where whites lived while allowing violations and exceptions where blacks lived. More recently, some suburban municipalities have practiced racial zoning indirectly, setting development standards so high that minorities were effectively priced out of the community. The post-World War II Levittowns, whose builder is reputed to have said he could address a housing problem or a race problem but not both, reflect the collective impact of these various practices.

Efforts to eliminate racial discrimination in housing have been relatively slow to take effect and only partially successful. The Fourteenth Amendment's equal protection clause made discriminatory actions by a public body unconstitutional, but the issue received little public attention until the civil rights movement of the 1960s. Then the fair housing provisions of the 1968 Civil Rights Act prohibited much public and private discrimination. Still, discriminatory actions by individual property owners, realtors, and lenders were often difficult to detect. Moreover, to gain relief from the discrimination—whether public or private—the individual affected had to pursue the matter through the court system. So despite the existence of legislation and regulations to prohibit racial discrimination in housing, the actions of individuals and public bodies have combined with historic housing patterns to greatly limit minorities' choice of residential location.

Conclusion

Housing in the United States has traditionally been viewed as a consumer good rather than a public or social good. Thus, the United States has relied on the private sector to provide it, with government intervention only to address the inequitable by-products of the market system. But the nature of housing itself—the cost to construct and maintain it and its relative inability once built to be transformed or relocated—has limited the impact of that intervention. The result has been continued segregation of people in the United States by race and class and aggravation of the problems that segregation has caused. (SEE ALSO: **Discrimination; Federal Government;** *Right to Housing; Section 608; Substandard Housing*)

—*Patricia Burgess*

Further Reading

Doucet, Michael J. and John C. Weaver. 1991. *Housing the North American City*. Montreal: McGill-Queen's University Press.

Friedman, Lawrence M. 1968. *Government and Slum Housing: A Century of Frustration*. New York: Arno.

Hirsch, Arnold R. 1983. *Making the Second Ghetto: Race and Housing in Chicago, 1940-1960*. New York: Cambridge University Press.

Jackson, Anthony. 1976. *A Place Called Home: A History of Low Cost Housing in Manhattan*. Cambridge: MIT Press.

Jackson, Kenneth T. 1985. *Crabgrass Frontier: The Suburbanization of the United States*. New York: Oxford University Press.

Lubove, Roy. 1962. *The Progressives and the Slums: Tenement House Reform in New York City, 1890-1917*. Pittsburgh: University of Pittsburgh Press.

Mitchell, J. Paul, ed. 1985. *Federal Housing Policy & Programs: Past and Present*. New Brunswick: Rutgers, State University of New Jersey, Center for Urban Policy Research.

Radford, Gail. 1997. *Modern Housing for America: Policy Struggles in the New Deal Era*. Chicago: University of Chicago Press.

Wright, Gwendolyn. 1981. *Building the Dream: A Social History of Housing in America*. Cambridge: MIT Press. ◄

► Hogan

Hogan is a Navaho word, translating roughly into "home place." It has meanings of shelter, house, and a place for family activities. The hogan is the traditional housing of the Navaho people, a tribe of American Indians located in the Southwestern region of the United States, including the states of Arizona, Colorado, New Mexico, and Utah. The hogan is a circular and dome house. In Navaho mythology, the first hogan was built by the Gods and Spirits for a newly born Navaho goddess, Changing Woman. Because of this beginning, the Navaho hogan is viewed as sacred. Therefore, it must be constructed in a certain manner. After construction is completed and before occupation, a Blessing Way ceremony must be conducted to ensure good

Figure 13. Hogan Interior
Photograph courtesy of Ed Kosmicki/University of Colorado.

fortune for the new hogan and the household that will occupy it. Often, this is an extended three-generation family.

Traditional hogans are windowless, one-room dwellings. The single room is never subdivided. Hogans are ordinarily circular. They may also have a polygonal shape. The usual floor plans are octagons. The traditional hogan is never rectangular. It is almost 16 feet in diameter with a domed ceiling of peeled logs and a smoke hole at the top. The perimeter walls are about 5 ½ to 6 feet high. On average, the interior height of the dome will reach 10 feet. The floor of old style hogans was about 24 inches below ground level. This allowed a natural bench to be created inside the hogan. Modern style hogans, which were widespread on Navaho reservations in the mid-1990s, do not have this bench. The walls are constructed of overlapping mud-covered logs or sandstone. The dome roof is a cribbed ceiling made also of overlapping mud-covered logs. The dome form reflects the sacred sky. This hogan form is typical and represents the female element. A female family member typically owns the hogan. Male hogans are short cone-shaped structures.

The circular hogan has an east-facing door and earthen floor. On the first day of construction, the hogan's single doorway is centrally oriented toward the rising morning sun. The first rays of light mark the central point of the doorway. The traditional doorway was covered with animal hides or by a woven ruglike covering. Contemporary hogans use wood with hinges for the doors. Light enters through the doorway and smoke hole. Contemporary hogans may allow a single window located on the south or west wall. The north wall may never have a window because the dead reside in the north. The traditional hogan's hearth is on the dirt floor under the smoke hole that is 1 1/12 feet east of the center of the dome roof. The hearth

is on the flat surface of the floor with a large flat stone set at one of the cardinal directions of the fire. Contemporary hogans have replaced the hearth with small wood burning stoves.

Within the hogan, spaces are culturally and religiously defined. Males sit on the north side and females must sit on the south. Visitors and medicine people sit on the west, facing the east. A hogan thus represents the four cardinal directions with an east-facing door, reflecting the Navaho cosmos and the center of their religious being.

The hogan's form and technique of construction have remained relatively unchanged for centuries. In 1989, the University of Colorado in Boulder, Colorado, undertook an experiment in blending traditional architectural design with appropriate technology. This experiment included the construction of three demonstration hogans on the university's campus, using solar passive technology. Renovations slated for 1998 include plans for the hogans to serve as a religious center, conference facility, art gallery, and office for the local American Indian Community. (SEE ALSO: *Cultural Aspects; Native Americans*)

—*Charles Cambridge* ◀

▶ Home

In casual language, the terms *home* and *house* are often used interchangeably. In some instances, their meanings do indeed overlap. However, just as not every neighborhood is a community, so also not every house is a home. A house is a physical construction. A home is a mental construct. A house is a tangible, concrete object. Home is an elusive, nebulous notion. Home signifies an emotional attachment. It connotes a relatively permanent place of refuge where people find comfort, a base from where they can safely explore the outer world. It is something to which people like to return. Leaving home can be a psychologically difficult process, unlike vacating a house, which involves more pragmatic issues.

Not all houses are homes to their occupants, nor are all homes found in houses. Partners in newly formed households may hold emotional allegiances to the different hometowns in which they grew up. Sometimes, immigrants feel homesick for their country of origin, and soulmates maintain homes in each other's hearts. In this sense, home is lived experience.

Insofar as houses are homes, their coincidence is shaped significantly by economic and cultural factors. Historians have shown how the evolution of the idea of home in Europe parallelled the emergence of a new form of household in the wake of the Middle Ages. This process of household formation did not occur uniformly throughout society but took place first in the large houses of households of means rather than in the cramped quarters of the poor. Lacking patrons among the religious and aristocratic establishments during the "Golden Age" of the 17th century, Dutch painters depicted household scenes in houses populated by a new bourgeoisie and merchant class whose affluence derived in no small part from the labor by and exploitation of others, less fortunate, whose stark living conditions in the same low countries or in far-off colonized regions are not featured nearly as much in pieces of the period.

Also historically, cultural norms have typically ascribed to women responsibility for transforming houses into homes. Still today, survey questions and bureaucratic forms bestow the role of homemaker upon women who are so designated when not gainfully employed in the official labor force.

The same observation applies to environments of the home. Cultural definitions have assigned roles in the public domain primarily to men, whereas women have more typically been confined to the domestic realm for the tasks of homemaking and child rearing. Thus, low-density residential environments, although gender neutral in themselves, in practice discriminated against women in the particular cultural and economic context that prevailed.

Use of the house as home is similarly governed by cultural and economic factors. For example, the social organization of relationships among the members of a household occupying a house produces a certain pattern of spatial and temporal usage of the interior space. In hierarchical arrangements, this has usually meant that women and children were relegated to smaller and less desirable parts of the house (e.g., a dark kitchen at the back or a small room with little privacy). The physical design of a house, in turn, reifies the household structure. This is clearly seen in the traditional Chinese courtyard dwelling whose physical layout has served to maintain and reinforce patriarchal dominance.

Housing as home implies a measure of control over its interior space and security of tenure. Both can be provided by a variety of formal and informal mechanisms. In many countries, private ownership is the prevailing legal arrangement. In the ways and to the extent that the experience of home is related to private ownership, it is linked as well to barriers impeding or preventing access to private ownership. The laws of certain countries, for example, prohibit women from owning property altogether, while, in other countries, women face extra hurdles in obtaining mortgage loans. Because of affordability problems and continuing discrimination, for many poor and minority households in the United States, the American dream of private home-ownership remains just that: a dream.

The differential meaning of home also manifests itself in other ways. Houses become homes, in part, by personalization—decoration of the interior, choice of furniture, extension of the porch, remodeling of the kitchen. Studies show how households often seek to use their homes as a means to give expression to their self-identity. This tendency may produce certain designs, colors, or floor plans characteristic of a particular ethnic or religious group and supportive of their customs and cultural preferences. It may also produce results patterned after occupational and class differences. Research has found, for example, that businesspeople tend to prefer living in ostentatious houses, making a symbolic statement about their status to the community, whereas those working in the helping professions are more attracted to more inward-oriented houses. Low-income-households, restricted to mass-produced, standardized

dwellings, have very limited opportunities for such self-expression. Likewise, researchers have found that at the level of the home environment, households that can afford it will "self-select" a neighborhood that supports their preferred lifestyle, while households with lower incomes are restricted to less choice areas.

Formal and informal rules for house design and official regulations are often not responsive to those who, by choice or otherwise, are not part of mainstream society. For example, occupancy standards of the Department of Housing and Urban Development are geared to the norms of white, middle-class households that aspire to separate spaces for preparing food, eating meals, and socializing and expect a particular assignment of bedrooms to children, none of which is uniformly shared by, for example, Hispanic households. Similarly, many Native Americans in reservations live in HUD housing built in disregard of long-standing and cherished tribal values. The new town of Chandigarh, India, remains one of the most vivid examples of a mismatch between houses provided by the experts, paid for by the powerful, and homes desired by the residents who live in them. Le Corbusier, brought in to help realize the (Eastern perception of) the Western ideal, supplied the telling words:

> Taking into account the wishes expressed by families? NO, I don't believe we can do that. We have to conceptualize, discern, and supply; put the question to the right person. . . . We have to build (the housing environment) and install the occupants, then Radiant City operates. (Chombart de Lauwe 1959, p. 201)

Fortunately, many of the entries in this encyclopedia offer positive examples of more enlightened approaches that emphasize design and planning of housing with greater participation by the households who will make it their home. (SEE ALSO: **Behavioral Aspects**; *Children; Cultural Aspects; Feminist Housing Design*)

—*Willem van Vliet*--

Further Reading

Altman, I. and C. M. Warner, eds. 1985. *Home Environments*. New York: Plenum.

Ardener, S., ed. 1981. *Women and Space: Ground Rules and Social Maps*. London: Croom Helm.

Benjamin, D. N. and D. Stea, eds. 1995. *The Home: Words, Interpretations, Meanings and Environments*. Aldershot, UK: Avebury.

Brolin, B. C. 1972. "Chandigarh Was Planned by Experts But Something Has Gone Wrong." *Smithsonian* 3(3):56-63.

Chombart de Lauwe, P. H. et al. 1959. *Famille et Habitation. Vol. 1: Sciences Humaines et Conceptions de l'Habitation*. Paris: Centre National de Recherche Scientifique.

Cooper Marcus, C. 1995. *House as Mirror of Self: Exploring the Deeper Meanings of Home*. Berkeley, CA: Conari.

Huizinga, J. H. 1953. *The Waning of the Middle Ages*. Garden City, NY: Doubleday Anchor.

Rapoport, A. 1969. *House Form and Culture*. Englewood Cliffs, NJ: Prentice Hall.

Rybszynski, W. 1986. *Home*. New York: Penguin.

Sandford, J. 1976. *Cathy Come Home*. London: Marion Boyars. ◄

▶ Home Care

Home care in the United States is a diverse and rapidly growing service industry. More than 19,000 providers deliver home care services to some 7 million individuals who require such services because of acute illness, long-term health conditions, permanent disability, or terminal illness. Annual expenditures for home care were expected to exceed $27 billion in 1995. Despite this substantial commitment of resources, there remains inadequate access to home care services, especially for individuals who need long-term care.

Home Care Providers

The first home care agencies were established in the 1880s. Their numbers grew to some 1,100 by 1963 and to nearly 19,000 in 1995. Home health agencies, home care aide organizations, and hospices are known, collectively, as "home care organizations."

The National Association of Home Care (NAHC), a trade organization located in Washington, D.C., that represents home care agencies, had identified a total of 18,874 home care agencies in the United States as of December 1995. This number consisted of 9,120 Medicare-certified home health agencies, 1,857 Medicare-certified hospices, and 7,897 home health agencies, home care aide organizations, and hospices that did not participate in Medicare.

Medicare-Certified Agencies

Home care agencies of various types have been providing high-quality, in-home services to people in the United States for over a century. However, Medicare's enactment in 1965 greatly accelerated the industry's growth. Medicare made home health services, primarily skilled nursing and therapy of a curative or restorative nature, available to the elderly and, beginning in 1973, to certain disabled younger people.

Between 1967 and 1980, the number of agencies certified to participate in the Medicare program nearly doubled, from 1,753 to 2,924. Between 1980 and 1985, the number of agencies nearly doubled again to 5,983.

The various types of home care agencies can be classified into two broad categories—freestanding and institution-based agencies.

FREESTANDING HOME HEALTH AGENCIES

- ▶ *Voluntary agencies.* These are visiting nurse associations (VNA) or similar nonprofit organizations governed by a board of directors and usually financed by tax-deductible contributions as well as by their earnings.
- ▶ *Public agencies.* These are governmental agencies operated by a state, county, city, or other unit of local government having a major responsibility for prevention of disease and for community health education. They are sometimes referred to as "official" or "governmental" agencies.
- ▶ *Proprietary agencies.* These are for-profit home health agencies and, as mentioned above, one of the fastest growing of all agency types.

- ▶ *Private nonprofit agencies.* These are privately developed, governed, and owned nonprofit home health agencies.

INSTITUTION-BASED HOME HEALTH AGENCIES

- ▶ *Hospital-based agencies.* These are agencies that are operated as an integral part (e.g., a department) of a hospital. Agencies that have working arrangements with a hospital, or perhaps are even owned by a hospital but operated as separate entities, are classified as freestanding agencies under one of the categories listed above.
- ▶ *Other institution-based agencies.* In addition to hospitals, a growing number of rehabilitation facilities and skilled nursing facilities are operating home health agencies.

Certified Hospices

Medicare added hospice benefits in October 1983, 10 years after the first hospice was established in the United States. Hospices provide palliative medical care and supportive social, emotional, and spiritual services to the terminally ill and their families, primarily in the patient's home. The core hospice interdisciplinary team includes physicians, nurses, medical social workers, and counselors, who coordinate an individualized plan of care for each patient and family.

Although the hospice concept dates to ancient times, the U.S. hospice movement did not begin until the 1960s. The first hospice in the country, the Connecticut Hospice, began providing in-home services in March 1974. The number of Medicare-certified hospices grew from 31 in January 1984 to 1,857 in December of 1995.

Noncertified Agencies

The 7,897 uncertified home health agencies, home care aide organizations, and hospices that remain outside Medicare do so for a variety of reasons. Some do not provide the kinds of services that Medicare covers. For example, home care aide organizations that do not provide skilled nursing care are not eligible to participate in Medicare. About 20% of these noncertified agencies are involved primarily in the provision of homemaker/home health aide services, about 25% are primarily providing hospice and/or bereavement services, and about 13% are primarily providing high-tech services. Most of the rest of these agencies offer a multiplicity of services.

Home Care Expenditures and Use

National Expenditures

National expenditures for personal health care totaled approximately $879 billion in 1995. Of this amount, nearly two-thirds goes for hospital care and physicians' services. Total home care spending is difficult to estimate, in part, because home care expenditures constitute only a small fraction of national health spending.

In annual estimates of national health care spending from both the Department of Health and Human Services and Congressional Budget Office, home care spending is only partially identified. Neither of these sources includes the home care delivered by noncertified home care agencies or hospital-based home health agencies. The most comprehensive estimate of national home care spending comes from the National Medical Expenditures Survey (NMES), which is conducted every 10 years by the Agency for Health Care Policy and Research. According to the 1987 NMES study, $11.6 billion was spent on home care in that year. More recently, the Health Care Financing Administration estimated home care spending of $34.6 billion in 1995. This represents roughly 4.0% of national health care spending.

Medicare is the largest single payer of home care services. In 1992, Medicare spending accounted for about a third of total home care expenditures. The second largest single payer of home care services is Medicaid. It accounts for about 25% on all home care spending. Several other public funding sources, including the Older Americans Act, Title XX Social Services Block Grants, the Department of Veterans Affairs, and Civilian Health and Medical Program of the Uniformed Services (CHAMPUS), account for less than 1% of total home care expenditures. Private insurance makes up about 5.5% of home care payments. About one-third of home care services are financed through out-of-pocket payments.

The Medicare Home Health and Hospice Benefits

HOME HEALTH CARE

When Congress enacted the Medicare program in 1965, it included coverage for home health care services. To be eligible, a beneficiary must (a) be homebound, (b) be under the care of a physician who establishes a home health plan of care, (c) need at least one of the qualifying services (intermittent skilled nursing, physical therapy, speech therapy, or continuing occupational therapy), and (d) receive medically reasonable, necessary care from a Medicare-certified home health agency. Eligible beneficiaries may receive Medicare-covered nursing and home health aide care on a part-time or intermittent basis as well as physical, occupational, and speech therapy services; medical social services; medical supplies; and durable medical equipment. There is no limit on the number of covered visits, and except for a 20% co-insurance for durable medical equipment, beneficiaries pay no deductibles or other copayment.

The home health benefit represents a relatively small proportion of Medicare spending—only about 9% of total benefit payments in 1995. About half of the estimated $180 billion Medicare benefit payments go to hospitals and nearly a quarter to physicians. Hospice payments account for less than 1% of total Medicare benefit payments.

In 1995, more than 37 million aged and disabled persons were enrolled in the Medicare program. About 3.6 million enrollees received home health services—nearly double the number of home health recipients in 1990. In the same period, Medicare home health expenditures more than tripled from $3.8 billion in 1990 to an estimated $16.2 billion in 1995. Most of the new spending occurred as a result of

visits, which increased from 63.7 million in 1990 to an estimated 252 million in 1995.

Growth in the Medicare home health benefit can be attributed to specific legislative expansions of the benefit and to a number of sociodemographic trends that have fostered growth in the program from the beginning and that will no doubt continue to do so for many years to come. These trends include the aging of the population, technological advances that have made it possible to substitute home care for hospital care for an increasing number of patients, and the increased public and professional awareness of home health care as an appropriate, safe, and often cost-effective setting for the delivery of health care services.

The first specific legislation expansions occurred in 1972, when Congress repealed the 20% co-insurance provision and extended Medicare coverage to disabled Social Security beneficiaries and to people with end-stage renal disease. Further expansions came in 1980 when Congress removed the 100-visit limit on home health services under both parts of the program, eliminated the three-day prior hospital stay under Part A, eliminated the applicability of the Part B deductible to home health services, and permitted unlicensed proprietary home health agencies to participate in Medicare.

The 1983 enactment of prospective payment for Medicare hospital patients brought another significant expansion of the home health benefit, because patients were discharged from the hospital "quicker and sicker." The percentage of all Medicare hospital patients discharged to home health care increased from 9% to 18% as a result of that legislation.

Several changes in claims processing practices occurred during the mid-1980s and, when combined with the changes in reimbursement methodologies, led to general confusion among agencies about what services would be reimbursed. The result was a dramatic increase in the number of claims denied as well as a so-called chilling effect in which some Medicare-covered claims were diverted to Medicaid and, regrettably, some patients went without care. This "denial crisis" led in 1987 to a lawsuit brought by a coalition led by Representatives Harley O. Staggers (D-WV) and Claude Pepper (D-FL), consumer groups, and NAHC. The lawsuit resulted in a rewrite of the Medicare home health payment policies to clarify the coverage rules.

Just as a lack of clarity and arbitrariness had depressed growth rates in the preceding years, the policy clarifications that resulted from the court case allowed the program for the first time to provide beneficiaries the level and type of services that Congress intended. The correlation between the policy clarifications and the increase in visits is unmistakable. The first upturn in visits (25%) came in 1989 when the clarifications were announced, and an even larger increase took place (50%) in 1990, the first full year the new policies were in effect. In the 1989 to 1992 period following the court ruling, the home health benefit increased at a rate that nearly doubled the 23% average experienced over the life of the Medicare program.

Projections from the Health Care Financing Administration (HCFA) actuaries indicate that the home health benefit has matured and that expenditure increases will fall to 12%

by 1998. However, even as the acute care benefit matures, new home care growth could occur as a result of comprehensive reform of the U.S. health care system or development of a federal long-term care program.

THE MEDICARE HOSPICE BENEFIT

Congress enacted legislation in 1982 that created a Medicare hospice program. Medicare-covered hospice services may be provided to terminally ill Medicare beneficiaries with a life expectancy of six months or less. The Medicare hospice benefit is divided into four benefit periods: two of 90 days, one of 30 days, and one of unlimited duration. The beneficiary must be recertified as terminally ill at the beginning of each benefit period. The following covered services are provided by the hospice as necessary to give palliative treatment for conditions related to the terminal illness: nursing care; services of a medical social worker, physician, counselor, home health aide, and homemaker; short-term inpatient care; medical appliances and supplies, including drugs and biologicals; physical and occupational therapies; and speech-language pathology services.

Medicare hospice expenditures have grown from $112 million in 1987, to $1.3 billion in 1994. A total of 221,849 beneficiaries received hospice services under Medicare in fiscal year 1994.

Medicaid

MEDICAID HOME HEALTH SERVICES

To receive federal matching funds, states must offer certain basic services in their Medicaid plans, including home health care. The home health benefit must include coverage for three mandatory home health services—part-time nursing, home health aide, and medical supplies and equipment—and one optional service category—physical therapy, occupational therapy, and speech pathology and audiology services.

Between FY 91 and FY 94, expenditures increased from $4.1 billion to $7.0 billion, an increase of nearly 72%. As in the case of Medicare, home health services represent a relatively small part of total Medicaid payments. Of the $108 billion in Medicaid benefit payments, more than half went for hospital and skilled nursing facility services. Home health services make up about 6.5% of the payments.

MEDICAID HOSPICE SERVICES

Hospice care is an optional Medicaid service, which in 1997 was offered by 42 states. Payments for hospice services were estimated at $198 million in FY 94.

Home Care Recipients

Estimates indicate that as many as 9 to 11 million people in the United Sates need home care services. Most will receive services from so-called informal caregivers—family members, friends, or others who provide uncompensated care. Some will also receive formal services—that is, purchased or compensated services—from home care providers.

The NMES findings indicate that 5.9 million individuals, roughly 2.5% of the U.S. population, received formal home

care services in 1987. Of these recipients, about half were over the age of 65, and the amount of home care they used tended to increase with age. About 40% of the home care recipients had functional limitations in one or more activities of daily living. Age and functional disability are likely predictors of the need for home care services.

Another survey, conducted by the National Center for Health Statistics, gathered information on client diagnoses. It found that more than 25% of home care patients admitted to home health agencies in 1992 had conditions related to diseases of the circulatory system. Persons with heart disease, including congestive heart failure, made up 49% of all conditions in this group. Stroke, diabetes, and hypertension were also frequent admission diagnoses for home care patients.

Many hospital patients are discharged to home care services for continued rehabilitative care. As hospital stays shortened in the early 1980s, the percentage of Medicare patients discharged to home health care increased from 9.1% in 1981 to 17.9% in 1985. The four diagnoses most likely to result in posthospital home health care use include patients who have had a stroke, heart failure, or major joint procedure and those with chronic obstructive pulmonary disease.

Caregivers

Informal Caregivers

Estimates indicate that almost three-quarters of severely disabled elders receiving home care services in 1989 relied solely on family members or other unpaid help. Eight of 10 informal caregivers provide unpaid assistance an average of four hours a day, seven days a week. Three-quarters of informal caregivers are female, and nearly a third are over the age of 65.

Formal Caregivers

Formal caregivers include those professionals and paraprofessionals who provide in-home health care and personal care services. Little information is available on the total number of formal caregivers. Neither the Bureau of Labor Statistics nor the major organizations that collect information on health care providers gather comprehensive information on the home care industry. The Bureau of Labor Statistics information, for example, does not include public home care agencies or hospital-based home care agencies. The HCFA gathers information only on Medicare-certified agencies. Rough estimates derived from these partial sources indicate that more than 600,000 individuals were employed by home care agencies in 1995.

Current Home Care Policy Issues

Improved Access

The desire to control costs while also improving access to care is the primary force driving health care policy in the United States. Total health expenditures were $988.5 billion in 1995, representing 13.6% of the gross domestic product. Despite this extraordinary commitment of resources, access to basic health care and long-term care services remains a problem for millions of people in the United States. Nearly one in four nonelderly Americans lacks access to basic health care services, either because of having no health insurance coverage or having insurance inadequate to meet their needs. Long-term care poses an even greater problem. Millions of people of all ages need long-term care services, and little help is available through federal or state programs.

Cost Containment

Cost containment strategies that seek improved access would have a direct impact on home care. For one, the pursuit of least costly care could result in an increased reliance on home-based care. In many cases, home care is a cost-effective service, not only for individuals recuperating from a hospital stay but also for those who, because of a functional or cognitive disability, are unable to take care of themselves. However, cost-effectiveness should not be the only rationale for home care. The best argument for home care may be that it is a humane and compassionate way to deliver health care and supportive services. Home care reinforces and supplements the care provided by family members and friends and maintains the recipient's dignity and independence, qualities that are all too often lost in even the best institutions.

Cost containment strategies that accelerate the trend toward managed-care delivery systems will also have a significant impact on home care. Managed-care plans create strong financial incentives to deny services, and home care's experience with managed care under the Medicare program has not been a positive one. Heavy-care patients, whose needs can be both expensive and lengthy, are especially vulnerable under managed-care plans that seek primarily to control costs. When plans fail to achieve sustained cost savings—and many payers have found that promised savings were experienced only as a one-time phenomenon—benefits are often reduced or eliminated altogether. Patients have sometimes been illegally denied home care services and had to sue insurers to get care. Managed care has not proven to be a panacea for controlling costs and ensuring access to appropriate services. Nevertheless, its use is likely to be encouraged as part of health care reform efforts. Therefore, stronger safeguards than those that exist under current plans are needed to ensure that cost savings are not achieved by denying needed services.

Medicare Cutbacks

In the 1990s, the substantial growth in Medicare expenditures for home health care caught the attention of policymakers. Under ever-increasing pressure to reduce federal spending, Congress repeatedly turned to large-ticket items such as Medicare. Although home health has remained a relatively small part of Medicare, it is among the fastest growing benefits. The sheer size of home health expenditure increases helped to make the program a federal budget target in 1997. The Balanced Budget Act of 1997 (P.L. 105-33) imposed new limits on home health agency reimbursement, which, as with other proposals for limiting growth (such as beneficiary copayments), is expected to result in slower spending growth at the cost of access to

services. (SEE ALSO: **Elderly**; *Foundation for Hospice and Homecare; National Association for Home Care*)

—*Jill L. White*

Further Reading

Bishop, Christine and Kathleen Carley Skwara. 1993. "Recent Growth of Medicare Home Health." *Health Affairs* (Fall):95-110.

Davis, A. Romaine, J. Boondas, and A. Lenihan. 1995. *Encyclopedia of Home Care for the Elderly*. Westport, CT: Greenwood.

Kaye, Leonard W. 1992a. *Home Health Care*. Newbury Park CA: Sage.

———. 1992b. *Home Health Care: Geriatric Case Practice Training*. Vol. 2. Washington DC: George Washington University Medical Center.

Nassif, Janet Zhun. 1985. *The Home Health Care Solution*. New York: Harper & Row.

National Association for Home Care. 1996. *Basic Statistics About Home Care*. Washington, DC: Author.

"New Developments in Home Care Services for the Elderly: Innovations in Policy, Program and Practice." 1995. *Journal of Gerontological Social Work* 24(3-4, Special issue).

Rivlin, Alice M. and Joshua M. Wiener. 1988. *Caring for the Disabled Elderly*. Washington, DC: Brookings Institution.

Spiegel, Allen D. 1983. *Home Healthcare*. Owings Mills, MD: National Health Publishing. ◀

▶ Home Energy Rating Systems

A home that is designed and built for a high level of energy efficiency costs more than one that is not. This difference in cost can vary from several hundred to several thousand dollars. The construction features responsible for the increased costs are usually not detectable to an untrained person, and despite decades of research and demonstration projects to educate consumers about the potential for energy efficiency, not all homebuyers ask questions about this aspect of a home when they consider a purchase. Reasons include lack of interest on the part of homebuyers. Other, more easily recognizable features, such as attractive kitchens and bathrooms, are often more important. Buyers may also assume that, because the technology for high levels of energy efficiency exists, all new homes are built to conserve energy.

Although many states have provisions in building codes that require specified levels of energy efficiency, these provisions usually result in *minimum* standards that do not reflect what can be achieved. Even in states where energy conservation construction codes do exist, they do not provide information to consumers that would allow them to estimate what their energy costs would be in homes they consider for purchase. Energy labeling systems, common for cars and appliances, allow consumers to estimate operating costs associated with fuels used to operate these items. Now these systems are gaining in popularity for homes, where they are referred to as home energy rating systems (HERS).

HERS programs have existed in some form since the 1950s, when General Electric developed the Gold Medallion home program to encourage the construction of affordable and efficient all-electric homes. Electric utility companies participated in this program and required these homes to meet specific insulation standards. The Gold

Medallion program continued until 1974; shortly after that time, other HERS programs began to appear, such as the Good Cents program implemented in 1976 by the Gulf Power Company. Since that time, more recognition has been paid to the effects of energy costs on housing affordability and the need for homebuyers to compare potential energy costs from one home to another.

The overall goal of HERS programs is to make comparisons of a home's level of energy efficiency as easily as already possible for cars and appliances. Although this objective seems straightforward and simple, it is more complicated than it seems. People may use energy differently in identical homes. A well-known study, conducted by Princeton University, found that the behavior of occupants in homes has more to do with energy consumption than do building features of these homes. This is because of individual preferences for thermostat settings, how often doors and windows are opened, and maintenance practices followed for furnaces and boilers.

HERS programs consist of prescriptive systems, which typically mandate levels of insulation and types of doors and windows, or performance systems, which specify broad objectives that can be achieved in different ways (e.g., alternative insulation or airsealing strategies). Homes are certified in some HERS programs through inspections that include specialized equipment, including blower door tests. These tests use a special fan that depressurizes a house and measures the rate of air infiltration.

Once a home meets requirements for a HERS program, it is registered as such. This registration can be used as a marketing tool. Even with advantages that HERS offer, barriers have existed to their widespread implementation. A formidable barrier has existed among lenders, who calculate debt-to-income ratios for potential buyers to determine maximum mortgage amounts. Because highly energy-efficient homes have a higher initial purchase price than conventional homes, many buyers have been unable to qualify for mortgages for the higher purchase costs, because they result in debt-to-income ratios that lenders consider too high. Progress has been made in overcoming this barrier through programs in the secondary mortgage market.

Agencies in the secondary mortgage market, such as Fannie Mae and the Federal Home Loan Mortgage Corporation, purchase mortgages from lenders and sell them to investors, thereby allowing lenders to have more cash available for more mortgages. These agencies now provide incentives to lenders to "stretch" debt-to-income ratios for homebuyers—in effect, by allowing homebuyers to qualify for the larger mortgages necessary for energy-efficient homes. This practice recognizes the contribution of energy costs to total shelter costs, which include principal, interest, taxes, and energy costs. With lower energy costs in efficient homes, buyers have more money available for principal and interest. Mortgages made in this way are referred to as energy-efficient mortgages (EEMs). Links are being forged between EEM and HERS programs, to make it easier for buyers to purchase energy-efficient homes.

Another barrier that has existed to widespread implementation of HERS programs is builder resistance. In the United States, the homebuilding industry is slow to change because of its fragmented and localized nature. Building codes that can vary from one municipality to the next contribute to this situation, as does the fact that homebuilding is a trade based in the crafts, not the sciences. This, too, is changing, as operators of HERS programs now provide builder training and cash incentives to construct homes that qualify for a home energy rating.

These advances are likely to make the construction of highly energy-efficient homes more a matter of routine, allowing consumers to compare energy conservation levels among homes and to make choices for more efficient use of energy resources, while at the same time making housing more affordable. (SEE ALSO: *Energy Conservation*)
—*Joseph Laquatra*

Further Reading

Hendrickson, Paul. 1988. "Home Energy Rating Systems: Information to Increase Efficiency." *Home Energy* 5(5):22-26.
Nisson, J. D. Ned. 1993. "New Council Begins Unifying Home Energy Rating Systems." *Energy Design Update* 13(5):1.
Sonderegger, Robert C. 1978. "Movers and Stayers: The Resident's Contribution to Variation across Houses in Energy Consumption for Space Heating." *Energy and Buildings* 1:313-24.
Vories, Rebecca. 1991. "Making 'HERS' a Household Word." *Home Energy* 8(5):30-35.
Wilson, Alex and John Morrill. 1996. *Consumer Guide to Home Energy Savings*. Washington, DC: American Council for an Energy Efficient Home. ◄

► HOME Investment Partnerships Act

The HOME Investment Partnerships Act is part of the Cranston-Gonzales National Affordable Housing Act, passed in 1990. The legislation emphasizes the role of state and local governments in overseeing affordable housing provision, and the HOME program channels funding to these local governments to use as they see fit in the provision of affordable housing. The program also emphasizes the roles of community housing developers and of the private sector, attempting to spur and to support public-private partnerships in the development and management of affordable housing. The act represents a "new paradigm for federal involvement in housing" that emphasizes local control.

HOME funds are allocated to state and local governments in the form of block grants with matching-funds requirements. Funds must be used to develop or to support affordable rental housing and homeownership programs. Governments can use the funds for acquisition or rehabilitation of affordable housing, to finance the costs and relocation expenses of displaced people, or for rental assistance. HOME funds can be used for construction of affordable housing only if (a) localities certify that construction is necessary for a particular neighborhood revitalization program, (b) the neighborhood is one of moderate- or low-income residents, and (c) housing will be built by a community development organization or a public agency. These criteria can be waived if construction will meet the needs of large families or people with disabilities or if the project is for single-room occupancy housing.

To encourage the work of community housing development organizations (CHDOs), HOME requires that at least 15% of funds allocated to a jurisdiction be reserved for these organizations; CHDOs can use the money to develop or buy affordable housing or for technical assistance. In addition, 90% of HOME funds used on a given project must assist families whose incomes are 60% or less of the median area income; for homeownership assistance, all funds must be used to assist low-income families. Funds are to be distributed to states and local governments, with 60% earmarked for local jurisdictions and 40% designated for states. Initially, matching requirements ranged from 25% to 50%, depending on the nature of the project; they were reduced to 25% to 30% in 1991 legislation. Funds are distributed into trust funds, with the stipulation that if money is not used within two years, it must be reallocated.

The HOME legislation also directed the Department of Housing and Urban Development (HUD) to develop model programs that local jurisdictions could adopt and to provide support to jurisdictions in their efforts to develop capacity to identify and meet affordable housing needs. By 1994, HUD program guides were available describing how HOME could be used for construction of affordable housing, rental-project financing, sweat equity programs, and first-time homeownership programs.

The HOME Act terminated several programs, including the rental rehabilitation and Housing Development Action Grants (HODAG) program within the 1937 Housing Act, the Section 312 rehab program, the 1987 Nehemiah homeownership program, the moderate rehabilitation program under Existing Housing Section 8, and the 1974 Urban Homesteading program. The Section 8 moderate rehab program had come under attack during a HUD scandal, unraveling as HOME legislation was being considered.

As part of the 1990 National Affordable Housing Act, HOME was a response by the federal government to the Reagan era, during which massive cuts in affordable housing programs had occurred, especially in construction. Critics blamed rising homelessness and affordability problems on these changes. As incoming president, George Bush had promised to work to restructure the nation's housing programs, with the help of his HUD secretary, Jack Kemp. The HOME program, however, was largely a Democrat-supported program, often competing against Kemp's pet program, HOPE, first for passage and then for funding. Republicans were successful in restricting the new construction component of HOME and of targeting the program as narrowly as possible to the very needy.

HOME received $1.5 billion for fiscal year 1991, $1 billion in fiscal year 1992, and $1.2 billion in fiscal year 1993, allocations higher than those requested by Bush. When Bill Clinton took office in 1993, only 2% of the allocated funds had been spent, of only 4% committed to specific jurisdictions. On his appointment as HUD secretary, Henry Cisneros promised to seek accelerated spending of these funds. Local governments and nonprofit developers testified to Congress in 1993 that changes in matching requirements, targeting requirements, use requirements, and definitions of community-based developers would facilitate use of the funds. For fiscal year 1994, the program received $1.3 billion, about $3 million less than Clinton had requested. In fiscal year 1997, the program received $1.4 billion. (SEE ALSO: *Community Development Corporations; Cranston-Gonzalez National Affordable Housing Act; Homeownership and Opportunity for People Everywhere; Single-Room Occupancy Housing*)

—Mara Sidney

Further Reading
Congress and the Nation. Vol. 7. 1993. Washington, DC: Congressional Quarterly Service.

Congressional Quarterly Almanac. Washington, DC: Congressional Quarterly. Published annually from 1948.

Hays, R. A. 1995. *The Federal Government and Urban Housing: Ideology and Change in Public Policy.* 2d ed. Albany: State University of New York Press. ◄

► Home Matching

The taking in of boarders and lodgers by families to supplement household income was commonplace in late 19th-century U.S. cities. The scarcity of rental housing and the limited incomes of farm-to-city transplants, European immigrants, and young, unmarried people made this an important housing alternative. By the 1920s, however, more available conventional housing and less favorable societal attitudes toward sharing living space contributed to its decline. The 1950s and 1960s witnessed an especially sharp drop in the incidence of this living arrangement, and interest in household sharing did not pick up again until the 1970s. The role played by community organizations in pairing home seekers with persons wishing to share their homes can be traced to at least as early as the 1950s, but the major growth of these matching services did not occur until the 1980s in response to the needs of elderly homeowners who saw the merits of home sharing. Whereas only about a dozen home-sharing agencies existed in the early 1980s, by the early 1990s, an estimated 278 agencies across the United States were involved in home matching. Most match-up programs linking older people with housemates are run by nonprofit organizations, although some are operated as programs by city governments or by public housing authorities. Their annual budgets are small (median amount is about $25,000). The most popular sources of funding are municipal and county governments, foundations, state governments, and private donors. Funding used by these entities may come from federal programs. Most home-matching programs are staffed by only one full-time employee and rely heavily on volunteers. There is, however, no single home-matching model. Most screen applicants and attempt to make appropriate matches, but others merely provide lists of persons interested in home sharing. Some (sponsored by housing-related programs) see their primary goal as the creating of affordable housing situations; others (sponsored by human service programs), provide living arrangements that offer companionship and personal assistance. Although most programs are intended to serve the elderly, they are increasingly addressing the home-sharing needs of younger populations.

Elderly persons are increasingly preferring to age in place (that is, remain as long as possible) in their own homes, even though sometimes they are overhoused (that is, living in larger and more expensive accommodations than needed), are living alone, and have disabilities and chronic health problems making independent living difficult or unsafe. Thus, they view home-sharing arrangements as a desirable alternative for several reasons. Home sharing offers a source of income, added physical and emotional security, companionship to combat loneliness, assistance with everyday chores or errands, and assistance with activities such as bathing, dressing, meal preparation, walking, and getting around. The housemates matched with elderly persons in turn have assumed a variety of roles, ranging from simply boarders or lodgers who provide only minimal household services to caregivers who provide hands-on personal care.

According to the latest decennial survey (1990) by the U.S. Bureau of the Census, only a very small percentage of U.S. older persons (about 2% of both males and females aged 65 and older) shared their households with someone who was unrelated to them. These statistics are misleading, however. In 1990, among persons over the age of 85, about 3.8% of men and 5.5% of women lived in these household arrangements. Moreover, in the 1970s and 1980s, this was one of the fastest growing living arrangements of elderly persons. The extent that shared household arrangements are arranged through an organization is unknown, although they probably make up only a small fraction of the total. Nonetheless, current knowledge about the causes and consequences of older people living with an unrelated person primarily derives from research on these formally arranged matchups.

Older people interested in home sharing are usually homeowners, women, living alone (and often widowed), and persons with no more than a high school degree. They can be divided into at least three distinctive groups. The first group, often referred to as traditional home sharers, includes relatively healthy, active, and younger elderly persons who use the home-sharing arrangement to supplement their monthly incomes, to avoid being alone, to secure assistance with property maintenance tasks, and to have someone around in case of an accident or medical emergency. The second most common type of match involves older people, usually at least in their 70s, who need assistance with housekeeping, cooking, shopping, and other errands because they have some physical impairment. In return for the assistance obtained from these matches, these older people often provide free room and board. A third category of matches involves physically or mentally very frail elderly who are almost completely unable to care of themselves or their homes. They require almost constant supervision and much assistance when performing their everyday activities. Because of the substantial time and emotional demands placed on the housemate, these matches are the most difficult for agencies to arrange and constitute only a small percentage of all matches. Usually, the older person must offer these housemates free room and board plus a monthly stipend.

Housemates seeking home-sharing accommodations are younger, often under age 30, have lower incomes, and are often in transitory periods of their lives (students, the temporarily unemployed, recent immigrants, the poor, displaced homemakers, and the recently divorced or separated). Those willing to provide personal care are somewhat older.

The obvious drawbacks of home sharing are a loss of privacy and autonomy by the older occupant and the possibility of personality and lifestyle conflicts with the housemate. These matches are also often short-term arrangements. Because of incompatibility, the transitory characteristics of the home seeker, or the fluctuating needs of the older person, the average match lasts nine months. Thus, older persons wanting a home-sharing relationship often have to tolerate the prospect of having a succession of housemates or live-ins. The extra monthly income received from the housemates of older residents may also jeopardize their eligibility for food stamps or Supplemental Security Income payments. Very restrictive zoning ordinance definitions of "family" in single-family neighborhoods may also act as an impediment.

The largest and most prominent organization that promotes home matching for elderly persons is the National Shared Housing Resource Center (NSHRC) in Baltimore, Maryland, established in 1981 by Gray Panther activist Maggie Kuhn. NSHRC is now a volunteer-led organization directed by 10 regional coordinators. Staff members offer training and technical assistance to shared-housing sponsors, and the organization serves as an information clearinghouse on shared-housing trends and issues. (SEE ALSO: Elderly; *Shared Group Housing*)

—*Stephen M. Golant*

Further Reading

Danigelis, Nicholas L. and Alfred P. Fengler. 1991. *No Place Like Home: Intergenerational Homesharing through Social Exchange.* New York: Columbia University Press.

Hemmens, G. C., G. J. Hoch, and J. Carp (Eds.). 1996. *Under One Roof: Issues and Innovations in Shared Housing.* Albany: SUNY Press.

Horne, Jo and Leo Baldwin. 1988. *Home-Sharing and Other Lifestyle Options.* Washington, DC: Scott, Foresman.

Jaffe, Dale J., ed. *Shared Housing for the Elderly.* Westport, CT: Greenwood.

Jaffe, Dale J. 1989. *Caring Strangers: The Sociology of Intergenerational Homesharing.* Greenwich, CT: JAI.

Schreter, Carol and Lloyd Turner. 1986. "Sharing and Subdividing Private Market Housing." *The Gerontologist* 26:181-86. ◀

▶ Home Mortgage Disclosure Act

Debates over redlining and reinvestment have historically focused on one central question: Do mortgage-lending practices and patterns reflect discrimination or risk? Lenders generally assert that as profit-making institutions their behavior is guided by rational assessment of risk, sound business practices, and the desire to lend money to anybody who is creditworthy. Industry critics contend that lending has long been influenced by subjective as well as objective considerations, resulting in discrimination against older urban communities and racial minorities throughout metropolitan areas. The debates were long carried out in the absence of public information on the geographic distribution of loans and the racial composition of applicants and borrowers. In 1975, the Home Mortgage Disclosure Act (HMDA) was passed as part of an effort to fill this information gap. In conjunction with the Community Reinvestment Act (CRA), passed two years later, requiring most lenders to affirmatively assess and respond to the credit needs of their entire service areas, including low- and moderate-income communities, HMDA has also served to encourage lenders to address more aggressively the credit needs of underserved areas.

HMDA is basically an information disclosure requirement under which most lenders are required to report the number, dollar amount, and type of mortgage loans made by census tract in metropolitan areas. They must also report the race, income, and gender of each applicant and the disposition of each application. The law has evolved and been amended several times since its inception.

From the time HMDA took effect in 1976 through 1989, the law required commercial banks, savings and loan associations, savings banks, and credit unions with assets of $10 million or more and offices in metropolitan areas to disclose publicly the number of mortgage loans they originated and the dollar volume of those loans by geographic location, usually census tracts. Lenders have been required to make this information available at an office within the metropolitan area, and there is a central repository where reports for all lenders within the metropolitan area can be obtained. Institutions are permitted to charge reasonable duplicating fees for their reports. Reports, as well as computer tapes and compact disks containing the raw data, can also be obtained from the Federal Financial Institution Examination Council in Washington, D.C. Information is also available via the Internet.

Initially, HMDA was a temporary law that was extended with each expiration date. Later, it became a permanent law. In 1987, it was amended to include the mortgage banking subsidiaries of bank holding companies and savings and loan service corporations. In 1989, it was amended again to include all mortgage-lending institutions that originate more than five mortgages in any metropolitan area. As part of the Financial Institution Reform, Recovery, and Enforcement Act (FIRREA—the savings and loan bailout bill) in 1989, mortgage bankers not previously covered by the act were included. More significantly, lenders now had to report the race, income, and gender of each applicant as well as the census tract of the property. Such information must be reported for each type of loan (conventional or government insured—Federal Housing Administration, Department of Veterans Affairs, or Fannie Mae) and the purpose of the loan (single family, multifamily, home improvement, and nonoccupant). FIRREA also requires lenders to report whether the loan was accepted or rejected and who purchased the loan if it was not retained in the lender's portfolio.

HMDA data have been widely used to track the flow of credit, to identify redlined areas and institutions that most egregiously avoided low-income and minority communities, and as the basis for several multimillion dollar reinvestment agreements that many lenders have signed since this law was enacted. More than $210 billion have been committed for reinvestment in agreements between lenders and community groups in more than 100 cities.

Important limitations of HMDA data have been noted. The fact that disclosure was restricted to aggregate data at the census tract level until FIRREA meant that experiences of individual applicants and borrowers could not be traced. This left open the possibility of the "ecological fallacy" by which conclusions about discrimination against individual borrowers could be erroneously inferred from findings drawn from aggregate data. Such data also provided no information on loan demand, thus raising the question as to whether low levels of lending resulted from discrimination by lenders, lack of demand by borrowers, or other causes. The individual applicant reporting requirements established by FIRREA eliminated some of these shortcomings. But HMDA still provides no information on factors such as the credit record or employment history of applicants or the condition of the properties they intended to purchase with the loans.

Despite these limitations, research based on HMDA data has proven quite valuable in addressing redlining issues. Many studies using HMDA data have found that more loans and more loan dollars are going to predominantly white and suburban communities than to older urban or predominantly nonwhite areas and that such relationships hold even after taking into consideration median family income, average age of housing, and other demographic factors presumably associated with risk. These findings have led researchers to move beyond HMDA data, to look at actual loan files of selected institutions and to send testers (pairs of individuals posing as mortgage loan applicants who are matched on every demographic characteristic except one—their race—to identify racial bias in preapplication screening), providing more definitive conclusions regarding racial bias. For example, the Federal Reserve Bank in Boston examined more than 3,100 applications at 300 area financial institutions and found that black applicants were rejected 60% more often than comparable white applicants as measured by 38 variables taken from the loan applications.

Another provision of FIRREA, which was stimulated at least in part by the findings of HMDA-based studies, is that it required regulators to publicly disclose the CRA rating and the rationale for the rating that they give to the institutions over which they have jurisdiction. (Under the CRA, institutions are periodically examined to assess their community reinvestment performance, and they are given one of four ratings: outstanding, satisfactory, needs improvement, or substantial noncompliance.) Consequently, community groups and the general public now have even more information on the mortgage-lending practices of financial institutions. As indicated above, such research has led directly to many community reinvestment initiatives. HMDA-based research has directly and indirectly led to more aggressive law enforcement practices by federal financial regulatory agencies and greater commitment by lending institutions to meeting the credit needs of low- and moderate-income families and racial minorities in the nation's metropolitan areas.

HMDA was never intended to be the sole piece of information on lending practices or credit needs. However, information that continues to be available from HMDA, along with additional data sources, is shedding new light on redlining problems and constitutes an essential tool in efforts to solve them. (SEE ALSO: *Community Reinvestment Act;* **Discrimination;** *Redlining*)

—*Gregory D. Squires*

Further Reading

Federal Financial Institutions Examination Council. 1996. *A Guide to HMDA Reporting: Getting It Right!* Washington, DC: Author.

Fix, Michael and Raymond J. Struyk. 1992. *Clear and Convincing Evidence: Measurement of Discrimination in America.* Washington, DC: Urban Institute Press.

Galster, George. 1991. "Statistical Proof of Discrimination in Home Mortgage Lending." *Review of Banking & Financial Services* 7:187-97.

Goering, John and Rob Wienk (eds.). 1996. *Mortgage Lending, Racial Discrimination, and Federal Policy.* Washington, DC: Urban Institute Press.

Munnell, Alicia H., Geoffrey M. B. Tootell, Lynn E. Browne, and James McEneaney. 1996. "Mortgage Lending in Boston: Interpreting HMDA Data." *American Economic Review* 86(1):25-53. ◀

▶ Home Owners' Loan Act of 1933

Signed into law on June 13, 1933, Public Law 73-43 was the first significant piece of New Deal housing legislation. Like other early New Deal legislation, its focus was on immediate relief from the tremendous problems presented by the Great Depression. In this case, the Roosevelt administration and Congress were responding to the high rate of foreclosures and potential foreclosures confronting homeowners and the mortgage industry. The primary purpose of the legislation was to help homeowners keep their homes and to prevent lenders from having to carry properties for which there was little demand. It directed the Federal Home Loan Bank Board (FHLBB), which had been created the previous year, to establish the Home Owners' Loan Corporation (HOLC). Federal Home Loan Banks had been authorized in the 1932 Federal Home Loan Bank System Act to make direct loans to homebuyers who could not obtain loans from any other source. This provision, however, had failed to stem the tide of the many foreclosures that were occurring at that time. HOLC, by providing long-term, complete amortization loans, with terms of up to 15 years and interest rates not to exceed 5%, was to salvage the many home mortgages that ran the risk of not being repaid. See the entry under Home Owners' Loan Corporation for more information on HOLC's accomplishments and impacts.

In addition, the act authorized the FHLBB to encourage the creation of federal savings and loan associations that would be examined and regulated by the FHLBB. Federal savings and loans were to make loans on the security of their shares or on loans made for homes and business property located within 50 miles of a federal savings and loan home office. A maximum of 15% of a federal savings and loan's assets could be loaned for amounts greater than $20,000 and further than 50 miles away. Such restrictions were established to make certain that savings and loans concentrated on home loans within their home territory.

Such restrictions lasted until the 1980s when, under the guise of deregulation, savings and loans were able to increase the proportion of their loans to commercial developments. Savings and loans subsequently made many bad loans, thereby prompting Congress to pass the Financial Institutions Reform, Recovery and Enforcement Act (FIRREA) in 1989 to bail out the savings and loan investment losses that had been insured by the federal government. (SEE ALSO: *Home Owners' Loan Corporation;* **Mortgage Finance**)

—*Charles E. Connerly*

Further Reading

Brueggeman, William and Jeffrey Fischer. 1977. *Real Estate Finance and Investments.* 10th ed. Homewood, IL: Irwin.

Jackson, Kenneth T. 1985. *The Crabgrass Frontier: The Suburbanization of the United States.* New York: Oxford University Press.

Keith, Nathaniel S. 1973. *Politics and the Housing Crisis Since 1930.* New York: Universe Books. ◀

▶ Home Owners' Loan Corporation

The Home Owners' Loan Corporation (HOLC) was created by the federal government in 1933 to refinance delinquent or defaulted single-family home mortgages during the Great Depression. By stemming the rising tide of home foreclosures, the HOLC stabilized local housing markets shaken by the economic downturn, restored the viability of financial institutions burdened by nonperforming real estate assets, and won the support of moderate-income homeowners for New Deal programs. The liberalized mortgage underwriting ratios and risk-rating system used by the HOLC were later adopted by the Federal Housing Administration (FHA), with the effect of catalyzing the growth of middle-class suburbia and accelerating the decline of older, inner-city neighborhoods.

By early 1933, during the depths of the Great Depression, more than 1,000 homes were being foreclosed each day. Shortly after ordering a national bank "holiday," President Roosevelt introduced legislation to establish the HOLC as a vehicle to bail out both small property owners and distressed lenders. In June 1933, Congress authorized the HOLC to issue $2 billion in government-backed bonds over a three-year period that would fund the purchase of nonperforming loans, finance short-term taxes and home repairs, and write new mortgages with long-term repayment schedules and low interest rates. Many owners of one-to four-unit buildings were unemployed and unable to renegotiate their outstanding debt on favorable terms with small local banks and thrifts weakened by the financial crisis. The HOLC provided 15-year loans for up to 80% of appraised value to these homeowners who lost their houses after 1930 or were unable to refinance. The maximum loan amount was $14,000, and the maximum appraised property value was $20,000, with monthly amortized repayments of principal at 5% interest. These underwriting terms were a significant departure from the practices of private lenders during the 1920s. In particular, high loan-to-value ratios and the concept of long-term amortization were important innovations that substantially advanced housing affordability for people in the United States.

By 1936, when it stopped making loans, the HOLC had refinanced 1 million homes totaling $3.1 billion, or 10% of the nation's nonfarm owner-occupied housing. The HOLC opened offices in 200 cities in each of the 48 states and the District of Columbia. Considerable local discretion was granted in the administration of the program, with loan approval controlled by state offices. One-fourth of all HOLC loans were made in Ohio, Michigan, and New York, with Detroit and Cleveland as two of the most active cities. In 1939, Congress extended the repayment term to as long as 25 years, and interest rates were lowered to 4.5%. When the HOLC was liquidated in 1951, it had foreclosed on 19% of its loans, acquiring and selling nearly 200,000 properties during its 18-year history.

The HOLC worked with banks and savings and loans to develop a risk-rating system that grouped neighborhoods into four categories: new, homogeneous, and "in demand" (A); "still desirable" (B); in the process of "declining" (C); and already declined (D). Color-coded maps were used to illustrate this geographic pattern of risk, with D neighborhoods colored red. When the FHA later adopted the HOLC ratings for its mutual mortgage insurance program, private lenders used these maps to "redline" neighborhoods that were defined as poor credit risks, exacerbating their decline. Although the HOLC approved only 54% of the loan applications it received, and few mortgages were refinanced in poor, densely populated black ghettos where the housing was predominantly rental, it did make loans in many lower-rated areas. In Chicago, 60% of HOLC loans were in C or D neighborhoods, and in a sample of HOLC lending in the New York region, 56% were in areas at least partially inhabited by immigrants or African Americans. Most important, HOLC helped to revive urban mortgage lending after the financial crash of the Great Depression and pioneered low-cost home financing techniques that the FHA used after World War II to increase the national homeownership rate and make suburban housing more affordable. (SEE ALSO: *Home Owners' Loan Act of 1933*)

—*John T. Metzger*

Further Reading

Cohen, Lizabeth. 1990. *Making a New Deal: Industrial Workers in Chicago, 1919-1939.* Cambridge, UK: Cambridge University Press.
Harris, C. Lowell. 1951. *History and Policies of the Home Owners' Loan Corporation.* New York: National Bureau of Economic Research.
Jackson, Kenneth T. 1985. *Crabgrass Frontier: The Suburbanization of the United States.* New York: Oxford University Press.
Schlesinger, Arthur M., Jr. 1959. *The Age of Roosevelt: The Coming of the New Deal.* Boston: Houghton Mifflin. ◀

▶ Homelessness

Definition

Homeless people lack regular, private access to conventional housing. This clearly includes people sleeping on sidewalks, in parks, in public places, and in emergency shelters. People temporarily sharing housing, receiving short-term custodial care (e.g., at a halfway house or detox center), renting a room by the week, or otherwise facing weekly shelter uncertainty do not always count as homeless. But most homeless come from this population, which sociologist Peter Rossi called "the precariously housed."

The Problem

Unexpected natural disasters, wars, and civil disruptions often produce wide-scale temporary homelessness. Refugees escaping floods or warfare suffer the loss of home and community and so become homeless. In contrast with such concentrated episodes of shared homelessness, which often arouse public sympathy and generous amounts of relief, dispersed and intermittent homelessness among the destitute poor seldom raises public alarm and care. In the United States in the early 1980s, the number of poor people sleeping in public places increased. The visibility of this privation evoked commentary and attention among citizens, analysts, and officials. Despite national economic improvement in the mid 1980s and 1990s, homelessness persists.

Identifying the Homeless

Counting the homeless aroused heated public debate. The Department of Housing and Urban Development (HUD) sponsored a national survey of shelter providers in 1984 that estimated the number of homeless people. HUD's estimate of 200,000 was substantially less than the estimate of three million that homeless advocates had made up for media consumption. Political disputes over the importance of the problem spilled over into methodological debates about the size of the population. Advocates pushed large estimates to support their claims about the importance of the problem and the need for public attention to take action and solve it. Analysts emphasized the importance of basing estimates on prudent enumeration methods unbiased by moral and political bias.

Still, disputes continue among analysts using different enumeration procedures. Studies measuring the proportion of the population who are homeless on a particular day (incidence) usually produce lower counts than studies that measure the proportion of the population who have ever experienced homelessness (prevalence). For instance, one recent nationwide study found that lifetime prevalence for all types of homelessness combined was 14.0% (26 million people) and 4.6% (8.5 million people) for the five years between 1985 and 1990. Lifetime "literal homelessness" (sleeping in shelters, abandoned buildings, bus and train stations, etc.) was 7.4% (13.5 million people). Five-year prevalence of self-reported homelessness among those who had ever been literally homeless was 3.1% (5.7 million people).

The Causes of Homelessness

Like many other low-status social groupings, homeless people do not possess the social power and standing to define their own identity. Classification categories reflect the relevant measures used by different analysts and caretakers to evaluate the characteristics, needs, motivations, resources, and conditions of homeless people. Inquiry into the causes, conditions, and prospects of the homeless follow different disciplinary pathways and so end up with different conclusions. Psychiatrists, psychologists, and social workers study mental illness and addictive behavior among the homeless. They focus on homeless people and ask the following kinds of questions: What characteristics or behaviors increase the prospects of becoming homeless? How might caretakers identify these and intervene to reduce their disabling effects? Community development officials, planners, low-income housing providers, and activists study the inefficiencies and inequalities of local labor and housing markets. They focus on the system of housing provision and ask the following types of questions: What government policies and market practices limit the affordability of decent housing? How might public and private housing providers increase the quantity and quality of affordable housing to the homeless?

Economic Distress

Homeless people usually come from households that experience chronic poverty. Few homeless persons have gone directly from middle-class prosperity onto the streets. However, why do some poor go homeless and others remain housed? First, homeless adults obtain less income for longer periods of time than do other poor people. They work less often, receive lower wages, and receive less public aid. Second, they tend to have less work experience and less education than do the poor who manage to avoid becoming homeless. Third, the number of low-wage employment opportunities for the unskilled has diminished in recent decades and thus reduced job availability and wage levels for the poor. Taken together, these changes have increased the share of destitute poor—people with incomes so meager that paying for food and other daily necessities leaves too little for shelter.

Mental Illness

Changes in health care practices for the mentally ill have greatly reduced the number living in state hospitals over the past 30 years. Outpatient treatment programs now serve the mentally ill who must find shelter in residential neighborhoods. Unfortunately, states do not budget sufficient funds to provide full-service residential care for the mentally ill in different community settings. Mentally ill poor people face greater risk of homelessness (than do those without such illness), and the experience of homelessness increases the risk of such illnesses among the destitute.

Drugs and Addiction

Alcohol addiction has long been associated with the urban homeless. The stereotype of the skid row bum exaggerated the scope of alcoholism but did describe a large category of the inner-city homeless. Drunkenness continues among large numbers of homeless who can afford inexpensive alcoholic beverages. The widespread availability of crack cocaine in the late 1980s exposed the inner-city poor to a powerful and relatively inexpensive drug. Crack addiction has contributed to increasing homelessness among the inner-city poor, especially among women and children.

Lack of Affordable Housing

Throughout the first six decades of this century, the destitute and transient poor (the old homeless) lived in the hotels and lodging houses located in urban skid rows. Condemned as obsolete and dangerous slum housing in the 1950s and 1960s, most of these single-room occupancy hotels (SROs) were demolished as part of downtown urban renewal efforts. Thousands of tiny (60 to 200 square foot), low-rent, single-room units were lost. Because local building codes outlawed new construction of SRO hotels or rooming and lodging houses, the destroyed units were never replaced.

For many poor households, rent increased faster than income in the early 1980s. This contributed to a growing housing affordability gap. But for the destitute poor, such increases were seldom cause for eviction. The homeless could not afford even the cheapest studio apartments. Shared housing increased as homeless adults who would have earlier moved into an SRO or lodging house doubled up with others. Government-subsidized housing was little help. Single homeless adults were not eligible for government-housing subsidy, and subsidized households were not allowed to shelter these adults.

Efforts to End Homelessness

Government policy debates throughout the 1980s tended to move back and forth between people-centered and housing-centered arguments about the causes of homelessness. The people-centered approach urged government to provide emergency shelter and a variety of social services to aid the homeless poor. The housing-centered approach emphasized the importance of providing long-term housing subsidies and removing regulatory barriers to the construction of a wide variety of shared-housing arrangements. Plans to care for the homeless usually combine both in a three-tiered scheme. First, emergency shelter and basic services take the homeless person off the street. Second, transitional housing offers social services, employment training, public aid, and from six months to two years of secure housing in a shared-living environment to equip the homeless person for independent living. Finally, in the third step, the shelter resident moves into his or her own low rent (preferably subsidized) rental dwelling, receiving occasional support services to cope with unexpected emergencies.

Most provisions and care for the homeless have come through the efforts of local government and not-for-profit community service agencies. Local philanthropic and caretaking institutions led the way in documenting the urgency of the homeless problem in the first half of the 1980s. Their resources could not keep pace with growing demand. It proved a hard sell. The federal government response to the homeless was initially belated and meager. Not until 1983 did the Reagan administration support congressional appropriations for an emergency food and shelter program administered by the Federal Emergency Management Agency (FEMA) that totaled about $140 million. During the same fiscal year, about $77 million dollars in HUD Community Development Block Grant funds were used by local governments to acquire and rehabilitate shelters for the homeless. An Interagency Task Force on the Homeless was also created in 1983 to find ways to use existing federal resources to help the homeless. Despite the small amount of funds involved, these federal dollars provided an important source of funds for local shelter providers and caretakers who used the federal funds to leverage additional contributions from state and local governments. Advocates and lobbyists for the homeless persisted and eventually succeeded in gaining legislative support for a bill that provided new federal funds for homeless shelters and services.

The 1987 Stewart B. McKinney Act marked the first major step by the federal government to address the problem of homelessness, not as a temporary emergency condition but as a serious social problem. The bill authorized more federal funds ($350 million) for the homeless in fiscal year 1987 than had been allocated in total over the previous four years. McKinney Act funds were appropriated to fund the local provision of food and shelter for the homeless while targeting services for the mentally ill homeless and those with substance abuse problems. The distribution of funds has shifted slightly over time with more money appropriated for transitional housing projects compared with emergency shelter projects and services for special populations.

The program elements of the legislation are based on a mobility model of individual effort and improvement. Proposals for shelters and service programs that target vulnerable population groups such as the elderly, youths, and women get funding priority. Almost a third of total funds go for specialized health projects that offer services and shelter for the physically and mentally handicapped homeless, as well as for those with substance abuse problems. The intent is to promote programs that will assist vulnerable and needy homeless individuals to recover their former independence and obtain permanent housing—hence, the increasing emphasis on funding transitional housing projects. But this does little to either prevent homelessness or increase the availability and affordability of housing for the poor.

The McKinney funds proved an important resource for shelter development and service provision. Local governments, but mainly nonprofit organizations, built a wide variety of emergency and transitional shelters. But the provision of affordable permanent housing has not kept pace. In 1994, a decade after its first report on the nation's homeless, HUD produced another more ambitious document outlining a plan to "break the cycle of homelessness." The plan urges adoption of emergency, transitional, and permanent supportive housing. The authors of this plan argue for a comprehensive approach that combines people-centered supports (e.g., addiction treatment) and permanent affordable housing (e.g., new SRO housing or subsidized rental units). This approach reframed the homeless problem as part of the larger problem of growing poverty at about the same time homelessness lost its standing as a popular public issue. A decade of reform efforts and study provided knowledge of what to do to remedy the problem, but the political will to put this knowledge to public use has languished.

HUD, facing deep congressional budget cuts and Republican threats of wholesale elimination in 1995, was forced to restructure. The McKinney Act homeless programs will be lumped into a categorical block grant allocated to state and local governments by a needs-based formula rather than by competitive funding. These governments will take responsibility for prioritizing and monitoring the use of the most likely smaller pot of funds. Meanwhile, HUD plans to develop new information systems that state and local governments can use to identify the homeless and coordinate the provision of shelters and services. In effect, homelessness has lost standing as a national problem. (SEE ALSO: **Affordability**; *Community for Creative Non-Violence; Hoovervilles; Interagency Council on the Homeless;* Journal of Social Distress and Homelessness; *National Alliance to End Homelessness; National Coalition for the Homeless; National Resource Center on Homelessness and Mental Illness; Partnership for the Homeless, Inc.; Single-Room Occupancy Housing; Stewart B. McKinney Homeless Assistance Act; Transitional Housing*)

—Charles J. Hoch

Further Reading

Burt, Martha. 1992. *Over the Edge: The Growth of Homelessness in the 1980's.* New York and Washington DC: Russell Sage and Urban Institute Press.

Hoch, Charles and Robert Slayton. 1989. *New Homeless and Old: Community and the Skid Row Hotel.* Philadelphia: Temple University Press.

Jencks, Christopher. 1994. *The Homeless.* Cambridge, MA: Harvard University Press.

Liebow, Elliot. 1993. *Tell Them Who I Am: The Lives of Homeless Women.* New York: Free Press.

Link, B. G., E. Susser, A. Stueve, J. Phelan, R. E. Moore, and E. Struening. 1994. "Lifetime and Five Year Prevalence of Homelessness in the United States." *American Journal of Public Health* 84(12).

Rossi, Peter. 1989. *Down and Out in America.* Chicago: University of Chicago Press.

U.S. Department of Housing and Urban Development. 1994. *Priority: Home! The Federal Plan to Break the Cycle of Homelessness.* Washington DC: Author.

Wolch, Jennifer and Michael Dear. 1993. *Malign Neglect: Homelessness in an American City.* San Francisco: Jossey-Bass. ◀

▶ Homeownership

Homeownership or, more precisely, owner occupancy, is part of the American dream. As an investment, it offers an opportunity to accumulate wealth. As a sign of personal achievement, it carries social status. Ideally, it takes the form of a single-family home. Even when bought from a speculative builder, this dwelling form is seen to offer abundant scope for the expression of individual tastes and creativity.

North America has one of the highest levels of homeownership in the world, especially considering that it is a mainly urban nation. Housing is usually more expensive in urban than in rural areas and most expensive in the largest cities. High land costs account for much of the difference, along with more elaborate municipal services and higher wage rates. As a result, a relatively low proportion of city dwellers can afford their own homes. This was especially true before the refinement and wide diffusion of condominium tenure, which has made it possible for households to acquire quite small and inexpensive dwellings in multiunit structures. For much of the 20th century, levels of urban homeownership have been about 30 percentage points lower in urban than in rural areas and lowest of all in New York City, the largest metropolitan area. As a result, it is most appropriate to compare the United States with countries that are urbanized to about the same degree. In terms of homeownership rates, the United States leads Europe, with the main exception (in recent years) of the United Kingdom; is broadly comparable with Canada; and lags slightly behind Australia.

Historically, the high level of urban homeownership in Canada and the United States has reflected a combination of circumstances. Cheap and abundant land was once very important, as was the early development of balloon frame technology in the third quarter of the 19th century. Balloon framing, employing milled lumber and nails, took advantage of cheap timber resources in North America. In contrast to the prevailing methods of wood and masonry construction in Europe, it requires less skilled and cheaper labor. Indeed, it is simple enough that many people are able to build their own homes. In Europe, building regulations and mandatory services helped to set a high minimum on the costs of urban housing, but in North America, many suburbs were unregulated and poorly serviced as late as the 1940s. As a result, many lower-income households were able to acquire their own homes—albeit modest in size and quality and sometimes on remote, ill-serviced sites.

In the past 60 years, mortgage credit has come to play a vital role in the growth of homeownership. In 1890, only 28% of all nonfarm owner-occupied homes were mortgaged. At that time, however, lenders required large down payments, loan terms were short (5 years was common), and the predominant "balloon" mortgage form (according to which the borrower paid only interest, the principal being due in full at the end of the term) did not greatly encourage homeownership. The major change came in the 1930s when the National Housing Act established the Federal Housing Administration (FHA), which introduced mortgage insurance and soon established the long-term amortized mortgage as the norm. Today, the affordability of homeownership depends heavily on the terms of mortgage credit. About two-thirds of all nonfarm owner-occupied dwellings are mortgaged, and more than four-fifths of all new homes are purchased with credit. These factors, coupled with rising real incomes, account for the growing level of urban homeownership over the past century. Except for the 1930s, the level has risen fairly steadily from about 33% in 1891 to 65% today.

Social Dimensions

The growing level of owner occupancy reflects not only that more people are able to own their own home but also that people own homes for a larger proportion of their lives. Owning a home requires a substantial income and also a down payment. For both reasons, young households are unlikely to own, or even to be buying, a home. In the 19th and early 20th centuries, it was not uncommon for owner-occupants to first acquire a home in their 40s. In many cases, this was the stage when older sons or daughters, still living at home, began to work and also to make a contribution to the household's income. Since World War II, the wide diffusion of higher ratio mortgages (with loan-to-value ratios of 80% and up) has reduced the need for substantial savings, and many people now acquire their first home in their 30s or even their late 20s. Even today, however, homeownership rates still increase steadily with the age of household heads, rising rapidly from 17.1% among those aged between 15 and 24, to 45% for those aged 25 to 34, to 66% for those aged 35 to 44, and peaking around retirement (80%, ages 55-64, and 79%, ages 65-74). The elderly are those who are most likely to own their homes outright.

People in the United States have long viewed homeownership as a mark of social status, to the point that some have suggested that acquiring a home makes a person, or a family, middle class. There is a certain element of truth to this in that the ownership of domestic property is an investment that can yield a long-term capital gain that changes the household's economic situation. As with any investment, there are risks, and for specific periods and in specific places, property values have fallen. Over most of the past half century, however, they have risen more rapidly than inflation. In addition, owner-occupants have been favored through the nontaxation of imputed rent, coupled with their ability to deduct mortgage interest from taxable income. As a result, buying a home has often been one of the best investments that most households are able to make, providing advantages to the owner that are denied the tenant.

Some writers have suggested that ownership of domestic property helps define social class, but the argument is controversial. More commonly, it is viewed as a reflection of income and status achieved elsewhere, usually in the labor market. In these terms, however, it makes little sense to see owner occupancy as a sign of the middle class. Historical research has indicated that the desire to own one's home has been especially strong among immigrant workers. Levels of homeownership among blue-collar workers have often been as high as among the professionals and managers, higher when income is controlled. Immigrants, many from rural cultures that attached great significance to the ownership of land, have been eager to put down roots in the New World. Workers, arguably, have sought to own property both as a source of security and also as a way of gaining control over some aspect of their lives. Far from being a middle-class trait, then, homeownership might be viewed as a feature of the immigrant working class.

Although immigrants in general have worked hard to acquire homes, cultural differences have often been invoked to explain ethnic and racial variations. Historically, there is evidence to suggest that some ethnic groups (e.g., the Irish) were especially eager to acquire property, whereas others (e.g., Jews) placed more emphasis on education. The case, however, has never been made convincingly, and evidence for Montreal, for example, indicates that it can be overstated. Throughout the late 19th and early 20th centuries, Montreal had a very low level of homeownership, which many observers attributed to the fact that French Canadians were not much interested in owning their own homes. Research, however, has shown that ownership levels among French and English Canadians in Montreal were virtually identical. Evidently, if there was a cultural influence at work, it was one of place, not ethnicity, and it rested on the historical development of a distinctive set of institutions—lending practices, building regulations, ways of using dwelling space—within the local housing market. Significant variations in local market conditions and in immigrant destinations make it difficult to generalize about the impact of ethnicity prior to World War II.

One of the most persistent contrasts in homeownership is that between native-born whites and African Americans. Throughout the 20th century, nonfarm ownership rates have always been about 20 percentage points lower among African Americans than among whites. In 1995, the rates were 68% and 43%, respectively. To some extent, this can be explained by lower incomes and family structures that have made owning a home more difficult. In recent years, the prevalence of female single-parent households has been especially important. To a still unknown degree, however, it is clear that discrimination has played a part. Early federal activity explicitly discouraged mortgage lending in racially mixed or black neighborhoods and, given the paucity of black-owned lending institutions, African Americans have found it especially difficult to acquire homes. A contributory factor has been the effect of exclusionary zoning, which for many years made it difficult for African Americans to move into the suburbs where homeownership has usually been most affordable.

Significance

The wide diffusion of homeownership has had extensive effects on individuals, communities, and the wider economy. The nature of some of these effects is clear, although none is entirely straightforward. One of the more direct effects of owner occupation is to reduce residential mobility. Mobility rates for owner-occupants are several times lower than for tenants. In part, this is because selling a house is always more costly and time-consuming than changing apartments, especially in a depressed market. To some extent, however, the direction of causality goes the other way too: Those who have families and comparatively stable jobs, and who do not expect to move soon, are the most likely to buy a home. One of the implications of this is that a neighborhood of homeowners is likely to be comparatively stable, not only in terms of residential turnover but also in its tenure and social composition.

On the average, owners also differ from tenants in having stronger attachments to a neighborhood, and not just because they are likely to have been residents for a longer period. Tenants also care about the livability of their dwellings and neighborhoods, but owners have a financial stake as well. For that reason, owners care more about what happens in the local neighborhood and are more likely to become involved in community politics. This is especially true in situations in which the character of the neighborhood—and hence, local property values—is threatened, for example, by proposed redevelopment. Even in areas that are mixed in their tenure composition, homeowners usually dominate residents' associations, and homeowner politics tend to be conservative. At the neighborhood level, it is associated with the NIMBY (not in my backyard) syndrome and, at the scale of whole suburban jurisdictions, with the exclusion of undesirable land uses and persons, the latter being defined in terms of income, race, or both. Homeowner interests, coupled with the strong tradition of local government in the United States, have played a significant role in the segregation and political fragmentation of metropolitan areas. They have also shaped significantly the nature of local democracy.

In economic terms, one of the effects of homeownership has been to underpin and promote growth. Single-family homes use up much more land than do other forms of housing. An extensive infrastructure of roads, pipes, wires, and so forth is required to sustain low-density suburban development. At the same time, single-family, owner-occupied homes promote an individualistic style of life in which every household is encouraged to buy its own array of appliances and automobiles. This form of development helps to sustain consumer demand and, hence, economic growth. This has been interpreted in various ways. To many, including a wide array of manufacturers and lending institutions, not to mention those involved in the real estate industry itself, it has been welcomed. Their argument is that consumers have bought what they wanted, thereby creating new jobs that have in turn helped to sustain consumer demand. This point of view has been more or less explicitly adopted by federal governments since at least the 1930s. Initially, housing programs were initiated to revive the construction industry, as well as to assist the manufacturers of building materials, furnishings, and appliances. Homeownership has been promoted in various ways, most notably through the tax system by allowing owner-occupants to deduct interest payments on mortgage debt. After World War II, implicit and explicit housing programs were complemented by a federal highway program that directly promoted the use of automobiles and, hence, led to suburban sprawl.

From at least the 1970s onward, however, the private and public sector proponents of suburban homeownership have come under increasing attack. Initially, Marxists (in particular) emphasized that growth has relied on the rapid and excessive extension of consumer credit, of which mortgages are by far the largest single form. This, arguably, has locked consumers into a treadmill while generating profits for financial institutions, as well as for the manufacturers of steel, rubber, lumber, and so forth. More recently, feminists have argued that individual homeownership in low-density suburbs has helped oppress women by creating excessive domestic work. Although, in principle, men might have shared equally in this work, in practice they still do not. Feminists, in particular, have also emphasized that homeownership is less affordable, and may in some respects be less appropriate, for "nontraditional" households, including singles and single parents. Some of these new housing needs have been met through condominium developments, but a number of writers have called for a more radical rethinking of ways in which people might be housed, ways that place less emphasis on individualized solutions and more on cooperation. Most recently, a new generation of scholars has raised concerns about the impact of low-density suburban development on the environment, both in terms of the use of natural resources, notably fossil fuels, and in the production of garbage, air pollution, and other wastes. In principle, not all of these criticisms apply to homeownership per se. However, the close association in the United States of this tenure form with a particular style of suburban living must raise questions about whether homeownership should be promoted as vigorously in the next century as it has been in the past. (SEE ALSO: *Homeownership and Opportunity for People Everywhere; Implied Warranties of Quality in the Sale of New Homes; Low-Income Housing Preservation and Resident Homeownership Act of 1990; National Homebuyers and Homeowners Association; National Low Income Housing Coalition; Owner Building; Owner Take-Back Financing; Shared Ownership; Taxation of Owner-Occupied Housing*)

—*Richard Harris*

Further Reading

Clarke, C. E. 1986. *The American Family Home, 1880-1960.* Chapel Hill: University of North Carolina Press.

Harris, R. 1990. "Working-Class Homeownership in the American Metropolis." *Journal of Urban History* 17(1):46-69.

Harris, R. and C. Hamnett. 1987. "The Myth of the Promised Land: The Social Diffusion of Home Ownership in Britain and North America." *Annals, Association of American Geographers* 77(2):173-90.

Hayden, D. 1984. *Redesigning the American Dream. The Future of Housing, Work, and Family Life.* New York: Norton.

"Homeownership for Low-Income Households: Benefits and costs for Residents and Communities." 1996. Paper presented at the Fannie Mae Annual Housing Conference, May 24, 1995. *Housing Policy Debate* 7(1, Special issue).

Hughes, James W. 1996. "Economic Shifts and the Changing Home-ownership Trajectory." *Housing Policy Debate* 7(2):293-325.

Megbolugbe, Isaac F. and Peter D. Linneman. 1993. "Home Ownership." *Urban Studies* 30(4-5):659-82.

Perin, C. 1977. *Everything in Its Place.* Princeton, NJ: Princeton University Press.

Pratt, G. 1986. "Housing Tenure and Social Cleavages in Urban Canada." *Annals, Association of American Geographers* 76:366-80.

Saunders, P. 1990. *A Nation of Home Owners.* London: Unwin Hyman. ◄

► Homeownership and Opportunity for People Everywhere

Enacted in Title IV of the landmark Cranston-Gonzalez National Affordable Housing Act (NAHA) of 1990, the Housing Opportunities for People Everywhere (HOPE) initiative resulted from the efforts of successive Republican administrations to legislate a national program of homeownership and self-management for the residents of public and publicly assisted rental housing. The public housing sales component of HOPE, in particular, sprang from the Public Housing Homeownership Demonstration of the Reagan Administration, which was based on the British ex-

periment of the 1970s and 1980s with selling publicly owned council housing. This strategy became a centerpiece of the housing policy promoted by the Bush administration and housing secretary Jack Kemp. The U.S. Department of Housing and Urban Development (HUD) administers the program.

Within the overall HOPE initiative were three main programs in Subtitles A, B and C, also known as HOPE I, II, and III. HOPE I and II support conversions of existing multifamily rental housing to homeownership, including cooperative ownership, and to resident self-management. HOPE I provides assistance for the conversion of public and Indian housing. HOPE II provides such assistance for residents in multifamily projects of five units or more that are Federal Housing Administration (FHA)-distressed or -insured or held by HUD, the U.S. Department of Agriculture (USDA), Resolution Trust Corporation (RTC), or state and local governments. HOPE III facilitates homeownership by single-family homebuyers in developments of one to four units, including scattered-site developments, owned or held by HUD, USDA, RTC, state or local governments, public or Indian housing authorities, and the Department of Veterans Affairs.

Legislation in 1992, the Housing and Community Development Act (HCDA), added a new subtitle, a Youthbuild program, to expand the supply of permanently affordable housing for homeless and low-income and very low income families by training and employing disadvantaged youth.

Under each HOPE program, HUD is authorized to award eligible applicants two kinds of grants to develop specific homeownership programs: planning and implementation grants. Applicants may be public agencies, housing authorities, resident councils, or nonprofits. With a few exceptions, the planning and implementation activities that may be undertaken, whether for multifamily or single-family housing, are the same. Planning grants offer up to $200,000 to fund the necessary preliminary activities leading to homeownership and include (a) training and technical assistance for applicants, (b) feasibility studies, preliminary architectural and engineering work, (c) homebuyer and homeowner counseling, (d) planning for economic development and job training that promote economic self-sufficiency, (e) development of security plans, and (f) preparation of an application for an implementation grant. HOPE I and II planning grants may also be used for development of resident management corporations and resident councils.

Implementation grants facilitate the actual conveyance of the property. Permissible activities include (a) architectural and engineering work; (b) legal fees; (c) acquisition of the property for the purpose of transferring ownership to eligible families; (d) rehabilitation; (e) temporary relocation during rehabilitation and permanent relocation of families that elect to move; (f) counseling and training of homebuyers and ongoing trainings after transfer; (g) economic activities that promote the economic self-sufficiency of homebuyers, residents, and homeowners; and (h) administrative costs not exceeding 15% of the grant. In addition, HOPE I and II grants fund the development of resident management corporations and councils, some project operating expenses and replacement reserves, and implementation of replacement housing plans.

To receive implementation grants, applicants must satisfy a matching requirement using nonfederal sources. For HOPE I applicants, the match is 25% of the grant amount; for HOPE II and III, the match is 33%. Federal low-income housing tax credits; HUD community development block grants; the value of land, real property, and infrastructure; fee waivers and deferrals; and in-kind contributions are examples of nonfederal match sources.

Eligible applicants for HOPE I and II planning and implementation grants are public and Indian housing authorities, other public organizations and bodies, resident management corporations, resident councils, cooperative associations, and private nonprofit organizations. Under HOPE III, private nonprofit organizations, cooperative associations, and public agencies in cooperation with private nonprofits are eligible applicants. The HCDA of 1992 established that mutual housing associations could also be eligible applicants under the HOPE II program and created a preference for public and Indian housing residents in the HOPE III program.

The definitions of eligible families differ by program. HOPE I program beneficiaries must be tenants in public or Indian housing at the time the implementation grant is approved, a low-income family, or a family or individual assisted under the housing programs administered by the USDA. For the purposes of HOPE II, eligibility is limited to tenants of the eligible property at the time of implementation grant approval or families or individuals whose incomes do not exceed 80% of the area median income adjusted for household size. HOPE III limits eligibility to first-time homebuyers with incomes at or below 80% of area median income.

Initial sales prices of units established under each of the HOPE programs (including principal, interest, insurance, taxes, and closing costs) should be affordable to eligible families at no more than 30% of adjusted income. Units transferred within 20 years of acquisition are subject to resale restrictions that limit the family's consideration for its interest in the property so that undue profits may not be realized. Within 6 years, families may withdraw their contribution to equity, an appreciation factor applied against equity and the value of any improvements. Within 6 to 20 years, HUD is authorized to recapture an amount equal to the remaining balance of the note on the property. Of the net sales proceeds that cannot be retained by the homeowner and that are recaptured by HUD, 50% are distributed to the entity that originally transferred ownership interests to the family. The other 50% is retained by HUD.

In the case of multifamily housing tenants, families that decide not to purchase a unit or are unqualified to do so are protected against eviction by reason of a homeownership program. They can continue to reside in the housing at affordable rents or, if electing to move, must receive a unit in other public housing or Section 8 rental assistance for use in other privately owned housing. (SEE ALSO: Af-

fordability; Homeownership; *National Affordable Housing Act of 1990*)

—*Robert Wiener*

Further Reading

Cranston-Gonzalez National Affordable Housing Act. 1990. Conference Report 101-922. 101st Cong., 2d sess. Washington, DC: Government Printing Office.

House Conference Report 101-943, Affordable Housing Act (P.L. 101-625), Committee of Conference, 101st Cong. 2d sess., reprinted in 1990 *U.S. Code Congressional and Administrative News Legislative History*, p. 6100.

House Report 102-760, Housing and Community Development Act (P.L. 102-550), House Subcommittee on Housing and Community Development of the Committee on Banking, Finance and Urban Affairs, 102d Cong. 2d sess., reprinted in 1992 *U.S. Code Congressional and Administrative News, Legislative History*, p. 3290.

U.S. Department of Housing and Urban Development. 1992. *1992 Programs of HUD*. Washington, DC: Author. ◀

▶ Hoovervilles

The Great Depression of the early 1930s left many people in the United States unemployed, without income, and eventually, without homes. Although a number of these homeless became transients who slept in various outdoor places, others—many with families—developed ramshackle, temporary communities. These communities came to be known as "Hoovervilles," named for the president at the time, Herbert Hoover, whom many blamed for the crisis.

By 1932, virtually every major city had a Hooverville. Typically, these shantytowns were built on empty lots and derelict lands on the outskirts of cities or towns. They consisted of informal arrangements of shacks built of whatever residents could find—scraps of wood and cardboard, fence posts, and frames of run-down cars and trucks.

Contrary to the common stereotype of the homeless as transient males or skid row bums, most of the residents of Hoovervilles were adult women and men who had been employed but lost their jobs and homes as a result of the Great Depression. Many had previously held blue-collar jobs. Residents varied in age from the very young to the elderly. Entire families settled in Hoovervilles, including numerous children.

Although life differed from place to place, many residents found odd jobs around the cities and worked in return for clothes, food, or occasionally, a little money. Some residents occupied their time by "rustling," collecting scraps from the streets and alleys to use or sell for cash. Children were often enrolled in public schools, although for many it meant a long walk. Food was found from scrounging, hustling, charity organizations, and philanthropists. The police often visited the Hoovervilles but seldom interfered with life there. Crime rates tended to be very low. Over time, many residents of Hoovervilles came together and organized social activities. For the most part, residents tried to maintain some semblance of life as it had been before.

However, sanitary conditions within Hoovervilles were not good. For example, in one Hooverville of more than 600 shacks, only 15 of the dwellings had any form of sanitary facilities. Although clean water was available to many from fire hydrants, most Hoovervilles had no running water or electricity. Roads were not paved or cobblestoned and, therefore, were muddy and not easily passable during rains.

Some of the more known Hoovervilles were located in St. Louis, Seattle, New York, and Washington, D.C. St. Louis's Hooverville was perhaps the largest. Located between railway tracks and the Mississippi River, at its peak, it had a population of more than 3,000 people. It included four churches, a welcoming center that distributed charity and hot meals, and a community center that held classes and provided entertainment.

New York was the site of several Hoovervilles. One, built atop a garbage dump, was home to approximately 600 people. New residents built their shacks on whatever unoccupied spot they could find. Supplies were plentiful from the dump, and water was available from fire hydrants. Some residents, able to move on, sold their homes to new residents for as much as $50.

Perhaps the most widely known of all Hoovervilles was established in Washington, D.C. In the summer of 1932, 25,000 unemployed veterans of World War I and their families congregated in Washington. They came to recover a soldier's bonus promised to them by Congress in the 1920s but not to be paid until 1945. They became known as the "Bonus Expeditionary Force" or the "Bonus Army." The largest encampment was alongside the Anacostia River. Other veterans put up tents and shacks on Pennsylvania Avenue, within view of the Capitol. Although peaceful, to some in the city, the mass of veterans appeared unsightly and terrifying. After several weeks, the army was called in by President Hoover to remove them. General Douglas MacArthur led troops who fired at the residents and burned their temporary homes. Two veterans were killed, hundreds were injured, and all of their tents and shacks were destroyed.

Most Hoovervilles were communities for people who had nowhere else to go. Owing to governmental assistance and the New Deal programs put into place by Franklin Delano Roosevelt's administration, by the mid-1930s, many Hooverville residents were able to find more stable work and, eventually, more permanent homes in established or new neighborhoods. Thus, Hoovervilles slowly diminished in size and number until the last one was demolished in the late 1930s. (SEE ALSO: **Homelessness**)

—*Heather C. Melton*

Further Reading

Crouse, Joan M. 1986. *The Homeless Transient in the Great Depression: New York State 1929-1941.* New York: State University of New York Press.

Miller, Henry. 1991. *On the Fringe: The Dispossessed in America.* Lexington, MA: Lexington Books.

Rossi, Peter H. 1989. *Down and Out in American: The Origins of Homelessness.* Chicago: University of Chicago Press.

Towey, Martin G. 1980. "Hooverville: St. Louis Had the Largest." *Gateway Heritage* 1(2):2-11.

Watkins, T. H. 1993. *The Great Depression: America in the 1930s.* Boston: Little, Brown. ◀

▶ Hospice Care

Since medieval times, there have been special guest houses for travelers in need of medical care. Wounded crusaders and pilgrims suffering from disease could go to a so-called hospitium where they would get meals, receive treatment, and have a place of relative comfort for recuperation.

In the early 1960s, British physician Cicely Saunders led the way for the modern hospice movement when she practiced and advocated the use of compassionate care for the dying. Her pioneering work resulted in 1967 in the establishment of St. Christopher's Hospice near London, England. However, today, hospice care is available not only in many institutional settings, including hospitals and nursing homes, but increasingly also in the community in people's own homes. This trend is in line with a demonstrated preference of many elderly for "aging in place."

Hospice care eschews the use of artificial life support systems and life-prolonging medical technology when there is no prospect for recovery. However, its purpose is not to shorten the final stage of life of those who are terminally ill. Rather, it aims to facilitate dying with dignity in a home-like environment, without pain and discomfort. To this end, it relies on a program of medical and social services designed to help terminally ill patients and their relatives and friends cope with the physical and emotional distress often associated with terminal illness. Although elderly persons dying of cancer have commonly been the recipients of hospice care during the last six months of life, there are growing numbers of beneficiaries among others facing life-ending circumstances, including those suffering from AIDS.

Most providers of hospice care meet a specified range of services and standard of care required for Medicare certification. They employ multidisciplinary teams that usually include a physician, a registered nurse, a psychologist and/or social worker, and a member of clergy. In addition, trained volunteers play a key role. In the late 1990s in the United States, over 70,000 volunteers contributed more than 5 million hours of service. The model underlying hospice care also stresses a participatory approach, encouraging involvement by relatives, friends, and the dying themselves. In 1998, there were about 2,600 hospices in the United States, averaging 25 clients. A total professional staff of more than 20,000 was estimated to serve over 340,000 clients annually, of whom two-thirds were older than age 65.

Research has found that hospice care is often less costly than conventional care during the final stages of life. Day-to-day care by volunteers and by the client's social network helps reduce costs, which are also lowered by use of less expensive technology and diminished overhead. Hospice care is a covered benefit under Medicare. Coverage is also available under the Medicaid programs of about 40 states and many private health insurance schemes. Financing policies are not uniform and vary from facility to facility. However, generally, hospices strive to accommodate clients on the basis of need rather than ability to pay. (SEE ALSO:

Acquired Immune Deficiency Syndrome; **Elderly;** *Home Care; Hospice Foundation of America*)

—*Diana E. Axelsen*

Further Reading

Buckingham, Robert W. 1996. *The Handbook of Hospice Care.* Amherst, MA: Prometheus.

Connor, Stephen R. 1997. *Hospice: Practice, Pitfalls, and Promise.* Bristol, PA: Hemisphere Publishing Corporation.

Delfosse, Renee. 1995. *Hospice and Home Health Agency Characteristics: United States, 1991.* PLACE: U.S. National Center for Health Statistics Division of Health Care Statistics.

Lattanzi, A. 1998. *The Hospice Choice.* New York: Simon & Schuster.

National Hospice Organization Ethics Committee. 1995. *Hospice Code of Ethics.* Arlington, VA: National Hospice Organization.

Saunders, Dame C. 1997. *Hospice Care on the International Scene.* New York: Springer.

Volicer, Ladislav, ed. 1998. *Hospice Care for Patients with Advanced Progressive Dementia.* New York: Springer.

► Hospice Foundation of America

The Hospice Foundation of America (HFA) was initially chartered in 1982 as the Hospice Foundation, Inc. to provide fundraising assistance to hospices operating in southern Florida. In 1990, HFA expanded its scope to the national level. A board of health policy experts guides the foundation's efforts to offer leadership regarding the entire spectrum of end-of-life issues, to serve as a resource on hospice issues and care of the terminally ill for policy makers, and to advocate for sound principles of hospice.

In 1996, HFA awarded over $186,000 to 51 organizations in 14 states. It distributes *Journeys,* a monthly newsletter, for recently bereaved people, and issues other publications of interest to the hospice world. As a not-for-profit public charity, HFA also conducts public education programs, sponsors research on ethical issues and the economics of terminal care, provides technical assistance to organizations and participates in the formulation and implementation of public policy. Financial support comes from private contributions, corporate grants, and other donations. For more information, contact HFA at 2001 South Street NW, Suite 300, Washington, DC Phone: (202) 638-5419 or 1-800-854-3402. (SEE ALSO: **Elderly;** *Home Care; Hospice Care*)

—*Diana E. Axelsen*

► House of Representatives, Committee on Banking and Financial Services

The Committee on Banking and Financial Services in the U.S. House of Representatives oversees legislation and regulation regarding the nation's financial system, its housing policy, and its urban affairs. Proposed housing policy is delegated to the Subcommittee on Housing and Community Opportunity, called Housing and Community Development before 1995. Committee priorities have shifted over time, as the committee responded to pressing issues

and presidential initiatives and as committee members and their policy priorities shifted.

The committee system evolved as a way for Congress to handle its diverse and growing workload. The House of Representatives' 19 standing committees, permanent committees with fixed jurisdictions, allow members to specialize in one or two policy areas. Committees have research staffs to aid their information-gathering and administrative tasks. Usually, the staff is organized on a partisan basis, with the committee chair overseeing hiring decisions; the minority party typically receives at least one-third of the committee staff. Representatives also have personal staffs, which they may use to supplement the committee staff.

Proposed legislation is routed to the appropriate committee, where members hold hearings on the proposal, debate its merits, and bargain among themselves, with the executive branch, and with interest groups. The committee chair and the relevant subcommittee chair decide whether and when the committee will consider a bill. Eventually, the committee either produces a resolution ready for debate and vote in the House or decides to abandon the proposal altogether.

The committee with jurisdiction over housing policy, originally named the Committee on Banking and Currency, was established in 1865. The committee had only nine members in 1865; by 1933, it had 25 members, and by 1989, membership reached its current level of about 50.

At the start of each Congress, representatives request committee assignments from the Committee on Committees for their political party. Current committee members may hold their assignments or request a transfer. Several criteria guide committee assignments, from member interest and expertise to geographical balance to electoral considerations such as how members might best serve their constituencies or improve reelection prospects. The House majority party will have a majority on every committee.

Typically, committee chairpersons are chosen based by seniority, meaning that the ranking member of the majority party will become the chair. Exceptions occur, however, and each party has rules that permit challenges to seniority-based assignments. Committee chairs appoint subcommittee members and chairs. Since the congressional reforms of the 1970s, subcommittees in the House have become more independent of their parent panel and subcommittee leaders have more control over proceedings, staff, and budget.

The committee's housing agenda has shifted over time as the president, interest groups, committee members, and other legislators developed the direction of federal housing policy. Thus, the committee has worked on legislation ranging from the federal government's first major intervention in the U.S. housing market during the Great Depression to the urban renewal and community development programs of the 1960s era to retrenchment and privatization under President Reagan. Groups lobbying the committee have typically wanted either more or less intervention from government, no matter which particular housing issue is under examination. In general, proponents of an expansive federal housing policy have included urban interests, such as the National League of Cities and the Conference of Mayors, the National Association of Housing and Redevelopment Officials, and advocacy groups for the poor, such as the National Low Income Housing Coalition. Opponents of government intervention typically include housing industry trade associations representing builders and realtors, along with the banking industry and business interests such as Chambers of Commerce.

Committee members themselves divide along party lines, where the split also occurs according to degree of support for government intervention with Democrats typically pressing for continued or increased intervention and Republicans seeking government restraint. Other splits that have emerged over the years among committee members include regional divisions and urban/rural splits.

The House committee has taken a more interventionist approach to housing policy than its Senate counterpart and than many recent presidents. Because the bodies often pass different versions of housing bills, compromises are usually reached in conference committees with members from the House and Senate. But even when a compromise has been reached, committee proposals can be altered on the House floor through the introduction of amendments. In addition, passage of a bill does not guarantee that the funding it authorizes will be appropriated. The House Appropriations Committee has become an increasingly significant policymaker, able to blunt the effect of a program authorization bill by funding it at a low level.

The committee's party orientation changed in 1994, when Republicans won the House majority for the first time since 1953. New congressional leaders changed the committee's name from Banking, Finance and Urban Affairs to Banking and Financial Services, reflecting a change in policy emphasis. In addition, the housing subcommittee's name changed from Housing and Community Development to Housing and Community Opportunity.

Although some bills pass through the committee and on to the House floor without much debate or opposition, others spur significant controversy. The bills that eventually became the Housing Act of 1949, authorizing a broad program of slum clearance and public housing, provoked at least four years of fighting among conservative and liberal legislators and lobbyists. Opponents called public housing "socialistic," whereas the Truman administration and congressional supporters called it a historic milestone that would advance the welfare of the American people. The bill did eventually win bipartisan support in Congress, and opponents shifted their focus toward the appropriations process.

Sometimes conflicts over proposed legislation pit congressional committees against the president or against each other. For example, President Reagan's early efforts to reduce government involvement in housing and to create incentives for private sector initiatives met opposition in the Democratic House. Although Reagan succeeded in reducing the scope of federal housing programs, congressional committees were sometimes able to temper the drastic reductions in housing allocation that Reagan sought. The House Banking Committee typically would increase funding authorization in the administration's proposed legislation, and compromises would be struck during conference committee sessions.

In 1981, for example, the House Banking Committee reported more generous housing measures to the floor than Reagan's proposals, but the Republican-controlled Senate was able to thwart these efforts in conference committee. In 1983, however, the House Banking Committee chair was able to win presidential and Senate support for a more generous housing bill by linking it to an unrelated bill that was important to Reagan.

Congress also functions to oversee the executive branch bureaucracy, and so the House Banking Committee in theory oversees implementation of housing legislation. Committee oversight techniques include informal communication between committee and agency staff, program evaluations by congressional support agencies such as the General Accounting Office, oversight and program reauthorization hearings, and review of proposed agency rules. In practice, owing to committees' heavy legislative workload, congressional oversight occurs sporadically, often when scandals or crises gain public attention. For example, in 1989, following a report by the Department of Housing and Urban Development's (HUD) inspector general, three congressional committees, including the House Banking Committee, began investigating influence peddling at the agency.

Committee hearings, reports (that include the text of proposed legislation), and summaries of activity for each congressional session are published by the Government Printing Office. They are available in libraries that are depositories of federal government publications. Committee staff can be reached at 2129 Rayburn House Office Building, Washington, D.C., 20515-6050; phone: (202) 225-7502. For hearings and schedule information, call (202) 225-7588. (SEE ALSO: **Federal Government**; *Senate Committee on Banking, Housing and Urban Affairs*)

—*Mara Sidney*

Further Reading

Congress and the Nation. Vols. 1-7. 1965-1993. Washington, DC: Congressional Quarterly Service.

Congressional Quarterly Almanac. Washington, DC: Congressional Quarterly. Published annually from 1948.

Congressional Quarterly Weekly Report. A weekly news magazine.

Rieselbach, Leroy N. 1995. *Congressional Politics: The Evolving Legislative System.* Boulder, CO: Westview. ◀

▶ Household

Household is a research term used to describe all persons who occupy the same housing unit. In its enumeration of the U.S. population and its living patterns, the Bureau of the Census considers a "housing unit" as "a house, an apartment, a mobile home, a group of rooms, or a single room that is occupied."

The Bureau of the Census uses household to track population trends, and the decennial census figures are used as a basis for program and policy decisions by government.

Many survey researchers and marketers also use census information for their academic or business purposes. Thus, the demographic and economic information relating to household behavior is used by both public and private sector decision makers to determine current needs and future directions.

Many ongoing national surveys, such as the Consumer Expenditure Survey (CES), the Current Population Survey (CPS), and the Survey of Income and Program Participation (SIPP), use relatively large samples of household units as part of the information they generate about consumer habits, population trends, and welfare status.

Furthermore, patterns of childbearing, the aging of the baby boom population, and trends such as people entering into marriages later in life, average family size, and teenagers leaving the home all are reflected in the household data generated by the census.

In the past, the Bureau of the Census used the term *head of household* to refer to the major income provider in a housing unit, or the oldest adult. Since the 1980 census, the term *householder* has been used to describe the person by whom the housing unit is owned or rented. In about 95% of the cases, the householder is requested to fill out the census questionnaire (who then usually returns the form by mail).

The Bureau of the Census divides households into two major categories—family and nonfamily households. Family households must have a householder present and at least one other family member related to the householder. A nonfamily household includes no relatives of the householder and, in most cases, signifies one person living alone.

Tracking the number of households in the United States reveals important information about population trends. There were approximately 63.4 million U.S. households identified by the 1970 census, about 80.8 million estimated by the 1980 census, and more than 91.9 million households identified by the sample survey of the 1990 census. The average number of persons per household has decreased over the years, from 3.67 persons per household in 1940 to 3.33 per household in 1960 to 2.73 persons per household in 1980 to 2.63 per household in 1990.

Researchers wishing to use census information are able to access these data by computer (through CD-ROM capabilities or other vehicles such as DIALOG information services), all of which are considered public information. With the information on household behaviors available through the census and other surveys, the household has become the major focus of survey research and consumer behavior. (SEE ALSO: *Household Composition; Household Size; Nontraditional Households*)

—*John S. Klemanski and John W. Smith*

Further Reading

Sardamoni, K. 1992. *Finding the Household: Conceptual and Methodological Issues.* Newbury Park, CA: Sage.

Sweet, James A. and Larry L. Bumpass. 1988. *American Families and Households.* New York: Russell Sage.

U.S. Department of Commerce, Bureau of the Census. 1992, March. *1990 Census of Population and Housing: Summary Population and Housing Characteristics: United States.* Washington, DC: Government Printing Office.

U.S. Department of Commerce, Bureau of the Census. 1992, November. *1990 Census of Population and Housing: Summary Social, Economic, and Housing Characteristics: United States.* Washington, DC: Government Printing Office. ◀

▶ Household Composition

The demand for housing is the result of three interrelated decisions. First is household formation, the decision by one or more individuals to leave an existing living arrangement and enter the housing market to obtain their own unit. Although traditionally thought of as a married couple with or without children, households also include single persons, single parents, and nonfamily groupings. Demographic, social, and economic factors determine the composition of households, and in recent years, the trend has been to nontraditional households.

Having decided to form a household, the second decision concerns tenure choice, the decision of whether to own or rent a housing unit. Third is the quantity decision of how much housing to obtain. Historically, more work has been devoted to understanding the determinants of these last two decisions than the household formation decision.

A household is virtually synonymous with an occupied housing unit. Excluded are persons living together in group quarters, such as dormitories, military barracks, and prisons, but unrelated individuals can constitute a household. The character of the housing unit or housing shortages can result in more than one household occupying the unit. Separate households exist if the unit has separate direct access to living quarters for different households. On the other hand, extended families including three generations would constitute a single household if in a single housing unit.

In recent years, the pattern of household composition has been changing dramatically. Historically, most households have been family households, defined as those that include relatives of the householder. Family households are primarily composed of a husband and wife with or without children. A majority of households have continued to be married couples, making up 56% of all households in 1996, and the number of such households has grown with the general increase in population and households. However, their percentage of total households has significantly declined since the 1960s, when they were almost 75% of all households (see Table 10). In contrast, a growing percentage of total households has been nonhusband wife family households. These households include single parents, generally mothers, with children and other family households, such as related individuals living together.

The most rapid percentage increases in households have been nonfamily households. Dominant among these types of households are single-person households, both male and female. Nonfamily households containing more than one member have been less important as a contributor to the

TABLE 10 Household Composition, by Percentage: 1996, 1990, 1980, 1970

Type of Household	1996	1990	1980	1970
Family households				
Married couple, with no own children under 18	28.8	29.9	29.9	30.3
Married couple, with children under 18	25.0	26.3	30.9	40.3
Mother with children	7.7	7.1	6.7	4.5
Father with children	1.6	1.2	0.8	0.5
Other family households	6.8	6.5	5.4	5.6
Nonfamily households				
Persons living alone	25.0	24.6	22.7	17.1
Other nonfamily households	5.1	4.6	3.6	1.7

SOURCE: U.S. Bureau of the Census. 1993. "Household and Family Characteristics, March 1992." *Current Population Reports.* P20-467. Washington, DC: Government Printing Office; U.S. Bureau of the Census. 1995. "Household and Family Characteristics, March 1994." *Current Population Reports.* P20-483. Washington, DC: Government Printing Office; U.S. Bureau of the Census. 1997. "Household and Family Characteristics, March 1996." *Current Population Reports.* P20-495. Washington, DC: Government Printing Office.

total number of households. although they have increased the most rapidly in relative terms. Other changes in household composition have been declining household sizes and increasing numbers of elderly households.

Changes in household composition are the result of demographic, social, and economic factors. Analysis of the population tends to occur in five-year age intervals. The 25- to 29-year-old and 30- to 34-year-old age groups are the prime household-forming age groups. These groups are leaving parents, group quarters, or other living arrangements to form new households. Persons of this age are likely to form nonfamily households. In the 1970s, these groups composed the post-World War II baby boom population in the United States and other countries that led to a surge in new household formations. The latter factor has decreased in importance as this age cohort has moved out of prime household-forming years and been replaced by the "baby bust" cohort that resulted from lower birth rates in the 1960s.

As the population lives longer and is healthier and thus able to maintain an independent lifestyle, an increasing percentage of householders, or persons who formerly were referred to as household heads, are over the age of 65. These households often consist of widows.

Changing fertility rates, delayed marriage, divorce rates, and an increase in the number of females having children outside of marriage cause changes in the number of married couples without children, single-person households, and single-parent families. As fertility rates have declined, households have become smaller. Delay in marriage may result in more single-person households being formed. The impact of marriage and divorce are not clear, however, because marriage may cause a decline if two single-person households are combining but may increase households if the partners are leaving family or other households.

Economic factors also contribute to changing household composition and may, in turn, contribute to social change. These factors operate through changes in age-specific householder rates, or the rate at which the population in a particular age group forms into households. The formation of a household is a result of the propensity and ability of a person or group to enter the housing market. Propensity relates to the desire for privacy. Ability is the economic wherewithal to act on privacy desires.

Propensity to form a household is in part a function of age, as children reach an age at which they desire independence from their parents. Tastes and preferences may change over time, influencing views regarding living arrangements, marriage, and divorce. Education may increase the desire for privacy through both the experience of attending school away from home and the knowledge and tolerance of lifestyle differences that include nonfamily households. Military service or other experiences living in group quarters and achieving a degree of independence increase the probability of establishing a household. Type of employment influences the likelihood of forming a household in response to career pressures and social culture. These factors may also affect the propensity to marry, possibly leading to a decision to delay marriage. Location of a job is also a factor in that migration to another location makes relatives unavailable as a potential living arrangement. Thus, an increasingly mobile labor force creates the need for additional households.

Primary among the economic factors affecting ability to form a household is income. As women entered the labor force in larger numbers, they increased their ability to form households as individuals or in a nonfamily household. This decision may be made by a never-married individual or as the result of a divorce. Increasing income for other individuals through changes in the economy or increases in government transfer payments such as welfare and Social Security allows the establishment or maintenance of independent households. Age at marriage is thus influenced by economic factors related to the ability of individuals to maintain separate households. Similarly, divorce may be an economic decision and may result in either an increase in households if both parties maintain separate households after a divorce or a decrease if both join existing households.

Households are individuals occupying a housing unit, so the availability and cost of housing influences the ability to form a household. High housing costs, particularly homeownership costs, can lead to both spouses in a marriage working to generate sufficient income to afford a house. Low housing costs can make it possible for an individual to acquire a housing unit. Finally, housing costs can enable an elderly individual or couple to maintain an independent household. Changes in housing affordability and income have led to delayed household formation in the younger population in recent years; in the early 1990s, almost one-third of unmarried persons between the ages of 25 and 29 lived with their parents.

Household composition also has a relationship to the housing stock. Certain types of households are more likely than others to occupy particular forms of housing. Young, newly formed households are likely to rent, childless couples and the elderly may be the best market for condominiums, and families with children desire large single-family homes. The composition of households is therefore partly dictated by the available housing stock, but the future housing stock is in turn built in response to expectations of future household formation patterns. As the baby boom population became young adults in the 1970s, it formed single- and two-person households and created a demand for apartments, condominiums, and similar housing units. As this group has matured, its housing needs have changed to include single-family units and more space to accommodate families.

Without consideration of the potential for changes in householder rates, the level of new household formations in the United States will decline into the next century because of the aging of the baby boom age cohorts and the smaller cohorts that will move into the prime household-forming years. In the peak years during the 1970s, an average of more than 1.7 million new households were formed per year; by the year 2000, that number may fall below 1 million as the total number of households in the United States will exceed 105 million. The projected rate of new household formation also has implications for household composition patterns, as the age structure of the population shifts to an older average and declining demand may lower housing prices, thereby increasing household formation and resulting in smaller households. Finally, as the baby boomers move into the 35- to 54-year-old age group, their incomes are expected to rise through their increased earning power in the labor force. (SEE ALSO: *Household; Household Size; Nontraditional Households*)

—*Marc T. Smith*

Further Reading

Joint Center for Housing Studies. *The State of the Nation's Housing.* Cambridge, MA: Harvard University, Author. Various years.

Masnick, George and Mary Jo Bane. 1980. *The Nation's Families: 1960-1990.* Boston: Auburn House.

Myers, Dowell, ed. 1990. *Housing Demography: Linking Demographic Structure and Housing Markets.* Madison: University of Wisconsin Press.

Sweet, James A. and Larry L. Bumpass. 1987. *American Families and Households.* New York: Russell Sage.

U.S. Bureau of the Census. *Current Population Reports.* Series P-25. Washington DC: Government Printing Office. Various issues. ◄

► Household Size

Because a household generally implies an occupied housing unit, the size of households has important implications for the level and nature of housing demand. The population could decline, but the demand for housing could remain unchanged if the size of households declined. Household size experienced a very rapid decline in the 1970s and has declined steadily throughout this century (see Table 11). The average number of persons per household, about 3.3 during the 1950s, is approaching 2.6 in the 1990s.

Persons who do not live in group quarters (which includes, for example, dormitories, prisons, military, nursing

TABLE 11 Average Household Size, 1790 to 1996

Year	Average Household Size
1850	5.55
1860	5.28
1870	5.09
1880	5.04
1890	4.93
1900	4.76
1910	4.54
1920	4.34
1930	4.11
1940	3.65
1950	3.34
1960	3.30
1970	3.14
1980	2.76
1990	2.63
1992	2.62
1994	2.67
1996	2.65

SOURCES: U.S. Bureau of the Census. 1975. *Historical Statistics of the United States, Colonial Times to 1970* Part 1, Series A-349-355. Washington, DC: Government Printing Office; Sweet, James A. and Larry L. Bumpass. 1988. *American Families and Households.* New York: Russell Sage; U.S. Bureau of the Census. 1992. "Household and Family Characteristics: March 1992." *Current Population Reports.* Series P20-467. Washington, DC: Government Printing Office; U.S. Bureau of the Census. 1997. "Household and Family Characteristics, March 1996." *Current Population Reports.* P20-495. Washington, DC: Government Printing Office.

homes, and the homeless) are divided into households, and for this population, a household refers to a person or group of people who occupy a housing unit. Households can be either family or nonfamily. Family households include married couples with children, married couples without children, and single-parent families. These households may include nonrelatives. Nonfamily households include single persons, the predominant type of such household, and unrelated persons who share a housing unit. About one-quarter of all households were single persons in 1990, compared with about 18% in 1970. Married couples have declined to fewer than 56% in 1994. A single household may include two families, such as two married couples or a family and an unrelated boarder, if their living quarters are not separated.

Several factors affect the average size of households. First is the birth rate or the number of children per household. Second is the proportion of households that are nonfamily households and, in particular, single-person households. Third is the divorce rate and the extent to which two households result from divorce. A related fourth factor is the number of single-parent families. Finally, the ability of the elderly to maintain an independent one- or two-person household affects average household size.

Household size is also related to householder rates, the rate at which the population in a particular age category forms itself into households. Average household size across the population can change as a result of the changing age structure of the population, with the best example being the post-World War II baby boom. The surge in births following the war created a large population cohort group that, as it moved into the prime household-forming years, generally ages 25 to 34 years, created a large number of new, small households. As this group has aged and moved into child-rearing years and smaller population groups have followed as a result of the decline in births following the baby boom, the demographic force creating smaller household size has moderated. The growth in elderly population also leads to an increase in one- and two-person households to the extent that this population can remain independent rather than merging with other households.

Households form because a person or group has the desire to form an independent household and is able to act on that desire. Desire is influenced by education, occupation, migration, and changing tastes and preferences, among other factors. Ability is a function of the income of the individual or group and of housing availability and cost. Housing cost may include the perception of expected housing price appreciation so that a household may form and purchase a house as an investment. Increased female labor force participation, government transfer payments, and perceptions of low relative housing prices are among factors that led to more single-person and nonmarried-couple households in the 1970s and 1980s. As smaller numbers move into the prime household-forming age groups in the 1990s and the following decade and because housing appreciation rates are expected to slow, the size of households may halt its downward trend. (SEE ALSO: *Household; Household Composition*)

—*Marc T. Smith*

Further Reading

Joint Center for Housing Studies. *The State of the Nation's Housing.* Cambridge, MA: Harvard University, Author. Various years.

Masnick, George and Mary Jo Bane. 1980. *The Nation's Families: 1960-1990.* Boston: Auburn House.

Myers, Dowell, ed. 1990. *Housing Demography: Linking Demographic Structure and Housing Markets.* Madison: University of Wisconsin Press.

Sweet, James A. and Larry L. Bumpass. 1987. *American Families and Households.* New York: Russell Sage.

U.S. Bureau of the Census. *Current Population Reports.* Series P-25. Washington DC: Government Printing Office. Various issues. ◀

▶ Housing Abroad: Canada

As a highly urbanized nation, Canada has high land and housing costs in the larger metropolitan areas. Canada's 30.5 million people live in 11.6 million households, 78% in urban areas. One-third of all Canadians live in the three largest metropolitan areas—Toronto, Montreal, and Vancouver (1998 figures). Urban as well as regional population concentration is a continuing trend. Ontario, the largest province, had 38% of the nation's population (5% more than three decades ago).

Demand for housing is fueled not only by this internal migration, leading to greater urban and regional concentration, but also by declining household size and high im-

migration levels. Average household size has slipped from four persons per household 1951 to 2.6 in 1996. During the same period, one-person households increased from 7% to 24% and two-person households from 21% to 32% of the total. Less than 5% of all households had five or more persons, compared with 33% in 1951. In addition, Canada accepts about 200,000 immigrants and refugees annually, most of whom settle in the three largest metropolitan areas. In 1998, 5.1 million Canadians (17% of the population) were immigrants and more than half (57%) lived in the Toronto, Montreal, and Vancouver metropolitan areas. The Toronto area alone is home to 1.5 million immigrants (38% of Toronto's population).

On average, Canadians spend 18% of their gross household income on housing (excluding utilities). About 20% of Canada's households spend more than 30% of household income on shelter. Of these, 58% were renters, 36% were owners with mortgages, and 6% were mortgage-free owners. Homeowners have about double the median household income of renters. Housing expenditures-to-income ratios are highly unequal owing to the large gap between rich and poor. The highest two-income quintiles spend 7% and 12% of household income on housing, the middle quintile spends 20%, and the lowest two spend 25% (the second lowest) and 36% (the lowest-income quintile).

Homeownership

Just under two-thirds of Canadian households are homeowners. The homeownership rate has fallen from 66% in the 1950s and 1960s to about 62% since the mid-1970s. There are significant regional variations: 75% in the Atlantic provinces, 66% in the prairie provinces, and 55% in Quebec. Ontario and British Columbia are close to the national average of 64%.

The typical house-building firm in Canada is small, building fewer than 10 houses per year. Even the largest firms, which may build up to 2,000 houses per year, are small in scale compared with the average firm in other goods-producing industries (e.g., automobiles, consumer appliances). Few house builders operate in more than one market area.

Until condominium ownership was introduced during the early 1970s, most homeowners occupied a detached or semidetached dwelling. About 57% of the housing stock consists of single, detached houses, a decline from about 70% in the 1950s. The introduction of condominiums allowed households in the more expensive urban markets to become homeowners. There are about 370,000 owner-occupied condominiums located mainly in the larger metropolitan areas (6% of the owner-occupied housing stock). Half of Canada's homeowners have paid off their mortgage. Property taxes and mortgage interest on owner-occupied houses are not tax deductible. The major tax advantage for homeowners is a tax exemption for capital gains on owner-occupied houses. In 1954, the federal government introduced mortgage insurance, protecting lenders from loss on insured loans. This not only made available greater private sector lending but also allowed Canada's banks to enter the mortgage-lending business. In the late 1960s, the practice of setting a National Housing Act (NHA) mortgage interest rate was discontinued.

Between the early 1970s and mid-1980s, a variety of short-term subsidy programs were introduced to assist first-time homebuyers and to stimulate home construction. They included a tax-subsidized down payment savings program, one-time cash grants to assist with the down payment, and mortgage interest rate subsidies. There has been no thorough evaluation of the impact of these programs, which were ad hoc responses to political pressure from the house-building industry and from middle-class voters facing high mortgage interest rates. Homeownership levels, however, did not increase as a result of these initiatives. Analysts tend to agree that these programs shifted demand for owner-occupied housing forward and that upwardly mobile middle-income households tended to be the beneficiaries. There had been no noticeable beneficial impact on the rental market in terms of relieving demand pressures. Interprovincial migration, immigration, and demographic change are major factors determining changes in demand in urban rental housing markets.

Private Renting

Canadians who cannot afford to own a house are renters in either the private rental sector (30.4% of households) or in the relatively small nonmarket social housing sector (5.6% of households). Private renting has increasingly become a residual sector. Anyone who can afford to buy a house or condominium tends to do so. Since the early 1970s, an increasing percentage of lower-income households are renters (from 45% in 1971 to 65% in 1996 for the lowest-income quintile), whereas fewer high-income households remain in the rental sector (a decline of 25% to 11% of the highest-income quintile households). This trend means that the private rental sector will continue to be dominated by households with incomes in the lowest two quintiles of the population, and consequently, the least economic resources to stimulate economic demand in the rental market. Growing social need rather than market demand is the result.

This polarization of Canadian households by tenure and income has already caused serious supply problems in the private rental sector. The rent levels required to cover the cost of building new units are well above the ability of a majority of renters to pay. Vacancy rates across Canada, especially in some major cities, have been extremely low since the early 1970s. Except for recessions, the average vacancy rate for the largest 25 metropolitan areas is about 2% over the past decades. Toronto and Vancouver typically have vacancy rates under 0.5%. These low vacancy rates have not led to much unsubsidized private investment in the construction of new rental housing. The continuation of a viable private rental market, as a source of new rental units, is in doubt owing to the upward cost pressures on supplying a rental unit and the downward trend in the income profile of renters.

Between 1974 and 1985, a number of federal and provincial rental housing supply subsidy programs were initiated, providing investors with either tax benefits or subsidized second mortgages for a 10- or 15-year period. The subsidies were not deep enough to assist those in need, and

the supply of new units was not adequate enough to have much impact on vacancy rates in the more expensive and tighter rental markets. The share of the rental sector composed of conventional apartment buildings continues to decline. The rental stock in the greater Toronto area (650,000 units in 1998), for example, consists of the following mix of unit types: 49% conventional private sector apartment buildings, 20% social housing units, 14% rented houses, 9% apartments in houses, and 8% rented condominium apartments. In 1994, Ontario became the first province to allow one self-contained apartment to be created by any homeowner, thus eliminating the illegal zoning status of some 100,000 apartments in houses.

Social Housing

About 625,000 rental units (5.6% of Canada's housing stock) is owned and operated outside the private sector. Subsidized housing policy and programs in Canada and the United States, once very similar, have developed along very different lines since the 1970s. Canada was one of the last Western nations to initiate a public housing supply program (in 1949). What is commonly known as public housing in Canada consists of federally subsidized housing owned and managed by public housing authorities with means test criteria that target 100% of the units for the very poor. There are about 205,000 public housing units, built mainly during the 1960s and 1970s (about 2% of the housing stock). Of the projects, 80% contain fewer than 50 units, although the 11% of the projects that have more than 100 units account for nearly half the total stock.

By the late 1960s, widespread dissatisfaction with public housing created the political will to experiment and innovate with improved means of supplying and managing nonmarket housing for low- and moderate-income households. The 1973 amendments to the NHA created a permanent stock of good quality nonprofit "social housing" along with a community-based housing development sector (commonly called the "third sector," in contrast to the private and state sectors). Responsibility for housing policy and programs is shared by the federal and provincial governments, although the federal government has played the major role. The Canada Mortgage and Housing Corporation (CMHC) is the federal government's housing agency that administers the NHA programs.

The public, private and cooperative versions of nonprofit housing as well as the rural and native nonprofit housing programs are together commonly called *social housing*. They are socially assisted (receive direct subsidies), and in contrast to the older public housing program, they house people from a broader social and income mix. The smaller-scale social housing projects, developed and managed by local groups, including the residents themselves, was viewed as a preferable option for tenants and for the communities being asked to accept them. Since 1973, Canada has built about 400,000 social housing units (which equals about 3.5% of the total housing stock). About 50,000 of these were financed by provincial governments (Ontario, Quebec, and British Columbia), with the rest financed jointly by federal and provincial governments.

The more innovative component of Canadian nonprofit housing, and as a result, the most closely watched and evaluated, is the nonprofit, nonequity cooperative housing program. After a review of housing and urban development policy in 1969, federal government housing officials agreed that socially mixed nonprofit housing cooperatives were an improved alternative to the large public housing projects that were often built in conjunction with urban renewal. The 1973 amendments to the NHA created a national co-op housing program. Unlike other types of nonprofits, members of housing co-ops own and manage their projects. It is a nonequity form of homeownership. In the mid-1980s, the federal government explicitly defined the objective of the cooperative housing program as providing "security of tenure for households unable to access homeownership." Co-op housing units cannot be sold or even passed on to a friend. When someone moves out, another household from the co-op's waiting list moves in. Because residents do not invest in the units, they take no equity when they leave. Canada's 2,000 housing co-ops (with 80,000 units) are democratically owned and managed subsidized housing. A majority (70%) of Canada's housing cooperatives are managed directly by the residents on a voluntary basis. About 30% of the cooperatives, usually the larger ones, retain full-time or part-time paid staff.

In 1986, the federal government and the cooperative housing movement agreed to experiment with a new mortgage instrument, the index-linked mortgage (ILM) rather than the equal-payment mortgage (EPM). Interest rates on ILMs are based on a fixed "real" rate of return—the rate of return the lender wants after inflation—plus a variable rate adjusted according to inflation. Therefore, no provision has to be built into the rate of interest to take account of risk—anticipated inflation—as there is in EPMs. This makes the initial payments of ILMs much more affordable to potential borrowers. To maintain the real rate of return that the lender wants, the interest rate is adjusted periodically based on the rate of inflation. The federal government's evaluation of the ILM after its initial five years in operation found that lower than real interest rates were realized by the ILM, making it "a more cost-effective mortgage instrument than the EPM," resulting in savings that makes the latest funding formula "a more cost-effective way to deliver co-operatives housing" than previously.

In the early 1990s, like the Reagan and Bush administrations in the United States, Canada's Conservative government made a decision to curtail much of its spending on social housing. Very few new projects are being built in the mid-1990s as a result. In the 1980s, the province of Ontario initiated its own unilateral social housing supply program to help address housing needs in that province, building about 45,000 units. A change in government in 1995 also resulted in the termination of Ontario's social housing supply program.

Conclusion

Canada's housing system is characterized by a reliance on the private sector to develop land, build new housing, and renovate existing housing. A small niche for nonprofit and cooperative housing emerged from a process of experimentation and steady growth through the 1970s and 1980s. By the mid-1990s, few subsidy programs remain in either the private or nonprofit sectors. The rate of homeownership is about the same as in the United States, even though there is no mortgage interest and property tax subsidy program in Canada.

The gap between housing costs and household incomes in Canada is a serious one, particularly for renter households. Many families and individuals have incomes that are inadequate to keep pace with the cost of housing in many parts of the country.

The Canadian experience with its innovative form of social housing supply and management shows that it takes time to build the capacity of the municipal and community-based nonprofit sector. A community-based nonprofit housing development capacity cannot emerge, mature, and sustain itself without longer-term stability in government commitment to social housing supply programs. Canada has spent almost two decades developing and investing in a community-based nonprofit housing capacity—with program delivery mechanisms that work reasonably well, with dependable, although recently declining, funding and increasingly experienced and sophisticated nonprofit developers and managers. The Canadian experience shows that the incubation process pays off down the road. In broad terms, what a successful nonprofit housing sector requires is a coordinated national system to undertake two goals: capacity building, to enhance the organizational capacity of the nonprofit sector at the local level; and development support, to provide subsidy resources for their housing development activities. This is what Canada began doing in 1973, although the federal government's commitment to this policy was being reassessed in the 1990s. (SEE ALSO: *Canada Mortgage and Housing Corporation; Cooperative Housing;* **Cross-National Housing Research**)

—*J. David Hulchanski*

Further Reading

Bacher, John C. 1993. *Keeping to the Marketplace: The Evolution of Canadian Housing Policy*. Montreal: McGill-Queen's University Press.

Bourne, Larry S. and David Ley, eds. 1993. *The Changing Social Geography of Canadian Cities*. Montreal: McGill-Queen's University Press.

Cooper, Matthew and Margaret Critchlow Rodman. 1992. *New Neighbours: A Case Study of Cooperative Housing*. Toronto: University of Toronto Press.

Dreier, Peter and J. David Hulchanski. 1993. "The Role of Non-profit Housing in Canada and the United States: Some Comparisons." *Housing Policy Debate* 4:43-81.

Fallis, George and Alex Murray, eds. 1990. *Housing the Homeless and Poor: New Partnerships Among the Private, Public, and Third Sectors*. Toronto: University of Toronto Press.

Hulchanski, J. David. 1988. "The Evolution of Property Rights and Housing Tenure in Post-War Canada: Implications for Housing Policy." *Urban Law and Policy* 9(2):135-56.

Miron, John R., ed. 1993. *House, Home, and Community: Progress in Housing Canadians, 1945-1986*. Montreal: McGill-Queen's University Press.

Pomeroy, Stephen P. 1995. "A Canadian Perspective on Housing Policy." *Housing Policy Debate* 6(3):619-54.

Prince, Michael J. 1995. "The Canadian Housing Policy Context." *Housing Policy Debate* 6(3):721-58.
Sewell, John. 1994. *Houses and Homes: Housing for Canadians.* Toronto: James Lorimer. ◄

► Housing Abroad: Western and Northern Europe

The housing systems of Western and Northern Europe are marked by considerable diversity. Economic, political, historical, and social influences have created different institutional, legal, financial, and organizational arrangements. Consequently, these countries have a different legacy of housing stock and built form and of tenure and ownership. The variety of arrangements is most commonly illustrated through the statistics of housing tenure, with Switzerland retaining a large private rented sector and Ireland and the United Kingdom very small private rented sectors, substantial nonprofit sectors in the Netherlands and Sweden, and large and recently expanded homeownership sectors in Belgium, Ireland, and the United Kingdom. In the latest phase of policy development, these very different systems have been exposed to very similar strains and pressures. Not only have their economies and financial systems been subject to the same global pressures with problems of economic recession, rising unemployment, and fiscal austerity, but these have interacted with changes in demographic structure (aging populations) and an aging housing stock to create new demands. Problems of homelessness have become more prominent everywhere, and increased immigration associated with changes in Eastern Europe, as well as in North Africa, has had a wide impact. Common responses to economic and social change have also been associated with the expansion and evolution of the European Union.

Common pressures and elements in changing systems have led some commentators to suggest that the housing systems in these countries have become more similar—a convergence thesis—but it is important not to understate very different starting points and differences in policy responses and the pace of change.

Over recent years, the major area in which housing innovations have occurred has related to privatization. This has been most prominent in the United Kingdom and Ireland where housing that is not provided through the market has mainly been provided directly by the State (through the local government system) rather than through autonomous or semiautonomous independent/voluntary organizations. In the United Kingdom, the major innovation of providing a right to buy to tenants of State housing is nearly 20 years old, and although there have been detailed changes in this policy area, the innovations in relation to privatization have involved additional policies. In the United Kingdom, a number of schemes have been designed to encourage demunicipalization or transfers of ownership away from the State. Legislation provides for tenants' choice of landlord, enabling tenants to choose (through secret ballot) an alternative landlord. An alternative route for tenants unable to take advantage of the right to buy has been a rent-to-mort-gage scheme enabling such tenants to become homeowners. In practice, neither of these schemes has had much effect, and there has been minimal response. The sale to sitting tenants remains the dominant mechanism for demunicipalization. This form of privatization has expanded most in Western and Northern Europe—but nowhere on the same scale as in the United Kingdom and Ireland. In the Netherlands, for example, government encourages privatization, but there is no right to buy granted to tenants. Rather, the nonprofit housing associations that dominate the Dutch social rented sector have the power to decide whether to sell or not.

These privatization schemes have operated in an environment of increasing social inequality and against a common background of concern to control and restrain public expenditure. Housing budgets have commonly been under pressure, and concern to target subsidy and other expenditures has grown. Rising rents and changing subsidy schemes have been common, and housing allowance systems have become more prominent as ways of assisting those in most need to meet rising rents. Although there has been considerable review of these schemes, there has also been increasing concern about the poverty trap for low-income households, and the increasing costs of housing allowance schemes have created a pressure for changes in social security arrangements. Rising rents have also made homeownership more attractive and encouraged people to take advantage of opportunities to buy as sitting tenants. In England, this combination of factors has led to a further privatization innovation–large-scale voluntary transfers. These involve the transfer of all or almost all of a local authority's stock to an independent housing association. Although this speeds demunicipalization, ironically, the motive has often been to set up an arrangement that will, in the long run, prevent all properties from being sold to homeowners under the right to buy.

Increasing social inequality and demand for rented housing in European countries has coincided with declining investment or subsidy for housing by the State and increasing problems on mass housing estates built in previous decades. In this environment, policy innovation has involved new schemes for the homeless, including increased provision of hostels and shelters in many major cities. More prominent have been policy initiatives to deal with the problem of run-down estates.

In France, a series of reports emphasized the social and economic problems of large estates (*grande ensembles*) and the need to develop new approaches with a strong participation by tenants, local communes, and landlords. Funding from government was linked to a formal agreement between all partners and was related to capital works and social programs. These programs had a considerable impact on landlords' general practice with a greater emphasis on participation. In Britain, a range of programs was designed to enable innovative approaches to be adopted in unpopular estates. These included the Priority Estate Project (PEP), Estate Action Initiatives, Housing Action Trusts (in England), and Partnerships (in Scotland). The PEP approach has matured over a period of more than 15 years, emphasizing local estate management and resident involvement.

Estate Action, between 1985 and 1995, has similarly emphasized local management and resident involvement as necessary elements in physical modification to estates. The emphasis on resident involvement has become a feature of initiatives, with government promoting the idea of a right to manage and tenant management organizations. In Denmark, the problems of less popular estates were addressed by devising a new financial framework that released money for capital expenditure and by reemphasizing cooperative structures and management. Increased emphasis was placed on tenants' role in management and on electing the majority to the board of social housing companies. Following this, almost all estates have functioning estate boards responsible for estate-level accounts.

A similar emphasis on tenant participation and empowerment has been a feature of initiatives relating to mass housing estates in other countries and has linked to debates about citizenship, governance, and social exclusion.

In these and other developments, there has been increasing concern about racial attacks and renewed demands resulting from immigration. These pressures have been particularly severe in Germany and less so in those countries farther to the north and west.

New approaches to housing for older people and for people with learning difficulties have also been prominent. A general reaction against institutional care has been strengthened by the fact that care in the community generally involves a lower cost. A range of innovations involve forms of community support and supported accommodation, day care centers, domiciliary workers, and a considerable variety of specialist support to enable groups previously consigned to institutions to flourish or survive within ordinary dwellings and neighborhoods. A related set of policy innovations has been concerned to support families and carers, by enabling them to have relief from continuous care roles. Finally, schemes to assist older people to maintain or adapt their existing homes so that they are able to continue to live in them if they so wish, relate to the same issue. These schemes have particularly been developed for homeowners and include subsidized agency services as well as assistance with the costs of work carried out on their properties. Awareness of the increasing significance in the population of older people and people with disabilities has also prompted new interest in designing homes that are barrier free or easily adaptable for lifetime needs. Incorporating appropriate elements in initial design increases costs of construction by a much smaller amount than the major adaptation costs incurred in converting a property designed without attention to lifetime changes. Awareness of lifetime costs associated with poor insulation and energy efficiency has also led to innovations in design and encouragement of this. The increased awareness of debates about the environment and sustainability have added to concerns about individual affordability in prompting innovative thinking.

A key feature of schemes relating to older people, community care, the revival of marginalized estates, homelessness, and the encouragement of increased private finance through privatization schemes is that of interagency working. The common agenda faced in recent years has led to

innovation in the ways of developing and carrying out policies as well as to innovation in policy itself. Thus, the language of policy in the 1990s is of partnership, joint funding, joint ventures, joint working, coordination, liaison, consultation, participation, and facilitation. There are major innovations in ways of working and ways of financing projects. This particularly involves State and nonprofit organizations working with the private sector. Public policy is increasingly developed in consultation and negotiation with potential partners and carried out in collaboration with them. The term *enabling* is increasingly used to signify a shift in the role of public agencies from a previous role as direct provider.

A final aspect of innovation in relation to Western and Northern Europe relates to the role of the European Union (EU). Under the treaties that have established the EU, European legislation applies in all member States of the community, and the European Court of Justice has the task of ensuring that the European treaties are interpreted and implemented in accordance with community law. Housing, however, was not included in these treaties and, significantly, has not been within the competence of the EU. Housing policy continues to be the responsibility of national and local authorities in individual member States. However, the agenda emerging from the EU in the early 1990s has a new emphasis on the generation of employment, economic regeneration, and the encouragement of partnerships involving employers, employees, and the State to achieve this. It is also concerned with problems of unequal economic development and lagging regions and with integration, cohesion, and combating social exclusion. These objectives are being pursued at a number of levels, but the key element from the housing perspective is the development of programs that target the same areas and communities that are central to the objectives of national housing policies. The new European agenda means that there are more funds and more programs for an increased proportion of the area and population of Europe. This is not directed at housing, but housing organizations are potential partners, especially if they are able to define their roles to embrace community development and regeneration. As a consequence of this, an increasing number of local initiatives involving housing are being developed with EU support. They reflect the language and priorities of the EU, and housing organizations are being influenced by the presence of European programs in developing new approaches and proposals and in organizing their activities. Innovations in housing in Western and Northern Europe are likely to be increasingly influenced by the policies and programs of the EU, and its expansion will extend this effect. (SEE ALSO: *Centrally Planned Housing Systems;* **Cross-National Housing Research;** *European Network for Housing Research; European Social Housing Observation Unit; Welfare State Housing*)

—*Alan Murie*

Further Reading

Boelhouwer, P. J. and H. M. H. van der Heijden. 1992. *Housing Systems in Europe: Part I. A Comparative Study of Housing Policy.* Delf, The Netherlands: Delft University Press.

Danermark, B. and I. Elander. 1994. *Social Rented Housing in Europe: Policy, Tenure and Design*. Delft, The Netherlands: Delft University Press.

Emms P. F. 1990. *Social Housing: A European Dilemma*. Bristol, UK: University of Bristol.

Lundqvist, L. 1992. *Dislodging the Welfare State? Housing and Priva-tisation in Four European Nations*. Delft The Netherlands: Delft University Press.

Malpass, P. and R. Means. 1994. *Implementing Housing Policy*. London: Open University Press.

Papa, O. 1992. *Housing Systems in Europe: Part II. A Comparative Study of Housing Finance*. Delft, The Netherlands: Delft University Press.

Power, A. 1993. *Hovels to High Rise*. London: Routledge.

Power, A. (forthcoming.) *Uneasy Estates*. London: Routledge. ◄

► Housing Act of 1949

The Housing Act of 1949 (P.L. 81-171) represents one of the most often cited pieces of federal housing legislation in history. Its long road to passage was filled with debate and controversy between conservatives and liberals. Conservatives consistently felt uncomfortable with the U.S. commitment to public housing and opposed the legislation. The real estate lobby and industry abhorred the growing presence of the federal government in housing. To them, the federal government was already too big and the legislation was seen to result in more centralization of power in Washington, D.C. The real estate industry felt it could meet the nation's housing needs with minimal interference from the federal government. Moreover, conservatives did not like that urban redevelopment was secondary to the primary goal of expanding the number of public housing units. To conservatives, public housing was not the tool to rebuild urban neighborhoods. They favored a more comprehensive urban redevelopment program.

Liberals countered the conservative views by stressing the social importance of housing. They supported public housing and its increasing numbers. The legislation included the authorization of the construction of 810,000 units of public housing over a six-year period. These units would replace those units destroyed by urban development. Liberals overcame conservative opposition and cheered the passage of the Housing Act of 1949. Eventually, the construction of these 810,000 additional units was relegated to secondary importance and took some 20 years to complete, because priority was given to redeveloping areas for commercial purposes.

A number of organizations participated in the lengthy debate over the passage of the Housing Act of 1949. Outspoken advocates included Catherine Bauer Wurster and W. Blucher. Among the groups supporting the legislation were the American Institute of Architects, National Urban League, National Housing Conference, League of Women Voters, AFL-CIO, and the U.S. Conference of Mayors. Among those opposing the legislation were the U.S. Savings and Loan League, Chamber of Commerce of the United States, National Association of Home Builders of the United States, National Association of Mutual Savings Banks, and the National Association of Real Estate Boards.

In passing the Housing Act of 1949, Congress declared the following broad national policy:

> The general welfare and security of the Nation and the health and living standards of its people require housing production and related community development sufficient to remedy the serious housing shortage, the elimination of sub-standard and other inadequate housing through the clearance of slums and blighted areas, and the realization as soon as feasible of the goal of a decent home and suitable living environment for every American family, thus contributing to the development and redevelopment of communities and to the advancement of the growth, wealth, and security of the Nation.

Title I of the act provides federal funding to assist localities with "slum clearance and urban redevelopment" efforts (which created what was later to be known as "urban renewal under the Housing Act of 1954). Cities, not the federal government, chose the areas to undergo urban renewal. Title II extended and authorized increased Federal Housing Administration mortgage insurance. Title III authorized the construction of more low-rent public housing units. Additional funding was also provided for housing research, technical services, farm housing research, a decennial census of housing, and for Washington, D.C., to participate in the Title I Slum Clearance and Community Development program.

President Harry S. Truman was deeply concerned that little progress had been made in improving the housing conditions of millions of U.S. families. To Truman, it was disgraceful that the private sector had failed to meet the nation's housing needs and that different localities could not eliminate slums themselves. As such, federal assistance was necessary to eliminate substandard housing and improve the living conditions of U.S. families. In a June 15, 1949, statement, President Truman proclaimed the Housing Act of 1949

> of great significance to the welfare of the American people. It opens up the prospect of decent homes in wholesome surroundings for low-income families now living in the squalor of slums. It equips the Federal Government, for the first time, with effective means for aiding cities in the vital task of clearing slums and rebuilding blighted areas.

—*Roger W. Caves*

Further Reading

Aaron, Henry J. 1972. *Shelter and Subsidies: Who Benefits from Federal Housing Policies?* Washington, DC: Brookings Institution.

Abrams, Charles. 1965. *The City Is the Frontier*. New York: Harper & Row.

Davies, Richard O. 1966. *Housing Reform During the Truman Administration*. Columbia: University of Missouri Press.

Fisher, Robert Moore. 1959. *Twenty Years of Public Housing*. New York: Harper & Brothers.

Gelfand, Mark I. 1975. *A Nation of Cities: The Federal Government and Urban America 1933-1965*. New York: Oxford University Press.

Hays, R. Allen. 1995. *The Federal Government and Urban Housing*, 2nd ed. Albany: State University of New York Press.
Phares, Donald, ed. 1977. *A Decent Home and Environment: Housing Urban America.* Cambridge, MA: Ballinger. ◄

► Housing Act of 1954

In 1953, the Eisenhower Advisory Committee on government housing policy and programs recommended that a broader program of urban renewal replace the narrow program of urban redevelopment that was found in the Housing Act of 1949. President Eisenhower supported the change, and one year later, the Housing Act of 1954 (P.L. 83-560) changed the program title of the Housing Act of 1949 from the Slum Clearance and Urban Redevelopment program to the Slum Clearance and Urban Renewal program. It followed President Eisenhower's proclamation, in a January 25, 1954, message to Congress, that "the national interest demands the elimination of slum conditions and the rehabilitation of declining neighborhoods." With housing problems continuing to plague the United States with millions of people living in slums, it became national policy to clear slums and blighted areas and, ultimately, to eliminate the causes of slums and blight. After signing the act on August 2, 1954, President Eisenhower indicated that its passage would "raise the housing standards of our people, help our communities get rid of slums and improve their older neighborhoods, and strengthen our mortgage credit system . . . strongly stimulate the nation's construction industry and our country's entire economy." It was now more than just a federal housing program. It stands as one of the most significant pieces of federal legislation ever passed that affected U.S. cities. It has also been controversial and criticized by some individuals for being a "people removal" program.

The new urban renewal program represented a more comprehensive tool that offered localities federal assistance for conservation and rehabilitation as well as assistance for slum clearance. The earlier "primarily residential" clause of the Housing Act of 1949 was broadened to allow urban sites to be developed for commercial and institutional developments, not just for housing. The previous requirement that cleared slum land had to be used for housing was strongly opposed by supporters of urban renewal.

To receive federal funding under the urban renewal program, localities were required to develop a "workable program." The workable program had to contain the following requirements:

1. Adequate codes and ordinances—standards for construction and standards for housing
2. A comprehensive community plan for land use and public capital development that would include a land use plan, a thoroughfare plan, a community facilities plan, a public improvement program, a zoning ordinance and map, and subdivision regulations
3. Neighborhood analysis to show the existence and extent of blight

4. Administrative organization adequate to an all-out attack on slums and blight
5. Program for relocation of displaced families
6. Citizen participation
7. Adequate financial resources for accomplishing the aforementioned requirements

The Housing Act of 1954 authorized several mortgage insurance programs. The Section 220 program authorized a Federal Housing Administration mortgage insurance program for rehabilitation and neighborhood conservation housing insurance. It was designed to assist the financing required for the rehabilitation of existing dwellings and the construction of new dwellings within urban renewal areas. The Section 221(d)(2) program offered mortgage insurance to increase homeownership opportunities for low- and moderate-income families, especially those displaced by urban renewal. The Section 221(d)(3) and Section 221(d)(4) programs provided mortgage insurance for low- and moderate-income families and for families displaced as a result of urban renewal activities. The 221(d)(3) program offered 100% mortgage insurance for nonprofit developers and 90% for profit-oriented developers. The 221(d)(4) program provided mortgage insurance for profit-oriented developers for moderate-income projects.

Assistance for facilitating comprehensive planning was also included in the act. Section 701 provided federal funding for the development of comprehensive planning at the state, regional, and local levels. More specifically, under Section 701, grants were made to

> state planning agencies for the provision of planning assistance (including surveys, land use studies, urban renewal plans, technical services and other planning work, but excluding plans for specific public works) to cities and municipalities having a population of less than 25,000 according to the latest decennial census. The Administrator is further authorized to make planning grants for similar planning work in metropolitan, or regional planning agencies empowered under State or local laws to perform such planning.

Many planning agencies owe their existence to Section 701 funding.

—*Roger W. Caves*

Further Reading

Anderson, Martin. 1964. *The Federal Bulldozer: A Critical Analysis of Urban Renewal.* Cambridge: MIT Press.
Davies, J. Clarence III. 1966. *Neighborhood Groups and Urban Renewal.* New York: Columbia University Press.
Flanagan, R. M. 1997. "The Housing Act of 1954: The Sea Change in National Urban Policy." *Urban Affairs Review* 33(2):265-87.
Greer, Scott. 1965. *Urban Renewal and American Cities.* Indianapolis, IN: Bobbs-Merrill.
Rossi, Peter H. and Robert A. Dentler. 1961. *The Politics of Urban Renewal.* New York: Free Press of Glencoe.
Rothenberg, Jerome. 1967. *Economic Evaluation of Urban Renewal.* Washington, DC: Brookings Institution.
Wilson, James Q., ed. 1966. *Urban Renewal: The Record and the Controversy.* Cambridge: MIT Press. ◄

► *Housing Affairs Letter*

Housing Affairs Letter is a weekly nationwide publication covering the housing markets, legislation, and regulations. First published in 1961. Price: $399 per year. First published in 1961, *The Housing Affairs Letter* contains current information on housing activity nationwide, including public, private, and subsidized housing; federal legislation; regulations; and housing finance, including secondary mortgage market developments. The subscription price is $409 per year for 50 issues. Address: C.D. Publications, 8204 Fenton St., Silver Spring, MD. 20910-2889. Phone: (301) 588-6380. Fax: (301) 588-6385.

—*Laurel Phoenix* ◄

► Housing after Disasters

Throughout the history of mankind, communities affected by earthquakes, floods, hurricanes, landslides, or fires had to deal with the problem of homelessness. However, only relatively recently has study of housing after disasters become a part of scholarly discourse. In most natural disasters, housing is particularly prone to the forces of destruction. For example, following the 1993 Killari (India) earthquake, about 250,000 stone masonry houses were destroyed, and in the 1995 Kobe (Japan) earthquake, more than 20,000 households lost their apartments and houses. To clarify the dynamics of housing delivery after disasters, this entry will first present the conceptual underpinnings that have emerged.

Four concepts can be distinguished. First, *emergency sheltering* refers to victims' refuge outside of their own homes for brief periods during an emergency. Spending several hours in a designated community shelter in an anticipation of a hurricane landing or staying for several days during a flood in a friend's house located on high ground are examples of emergency sheltering. *Temporary sheltering* extends beyond the emergency period, and in addition to refuge, it also includes activities such as feeding and auxiliary services. *Public sheltering*, the least favored of all sheltering options, is typically the most extensively planned of all postdisaster housing alternatives. Nonmass temporary sheltering arrangements (e.g., staying with friends or relatives) are preferred by the victims, but are less well understood and studied. In contrast to sheltering, both *temporary* and *permanent housing* involve the resumption of everyday household activities. In the case of the former, permanent housing will eventually be obtained, and in the case of the latter, the units obtained represent replacement housing. For a student of housing policy, the processes that surround the delivery of permanent housing after disasters are of central interest.

Following major disasters, housing recovery policies differ across societies depending on their cultural, economic, and political characteristics. These policies are determined by the link between the circumstances surrounding the disaster itself and the manner of housing delivery in nondisaster times. Thus, early postdisaster housing decisions have implications for long-term recovery. Early decisions to opt for full-fledged temporary housing, on the one hand, can escalate the cost of reconstruction and prolong the time for permanent housing completion but, on the other hand, can facilitate a proper rehabilitation process, as illustrated in Northern Italy after the 1976 Friuli earthquake. In general, reconstruction policies after disasters aim to improve the performance of structures in a next disaster, by addressing the four principal sources of vulnerability: siting, buildings, amenities, and resources.

Siting

Rapid and uncontrolled growth of urban populations puts a severe strain on cities already struggling with a shortage of available land. Especially in developing countries, the choice of hazardous locations contributes to housing vulnerability: High-risk sites such as hillsides and riverbanks often house high-density squatter settlements, owing primarily to severe housing shortages (e.g., in Caracas and Bombay) and to residents' low incomes.

Housing policy response to hazardous sites is determined by several factors. High land values and the advocacy of occupants of hazardous locations may prompt affluent societies to resort to site improvements and sophisticated technological solutions, which is typically the case in the United States and Japan. In contrast, in most developing countries, the traditional official reconstruction policy has been to abandon the site and relocate the people to a different area, presumably less vulnerable to earthquake hazards. However, the three highest housing priorities of low-income populations in developing countries are location (i.e., close proximity to jobs and services), security of tenure, and community services. Hence, in these countries, relocation after disasters rarely succeeds in the long term. This has been the fate of resettlement following numerous earthquakes in Turkey, which invariably resulted in costly reconstruction efforts—and in the reoccupation of old sites. Recently, personal and community attachment to a particular place called "home" was better understood by the decision makers in Mexico City after the earthquake of 1985, producing rehabilitated housing, to higher seismic safety standards, in the same locations.

Buildings

The residents' exposure to the effects of disasters is further compounded by the hazardous structures they live in. For example, overcrowding, nonexistent or malfunctioning infrastructure, and the lack of maintenance owing to strict rent control, were among the primary causes of failure of substandard urban structures in Mexico City following the 1985 earthquake. In the 1993 earthquake in Killari, India, poor construction quality and inadequate traditional building techniques contributed to enormous housing losses.

Reconstruction policies focus on solving the problems of seismic safety through several approaches, including introducing new construction standards, new building technologies, training local artisans, and building demonstration houses. In India, local artisans have been trained in aseismic building techniques, using local construction materials such as stone and brick. An engineer stationed in the field develops a solution to repair and strengthen each damaged house. Verification of reconstruction quality is a prerequisite for the release of the aid installments to which beneficiaries are entitled.

By focusing on improving the most visible effects of a disaster—repairing damaged houses and replacing those no longer habitable—reconstruction frequently overlooks the buildings with minor or no damage, although they may be the primary elements of vulnerability to future disaster. Moreover, economic constraints on reconstruction sometimes trade off cultural acceptability of new structures for lower cost. For example, poor thermal performance of prefabricated buildings in Turkey and Guatemala severely affected the comfort and the lifestyle of disaster victims and led many of them to abandon their new houses.

Recent examples, however, show that, increasingly, reconstruction policies include these considerations. In Mexico City, housing conditions were greatly improved, dwelling size was increased, new facilities were provided, and at the same time, the cultural values and identity of individual neighborhoods were preserved. Similarly, the reconstruction in Datong-Yanggao, China, following the 1989 earthquake was inspired by local architectural styles with a special emphasis on local cultural tradition.

Amenities

Lack of basic services such as water, drainage, and sewer facilities in settlements may also increase the disaster vulnerability of residents. For example, constant saturation of foundations because of lack of surface drainage contributes to these structures' poor performance in earthquakes. Housing reconstruction policies, therefore, must not neglect at least the most basic of infrastructure services— water, sewer and electricity. However, when in their quest for earthquake safety the planners select geologically "safe" sites, these most frequently are unconnected to infrastructure networks. The limited financial resources for reconstruction and the prohibitive cost of servicing the land often mean that the relocation sites will receive new buildings and residents but not water and sewers. Such was the case in Turkey following the Gediz and Lice earthquakes of 1970 and 1975, respectively, where fully equipped houses were built on unserviced land. However, more recently, the experiences of Mexico City and India show how economic and cultural considerations successfully guided postdisaster rehabilitation.

Resources

Housing reconstruction after disaster is always a massive endeavor requiring mobilization of large financial and human resources. At the same time, it is a rare opportunity to engage such formidable force for the tasks of development. Although assistance from outside the community is often welcome and sometimes necessary as a "catalyst" for successful reconstruction, it must not take on a leading role.

Many people participate in planning and implementation of reconstruction after a disaster, but the victims are often left out. Reconstruction is thus deprived of local skills, experience, labor, institutions, and sometimes significant funds for rebuilding. In the long run, the community is also robbed of the positive values of involvement in the rebuild-

ing and runs the risk that its traditional way of life may be changed beyond recognition. External decision makers often bring ready-made solutions and foreign technologies inappropriate to communities whose residents are excluded from meaningful participation in reconstruction. Thus, not only the disaster itself but also the exclusion of local community members from the decision-making and rehabilitation processes combine to undermine the process of redevelopment.

Housing reconstruction in India after the 1993 Killari earthquake involved massive victim participation in housing rehabilitation, affecting about 250,000 households: Residents of the relocation villages took active part in the design of houses and the actual layout of the 52 villages. The bulk of the program, however, was the in situ repair, strengthening, and reconstruction of about 210,000 houses scattered in 2,500 villages. The program involved active participation by the beneficiaries, who—assisted by the advice of engineers and the know-how of local artisans trained in earthquake-resistant building techniques—initiated repair or rebuilding of their houses. The government acted only as a facilitator, by providing the training and field assistance through 700 engineers deployed in the villages. This reconstruction process brought together the special local cultural needs and the technical outside skills needed for a successful recovery. (SEE ALSO: *Environmental Hazards: Earthquakes; Environmental Hazards: Flooding; Environmental Hazards: Hurricanes*)

—Jelena Pantelic

Further Reading

Berke, Philip R., Timothy Beatley, and Clarence Feagin. 1993. "Hurricane Gilbert Strikes Jamaica: Linking Disaster Recovery to Development." *Coastal Management* 21(1):1-23.

Cuny, Frederick C. 1983. *Disasters and Development*. New York: Oxford University Press.

Diacon, Diane. 1993. "Typhoon Resistant Housing in the Philippines: The Core Shelter Project." *Disasters* 16(3):266-271.

Kreimer, Alcira. 1991. Reconstruction after Earthquakes: Sustainability and Development. *Earthquake Spectra* 7:97-104.

Oliver-Smith, Anthony. 1990. "Post-Disaster Housing Reconstruction and Social Inequity: A Challenge to Policy and Practice." *Disasters* 14(1):7-19.

Pantelic, Jelena. 1991. "The Link Between Reconstruction and Development." *Land Use Policy* 8:343-47.

Quarantelli, Enrico L. 1984. "The Phenomena of Sheltering and Housing" and "General Observations and Conclusions." Pp. 1-5 and 73-80 in *Sheltering and Housing after Major Community Disasters: Case Studies and Observations*, by E. L. Quarantelli. Newark: University of Delaware, Disaster Research Center. ◀

▶ Housing Allowances

Housing allowances are economic assistance to lower-income households who require help to pay rent for a dwelling unit that meets certain minimum physical standards. A housing allowance is the remedy for the rent gap, or the difference between how much a household can afford to spend on rent as a percentage of income and what the market requires it to spend. Housing allowances are paid either directly to the landlord or to the program participant.

A housing allowance is a demand-side (consumer) subsidy and reflects the reality that affordability is the major housing problem for lower-income households. However, it should be noted that in some housing markets, shortages of affordable units still exist, a deficiency that suggests the need for a complementary supply-side (producer) subsidy wherein additional low-cost units should be built for a lower-income target population. A careful analysis of market conditions will suggest whether one or both of these approaches will be required to ensure adequate housing opportunities for the poor at affordable prices.

Background

During the 19th and early 20th centuries, major cities in Europe and the United States experienced critical housing shortages and tremendous population pressures as migrants were attracted by the economic opportunities of the Industrial Revolution. In addition, urban migration increased as workers were displaced by decline in agricultural employment or by political or religious oppression. Another constraint on housing supply stemmed from the installation and upgrading of infrastructure in the latter part of the 19th century as a component of the retrofit of Europe's major cities, such as Paris under the aegis of Baron Georges Eugène Haussmann. As a result, displaced households were crowded into slum dwellings of subminimal physical standards amid crushing population densities.

From 1890 to 1940, slum clearance gained importance as a major policy objective, which further exacerbated housing shortages. In response, philanthropic housing enterprises, efforts by cooperatives and trade unions, and government-sponsored housing initiatives combined to reduce the very serious shortage of affordable, physically sound dwelling units. This trend continued after World War II and its widespread devastation. Gradually, however, the postwar era witnessed the amelioration of housing opportunities so that, eventually, supply shortfalls were no longer the chief determinant of public sector housing market operations.

As societies progressed toward a housing market policy based on supply-and-demand equilibrium, economic rents, and decreasing government intervention, it became necessary to subsidize rents for those not able to participate in the economic recovery. The need for this demand-side subsidy stimulated considerable thought concerning appropriate strategies that now define housing allowance policies.

In the United States, the postwar high-production strategies that resulted in entire suburban neighborhood units built by large-scale merchant builders, the acceleration in subsidized housing production following the Housing and Urban Development Act of 1968, and the accompanying loss of population in many major cities all combined to create an adequate housing supply in most parts of the country by 1980. As a result, the Final Report of the President's Commission on Housing declared in 1982 that affordability was the primary housing problem of the poor. The Commission recommended that the federal government's major housing initiative focus on a housing payments program to be supplemented by Community Development Block Grant funding to provide housing supply assistance in locations where it was warranted. This conclusion was supported by the findings of the Department of Housing and Urban Development's (HUD's) $150 million Experimental Housing Allowance Program (EHAP), a carefully designed research effort. EHAP's demand experiment examined individual household responses to housing allowances, the supply experiment looked into the market effects of housing allowances and their impacts on housing price and supply, and the administrative agency experiment provided insights concerning the impact of housing allowances on local public housing authorities.

EHAP results convincingly supported a housing allowance program on the grounds that it avoided the cost inefficiencies of subsidized new construction while at the same time demonstrating that housing allowances did not result in market distortions.

Considerations in the Formulation of a Housing Allowance

Any housing allowance policy must address concerns relevant to economic and physical standards. A review of these follows.

Economic Issues

How much can a household afford to pay for housing? If the economic cost of housing is in excess of what a household can afford, how much of that rent gap should be covered by the public sector? Up to what income levels should households qualify for housing allowance assistance? What is the maximum rent that can be subsidized by a housing allowance? Finally, should every eligible household receive a housing allowance?

Housing Affordability and the Rent Gap. Virtually every advanced market economy has developed a consensus as to what percentage of income a household should expend for shelter. In the United States, "a week's wages for a month's rent" has been a generally accepted rule of thumb since the 19th century, whereas European thought has tended to converge on a figure of about half that size. The rent gap is the difference between what a family should spend on housing, based on agreed-on percentage of income, and what the market requires it to allocate for housing. Once the magnitude of this rent gap has been established, the next question is, how much of the gap should be bridged by a housing allowance? In general, the larger the rule-of-thumb percentage, sometimes termed a *housing-income ratio,* the fuller the coverage of the rent gap by the housing allowance. The U.S. Section 8 Existing Housing Program, which mandates a housing-income ratio of 30%, provides 100% coverage of the rent gap. In Europe, this coverage is often less, usually between 50% and 85%, but then so is the required housing-income ratio. Furthermore, a much greater percentage of eligible European households receive a housing allowance than do U.S. households.

Income Levels. Virtually all housing allowance programs mandate a household income ceiling, but its method of calculation varies. Often, it is a combination of factors, centering on household size and income as a percentage of a given standard—for example, national median income.

Sometimes, variables such as rent and family assets also enter into the calculation.

Maximum Rents. Any housing allowance should be sufficient so that the recipient has access to housing that meets minimum physical requirements, but it should not be so generous so that above-average or luxury units qualify. In the United States, the Section 8 Existing Program's Fair Market Rent standard specifies units should be "of modest design with suitable amenities." A mid-1980s Australian housing allowance proposal called for "adequate accommodation in an average suburb," and other approaches involve estimates of a probable market rent when supply-and-demand characteristics are adequately balanced. This invariably involves regional adjustments for high-cost areas, such as San Francisco Bay or Greater London.

Eligibility and Equity. Should a housing allowance be an entitlement to any qualifying household? With a supply-side subsidy system, only a limited number of households can be sheltered in assisted units specifically built for that purpose. Moreover, once a family occupies a public housing unit, it tends to remain there despite increases in income and decreases in household size. This scenario is widespread in Europe. As a result, the concept of horizontal equity, where all households of similar income are treated alike, cannot be implemented. Under a demand-side subsidy, such as a housing allowance, all households of qualifying income can be served if the political will exists to fund this subsidy. In practice, however, horizontal equity considerations are difficult to implement. In the United States, well under 20% of all eligible households receive rental assistance housing of any type. In Europe, where income tax policy tends to be more highly redistributive, the cost of a housing allowance on a per capita basis is less than in the United States. As a result, a high degree of horizontal equity can be achieved.

Physical Standards

Can housing allowances improve the physical standards of accommodation for program participants? If increased administrative costs are considered an acceptable trade-off, the quality of the dwelling unit and overcrowding can be monitored. Performance-based housing quality standards have been implemented in the U.S. Section 8 Existing Housing Program, and inspections occur either annually or prior to a new tenancy; the same applies to checks on household income. If deficiencies are found, the rent gap subsidy paid to the landlord is withheld until the unit passes a reinspection. Because the Section 8 Existing Program is targeted toward households with incomes 50% or less of area median, the subsidy payment to the landlord is usually substantial, so every incentive exists to comply with inspection directives with respect to repairs.

Evaluation

The question, How much for housing? involves not only individual levels of expenditure but the allocation of national resources as well. For societies that perceive shelter as a basic entitlement, both housing allowances and subsidized new construction programs can work together to provide rent gap remedies and alleviate shortages of afford-able units. In the United States, whose Supreme Court has ruled that housing is not a fundamental right protected by the Constitution and where large budget deficits have sharply reduced domestic spending since the early 1980s, policy has shifted to limiting program growth, reducing benefits, and targeting eligible populations more sharply. Under these conditions, the cost-effectiveness of demand-side subsidies such as housing allowances has emerged as a superior policy choice. These subsidies promote greater locational options, tenant self-reliance, and market search skills. In addition, the existing housing stock is effectively used and administrative oversight can ensure satisfactory levels of maintenance.

Although many believe that a housing allowance is more of an income transfer policy than a housing policy, the complexities of integrating it with a general income maintenance program have proven formidable. The political appeal of housing assistance far transcends the difficult realities of any further expansion of the welfare state. (SEE ALSO: **Affordability;** *Experimental Housing Allowance Program;* **Subsidy Approaches and Programs**)

—*Daniel J. Garr*

Further Reading

Bradbury, Katharine L. and Anthony Downs, eds. 1981. *Do Housing Allowances Work?* Washington, DC: Brookings Institution.

Fallis, George. 1993. *On Choosing Social Policy Instruments: The Case of Non-Profit Housing, Housing Allowances or Income Assistance.* New York: Pergamon.

Friedman, Joseph and Daniel H. Weinberg, eds. 1983. *The Great Housing Experiment.* Beverly Hills, CA: Sage.

Howenstine, E. Jay. 1986. *Housing Vouchers: A Comparative International Analysis.* New Brunswick: Rutgers, the State University of New Jersey, Center for Urban Policy Research.

Priemus, Hugo. 1984. *Housing Allowances in the Netherlands.* Delft, The Netherlands: Delft University Press.

Stahl, Konrad and Raymond J. Struyk. 1985. *U.S. and West German Housing Markets: A Comparative Economic Analysis.* Washington, DC: Urban Institute Press.

Struyk, Raymond J., Neil Mayer, and John A. Tuccillo. 1983. *Federal Housing Policy at President Reagan's Midterm.* Washington, DC: Urban Institute Press. ◀

▶ Housing and Community Development Act of 1992

The Housing and Community Development Act of 1992 has 16 parts, known as titles, that deal with a variety of topics. The following briefly describes the main points of each title, except Title X (the Residential Lead-Based Paint Hazard Reduction Act of 1992), which is discussed in detail elsewhere in this volume. Although the act creates a number of goals, the actual implementation of these goals is subject to appropriation of funds.

Title I—Housing Assistance

This title authorizes funding for various categories of federally assisted housing programs, including public housing, Section 8 housing, Indian housing, and homeownership programs. Major reconstruction of obsolete public and Indian housing projects, reform of public housing management, public housing operating subsidies, public housing homeownership, and revitalization of severely distressed public housing are also discussed. This title gives the U.S. Department of Housing and Urban Development (HUD) authority to make grants to enable applicants to develop and carry out Youthbuild programs, which are intended to enable economically disadvantaged youth to obtain employment skills and expand the supply of affordable housing.

Title II—Home Investment Partnerships
This title authorizes funding of portions of the Cranston-Gonzalez National Affordable Housing Act. It extends the meaning of "affordable housing" to include permanent housing for disabled homeless persons, transitional housing, and single-room occupancy housing.

Title III—Preservation of Low-Income Housing
This title authorizes funding of prepayment of mortgages insured under the National Housing Act. It establishes a preservation program and authorizes HUD to provide it with technical assistance and capacity building.

Title IV—Multifamily Housing Planning and Investment Strategies
The owners of covered multifamily housing and covered multifamily housing for the elderly (e.g., funded by Section 202 of the Housing Act of 1959, Section 221 of the National Housing Act, etc.) submit needs assessments to HUD. These include descriptions of financial assistance and other assistance needed to maintain these properties in livable condition. Assessments for all such properties will be submitted by 1995.

Title V—Mortgage Insurance and Secondary Mortgage Market
This title sets maximum mortgage amounts covered by Federal Housing Administration (FHA) mortgage insurance. It defines limitations on Government National Mortgage Association (GNMA) guarantees of mortgage-backed securities. It directs HUD through the FHA to demonstrate the effectiveness of providing new forms of federal credit enhancement for multifamily loans.

Title VI—Housing for Elderly Persons and Persons with Disabilities
This section authorizes funding for supportive housing for the elderly and persons with disabilities. It directs HUD to conduct a demonstration to determine the feasibility of purchasing and installing elder cottage housing units adjacent to existing one- to four-family dwellings. It directs HUD and the Department of Agriculture to issue proposed interim regulations implementing Section 802 of the Cranston-Gonzalez National Affordable Housing Act, with respect to eligible federally assisted housing. It gives authority to public housing agencies to provide designated public housing and assistance for disabled families. Owners of federally assisted housing are required to select tenants on the basis of criteria established by HUD. These criteria will enable owners to make reasonable accommodations for persons with disabilities and comply with civil rights laws. Owners of covered Section 8 housing projects may give preference to elderly families and shall reserve units for disabled families. Owners are required to provide service coordinators to assist these tenants.

Title VII—Rural Housing
This title authorizes funding to insure and guarantee loans for low- and moderate-income borrowers for rural housing.

Title VIII—Community Development
This title authorizes funding of Community Development Block Grants. It requires HUD to develop guidelines for the evaluation, selection, and review of economic development projects, which can include enterprise zones and colonias. It authorizes funding for the Neighborhood Reinvestment Corporation. It directs HUD to carry out a five-year demonstration program to determine the feasibility of facilitating partnerships between institutions of higher education and communities to solve urban problems through research, outreach, and the exchange of information. It directs HUD to establish a demonstration program to determine the feasibility of assisting states and local governments to develop computerized databases of community development needs. It directs HUD to carry out a demonstration program to test new models for bringing credit and investment capital to targeted geographic areas and low-income persons in such areas. It authorizes funding of a nonprofit community-based public benefit corporation created in response to the civil disturbances of April and May 1992, in Los Angeles, California.

Title IX—Regulatory and Miscellaneous Programs
This title states that the policy of Congress is to direct employment and other economic opportunities generated by federal financial assistance toward low-income persons and recipients of government housing assistance. It authorizes funding for HUD research and development. It authorizes HUD to use private nonprofit fair housing organizations to investigate violations of the Civil Rights Act of 1968 and to enforce remedies to such violations.

This title directs HUD to develop a new standard for hardboard panel siding on manufactured housing. It authorizes funding of the National Commission on Manufactured Housing to examine the feasibility of expanding standards governing manufactured home sales. It directs HUD to establish the Solar Assistance Financing Entity to assist in financing solar and renewable energy capital investments.

This title modifies the separate capitalization rule for savings association's subsidiaries engaged in activities that are not permissible for national banks. Some modifications of other banking rules are also included.

Title XI—New Towns Demonstration Program for Emergency Relief of Los Angeles
The purpose of this title is to provide for the revitalization and renewal of inner-city neighborhoods in Los Angeles, California, that were damaged during the civil disturbances of April and May 1992 and to demonstrate the effectiveness

of new town developments for accomplishing this goal. New town demonstration areas are geographic areas defined in new town plans. These areas have pervasive poverty, unemployment, and general distress. The plans provide for housing units, jobs, social services, supplemental resources, contractors and developers, and financing for homebuyers. New towns are eligible for federal mortgage insurance, secondary soft mortgage financing, and community development assistance.

Title XII—Removal of Regulatory Barriers to Affordable Housing

A "regulatory barrier" to affordable housing is any public policy, including statutes, ordinances, regulations, or administrative procedures or processes, required to be identified by a jurisdiction in connection with its comprehensive housing affordability strategy, which is mandated by the Cranston-Gonzalez National Affordable Housing Act. This title authorizes grants to state and local governments that encourage them to identify and remove these regulatory barriers. It also establishes a regulatory barrier clearinghouse that collects information regarding the barriers.

Title XIII—Government Sponsored Enterprises

This title is also known as the Federal Housing Enterprises Financial Safety and Soundness Act of 1992. It protects the taxpayers against liability for the activities of the enterprises known as the Federal Home Loan Mortgage Corporation, Fannie Mae, or the Federal Home Loan Banks. Any such enterprise or bank is *not* backed by the full faith and credit or the United States.

The title establishes the Office of Federal Housing Enterprise Oversight within HUD. It gives HUD authority to establish housing goals, monitor and enforce housing goals, and approve any new programs of each enterprise mentioned above. It requires each enterprise to determine the amount of total capital that is sufficient for the enterprise to maintain positive capital during 10-year "stress" periods. The director of the Office of Federal Housing Enterprise Oversight is authorized to enforce this provision.

Title XIV—Housing Programs under the Stewart B. McKinney Homeless Assistance Act

This title is also known as the Stewart B. McKinney Homeless Assistance Act of 1992. It authorizes a shelter grants program and a supportive housing program. The purpose of the latter is to assist homeless persons in the transition from homelessness and provide supportive housing that enables independence. This title also authorizes HUD to make grants to applicants to demonstrate the feasibility of providing low-cost housing to homeless persons who are unwilling or unable to participate in mental health treatment programs or receive other supportive services. This housing, known as "safe havens," is to demonstrate whether and on what basis eligible persons choose to reside in them and whether safe havens are cost-effective in comparison to other alternatives. It will also show the extent to which residents become willing to participate in various treatment programs after a period of residence in a safe haven and to move to more traditional forms of permanent housing.

Title XV—Annunzio-Wylie Anti-Money Laundering Act

This title authorizes the comptroller of the currency to revoke the charter of federal depository institutions that have been convicted of money laundering or cash transaction reporting offenses. Likewise, credit unions will lose their organization certificate and state depository institutions will lose their insurance for these offenses. Nonbank financial institutions are prohibited from transmitting illegal money. Improvements in enforcement of laws prohibiting money laundering are also prescribed.

Title XVI—Technical Corrections of Banking Laws

This title is also known as the Federal Deposit Insurance Corporation Improvement Act of 1991. It provides technical corrections relating to Title I of the Federal Deposit Insurance Corporation Improvement Act of 1991.

—*Barbara A. Haley* ◀

▶ Housing and Development Law Institute

The Housing and Development Law Institute (HDLI) is a nonprofit organization founded in 1984 that provides legal and educational services for the public affordable housing sector. Membership is composed of 250 attorneys nationwide who are sponsored by public housing agencies and private law firms. The HDLI sponsors educational seminars for its members, produces public housing law manuals, and publishes several periodicals, including *The Authority,* a quarterly summary of law cases affecting housing agencies; *The HUD Regulator,* a monthly summary of the Department of Housing and Urban Development regulations with commentary; and *HDLI Networking and Feedback.* Staff size: 4. Contact person: William F. Maher, Executive Director of Counsel. Address: 630 Eye Street, Washington, D.C. 20001. Phone: (202) 289-3400. Fax: (202) 289-3401.

—*Caroline Nagel* ◀

▶ *Housing and Development Reporter*

The *Housing and Development Reporter* is a weekly, loose-leaf national publication and reference service providing comprehensive coverage of housing and urban development programs and management issues, community and economic development, and related business, finance, and tax issues. It was first published in 1973. *HDR Current Development Reports* cover action from Congress, the courts, and the state and federal agencies that affect housing, community and economic development, and real estate business, finance, and taxation. *HDR Reference Files* include agency regulations and instructions, full text of all significant housing and development legislation, program descriptions, and directories of congressional committees as well as of housing and development organizations and agencies. Price is $860 per year. Editors: Bruce S. Lane and Charles L. Edson. Address: Warren Gorham Lamont, 31 St. James Ave., Boston, MA. 02116-4112. Phone: 1-800-950-1214 or (617) 423-2020. Fax: (617) 423-2026.

—*Laurel Phoenix* ◀

► Housing and Home Finance Agency

The Housing and Home Finance Agency (HHFA) was the predecessor to the U.S. Department of Housing and Urban Development (HUD). It existed from 1947 to 1965. The HHFA was a weak, coordinating umbrella organization for some, but not all, federal housing programs.

Creation

The HHFA was created at the termination of the National Housing Agency (NHA). The NHA was created in 1942 to oversee wartime housing activities. It included public housing (U.S. Housing Authority), the Federal Housing Administration (FHA), and the Federal Home Loan Bank Board (FHLBB). The NHA was caught in the battle between the public housing and urban lobby and the private sector (especially the realtors and lenders) over the shape of post-World War II housing and redevelopment policy. The former sought a sweeping urban slum clearance and public housing program to be led by the NHA. The latter triumphed in 1946 after the Republicans won control of the U.S. Congress when the NHA was denied permanent status. President Truman then appointed a housing expediter to coordinate an emergency housing program for veterans.

However, faced with strong conservative opposition, the housing expediter resigned in 1947. Truman then created the HHFA to coordinate the public housing program and the FHA. The HHFA's administrator had no direct control over either agency, because their heads were presidential appointees also. The HHFA was further weakened because the veterans' and farmers' housing programs and the Federal Home Loan Bank Board were not included under the HHFA umbrella.

The HHFA under Truman

The HHFA's future was tied to President Truman's presidential fortunes. With Truman's surprise victory in 1948, Congress finally passed the much-debated Taft-Wagner-Ellender housing legislation in 1949. It created a new urban redevelopment program. Opposition to public housing was strong, but the Truman administration won authorization for 810,000 new units over a six-year period.

However, with the almost immediate onset of the Korean conflict in 1950 and Truman's decision not to seek reelection, the HHFA's future direction awaited the outcome of the 1952 presidential election.

The HHFA under Eisenhower

Eisenhower's administration was conservative. He appointed Albert Cole as HHFA administrator, a former Kansas Republican Congressman who had opposed the NHA, the HHFA, and public housing. However, public housing survived, although its maximum construction authorization was reduced to only 35,000 units annually. In 1954, Eisenhower's administration rejected Cole's request to give the HHFA administrator full control over its subordinate agencies, which now included the Urban Renewal Administration. The 1954 Housing Act renamed Urban Redevelopment as "Urban Renewal" and authorized rehabilitation,

in addition to slum clearance. It emphasized federal support for code enforcement and planning.

The Eisenhower administration rejected the agency for cabinet status in 1957. It opposed broadening the HHFA's mandate to urban affairs. As the second term of his administration ended, President Eisenhower twice vetoed 1959 congressional legislation increasing federal housing aid, before finally relenting.

The HHFA under Kennedy and Johnson

President John Kennedy appointed veteran New York housing official Robert Weaver as his HHFA administrator. This was noteworthy because Weaver became the highest-ranking black official in the Kennedy administration. Housing legislation increased HHFA funding in 1961.

That same year, Kennedy tried to elevate the HHFA into a cabinet-level Department of Urban Affairs and Housing, but this was defeated in Congress. His successor Lyndon Johnson was able to create the Department of Housing and Urban Development (HUD) as the successor to the HHFA in August 1965, following his 1964 landslide electoral victory. In January 1966, Robert Weaver became the first African American Cabinet member. (SEE ALSO: **Federal Government**)

—*W. Dennis Keating*

Further Reading

Gelfand, Mark I. 1975. *A Nation of Cities: The Federal Government and Urban America, 1933-1965.* New York: Oxford University Press.

Keith, Nathaniel S. 1973. *Politics and the Housing Crisis Since 1930.* New York: Universe Books. ◄

► Housing and Society

Housing and Society is a triannual journal concerned with the interests of architects, interior designers, researchers, realtors, and educators in the housing field. First published in 1973 as *Journal of Housing Educators,* its articles cover a broad range of topics, including housing alternatives, design, utilities, interiors, energy, policy, and behavioral aspects of housing. Circulation: 450. Price: $50 per year. Editors: Julia Beamish, Kathleen Parrott, Rosemary Goss, Michael Johnson, Rebecca Lovingood. Address: College of Human Resources, 240 Wallace Hall, Virginia Tech University, Blacksburg, VA 24061-0424. Phone: (703) 231-8881. Fax: (703) 231-3250.

—*Caroline Nagel* ◄

► Housing and Urban Development Act of 1968

The Housing and Urban Development Act of 1968 (P.L. 90-448) represented a landmark piece of federal housing legislation. Broadly expanding earlier housing legislation, this act declared a numerical national goal of constructing or rehabilitating 26 million housing units over a 10-year period with 6 million of the units devoted to low- and moderate-income families—a goal recommended by the President's Committee on Urban Housing (the Kaiser Commission). President Johnson, after signing the act on August 1, 1968, proclaimed it as the "Magna Carta to liberate our cities."

The Housing and Urban Development Act of 1968 contained a wide array of programs designed to help lower-income families obtain decent housing. Congress recognized that the Housing Act of 1949's broad national goal of "a decent home and suitable living environment for every American family" had not been realized. Under Section 2, Congress also indicated that this failure to achieve the goal "is a matter of grave national concern; and that there exist in the public and private sectors of the economy the resources and capabilities necessary to the full realization of this goal." The act contained 17 titles: lower-income families, rental housing for lower-income families, Federal Housing Administration insurance operations, new community land development, urban renewal, urban planning and facilities, urban mass transportation, secondary mortgage market, national housing partnerships, rural housing, urban property protection and reinsurance, District of Columbia Insurance Placement Act, national flood insurance, interstate land sales, nonprofit hospitals mortgage insurance, 10-year housing goal, and miscellaneous activities.

Two housing assistance programs were established under the legislation. A Section 235 program was established to help low- and moderate-income families achieve home-ownership by providing them with an interest rate subsidy. The federal government would share some of the burden if a low-income family could not afford the full cost of buying a home. The program would work hand in hand with achieving the national goal of 6 million low- and moderate-income family housing units. A rental assistance program, Section 236, was also created that provided subsidies for the construction or rehabilitation of rental and cooperative housing.

The legislation also encouraged the development of entire new towns or new communities. Among the activities authorized under the legislation included the ability of the secretary of the Department of Housing and Urban Development to guarantee various financial obligations to help finance new town development and grants to assist new communities finance basic infrastructure projects.

Urban renewal activity funding increased under the legislation. Funding was also authorized for neighborhood development programs, urban renewal projects in model cities areas, and projects to demolish nonresidential structures infested with rats.

Title VIII of the legislation separated the Federal National Mortgage Association into two corporations: the Government National Mortgage Association (GNMA) and the Federal National Mortgage Association (FNMA). GNMA, a government corporation located within HUD, supports programs that add liquidity to the secondary mortgage market, thus allowing more money available for new mortgages. FNMA, a government-sponsored private corporation, also performs in the secondary mortgage market by purchasing government-insured or -guaranteed mortgages and later reselling them as the demand for mortgages decreases.

A number of additional programs were expanded in the legislation. For instance, additional funding for the model cities program was authorized as was funding for assisting states in providing urban information and technical assistance to communities of less than 100,000 population. (SEE ALSO: *New Towns*)

—*Roger W. Caves*

Further Reading

Hays, R. Allen. 1985. *The Federal Government & Urban Housing.* Albany: State University of New York Press.

Hendershott, Patric H. and Kevin E. Villani. 1978. *Regulation and Reform of the Housing Finance System.* Washington, D.C.: American Enterprise Institute for Public Research.

The President's Committee on Urban Housing. 1969. *The Report of the President's Committee on Urban Housing: A Decent Home.* Washington, D.C.: Government Printing Office.

Weicher, John C. 1980. *Housing: Federal Policies and Programs.* Washington, DC: American Institute for Public Policy Research. ◀

▶ Housing Assistance Council

The Housing Assistance Council (HAC) is a nonprofit corporation that fosters the development of low-income housing in the rural United States through technical assistance, seed money loans, research projects, housing programs, policy assistance, and training and information services. HAC offers several publications, including the *HAC News,* a free biweekly guide to federal rural housing programs; *State Action Memorandum,* a free bimonthly newsletter focusing on state rural housing programs; and various research reports and technical manuals published as needed. Founded in 1971, the council operates with a staff of 33 on an annual budget of $2.5 million (1993). It is funded primarily through grants and contracts. Besides a national office in Washington, D.C., HAC has three regional offices located in California, New Mexico, and Georgia. Executive Director: Moises Loza. Address: 1025 Vermont Avenue, NW, Suite 606, Washington, DC 20005. Phone: (202) 842-8600. Fax: (202) 347-3441. (SEE ALSO: *Rural Housing*)

—*Caroline Nagel* ◀

▶ Housing Bonds

Government has an interest in seeing to it that all households are adequately housed. When there is a shortage of housing or banks are unwilling to lend money for housing, one of the methods by which government can stimulate the housing market is to assume the role of lender. Rather than contract for the housing to be built for public ownership, such as with public housing, it may assist private builders and developers by making housing development loans available on favorable terms. Alternatively, the government may make loans to homebuyers so that they may purchase new or existing housing.

Government can raise the funds for this effort either through taxation or through its capacity to borrow money in the bond markets. Using the bond market is advantageous in that no tax revenues are needed; instead, the government can repay the bond buyers using only the revenues from the housing that is financed by the bond proceeds.

Government bonds can be sold on the bond market with or without a general obligation on the part of the issuing government. A *general obligation* means that the government is obligated to raise taxes as might be necessary to ensure repayment to the bond buyer of the principal and interest on the bonds. An alternative to this general obligation bond is referred to as a revenue bond or "moral obligation" bond. Housing bonds fall into this category. The unit of government that is issuing the bond is not pledging its full faith and credit; that is, it is not pledging to raise taxes to repay the bondholders. Rather, it is pledging only the revenues to be raised from the housing that will be developed or purchased using the proceeds from the sale of the bonds. If the housing project does not succeed, it will be sold to cover the bond obligation. The government has a moral obligation to use other funds to cover losses only in the rare event that all of the bond debt cannot be repaid from the sale of the property. This moral obligation is only an informal understanding that the government would not allow the issuing agency, usually a state housing finance agency, to default on the obligation to repay the bondholders.

Bonds may be either tax exempt or taxable. Tax exempt means that the interest paid to the bondholders is exempt from federal income taxes. Taxable means that this is not the case. If the bonds are tax exempt, the issuing agency becomes subject to a great many rules imposed by the federal government. They restrict the total dollar value of housing bonds that can be issued during any one year. They place limits on the types of housing and the occupancy of the housing that can be financed. They place constraints on the use of the funds between the time the bonds are sold and the time that the housing is developed. Finally, they place severe limits on the interest rates that can be charged to the owner of the housing that borrows the money. By being exempt from taxation, the bond buyers are willing to accept a lower interest rate than they would accept on a bond with taxable interest. This lower interest rate can be passed along to the owner of the housing being financed by the bond proceeds. With the lower interest rate, a home-buyer is able to better afford a home or a developer can charge lower prices or rents on the housing produced, making it more affordable to needy households.

The bonds may be used to finance either rental housing, usually multifamily developments, or owner-occupied housing. If it is a rental development, the bond proceeds are loaned to a property owner who must agree to operate the rental units in a manner that conforms with the applicable laws governing the use of housing bond proceeds. These rules generally govern the maximum income that a household may have to occupy a unit and the maximum rent that can be charged for the unit. Such loans are typically made by a state housing finance agency, although local units of government have issued municipal housing bonds in the past.

If the bonds are to finance owner-occupied housing, the same type of restrictions applies. Typically, the household buying the home must meet income limitations, and the price of the home must fall below a specified maximum amount. Again, a state housing finance agency may originate the loan, or it may transfer the bond proceeds to conventional lenders who will make the loans to eligible homebuyers on terms dictated by the agency. Tax-exempt-housing bonds issued for owner occupancy are referred to as *mortgage revenue bonds*.

If the bonds are taxable, relatively few limits are imposed on the bonds, but being taxable, the interest rates paid on the bonds are similar to the rates that would be charged on loans borrowed from conventional lenders. States may use taxable bonds from time to time to provide funds for housing development when the conventional lenders are not making funds available or when federal limits on tax-exempt bonds apply.

Housing bonds provide a means for state or local government to use its power to raise money in the bond market to augment the supply of credit for housing. The additional credit can foster housing development that would not occur in the market or that would not be affordable without the assistance provided though the housing bond system. Given their low cost to state and local government, housing bonds will continue to be popular, especially during periods when private sector lenders are unwilling or unable to supply as much credit as the marketplace requires. (SEE ALSO: *Local Government; Mortgage Revenue Bonds; State Government*)

—*Kirk McClure*

Further Reading

Kaufman, George, ed. 1981. *Efficiency in the Municipal Bond Market: The Use of Tax-Exempt Financing for "Private" Purposes*. Greenwich, CT: JAI.

Peterson, George and Brian Cooper. 1979. *Tax Exempt Financing of Housing Investment*. Washington, DC: Urban Institute Press.

Zimmerman, Dennis. 1991. *The Private Use of Tax-Exempt Bonds*. Washington, DC: Urban Institute Press. ◀

▶ Housing Codes

Municipalities seek to ensure housing quality through the enforcement of building and housing codes. Building codes establish standards for new construction and remodeling. Housing codes emphasize standards of maintenance. Although there is overlap between these codes (e.g., they may both contain minimum plumbing requirements), housing codes are unique in their emphasis on how the structure is used and maintained. Housing codes typically set maximum limits on the number of persons per room as well as the minimum square feet per bedroom (related to the number of people who will use it). They often prohibit habitation of cellar areas. They set forth minimum conditions for property maintenance, including, but not limited to, sanitation (prohibiting the accumulation of garbage); structural safety of stairs, porches, and railings; operability of windows and heating units; and avoidance of dampness and mold growth.

Modern housing codes have their origin in the New York Tenement Housing Law of 1867, which legitimated the railroad flat and prohibited the construction of anything worse. The railroad flats were long, narrow apartments built four to a floor in five- to seven-story, common-wall, walk-up structures on 25- × 100-foot lots. The only ventilation was from windows in the front and back of the building.

The immediate impetus for housing reform was the 1865 cholera epidemics in Europe preceded by various outbreaks of smallpox, dysentery, tuberculosis, and other diseases in New York City's slums. The draft riots of 1863, a violent, destructive reaction to discriminatory conscription, caused widespread public fear of the social and moral condition of the immigrant. The New York Association for Improving the Condition of the Poor, established in 1843, raised the specter of the city's being overrun by beggars and thieves. Better housing was seen as a means to protect the existing order as well as to impose middle-class manners and morals on immigrant working-class people.

Among the requirements of the 1867 housing act were proper fire escapes, a toilet or privy for every 20 occupants, and a water faucet on each floor. These standards were expanded in 1879 through the "Old Law," which specified air shafts between adjacent buildings, window openings into each room, and two toilets for each floor.

Housing conditions in New York City continued to be deplorable. Lawrence Veiller, founder of the National Housing Association, led the campaign for reform, authoring the New York Tenement House Law of 1901 (the "New Law"), which established new construction standards, mandated adaptations to existing structures, and specified maintenance requirements. Enforcement was to be accomplished through construction and remodeling permits, an inspection system, and penalties for noncompliance. The task was enormous. At the turn of the century, New York City had more than 80,000 tenements, three-quarters of which had been built under the old law. Among these tenements were 350,000 interior rooms with no windows. Thousands of the tenements were still making use of the school sink, a trough of water placed below the seat of a latrine.

Although New York City had by far the worst housing conditions of any city in the United States, many others experienced problems of dilapidation, fire danger, and poor sanitation. The nature of the problem varied by city. Chicago, with predominantly one- and two-family dwellings, found the rear house to be a source of blight. Houses built on narrow alleys in Philadelphia and Baltimore often lacked water or sewer connections, and the alleys themselves were filled with dirt and refuse. Kansas City had dilapidated shanties and cellar dwellings.

By 1920, 11 states had adopted housing legislation, most of them drawing on the New York Tenement Act of 1901 or Veiller's books: *A Model Tenement House Law* (1910) and *A Model Housing Law* (1914). By that time, nearly 20 cities had new housing codes; 20 more addressed housing concerns through their building and health ordinances.

The emphasis on restrictive housing codes lost favor among housing reformers in the 1930s. It was realized that the cost burden of required improvements could price many tenants out of the market. Inspectors had to confront the inherent difficulties in enforcing certain standards, such as laws against overcrowding; landlords taking in lodgers served an important function in helping immigrants to adjust to a new culture. A restriction against encumbering fire escapes was especially difficult to enforce because tenants used this area for additional living and storage space.

Public policy began to place greater emphasis on the construction of public housing. Consequently, housing codes as an implementation tool received limited attention for the next three decades. This changed with the federal Housing Act of 1949, which set forth as an allocation criterion for federal urban redevelopment resources the adoption and modernization of local codes establishing adequate dwelling standards addressing health, safety, and sanitation. To receive federal funding, the Housing Act of 1954 required cities to have a "workable program" of urban renewal; a housing code was specifically mentioned as a possible element for this program. The Housing and Home Finance Agency (predecessor to the U.S. Department of Housing and Urban Development [HUD]) subsequently made housing codes mandatory, and the statute was amended in 1964 to reflect this change. The number of cities with housing codes grew from 56 in 1955 to 736 by July 1963. According to HUD's 1970 Annual Report, by the end of 1970, $261.3 million in federal funds were reserved for code enforcement and 163,536 housing units had been brought into compliance.

The current number of municipalities with housing codes is not known because no one is collecting this information. Housing codes are more common east of the Mississippi, although an increasing number of western cities are adopting these codes, particularly in California. Housing codes are found primarily in larger cities where housing problems warrant establishing maintenance standards and resources are sufficient to support code enforcement. An important exception is New York State; housing maintenance standards are included as a section of the mandatory fire prevention and building code.

Many housing codes are based on a model developed by the American Public Health Association in conjunction with the Centers for Disease Control. There is no strong constituency in support of uniformity in housing codes (in contrast to the building codes), and as a result, there is more variability, with the regulations being adapted to address local needs.

Although housing code enforcement tends to be complaint driven, some municipalities use the regulations to target problem areas or even specific landowners who have a history of violations. These efforts are most successful when there is community support, resources are available for needed repairs, and the housing market can absorb higher costs resulting from the improvements.

Although housing codes are not a panacea, used in conjunction with other policies and programs, the codes can be an effective tool for maintaining and improving housing quality. (SEE ALSO: *Building Codes; Code Enforcement; Health Codes; Housing Occupancy Codes*)

—*Deborah A. Howe*

Further Reading

DeForest, Robert W. and Lawrence Veiller. 1903. *The Tenement House Problem.* Vols. I and II. New York: Macmillan.

Friedman, Lawrence M. 1968. *Government and Slum Housing: A Century of Frustration.* Chicago: Rand McNally.

Lubove, Roy. 1962. *The Progressives and the Slums.* Pittsburgh, PA: University of Pittsburgh Press.

Meier, Ron B. 1983. "Code Enforcement and Housing Quality Revisited: The Turnover Case." *Urban Affairs Quarterly* 19(2):255-73.

Plunz, Richard. 1990. *A History of Housing in New York City.* New York: Columbia University Press. ◄

► Housing Completions

Housing completions are widely used as an indicator of business activity. In the methodology of the U.S. Bureau of the Census, a newly constructed one-unit structure is deemed completed when all finish flooring—carpeting, if used in place of finish flooring—has been installed or when first occupied, if earlier. In buildings with two or more housing units, all units in the structure are deemed completed when 50% or more of the units are occupied or available for occupancy. In counting completions, the bureau does not include group quarters (such as dormitories and rooming houses), transient accommodation (such as transient hotels, motels, and tourist courts), moved or relocated buildings, mobile homes, or housing units created in an existing residential or nonresidential structure.

Furthermore, the census bureau excludes publicly owned housing units (contract awards) in their measure of completions, but they do include units in structures built by private developers with public subsidies or that, on completion, are offered for sale to local public housing authorities under the Department of Housing and Urban Development (HUD) "Turnkey" program.

In addition, the census bureau excludes mobile homes (those that have transportation gear as an integral part of the unit and can be towed from site to site). Mobile homes are an important source of new housing. The bureau estimated, for example, that manufacturers shipped 319,700 mobile homes during 1996.

The U.S. Bureau of the Census estimates housing completions using the monthly Survey of Construction. Bureau representatives draw random samples of building permits in about 840 jurisdictions where building permits are required for construction. Recipients of building permits are surveyed to determine whether dwellings under construction have been completed, and sample data are then blown up to form national estimates for permit-issuing areas. Upward adjustment of these housing completions is made to account for starts without permit authorization. The bureau also estimates completions in places where no building permit is required; in this case, national estimates are blown up from a random sample of about 130 places that the bureau searches intensively for housing units completed.

TABLE 12 Annual Net Privately Owned Housing Units (shown in thousands) Completed, United States, 1968 to 1996

Year	All Starts	One Unit[a]	Two Units	Three or Four Units	Five or More Units
1968	1,320	859	44	33	384
1969	1,399	808	44	35	512
1970	1,418	802	43	42	532
1971	1,706	1,014	51	55	586
1972	2,004	1,160	54	65	725
1973	2,101	1,197	60	64	780
1974	1,729	940	44	52	693
1975	1,317	875	32	29	382
1976	1,377	1,034	41	37	266
1977	1,657	1,258	49	46	304
1978	1,868	1,369	59	57	382
1979	1,871	1,301	61	64	445
1980	1,502	957	51	67	426
1981	1,266	819	49	62	336
1982	1,006	632	30	51	293
1983	1,390	924	37	55	374
1984	1,652	1,025	35	77	515
1985	1,703	1,073	36	61	534
1986	1,756	1,120	35	51	550
1987	1,669	1,123	29	42	475
1988	1,530	1,085	24	33	389
1989	1,423	1,026	24	35	338
1990	1,308	966	17	28	297
1991	1,091	838	17	20	217
1992	1,158	964	15	21	158
1993	1,193	1,039	10	17	127
1994	1,347	1,160	12	20	155
1995	1,313	1,066	15	20	212
1996	1,413	1,129	14	20	141

SOURCE: Data made available by the Construction Starts Branch, Construction Statistics Division, U.S. Bureau of the Census.

a. One-unit structures include all fully detached and certain attached (semidetached, row house, and town house) dwellings, except for attached structures in which heating, air-conditioning, water supply, power supply, or sewage disposal facilities are shared.

Table 12 summarizes total annual housing completions in the United States since 1968. In general, completions trough before the trough in the business cycle, then rise quickly during the first couple of years of the trough-to-peak of the business cycle; starts peaked in 1973, 1978, and 1986. Also evidenced in Table 12 is the postwar baby boom, which contributed to the surge in housing completions in the early 1970s. The great apartment building boom of the late 1960s and early 1970s is also apparent. (SEE ALSO: *Building Cycle; Housing Starts*)

—*John R. Miron*

Further Reading

Topel, R. and S. Rosen. 1988. "Housing Investment in the United States." *Journal of Political Economy* 96(4):718-40.

U.S. Department of Commerce, Bureau of the Census. 1995. *Current Construction Reports: Housing Completions.* Washington, DC: Author. ◄

► Housing Construction Process

The process by which houses are built has changed over the centuries, but it still involves certain fundamentals, including a site, production and organization of materials, skilled workers, and the actions of clients and users. A site is found; the earth is shaped; public utilities are brought in; building permission is sought; materials are brought to the site, manipulated, and assembled; and a household brings in personal possessions and makes a home in the building. Later, the house is modified to suit changed expectations of the household, to replace obsolete or defective technical parts, or to meet preferences of a new household.

Perhaps one of the most significant distinctions in housing construction processes is between dwellings built because a household has acted to set the process in motion and houses built when a builder takes initiative. In the former, a buyer may approach a builder and "order" a house to be built, using standard plans or plans drawn to suit the buyer's preferences. In the case of a party wishing to occupy an existing multifamily building, an interior contractor may be hired to "fit-out" the space according to the new occupant's requirements. There is also the case of a house constructed without any household acting, to be sold on the market. Such speculative construction, in anticipation of use, shifts risk to the maker.

How a Conventional House Is Built

The construction of a conventional light wood frame house begins with a decision to build. A contract is made between a party who wants the house and the party who will build it. Financing is arranged for the construction, and usually at the same time, a commitment is obtained from a lending institution for the permanent financing once the ownership of the house changes hands to the new owner. If a site has already been purchased, a building permit is obtained from the local regulatory jurisdiction after plans have been reviewed and approved. Arrangements are made for installation of services such as water, sewer, gas, electricity, and telephone and for temporary electrical service for the use of power tools during construction. These services will later be permanently attached to the house.

The next step is to "lay out" the site, indicating the exact location of the building and other improvements in preparation for excavation, grading, and other site work. An excavation is made for the subgrade structure, called a *foundation*. This is normally made of poured-in-place concrete and/or concrete masonry construction. Around the perimeter of the foundation wall in the case of a house with a basement, a drainage line is placed in a gravel bed at the foot of the foundation wall. The outside of the foundation wall is waterproofed. When no basement is planned, but the house will be built on a concrete slab, piping, electrical conduit, and sometimes heating ducts are positioned, after which gravel is laid and covered by a vapor barrier and concrete is poured for the floor.

Next, when there is a basement, a wooden floor platform is built on top of the foundation walls, supported by the perimeter foundation walls and, if necessary, steel or wooden beams placed as needed to reduce the floor joists'

spans. Plywood or a substitute sheet material is glued and nailed over the joists, providing a working floor surface for the next stages. Wall frames of wooden studs are made flat on this platform and tilted up into place and braced. When completed, these walls support another platform like the first one, itself prepared for the next series of walls above it in a multistory building. This process continues for the correct number of floors, usually not more than four in wooden construction. The roof is then built, using flat or sloped wooden roof rafters or prefabricated wood trusses made off-site and lifted into place by a crane. The roof is braced and covered with sheathing material, after which the waterproof surface and gutters for draining away water are installed. Walls are permanently braced against horizontal wracking. Windows and doors are then installed in prepared openings and joints around them sealed. Then the exterior walls are sheathed with insulating boards or other sheet materials prior to the application of the exterior cladding. This cladding can be a layer of wood or substitute siding, masonry, or stucco. The house is now secured against degradation by weather and from pilfering.

Some "rough" interior work on wiring, plumbing, heating, and ventilating can be done at the same time that the exterior finish work is under way. In houses with basements, piping is positioned in accord with the overall plumbing design, a vapor barrier is laid, and a concrete basement slab is poured. Insulation materials, normally fiberglass "batts" manufactured to standard dimensions corresponding to the spacing between wooden framing members, are put in the wall and ceiling cavities after wiring, ductwork, and plumbing are installed, to meet the specified energy conservation standard adopted for the governing jurisdiction. This is followed by a vinyl vapor barrier on the warm side or inside of the insulation. Drywall, sheetrock, or other paneling is then installed and joints are taped and sanded, ready for paint or wallpaper. Doors and door frames are installed as well as other woodwork such as baseboards, fireplace mantels, stair balustrades, and accessories such as cabinet shelves, rods, and medicine cabinets. The walls and ceilings are spray painted or otherwise finished with wall covering. Kitchen and bath cabinets are placed, and plumbers and electricians return to complete their work on wiring, plumbing fixtures, and heating and air-conditioning equipment. Finishes are then laid on floors and walls and final baseboard trim is installed. Landscaping is completed, and the house is ready for the occupant's decorative features, window coverings, and personal possessions.

Time to construct a single house varies with its size and complexity and with the speed with which decisions—including governmental reviews—are made and supplies arrive on the site. A small and simple house can be completed in less than a month, particularly when prefabricated parts are used; more complex houses in the range of 3,000 square feet can take as much as six to eight months or more to complete. In general, the weather-tight shell of the house, making up approximately 20% to 30% of the total value of the house, is erected in a matter of weeks for an in situ building. The portion of work that takes longest is the interior construction, finishes, and equipment. Here, the complex interweaving of resource systems in walls and floors requires an intricate choreographing of trades. The rapid increase in the number of resource systems now found in houses has made this portion of work both the most expensive and also the most troublesome to coordinate.

Regulatory Review Process

Each jurisdiction in which a house is built has a regulatory review process intended to protect the public health and safety. This process starts with a preliminary review of the technical drawings, to make sure that they comply with the requirements for structural, electrical, and plumbing systems design. The drawings are reviewed in the local building inspection department. If the drawings meet the requirements of the building code, a permit is issued following payment of a fee.

Once the permit is granted, building may commence. At specified intervals, a building inspector will visit the construction site and review work in progress, making sure that it complies with the approved drawings. Any problems or discrepancies are discussed and solutions approved and "signed off" by the inspector. If work has been done that does not comply, it may have to be removed or corrected before work can continue.

Several inspections take place. The first occurs before the footings and foundation walls are poured, to check on the correctness in the depth of the excavation and on other preparations, including placement of reinforcing.

The next inspection usually occurs after the framing of the house is completed and all rough wiring and plumbing is in place. This must be done before the drywall is installed. Water supply lines are pressure tested to assure that there are no leaks, and the electrical panel box is inspected to be sure that the electrical circuits are installed as approved. If the jurisdiction has stringent energy efficiency and performance requirements, the thermal barriers are inspected while still visible.

The last inspection is made after all the major interior work is completed, the electrical service and plumbing fixtures are installed, and the heating and air-conditioning systems are in place. When all regulatory provisions are deemed to have been met, a certificate of occupancy is granted, indicating that the building department has given final approval of the construction. In areas with special regulatory provisions, such as historic district designation, proximity to a floodplain, or other special conditions, further reviews will be required.

This process is virtually the same in all U.S. jurisdictions. Variations may be found, however, because of the particular building code in force, because of the reputation of the builder, or because of the complexity of the design. In multifamily buildings, more visits are made, but the same basic steps apply. (SEE ALSO: *Construction Industry*)

—*Stephen Kendall*

Further Reading

Bender, Richard and Forrest Wilson, eds. 1973. *A Crack in the Rearview Mirror.* New York: Van Nostrand.

Dietz, Albert G. H. 1979. *Dwelling House Construction.* Cambridge: MIT Press.

Eichler, Ned. 1982. *The Merchant Builders.* Cambridge: MIT Press.

Kaiser, Edgar F. 1968. *The Report of the President's Committee on Urban Housing: A Decent Home.* Washington, DC: Government Printing Office.

Mason, Joseph B. 1982. *History of Housing in the U.S., 1930-1980.* Houston, TX: Gulf. ◀

▶ Housing Costs

Housing costs represent the sum total of various factors, including property acquisition and development, maintenance, and financing. Policymakers and activists who are concerned with providing affordable housing need to understand these cost components and the extent to which they can be influenced.

A 1995 survey of U.S. developers conducted by the National Association of Home Builders (NAHB) found the following average distribution of costs for a standard house, defined as 2,000 square feet with two and a half baths and three or four bedrooms. The average selling price for such a house was $183,585:

Total construction costs	53.2%
Finished lot (including financing)	24.4%
Financing (construction)	2.0%
Overhead and general expenses	5.8%
Marketing (including homebuyer financing)	2.2%
Sales commission	3.3%
Profits	9.1%

Although the relative and absolute values of each of these components will vary depending on the particular developer, the local housing market, and the regulatory framework, this list forms a useful basis for further consideration.

Total construction costs include the cost of building permits. Many municipalities also impose impact fees to help cover the costs of public infrastructure such as parks, schools, and roads. Impact fees of $3,000 per dwelling are not uncommon; they have been documented as high as $45,000. The increasing use of impact fees and municipal efforts to ensure that building fees cover the full cost of inspections suggest that the relationship between these fees and total housing costs is likely to grow.

Construction costs also include both labor and materials. In general, labor constitutes about one third of total construction costs. Material costs are emerging as a somewhat volatile part of housing construction. When Hurricane Andrew destroyed more than 80,000 homes in Florida in August of 1992, it caused an increase in plywood prices throughout the United States. As a result, the cost of a 1,700-square-foot home rose nearly $600 in Orlando and $173 in Boston. As another example, from October 1992 to March 1993 lumber prices rose more than 90% adding $4,500 to new home prices. This situation was largely attributed to federal restrictions on the lumber supply in an effort to protect the northern spotted owl.

According to the NAHB's 1995 standard housing data, construction costs in the United States averaged $48.93 per square foot. There can be a wide range in this cost due to economies of scale realized through construction of multiple houses at one time or construction of a large house rather than a small house. Higher-amenity houses that make use of more expensive materials and skilled labor will also cost more per square foot.

A *lot* refers to a parcel that is fully serviced according to local subdivision requirements. Typically, this reflects the cost of roads, water, and sewage services as well as sidewalks in more urbanized areas. Site development is expensive. For example, one Oregon developer reported costs for a subdivision completed in 1997: Streets, curbs, sidewalks, stormwater drainage, trenches for utilities, and installation of street lights ran $11,795 per lot with an average cost of $203 per lineal foot. Water lines cost $45 per foot and sanitary sewers cost $37 per foot.

Municipalities establish subdivision standards to avoid the future need to expend public funds for road improvements and the provision of public amenities. As a result, the costs of the infrastructure are "up-fronted" and subsequently incorporated in the initial asking price of a house. This explains why in 1949, when very few subdivision regulations were in place, lot costs constituted only 11% of total housing costs compared with 24% in 1995.

Lot costs vary dramatically from one housing market to the next; they can also change with remarkable speed, rising or falling 30% in a span of six months to a year in particularly volatile markets. As an example of costs, 6,000- to 6,500-square-foot lots in the western part of the Portland, Oregon, metropolitan region averaged $42,000 to $45,000 in 1994. Similar lots in southern California would have run $75,000 before the recession of the early 1990s.

Construction loans are typically .25% above mortgage rates for owner-occupied housing. This financing in combination with property taxes can be particularly costly for developers if there are excessive delays in the development process, as can happen with protracted decision making at the local level.

The combination of lots and building/impact fees can constitute, in extreme cases, as much as 35% to 45% of total costs. This cluster of costs is most affected by public policy. A focus on regulatory reform can identify various opportunities where different levels of government can modify regulations and streamline decision making so as to reduce cost impacts on housing.

As an example, one affordable housing initiative in Hyde Park, New York, found that the zoning ordinance requirement for curbs throughout an apartment complex property translated into a $7 cost per month for each of 77 apartments. The town ultimately amended the zoning ordinance to remove those regulations, which added costs but provided questionable benefits.

The clustering of housing units on small lots can realize savings in infrastructure costs (and provide commonly held open space) in comparison with the same number of more widely dispersed units on larger lots. One of the most significant opportunities for reducing costs is to build at higher densities; this can be encouraged if not mandated through zoning and related development regulations. Zoning should also allow for less costly housing alternatives, such as manufactured housing or attached single-family

dwellings. Some municipalities give density bonuses to developers who make a portion of their housing available at lower cost.

Sponsorship of housing development by not-for-profit organizations, specifically community development corporations (CDCs), is an important means of creating affordable housing. Habitat for Humanity, Inc. is particularly well-known for its ability to draw on donations of materials and volunteer labor to build low-cost housing throughout the United States and the world. CDCs can take advantage of grants and other resources as well as tax benefits that are not normally available to private developers.

Those who are concerned with providing affordable homeownership need to consider not just the selling price of the house but other expenses that can be equally important in determining whether a particular household can maintain ownership. These include financing, energy costs, insurance, and property taxes. In determining the amount of housing debt a household can assume, financing institutions typically use a maximum of 28% of gross income for all housing costs. The various factors of housing cost can, therefore, represent serious limitations with respect to what a household can afford to buy.

Changes in mortgage rates have a dramatic impact on monthly costs. On an $80,000 mortgage amortized over 30 years, 6% versus 9% translates into a cost difference of $164.05 per month. Thus, the higher rate requires an additional $7,000 in annual income for the same mortgage.

Building codes mandate strong energy conservation measures, thus lessening the impact of energy costs on total monthly costs, although these can still be in excess of $70 per month. The cost of energy measures is, however, incorporated into the total cost of the house. Therefore, the purchaser has to pay more for the house to save on future energy costs.

Insurance rates are related to housing value. Basic coverage in the late 1990s for an $80,000 house begins at $250 per year. Property taxes represent a much more significant cost, often adding hundreds of dollars per month.

Maintenance becomes a significant cost factor in older housing stock. If housing values are low owing to the lack of demand, necessary maintenance and repairs may be neglected because the homeowner realizes that the value of the work would not be recouped on sale of the property. Deferred maintenance also occurs when an owner has a limited income.

As the value and quality of older housing stock declines relative to new construction, these dwellings become more affordable to lower-income households. The better built and historically interesting units may ultimately become a source of low-cost housing for households with limited resources but a willingness to use their own labor. This can lead to the process of gentrification, which ultimately converts low-cost housing into relatively high-cost housing.

Issues of housing cost are somewhat different when it comes to rental housing. Rental rates need to cover all the costs associated with homeownership as well as management expenses and a margin of profit. Renters are particularly vulnerable to the buy/sell cycles of the housing market. Unlike homeowners, renters do not benefit through the personal realization of capital gains. Renters' housing costs often increase on the sale of their units as rents are raised to cover the costs of refinancing. One way of addressing this situation is through the development and long-term ownership of rental housing by CDCs.

Energy costs are of particular concern in older rental housing stock that lacks effective conservation measures, such as adequate insulation, storm windows, and weather stripping. Various programs sponsored by community action agencies and utility companies have focused on providing energy conservation services to low- and moderate-income households as a means of reducing energy (and hence, housing) costs and increasing comfort. These measures are typically rated with respect to their payback period—that is, the number of years before the value of the energy savings will equal the cost of the conservation measure.

An analysis of housing costs is a complex subject because it involves many different considerations. For any particular housing initiative, it is necessary to detail actual housing costs and revenues and to systematically explore the implications of changing the values of different variables. A pro forma is used for this purpose. The ultimate goal is to ensure that revenues will exceed costs. Those concerned with affordable housing will have the added challenge of designing housing initiatives so that the end product, in all its dimensions, is available at a cost that can be afforded by households in the targeted income range. (SEE ALSO: **Affordability**; *Housing Investment; Housing Price*)

—*Deborah A. Howe*

Further Reading

Hershey, Stuart S. 1987. *Streamlining Local Regulations: A Handbook for Reducing Housing and Development Costs.* Washington, DC: Department of Housing and Urban Development.

Kemp, Peter. 1990. "Income-Related Assistance with Housing Costs: A Cross-National Comparison." *Urban Studies* 27(6):795-808.

Mallach, Alan. 1984. *Inclusionary Housing Programs.* New Brunswick: Rutgers, the State University of New Jersey, Center for Urban Policy Research.

National Association of Home Builders. Address: 1201 15th Street, NW, Washington, DC: 20005-2800. Phone: 1-800-368-5242. Fax: (202) 822-0377.

"Not in My Back Yard": Removing Barriers to Affordable Housing. 1991. Washington, DC: Advisory Commission on Regulatory Barriers to Affordable Housing.

Real Estate Research Corporation. 1974. *The Costs of Sprawl.* Washington, DC: Government Printing Office. ◄

► Housing Courts

Specialized courts designed to handle legal disputes between landlords and tenants—known as housing courts or landlord-tenant courts—became a popular idea in urban areas throughout the United States in the 1960s and 1970s. The impetus for housing courts came from widespread dissatisfaction with existing dispute-resolution mechanisms on the part of both tenants and landlords. Proponents of the idea maintained that housing courts would, through expertise and economies of scale, be able to resolve disputes both expeditiously and equitably. By the time enthusiasm for the housing court idea began to wane in the 1980s, many of the nation's largest municipalities, including New York City, Los Angeles, and Boston, had established housing courts.

Landlord-tenant relations have always been intense and fractious. Each party to the landlord-tenant relationship seeks something from the other. A tenant seeks the best quality home at the most affordable price. A landlord seeks to maximize economic return at minimal expense. Because of this tension, disputes between landlords and tenants are commonplace. Yet because the balance of power in the legal relationship has rested firmly with landlords, historically, legal issues were few and methods for dealing with disputes were relatively simple.

The most common legal disputes between landlords and tenants are eviction proceedings. Since the early 19th century, all jurisdictions in the United States have had "summary proceedings" statutes to govern the eviction process. Summary proceedings are accelerated legal proceedings designed to move toward judgment in a far shorter time frame than ordinary civil proceedings. By providing an expedient means to obtain court-ordered evictions, summary proceedings were supposed to encourage landlords to seek eviction in court rather than use self-help to remove tenants. For the most part, summary proceedings succeeded in bringing the eviction process into the realm of the courts.

Until the 1960s and 1970s, in most jurisdictions, tenants had few legal defenses to eviction proceedings. However, during that period, in response to tenant activism related to the civil rights movement, most states adopted the *implied warranty of habitability:* the notion—novel at the time—that a tenant's obligation to pay rent depended on a landlord's providing a habitable dwelling. In addition, housing codes; federal, state, and local housing subsidy programs; rent regulation (in some jurisdictions); and new procedural safeguards against eviction had, along with the warranty of habitability, made landlord-tenant law into an increasingly complex body of law. In large municipalities such as New York City, eviction proceedings, proceedings brought by tenants against landlords for repairs, and municipal lawsuits against landlords for code violations were all brought in separate courts with jurisdiction over a wide variety of lawsuits in addition to housing matters. Although tenants began to have more substantial defenses to landlords' eviction claims and legal remedies for landlords' misfeasance, these rights remained more theoretical than real, because it was virtually impossible for tenants to assert their rights on their own in these separate forums. Landlords were also dissatisfied because these new rights had made summary eviction proceedings less summary.

Housing courts were established as specialized courts that exclusively adjudicate housing disputes to address these landlord and tenant concerns and to accommodate the evolution in landlord-tenant law. Although this factor is common to all housing courts, the structure and approach of housing courts differ from jurisdiction to jurisdiction. Some housing courts have jurisdiction over rent and eviction cases only, some over code enforcement cases only, and some have broad jurisdiction over all cases that directly or indirectly affect housing conditions in the locality. Some courts are more informal than others. Informal courts use hearing officers instead of judges and do not apply technical rules of evidence; some require parties to appear without lawyers. The powers of the housing courts also differ from jurisdiction to jurisdiction. In some housing courts, the powers of judges are fairly limited, permitting them to render judgments in simple eviction cases and no more; in other housing courts, they are quite broad: Boston's housing court, for example, granted a quarter million dollar negligence award for lead paint poisoning, and although ultimately reversed, New Haven's housing court ruled that the State of Connecticut has a responsibility to provide emergency shelter beyond the state-imposed 100-day limit. Innovations that have been tried in housing courts include permitting private parties to act in a prosecutorial-type role and bring criminal proceedings against landlords for code violations (Boston); having nonattorney advisers in court to assist tenants and landlords (Boston); having a legal assistance office in court (Detroit); and equipping courtrooms with computer terminals that have online, read-only access to municipal records of housing code violations and tax arrears (New York City).

Those who use them do not generally perceive housing courts to have fulfilled their promise of efficient, equitable resolution of landlord-tenant disputes. Many housing courts have come under severe criticism; few have staunch defenders. The large majority of tenants in urban areas are poor people, and housing courts have been plagued by the problems that face most courts in which poor people are forced to litigate, such as inadequate, overcrowded facilities, a high volume of cases, and insufficient time for each case. And because of the close nexus between poverty and race in the United States, these inadequacies most severely affect members of minority groups. The majority of defendants in eviction proceedings cannot afford counsel and cannot obtain free counsel because there are not enough attorneys available who provide free representation. Although there is a growing body of tenant protection legislation on the books and housing court judges have greater expertise in housing law than other judges, most tenants are unaware of or unable to articulate the legal defenses they may have to eviction proceedings. A study of Detroit's housing court, for example, found that changes in the law such as the advent of the warranty of habitability were less significant for tenants than the availability in the court of clear, understandable information on tenant rights. Moreover, although the articulated purpose of most housing courts was to consolidate all types of proceedings in a single forum, housing courts have been criticized by tenants as placing undue emphasis on rent collection and eviction to the exclusion of tenant claims, such as needed repairs and rent overcharges. Landlords have also vociferously criticized housing courts and have, for example, initiated lawsuits in Washington D.C. and New York City that complained of delays in processing evictions and protenant bias on the part of the tribunal.

More housing cases are litigated than any other type of civil litigation. The stakes in housing courts are extremely high, the issues housing courts address are increasingly complex, and the litigants are highly emotional. Localities have established housing courts in response to the evolving nature of landlord-tenant law in an attempt to placate both sides of the landlord-tenant divide. Neither side seems to have been satisfied. (SEE ALSO: *Eviction; Right to Housing*)

—*Andrew Scherer*

Further Reading

American Bar Association, Special Committee on Housing and Urban Development. 1981. *Specialized Courts: Housing Justice in the United States—Executive Summary.* Washington, DC: Department of Housing and Urban Development.

DeGraffe, Luis Jorge. 1990. "The Historical Evolution of American Forcible Entry and Detainer Statutes." *Seton Hall Legislative Journal* 13:129.

Keating, W. Dennis. 1987. "Judicial Approaches to Urban Housing Problems: A Study of Cleveland Housing Court." *The Urban Lawyer* 19(2):345-65.

Legal Assistance Foundation of Chicago, National Lawyers Guild and the Chicago Council of Lawyers. 1977. *Judgment Landlord: A Study of Eviction Court in Chicago.* Chicago: Authors.

Miller Mosier, Marilyn & Richard A. Soble. 1973. "Modern Legislation, Metropolitan Court, Miniscule Results: A Study of Detroit's Landlord-Tenant Court." *Journal of Law Reform* 7(9):66.

Washington University School of Law. 1979. "Housing Courts." *Urban Law Annual* 17(Special issue). ◄

► Housing Credit Access

The improvement of housing conditions in developed and underdeveloped countries requires access to credit. Individuals leverage their savings through a down payment to secure long-term amortized financing for homeownership; homeowners can obtain smaller loans for making structural repairs; and developers leverage their own funds or equity ownership position to finance the construction, acquisition, and rehabilitation of residential property. Barriers to credit access, such as low incomes or racial, gender, and locational discrimination, are among the most important obstacles faced by households seeking affordable, quality dwelling units. Efforts to expand access to housing credit have led to reforms in the lending practices of mainstream financial institutions and to the creation of alternative credit institutions and instruments. For example, credit access can be expanded through the use of low-interest loans; relaxed minimum loan requirements; waivers of closing costs, loan points, or the mortgage insurance requirement; reduced down payments; sweat equity; credit counseling; multiple or nontraditional sources of income; flexible debt-to-income qualifying ratios; lease-purchase arrangements; and deferred second mortgages.

In the United States, federal intervention in housing finance (through government mortgage insurance, government-sponsored enterprises in the secondary mortgage market, and the Federal Home Loan Bank system) has emphasized macroeconomic goals over social housing concerns. Since the 1960s, the economic power of financial institutions in determining spatial patterns of growth and decline within metropolitan areas as well as the distribution of individual wealth (in 1993, home equity and rental property accounted for one-half of the net worth of U.S. house-holds) has increasingly been recognized by policymakers and housing advocates, leading to the passage of several laws and regulations and the creation of public-private housing finance programs designed to expand access to credit for areas and groups that are underserved by the mortgage markets.

The Fair Housing Act of the Civil Rights Act of 1968 (and subsequent amendments) prohibits discrimination in the sale, rental, or financing of housing by race, color, religion, gender, handicap, familial status, or national origin. The Equal Credit Opportunity Act of 1974 (and subsequent amendments) prohibits discrimination in credit transactions by gender, marital status, race, color, religion, national origin, age, or receipt of public assistance income. To combat the problem of geographic discrimination, known as *redlining,* the Community Reinvestment Act of 1977 (CRA) requires federally regulated depository financial institutions to serve the credit needs of local communities in areas where they are chartered to conduct business, through lending activities, community development investments, and basic financial services.

In some states and localities, housing finance agencies use tax-exempt revenue bonds to fund below-market-rate mortgages to first-time homebuyers and developers of low- and moderate-income multifamily rental housing. A national network of alternative community development financial institutions (CDFIs), made up of locally based community development banks, credit unions, and nonprofit revolving loan funds, is combining philanthropic and government funds with other investment sources to stimulate access to credit in economically disadvantaged urban and rural areas. All of these conventional and alternative financial institutions rely on federal programs such as the low-income rental housing tax credit, Section 8 rental assistance, the Community Development Block Grant, and the HOME housing block grant, to preserve housing affordability.

The role of the federal financial regulatory agencies and the U.S. Department of Housing and Urban Development (HUD) in regulating the U.S. housing finance system has been strengthened since the late 1980s by (a) congressional amendments to the CRA, the Home Mortgage Disclosure Act, and the Fair Housing Act; (b) legislative reform of the Federal Home Loan Bank system; (c) expanded oversight of the government-sponsored enterprises (GSEs) in the secondary mortgage market; and (d) revisions to the CRA enforcement regulations. These agencies now collect and analyze a wealth of data on an annual basis, describing local activity in the primary mortgage market (the origination of home loans by lending institutions), the secondary mortgage market (the purchase of home loans by investors), the private mortgage insurance industry, and other financial services.

Sources of Data

The Home Mortgage Disclosure Act of 1975 (HMDA) requires federally regulated financial institutions to disclose their mortgage lending on an annual basis by the census tract location of the property for which financing is being requested. Until 1990, HMDA collected data only on home loans that were actually originated by commercial banks, savings institutions, and credit unions (which are not regulated by CRA), indicating the amount, purpose, and type of each loan. In 1980, when HMDA was renewed by Congress for five more years, federal regulators agreed to make the lending reports available to the public in metropolitan area depositories (until then, the reports were available only on request from each institution) and to computerize the national database.

Congressional amendments in 1987 and 1989 renewed HMDA permanently and expanded the coverage of institutions to include mortgage-banking subsidiaries of banks and thrifts and independently owned mortgage companies (mortgage bankers are not regulated by CRA). Also starting in 1990, the loan application registers (LARs) of each reporting institution were made available to the public in summarized and detailed format, describing the disposition of each home loan application (including loans purchased and refinanced) and the race, gender, and income of every loan applicant. The institutional threshold for HMDA reporting (at least $10 million in assets with an office in a metropolitan area) expanded in 1993 to include smaller lenders that originate more than 100 home purchase loans or refinancings annually.

In 1997, the asset-based reporting threshold for depository institutions was increased from $10 million to $28 million (with future annual adjustments in this exemption based on the consumer price index). Commercial banks and savings institutions with assets of at least $250 million are required to disclose HMDA records for the entire United States, not just the metropolitan areas where they maintain offices. The HMDA data since 1992 are available in computerized CD-ROM format from the Federal Financial Institutions Examination Council (FFIEC) and can be retrieved on the Internet through the Right-to-Know Network of the Unison Institute (www.rtk.net).

The type of information available through HMDA is listed in Table 13. National statistics on HMDA-reported lending to underserved groups and areas are summarized in Table 14. In general, the absolute number of home mortgage loan originations to underserved groups and areas, and these loans as a percentage of the total, has been increasing since 1990. Table 14 verifies this trend of successful social credit allocation for conventional home purchase loans, home improvement loans, and Federal Housing Administration (FHA)/Veterans Administration (VA) home purchase loans (except for a declining proportion of FHA/VA loans in central cities). For example, the number of conventional home purchase loans to black borrowers tripled between 1990 and 1995, and as a percentage of the total number of conventional home purchase loans, these loans increased from less than 3% in 1990 to 5% by 1995. The number of conventional home purchase loans to low- and moderate-income borrowers has more than tripled since 1990, and as a percentage of the total number of conventional home purchase loans, these loans increased from just 11% in 1990 to 23% by 1994. Denial rates among underserved groups and in underserved areas for conventional home purchase and home improvement loans have been increasing, however, because of the expanded number

TABLE 13 Sources of Data on Housing Credit Access in the United States

Type of Information Available on an Annual Basis for Each Loan or Insurance Application and Each Loan Purchase	Primary and Secondary Mortgage Markets FFIEC/HMDA[a]	Private Mortgage Insurance FFIEC/MICA[b,c]	Government-Sponsored Enterprises HUD Public Use Data
Lending institution	x		
Type of lending institution			x
Mortgage insurance company		x	
Federal regulatory agency	x		x
Loan amount	x	x	x[d]
Loan-to-value ratio			x[c,d]
Loan purpose	x	x	x
Federal mortgage insurance/guarantee	x		x
Owner-occupancy	x[c]		x[c]
Number of units			x
Income mix of units			x[d,e]
Affordability level of rental units			x
Number of bedrooms			x[e]
Census tract location of property	x	x	x[f]
Race/national origin of applicant/borrower	x[g]	x	x[c]
Gender of applicant/borrower	x[g]	x	x[c]
Annual income of applicant/borrower	x[c,g]	x	x[c]
Number of applicants/borrowers	x[g,h]		x[c]
Age of applicant/borrower			x[c]
First-time homebuyer			x[c]
Action taken on application	x	x	
Reasons for denial of application	x[i]		
Government-sponsored enterprise (GSE) purchasing agency	x		x
Type of non-GSE purchasing institution	x		

a. Federal Financial Institutions Examination Council/Home Mortgage Disclosure Act.
b. Federal Financial Institutions Examination Council/Mortgage Insurance Companies of America.
c. Single-family (1-4 unit) loans only.
d. Data grouped into categories; exact number not available (except for loan amount of single-family loan purchases under $200,000).
e. Multifamily (5+ unit) loans only.
f. Not available for correlation with type of lending institution (single-family loans only), loan-to-value ratio, loan purpose, federal mortgage insurance/guarantee, owner-occupancy, number of units, income mix of units, affordability level of rental units, and number of bedrooms.
g. Reporting is optional for loans purchased.
h. Up to two co-applicants.
i. Reporting is optional.

of applications as well as nondiscretionary "credit scoring" policies that have recently been adopted by financial institutions at the urging of the government-sponsored enterprises in the secondary mortgage market. The national denial rate for conventional home purchase loan applications by blacks increased from 33% in 1994 to 49% by 1996.

More than one-half of all insured single-family home mortgage loans in the United States are insured by private mortgage insurance (PMI) companies. Homebuyers often must acquire private mortgage insurance as a prerequisite for receiving a mortgage loan, but the HMDA does not provide any data on PMI companies or the application process for PMI. Since 1993, the eight PMI companies that dominate the industry (comprising the trade association membership of the Mortgage Insurance Companies of America [MICA]) have disclosed their insurance application records on an annual basis through the FFIEC. Similar to HMDA, the records disclosed by each company include information describing each insurance application: the census tract location of the property; the race, gender, and income of the applicant; the amount and purpose of the mortgage loan; and the disposition of the application (see Table 13). Analysis of these PMI data by the Federal Reserve indicates that insurance applicants who are nonwhite, low- and moderate-income or buying a home in a census tract that is predominantly minority, low- and moderate-income or located in a central city are more likely to be denied insurance.

More than one-half of all outstanding single-family mortgage debt is securitized in the secondary mortgage market, including almost 9 of 10 FHA- or VA-backed home loans. Approximately 90% of all securitized single-family mortgage debt is held by the government-sponsored secondary market entities, principally Fannie Mae (formerly the Federal National Mortgage Association) and the Federal Home Loan Mortgage Corporation (Freddie Mac). Fannie Mae and Freddie Mac are both regulated by HUD. The influence of these government-sponsored enterprises on home mortgage underwriting procedures has been growing

TABLE 14 Residential Mortgage Lending in the United States by Financial Institutions Reporting under the Home Mortgage Disclosure Act, 1990 to 1996

| | Year | Number of Loans | Percentage of All FHA/VA/FmHA Home Purchase Loans (%) | FHA/VA/FmHA Market Share (%) | Loan Applications | | |
					Originated (%)	Denied (%)	Other (%)
FHA/VA/FmHA Home Purchase Loans							
Borrowers							
Black	1990	47,579	8.3	50.3	60.1	26.2	13.7
	1993	81,057	9.9	49.9	65.1	22.2	12.7
	1996	111,748	12.7	45.1	70.3	15.5	14.2
Hispanic	1990	30,474	5.3	30.5	66.7	18.4	15.0
	1993	66,089	8.1	42.0	69.8	14.6	15.7
	1996	109,343	12.4	44.6	76.1	10.2	13.7
Low and moderate income	1990	110,162	23.7	42.4	70.8	18.0	11.2
	1993	260,387	37.2	39.0	75.6	14.4	9.9
	1996	310,788	39.7	35.8	77.4	11.2	11.3
Census Tracts							
Minority (50%-100%)	1990	30,095	7.0	26.9	63.1	22.1	14.8
	1993	73,433	10.7	37.0	69.7	17.5	12.8
	1996	97,036	12.7	37.5	71.9	12.5	15.6
Low and moderate income	1990	57,195	13.2	30.0	68.7	17.7	13.6
	1993	107,348	15.5	36.7	71.9	15.8	12.3
	1996	133,729	17.3	34.4	74.3	11.6	14.0
Central city	1990	215,423	49.0	31.8	73.7	14.6	11.6
	1993	318,944	45.9	31.1	76.1	13.2	10.7
	1996	355,567	45.4	28.2	77.9	10.0	12.1

| | Year | Number of Loans | Percentage of All Conventional Home Purchase Loans (%) | Conventional Market Share (%) | Loan Applications | | |
					Originated %	Denied (%)	Other (%)
Conventional Home Purchase Loans							
Borrowers							
Blacks	1990	47,045	2.9	49.7	50.8	33.6	15.6
	1993	81,322	3.4	50.1	51.5	34.0	14.5
	1996	135,944	4.6	54.9	35.1	48.8	16.1
Hispanic	1990	69,548	4.3	69.5	61.2	21.4	17.4
	1993	91,345	3.9	58.0	59.6	25.1	15.3
	1996	135,683	4.6	55.4	46.8	34.4	18.8
Low and moderate income	1990	149,930	11.4	57.6	61.7	25.8	12.5
	1993	407,059	21.6	61.0	67.0	21.5	11.5
	1996	558,162	23.2	64.2	50.5	34.2	15.2
Census Tracts							
Minority (50%-100%)	1990	81,869	6.5	73.1	61.1	21.1	17.8
	1993	124,891	6.7	63.0	62.6	23.0	14.4
	1996	161,809	6.8	62.5	50.3	30.7	19.0
Low and moderate income	1990	133,554	10.7	70.0	63.4	20.1	16.6
	1993	185,014	9.9	63.3	65.0	21.8	13.2
	1996	255,204	10.8	65.6	50.9	31.7	17.4
Central city	1990	461,325	36.3	68.2	70.8	14.7	14.5
	1993	708,088	37.7	68.9	74.2	14.1	11.7
	1996	903,289	37.6	71.8	63.5	21.1	15.4

(Continued)

TABLE 14 (Continued)

	Year	Number of Loans	Percentage of All Home Improvement Loans (%)	Loan Applications		
				Originated (%)	Denied (%)	Other (%)
Home Improvement Loans						
Borrowers						
Black	1990	41,307	5.6	54.7	36.9	8.4
	1993	59,872	6.8	51.4	38.1	10.4
	1996	79,409	8.5	43.7	42.8	13.4
Hispanic	1990	23,794	3.2	56.1	32.5	11.4
	1993	51,308	5.8	55.7	35.0	9.3
	1996	64,075	6.8	48.0	40.6	11.4
Low and moderate income	1990	147,370	24.8	59.1	31.9	9.1
	1993	211,870	30.6	55.2	34.8	10.0
	1996	262,848	29.0	43.0	42.2	14.7
Census Tract						
Minority (50%-100%)	1990	52,779	9.2	49.5	38.5	11.9
	1993	80,024	12.0	47.3	40.5	12.1
	1996	108,692	12.4	35.4	46.9	17.7
Low and moderate Income	1990	74,200	13.0	54.4	34.9	10.7
	1993	115,304	16.9	52.2	36.8	11.0
	1996	153,628	17.3	40.2	43.7	16.1
Central city	1990	227,140	38.9	63.0	26.4	10.6
	1993	291,105	42.1	60.6	28.7	10.7
	1996	371,099	40.9	47.4	36.0	16.5

SOURCE: Federal Financial Institutions Examination Council.

NOTE: The number of reporting institutions increased by 6.4% between 1992 and 1993 due to expanded coverage of mortgage bankers. In 1990, residential mortage lending was reported for metropolitan areas only. In 1996, commercial banks and savings institutions with assets of $250 million or more also reported mortgage loans made in nonmetropolitan areas. FHA/VA/FMHA and conventional home purchase loans and home improvement loans finance one- to four-unit properties. For depository institutions with assets of $30 million or less, reporting the race or national origin, gender, and gross annual income of the loan applicant is optional. For applications taken by mail, reporting the race or national origin and gender of the loan applicant is not required if the applicant does not provide it. For phone applications, reporting this information is optional. Low and moderate income is defined as a loan applicant with a gross annual income at or below 80% of the median family income of the metropolitan area or a census tract with a median family income that is equal to 80% or less of the median family income for the metropolitan area. Census tract boundaries for 1990 are from the 1980 census. Census tract and metropolitan area family income data for 1990 is an inflation-adjusted update from the 1980 census. Census tract and metropolitan area family income data for 1993 and 1996 are updates from the 1990 census, based on estimates from HUD. The market share for FHA/VA/FMHA and conventional home purchase loans refers to the percentage of all home purchase loans in the particular category that were either insured or guaranteed by the federal government (through the Federal Housing Administration, the Department of Veterans Affairs, or the Farmers Home Administration), or were "conventional"" (not government backed). Actions taken on loan applications are grouped into three categories: (a) approved and originated, (b) denied, and (c) other. The category "other" includes applications approved but not accepted, applications withdrawn, and application files closed for incompleteness.

since the 1970s, especially with the decline of the savings and loan industry. Title XIII of the Housing and Community Development Act of 1992 gave HUD the authority to establish annual goals for mortgage loan purchases made by Fannie Mae and Freddie Mac and to disclose information describing these loan purchases (see Table 13).

In 1996, HUD required that 40% of the dwelling units financed by Fannie Mae and Freddie Mac be occupied by families with an income at or below the median level for the metropolitan area, increasing to 42% for 1997 to 1999. HUD required that 21% of all mortgages purchased by Fannie Mae and Freddie Mac in 1996 be located in census tracts with either (a) a median family income of 90% or less of the area median or (b) a minority population of 30% or more combined with a median family income of 120% or less of the area median. This goal was increased to 24% for 1997 to 1999. Finally, HUD required that 12% of all

units financed by Fannie Mae and Freddie Mac in 1996 be occupied by owners or renters with an income at or below 80% of the area median income *and* located in census tracts with a median family income of 80% or less of the area median or by owners or renters with an income at or below 60% of the area median income with no restriction on census tract location. This goal was set at 14% for 1997 to 1999. Within this category of mortgage purchases, Fannie Mae and Freddie Mac must buy multifamily housing loans equal to at least 0.8% of the total dollar volume of their mortgage purchases in 1994.

Uses of Data

The data on housing credit access outlined in Table 13 can be used to (a) assess community credit needs through the Community Reinvestment Act (CRA), (b) prepare local housing and community development plans, (c) enforce fair lending and fair housing laws, and (d) monitor and evaluate public-private housing finance programs. Community-based organizations and sometimes local governments have analyzed HMDA data to define the problem of redlining, demonstrate the failure of lending institutions to comply with CRA, and negotiate lending agreements targeted to underserved areas and groups. This HMDA research can be effective in building a coalition for negotiations, collectively designing and implementing a reinvestment program, and generating media attention. Under the revised CRA regulation, the federal financial regulatory agencies must use HMDA data as the primary source of information for evaluating CRA lending performance, and HMDA analysis is an important component of CRA strategic plans that financial institutions have the option of preparing, in collaboration with community-based organizations and the regulatory agencies. Also, market studies prepared by specialized community development financial institutions increasingly rely on HMDA analysis to demonstrate the need for financial services in distressed communities.

Research on housing credit access can be used by local governments to support the planning requirements of the federal Community Development Block Grant and the federal HOME housing block grant and to prepare the "analysis of impediments" study mandated by the local planning requirements of the Fair Housing Act. This research can also be used in the planning process for federal empowerment zones. State housing finance agencies can use data on primary and secondary mortgage markets to create financing partnerships that leverage state government funds. Analyses of mortgage-lending patterns in metropolitan areas can also support growth management programs by informing strategies to control the spatial pattern of investment and expand the supply of affordable low- and moderate-income housing, through inclusionary or "fair share" housing programs. Finally, HMDA analysis may indicate a possible pattern or practice of discrimination by a lending institution (such as differential treatment based on race) and lead to investigations of discrimination in mortgage loan underwriting and marketing. The high-profile investigations of Decatur Federal Savings and Loan and Shawmut Mortgage Company during the early 1990s were prompted by HMDA studies conducted by the *Atlanta Journal-Constitution* (Decatur) and the Federal Reserve Bank of Boston (Shawmut).

Compared with U.S. census information, which is gathered every 10 years, HMDA data describes on an annual basis in each census tract who is applying for housing credit (and being approved or rejected) from which institutional lenders. In this way, HMDA analysis is a tool to monitor local residential real estate markets. To determine local credit needs, mortgage-lending data are correlated with census tract information on race, income, population, households, and housing units and mapped spatially through the use of geographic information systems, to analyze disparities by race, gender, income, and location. Market share analysis is used to compare the volume of loan applications, approvals, and purchases among competing institutions in defined markets and submarkets.

Housing credit access in the United States is increasingly being impaired by larger trends in financial services, such as the computerization of mortgage loan origination, the rise of "subprime" lenders and finance companies that charge high interest rates to borrowers with low credit scores, and the growth of "predatory" lenders in low-income and urban areas where bank branches have closed and financial services are unavailable or inadequate. HMDA data do not include the interest rate charged on mortgages, and reporting the race and gender of the borrower for applications taken by phone is optional. The revised CRA regulation is making available additional information on bank branch locations, recent branch closings, and small-business lending.

In India, southeast Asia, and South Africa, community-based organizations—also known as nongovernmental organizations (NGOs)—are reforming the practices of national and international development finance institutions to expand access to credit for low-income housing and neighborhood development. "Peer group" lending and savings models pioneered in developing countries are being replicated in the United States through strategies such as individual development accounts (IDAs), designed to stimulate savings by the poor for housing expenses, education, and enterprise development. (SEE ALSO: **Affordability;** *Community Reinvestment Act;* **Discrimination;** *Fannie Mae; Home Mortgage Disclosure Act; Housing Finance; Lending Institutions; Neighborhood Housing Services of America; Redlining; Section 237; Section 502; Uniform Residential Landlord and Tenant Act*)

—*John T. Metzger*

Further Reading

Fish, Gertrude Sipperly, ed. 1979. *The Story of Housing.* New York: Macmillan.

Goering, John, ed. 1996. "Race and default in credit markets." *Cityscape* 2(1).

Goering, John and Ron Wienk, eds. 1996. *Mortgage Lending, Racial Discrimination, and Federal Policy.* Washington, DC: Urban Institute Press.

Hudson, Michael. 1996. *Merchants of Misery: How Corporate America Profits from Poverty.* Monroe, ME: Common Courage Press.

Oliver, Melvin L. and Thomas M. Shapiro. 1995. *Black Wealth/White Wealth: A New Perspective on Racial Inequality.* New York: Routledge.

Parzen, Julia Ann and Michael Hall Kieschnick. 1992. *Credit Where It's Due: Development Banking for Communities.* Philadelphia: Temple University Press.

Squires, Gregory D., ed. 1992. *From Redlining to Reinvestment: Community Responses to Urban Disinvestment.* Philadelphia: Temple University Press.

Yinger, John. 1995. *Closed Doors, Opportunities Lost: The Continuing Costs of Housing Discrimination.* New York: Russell Sage. ◄

► Housing Demand

Housing demand is the preference for accommodation expressed subject to the constraints of income and price. Economists usually presume that each consumer has preferences that he or she uses to rank choices from most to least preferred. These rankings are generally thought not to vary over time or with respect to income and price. Faced with a set of prices, consumers then choose the highest ranked from among choices within their budget. Demand is also usefully seen as the outcome of strategies used to cope with a costly part of the consumer's budget. Coping strategies include choosing less adequate housing (e.g., smaller dwelling, housing in poor repair, or housing with few amenities), moving to a remote location where housing is cheaper, sharing accommodation with others, taking greater risks in financing, rent payments in-kind, and use of sweat equity (e.g., buying a dilapidated home to fix up with one's own labor).

Demand is commonly viewed as a list of prices for housing and the specific quantity of housing chosen or demanded at each price—with income, other prices, and preferences held constant. A demand curve portrays this price-quantity relationship in graphical form. For many commodities, demand can be readily measured as a quantity—for example, fifty loaves of bread annually. In the case of housing, the quantity demanded is usually just one dwelling. However, the housing stock includes much variety; dwellings differ in attributes such as floor area, number of rooms, number of baths, quality of construction, design, built-in features (e.g., central air-conditioning), contract conditions and property rights, access to local public services, and other neighborhood amenities, including safety, social mix, and proximity to work, schools, and shopping. Here, consumer demand is usually measured in terms of the bundle of attributes (generically referred to as "quality") of the dwelling.

In view of this widely observed variety of stock, some scholars characterize housing in terms of presence, absence, or amount of one or a few important attributes. This includes studies in which a single measurable feature of the dwelling is used to proxy quality: for example, number of rooms, number of baths, lot size, or age of dwelling. Related to these are studies based on discrete choice—typically, the presence or absence of a specified dwelling attribute (e.g., brick sheathing or central air-conditioning). Still other scholars combine these two and model demand simultaneously as a discrete choice (e.g., tenure) and as continuous variable (e.g., floor area).

Many economists use a different approach in which they attempt to estimate, for each dwelling, an unobservable flow of "housing service" (i.e., a quality of accommodation). Underlying this approach is the assumption that producers in a competitive market will, over time, vary the amounts of housing attributes supplied until each attribute is equally profitable; hence, the expenditure on housing as a consumption good should be directly proportional to the flow of housing services. This perspective has led some scholars to measure demand simply as housing expenditure. However, expenditure can be thought of as the product of the quantity of housing service and the price of a unit of service. This concern over quantity-price decomposition has led other scholars to estimate the demand for housing by a consumer by dividing housing expenditure by the typical expenditure for a dwelling of standard quality. Hence, demand is measured as the number of dwellings of standard quality that a given expenditure could purchase.

That housing stock is a capital good with a long life complicates the analysis of housing demand. As a thought experiment, imagine that all housing was made of cardboard and quickly became run-down; each consumer would have to return to the market frequently to purchase new cardboard stock from homebuilders. In that case, the suppliers of housing stock would be largely the new homebuilders; there would be little demand for resales of cardboard stock. In the real world, however, housing is durable, and consumers are more likely to sell their dwellings at some point in their lives. They then become suppliers of existing stock and are mindful of the capital gain, or loss, realized on disposition. A high resale price downstream reduces the effective price of homeownership and hence increases the demand for housing today.

Many studies of housing demand focus on mode of tenure. Most of these studies look at only two alternatives (private sector rental and homeownership) and ignore others (e.g., leasehold, equity cooperatives, nonequity cooperatives, public housing, collective housing, or reservation housing) or blur distinctions (e.g., condominium versus freehold ownership, long-term vs. short-term leases). In such studies, the demand for homeownership is sometimes seen as a portfolio choice problem (how the household holds its financial assets) but more often is seen in terms of the relative costs of homeownership vis-à-vis renting. Here, aspects of income taxation as they relate to the imputed rents of homeowners, capital gains (or losses) on disposition of property, and deductions for mortgage interest, property tax, and depreciation are commonly found to be important.

In an important sense, there are actually two kinds of demand for housing. One is the demand for housing stock by homeowners and landlords. This is the demand that leads to investment in new construction and to new housing starts and completions. The other is the demand for housing services by consumers (whether homeowners or renters). Landlords typically combine housing stock with property management services, utilities, security, and maintenance to supply the accommodation that renters demand. Homeowners typically combine stock with utilities, security, maintenance, and their own management skills to supply accommodation to themselves.

Suppose that in a local housing market over a short period of time, we observe the number of dwellings sold or newly rented and the corresponding price paid. Let us refer to these consumers as "new" homeowners or renters, in contrast to "sitting" consumers (those who acquired or last relet their premises before the period of time under study). In a market in competitive equilibrium, no new consumer willing to pay the market price forgoes consumption, and hence, the amount of housing service exchanged is the amount demanded at that price. If we further assume that sitting consumers are utility maximizing and have no transaction costs (e.g., associated with lease or contract termination, moving, or property transfer), then the housing services provided by their dwellings can be added to the amount purchased by new consumers. To the extent that sitting consumers have such transaction costs, a discrepancy can rise between the demand for housing over the short term and over a longer term. (SEE ALSO: *Housing Investment; Housing Markets; Housing Supply,* **Tenure Sectors**)

—*John R. Miron*

Further Reading

Formby, J. P. and W. J. Smith. 1996. The Income Elasticity of Demand for Housing: Evidence from Concentration Curves." *Journal of Urban Economics* 39(2):173-92.

Pozdena, R. J. 1988. *The Modern Economics of Housing: A Guide to Theory and Policy for Finance and Real Estate Professionals.* New York: Quorum.

Sternlieb, G. and J. W. Hughes. 1986. "Demographics and Housing in America." *Population Bulletin* 41(1):3-34. ◄

► Housing Distribution Mechanisms

Processes of housing distribution are fundamental to patterns of social and spatial segregation, residential mobility, and the operation of housing choice. These processes link the physical housing stock (newly built dwellings as well as the existing stock of dwellings) with users of the stock. Essentially, housing distribution mechanisms engage when dwellings are available for use, and they serve in determining who comes to use these dwellings.

Housing distribution mechanisms cannot be assumed to operate efficiently or fairly and cannot be taken for granted as reflecting various features of the housing market. Nor, however, do they operate in a wholly independent or autonomous fashion. Rather, they relate to the structure and operation of other processes (production, finance, consumption) but introduce additional factors that cannot always be read off from these processes. For example, the fact that a dwelling has been built by a private developer and builder does not determine what the distribution mechanism will be after construction is completed. The dwelling could have been built for a variety of different owners or in different locations that will affect how distribution is managed.

Discussion of housing distribution processes, nonetheless, cannot be wholly divorced from supply or the features of the dwelling. The stock available for distribution (newly built or secondhand dwellings becoming vacant) changes over time and differs between locations. Potential users are in competition for the supply that is available, and those mediating this competition will adjust their policies and practices in the light of changes in supply as well as demand.

The most appropriate framework for considering the operation of housing distribution is one that starts with competition between individuals (or households) for housing resources. Because dwellings differ in terms of building type, size, age, location, quality, and condition, competition will be more or less intense. This does not assume uniform values: Different households will have different requirements for housing, depending on health, age, family size and structure, type and location of employment, schools and other services, and lifestyle factors. In a pure market system, the price mechanism would be the key way of resolving this competition. However, no housing systems

work in this way. In many countries, there are significant nonmarket sectors. In addition, the structure of legislation and regulation affects the operation of market sectors and the particular features of housing distinguish it from other commodities. These features mean that more complex and elaborate arrangements are required to sustain flows of information about what is available in terms of properties and financial arrangements for paying for housing. Market failure is an intrinsic feature of the housing sector, and arrangements to prevent and compensate for this have been widely developed.

The conventional framework for the analysis of how housing competition is managed in advanced economies (and more widely) distinguishes between different distribution mechanisms in different tenures. Such mechanisms exist in all tenures, but they may be very different in different tenures. The conventional approach distinguishes between mechanisms associated with house purchase and renting. Within each of these initial categories are important subdivisions. These are most evident in the rented sector, where processes in private (market) sectors are generally distinguished from those in nonmarket or nonprofit sectors (e.g., public sector housing, voluntary or charitable organizations, and cooperatives) where distribution mechanisms are labeled as bureaucratic rather than market.

Where owners of housing seek to distribute or allocate their housing on criteria other than ability to pay a market price, they are involved in nonmarket bureaucratic rationing processes. In the housing research literature, these processes have been most fully investigated for large public sector and nonprofit landlords. In this literature, a number of separate elements in bureaucratic processes are apparent:

1. Primary eligibility criteria are generally laid down in policy to determine who can obtain housing. Such criteria may involve specific tests of income (as in the United States) and of residence. They may also refer to age and family structure (e.g., married couples or families with children). Where households would otherwise be excluded, there may be specific exemptions, for example, for veterans, people with family ties to the area, or those with particular disabilities.

2. Secondary criteria that determine how households are selected from among those qualifying in terms of primary criteria. These criteria will determine whether people are offered a house at all (in a situation of overall shortage) or what property is offered (who gets the best houses and who gets the worst).

3. This secondary allocation process may be operated by managers in an unaccountable fashion. However, it is more likely to be the case that clear published rules and systems are used. For example, in the large professional public and nonprofit housing sectors in the democracies of Western and Northern Europe, various rules are set out. These may involve points schemes (prioritizing applicants in terms of specific circumstances, such as overcrowding, lack of adequate amenities, sharing, housing condition, health status), date order schemes (first come, first served), or a combination of these. Other bureaucratic schemes have placed great importance on occupational or other merit

categories. Such merit- or status-based schemes are regarded as typical of those operating under state socialist systems in much of Eastern Europe where they were designed to reward politically or economically favored groups.

4. Primary and secondary allocation processes determine who is at the head of waiting lists for different types of dwellings. There is not a single list but, rather, lists that reflect rules about the fit between household and dwelling size (avoiding both overcrowding and "underoccupation") and expressed preferences about neighborhood or location. In some cases, landlords have also graded their properties and applicants as suitable for different grades of property.

5. The final element in bureaucratic allocation involves "matching" dwelling and household. In this final phase, the manager has a property available to let and a waiting list to draw from. In many cases, the person at the very top of the list may not be offered the property or may refuse it and the manager's task is to allocate the property as quickly as possible (to minimize rent loss) to a suitable applicant from among those at the top of the relevant list. The most detailed studies of this matching process emphasize the different pressures on those operating the process—from managers wanting minimum vacancy periods and others wanting people who will fit in. As a result, labeling and stereotyping influences judgments and decisions. The outcomes are also influenced by applicants' decisions to accept or refuse offers and by the rules related to such offers. For example, in the British system, certain households, notably the homeless, may be eligible for only one "reasonable offer." Others in the housing waiting list may be able to refuse more properties without damaging the chance of a further offer.

A number of comments can be made about how these bureaucratic schemes operate. First, they can be based on coherent and reasonable rules that are known to applicants and are not secret. These rules can allow for choice and do not represent command systems unresponsive to the applicant. Second, and following from the first point, applicants have differential bargaining power. Some of this derives from knowledge, some from the demand category and its status, and some from ability to wait for a better offer. Thus, it is argued that those in the most desperate housing situations are often formally in the lowest currency category (e.g., homeless) but in any case are less likely to be able or willing to wait and more likely to accept the first house offered even if it is less desirable. The outcome of all the processes outlined above can therefore appear to be very similar to a market process. Those with the most limited resources or bargaining power are most likely to be housed in the housing for which there is least demand. Third, these general patterns are least likely to work for those from minority ethnic groups. In some cases, this has been a result of institutional discrimination or rules that systematically disadvantage such groups (e.g., residence, ethnic mix), but even where policies have been designed to avoid discrimination, equal treatment of all races has not been achieved. A range of factors influences this, including the application of codes and stereotypes that grow out of the wider society rather than the housing allocation process itself.

Bureaucratic allocation systems succeed to some degree in breaking the association between what housing people obtain and what their income is. Even when policy and practice are most carefully constructed to be fair and provide equal treatment, however, there is ranking and stratification and unequal treatment. If the same questions are asked about distribution mechanisms in the private sector, some of the same issues arise. Thus, in the private rented sector, the research evidence shows that a variety of processes cast doubt on the assertion that ability to meet the rent is the only consideration in selecting tenants. There is a history of explicit and overt discrimination against particular households (as well as of action to illegally evict the most vulnerable).

The practice of insisting that properties are no longer available for rent when certain applicants apply has the same effect. This has applied most evidently to minority ethnic groups. Even when such explicit discrimination does not operate (partly because of legal measures), a range of practices used by landlords and their agents make access to private rented housing difficult for some who could afford the rent. These most notably relate to the payment of deposits, key money, and rent in advance, all of which can require that households have substantial funds before they can obtain tenancies. The large number and different orientation of private landlords also affect the operation of this sector of the housing system. Smaller landlords, in particular, may vary considerably in what procedures and selection criteria they use. As a result of this range of considerations and different forms of regulation affecting the sector, it is widely regarded as providing both the most open access and the least consistent and transparent distribution mechanisms.

The dominant tenure in most market economies is now private homeownership, and the mechanisms are, consequently, those affecting who buys dwellings available for purchase. Initially, there are, as in other sectors, questions about information and knowledge of the market. Agencies involved in this process may adopt practices that selectively provide information based on racial and other prejudices and stereotypes.

The management of the homeownership sector involves other gatekeepers, including lawyers, valuers, mortgage and insurance brokers, banks, and other institutions providing loans for house purchase. The practices of all of these agencies may affect the decisions of households. False or misleading information and advice may apply in some cases. In others, practices in relation to assessment of risk are crucial. This is best documented in relation to decisions to provide mortgage finance. Evidence from both Europe and North America show that financial institutions have, at times, red- or bluelined areas on which they will not lend or will lend only on unfavorable terms. The operation of the market in these areas (of buying and selling) is affected by such practices, which do not relate to accurate assessment of the type, quality, or condition of individual dwellings.

Similarly, decisions on the creditworthiness of applicants for loans may relate to a variety of age, gender, income, and occupational factors and involve social stereotyping similar to that identified in bureaucratic rationing systems. Applications for loans from people with the same incomes or in the same age or occupational categories may be treated differently. The distribution of properties for homeownership is normally through a market process (except where the opportunity to buy is exclusively available to one buyer, as in some privatization schemes). In this situation, access to loans is crucial. However, other sources of funds (the extended family or community, inheritance, proceeds from previous sales of homes) also affect what is affordable or obtainable, when such funds exist, and when the lenders' loan is less in relation to the value of the property, loans are likely to be easier to obtain.

Although research evidence has been most effective in analyzing the different elements involved in the distribution of housing outside the market sector, it is apparent that a range of actors equally affect processes in the private sectors. In all tenures (including cooperatives, accommodation provided by employers, shared ownership, and a range of initiatives designed to enhance access to housing for particular groups), there are a series of separate processes involved in bringing potential users of housing into occupation of that dwelling. These processes involve the application of formal rules and policies but also involve the management of information and competition. Just as bureaucratic processes may produce outcomes that bear similarities to market processes, these market processes are managed and implemented by bureaucracies that are affected by wider societal pressures and can result in outcomes that would not be predicted by market models. Bargaining power and distribution mechanisms in all tenures involve a range of social attributes and relate to the nature of the wider society. (SEE ALSO: *Centrally Planned Housing Systems;* **Cross-National Housing Research;** *Welfare State Housing*)
—*Alan Murie*

Further Reading

Bassett, K. and J. Short. 1980. *Housing and Residential Structure: Alternative Approaches.* London: Routledge & Kegan Paul.

Grigsby, W. G. 1963. *Housing Market and Public Policy.* Philadelphia: University of Pennsylvania Press.

Henderson J. and V. Karn. 1987. *Race, Class and State Housing.* Brookfield, VT: Gower.

Johnston R. J. 1971. *Urban Residential Patterns.* London: G. Bell.

Kemp, Peter. 1990. "Income-Related Assistance with Housing Costs: A Cross-National Comparison." *Urban Studies* 27(6):795-808.

Murie A., P. Niner, and S. Watson. 1976. *Housing Policy and the Housing System.* London: Allen & Unwin.

Rex J. and R. Moore. 1967. *Race, Community and Conflict.* Oxford, UK: Oxford University Press. ◄

► *Housing Finance*

Housing Finance tracks the housing market in the United Kingdom. It is published quarterly, in February, May, August, and November, by the Council of Mortgage Lenders. Each issue includes a commentary on housing market trends during the preceding quarter. In addition, statistical tables illustrate trends in housing tenure and construction, mortgages, housing prices, property transactions, savings, and building societies. Feature articles cover particular segments of the housing market, such as women or first-time buyers and particular geographic locations. Articles also explore the implications of the housing market's changing social and economic context for housing finance mechanisms. *Housing Finance* is available for £20.00 per copy from the BSA CML Bookshop, 3 Savile Row, London W1X 1AF. (SEE ALSO: *Collateral Mortgage Obligation; Housing Trust Funds; Impact Fees; Linkage; Mortgage Finance; Syndication; Tax Increment Financing*)
—*Mara Sidney* ◄

► Housing Finance Agencies

Government frequently finds it necessary to intervene in the housing market to correct problems that it finds. Among the problems frequently encountered is the absence of lenders willing and able to make loans. These loans are necessary to permit developers to acquire, rehabilitate, or build multifamily housing, and they are necessary to permit households to purchase a home, whether a new home or an existing one. Government can resolve these problems by becoming a lender. A common method through which government fills this role is the creation of a housing finance agency (HFA). HFAs are, simply, a type of government-sponsored bank providing funds for housing purchase and development.

Over time, government at various levels has recognized a need to form HFA's as alternatives to the private market lenders. The supply of capital for the development of multifamily housing rises and falls periodically as part of normal business cycles in the economy, but over the long term, the supply of capital for multifamily housing development has decreased, leaving the development industry with a shortage of financing for otherwise needed projects. The secondary market for multifamily housing, especially smaller, affordable housing is also very limited, creating a shortage of financing for development. Finally, private financing for the development of affordable housing is frequently in short supply, especially in geographic areas experiencing a high incidence of poverty or physical deterioration. HFA's generally operate as a source of funding to resolve these problems.

HFAs exist as entities created by state-enabling statues. The statutes dictate the scope of powers of each HFA as well as the organizational form of the agency. Usually, the agencies operate under the review of a board of directors who set policy. An elected official (such as a governor for a state-sponsored HFA) generally appoints these board members. Some of the members of the board are often public officials for whom service on the HFA board is part of their duties. This board appoints an executive director who hires other professional staff as needed to implement the programs with which the HFA is charged.

Generally, HFAs are entities serving a state, but many agencies exist that serve other jurisdictions. Every state, except for Arizona and Kansas, has a state HFA. Arizona performs these government-sponsored lending duties through its state Department of Commerce. Some large cities, such as New York City, have their own HFA. Other HFAs exist that serve counties or metropolitan areas, such as those in the Baltimore and San Diego areas.

HFAs generally have the power to sell housing bonds on behalf of the state within which they are located. Proceeds from the sale of these bonds are used to finance housing that may be either multifamily or single family and may be for rental property or for owner occupancy. Interest paid on these bonds is, in most, although not all cases, exempt from federal income taxes. This exemption enables the HFA to loan the proceeds from these bond issues at interest rates that are lower than could be obtained from private lenders. The reduced interest charges help to make the housing more affordable to the end user, either through lowering rents or lowering the monthly payments of an owner occupant.

An HFA may function as a primary lender, making loans directly to borrowers, such as developers of affordable multifamily housing. HFAs may also serve as secondary market lenders, making loans to a set of private banks who will make loans to the end borrowers. The private banks will then service the loans, collecting payments from the borrowers and sending periodic payments back to the HFA for it to repay its bond buyers.

HFAs frequently perform other auxiliary tasks—for example, administering a variety of state and federal housing programs such as the Low-Income Housing Tax Credit program. They function in nearly all parts of the United States and have become an important part of the development of affordable housing. They are especially crucial to the developers of affordable rental housing using the federal low-income housing tax credits and to first-time homebuyers using mortgage revenue bonds. (SEE ALSO: *State Government*)

—*Kirk McClure* ◄

► *Housing Finance International*

Housing Finance International is a quarterly journal of the International Union of Housing Finance Institutions. First published in 1986, it presents nontechnical articles relating to the housing finance field. Circulation: 1,000 worldwide. Price: $75 for two issues. Address: Secretary General: Mr. Don Holton, International Union for Housing Finance, 111 East Wacker Drive, Chicago, IL 60601-3704. Fax: (312) 946-8202. Editor: Dr. Michael Lea, Cardiff Consulting Services, 2207 Via Tiempo, Suite 100, Cardiff, CA 92007. Fax: (619) 942-5107.

—*Caroline Nagel* ◄

► *Housing Finance Review*

A now defunct quarterly journal, *Housing Finance Review* covered all aspects of housing finance, including portfolio management by private financial institutions, the operations and policies of official institutions involved in housing finance, and various types of mortgages and mortgage-related instruments. It also reviewed the structure of primary and secondary mortgage markets, the effects of housing finance on construction and on general economic activity, housing subsidies, and consumer protection issues related to housing and housing finance markets.

—*Laurel Phoenix* ◄

► Housing-Income Ratios

Housing-income ratios are indicators that describe housing cost characteristics in the context of household income. Price-to-income and value-to-income ratios show the relationship between the price or value of a home and owner income. Payment-to-income ratios describe the relationship between mortgage or rent payments and income. Certain housing-income ratios have become accepted as "rules of thumb" or standards for determining an appropriate amount of housing expense. For example, homebuyers may be advised to shop for homes priced at twice to 2.5 times their gross annual income. Another rule of thumb asserts that a household today should pay no more than 30% of its total income for housing. The 30% standard evolved from late 19th-century experience, which allowed "a week's pay for a month's rent." Early in this century (i.e., 1920s and 1930s), mortgage lenders used a 25% rule of thumb, which predominated for more than 50 years; the 25% rule was based on home loan repayment patterns over that period. During the 1980s, higher consumer expenditures for housing provided a rationale for a 30% standard. Rules of thumb provide convenient but not always accurate guidelines for budgeting and policy-making decisions. Taken individually, some households spend more, and others spend less, than 30% of their income on housing.

Mortgage analysts study housing-income ratios to learn points at which borrowers become overloaded with debt. The evaluation of existing loans provides a statistical basis for setting lending standards to avoid mortgage delinquency and default. Following are two examples of ratio-based loan qualification criteria:

> ► *Payment-to-income ratio:* Monthly payments for mortgage principal, interest, taxes, and insurance (PITI) should not exceed N% (e.g., 25%-28%) of monthly income.
> ► *Debt-to-income ratio:* Monthly PITI payments plus other payments for long-term debt (e.g., auto loans, student loans, revolving credit) should not exceed N% (e.g., 33%-36%) of monthly income.

Because housing costs vary over time and from place to place, housing-income ratios are a way to compare relative costs rather than dollar values. Policymakers use ratio-based standards of need to implement housing assistance programs nationally. For example, residents in federally assisted rental housing currently pay 30% of their household income toward rent. Federal programs have generally followed rules of thumb established with conventional housing market data. However, consumer advocates have asserted that market-based rules are not accurate indicators of what lower-income households can or should spend on housing. Although housing-income ratios can be a good way to predict outcomes based on existing data, housing experts caution against applying common rules to dissimilar populations. Factors of income, tenure (i.e., rent or own), race, welfare status, family composition, lifestyle, and age as well as external factors (e.g., discrimination, housing shortages) have all been linked to deviations from established housing-income ratios. Moreover, ratios have been found to vary from country to country, with significant differences between market and nonmarket economies. To illustrate, average costs in market economies may range from 15% to 30% of income and up, depending on the country, year, housing type, and population subgroup. Furthermore, the 30% standard for subsidized housing in the United States contrasts sharply with housing-income ratios for state-provided housing that prevailed in the late 1980s in the following formerly socialist countries (Renaud 1992, p. 882):

Country	Year	Rent (% of income)
Bulgaria	1988	12.1
Hungary	1987	8.7
Poland	1986	4.4
Romania	1989	4.4
Yugoslavia	1988	9.3
USSR	1989	2.5

Program-to-program and country-to-country comparisons must also account for (a) the extent to which both rent payments and government funds support the actual cost of housing, (b) variations in income tax regulations and lending mechanisms, and (c) the presence or absence of noncash income supplements (e.g., rent control, food stamps, health care benefits, etc.). (SEE ALSO: **Affordability**; *Affordability Indicators; Affordable Housing Indices; Brooke Amendments; Housing Costs; Rent Burden*)

—*Marjorie E. Jensen*

Further Reading

Daye, Charles et al. 1989. *Housing and Community Development: Cases and Materials.* 2d ed. Durham, NC: Carolina Academic Press.

Feins, Judith D. and Terry Saunders Lane. 1983. "Defining the Affordability Issue." In *Housing Supply and Affordability,* edited by Frank Schnidman and Jane A. Silverman. Washington, DC: Urban Land Institute.

Renaud, Gertrand. 1992. "The Housing System of the Former Soviet Union: Why Do the Soviets Need Housing Markets?" *Housing Policy Debate* 3:877-99.

Stone, Michael E. 1993. *Shelter Poverty: New Ideas on Housing Affordability.* Philadelphia: Temple University Press.

Van Vliet--, Willem and Yosuke Hirayama. 1994. "Housing Conditions and Affordability in Japan." *Housing Studies* 9:351-67. ◀

▶ Housing Indicators

In October 1990, the World Bank launched the Housing Indicators Program, financed in part by the U.N. Centre for Human Settlements (Habitat), as an essential step in the implementation of the Global Strategy for Shelter to the Year 2000, endorsed by the U.N. General Assembly in 1988. Both developing countries and more developed economies participate in the Housing Indicators Program. At the end of January 1992, 52 countries were involved in the program.

The program has three aims:

1. To provide a comprehensive conceptual and analytical framework for monitoring the performance of the housing sector
2. To provide important new empirical information on the high stakes of policy making in the housing sector for societies and economies
3. To initiate new institutional frameworks that will be more appropriate for formulating and implementing future housing policies

The Global Strategy for Shelter is based on the assumption that government does not provide housing but plays an enabling role. Governments should thus facilitate, energize, and support the activities of the private sector, both formal and informal.

Questions Posed by the Program

Governments require instruments to see housing policy from a more global and comparative perspective. Only then can the lessons learned in one country become relevant to another. The Housing Indicators Program responds to this need. The program seeks to answer three fundamental questions:

1. Can informative, robust, reliable, and cost-effective techniques be developed to do the following:

   a. measure key aspects of housing sector performance;
   b. establish the links between the socioeconomic and policy environment and key housing sector outcomes, and
   c. establish the links between housing sector outcomes and broad social and macroeconomic performance?

2. How should the use of key indicators of housing sector performance be integrated into the formulation of national shelter strategies and international development assistance to the housing sector?
3. What institutional developments can be initiated to ensure that housing indicators will be used effectively in informing housing sector policy?

Indicators and Modules

The World Bank developed an extensive series of indicators. It distinguishes key, regulatory, and alternate indicators. Key indicators are the most powerful indicators of housing sector performance across countries and time. Alternative indicators provide a different way of measuring the same thing as a key indicator but with more readily available data.

The first version of the Housing Indicators Program used 25 key housing indicators, 10 regulatory indicators, and 10 socioeconomic impact indicators, grouped into six modules:

1. The Housing Affordability Module, which deals with house prices, rents, and household incomes
2. The Housing Finance Module, which deals with mortgages, credits, and interest rates
3. The Housing Quality Module, which deals with key attributes of housing quality
4. The Housing Production Module, which deals with housing production and investment
5. The Housing Subsidies Module, which deals with subsidies
6. The Regulatory Audit Module, which deals with regulations affecting the exchange of land and housing, land registration and ownership, housing finance regulation, rent control, administrative delays, land use, land development controls, and property taxation

All of the key indicators and alternate indicators are numbers, percentages, or ratios. Several of the regulatory indicators are composite indicators, formulated from responses to a large number of simple yes-or-no questions concerning the regulatory and institutional environment of the housing sector.

In each of the 52 participating countries, a housing indicators consultant was recruited. The consultant's task was to quantify the indicators for a particular city (usually the capital) in that country. In 1990, the consultants conducted an extensive survey of the housing sector, aimed at obtaining values for 25 key indicators of housing sector performance, 10 alternate indicators, 10 regulatory indicators designed to quantify the regulatory and institutional framework within which the housing sector operates, and 10 alternate regulatory indicators.

The survey aimed to do the following:

1. Create a basic set of *indicators* of the housing sector
2. Obtain *current estimates* for these indicators in 40 to 50 countries
3. Establish *key relationships* between these indicators and between them and key indicators of social and economic development

Additional, more practical, aims were these:

1. Provide an *analytical tool for governments* for measuring the performance of the housing sector in a comparative, consistent, and policy-oriented manner
2. Establish *baseline data* for new national shelter strategies and new housing sector loans
3. Create a *framework for comparing housing sector performance* between cities, countries, and different time periods

4. Help establish a *new institutional framework* for formulating and implementing national sectorwide housing policies
5. Work toward the creation of an *international network of experts and institutions* capable of overseeing the development of the housing sector

Conceptual Framework

The conceptual framework of the Housing Indicators Program is market oriented. Market prices are determined by supply-and-demand factors. Housing supply is affected by the availability of resource inputs, such as residential land, infrastructure, and construction materials. The organization of the construction industry, the availability of skilled and productive construction labor, and dependence on imports also affect it. Housing demand is determined by demographic conditions (e.g., rate of urbanization, new household formation) and macroeconomic conditions affecting household incomes. Demand is also influenced by the availability of housing finance and by government policies: taxation and subsidies, particularly subsidies targeted at the poor.

Both supply and demand of housing are affected by the regulatory, institutional, and policy environment. The conceptual framework also assumes that housing policies and housing outcomes may in turn affect broader socioeconomic conditions, such as the mortality rate for children under the age of five, the rate of inflation, the household savings rate, manufacturing wage and productivity levels, capital formation, and the balance of payments of the government budget deficits.

Normative View

The housing indicators team presented a normative view of how the housing sector should work. Policy making in the housing sector must be based on such a normative view. Therefore, it is necessary to look at the performance of the housing sector from a number of different perspectives: those of housing consumers, housing producers, housing finance institutions, local governments, and central governments. Each of these five perspectives focuses on different aspects of the sector and on different qualitative norms.

Examples of qualitative norms regarding housing consumers are these:

► Everyone is housed.
► There is a separate dwelling unit for every household.
► Housing expenditures do not take up an undue portion of household income.
► House prices are not subject to undue variability.
► Living space is adequate.
► Structures are safe, providing adequate protection from the elements, fire, and natural disasters.
► Infrastructure services and amenities are available and reliable.
► Location provides good access to employment opportunities.
► Tenure is secure and protected by due process of law.

▶ Households may freely choose between different housing options and different housing tenures (owning vs. renting).

▶ Housing finance is available to smooth housing consumption over time and allow households to pursue desired patterns of saving and investment.

▶ Adequate information on housing options is available and affordable, to ensure efficient choice.

Qualitative norms are similarly formulated for housing producers, housing finance institutions, local governments, and central governments. In this way, the Housing Indicators Program provides a normative overview of a well-functioning housing sector from various key perspectives.

Qualitative Regulatory Norms

The government must provide an enabling legal framework for the entire cycle of housing development, transaction, use, maintenance, and replacement. Qualitative norms can be embodied by a limited number of concrete policy goals and measured by quantitative indicators.

The housing indicators team produced qualitative regulatory norms related to housing market development, land market development, housing finance development, public sector involvement, low-price distortions, bureaucratic bottlenecks, affordable standards, compliance, squatter tolerance, and housing as a local tax base

Research Outputs

The Housing Indicators Program resulted in indicator values for a set of 25 key indicators, the 10 alternate indicators, the 10 regulatory indicators, and the 10 alternate regulatory indicators, for 40 to 50 cities in selected countries.

A World Bank monograph introduced the conceptual framework for using housing sector indicators to measure sector performance and to guide policy and discussed the indicators and the relationships between them. Its second part discussed methods for data collection, data processing and analysis, institutional arrangements for creating and using indicators, and comparative costs of alternative approaches. Papers by country-based consultants used comparative data on the indicators to discuss national housing policy and future monitoring of the housing sector.

Conclusions

The housing indicators have already been updated and amended several times. Further changes are foreseen. If the indicators are maintained for a few years, the raw data can be used not only for cross-sectional analysis but also for time-series analysis. This could make a substantial contribution to the utility of housing indicators.

In the mid 1990s, the housing indicators team was comparing all 52 participating countries. A more narrowly focused study might apply the indicators to developing countries. The results could serve as a basis for a comparison with West European countries, which are undergoing a gradual harmonization of housing policy. Or they could provide a basis for comparison with countries in Central and Eastern Europe, which are reorienting their housing markets and policies. The Housing Indicators Program is a promising instrument for comparative housing research and policies. (SEE ALSO: **Cross-National Housing Research;** *Global Strategy for Shelter; Housing Markets World Bank*)

—*Hugo Priemus*

Further Reading

Mayo, Steve. 1991. *Housing: Enabling Markets to Work.* World Bank Policy Paper. Washington DC: World Bank.

Priemus, H. 1992. "Housing Indicators: An Instrument in International Housing Policy?" *Netherlands Journal of Housing and the Built Environment* 7(3):217-38.

World Bank. 1993. *The Housing Indicators Program.* Vols. I-IV. Washington DC: World Bank. ◀

▶ Housing Inequity

Inequity in housing has both theoretical and practical dimensions. As part of political economy, housing has association to social theory, to the philosophy of social science, and to political choice. Housing also has large significance as one of life's basic needs. The visual presence of varying standards, ranging from palaces to insanitary squatter settlements, arouses moral sensitivities about inequity, and it arouses a sense of injustice. Some prima facie claims of injustice have a strong imperative for urgent action for corrective responses. In many squatter settlements in developing countries, the matters at stake are life and death, involving high rates of child mortality and appallingly unsatisfactory conditions of life.

Housing issues link ethics, political economy, the pragmatic issues in social inequity, and urgency for reform. Links can occur through the way property rights are expressed in housing, in the institutional conditions that prevail in housing systems, and in the role that "home" plays in the local culture.

Housing systems can be understood as comprising production, consumption, allocations to occupiers, maintenance, finance, and various industries that supply resources for housing and the occupancy of housing. The character and operation of housing systems differ according to whether housing operates in subsistence economies, in state socialist economies, in market economies, or in other economic systems. By the 1900s, most economies in the world had become capitalist or market based, and their housing systems involve a mix of state resourcing, market allocation, voluntarist expression (e.g., nongovernment housing organizations [NGOs]), and household self-help in saving, construction, maintenance, and so on. The particular institutional settings, and especially the way financial conditions are structured, largely determine price-access to housing. Price-access has a relationship to the distribution of household income.

In entering a submarket or tenure sector in a housing system, households face opportunity sets that define and constrain their choices. Feasible choices are those that are "affordable." Affordability involves three interacting factors. First is the income of the household as its capacity to save. This will be conditioned by economic growth rates and distributions of income growth in the wider economy.

Economies with stable growth rates and patterns that improve the absolute and relative positions of households in the bottom 40% of the distribution of income are likely to have fewer affordability problems than those with low growth and patterns that increase inequality and the numbers of households in poverty. A second factor is the way housing finance is provisioned and structured. This includes mortgage and credit instruments, levels and designs of subsidies in housing, and their ultimate results in rent-to-income and housing payment-to-income ratios. A third factor is the medium-term trend in housing costs. These may be conveniently expressed as an index in comparison with indices in retail prices in general.

The three factors influence the opportunity set and price-access of any particular household in the distribution of household income. Affordability is a relationship between income distribution, a household's income, and the costs and financial conditions in housing submarkets. The degree of inequality and inequity in the *prior* distribution of income—including volumes and intensities of poverty among the lower parts of distribution—will have a large determining influence on current housing inequalities and inequities. The structuring of housing systems by government policies variously endorses the prior distribution, increases the inequalities, or moderates the inequalities and the inequities. Generally, housing policy has limited, although important, potential to reduce antecedent inequality.

The appropriate policies for reducing poverty and inequality vary according to context. For example, in an industrialized country without gross housing deficits in most of its submarkets, housing allowances can be used to close the housing (affordability) gap for some poor households. In the grossly undersupplied housing of the formerly socialist countries of Central and Eastern Europe in the 1950 to 1989 period, the release of market forces would have improved housing opportunities among the poor. In developing countries in the 1990s, low-income housing can be improved by income generation schemes, providing basic utilities and services, and facilitating self-help in housing for those living in squatter settlements.

Appearances alone do not provide good diagnoses or foundations on which to formulate policies. For example, in the former socialist countries of Central and Eastern Europe, rent-to-income ratios were very low, suggesting that household outlays on housing would also be low. However, under the full consideration of housing's connectedness to socialist economics, this was not the case. Housing outlays were high and paid in hidden ways, by implicit taxes on wages, by cost recovery in the prices of goods produced by enterprises that provided housing for employees, and by the tax transfer system in public budgets. Taking another example, some forms of subsidy in industrialized countries (e.g., subsidized rates of interest and grants to providers of social housing) that apparently improve price access for low-income households can lead to "horizontal" and "vertical" inequity. Horizontal inequity occurs where households with similar incomes and needs pay unequal amounts for their housing or have significantly differing standards of housing. Vertical inequity happens when higher income groups pay less in their housing outlays than lower income groups. Vertical and horizontal inequity occurs in most housing systems. It is partly derivative from prior inequalities in income and from badly designed policies in finance and allocation, including some income support schemes. But some of it is inevitable in a "here and now" perspective, because housing choices and allocations occur over longer perspectives than in the here and now. When *lifetime* incomes and housing outlays are considered, some of the here and now inequity disappears; but for some, inequity increases, especially when comparing outlays among renters and homeowners. By pensionable age many renters will have spent more on housing outlays than more affluent homeowners.

In housing, inequity and inequality are closely aligned with each other. However, principles of equity are wider than those of inequality; they include considerations of *merit, productive contribution,* and *effort* as well as claims based on *need.* The post-1960 social science literature often suggests that ideas from ideology, welfare, and good policy entirely coincide. They do not. Although they have overlap, ideology (e.g., market, socialist, and welfare state advocacy) and good policy have some independence from each other. It is important to distinguish among policy objectives of antipoverty, egalitarian, and housing poverty alleviation principles. Egalitarianism is about the "goodness" of the whole of the distribution of income and the way housing is related to it. Antipoverty is about the bottom sections of the distribution of income. Housing poverty goes beyond rent-to-income and similar ratios and also includes considerations of substandard conditions, crowding, and housing-related aspects of health.

Inequity in housing centrally relates to economic conditions, to institutional frameworks, and to the nature of property rights. Property rights refer to cultural, social, and institutional understandings as well as to legally defined attributes. Hence, inequity in housing includes tenure, use rights, occupancy rights, transfer rights, financial obligations in credit, and the conditions in rental markets. These rights will be articulated in state, market, and social processes in society. Ultimately, what is deemed to be socially acceptable in inequality and inequity in housing in a given society is a question of social values in that society, values on which there may or may not be universal accord. (SEE ALSO: **Tenure Sectors**)

—*Cedric Pugh*

Further Reading

Ball, M., M. Harloe and M. Martens. 1990. *Housing and Social Change in Europe and the USA.* London: Routledge.

Eggertsson, T. 1990. *Economic Behaviours and Institutions.* Cambridge, UK: Cambridge University Press.

Furubotn, P., ed. 1974. *The Economics of Property Rights.* Cambridge, MA: Ballinger.

Pugh C. 1990. *Housing and Urbanisation: A Study of India.* New Delhi: Sage.

Rescher, N. 1966. *Distributive Justice.* New York: Bobbs-Merrill.

Sa-aadu, J., James D. Shilling, and C. F. Sirmans. 1991. "Horizontal and Vertical Inequities in the Capital Gains Taxation of Owner-Occupied Housing." *Public Finance Quarterly* 19(4):477-85.

Turner, B., J. Hegedus, and I. Tosics, eds. 1992. *Housing and Housing Policy in Eastern Europe and the Soviet Union.* London: Routledge. ◄

► Housing Investment

The wealth of a nation is derived from its stock of physical and human capital (e.g., education and job training). The physical capital stock includes machines, tools, equipment, factories, dams, highways, consumer durables, and housing. For the United States, the housing capital stock accounts for approximately 38% of the total physical capital stock.

At any point in time, a nation's housing capital stock is fixed and thus a country's average housing standard is fixed. Over time, however, the housing stock and the associated flow of consumer benefits from this stock, called "housing services," can be altered through various kinds of housing investment activities. Three basic investment activities are (a) new housing construction, (b) maintenance of the existing housing stock, and (c) housing stock conversions. Each of these are discussed below.

Changes in the housing capital stock depend in part on the scale of new residential construction. For wealthy countries, new housing construction expenditures account for 4% to 6% of total national production in a given year; in extremely poor countries, usually less than 3%.

What, then, are the primary determinants of the rate of new housing construction? Economists are in general agreement that in the long run, the fundamental determinants of the housing stock are inflation-adjusted incomes, interest rates, construction costs, and demographics. Consequently, if a high housing standard is a key social goal, the basic policy implication is that government policies must promote low interest rates and a rapid rise in the standard of living.

In the short run, however, housing construction generally exhibits large year-to-year fluctuations, and there is considerable debate concerning the reasons for this cyclical instability. In the housing literature, the debate has focused on the supply of mortgage credit relative to its cost in explaining short-run construction cycles. Given the institutional structure and regulation of the U.S. housing finance system up until the 1980s, the consensus view was that short-run variations in the supply of mortgage credit was the dominant cause of cyclical instability in housing construction. Consequently, since the early 1970s, a major U.S. housing policy goal has been both to reduce fluctuations in the supply of mortgage credit as well as to increase the total *flow* of mortgage credit. Empirical studies suggest that U.S. policy measures have reduced fluctuations in mortgage credit somewhat but that increased government loans have increased the supply of mortgage credit substantially less than dollar for dollar. It appears that government loans increase total mortgage credit only 20% to 35% of the government loan amount in the short run and less than this in the long run.

Since the early 1980s, however, regulatory changes and the rapid development of the secondary mortgage market have led to better functioning capital markets. These institutional changes should reduce the role of mortgage credit availability as the major explanation for cyclical instability in housing construction and focus most emphasis on fluctuations in interest rates.

Concomitantly, although inflation-adjusted economic variables guide rational economic decision making, an important exception can occur with respect to housing investment. Most housing investment must be financed through borrowing. Increases in the expected rate of inflation tend to increase the market rate of interest. Under a fixed-level payment mortgage contract, however, the level of market interest rates can influence housing demand.

Specifically, a rise in the market interest rate raises the actual finance costs in the early years of the repayment period while lowering actual costs in the latter years. The key problem in this case is that the rise of inflation-adjusted payments in the early years of the loan period can be difficult for households to meet out of current income, causing most households cash flow problems in the financing of their housing. This finance problem is often referred to as the "tilt" problem, because inflation in this case alters the time profile of the inflation-adjusted payments stream over the repayment period. Consequently, in an inflationary environment, alternatives to the fixed-level nominal payment mortgage contract emerge to more evenly distribute the actual financing costs over time. A common response in this respect is offering different kinds of variable-rate mortgages.

Another important determinant of housing demand, and thus housing investment, is the price of housing relative to other goods. With respect to capital goods, defining the relevant "price" is no simple matter. In recent years, a user cost of capital approach has been applied to the pricing of housing capital. Specifically, the cost of housing to a consumer per month or year (i.e., its rental price) depends on several factors, such as the market interest rate, the expected rate of inflation, the income tax rate, and expected capital gains (or losses) from owning a dwelling. Of particular importance to note here is that inflation, in combination with specific features of the income tax code, can significantly change the inflation-adjusted, after-tax mortgage interest cost faced by consumers. Consequently, the rental price of housing can fall, increasing the amount of housing demanded, even in periods when market interest rates and inflation-adjusted construction costs are rising. In the United States, the surge in housing demand in the late 1970s seems to be explained in large measure by significant decreases in inflation-adjusted, after-tax mortgage rates.

In contrast to demand-side theories, some analyses have focused on the supply side. One major empirical study found that new residential construction is highly responsive to changes in the asset price of housing, with most of the long-run response occurring within one year.

Changes in housing supply can also occur as a function of changes in the existing stock of dwellings. This kind of investment activity is more important than generally recognized. By the 1980s, investment in the maintenance and restoration of the U.S. housing stock accounted for approximately 30% of total housing investment per year. Investment in the existing housing stock can take several forms. First, because of weathering and use, housing capital physically deteriorates overtime without repair and maintenance. Through maintenance, physical deterioration can be greatly slowed and the life of the housing stock extended.

Under existing economic incentives, the U.S. housing stock depreciates slowly, about 1% per year.

In addition to repair and maintenance, the housing stock can be altered by conversions. Dwellings can be significantly raised in quality by expenditures that physically upgrade a dwelling structure in terms of its internal layout, equipment standard, aesthetic quality, and size. Likewise, such changes can be reversed to lower the structural quality of dwellings. Relatedly, dwelling units can be converted to alternative nonhousing uses (e.g., commercial or retail use, parking lots, etc.).

It is important to note here that under market incentives, property owners are guided in their investment behavior by the differential profit opportunities presented to them. For example, basic supply-and-demand changes as well as government policies (e.g., various tax and subsidy policies) that increase the profitability of specific kinds of real estate investments will induce greater investment in these areas.

An important kind of housing investment activity in many countries is the improvement investments made by owners of single-family housing. Given the complex nature of the household's choice in this instance, few unambiguous theoretical predictions can be made, thus necessitating careful empirical studies. Research suggests that the likelihood and the amount of improvement investments increase with household income and dwelling age. In addition, when the cost of homeownership is rising, the cost of improvement is relatively low, thus giving incentive to increase improvement expenditures. Finally, high property tax rates and low expectations of housing price appreciation tend to decrease the demand for improvement expenditures.

A basic implication from the above discussion of housing investment is that the supply of additional housing under market allocation is much more responsive to economic incentives than generally recognized. In any given year, the supply of additional housing units is derived from four sources: new housing construction, decreases in vacancies, smaller housing stock losses, and additions to the housing stock from remodeling and renovations. Most housing analysts have tended to focus almost exclusively on new construction, implicitly suggesting that it is virtually the sole source of new additions to the housing stock. U.S. data for the period 1961 to 1985 show that net additions play a major role in the short-run adjustment of the housing market. For the U.S. case, on average, one-half of a rise in housing demand from increased household formations is met by reduced losses from the housing stock or nonnew construction additions during the concurrent year.

At this point, it seems important to ask, In what way, if any, do market incentives cause less than optimal maintenance of the housing stock? The primary argument pointing to inappropriate market incentives relates to the presence of external effects in housing markets. These external (neighborhood) effects may give rise to inefficiency in housing markets, particularly in highly dense urban slums. In such markets, a less than socially optimal level of maintenance is likely as some of the benefits of such investment accrue to neighboring property owners. Although a valid

theoretical point, there is little empirical evidence of significant undermaintenance in this respect. If policymakers, however, believe that significant undermaintenance exists, then the basic policy response is some form of subsidy or tax incentive for the purpose of increasing maintenance.

It is important to note that because of major conceptual problems and data limitations, economists' understanding of housing investment is far from complete. Moreover, the relatively few econometric estimates of housing supply parameters are not definitive. Nevertheless, both theory and evidence indicate that housing investment is responsive to market incentives—namely, the real price of housing and real interest rates. Likewise, both theory and evidence indicate that increases in real incomes (i.e., real economic growth) will increase housing investment. In contrast, there seems a credible basis for believing that housing price controls, at least the kind generally implemented by governments, adversely affect housing investment, in particular, the maintenance and conservation of the housing stock.

Under market allocation, builders face strong pressures to supply the kinds of housing that consumers desire, strong incentives exist for the conservation of the housing stock, and competitive forces tend to strongly promote least-cost construction. This is not to suggest, however, that the present U.S. housing markets are in any sense "perfect." Clearly, various kinds of discriminatory practices against various racial and ethnic groups exist. Moreover, housing is a good that can be produced only at high resource cost. During the 1960s and early 1970s, there were high hopes that industrialization of the building process would lead to substantially lower construction costs. However, little credible evidence exists that industrial building systems have achieved significant economies. Consequently, given the inherently high cost of housing, the poor and others (e.g., large families) will often face major affordability problems with respect to housing, in particular housing of a good standard.

On the other hand, East and West European countries have relied on long-term housing price controls, and various forms of economic planning of the housing sector have encountered major efficiency problems. Specifically, it often appears that housing production has not been well matched to consumer preferences, the housing stock has been undermaintained, and there is considerable doubt that construction has been carried out at least cost. (SEE ALSO : *Housing Costs; Housing Demand; Housing Markets; Housing Starts*)

—*Thomas S. Nesslein*

Further Reading

Ball, Michael and Andrew Wood. 1996. "Does Building Investment Affect Economic Growth?" *Journal of Property Research* 13(2):99-114.

Hendershott, Patrick and Marc Smith. 1988. "Housing Inventory Change and the Role of Existing Structures." *American Real Estate and Urban Economics Association Journal* 16:364-78.

Hughes, Jonathan and George Sternlieb. 1987. *The Dynamics of America's Housing*. New Brunswick: Rutgers, State University of New Jersey, Center for Urban Policy Research.

Mills, Edwin and Bruce Hamilton. 1989. *Urban Economics*. 4th ed. Glenview, IL: Scott, Foresman.

Montgomery, Claire. 1992. "Explaining Home Improvement in the Context of Household Investment in Residential Housing." *Journal of Urban Economics* 32:326-50.

Nesslein, Thomas. 1998. "Housing: The Market Versus the Welfare State Model Revisited." *Urban Studies* 25:95-108.

Rothenburg, Jerome et al. 1991. *The Maze of Urban Housing Markets*. Chicago: University of Chicago Press.

Smith, Lawrence, Kenneth Rosen, and George Fallis. 1998. "Recent Developments in Economic Models of Housing Markets." *Journal of Economic Literature* 26:29-64.

Topel, Robert and Sherwin Rosen. 1988. "Housing Investment in the United States." *Journal of Political Economy* 96:718-40. ◀

▶ Housing Markets

Definition

Housing is a complicated concept that can best be described as a collection of attributes that come bundled together—a physical structure, neighbors and neighborhood, accessibility to work and urban amenities, private and public rights, income and investment opportunities. The joint consumption of these attributes by a household determines its economic and social well-being.

The urban housing market is the set of institutions and procedures for distributing households in a given urban area across housing units. Although housing prices, household incomes, and household preferences are major determinants of housing allocation, the actual distribution of households is largely influenced by the behavior of various housing market participants, including builders and developers, real estate agents, and financial institutions, each of which imposes its own political, institutional, and cultural framework and interests on one another. Private market mechanisms, although sufficient for the allocation of households who can afford market prices, have to be either regulated or complemented by public sector mechanisms to assist those whose incomes are inadequate or those who may be subject to discrimination based on characteristics such as race, ethnicity, or national origin.

Structure

The urban housing market is composed of a residential stock and a collection of households. The residential stock consists of housing units differentiated primarily on the basis of tenure (owner or renter occupied), ownership (private or public), type of structure (single family or multifamily), design (colonial, rambler, or town house), location, and the general condition of the structures. The occupancy of a given housing unit by a household offers a combination of attributes that come bundled together that, for analytical purposes, are referred to as "housing services." Households, who are differentiated on the basis of socioeconomic characteristics and lifestyles, demand a certain level and quality of housing services. Housing prices and rents are the outcome of the interactions of household demand for housing services and the supply of housing services that is offered by the urban residential stock.

There are two origins of demand for housing. The first is investment demand for the residential stock—that is, the desire of investors (landlords, speculators, and homeowners) to invest in residential stock relative to other investment opportunities. The second is the demand for "housing services," the consumption of which provides occupants with a certain level of residential satisfaction. Some services are internal to the structure, such as the number of bedrooms and the condition of the structure; others are external and result from the geographic location of the structure, including accessibility, safety, neighborhood characteristics, and the level and quantity of public services. The latter may be termed *externalities,* and these services are analytically difficult to quantify in terms of their contribution to household residential satisfaction and to resulting house prices.

Both the housing stock and households are dynamic elements; that is, their size, quality, and composition change over time, resulting from internal forces, such as aging and household formation, or in response to changes in labor market conditions or national economic and population growth trends. Any change in the household characteristics will redefine the level and nature of housing services demanded from the residential stock. In addition, changes in household tastes and preferences may also lead to changes in housing services demanded by households. The stock responds to modified housing demand through either new construction or modification of the existing stock—for example, mergers and subdivisions or conversions of property from nonresidential to residential use. The speed and success of supply adjustments to changing housing demand determines the efficiency of housing market operation in any local housing market. The extent of the discrepancy between household demand and the housing supply determines both the rate of housing transactions and the flow of population into and out of a particular market. These shifts have important implications regarding occupancy patterns, resulting neighborhood change, and metropolitan development.

Operation

Housing market outcomes are the result of interactions between demand and supply of housing services offered by the existing residential stock. Households choose different bundles of housing services according to their preferences and financial constraints. House prices are the major mechanism by which households are allocated to various units in the residential stock by tenure, price, quality, and location.

At any point, it is possible to envision a match between the *supply of housing services* from the existing residential stock and the *demand for housing services* among existing households. Households are assumed to be willing to pay for housing services in such quantities to maximize their residential satisfaction subject to their income and preferences. From the perspective of the consumer, the selection of a housing unit with particular attributes involves a series of trade-offs, given household preferences and income. On the basis of these trade-offs, households offer bid prices for available units. From the perspective of the producer, an asking price is based on the cost of housing production and profit expectations. Thus, the allocation of households to housing units is a competitive bidding process. The sale price (or rent) of a given unit is therefore the "market clearing" price at which households' bid prices equate with the producers' asking prices.

An important element of the urban housing market is how the housing stock and the demand for housing both reach equilibrium over time. Housing supply decisions made at the local level by landowners, developers, financial institutions, and local government are part of a larger residential development process that is heavily regulated by local land use and building regulations. A major component of this process is the availability of mortgage credit. Accordingly, the housing supply is highly sensitive to cyclical economic conditions at both the national and local levels. In addition, the changing nature of the local building industry in terms of the size of building firms and the available building technology affects the cost, quality, and location of the housing supply. On the demand side, the ability of households to adjust their housing consumption is limited by settlement costs and the high costs involved in making a physical move. This limitation is necessarily more pronounced for lower-income households, some of which might face constraints because of race, ethnicity, and lifestyle.

Although competitive market mechanisms are quite satisfactory in responding to housing consumption decisions of households with sufficient incomes, residential market outcomes for households with lesser means are affected greatly by the behavior of various actors—real estate agents, landlords, and financial institutions—because they affect credit availability and racial discrimination. Specifically, supply, maintenance and repair, financing, and consumption decisions in a locality are based on the benefits of housing investment relative to the returns on investment in other business opportunities in the capital markets and the performance of the local labor market. Therefore, the magnitude and the quality of housing services that households with modest and low incomes (usually renters) can obtain is closely linked with the investment decisions of landlords and financial institutions. This necessitates the imposition of certain rules and regulations on institutions involved in housing, further complicating the operation of urban housing markets. These several factors hinder the notion of a purely competitive market system as the backbone of housing market operations.

Market Failures and Externalities

Two unique qualities of housing are its spatial fixity and durability, both of which make it very difficult to alter the level and quality of housing services attached to residential units. This may lead to scarcity for housing with characteristics for which there is inelastic consumer demand, such as single-family detached structures or school quality. Hence, the degree of substitution between houses characterized by a given set of attributes is limited. The outcome of relatively inelastic demand and supply for certain housing characteristics is the segmentation of the urban housing stock into distinct geographic submarkets. Differential access to housing because of discrimination, limited information, and institutional constraints are added factors that can lead to a degree of geographic segregation.

Neighborhood effects are an integral component of housing market operations at both the individual and aggregate levels. Neighborhood effects can be defined as all attributes that enter into the housing bundle as a consequence of the housing unit's geographic location—both its absolute location and the neighborhood in which the unit is located. Thus, in addition to a housing unit's structural attributes, the array of locational characteristics (neighbors, accessibility, and public service provision) enters into both individual housing demand-and-supply decisions and resultant market-level behavior. Therefore, one important concern that has evolved is whether neighborhood differentials are reflected in housing prices as premiums (or discounts). If, for example, race acts as an externality, then households of that race may encounter premiums that lead to an inequity in the price paid for housing services as well as to subsequent social inequities. The measurement of the impact of negative (or positive) externalities on housing prices is therefore important for guiding public policy and regulatory requirements.

Policy Response

Housing is important to both the national and local economies. At the national and local levels, housing is a major source of employment in many industries. Residential real estate is a significant component of national wealth. Housing is important to local economies as the greatest consumer of urban space and a large tax revenue generator. It is therefore not surprising that the functioning of urban housing markets is heavily regulated by public mechanisms. Policy response involves both indirect involvement in the operation of private market mechanisms and direct involvement in housing production and allocation for the provision of adequate and affordable housing for the urban poor. The public allocation mechanism is based primarily on housing needs relative to established standards set by society regarding what is adequate housing and how much households should pay for housing given other necessities. Some radical social scientists argue that government involvement will not be effective in dealing with housing problems as long as housing is treated as an economic commodity and participants in the housing production process serve the interests of private capital. (SEE ALSO: *Filtering; Housing Demand; Housing Indicators; Housing Investment; Housing Shortage; Housing Supply; Market Segmentation; U.S. Housing Market Conditions*)

—*Isaac F. Megbolugbe and Ayse Can*

Further Reading

De Leeuw, Frank. 1975. *The Web of Urban Housing: Analyzing Policy with a Market Simulation Model.* Washington, DC: Urban Institute.
Myers, Dowell. 1990. "Systematic Biases in Housing Market Analyses." Pp. 107-23 *Research in Real Estate*. Vol. 3, edited by C. F. Sirmans. Greenwich, CT: JAI.

Stokes, Charles J. 1976. *Housing Market Performance in the United States.* New York: Praeger.

U.S. Department of Housing and Urban Development, Office of Policy Development and Research. *U.S. Housing Market Conditions Quarterly* (see each issue). ◀

▶ Housing Norms

Housing norms are the social pressures on individuals and households to live in housing with prescribed characteristics. Norms are not characteristics of households; they are characteristics of societies and segments within societies. Nevertheless, to be effective in influencing behavior, norms should be internalized by significant portions of the population. Housing norms are societal phenomena but are implemented by households.

When a household faces a decision about housing, there are two important influential forces. The first includes the norms that suggest to the household members what kind of dwelling they "should" live in. The second includes constraints (such as economic resources) that limit the kind and quality of dwelling. Neither the norms nor the constraints directly cause behavior. Rather, households make a mental calculation that weights current housing conditions, the housing norms, relevant values, and the constraints impinging on them and develop preferences. Preferences indicate the type, quality, and other characteristics of housing that households are willing and able to obtain.

It is crucial to the understanding of housing norms to distinguish preferences from norms. This distinction is especially important because preferences can differ enormously among segments of the society, whereas the norms that apply to specific households are quite homogeneous.

It may seem bizarre to say that poor people prefer low-quality housing. The norms certainly do not prescribe poor-quality housing for poor families. The income constraint and the family's values lead the family to reject going hungry to obtain high-quality housing. In light of this constraint, the poor household prefers housing that is of poor quality because it is inexpensive. Preferring and choosing less expensive housing leaves money for food and clothing. Such preferences do not suggest a positive desire to live in poor housing. It is important not to confuse preferences and norms in household decision making.

Housing norms in the United States and in most if not all Western cultures prescribe homeownership, single-family dwellings, sufficient numbers of rooms (especially sleeping spaces) for the number of household members of each age and sex category, and at least some private outdoor space. Obviously, there are additional, more specific and sometime divergent norms that differ significantly among social categories within and among societies, but those are the most widely shared norms.

Even housing conditions that would be seem to reflect unique norms in a particular society may be conditioned by factors such as the level of technology and the availability of certain kinds of building materials. For example, flat-roofed dwellings require low rainfall or a high level of technology. Adobe houses seldom exist if the needed type of soil is absent. The existence of particular housing conditions should not be used to infer that norms prescribing those conditions exist.

Given the amazing variation in housing conditions among (a) segments within specific societies and (b) among societies, the claim that certain housing norms are nearly universal may seem strange. This seemingly bizarre conclusion will become understandable in later parts of this entry.

Housing norms can be expressed informally as in the pressures exerted by friends and family, or they can be manifested formally as in the laws and regulations about housing occupancy and development. Those expressions of norms are never exactly representative of the norms as internalized by the members of the society. Indeed, when the national policymakers and the population differ in their ideas about housing, the formalized norms may be very different from the norms as internalized by the members of the population.

The expression of norms involves social, psychological, and political processes that ensure a certain amount of "error." The housing codes and regulations adopted into law typically reflect not only norms but also the resources and other constraints affecting the society. For example, housing regulations in the Netherlands can be expected to reflect not just families' desires for spacious housing but the constraints of land and other inputs into the housing development process.

One should not use the existence of particular behaviors or social conditions to infer that there are norms that prescribe them. The absence of certain conditions or behaviors also should not be used to infer that norms favoring them do not exist. Very often, individuals and groups behave (or fail to behave) as they do because they cannot do otherwise. Failure to conform to norms either by omission or commission very often is caused by constraining factors in the situation. Therefore, inferences about the existence of norms from the existence or absence of norm-conforming behavior are fallacious.

There are at least two reasons that many households do not live in housing that meets widely shared norms (renters, residents of apartments, people in crowded housing, those with no backyard, etc.). First, constraints that can make it impossible include the following:

1. Resources (income, wealth, information, skills, time)
2. Family organization (the household's ability to effectively make and implement decisions about its housing)
3. The housing market (prices, supplies of housing, building materials, mortgage money)
4. Predispositions (psychological characteristics of the household—apathy, ambition, etc.)
5. Discrimination (because of race, ethnicity, sex, age, disability, social class)

The second reason is the value system of the household. A household's values may be such that other things are held to be more important than housing, leading it to live in housing that does not meet the norms so it can have something else in greater quantity or quality.

Households may experience negative sanctions (punishments) for living in nonnormative housing or positive sanctions (rewards) for living in housing prescribed by the norms. There are many ways in which sanctions are administered. They may be administered informally, as when friends at work subtly ask when one is going to buy a house, or formally, when it is easier to get a car loan if one is a homeowner than if a renter.

Another indicator of the importance of housing norms is that households living in nonnormative housing usually are dissatisfied with it. Dissatisfied households usually want either to move or to make alterations. Moving and alterations are forms of housing adjustment. If the constraints and values make it impossible for households to move or alter the present dwelling, they may engage in adaptation. Housing adjustment involves making changes in the dwelling. Adaptation involves making changes in the household.

There are two basic concepts in adaptation: (a) needs reduction and (b) constraint reduction. The preferred method of dealing with housing problems or deficits is adjustment behavior through which deficits are removed. Because of the severity of the constraints, some households are unable to adjust. They are forced to consider one or both forms of adaptation.

Alternatively, when faced with severe constraints, a household can decide that it does not have to try to meet the norms. Then it can develop a modified version that is easier to meet. Another method of needs reduction is to change the composition of the household. If a family is getting crowded perhaps an older child will move into an apartment to release some space for the younger children. If the house is too big and expensive, another household might move in and the two could share housing expenses.

Of the five types of constraints, only a few are susceptible to reduction by the individual household. Discrimination, for example, is not under the control of the victim. Nor are the prices and supplies of housing and related goods and services under the control of individual households.

Changeable constraints include those on resources, which can sometimes be removed by, for example, investing in training or education to improve income. Reorganizing the household could reduce household organizational constraints. Sometimes the member of the household who handles the money and makes economic decisions is not the most qualified. If it is having trouble obtaining satisfactory housing, such a household could reorganize and put someone else in charge of money. Then it could perhaps remove housing deficits by more effective management.

Housing norms are the central concept in housing adjustment theory. The motivation for households to engage in adjustment or adaptation results from a perceived normative deficit or imbalance in their housing. Norms indicate the kind of housing households "should" have. If they do not have such housing, we conclude they have a deficit or a normative imbalance. If housing is important to households because of their values, the presence of a deficit is likely to cause them to be dissatisfied. If they are dissatisfied, they are likely to be motivated to correct the deficit. If they

are motivated, they are likely to engage in adjustment. If they cannot adjust, they are likely to try adaptation.

All of those "if statements" are linked to the presence or absence of constraints. Housing adjustment theory thus represents a causal chain from housing conditions to dissatisfaction to adjustment behavior to adaptive behavior. Progress through the chain depends on the household's ability to complete the housing adjustment process. That ability depends on the strengths of the various constraints.

One of the forces in the changes in specific content of housing norms is technology. Economic development can also be a source of change in housing norms. It is important, however, not to confuse the release of constraints, which could accompany economic growth, with true change in norms. (SEE ALSO: *Residential Mobility; Residential Preferences; Residential Satisfaction*)

—*Earl W. Morris and*
Mary Winter

Further Reading

Cho, J., E. W. Morris, and M. Winter. 1990. "Removing Housing Deficits in the Transition from Rental to Ownership." *Housing and Society* 17(2):45-59.

Crull, S. R., M. Eichner, and E. W. Morris. 1991. "Two Tests of the Housing Adjustment Model of Residential Mobility." *Housing and Society* 18(3):53-64.

Lodl, K. A. and E. R. Combs. 1989. "Housing Adjustments of Rural Households: Decisions and Consequences." *Housing and Society* 16(1):13-22.

Morris, E. W. and M. Jakubczak. 1988. "Tenure-Structure Deficit, Housing Satisfaction, and the Propensity to Move: A Replication of the Housing Adjustment Model." *Housing and Society* 15(1):41-55.

Morris, E. W., M. Winter, M. B. Whiteford, and D. C. Randall. 1990. "Adjustment, Adaptation, Regeneration and the Impact of Disasters on Housing and Households." *Housing and Society* 17(1):1-29. ◀

▶ Housing Occupancy Codes

Occupancy standards define the maximum number of people permitted to occupy a dwelling. The official purpose of the standard is to promote health and safety by eliminating undue overcrowding. Although there is general consensus that minimum enforceable standards are necessary, debates have raged for over a century about how to ascertain the maximum number of people who may share a space before it becomes unhealthy. Current U.S. standards, starting in the second half of the 19th century, have their origins in the public health movement. Since the 1950s, the building industry has increased its influence in detailing the specifics of standards, inscribing them in the housing code section of their model building codes. Such model codes are often adopted and adapted by local, state, and federal agencies, such as the Department of Housing and Urban Development (HUD). The American Public Health Association (APHA), in its 1986 model code, *Housing and Health,* suggests that occupancy standards championed by the building industry represent a shift from the primacy of social interests and the desire to create healthy environments to the primacy of financial interests in housing design and use.

Current occupancy standards are measured two ways. One is by the number of people per bedroom, regardless of bedroom size. The second is by number of people per square feet, calculated by both the overall person-to-dwelling size and person-to-bedroom size. For determining overall dwelling size, codes generally include only habitable rooms (excluding bathrooms, hallways, storage areas, etc., and sometimes kitchens). In some codes, measurements are wall-to-wall; in others, the area taken up by built-ins that extend into the room are discounted.

Housing codes also set design criteria that affect occupancy levels, including (a) ventilation and lighting, (b) minimum allowable room size per person, (c) the requirement that one room be a certain minimum size (frequently 120 or 150 square feet), and (d) often, prohibiting a room from being counted as a bedroom if it is a pass-through to another room.

Affordable housing advocates, public health officials, the building industry, and policymakers at different levels of government bring different perspectives to the design of occupancy standards. For such a small code—it tends to be under a page in length—it has a disproportionately large effect on the ethnic, racial, social, and economic structure of cities: How one calculates the acceptable number of people-per-unit directly affects the availability and affordability of housing, and by extension, homelessness, coercive racial and ethnic segregation, and access to services.

In some cases, restrictive occupancy codes have been created explicitly to maintain a particular status quo in a community. Although appearing neutral, codes may be used to discriminate illegally against people on the basis of race, national origin (ethnicity), families with children, disability, and other protected categories under the Fair Housing Act. The primary ways occupancy standards are used for discriminatory purposes are these: uneven enforcement, targeting certain groups; setting standards biased against minority racial, ethnic, or religious groups in a community; and writing zoning policies that restrict the number of unrelated people who may share a unit.

Historically, the 1870 San Francisco Lodging House Ordinance ("cubic air" ordinance) was enacted at the request of the Anti-Coolie Association and required a minimum of 500 cubic feet of air space per person. It was disproportionately enforced in Chinatown where low-paid, single, working Chinese men shared rooms with less air space each than mandated. In 1876, California made this minimum a statewide law (other states also used this standard, some more consistently enforcing them across the population than others). Since the enactment of the 1968 Fair Housing Act, legal recourse against discriminatory occupancy standards and the differential enforcement of standards has improved. For instance, in 1994, the U.S. attorney general successfully sued the City of Wildwood, New Jersey, for imposing an "unreasonable" occupancy standard that was more restrictive than the state's and that would have disproportionately affected families with children and local Latino households.

Discussions about whether the standards should be more or less restrictive inevitably incorporate questions of what constitutes good housing, whether good housing should be a right or a privilege, and whether business interests and profit or the availability of decent affordable housing should be more heavily weighted. For example, should a family of five be permitted to rent a two-bedroom dwelling, or does the owner have sufficient business necessity to require that they pay for a three-bedroom dwelling? Occupancy standards thereby restrict where a household may live. Intended or unintended consequences ensue, segregating low-income households in run-down neighborhoods with poor services, where larger homes are available at a lower rent or where property owners permit higher densities. (Property owners may allow less restrictive standards, but renters have no recourse in law if denied a unit in which the number of people exceeds the approved code.)

History

Nineteenth- and early 20th-century slum conditions led to public health intervention in housing. Because of high rent and low pay, households, many of whom were immigrants, were forced to move large numbers of people into one- to three-room apartments notorious for their inadequate ventilation, lighting, and maintenance. This environment created ideal breeding grounds for vermin and the transmission of contagious diseases, resulting in the housing codes coming under the purview of sanitation experts, while remaining a preoccupation of social reformers.

Lawrence Veiller, a major force in the progressive housing reform movement and secretary of the New York Tenement House Commission, was instrumental in establishing the 1901 New York Tenement Act. This legislated important new building code safeguards (related to the physical structure of buildings) as well as new housing codes (related to living conditions). In his influential 1910 book, *Housing Reform,* Veiller dismissed some contemporary beliefs that certain "foreign elements," such as Poles, Slavs, Italians, Hungarians and Jews, predominated in northern urban tenement slums because of "faults . . . in the blood, hereditary traits and instincts impossible to overcome." Instead, he argued that better housing would lead to physically and sociologically healthier people because it was not the intrinsic character of people that led to poverty and its ills but the physical environment in which the newcomers lived.

Housing reformers believed that decent, safe, and sanitary housing should be a right, that everyone should have what they, the reformers, considered a reasonable amount of space. Although generous in intent, they were highly assimilationist and paternalistic in practice. They attempted to assimilate low-income immigrants by rearranging their domestic social and spatial relations to conform to the reformers' sense of moral order, emphasizing individualism and a belief in the necessity of privacy within the home for physical, psychological, and moral health. They explicitly decried intergenerational sharing of sleeping spaces. Although immigrants from many cultural backgrounds did not agree with the reformers' emphasis on personal privacy, they did agree with the emphasis on cleaner, larger homes. Remedial legislation designed to provide more individual space was geared toward limiting the number of people in an apartment, not toward limiting the ever-increasing, and sometimes usurious, rents that forced people to share small units.

Starting with its 1939 publication, *Basic Principles of Healthful Housing*, the APHA issued influential model housing codes that explicitly tied definitions of crowding and the appropriate combination of occupants and unit size to a public health rationale. In its 1950 volume, *Planning the Home for Occupancy*, the rationale is also explicitly class correlated: The minimum standards considered necessary to attain the goal of "healthful housing . . . closely approximates actual practice in the high-income groups." The APHA relied heavily on the 1935 English law on overcrowding, which limited occupancy to two people per room. The APHA ideal was more restrictive at one person per bedroom, to enable "isolation" from the "intrusion" of other householders. Realizing this standard would be hard to uphold, it stated "we can at least insist on a room shared with not more than one other person."

Like housing reformers before, the APHA conflated socially and culturally produced preferences with universal health and necessity. Cross-cultural and historical analyses provide evidence that concepts concerning the preference to share sleeping and living spaces often relate to deeper core values, such as the emphasis on individualism or communality. In contrast to the U.S. model, middle-class households in countries as different as Mexico and Japan commonly choose intergenerational sleeping arrangements and more than two people per bedroom, while leaving other bedrooms unused. Current psychological literature does not substantiate the argument that a less restrictive occupancy standard would have a negative psychological or physical impact.

Thus, in legislating occupancy codes, it is essential to understand the sociocultural and historical basis of a preferred system. It is equally important to question whether the nuclear family is being privileged over other household configurations, such as extended families or two single-parent households sharing a home, and whether the justifications are indeed health and safety based or if they are more closely aligned to particular cultural and economic preferences.

Housing occupancy codes are intended to protect renters from unhealthy living conditions and unscrupulous property owners and to protect the investment of owners. What continues to be contested is the balance among these interests, the potential discriminatory implications for protected categories of people under the Fair Housing Act, and the larger policy implications for the structure of cities. (SEE ALSO: *Certificate of Occupancy; Doubling Up; Housing Codes; Space Standards*)

—*Ellen-J. Pader*

Further Reading

Mood, Eric. 1986. *APHA-CDC Recommended Minimum Housing Standards*. Washington DC: American Public Health Association.

Myers, D., W. C. Baer, and S-Y Choi. 1996. "The Changing Problem of Overcrowded Housing." *Journal of the American Planning Association* 62(1):66-84.

National Multi Housing Council. Homepage on the Internet (www. nmhc.org).

Pader, E. J. 1994. "Spatial Relations and Housing Policy: Regulations That Discriminate against Mexican-Origin Households." *Journal of Planning Education and Research* 13(2):119-35.

U.S. Department of Housing and Urban Development. 1990. "Occupancy Requirements of Subsidized Multi-Family Housing Programs." Handbook 4350.3. Washington, DC: Office of Housing.

Wright, G. 1981. *Building the Dream: A Social History of Housing in America*. Cambridge: MIT Press. ◀

▶ Housing Policy Debate

A scholarly journal published quarterly by the Fannie Mae Foundation (previously by Fannie Mae Office of Housing Research), *Housing Policy Debate* provides insightful discussion of and original research on a broad range of housing issues. Recent articles explored affordable housing shortages, housing policies for distressed urban neighborhoods, new research on homelessness, and innovations in public housing. First published in 1990. Circulation: 7,000 internationally. Editor: James H. Carr; Fannie Mae Foundation; 4000 Wisconsin Avenue, NW; North Tower, Suite One; Washington, DC 20016-2804. Phone: 202-274-8000. Fax: 202-274-8100. E-mail: fmfpubs@fanniemaefoundation.org. World Wide Web: http://www.fanniemaefoundation.org.

—*James H. Carr* ◀

▶ Housing Price

The principal purposes of housing price measurement are (a) to compare the cost of living among cities, neighborhoods, or groups of consumers; (b) to describe changes in the affordability of housing over time; and (c) to measure access to homeownership. These are different purposes that sometimes call for different ways of measuring the price of housing. However, their essence is that we typically want to speculate how much a dwelling might reasonably cost consumers if they were to live elsewhere within the private sector.

To almost anyone but an economist, price is simply the amount spent by a typical consumer either to purchase a dwelling or to rent it. Although conceptually simple, this definition is complex in practice. Consider the case of renting: Here, we might think of "housing price" as the amount paid by a tenant under a standard lease to a for-profit landlord for premises used for accommodation. Price typically includes the monthly rent payment. However, rent sometimes includes parking, heat, utilities, property taxes, use of on-site recreation facilities, cable TV, and other services. To compare rents among dwellings, neighborhoods, or cities, we must specify the set of services included. If our interest is in comparing the rents of equivalent dwellings, the rent of a tenant who does not receive a complete set of services must be "grossed up" by the amount spent to purchase the services separately. Because some tenants cause excessive wear and tear, are noisy or disruptive, or are in other ways costly to serve, housing price may also vary among categories of consumer; thus, the measure of housing price must be specific to a particular kind of tenant. Similarly, rent may also vary with the length of a lease and other lease conditions (including instances in which a landlord is also the employer).

In the case of homeowners, there are two kinds of housing price measures. In some instances, price is measured simply as the cost of purchasing a dwelling. Typically, this consists only of the purchase price itself, but sometimes it may also include property transfer taxes as well as related legal, insurance, and brokerage fees. Alternatively, for either renter or owner, housing price can be measured in terms of the operating cost of a dwelling. There are two widely used measures of operating cost. One is cash flow price: a price measure based on out-of-pocket expense that is often used to evaluate the affordability of homeownership. Cash flow price includes, for instance, mortgage interest payments; other components are listed in Table 15. At the same time, actual interest paid may be misleading because some homeowners are mortgage free and those that have mortgages may pay differing interest expenses depending on the amount, type, and starting or renewal date of their mortgage. In contrast, measures of access to homeownership, for example, typically assume that the consumer is able to make a specified down payment and finances the remainder with a conventional mortgage.

The second measure of operating cost is user cost, which measures the opportunity cost of homeownership. Unlike cash flow price, user cost includes foregone interest income on owner's equity in the home, property depreciation, and expected capital loss after sale during the period (negative if capital gain) and excludes mortgage principal repayment (see Table 15). An after-tax user cost measure is also used sometimes because foregone interest income on homeowner equity (the opportunity cost of their equity participation to homeowners), mortgage interest expense, depreciation, and capital gains (not to mention the implicit rents that could be earned had homeowners offered their dwellings for rent) are treated differently in income taxation from other kinds of household income or expenditure.

However, dwelling rents and selling prices will vary substantially from one part of a city to the next, even among otherwise comparable dwellings. One principal reason is the difference in commuting costs. Households are willing to pay more for a dwelling that gives them a faster or less expensive journey to work. In other words, households are willing to pay more for housing to save transportation costs. This argument has led some researchers to calculate a gross price for housing that sums shelter cost and commuting cost.

This raises the broader issue of variations in housing quality. Dwellings differ in terms of floor area, number of rooms, garage spaces, architectural design and layout, quality of construction, state of repair, and outdoor private space as well as location. To compare whether housing in one market is more costly than in another, we usually want to control for differences in housing quality. There are two commonly used approaches here. One is to select a popular type of dwelling and then compare how the price of that

TABLE 15 Expense Components Included in Cash Flow Price and User Cost Approaches to Housing Price

Expense Component	Cash Flow Price	User Cost
Mortgage interest paid	yes	yes
Imputed interest on owner equity	no	yes
Mortgage principal repayment	yes	no
Depreciation	no	yes
Utilities (heat, light, water)	yes	yes
Property taxes	yes	yes
Condominium fee (if applicable)	yes	yes
Land rent (if applicable)	yes	yes
Other occupancy charge (if applicable)	yes	yes
Maintenance expense	yes	yes
Homeowner's insurance	yes	yes
Capital loss/gain	no	yes

dwelling varies from one market to the next; the Coldwell Banker housing price series, for example, uses a standard three-bedroom house of 2,000 square feet in neighborhoods where corporate transferees would tend to locate.

The second approach is to see that the local housing stock is varied and that the housing market acts to price each dwelling as a weighted sum of that dwelling's characteristics and thus to estimate a hedonic (marginal) price equation (a statistical relationship that ties the selling price of an owned home—or rent, in the case of rented dwellings—to the physical characteristics of the dwelling and its neighborhood setting) for each market that replicates how housing price varies with dwelling characteristics. A price index can then be calculated by looking at how the average price of a given portfolio of housing units has changed over time. This is the approach used by the Bureau of the Census (U.S. Department of Commerce) in its price index for new one-family houses sold (see Table 16). The housing characteristics included in the index include size of dwelling, lot area, number of stories, number of bathrooms, and presence of central air-conditioning, garage, basement, and fireplace. The right-hand column of Table 16 shows how selling prices of new houses have changed between 1977 and 1996; price here varies in part because of the changing mix of dwellings sold. The left-hand column shows what the average selling price in the same year would have been for a given mix of dwellings (the mix sold in 1992). Because the given mix would have sold for more in 1977 than the average sale price at the time, we know that the mix changed in favor of more costly units between 1977 and 1992. (SEE ALSO: **Affordability;** *Housing Costs*)

—*John R. Miron*

Further Reading

Bryan, T. B. and P. F. Colwell. 1982. "Housing Price Indexes." *Research in Real Estate* 2:57-84.

Devaney, F. J. 1994. "Tracking the American Dream: 50 Years of Housing History from the Census Bureau: 1940-1990." Pp. 44-46

TABLE 16   Average Sales Price of Portfolio (New Single-Unit Houses Sold in 1987) Compared with Average Price of Houses Actually Sold, United States, Selected Years, 1977 to 1992

Year	Average Price (in dollars) of Portfolio According to Price Index	Average Price (in dollars) of Houses Actually Sold in That Year
1977	67,400	54,200
1982	108,400	83,900
1987	127,700	127,200
1992	144,100	144,100
1993	150,300	147,700
1994	157,500	154,500
1995	161,900	158,700
1996	165,100	166,400

SOURCE: U.S. Department of Commerce, Bureau of the Census. 1997. *Current Construction Reports: Characteristics of New Housing, 1996.* Series C25. Table 25. Washington, DC: Author.

in *Current Housing Reports.* H121/94-1. Washington, DC: U.S. Department of Commerce, Bureau of the Census.

Green, R. and P. H. Hendershott. 1996. "Age, Housing Demand, and Real House Prices." *Regional Science and Urban Economics* 26(5):465-80.

Megbolugbe, Isaac F., ed. 1995. "House Price Indices: Research and Business Uses." *Journal of Housing Research* 6(3, Special Issue).

Megbolugbe, Isaac F., ed. 1996. "House Price Indices: Policy, Research, and Business Applications." *Journal of Housing Research* 7(2, Special issue).

Pozdena, R. J. 1988. *The Modern Economics of Housing: A Guide to Theory and Policy for Finance and Real Estate Professionals.* New York: Quorum (see Chapter 6).

Steele, M. 1993. "Incomes, Prices, and Tenure Choice." Pp. 41-63 in *House, Home, and Community: Progress in Housing Canadians, 1945-1986,* edited by J. R. Miron. Montreal: McGill-Queen's University Press.

U.S. Department of Commerce, Bureau of the Census. 1995. *Current Construction Reports: Characteristics of New Housing: 1995.* Washington, DC: Author.

U.S. Department of Commerce, Bureau of the Census. 1995. *Current Construction Reports: Housing Completions.* Washington, DC: Author. ◀

▶ Housing Shortage

The term *housing shortage* has both economic and normative interpretations. In a positive economic theory, a housing shortage is the amount by which the demand for housing at a given price exceeds the supply of housing. In a normative interpretation, a housing shortage is the amount by which the need for housing exceeds the available supply.

In the former interpretation, we should imagine a demand curve and a supply curve for housing in each local housing market. The demand curve is a list of housing prices showing the total amount of housing demanded in the market at each price point: *Price* here can mean either the selling price of an owner-occupied dwelling or the monthly carrying cost of a rented or owned dwelling. The supply curve is another list of house prices showing the total amount of housing supplied at each price point. Although it may be true that many consumers want more housing in principle (whether *more* is defined in terms of quantity or quality), the positive economic view is that an equilibrium price exists, at which there is no excess demand. However, this does not entirely exclude the possibility of a housing shortage. Two cases of housing shortage can be envisaged. One case is a situation in which the market is not in equilibrium for some reason. Sometimes, this happens because the housing market is imperfect: For example, externalities exist (costs of a property not borne directly by the buyer or seller) or information is incomplete. Other times, public action prevents the housing market from reaching equilibrium. One common example is a rental market under rent control; rents are held below market, and hence the amount of housing demanded is greater than the amount of housing that landlords are prepared to supply.

Another common example is public housing, for which rents are subsidized and dwellings are allocated by merit rather than market; here, there are housing shortages because more consumers want to live there than the system is prepared to supply. Still another example would be local restrictions on building that prevent developers from offering housing that some consumers want. The second case draws a distinction between supply in the short run versus the long run. In the short run, the supply of housing is relatively fixed. A sudden increase in demand leads to a run-up in the price of housing. However, over the longer term, housing suppliers respond by building more housing and converting other structures to housing, thus driving price back down. A housing shortage here is seen to be the difference between demand at the currently high short-run equilibrium price and the demand that would be met if only price had fallen to its long-run equilibrium level.

Now, consider the second, normative view of housing shortage. To begin, a policy planner would (a) identify categories of potential consumers having similar housing needs, (b) enumerate these categories, and (c) specify minimum standards of housing for each kind. Potential consumers, as used here, constitute more than just the set of households currently observed. Some households consist of two or more consumers who are presently doubled-up but for whom separate accommodation is preferable. The homeless are also potential consumers. Standards of housing adequacy for each category of potential consumer would be based on perceived needs, arise from a perspective that sees housing as a necessity of life, and presumably describe the kind of housing that consumers would choose if only they were able to afford it. In this sense, housing standards are not independent of the planner's notion of a minimum standard of living. The planner then measures housing shortage as the number of consumers who are not presently in adequate housing.

Standards of housing adequacy can be defined along several dimensions. One is physical adequacy of the dwelling. The Canadian government uses what it calls the "core need" approach to measure physical adequacy. This measure rates a dwelling as adequate if it requires only regular upkeep or, at most, minor repairs and if it possesses hot and cold running water, an inside toilet, and an installed bath or shower. Across Canada's 10 provinces, 904,000 households lived in inadequate dwellings in 1995.

A second dimension is suitability of the dwelling with respect to the persons living there, taking into account that needs differ: for example, by gender, activity limitation, workforce participation, and ability to communicate.

A third dimension of adequacy is accessibility to community facilities and services, jobs, shopping, and other amenities. Related to this is the importance of neighborhood social mix; some planners and housing experts argue that neighborhoods should mix incomes and household types.

A fourth dimension of adequacy has to do with the relationships of housing to the physical environment. Within this, adequacy may be defined in terms of the presence of harmful chemicals in or near housing (e.g., methane and radon gases, urea-formaldehyde foam insulation, soil contamination, and PCB and other toxic waste storage and disposal) and contamination within the home arising from supertight construction techniques and synthetic building materials (the "sick building syndrome"). In a similar vein, adequacy can be assessed in terms of the appropriateness and efficiency of the residential structure and its mechanical components as seen from a sustainable development, perspective: for example, construction waste disposal.

A fifth dimension of adequacy is empowerment and personal safety within the home and neighborhood. In the context of the private home, this would include mechanisms for protecting against domestic violence. In the case of communal buildings, this includes consideration of physical security in basement garages, stairwells, elevators, and walkways. For all kinds of dwellings, physical safety in the neighborhood, sense of control, and sense of community would be aspects of concern in assessing adequacy.

From a normative view, is housing inadequacy simply indicative of low income? Put differently, if every consumer were sufficiently affluent, would housing adequacy no longer be an issue in public policy? The answer would have to be largely, but not entirely, yes. An affluent consumer could afford to purchase or rent a dwelling that met many of the requisite standards for adequacy. However, in some respects, adequacy is also an outcome of municipal provision of services and amenities, and these are typically out of the hands of the consumer. (SEE ALSO: *Housing Markets*)

—*John R. Miron*

Further Reading

Hallett, G. 1993. *The New Housing Shortage: Housing Affordability in Europe and the USA*. London: Routledge.

Hays, R. Allen. 1995. *The Federal Government and Urban Housing: Ideology and Change in Public Policy*. Albany: State University of New York Press.

Salins, P. D. and G. C. S. Mildner. 1992. *Scarcity by Design: The Legacy of New York City's Housing Policies*. Cambridge, MA: Harvard University Press.

Stegman, Michael A. and Michael I. Luger. 1993. "Issues in the Design of Locally Sponsored Home Ownership Programs." *Journal of the American Planning Association* 59:417-32.

Wolch, Jennifer, and Michael Dear. 1993. *Malign Neglect: Homelessness in an American City*. San Francisco: Jossey Bass. ◄

► Housing Starts

Housing starts are widely used to indicate business activity and consumer confidence. The U.S. Bureau of the Census defines the start of construction as when excavation begins for the footings or foundations of a privately owned building containing any housing units. Construction of every housing unit in a multi-unit building is assumed to start with excavation for the building.

In counting housing starts, the bureau does not include moved or relocated buildings, housing units created in an existing residential or nonresidential structure, or housing units that form group quarters. Furthermore, the bureau excludes publicly owned housing units (contract awards) in its measure of housing starts, but it does include units in structures built by private developers with public subsidies or units that, on completion, are offered for sale to local public housing authorities under the Department of Housing and Urban Development (HUD) "Turnkey" program. In addition, the bureau excludes mobile homes (those that have transportation gear as an integral part of the unit and can be towed from site to site), although they do include excavations for prefabricated, panelized, componentized, sectional, and modular units as housing starts.

The Bureau of the Census estimates housing starts using the monthly Survey of Construction. Bureau representatives draw random samples of building permits in about 840 jurisdictions where building permits are required for construction. Recipients of building permits are surveyed to determine whether excavation has begun, and sample data are then blown up to form national estimates for permit-issuing areas. Upward adjustment of these housing starts is made to account for starts without permit authorization or prior to permit issuance. The bureau also estimates housing starts in places where no building permit is required; in this case, national estimates are blown up from a random sample of about 130 places wherein the bureau searches intensively for all housing units started.

Table 17 summarizes annual counts of housing starts in the United States since 1959. In general, starts rise quickly in the first two years of the trough-to-peak of the business cycle; starts peaked in 1963, 1972, 1978, and 1986. Also evidenced in Table 17 is the postwar baby boom, which contributed to the surge in housing starts in the early 1970s. (SEE ALSO: *Building Cycle; Housing Completions; Housing Investment*)

—*John R. Miron*

Further Reading

Cammarota, Mark T. 1989. "The Impact of Unseasonable Weather on Housing Starts." *American Real Estate and Urban Economics Association Journal* 17(3):300-13.

Puri, Anil K. and Johannes Van Lierop. 1988. "Forecasting Housing Starts." *International Journal of Forecasting* 4(1):125-34.

U.S. Department of Commerce, Bureau of the Census. 1995. *Current Construction Reports: Housing Starts*. Washington, DC: Author.

TABLE 17 Annual Net Privately Owned Housing Units Started (in thousands), United States, 1959 to 1992

Year	All Starts	One Unit[a]	Two Units	Three or More Units	Three or Four Units	Five or More Units
1959	1,517	1,234	56	227		
1960	1,252	995	44	214		
1961	1,313	974	44	295		
1962	1,463	991	49	422		
1963	1,603	1,012	53	538		
1964	1,529	971	54		55	450
1965	1,473	964	51		36	423
1966	1,165	779	35		27	325
1967	1,292	844	41		30	376
1968	1,508	899	46		35	527
1969	1,467	811	43		42	571
1970	1,434	813	42		42	536
1971	2,052	1,151	55		65	781
1972	2,357	1,309	67		74	906
1973	2,045	1,132	54		64	795
1974	1,338	888	33		35	382
1975	1,160	892	35		30	204
1976	1,538	1,162	44		42	289
1977	1,987	1,451	61		61	414
1978	2,020	1,433	62		63	462
1979	1,745	1,194	56		66	429
1980	1,292	852	49		61	331
1981	1,084	705	38		53	288
1982	1,062	663	32		48	320
1983	1,703	1,068	42		72	522
1984	1,750	1,084	39		83	544
1985	1,742	1,072	37		56	576
1986	1,805	1,179	36		48	542
1987	1,621	1,146	28		38	409
1988	1,488	1,081	23		35	348
1989	1,376	1,003	20		35	318
1990	1,193	895	16		21	260
1991	1,014	840	16		20	138
1992	1,200	1,030	12		18	139
1993	1,288	1,126	11		18	133
1994	1,457	1,198	15		20	224
1995	1,354	1,076	14		19	244
1996	1,477	1,161	16		29	271

SOURCE: Data made available by the Construction Starts Branch, Construction Statistics Division, U.S. Bureau of the Census.

a. One-unit structures include all fully detached and certain attached (semidetached, row house, and town house) dwellings, except for attached structures in which heating, air-conditioning, water supply, power supply, or sewage disposal facilities are shared.

U.S. Department of Commerce, Bureau of the Census. 1987 to present. *Construction Reports: New Residential Construction in Selected Standard Metropolitan Statistical Areas.* C21. Washington, DC: Government Printing Office. Quarterly with annual cumulations.

U.S. Department of Commerce, Bureau of the Census. Construction Statistics Division. 1976 to present. *Construction Reports: Characteristics of New Housing.* C25. Washington, DC: Government Printing Office. Annual.

U.S. Department of Housing and Urban Development. *U.S. Housing Market Conditions.* Washington, DC: Office of Policy Development and Research. Published quarterly. ◄

► *Housing Studies*

A quarterly, *Housing Studies* is a refereed international journal presenting original research on all aspects of housing studies and housing policy. The first volume in was published in 1986. Subscribers include researchers and institutes in the field of housing as well as academic libraries worldwide. Managing Editor: Moira Munro, School of Planning and Housing, Heriot Watt University, Edinburgh College of Art, 79 Grassmarket, Edinburgh EH1 2HJ, Britain. Phone: 44 131 221 6162. Fax: 44 131 221 6163. North American editor: J. David Hulchanski, Center for Applied Research, Faculty of Social Work, University of Toronto, 246 Floor St. West, Toronto, Ontario, M5S 1A1, Canada. Phone: (416) 978 1973. Fax: (416) 978 7072.

—Caroline Nagel ◄

► Housing Subsidies in the United States and Western Europe: History and Issues

Housing subsidies provide financial support by government to stimulate production and consumption of housing. Direct government appropriations discussed in this entry support both rental and owner-occupied housing through either government provision (social housing) or provision by private for-profit or not-for-profit owners or sponsors. Direct appropriations are usually limited to support of housing for households with low or moderate incomes or with special needs, such as the elderly or single-parent families. Other less direct forms of subsidy, either through the tax code or through housing finance intermediaries, which are treated more fully in other entries, are usually far less restrictive with respect to populations served and are more likely to aid owner-occupied rather than rental housing.

The Fundamental Problem

The objective of housing subsidies discussed in this entry is to help provide adequate housing to households with insufficient means. There are two basic ways to close the gap between the cost of decent housing and what households can afford to pay. First, subsidies can lower the capital cost of the producer—be it a central government agency, a local authority, a not-for-profit housing organization, a housing cooperative, or a private enterprise—so that the occupant can be charged an affordable rent. This is a producer's—or supply-side—subsidy. Second, subsidies can supplement household income so that, by paying a reasonable percentage of income toward rent, a household can obtain decent housing and yet have adequate resources for essential non-housing expenditures. This is a consumer's—or demand-side—subsidy. Housing policy rhetoric and, to a lesser extent, policy itself have tended to move away from producer subsidies toward consumer subsidies. It is argued that consumer subsidies are more efficient because they provide benefits to consumers at a lower cost to government.

This entry reviews the evolution of housing subsidy policy among the highly industrialized countries of Western

Europe and the United States and then poses some central issues in contemporary housing subsidy policy.

The Evolution of Housing Subsidies

Early Policy

Housing subsidies find their origins in the Industrial Revolution of the Western world. During the 19th and early part of the 20th century, the transfer of production and commerce from agriculture to towns and cities—notwithstanding all the great gains in national productivity—had in many ways a devastating impact on the lives of workers and their families. The security of the farm and village gave way to the insecurity of the urban slum as overcrowding, insalubrity, and poor construction eroded the quality of home life.

The workers' housing problem became a concern of governments of European countries long before it was noticed in the United States, mainly because of three significant differences in the two environments. First, during the 19th century and the beginning of the 20th century, the European economic environment was relatively closed in comparison with the North American environment. The vast undeveloped American continent provided almost unlimited opportunities for individual initiative, freedom, and independence. By contrast, European workers were hemmed in geographically and by a more rigid social and economic structure. As compensation for weakness in the market, European workers sought economic empowerment through the formation of trade unions and the organization of socialist political parties in a manner quite foreign to the American experience.

Second, there was a widespread European view that social justice and human welfare for low-income families could be achieved only by cutting out for them a bigger piece of the national economic pie, which was believed to be of more or less fixed size. In North America, the feeling tended to be that, by the country's pushing westward, the size of the pie was constantly increasing. The modern welfare state had its European origins in the 1880s under Chancellor Otto von Bismarck in Germany and became fairly well codified throughout most of Western Europe under conventions developed by the International Labour Organization (established under the Versailles Treaty in 1919) long before the United States turned its attention to developing national social policies during the Great Depression of the 1930s.

Third, from the above considerations, European nations developed a strong value judgment that housing constituted a basic element in the worker's standard of living and that government had a responsibility to ensure that workers and their families had access to housing that met a minimum acceptable standard. Workers' housing came to be widely regarded as a social service comparable in many ways to education and health care.

In European countries, the seeds of housing subsidy policy were sown during the latter part of the 19th century. Progressive employers (e.g., the Cadburys and Rowntrees in Great Britain and the Guiness Trust in Ireland) provided subsidized housing for their workers, and well-meaning philanthropists embarked on subsidized housing schemes for the poor. Cooperative housing dates back to the Rochdale Cooperative Movement launched by Robert Owen in England in 1823, inspiring the first German cooperative housing society in 1849. In the United States, a philanthropic housing movement, centered in New York, developed in the latter part of the 19th century.

In the realm of public policy, regulatory measures such as building and occupancy codes, aimed at establishing minimum standards of safety and health, were adopted in many communities. National legislation was adopted in Ireland in 1876 and in France in 1894. However, relatively little impact on the national housing market was apparent before World War I.

With little new construction during World War I, serious housing shortages developed. Rent controls were introduced to protect workers and to prevent profiteering and inflation. Prime Minister Lloyd George in Great Britain promised returning veterans "houses fit for heroes to live in." One of the major housing objectives of governments became the provision of social housing—that is, minimum-standard housing owned and managed wholly or in part by government or one of its agencies at a rent that low-income workers could afford to pay.

Public financial assistance was mainly in the form of subsidies to produce housing, usually with the central government loaning the bulk, if not all, of the capital cost of the housing to a local agency. In some cases (e.g., Great Britain), the local body was a public instrumentality such as the local housing authority. In other countries (e.g., the Netherlands), the local body tended to be a nonprofit, quasi-public housing association. In still other countries, such as Germany and Sweden, the local body frequently took the form of housing cooperatives that were largely dominated by trade unions. Subsidies usually took the form of low-interest rates on capital loans. The central government determined what a reasonable rate of interest should be on its loan and then shouldered the difference between that rate and the market rate at which it had to borrow. During the first half of the 20th century, the reasonable rate ranged from 1% to 4%, but on occasion was even zero. In addition, public financing often involved the creation or strengthening of government-owned, although semiautonomous, housing mortgage banks.

The extent of government financial assistance differed from country to country as well as over time. In Germany, public financing accounted for more than half of total new housing investments between 1924 and 1930. In Great Britain, local authority projects provided 61% of all housing units built between 1920 and 1923, but with a switch to conservative political power, the local authority share fell to only 34% between 1924 and 1930. In Belgium and the Netherlands, about half the dwellings built in the 1920s obtained government financial assistance. Sweden dropped its housing subsidy program after 1924 while continuing public support to mortgage credit institutions but resumed subsidies 10 years later as an antidepression measure.

In the United States, direct federal involvement in housing production commenced during the Great Depression of the 1930s. In 1937, the federal government adopted a national housing program that provided capital subsidies to local housing authorities, which were expected to construct housing and charge rents that would cover operation and maintenance costs. The public housing program was conceived primarily as a measure of attacking unemployment and providing jobs and only secondarily as a means of meeting people's housing needs. In fact, the program was viewed as providing temporary housing for the working or "deserving poor," who could not find decent housing on the private market; households with little or no means were excluded by an income floor. Not until the late 1960s, with the passage of the Brooke Amendment, did public housing evolve into a program for the very poor, when operating subsidies were introduced to allow public housing to serve low-income families.

The Impact of the War and Postwar Policy: 1939 to 1970

World War II witnessed a catastrophic worsening of housing conditions. Most countries entered the war with a backlog of housing needs inherited from the Great Depression. During the war, below-normal housing construction persisted as resources were diverted to military purposes; virtually all countries adopted rent controls to protect tenants from scarcity-based rent increases and to promote price stabilization.

After the war, most involved countries required an all-out housing effort to heal the ravages of war or to meet the new demands created by major demographic changes associated with refugee movements in Europe, postwar urbanization, and new household formation. The responses varied considerably, affected by both the level of direct involvement in the war and established political philosophies. Countries such as Australia, Belgium, Canada, Japan, New Zealand, Norway, Switzerland, and the United States placed high priority on private housing, whereas Denmark, Finland, France, Germany, Ireland, the Netherlands, Sweden, and Great Britain developed a considerable amount of social rental housing.

Building on their prewar experience, European governments moved energetically to tackle the national housing shortage with a wide range of policies. In addition to producer subsidies developed earlier, Austria, Belgium, Denmark, Germany, France, Finland, Ireland, Italy, Japan, and the Netherlands developed contractual savings-for-housing plans offering economic incentives to people willing to enroll in a regular homeownership savings plan. Such plans mobilized substantial capital insulated from financial instabilities that had bedeviled housing construction in many countries.

In the United States, the 1949 National Housing Act made a commitment to "a decent home and suitable living environment" for every American. However, the responses to the nation's housing needs were carefully designed not to compete with private initiatives. Public housing income ceilings were limited so that public provision did not compete with the private market. By the end of the 1950s, the United States had initiated the first of a number of supply-side subsidy programs in the form of below-market interest rate mortgages for rental and owner-occupied housing. The programs were targeted initially to the elderly, to house-

holds of moderate means, and eventually, with the introduction of "rent supplements," to low-income households. In 1974, the so-called Section 8 program became the major low- and moderate-income housing initiative providing support for new production, rehabilitation, and a rent certificate program to provide assistance directly to households.

Western European nations also subsidized private rental housing. In Germany, Spain, and Switzerland, subsidies to for-profit builders did not differ from those to nonprofit builders, except that considerable personal investment was customary. In Denmark, annual subsidies to for-profit builders were smaller than those to nonprofit builders. France provided annual subsidies and concessional interest rates; the Netherlands provided lump sum grants.

Finally, some governments placed important financial responsibility for workers' housing on employers. Both France and Italy adopted laws requiring employers to pay into funds for financing workers' housing, with any subsidies qualifying as tax deductions. Other countries, including Austria, Belgium, Finland, Germany, Japan, the Netherlands, and Sweden extended tax incentives to employers constructing or making loans for workers' housing.

By 1957, in most of highly industrialized Western Europe, more than half of all housing units received some form of public financial assistance; in fact, in France, Ireland, the Netherlands, and Sweden, it was more than 90%. In the United States, however, during the 1950s and 1960s, subsidized units accounted for less than 5% of annual additions to the housing stock. In 1968, the United States established national housing goals for the next decade, proposing construction or rehabilitation of 26 million units, 6 million to be subsidized by government for low- and moderate-income families. By 1997, the federally subsidized public and private housing stock totaled about 5.7 million units, including those occupied by tenants receiving vouchers and rent certificates.

Housing policies had social objectives other than simply supplying housing. They were often designed to help stabilize employment and the economy as a whole. In European countries, seasonal winter unemployment as well as cyclical unemployment were policy concerns. Another objective was to conserve the existing housing stock that was often in a state of decay, especially important in European cities with masonry construction possessing high architectural quality. Finally, in Europe, housing subsidies were a tool to promote labor mobility.

The 1970s: Inflation, Stagnation, and a Slowdown in Subsidies

By the early 1970s, a variety of factors brought about significant changes in the orientation of national housing policies. First, the "crude housing shortage" inherited from the war had generally been overcome. In fact, in the early 1970s, Denmark, France, Germany, Sweden, Switzerland, and Great Britain had some difficulty renting newly produced social housing units. Western European countries had started to think in terms of improved living standards, including automobiles, more living space, and homeownership. These social demands promoted a shift away from social housing and toward homeownership—even in Sweden, the Western European country with the most highly developed welfare state.

However, postwar affluence soon began to fade. Disruptions in the international oil market had a powerful impact on costs in the non-oil-producing world. Higher housing subsidy burdens on governments led to disenchantment with public expenditures on housing. Recurring balance-of-payments difficulties and strong inflationary pressures, particularly in Denmark, the Netherlands, and Great Britain—all heavily dependent on foreign trade—brought frequent credit restrictions resulting in "stop-and-go" construction policies that upset the financing of low-cost housing. Finally, the continuation of rent controls adopted during World War II, primarily in Western Europe, created major inequities in rent structures, resulted in a misallocation of space, and adversely affected the quality and quantity of private rental housing.

As a consequence of these changes, major shifts in national housing subsidy policy emerged. First, Western European governments have moved away from general housing subsidies to more emphasis on households with the greatest need—those with low incomes or other specific attributes, such as the elderly, the handicapped, large families, and single-parent families. U.S. housing subsidy policy has also increasingly concentrated on the most needy families. Second, consumer (demand-side) subsidies emerged as the preferred tool for dealing with some of the new issues, such as equalization in the rent structure, redirecting subsidies to the neediest households, and cutting back on total expenditure for housing subsidies.

Finally, for a number of reasons, homeownership subsidies were given a higher priority. The expansion of homeownership—generally to households with moderate incomes or better—bypassed the need for large rental subsidies to poorer families and transferred housing maintenance costs from the public to the private sector; it increased the total supply of savings and capital for the housing sector; it provided a hedge for individuals against inflation; and it responded to the growing desire of an affluent society to devote more of its income to higher-quality living. The condominium system of ownership was a notable development in both Canada and the United States.

Housing Subsidies Since 1980

In the 1980s and 1990s, changing supply-and-demand factors produced new pressures on housing affordability. On the demand side, rising income disparities became increasingly apparent. Wage rates and job opportunities in central cities were being jeopardized by the flight of industry and competitive pressures from globalization of the economy. Average unemployment rates rose from 2% to 3% during the 1960s to 8% to 10% in subsequent decades.

On the supply side, fiscal austerity to cope with rising national deficits, inflation, and high interest rates led to significant reductions in subsidies for housing production and maintenance. Many high-rise projects—especially those located in areas of high industrial unemployment (e.g., Glasgow and St. Louis)—encountered design defects, technical obsolescence, and progressive physical deterioration. Serious economic and social problems compounded

the situation as these projects became increasingly occupied by poor single-parent families, minorities, and the unemployed. Vandalism, drug trafficking, and other criminal activities also took their toll. Accelerated decay hit many inner-city areas, including older, privately owned rental housing whose landlords had insufficient incentives to maintain their property. As part of fiscal austerity, many governments in Western Europe and the United States embraced privatization and decentralization. Social rental housing was sold to either tenants or private landlords. In Great Britain, nearly 1.6 million units of "council housing" (25% of the 1981 total of 6.4 million) were sold; other countries, including the United States, had smaller programs. Central governments devolved spending responsibilities to subnational units, whose resources, however, were far more limited. At the same time, not-for-profit housing organizations helped fill some of the gap in the provision of social housing. (Not-for-profits had long been a salient feature of national housing policy in Canada, Germany, and the Netherlands, where they have owned more than half the social housing stock.)

On the demand side, housing allowances increasingly provided direct assistance to households by meeting the gap between the cost of adequate housing and a percentage of income that lower-income households could reasonably be expected to pay. In Western European housing allowance systems, rent-to-income ratios averaged 10% to 15%, with many including sliding scales to take account of family size and income. In Austria, France, and Great Britain, for very low levels of income, the ratio was zero. In the United States, historically, a 25% rent-to-income ratio was regarded as reasonable, but in the early 1980s, the ratio was raised to 30% because government was not prepared to increase appropriations for federal housing subsidies. In most Western European systems, the housing allowance was an entitlement, meaning that anyone satisfying the eligibility criteria could apply for and receive financial assistance. In the United States, rent certificates and, subsequently, housing vouchers were not entitlements; budgetary limits restricted coverage to only a fraction of eligible households.

Support for homeownership—especially for first-time homebuyers—has come increasingly through subsidized loans and outright grants, as well as through improved mortgage instruments: for example, low down payments, deferred payments, index-linked mortgages (adjusted as income rises), and equity or shared appreciation mortgages (in which the lender shares in future capital gains in return for a lower down payment). Bank deregulation has increased the supply of housing funds but has led to higher real mortgage rates (adjusted for inflation).

At the beginning of the 1980s, along with decentralization, local governments in Europe and North America shifted toward area-based approaches to urban regeneration, with important new elements: (a) tenant empowerment in the form of responsibility for housing management; (b) partnerships among public, private, and not-for-profit agents; and (c) renewed attention to economic development as a tool for job creation.

As the United States phased out direct producer subsidies and made dramatic reductions in tax subsidies to rental housing, it did introduce in the late 1980s one important new initiative to promote private housing production—a tax credit for low-income housing, which usually requires partnerships between for-profit firms and not-for-profit housing providers. Unlike the homeowners' mortgage interest tax deduction, which is a virtual entitlement, the amount of activity permissible under the low-income tax credit is subject to budgetary limits.

Contemporary Issues

The following seven issues are central to current policy debates and will shape housing policy into the future:

1. *Homelessness.* There are many strongly debated points. First, estimates of the size of the homeless problem vary greatly. Second, there are great differences concerning public responsibility. One view is that many are homeless by choice; therefore, government has little, if any, special responsibility. Another view is that many people have become homeless for reasons beyond their control; thus, government has a responsibility to help. Third, a variety of solutions are proposed, ranging from the mere provision of shelters to comprehensive support services to help reintegrate the homeless into the economy and society. Fourth, increasing emphasis is placed on prevention, ranging from short-term emergency measures to fundamental social and economic reform to eliminate poverty.

2. *Social Housing Versus Private Housing.* In many countries, particularly Great Britain and the United States, public housing has been deeply questioned as a result of social, economic, technical, and management problems encountered in many projects, especially in the inner city. In other countries (e.g., the Netherlands and Scandinavian nations), social housing has maintained high standards and support. There is strong debate concerning three major alternatives: rebuilding public housing programs, expanding not-for-profit housing organizations' initiatives, and relying on private housing markets, including, where necessary, direct subsidies to rental housing.

3. *Producer Subsidies Versus Consumer Subsidies.* What is the best way to close the gap between a household's ability to pay and the cost of housing? In the past two decades, the rhetoric has favored consumer subsidies on the grounds that they are more efficient; not only do they allow for greater consumer choice, but the average cost per household can be half that of producer subsidies. Nevertheless, nations have continued to provide new producer subsidies, partly because of localized shortages (associated with rural-urban migration, economic growth, and households with special needs) and partly as a result of political pressure from private and not-for-profit housing providers.

4. *What Is a Reasonable Rent-to-Income Burden?* The wide variations, from zero to 30%, were noted earlier. "Reasonableness" is a function not just of the circumstances of households in need but also of the constraints placed on any program by the willingness of taxpayers to bear the cost.

5. *Is Socioeconomic Integration Viable?* There is strong agreement within the policy community that it is desirable. As governments have focused support on the neediest families, production subsidies have tended to concentrate the poor both geographically and within buildings or projects. In the United States, the siting of low-income housing in better-off neighborhoods has met with community resistance. Housing allowances, which enable recipients to find their own housing, help promote a desirable social and economic mix within buildings and across communities, although the level of financial support provided may limit opportunities—as has been the case with the U.S. housing voucher program. Canada has been a successful innovator in promoting economic integration by requiring nonprofit and cooperative housing organizations receiving interest subsidies to accommodate up to 30% of tenants from low-income groups. Similarly, the low-income housing tax credit in the United States requires that at least 20% of housing units be allocated to low-income families.

6. *Is Tax Policy a Fair and Efficient Way to Subsidize Housing?* There are two major controversies: First, should homeownership subsidies (e.g., low-interest loans, grants, and tax-deductible mortgage interest payments) be limited to low- and moderate-income households? Practically all nations have placed some restrictions on tax deductibility of mortgage interest. Australia, Canada, and New Zealand have abolished it. However, in the United States, restrictions are minimal, and the magnitude of this virtual entitlement is enormous. Such tax expenditures rose from $2.9 billion in 1969 to $40 billion in 1996, with $28 billion (70%) of these benefits being received by taxpayers with incomes of $75,000 or more. In contrast, total outlays by the U.S. Department of Housing and Urban Development for subsidized housing in 1996 were $21 billion. (Substantial government funds also flow indirectly to private, mostly rental, housing through income support programs for the poor.) The second issue is whether tax policy should be neutral with respect to tenure. Notwithstanding some strong arguments justifying government support of homeownership mentioned earlier, there are widely held opposing views that not only does such treatment discriminate against renters, but it may also result in a misallocation of resources with too many going to homeownership and too few to other forms of housing and nonhousing investment. The problem is the difficulty of taking away a benefit from a politically and economically powerful group of homeowners.

7. *Do Citizens Have a Right to Decent Housing, Similar to Education and Health Care?* European governments have long embraced the objective of decent housing for all. The United States did so in 1949. Nevertheless, the right to decent housing was heatedly debated at the second U.N. Conference on Human Settlements (Habitat II) in 1996. The European Union supported explicit and unqualified recognition of a separate, distinct, and existing "human right to adequate housing." In contrast, the United States, Japan, and a number of Latin American countries asserted that housing is an important component of an existing "right to an adequate standard of living" already set out in the U.N. Universal Declaration of Human Rights. The United States further maintained that this derivative housing right—along with the right to food, clothing, and other basic necessities—should be realized progressively based on availability of resources. The compromise reached

embraced both the U.S. belief that the right to decent housing is an important component of the right to an adequate standard of living and the European Union's concern that governments have a recognized obligation to protect, promote, and ensure this right.

However, whatever form the rhetoric may take, the true measure of a country's commitment is how well it translates an abstract right into reality. Do homeless families and those living in substandard conditions have access to adequate housing? Is the right to adequate housing legally enforceable? Historically, Western European countries have gone a great distance in making this right a reality through comprehensive social housing programs and housing allowance entitlements. Public housing authorities in the United Kingdom have a statutory duty to house homeless families. In New York State, the right to shelter for the homeless has been upheld in the courts. But generally, citizens do not have a clear, legally enforceable right to adequate housing. Moreover, during an era in which fiscal austerity has constrained housing programs in Western Europe and the United States, the collective will to honor this right has—at least temporarily—been seriously eroded. The right to adequate or decent housing will likely remain a contentious issue well into the 21st century. (SEE ALSO: **Homelessness;** *Privatization;* **Public Housing;** *Rent Control; Right to Housing; Section 8;* **Subsidy Approaches and Programs;** *Taxation of Owner-Occupied Housing*)

—*E. Jay Howenstine and Elizabeth A. Roistacher*

Further Reading

Best, Richard. 1996. "Successes, Failures, and Prospects for Public Housing Policy in the United Kingdom." *Housing Policy Debate* 8(3).535-63.

Hallett, Graham, ed. 1993. *The New Housing Shortage: Housing Affordability in Europe and the USA.* London: Routledge.

Howenstine, E. Jay. 1986. *Housing Vouchers: A Comparative International Analysis.* New Brunswick: Rutgers, State University of New Jersey, Center for Urban Policy Research.

Organization for Economic Cooperation and Development. 1986. *Tax Policies and Urban Housing Markets.* Paris: Author.

Organization for Economic Cooperation and Development. 1987. Urban Housing Finance. UP/H (87)1. Paris: Author.

Organization for Economic Cooperation and Development. 1994. *The Multi-Sector Approach to Urban Regeneration: Towards a New Strategy for Social Integration, Housing Affordability and Livable Environment.* ENV/UA/H (94). Paris: Author.

Roistacher, Elizabeth A. 1987. "Housing and the Welfare State in the United States and Western Europe." *Netherlands Journal of Housing and Environmental Research* 2(2):143-75.

Stegman, Michael A. 1996. "The Right to Housing Debate." *The Urban Age* 4(2):9.

U.N. Economic Commission for Europe. 1980. *Major Trends in Housing Policy in ECE Countries.* ECE/HBP/29. Geneva: Author.

Van Vliet--, Willem and Jan van Weesep, eds. 1990. *Government and Housing: Developments in Seven Countries.* London: Sage. ◄

► Housing Supply

Housing supply is a term commonly used in two distinct ways. One is in reference to the amount of housing stock; this is the *stock supply.* The other usage is in reference to the flow of housing services, or the quality of accommodation flowing from that stock: This is the *flow supply.*

In its simplest sense, we can think of housing stock supply as a count of dwellings available for human habitation. In practice, this is an elusive concept to measure. The census, for example, counts all dwellings in which people normally reside (that is, 102.3 million dwellings in the United States in 1990); however, this is only part of the supply. Vacant and seasonal dwellings that can be occupied year-round are also part of the stock supply. So, too, might other structures that can be occupied readily but that are presently vacant (e.g., houseboats, trailers). If we were to go so far as to add tents to this list, we immediately see that we cannot enumerate a stock without first defining the concept of dwelling. Also not transparent is the case of other structures (e.g., vacant office buildings, garages, hotels, or warehouses) or spaces within existing dwellings that could be converted, with modest effort, into separate habitable units.

This stock supply changes over time. On the positive side, developers and self-builders construct new buildings, mobile homes, and other habitable structures. Within existing structures, owners may reconfigure the layout to add another dwelling unit. In addition, landlords may convert space within nonresidential structures into new dwellings. These all constitute additions to the housing stock. On the negative side, some buildings are demolished by their owners; others collapse with age and neglect or are condemned; still others burn down or are otherwise destroyed by accident; and some are converted to nonresidential uses. Within a given building, the number of dwellings may also drop because of conversion activity (e.g., converting a triplex back to a single-unit structure). The housing stock supply grows on net if additions exceed such losses. Data on recent private sector additions to the U.S. housing stock are summarized in Table 18.

As in any market, housing stock supply can be thought to change in response to price. Viewed this way, housing supply is actually a list of housing prices showing the quantities supplied at each price. Over the short term (that is, a day, week, or perhaps month), stock supply is generally thought to be only weakly responsive to price, because property owners need time to alter the amount of stock. Over the longer term, however, developers have time to construct more housing as additions to the stock become profitable. Profit-maximizing developers will look at costs of different ways of producing dwellings using existing structures (as in conversions) or available land (as in new construction), building materials, labor, and financing as well as building code and other regulatory restrictions in determining if and how to respond to changes in market price.

To this point, we have thought of dwellings as homogeneous stock. In fact, dwellings come, figuratively speaking, as "apples and oranges." Some dwellings are large, others small; some dwellings have municipal water and sanitary sewer connections, and others do not; some are basement apartments, and others are penthouse suites; and so on. In the United States in 1990, for example, 65.8 million dwellings were in one-unit structures, 28.0 million were in multi-

TABLE 18 Characteristics of New Privately Owned
Dwellings Completed (in thousands), United
States, 1989 to 1996

	1989	1991	1993	1995	1996
All dwellings	1,423	1,091	1,193	1,313	1,413
In structures with one unit					
For sale	661	481	642	682	746
For owner occupancy on owner's land	335	335	375	350	350
For rent	31	21	23	33	33
In structures with two or more units					
For sale	90	58	44	51	50
For rent	307	197	109	196	234

SOURCE: U.S. Department of Commerce, Bureau of the Census. 1994. *Characteristics of New Housing: 1993.* C25/93-A. Table 1. Washington, DC: Author. U.S. Department of Commerce, Bureau of the Census. 1997. Characteristics of New Housing: 1996. C25/96A. Table 1. Washington, DC: Author.

unit structures, and 8.5 million in mobile homes or other. To simply enumerate dwellings in the stock is to miss important differences among housing units. An alternative measure of housing stock would take into account variations in housing quality, as reflected in differences in the cost of producing (or the price of purchasing) each unit of housing.

Economists argue that we should think of housing supply not in terms of homogeneous units but, rather, as a varied stock of physical capital (that is, floors and roofs, walls, doors and windows, locks, plumbing and electrical systems, and so on) that is combined with heat, utilities, built-in appliances, cleaning and maintenance supplies, other purchased inputs, and labor to produce a level of dwelling accommodation. This approach to housing supply, which conceptualizes housing in terms of the quality of accommodation "flowing" from it, is called the flow supply of housing. In the rental market, landlords produce levels of accommodation for sale to tenants in an explicit market. Homeowners, on the other hand, each produce a level of accommodation to meet their own needs—an implicit market. Economists argue that we can think of each dwelling as providing a level of accommodation that can be systematically varied to meet consumer needs in a competitive market. For example, suppose landlords find that an on-site swimming pool allows them to obtain a higher rent. Landlords as a group will have an incentive to add swimming pools to their apartment projects up until the market is satiated—that is, when the present value of all future rent increments attributable to a pool falls until it just equals the discounted cost of installing and maintaining the pool. In a competitive market for housing, we might therefore expect the flow value of a unit of housing stock to be decomposable into the presence of various dwelling characteristics, each times the marginal, or hedonic, price (a statistical relationship that ties the selling price of an owned home—or rent, in the case of rented dwellings—to the physical characteristics of the dwelling and its neighborhood setting) of that characteristic.

Just as with housing stock supply, we can think of flow supply as responsive to price. Landlords, for example, will alter the stock supply of housing and alter the level of accommodation provided in existing housing in response to opportunities for increased profit. Flow supply then can be thought of as a list of prices for housing service showing quantities supplied at each price—just as we have done above with stock supply earlier. Over the shorter term, flow supply will generally be less responsive to price, because property owners need time to alter stock supply and the other inputs used to produce a particular level of accommodation. Over the longer term, however, developers have time to build and adapt their housing.

Stock supply and flow supply are each useful concepts in understanding housing markets. Stock supply, with its focus on number of dwellings, tells us how many households can be accommodated on net by new construction and other sources of stock change. Flow supply, with its focus on level of accommodation, tells us about changes in dwelling quality, size, and amenities. (SEE ALSO: *Housing Demand; Housing Markets*)

—*John Miron*

Further Reading

Bramley, G. 1993. "Planning, the Market and Private Housebuilding." *Planner* 79(1):14-16.

Eatwell, J., M. Milgate, and P. Newman, eds. 1987. *The New Palgrave: A Dictionary of Economics.* Vol. 4. London: Macmillan (see pp. 506-09, 554-56).

Fallis, G. 1993. "The Suppliers of Housing." Pp. 76-93 in *House, Home, and Community: Progress in Housing Canadians, 1945-1986,* edited by J. R. Miron. Montreal: McGill-Queen's University Press.

Joint Center for Housing Studies of Harvard University. *The State of the Nation's Housing.* Cambridge, MA: Author. Annual publication.

McKellar, J. 1993. "Building Technology and the Production Process." Pp. 136-54 in *House, Home, and Community: Progress in Housing Canadians, 1945-1986,* edited by J. R. Miron. Montreal: McGill-Queen's University Press.

Pozdena, R. J. 1988. *The Modern Economics of Housing: A Guide to Theory and Policy for Finance and Real Estate Professionals.* New York: Quorum (see Chapter 3).

U.S. Department of Commerce, Bureau of the Census. 1995. *Current Construction Reports: Characteristics of New Housing: 1995.* Washington, DC: Author. ◀

▶ Housing Trust Funds

Housing trust funds are distinct accounts, typically established by city, county, or state governments, with dedicated ongoing revenue committed exclusively to support housing affordable to lower-income households.

By 1997, more than 115 housing trust funds had been created by local and state governments to secure funds for needed housing programs—more than six times as many as existed in 1985. As the federal government reduced its commitment to meeting lower-income housing needs in the 1980s, local and state governments were forced to search for needed funds to address the growing problems of inadequate affordable housing and homelessness.

Because housing trust funds have a secure stream of revenue, they remove the need to address affordable housing during annual budget battles, giving the right to housing a stature it has not enjoyed in the United States. Although housing trust funds have not been able to obtain funding sufficient to address all of the housing needs that exist for poor residents, they have established a precedent for recognizing the importance and advantages of addressing long-term funding needs.

Revenue Sources

The key characteristic of a housing trust fund is that it receives ongoing revenues from dedicated sources of funding, such as taxes, fees, or loan repayments. Revenue sources for housing trust funds vary widely and differ for city, county, or state funds. Typically, legislation or an ordinance is passed that either (a) increases an existing revenue source, such as a real estate transfer tax, and commits the increase to the housing trust fund on an annual basis or (b) commits all or a portion of a revenue source that is not currently supporting affordable housing, such as the repayments from an urban development action grant loan. In this sense, housing trust funds have been able to secure new funds for housing. Most housing trust funds commit the revenue they receive annually to eligible projects, and although some repayments into the fund occur through the loans made, they do not solely rely on interest and income earned or loan repayments for their continued source of funding. The ongoing nature of the committed revenue stream makes housing trust funds unique.

To support their housing trust funds, cities have relied on in-lieu fees from inclusionary zoning ordinances, linkage programs requiring an exaction from office and commercial developers to offset the impact of additional employees on the supply of housing, preservation ordinances that place a penalty on the demolition or conversion of rental units, tax increment districts, repayment of government loans (such as urban development action grants), program income from government programs (such as the community development block grant program), hotel and motel taxes, or other local taxes and fees.

County housing trust funds have relied on document-recording taxes or fees, sale of county-owned land, property taxes, or other sources. States have been able to commit a wide variety of revenue sources to their housing trust funds, including real estate transfer taxes, interest from title insurer accounts, document-recording fees, interest from real estate escrow accounts, escheat funds, mortgage revenue bond reserve or surplus funds, and interest from tenant security deposits, among others.

Although a handful of housing trust funds are able to commit around $50 million a year or more in revenue, most enjoy between $5 and $15 million in annual revenues. Some housing trust funds operate on less than $500,000 a year. Historically, housing trust funds have been able to leverage for every housing trust fund dollar committed at least an additional five dollars from private or other government funds. Thus, housing trust funds bring in far more dollars to support needed housing than would otherwise probably be available. In many instances, housing trust funds provide matching funds for programs whose available dollars might otherwise go unused, such as the federal McKinney program or funding from the National Affordable Housing Act through the HOME program.

Administration

Typically, enabling legislation or ordinances describe the purpose of a housing trust fund, how it will be administered, how the funds can be spent, what broad requirements must be met, the revenue sources dedicated to the fund, and other procedural requirements, such as monitoring and evaluation. Administrative rules and program guidelines are usually developed subsequently by the governing board and staff to guide the implementation of the fund.

A governing board, appointed by the mayor, governor, or other elected body usually administers housing trust funds. The enabling legislation typically requires broad representation on this board, and members of the board usually represent government, private and nonprofit developers, realtors, financial institutions, corporations, apartment owners and managers, housing advocates, low-income tenants, and the public at large. Responsibilities of the board range from advisory to oversight of the housing trust fund, including establishing policy and program guidelines and determining which applicants are awarded funds.

Although a few housing trust funds have been structured as independent nonprofit entities or are administered by an existing nonprofit organization or foundation, most housing trust funds are staffed through local departments of housing and community development or other government agencies. Several state housing trust funds are administered by state housing finance agencies, others by departments of housing, development, or commerce. Most local housing trust funds are administered by whatever department or agency customarily works with housing programs; a few are administered by redevelopment agencies or housing authorities.

Programs Funded

Some housing trust funds are used to fund programs that are already underway or administered by the agency or department administering the housing trust fund. In these cases, the funds go to support these housing programs and are often combined with other available housing dollars. Other housing trust funds establish specific programs, such as a rental rehabilitation program, a first-time homebuyers program, or a homeless shelter assistance program that will each receive portions of the available revenue.

It is more common, however, for housing trust funds to make their funds available to eligible applicants through periodic requests for proposals that outline eligible activities, application procedures, review criteria, and other priorities and requirements. Although some housing trust funds may restrict the use of their funds, they more typically provide gap financing for eligible projects either as the first money in to enable the project to get under way or as the last funds committed to ensure a project's feasibility.

Although housing trust funds vary in the types of housing activities they will support, they have provided funding for every phase of housing production and all types of housing activities. Typically, funds are awarded as loans or grants to support the acquisition, construction, or rehabilitation of housing. Housing trust funds that encourage projects that serve the very lowest incomes usually rely on grants or forgivable loans as a necessary way to make these projects work. Almost all housing trust funds support the hard costs of project development, and some support housing-related activities, such as counseling, child care, and the like. Although most housing trust funds encourage the provision of permanent housing, some also support transitional and emergency housing projects. Some housing trust funds make funds available for rental assistance, and a few attempt to use some funding to support operational costs of projects to help keep units affordable to very low incomes.

Most housing trust funds allow private and nonprofit developers, government agencies, individuals, and, at times, housing authorities to apply for funding. Increasingly, housing trust funds are recognizing that nonprofit development corporations play a very important role in ensuring that the goals of providing long-term affordable housing for very low income households are met. To assist nonprofit organizations in developing or expanding their ability to compete for funds by creating good projects, housing trust funds have created special-capacity building programs to help nonprofit development organizations develop their skills and resources. Some funds also make predevelopment funds available, which are often especially difficult for nonprofits to obtain.

Virtually all housing trust funds are established to provide funding for housing affordable to low- and moderate-income households, although it is common for funds to target resources specifically to low-income and very low income households. Some housing trust funds focus on programs and services for the homeless.

Creating a Housing Trust Fund

Housing coalitions are often responsible for the development of housing trust funds. These coalitions may come together for the express purpose of creating a housing trust fund, or an existing coalition may identify this as a priority on their agenda. It is not unusual for it to take a year or two to create a housing trust fund. The campaign often involves identifying housing needs, researching possible revenue sources, working with government agencies to determine an administrative structure, developing priorities for funding, and working to get the proposal passed by city council, county commissioners, or the state legislature.

The campaign can be very controversial. It requires elected officials to make a commitment to address housing needs and to do so, often, under the constraints of tight fiscal conditions. Increasing or creating a tax or fee and committing it to a housing trust fund require legislative leadership sufficient to convince elected officials that making these kinds of tough decisions is in the overall interest of their constituency. This often requires widespread public campaigning and support. Thus, coalitions are often involved in education and lobbying activities to make the housing trust fund campaign a success.

Opposition to the creation of housing trust funds is common. Not only will such opposition come from those who do not support low-income housing, but it comes in particular from those who are most closely tied to the revenue sources that are sought after for the housing trust fund.

Thus, housing trust fund advocates must develop convincing arguments to support the creation of these funds. Housing trust funds are widely regarded as beneficial to local economies because of the resultant construction activities, increased taxes from improved property, sales taxes, and income generated. Housing advocates support housing trust funds because they so often work well with community-based nonprofit development organizations, provide flexible funds for projects, and support housing for lower-income households too long ignored by other government programs. After more than 15 years of experience with housing trust funds, it is clear that these funds have been successful in helping address the housing needs of low-income and very low income households through their flexible programs and the availability of ongoing funds. (SEE ALSO: *Housing Finance*)

—*Mary E. Brooks*

Further Reading

Brooks, Mary E. 1989. *A Citizens Guide to Creating a Housing Trust Fund*. Washington, DC: Center for Community Change.

Brooks, Mary E. 1992. "Part I: Housing Trust Funds: What Makes Them Work"; "Part II: Victorious Housing Trust Fund Campaigns"; and "Part III: Housing Trust Funds: How They Work." *Shelterforce* (May/June, July/August, and September/October):6-10, 6-9, 10-13+.

Brooks, Mary. 1997. *A Status Report on Housing Trust Funds in the United States*. Washington, DC: Center for Community Change.

Connerly, Charles E. 1989. *A Guide to Housing Trust Funds: Tools for Community Development*. Washington, DC: Neighborhood Reinvestment Corporation.

Petherick, Glenn. *State Housing Trust Funds: Targeted Supports for Affordable Shelter*. Washington, DC: National Council of State Housing Agencies.

Rosen, David. 1987. *Housing Trust Funds*. Chicago: American Planning Association. ◀

▶ HUD Minimum Property Standards

The Department of Housing and Urban Development (HUD) Minimum Property Standards are intended to provide a sound basis for all housing units constructed under HUD mortgage insurance and low-rent public housing programs. They comprehend material standards developed by the housing industry and accepted by HUD as well as technical suitability standards adopted by HUD for materials and products for which there are no industry standards acceptable to HUD. In addition, these housing units must also conform to state or local building codes if the HUD secretary has approved the code in question. Each HUD regional office maintains a current list of jurisdictions with both accepted and partially accepted building codes.

For multifamily and care-type properties, the following model national building codes have been incorporated by reference into HUD Minimum Property Standards: the

Building Officials and Code Administrators (BOCA) Basic/ National Building Code; the Standard Building Code; the National Electrical Code; and the Uniform Building, Plumbing and Mechanical Codes. These codes provide standards for crucial construction functions such as fire safety, light and ventilation, structural loads (including seismic safety), foundation systems, materials standards, construction components, glass, mechanical systems (including heating, cooling, and ventilation), plumbing, electrical systems, and elevators. In addition, certain other technical standards pertain and are published in HUD Handbook No. 4910.1.

For single-family detached homes, duplexes, triplexes, and dwelling units in a structure characterized by a side-by-side town house arrangement, the Council of American Building Officials (CABO) One and Two Family Dwelling Code and the Electrical Code for One and Two Family Dwellings have been incorporated by reference into HUD Minimum Property Standards. With the exception of elevators, the same standards for crucial construction functions apply to single-family homes, duplexes, triplexes, and so on that apply to larger properties.

In addition, general acceptability criteria have been established for single-family homes, duplexes, and triplexes, including regulations pertaining to site conditions (especially toxic materials, ground movement, drainage, and erosion), to vehicular and pedestrian access, and to utilities and water supply. HUD has also developed standards for factory-produced (i.e., modular or panelized) components, which have been published in Handbook No. 4950.1.

Finally, minimum property standards concern themselves with thermal requirements relevant to structural insulation. In addition to conventional criteria in this area, passive solar energy and the related storage and reradiation capacity of masonry, water, and other masses may be used if they are cost-efficient and do not compromise structural safety and durability. Changes in minimum property standards are inevitable over time and notices of such changes are published in the *Federal Register*. (SEE ALSO: *Building Codes*)

—*Daniel J. Garr*

Further Reading

Code of Federal Regulations. Vol. 24, Part 882.109, 1996, pp. 103-108. Washington, DC: Government Printing Office. ◄

► *HUD Statistical Yearbook*

This was housing's equivalent to the U.S. Bureau of the Census *Statistical Abstract* because it presented in one volume a universe of data related to housing and urban development. The Department of Housing and Urban Development (HUD) published it annually from the late 1960s through 1980, with 1979 the last year covered. In the *1979 Statistical Yearbook,* each office in HUD reported statistics on its programs. Hence, Community Planning and Development reported data on the following programs: Community Development Block Grant, Urban Development Action Grant, Rehabilitation Loans, Reloca-

tion, Urban Renewal, and Urban Homesteading. Although some of the tables were rather mundane, others were quite revealing. For example, in FY 1979, there were 11,937 families and individuals displaced under HUD-funded programs, of which 51% were black.

In addition, the HUD *Statistical Yearbook* also reported general statistics related to housing and urban development, thereby making this information conveniently available to researchers. Reflecting the diversity of the data collected were the following: composition of households, 1960 to 1978; new private and public units started, 1969 to 1979; mobile home shipments and estimated retail sales, 1969 to 1979; value of new construction put in place, private and public, 1969 to 1979; indices of dwelling unit construction cost, 1972 to 1979; foreclosures on Federal Housing Administration (FHA) and Veterans Administration (VA) home mortgages, 1969 to 1979.

The *HUD Statistical Yearbook* became a victim of Reagan administration cutbacks; the last issue appeared the month that Ronald Reagan defeated Jimmy Carter. (SEE ALSO: *U.S. Bureau of the Census*)

—*Charles E. Connerly* ◄

► HUD User

HUD Clearinghouses

The U.S. Department of Housing and Urban Development (HUD) sponsors several clearinghouses that disseminate information on housing and community development issues. The mandates of these clearinghouses vary, as does the type and subject matter of the information they provide.

The oldest of HUD's clearinghouses—and the broadest in scope—is HUD USER (P.O. Box 6091, Rockville, MD 20850; 1-800-245-2691), a research information service sponsored by the department's Office of Policy Development and Research (PD&R) since 1978. HUD USER collects, develops, and disseminates research information on key topics as diverse as public and assisted housing, building technology, community and economic development, homelessness, fair housing, local land use regulation, and lead-based paint abatement. The clearinghouse distributes HUD-sponsored research reports, including hundreds of new and out-of-print titles, as well as audiovisual materials and housing-related computer packages. It also makes available the data sets from many of HUD's most important research initiatives, including the nationwide Housing Discrimination Study and a detailed assessment of the physical and financial condition of the HUD-insured multifamily housing stock.

Each year, HUD USER responds to thousands of information requests from researchers, policymakers, individuals, and governments by providing documents, referrals, and searches of its extensive bibliographic database. To increase public awareness of housing research and policy issues, HUD USER publishes a free monthly newsletter, *Recent Research Results,* and the standard reference work, *Directory of Information Resources in Housing and Community Development.* Its other publications include resource guides and briefs that discuss research on a variety of housing issues.

After the mid-1980s, HUD has established several other information clearinghouses designed to support particular Federal programs or policy priorities. These included the following:

- ▶ *Fair Housing Information Clearinghouse* (P.O. Box 6091, Rockville, MD 20850; 1-800-434-3442), sponsored by HUD's Office of Fair Housing and Equal Opportunity (FHEO), supports the department's fair housing mission by disseminating reports, regulations, brochures, and public service announcements on topics such as fair housing enforcement techniques, accessibility issues, mortgage and insurance discrimination, and fair housing education.
- ▶ HUD's *Regulatory Reform for Affordable Housing Information Center* (P.O. Box 6091, Rockville, MD 20850; 1-800-366-4629) was established in 1992 following the widely publicized report of the Advisory Commission on Regulatory Barriers to Affordable Housing convened by Jack Kemp, President Bush's HUD secretary. This clearinghouse offers reference, referral, and technical assistance services to state and local governments, the housing industry, planners, and advocacy groups. It provides information on exemplary state and local regulatory reform initiatives in areas such as inclusionary zoning, growth management, manufactured housing, streamlining the local development permit process, and other relevant topics.
- ▶ Two clearinghouses established by HUD's Office of Resident Initiatives serve the information and technical assistance needs of public housing residents and staff. The *Drug Information & Strategy Clearinghouse* (P.O. Box 6424, Rockville, MD 20850; 1-800-578-3472) provides information on drug abuse prevention, intervention, treatment, and education as well as on community policing, youth programs, and other topics. It also offers assistance to applicants for HUD's Public Housing Drug Elimination Program grants and

referrals to technical assistance providers. The *Resident Initiatives Clearinghouse* (P.O. Box 6424, Rockville, MD 20850; 1-800-245-2691), an outgrowth of then-Secretary Kemp's commitment to "empowering" public housing residents, disseminates information, referrals, and publications to individuals and organizations interested in resident management, economic development, supportive services, and homeownership for residents of public and Indian housing.

Most recently, other HUD offices have developed clearinghouses intended to improve communication with local participants in their programs, providing information such as program regulations, notices, announcements, and case studies:

- ▶ *American Communities* (P.O. Box 7189, Gaithersburg, MD 20898-7189; 1-800-998-9999) is the information center of HUD's Office of Community Planning and Development (CPD), which administers many of HUD's most visible and popular grant programs. It offers information, referrals, and documents on programs such as Community Development Block Grants (CDBG), HOME, HUD's homelessness and special needs assistance programs, Empowerment Zones/Enterprise Communities, and other CPD activities.
- ▶ *Multifamily Housing Clearinghouse* (P.O. Box 6424, Rockville, MD 20850; 1-800-955-2232), recently established by the Office of Multifamily Housing, provides informational and resource support related to various aspects of HUD-insured or HUD-assisted multifamily housing, including supportive housing for the elderly (Section 202) and the disabled (Section 811), as well as multifamily development, management, preservation, property disposition, and resident initiatives. (SEE ALSO: *U.S. Housing Market Conditions*)

—*Laurel Phoenix*

I

▶ Igloo

The igloo is one of the house types of the Eskimo people. Eskimo culture developed along the shores of the Baffin Land and the northern parts of Hudson Bay in latitudes 65 to 70 and has been there for more than 2,000 years. The climate is arctic, with extremely cold winters and short cool summers. During the winter, daylight hours are short and winds severe, making it difficult to leave shelter for days. The Eskimo have adapted to these harsh conditions by developing ingenious technology and tools. They have also evolved cognitive capabilities that allow them to draw maps of large territories with accuracy and social behaviors that ensure survival through cooperation and demographic checks and balances.

Eskimo social structure consists of bands of families, typical among hunters and gatherers who live in marginal environments. Technology evolved in response to limitations in locally available materials: snow-ice, skin, bone, and stone. The igloo is the special response of the Eskimo who congregate early in winter in settlements along the shore on the floe ice. Here, seals keep several breathing holes open and visit them during the day. Hunters wait by all of the holes to guarantee a successful hunt. Hence, the rationale for igloo communities consisting of several distantly related extended families.

Winter is the season for social life and ceremonial activities. Several extended families occupy a single camp and hunt over an established territory. They form a cohesive social unit, with several camps strung out along the shores 10 to 30 miles apart. Visits among these villages are frequent. Beyond it, traditional social organization does not function as a political unit. In the spring, when the seals come out, the winter camp breaks up into smaller units and extended families set up tents at the shore. As the ice melts, the kayak takes the place of the dogsled. The families move inland into the tundra, where they hunt caribou and musk ox and fish for salmon in the summer and fall. Summer camps comprise small seal or caribou tents with a ridgepole and a semiconical rear. Some groups occupy permanent stone and earth houses. These are oval or rectangular, three or four yards across with a long, narrow entrance passage.

These houses are excavated into the ground. However, the floor of the main chamber is a foot higher than the passage to prevent draft. The bedding is laid out at the back, and the sides are for cooking and storage for two families who occupy the house. The rafters of whalebone or driftwood are covered with a double layer of seal skins with moss in between.

The igloo follows the plan of the stone house. Large blocks of snow are cut from compacted snow, using an ivory knife, and laid in a spiral, sloping inward to build a dome. The final block is lowered into position from the outside. The cracks are tightly packed with snow. During the winter, the structure becomes more solid as the inner walls melt and freeze. Side ledges and a rear platform are built of snow and covered with moss and skins to prevent excessive melting. Ice or gut skin windows are set above the exit tunnel, which is subdivided into domed or vaulted sections for storage. The exit has a sharp bend to help reduce the inflow of cold air. The main chamber is lined with skins held in position by sinew cords passing through the walls of the dome and held by toggles. A considerable air space is left between the snow roof and the skin ceiling for insulation. Therefore, a temperature of 10° to 20°F above freezing can be maintained without melting the igloo. Where several families are camping together, a large chamber is often built as a meeting place for singing, dancing, and sorcerers' shows. Dwellings are connected to it by galleries. During journeys, the Eskimo also build temporary igloos about two yards in diameter in an hour or so for camping through the night.

How is the life of people under marginal conditions, such as the Eskimo, relevant for contemporary societies? The technological adaptations (in the form of dwelling and tools), social adaptations (in terms of flexibility by temporary fission of coresidential groups acting as production and distribution units), and extensive knowledge of natural processes and terrain required for survival are all examples of sustainability. However, our learning from the Eskimo is limited by a widespread negative view of "subsistence economies." Such judgment arises from a perspective of affluence and a belief that resources are infinite. An alternative perspective would be that our material ends are few and finite and our ways of obtaining them are adequate. Then, there is plenty to learn from the Eskimo: how to build with local materials, how to plan dwelling complexes that encourage cooperation, and how to maintain a minimum ecological footprint. (SEE ALSO: *Cultural Aspects*)

—*Fahriye Hazer Sancar*

Further Reading

Balikci, A. 1968. "The Netsilik Eskimos: Adaptive Processes." In *Man the Hunter,* edited by R. B. Lee and I. DeVore. Chicago: Aldine.

Edwards, Clint. 1996. *The Igloo.* Fredricksburg, VA: Igloo Publishing.

Forde, C. Daryll. 1963. *Habitat, Economy and Society.* New York: E. P. Dutton.

Houston, James A. 1995. *Confessions of an Igloo Dweller.* Boston: Houghton Mifflin.

National Film Board of Canada. 1951. *How to Build an Igloo.* Ottawa: Author.

Pioneers of Alaska. 1988. *Constitution and By-Laws of the Grand Igloo, Pioneers of Alaska and the Constitution for the Subordinate Igloos and Auxiliaries.* Anchorage, AK: Author.

Smutek, Ray. 1975. *Igloo: Building Eskimo Snowhouses.* Renton, WA: Off Belay.

Steltzer, Ulli. 1995. *Building an Igloo.* New York: Henry Holt.

Yue, Charlotte. 1993. *The Igloo.* Orlando, FL: Harcourt Brace Jovanovich. ◀

▶ Immigration and Housing

Sudden massive immigration, including entry of refugees, asylum seekers, and illegals, means expanded demand for housing and a strain on existing supply. Demand may vary by characteristics of immigrants, such as household composition. Many single men use hotels, hostels, or employer-provided units, whereas families need apartments or houses. Those from countries with norms of extended-family living or families having many children need large units. Type of demand is also influenced by immigrants' income, usually meager, but in some cases, it is considerable if they are businessmen or professionals (e.g., Hong Kong investors, Indian engineers, Iranian upper class). The former need cheap rentals, whereas the latter may buy suburban homes, as in the Chinese community in Vancouver, Canada. Ethnic groups inclined to doubling up, such as the Vietnamese, may afford a more standard apartment or house because they may have several wage earners per household. Those sending money home may wish to save on apartment costs, whereas others with families present incur greater housing costs, often paying a high proportion of income for rent.

Occupation may affect housing demand and location outside of an ethnic enclave. Immigrants often live in their business premise (Korean green grocers, Indian motel operators) or at times in company housing. Whether immigrants are first or second generation can also influence location. Spatial mobility often follows socioeconomic mobility. Immigrant status may determine initial housing type and location. Refugees in Germany often started in hostels and barracks, whereas in the United States many Vietnamese were resettled in dispersed locations.

Type of tenure may reflect the stage of the immigrant's housing career. For a single person, a hotel, a rented room,

or employer housing may suffice. When a family comes, housing choice can change to a private apartment or social housing, especially in Europe, and then to homeownership. In the United States, immigrants are seen as the largest untapped potential in the homebuyers market. However, South Asian immigrants in Britain and Moroccans in the Netherlands turned to homeownership early because it offered the only available housing.

Accessibility to certain housing is a major concern, including access to public housing and housing assistance (vouchers, Section 8, rent rebates). By the 1980s a number of immigrants to the United States were using Section 8. In Europe, barriers to social housing were decreasing, especially in hard-to-rent units unpopular with locals. These were often high-rise units or remote housing at urban peripheries in Britain, Sweden, France, the Netherlands, and Germany.

Discrimination and segregation by government and private parties have produced barriers to accessibility. Institutional racism by lenders and realtors and the lack of enforcement of fair housing legislation have been extensively documented for the United States and Britain. Negative resident attitudes toward minorities and racial harassment have also been a focus of research. Other research has focused on the development of ethnic enclaves and the segregation of ethnic groups. Some governments have adopted dispersal policies (Sweden, the Netherlands, U.S. resettlement of Vietnamese refugees). Related attempts concern the specification of unit quotas (percentage of units mandated to be reserved for immigrant minorities) and limits on proportion of immigrants in an area (Berlin). Housing rights of immigrants are often a function of a nation's civil rights, city ordinances, and newcomers' status as guest workers on contract, permanent immigrants, or refugees. Accessibility to public housing in Europe has been related to residence in an urban renewal area, causing displacement of immigrants and creating a need to rehouse them.

More generally, some have expressed concerns about the broader effects of housing immigrants, including the effect on the existing supply of low-cost private and public housing and associated competition between poor indigenous households and newcomers, and the attendant social tensions, violence and political conflict observed in, for example Germany, France, and Sweden. (SEE ALSO: *Farmworker Housing; Segregation*)

—*Elizabeth Huttman*

Further Reading

Alba, Richard D. and John R. Logan. 1993. "Assimilation and Stratification in the Homeownership Patterns of Racial and Ethnic Groups." *International Migration Review* 26(4):1314-341.

Daly, Gerald. 1996. "Migrants and Gate Keepers: The Link between Immigration and Homelessness in Western Europe." *Cities* 13 (1):11-23.

Moghadam, Fathali M., Donald M. Taylor, and Richard N. Lalonde. 1989. "Integration Strategies and Attitudes Toward the Built Environment: A Study of Haitian and Indian Immigrant Women in Montreal." *Canadian Journal of Behavioural Science/Revue Canadienne de Sciences du Comportment* 21(2):160-73.

Myers, Dowell and Seong Woo Lee. 1996. "Immigration Cohorts and Residential Overcrowding in Southern California." *Demography* 33(1):51-65.

Pader, Ellen-J. 1994. "Spatial Relations and Housing Policy: Regulations that Discriminate against Mexican-Origin Households." *Journal of Planning Education and Research* 13(2):119-35.

Ray, Brian K. and Eric Moore. 1991. "Access to Homeownership among Immigrant Groups in Canada." *Canadian Review of Sociology and Anthropology—Revue* 28(1):1-29.

Van Weesep, Jan. 1993. "Housing the 'Guest Workers.'" Pp. 167-94 in *Exploring the Margins of the New Europe,* edited by Costis Hadjimichalis and David Sadler. London: Wiley. ◄

► Impact Fees

Until about the mid-1970s, local, state, and federal government generally provided the capital improvements and public services, known as off-site infrastructure, to serve new housing developments. This infrastructure was provided at no charge to the developer of the subdivision, the builder of the individual housing units, or the owners of the dwellings. Typically, costs of off-site infrastructure were borne by the community through property taxes or funded by federal grants. Recently, however, fiscal pressure at all levels of government and public resistance to increasing property taxes have caused government to require housing providers to pay some of the cost of infrastructure serving new development. One of the proposed solutions to the infrastructure funding problem is *impact fees,* or *development exactions,* as they are alternatively known.

Impact fees are cash assessments levied against builders, usually at the time a building permit is issued, that are used to pay for the off-site infrastructure necessitated by growth and development. Impact fees are variously used to provide capital improvements—such as sewers, water supply and storm drainage facilities, parks, and roads—and public services—such as police and fire protection. Fee structures are usually based on the marginal cost of providing infrastructure to new developments, although, in reality, they are more likely to be based on the average marginal, or incremental, cost of serving units in a new subdivision. Occasionally, liaised fee schedules are encountered that are designed to channel or direct growth. For example, developers opting to develop land closer to established roads will pay a lower highway impact fee than developers electing to develop land farther removed from established roads.

Critics have challenged the validity of impact fees on the basis that they are taxes and, thus, not a legitimate exercise of the police power. Case law, however, has made it clear that impact fees are a permitted use of the police power, provided certain criteria are met. These criteria are often referred to as the test of *rational nexus.* The concept of rational nexus holds that (a) new development must require that existing systems of providing public services be expanded, (b) the fees charged to new users must be no more than what would have been charged in the absence of impact fees, and (c) any fees charged must be used expressly for the purpose for which they were levied. In other words, to be legally valid, impact fees must bear some proportionate relationship to the cost of providing infrastructure to new developments and be spent for the purpose for which they were charged.

Economic issues surrounding the use of impact fees primarily focus on who actually pays the fee: the developer, the homebuyer, or the landowner from whom the land was purchased. The answer to this question can be determined only if the structure of the market where the fee is assessed is known. Specifically, depending on the price elasticity of supply and demand, it can be shown that any or all of the parties listed above will pay some or all of the fee. Empirical research in this area has concluded, however, that in situations in which impact fees are used, the price elasticity of supply and demand is such that the housing supplier is most always able to pass the cost forward to the homebuyer and very little, if any, of the cost is borne by the builder, developer, or the original landowner. These are markets where, typically, demand is quite inelastic (unresponsive to changes in price) and supply is very elastic (responsive to changes in price). Furthermore, it has been shown by housing researchers that developers are able to pass along not only the cost of the impact fee but an additional amount that reflects the increased costs of financing and marketing properties that were constructed subject to impact fees. Related work has shown that impact fees affect not only new housing prices but may affect the prices of existing dwellings as well, to the extent that these are competitive with new housing units.

Impact fees will not be the sole answer to solving public financing problems of infrastructure provision, but certainly they will constitute a portion of the solution. With more cities and counties adopting impact fee ordinances, it is a reasonable assumption that their use will continue to grow. (SEE ALSO: *Density Bonus; Housing Finance; Linkage*)

—*Charles J. Delaney*

Further Reading

Juergensmeyer, J. and R. Blake. 1981. "Impact Fees: An Answer to Local Government Capital Funding Dilemma." *Florida State University Law Review* 9(3):416-45.

Ross, Dennis H. and Scott Ian Thorpe. 1992. "Impact Fees: Practical Guide for Calculation and Implementation." *Journal of Urban Planning and Development* 18(3):106-18.

Singell, L. D. and J. H. Lillydahl. 1990. "An Empirical Examination of the Effect of Impact Fees on the Housing Market." *Land Economics* 66(1):100, 134.

Snyder, T. and M. Stegman. 1986. *Paying for Growth: Using Development Fees to Finance Infrastructure.* Washington, DC: Urban Land Institute.

Weitz, S. 1985. "Who Pays Infrastructure Benefit Charges: The Builder or the Home Buyer?" In *The Changing Structure of Infrastructure Finance,* edited by J. C. Nicholas. Cambridge, MA: Lincoln Institute of Land Policy. ◄

► Implied Warranties of Quality in the Sale of New Homes

New homes, like other products, sometimes contain flaws; for example, a pipe or window may leak. What rights do buyers have against the sellers of defective new homes? In most states, the law provides that sellers guarantee that newly constructed homes will be free of certain defects for some time, and, therefore, buyers who purchase homes containing such defects have rights against the sellers. These warranties provided by law are called *implied warranties*. The law generally does not require sellers to inform buyers about implied warranties, but some sellers have voluntarily chosen to do so, and indeed, some sellers have elected to provide even more than the law requires by telling buyers that they provide additional warranties, called *express warranties*. The law has created implied warranties in sales of new homes in only the last few decades, and consequently, the rules governing them are still developing.

History

Until the 1960s, under the doctrine of caveat emptor, buyers of new houses had virtually no rights against their sellers if the homes turned out to be defective, except in the rare case in which the seller had made false statements to the buyer before the sale occurred. The courts adopting these rules reasoned, first, that buyers could protect themselves against incompetent or dishonest sellers by inspecting the house while it was constructed to verify that construction was proceeding properly and, second, that buyers could withhold payment until errors were corrected. In the 1960s, however, a number of law professors criticized this rule of caveat emptor, observing that the law provided that sellers of goods, even inexpensive goods such as pens and lightbulbs, impliedly warranted that their products were fit for the ordinary purposes for which such products were used. It seemed odd that homes—which are likely to be among the most expensive and important purchases ever made by a consumer—did not also carry implied warranties with them. These professors also observed that sellers were in a better position to ensure that houses were constructed properly than buyers who were less sophisticated about construction or who were not able to observe the house at the time it was being built.

As a result of this new thinking, courts in the 1960s began ruling that sellers of new homes automatically warranted that the house was free of certain defects. For example, when the walls of a new home began to crack four months after the sale, the builder was held liable to the buyer. These warranties have been described in different ways: They have been called warranties of habitability or fitness for intended purpose or workmanlike or skillful construction, among other things. The court decisions have generally been somewhat vague about exactly what is covered by the warranty, and different states have provided different amounts of protection to homebuyers.

More recently, a handful of states have enacted statutes regulating these warranties. Some of these statutes provide significant detail about the nature of the warranty and the seller's obligations.

In addition, a number of builders have chosen to provide express warranties, in which they state the precise scope of the warranty. The National Association of Home Builders has created the Home Owners Warranty Program. The Home Owners Warranty provides a warranty against various defects for periods ranging from 1 year to 10 years.

Builders who wish to offer the Home Owners Warranty must meet certain standards. Under the Home Owners Warranty Program, when a defect manifests, the builder must repair, replace, or pay for repair or replacement of the defect. The program also requires builders to ensure against the possibility that they will not meet their obligations under the warranty.

Why Can't the Market Solve the Problem?

One fundamental issue arising in connection with the implying of warranties is whether the law should intervene at all. The market for new homes is somewhat different from the market for most consumer items. Many consumers retain counsel when negotiating for the purchase of a new home; it can be expected that competent attorneys would bargain for express warranties, and thus, the law need not require sellers to supply implied warranties. However, many consumers do not retain attorneys when purchasing new homes. In addition, it appears that even those consumers who retain lawyers may not be well-advised on warranties. Many lawyers do not become involved in home purchase transactions until too late for negotiations. Others may not seek to negotiate on warranty issues. Thus, society cannot depend on attorneys to protect the interests of homebuyers where warranties are concerned.

Many consumers suffer from limitations that may impair their ability to seek adequate warranty protection when purchasing a new home. Studies show that many shoppers for new homes do little house hunting before settling on a choice. These studies have found that the median number of homes visited by prospective home purchasers ranges from two or three to six or seven. It appears that many homebuyers put little effort into their search and are not well informed about the choices confronting them.

Studies have also shown that consumers suffer from information overload when confronted with too many choices or too complex choices; that is, consumers may feel overwhelmed by information. Consumers may respond to information overload by choosing one of the first houses they visit, thus ending the selection process, or by focusing on only some characteristics and so settling on a house without taking warranties into account.

In addition, it is doubtful that consumers are able to determine accurately the value of a warranty. A warranty is a type of insurance against the risk that a defect will occur. The value of the warranty depends on the likelihood that the defect will be present in a house. Consumers who lack the ability to predict with accuracy the probability that a defect will occur will not possess the ability to determine the value of the warranty. Studies show that consumers are not very skilled at assessing probabilities. Many consumers habitually underestimate the likelihood that a problem will occur and so do not protect themselves against the occurrence of the problem. For example, many consumers do not purchase adequate flood and earthquake insurance, refuse to wear seat belts, and underestimate the number of consumer problems they experience, relative to other consumers. Consumers who engage in these behaviors are also likely to underestimate their need for a warranty against housing defects.

In short, many consumers lack the ability to make a competent judgment about whether or not they need a warranty. Consequently, many states have chosen to require sellers to provide them.

Unsettled Issues

Because the development of implied warranties in new homes is so recent, many issues concerning these warranties are still unsettled in many states. One such issue is whether sellers should be permitted to disclaim implied warranties; that is to say, under what circumstances should a seller be able to avoid providing the warranty supplied by law? At least three states do not permit sellers to disclaim warranty protection. New York, Louisiana, and Indiana all require that builders provide some warranty protection, even if the buyer does not wish it. But few states have adopted that rule. A number of states have permitted sellers to disclaim warranties if the disclaimers are clear, specific, and conspicuous to the buyer so that the buyer can understand that he or she is not receiving warranty protection.

Another issue is whether subsequent purchasers of the house should be able to take legal action against the original builder if the defect appears during the warranty period. Often, people sell houses to others. What rights do later purchasers have on the warranty? Some courts have held that the second or even third purchaser can recover, on the theory that the rationale for creating the warranty in the first place argues for extending it to all purchasers. Other courts have denied protection to subsequent purchasers, reasoning that they did not enjoy a contractual relationship with the builder.

Conclusion

Implied warranties of quality in new homes are an important development in protecting the unstated expectations of consumers who are unable to protect themselves. Although a number of issues covering implied warranties remain to be resolved, courts and legislatures are working to fill in the gaps. (SEE ALSO: **Homeownership**)

—*Jeff Sovern*

Further Reading

McDaniel, Amy L. 1990. "The New York Housing Merchant Warranty Statute: Analysis and Proposals." *Cornell Law Review* 75:754-82.
Meeks, Carol B. and Eleanor H. Oudekerk. 1981. "Home Warranties: An Analysis of an Emerging Development in Consumer Protection." *Journal of Consumer Affairs* 15(2):271-89.
Pridgen, Dee. 1995. *Consumer Protection and the Law.* Deerfield, IL: Clark Boardman Callaghan (see Chapter 18).
Roberts, Ernest F. 1967. "The Case of the Unwary Home Buyer: The Housing Merchant Did It." *Cornell Law Review* 52:835-70.
Sovern, Jeff. 1993. "Toward a Theory of Warranties in Sales of New Homes: Housing the Implied Warranty Advocates, Law and Economics Mavens, and Consumer Psychologists under One Roof." *Wisconsin Law Review* 1:13-103. ◄

standards of fairness and justice. One of the areas of concern is the tax status of the housing services available to homeowners and tenants. Under the U.S. federal income tax code, the value of housing services consumed by homeowners is not subject to tax liability. On the other hand, the payment of rent by the tenant to the owner is taxable as income. The market value of the housing services consumed by homeowners that avoids federal taxation is called the *imputed rental income.*

Imputed rental income is often estimated by comparing what the contract rent for the housing services would be in the private rental sector. It is important to hold constant the rights of homeowners and tenants when making such comparisons. Rents are said to be *imputed* because they are estimated based on evidence of what would be rental income if rents were paid.

As indicated above, the U.S. income tax code does not treat homeowners and tenants identically on the basis of taxation of housing services. Although this is not the only instance of preferential treatment provided to homeowners (i.e., others include the deductibility of property taxes, mortgage interest deductions, special once-a-lifetime exemptions from capital gains taxes—all in support of homeownership), immunity of imputed rental income to federal (and state) income tax liability is a major element in the U.S. tax system. Estimates on the magnitude of this policy choice indicate that the benefits from this policy alone exceed all of the other preferential policies for homeowners in the United States.

Some countries tax imputed rental income to provide a more even balance between homeowners and tenants. There are problems with choosing this policy, such as measuring the size of the benefits in the absence of reportable evidence. Also, if imputed rental income is to be taxable, it is important to keep the benefits comparable for similar households. Nonetheless, most public finance economists agree that income taxes should be as broad based as possible; failing to tax imputed rent moves away from this theoretical ideal. (SEE ALSO: *Taxation of Owner-Occupied Housing*)

—*Austin J. Jaffe*

Further Reading

Blankenship, Frank J. 1989. *The Prentice Hall Real Estate Investor's Encyclopedia.* Englewood Cliffs, NJ: Prentice Hall.
Hein, Scott E. and James C. Lamb, Jr. 1981. "Why the Median-Priced Home Costs So Much." Federal Reserve Bank of St. Louis. *Review* 63(June/July):11-19.
Jacobus, Charles J. 1996. *Real Estate Principles.* Upper Saddle River, NJ: Prentice Hall.
McKenzie, Dennis J. 1996. *Essentials of Estate Economics.* Upper Saddle River, NJ: Prentice Hall.
Peiser, Richard B. 1992. *Professional Real Estate Development: The ULI Guide to the Business.* Washington, DC: Urban Land Institute,
Shim, Jae K. 1996. *Dictionary of Real Estate.* New York: John Wiley.
Tosh, Dennis S. 1990. *Handbook of Real Estate Terms.* Englewood Cliffs, NJ: Prentice Hall. ◄

► Imputed Rental Income

When comparing tenures in housing markets, tax authorities are faced with the difficult challenge of maintaining

► In Rem Housing

Housing abandonment became a dramatic phenomenon in many U.S. cities during the late 1960s and early 1970s. *In rem housing,* a term most frequently used in New York City, constitutes one form of landlord abandonment and connotes transfer of ownership to city government in lieu of tax payments. Other forms of abandonment include cessation of provision of services and mortgage foreclosure, both of which may accompany nonpayment of taxes. Many buildings that end up in rem have experienced a long period of disinvestment and cutbacks in services prior to being taken. Defaults on federal- and state-sponsored mortgage guarantees as well as takeovers of poorly managed public housing authorities have resulted in a large analogous class of publicly owned housing.

In New York City, the increased visibility of in rem housing accompanied large out-migration of middle- and upper-class white populations to the suburbs and other parts of the United States, leaving poorer, frequently minority, residents as the prime consumers of rental housing. The city continued to auction buildings to the highest bidder only to see the majority return quickly to city ownership. Between 1965 and 1968, 38,000 units of housing were lost. In an effort to force earlier payment of tax arrears and to prevent further deterioration and displacement prior to city ownership, Local Law 45 was passed in 1976 to allow the city to vest property after one year of tax delinquency instead of three. This ruling resulted in city ownership of an even larger proportion of housing than previously. During the late 1970s, abandonment in New York City was estimated to lead to the loss of more than 31,000 units a year. Although large numbers of these buildings created landscapes of complete abandonment reminiscent of cities bombed in wartime, at any given time, up to 50,000 of the units were occupied.

The policy focus has been on returning the in rem housing to private ownership by community organizations, tenant-owned limited-equity cooperatives, or new private landlords through a variety of programs. These programs have included subsidies for renovation, primarily from New York City's share of federal Community Development Block Grants, and direction of federal rent subsidies to qualified households in buildings sold through these programs. Evaluations have documented the critical role of this housing stock in sheltering the poorest New Yorkers, whose incomes average only about half that of public housing residents and a third or less of the median household income. Limited-equity tenant co-ops have demonstrated a strong, replicable ability to provide good, largely affordable, housing conditions and high levels of security even in very deteriorated neighborhoods. None of the privatization programs have been able to house as many unsubsidized, very low income residents. Vacant in rem buildings have become the primary source of new low-income housing, redeveloped, often with tax credit financing or by public-private partnerships, as new housing for low- and moderate-income households. The current and former in rem stock has been shown to both generate the most incidences of homelessness and provide one of the main resources for rehousing homeless households.

Pressure on city finances has led to repeated efforts to divest itself of this stock. These efforts have failed because of the steady flow of new in rem buildings, as well as inability to dispose of all existing holdings. In the mid 1990s, a new Republican administration halted vesting of new property and developed new programs that would bring private and tax credit financing to disposition of the occupied stock. Threats to the federal portion of renovation and operating subsidies raise questions about the viability of the new approach and the future of buildings sold through previous programs that relied on federal rent subsidies. Vacant buildings have come to make up a smaller proportion of in rem properties, reducing the potential supply of new units available for low-income populations and increasing the threat of displacement and higher levels of homelessness among residents of occupied in rem buildings. (SEE ALSO: **Abandonment; Urban Redevelopment**)

—*Susan Saegert*

Further Reading

Bratt, R. G., C. Hartman, and A. Meyerson, eds. 1986. *Critical Perspectives on Housing.* Philadelphia: Temple University Press.

Leavitt, J. and S. Saegert. 1990. *From Abandonment to Hope: Community Households in Harlem.* New York: Columbia University Press.

Stone, M. E. 1993. *Shelter Poverty: New Ideas on Housing Affordability.* Philadelphia: Temple University Press. ◄

► In Situ Construction

Housing built on-site is sometimes called *in situ house construction.* This is the dominant method of residential construction, including multifamily housing. This approach refers to building in which a greater proportion of work on the final house assembly is accomplished on-site compared with work done off-site. Since the light wood frame was first used in the 1830s, a gradually increasing percentage of value added to elements used in housebuilding has migrated off-site. Today, a site-built house may have prefabricated floor and roof trusses, windows, fireplaces, and kitchen and bathroom elements. Although foundations remain largely site operations, new products are being introduced to speed foundation work. Developments in mass-produced items, such as new exterior and interior finishes, structural elements, glazing, prehung doors, fasteners and sealants, and heating, ventilation, and air-conditioning systems, have also helped to increase the efficiency of site building.

There are basically two classes of site builders. One is "production homebuilders," large organizations that may also have modular and panelized division and may operate in multistate or national markets. The other is "small-volume homebuilders," who build, on average, less than 50 houses each year and who focus on local markets. A large number of single-family houses are built by contractors, sometimes called "pick-up-truck" builders, who build one house at a time on a speculative basis, operating from their home offices with very low overhead and no permanent construction crew.

Both types of site builder have adopted rationalization procedures to increase their efficiency in competition with builders using the output of modular and panelized producers. This has included improvements in scheduling and management, arrangements with suppliers, computer-assisted design and accounting, and the adoption of new tools and materials that speed work, reduce the labor component, and improve quality.

Also part of in situ building is remodeling activity, which accounted for more than $112 billion in expenditures in 1996. This activity includes both rented and owned units, single-family and multifamily units, and construction improvements and upkeep.

In situ residential construction volume is difficult to quantify because it uses both prefabricated and mass-produced elements and may thus be counted as panelized housing or "industrialized housing" in some documentation. (SEE ALSO: **Industrialization in Housing Construction**)

—*Stephen Kendall*

Further Reading

Berg, Rudy, G. Z. Brown, and Ron Kellet. 1990. *An Analysis of U.S. Industrialized Housing.* Eugene: University of Oregon, Center for Housing Innovation.

Dietz, Albert G. H. 1979. *Dwelling House Construction.* 4th ed. Cambridge: MIT Press.

Eichler, Ned. 1982. *The Merchant Builders.* Cambridge: MIT Press. ◀

▶ Inclusionary Zoning

Inclusionary zoning is a body of zoning techniques designed to encourage the development of affordable housing for lower-income groups. Courts and legislatures have created or imposed these techniques to correct the social and economic inequities of exclusionary zoning.

Exclusionary zoning, by contrast, is a body of zoning techniques that have the effect of driving up the cost of housing and that therefore exclude lower-income households from a community. For example, by restricting the development of multifamily housing or of rental units with large numbers of bedrooms, communities limit the supply of housing on which lower-income households must rely. The result is fewer and costlier housing opportunities for households of limited means.

Since the beginning of the 1970s, exclusionary zoning has come under increasing attack. Where the courts find that local zoning improperly excludes housing for lower-income groups, they may impose a "builder's remedy": a court order directing local authorities to issue a building permit for an excluded housing project or development. But courts and legislatures have also developed a body of inclusionary zoning techniques that go beyond the facts of particular disputes about particular projects or developments and reshape a community's zoning program. In some instances, courts have ordered local authorities to adopt such measures. In other instances, state legislatures have superimposed these inclusionary zoning measures on local zoning ordinances. But in all instances, these inclusionary zoning techniques are remedial measures meant to correct the effects or reverse a history of exclusionary zoning.

New Jersey courts have led the way in the attack on exclusionary zoning and in the development of inclusionary zoning remedies. The New Jersey Supreme Court in *Southern Burlington County NAACP v. Township of Mt. Laurel* (1975) (*Mt. Laurel I* N.J. 36 F. 2d 713) declared that every "developing" municipality in the state must provide a "realistic opportunity for the construction of its fair share of the present and prospective regional need for low and moderate income housing." Eight years later, the New Jersey Supreme Court revisited its (now celebrated) "fair share" doctrine, in *Southern Burlington County NAACP v. Township of Mt. Laurel* (1983 N.J. 456 F. 2d 390) (*Mt. Laurel II*). Reviewing intervening "Mt. Laurel" litigation in other communities as well as Mt. Laurel's own efforts to rewrite its zoning ordinance, the court in *Mt. Laurel II* restated and refined the fair share doctrine, imposed the fair share responsibility on all communities (not just "developing" communities) and enumerated the inclusionary zoning techniques that communities may adopt to provide a "realistic opportunity" for low- and moderate-income housing that *Mt. Laurel I* required.

According to *Mt. Laurel II*, removing exclusionary zoning devices from their zoning ordinances is "the very least" that communities must do. But unless the removal of such regulatory barriers to affordable housing makes it "realistically possible for lower income housing to be built," communities must also adopt "affirmative measures" to encourage the building of such housing. Among the "affirmative measures" enumerated by the court were four inclusionary zoning techniques.

The first is incentive zoning. Under incentive zoning, developers earn a density bonus if they construct lower-income housing. The bonus may be fixed so that a developer earns a predetermined density bonus for participating in a lower-income housing program. Alternatively, the bonus may be on a sliding scale so that the permitted density rises with the number of lower-income housing units a developer provides. Relying on an empirical study of incentive zoning, the court in *Mt. Laurel II* observed that developers are reluctant to take advantage of incentive zoning. The court concluded that an effective inclusionary zoning program cannot rely on incentive zoning alone. But the density bonus remains an attractive inclusionary zoning technique. Some states, such as California, have adopted legislation mandating a density bonus if a developer provides affordable housing.

The second inclusionary zoning technique described by the court in *Mt. Laurel II* is the mandatory set-aside. Typically, mandatory set-aside programs work by requiring that developers agree to dedicate a certain percentage of their developments to low- or moderate-income housing before getting the site plan approvals or rezonings they require. The court in *Mt. Laurel II* cautioned local authorities not to let developers defer construction of low- and moderate-income units until the final phase of their projects. The court explained that developers could easily renege on their commitment to build those units, if their construction was deferred to the last phase of a project. According to the court in *Mt. Laurel II*, if a developer agrees to set aside 20% of a development for lower-income housing, that developer should set aside 20% of each phase or stage of the development for such housing "to the extent this is practical."

The mandatory set-aside is a widely adopted inclusionary zoning technique. Many communities outside New Jersey have enacted mandatory set-aside ordinances.

The third inclusionary zoning technique recommended by the court in *Mt. Laurel II* is zoning for mobile homes. The court reasoned, "As the cost of ordinary housing skyrockets for purchasers and renters, mobile homes become increasingly important as a source of low cost housing." Prohibitions on the exclusion of mobile homes or manufactured housing or requirements that communities afford opportunities for such housing are now common inclusionary zoning techniques.

Fourth, and finally, the court in *Mt. Laurel II* recommended what it called "least cost" housing, but only if communities cannot meet their "fair share obligation" by using any other inclusionary zoning technique. "Least cost" housing, as the court explained it, is not lower-income housing but, rather, middle-income housing built at the least cost possible (and according to the court's "stringent" guidelines). Such housing, the court theorized, is most likely to "filter down" to the poor as housing is recycled among the classes of society.

In 1985, New Jersey enacted the New Jersey Fair Housing Act. The act provoked controversy because it transferred initial control over suburban zoning and housing from the judiciary to an administrative Council on Affordable Housing. But the act was consistent with the *Mt. Laurel* decisions in requiring that all New Jersey communities (a) adopt master plans before they adopt or amend zoning ordinances, (b) include a "housing plan element" in their master plans, and (c) provide for low- and moderate-income housing in their housing plan elements, consistent with their fair share obligations.

Several other states have also enacted legislation requiring communities to adopt comprehensive plans that include housing elements or that at least include some provision for lower-income housing. Examples include Arizona, California, and Oregon. Comprehensive planning is mandatory in some of these states.

But legislative action to encourage or mandate inclusionary zoning practices takes other forms as well. Several states have adopted legislation requiring communities to zone for manufactured housing, including mobile homes. In New Hampshire, the legislature requires all municipalities to "afford reasonable opportunity for the siting of manufactured housing" and prohibits the complete exclusion of such housing from any community.

Some New England states have adopted legislation that authorizes a state agency or a court to overrule local zoning decisions that exclude affordable housing. In Connecticut, for example, a court appeal is authorized if local zoning restrictions have a substantial impact on housing affordability. The municipality has the burden in court to justify its decision with health, safety, and other legal matters that outweigh the need for affordable housing.

It is not certain that inclusionary zoning techniques will always lead to the provision of affordable housing. A study of mandatory set-aside programs in New Jersey in the five years after *Mount Laurel II* suggests that multifamily housing projects with a percentage of units let at below-market rates may not be profitable. The same study reveals a strong developer preference for ownership units over the rental units customarily sought by lower-income households. This experience suggests that inclusionary zoning techniques alone will not solve the problem of affordable housing.

However, the court in *Mt. Laurel II* acknowledged that inclusionary zoning techniques were not the only "affirmative measures" that communities might have to take to provide a "realistic opportunity for low and moderate income housing." According to the court there are "two basic types of affirmative measures." One is inclusionary zoning. But the other is encouraging or requiring the use of housing subsidies offered by state and federal governments. (SEE ALSO: **Discrimination**; *Exclusionary Zoning; Zoning*)

—*Daniel R. Mandelker and Harold A. Ellis*

Further Reading
Bergman, E. M. 1974. *Eliminating Exclusionary Zoning: Reconciling Workplace and Residence in Suburban Areas.* Cambridge, MA: Ballinger.

Mandelker, D. R. 1993. *Land Use Law.* 3d ed. Charlottesville, VA: Michie.

Mandelker, D. R., R. A. Cunningham, and J. M. Payne. 1995. *Planning and Control of Land Development: Cases and Materials.* 4th ed. Charlottesville, VA: Michie.

Merriam, D., D. J. Brower, and P. D. Tegeler, eds. 1985. *Inclusionary Zoning Moves Downtown.* Washington, DC, and Chicago: Planners Press, American Planning Association.

Salsich, P. 1991. *Planning, Zoning, Subdivision Regulation, and Environmental Control.* Colorado Springs, CO: Shepard's/McGraw-Hill.

Steinberg, M. 1989. *Adaptations to an Activist Court Ruling: Aftermath of the Mt. Laurel II Decision for Lower Income Housing.* Cambridge, MA: Lincoln Institute of Land Policy.

Tucker, W. 1991. *Zoning, Rent Control, and Affordable Housing.* Washington, DC: Cato Institute.

White, S. Mark. 1992. *Affordable Housing: Proactive and Reactive Planning Strategies.* Planning Advisory Service Report No. 441. Washington, DC, and Chicago: American Planning Association. ◄

► Incumbent Upgrading

The terms *community conservation, neighborhood preservation,* and *incumbent upgrading* refer to the same phenomenon: efforts to improve the physical conditions of inner-city areas with the existing population remaining in place. Interest in incumbent upgrading dates back to the 1970s when growing numbers of community activists, scholars, and policymakers advocated for the conservation of the urban housing stock and the preservation of urban neighborhoods. This interest grew out of recognition of the economic value of the central-city housing stock, an awareness of the need to maintain the viability of the central city, and recognition of the psychological attachments of existing residents to older areas.

It is important to distinguish incumbent upgrading from gentrification, a second type of neighborhood revitalization. Gentrification is a form of neighborhood improvement associated with the replacement of a low-income population by a more affluent one. Gentrifying and incumbent-upgrading communities are usually easy to distinguish from one another. Gentrifying neighborhoods are more likely to be located close to downtown and tend to be located closer to museums, parks, and other marketable amenities. In contrast, incumbent-upgrading neighborhoods tend to be located some distance from downtown and lack architecturally distinctive housing and other amenities. Local government plays a far more significant role in incumbent upgrading of neighborhoods.

Urban renewal was the federal government's main response to urban problems from the end of World War II through the early 1960s. During the 1960s, federally assisted code enforcement programs were a major force in shifting the urban renewal program toward the conservation and rehabilitation of neighborhoods. This paved the way for the Community Development Block Grant (CDBG) program of the 1974 Community Development Act under which local communities received federal grants to carry out community development activities. In the first five years under the CDBG program, funds spent for conservation of the housing stock rose from 16% to 31% of the total, second only to elimination of slums and blight.

During the 1970s, incumbent-upgrading programs such as Neighborhood Housing Service (NHS) and the Urban Homesteading Demonstration (both national efforts), broadened their scope to deal with neighborhood problems as well as with substandard housing conditions. The NHS model is perhaps the best known program developed during this period. The NHS model was originally created in Pittsburgh as a result of efforts by community activists and lenders. The Urban Reinvestment Task Force (composed of representatives of HUD, the Federal Home Loan Bank Board, the Federal Reserve System, the Federal Deposit Insurance Corporation, and the Comptroller of the Currency) has since initiated NHS programs throughout the United States. Key components of NHS include (a) a high degree of resident involvement in the operation of the program, (b) local government participation through capital improvements and code enforcement programs, (c) the agreement of financial institutions to reinvest in the community by making market rate loans to qualified buyers and through contributions to the NHS to support operating costs, and (d) a high-risk loan fund to families who cannot meet usual credit risk standards.

The objective of neighborhood incumbent-upgrading efforts is to alter market conditions so that the private market will continue to finance housing purchase and rehabilitation in the neighborhood without subsidy dollars after the specific program (such as NHS) is ended. For these programs to be successful within a limited time period, they must produce neighborhood spillover effects. That is, visible housing improvements (resulting from governmental loans and grants to program participants) should make nonparticipating neighbors more confident about the future of the area, thereby making them more likely to stay and invest in housing improvements.

Stimulating such spillover effects and promoting greater confidence is no easy matter. The assumption by many economists that the aging of the housing stock is the key factor in explaining neighborhood decline is clearly unrealistic and has led to an overemphasis on cosmetic housing improvements as the main thrust for neighborhood upgrading efforts. Some neighborhoods have experienced more rapid physical and social decline—measured by housing abandonment, property value declines, and increases in crime—than would be expected in terms of the age of the housing alone. In many of these areas, decline has racial and economic aspects. To be successful, neighborhood upgrading efforts must also account for demographic change.

Scholars have proposed a number of approaches to promote neighborhood upgrading but have tended to ignore the need for demographic stabilization, thereby providing an overly rosy picture concerning the prospects for upgrading. Anthony Downs proposes physical programs such as code enforcement and street improvements for areas in the early stages of decline; Rolf Goetze recommends government efforts to influence the public image of marginal communities through posters, "positive" news stories, and television programs; finally, Roger Ahlbrandt and James Cunningham recommend city-funded community organizing efforts to improve the social fabric of urban neighborhoods.

Evaluations of neighborhood upgrading programs have added to the overly optimistic assessments of the prospects of these efforts. Assessments of these programs have generally been positive, but the methodologies tend to be seriously flawed. Most of the studies rely on aggregated data from the census or other sources, whereas survey analysis at the individual household level is needed to test conclusively for the existence of spillover effects.

A 1982 study of the South Shore Bank, Chicago, a widely publicized program that provided the basis for President Clinton's proposal for community banks, illustrates the generally weak evaluations. The bank has sought to provide increased funds for creditworthy single-family and multi-family building owners and has also attempted to be a catalyst for neighborhood revitalization activities. Study results based on aggregated data seemed to indicate success. Property values, which had declined, increased again during the late 1970s. The increases were greatest in the areas with the largest number of loans. The study failed, however, to identify spillover effects. It is quite possible that most of the improvement in multifamily buildings occurred as a result of the South Shore bank loans and that few other building owners improved their properties. Furthermore, although crime was a serious issue in the community, it was virtually ignored in the report.

A 1986 secondary analysis of the Urban Homesteading Demonstration (UHD) surveys data set is one of the handful of evaluations to focus on program impacts at the household level. The federal program permitted HUD to transfer to cities one- and two-family homes, which were sold to families for a dollar under the condition that buyers rehabilitate their homes and remain in them for at least three years. The program was expected not only to help homesteaders by providing them with good housing values but also to stimulate improvement in the surrounding neighborhoods. There was little evidence that such improvement occurred. Householders living closest to homesteading properties were no more likely to remain at their locations or to invest in improvements than were others when background characteristics were controlled. It was impossible to detect spill-overs because the density of homesteading properties was too low (usually about 1% of the housing stock) to affect investment or mobility behavior in the larger area. In addition, many of the UHD neighborhoods were also experiencing significant racial and income changes at the same time. The demonstration project was too narrowly focused on housing and other physical improvements to achieve demographic stabilization and to ensure that the housing improvements would be maintained over time.

The prospects for local neighborhood revitalization programs such as UHD and NHS vary considerably by community type. The outlook is poor in unstable communities undergoing rapid change in social class, income, and racial composition and characterized by crime and other social maladies associated with high population turnover. The prospects for achieving successful upgrading are much better in stable communities because the program can focus on the more attainable goal of reducing physical deterioration without also having to deal with poverty, discrimination, and the more intractable problems associated with demographic stabilization.

In summary, the neighborhood upgrading programs implemented since the 1970s have done much good: providing lower- and moderate-income families with good housing values, helping to conserve the older housing stock, bringing bank loans into areas redlined in the past, and empowering residents through programs such NHS. However, even the best of these programs are ill equipped to deal with demographic stabilization. (SEE ALSO: *Gentrification; Historic Preservation;* **Urban Redevelopment**)

—*David P. Varady*

Further Reading

Ahlbrandt, Roger S. Jr. and James V. Cunningham. 1979. *A New Public Policy for Neighborhood Preservation.* New York: Praeger.

Byrum, Oliver E. 1992. *Old Problems in New Times: Urban Strategies for the 1990s.* Chicago: Planners Press.

Clay, Phillip L. and Robert M. Hollister. 1983. *Neighborhood Policy and Planning.* Lexington, MA: Lexington Books.

Downs, Anthony. 1982. *Neighborhoods and Urban Development.* Washington, DC: Brookings Institution.

Goetze, Rolf. 1976. *Building Neighborhood Confidence: A Humanistic Strategy for Urban Housing.* Cambridge, MA: Ballinger.

Nenno, Mary and Paul Brophy. 1982. *Housing and Local Government.* Washington, DC: International City Managers Association.

Schoenberg, Sandra Perlman and Patricia Rosenbaum. 1980. *Neighborhoods That Work: Sources for Viability in the Inner City.* New Brunswick, NJ: Rutgers University Press.

Varady, David P. 1986. *Neighborhood Upgrading: A Realistic Assessment.* Albany: State University of New York Press. ◄

► Index of Isolation

The Index of Isolation measures the degree of spatial isolation between any two nonoverlapping groups (say between Koreans and Hispanics or between college graduates and those with only a high school degree or between blacks and the remainder of the population). These indices are normally applied to urban areas and were originally designed for racial and ethnic groups. However, they can also be applied in other contexts as, for example, segregation in a school system or gender segregation by occupation. The index is easily computed and ranges from 0 to 1.0 in a manner that has a straightforward operational interpretation.

The Index of Isolation (II) was proposed by Shevky and Williams in 1949 and later modified by Bell in 1954. Bell's modification is the quotient of two measures: "The numerator is . . . the probability P that the next person a random individual from group 1 will meet is also from group 1. The denominator is the hypothetical probability . . . assuming group 1 is homogeneously mixed in all the census tracts of the city" (pp. 357-58). In other words, the actual level of isolation was "standardized" by dividing by the composition of the entire city. (This revised index is identical with the square of a well-known statistic, *eta.*) Lieberson dropped the denominator of the Bell index and reinterpreted the numerator to provide a straightforward summary index of the actual isolation of the group(s). If i and j refer to any two different ethnic and racial groups with no overlap in membership, then $_iP^*_j$ gives the isolation of members of Group i from members of Group j; $_iP^*_i$ indicates the isolation of members of Group i from all others in the population. These measures take into account the spatial dissimilarity of Group i from either Group j or from the entire non-i population *and* the ethnic and racial composition of the population. By comparison, the widely used measure of segregation described by Duncan and Duncan, *ID,* controls for composition by comparing the spatial distribution of two or more groups in terms of their percentage distributions across spatial units. Computation and interpretation of the P*-type measures is straightforward (see Lieberson 1980, pp. 253-257; Lieberson 1981, pp. 67-70).

It is important to recognize that P* indices of isolation are not substitutes for other measures of segregation. Rather, each index is responsive to different dimensions and appropriate for particular applications—indeed, in some cases, several measures may be necessary (e.g., see the discussion of hypersegregation by Massey and Denton). These authors examine isolation, unevenness or settlement, clustering of segregated neighborhoods, and the proximity of segregated neighborhoods to the urban core; a high score on four of the five measures is defined as hypersegregation. In evaluating the utility of P*, its asymmetry is a particularly important feature, reflecting the fact that the population composition of the city is taken into account. As a consequence, the isolation between two groups, i and j, is experienced differently by the two groups because it is affected by the relative numbers of the two populations. This also means that changes in composition over time will have consequences for the relative level of potential interaction between the groups. Thus, for example, a growth in the black component of a city will alter the frequency of contact among blacks, among whites, and between whites and blacks—even if the pure pattern of segregation, as measured by *ID,* remains constant. On the other hand, the inclusion of population composition in the computation of P* means that the value of the isolation index can be largely a function of composition rather than any differences in spatial patterns of the two groups. Of course, often, this will reflect a reality that is pertinent to a specific problem under consideration. The value of P* in excess of that owing to composition can be readily determined, as was the goal in the proposal by Bell. Also, there are formulas for determining the minimum and maximum P*, given the level of *ID* index and population composition (see Lieberson 1981, pp. 76-79). (SEE ALSO: *Segregation; Segregation Index*)

—*Stanley Lieberson*

Further Reading

Bell, Wendell. 1954. "A Probability Model of the Measurement of Ecological Segregation." *Social Forces* 32:357-64.

Duncan, Otis Dudley and Beverly Duncan. 1955a. "A Methodological Analysis of Segregation Indexes." *American Sociological Review* 20:210-17.

———. 1955b. "Residential Distribution and Occupational Stratification." *American Journal of Sociology* 60:493-503.

Lieberson, Stanley. 1980. *A Piece of the Pie: Blacks and White Immigrants since 1880.* Berkeley: University of California Press (see Chapter 9).

———. 1981. "An Asymmetrical Approach to Segregation." Pp. 61-82 in *Ethnic Segregation in Cities,* edited by Ceri Peach, Vaughan Robinson, and Susan Smith. London: Croom Helm.

Lieberson, Stanley and Donna Carter. 1982a. "A Model for Inferring the Voluntary and Involuntary Causes of Residential Segregation." *Demography* 19:511-52.

———. 1982b. "Temporal Changes and Urban Differences in Residential Segregation: A Reconsideration." *American Journal of Sociology* 88:296-310.

Massey, Douglas S. and Nancy A. Denton. 1989. "Hypersegregation in U.S. Metropolitan Areas: Black and Hispanic Segregation along Five Dimensions." *Demography* 26:373-93.

———. 1993. *American Apartheid: Segregation and the Making of the Underclass.* Cambridge, MA: Harvard University Press.

Shevky, Eshref and Marilyn Williams. 1949. *The Social Areas of Los Angeles.* Berkeley: University of California Press. ◀

▶ Industrialization in Housing Construction

Industrial production is the manipulation of materials to make many of a given product or part, finished and ready to use. The industrial revolution of the late 18th and early 19th centuries was significant because, for the first time, machines and organizational methods were harnessed by those who wanted to produce great quantities of articles and, by using standard methods, reduce cost and improve quality. Industrial production has been remarkably successful in making uniform products. However, with computational support, manufacturers can now customize production to a much greater extent. Industrial producers can, therefore, now make available a great variety of high-quality products to be used in the construction of individual buildings. House construction begins where industrial production stops.

Today, industrialized housing refers to the production of large building elements made in factories, such as panels, modular units, or trusses, using sophisticated machines and equipment to reduce labor, control waste, and improve quality. These products, being large, expensive, and complex, are generally made by companies that specialize in their production and are specified for a particular site or purchaser. That is, they are ordered by a party who will use them to make a house on a site. The products are then transported to a building site and brought together with other products, the site, and utilities and, in the context of community regulations and traditions, made into a building or house. Sometimes this kind of production is called *prefabrication*.

Industrial production more generally refers to the manufacture of uniform finished products that are of general use, available for use on any site, and made, unlike modular units or roof trusses, at the initiative of the manufacturer. These include products such as plumbing fixtures, aluminum profiles, sheet materials of all kinds, dimension lumber such as "2 × 4s," packaged heating and air-conditioning systems, cabling and other electrical apparatus, piping, and other general products. These products are the mainstay of house building, each seeing many steps of specification and manipulation on its way to a final assembly in house construction.

The power of modern industrial means lies in the ability to produce large numbers of an object faster and with more uniformity than was previously possible. Industrial mass production allows lower unit prices, and assembly line production techniques allow for maximum efficiency, uniformity, and accuracy of physical characteristics, such as dimensions, surface finishes, and weight, with reduced dependency on human labor. Thus, when a party identifies a market for large numbers of identical products, the powers and principles of industrial mass production are essential to success.

This process has demonstrated its effectiveness in making available a vast range of commodity and consumer products, including a very large and growing repertoire of building products. Although the dominant house-building system, the light wood frame, remains a constant, no part of it has remained unchanged. This is independent of whether it is used in on-site or in modular, panelized, or other production methods, by large corporate production builders or small "pick-up-truck" homebuilders. Each part of the system is mass produced in increasingly sophisticated plants and distributed in worldwide supply chains.

From early in the 20th century, interest in the United States and elsewhere in applying industrial production to solve housing problems has remained strong. In the United States, there was a decided acceleration of federal and private sector interest and investment in this effort between 1930 and 1970. The increased interest in applying industrial production methods to housing coincided with increased confidence in the application of scientific management to broader social problems.

The general goal was the large-scale production of houses in factory settings. Experience with automobile mass production was often used as the argument for mass producing houses. This belief naturally gave rise to the search for the "ideal plan" or "ideal house types," although modern mass production and building technology by themselves do not demand uniformity and repetition in floor plans or building design. Rather, uniform or standard plans are the result of hierarchical, centralized decision making rather than of user participation.

Houses are much more complex, both technically and culturally, than other mass-produced consumer products. They also are subject to local community decisions and regulations, unlike other objects derived from industrial mass production. Thus, the assumption that mass producing housing would be similar to mass producing other products is false. Companies that tried to introduce new housing systems suited to factory mass production met with failure.

In contrast, the factory production of mobile homes, using a combination of wood and steel or aluminum framing and other light weight materials and techniques, is similar to the production of modular houses: large, complex assemblies are made in a factory, taken to a building site, and constructed onto a foundation and attached to public services. Those that succeeded in moving house production into factories did so by using conventional technology and organizational arrangements and by moving some percentage of site construction into protected environments for later transport to a building site. (SEE ALSO: *Construction Technology; Flexible Housing; In Situ Construction; Manufactured Housing; Modular Construction;* Open House International; *Operation Breakthrough; Panelized Housing; Prefabrication; Support-Infill Housing*)

—*Stephen Kendall*

Further Reading

Bender, Richard and Forrest Wilson, eds. 1973. *A Crack in the Rear View Mirror: A View of Industrialized Building.* New York: Van Nostrand.

Bruce, Alfred and Harold Sandbank. 1972. *A History of Prefabrication.* New York: Arno (originally published in 1944).

Finnimore, Brian. 1989. *Houses from the Factory: System Building and the Welfare State.* London: Rivers Oram.

Habraken, N. John. 1988. *Transformations of the Site.* Cambridge, MA: Awater Press and MIT Department of Architecture.

Herbert, Gilbert. 1984. *The Dream of the Factory-Made House: Walter Gropius and Konrad Wachsmann.* Cambridge: MIT Press.

Hounshell, David A. 1984. *From the American System to Mass Production, 1800-1932.* Baltimore: Johns Hopkins Press.

Kaiser, Edgar F., chairman. 1968. *A Decent Home: The Report of the President's Committee on Urban Housing.* Technical Studies, Vol. II. Washington, DC: Government Printing Office.

Kelly, Burnham. 1959. *Design and the Production of Houses.* New York: McGraw-Hill.

Russell, Barry. 1981. *Building Systems, Industrialization, and Architecture.* John Wiley.

"The Twentieth Century Fund." 1944. In *American Housing: Problems and Prospects,* Miles L. Colean, research director. New York: Twentieth Century Fund. ◀

▶ Infill Housing

Infill housing refers to new construction on existing, vacant parcels of land amid existing buildings in developed neighborhoods. The land on which infill housing is built was either bypassed in earlier phases of construction or cleared in central cities or suburbs to encourage infill development. *Infill* is sometimes synonymous with *odd-lot development. Sprawl* is the opposite of infill and is construction on raw, vacant parcels without public infrastructure such as roads, sewers, and schools.

Infill examples include detached, secondary units such as rear houses, attached accessory units, or entirely new single-family houses and apartments. Infill can be new construction built on-site or factory built. Infill projects emphasize compatibly designed structures blending with an already developed environment. Special design complications arise when builders attempt to develop infill within historical districts.

Infill housing can revitalize urban neighborhoods along with urban renewal, loft conversion, gentrification, urban homesteading, and techniques such as bringing properties up to building code. Arguments for revitalizing infill include increased direct savings in infrastructure costs compared with the infrastructure costs associated with fringe development. Infill also preserves open space at the urban fringe and stabilizes older neighborhoods, thus enhancing community values. Infill occurs not only in low-value districts but also on high-value land by builders seeking profits. Often, this is done without regard to the surrounding area, thereby changing the character of the neighborhood and evoking community protest.

However, not all urban developed vacant land is available for infill. The land may be inaccessible, in a flood plain, or on too steep an incline or may have been a landfill or otherwise contaminated. Often, infill housing is perceived as a high-risk venture with low rates of return. In many central cities and inner-ring suburbs, developers do not rebuild large blocks of developed vacant land because these sites are seen as low-profit areas compared with the more outlying areas. Single-lot owners can initiate infill projects. Projects of a dozen units or more can be undertaken by inexperienced builders in the process of establishing themselves. Not-for-profit corporations can create infill in close cooperation with city and county planning departments that offer incentives such as fee waivers, cancellation of impact fees, or even subsidization of connection fees. In addition, government agencies, such as housing authorities, can sponsor infill by targeting areas prime for redevelopment. Infill can be the product of public-private joint ventures. The public contribution typically consists of providing land with clear title, preliminary design, and/or batch-processed site plan review.

Infill adds to the local tax base, eliminates unkempt parcels, and reduces road wear by decreasing the daily urban commute. Local communities encourage infill where parcels are taxed at full assess valuation so that owners have incentives to sell or build. However, cities discourage infill when parcels have uncleared tax liens or when government promotes new fringe growth by expanding public infrastructure. (SEE ALSO: *Floor-Area Ratio; Historic Preservation; Land Value Taxation; New Urbanism; Temporarily Obsolete and Derelict Sites;* **Urban Redevelopment**)

—*John W. Smith and John S. Klemanski*

Further Reading
Beasley, Ellen. 1988. *Design and Development: Infill Housing Compatible with Historic Neighborhoods.* Information Series #41. Washington, DC: National Trust for Historic Preservation.
Frank, James E. 1989. *The Costs of Alternative Development Patterns: A Review of the Literature.* Washington, DC: Urban Land Institute.
U.S. Department of Housing and Urban Development. 1980. *Urban Infill: The Literature.* Washington, DC: Government Printing Office.
Wentlin, James W. and Lloyd W. Bookout, eds. 1988. *Density by Design.* Washington, DC: Urban Land Institute. ◄

► Institute of Real Estate Management

Founded in 1933, the Institute of Real Estate Management (IREM) provides education and certification for real estate management professionals—managers of residential, commercial, and industrial properties; asset managers and site managers; property supervisors; and management company owners. IREM confers certification as a CPM® (Certified Property Manager®) and runs the ARM® (Accredited Residential Manager®) and AMO® (Accredited Management Organization®) programs. It offers more than 40 publications. Current membership: 10,000. Staff size: 90. Executive Vice President: Ronald Vukas. Address: 430 N. Michigan Ave., Chicago, IL 60611. Phone: (312) 329-6000. Fax: (312) 661-0217.

—*Laurel Phoenix* ◄

► Interagency Council on the Homeless

The Interagency Council on the Homeless is a federal agency established in 1987 by the McKinney Homeless Act to provide governmental leadership for activities to assist homeless individuals and families. The council (a) plans and coordinates the federal government's homeless assistance programs, (b) recommends or makes policy changes to improve assistance programs, (c) monitors and evaluates homeless assistance provided by all levels of government and the private sector, (d) provides technical assistance to community homeless assistance projects, and (e) disseminates information on federal resources available to serve the homeless population. Council publications include a variety of brochures on federal assistance programs, funding sources, and volunteer opportunities in homeless projects; fact sheets on the demographic characteristics of the homeless; and manuals on initiating community homeless assistance programs.

Members of the council include the heads (or their designees) of 12 cabinet-level departments in the federal government, the Office of Management and Budget, and five independent agencies: the Corporation for National Service, the Social Security Administration, the Federal Emergency Management Agency (FEMA), the General Services Administration, and the U.S. Postal Service. Council activities are developed and carried out by a policy group composed of subcabinet representatives from each member agency. The group provides a forum for coordinating policies and programs and for preparing recommendations for consideration by the full council. In addition to its Washington, D.C., staff, the council has a full-time regional coordinator provided by the Department of Housing and Urban Development in each of the 10 federal regions. Council funding is appropriated by Congress. Address: 451 Seventh Street, SW, Room 7274, Washington, DC 20410. Phone: (202) 708-1480. Fax: (202) 708-3672. (SEE ALSO: **Homelessness**)

—*Caroline Nagel* ◄

► Interest Rates

Interest rates reflect the compensation due to the entity that postpones consumption (the lender) and the cost to the entity that accelerates consumption by exercising a demand for capital (the borrower). The lender demands a profit on the funds made available to the borrower, based on (a) a real return, adjusted for inflation; (b) another increment of return based on prevailing perceptions of inflation; (c) an additional premium to compensate for potentially incorrect perceptions of inflation at the time the transaction was made; (d) a further premium to offset the risk of a borrower default; and (e) the cost of capital incurred by the lender, such as returns on deposits paid by banks to depositors.

There are many determinants of interest rates, especially since financial markets have become globalized. A primary factor is the budget established by the national government. In the United States, it has been the practice to determine spending priorities and then to try to finance them, primarily through tax revenues. This method has resulted in the federal government's operating at a deficit every year since 1960 with the sole exception of a balanced budget in 1969. The government's escalating borrowing needs have reduced the amount of credit available for private investment, which "crowds out" other borrowers and increases interest rates. Not surprisingly, efforts are made periodically to control federal spending or even to demand a balanced federal budget. In contrast, modern proponents of the theories of John Maynard Keynes (1883-1946) have argued that deficit spending stimulates the economy more than high interest rates limit economic growth. However, although Keynes's theories addressed the deflationary economy of the 1930s, he later warned of the dangers of deficit spending, especially when inflation poses a threat.

Monetary policy is a second variable that influences interest rates. The Federal Reserve Board of Governors regulates the amount of money in circulation, which it monitors by various measures, such as currency and funds held in savings and checking accounts. When the Federal Reserve slows the growth of the money supply, it creates an upward pressure on interest rates. Sometimes it can dramatically contract monetary growth as a response to inflation, an action that produced the extremely high interest rates of 1981 and 1982. Likewise, it can accelerate growth in the money supply if inflation is under control and economic growth has slowed.

A third determinant of interest rates is the perception of future inflation. A recessionary economy with relatively high unemployment, lagging demand for durable goods (such as housing and automobiles), a low rate of factory use, and falling consumer confidence signals that the Federal Reserve would adopt a stimulus-oriented monetary policy. In contrast, rising consumer prices, high levels of factory use, a rate of unemployment approaching a point where insufficient supplies of labor propel wages upward, and higher commodity prices, such as those for oil, forest products, and industrial metals, would suggest rising inflation. In a sense, financial markets welcome "bad news" such as increasing unemployment, for it indicates the potential for lower interest rates, whereas a strong economy and low unemployment—"good news"—may compel the Federal Reserve to raise interest rates.

Federal Reserve actions have a direct impact on short-term interest rates, especially the closely watched federal funds rate, which financial markets monitor on a daily basis. However, the Federal Reserve's actions have only an indirect affect on the longer end of the maturity spectrum, especially the bellwether 30-year treasury bond. Long-term rates are subject to myriad influences relating to the globalization of financial markets. If raising short-term rates meets with the approval of world markets, it could result in decreased interest rates on the 30-year bond. Or if instability in emerging economies precipitates a "flight to quality," long-term U.S. interest rates could decline owing to the demand for the dollar as an economic "safe haven." In contrast, the reunification of Germany precipitated a large demand for capital, resulting in higher worldwide interest rates. Earthquakes requiring reconstruction in densely populated Japan could have a similar effect.

Mortgage financing is susceptible to all these trends. Fixed-rate, 30-year mortgages are closely linked to yields on the 30-year bond. However, adjustable-rate mortgages (ARMs) would be more sensitive to interest rate movements on the shorter part of the maturity spectrum, especially federal funds. The housing consumer would have to evaluate these factors in choosing a fixed-rate mortgage or an ARM. Because trends in world interest rates fluctuate over time, many homeowners must consider when to refinance their properties as well as other debts. As interest rates fall, consumers might refinance fixed-rate loans to take advantage of declining interest rates by switching to ARMs, whose costs will drop in a parallel manner. As the interest rate cycle bottoms out, consumers will refinance ARMs with fixed-rate loans to "lock in" the certainty of a low and predictable mortgage payment.

The direction of future interest rates is always extremely difficult to judge. Economic fluctuations (both global and domestic), tensions and instability in sensitive geopolitical locations such as the Middle East, investor perceptions and psychology, and fluctuating international currency valuations, to name a few, are all ingredients in this complex mix. (SEE ALSO: **Mortgage Finance;** *Usury Laws*)

—*Daniel J. Garr*

Further Reading

Blankenship, Frank J. 1989. *The Prentice Hall Real Estate Investor's Encyclopedia.* Englewood Cliffs, NJ: Prentice Hall.

Jacobus, Charles J. 1996. *Real Estate Principles.* Upper Saddle River, NJ: Prentice Hall.

McKenzie, Dennis J. 1996. *Essentials of Estate Economics.* Upper Saddle River, NJ: Prentice Hall.

Peiser, Richard B. 1992. *Professional Real Estate Development: The ULI Guide to the Business.* Washington, DC: Urban Land Institute. white

Shim, Jae K. 1996. *Dictionary of Real Estate.* New York: John Wiley.

Tosh, Dennis S. 1990. *Handbook of Real Estate Terms.* Englewood Cliffs, NJ: Prentice Hall. ◀

▶ International Association for Housing Science

The International Association for Housing Science (IAHS) is an international, interdisciplinary nonprofit organization that promotes research on housing science, disseminates research findings to members of the public and private sector housing fields; and sponsors international conferences to develop innovative ideas and methods to solve worldwide housing shortages. The IAHS publishes the *International Journal for Housing Science and Its Applications.* A member of the United Nations, New York, as ECOSOC-NGO, it has organized 23 world congresses and published 22 volumes of Housing Congress Proceedings. It has more than 200 members from 41 countries. Members work in a variety of fields related to human habitat, including engineering, architecture, law, sociology, finance, labor organizing, and land development. Founded in 1972. President: Dr. Oktay Ural. Address: P.O. Box 340254, Coral Gables/Miami, FL 33134. Phone: (305) 348-3171. Fax: (305) 348-3797.

—*Caroline Nagel* ◀

▶ International Association for People-Environment Studies

Founded in 1981, the objectives of the International Association for People-Environment Studies (IAPS) are to (a) facilitate communication among those concerned with the relationships between people and their physical environment, (b) stimulate research and innovation for improving human well-being and the physical environment, and (c) promote the integration of research, education, policy, and practice. IAPS offers its international members conferences, symposia, and seminars in English and French; promotes ongoing topical study networks (e.g., housing, landscape, cultural aspects of design); and networks with other international organizations devoted to this topic (Environmental Design Research Association [EDRA]—U.S.; Man Environment Relations Association [MERA]—Japan; People and Physical Environment Research [PAPER]—

Australasia). Publications include a newsletter, a membership directory, and conference and seminar proceedings. Membership dues: Regular—$50; students/retired/unemployed—$25. Treasurer: Roderick Lawrence. Address: CUEH, University of Geneva, 102 Boulevard Carl-Vogt, 1211 Geneva 4, Switzerland. Phone: 41-22-7058174. Fax: 41-22-7814100.

—Laurel Phoenix ◀

▶ International Federation for Housing and Planning

Founded in 1913 by Sir Ebenezer Howard as the "International Garden Cities and Town Planning Association," International Federation for Housing and Planning's (IFHP) purpose is to provide easy access to internationally available sources of knowledge in housing, urban and regional planning, the environment, and related fields to improve housing and planning practice throughout the world. A major activity is organizing international congresses in the field of housing and planning to disseminate information and provide opportunity for professionals to meet colleagues. Publications include congress reports and a bimonthly newsletter. IFHP has individual and organizational members worldwide. Staff: Five. Secretary General: Mrs. E. E. van Hylckama Vlieg. Address: Wassenaarseweg 43, 2596 CG The Hague, the Netherlands. Phone: 31 70. 324 4557. Fax: 31 70. 328-2085.

—Laurel Phoenix ◀

▶ *International Journal of Urban and Regional Research*

The *International Journal of Urban and Regional Research* is a quarterly publication covering all aspects of urban and regional development research, including housing. The audience is primarily academic. Circulation: 1,200 worldwide. First published: 1977. Price: Institutions in the United Kingdom and Europe, £109; in the United States, $201; in the rest of the world, £121. Individuals in the United Kingdom and Europe, £43; in the United States, $49; in the rest of the world £49. Reduced rates available (IBG, ISA, BSA, ECPR, KNAG). Editor: Patric Le Galès, CEVIPOF (FNSP, CNRS), Paris; Events and Debates Editor: Susan Fainstein, Rutgers University; Reviews Editor: Linda McDowell, University of Cambridge. Submissions piolicy: anonymously refereed, submit four paper copies, 8,000-word limit, and a 200-word abstract to Assistant Editor: Terry McBride, Flat 3,

19 Walpole Terrace, Brighton BN2 2ED, UK. Phone and fax: 44 (0) 1273 676435. E-mail: tmcbride@mistral.co.uk.

—Laurel Phoenix ◀

▶ International Sociological Association Research Committee on Housing and the Built Environment

The International Sociological Association (ISA) Research Committee on Housing and the Built Environment is an international association founded in 1978 to provide a forum for promoting research and communication among housing researchers. The Research Committee had its origins in 1974 and has organized sessions on housing and related matters at each subsequent World Congress of Sociology, including Mexico City, New Delhi, Madrid, and Bielefeld, Germany. In addition, it conducts biennial international conferences. Past venues have included Amsterdam, Paris, Montreal, and Beijing. It has also sponsored a number of smaller, regional meetings in places such as Budapest, Prague, Kobe, Nairobi, and Hamburg. Results of these meetings, organized directly by the group or under its auspices in collaboration with host institutions, have been widely disseminated through a steady flow of publications. Conference papers cover methodological, theoretical, and policy issues. The committee's international membership provides a base for cross-national reports and studies of a range of issues characteristic to given societies. Members receive *Notes,* a newsletter published three times a year covering recent publications, current research, and upcoming conferences, and a membership directory. Membership subscriptions ($20 for 2 years) provide the committee's budget. In 1993, there were 337 members from approximately 30 countries. The United States accounts for the largest proportion of members, but substantial numbers come from Western and Eastern Europe, Scandinavia, and the Third World. Although remaining loosely affiliated with the International Sociological Association (ISA), a majority of the membership represents disciplines other than sociology, including geography, political science, economics, planning, and policy analysis. Contacts between group members have led to mutual visits, guest lectures, and collaboration in research and practice. Secretary: Patricia Edwards. Address: College of Architecture and Urban Studies, 202 Cowgill Hall, Virginia Polytechnic Institute and State University, Blacksburg, VA 24061-0205. Phone: (703) 231-6386. Fax: (703) 231-9938.

—Leslie Kilmartin

J

▶ Joint Center for Housing Studies

The Joint Center for Housing Studies is an interdisciplinary research institute affiliated with the Graduate School of Design and John F. Kennedy School of Government at Harvard University. The purpose of the center is to bring together members of the academic community with housing industry executives and government officials to study trends in design, production, and consumption of housing and to analyze public and private efforts to meet affordable housing needs in the United States. Its research has been instrumental in shaping housing legislation at the local, state, and federal levels and in informing private groups of various aspects of housing policy. The Joint Center has released several books, papers, and articles on housing issues for both academic audiences and the larger audience of public policy decision makers and housing industry analysts. Included in the Joint Center's publications is *The State of the Nation's Housing,* a series of reports providing a periodic assessment of the nation's housing situation. Founded in 1959 as the Joint Center for Urban Studies, the center's research program is supported by private industry, governmental bodies, nonprofit organizations, and trade associations. Director: Nicholas P. Retsinas. Address: Harvard University, 79 John F. Kennedy Street, Cambridge, MA 02138. Phone: (617) 495-7908. Fax: (617) 496-1722.

—*Caroline Nagel* ◀

▶ Joint Tenancy

One of the features of Anglo-American law is the ability to own property concurrently (i.e., more than one owner of the same property can exist as co-owners). The most common form of concurrent ownership is joint tenancy. Under this form of ownership, co-owners own undivided, equal shares as partners.

To establish a joint tenancy, four tests must be met. These are traditionally known as the "four unities." All cotenants must have entered into the agreement at the same time (unity of time), with each tenant owning an undivided interest in the property (unity of title), with all interests for the same period and of the same type (unity of interest), and that the ownership form was created by a single conveyance (unity of possession). If any of these tests cannot be met, the organization will not be held to be a joint tenancy.

The ownership form most often compared with joint tenancy is called tenancy in common. Unlike joint tenants, tenants in common may hold unequal interests, acquired over a period of time. In addition, tenants in common may also convey their interests without disturbing the other tenants. A tenant in common owns interests such as common stock; joint tenants are more like strict, joint partners.

A similar ownership form to joint tenancy is called tenancy by the entirety where permitted by law. In this form, a fifth unity (unity of person) is required. During a marriage, co-ownership under tenancy by the entirety views a husband and wife as a single legal person. In the event of the termination of the marriage, the tenancy is dissolved.

If any of the unities are broken, the joint tenancy is automatically terminated and a tenancy in common is then said to exist. Courts have raised the standards necessary to meet the unity tests required for joint tenancies. Because of difficulties associated with the concurrent ownership forms, although once being relatively easy to satisfy the requirements for joint tenancies, in recent years, it has become more difficult. (SEE ALSO: **Tenure Sectors**)

—*Austin J. Jaffe*

Further Reading

Blankenship, Frank J. 1989. *The Prentice Hall Real Estate Investor's Encyclopedia.* Englewood Cliffs, NJ: Prentice Hall.

Jacobus, Charles J. 1996. *Real Estate Principles.* Upper Saddle River, NJ: Prentice Hall.

Jaffe, Austin J. and C. F. Sirmans. 1995. *Fundamentals of Real Estate Investment.* 3d ed. Englewood Cliffs, NJ: Prentice Hall.

McKenzie, Dennis J. 1996. *Essentials of Estate Economics.* Upper Saddle River, NJ: Prentice Hall.

Peiser, Richard B. 1992. *Professional Real Estate Development: The ULI Guide to the Business.* Washington, DC: Urban Land Institute.

Shim, Jae K. 1996. *Dictionary of Real Estate.* New York: John Wiley.

Tosh, Dennis S. 1990. *Handbook of Real Estate Terms.* Englewood Cliffs, NJ: Prentice Hall. ◀

► Journal of Housing and Community Development

The *Journal of Housing and Community Development* is a bimonthly publication on the needs and concerns of public housing and redevelopment officials. Contents include direct administrative, organizational, and operational reports; articles on legal, governmental, and technical development and articles on the development of national trends as they affect housing and community development programs at national, state, regional, and local levels. The audience is the housing and community development industry. Circulation: 13,500 nationwide. First published in 1944. Price: $4 per copy. Submissions policy: Query first, indicating subject, approach, development, and conclusions, along with professional credentials and experience of the author. Duplicate hard copy manuscripts submitted in double space form, no longer than 12 pages. Submission Editor: Terence Cooper. Address: 630 Eye St., NW, Washington, DC 20001-3736. Phone: (202) 239-3500. Fax: (202) 239-8181.

—*Laurel Phoenix* ◄

► Journal of Housing Economics

The *Journal of Housing Economics* is a quarterly publication focusing on economic research related to housing and targeted toward academics and professionals. Topics include international comparisons, impact of the local public sector, and construction industry studies. Distributed internationally, the journal's purpose is to encourage analysis of public policy issues regarding housing. First published in 1991. Submission editor: Henry O. Pollakowski. Address: Joint Center for Housing Studies, Harvard University, 79 John F. Kennedy St., Cambridge, MA 02138. Phone: (617) 495-3049. Fax: (617) 496-1722.

—*Laurel Phoenix* ◄

► Journal of Housing for the Elderly

The *Journal of Housing for the Elderly* is a quarterly journal focusing on the interests of architects, urban planners, managers, and policymakers specializing in seniors' housing. Published since 1983, the journal seeks to enhance the residential environment for the elderly through the rapid publication of original, cross-disciplinary research in the housing and aging fields. Articles cover a broad range of issues, including housing legislation, managerial and financial strategies, housing-related service delivery, housing alternatives, energy conservation, and privacy needs of tenants. Price: $85 per year for institutions; $40 per year for individuals. Editor: Dr. Leon Pastalan. Address: University of Michigan, College of Architecture and Urban Planning, 2000 Bonisteel Boulevard, Ann Arbor, MI 48109-2069. Phone: (313) 763-9560. Fax: (313) 763-2322. (SEE ALSO: **Elderly**)

—*Caroline Nagel* ◄

► Journal of Housing Research

A scholarly journal published twice a year by the Fannie Mae Foundation (previously by the Fannie Mae Office of Housing Research), *Journal of Housing Research* presents theoretical and empirical research on housing and finance issues. Recent articles have examined homeownership, housing demography, disparities in mortgage lending, residential segregation, and housing affordability. First published in 1990. Circulation: 4,800 internationally. Editor: James H. Carr. Address: Fannie Mae Foundation; 4000 Wisconsin Avenue, NW; North Tower, Suite One; Washington, DC 20016-2804. Phone: 202-274-8000. Fax: 202-274-8100. E-mail: fmfpubs@fanniemaefoundation.org. World Wide Web: http://www.fanniemaefoundation.org.

—*James H. Carr* ◄

► Journal of Property Management

The *Journal of Property Management* is a bimonthly journal focused on asset and property management of investment-grade real estate. First published in 1934. Circulation: 21,000 in the United States and Canada. Price: $36.95 per year. Editor: Mariwyn Evans. Address: 430 N. Michigan Avenue, Chicago, IL 60611. Phone: (312) 329-6058. Fax: (312) 661-0217.

—*Caroline Nagel* ◄

► Journal of Real Estate Research

The *Journal of Real Estate Research* is a quarterly journal that focuses on business decision-making applications of scholarly real estate research. The journal is especially interested in research that can be useful to business decision makers in areas such as development, finance, investment, management, market analysis, marketing, and valuation. The official publication of the American Real Estate Society, its audience is primarily real estate academics and professionals. Circulation: 1,600 internationally. First published in 1986. Price: Individual $85 in the United States; $95 internationally. Includes the *Journal of Real Estate Literature, Journal of Real Estate Portfolio Management, Real Estate Research Issues,* an annual monograph, *Real Estate Capital Markets Report,* and *ARES Newsletter.* Submission policy: Anonymously refereed, authors should submit four copies of an original, double-spaced, typed manuscript. Submission editor: G. Donald Jud. Address: University of North Carolina at Greensboro, Bryan School of Business and Economics, Department of Finance, Greensboro, NC. 27412. Phone: (910) 334-3091. Fax: (910) 334-5580. E-mail: JUDDON@IAGO.UNCG.EDU.

—*Laurel Phoenix* ◄

▶ Journal of the American Real Estate and Urban Economics Association

The *Journal of the American Real Estate and Urban Economics Association* is a quarterly publication on scholarly research in real estate issues. Its purpose is to facilitate communication among academic researchers and industry professionals and to improve decisions related to real estate. The audience is primarily academic and professional researchers. Circulation: 1,300. First published in 1971. Price: $50 per year. Manuscripts are both solicited and invited. Submission editors: Jim Shilling and Kerry Vandell. Address: School of Business, University of Wisconsin, 5262 Grainger Hall, Madison, WI 53706. Phone: (608) 262-8602. Fax: (608) 265-2738.

—Laurel Phoenix ◀

▶ Journal of Social Distress and the Homeless

The *Journal of Social Distress and the Homeless* is a quarterly journal that provides an international, interdisciplinary forum for original, peer-reviewed papers on psychosocial distress. Articles explore contemporary and historical issues related to education, health care, criminal justice, economics, and the family, treating topics such as homelessness, urban violence, and racial tension. The journal publishes results from experimental research, clinical papers, theoretical work, literature reviews, short notes, book reviews, and from time to time, special issues. Its audience consists of professionals in a wide range of disciplines in the social, behavioral, and medical sciences and in the humanities. Yearly subscriptions: $165 (domestic) and $195 (foreign) for institutions; individual subscriptions: $41 (domestic) and $48 (foreign). Submissions should be sent in quadruplicate to the editor: Dr. R. W. Rieber, John Jay College of Criminal Justice, City University of New York, 445 West 59th St., New York, NY 10019. (SEE ALSO: **Homelessness**)

—Laurel Phoenix ◀

▶ Journal of Urban Economics

The *Journal of Urban Economics* is a bimonthly publication focusing on urban economics. The journal also has brief notes commenting on published work, new information, or new theoretical advances. The audience is economists. First published in 1974. Price: $283 per year in the United States and Canada; $349 per year in all other countries. Submission policy: Submit original, double-spaced manuscript in triplicate. Submission editor: Professor Jan K. Brueckner. Address: Department of Economics, University of Illinois, 1206 S. Sixth St., Champaign, IL 61820. Phone: (619) 699-6388. Fax: (619) 699-6800.

—Laurel Phoenix

► Land Value Taxation

Land value taxation has emerged in recent decades as a viable alternative to the conventional property tax. By 1995, 16 U.S. cities were using the *two-rate property tax*.

What It Is

In contrast to the typical property tax, which imposes a single tax rate on the *combined* value of land and buildings, the two-rate tax imposes a low rate on buildings and a higher rate on land values. Higher land taxes discourage land speculation; lower building taxes generate construction, easing shortages of decent affordable housing.

Background

American economist Henry George (1839-1897), building on ideas of Adam Smith and David Ricardo, favored taxing socially created land values instead of private production. Early instances of untaxing housing occurred in two "single-tax" communities that still flourish—Fairhope, Alabama (1893) and Arden, Delaware (1900). Municipal corporations own the land. Residents lease their lots, paying rents based solely on land value, regardless of the size or quality of their homes.

California's Modesto and Turlock irrigation districts applied this principle in the 1920s. Taxing owners according to their land values, reflecting the advantages of proximity to water, was credited with breaking up huge estates and fostering smaller intensive farms.

Before the advent of extensive federal grants, many U.S. cities funded street and utility extensions through public improvement districts. Beneficiaries—adjacent landowners—were assessed per front-foot, a rough land value basis.

Pennsylvania Story

Pittsburgh and Scranton pioneered the two-rate tax in 1913. After legislators gave all home rule cities this option, Harrisburg adopted it in 1975 and others followed suit: Aliquippa, Clairton, Coatesville, Connellsville, Du Bois, Duquesne, Lock Haven, McKeesport, New Castle, Oil City, Titusville, Allentown, and Washington. Tax rates vary from 2 to 16 times higher on land than on buildings. School districts, which have taxing power in the state, won permission to use this reform in 1993; several opted for it by 1995.

Other Developments

In 1987, Peoria, Illinois, abated taxes on any increased building values within its enterprise zone while continuing to tax land value increases. The blighted zone subsequently enjoyed a higher construction rate than the rest of Peoria.

Many variations are used in other countries. In Australia and New Zealand, a majority of local jurisdictions tax property on land value only; housing and other buildings are tax free. For three decades, Taiwan has preserved scarce farmland and induced intensive use of urban sites with a land value tax plus a progressive land increment tax.

Effects

Two-rate taxing cities in the United States have experienced the following results:

- ► *Lower residential taxes*. Most owners benefit. Seniors are not driven from their homes by excessive taxes.
- ► *Affordable housing*. Construction and rehabilitation are stimulated. The larger housing supply holds prices and rents in check.
- ► *Central business district revitalization*. Higher charges on idle sites and lower building taxes induce *private* urban renewal of central business districts.
- ► *Efficient land use*. More intense use of in-city sites reduces sprawl, avoiding costly duplication of infrastructure.

Restraining Factors

Given this positive record, why is the reform not spreading more rapidly? Explanations include the following:

- ► Local officials are wary of changing their major funding source.
- ► Installing two-rate taxes often requires amending state constitutions, which is difficult and time-consuming.
- ► Opposition is mounted by slumlords and land speculators who profit from the current system.

TABLE 19 Average Annual Value of Building Permits (Values in Millions of Constant 1982 Dollars)

	1960 to 1979	1980 to 1989	Percentage of Change
Pittsburgh	$181.734	$309.727	+70.43
Akron	134.026	87.907	–34.41
Allentown	48.124	28.801	–40.15
Buffalo	93.749	82.930	–11.54
Canton	40.235	24.251	–39.73
Cincinnati	318.248	231.561	–27.24
Cleveland	329.511	224.587	–31.84
Columbus	456.580	527.026	+15.43
Dayton	107.798	92.249	–14.42
Detroit	368.894	277.783	–24.70
Erie	48.353	22.761	–52.93
Rochester	118.726	82.411	–30.59
Syracuse	94.503	53.673	–43.21
Toledo	138.384	93.495	–32.44
Youngstown	33.688	11.120	–66.99
15-city average	167.504	143.352	–14.42

Dun and Bradstreet figures (Oates and Schwab 1995, Table 3).

► Popular and professional journals have largely ignored the reform; whereas California's Proposition 13 captured national headlines, Pittsburgh's expanding two-rate tax was virtually unpublicized; yet Pittsburgh went on to surpass most cities in downtown construction and affordable housing (see Table 19).

In 1979, Pittsburgh significantly shifted taxes from buildings to land values. Other cities in the comparison use the conventional method, taxing buildings and land values at the same rate.

Outlook

Expansion of the reform nevertheless seems likely. Seven Nobel economists—James Buchanan, Milton Friedman, Franco Modigliani, Paul Samuelson, Herbert Simon, Robert Solow, and James Tobin—have endorsed land value taxation. Advocates are promoting its adoption, among other places, in Missouri, New Hampshire, West Virginia, Minnesota, Maryland, Wisconsin, and the District of Columbia.

High land values, often blamed for causing urban housing problems, may—if taxed more rationally—become a major key to their solution. (SEE ALSO: *Brownfields; Infill Housing; Property Tax; Speculation*)

—*Walter Rybeck*

Further Reading

Andelson, Robert V., ed. 1997. *Land Value Taxation around the World.* 2d ed. New York: Schalkenbach Foundation.

Gaston, Paul M. 1993. *Man and Mission: E. B. Gaston and the Origins of the Fairhope Single-Tax Colony.* Montgomery, AL: Black Belt Press.

George, Henry. [1879] 1993. *Progress and Poverty.* New York: Schalkenbach Foundation.

Harriss, C. Lowell, ed. 1983. *The Property Tax and Local Finance.* New York: Academy of Political Science.

Oates, Wallace E. and Robert M. Schwab. 1995. *The Impact of Urban Land Taxation: The Pittsburgh Experience.* Baltimore: University of Maryland Press.

Rybeck, Walter. 1991. "Pennsylvania's Experiments in Property Tax Modernization." *NTA Forum* (Spring):1-5. ◄

► Lease

A lease is an implicit or explicit contract between two parties—the lessor (typically the landlord) and the lessee (typically the tenant)—that stipulates the mutual obligations and consideration to be paid (e.g., rent) by the lessee in return for specified rights to use the property. Although varying widely in form, typical lease conditions in a residential tenancy might include (a) the length of the contract, (b) options for renewal, (c) initial rent (or other compensation) and mechanisms for change, (d) allocation of other costs, (e) performance standards to be met by landlord or tenant, (f) subletting, (g) penalties for failure to perform, and (h) conditions for termination.

At its heart, a lease is a contract to govern contingencies. In general, neither landlord nor tenant can perfectly anticipate what might happen over the life of a lease. A well-designed lease can reduce uncertainty by means of mutual restrictions and guarantees with a view to ensuring the well-being of both landlord and tenant. For landlords, the principal risks are that (a) their costs may increase over the duration of a lease in ways that had not been anticipated, (b) market conditions may change in a way that makes some other use of the property more valuable than this tenancy, or (c) the tenant (or subtenant) will subsequently prove to be costly: for example, waste heat or energy, vandalize or damage the property, be tardy in paying rent, or be noisy, unsanitary, dangerous, unruly, or disruptive. For their part, tenants cannot always be confident that the landlord will behave appropriately. Presumably, a residential tenant wants "quiet enjoyment" of the property (e.g., unrestricted access to the rented premises; adequate maintenance and quick repairs by the landlord; timely provision of heat, water, and other utilities; freedom from undue eviction; protection from noisy or unruly neighbors; and control over access to the rented premises by the landlord), protection against undue rent increases during the lease, and adequate notice of any changes in lease conditions. Even if the tenant has confidence in the current landlord, the dwelling might be sold at some future date to a new landlord who might behave differently.

In principle, landlord and tenant are free to draft the terms of the lease, subject to the provisions of government legislation and common-law practice. However, critics argue that many landlords impose lease conditions that tenants must either take or leave (that is, cannot freely negotiate). In recent decades, many jurisdictions in North America and elsewhere have sought to remedy such imbalances by assigning property rights to residential tenants through legislation that supersedes lease provisions. Such packages of rights are generically termed *security of tenure,* although the packages have changed over time and differ among jurisdictions. Security of tenure provisions typically include things such as (a) the maintenance responsibilities of the landlord; (b) the right of tenants to sublet; (c) conditions for seizure of premises or contents by the landlord; (d) access to rented premises by landlords; (e) restrictions on damage deposits, key money, and other landlord charges; and (f) protection against unjustified eviction. In general, security of tenure legislation can be thought to arise from the same concerns that give rise to consumer protection legislation generally. In some jurisdictions, rent regulation has also been introduced—in part, to protect security of tenure. Although rent regulation is also used to achieve other social goals, one important benefit has been to ensure that a landlord cannot otherwise evict a tenant simply by proposing an unjustified rent increase—what has been referred to as an economic eviction. (SEE ALSO: *Eviction; Private Rental Sector; Security Deposit*)

—*John R. Miron*

Further Reading

Blankenship, Frank J. 1989. *The Prentice Hall Real Estate Investor's Encyclopedia.* Englewood Cliffs, NJ: Prentice Hall.

Jacobus, Charles J. 1996. *Real Estate Principles.* Upper Saddle River, NJ: Prentice Hall.

McKenzie, Dennis J. 1996. *Essentials of Estate Economics.* Upper Saddle River, NJ: Prentice Hall.

Miceli, T. J. 1992. "Habitability Laws for Rental Housing." *Urban Studies* 29(1):15-24.

Miron, J. R. 1990. "Security of Tenure, Costly Tenants and Rent Regulation." *Urban Studies* 27(2):167-83.

Peiser, Richard B. 1992. *Professional Real Estate Development: The ULI Guide to the Business.* Washington, DC: Urban Land Institute.

Shim, Jae K. 1996. *Dictionary of Real Estate.* New York: John Wiley.

Tosh, Dennis S. 1990. *Handbook of Real Estate Terms.* Englewood Cliffs, NJ: Prentice Hall. ◄

► Leased Housing

The leased housing program was a variant of the public housing program created in 1965. It is also known as the Section 23 program. The leased housing program is generally viewed as the forerunner to the housing allowance-type programs that followed—the Section 8 existing housing and voucher programs.

In the conventional public housing program, local public housing authorities (PHAs) own and manage developments and then rent units to eligible households. In the leased housing program, PHAs entered into rental contracts with private landlords. The PHA was responsible for selecting tenants who paid 25% of their income for the housing, and the PHA paid the remainder of the contract rent to the landlord.

With the enactment of the Housing and Community Development Act of 1974, the Section 8 existing housing program, which also enabled low-income households to rent private apartments, replaced the leased housing program. A major difference between the two programs was

that in the leased housing program, PHAs would locate the units and enter into a contract directly with the landlords, whereas in the Section 8 existing program, individual households were responsible for locating their own units on the private rental market. Tenants are required to pay 30% of income for their units (changed from 25% in 1981), and the PHA, using federal funds, pays the difference between that amount and the full rental. In the Section 8 existing program, tenants cannot rent units that exceed the fair market rental amount for their area, and their contributions cannot exceed 30%. In a variant of this program, the voucher program, tenant contributions may exceed 30% of income, but the allotted subsidy is based on the difference between 30% of income and the fair market rental. A total of about 100,000 units were financed through the leased housing program until its demise in 1974. (SEE ALSO: **Public Housing;** *Public Housing Authority; Section 8 program; Turnkey Public Housing*)

—*Rachel G. Bratt*

Further Reading

Hartman, Chester W. 1975. *Housing and Social Policy.* Englewood Cliffs, NJ: Prentice Hall.

National Association of Housing and Redevelopment Officials. 1990. *The Many Faces of Public Housing.* Washington, DC: Author.

Solomon, Arthur P. 1974. *Housing the Urban Poor.* Cambridge, MA: MIT Press. ◄

► Lending Institutions

The history of banking and financial services in the United States is a complex story of institutional change, evolution, and specialization. In general, financial institutions that directly extend credit can be categorized as either (a) depository, such as commercial banks, thrift institutions, and credit unions, in which loans are made from deposits; (b) nondepository, such as mortgage banks and finance companies, in which loans are made from outside sources of funds not directly controlled by the institution; or (c) fiduciary, such as life insurance companies and pension funds, in which loans are made from the proceeds of long-term vested obligations.

Thrift institutions, composed of savings and loan associations and mutual savings banks, originated as a source of funds for the construction and purchase of homes. Individuals pooled their savings into "share accounts" held by local, cooperative "building and loan" associations. These state-chartered, tax-exempt institutions relied on long-term savings and time deposits and specialized in mortgage lending, unlike commercial banks that took demand deposits and made shorter-term loans to businesses. Until 1920, mutual savings banks were the leading institutional source of real estate credit in the United States, particularly for multifamily housing (apartment buildings), largely because of their geographic concentration in the older, densely populated, built-up areas of the Northeast. State laws restricted the expansion of savings banks as the country grew, and during the 1920s boom they were surpassed by savings and loans associations that rapidly proliferated to finance single-family housing (one- to four-unit dwellings). These institutions tended to be more geographically dispersed and less concentrated in the major banking centers, with the exception of Philadelphia, where the first savings and loan was formed in 1831.

Life insurance companies also hold long-term savings and make investments into real estate. Insurers, however, were fewer in number than thrifts in their heyday, less encumbered by regulation, and often purchased loans originated by others. In addition, life insurance companies were more likely to have a diverse investment portfolio, with loans for single-family and multifamily housing as well as for commercial and industrial real estate. With fewer geographic restrictions than depository institutions, life insurers and (later) pension funds could easily penetrate multiple markets and often maintained only indirect connections to local communities through "correspondent" relationships with mortgage originators.

Commercial banks were started during the mercantile era by merchants who pooled funds to make loans to each other that were secured by their goods of exchange. Banking expanded to serve the growing concentration of business activity within urban centers, and after the Civil War, deposits drawn down by checks replaced the circulation of bank notes as the principal means of credit provision, facilitating a greater volume of bank transactions as well as increased commerce between and within cities. In some states, banks could finance real estate in addition to business development and expand beyond their central headquarters location by opening branches in outlying commercial districts.

In Illinois, where branch banking was prohibited, neighborhood banks were established to serve community areas within Chicago outside of the downtown Loop district. Cities such as New York became money centers as the largest banks—often organized into clearinghouse associations—held and managed reserves and correspondent balances from banks in other cities, which in turn became depositories of these funds from "country banks" located in rural areas. The Federal Reserve Act of 1913 created 12 regional Reserve Banks and the Federal Reserve Board to supersede this existing system, supplying funds to all banks with federal charters—known as national banks—and some state-chartered banks. Commercial banking grew rapidly and became increasingly urbanized during the first three decades of the 20th century, especially in the Northeast and Midwest, where most of the nation's population and biggest cities were concentrated. National banks were granted expanded powers to invest in real estate and home mortgages, and in the late 1920s, "group banking" gained wider acceptance as a form of corporate organization in which multiple bank units were controlled by a single holding company, usually based in an urban center.

The decentralization of urban centers that accelerated during the 1920s yielded profitable opportunities for mortgage lenders in new, suburban regions that were developed and marketed to an emerging middle class. These suburbs were governed by zoning and subdivision regulations designed to protect the long-term stability of land uses and property values, as well as to exclude African Americans

and occasionally other groups such as Jews. Lenders avoided the most blighted and transient areas of mixed land use near the urban core, where deteriorating tenement structures stood side by side with small factories, warehouses, and abandoned commercial buildings. The continued influx of foreign immigrants from Central and Eastern Europe and the increased migration of poor African Americans from the rural South into the crowded cities of the Northeast and Midwest created demand for locally based financial institutions owned by and marketed to these disadvantaged groups, who typically identified by ethnic, national, and racial origin and by religious affiliation. Mainstream lenders perceived working-class immigrants and especially low-income African Americans descended from slaves as negative influences on property values. During the 1920s, the inner-city neighborhoods where they settled and often became segregated were viewed as slum districts with substantial credit risk. Workers and merchants in these communities who saved money and sometimes moved to nearby areas within the city often pooled their funds to create small banks and thrifts to serve these neighborhoods that larger institutions would not. Ethnic banking became a vehicle for economic advancement in cities such as Chicago.

The economic collapse of the Great Depression changed both the structural and spatial dynamics of financial services. In the cities, many of the smaller neighborhood banks and branch facilities closed as real estate values plunged, and depositors lost the money in their accounts. In response to the wave of bank and thrift closings, the federal government greatly expanded its regulatory oversight and support of depository institutions. The most far-reaching change was the introduction of federal deposit insurance, which stopped the withdrawal of funds during the crisis and became a competitive necessity for many banks and thrifts seeking a "safe and sound" image. By 1934, 90% of the nation's commercial banks—and all members of the Federal Reserve—were insured through the new Federal Deposit Insurance Corporation (FDIC). The Federal Home Loan Bank Act of 1932 created a national system of liquidity and control for thrifts and insurance companies making long-term mortgage loans that was somewhat parallel to the Federal Reserve network open to eligible commercial banks. But the availability of funds through 12 regional banks governed by a centralized Federal Home Loan Bank Board (FHLBB) did not stop the growing number of insolvent institutions and the loss of deposits to insured commercial banks. Worried savings and loan executives lobbied successfully for the creation of the Federal Savings and Loan Insurance Corporation (FSLIC) under the National Housing Act of 1934. Federal operating charters, which had been used to confer national bank status on larger banks since 1863, were also made available to thrifts for the first time through the FHLBB.

In addition to this new regulatory system, the federal government took action to revive local real estate markets burdened by the rising tide of home foreclosures and plummeting land values. The Home Owners' Loan Corporation (HOLC) was formed to refinance the huge number of mortgage loans falling into delinquency and default and served to bail out both working-class homeowners and the financial institutions where they had borrowed money. The Federal Housing Administration (FHA) was created in 1934 as an economic stimulus for homebuilding and mortgage lending. FHA mutual mortgage insurance became a policy tool for mitigating the risk of lending, reducing the down payment amount and monthly debt service needed for homeownership, creating industrywide standards for loan underwriting and property appraisal, and enforcing site planning requirements in new residential subdivisions. FHA insurance, for which multifamily housing and home improvement loans were also eligible, was used by mortgage companies and commercial banks to expand their activities in housing finance. In particular, FHA-insured loans became the core product of mortgage bankers, who began to significantly expand their originations of single-family home loans. Mortgage bankers worked closely with realtors, large-scale suburban "community builders," and national investors such as life insurance companies, all of whom valued their expertise in working with FHA programs. For the mortgage companies, FHA-insured lending and loan servicing became a reliable source of fee income, and the standardized mortgages were easy to package for sale to the secondary market.

The policies and administration of the FHA were strongly influenced by the prevailing attitudes and practices of financial institutions and real estate developers, which resulted in a built-in spatial bias favoring suburban areas over central cities. Under pressure during the Great Depression to operate in a businesslike manner and minimize potential losses, the FHA adopted a risk-rating system that directed mortgage insurance to newer middle-class neighborhoods planned for a racially and ethnically homogeneous population but refused to underwrite loans in older, poor and working class inner-city areas. This neighborhood rating system, accompanied by color-coded maps that highlighted the locations of greatest and least perceived risk, was patterned after the methods used by the HOLC, which downgraded low-income urban districts but nevertheless refinanced many loans there. The HOLC itself had simply adopted the neighborhood risk ratings that were commonly found in private industry, as indicated in the procedures recommended by the leading real estate appraisal textbooks.

The repeal of branching restrictions in the 1950s and 1960s enabled banks and thrift institutions to expand and sometimes relocate through mergers and acquisitions into suburban areas and growing "Sunbelt" regions. This trend was perhaps best exemplified by the rise of the powerful savings and loan industry in California. Periodic interest rate "shocks" and credit "crunches" during the 1960s, resulting from the globalization of the economy and government spending for the Vietnam War and social policy initiatives, caused further change in the U.S. financial system. The Federal National Mortgage Association (now known as Fannie Mae), which was formed in 1938 to support the FHA program through the purchase of government-insured home loans, was joined by the Federal Home Loan Mortgage Corporation (Freddie Mac), established in 1970 to buy conventional, privately insured loans from FHLBB thrifts and by the Government National Mortgage Association (Ginnie Mae), created by the Housing Act of 1968 to package and sell investment securities backed by FHA- and Veteran Administration-guaranteed (VA) loans. After 1970, both Fannie Mae and Freddie Mac gained new powers as privatized government-sponsored enterprises, enabling Fannie Mae to purchase conventional instead of just FHA- or VA-backed loans, Freddie Mac to purchase conventional mortgages originated by commercial banks and mortgage companies instead of just thrifts, and allowing both entities to issue their own mortgage-backed securities. These secondary market agencies gained considerable influence by establishing their own loan underwriting guidelines for primary lenders to use in making loans that are then sold, and they stimulated the rapid growth of the mortgage banking industry. By 1986, Fannie Mae, Ginnie Mae, Freddie Mac, life insurance companies, pension funds, and other financial institutions and investors were buying three out every four mortgages originated.

As interest rates grew with inflation during the 1970s, savings and loans began to experience "disintermediation" resulting from increased institutional competition for deposits. This created a mismatch between their traditional long-term, fixed-rate mortgage liabilities and their changing asset base of liquid, short-term, high- and variable-rate deposit accounts. Disintermediation weakened the position of savings and loans as a source of funds in housing, and the industry collapsed after further deregulation in the 1980s, which led to an even stronger role for the secondary market and mortgage bankers in housing finance. Commercial banking continued to grow and consolidate with regional interstate expansion during the 1980s and the passage of a national interstate branching law in 1994.

The dollar volume of single-family home mortgage originations by thrift institutions declined from one-half of the annual total in 1980 to only one-fifth by 1994, whereas the market share held by banks and mortgage companies grew from 43% to nearly 80% over the same period. Multifamily mortgage lending by commercial banks also increased rapidly during the 1980s, from only 10% of the dollar volume of all originations in 1980 to nearly two-thirds by 1994. These lending institutions are now subject to a new financial regulatory framework designed to prohibit geographic "redlining" and racial discrimination in home mortgage lending. Community-based housing organizations and local governments have worked with primary lenders and secondary market agencies to create programs that target mortgage loans to older cities, working-class and minority borrowers, and first-time homeowners. In addition, specialized financial institutions have been chartered, such as community development banks and credit unions, to serve economically disadvantaged urban and rural areas. (SEE ALSO: **Mortgage Finance**; *Resolution Trust Corporation*; *Savings and Loan Industry*)

—*John T. Metzger*

Further Reading

Dymski, Gary A., Gerald Epstein, and Robert Pollin, eds. 1993. *Transforming the U.S. Financial System: Equity and Efficiency for the 21st Century*. Armonk, NY: M. E. Sharpe.

Eichler, Ned. 1988. *The Thrift Debacle*. Berkeley: University of California Press.

Florida, Richard L., ed. 1986. *Housing and the New Financial Markets*. New Brunswick: Rutgers, State University of New Jersey, Center for Urban Policy Research.

Goldsmith, Raymond W. 1958. *Financial Intermediaries in the American Economy Since 1900*. Princeton, NJ: Princeton University Press.

Kim, Sunwoong and Gregory D. Squires. 1995. "Lender Characteristics and Racial Disparities in Mortgage Lending." *Journal of Housing Research* 6(1):99-114.

Mayer, Martin. 1974. *The Bankers*. New York: Weybright & Talley.

Taub, Richard P. 1994. *Community Capitalism*. Boston: Harvard Business School Press. ◄

► Levittowns

Changes in government housing policies in the years following World War II altered both the availability and the nature of housing for lower-income U.S. workers. Between 1947 and 1967, millions of low-cost, rudimentary, single-family houses were constructed in suburban housing developments under the terms of the Veterans Emergency Housing Program, the Federal Housing Acts, and the GI Bill. These developments are typified by Levittown, the 17,500-unit subdivision constructed on Long Island by the firm of Levitt and Sons under the management of William Levitt.

During the war, Levitt had built worker housing for defense plants at Norfolk, Virginia. The units were designed to provide the bare essentials of shelter and were cheaply and rapidly constructed using standardized mass-produced materials and innovative design. Among the innovations were the use of homosote, plywood, and sheetrock in 4' × 8' components and the introduction of concrete slab foundations, which speeded construction by eliminating the need for basements.

Using the innovations developed during the war, Levitt began construction of the first of the three postwar subdivisions that he would call Levittown (Island Trees, Long Island, 1947-1951; Bucks County, Pennsylvania, 1951-1955; and Willingboro, New Jersey, 1958-1965).

Levittown, Long Island

The first Levittown was built in the hamlet of Island Trees on Long Island. Construction began in late spring of 1947. The first 2,000 houses were completed and occupied in October. Over the next year, an additional 4,000 houses would be added to the subdivision. All of the first 6,000 houses were Cape Cods, a style that the Federal Housing Authority (FHA) had approved for funding under its programs. All were rental units and rented for $52 per month, which was roughly one week's pay for the target constituency, who had an income of between $2,500 and $3,000 per year.

The platform-framed houses were a reduction of the middle-class house of the prewar period to an affordable minimum: a 750-square-foot, four-room house with a kitchen and living room, a single bath, and two bedrooms. The unfinished attic could be converted into two additional bedrooms. Every house came equipped with basic, con-struction-grade kitchen appliances; enameled metal cabinets; a refrigerator; and stove. In addition, the builder included a washing machine. The floor plan was based on four-foot units, which capitalized on the rising availability of standardized components in sheetrock, windows, and plywood. Other economies were based on technologies developed during the war for military housing, such as asbestos siding, asphalt tiles, and latex paint. The houses were set on concrete slab bases with radiant heating provided by in-the-floor hot water coils. This innovation required a modification of the local building ordinance, which was granted. Levitt also requested permission to substitute 5/16" plywood for the traditional 3/4" sheathing. This modification was not approved until after the completion of the Island Trees development.

The construction at Levittown altered the traditional relationship between worker and product and attempted to bring the construction industry into the industrial age by treating the workers and the houses as if they were units on an assembly line. The building process, which Levitt called the "on-site factory," was broken up into some 30 operations, each of which was a complete entity for which a work crew was trained. Like the inverse of the factory assembly line, the crews moved from house to house completing a single operation before moving on to repeat it at the next site.

In the final months of 1948, the FHA, concerned about market saturation, questioned the viability of Levittown-type Cape Cod style rental projects and reduced its funding commitments for them. In response, Levitt redesigned the basic model for his houses and made them available solely for the proprietary market. The facade on the new model—the "49er" or Ranch—was more modern and contained an additional 50 square feet of floor space. There were four exterior design variations on a single floor plan, which featured a kitchen at the front of the house and a living room in the rear, facing the garden. The four-room layout included the kitchen appliances and washing machine, a fireplace, and a picture window as standard equipment. Here too, materials were durable but basic; the kitchen cabinetry was enameled steel; the floors, asphalt tile; and the outer walls, asbestos shingles.

With his ranches, Levitt introduced the idea of an "annual model"—a concept borrowed from the automotive industry—into the field of housing. As with automobile design, the annual model changes in the houses offered no substantive change in overall design. The more radical changes came in three- to four-year intervals and, for Levitt, were represented by moves to new locales as the available land in each area was reduced.

Although in form it was essentially the same as the 1949 models, the 1950–1951 Ranch included a finished third bedroom in the attic, a built-in television, and a carport. A fifth variation in the elevation was also offered. The 1949 model had sold initially for $7,990, but real estate inflation increased the cost of succeeding models. In all, 11,500 ranch models were built between 1949 and 1951 when the subdivision was completed. At completion, the development also provided seven village greens with stores and playgrounds, nine swimming pools, and a community meeting hall.

The first Levittown was subjected to considerable scrutiny from a variety of critics and fields. Architects, social scientists, and journalists levied judgment on the size, scale, and economics of the houses as well as on the quality of life the development offered to its residents. The critics claimed that the economic and demographic stratification of the population, along with the repetitive design of the houses, resulted in a lifestyle that was culturally stifling. The impact of low-density housing combined with increased distance from major urban areas suggested to the critics that these housing developments would degenerate into suburban ghettoes in which woman and children—particularly those of adolescent age—would be without meaningful activity. Many of the criticisms would have applied equally to other settings, but the tendency of the analysts to focus on the new suburbs led them to imply a causal relationship between the low-cost houses and the quality of the lives of their inhabitants.

Other critics expressed the fear that the low cost and reduced amenities of Levittown would result in a slum. The criticisms not only stung but reduced the marketability of the Levittown product. The Levitts responded by redesigning their product for the successive Levittowns and incorporating more of the amenities being offered by the local builders.

Despite the prognoses of most of the observers, the community was well received by those in the target market. New, the houses sold as quickly as they could be constructed; used, they continued to sell as the original residents moved on. Subsequent titles in the federal housing policies made home renovation very attractive, and the homeowners in Levittown took advantage of the terms to expand and remodel their houses into more commodious homes. As a result, the houses continue to withstand the test of time, and the community has flourished.

Levittown, Pennsylvania

The success of the Long Island Levittown led, in 1951, to the construction of a second 17,000-unit development just outside Philadelphia in Pennsylvania's Bucks County. Although the builder had abandoned the traditional Cape Cod model in favor of a house that was more modern in flavor, the Bucks County Levittown shared many of the elements of the Long Island community. Several exterior variations or "models" were offered each year with little or no variation in the interiors. The houses were clustered in neighborhoods in which the houses represented a single annual model, and land was set aside for schools and churches. Unlike the Long Island development, the Bucks County project did not offer internal "village greens" with small shops and playgrounds. The village green concept, with its emphasis on a walk-in trade, was already outmoded in the automobile culture of the 1950s, and most of the small shops had failed. Instead, Levitt built a large shopping center at the outer edge of the community, which was more viable.

Although the veterans no longer had an exclusive claim on the first choice of the houses, the continuation of the GI Bill as the primary funding mechanism for purchasing the houses meant that the target market remained similar to that of the Long Island development—blue-collar workers and the lower end of the white-collar middle class. The great majority of the houses were designed to sell for about $10,000, although a few sold for $17,000. The more expensive models were intended for those who might move in at the low end of their earning capacity but remain as their earnings increased. In addition, a percentage of the units were rented for $65 a month. This variation in pricing addressed one of the major criticisms of those who had objected to the socioeconomic stratification of the earlier Levittown.

Levittown, New Jersey

By 1955, the market had begun to shift. Levitt planned the third Levittown at Willingboro, New Jersey, for a market that was somewhat more affluent. To attract that market, he offered a variety of models with three to four bedrooms and a footprint of more than 1,200 square feet. The Willingboro houses were built in a more traditional "neo-Colonial" style and cost from $11,500 to $14,500, depending on the number of rooms. In 1964, a five-bedroom model was introduced at a commensurately higher selling price. The increase in selling price reflected the increase in space, rather than an upward shift in quality. Despite the inflation, Levitt had kept the cost per house at a constant level of about $10 per square foot.

A major innovation in this Levittown was the decision to mix the models within the neighborhoods. Whereas in the earlier Levittowns each neighborhood was limited to the variations on that year's annual model, in Willingboro, the houses were introduced in a variety of styles and models each year and were interspersed throughout the development. This change addressed one of the most persistent of the anti-Levittown criticisms by reducing the tract house appearance and implying the more traditional, organic, suburb of mixed styles and periods. Nevertheless, in reaction to the critics' repeated use of the term *Levittown* to signify the problems of the postwar subdivisions, the residents of the New Jersey Levittown voted in the late 1960s to change the name of their community to the name of the original hamlet, Willingboro.

Although the Levitt firm continued to build suburban housing, not only on the U.S. mainland but as far away as Puerto Rico, France, and Israel, the Levittown name was not revived. (SEE ALSO: *Department of Veterans Affairs; Suburbanization*)

—*Barbara M. Kelly*

Further Reading

Buhr, Jean Dieter. 1988. "The Meaning of Levittown." Ph.D. dissertation, Department of Geography, Rutgers University, New Brunswick, NJ.

Dobriner, William M. 1981. *Class in Suburbia*. 2d ed. Westport, CT: Greenwood.

Gans, Herbert. 1982. *The Levittowners: Ways of Life and Politics in a New Suburban Community*. 2d ed. New York: Columbia University Press.

Keller, Mollie. 1990. "Levittown and the Transformation of the Metropolis." Ph.D. dissertation, Department of History, New York University.

Kelly, Barbara M. 1993. *Expanding the American Dream: Building and Rebuilding Levittown*. Albany: SUNY Press.

Liell, John. 1952. "Levittown: A Study in Community Planning and Development." Ph.D. dissertation, Department of Sociology, Yale University, New Haven, CT. ◀

▶ Limited-Dividend Development

From the early 1960s to the mid-1980s, the limited-dividend approach was one of the major ways the federal government stimulated private investment capital for the production of low- and moderate-income rental housing. Thousands of units were developed with capital from investors who agreed to take a limited, fixed return or dividend on their investments paid out of the distributions from project rents. In exchange, they received other financial benefits, the principal one being substantial tax savings from passive losses on property depreciation. An estimated 330,000 units were produced with such capital infusions through the U.S. Department of Housing and Urban Development (HUD) Section 221(d)(3) and Section 236 multifamily programs. Another 300,000 units were developed via the Farmers Home Administration (FmHA—now Rural Housing Service [RHS]) Section 515 program within the U.S. Department of Agriculture.

Up until 1986, investors in real estate were permitted a special exemption from an "at-risk" provision in federal tax law that allowed them to deduct from taxable income more in losses from property depreciation than their actual cash investment. In contrast, deductions allowed for losses in most non-real estate investments were limited to the actual cash investment or the amount for which direct investor liability existed. Moreover, real estate investors who engaged in new residential construction were favored by additional provisions allowing the highest rates of accelerated property depreciation. The Economic Recovery Tax Act (ERTA) of 1981 created an even more advantageous depreciation benefit by reducing the period over which losses could be accelerated, favoring low-income housing with higher rates of acceleration than market rate housing and applying that rate to both new and existing properties.

Compared with other real estate investments, government-assisted housing provided among the greatest tax benefits. The depreciation deduction was based on the full development cost of the property, including the portion financed. Because these developments were so highly leveraged, requiring an equity contribution of only 10% or less of development cost, investors could buy in with a minimal up-front outlay and therewith reduce their personal income tax liabilities by offsetting ordinary income, such as salary, interest, and dividends, against property depreciation in dollar amounts over time that well exceeded the amounts of their original cash investments.

Depreciation losses on these properties were usually accelerated over a 20-year period. Investors anticipated that the properties would be sold at the end of that period when the tax shelter had been fully exhausted. Although postponement of taxes due in each year of the investment increased the amount due upon sale, most of the gains from sale would be taxed at the lower capital gains rate. Moreover, prepayment provisions in HUD and FmHA (RHS) financing enabled current or prospective owners to prepay the underlying government loan, remove all low-income use restrictions, and convert these properties to market rate operations at substantially enhanced market values. This problem first began to emerge in the late 1970s and 1980s.

To take advantage of these tax benefits, project developers, often acting in the capacity of general partner, organized groups of investors, sometimes called syndicates, by selling or syndicating units of limited partnership. The general partner received up-front revenues through the sale of partnership units to the limited partners, ongoing revenues from management fees and other fees payable from the limited-dividend distribution and some gain upon sale or refinancing. The limited-dividend partners received the tax benefit and the prospect of future gains upon sale, while having their actual out-of-pocket cash liability limited to the amount invested. Other types of limited-dividend investors in this housing were nonsyndicated profit entities organized as individuals, partnerships, or corporations.

In return for the shelter, investors agreed not only to accept a limited cash flow from rents (initially, 6% of equity per annum) but to methods of operation, income controls for admissions, and rental charges that would ensure occupancy by the targeted income groups. The rents were made affordable to low-income and very low income households through highly favorable loan terms to developers and, sometimes, additional project-based rental subsidies.

Many of the units that were financed in earlier years and had realized most or all of their depreciation losses were resyndicated in the mid-1980s under the new rules created in ERTA. General partners purchased and resold or resyndicated the properties to new limited-dividend investors who received the depreciation benefits. The schedule for depreciating properties was restarted based on the current market value.

The Tax Reform Act of 1986 removed the possibility of passive losses through investment in rental housing, eliminated preferential capital gains rates, lowered marginal tax rates, and substantially lengthened depreciation periods. With substantially reduced tax benefits remaining, existing owners have experienced diminished returns and greater incentives to liquidate their investments. Prospective investors in low-income housing were offered a tax credit in lieu of passive losses from depreciation. The credit has become the major incentive for private investors still willing to invest in low-income housing and accept limited dividends on their investments. (SEE ALSO: *Depreciation of Property; Section 221[d][3] and [4]; Section 236; Section 515; Tax Incentives*)

—*Robert Wiener*

Further Reading

Blankenship, Frank J. 1989. *The Prentice Hall Real Estate Investor's Encyclopedia*. Englewood Cliffs, NJ: Prentice Hall.

Jacobus, Charles J. 1996. *Real Estate Principles*. Upper Saddle River, NJ: Prentice Hall.

McKenzie, Dennis J. 1996. *Essentials of Estate Economics*. Upper Saddle River, NJ: Prentice Hall.

National Low Income Housing Preservation Commission. 1988. *Preventing the Disappearance of Low Income Housing.* Washington, DC: National Corporation of Housing Partnerships.

Peiser, Richard B. 1992. *Professional Real Estate Development: The ULI Guide to the Business.* Washington, DC: Urban Land Institute.

Shim, Jae K. 1996. *Dictionary of Real Estate.* New York: John Wiley.

Tosh, Dennis S. 1990. *Handbook of Real Estate Terms.* Englewood Cliffs, NJ: Prentice Hall.

► Limited-Equity Cooperatives

Limited-equity cooperatives are housing cooperatives that restrict the resale price of shares by limiting the equity returned to departing members. When residents leave, they generally receive the sum of their initial share price and their contribution to amortizing the cooperative's principal. Sometimes, the equity is adjusted for interest, inflation, or improvements made to the occupied unit. Cooperatives emphasizing affordability minimize such adjustments or cap transfer values to fix resale prices over time and to retain the value of any subsidies or property appreciation. Nationwide, there are an estimated 224,000 limited-equity co-op units, of which 60,000 are sponsored by New York city and state governments.

These cooperatives are affordable because, as self-governing nonprofit corporations, shareholders are not personally liable for the financial obligations of the cooperative. This makes it possible for low-income households to buy cooperatives even if they do not qualify for an individual mortgage. If a member defaults, the collective mortgage remains intact. Initial share prices are usually restricted to a month's rent, about the same as a security deposit on a rental unit. However, the limited-equity provisions guarantee that the share prices do not rise even when the value of the property does. New shareholders pay about the same amount as departing members. Because the cooperative holds a single mortgage regardless of turnover, transaction costs are eliminated and monthly payments are fixed over time. However, prolonged vacancies can erode reserve funds and, eventually, the cooperative's ability to meet its obligations.

Limited-equity cooperatives are designed for the use, not the profit of the members, who are usually low- and moderate-income households. Each household receives one share and one vote. Members have the right to occupy their units in return for monthly payments. Low-income cooperative residents have more security than they would have as tenants. Limited-equity cooperatives reduce the risk of displacement from escalating rents and maintain the long-term affordability of housing in booming markets. Unlike tenants, members can also control their housing conditions. They participate in selecting their neighbors, maintaining the property, and making decisions.

Some limited-equity cooperatives, such as those in New York, are affiliated with mutual housing associations (MHAs). MHAs are democratically controlled nonprofits, which themselves may be cooperatives, that develop and provide services to multiple affordable, resident-controlled housing projects. Low-income cooperatives often need professional, technical, and financial assistance because they are so complex and residents are more likely to lack the necessary skills, experience, education or, most typically, money to address unexpected problems. Many limited-equity cooperatives have been formed in abandoned or older buildings where structures and systems slowly deteriorate and require major infusions of capital to repair. MHAs provide cooperatives with a safety net in such circumstances.

Limited-equity cooperatives qualify for some subsidies unavailable to homeowners. Under Section 216(b)(1), nonprofit limited-equity cooperatives are eligible for tax-exempt private activity bond financing and are treated as rental projects if they have significant proportions of residents with low incomes. Receiving tax-exempt bond financing disqualifies residents from deducting mortgage interest and property taxes from their personal income taxes, but in the case of low-income households, these deductions are not worth much anyway. The same logic applies to syndication. Cooperatives rarely have incomes to offset, so they can raise capital through limited partnerships that receive Low-Income Housing Tax Credits. Nevertheless, cooperatives that accept federal subsidies lose some of their autonomy. In the case of low-income cooperatives, this may entail accepting residents on a first-come, first-served basis or on the basis of housing need, instead of selecting members committed to cooperative living. In racially and ethnically diverse low-income neighborhoods, this has caused friction.

The first limited-equity cooperatives in the United States were socially homogeneous. Among the best known is the Amalgamated Houses of the Amalgamated Clothing Workers Union. This low-rise neo-Tudor apartment complex was built adjacent to a park in the Bronx, New York, had extensive gardens and community facilities, and mostly served Jewish workers in the garment industry who escaped the slums of lower Manhattan. Although Amalgamated survived the Great Depression, most of the early working-class cooperatives could not carry their unemployed residents and eventually, failed. In the 1950s, most cooperatives were for middle-income households. However, the United Housing Foundation, under the direction of Abraham Kazan, who had been involved with the Amalgamated cooperatives, continued to convert or build limited-equity cooperatives. In addition, about one-sixth of the units in New York's Mitchell-Lama cooperatives went to low-income households. In the 1980s, as property values boomed and 20-year restrictions on mortgage prepayments expired, some Mitchell-Lama cooperatives sought to convert from limited-equity to market rate, which could have potentially displaced low-income residents. However, partly because the real estate market later cooled down, only one cooperative had left the Mitchell-Lama program as of 1995.

Most federal cooperative programs of the 1960s had limited-equity restrictions, but their subsidies were not deep enough to reach low-income households. That began to change in the 1970s, when Section 8 subsidies were made available to some cooperative members. By the 1980s, virtually the only cooperative subsidies were for low-income households, including public housing tenants. Nevertheless, most limited-equity cooperatives today must combine federal assistance with other subsidies from state and local governments, private lenders, and nonprofits to produce housing affordable to low-income residents. (SEE ALSO: **Community-Based Housing**; *Cooperative Housing*)

—*Hilary Silver*

Further Reading

Dolkart, Andrew. 1993. "Homes for People: Non-Profit Cooperatives in New York City—1916-1929." *Cooperative Housing Journal* (January/February):13-22.

Levy, Herbert. 1992. "An Introduction to Limited Equity Cooperatives." *Cooperative Housing Journal* (Annual Issue):10-11.

Miceli, Thomas J., Gerald W. Sazama, and C. F. Sirmans. 1994. "The Role of Limited Equity Cooperatives in Providing Affordable Housing." *Housing Policy Debate* 5(4):469-90. ◀

▶ Linkage

Linkage refers to the practice by local governments of making land use and zoning approval for nonresidential (usually commercial) development contingent on the provision of affordable housing. In linkage programs, nonresidential developers are required to provide affordable housing as a condition of permit approval. Linkage emerged during the 1980s at a time when many cities were experiencing a robust downtown office economy at the same time they were suffering from shortages of affordable housing. A number of cities, generally on the East and West Coasts, have instituted linkage programs since 1980. At a time when the U.S. federal government was withdrawing support for low- and moderate-income housing, linkage was seen as a means of generating capital for affordable housing development at the local level.

The logic of linkage is based on the connection between nonresidential development and increased demand for affordable housing. As commercial and industrial development introduce new employment opportunities and new households into a local market, the competition among residents, old and new, for housing opportunities increases. In a tight housing market, with few vacancies, the competition leads to price escalation. Under these conditions, higher-income households outbid lower-income families, causing displacement. Without an increase in the housing stock to accommodate the increase in demand generated by nonresidential development, low-income families are forced out of the housing market. Therefore, linkage programs require nonresidential developers to provide affordable housing to mitigate the potential housing market problems produced by their developments.

History of Linkage

Linkage programs emerged at the historical confluence of three trends in the political economy of U.S. cities. The first trend was the phenomenal growth in the office-based economy of major cities. From the late 1970s through most of the 1980s, cities across the country experienced a huge increase in the amount of commercial office space in their downtown cores. These high-rises were home to the expanding business service industries and corporate management facilities that have come to dominate U.S. urban areas since the mid 1970s. Inner-city neighborhoods directly surrounding downtown cores were often transformed from centers of very low income housing and low-density, light industrial/warehousing functions to corporate centers of finance and management and upscale retail and commercial nodes built to accommodate a new, more professionalized workforce.

At the same time in the United States, many urban areas were experiencing a severe housing affordability crisis. While affordable units were demolished to make way for offices, inflation in housing costs was significantly outpacing incomes. Hundreds of thousands of persons became homeless, and millions of other families were paying high percentages of their incomes for shelter. Gentrification in inner-city neighborhoods also eliminated affordable housing for lower-income people. The crisis of affordability, furthermore, seemed most acute in those cities experiencing economic transformation to corporate service and management functions.

The final trend that led to local housing solutions such as linkage, was the almost complete withdrawal of the federal government from housing assistance during the 1980s. Under the Reagan administration, the HUD budget authorization fell by more than 80% compared with its high mark in 1978. So at the same time that housing conditions were worsening in inner cities, the federal government was dramatically reducing its contribution to assisted housing for low-income persons. This led local officials as well as low-income advocates to look at local sources of housing assistance. The healthy commercial office space markets in many cities provided an opportunity to raise needed housing capital. Linkage programs are attractive to local officials because they allow local governments to raise capital for housing assistance without raising taxes, incurring debt, or using general revenues.

The City of San Francisco was the first major city to adopt a housing linkage program. San Francisco's program came about as the result of community activists concerned about the deleterious effects of rapid downtown office space construction. These activists were also housing advocates, and they proposed to link an exaction on office developers to the production of affordable housing. Their initial expectation was that the exaction would significantly reduce the rate of growth in downtown construction. In fact, San Francisco office developers were not deterred in any way by the linkage fee. The rate of office construction continued unabated and the program generated close to $30 million in revenues over the first five years.

The success of the program led other municipalities to try linkage programs of their own. Boston and Miami adopted linkage programs in 1983, Seattle followed suit in 1984, and other cities added programs throughout the rest of the decade. By the late 1990s, 10% to 15% of medium- to large-sized cities practice some form of linkage. This includes cities that do not have a formal program but nevertheless negotiate linkage fees on a project-by-project basis. Most of the cities adopting linkage were characterized by strong nonresidential development markets in the 1980s and are located on either coast, with California, Florida, and New Jersey having a disproportionate number of cities using linkage.

The two largest programs have been in San Francisco and Boston. In San Francisco, the program assisted more than 5,000 units between 1981 and 1989. In Boston, developers committed more than $46 million that assisted almost 2,500 units. In other cities, the programs have generated more modest activity.

How Linkage Programs Work

There are many variations in the way linkage programs can be designed. Generally, the enabling legislation for linkage is provided in local governments' zoning authority. Thus, typically, linkage is implemented as part of the permit approval process for nonresidential development. In most places, the linkage fee is required as a condition of permit approval and must be in place by the time the certificate of occupancy is issued for the nonresidential development.

Most linkage programs require commercial developers to provide affordable housing. This is especially true of the earlier programs—those adopted prior to 1986. As the decade of the 1980s progressed, local officials expanded the concept to deal more inclusively with other types of nonresidential developers. A review of 15 linkage programs revealed that 8 of them applied not only to commercial developers but also to industrial developers or, in some cases, to all nonresidential development.

Nonresidential developers can fulfill their housing obligations in a variety of ways. Some jurisdictions allow only a contribution of fees to a locally administered trust fund. Others allow the developer to build the housing as well as pay into a fund. One program, in Cambridge, Massachusetts, allows the developer to donate land.

Other program design choices include (a) whether or not the trust fund revenues will be used exclusively for affordable housing or whether some market rate housing is allowed, (b) whether linkage applies only to downtown development or to development citywide, (c) whether there are spatial limitations on where the assisted housing can be built (i.e., in proximity to the nonresidential development or citywide), and (d) whether the program is a mandatory requirement, a voluntary program, or incorporated into an incentive zoning package (e.g., in return for a density bonus).

Implementation Issues

The main concern with linkage programs is that they will discourage nonresidential development. Local officials must consider the possibility that imposing linkage requirements might drive nonresidential developers to locate their projects in other municipalities that do not have linkage requirements. If this impact were large enough, linkage might have the perverse effect of short-circuiting the very economic growth that serves as its basis.

Most of the evidence, however, suggests that this has not happened. One study found that most developers simply planned to pass the added cost created by linkage on to their office space tenants. Another study showed that the cost of compliance in San Francisco represented between 1% and 2% of total nonresidential development costs. In the cities that have adopted linkage, these costs are too low and too easily passed on to deter nonresidential developers.

The legal issues related to linkage are a greater concern for local officials. The first legal question is whether the exaction constitutes an illegal taking of private property. The taking issue hinges on whether the amount of the fee is reasonable and proportional to the impact of the development. In practice, most municipalities have been quite conservative in the computation of the amount of the housing exaction. This has made the "taking" issue moot.

The most important legal issues are related to constitutional guarantees of "due process." A procedural due process violation can be avoided by implementing linkage through a city ordinance rather than through administrative rules. A substantive due process issue relates to how well the exaction is linked to the nonresidential development—that is, how clearly it has been shown that nonresidential development affects the availability of low-income housing. The *Nollan v. California Coastal Commission* decision handed down by the U.S. Supreme Court in 1987 requires that a direct relationship be established between the exaction and the need created by new development. This "rational nexus" ruling requires cities to have studied and documented the link between nonresidential development and affordable housing needs.

Some argue that linkage has been overemphasized as a solution to local housing problems. First, there is the problem of its suitability for many cities. Without a strong nonresidential development market, a city cannot make use of the program. It seems unlikely, therefore, that linkage will ever be widely adopted as a means of generating local housing revenues. Similarly, some argue that linkage is a poor substitute for greater federal responsiveness to housing issues and, by itself, is an insufficient base on which to build a local housing strategy. Even in San Francisco and Boston where linkage has generated millions of dollars in housing capital, these resources have not been enough to fully address local housing needs. These arguments seem somewhat unfair, however, because linkage has never been presented either as a solution for all local governments or as a complete replacement for federal funds.

A related criticism of linkage is that it provides nonresidential developers with a means of avoiding further mitigation for the unequal social impacts of their development. Some, for example, argue that once developers pay the linkage fee, opposition groups lose their grounds for demanding further responsiveness from developers on social issues. That is, "linkage amounts to a license to externalize the social costs of development." This critique suggests that linkage is only a partial mitigation of the negative externalities of urban development and that it may serve to impede full mitigation. The example of San Francisco, however, where housing mitigation has been combined with a transportation exaction, day care requirements, and development limits suggests that this is not necessarily the case.

Since these programs emerged in the early 1980s, linkage has provided local officials and activists in some cities with an effective means of generating housing subsidies in a decade in which the U.S. federal government was drastically reducing its commitment to housing. Linkage is best considered as part of a larger strategy to address urban housing needs. (SEE ALSO: *Density Bonus; Housing Finance; Impact Fees*)

— *Edward G. Goetz*

Further Reading

Crook, A. D. H. 1996. "Affordable Housing and Planning Gain, Linkage Fees and the Rational Nexus: Using Land Use Planning Systems in England and the USA to Deliver Housing Subsidies." *International Planning Studies* 1(1):49-72.

Dreier, Peter and Bruce Erlich. 1991. "Downtown Development and Urban Reform: The Politics of Boston's Linkage Policy." *Urban Affairs Quarterly* 26:354-75.

Goetz, Edward G. 1988. "Office-Housing Linkage Programs: A Review of the Issues." *Economic Development Quarterly* 2:182-96.

———. 1989. "Office-Housing Linkage in San Francisco." *Journal of the American Planning Association* 55:66-77.

Herrero, Teresa R. 1991. "Housing Linkage: Will It Play a Role in the 1990s?" *Journal of Urban Affairs* 13:1-20.

Keating, W. Dennis. 1986. "Linking Downtown Development to Broader Community Goals." *Journal of the American Planning Association* 52:133-41.

Porter, Douglas, ed. 1985. *Downtown Linkages*. Washington, DC: Urban Land Institute.

Smith, Michael Peter. 1989. "The Uses of Linked-Development Policies in U.S. Cities." In *Regenerating the Cities*, edited by M. Parkinson, B. Foley, and D. R. Judd. Glenview, IL: Scott, Foresman. ◄

► Local Government

Housing is the physical, economic, and social backbone of local community structure. Although direct housing activity was generally regarded in the United States as a private enterprise rather than a municipal government function, the general powers of local governments have had a powerful influence on both housing and development for a long period of time. The two principal mechanisms used by municipalities since the 1920s to control land use are zoning and subdivision regulations. Building codes are another significant area of municipal influence. However, only after disastrous fires in Chicago (1871) and San Francisco (1906) did governments give serious attention to local comprehensive building codes; this was a slow-growing movement, beginning with the model codes prepared by local building officials beginning in 1915. The evolutionary adoption of municipal housing occupancy codes, which governed health and safety conditions, also influenced housing and development. A major expansion in zoning, building, and occupancy codes came after 1954 when the federal government required that cities and other local public bodies adopt housing, zoning, building, and other local codes and regulations as a condition of receiving federal urban renewal assistance. Known as the "workable program," this action resulted in the adoption of building codes in more than 3,000 localities. Still another area of municipal activity affecting local housing and development is property taxes, a dominant source of municipal revenue; increasingly, local governments have learned to use taxation to promote housing and to guide development.

A 1991 analysis of implementing land use regulations to promote housing (housing linkage, preservation, replacement, inclusionary zoning, and rent control) docu-

mented that in 133 cities with populations of more than 100,000, from 12% to 19% used these mechanisms. Along with individual functional activities, there has been an evolutionary growth in more comprehensive local planning instruments that set forth long-term policy goals for community development.

More aggressive action by the federal government in the 1930s in providing assistance for slum clearance and the development of low-income housing brought local governments into a new arena of direct local housing—and ultimately, community revitalization—activity. Initially, this federally assisted activity was administered by local housing and redevelopment agencies separate from general-purpose local governments. Beginning with the Housing and Community Development Act of 1974, local governments assumed the administrative role under Community Development Block Grants.

The decade of the 1980s brought an array of new locally generated housing and community development initiatives and a new range of participants in housing and urban development activity. It also witnessed more sophisticated efforts to merge traditional local governmental powers (land use, zoning, building codes, housing occupancy codes, taxation) with the more direct housing and community development activity supported by federal assistance. By the late 1990s, a more comprehensive local housing and urban development function was in formation.

New Initiatives in the 1980s

The long evolution of the housing and urban development function in local government reached a new stage of expansion in the 1980s, propelled by changes in local economic structure, changing demographics, a decline in neighborhood conditions, new areas of low-income housing need, and a decline in federal housing and urban development assistance. These new interventions took a variety of forms: public-private partnerships with new support for neighborhood community development organizations, housing trust funds, linkage to private development, use of tax increment proceeds, and affordable housing financing. At present, these interventions are a dynamic, diverse mixture of activity but with few comprehensive strategies to link them in a total community effort.

Public-Private Partnerships

Local public-private partnerships take a variety of forms: Some are focused on a single development or project. In 1989, the U.S. Conference of Mayors identified 127 examples of individual partnership projects, including homeownership, rental housing, home improvement and rehabilitation, and housing for special needs (elderly housing, transitional housing, single-room occupancy housing). Using a broad definition, the National Association of Housing Partnerships catalogued 26 partnerships in 1990, 43 in 1991, 64 in 1992, and 73 in 1993.

The more structured permanent partnerships establish an ongoing "program" of activities going beyond single projects, covering a communitywide base and extending over time. In general, such partnerships are based on agreements between a locality's corporate and financial institutions, the city government, and neighborhood-based nonprofit organizations. Typically, the private entities and the city government provide investment capital to undertake an agreed-on agenda of neighborhood improvement activities, usually carried out by neighborhood organizations. The glue holding the partnership together is a cadre of financial and development specialists who structure the investment package and administer the assistance; often, the investment incentive for the private investor is a tax benefit under the federal low-income housing tax credit program. The growth of local partnerships has been stimulated by foundations and private corporate funding through the Enterprise Foundation, the Local Initiatives Support Corporation, the Neighborhood Reinvestment Corporation, and Fannie Mae (formerly, the Federal National Mortgage Association). A number of early partnerships, particularly in Chicago (1985), Cleveland (1986), and Boston (1988) became models for other cities.

Local Housing Trust Funds

The concept of local housing trust funds, often using dedicated sources of local revenue, was a product of the late 1970s and early 1980s in response to cuts in federal housing assistance and increasing local pressures to provide low-income housing. In 1977, the Community Redevelopment Agency of Los Angeles used tax increment revenues received from three commercial redevelopment projects to create five different neighborhood trust funds for new low-income housing. Early pioneers in trust funds were Dade County (Florida) and Montgomery County (Maryland). In 1988, the National Association of Housing and Redevelopment Officials documented 12 local housing trust funds defined as a pool of money, either from dedicated revenues or appropriations or a combination of these, that was earmarked for specific purposes, usually the financing, construction, acquisition, or rehabilitation of housing for low- or moderate-income families. In 1994, the Housing Trust Fund Project of the Center for Community Change reported housing trust funds in 32 localities and 15 counties, using diverse sources of funding: Among the most common sources are (a) housing linkage fees and inclusionary zoning (particularly in California), requiring private commercial developers to include low-income housing in their development plans; (b) revenues from government programs, including the federal Community Development Block Grant (CDBG) and HOME programs; and (c) city appropriations and real estate sources (particularly real estate transfer taxes). These sources fluctuate with economic conditions. Local funds are sometimes administered by independent nonprofit organizations with members appointed by the city government; in many cities, funds are administered through local government agencies.

Linkage to Private Development

Increasingly, local governments have tapped the private development process for housing through linkage agreements, density bonuses, and inclusionary zoning ordinances. Linkage agreements usually involve a contribution for housing by a commercial or office developer or an agreement to build replacement housing. The rationale for linkage exactions is the nexus between the development and its impact on the local housing market. Linkage programs are generally described as "mandatory" exactions (Boston, San Francisco, Santa Monica) or "voluntary" and linked to density bonuses for the developer (Seattle and Miami). Increasingly, cities negotiate linkage agreements on an individual project basis rather than requiring a mandatory housing contribution based on the square footage in the new development. Boston's parcel-to-parcel program and New York's Battery Park bond float represent still further approaches to linkage that tie access to prime development sites with neighborhood sites in need of reinvestment.

Inclusionary zoning has been used primarily in suburban counties and small cities. The philosophy of inclusionary housing programs is to make development of lower-income housing an integral part of other development taking place in the community by setting either mandatory or voluntary objectives for the inclusion of lower-income housing in prospective market rate developments, coupled with incentives, such as density bonuses, to facilitate these conditions or objectives. It is estimated that there are more than 50 inclusionary zoning ordinances, mostly in California. Although linkage has been described as an emergency-type solution for cities dependent on a rather narrow range of potential investors in high-profit, "hot market" situations, it is a useful tool in appropriate situations: Between 1983 and 1989 in Boston, under its mandatory exaction program, 41 developers committed $76 million to housing programs. Linkage housing programs have also forcefully raised the issue of making lower-income housing an integral part of total community development.

Tax Increment Financing

The use of increased local taxes received from redevelopment areas for other housing and community development needs—tax increment financing—has been authorized in at least 35 states but is used widely in only two—California and Minnesota. California law requires that 20% of tax increment revenues be used for housing, unless a particular redevelopment project is exempted. As noted above, in Los Angeles, five housing trust funds were financed from three redevelopment projects, with activity totaling about 15,000 low-income housing units. Duluth, Minnesota, used tax increment monies to develop four downtown development ventures, including 420 units of housing.

Affordable Housing Financing

Perhaps the most widespread mechanism used by local governments to assist the development of affordable housing is the issuance of tax-exempt mortgage bonds that lower the development cost through below-market interest rates. Often, these issuances are made in conjunction with projects developed by public-private housing partnerships (see above). CDBG funds are usually included in these financing packages are. The Association of Local Housing Finance Agencies (ALHFA) reported in 1987 a total of 129 full-fledged local housing finance agencies consisting of one-third operating divisions of city or county government, one-third single-purpose local housing finance agencies, and one-third city or county housing authorities (financing for other than federally assisted public housing). Local agency issuances of mortgage revenue bonds grew from

$1.54 billion in 1981 to $4.52 billion in 1985. Typically, tax-exempt, bond-financed interest rates are 1.5% below the conventional market rate. The federal Tax Reform Act of 1986 (P.L. 99-154) imposed restrictions on tax-exempt mortgage bonds, which resulted in a reduction in issuances. Local mortgage bonds were made more viable, however, by the authorization of Low-Income Housing Tax Credits (LIHTCs) for housing investors in this same act. In 1991, local housing finance agencies issued $2.14 billion in mortgage revenue bonds. Further restrictions in legislation of 1988, 1989, and 1990 reduced the authorized volume of bond issuances and placed tight targeting restrictions on LIHTCs. Since 1995, powerful conservative political forces in Congress have sought to eliminate the LIHTC but as of 1998, without success.

Future Issues: Structures and Strategies

Expanding housing and community development activity by local public agencies in the decade of the 1980s has had significant impacts on the local administrative structures of housing and community development and raised new issues about the coordination of these efforts under common goals or community improvement strategies. The diversity and complexity of current activity has yet to be completed and evaluated.

From an evolutionary perspective, local administrative structures were established to carry out zoning, building, and housing codes functions, usually by departments of local government. Beginning in the 1930s, new structures in the form of independent local housing and redevelopment authorities were established to administer federally assisted public housing and urban redevelopment programs. In 1974, the general-purpose city government was given authority to administer a consolidated CDBG program, not including public housing. Local administration of CDBGs has taken a number of forms: Larger cities have tended to incorporate it into larger departments of housing, community development, or both; smaller communities have usually established "offices of community development" under the office of the mayor. Local housing authorities still carry out the responsibilities of federally assisted public housing and, since 1974, local responsibility for the administration of the federally assisted Section 8 rent certificate program; local housing authorities have also initiated new housing development or rehabilitation programs without federal assistance.

Local redevelopment authorities (or combined housing and redevelopment authorities) have tended to concentrate on longer-term community revitalization efforts using linkages with private development or tax increment proceeds from redevelopment projects. Beginning in the 1960s, there was some movement to consolidate local organizational structures for housing and community development, most notably in New York City (the Department of Housing Preservation and Development), Baltimore (the Department of Housing and Community Development), and Dade County, Florida (the Department of Housing and Urban Development). Most localities, however, have not undertaken full consolidation of housing and community development functions, retaining the original organizations designed to respond to federal assistance and adding new

locally generated activities under this organization. Each locality develops its own organizational structure—there is no uniform pattern because there is no uniform pattern of local housing initiatives. Larger cities have developed coordination mechanisms usually centered in the mayor's office; in the past, in cities such as Baltimore, the mayor convened a "cabinet" of department heads with responsibilities affecting housing and community development to effect coordination of efforts.

A survey of the "primary affordable housing departments" by the University of Illinois in the 12 largest cities in 1991 identified a range of functions carried out by these agencies, with the most prominent function (67%) being housing. Two cities (Baltimore and Indianapolis) also include community development, economic development, planning, and code enforcement. Two cities (Detroit and Memphis) include economic development along with housing. Some cities have created quasi-public corporations to administer or oversee long-term development initiatives. Since 1991, the Baltimore Department of Housing and Community Development has added a significant human services component to its organizational structure and has separated the public housing program as an independent authority.

One of the major issues facing local governments in the future is how to coordinate the multiple housing activities—public, private, and nonprofit—now taking place within their political boundaries and how to achieve a political consensus on a comprehensive strategy to shape the total development of the city. This is particularly true in relationship to the expanding number and influence of nonprofit neighborhood community development corporations. The existing instruments, such as the federally required comprehensive housing affordability strategy (CHAS), make up only one aspect of the total strategy that is required. Although a few cities are taking action to devise comprehensive strategies, it is not a widespread pattern. Bringing together the fragmented organizational structure and developing comprehensive city development strategies are critical issues for the future. (SEE ALSO: *Density Bonus; Federal Government; Housing Bonds; Inclusionary Zoning; Linkage; State Government*)

—*Mary K. Nenno*

Further Reading

Mallach, Alan. 1984. *Inclusionary Zoning Programs: Policies and Practices.* New Brunswick: Rutgers, the State University of New Jersey, Center for Urban Policy Research.

Nenno, Mary K. 1996. *Ending the Stalemate: Moving Housing and Urban Development into the Mainstream of America's Future.* Lanham, MD: University Press of Amatch.

Nenno, Mary K. and George C. Colyer. 1988. *New Money and New Methods: A Catalog of State and Local Initiatives in Housing and Community Development.* Washington, DC: National Association of Housing and Redevelopment Officials.

Stegman, Michael A. and J. David Holden. 1987. *Nonfederal Housing Programs: How States and Localities Are Responding to Federal Cutbacks in Low Income Housing.* Washington, DC: Urban Land Institute.

Suchman, Diane R. with D. Scott Middleton and Susan L. Giles. 1990. *Public-Private Housing Partnerships.* Washington, DC: Urban Land Institute. ◀

▶ Local Initiatives Support Corporation

The Local Initiatives Support Corporation (LISC) was formed in 1979 to provide low-cost financing and technical support to locally based community development corporations (CDCs) across the United States. LISC is a national nonprofit "intermediary" (employing a staff of 500) that funnels grants, loans, and equity investments from corporations, private investors, and foundations to CDC housing, real estate, and business development projects in low-income areas. Over 1,400 CDCs have used $2.4 billion from LISC to leverage $3.1 billion in additional financing, resulting in the development of 73,000 housing units and 10 million square feet of commercial space in distressed urban and rural communities.

The New York-based LISC was created by the Ford Foundation, which helped to establish the first urban CDCs during the 1960s. Ford provided matching funds for the initial $10 million capitalization that included financial commitments from Aetna Life and Casualty, Prudential Insurance Company, International Harvester, Levi Strauss, the ARCO Foundation, and Continental Illinois Bank. This was quickly followed by local fundraising efforts to start "areas of concentration" in the South Bronx and several other cities where LISC would operate offices, provide focused assistance to CDCs, and attract additional public and private support. LISC programs now serve the nation's seven largest cities and 30 other areas. More than 1,600 corporations, philanthropies, and investors have contributed capital and otherwise provided leadership for LISC's work.

The importance of LISC has grown as federal funding for housing and community development was reduced during the 1980s. LISC makes grants, low-interest loans, loan guarantees, and other forms of subsidized and specialized financing to assist CDCs in identifying and developing project proposals, and bridging the "gap" to secure firmer, long-term commitments of public and private funds for project implementation. Local LISC program officers also provide technical assistance to CDCs with the goal of building their organizational and project development capacity. LISC created the National Equity Fund (NEF) in 1987 to syndicate federal Low-Income Housing Tax Credits. NEF has raised $2 billion in equity investments for 37,000 units of nonprofit rental housing. LISC then formed the Local Initiatives Managed Assets Corporation (LIMAC) in 1988 to establish a secondary market for low-income housing and community development lending. LIMAC purchases loans made by LISC, community development financial institutions, and commercial banks, pooling them into mortgage-backed securities for sale to Freddie Mac and other institutional investors. To encourage commercial development in neighborhoods where these housing investments are targeted, LISC started the Retail Initiative in 1992 to provide equity capital for inner-city supermarkets and other related retail services developed by CDCs. Since 1994, LISC has allocated financial and technical assistance to CDC social, employment, and public safety programs, and rural CDCs.

During the mid-1980s, LISC played a key role in creating the Boston Housing Partnership, the Chicago Housing Partnership, and the Chicago Equity Fund, all of which are local programs for the nonprofit rehabilitation of low-income multifamily housing, with critical support from state and municipal government. LISC joined with the Enterprise Foundation, another national community development intermediary, to establish the New York Equity Fund in 1987 and expand the activities of the Cleveland Housing Network. LISC and Enterprise also distribute loans and grants to CDCs in selected cities through the National Community Development Initiative, a funding consortium of major foundations, corporations, and the U.S. Department of Housing and Urban Development. To strengthen the capacity of CDCs, LISC launched local efforts such as Boston's Neighborhood Development Support Collaborative, and in 1992 organized the five-year Campaign for Communities to raise more private capital. LISC's work in states and cities during the 1980s became a model for replication through the new HOME program under the National Affordable Housing Act of 1990, resulting in increased federal funding of LISC and its affiliates. In 1997, LISC joined Enterprise, the National NeighborWorks Network, and Habitat for Humanity International to form the Community Compact, which will invest $13 billion over four years to produce 200,000 housing units across the United States. Contact: Paul Grogan, President and Chief Executive Officer, Local Initiatives Support Corporation, 733 Third Avenue, 8th Floor, New York, NY 10017, phone (212) 455-9800. (SEE ALSO: *Enterprise Foundation; Neighborhood Reinvestment Corporation*)

—*John T. Metzger*

Further Reading

Local Initiatives Support Corporation. 1997. *A Partnership of Progress: 1996 Annual Report*. New York: Author.

Mayer, Neil S. 1990. "The Role of Nonprofits in Renewed Federal Housing Efforts." In *Building Foundations: Housing and Federal Policy*, edited by Denise DiPasquale and Langley C. Keyes. Philadelphia: University of Pennsylvania Press.

Roberts, Benson F. and Fern C. Portnoy. 1990. "Building a New Low-Income Housing Industry: A Growing Role for the Nonprofit Sector." In *The Future of National Urban Policy*, edited by Marshall Kaplan and Franklin James. Durham, NC: Duke University Press.

Vidal, Avis C., Arnold M. Howitt, and Kathleen P. Foster. 1986. *Stimulating Community Development: An Assessment of the Local Initiatives Support Corporation*. Research Report R86-2, 1986. Cambridge, MA: Harvard University, John F. Kennedy School of Government, State, Local, and Intergovernmental Center. ◀

▶ Local Urban Homesteading Agency

Local urban homesteading agency (LUHA) was the term used to refer to entities responsible for the administration of the federal urban homesteading program authorized by Section 810 of the Housing and Community Development Act of 1974. States and units of local government applied to the Department of Housing and Urban Development (HUD) for approval to operate urban homesteading programs and to designate LUHAs. The LUHA could be the state, a unit of local government, another public agency, or a qualified nonprofit agency. A majority of LUHAs were local governmental agencies, and a minority were state government agencies or nonprofit designees.

LUHAs were funded either from general revenues available to the city, county, or state and/or from Community Development Block Grant (CBDG) funds. The Section 810 dollars did not cover administrative costs of the programs. CDBG monies composed the principal source of administrative support for most local programs.

LUHAs used unoccupied one- to four-family properties. The procedure involved two steps. First, the properties were transferred from the federal agency to the LUHA and then from the LUHA to the homesteader. The transfer of properties from the federal agency was essentially a paper transaction in which the funds allocated for the LUHA's urban homesteading program were reduced by the value of the property and given to the agency to pay for the properties. The LUHAs never actually received the funds themselves.

The LUHAs would subsequently transfer the properties to eligible homesteaders. Eligibility was established by each individual program. Selection usually involved a lottery process because the number of applicants often exceeded the number of properties. The homesteaders received possession of the properties at no or nominal cost with a conditional deed. After the conditions of the urban homesteading program were met—repairs made within three years and residence for a minimum of five years—the homesteader obtained fee simple final title to the residence.

Properties from the inventories of HUD, the Department of Veterans Affairs, and the Farmers Home Administration were used in the programs. Aside from the national inventory, LUHAs sometimes used locally owned properties that had been abandoned and acquired by local governments. HUD operated a Local Property Demonstration Program, which concluded in 1987, in which 11 cities homesteaded locally acquired properties. In this demonstration, Section 810 funds were used to compensate city agencies for the value of properties acquired. In addition, some communities used other funds to acquire federal properties for use in urban homesteading efforts.

The responsibilities of the LUHA also included ensuring homesteaders' compliance with program criteria. After receiving the properties, the LUHAs proceeded with steps in the urban homesteading process, which included selection of homesteaders, conditional conveyance of title, occupancy and rehabilitation of property by homesteaders, and final conveyance of title. Final conveyance occurred after all program requirements had been met, including completion of all rehabilitation and residence by the homesteader for five years. LUHAs had to monitor the repair of properties in keeping with program requirements. Federal law did not require the LUHA itself to provide funding for the urban homesteading program, although some used Section 312 rehabilitation program funds and CDBG program funds to assist buyers with the necessary rehabilitation.

LUHAs that did not receive properties during a program year were categorized as inactive. The legislation did not require LUHAs to homestead a specific number of properties, but as the program evolved, HUD did encourage them to plan on homesteading at least five properties per year to be cost-effective and to have a discernible neighborhood impact. The number of properties acquired annually varied. In 1987, for example, properties received by individual LUHAs ranged from none to 51.

When funding for the Section 810 urban homesteading program ended in 1991, LUHAs continued to process outstanding homestead properties. LUHAs that received properties in 1991 were given an additional five years to convey clear title to the selected homesteaders. For practical purposes, LUHAs ceased to exist with the termination of the urban homesteading program and the designation—LUHA—is no longer in common usage. Its homeownership objectives were superseded by HUD's HOME Program. (SEE ALSO: *Urban Homesteading*)

—*Mittie Olion Chandler*

Further Reading

U.S. Department of Housing and Urban Development. 1989. *Report to Congress on Community Development Programs*. Washington, DC: Author. ◄

▶ Locally Unwanted Land Use

Industrialized societies face a large, distinct, and fast-expanding number of development projects that are regionally or nationally needed or wanted but objectionable to many people who live near them. Examples of such "pariah" land uses may be low-income housing, halfway houses, hazardous waste facilities, power plants, airports, prisons, and highways. These projects create a political tension: Societies want them, but individuals—and often communities—do not want them close by. The projects are locally unwanted land uses, or LULUs.

Few people want to live near a LULU, but they are frequently compelled to do so and become resigned to the prospect, finding ways to adapt because they have no choice. A LULU is considered dangerous (hazardous waste facilities), noisy (airports), ugly (power plants), smelly (many factories), polluting (all of the above), or perhaps, occupied by potentially disruptive residents (low-income housing), depressing residents (old-age homes), or otherwise unwelcome persons (halfway houses).

A LULU may offend its neighbors because of its innate features—for instance, its technology or its occupants. It may offend because of its consequences—the traffic, industrial by-products, or additional development problems it creates or the difficulties its management could pose.

LULUs strain the sense of equity; they gravitate to disadvantaged areas such as slums, industrial neighborhoods, and poor, minority, sparsely populated, or politically underrepresented places that are unable to fight them off and so become worse places after they arrive. For instance, in New Jersey, a notorious producer of LULUs, most cluster within 10 miles of the central and northern portions of the New Jersey Turnpike (itself a LULU); around the chemical-processing towns on or near the Hudson River (such as Bayonne or Elizabeth); or in Newark, Jersey City, Paterson, Trenton, and Camden, the state's most distressed large cities. LULUs tend not to appear in the high-income suburbs of New York City or Philadelphia; southern agricultural or coastal recreational New Jersey; or nature preserves such as the Pinelands or the Delaware Water Gap.

A LULU offers regional benefits, perhaps national ones. It satisfies a nonlocal public need or private demand. The problem is that its economic and environmental costs to third parties—what economists call its negative externalities—fall heavily on its locality. The locality shares in the LULU's benefits but gets more of its costs. So a LULU may pit a nearby minority (the LULU's neighbors, who mainly and intensely feel its costs) against a relatively apathetic, more distant majority (the rest of the relevant population, who mainly feel the LULU's benefits, but not intensely). The locality and the region disagree about the LULU because they experience it differently.

Thus, LULUs pose the dilemma of having in society's midst big projects no one wants nearby. The United States typically deals with the difficulty through land use regulations, pollution controls, environmental impact statements, citizen participation, eminent domain, preservation areas, and lawsuits. But the task is getting harder, and the old approaches no longer suffice. In recent years, the U.S. public has become more sensitive to the deleterious environmental and economic effects of LULUs. It is less willing to see them located or operated indiscriminately. The result, reflecting the NIMBY ("not in my back yard") syndrome, has often been LULU blockage. (SEE ALSO: *Brownfields; Environmental Contamination: Toxic Waste; Not in My Back Yard; Temporarily Obsolete Abandoned Derelict Sites*)

—*Frank J. Popper*

Further Reading

Brion, Denis J. 1991. *Essential Industry and the NIMBY Phenomenon.* New York: Quorum.

DiMento, Joseph and LeRoy Graymer, eds. 1991. *Confronting Regional Challenges: Approaches to LULUs, Growth, and Other Vexing Governance Problems.* Cambridge, MA: Lincoln Institute of Land Policy.

Kaufman, S. and J. Smith. 1997. "Implementing Change in a Locally Unwanted Land Use." *Journal of Planning Education and Research* 16(3):188-200.

Piller, Charles. 1991. *The Fail-Safe Society: Community Defiance and the End of American Technological Optimism.* New York: Basic Books.

Popper, Frank J. 1992. "The Great LULU Trading Game." *Planning* 58(5):15-17.

Portney, Kent E. 1991. *Siting Hazardous Waste Treatment Facilities: The NIMBY Syndrome.* New York: Auburn House. ◀

▶ Lodging Accommodation

People in the United States have long believed that every nuclear family should have its own self-contained living quarters. This view was implied in the famous rhetoric of the 1949 Housing Act, with its assertion that every family should have a decent home. It dates back, however, to the 19th century when reformers attacked the widespread practice of taking lodgers into private homes. As incomes rose, the notion expanded to include single persons too. Some apartments intended for single individuals were built in the 1920s, and these became common by the 1960s. Today, most people who can afford to own or rent their own dwelling unit do so. Only those with low incomes or who are for some reason unable to care for themselves live in lodgings.

Lodging accommodation may be distinguished from homes on the one hand and from shelters on the other. In contrast with a home, lodgings are not self-contained. Typically, a kitchen or bathroom, and sometimes other living spaces, are shared. At the extreme, a very fine line separates a lodger from a tenant. For example, in Toronto until at least the 1950s, many single-family dwellings had been converted into two or more units, each with its own kitchen and bathroom. Because of the configuration of the narrow, semidetached dwellings, these units often shared a single entrance to the street. For this reason, the Canadian census in 1951 treated these as single dwellings that contained lodging families, although most occupants probably thought of themselves as tenants. Lodgers, even more than tenants, tend to be mobile. They rent space by the week or, at most, the month. Even so, lodging accommodation can be distinguished from shelters by its relative permanency. Shelters are available for single nights. Extended stays are often discouraged, and the shelter itself is often off-limits during the day. In contrast, lodgers have the right to use their space 24 hours a day. On these criteria, some emergency shelters might more properly be thought of as lodging accommodation. Here, it is clear that, again, one type of accommodation can shade into the other.

Lodging accommodation takes a variety of forms and entails a range of experience. Most accounts stress the anonymity of lodgers. There is some truth to this, but the point can easily be overstated. The largest cities were indeed the first to develop dedicated lodging houses, but because they brought together people with at least some life experiences in common, these were not necessarily impersonal. Moreover, they were typically clustered into lodging-house districts where a variety of commercial institutions sprang up to serve the lodgers' needs: restaurants, laundries, pool halls, and so forth. More important, even in the largest cities, most lodgers lived in private homes. At the turn of this century, a national survey yielded the estimate that about one quarter of all urban households contained at least one lodger or lodging family. At that time, the number of lodgers in private homes would have outnumbered those living in dedicated lodging houses by a ratio of perhaps 10 to 1. Many lodgers were treated as one of the family, but even when this was not the case, the lodging relationship was more personal than that of the typical tenant.

Increasingly, even this limited intimacy was shunned. Typically, households took in lodgers when they were pressed for money and when they had some available space. Space was most limited in households that contained children, so lodging was commonest in older households where children had left home. In the late 19th and early 20th centuries, the practice of lodging was strongly criticized for compromising family privacy and morals. Such criticisms probably had little effect on those who needed the additional income that lodgers provided, but there is evidence to suggest that privacy was a general concern. After about 1900, the typical form of lodging changed. Hitherto, it had been usual for the lodger to board, paying not only for accommodation but also for services such as meals and laundry. Over the course of this century, however, *boarding* gave way to *rooming,* in which the lodger simply rented a room and sought commercial establishments for other needs. Although neither contemporaries nor later scholars have been consistent in their usage of the key terms, it makes sense to distinguish roomers and boarders in the above fashion and to treat each as specific types of the more general category "lodgers."

In most cases, hosts took in lodgers as a means of supplementing income from other sources. The work involved usually fell to the woman of the household, and indeed, taking in lodgers was one of the most common ways in which women earned income until at least the 1920s. For a significant minority of people, including many widows, lodgers became the main income source. Lodging-house keepers adapted their existing homes or bought larger dwellings for this purpose. In neither case was it usually necessary to make major structural alterations, especially if the owner provided meals from a single kitchen. Houses converted in this fashion are still quite a common source of lodging accommodation, usually being concentrated in central, and often rather run-down, districts of the city.

Beginning in the largest cities around the turn of the century, a variety of organizations began to address the demand for cheap lodgings. Faced with high labor turnover or with the difficulty of attracting and keeping young women, some employers had already been building boarding houses for their workers. Then, others began to erect lodging houses, cheap residential hotels, or both for people who otherwise would have lodged in private homes. Many were commercial operations, although a good number of service and philanthropic organizations (e.g., the YWCA) built and operated hostels that housed specific types of people on a continuing basis. These hostels, especially those housing young women, sometimes came with strict supervision and house rules as well as recreational programs. Resented or welcomed, these hostels provided a sort of home away from home so that, again, the lodging experience was not one of complete anomie.

Over the course of the 20th century, lodging has become socially marginalized. There are three main dimensions to this process, which, although related, have occurred in distinct phases. First was the shift from boarding to rooming. Second came an absolute decline in the numbers of people seeking lodgings. This decline occurred in stages. In the United States, the 1920s saw a substantial decline in lodging as the first nationwide apartment boom created fairly inexpensive accommodation for young couples and singles, the latter often sharing a flat with another person of the same sex. The Great Depression reversed this trend, however, for many people lost their jobs and were unable to maintain their own households, and during World War II severe housing shortages compelled many people to "double up." It was only in the late 1940s that lodging went into a sustained, and apparently permanent, decline.

The third long-term shift has been in the social character of those who are involved in the practice of lodging, either as lodgers or as hosts. In the 19th century, all kinds of people were involved. There was little or no stigma to the practice. Affluent families had more living space than workers, and many rented it out. At the same time, businessmen and professionals often used the better type of boarding houses as an alternative to hotels. Criticism stigmatized lodging, but as late as the 1940s, many middle-class families still

took in roomers, who were typically workers engaged in a variety of respectable manual or clerical occupations. At that time, it was still quite common for people to take in lodgers, not from absolute necessity but to be able to save for the down payment on a home. The prosperity of the postwar years, however, coupled with the diffusion of low-ratio amortized mortgages, soon eroded the practice. Increasingly, those who take in lodgers have done so because they had no choice, and only those who cannot afford any other type of accommodation choose to lodge. The main exception is among certain ethnic communities. Short-term lodging has long been viewed as an accepted mechanism of adjustment for recent immigrants, and in many cases, lodgers have been treated as extended kin. Immigrant hosts, more than most, seem to have viewed taking in lodgers as a means of saving for, or paying for, a home. This is still true.

Although for much of the past century lodging was condemned as morally dangerous, in recent years scholars have generally interpreted it as a legitimate means whereby families and individuals are able to manage on low or irregular incomes. Indeed, some have stressed the value of lodging as part of the necessary malleability of household structures in the face of economic uncertainty. The corollary is that, for as long as it was common, lodging provided a significant element of flexibility within the housing market. Housing markets are local. They respond not only to general economic booms and recessions but also to more local movements of capital and people. Lodging in private homes enables a local housing market to handle rapid population growth. Today, this source of market flexibility has largely been lost, thereby putting greater pressure on those private and public agencies responsible for building new homes. (SEE ALSO: *Doubling Up; Single-Room Occupancy Housing*)

—*Richard Harris*

Further Reading

Harris, R. 1992. "The End Justified the Means: Boarding and Rooming in a City of Homes, 1890-1951." *Journal of Social History* 26(2):331-58.
———. 1994. "The Flexible House: The Housing Backlog and the Persistence of Lodging, 1891-1951." *Social Science History* 18(1):31-53.
Modell, J. and T. Hareven. 1973. "Urbanization and the Malleable Household." *Journal of Marriage and the Family* 35:467-79.
Peel, M. 1986. "On the Margins: Lodgers and Boarders in Boston, 1860-1900." *Journal of American History* 72(4):813-34. ◄

► Loft Living

Loft living represents an opportunity to recapitalize old manufacturing buildings by converting them to residential use. Like gentrification, loft living brings new residents into old buildings and neighborhoods. It also helps infuse a middle-class and upper-middle-class concern with cultural consumption—mainly in art galleries, restaurants, and specialized boutiques—into attempts to redevelop downtowns. As a real estate market, loft living responds to a specific taste for center-city living when housing prices are high. It has been most successful when and where there is high demand for and low production of "luxury" housing: in New York City, primarily beginning in the 1960s, and also in Boston, Chicago, Philadelphia, Los Angeles, and San Francisco. Loft living has also been used as a cultural or "lifestyle" model and a stimulus to residential reinvestment in historic buildings in older cities outside the United States, notably in London's Docklands and arrondissements north of the Seine in Paris. It has spread most widely, however, in North America. This may reflect a greater openness to new spatial expressions of middle-class urban needs, a perennial desire for large housing spaces, and a specific history and geography of deindustrialization and cultural production.

Loft living cannot be understood apart from the deindustrialization of major U.S. cities. Decentralization of manufacturing to suburbs and exurbs, from the 1950s on, reflected modern industries' needs for large, continuous floor space, efficient loading and unloading facilities, and access to highways. Late 19th-century loft buildings on crowded city streets did not meet these needs. But deindustrialization also reflected the desires of property-owning elites, connected to both financial institutions and city planning commissions, who wanted to enhance the image of a city and upgrade the economy from one based on manufacturing to one based on services. To reach these goals, elites, planners, and elected officials designed and invested in downtown office centers, revised zoning maps to restrict manufacturing to smaller areas, and rejected street and rail improvements that would have made freight handling easier. Although they did not plan to convert loft buildings to housing—demolition and new construction being much preferred to preservation—local business and political leaders had long talked about the possibility of attracting more middle-class residents to areas surrounding central business districts.

Loft living is also inseparable from demographic and labor market changes that began to be observed during the early 1960s. Young college graduates, living alone or in couples, sought inexpensive housing close to jobs and entertainment in center cities. At the same time, an increase in college-trained artists and graduates of art institutes expanded the labor force of cultural producers who required access to art markets that were generally located in the largest cities with the most diversified commercial economies. These expansions of the middle class occurred outside the double structural shift of the post-World War II period—suburbanization and upward social mobility—so that, in a sense, a younger generation "returned" to the cities their parents had left behind. Like the larger phenomenon, gentrification, of which it is a part, loft living spurred a discussion of whether housing choices were related to a more comprehensive change of social values.

Indeed, living in a loft evokes both the simplicity of doing without the usual amenities of passenger elevators, neighborhood schools, and dry cleaners and the hedonism of large spaces, skylights, and plant-laden, marble-clad interior design. The variation to some extent reflects the different income levels of people who buy or rent lofts, as well as the type of lofts—improved or "raw" space—they choose to inhabit. It also reflects the different periods in which the living-loft market develops: from the raw space taken and rehabilitated by "pioneers," often artists seeking live-in studios, to the finished, and quite possibly small, "loft-apartments" offered in subdivided lofts by professional real estate developers.

The first loft dwellers in New York City were artists who needed cheap living quarters and did not mind living in their studios. Although living in most of these work spaces violated local building codes and zoning laws, local authorities were either ineffective or benignly unconcerned about removing illegal residents. Museum directors, gallery owners, and art collectors often visited these artists to examine their work. So artists' lofts acquired a certain cachet among cultural elites. By the late 1960s, the lack of facilities and amenities, including lack of full-time heat, hot water, bathrooms, and kitchens, represented an opportunity for the self-expression of designers, who were either professional interior designers and architects or loft dwellers engaged in artistic occupations. Unimproved lofts provided ideal space for self-expression because they typically offered spaces as large as 2,000 to 5,000 square feet, often with oversize, although grimy, windows. Continued illegal residence did not dissuade artists and others with more money from investing in their lofts. However, the emergence of public and elite support for arts activities as an indicator of urban revitalization legitimized loft living as a necessary support of a downtown arts establishment. Moreover, the rise of a movement for historic preservation of architecturally significant old buildings gave a significant boost to loft dwellers' efforts to change their legal status. With social recognition of both artists and old, cast-iron manufacturing buildings, loft living slowly won legal acceptance.

The most dramatic, and most closely watched battles, over the legality of loft living were fought in Lower Manhattan from the early 1960s to the mid 1970s. First, in 1964, artists won the right to live in lofts in areas zoned for manufacturing. The city government established a procedure to certify artists, based on their sources of income; this certification entitled them to become legal residents of manufacturing lofts. A sign, AIR—Artist in Residence, was posted on the exterior of the building to alert firefighters who might be called to an emergency. At this stage, loft dwellers did not have the right to municipal services such as garbage pickups. Neither did they have tenants' rights vis-à-vis amenities and upkeep, rent controls, or eviction. A second, indirect step toward legalization of loft living was taken with the passage of a Landmarks Preservation Law in 1965, which signaled the city government's readiness to save the old manufacturing neighborhoods surrounding Wall Street from demolition and urban renewal. The next step, taken in 1971 after much lobbying by artists' organizations, was the creation of an artists' residential district within a Lower Manhattan manufacturing zone now known as SoHo ("south of Houston Street"). Two years later, the cast-iron facades of most manufacturing buildings in SoHo won the core of the area designation as a historic district. Although artists' lofts were scattered over a wider area, the initial legal focus on SoHo, as well as its central location, led to an increasing concentration of art galleries and a reputation as an artists' quarter. This in turn spurred tourism, the opening of clothing shops, and more residential use of manufacturing lofts. Pressed by manufacturing tenants whose landlords now preferred higher residential rents, the local community board demanded that building owners make all reasonable efforts to retain manufacturing uses. The artists' district, in fact, prohibited nonmanufacturing uses on ground floors and first floors unless landlords demonstrated they could not find a manufacturing tenant. Nevertheless, SoHo manufacturers, mainly at the low end of the garment industry, were neither strong nor numerous enough to compete with potential residential and commercial tenants.

In 1975, in a recession for housing construction, the New York City Council passed a law that offered incentives to real estate developers for large-scale residential conversions. Shortly after, the artists' district of SoHo was enlarged and two new artists' districts were designated in adjacent loft neighborhoods. The entry of professional developers into the loft market effectively removed barriers against nonartists living in lofts. It also legitimized residential lofts in the eyes of banks and other lending institutions that had refused to grant mortgages or construction loans to a basically illegal form of housing.

By the end of the 1970s, loft living was established as a regular, although still somewhat eccentric, housing market. A New York City Loft Board was established as a city agency in 1983 to oversee the legalization of loft buildings as multiple dwellings. However, because landlords retain the power to deny residential tenants' petition to convert the building legally to an apartment house—and to make the required improvements—very few buildings have passed through the Loft Board's certification procedures. Most landlords also wish to circumvent the Loft Board's strictures on rents. Moreover, during the 1980s, two developments in the larger financial and real estate communities limited the expansion of loft living. From 1983, office construction increased both downtown and midtown as New York City got caught up in an expansion of global business activities. At the same time, many owners of residential buildings, including loft buildings, sold out to tenants in co-op or condominium conversions. These changes shifted developers' attention to commercial real estate markets and diffused loft dwellers' interests.

In cities with a smaller labor force of cultural producers, living lofts have been marketed as another form of gentrification. They are a return to downtown, to gracious living in large, open spaces, and to a housing market that emphasizes individual choice over government-sponsored urban renewal. Despite the significant role of government in forming the living-loft market in New York City, lofts are seen as a sign of the vitality of private urban housing markets. (SEE ALSO: *Gentrification; Historic Preservation*)

—*Sharon Zukin*

Further Reading

Hudson, James R. 1988. *The Unanticipated City: Loft Conversions in Lower Manhattan.* Amherst: University of Massachusetts Press.
Jackson, Peter. 1985. "Neighborhood Change in New York: The Loft Conversion Process." *Tijdschrift voor Economische en Sociale Geografie* 76(3):202-15.
Simpson, Charles R. 1981. *SoHo: The Artist in the City.* Chicago: University of Chicago Press.
Stratton, Jim. 1977. *Pioneering in the Urban Wilderness.* New York: Urizen.
Vance, Mary. 1988. *Loft Buildings: Recent References.* Monticello, IL: Vance Bibliographies.
Zukin, Sharon. 1989. *Loft Living: Culture and Capital in Urban Change.* 2d ed. New Brunswick, NJ: Rutgers University Press. ◀

▶ Low-Income Housing Preservation and Resident Homeownership Act of 1990

To address the "prepayment" and loss of hundreds of thousands of rental units financed by the Department of Housing and Urban Development (HUD), Congress fashioned the Low-Income Housing Preservation and Resident Homeownership Act (LIHPRHA) in 1990. Located in Title VI of the omnibus Cranston-Gonzalez National Affordable Housing Act (NAHA) of 1990, this landmark legislation repealed earlier emergency legislation passed in Subtitle B of the Emergency Low-Income Housing Preservation Act (ELIHPA) of 1987. The Title VI program, as it is commonly called, created a detailed procedural scheme for treating all future prepayment-eligible HUD properties. Some properties that were already subject to ELIHPA (Title II) were permitted the option of proceeding under either the "old" or "new" programs.

The HUD prepayment problem arose, beginning in the mid-1980s, as privately owned rental units built through the Section 221(d)(3) and Section 236 programs in the 1960s and 1970s began to reach the 20-year date when they could be prepaid. Under both programs, eligible borrowers were nonprofit and for-profit sponsors who received low-interest loans and agreed to pass their interest savings on to low- and moderate-income households in the form of lower rents. In some cases, additional project-based Section 8 assistance was provided for lower-income tenants who were paying more than 25%, later increased to 30%, of income for rent.

The low-income use of the projects was controlled through regulatory agreements that were coterminous with the 40-year term of the loans. However, HUD granted for-profit owners, unlike nonprofits, the ability to prepay their loans after only 20 years, which would effectively remove all regulatory restrictions on admissions and rents. At the time, neither Congress nor HUD realized the incentive for-profit owners would have to actually exercise their prepayment options in the future. For-profits were primarily drawn to invest in such properties by the prospect of offsetting ordinary income against property depreciation in amounts that were many times their original cash investment. Therefore, when tax shelters were exhausted after 10 to 20 years and many projects came to be worth much more as market rate than as subsidized properties, the pressures for prepayment became irresistible. The size of the inventory built by or later transferred to such owners has been subject to some dispute. According to a report by the National Housing Trust in 1992, 3,874 properties with 468,874 units were eligible for prepayment under the Title VI or Title II programs or both. Other sources, such as the National Low Income Housing Preservation Commission in 1988 and HUDs Office of Inspector General in 1995, have reported total unit numbers in the range of 350,000 to 400,000.

Title VI represented a historic compromise between the interests of owners and low-income tenants. In essence, what Congress did was to greatly restrict the conditions under which owners could prepay and, in exchange, offer owners fair market compensation in the form of a basket of financial incentives to continue operation of the housing in low-income use or transfer the housing to certain qualified purchasers who agreed to keep the units affordable. Congress also provided the occupants of these properties distinct rights of notification and participation, including the opportunity to become homeowners or participants in the new ownership.

To qualify for the program, a prepayment-eligible building had to be within 2 years of its prepayment date, in other words, at least 18 years into the term of its original mortgage. The process started when an owner filed a notice of intent with both HUD and the tenants indicating which one of three options the owner pursue: (a) to take financial incentives from HUD and continue operating the housing in low-income use for the remaining useful life of the property, (b) to transfer the property to a qualified purchaser who will agree to extend affordability for the same term, or (c) to prepay and terminate all use restrictions. The remaining useful life was defined as at least 50 years from the date of the agreement to extend affordability.

In most instances, owners were unable to justify prepayment because they could not demonstrate that such action would not materially affect the housing opportunities of minorities and cause displacement or adversely affect the supply of safe, decent, and affordable housing.

Once the property's value was determined, the owner had to notify HUD whether the final intent was to continue in the program or transfer the housing. If the decision was to transfer, a 15-month marketing period commenced wherein an owner could accept an offer to purchase the housing only from certain qualified buyers at the appraised market value. During the first 6 months, offers could be accepted only from an organization of the project's tenants or community-based nonprofit organizations endorsed by a majority of the tenants. For the next 6 months, offers could be accepted from a tenant organization, any nonprofit organization, or a public agency. In the last 3 months, the owner could accept an offer from any of the above buyers and, in addition, from a for-profit buyer, provided it agreed to the same terms.

After 15 months, if a bona fide purchase offer was not received, owners were free to prepay the mortgage. Qualifying low-income tenants received portable Section 8 assistance, and certain tenants were provided protected occupancy for three years, but thereafter the units were permanently deregulated as low-income housing.

Under either a sales or "stay-in" scenario, owners were required to prepare a "plan of action" that described any proposed changes in the status of the mortgage and regulatory agreement, the low-income affordability restrictions, and the current ownership. Plans of action also had to include a proposed management and capital improvements plan. Authorized incentives for a plan to stay in included HUD Section 241(f) mortgage insurance for a 70% equity loan and capital improvement loan, additional Section 8 rental assistance and other incentives sufficient to ensure the owner a fair return on investment, set at 8% of equity. Authorized incentives for a plan to transfer included Section 241(f) mortgage insurance for 95% of the difference between the market value of the housing and the balance of the existing HUD loan (which is assumed), a loan for rehabilitation, additional Section 8 assistance, and a small project asset management fee.

For transfers in which a majority of tenants elected to participate in a HUD homeownership strategy, HUD was authorized to offer additional loans and grants to ensure that no occupant of the housing paid more than 35% of income. Tenants had to first form a HUD-approved "residents council" and affiliate with a nonprofit or public agency.

The experience of Title VI and its predecessor, Title II, showed that the great majority of owners, when presented with the option of selling or staying in with financial incentives, was motivated to hold the property. This was primarily because the after-tax consequences of a transfer, given federal capital gains laws and the market for such properties, makes the sales option less attractive than a stay-in, which was not a taxable transaction. Starting in FY 1995, Title VI encountered growing political opposition amid court challenges by owners, reports of high costs, and a general movement in Congress to constrain HUD spending. In FY 1995, unobligated program funds were targeted by Congress for recision. In FY 1996 and FY 1997, both the administration and key members of Congress sought to completely eliminate the program. Tenants' advocates, however, were able to convince Congress to extend the life of the program, albeit with significantly reduced funding and program changes.

Several of these changes greatly altered the character of the original program. First, with limited funds and great demand, HUD decided in FY 1995 to give priority to sales, effectively removing the stay-in option. Owners with plans of actions to stay in were allowed, during a short window of time, to switch to the sales option. Second, in FY 1996, HUD decided to provide capital grants to finance sales and rehabilitation rather than equity loans. By offering straight grants, HUD allowed buyers to freeze unit rents at presale levels and avoid the need for long-term commitments of project-based Section 8 funds to support the higher rents that would have resulted had additional debt been layered on the property. Third, in March 1996, Congress restored the ability of owners to prepay without restriction. Qualified tenants facing rent burdens after prepayment were to be provided tenant-based Section 8 vouchers or certificates to stay in their homes or move to other housing.

According to the National Housing Trust, through FY 1997, about 770 properties with 94,621 units nationwide were preserved under the Title VI and Title II programs. In the FY 1998 HUD Appropriations Act, however, Congress finally denied funding for the program. Some financial restitution was authorized for owners who had incurred uncompensated transactional costs. Funds were also authorized for provision of prepayment Section 8 vouchers and certificates. Currently, existing or new owners of properties who prepay and raise rents are obligated to accept

Section 8 vouchers and certificates from tenants who occupied the housing at the time of prepayment. (SEE ALSO: **Affordability; Homeownership**)

—*Robert Wiener*

Further Reading

National Low Income Housing Preservation Commission. 1988. *Preventing the Disappearance of Low Income Housing*. Washington, DC: National Corporation of Housing Partnerships.

Wiener, Robert. 1991. Congress Challenges Cities: Hang on to Low-Income Housing. *Western Cities* (January). ◀

▶ Low-Income Housing Tax Credits

Low-Income Housing Tax Credits are provided through the U.S. federal tax code to investors in the construction and rehabilitation of low-income rental housing. Since the low-income housing tax credit program was initiated in 1986, it has emerged as the primary tax incentive for stimulating the development of low-income housing in the United States. By 1995, almost $2.5 billion in tax credits had been allocated to develop about 739,000 units across the country.

The low-income housing tax credit program differs from most other federal housing programs because it is operated through the tax code. Consequently, it does not require an annual federal appropriation. Rather than a direct outlay of funds, the tax credit program results in a "tax expenditure," or a reduction in individual and corporate federal income tax liabilities. The program's cost to the federal government can be measured by estimating the reductions in the tax liabilities of individuals and corporations that claim the tax credit on their tax returns. In 1995, the program was estimated to cost the federal government $2.2 billion in foregone tax revenue.

The Origin of Low-Income Housing Tax Credits

Low-Income Housing Tax Credits were initially established by the Tax Reform Act of 1986. They were created to replace other tax incentives, such as special accelerated depreciation, that were eliminated by the 1986 act. Congress hoped that replacing these other tax incentives with Low-Income Housing Tax Credits would more efficiently target the housing produced to those most in need of assistance. The earlier tax incentives were criticized for not clearly linking the amount of a project's subsidy with the percentage of units in the project serving low-income households. In addition, households with incomes as high as 80% of the area's median income qualified to live in the units produced with these earlier incentives. Furthermore, these incentives did not limit the rents that low-income tenants could be charged. The tax credit program was designed to address each of these concerns.

The tax credit program was authorized for only a 3-year period by the 1986 act. However, the program quickly gained widespread support. Low-income housing advocates, concerned about declining federal appropriations for the development of low-income housing during the early 1980s, were enthusiastic about the additional housing that the tax credit program could help produce. Other supporters believed the program was effective because it enticed private capital into helping meet the needs of the nation's low-income households. After it was temporarily extended several times, the tax credit program was made a permanent part of the federal tax code by the Revenue Reconciliation Act of 1993.

Administration of the Low-Income Housing Tax Credit Program

The low-income housing tax credit program is administered by the Internal Revenue Service of the U.S. Department of the Treasury and by state tax credit allocation agencies. Most states have selected their state housing finance agencies to function as their tax credit allocation agencies.

Unlike some other tax incentives, such as the home mortgage interest deduction, the federal government limits its tax expenditure for Low-Income Housing Tax Credits by limiting the amount of credits that may be claimed each year. Each state allocation agency receives an annual allocation of tax credits on the basis of the state's population. As of 1995, the states received annual tax credit allocations of $1.25 per resident. Credits unused in previous years may be carried over and used by state allocation agencies.

Each state allocation agency reviews applications submitted by developers and awards tax credits on the basis of federal requirements and the state's special housing needs. The agency is required to document the state's housing needs and priorities in a "qualified allocation plan."

State allocation agencies are required to monitor compliance with the federal requirements for projects funded with tax credits and to report cases of noncompliance to the Internal Revenue Service. Noncompliance with federal requirements, such as serving too few low-income households, can result in recapture of a portion of the tax credits that a developer or investor has already received.

Using Low-Income Housing Tax Credits

The program provides a 10-year tax credit to project investors for eligible projects in which units are set aside for at least 15 years of use by low-income households. A developer can use tax credits for many types of low-income rental housing, including constructing new units, rehabilitating units he or she already owns, and acquiring and rehabilitating existing properties. The number of units in a project aided by tax credits can range from one to hundreds.

Projects funded with Low-Income Housing Tax Credits must meet federal guidelines pertaining to the income and rents of tenants living in the project. A minimum percentage of a project's units must be set aside for low-income tenants. Generally, developers may choose between setting aside 20% of the project's units for tenants with incomes no higher than 50% of the area's median income or setting aside 40% of the units for tenants with incomes no higher than 60% of the area's median income. The rents charged for these low-income units may not be greater than 30% of this maximum income level. Rents are also adjusted to reflect the number of bedrooms in the unit.

Both for-profit and nonprofit housing developers can use Low-Income Housing Tax Credits. To convert their tax credit allocations into cash for developing the units, many developers form limited partnerships, often through syndicators, with corporations or individual investors who have tax liabilities that could be reduced through a tax credit. When the partnership is formed, the investor contributes cash in return for the tax credits and other associated tax benefits that the project is expected to provide. In a limited partnership, the developer typically assumes the role of the "general partner" and is responsible for developing and managing the units; the investors are the "limited partners" and are responsible for providing up-front cash in return for annual tax benefits.

As limited partners, the investors in tax credit projects share in any profit or loss incurred while operating the project. Accordingly, limited partners can deduct project losses, in addition to claiming the tax credit on their income tax returns. Although corporate investors can claim an unlimited amount of these deductions, the amount claimed by individual investors is subject to limitations if they have no income to claim from other rental property.

After they receive the last of their annual tax credits and the project has provided low-income housing for 15 years, investors may sell or donate the units to a new owner. The tax credit program requires that investors try to ensure that the units will be used for low-income households for an additional 15 years. However, if investors trying to sell the units cannot find a buyer willing to continue to use the units to serve low-income households, the units may be converted to another use.

Amount of Funding Provided Through Tax Credits

The size of a project's tax credit allocation depends primarily on its development costs, whether it is receiving other types of federal assistance, and the type of activity involved (that is, new construction, rehabilitation, or acquisition). However, state allocation agencies are also required to evaluate the reasonableness of a project's costs and to adjust the allocation to ensure that no more tax credits are awarded than necessary to make the project viable. The amount of funding that a tax credit allotment ultimately provides for paying development costs depends on the amount that investors are willing to pay for the credits. Some of the cash raised from investors is used to pay administrative costs, such as syndication fees. The remaining funds are available to apply toward the costs of developing the low-income units.

For newly constructed or substantially rehabilitated housing that does not receive certain other federal subsidies, the total amount of credits awarded to the project can have a present value of up to 70% of the eligible cost of developing the low-income units. Generally, each year for 10 years investors in these projects would take a credit on their income taxes equal to about 9% of the eligible development costs. For housing that is also receiving certain other federal subsidies and for the purchase of housing that is undergoing rehabilitation, the total amount of credits awarded to the project can have a present value of up to 30% of the eligible development costs. Investors in these projects would typically take a credit each year for 10 years of about 4% of the eligible costs.

Combining Tax Credits with Other Sources of Funds

Because tax credits rarely cover the full cost of developing the units in a project designated for low-income households, most developers must obtain supplemental funds, such as conventional or federally insured loans and state subsidies. Some developers also rely on supplemental funds, such as Department of Housing and Urban Development (HUD) Section 8 rental assistance, to help defray operating costs and debt service on loans when low-income tenant rents do not cover monthly costs.

Although many developers combine the tax credit with other subsidies because they want to serve very low income households, others have done so to receive unusually high profits. Cases of high profits were found among projects combining tax credits with HUD assistance as well as projects combining tax credits with Farmers Home Administration (now Rural Housing and Community Development Services [RHCD]) assistance. Many projects funded through the Farmers Home Administration's rural rental housing program receive tax credits.

In an attempt to ensure that the amount of assistance provided through the tax credit is not more than necessary to make a project viable, both HUD and the National Council of State Housing Agencies established cost guidelines for state allocation agencies to use when reviewing tax credit applications. The HUD guidelines issued in 1994 establish standards for evaluating the reasonableness of builders' profits, developers' fees, and syndication costs for projects combining tax credits with HUD and certain other subsidies. The voluntary guidelines established by the National Council of State Housing Agencies in 1993 are intended to be used during a state allocation agency's review of each tax credit application, regardless of whether the project receives other federal subsidies.

The low-income housing tax credit has been widely used in the United States to develop additional housing for low-income households. Supporters of low-income housing development argue that housing needs of some low-income households in certain parts of the country cannot be met through the existing housing stock. As long as shortfalls in the housing available in some communities and the lack of other sources of federal funding for low-income housing development persist, the tax credit is likely to remain a key tool of U.S. housing policy. (SEE ALSO: *Tax Expenditures; Tax Incentives*)

—*Leslie Black-Plumeau*

Further Reading

Guggenheim, Joseph. 1994. *Tax Credits for Low Income Housing: Opportunities for Developers, Non-Profits, and Communities under Permanent Tax Provisions*. Glen Echo, MD: Simon.

U.S. Congress. Joint Committee on Taxation. 1994. *Estimates of Federal Tax Expenditures for Fiscal Years 1995-1999*. Washington, DC: Government Printing Office.

U.S. General Accounting Office. 1993. *Public Housing: Low-Income Housing Tax Credit as an Alternative Development Method*. GAO/RCED-93-31. Washington, DC: Author.

U.S. General Accounting Office. 1997. *Tax Credits: Opportunities to Improve Oversight of the Low-Income Housing Program*. GAO/GGD/RCED-97-55. Washington, DC: Author.

M

► Maintenance

Housing maintenance describes the processes through which homeowners, landlords, and tenants offset physical deterioration in their dwellings and obsolescence arising from changes in technology and taste. Maintenance activity is an input (the expenditure and effort by the homeowner, landlord, or tenant) whose effect can also be described as an output (the condition or state of repair of the dwelling). In principle, the former includes steps taken to maintain either the physical capital (that is, the dwelling or parts thereof) or the human capital (that is, the skills and effort) required to operate it. In practice, housing maintenance is difficult to measure. For one thing, it is hard to measure the amount of time spent in do-it-yourself repair work, and data on skill levels are scarce. For another, it is difficult to separate maintenance expenditure from other costs of housing operation. For example, a regular expenditure on prevention (floor wax, cleaning and janitorial services, lawn maintenance, trash and snow removal) can prolong the useful life of a dwelling and its fittings, but these are generally not counted in maintenance expenditure.

In practice, information about maintenance takes the form mainly of a narrow range of expenditure data. A good example is a U.S. Department of Commerce series titled *Expenditures for Maintenance and Repairs* (*EMR*) available quarterly (see Table 20). *EMR* covers only out-of-pocket expenses by property owners (expenses by tenants are ignored) connected to items that are permanently attached or firmly affixed to some part of the house. The dwelling universe of *EMR* is broad; it includes both single and multi-unit structures, private and publicly owned structures, non-farm and farm properties, occupied rental or owned dwellings, and vacant dwellings. It excludes only properties that are primarily nonresidential, institutional accommodation, and unusual living quarters (such as tents or boats). However, *EMR* is narrow in terms of scope of expenditures. It does not count repair or replacement of many household appliances, nor furniture, carpets, draperies, and other furnishings. Expenditures for landscaping of the property (planting of flowers, trees, and shrubs, for example) are

also excluded from *EMR* (although expenditures for grading, draining, fencing, and paving are included). *EMR* does count expenditures for incidental maintenance and repairs that keep a property in ordinary working condition rather than an additional investment in the property. Under "maintenance," *EMR* includes, for example, expenses for painting, wallpapering, floor sanding, and furnace cleaning. Under "repairs," *EMR* also comprises expenditures for heating, plumbing, electrical work, and other activities that involve upkeep of residential properties. Repairs also mean replacement of parts and of whole units. However, not included here are "major replacements" (complete furnace or boiler, entire roof, central air conditioner, all siding, water heater, entire electrical wiring, doors, plumbing fixtures, all water pipes, windows, septic tank or cesspool, sink or laundry tub, complete walks or driveways, garbage disposal unit) or "construction improvements" (includes new additions and alterations to the property, down to items as small as a new electrical socket, wall switch, or shelf). That *EMR* measures preventive maintenance rather than expenditure for major repairs is also evidenced implicitly in Table 20. Under normal circumstances, we might expect homeowners to first run into major repairs about two decades after construction. At this point, the roofing, furnace and central air conditioner, and major appliances begin to reach the stage where normally they would be replaced. However, Table 20 shows that, in 1993, homeowners of dwellings built in the 1960s spent about the same amount in total as owners of homes built in the 1970s and 1980s. Because the number of dwellings constructed in each decade was similar, *EMR* shows no such "bump."

Which factors determine how much maintenance activity is undertaken directly by a household? One is tenure. Renter households generally spend little on maintenance directly; however, part of the monthly rent is spent by the landlord undertaking maintenance. Households in condominiums and cooperatives may spend more money directly on maintenance but similarly contribute to buildingwide repairs through condominium fees and occupancy charges.

Economic circumstances also influence housing maintenance. Consumers often put off maintenance until they can afford to undertake it, or doing repairs themselves rather

TABLE 20 Annual Expenditures on Maintenance and Repair (EMR), in Millions of Dollars, United States, 1991 to 1995

	1991	1992	1993	1994	1995
	($m)	($m)	($m)	($m)	($m)
All properties	49,840	45,154	41,699	45,154	42,077
Owner-occupied one-unit properties	23,645	23,802	21,175	24,241	25,076
Built since 1990	n.a.	n.a.	901	702	1,740
Built 1980-89	n.a.	n.a.	3,496	3,301	3,973
Built 1970-79	n.a.	n.a.	3,503	3,928	4,379
Built 1960-69	n.a.	n.a.	3,444	4,502	3,984
Built before 1960	n.a.	n.a.	7,975	10,096	9,469
Not stated	n.a.	n.a.	1,856	1,711	1,532
Other properties	26,195	21,352	20,524	20,913	17,001

SOURCE: U.S. Department of Commerce, Bureau of the Census. Various years. Current Construction Reports: Expenditures for Residential Improvements and Repairs. C50. Washington, DC: Author.
NOTE: n.a. = data not available.

than using a contractor reduces expenditures. In either case, repair expenditures decline when times are tough. The effect of the early 1990s recession is evidenced in Table 20, with the sharp drop in maintenance after 1991: first in multi-unit and rental properties, then later in owner-occupied, one-unit properties. In a similar vein, homeowners cope with lower incomes in old age in part by undertaking less maintenance.

Housing maintenance is also related to the household division of labor. Persons living alone and households made up of ability-impaired persons may be less able to undertake maintenance and hence may simply forego repairs.

The state of the local housing market forms another factor impinging on maintenance activity. If house prices are low, homeowners have less incentive to maintain their dwellings. Put differently, reduced maintenance in the presence of physical deterioration and obsolescence is one way for households to disinvest in their housing.

Households vary in risk taking. Households with a greater aversion to risk are more likely to adopt a regime of regular maintenance. In this sense, maintenance is like an insurance premium against a major failure in dwelling equipment or structure. Other households might prefer not to spend on maintenance and instead take the risk of major failure.

Finally, there is the question of maintenance versus major repair. Housing maintenance is often thought to be subject to economies of scale: For example, replacing a copper pipe is less expensive if at the same time one replaces the wall within which the pipe is buried. Such economies of scale differ from one dwelling to the next because of differences in design and technology. Therefore, the choice between maintenance and major repair may be substantially constrained by the kind of dwelling in which the consumer lives. (SEE ALSO: Obsolescence)

—John R. Miron

Further Reading

Albon, R. P. and D. C. Stafford. 1990. "Rent Control and Housing Maintenance." Urban Studies 27(2):233-40.
Chinloy, P. 1980. "The Effect of Maintenance Expenditures on the Measurement of Depreciation in Housing." Journal of Urban Economics 8:86-107.
Kutty, Nandinee K. 1996. "The Impact of Rent Control on Housing Maintenance: A Dynamic Analysis Incorporating European and North American Rent Regulations." Housing Studies 11(1):69-88.
Littlewood, Amanda and Moira Munro. 1996. "Explaining Disrepair: Examining Owner Occupiers' Repair and Maintenance Behavior." Housing Studies 11(4):503-26.
Margolis, S. E. 1991. "Depreciation and Maintenance of Houses." Land Economics 57(1):91-105.
Read, C. 1991. "Maintenance, Housing Quality, and Vacancies under Imperfect Information." American Real Estate and Urban Economics Association Journal 19(2):138-53.
Spivack, Richard N. 1991. "The Determinants of Housing Maintenance and Upkeep: A Case Study of Providence, Rhode Island." Applied Economics 23:639-46.
Sweeney, J. L. 1974. "Housing Unit Maintenance and the Mode of Tenure." Journal of Economic Theory 8(2):111-38.
U.S. Department of Commerce, Bureau of the Census. 1994. Current Construction Reports: Expenditures for Residential Improvements and Repairs. Washington, DC: Author. ◄

► Manufactured Housing

Manufactured housing—more commonly referred to as house trailers, mobile homes, or just trailers—provides housing for 1 of every 16 persons in the United States and accounts for approximately one-quarter of all new homes sold annually. The popularity of this form of housing is strongly associated with its relatively low price. At less than half the construction cost per square foot of conventionally built dwellings, manufactured housing is the primary form of unsubsidized affordable housing in the United States. However, because manufactured homes are typically associated with lower-income households, many communities effectively exclude them. More extensive use of manufactured housing depends largely on reducing discriminatory zoning practices.

Development of the Industry and Market

The manufactured housing industry had its origins in the 1920s, when millions of people in the United States began taking to the road to enjoy automobile camping in the countryside. The travel trailer developed as a camping accessory. Their owners built three-quarters of the approximately 250,000 trailers in use by the end of the 1930s. But manufacturers, the largest of which used assembly line techniques, produced the remainder.

Although welcomed at first by communities as a source of tourist dollars, by the late 1930s trailer users were seen by many municipal officials as undesirable. Fearing that trailers would spawn year-round squatter towns, local regulations restricting trailer use proliferated. The industry attempted to counter this development by promulgating a model code that permitted the use of trailers for recreation but prohibited them as a form of permanent housing.

Figure 14. Modern manufactured housing is designed to be compatible in appearance to site-built homes.
Photograph courtesy of Manufactural Housing Institute, Arlington, Virginia.

Figure 15. Manufactured housing used in a HUD-sponsored infill project in the South Bronx, New York.
Photograph by A. Wallis.

World War II brought gasoline rationing and a decline in autocamping. To gain federal government acceptance of the product as temporary, portable shelter, the industry now found it necessary to reverse itself and promote the idea of trailers as housing. This effort achieved partial success, and during the war, the federal government purchased 90% of the more than 50,000 house trailers manufactured. By the middle of the war, however, the government decided to cease use of trailers for fear they would degenerate into postwar slums. Sales withered and many manufacturers went out of business.

After the war, a severe housing shortage, combined with demand for portable housing by military personnel and itinerant construction workers, resulted in record-high trailer sales to households requiring year-round housing. By the mid-1950s, it was clear that the industry had two distinct markets: one requiring year-round but mobile housing and another desiring shelter for automobile camping and cross-country touring. The latter market eventually distinguished its product as recreational vehicles or RVs, whereas the house trailer became known as the mobile home.

By the 1960s, the mobile home market was shifting from buyers seeking portable, year-round shelter to households for whom affordability was a principal concern. The declining importance of mobility allowed manufacturers to market units that were longer and wider but that required towing by a commercial mover. These larger units were designed to look more like site-built houses, featuring conventionalized floor plans.

In a 1970 message to Congress on the nation's housing, President Nixon recognized the importance of mobile homes to the supply of affordable housing. Two years later, mobile home production reached a historic high of 576,000 units shipped. But the growing market also strengthened demands that mobile home construction practices be regulated. In 1974, Congress passed the Mobile Home Construction and Safety Standards Act, more commonly referred to as the "HUD Code." In 1981, the act was revised, formally recognizing the industry's new name of *manufactured housing*.

In 1995, there were approximately 18 million people living full-time in 8 million manufactured homes in the United States, constituting about 6% of all occupied housing units. The housing type continues to appeal to households seeking affordable housing, but other important attractions include ease of maintenance, design and construction quality, and improved financing options.

Historically, the manufactured housing population has been bimodal, with heavier representations of younger and older households than the U.S. population in general. This pattern has been changing, becoming more characteristic of the general population. A major industry survey found that from 1981 to 1993, the average age of the head of household owning a manufactured home increased from 46 to 51 years of age. This age shift reflects greater representation of households where the head is 40 to 59 years of age, increasing from 23% to 32%. By contrast, the number of household heads that are 60 years of age or older grew at a slower rate, 28% to 35%, and the percentage of households in which the head was 30 years of age or less, declined from 24% to 8%.

Corresponding to this age shift, the average household in 1993 had a higher income adjusted for inflation, even though 28% of heads of household were retired. This increase in household income is more pronounced for new buyers. Nevertheless, the average manufactured-home household still has a lower income than the U.S. average and a lower level of educational attainment.

Although it remains an important form of affordable housing, minorities are only half as prevalent in manufactured housing as in the population in general. However, the minority population living in manufactured homes has increased from 2% in 1970 to 10% in 1990.

Communities and Placement

Early travel trailer users, or "trailerites," often parked along a country road for the evening. Commercial and municipal trailer parks began appearing in the 1920s, frequently mixing trailers with tents. Although municipal parks soon disappeared, commercial parks, often run as "mom-and-pop" businesses, flourished. During the war, the federal government developed trailer parks, primarily for the use of war workers living in trailers designed and built under government contract.

With the development of more stationary mobile homes in the 1950s, a wide variety of parks or mobile home communities emerged to provide rental sites and facilities for more permanent living. The typical park had about 100 homesites and was still frequently run as a family business. Luxury parks also developed, catering to higher-income households desiring a seasonal home or a retirement community.

Despite the improved quality of mobile homes and mobile home communities, many local governments continued to discriminate against their use. Mobile homes were excluded from residential zones and often confined to less desirable areas, including those that were flood prone. Frequently, to avoid such restrictions, parks and individually sited mobile homes were located outside of municipalities. However, as municipalities expanded, they became "grandfathered" into the allowable land use.

Today, purchasers of manufactured homes are more likely to locate on a private lot than in a park, or "land lease community" as they are now called. Whereas in 1981, 49% of new purchasers responding to an industry survey located in a land lease community, in 1996 only 32% chose that option. By comparison, those locating on private property rose from 45% to 68%. The ability to finance home and land together through a single mortgage has made the private-siting option more attractive. In addition, development of land lease communities continues to encounter local zoning restrictions, thereby limiting the supply of new rental sites.

There are about 25,000 manufactured housing communities in the United States. Communities being developed today exhibit most of the same design standards as subdivisions for site-built homes. Communities have several hundred home sites and are professionally managed. The homes themselves are larger, oriented parallel to the street, and placed on larger lots. They look like conventional homes and often have such site-built additions as an attached garage or enclosed patio. Community amenities, such as a golf course and recreation center, remain attractions for locating in a community development, but these are offered to appeal to the upper end of the market. By contrast, older, more affordable, basic housing communities are disappearing as land appreciation induces owners to convert their property to other uses.

Manufacturing

Manufactured housing is produced in enclosed factories using assembly line techniques. According to the Manufactured Housing Institute, the industry's primary association, in 1992, 85 companies produced homes in 234 factories throughout the United States. Approximately 5,500 retailers sold manufactured homes. An increasing amount of this retail now takes place within land lease communities where purchasers select from model homes displayed on site.

The relatively low cost of manufactured housing is achieved through a combination of large-quantity purchase of materials, mass production assembly techniques, and the use of low-skilled and often nonunionized labor. The typical home can be constructed in a single day with a fraction of the labor required of conventional on-site construction. The average factory produces more than 700 homes a year.

Whereas the average site-built house, excluding land, cost $58.11 per square foot to build in 1996, the average manufactured home cost $27.83 per square foot. In 1996, the average site-built home cost $124,650 and had 2,145 square feet, whereas the average mobile home cost $38,400 and had 1,380 square feet of space. The product mix of manufactured homes was 52% multisection and 48% single section. Although multisection homes are more expensive on a per square foot basis, their percentage of the market has been growing steadily since they were first introduced in the early 1960s.

Manufactured homes are delivered from the factory, usually completely furnished, requiring only the hook up of utilities at the site. The average manufactured home is shipped within 300 miles of the factory in which it was produced. It is expected that of the homes being delivered today, only about 1% will ever be relocated.

Legal Status of Mobile Homes

The manufactured home is clearly a hybrid between vehicle and house, and as such, it presents considerable ambiguity regarding its proper legal treatment. This ambiguity extends to owners who have purchased a manufactured home but live in a land lease community.

Construction and Safety Regulation

During the first several decades of its development, the manufactured home was regulated as a vehicle rather than as a dwelling. Construction quality varied. In 1963, the industry developed a voluntary safety and construction code. Over the next decade, several states adopted this code as a mandatory requirement. The HUD Code, placed in effect in 1976, incorporated most of the industry's voluntary code and added additional standards.

Manufactured housing is the only privately marketed building type regulated by a federal code. The HUD Code preempts state and local codes, which usually govern building construction. Consequently, manufactured homes produced in one state can be shipped to other states under a uniform construction standard.

Although manufactured homes are still often regarded as poorly built and prone to fire and high-wind hazards, the HUD Code assures a dwelling that meets safety standards equivalent to those required of site-built homes. In addition, the code sets special standards for thermal, high-wind, and heavy-snow-load zones. Homes must be built to standards that apply to the zone where they are to be located.

Taxation

The taxation of mobile homes is established by state law and varies from state to state. In some states, manufactured homes are treated as personal property, similar to taxation of an automobile. Usually a home permanently affixed to a privately owned site is treated as real property. Many states use both a personal and a real property tax treatment, depending on how a unit is affixed to its site and whether the unit is on a lot owned by the same party.

The basic concern of state and local governments regarding the taxation of manufactured homes is whether sufficient revenue can be collected from their occupants to offset the cost of public services, especially schools, provided to them. To help compensate for lower tax revenues, some states charge separate fees to manufactured-home households living in land lease communities.

Financing

Although manufactured homes are significantly less expensive than site-built homes, they do not provide the same financing advantages. Traditionally, manufactured-home loans have been treated like automobile loans rather than home mortgages, even when the home was permanently affixed to an owner-occupied site. Consequently, the interest rate paid on manufactured homes has been significantly higher than that available to buyers of conventional homes. Moreover, because the method of financing is essentially treated as a consumer loan, the interest is not tax deductible.

Today, a number of federally guaranteed programs offer conventional mortgages for a manufactured home purchased with land. These mortgages are offered at rates close to but still slightly higher than those for site-built houses. Although the rates are more favorable, these mortgages are used by only a small percentage of mobile home purchasers because many banks and mortgage agencies do not promote their use.

Another financial aspect of owning a manufactured home concerns appreciation of value. A manufactured home not affixed to its own site usually depreciates rapidly, much like an automobile. But a mobile home permanently attached to a site and sold as a single unit can appreciate in value, although usually not as rapidly as a site-built house. Variations in financing, tax treatments, and appreciation in value make it difficult to determine the long-term comparative cost advantages of owning a manufactured home.

Discrimination

Throughout its history, from trailer to manufactured home, this form of housing has consistently encountered discrimination. As a building type, it has been perceived as a threat to existing housing values. Its occupants have been stereotyped as belonging to a lower socioeconomic class. But manufactured housing has also consistently been one of the most affordable housing options available.

Beginning in the mid-1970s, the legislatures, courts, and attorneys general of many states have acted to prohibit the use of local zoning codes to exclude manufactured homes. These efforts were motivated by a concern that communities make affordable housing options available. For example, California allows manufactured homes in all residential areas as long as the homes are compatible with the neighborhood. Colorado requires that all communities provide zoning for the single-lot placement of manufactured homes. The New Jersey Supreme Court ruled that local zoning

prohibitions against manufactured homes was discriminatory and an unconstitutional application of local zoning authority. Despite these efforts, significant barriers against the use of manufactured housing remain.

Tenants' Rights

The rights of people who own a manufactured home but live in a land lease community are often ambiguous. In many states, such households have fewer rights than those living in rental housing, because they are renting only land not a dwelling. They can be evicted without notice and restricted in the length of time guests may live in their home. Leases may not be available, and rent increases can be extreme. They may be required to pay a fee if they sell the home to someone who wishes to keep it in the same community, and their landlords can sell the entire community or convert it to a different use, forcing all tenants to leave.

To remedy abuses arising from this situation, 32 states have now enacted laws defining the rights of land lease community tenants. In addition, many municipalities with a very limited supply of affordable housing have placed moratoria on the conversion of land lease communities to other uses. Nevertheless, state and municipal laws vary widely in their protections, and rental community residents still lack many of the rights commonly accorded tenants of conventional rental housing.

The Future

The number of manufactured homes produced rose sharply throughout the 1960s, reaching historic highs in the early 1970s. Then sales dropped off precipitously, at first owing to a general recession in the economy, but not recovering to anywhere near market highs even after that. In 1991, the industry experienced a 30-year low, with fewer than 172,000 units shipped. However, in the mid-1990s, it enjoyed healthy growth, with 363,411 shipments reported in 1996—more than twice the number shipped in 1991.

In part, the decline in units shipped can be attributed to increased sales of modular housing. Combined manufactured and modular housing sales account for about a third of all new single-family housing produced in an average year. Modular housing is built with technology virtually identical to that employed for manufactured housing, and many factories produce both types of units.

The principal differences between manufactured and modular housing are legal and structural. By law, manufactured homes must be built with a permanently affixed chassis and axles. Modular homes are built on a wood platform. Their construction is regulated by state codes that vary across the United States.

In the future, the distinction between manufactured and modular homes is likely to disappear. As both products assume more of a traditional houselike appearance, they are likely to achieve greater acceptability with local zoning officials. However, because these modifications generally increase cost, the real challenge will come in retaining affordability in the pursuit of acceptability. (SEE ALSO: *Construction Technology; ECHO Housing;* **Industrialization in**

Housing Construction; *Manufactured Housing Construction and Safety Standards Act*)

—*Allan D. Wallis*

Further Reading

Boehm, Thomas P. 1995. "A Comparison of the Determinants of Structural Quality between Manufactured Housing and Conventional Tenure Choices: Evidence from the American Housing Survey." *Journal of Housing Economics* 4:373-91.

Geisler, Charles C. and Hisayoshi Mitsuda. 1987. "Mobile-Home Growth, Regulation, and Discrimination in Upstate New York." *Rural Sociology* 52(4):532-43.

Manufactured Housing Institute. *Quick Facts*. Annual pamphlet on the manufactured housing industry, available from the Manufactured Housing Institute, 2101 Wilson Blvd., Suite 610, Arlington, VA 22201.

National Commission on Manufactured Housing. 1994. *Final Report*. Washington, DC: Government Printing Office.

Sanders, Welford. 1993. *Manufactured Housing Site Development*. Planning Advisory Service Report No. 445. Chicago: American Planning Association.

Sheldon, Jonathan and Andrea Simpson. 1991. *Manufactured Housing Park Tenants: Shifting the Balance of Power*. Washington, DC: American Association of Retired Persons.

Wallis, Allan. 1992. *Wheel Estate: The Rise and Decline of Mobile Homes*. New York: Oxford University Press.

White, Mark. 1996. "State and Federal Planning Legislation and Manufactured Housing: New Opportunities for Affordable, Single-Family Shelter." *Urban Lawyer* (Spring):263.

Wilden, Robert W. 1995. "Manufactured Housing: A Study of Power and Reform in Industry Regulation." *Housing Policy Debate* 6(2):523-37. ◀

▶ Manufactured Housing Construction and Safety Standards Act

The Mobile Home Construction and Safety Standards Act, passed by Congress in 1974 and implemented in June 1976, authorizes the U.S. Department of Housing and Urban Development (HUD) to establish and enforce a code for the manufacture and safety performance of mobile homes. The act, commonly referred to as the "HUD Code," was amended in 1980 with the term *manufactured housing* replacing *mobile home*. Although national in scope, the HUD Code divides the United States into several regions requiring specific insulation and structural requirements.

Prior to 1974, federal regulation of mobile homes was limited, falling within the purview of the U.S. Department of Transportation, which was concerned with its construction and safety only insofar as mobile homes functioned as vehicles. However, their safety as dwellings was an issue of considerable and growing concern.

Fearful that individual states might adopt their own safety and construction standards, in 1963 the Mobile Home Manufacturers Association (MHMA) developed a voluntary code. Manufacturers subscribing to the code agreed to in-factory inspections, and they could affix a seal to their units certifying compliance. By 1973, 45 states had adopted the MHMA code, making it virtually mandatory for all mobile home sales.

Although the MHMA code served as a de facto national construction standard, critics of the industry continued to raise concerns over lax inspections and dwelling safety hazards. A new, more restrictive wave of state codes appeared to be imminent when Congress, with encouragement from most of the manufactured housing industry, passed the HUD Code.

There are two unique aspects to the code. First, it makes manufactured housing the only privately built and owned building type whose construction is regulated by the federal government. Second, it preempts state and local codes regulating the safety and construction of manufactured housing. Foundation and land development standards, however, remain issues of local control.

The federal code has generally proven beneficial to the industry and consumers. Uniform construction standards have allowed for careful testing for durability and safety. The code has been periodically modified to improve safety, most notably in the improvement of built-in tie-downs—straps that wrap around the unit and are secured to a foundation or anchors to help prevent units from rolling over in high winds.

With assured safety and durability, federal mortgage-lending agencies have extended loan periods and amounts made available on manufactured home purchases (see Manufactured Housing). Moreover, interest rates on manufactured housing loans have fallen closer in line with site-built rates, whereas previously, they reflected automobile rates. The preemptive code has also been used to reduce zoning discrimination against the siting of manufactured homes, especially on single lots outside of parks. Despite these advances, discrimination against manufactured housing, especially in single-family housing zones, continues to be practiced in many communities. In most cases, communities that exclude manufactured housing ignore or attempt to circumvent the preemptive authority of the HUD Code, leaving it to the courts and state legislatures to pursue antidiscrimination actions. (SEE ALSO: *Manufactured Housing*)

—*Allan D. Wallis*

Further Reading

Sellman, Molly A. 1988. "Equal Treatment of Housing: A Proposed Model State Code for Manufactured Housing." *Urban Lawyer* (Winter):73-101. ◀

▶ Market Equity Cooperatives

Definition

Market equity cooperatives (MECs) are housing cooperatives in which the value of member equity shares changes according to changes in the market value of the housing owned by the cooperative. An MEC includes the following characteristics:

▶ Ownership of a share in the cooperative, which entitles a member to residency in one of the co-op's living units

► Free and voluntary membership, which entitles each household to one vote

► Membership control, which includes membership participation in the basic decision making of the co-op and in the appointment of the co-op management

► A form of "profit sharing," which means that members receive the economic benefits and losses from changes in both co-op operating costs and general market conditions

► Membership limited to those who have been approved for membership by the co-op's board of directors according to criteria consistent with governmental fair housing legislation

History

The first housing cooperatives were formed in 1720 in Rennes, France. By the early 20th century, housing cooperatives were found throughout Europe. The housing cooperative movement was especially strong in the Scandinavian countries. Although the first housing cooperative in the United States was established in New York City in 1876, they did not become well established in the United States until after World War I. By 1925, housing cooperatives were found in 16 cities, and by 1930, their value exceeded $500 million.

During the Great Depression, most of luxury cooperative apartments failed because of excessive mortgages and promoter profits, but also because of financial factors inherent in housing cooperatives (see the "Financing" section below). By 1934, only two of New York City's luxury co-ops remained, and more than 75% of the housing co-ops in New York and Chicago had gone bankrupt. The low- and moderate-income cooperatives did significantly better than the high-income ones, because they had higher equity-debt ratios, stronger financial reserves, and support from the unions and ethnic groups that sponsored these co-ops.

During World War II, a combination of rent controls and tax law changes that permitted the deduction of co-op mortgage interest and property tax payments by individual co-op members promoted a resurgence of middle-income co-ops in New York City. As a result of the New York State experience, co-ops became eligible for funding under the federal government's FHA mortgage insurance program. By 1958, about 48,000 of 100,000 co-op units then existing were financed by federally insured mortgages. During the 1960s and 1970s, there was a moderate rate of development of MECs. The strongest growth period for MECs was simultaneous with the growth of condominiums during the 1980s real estate boom. This boom resulted in the conventional conversion of approximately 550,000 units in existing apartment buildings into MECs by profit-motivated owners and brokers, mostly in New York City and mostly for above average-income families.

According to the National Association of Housing Cooperatives, in 1995 there were approximately 1 million living units owned by housing cooperatives in the United States. Approximately 700,000 of these units were organized as MECs and 300,000 of them as limited-equity cooperatives. Housing cooperatives represent 2.4% of the total stock of multi-unit housing nationwide and 5.7% in the Northeast. Housing cooperatives are also concentrated in the upper Midwest, California, and Georgia.

Financing

An MEC is financed by the cooperative's taking out a mortgage on the whole building. Co-op members pay a monthly carrying charge that covers a prorated share of the co-op's principal and interest payments on its mortgagee plus the operating and maintenance costs of the co-op's buildings and grounds. This kind of financing differs from that of a condominium in which each household takes out a mortgage on its living unit and then agrees to pay fees to the condominium association for the operation and maintenance of common areas.

Some of the financial factors inherent in housing cooperatives are demonstrated by the relative change value of MEC equity shares in the New York City market during the early 1990s. MEC share values were increasing more slowly than the value of individual condominium units because of the following financial factors:

► The market felt that the risk of owning an MEC equity share was greater than the risk of owning a condominium, because of the possibility that the whole co-op could go into default because of vacancy or late monthly payments in some of the units, whereas with a condominium, only the specific unit goes into default.

► The length of time to buy into a co-op is longer than that to buy a condominium, because in a co-op new members need to be approved of by the resident selection committee, whereas that step is not necessary for a condominium.

► There are more restrictions on subletting an MEC unit than a condominium unit, because of the desire to foster a cooperative spirit among the owner-residents of an MEC.

► With an MEC, the individual does not directly own his or her unit but only a share in the co-op that owns the building, whereas with a condominium, an individual directly owns his or her unit. Therefore, bankers feel that they have less direct collateral when financing an individual's purchase of an MEC share.

Economic and Intangible Advantages of MECs

Economic theory suggests that housing cooperatives have lower intertenant negative externalities than other ownership forms of multifamily housing. This means that residents of housing cooperatives are more likely to help in maintaining a low cost and neighborly living environment than are residents of other multifamily units. Because the members own the co-op, (a) more intertenant cooperation occurs, (b) members have more of a direct say in the operation of the co-op, and (c) they are able to experience directly the benefits of cooperation through lower monthly carrying charges and higher equity share values.

There are also fewer potential problems with "free riders" in a housing co-op. A free rider is a resident who tries to benefit from the actions of others without contributing his or her fair share of the work effort or costs. Because members of the co-op see themselves as owners of the co-op, there is usually a clear acceptance of the commonality of interests of all group members and a concomitant group pressure for all to contribute to a healthy social and physical environment for the co-op. (SEE ALSO: *Condominium; Cooperative Housing*)

—*Gerald W. Sazama*

Further Reading

Bandy, Dewey. 1993. *Characteristics and Operational Performance of California's Permanent Housing Cooperatives.* Davis: University of California, Center for Cooperatives.

International Labor Office. 1964. *Housing Cooperatives.* Studies and Reports, New Series, No. 66. Geneva: La Tribune de Geneve.

Miceli, Thomas, Gerald Sazama, and C. F. Sirmans. 1994. "Role of Limited Equity Cooperatives in Providing Affordable Housing." *Housing Policy Debate* 5(4):469-90.

National Association of Housing Cooperatives. 1991. *Summary of Housing Cooperatives in the United States.* Alexandria, VA: Author.

Siegler, Richard and Herbert J. Levy. 1986. "Brief History of Cooperative Housing." *Cooperative Housing Journal* (Annual issue): 12-20. ◄

► Market Segmentation

Market segmentation is a condition in which a housing market can be thought to consist of two or more submarkets between or among which there is little interaction. In other words, two submarkets are segmented if a shift in the demand or supply curve in one submarket, causing price and quantity of housing transacted in that submarket to change, has little or no effect on price and quantity in another submarket. Market segmentation may give rise to a condition in which the difference in price between two submarkets can be greater than can be explained in terms of the underlying difference in dwelling quality.

The possibility of market segmentation has been considered in a variety of contexts. One is geographic. Suppose that a city consists only of renters living in two geographically separate residential areas: A and B. Imagine a short-run scenario in which each market is in equilibrium and the price of a standard unit of housing is the same in both markets. Now imagine that the demand in Region A increases so that the equilibrium price in Region A increases. There would now be an incentive for renters to relocate from A to B; however, suppose that for some reason renters will not move (e.g., the cost of commuting from B is too high). Put differently, consumers at A have a zero elasticity of substitution between A and B. In that case, the price difference would persist and Markets A and B would be segmented (at least in the short run). Over the long run, landlords in Market A might be encouraged by the high rent to construct more housing; if so, rents in A would be driven down. In other words, the price difference would persist and A and B would continue to be geographically segmented only if, for some reason, landlords in A were to choose not to increase supply there. To summarize, in this example, it is the consumer's zero elasticity of substitution between A and B that gives rise to geographic segmentation

in the short run, and that price inelasticity of supply is also needed to sustain segmentation in the long run.

Consider a second fictitious example: A city has a rental submarket of three-bedroom bungalows painted yellow wherein there is an equilibrium price (P_y) characterized as the amount that (a) the marginal tenant is willing to pay for a yellow bungalow, (b) the marginal landlord is willing to accept, and (c) clears the market. Imagine now a second rental submarket, for three-bedroom bungalows painted brown, wherein the equilibrium price is P_b. Suppose now that yellow and brown bungalows are transacted in the marketplace until equilibrium is established in each submarket. Note that the only difference between dwellings here is the color of paint. If consumers were indifferent to color, then $P_b = P_y$. If consumers prefer yellow houses and the supply curve for each type of dwelling is upward sloped, then $P_y > P_b$. However, we might expect that because landlords can easily repaint their dwellings, at most, the price difference would be no larger than the cost of repainting a brown house. If the observed price difference were larger than this in equilibrium, we would conclude that the markets are segmented. Segmentation might arise, for example, if a municipal regulation prevented landlords from repainting their brown houses.

Another example serves to highlight the importance of neighborhood externalities. Suppose that a high-income consumer does not want to live in an otherwise comparable dwelling that is located in a low-income neighborhood. Put differently, "low-income neighborhood" is an externality effect that deters some consumers even if the price of accommodation in that neighborhood dropped suddenly.

In practice, are housing markets segmented? This is an empirical question. Scholars have found evidence that housing markets are segmented: for example, across regions (that is, between cities) of Great Britain, across race (among African American, Hispanic, and others) in New York City, between immigrants and nonimmigrants in Stockholm, and among lifestyle groups of homebuyers in Cincinnati.

Market segmentation has also been studied in the context of social housing. The reference to market here is not to a market allocation as described above, because social housing is allocated by a nonmarket mechanism. Instead, *market* is used here in reference to the groups in a social housing project: for example, immigrant, visible minority, elderly, ability-impaired, female, lone parent. Although the encouragement of social mix is a commonly stated goal in social housing policy, it is not uncommon for any one social housing project to contain one or only a few social groups. Social housing that can be characterized this way is socially segmented. The central question is whether public housing managers are discriminating among social groups or whether social groups choose to live in projects where they can have similar neighbors. (SEE ALSO: *Housing Markets; Residential Mobility*)

—*John R. Miron*

Further Reading

Eatwell, J., M. Milgate, and P. Newman, eds. 1987. *The New Palgrave: A Dictionary of Economics.* Vol. 4. London: Macmillan (see pp. 285-87).

Feitelson, E. 1993. "An Hierarchical Approach to the Segmentation of Residential Demand: Theory and Application." *Environment and Planning A* 25(4):553-69.

Galster, G. and J. Rothenberg. 1992. "Filtering in Urban Housing: A Graphical Analysis of a Quality-Segmented Market." *Journal of Planning Education and Research* 11(1):37-50.

Hamnett, C. 1989. "The Spatial and Social Segmentation of the London Owner Occupied Housing Market: An Analysis of the Flat Conversion Sector." Pp. 203-18 in *Growth and Change in a Core Region: The Case of Southeast England*, edited by M. Breheny and P. Congdon. London: Pion.

Huttman, E. D., W. Blauw, and J. Saltman, eds. 1991. *Urban Housing Segregation of Minorities in Western Europe and the United States*. Durham, NC: Duke University Press.

Palm, Risa. 1985. "Ethnic Segmentation of Real Estate Agent Practice in the Urban Housing Market." *Annals of the Association of American Geographers* 75(1):58-68.

Rothenberg, J., G. C. Galster, R. V. Butler, and J. Pitkin. 1991. *The Maze of Urban Housing Markets*. Chicago: University of Chicago Press.

Varady, D. P. 1991. "Segmentation of the Home Buyer Market: A Cincinnati Study." *Urban Affairs Quarterly* 26(4):549-66. ◀

▶ Migration

Issues of how, where, and why people move are strongly linked to housing considerations. Indeed, the definitions of mobility and migration are housing or residence based. Mobility is usually taken to mean any change in residence—across the street, to a different neighborhood, to a new city, or even abroad. Migration is a more restricted concept involving a long-distance change in residence, crossing a county boundary in the process. In theory, moves between counties entail a significant change in people's life patterns (where they work, shop, attend church, obtain medical care, visit family and friends, etc.) and their sense of community identity. Local mobility, on the other hand, does not involve a major change in these activities and affiliations.

Migration specialists have long believed that the causes of local mobility differ from those of long-distance migration. Housing and family considerations dominate the reasons why people make short-distance, local moves. Local moves arise from the need to establish an independent household, to marry or divorce, to accommodate more or fewer children, for a larger or higher-status home, to reduce rent or maintenance, to become a homeowner or alternatively to shift from ownership to renting, and for a host of other reasons related to family dynamics, changing living arrangements, and housing needs.

Conventional wisdom links longer-distance migration to jobs and other employment-related considerations. In theory, migrants flow from regions of low to high employment growth, from high to low unemployment rates, and from low to high income. But because some people derive "psychic income" or "utilities" from places that provide a pleasant climate, environmental and cultural amenities, and nearness to family and friends, they trade off economic for noneconomic benefits. Thus, the ultimate map of migration patterns reflects both economic and noneconomic causes of migration.

Migration selectivity factors are characteristics that distinguish those who move from those who do not. Migration selects out certain individuals on the basis of their age (people in their 20s have the highest probability of moving), education (people with high levels of education are especially prone to long-distance moving), race (white-Anglos tend to make more long-distance moves than blacks or Hispanics), and housing attributes. Between 1990 and 1991, renters made 62% of all long-distance (intercounty and abroad) moves in the United States but constituted only 33% of the total population. In part, this is because renters disproportionately fall into the highly mobile ages between 20 and 30, but it is also because renters are more likely than owners to be migrants, even when the effects of age, education, and income are taken into account. Weaker financial, social, and emotional attachments to rented compared with owned residences facilitate long-distance migration.

In Great Britain, the public housing sector accounted for a sizable share of the housing market, reaching a peak in 1976 of 31.4%. This proportion dropped to 22.8% in 1990 after the conservative governments of Margaret Thatcher and John Major stressed the privatization of public housing or council housing as it is known in Britain. The Thatcher and Major governments offered generous discounts to encourage tenants to buy their units.

One rationale underlying the privatization of council housing was to increase long-distance migration. During the late 1970s and early-1980s, scholars, economic analysts, and government officials asked why people in Britain did not move more often for job-related reasons. They argued that low rates of long-distance migration (only one-third of U.S. levels) did not allow the labor market to respond effectively and efficiently to the geographically uneven distribution of new employment opportunities. The inflexibility of the labor market was manifest in high rates of unemployment in the North and a labor shortage in the Southeast.

Heavy dependence on council housing was seen as a major barrier to migration. Britons seeking access to council housing were forced to place their names on lengthy waiting lists and, once having obtained housing, they waited again for a transfer to better, bigger, or more well situated units. The cumbersome, inflexible, and overly bureaucratic nature of the housing system was thought to dampen the migration tendencies of council tenants. Between 1971 and 1981, the incidence of migration was 32.2% among owners compared with 23.8% for council tenants. The obvious implication, from a housing policy perspective, was that shrinking the size of the council sector would stimulate more migration.

Spatially varying housing price inflation and the geography of housing affordability are especially relevant to the footloose elderly who are no longer place-bound by work. Elderly living in booming housing markets can "cash out" the equity in their homes by migrating to a less expensive housing market. In the United States, states with high home prices send more elderly migrants to Arizona than those with lower home prices, supporting the view that housing affordability helps to explain the map of elderly migration in the United States.

Typically, a migrant is a person making a permanent move between one spatially fixed residence and another. Excluded from this conceptualization are the homeless, itinerant workers, owners of recreational vehicles, second-homeowners, elderly "snowbirds," and college students. Among these special populations, movement and housing conditions are inextricably linked. The higher mobility rate among these groups can result in less permanent housing forms or in the nontraditional uses of more permanent housing types.

In the United States, a sizable and growing number of elderly persons are seasonal migrants, summering in the North and wintering in the South. These so-called snowbirds occupy a range of housing types depending on financial resources and personal preferences. Very few can afford to or, indeed, want to replicate their primary residence, usually located in North. One segment of snowbirds buys condominium apartments or mobile homes in a Sunbelt community, thereby establishing a sense of permanency and commitment to the community albeit on a seasonal basis. A second segment maintains weaker ties to the warm-climate community by renting an apartment for anywhere between one and six months. A recent study in Mesa, Arizona, a southwestern mecca for winter visitors, estimates that 7% of individual apartment units are rented to snowbirds. The influx of winter visitors generates a high degree of seasonality in the local apartment rental market. Rents are higher and vacancy rates are lower in the winter than in the summer.

A third and large component of the snowbird population lives in resorts especially designed to accommodate recreational vehicles (RVs). This segment is, in turn, divided about equally into those who live in so-called park units (permanent, factory-built units with the axles removed containing around 400 square feet of space) and those whose housing unit migrates with them between North and South. Some members of the latter group remain in one spot for the entire winter season, whereas others circulate among many destinations, staying two or three weeks at each resort. A subset of this latter group, estimated nationally to be as high as 1 million people, is referred to as "full-timers"—that is, persons who do not maintain a permanent residence anywhere but who live full-time on the road in their RVs. For them, home is on wheels, neighborhoods and communities are ephemeral, and social relationships are short term but intense.

Another snowbird population lives in RVs but disdains residence in RV resorts, preferring instead to camp at no charge anywhere from parking lots to undeveloped public land. This form of residence is called "boondocking." One extremely popular draw for boondockers in February is the Quartzite Rock and Gem Show, which draws several hundred thousand snowbirds to this desert outpost of fewer than 2,000 permanent residents in southwestern Arizona. An almost completely temporary community forms on public land, complete with banks, shops, recreational centers, and, of course, the Rock and Gem Show. Snowbirds from across the nation motor in to form temporary but highly functional neighborhoods. Socializing, which takes the form of card games, square dances, and television watching, is especially important in view of the RV's dearth of personal space.

From the other end of the world, both geographically and socially, comes another example of the unique housing accommodations of a highly migratory segment of the population. In South Africa, gold miners are migrant workers from rural parts of the country from neighboring nations like Mozambique, Swaziland, Botswana, and Lesotho. An estimated 98% of the gold industry's black workforce of 500,000 lives in highly regimented, male-only barracks adjacent to the mines. This system has forced miners to establish a rural base of landownership, family ties, and community support and to move back and forth between this rural base and the mine compound.

During the 1980s, changes in technology and the political economy of the gold mining industry forced management to look for alternatives to migrancy and compound housing for black workers. The need for more semiskilled and skilled black workers necessitated a more stable workforce. Greater stability, in turn, led to strategies for more owner-occupied family housing near the mines. However important as symbols of fundamental change in the industry's long-standing and almost exclusive dependence on the migrant labor system, efforts at increasing on- or near-site housing for black miners have been limited in scope. The cost of owner-occupied housing is prohibitive for many workers, and foreign workers (representing approximately 40% of the workforce) are barred from owning property in South Africa.

In conclusion, a number of significant questions in contemporary migration research involve rethinking long-held beliefs about migration and housing. The overly simplistic notion that local mobility responds to housing, whereas long-distance migration responds to jobs does not well reflect current realities. Finally, the definition of migration as a permanent move from one fixed residence to another has led migration scholars to fixate on one, very conventional life course trajectory and conventional forms of housing. As more and more working-age people move between first and second homes for pleasure and work, as more children move between two sets of divorced parents, and as the elderly move between fixed housing units or among places in mobile housing units, definitions of what is migration and, indeed, what is a home become blurry. (SEE ALSO: *Residential Mobility*)

—*Patricia Gober*

Further Reading

Champion, T. and T. Fielding, eds. 1991. *Migration Processes and Patterns.* London and New York: Belhaven (see Part II).

Clark, W. A. V. 1992. "Comparing Cross-Sectional and Longitudinal Analyses of Residential Mobility and Migration." *Environment and Planning A* 24:1291-1302.

Crush, J. and W. James. 1991. "Depopulating the Compounds: Migrant Labor and Mine Housing in South Africa." *World Development* 19:301-16.

Gober, P. 1993. "Americans on the Move." *Population Bulletin* 48(3): 1-40.

Hamnett, C. 1991. "The Relationship between Residential Migration and Housing Tenure in London, 1971-81: A Longitudinal Analysis." *Environment and Planning A* 23:1147-62.

Long, L. 1988. *Migration and Residential Mobility in the United States.*
     New York: Russell Sage.
Steinner, D. N. and T. D. Hogan. 1992. "Take the Money and Sun:
     Elderly Migration as a Consequences of Gains in Unaffordable
     Housing Markets." *Journal of Gerontology* 47:S197-S203. ◄

► Military-Related Housing

The military has an obligation to house its personnel. Although the housing of troops has long been an essential part of the military mission, providing housing for the families of U.S. military personnel has been a relatively modern development. Military family housing is a form of public housing not generally recognized as such, with some similarities to, as well as important differences from, the traditional public housing program.

As part of the basic benefits package, the housing needs of service personnel are addressed through direct provision of quarters or through an allowance to cover or supplement costs of housing in the private sector. It is thus an entitlement, albeit limited to those service members (and their dependents) above a certain pay rank (currently, E4).

The services (Army, Navy, Air Force, Marines, plus several tiny "services," such as the Defense Mapping Agency, Defense Logistical Agency, National Security Agency, Defense Intelligence Agency) limit the amount of housing they build, preferring to rely on the private housing supply in the community surrounding an installation. Only if such housing is unavailable or inadequate is a process initiated to build or lease family housing directly for the military.

The military has an inventory of some 400,000 family units, on-base and on nearby sites (although the base closings mandated by Congress and the Department of Defense [DOD] in the late 1980s and early 1990s are reducing that number), making the DOD by far the nation's largest landlord. The actual number of family units ever built by the military is far higher, because many units, temporary or permanent, have been disposed of or demolished.

About two-fifths of this stock is for the Army, one-third for the Air Force, one-sixth for the Navy, and a small fraction for the Marines. About one of every three military family housing units is outside the United States—primarily in Germany, Japan, Italy, and the United Kingdom. About one-third of military personnel with dependents live in DOD family housing; a considerably lower proportion of officers, as opposed to enlisted personnel, live in these quarters. At most bases, especially at installations in attractive, populous, and high-priced coastal areas, there are long waiting lists for these units. Base closings are adding to the shortage situation.

Programs

Military family housing has been funded and built under a variety of programs. During the New Deal 1930s, appropriations from the Works Progress Administration and Public Works Administration were used. Prior to and following World War II, various emergency "defense housing" programs, such as the Lanham Act, helped provide housing for servicemen, shipyard workers, and defense plant workers.

The first serious attempt to fashion large-scale military family housing began in 1949 with passage of the Wherry Housing Program (named for its Senate sponsor, Senator Kenneth Wherry of Nebraska). Under this program, housing was constructed by private developers on government-owned land or on land adjacent to military installations, with private FHA-insured financing for up to 90% of construction cost. Developers owned and operated the units, which were rented to military families. Between 1950 and 1955, approximately 84,000 units were built. Serious problems arose as a result of construction cost ceilings and poor maintenance by the developers-owners. Most of these units were subsequently purchased by the federal government under the successor Capehart program, passed in 1955 (and named after its Senate sponsor, Senator Homer Capehart of Indiana).

The Capehart Program, responding to the demonstrated failure of the private sector to operate decent housing, differed from the Wherry Program in two important aspects. FHA-insured private financing covered 100% of construction costs, and on completion, title was ceded to the individual services, which then took over the mortgage and management responsibilities. Generally, Capehart housing was built to higher standards than Wherry housing, although construction cost ceilings presented problems again. Some 115,000 units were built between 1956 and 1962. High mortgage costs, compared with the alternative of building housing with directly appropriated funds, led Congress to cancel the program.

Throughout this period, military family housing was also built with specifically appropriated funds in overseas locations (where FHA was not permitted to operate) and in high-cost areas of the United States (where cost ceilings precluded other programs). Since termination of the Capehart Program, military family housing has been built largely with appropriated funds—in effect, capital grants furnished via the DOD's congressionally approved military construction budget. Approximately 150,000 units have been built with direct financing.

At various times, the military has leased existing privately owned units for its personnel. But more recent introduction of housing allowances (see below) has virtually eliminated this practice.

Design and Management

Military housing developments range widely in size. Some have more than 2,000 units; others not more than a dozen units. A variety of housing types have been built. One- and two-story duplexes and row houses are most common. Single, detached units are built only for higher-ranking officers. High-rise apartments are rare, except in Japan. The majority of developments are located in small towns or on the outskirts of cities, where few if any community facilities exist, so many of the larger developments include such amenities.

Quality problems exist in military family housing, although the exact extent is not fully known, because the military does not undertake the same kinds of regular housing surveys carried out by the Bureau of the Census and the Department of Housing and Urban Development for the civilian stock, nor are there definitive DOD standards for housing adequacy. Although mid-1980s DOD reports stated that only 4% of such units were officially classified as "inadequate," because of substantial defects or inadequate space standards, by the mid-1990s, the secretary of defense was lamenting that as much as 60% of the military's housing stock had problems such as peeling lead-based paint, hazardous asbestos, cracked foundations, or faulty heating and cooling systems, and Congress was asked for billions of rehabilitation and replacement dollars—although DOD budgetary politics in the post-Cold War era suggest possible hyperbole. The individual services are responsible for management of these units, which often is carried out by civilian employees of the military. Large-scale modernization and renovation is done through contracts with private firms.

Generally, housing assignment is by pay grade (rank) and family size. Compounds are divided into groupings by rank. Upper-level noncommissioned officers and officers are assigned large units regardless of household size; for others, family size determines unit size. The different treatment ostensibly is rationalized by entertainment responsibilities of senior personnel but clearly has more to do with the status and privilege reflected in military rank.

Although racial discrimination and prejudice are not unknown in the military, it is one of the most integrated institutions in the United States, as manifested in these housing developments. Racial segregation per se does not exist there (although the assignment of housing by rank builds in a quite different race and gender mix among different housing areas, a phenomenon more obvious at the upper levels, where there are few minority or female admirals, colonels, etc.).

An important, somewhat unique factor in the management of these units is the control exerted over occupants by the employer-landlord. Requests to cease or modify behavior or improve property care are more likely to produce results when issued by a higher-ranking neighbor or via official complaint.

Because provision of quarters or a housing allowance is part of the basic compensation to military personnel, rents and utility costs are not charged to occupants. Eviction actions for nonpayment or for rent delinquency also do not arise.

When a base is closed or reduced in size, the unneeded units may be transferred to another federal agency, the local government, or to the private sector. Some are converted to low-rent public housing. More recently, arrangements have been made to use these facilities to house and provide services for homeless persons.

An alternative available to military families is owning or renting in the private market. Military personnel who do not live in rent-free DOD housing are entitled to a basic allowance for quarters (BAQ) in addition to their base pay. The BAQ, established in 1949, is adjusted periodically but has not kept pace with the rising cost of housing, leading to enormous affordability and housing quality problems for tens of thousands of families. In 1980, Congress created the variable housing allowance, for homeowners and renters, to supplement the BAQ in high-cost areas.

Unique Features

Several features stand out about the military family housing program, some of which could provide useful models for the civilian sector. First is the use of direct appropriations or capital grants for construction. In the private sector, the cost of credit is central to the cost of housing, creating an affordability crisis that manifests itself in lopsided rent-income ratios, increased crowding, homelessness, and high foreclosure rates. As an alternative to these pitfalls, housing programs that decrease or end reliance on debt financing might be given serious consideration. The fact that a government housing program has existed in the United States and has produced tens of thousands of units through such a financing system should be recognized and further studied.

A second unique feature of the military family housing system is that it is an entitlement program: Everyone who meets the rank and length of service criteria has a right to free housing or to a housing allowance. Nowhere else in the U.S. housing system does the concept of entitlement exist (except the indirect subsidy provided by the tax system's homeowner deduction, which is highly regressive and wasteful). A right to adequate, affordable housing is something U.S. society has neither recognized nor implemented. This situation endures despite the 1949 Housing Act that proclaimed a National Housing Goal of "a decent home and suitable living environment for every American family." Although a universal right or entitlement to decent, affordable housing seems unachievable in the near future, it is useful to know that something resembling this exists for a substantial part of the population.

A third feature worth emphasizing is the successful abolition of racial segregation in military family housing. Given the increasing patterns of racial segregation in society and the segregated nature of the low-rent public housing program, it is inspiring to see examples in which segregated living patterns have been overcome. (SEE ALSO: **Affordability; Public Housing;** *Racial Integration; Section 222*)

—*Chester Hartman*

Further Reading

Graham, Bradley. 1995. "The New Military Readiness Worry: Old Housing." *Washington Post,* March 7, p. A1.

Hartman, Chester and Robin Drayer. 1990. "Military-Family Housing: The Other Public Housing Program." *Housing and Society* 17(3):67-78.

Institute for Policy Studies Working Group on Housing. 1989. *The Right to Housing: A Blueprint for Housing the Nation.* Washington, DC: Author.

Hershfield, David C. 1985. "Attacking Housing Discrimination: Economic Power of the Military in Desegregating Off-Base Rental Housing." *American Journal of Economics and Sociology* 44(1): 23-28. ◄

► Mixed-Income Housing

Although mixed-income housing has emerged as a new development goal, the idea in fact describes how cities' housing stock was organized in the 19th and early 20th centuries. At that time, cities had a variety of housing types, densities, and monthly costs that catered to a full spectrum of affluent, middle-class and lower-income families.

In modern industrial cities, which flourished in the mid-1800s, most people lived relatively close to the workplace or close to a trolley line that could bring them to work. The city was laid out in such a way that fancy townhomes, moderate- and lower-priced apartments, and "cold water" tenements were located near to one another, compared with the current arrangement of most U.S. cities.

In the early 1920s, larger cities in the United States began to adopt zoning regulations that defined the allowed land uses in any one area, as well as allowable density, setback regulations, open space requirements, and other health and safety requirements (e.g., minimum width of stairways, number of exitways, building, materials, etc.). These zoning codes had the effect of separating out new residential uses from office, retail, and industrial activities. Over time, the concept of "highest and best" use emerged, which defined for any parcel the maximum amount and type of development that could be constructed under the zoning code regulations. The combination of separating land uses from one another and defining the land price as a relationship of the allowable building density encouraged property owners and investors to build a single type of housing form. For example, in a medium-density zone, a property owner could erect a building containing between 10 to 14 units per acre. A high-density zone would allow buildings that were designed between 15 to 20 units per acre. These numbers correspond to different housing types, such as single-family detached homes (4-12 dwelling units [DU]/acre), town homes (12-16 DU/acre), garden apartments (18-24 DU/acre), and high-rise towers (25-120+ DU/acre).

The response to zoning was to build the maximum number of units allowed and to repeat the same building forms multiple times to reduce costs and gain economies of scale. From 1940 to 1980, the suburbs produced whole neighborhoods with virtually identical housing types and densities. This resulted in an economic stratification that clustered families with similar incomes and, oftentimes, similar ethnic and racial backgrounds. The reliance of workers on commuting by automobile added to this residential stratification by enabling wealthier workers to live in nicer places farther away from the workplace. More recently, middle-class families have been commuting 30 to 90 minutes to and from work in order to buy less costly housing on inexpensive land located in the periphery of metropolitan areas.

Beginning in the early 1990s, planners and developers in the Bay Area established the concept of mixed-income residential projects, with several underlying assumptions:

▶ *Mixture of housing types:* To attract a mix of different incomes and to relate to existing residential and commercial areas, these projects would offer rental apartments, condominiums, and town homes, as well as higher-priced detached dwelling units.
▶ *Mixture of densities:* Building at an average density higher than the existing residential neighborhoods, these projects would offer a range of densities from 12 DU/acre for small-lot detached units and up to 50+ DU/acre for apartment and podium type development with parking under the units.
▶ *Mixture of family incomes:* The concept is to provide a combination ranging from low-income and very low income families, typically at less than 80% of the area medium income (80%-100% AMI), middle-income (100%-120% AMI) and market rate (120%+ of AMI) rental and for-sale units. Translated into monthly housing costs, this means that a single project might offer housing that ranges from under $350 per month for a two-bedroom apartment to a sales price of $240,000 for a single-family home.
▶ *Located near transit and retail:* To alleviate the need to use the car to drive to work and shopping, another key ingredient is to situate the development within walking distance to a transit station, bus stop, and grocery store. Typically, a quarter mile, or about four blocks, is considered convenient for people to walk. This arrangement may also permit a reduction in the need for parking spaces.
▶ *Partnership between city, developer, and not-for-profit organizations:* Producing a variety of housing types typically requires that the city, landowner, developer, and nonprofit housing organization need to cooperate to secure an option on the land, obtain land entitlements, assemble different sources of financing, and manage the rental portion of the project.
▶ *Use of special zoning and permit waivers:* A higher-density zoning category permits projects to be built at a higher average density, allows a greater variety of housing types, and lowers the land cost component per household.
▶ *Use of private lenders, local public dollars and state and federal subsidies:* Producing mixed-income housing necessitates securing different sources of financing, geared to fund a particular aspect of the development. For example, a low-income rental apartment might have special housing funds to subsidize rental rates. Tax credits are earmarked for rental construction for units targeted for families earning under 60% of AMI. Redevelopment agencies may loan funds for site acquisition and be repaid from private construction lenders.

The benefits of mixed-income projects include the following:

▶ Greater diversity of families living nearby one another
▶ A better use of decaying infill sites
▶ Less economic "ghettoization" of certain building types
▶ A better design fit with existing neighborhoods
▶ Improved transit ridership
▶ Shared infrastructure and land costs between market rate andsubsidized projects

The obstacles to providing mixed-income housing include the potential negative response from (a) adjoining property owners who do not want to see higher-density development, (b) city planners and elected officials who do not want to establish different housing patterns, (c) lending institutions who believe their investments may be less secure because of the proximity of different housing types and more than one management entity, and (d) residents who generally wish to protect property values and who want to limit socioeconomic diversity in their community. (SEE ALSO: *New Urbanism; Tent City*)

—*Will Fleissig*

Further Reading

Bernick, Michael S. 1996. *Transit Villages in the 21st Century.* New York: McGraw-Hill.

Hall, Kelvin Brian. 1996. *A Study of Environmental Semiotics in the Production of a Mixed-Income Housing Complex.* Master's thesis, Architecture, Rice University, Houston, TX.

Hoffman, Morton. 1995. "Mixed-Income Housing: A New Direction in State and Federal Programs." *Real Estate Issues* 20(2):40. ◀

▶ Model Cities Program

More than a full generation has passed since Lyndon Johnson proposed, and a somewhat reluctant but still responsive Congress enacted, the Model Cities Program. Yet its success or failure as concept, law, or program remains ringed by controversy. It was the centerpiece of the Housing and Urban Development Act of 1966 and, together with the 1968 Housing Act, constituted the historic high watermark in the nation's commitment to the well-being of its cities. Nonetheless, since its enactment, its impact economically, socially, and politically on cities has often been described as a failure.

As an urban strategy, the program was a first. Before Model Cities, major housing and urban assistance programs were, for the most part, bootlegged into law under the guise of a "greater" national purpose: (a) to stimulate economic recovery in the home mortgage and public housing programs of the Roosevelt years, (b) to reward returning veterans after World War II by more guarantee programs, or (c) to rescue downtown merchants through urban renewal.

Over time, these programs had uneven and often damaging consequences for central cities. They often hurt neighborhoods, displacing their poor, and left central business districts pockmarked with vacant sites. Even the stirrings of urban policy in the Kennedy administration, which advocated cabinet status for the assortment of programs under the agency umbrella of the Housing and Home Finance Agency and secured the first subsidies to private builders for low- and moderate-income housing, were tentative and incomplete. It remained for President Johnson's Great Society to identify cities as one of the three major domestic issues, equal in status to health and education. So the preamble of the 1966 act reads, "The Congress hereby finds and declares that improving the quality of urban life is the most critical domestic problem facing the United States."

Why and how Model Cities came to the top of the national agenda, why the conservative backlash of the 1970s and 1980s pushed urban issues off the agenda, and what evaluations of its contributions one makes 30 years later are central to understanding the process of urban public policy in the United States.

First, critics must place the Model Cities Program in the context of the times. What concerns triggered it? Why did it emerge?

Politically, it was the determination of a president to make a mark in domestic policy comparable to that of his early mentor, Franklin Roosevelt. The Eisenhower years witnessed just two major national initiatives—the 1956 interstate highway program and 1954 urban renewal. Both were destructive in their impact on large cities, especially in the wholesale clearance of poor neighborhoods. New concern for the poor surfaced in Michael Harrington's *The Other America,* the fledgling initiatives of President Kennedy's Committee on Juvenile Delinquency and Youth Crime, and a growing scholarly literature. Next, the urban cause found powerful political impetus in the civil rights movement, to which Lyndon Johnson gave priority in securing both the Civil Rights Act and the Economic Opportunity Act in 1964. (The latter's Community Action Program preceded Model Cities with a more radical version of empowerment.) Finally, the Watts riots in 1965 provided the spark to give urban issues a momentum that had never existed before, and the recurrent riots of 1967 continued the focus on the increasingly desperate straits of the nation's urban areas.

President Johnson relied on an outside task force to fashion Model Cities, an innovation that the Kennedy-Johnson White Houses used frequently. These task forces were composed largely of outside experts, often academics, and they drafted the major proposals for the Great Society. In 1964 and 1965, two urban task forces shaped the first presidential message on cities and conceived of the Model Cities Program (then called "Demonstration Cities").

The program's aim was to orchestrate the physical, political, and social dimensions of urban aid. Its two distinctive features were (a) *selectivity* (only 150 cities were chosen in the first phase compared with the 1,200 with active public housing and urban renewal programs at the time) and (b) a *metropolitan-wide effort* to link suburbs and central cities in common planning and community development programs.

Later on, critics argued that Model Cities was "politicized" by expanding the original number recommended by the task force and that key features of the metropolitan development component were lost in Congress. Although the program did accommodate the political process by showing deference to the constituencies of key congressional committees, the large majority of congressional districts did not receive grants, and selectivity remained a distinctive legislative innovation. Moreover, major provisions of the metropolitan component were enacted, providing for the review of local government projects by regional agencies and establishing regional councils of government (COGs). In 1968, national urban aid programs were greatly expanded when that year's act made a unique quantitative commitment to build 26 million housing units in 10 years, 6 million targeted for the poor.

The Model Cities program did not end with the 1968 elections, for President Nixon rejected the advice of his urban task force to abolish it, and Department of Housing and Urban Development (HUD) Secretary George Romney supported the commitments made in the 1960s. But neither did the program flourish. It was folded into the 1974 Housing and Community Development Act and renamed and redirected. In these years, a torrent of academic and media criticism fell on Model Cities, going beyond the early critique that it had been politicized. Critics charged that Model Cities was too little and too late, strangled by the Vietnam War; that it was too bureaucratic with too many rules and regulations; and that it was too "theoretical," a product of academic elites and not the communities themselves. Thus, a consensus emerged throughout the 1970s from conservatives and liberals alike that, in reality, Model Cities was chiefly a symbolic gesture, too small to have economic impact or change the pattern of urban politics.

Although the program continued *sub rosa,* President Nixon declared that the urban crisis had ended and focused on environmental concerns. President Carter, constrained by a faltering economy, undertook few new urban initiatives. Under the Reagan and Bush administrations, HUD resources were drastically reduced—more than any other federal department or agency. There also emerged a pattern of corruption and criminal conviction unlike any HUD had experienced before. The chief new initiative was the concept of "enterprise zones," which by the early 1990s had no track record and no financing.

Against this historical background and controversial evaluation, what responsible judgment can be struck about the program in the mid-1990s? First, as a concept emphasizing selectivity and competition among cities, both in demonstrating participatory and managerial competence, it broke new ground. It joined together, however uneasily, an alliance between city hall and community activists. Second, in company with other Great Society programs, it really helped the poor. The proportion of the U.S. population below the poverty line fell from 19% in 1964 to 12% in 1979. Politically, over the years, the program proved to be an effective political training ground for future minority officeholders, who captured mayoralties and other urban leadership posts in the 1970s and 1980s. If it did not overcome the centrifugal forces then dispersing jobs and households to the suburban fringe, it did provide new structures and processes by which minorities and new immigrants gained access to the political life of cities. Systematic studies in the mid-1980s demonstrated that, over time, Model Cities at the local level were powerful forces in the drive for political equality and community self-determination. As the 1992 Los Angeles riots make plain, that process, after long years of studied neglect, cannot alone keep the urban peace or ensure urban prosperity. But in company with civil rights and the other Great Society programs, it began the process of empowerment for the disenfranchised and the emergence of community-based organizations on which present housing, employment and educational programs increasingly rely. (SEE ALSO: **Urban Redevelopment**)

—*Robert C. Wood*

Further Reading

Browning, Rufus, Dale Rogers Marshall, and David H. Tabb. 1984. *Protest Alone Is Not Enough.* Berkeley: University of California Press.
Carmon, Naomi, ed. 1990. *Neighborhood Policy and Programs—Past and Present.* London: Macmillan.
Frieden, Bernard J. and Marshall Kaplan. 1975. *The Politics of Neglect: Urban Aid for Model Cities to Revenue Sharing.* Cambridge: MIT Press.
Harrington, Michael. 1971. *The Other America.* New York: Penguin.
Kaplan, Marshall and Franklin James, eds. 1990. *The Future of National Urban Policy.* Durham, NC: Duke University Press.
Schwartz, John E. 1983. *America's Secret Success: A Reassessment of Twenty Years of Public Policy.* New York: Norton.
Wood, Robert C. 1993. *Whatever Possessed the President? Academic Experts and Presidential Policy, 1960-1988.* Amherst: University of Massachusetts Press. ◄

► Model Codes

When local governments face the prospect of controlling building construction, they have the option of choosing between developing their own standards or adopting a model code. The first requires an expertise in building construction that is often beyond the capacity of local government. Adopting a model code is a much simpler alternative because such models represent the sum total of a considerable amount of experience and research on the part of the sponsoring organization. Over time, clear arguments in support of uniform construction standards and the validity of model codes have prevailed. A Federal Trade Commission staff report found that by the late 1980s, 97% of U.S. building codes were based on one of three model codes or a state code that was itself a derivative of these models.

The three models have a regional orientation. The Uniform Building Code, sponsored by the International Conference of Building Officials (ICBO), is used in western states. The National Building Code, sponsored by the Building Officials Conference of America (BOCA), is used in eastern states. The Standard Building Code, sponsored by the Southern Building Code Congress International (SBCCI), is used in southern states. In addition to the model building codes, other codes address more specific aspects of building construction. These include electrical codes, elevator codes, fire prevention codes, housing codes, and plumbing codes. The American Public Health Association in conjunction with the Centers for Disease Control has developed a model housing code that specifies maintenance standards.

As an example of a sponsor, the ICBO is a nonprofit organization founded in 1922. More than 1,600 city and county building departments and state agencies in the United States participate in drafting and approving its model. ICBO also supports research. The Uniform Building Code and its revisions are republished every three years as recommendations.

The three model building code organizations collaborate on issues of mutual concern, such as professional certification and federal policy initiatives under the auspices of the Council of American Building Officials (CABO), formed in 1972. In response to pressure from the National Home Builders Association, CABO sponsored development of the CABO One and Two Family Dwelling Code. This model code directly incorporates plumbing, mechanical, and electrical standards. Also in response to the Home Builders, the model organizations have take the bold step of abandoning their more regionally focused codes and working together in developing what will be known as the International Building Code (IBC). It is anticipated that the IBC will enable builders to be more globally competitive.

Despite the widespread adoption of model building codes, there is still resistance to their use. The state of Wisconsin has its own code, as do the cities of New York, Chicago, and Denver. Many states do not mandate the use of a building code, leaving this to the discretion of local governments. (SEE ALSO: *Building Codes*)

—*Deborah A. Howe*

Further Reading

Blankenship, Frank J. 1989. *The Prentice Hall Real Estate Investor's Encyclopedia*. Englewood Cliffs, NJ: Prentice Hall.

Hageman, Jack M. 1990. *Contractor's Guide to the Building Code Revisited*. Carlsbad, CA: Craftsman.

Jacobus, Charles J. 1996. *Real Estate Principles*. Upper Saddle River, NJ: Prentice Hall.

Joint Committee on Building Codes. 1977. *Report on Model Codes*. 1966 Series: United States. National Commission on Urban Problems. Background paper, No. 22. Washington, DC: Author.

Korman, Richard. 1989. "A Much-Misunderstood Contraption." *Engineering News-Record (ENR)* 222:25, 30-36.

McKenzie, Dennis J. 1996. *Essentials of Estate Economics*. Upper Saddle River, NJ: Prentice Hall.

Peiser, Richard B. 1992. *Professional Real Estate Development: The ULI Guide to the Business*. Washington, DC: Urban Land Institute.

Shim, Jae K. 1996. *Dictionary of Real Estate*. New York: John Wiley.

Tosh, Dennis S. 1990. *Handbook of Real Estate Terms*. Englewood Cliffs, NJ: Prentice Hall. ◄

► Modular Construction

As the building industry has become more complex, with an increase in the number of available parts and decision makers, agreement on modules, or basic units of size of building parts, has become necessary. These agreements constitute modular coordination.

The use of modules to coordinate the various dimensions of buildings was known in classical Greece and Rome and in Eastern cultures. But not until the period of Western industrialization did the adoption of dimensional modules became widespread. In the United States, modular coordination started in the 1920s. Albert Bemis proposed the four-inch cubical module in 1936. The first voluntary national dimensional standards for building modularity in the United States were adopted in 1945. International agreements on dimensional or modular standards in the building industry were widely discussed in the 1950s, principally designed to foster international trade in both products and services.

The objective of a dimensional module is to coordinate and reduce the variety of sizes of elements to simplify their production and assembly with other parts. Dimensional agreements among competitive producers also enable users of the products to consider options based on quality and performance, not on whether the item will fit with other parts. This principle applies to commodities such as masonry and wood products, sheet materials, various finish materials, and more complex products such as doors and windows, cabinets, and plumbing fixtures. All U.S. producers of masonry units (brick and concrete block) have agreed on a four-inch module. This dimension (approximately 10 cm) is also a basic module used in voluntary international standards. Likewise, pipes of different manufacturers can fit together because their producers agree on certain dimensional standards. However, to keep out competition and retain and enlarge an established market for their products, other manufacturers decide not to agree on dimensional harmonization.

The term *module* is also associated with housing production. In this case, it refers to the off-site fabrication of three-dimensional parts of buildings, which are then assembled on-site to make a complete structure. This is called *modular construction*. It is different from modular coordination in that it constitutes a specific way of organizing building construction, whereas modular *coordination* is a method of assuring that building components are dimensionally related.

Modular construction has a long history in the United States. Lithographs depicting the 19th-century settlement of the American West show complete two-story houses and sections of houses being delivered to new towns on flatbed rail cars crossing the prairies.

Modern modular construction involves the production, in factorylike settings, of building sections 12 to 14 feet wide, up to 60 feet long, and normally 10 feet high, complete with all exterior and interior finishes, cabinets, mechanical and plumbing equipment, and fixtures. The level of completion prior to delivery to the site varies, but most modular units are 90% to 95% complete and are ready to occupy after utility connections and final inspections by local building regulators.

A building may be made of two or more structurally independent modules, which may be stacked to make two- or three-story buildings, in a wide range of configurations. Uses of these buildings—using single or multiple modules—include for-sale and rental housing, offices, motels, retirement communities, schools, and clinics.

Most modular house construction uses wood frames. A few producers make concrete boxes, and some make steel-framed units. In the late 1960s and in the 1970s, several companies used modules made from concrete and other materials to construct large residential structures, in part stimulated by the federal government's Operation Breakthrough program.

Most producers employ some construction materials and methods different from site builders. For instance, materials that would degrade in wet weather construction can be used in the factory environment. In addition, devices are used in the production facilities to align and assemble large elements with automated gluing and fastening, and computers are used to manage inventory and waste materials reuse, as well as the cutting and shaping of materials.

Modular units are normally transported to the site by truck on temporary chassis. Production plants are concentrated in nonurban areas of the United States within approximately a 300-mile range of large, urban markets. Pennsylvania is one of the states making large numbers of modular buildings. Most plants hire nonunion workers and are subject to OSHA (Occupational Safety and Health Administration) safety requirements.

Commonly, modular houses are regulated by codes set at the state level. This category includes sectional or modular housing but does not include mobile or manufactured homes, which are regulated by a national code. Reciprocity agreements between some states allow a producer in one state to sell units in another. Where preemptive state codes do not exist, modular houses must comply with local codes. Appraisal values are comparable with site-built construction, and all qualifications for traditional mortgage financing are also met.

Modular house construction contributes an estimated 6% of total U.S. housing starts, with some production reaching 10% in the mid-Atlantic region in 1996. In 1997, modular house starts in the United States accounted for about 124,000 units. (SEE ALSO: *Flexible Housing;* **Industrialization in Housing Construction**)

—*Stephen Kendall*

Further Reading

Bemis, Albert Farwell. 1936. *The Evolving House. Vol. III: Rational Design.* Cambridge: MIT, Technology Press.

Bruce, Alfred and Harold Sandbank. [1944] 1972. *A History of Prefabrication.* New York: Arno.

Carreiro, Joseph et al. 1986. *The New Building Block: A Report on the Factory-Produced Dwelling Module.* Research Report No. 8. Ithaca, New York: Cornell University, Center for Housing and Environmental Studies.

Hallahan Associates. 1989. *Factory Built Housing in the 1990s: A Study on Modular and Panelized Housing from 1984 to 1994.* Washington, DC: Building Systems Council of the National Association of Home Builders.

Strategic Issues for Homebuilders, 1990-1992 and Beyond. 1990. Washington, DC: National Association of Home Builders, Department of Economics and Housing Policy. ◀

▶ Moratorium on Federally Assisted Housing Programs

On January 8, 1973, outgoing HUD Secretary George Romney announced a moratorium on all new commitments for federally subsidized housing programs, including Section 235, Section 236, and public housing. In addition, the freeze extended to sewer and water grants, open-space land programs, public facilities loans, urban renewal, and Model Cities funding, as well as to smaller programs under the jurisdiction of the Department of Agriculture's Farmers Home Administration. In this bombshell announcement, delivered to the annual convention of the National Association of Home Builders, Romney justified the moratorium, citing an "urgent need for a broad and extensive evaluation of the entire Rube Goldberg structure of our housing and community development statutes and regulations." As a replacement, he promised a "program of community development special revenue sharing of which these programs would become a part." This would materi-

alize in the Housing and Community Development Act of 1974.

Background

The program structure to which Romney referred originated in the 1930s as a New Deal response to the Great Depression. After World War II, the federal government sought to expand housing programs with its mandate for "a decent home and suitable living environment for every American family," a phrase that has persisted long after the passage of the Housing Act of 1949. That same act further expanded the federal role in cities with the controversial Urban Redevelopment program; its successor, Urban Renewal, continued that intervention in the Housing Act of 1954. Further scope was added to federal policy with broadened initiatives for FHA mortgage insurance, including actuarially risky "special purpose" programs such as mortgage insurance for multifamily housing.

The election of John F. Kennedy brought to power an administration that valued public-private cooperation in addressing social problems. As a result, the Housing Act of 1961 encouraged private lenders to generate capital for rental housing coupled with federal subsidies to reduce the interest rate and enhance affordability. These below-market-interest-rate (BMIR) programs, such as Section 221(d)(3), were characterized by high loan-to-value ratios for which FHA mortgage insurance was required. BMIR programs required treasury buydowns of the interest rate for as long as 40 years, a feature that many believed to be too costly an obligation.

As federal programs proliferated, earlier assisted efforts, most notably public housing, continued on an increasingly rocky course marked by escalating operating costs, crushing burdens of deferred maintenance, and ossified patterns of racial segregation. In 1972, the spectacular, nationally televised demolition of the immense Pruitt-Igoe project in St. Louis, barely 15 years old at the time, catalyzed public attention on all subsidized housing programs.

However, it was the Housing Act of 1968 that finally brought federally assisted housing to the brink. The act's Section 235 and Section 236 programs indelibly imprinted the aura of failure on both public opinion and on political institutions. The Housing Act of 1968 was the outgrowth of that decade's civil disorders, a phenomenon exhaustively studied by three federal commissions, Kaiser, Kerner, and Douglas. All of these reached the conclusion that a major cause of this unrest was inadequate housing, and the ensuing Housing Act of 1968 called for the construction, rehabilitation, or both of 26 million units in the 1969 to 1978 decade, 6 million of which were for low- and moderate-income households. On an annual basis, this represented an increase of about 70% over total housing starts for the period 1961 to 1968. The question, then, was could the federal government effectively manage this dramatic escalation in housing market intervention?

The Moratorium: A Study in the Irony of Presidential Politics

Before this question can be considered, we must develop an appreciation for the irony of the 1968 presidential election. In March of that year, Lyndon Johnson announced that he would not seek reelection, a decision dictated by the growing strength of the anti-Vietnam War movement. The question, then, was how would the next, conservative Republican President, Richard Nixon, whose main concern lay with foreign affairs, manage a broad liberal domestic mandate (the Housing Act of 1968) inherited from Johnson's Great Society? The Nixon administration realized that it could not reverse this Democratic social policy, especially because it had the support of Republican allies such as the National Association of Home Builders and, to a lesser extent, the Mortgage Bankers Association. Furthermore, the appointment of George Romney as HUD secretary brought a pro-urban, pro-housing advocate to the administration. In addition, Daniel Patrick Moynihan, later a Democratic senator from New York, provided a liberal balance to Nixon's more conservative advisers.

Nevertheless, Nixon soon placed his stamp on domestic policy with his "New Federalism" concept, a program that sought to simplify layers of Democratic programs and shift control over them to states and localities. At the same time, the high production strategies of the Housing Act of 1968 were gearing up as clouds developed over public housing. By 1970, widespread perceptions had arisen that these programs were in serious trouble. In December of that year, the staff of the House Committee on Banking and Currency reported,

> The Department of Housing and Urban Development and its Federal Housing Administration may well be on its way toward insuring itself into a national housing scandal. This conclusion has been reached because of the role that FHA has played in the operation of the section 235 and other programs. . . . FHA has allowed real estate speculation of the worst type to go on in the 235 program and has virtually turned its back to these practices. In many areas of the country, the 235 program is "carrying" the real estate market. In one county in the State of Washington, 80% of the real estate transactions in 1970 were made up of 235 houses. . . . In some areas, 235 purchasers are either "walking away" from their homes, or . . . are turning the houses back to (FHA).

The Section 236 program had also begun to experience difficulties early on, including a large number of foreclosures. In tandem with the discrediting of public housing as a result of the publicity surrounding the Pruitt-Igoe demolition, it was clear that all federally assisted housing programs were in serious political difficulty. Emboldened by his landslide reelection in 1972, President Nixon finessed earlier struggles with Congress over executive impounding of previously appropriated funds and declared a moratorium on all new HUD commitments.

In its wake, a research team led by Michael H. Moskow, HUD Assistant Secretary for Policy Development and Research, and new HUD Secretary James Lynn, concluded that subsidy programs were too costly and too inefficient. Their study, published in 1973 as *Housing in the Seventies,* constituted a broad indictment of past policies. It called for a new approach for the provision of housing subsidies, the

most important of which was the concept of a housing allowance that would emerge as Section 8 of the Housing and Community Development Act of 1974. (SEE ALSO: **Federal Government;** *Section 8; Section 101; Section 235; Section 236*)

—Daniel J. Garr

Further Reading

Boyer, Brian D. 1973, *Cities Destroyed for Cash.* Chicago: Follett.

Hays, R. Allen. 1995. *The Federal Government and Urban Housing.* 2d ed. Albany: State University of New York Press.

Meyerson, Martin and Edward C. Banfield. 1955. *Politics, Planning and the Public Interest.* Glencoe, IL: Free Press.

U.S. Department of Housing and Urban Development. 1973. *Housing in the Seventies.* Hearings before the Subcommittee on Housing of the Committee on Banking and Currency, House of Representatives, 93rd Cong. 1st sess., Part 3. Washington, DC: Government Printing Office. ◀

▶ Mortgage-Backed Securities

During the early 1970s, the process of securitization began in financial markets in the United States. Securitization transforms assets and liabilities that are illiquid and costly to sell into capital market instruments. This process enables owners of assets and debts to market their claims saving considerable amounts of resources. Similarly, this process permits investors to purchase interests in various types of assets in smaller shares, removing one of the traditional investment disadvantages ("lumpiness") of real asset markets. Securitization is also best viewed as a mechanism for providing liquidity to markets in which liquidity has traditionally been a problem.

One of the major forms of securitization in real estate financial markets has been the use of mortgage-backed securities (MBSs). These instruments were supported by the Government National Mortgage Association (GNMA or Ginnie Mae) and later by the Federal Home Loan Mortgage Corporation (FHLMC or Freddie Mac) and by Fannie Mae (formerly, the Federal National Mortgage Association). In effect, these agencies have assisted in the production and development of a national market for institutional-grade securities based on the characteristics and experience of the individual borrowers who make up the pool of mortgage holders underlying the security. By the late 1980s, more than $300 billion in MBSs were created.

Conceptually, the creation of an MBS is straightforward: A number of primary mortgage loans are purchased or pooled. Each loan is evaluated for its investment attributes: (a) expected interest and principal payments and (b) the probability of prepayments (virtually all modern U.S. mortgages extend the option to prepay to the borrower). Once the mortgages are "pooled," one or more securities are formed. The typical securities in U.S. mortgage markets have been mortgage pass-throughs, collateralized mortgage obligations (CMOs), mortgage-stripped securities, and real estate mortgage investment conduits (REMICs). Each of these has different forms and various characteristics.

MBSs are frequently based on liabilities (debt) claims to form investment securities. After all, a promise to pay represents an asset to the lender and thus, has value. The pooling of these assets is the crux of providing liquidity to investors.

Other assets in addition to MBSs have also been introduced. For example, there is a growing market in securities based on credit card debt, short-term receivables, and other types of liabilities. For firms, these developments provide marketability for these assets that previously did not exist. For financial institutions, MBSs provide the same type of marketability that enables banks and savings and loan associations to serve their primary function as financial intermediaries rather than as investors in the mortgage market.

MBSs are subject to risks, just like any other type of investment. Like bonds, the values of MBSs rise or fall inversely with the level of interest rates. Because the payment of interest is based on a given interest rate, increasing interest rates reduces the values of MBSs to keep this instrument competitive in the market. Similarly, a reduction in interest rate levels raises the values of fixed-income securities, including MBSs.

In addition, there is another, important source of risk associated with MBSs. Because borrowers have the right to prepay the remaining balance of the loan at their discretion, the owner of an MBS is uncertain how long the payments will continue. For example, declining interest rate levels will make prepayment attractive for borrowers. For MBS investors, declining interest rates means that the probability of prepayment increases.

Also, prepayment may occur for reasons unrelated to changing interest rates. The decision to move locations often requires prepayment of an existing mortgage. Changes in the family structure (i.e., divorces, marriages, deaths, etc.) may require a rearranging of financial positions. Retirement or changes in regional economies may also forebode prepayment. Therefore, it is not surprising that prepayment risk is a major consideration for investors in MBSs.

In the years to come, we should expect to see MBSs becoming increasingly a part of the U.S. financial system. MBSs for commercial mortgages have made great strides as well. MBSs of the future will be increasingly sophisticated, with various types of specialized features and options. MBSs are also spreading to other countries' financial systems. The resistance to MBSs appears to be coming from some governments who are apprehensive about financial innovations. They view MBSs as vehicles that increase the volatility of their financial systems. However, the riskiness of uncertain economic outcomes has always existed; MBSs and other mechanisms are modern instruments for risk management that in former times, were simply unavailable. (SEE ALSO: **Mortgage Finance;** *Neighborhood Housing Services of America*)

—*Austin J. Jaffe*

Further Reading

Archer, Wayne R. and David C. Ling. 1995. "The Effect of Alternative Interest Rate Processes on the Value of Mortgage-Backed Securities." *Journal of Housing Research* 6(2):285-316.

Chinloy, Peter. 1995. "The Market for Mortgage-Backed Securities." *Journal of Housing Research* 6(2):173-96.

Fabozzi, Frank J., ed. 1992. *The Handbook of Mortgage Backed Securities.* 3d ed. Chicago: Probus.

Fabozzi, Frank J. and Franco Modigliani. 1992. *Mortgage and Mortgage-Backed Securities Markets.* Boston: Harvard Business School Press.

MacDonald, H. 1996. "The Rise of Mortgage-Backed Securities: Struggles to Reshape Access to Credit in the USA." *Environment and Planning A* 28:1179-98.

McNulty, James E. 1988. "Mortgage Securities: Cash Flows and Prepayment Risk." *Real Estate Issues* (Spring/Summer):10-17.

Shilling, James D. 1995. "Rates of Return on Mortgage-Backed Securities and Option-Theoretic Models of Mortgage Pricing." *Journal of Housing Research* 6(2):265-84.

Smith, Stephen D. 1991. "Analyzing Risk and Return for Mortgage-Backed Securities." Federal Reserve Bank of Atlanta. *Economic Review* 76(January/February):2-11. ◄

▶ Mortgage Bankers Association of America

A national organization of corporations involved in real estate finance and related fields, Mortgage Bankers Association of America, founded in 1914, represents the real estate finance industry before Congress and federal regulators and provides members with educational and networking services, including correspondence courses, seminars, and conferences. MBA publications include *Real Estate Finance Today,* a biweekly newspaper dealing with residential and commercial lending, and *Mortgage Banking,* a monthly magazine that includes in-depth analyses of the real estate finance industry. Membership: 2,800 corporations nationwide. Staff size: 133. Contact person: Brian Chappelle, Senior Vice President, Corporate Relations and Membership. Address: 1125 15th Street, NW, Washington, DC 20005. Phone: (202) 861-6500. Fax: (202) 429-1672.

—*Caroline Nagel* ◄

▶ Mortgage Credit Certificates

Government, especially state and local levels of government, strives to help low- and moderate-income renter households become homeowners. Mortgage credit certificates offer a reduction in federal income taxes to eligible households, leaving more after-tax income for the purchase of a home. The certificates are good for as long as the household lives in the unit, but the household may have to repay some of the credit if it sells the home prior to living there for 10 years. The program is not an entitlement; the government annually allocates to each state only a limited number of mortgage credit certificates. Each state must devise a system to select participants for the program from among the eligible households living in the state. Eligible households are those homebuyers who (a) have a household income less than 100% of median household income in the metropolitan area, (b) are buying a home priced less than 90% of the area average home sales price, and (c) have not been a homeowner at any time during the prior three years. These restrictions are less stringent if the home is located inside a target area with very low income levels and chronic economic distress.

The credit amount is the interest paid during the year multiplied by the credit rate set by the state. It may be no less than 10% and no more than 50%, but independent of the credit rate, a household may not claim a credit of more than $2,000 per year. Selection of a credit rate poses a problem for states. If a state selects a high credit rate, more assistance is given to each individual household, making their housing more affordable, but fewer households can be served. If a state selects a low credit rate, more households can be assisted, but the assistance may be minimal, leaving the households with some level of housing affordability hardship.

The program was created as part of the Deficit Reduction Act of 1984 as an alternative to mortgage revenue bonds. Mortgage credit certificates were seen as more efficient in that they could provide benefits to households meeting the same eligibility standards but at a lower cost to the government.

The Deficit Reduction Act limits the total dollar amount of mortgage revenue bonds that any one state can issue during a single year and permits states to forego issuing mortgage revenue bonds using mortgage credit certificates instead. The exchange rate of certificates for bonds has varied over time to ensure a net savings to the federal government.

The mortgage credit certificate program is administered by each state, usually a state housing finance agency, or by a lower unit of government authorized by the state for this activity. Each year, states must decide how many mortgage revenue bonds to issue and how many to exchange for mortgage credit certificates. The decision may be based on credit market conditions. During some periods, the difference may narrow between the interest rate charged on conventional home purchase loans and the interest rate charged on home purchase loans funded from the proceeds of mortgage revenue bonds. When this spread becomes too narrow, mortgage revenue bonds have no advantage over conventional, unsubsidized loans, and the state turns to mortgage credit certificates. Ease of administration may also cause states to favor mortgage credit certificates. They can be put in place quickly and require only a small staff to administer compared with the large staff required to prepare, issue, and monitor mortgage revenue bonds.

Mortgage credit certificates have not been widely used, however; mortgage revenue bonds continue to be favored by most states. In 1990, only 17 states used the program, helping about 10,000 homebuyers, fewer than 500 homebuyers each for most states. Some of the lack of popularity of mortgage credit certificates has to do with the form of the benefits. Mortgage revenue bonds provide lower monthly mortgage loan payments to the eligible household. Mortgage credit certificates require the eligible household to make full, market-determined, monthly loan payments and to obtain its subsidy through a tax deduction. This indirect means of assistance is troublesome, especially if the household is unable to have its income tax withholding amount adjusted in accordance with the value of the credit. Some of the lack of popularity results from the fact that

they are nonrefundable. The credit may be deducted from the household's federal income tax liability only to the extent that the household has income tax liability. If the credit is greater than the household's income tax liability, the excess may be claimed during the next three years but is lost after that time.

Mortgage credit certificates assist needy households in buying a home through a reduction of their income tax liability. Although administratively easier to operate than mortgage revenue bonds, mortgage credit certificates have not been widely used owing to the indirect manner in which the subsidy is delivered to the needy household. As such, these mortgage credit certificates are not expected to become widely used, but they become relatively more popular during periods when the interest rates fall to a level close to the rate charged on mortgage revenue bonds. (SEE ALSO: *Low-Income Housing Tax Credits;* **Mortgage Finance;** *Mortgage Insurance; Mortgage Pass-Through Securities*)

—*Kirk McClure*

Further Reading

Blankenship, Frank J. 1989. *The Prentice Hall Real Estate Investor's Encyclopedia.* Englewood Cliffs, NJ: Prentice Hall.

Durning, Danny W., ed. 1992. *Mortgage Revenue Bonds: Housing Markets, Home Buyers and Public Policy.* Boston: Kluwer Academic.

Jacobus, Charles J. 1996. *Real Estate Principles.* Upper Saddle River, NJ: Prentice Hall.

McClure, Kirk. 1993. "Mortgage Credit Certificates as an Aid to Affordable Housing." *Journal of Planning Education and Research* 12(2):127-37.

McKenzie, Dennis J. 1996. *Essentials of Estate Economics.* Upper Saddle River, NJ: Prentice Hall.

Peiser, Richard B. 1992. *Professional Real Estate Development: The ULI Guide to the Business.* Washington, DC: Urban Land Institute.

Pryke, M., ed. 1994. "Property and Finance." *Environment and Planning A* 26:167-264.

Shim, Jae K. 1996. *Dictionary of Real Estate.* New York: John Wiley.

Tosh, Dennis S. 1990. *Handbook of Real Estate Terms.* Englewood Cliffs, NJ: Prentice Hall. ◄

► Mortgage Finance

Mortgage finance in the United States has evolved from an informal, unregulated process based on personal savings and short-term borrowing from individuals or local building and loan societies to a highly sophisticated national system of mortgage instruments and financial intermediaries supported by a secondary market of Wall Street investors. In summary, mortgages are a form of debt that enable consumers and developers to finance the purchase, construction, or rehabilitation of housing over an extended period of time, with minimal up-front, "out-of-pocket" expenditures for down payments or equity.

Until the financial crisis of 1929 precipitated the Great Depression, homes and apartment buildings were typically bought and sold on a cash basis or through short-term "balloon payment" mortgages made by the seller, wealthy individuals or loan brokers, or by institutions such as thrifts and life insurance companies. These mortgages required down payments of up to one-half or more of the entire cost, with a balloon payment of the total principal balance due after the borrower repaid interest costs over a term lasting usually no more than five years. Apartment buildings, which enjoyed a construction boom during the 1920s, were often financed through risky mortgage bonds issued by investment houses that exaggerated property values and overestimated demand to boost sales of these securities. The economic collapse of the Great Depression led to an unprecedented increase in home foreclosures and decline in land values.

The Home Owners' Loan Corporation (HOLC) was established by the federal government in 1933 to refinance the huge number of mortgage loans falling into delinquency and default, and served to bail out both working-class homeowners and the financial institutions that had borrowed money. Many owners of one- to four-unit buildings were unemployed and unable to renegotiate their outstanding debt on favorable terms with small local banks and thrifts struggling to survive the depths of the crisis. The HOLC provided 15-year loans for up to 80% of appraised value to these homeowners who lost their houses after 1930 or were unable to refinance. These underwriting terms were a significant departure from the practices of private lenders during the 1920s. In particular, high loan-to-value ratios and the concept of long-term loan amortization were important innovations that substantially advanced housing affordability for Americans and were subsequently adopted by the Federal Housing Administration (FHA), created in 1934.

FHA-insured home loans were mostly originated by mortgage companies and commercial banks (who had gained regulatory permission to grant mortgages during the 1920s), and did not gain widespread acceptance until after World War II, with the rapid growth in housing demand and a stronger secondary market through Fannie Mae (formerly, the Federal National Mortgage Association). Nevertheless, the FHA mutual mortgage insurance program became a competitive alternative to conventional, non-FHA-insured mortgage loans for buyers of low- and moderately priced housing, offering a statutory interest rate that usually remained below prevailing market rates. In addition, the low down payments, high loan-to-value ratios, and long-term amortized repayment schedules of the FHA forced the savings and loan industry, the nation's leading provider of home mortgage credit, to change its conventional underwriting methods. The property appraisal standards of FHA mirrored those found in private industry, however, resulting in a bias that favored new construction in homogeneous suburban subdivisions over existing housing in older urban neighborhoods with a heterogeneous or minority population. The FHA program, along with the Veterans Administration (VA) home loan guarantee, which required no down payment, greatly improved the upward mobility prospects of the white working class during the 1950s and 1960s; but FHA policies also exacerbated the uneven development of metropolitan areas and regions, leading to the spatial segregation of communities by race and class.

Periodic interest rate "shocks" and credit "crunches" during the 1960s, resulting from the globalization of the economy and government spending for the Vietnam War and social policy initiatives, caused further systemic change in U.S. housing finance. The FHA program together with an expanded, government-supported secondary market was increasingly used to stimulate and regulate the housing and financial sectors of the economy, as well as to improve socioeconomic conditions in distressed central cities. Fannie Mae, which was formed in 1938 to buy FHA-insured mortgages and later VA-guaranteed loans, was joined by the Federal Home Loan Mortgage Corporation (Freddie Mac), established in 1970 to purchase conventional, privately insured loans from thrift institutions in the Federal Home Loan Bank system, and the Government National Mortgage Association (Ginnie Mae), created by the Housing Act of 1968 to package and sell investment securities backed by FHA and VA loans.

After 1970, both Fannie Mae and Freddie Mac gained new powers as privatized government-sponsored enterprises, enabling Fannie Mae to purchase conventional instead of just FHA- or VA-backed loans, enabling Freddie Mac to buy conventional mortgages originated by commercial banks and mortgage companies instead of just thrifts, and allowing both entities to issue their own mortgage-backed securities. The FHA assumed additional responsibilities such as insuring loans in urban renewal areas, to low-income homeowners and developers of elderly and handicapped housing, and for new product types such as condominium units. Poor administration of the FHA program along with growing fiscal conservatism within government led to the resurgence of private mortgage insurance (PMI), which had been out of favor since the Great Depression. By 1980, PMI accounted for more than 40% of all insured mortgage originations. Mortgage insurance companies worked closely with savings and loans, mortgage bankers, Fannie Mae, Freddie Mac, and private secondary market firms to increase the volume of conventional, privately insured home mortgage lending, whereas the reoriented FHA focused on moderate-income and disadvantaged borrowers and previously neglected minorities.

As interest rates grew with inflation during the 1970s, savings and loans began to experience "disintermediation" resulting from increased institutional competition for deposits. This created a mismatch between their traditional long-term, fixed-rate mortgage liabilities and their changing asset base of liquid, short-term, high- and variable-rate deposit accounts. Disintermediation weakened the position of the savings and loan industry as a source of funds in housing finance and strengthened the role of the secondary mortgage market agencies, who already exercised considerable power and influence over the underwriting methods of primary, originating mortgage lenders. Alternative mortgage instruments, such as variable-rate mortgages, adjustable-rate mortgages (ARMs), and graduated-payment mortgages, were developed in response to the continued volatility of the interest rate environment, which by the early 1980s had significantly inflated the long-term cost of purchasing a home with fixed, market rate financing.

During 1984, when mortgage interest rates averaged 12%, ARMs accounted for more than one-half of all conventional single-family home loan originations. In addition, state and local governments were authorized to issue tax-exempt mortgage revenue bonds, enabling their housing finance agencies to make available below-market-rate home mortgages funded by the proceeds of bond sales. These programs gained popularity during the era of rising interest rates, but loans are usually targeted to low- and moderate-income areas and first-time homebuyers, and the annual volume of tax-exempt bond issues is capped to minimize federal revenue losses. From 1987 through 1992, mortgage revenue bonds financed more than 1.2 million housing units.

The dollar volume of outstanding single-family mortgage debt in the United States soared from $13 billion in 1925 to $45 billion in 1950, $280 billion in 1970, and $3.9 trillion by 1996. But FHA-insured and VA-guaranteed mortgages declined from 42% of outstanding single-family housing debt in 1950 to 35% in 1970 and only 15% by 1991, indicating the growth of conventional mortgage lending. In 1996, FHA and VA loans represented only 14% of new single-family home mortgage originations. Nearly one-half of outstanding single-family mortgage debt is now securitized through the secondary market. By 1994, this figure amounted to 49% for conventional single-family mortgages and 90% for FHA- and VA-backed single-family loans. During 1996, $371 billion of mortgage-backed securities were issued by Fannie Mae, Freddie Mac, and Ginnie Mae.

A continuing problem in the housing finance system is the existence of discriminatory biases that prevent disadvantaged minorities such as African Americans from achieving homeownership rates or accumulating home equity comparable to that for whites, even among households within the same income group. In addition to reforms in the FHA program designed to eliminate racial discrimination and geographic "redlining" in the approval of government home mortgage insurance, federal legislation since the 1960s has prohibited private discrimination against minorities seeking housing.

The Fair Housing Act of 1968 forbid racial discrimination by realtors and mortgage lenders in the sale or rental of private housing, and the Equal Credit Opportunity Act of 1974 extended this protection to other forms of consumer and commercial credit and to women. The Home Mortgage Disclosure Act (HMDA) of 1975 required federally regulated private depository institutions to disclose the geographic distribution of their home mortgage loans by census tract on an annual basis. The Community Reinvestment Act (CRA) of 1977 prohibited mortgage redlining and required federally regulated private lenders to affirmatively assess and meet the credit needs of communities within their geographic service area, subject to citizen and regulatory performance review. In 1989, both HMDA and CRA were strengthened through congressional amendments that expanded the disclosure requirements to include mortgage bankers as well as additional information on the race, gender, and income of all loan applicants and the disposition of loan applications. The public disclosure of CRA performance ratings was also authorized.

In 1992, federal legislation enhanced the regulatory and oversight powers of the U.S. Department of Housing and Urban Development (HUD) over the government-sponsored secondary market enterprises, Fannie Mae and Freddie Mac. The growing influence of both entities in mortgage finance was recognized, as HUD now monitors their compliance with annual regulatory goals for making targeted loan purchases in central cities, low- and moderate-income census tracts, and among low- and moderate-income homebuyers.

Community-based housing organizations and local governments have used this new financial regulatory framework to create mortgage-lending partnerships with banks, thrifts, secondary market agencies, and sometimes private mortgage insurers. These programs have used flexible underwriting criteria to make housing credit available in older cities and for working class and minority borrowers and first-time homeowners. In particular, down payment requirements have been reduced to less than 5% of the total purchase price, subsidies are used to diminish transaction closing costs, and debt-to-income qualifying ratios have been increased to more than 30% for front-end mortgage debt and sometimes more than 40% for back-end total consumer household debt. For example, Fannie Mae, which announced a $1 trillion commitment in 1994 for financing targeted housing needs through the end of the century, purchases mortgages through its Community Home Buyer's Program with loan-to-value ratios as high as 97%, down payment requirements as low as 3% combined with a 2% grant, and debt-to-income qualifying ratios of up to 33:38. By the mid 1990s, these initiatives had begun to produce results. According to an analysis of HMDA data by the Federal Reserve, conventional mortgage lending to low-income borrowers has increased by 37% since 1993, to African Americans by 36% (74% for low-income blacks), and to Hispanics by 25% (94% for low-income Hispanics). To manage risk, Fannie Mae and Freddie Mac have encouraged mortgage lenders to use credit scoring techniques that are found in consumer finance. This has led to the controversial practice of "subprime" lending, whereby borrowers perceived as credit risks are charged higher interest rates.

Financing for newly built or rehabilitated multifamily rental housing, both construction and long-term, is perceived by mortgage lenders as a greater risk than single-family home loans, because of the somewhat speculative nature of income from rental apartments, occasional opposition from communities that seek to exclude multifamily structures as an incompatible land use, and the negative ratings assigned by financial regulators that evaluate the condition and performance of institutional mortgage portfolios. Multifamily loan risk can be mitigated by FHA mortgage insurance, special secondary market programs, mortgage revenue bond financing, risk-sharing and credit enhancement arrangements, and the syndication of equity investments using federal tax credits for low-income rental housing. Other nonbank investors, such as pension funds and life insurance companies, have been important sources of financing for multifamily housing. Outstanding mortgage debt on multifamily dwellings (five or more units) has increased from $4 billion in 1925 to $10 billion in 1950, $58 billion in 1970, and $310 billion by 1994. Thirty-two percent of the 1996 amount was securitized through the secondary market, and 20% was insured by FHA. But new originations of FHA-insured multifamily loans have dropped from 31% of total originations to only 3% by 1994.

Mortgage lending is a vital element of the U.S. capitalist economy, providing funds for the housing sector and a reliable and predictable government-backed outlet for institutional investment, as well as a means to achieve home-ownership and accumulate household equity through the acquisition of a fixed asset that is likely to increase in value over time. Consumers can use their mortgage indebtedness to secure home improvement loans, finance household needs through a home equity line of credit, or capture the value of their owned home after retirement through "reverse equity" mortgages. In recent years, macroeconomic policy has paid considerable attention to the effect of interest rates on national growth and economic well-being. This has prompted various innovations in mortgage loan products and underwriting techniques designed to lower the short-term and long-term costs of borrowing for prospective homebuyers. The National Association of Realtors estimates that the monthly principal and interest payments for a median-priced single-family house in the United States fell from 36% of the median family income in 1981 to 19% by 1993, when declining interest rates caused a mortgage refinancing boom. Mortgage lending has become a public policy instrument to encourage housing production, support home sales activity and local real estate markets, stabilize the changing financial system, and regulate credit flows in the economy. Housing finance is also increasingly used as a vehicle to remedy historic patterns of racial discrimination and socioeconomic inequity and to stimulate the revitalization of declining areas. (SEE ALSO: *Alternative Mortgage Instruments; Amortization; Assumption of Loan; Balloon Mortgage; Chattel Mortgage; Federal Home Loan Bank System; Fixed-Rate Mortgage Loan; Government National Mortgage Association; Home Owners' Loan Act of 1933;* **Housing Finance;** *Interest Rates; Lending Institutions; Mortgage-Backed Securities; Mortgage Credit Certificates; Mortgage Insurance; Mortgage Pass-Through Securities; Mortgage Revenue Bonds; Mortgage-Stripped Securities; Negative Amortization; Owner Take-Back Financing; Principal; Refinancing; Reverse-Equity Mortgage; Rollover Mortgage; Sale-Leaseback; Second Mortgage; Secondary Mortgage Market*)

—*John T. Metzger*

Further Reading

Brueggeman, William B. and Jeffrey D. Fisher. 1996. *Real Estate Finance and Investments.* 10th ed. Homewood, IL: Irwin.

Canner, Glenn B. and Wayne Passmore. 1995. "Home Purchase Lending in Low-Income Neighborhoods and to Low-Income Borrowers." *Federal Reserve Bulletin* 81(2):71-103.

Dennis, Marshall W. and Michael J. Robertson. 1994. *Residential Mortgage Lending.* 4th ed. Englewood Cliffs, NJ: Prentice Hall.

Florida, Richard L., ed. 1986. *Housing and the New Financial Markets.* New Brunswick: Rutgers, State University of New Jersey, Center for Urban Policy Research.

Squires, Gregory D., ed. 1992. *From Redlining to Reinvestment: Community Responses to Urban Disinvestment.* Philadelphia: Temple University Press.

Weiss, Marc A. 1989. "Marketing and Financing Home Ownership: Mortgage Lending and Public Policy in the United States, 1918-1989." *Business and Economic History* 18:109-18. ◀

▶ Mortgage Insurance

Mortgage insurance is an insurance policy paid for by the borrower for the protection of the mortgage lender. In the United States, the Federal Housing Administration (FHA) provides government-supported mortgage insurance and private companies sell mortgage insurance in the private market. The overall impact of mortgage insurance is to allow people with relatively small amounts of savings to purchase housing with a mortgage.

The costly nature of housing means that most home purchasers need a mortgage loan. Mortgage lenders prefer a prospective buyer to have at least 20% of the purchase price for a down payment. Obtaining even 20% of the price of a house is prohibitive for many people. Mortgage insurance helps to overcome the gap between what lenders would like and what borrowers are able to achieve. If a borrower qualifies in every other way for a mortgage loan but has less than a 20% down payment, the lender will usually require mortgage insurance. In this case, the borrower pays premiums to protect the lender in case the borrower defaults.

Government Mortgage Insurance

The concept of mortgage insurance started with the FHA in the 1930s. The FHA showed that small premiums paid by a great many borrowers could be pooled to provide the money necessary to pay off the loans of those few who defaulted. The FHA programs require a down payment of between 3% and 5% of the purchase price of the house (the exact amount is determined by formula). The buyer pays the mortgage insurance premium, and the lender agrees to make the loan. The FHA does not make the loan; it simply provides insurance. If the borrower defaults, the FHA pays the lender the whole loan amount and the property belongs to the FHA. The borrower is not protected in any way. The borrower does benefit, however, from the FHA's standardized terms and from receiving a mortgage loan with a very low down payment. There are maximum price limits so that FHA insurance cannot be used for the purchase of more expensive housing. It primarily assists low- to moderate-income homebuyers.

The federal mortgage insurance programs have allowed a great many more people to qualify for mortgages. They have contributed to the relatively high proportion of U.S. households that own their own homes and the relatively large amount of U.S. debt tied to housing. The federal programs also have had impacts on the spatial distribution of housing, on the rapid expansion of the U.S. housing stock, and on racial segregation and central-city decline.

Private Mortgage Insurance

The private mortgage insurance (PMI) industry in the United States came into being after the FHA demonstrated that mortgage insurance schemes could be profitable. There are now more than a dozen private firms in the industry. They insure borrowers who have down payments ranging from 5% to 20% of the property price. An insurance policy purchased through a PMI company does not insure the entire amount of the loan (as does FHA). It insures only 20% or 25%. Consequently, PMI is generally less expensive than FHA insurance. In addition, PMI does not have to follow the FHA's price limits. The private companies respond more quickly to applications than the FHA does. All of these characteristics mean that PMI companies have skimmed off the most profitable part of FHA's business, leaving it with the lower-dollar-amount loans and borrowers with the lowest proportional down payments.

If, after paying on the loan for awhile, the borrower believes that he or she now has at least 20% equity in the property (that is, the remaining loan balance is less than 80% of what the property is worth), the financial institution can be asked to drop the requirement for mortgage insurance. This is at the discretion of the financial institution and will require at least a new appraisal to provide evidence. If the appraisal indicates that the remaining principal amount is now less than 80% of the value of the property, the financial institution may agree to drop the insurance, thus lowering the monthly cost to the borrower.

Private mortgage insurance has an effect similar to, although not as pronounced as, the effect of FHA mortgage insurance. It allows people who otherwise would not be able to buy a home to qualify for a loan and thus increases the proportion of homeowners in the United States and the country's reliance on debt financing for housing. (SEE ALSO: **Mortgage Finance;** *Section 203[b] and [i]; Section 203[k]; Section 221[d][2]; Section 221[d][3] and [4]; Section 222; Section 223[3]; Section 223[f]; Section 231; Section 232; Section 234; Section 237; Section 241; Section 245; Section 251; Section 255; Section 608*)

—*Hazel A. Morrow-Jones*

Further Reading

ABA Standing Committee on Lawyers' Title Guaranty Funds. 1991. *Buying or Selling Your Home: Your Guide to: Contracts, Titles, Brokers, Financing, Closings.* Chicago: American Bar Association.

Blankenship, Frank J. 1989. *The Prentice Hall Real Estate Investor's Encyclopedia.* Englewood Cliffs, NJ: Prentice Hall.

Brueggeman, William B. and Jeffrey D. Fisher. 1993. *Real Estate Finance and Investments.* 9th ed. Homewood, IL: Irwin.

Jacobus, Charles J. 1996. *Real Estate Principles.* Upper Saddle River, NJ: Prentice Hall.

Lederman, Jess. 1992. *The Secondary Mortgage Market: Strategies for Surviving and Thriving in Today's Challenging Markets.* Chicago: Probus.

McKenzie, Dennis J. 1996. *Essentials of Estate Economics.* Upper Saddle River, NJ: Prentice Hall.

Peiser, Richard B. 1992. *Professional Real Estate Development: The ULI Guide to the Business.* Washington, DC: Urban Land Institute.

Shim, Jae K. 1996. *Dictionary of Real Estate.* New York: John Wiley.

Tosh, Dennis S. 1990. *Handbook of Real Estate Terms.* Englewood Cliffs, NJ: Prentice Hall. ◀

▶ Mortgage Pass-Through Securities

A pass-through is a type of derivative mortgage security. It is formed by an issuer collecting, either through origination or purchase, a pool of homogeneous individual mortgages and selling the right to receive the cash flows from those mortgages. Typically, the issuer guarantees timely interest and principal repayments to the pass-through investors. The issuer (or a contracted agent) also performs the task of collecting and distributing the monthly payments to the pass-through investors. For providing the guarantee and servicing the mortgages, the issuer (or agent) receives a small fee, deducted from the monthly cash flows. The remaining cash flows are distributed to the investors based on their proportional ownership of the pool. This includes principal and interest payments as well any prepayments. If a homeowner defaults, the issuer/guarantor pays the investors the remaining balance owed as well as any outstanding interest. The issuer then seeks to recover payment from the homeowner.

The largest issuers of mortgage pass-throughs are the Federal Home Loan Mortgage Corporation (FHLMC, or Freddie Mac) and Fannie Mae (formerly, the Federal National Mortgage Association). These companies are government-sponsored enterprises that the market assumes are implicitly backed by the U.S. government. The Government National Mortgage Association (GNMA, or Ginnie Mae) does not directly issue mortgage pass-throughs but, rather, insures pass-through securities issued by private firms such as banks, mortgage companies, and homebuilders. Ginnie Mae is an agency under the Department of Housing and Urban Development and is backed by the full faith and credit of the U.S. Treasury. Although some companies have issued pass-throughs without the involvement of Fannie Mae, Freddie Mac, or Ginnie Mae, they have had little impact on the market.

The creation of mortgage pass-throughs in the early 1970s was a revolutionary event for the secondary mortgage markets. Combining many individual mortgages into a single pool provided economies of scale and diversification gains that allowed nontraditional housing lenders to enter the market. This has increased both the size and liquidity of the secondary mortgage market, which has, in turn, increased the availability of funds in the primary market. Pass-throughs also serve as collateral for a range of other derivative products, including collateralized mortgage obligations (CMOs) and interest-only/principal-only (IO/PO) strips.

More so than with other debt instruments, mortgage pass-throughs are extremely sensitive to changes in interest rates. When rates rise, the value of mortgage pass-throughs falls. Because of the borrowers right to prepay the mortgage, however, the value of a pass-through can fall even when the interest rates fall, a phenomenon known as *negative convexity*. Because of this additional interest rate risk, yields for pass-throughs are higher than for other government-insured debt. (SEE ALSO: *Mortgage Credit Certificates*)

—*Richard J. Buttimer Jr. and
Ronald C. Rutherford*

Further Reading

Bartlett, William W. 1994. *The Valuation of Mortgage Backed Securities.* Homewood, IL: Irwin.

Blankenship, Frank J. 1989. *The Prentice Hall Real Estate Investor's Encyclopedia.* Englewood Cliffs, NJ: Prentice Hall.

Brueggeman, William B. and Jeffrey D. Fisher. 1993. *Real Estate Finance and Investments.* 9th ed. Homewood, IL: Irwin.

Jacobus, Charles J. 1996. *Real Estate Principles.* Upper Saddle River, NJ: Prentice Hall.

Lederman, Jess. 1992. *The Secondary Mortgage Market: Strategies for Surviving and Thriving in Today's Challenging Markets.* Chicago: Probus.

McKenzie, Dennis J. 1996. *Essentials of Estate Economics.* Upper Saddle River, NJ: Prentice Hall.

Peiser, Richard B. 1992. *Professional Real Estate Development: The ULI Guide to the Business.* Washington, DC: Urban Land Institute.

Shim, Jae K. 1996. *Dictionary of Real Estate.* New York: John Wiley.

Tosh, Dennis S. 1990. *Handbook of Real Estate Terms.* Englewood Cliffs, NJ: Prentice Hall. ◀

▶ Mortgage Revenue Bonds

Some households are not able to buy a home because conventional mortgage loans are not available or are available only on terms that low-income households cannot afford. To resolve this problem, government may either become a lender to these households, making loans available on more favorable terms, or it may make funds available to conventional lenders on the agreement that the lenders use the funds to make loans to low-income households. Rather than raise taxes to generate the funds for these loans, a state or local government can sell mortgage revenue bonds. The proceeds from the sale of these bonds can then be used to fund the mortgage loans. The homebuyer who accepts one of these loans repays the loan on a monthly basis, with the payments passed back to the bond buyers.

This approach uses the borrowing capacity of government to raise funds. The funds obtained in this manner can be lent to eligible homebuyers on favorable terms. The favorable terms result from a specific provision of the federal income tax codes that exempt from federal income taxes the interest payments received by bondholders. As such, investors are willing to accept a lower interest rate on these tax-exempt bonds than they would accept on a comparable alternative investment that generates taxable income. This lower interest rate is passed on to the eligible borrower in that federal law requires all proceeds from the sale of the bonds, net a limited amount of allowable costs for issuing the bonds, be loaned to eligible borrowers for the purchase of a home. Generally, this interest rate will fall significantly below the interest rate charged by a conventional mortgage lender. The lower interest rate leads to lower monthly payments, which should reduce the burden of house payments on the income of the borrower. This may, in turn, help low-income homebuyers purchase more

housing than would otherwise have been possible, or it may help them surmount the price barrier that has kept them from entering into homeownership.

The use of mortgage revenue bonds began as an experiment during the early 1960s. Several states, led by New York, created state housing finance agencies to provide funds for private housing development. They created mortgage revenue bonds and, with few federal controls existing at that time and with their low cost to the issuing government, their popularity rose. This practice spread to local governments during the 1970s. From 1974 to 1990, more than $60 billion in mortgage revenue bonds were issued, assisting more than 1 million households.

Over time, the federal government has imposed a set of limitations on the issuance of these bonds to ensure better use of them and to protect the federal government from the unregulated demands that mortgage revenue bonds were imposing on the Treasury. The borrower must have a household income that is no more than 100% of the median household income found within the metropolitan area. The home that is purchased must be priced at no more than 90% of the average home price found in the area. Finally, the household must not have owned a home at any time during the three preceding years. These restrictions can be relaxed for large families or if the home purchased is located within a target area suffering from chronic economic distress. Beyond limiting the income and purchase price paid by the borrower, the federal government has imposed limits on the total dollar value of all mortgage revenue bonds issued within a state during any one year. These limitations have, in many cases, caused states to curtail their use of these bonds.

Mortgage revenue bonds have met with some controversy over the years. It has been charged that many of the households who eventually obtain a home purchase loan through the program did not actually need the assistance. In many cases, it has been found that the program participant could actually have afforded to purchase the home without any assistance.

A second controversy has arisen over inflated prices of the homes sold for participation in the program. Potential homebuyers are often unaware of the mortgage revenue bond program and are informed of the subsidy mechanism by a real estate broker or housing developer who is able to steer the buyer through the lending process. This provides the opportunity for the broker or developer to inflate the price of the home above its actual market value—in effect, capturing some of the benefits intended for the low-income buyer. To protect against this misuse of the program, states have found that they must advertise the program widely, make the funds available on a first-come, first-served basis, and not permit brokers or developers to reserve any funds for their exclusive use.

Finally, mortgage revenue bonds have been criticized as being administratively complex. The General Accounting Office found that only 12 to 45 cents of every federal dollar of income taxes foregone actually benefits homebuyers. The remainder is lost to high administrative costs and waste. Alternative subsidy mechanisms, such as mortgage credit certificates, avoid almost all of these costs.

Mortgage revenue bonds can provide a source of funds to assist first-time homebuyers. The program is costly to the federal government, although it is almost cost free to units of state and local government that use the program. Administration of the program is complex, and the program is subject to abuse if not well administered. The authority to issue these bonds is now highly regulated, and even the total amount that may be issued is subject to periodic budgetary decisions by the federal government. The use of this program is expected to continue with increased targeting of households needing assistance. The program will be most popular during periods when the interest rates charged on conventional loans are high relative to that charged on mortgage revenue bonds. This frequently occurs during periods of rapidly increasing inflation. (SEE ALSO: *Housing Bonds;* **Mortgage Finance**)

—*Kirk McClure*

Further Reading

Blankenship, Frank J. 1989. *The Prentice Hall Real Estate Investor's Encyclopedia.* Englewood Cliffs, NJ: Prentice Hall.

Dowall, David E. 1990. "The Public Real Estate Development Process." *Journal of the American Planning Association* 56:504-12.

Durning, Danny W., ed. 1992. *Mortgage Revenue Bonds: Housing Markets, Home Buyers and Public Policy.* Boston: Kluwer Academic.

General Accounting Office. 1988. *Homeownership: Mortgage Bonds Are Costly and Provide Little Assistance to Those in Need.* Washington, DC: Author.

Jacobus, Charles J. 1996. *Real Estate Principles.* Upper Saddle River, NJ: Prentice Hall.

McKenzie, Dennis J. 1996. *Essentials of Estate Economics.* Upper Saddle River, NJ: Prentice Hall.

Peiser, Richard B. 1992. *Professional Real Estate Development: The ULI Guide to the Business.* Washington, DC: Urban Land Institute.

Pryke, M., ed. 1994. "Property and Finance." *Environment and Planning A* 26:167-264.

Stegman, Michael A. and Michael I. Luger. 1993. "Issues in the Design of Locally Sponsored Home Ownership Programs." *Journal of the American Planning Association* 59:417-32. ◀

▶ Mortgage-Stripped Securities

With the development of mortgage-backed securities (MBSs), certain innovations have taken place. One of the more important developments has been the introduction of mortgage-stripped securities (or simply "mortgage strips.") Mortgage-stripped securities are derived from mortgage pools used to form all MBSs.

The most common form of mortgage-stripped security is the division of interest and principal payments into two separate securities. As borrowers make payments on their original mortgages, some of the payment is interest (repayment *on* the note) and some of the payment is principal (repayment *of* the note). The amortization process specifies the distribution of the debt service payment between interest and principal. At the MBS level, any interest received goes to the holder of one security; any principal received goes to another investor. Therefore, the investor holding the security who receives interest payment is said to have

an "interest-only" mortgage strip (or IO) and the investor who receives the principal payments is said to own a "principal-only" mortgage strip (or PO).

Clearly, there are differences between IOs and POs. Because interest owed is taken out of debt service payments first, the holder of an IO receives cash flow early in the life of the MBS compared with the owner of the PO. In fact, the division of the MBS between IO and PO splits the cash flow dramatically over time. Investors in IOs receive most of the interest early during the repayment life of the mortgages, and it declines sharply over time. Investors in POs receive very little during the early years, and the payments increase dramatically in the later years.

The key to valuation of IOs and POs is that the volatility of these securities is fundamentally different. Because most of the cash flow of POs is expected late in the repayment lives of the underlying mortgages, POs contain considerable risk. Thus, they promise a higher return to investors who bear the risk. This volatility is also due to interest rate and prepayment risk, which are also characteristic of mortgage-stripped securities.

The benefits of these instruments are substantial to institutional investors who have hedging objectives, given their other investments as well as particular investment needs. Strips have also proven to be quite successful in the Treasury bill market where liquidity is excellent and where default risk does not exist. (SEE ALSO: **Mortgage Finance**)

—*Austin J. Jaffe*

Further Reading

Fabozzi, Frank J., ed. 1992. *The Handbook of Mortgage Backed Securities.* 3d ed. Chicago: Probus.

Fabozzi, Frank J. and Franco Modigliani. 1992. *Mortgage and Mortgage-Backed Securities Markets.* Boston: Harvard Business School Press.

McNulty, James E. 1988. "Mortgage Securities: Cash Flows and Prepayment Risk." *Real Estate Issues* 13(Spring/Summer):10-17.

Sellon, Gordon H., Jr., and Deana VanHahman. 1988. "The Securitization of Housing Finance." Federal Reserve Bank of Kansas City. *Economic Review* (July/August), 3-20. ◀

▶ Mt. Laurel

Mt. Laurel refers to two landmark New Jersey State Supreme Court cases known officially as *South Burlington County NAACP v. The Township of Mount Laurel, 67 NJ 151 (1975),* and *South Burlington County NAACP v. Mt. Laurel, 92 NJ 158 (1983).* Generally, the cases are referred to as *Mt. Laurel I* and *Mt. Laurel II,* respectively. In both of these suits, the New Jersey Supreme Court found in favor of the National Association for the Advancement of Colored People and determined that the prevailing zoning in Mt. Laurel, and by extension, in virtually every suburban municipality in the state, excluded the poor because low-density requirements and costly housing standards precluded the possibility of low- and moderate-income families being able to afford to live in the community. Constitutionally, the court found that the power to zone is derived from the general police powers granted government. Those police powers must be exercised for the good of all of the people.

In *Mount Laurel I,* the court found that every municipality in a growth area (essentially the suburbs) had a constitutional obligation to provide, through its land use regulations, for a fair share of the present and prospective regional need for low- and moderate-income housing. In *Mount Laurel II,* the court expanded its ruling requiring municipalities to take affirmative actions so as to encourage the development of housing affordable to low- and moderate-income households. However, the court also suggested that it would be best if the elected branches of government established a process through which municipalities would understand the amount of affordable housing they would be responsible for within their region and the types of affirmative actions municipalities could take to fulfill their responsibilities. Not surprisingly, the *Mt. Laurel* decisions were not popular with municipalities. The court foresaw the politically expedient need for a process whereby municipalities could have a participatory role in policy-making and so requested that the legislature take over administration of *Mt. Laurel* from the courts.

In response, the state legislature passed the Fair Housing Act (P.L. 1985 c. 222 S2:27D—301, et. seq.) in 1985 to comply with the supreme court's direction. The Fair Housing Act created the Council on Affordable Housing (COAH) and gave it primary jurisdiction for the administration of housing obligations for each municipality in the state. COAH established fair share allocations for every municipality, identified housing regions, and established standards regarding how municipalities could affirmatively meet their obligations. Moreover, the Fair Housing Act gave municipalities the option of coming to COAH for substantive certification of a housing plan that explained how the municipality was going to affirmatively meet its fair share obligations. If a community received certification, the municipality's zoning ordinance was to be judged by the court to be in compliance with the court's mandates and thus exempt from legal challenges by builders and others who might seek to overturn a local zoning ordinance. Municipalities that did not seek or receive certification were still liable to having their local zoning ordinances challenged in court.

The COAH process accomplishes the following:

▶ Establishes the gross number of housing units each municipality must provide to meet their fair share
▶ Identifies the permissible methods to meet that obligation
▶ Provides technical assistance to communities in preparing a fair share plan to meet the obligation
▶ Mediates disputes arising from the fair share plan
▶ Adjudicates those disputes or transfers disputes to an appropriate body for resolution

COAH permits municipalities to meet their fair share obligations in the following ways:

▶ Granting higher-density zoning to a private developer if that developer agrees to provide at least 20% of all units on site as affordable (inclusionary development)

▶ Zoning vacant land at higher densities within a municipality even if no developer has proposed such a project for that site
▶ Building or causing to be built affordable rental projects, resulting in a municipality's being rewarded with credits in excess of the number of affordable units actually built (e.g., building one low-income rental unit counts as two affordable units against the municipality's fair share allocation)
▶ Providing zoning for municipal programs that encourage development of limited affordable age-restricted developments, accessory apartments, permanent group homes, and other alternative-living arrangements
▶ Paying another community, usually one with a larger low-income population, to build up to half of the paying or "sending" community's obligation within the "receiving" community's borders (known as a Regional Contribution Agreement or RCA), with the stipulation that the housing built or rehabilitated must be affordable to low- and moderate-income families

To encourage development of affordable housing in configurations other than within inclusionary developments, the state also established the Balanced Housing program that provides capital grant funds to subsidize the construction of new or rehabilitated units. The Balanced Housing program is financed by the state's real estate transfer tax. At the height of the building boom in New Jersey in the mid-1980s, the Balanced Housing program received revenues of close to $30 million per year. In fiscal year 1992, Balanced Housing received revenues of approximately $13 million.

In recent years, COAH and its staff have begun to change its focus from a regulatory and adjudicative body to an incentive-based body that assists municipalities in complying with the law. Since the late-1980 downturn in the state's economy, and in particular, the building industry, there is less pressure on municipalities to comply as a result of real or threatened builder law suits. To this end, COAH adopted new rules in 1993 for the following:

▶ Reformulation of the estimate of statewide affordable housing need over which COAH has jurisdiction from 145,000 units to 83,000 units, thus lowering municipal obligations
▶ Creation of richer incentives for production of rental, as opposed to for-sale housing
▶ Raising the minimum RCA payment (or subsidy) per unit to $20,000
▶ Permitting a wider range of densities in inclusionary developments so builders can better respond to market conditions
▶ Reducing the target affordability of both rental and sale projects to place units in closer reach of lower-income families

Implementation of the Fair Housing Act remains a highly contentious and litigious process, in the legislature, in communities, and in the courts. Emerging *Mt. Laurel* issues for COAH include the following:

- ▶ Enforcement of fair housing and affirmative action in the selection of homeowners and tenants by municipal and private housing sponsors
- ▶ Changing the focus from zoning to actual production of housing units
- ▶ Continued defense of the constitutionality of the *Mt. Laurel* decisions
- ▶ Creation of either regulatory, incentive-based, or if necessary, punitive approaches to municipalities that "drag their feet" or refuse to comply with the law

(SEE ALSO: *Exclusionary Zoning; Fair Share Housing*)
— *Nancy L. Randall and Daniel Hoffman*

Further Reading

Hughes, Mark A. and Therese J. McGuire. 1991. "A Market for Exclusion: Trading Low-Income Housing Obligations under Mount Laurel II." *Journal of Urban Economics* 29:207-17.

Hughes, Mark A. and Peter M. Vandoren. 1990. "Social Policy through Land Reform: New Jersey's Mount Laurel Controversy." *Political Science Quarterly* 105(Spring):97-111.

Huls, Mary Ellen. 1985. *Exclusionary and Inclusionary Zoning: A Bibliography*. Monticello, IL: Vance Bibliographies.

Oross, Marianne. 1989. "The Examination of the Social, Political and Economic Conditions Associated with Exclusionary Zoning Practices in Mount Laurel, NJ." PhD. Thesis, Graduate Program in Social and Philosophical Foundation in Education, Rutgers University, New Brunswick, NJ.

Payne, John M. 1987. "Rethinking Fair Share: The Judicial Enforcement of Affordable Housing Policies." *Real Estate Law Journal* 16(Summer):20-44.

Rose, Jerome G. 1977. *After Mount Laurel: The New Suburban Zoning*. New Brunswick: Rutgers, State University of New Jersey, Center for Urban Policy Research.

Steinberg, Marcia. 1989. *Adaptations to an Activist Court: The Aftermath of the Mount Laurel II Decisions for Lower-Income Housing*. Cambridge, MA: Lincoln Institute for Land Policy.

Williams, Norman, Jr. 1984. "The Background and Significance of Mount Laurel II (1983 New Jersey Supreme Court Decision Outlawing Exclusionary Zoning)." *Washington University Journal of Urban and Contemporary Law* 26:3-23. ◄

▶ Multifamily Housing

Based purely on design, a multifamily unit is one contained within a single building that houses several units. In most instances, units in a multifamily building are stacked atop each other and, therefore, are physically attached to three or more other units. For purposes of housing data collection, the U.S. Bureau of the Census distinguishes multifamily from other forms of housing as five or more units contained within a single building. Multifamily units can be either rental apartments or individually owned units, such as condominiums or cooperatives.

Of the total housing units in the United States, The U.S. Bureau of the Census estimates that in 1995, 12% are multifamily (see Figure 16). In comparison, 60% of the total units are single-family detached, indicating U.S. hous-

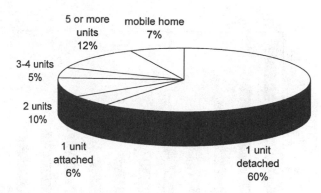

Figure 16. Percentage of Distribution of Housing Stock

ing consumers' strong preference for single-family houses. In most other countries, multifamily units constitute a much higher percentage of the total housing stock, owing to affordability, land availability, transportation systems, urban development patterns, and cultural preferences.

Forms of Ownership

The term *multifamily* can be confusing because many developers and real estate professionals use it to refer generally to all forms of rental housing. Although most rental units are in fact contained in multifamily buildings, the ownership of the units is not necessarily related to the size and design of the buildings. Particularly since the early 1980s, myriad multifamily building types have emerged specifically to house for-sale units.

Multifamily units developed for sale to individual homeowners are usually sold as condominiums or, less common, as cooperatives. Condominiums may be structured so that a unit owned is defined as an "airspace" or as a portion of real property, such as walls and land. All parts of the condominium development not specifically owned by an individual, as described on the deed, are owned in common by all of the development's owners. Ownership of common elements (which typically include outdoor areas, lobby, and hallways) is generally defined as a percentage proportionate to the square footage of each unit; the percentage interest in common area is granted to each unit owner by deed.

In the cooperative form of ownership, a single property is divided into units (or "use portions"), with each unit user owning shares of stock in the corporation that owns the building. Cooperative ownership is most commonly employed in multifamily mid- and high-rise buildings located in major urban markets such as New York City.

Building Types

Multifamily buildings can be categorized into three basic types based on the size and height of the buildings: (a) garden density (sometimes called garden apartments), (b) midrise buildings, and (c) high-rise buildings. The historical precedent for garden density multifamily buildings is the garden apartment, built widely in suburban areas during the postwar housing boom. This housing type was built typically for rental purposes. Garden density buildings are two or three stories, do not usually contain an elevator, and

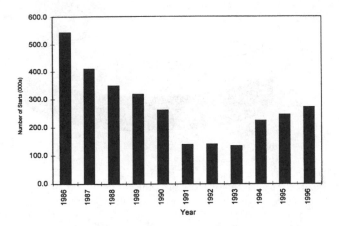

Figure 17. New Privately Owned Multifamily Housing

TABLE 21 New Privately Owned Multifamily Housing Units Started (in thousands)

Year	Total Number of Starts	Total Number of Multifamily Starts	Multifamily Starts as a Percentage of the Total
1986	1,805.4	542.0	30%
1987	1,620.5	408.7	25%
1988	1,488.1	348.0	23%
1989	1,376.1	317.6	23%
1990	1,192.7	260.4	22%
1991	1,013.9	137.9	14%
1992	1,199.7	139.0	12%
1993	1,287.6	132.6	10%
1994	1,457.0	223.5	15%
1995	1,354.1	244.1	18%
1996	1,476.6	270.8	18%

have 10 or more units within a single structure. Buildings are placed around a site to allow generous areas for landscaping and surface parking lots, yet a density of 10 to 20 units per acre is achieved. Over the past 15 to 20 years, garden density buildings have evolved into more complex and varied forms that may include elevators, underbuilding parking, varied rooflines, and a combination of stacked flats and townhouses. In addition, units contained in garden density buildings are more commonly built for sale to individual unit owners.

A midrise multifamily building is generally from four to eight stories high and includes elevators. Unlike garden density buildings, which are built with wood frame construction, midrise buildings are constructed with steel or reinforced concrete. High-rise multifamily buildings have construction characteristic similar to midrise buildings except that they feature eight or more floors.

Construction Activity

As shown in Figure 17 and Table 21, new privately owned multifamily housing starts have decreased sharply since peaking in the early to mid-1980s. These data indicate starts for all privately owned rental and for-sale housing (but do not include public housing developments undertaken primarily for low-income households). Privately owned multifamily construction peaked in 1985 at more than 576,000 new starts, which constituted 33% of the total housing starts in that year. In 1996, multifamily starts had fallen to only 270,800 units and constituted only 18% of the total housing starts.

This steady drop in multifamily housing starts can be attributed to several factors, including overbuilding during the early 1980s, shifts in market demand, and decreased financing sources for multifamily construction. Ready financing and tax advantages that existed before 1986 contributed to a substantial oversupply of multifamily units throughout all regions of the United States. The Tax Reform Act of 1986, coupled with increasing vacancy rates and dropping rents, led to a sharp decline in multifamily construction activity. The demand for new multifamily units

also softened due, in part, to demographic shifts—most notably, the aging of the baby boom generation and their desire to move into single-family houses to accommodate growing families.

By the mid-1990s, multifamily supply and demand reached a near equilibrium in most regions of the United States. This balancing was evidenced by decreased vacancy rates and modest rent increases. The outlook for multifamily housing construction for the future is favorable, although starts are not expected to approach the levels realized during the peak years of 1984 to 1986. The regions forecasted to experience the strongest gains are the South and West.

Financing Sources

Another reason for the sharp decrease in multifamily construction starts has been the unavailability of financing sources. Driven by favorable tax programs prior to 1986, syndicators compiled large flows of equity to finance rental projects. Changes in tax laws and a decrease of traditional lending sources have worked to discourage multifamily development since 1986. New sources of financing, however, began to emerge in the early 1990s, especially in the form of Real Estate Investment Trusts (REITs) funded by Wall Street investors.

Also important to the development of multifamily housing since 1986 has been the federal government's Low-Income Housing Tax Credit (LIHTC) program. This program grants tax credits to developers or investors who agree to set aside a percentage of a project's units for low-income households. The tax credits have worked effectively to attract capital to projects that otherwise might not have been able to secure financing. In 1995, 34.5% of all multifamily units started in the United States received LIHTCs (see Figure 18).

Market/Consumer Niches

Although the contemporary "American dream" still includes a single-family house in the suburbs, multifamily housing is the preferred housing option for millions of

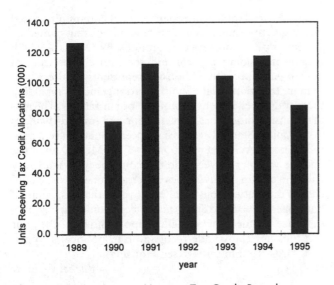

Figure 18. Low-Income Housing Tax Credit Rental

households. Households that have traditionally favored multifamily living include singles, unmarried couples, and the elderly. In 1996, census estimates show that the greatest number of renters were in the 25 to 29 age bracket, comprising 15.9% of all renters; renters over the age of 65 composed 13% of all renters. In the coming decade and beyond, the number of households with persons over the age of 65 will increase dramatically because of increased life spans and the aging of the baby boom population. This demographic shift is likely to increase long-term demand for multifamily housing.

Another factor influencing the demand for multifamily housing is affordability. Households that might otherwise prefer to live in single-family housing have opted for apartments or condominiums as a matter of financial necessity. In general, multifamily housing is less expensive to rent or purchase than single-family housing because of its higher density. This is not a hard-and-fast rule, however. Luxury multifamily housing is also built for high-income purchasers and renters who are attracted to this form of housing because of security, amenities, location, and lifestyle. (SEE ALSO: *Children; Condominium; High-Rise Housing; Section 201; Section 221[d][3] and [4]; Section 234; Section 241*)

—Lloyd W. Bookout

Further Reading

Bogdon, Amy S. and James R. Follain. 1996. "Multifamily Housing: An Exploratory Analysis Using the 1991 Residential Finance Survey." *Journal of Housing Research* 7(1):79-116.

Bookout, Lloyd W. et al. 1991. *Residential Development Handbook.* 2d ed. Washington, DC: Urban Land Institute.

Cagann, Robert A. 1994. *Rehabilitating Apartments: A Recycling Process.* Chicago: Institute of Real Estate Management of the National Association of Realtors.

Follain, James R. 1994. "Some Possible Directions for Research on Multifamily Housing." *Housing Policy Debate* 5(4):533-68.

Housing America's Population: The Outlook for Multifamily Housing. 1993. San Francisco, CA: Ernst & Young.

McKenna-Harmon, Kathleen and Laurence C. Harmon. 1993. *Contemporary Apartment Marketing: Strategies and Applications.* Chicago, IL: Institute of Real Estate Management of the National Association of Realtors.

"1994 Multi-Family Greenbook: A Comprehensive Guide to the Leaders in Multi-Family Property." *Commercial Property News,* November 1(Supplement).

U.S. Department of Housing and Urban Development. 1994. *Occupancy Requirements of Subsidized Multifamily Housing Programs.* Handbook 4350.3. Washington, DC: Office of Housing.

Wentling, James W. and Lloyd W. Bookout, eds. 1988. *Density by Design.* Washington, DC: Urban Land Institute. ◄

► Mutual Housing

The term *mutual housing* is used by many types of nonprofit housing organizations, and it can have a variety of different meanings. Mutual housing is a model of housing provision in which the residents belong to and control a membership organization that, in turn, owns and manages their housing. Many mutual housing associations (MHAs) in the United States are modeled after the former West Germany's Mutual Housing Association's ideals, which include (a) continuing production of long-term affordable housing, (b) ensuring resident control, (c) providing professional support, and (d) involving the community in the process. Mutual housing generally is sponsored by neighborhood development organizations, consumer cooperatives, nonprofit developers, or technical assistance providers.

There are three more or less distinct forms of MHAs. The *integrated-consumer cooperative* owns and controls all the housing it develops. Residents form the board and hold shares in the entire development rather than in their individual buildings. In the *dualistic* model, there is an MHA that both develops housing and provides technical assistance to a resident cooperative organization that owns and manages their building. The third type of mutual housing is the *federation,* a network of independently incorporated cooperatives joined under an MHA.

The Neighborhood Reinvestment Corporation, a congressionally chartered public nonprofit corporation committed to revitalizing declining neighborhoods for the benefit of current residents, provides funding and technical assistance to nonprofit community-based organizations across the United States. Since the mid-1980s, the Neighborhood Reinvestment Corporation has focused considerable resources on the development of MHAs.

The idea behind the Neighborhood Reinvestment Corporation's interest in mutual housing is credited to Congressman Jonathan Bingham of New York. On the basis of a trip to the former West Germany, Congressman Bingham sponsored legislation that would have created MHAs in the United States, modeled after those in Germany. Although this legislation failed to pass the 96th Congress, the Housing and Community Development Act of 1980 encouraged the development of MHAs, and the Neighborhood Reinvestment Corporation was asked to help develop a pilot program.

In Germany, "public benefit housing" has provided permanent affordable housing since the late 19th and early 20th centuries. As of the mid-1980s, some 4 million dwelling units were owned by social reform groups, churches, trade unions, tenant groups, and philanthropic associations.

The model advocated by the Neighborhood Reinvestment Corporation is closest to the integrated-consumer cooperative model, described above. It emphasizes the following: (a) the election of a board of trustees, a majority of whom must be residents of the mutual housing development or individuals waiting to move in; (b) security of tenure; (c) resident involvement in maintenance and management decisions; and (d) ability to nominate a qualified family member for immediate consideration for occupancy of the unit when the resident member gives up the unit or dies.

Mutual housing has many characteristics in common with other forms of nonprofit owned community-based housing. They all emphasize the role of citizens (both residents and community members), provide a high degree of security of tenure for occupants, and strive to provide decent-quality housing over the long term. One of the distinguishing features of MHAs is their commitment to continue to produce a significant supply of affordable housing until a threshold number of units is reached that allows for economic self-sufficiency. This goal is also shared by many, but not all of, other types of nonprofit housing providers.

Mutual housing also places great importance on trying to capture some of the key features of the homeownership experience, such as direct involvement with management decisions, security of tenure, investment of equity, and ability to pass the property on to heirs. In a survey of residents of a Baltimore MHA conducted in 1988, nearly two-thirds of the 39 respondents indicated that they think of themselves more as owners than as renters, and they indicated that some of the best features about the association are those closely related to the homeownership experience—financial and personal security.

In addition to the associations developed with the assistance of the Neighborhood Reinvestment Corporation, a number of other independent organizations have been responsible for developing thousands of units of mutual housing. For example, Cooperative Services, Inc., located in Oak Park, Michigan, was created in the 1930s and currently owns and manages about 5,000 units of housing, mostly for the elderly, in locations all across the United States.

MHAs are a far less prevalent form of community-based housing than resident cooperatives and housing owned by community development corporations. By providing a high degree of resident control and some of the benefits of homeownership, mutual housing represents an alternative to more traditional rental housing arrangements, including those sponsored by other nonprofit organizations. The mutual housing model has also encountered some difficulties in the United States. Lenders have been somewhat reluctant to finance developments sponsored by MHAs, and it is unclear whether lower-middle income people, who may have the option to buy homes on a traditional basis, would choose the mutual housing alternative. In that case, mutual housing in the United States may succeed only in providing subsidized homeownership-like opportunities for the poor and thus not enjoy wide recognition or political support. (SEE ALSO: **Community-Based Housing**)

—*Rachel G. Bratt*

Further Reading

Barnes, K. R. 1982. "Housing Associations: European Ideas, American Approaches." *Journal of Housing* 39(1):10-13.

Bratt, Rachel G. 1990. *Neighborhood Reinvestment Corporation-Sponsored Mutual Housing Associations: Experiences in Baltimore and New York.* Washington, DC: Neighborhood Reinvestment Corporation.

Bratt, R. G. 1991. "Mutual Housing: Community-Based Empowerment." *Journal of Housing* (July/August):173-80.

Cheuvront, Beverly. 1986. "Paving the Way for Mutual Housing." *City Limits* (August/September):20-21.

Goetze, Rolf. 1985. *The Mutual Housing Association: An American Demonstration of a Proven European Concept.* Washington, DC: Neighborhood Reinvestment Corporation.

Kinsky, John and Sarah Houde. 1996. *Balancing Acts: The Experience of Mutual Housing Associations and Community Land Trusts in Urban Neighborhoods.* New York: Community Service Society of New York.

Schwartz, Lisa. 1986. "Mutual Housing Associations: Promise for Affordable Housing." *Urban Land* (May):17-21.

Schwartz, Lisa. 1988. *Mutual Housing Associations: Lessons from the U.S. Experience.* Washington, DC: National Mutual Housing Network.

Van Ryzin, G. G. 1994. "Residents' Sense of Control and Ownership in a Mutual Housing Association." *Journal of Urban Affairs* 16: 241-53.

N

▶ National Alliance to End Homelessness

The National Alliance to End Homelessness is dedicated to the prevention and eradication of homelessness. It was founded in 1983 as the Committee for Food and Shelter. The alliance serves as a source of information on homeless issues and low-cost housing and as a network for homeless advocates in the private, public, and nonprofit sectors. The alliance offers a general membership as well as additional memberships in the Single Room Occupancy (SRO) Housing Network, the AIDS Housing Network, the Employment Programs Network, the Services Network, and the Transitional Housing Network. All members are entitled to a free monthly newsletter, discounts on alliance publications and alliance conferences, and participation in the Alliance Grassroots Network. The alliance is funded by corporate, foundation, and private donations and operates on an annual budget of $1 million (1994). Current membership: 2,800 nationally. Staff size: 9. President: Thomas L. Kenyon. Address: 1518 K Street, NW, Suite 206, Washington, DC 20005. Phone: (202) 638-1526. Fax: (202) 638-4664. (SEE ALSO: **Homelessness**).

—Caroline Nagel ◀

▶ National American Indian Housing Council

A national association founded in 1974, the National American Indian Housing Council (NAIHC) seeks to increase low-income housing opportunities for more than 500 Indian tribes and Alaska Native villages on reservation and trust lands and provides a national forum for tribal leaders and Indian housing professionals. The council conducts federal legislative advocacy efforts, provides educational training and on-site technical assistance to housing agencies, maintains an Indian Housing Resource Center, convenes an annual convention and legislative conference, and builds strong coalitions with other national organizations that inform and educate others about the needs in Indian housing. Publications include *NAIHC Pathway News,* a quarterly newsletter, and the *Washington Update,*

a legislative update. NAIHC is funded through federal, foundation, corporate, and private sector funds. Current membership: 192 Indian Housing Authorities. Staff size: 16. Executive Director: Christopher Boesen. Address: 900 Second St., NE, Suite 007, Washington, DC 20002. Phone: (202) 789-1754. Fax: (202) 789-1758. (SEE ALSO: *Native Americans*)

—Laurel Phoenix ◀

▶ National Apartment Association

The National Apartment Association (NAA) is the largest full-service trade association in the United States serving the specific needs of the multifamily housing industry. NAA is a federated trade association operating through over 150 affiliates nationwide and representing the interests of property owners, managers, lenders, builders, and suppliers. NAA offers its members enhanced professional development through educational programs, including certification programs for property supervisors, property managers, leasing agents, and maintenance personnel; free participation and listings in the only national database of handicapped-accessible apartment units; and representation on legislative and regulatory issues on all levels of government. Members are entitled to the association's bimonthly magazine, *Units,* and its free research services. Founded in 1939. Membership: 26,000. Staff size: 26. Contact person: Amy Dozier, Executive Vice President. Address: 201 N. Union Street, Suite 200, Alexandria, VA 22314. Phone: (703) 518-6141. Fax: (703) 518-6191. E-mail: information@naahq.com.

—Caroline Nagel ◀

▶ National Association for Home Care

The National Association for Home Care (NAHC) is a national trade association founded in 1982 through a merger of the National Association of Home Health Agencies and the Council of Home Health Agencies/Community Health

Services. It represents the nation's home care providers, including home health agencies, home care aide organizations, hospices, and the individuals they serve. NAHC works to ensure availability of humane, cost-effective, high-quality home care services by advocating home care's interests before Congress, the administration, and various federal and state regulatory agencies. In addition, NAHC provides educational programming and publications for home care providers. Its members include provider agencies, individual caregivers, associate organizations, and allied organizations. Its publications include books, periodicals, and brochures covering topics in administrative management, Medicare and Medicaid, political issues, and consumer information. Periodicals include *Caring Magazine,* a monthly magazine for the home care field ($45 per year); *Homecare News,* a monthly newspaper covering legislative and regulatory issues ($18 per year); *NAHC Report,* a weekly policy newsletter ($325 per year); *Hospice Forum,* a bimonthly newsletter of the Hospice Association of America ($105); and *Caring People Magazine,* a quarterly magazine about people who give their time to others ($18 per year). NAHC operates on a budget of $14 million (1995) funded through membership dues, conferences, publication sales, and advertising. Current membership: 6,000. Staff size: 100. Vice President for Research, Regulatory Affairs, and Education: Karen Pace. Address: 223 Seventh Street, SE, Washington, DC 20003-4305. Phone: (202) 547-7424. Fax: (202) 547-9559. (SEE ALSO: *Home Care*)

—*Laurel Phoenix* ◀

▶ National Association of Affordable Housing Lenders

Founded in 1988, the National Association of Affordable Housing Lenders' (NAAHL) mission is to build bridges between capital markets and U.S. neighborhoods. It is dedicated to delivering credit to low- and moderate-income families, first-time homebuyers, and affordable rental housing. NAAHL promotes partnerships among lenders, localities, and investors. Executive Director: Peter Bell. Address: 1726 18th St., NW, Washington, DC 20009. Phone: (202) 328-9171. Fax: (202) 265-4435.

—*Laurel Phoenix* ◀

▶ National Association of Housing and Redevelopment Officials

The National Association of Housing and Redevelopment Officials (NAHRO), founded in 1933, provides advocacy for affordable housing and strong, viable communities. NAHRO enhances the professionalism and effectiveness of its members through professional development training, conferences, and publications. The association produces numerous publications pertinent to housing and community development, including the *Journal of Housing and Community Development,* a bimonthly for $24 per year,

and the *NAHRO Monitor,* a twice-monthly newsletter sent to members only. In addition to publications, NAHRO also provides professional development training and certification programs and seminars; three national conferences each year; NAHRO's Legislative Network; expert technical and consultant services; NAHRO-NET—NAHRO's computer online information service; and Agency Awards of Merit and Excellence in Housing and Community Development. NAHRO operates on an annual budget of $6 million. Membership: 9,000 housing and community development agency officials who administer Department of Housing and Urban Development programs at the local level. Staff size: 39. Executive Director: Richard Y. Nelson, Jr. Address: 630 Eye Street, NW, Washington, DC 20001-3736. Phone: (202) 289-3500. Fax: (202) 289-8181.

—*Laurel Phoenix* ◀

▶ National Association of Housing Cooperatives

The National Association of Housing Cooperatives (NAHC) is a nonprofit federation of housing cooperatives, professionals, organizations, and individuals that promotes the interests of cooperative housing in the United States. The NAHC provides a network for those involved in the cooperative housing field, technical assistance on the development and operation of cooperatives, representation on issues of concern before Congress and federal agencies, and consulting and training services. In addition, the NAHC publishes the bimonthly *Cooperative Housing Bulletin* and the annual *Cooperative Housing Journal* for its members. Founded in 1950, the NAHC has 1,000 organizational and individual members nationwide. It operates on an annual budget of $510,000 (1997) and is funded through conference fees, membership dues, sales, grants, and donations. Staff size: 8. Executive Director: Herbert J. Cooper-Levy. Address: 1614 King Street, Alexandria, VA 22314. Phone: (703) 549-5201. Fax: (703) 549-5204. (SEE ALSO: *Cooperative Housing*)

—*Caroline Nagel* ◀

▶ National Association of Housing Partnerships

The National Association of Housing Partnerships (NAHP) has served as an advocacy and information-sharing organization for public and private nonprofit housing partnerships across the United States since 1989. NAHP also works to develop programs and policy related to the expansion of affordable housing opportunities and the revitalization of inner-city, suburban, and rural communities. Of NAHP's 65 members, 35 are local housing partnerships that bring together state or local governments and the private business sector in a state, metropolitan area, or city. Members receive a quarterly newsletter, a catalog of housing partnerships, and periodic updates via e-mail and fax. NAHP operates on an annual budget of $115,000 (1995) derived from foundation grants and membership fees. In August 1995, NAHP was awarded an $800,000 grant from HUD to operate a housing counseling program, with particular emphasis on first-time homeownership, through 15 local housing partnerships in 12 states. The Metropolitan Boston Housing Partnership currently administers all of NAHP's operations. Membership: 65 individuals and organizations nationwide. President: Robert B. Whittlesey. Contact person: Frank Shea. Address: 153 Milk St., Suite 300, Boston, MA 02109. Phone: (617) 946-3333. Fax: (617) 345-0123. E-mail: muelle@mbhp.com. (SEE ALSO: *Public/Private Housing Partnership*)

—*Mara Sidney* ◄

► National Association of Realtors

The National Association of Realtors (NAR) represents residential and commercial real estate brokers, salespeople, and property managers, appraisers, and counselors before Congress and promotes the right to own, transfer, and use real property. NAR provides members with educational programs, informational services, and research opportunities designed to enhance their businesses and the realty industry as a whole. In addition, NAR distributes a print and an online publication, *Today's Realtor®* and maintains a website at realtor.com. Founded in 1908, this organization maintains a membership of approximately 720,000 people throughout the United States and its territories. Staff size: 300. Contact person: Stephen D. Driesler, Senior Vice President/Chief Lobbyist. Address: 700 11th Street, NW, Washington, DC 20001. Phone: (202) 383-1238. Fax: (202) 383-1134.

—*Caroline Nagel* ◄

► National Center for Home Equity Conversion

National Center for Home Equity Conversion was founded in 1981 to promote the development of sound home equity conversion opportunities for older homeowners. Since its inception, the center has conducted seminars in 46 states and has trained more than 3,000 loan counselors nationwide. It also publishes *Your New Retirement Nest Egg: A Consumer Guide to the New Reverse Mortgages*. Director: Ken Scholen. Address: 7373 147th Street West, Suite 115, Apple Valley, MN 55124. Phone: (612) 953-4474. (SEE ALSO: *Reverse-Equity Mortgage*)

—*Caroline Nagel* ◄

► National Center for Housing Management

Founded in 1972, the National Center for Housing Management (NCHM) is a nonprofit organization created by Executive Order 11668 of the President of the United States to "strengthen and professionalize" public and privately owned federally assisted housing. NCHM provides

training and certification programs, a toll-free hotline service for certified professionals, and technical services for management improvement, Section 504/Fair Housing compliance, and management audits. NCHM also has a branch office in San Juan, Puerto Rico. Staff: 46. President: W. Glenn Stevens. Address: 1010 Massachusetts Avenue, NW, 4th Floor, Washington, DC 20001. Phone: (202) 872-1717. Fax: (202) 789-1179. Word Wide Web: www.nchm.org.

—*Laurel Phoenix* ◄

► National Center for Lead-Safe Housing

Founded in 1992, the National Center for Lead-Safe Housing's mission is to bring the public health, housing, and environmental communities together to combat childhood lead poisoning while preserving affordable housing. The center sponsors research on methods to reduce lead hazards and scientifically assess risks. It also promotes policies that clarify standards of care for landlords and builds the national coalition of health care providers, housing professionals, environmentalists, bankers, homeowners, renters, insurers, labor, and government needed to address the problem. Publications cover topics such as Department of Housing and Urban Development (HUD) reports on lead-based paints in housing, sampling methods for measuring settled lead dust, insurance liability, and mitigation costs for housing with lead exposure hazards. A training video, focusing on HUD's Lead-Based Paint Hazard Control Grant Program, covers paint chips, dust and soil sampling, and interviewing methods. Staff: 12. Budget: $2 million annually received from foundations and federal agencies. Contact person: Walter G. Farr. Address: 10227 Wincopin Circle, Suite 205, Columbia, MD 21044. Phone: (410) 992-0712. Fax: (410) 715-2310. (SEE ALSO: *Environmental Contamination: Lead-Based Paint Hazards*)

—*Laurel Phoenix* ◄

► National Coalition for the Homeless

The National Coalition for the Homeless (NCH) promotes affordable housing, health care, and benefits for the homeless through policy advocacy, public education, and grassroots organizing. The NCH's 70 board members serve as links between the coalition and homeless advocacy organizations on the state and local levels by providing technical assistance and informational materials. The NCH merged in 1992 with the Homelessness Information Exchange, which offers databases on programs, research, and personnel in the area of housing and homelessness. Coalition periodicals include *Safety Network Newsletter* (nine issues per year) and *Homewords* (four issues per year). Other current publications include *A Place Called Hopelessness: Shelter Demands in the '90s, Addiction on the Streets,* and *Heroes Today, Homeless Tomorrow?* The NCH operates on an annual budget of $550,000 (1993) and is funded by individual contributions and grants from foundations and corporations. Founded in 1983. Membership: 700 individuals and organizations; 12,000 affiliates nationwide. Staff size: 9. Executive Director: Fred Karnes, Jr. Address: 1612 K Street, NW, #1004, Washington, DC 20006. Phone: (202) 775-1322. Fax: (202) 775-1316. (SEE ALSO: **Homelessness**)

—Caroline Nagel ◀

▶ National Commission on Severely Distressed Public Housing

The National Commission on Severely Distressed Public Housing was created by Congress as part of the Department of Housing and Urban Development Reform Act of 1989. According to the legislation, the commission was established to (a) identify public housing developments in a severe state of distress, (b) assess the most promising strategies to improve the condition of distressed housing projects, and (c) develop a national action plan to eliminate by the year 2000 unfit living conditions in severely distressed developments.

As authorized by the legislation establishing the commission, the secretary of the U.S. Department of Housing and Urban Development (HUD) and the chairs of various committees in the Senate and the House of Representatives selected 18 commissioners. These commissioners included a variety of state and local government officials, heads of public housing authorities (PHAs), members of the business community, and tenant leaders. The members elected two cochairmen for the commission: The Honorable Bill Green, a member of the House of Representatives from New York City, and Vincent Lane, Chairman of the Chicago Housing Authority.

In preparation for writing its report and recommendations, the commission and its staff visited more than 25 cities and held 20 hearings. In addition, consultants were hired to analyze the physical and social problems facing public housing as well as to develop case studies of successful initiatives to alleviate these problems.

In its final report, the commission defined "severely distressed" public housing as housing that exhibited one or more of the following conditions:

▶ Families living in distress
▶ High rates of serious crimes in the development
▶ Barriers to managing the environment
▶ Physical deterioration of buildings

This definition was further elaborated by a detailed rating system for identifying which developments had the most severe problems. On the basis of its definition, the commission estimated that 86,000 of the nation's 1.4 million units of public housing were severely distressed, equal to approximately 6% of the entire public housing stock.

The commission's National Action Plan to eradicate severely distressed public housing by the year 2000 included a number of proposals to deal with the needs of the tenants, buildings, and management of public housing developments. Among the initiatives to assist people were direct

funding of resident management councils, the coordination of social and support services for residents, and the promotion of economic development through enterprise zones and hiring preferences for public housing residents.

With respect to improving the physical condition of the 86,000 units of severely distressed public housing, the commission recommended that Congress appropriate 7.5 billion dollars over 10 years and establish a separate administrative unit in HUD to coordinate programs. It further proposed that HUD grant waivers to PHAs to enable them to spend more to renovate severely distressed buildings than they would have been able to spend constructing new public housing developments. The commission also suggested that PHAs be permitted to build replacement units in areas where they would otherwise be forbidden because of high proportions of minority households.

In addition, the commission recommended a number of changes that would affect the management and operation of public housing, such as specific funding of PHA management improvements, accreditation for PHA employees, improved monitoring and data collection, and incentives for PHAs to provide for tenant representation on their boards. Experiments with profit-motivated and nonprofit management as well as partnerships with the private sector were also put forward as suggestions.

The commission also argued in favor of several changes in federal law that would promote income mixing in public housing developments such as setting maximum rents based on local market conditions and increasing the proportion of residents who could earn between 50% and 80% of area median income. Additional proposals to enhance livability at public housing developments included permitting PHAs to reserve certain developments exclusively for the elderly and improving projectwide security.

Several of the Commission on Severely Distressed Public Housing's recommendations were incorporated by Congress into federal law, including the creation of a new Office of Severely Distressed Public Housing Recovery within HUD. Two statutes passed in 1992 (the Housing and Community Development Act of 1992 and the Departments of Veterans Affairs and HUD and Independent Agencies Appropriations Act) contained several new programs designed to improve the management and condition of severely distressed public housing. Independent assessment of management practices was mandated. In addition, tenants of distressed public housing developments in "troubled" housing authorities were given the right to petition HUD to transfer management functions from PHAs to alternative management companies. Congress also appropriated additional money for PHAs to renovate severely distressed public housing. In accordance with the commission's suggestions, PHAs were authorized to spend more than the HUD-prescribed total development costs for refurbishing distressed public housing. A new federal program, the Urban Revitalization Demonstration (HOPE VI) was also enacted to provide funds for PHAs to renovate severely distressed or obsolete public housing.

Congress's response to the commission's recommendations on achieving a greater mix of incomes, however, was limited. The 1992 legislation removed the five-year time limit on ceiling rents and authorized PHAs to "skip" very low income households on their waiting lists to offer accommodations to other households under their own local preference systems. Nevertheless, although legislative proposals to relax income limitations were being debated in 1997, Congress had not yet increased the overall proportion of households living in public housing who could earn more than 50% of area median income beyond 25% (Schill 1993). (SEE ALSO: *Multifamily Housing;* **Public Housing**)

—*Michael H. Schill*

Further Reading

National Commission on Severely Distressed Public Housing. 1992a. *Case Studies and Site Examination Reports.* Washington, DC: Government Printing Office.

National Commission on Severely Distressed Public Housing. 1992b. *The Final Report.* Washington, DC: Government Printing Office.

Schill, Michael H. 1993. "Distressed Public Housing: Where Do We Go From Here?" *University of Chicago Law Review* 60:497-554.

Vale, Lawrence J. 1993. "Beyond the Problem Projects Paradigm: Defining and Revitalizing 'Severely Distressed' Public Housing." *Housing Policy Debate* 4(2):147-74. ◄

► National Commission on Urban Problems (Douglas Commission)

The National Commission on Urban Problems, popularly known as the Douglas Commission, was appointed by President Lyndon Johnson and made its report to him on December 12, 1968. Paul H. Douglas, former economics professor and U.S. Senator from Illinois, 1948 through 1966, chaired the commission. The commission's report was one of three major reports issued during the Johnson years that focused on the problems of the city; the other two reports were the Kaiser Committee report (President's Committee on Urban Housing) and the Kerner Commission report (National Advisory Commission on Civil Disorders). Like the Kaiser Committee report, the Douglas Commission report was issued in the closing days of the Johnson administration and therefore had no direct impact on public policy. Its chief historical importance lies in its role as an indicator of the nation's concern during the 1960s for the problems of the cities, in the indication of possible directions that the nation might have moved toward had the Democrats retained their control of the presidency, and in the significant effort it represented in collecting information on the problems of cities at that time.

In general, the Douglas Commission report was a blueprint for urban reform, and in addition to attention to housing, the report examined the federal urban renewal program, overall urban development policy, building and housing codes, development standards, urban government, and urban public finance. In terms of housing policy, the commission recommended (a) expansion of federal housing programs, while also calling for their simplification and consolidation within the Department of Housing and Urban Development, and (b) that the president deliver an annual housing message. Consistent with recommendations of the Kerner Commission, the Douglas Commission

placed emphasis on recommendations that would enable disbursement of affordable housing throughout metropolitan areas. Among its recommendations for the urban renewal program, the Douglas Commission argued for a federal uniform relocation policy that was eventually realized in the 1970 Uniform Relocation Assistance and Real Property Acquisition Act.

Regarding development policy in general, the commission took the approach of recommending that various federal programs and policies be used in a fashion to rationalize state and local planning. For example, the commission called on states to deny land use regulatory powers to local governments that lacked "development guidance programs" and urged that Federal 701 planning assistance funds be made available to states on the condition that they have adopted such legislation. The commission called for adoption of a grant program that would aid states that have adopted a statewide mandatory housing code. The commission also recommended that federal water and sewer grants be awarded only to jurisdictions whose building codes were not more restrictive than nationally recognized standards.

In terms of urban public finance, the commission called for a broadening of the tax base for metropolitan areas, recommending that less emphasis be placed on local property taxes for funding urban government expenditures and that fiscal inequities existing between poor and affluent jurisdictions be reduced, for example, through the geographic broadening of school taxes. Among its recommendations, the commission included a proposal for establishing a federal revenue-sharing program, which was eventually realized in the Federal Revenue Sharing and Community Development Block Grant Programs. Among the more controversial public finance issues taken up by the commission was the issue of land value taxation, by which taxes are levied on the increment in land values. Although, the full commission recommended that the U.S. Treasury study this issue, a minority report issued by four members, including Chairman Douglas, called for actual adoption of land value taxation as a mechanism by which property taxes could be shifted to landowners who benefit from increased property values associated with rising demand for their land. (SEE ALSO: *Building Codes;* **Federal Government;** *President's Committee on Urban Housing; Uniform Relocation Assistance and Real Property Acquisitions Policies Act of 1970;* **Urban Redevelopment**)

—*Charles E. Connerly* ◄

► National Community Reinvestment Coalition

The National Community Reinvestment Coalition (NCRC) battles discrimination through its dedication to increasing fair and equal citizen access to credit and banking services and products. A nonprofit organization founded in 1990, NCRC's membership includes local, regional, and national organizations. NCRC provides regulatory and legislative updates on community reinvestment issues and networking with legislators, academicians, the press, and community reinvestment leaders. It also offers workshops, conferences, and Community Reinvestment Act (CRA) training to members. In addition, it provides a library and newsclipping service, and members receive a bimonthly newsletter, the *NCRC Reinvestment Compendium;* a triannual newsletter, *Reinvestment Works;* and quarterly CRA ratings. The coalition is funded by foundations and government contracts. Membership dues are determined by size of member agency. Address: 733 15th Street, NW, Suite 540, Washington, DC, 20005-2112. Phone: (202) 628-8866. Fax: (202) 628-9800.

—Laurel Phoenix ◀

▶ National Consortium of Housing Research Centers

The National Consortium of Housing Research Centers (NCHRC) is a consortium of 18 research institutions with interests in housing and community development. Founded in 1987, the consortium is affiliated with the National Association of Home Builders. The mission of the consortium is to operate a national cooperative network of housing-related research centers to serve the residential and light-commercial construction industry, to facilitate the adoption of technological innovations by the industry, and to improve the quality of building science in the United States. Member institutions publish a variety of technical notes and research reports on a periodic basis. Contact person: Mark Nowak, Consortium Coordinator. Address: c/o National Association of Home Builders Research Center, 400 Prince Georges Blvd., Upper Marlboro, MD 20774. Phone: (301) 249-4000. Fax: (301) 249-0305.

—Robert C. Stroh ◀

▶ National Council of State Housing Agencies

The National Council of State Housing Agencies (NCSHA) is a nonprofit organization created in 1970 to assist its members in advancing the interests of lower-income and underserved people through the financing, development, and preservation of affordable housing. NCSHA's members are housing finance agencies (HFAs) with statewide authority. Member agencies operate in every state as well as in the District of Columbia, Puerto Rico, and the U.S. Virgin Islands. HFAs operate hundreds of affordable housing programs, including the Mortgage Revenue Bond (MRB) program, HOME Investment Partnerships Program (HOME), and the Low Income Housing Tax Credit (Housing Credit) program. NCSHA represents its member HFAs before Congress; the Administration; federal agencies concerned with housing, such as the Department of Housing and Urban Development (HUD) and the Treasury Department; and with other advocates for affordable housing.

NCSHA hosts conferences and provides members with regular updates on legislative and regulatory developments. In addition, NCSHA publishes a variety of reference manuals on using housing credits, housing trust funds, state equity funds, and related topics, which are available to the public. Members receive NCSHA's quarterly journal—*State Housing Finance*—its newsletters—*Capital Report* and *HFA Executive Directors' Advisory*—and other publications. For more information: National Council of State Housing Agencies, 444 North Capitol St., NW, Suite 438, Washington, DC 20001. Phone: (202) 624-7710. Fax: (202) 624-5899. URL: www.ncsha.org. (SEE ALSO: *State Government*)

—Linda Redmond ◀

▶ National Fair Housing Alliance

The National Fair Housing Alliance (NFHA) is a nonprofit organization dedicated to achieving fair, nondiscriminatory housing in the United States by (a) educating housing consumers and housing industry officials, (b) promoting the enforcement of fair housing and fair lending laws, and (c) researching the causes and effects of housing discrimination. Funded by grants from the U.S. Department of Housing and Urban Development, the NFHA also provides technical and program assistance to private and public fair housing agencies and consults for the housing and mortgage-lending industry. Founded in 1988, the alliance has 150 individual and (nonprofit) organizational members nationwide. Staff size: 4. Contact person: Shanna L. Smith, Executive Director. Address: 1212 New York Ave., NW, Suite 525, Washington, DC 20005. Phone: (202) 898-1661. Fax: (202) 371-9744. (SEE ALSO: **Fair Housing Amendments Act of 1988**)

—Caroline Nagel ◀

▶ National Foundation for Affordable Housing Solutions

The National Foundation for Affordable Housing Solutions, Inc., was founded in 1990 to preserve the existing stock of affordable housing, create new affordable housing, and link social services to housing needs in poor communities. The foundation works to develop affordable housing models and provide technical assistance in property acquisition by other nonprofit organizations. Through the end of 1997, supporting organizations to the foundation had acquired more than 1,600 units of affordable housing in 14 apartment communities. Through ownership, the foundation can implement its social service strategies. The foundation is funded entirely by private donations, technical services fees, and asset-based cash flow. Contact person: Martin C. Schwartzberg. Address: 11200 Rockville Pike, Suite 220, Rockville, MD 20852-3103. Phone: (301) 998-0498. Fax: (301) 998-0420. (SEE ALSO: **Affordability**)

—Caroline Nagel ◀

► National Homebuyers and Homeowners Association

Founded in 1972, the National Homebuyers and Home-owners Association (NHHA) provides educational publications to assist prospective homebuyers and testifies on Capitol Hill on behalf of homebuyers and homeowners. The association publishes the *Home Buyer's Checklist*, available in English or Spanish for $8.00. The NHHA operates on a small budget and has no membership. Staff size: 1. General Counsel: Benny L. Kass. Address: 1050 17th St. NW, Suite 1100, Washington, DC, 20036. Phone: (202) 659-6500. Fax: (202) 293-2608. (SEE ALSO: **Homeownership**)

—Laurel Phoenix ◄

► National Housing Act of 1934

Although the Home Owners' Loan Act of 1933 was the primary emergency *relief* act affecting the homeownership finance market, the National Housing Act of the 1934 (P.L. 73-479) was the primary legislation for *reforming* the housing finance market. The National Housing Act, signed into law on June 27, 1934, is best known for creating the Federal Housing Administration (FHA) and the mortgage insurance programs that made the purchase of a home easier to afford and safer to finance. The act established mortgage insurance programs for one- to four-family homes (Section 203), housing renovation and modernization (Title I), and multifamily dwellings (Section 207). These provisions had a major impact on the type of mortgages that were available, enabling many U.S. homebuyers to obtain long-term (20 to 30 years), self-amortizing home loans. In addition, the creation of FHA led to the development of national standards for mortgage underwriting, construction, and subdivision development.

Consistent with the legislation creating FHA, the National Housing Act also amended the Federal Home Loan Bank Act to authorize advances by Federal Home Loan Banks to savings and loans for mortgages insured by FHA.

Title III of the National Housing Act authorized the creation of a secondary market for home mortgages through the purchase and sale of mortgages. Under this authority, the Fannie Mae (formerly, the Federal National Mortgage Association) was chartered by the FHA on February 10, 1938.

Title IV of the National Housing Act established insurance for savings and loan deposits through the creation of the Federal Savings and Loan Insurance Corporation (FSLIC). The purpose of this title was to further shore up the home loan industry by encouraging savers to place their money in savings and loans.

—Charles E. Connerly

Further Reading

Federal Home Loan Bank Board. n.d. *Legislative History of National Housing Act, P.L. 73-479, [H.R. 962], June 27, 1934.* Washington, DC: Author.

Hays, R. Allen. 1995. *The Federal Government and Urban Housing: Ideology and Change in Public Policy.* Albany: State University of New York Press.
Mitchell, J. Paul. 1989. *Federal Housing Policy and Programs: Past and Present.* New Brunswick, NJ: Prentice Hall. ◄

► National Housing and Rehabilitation Association

The National Housing and Rehabilitation Association (NH&RA) is a national membership organization founded in 1971 that promotes partnerships among professionals in the affordable multifamily housing field. It represents its membership before Congress as well as at the Department of Housing and Urban Development (HUD) and other national forums. NH&RA disseminates information through five conferences per year as well as through publications, including *First of the Month,* a monthly update of new HUD rules, available housing funds, and key documents and how to obtain them, and *Multifamily Advisor,* a quarterly update on the multifamily industry. NH&RA serves its 320 members with a staff of 5. Executive Director: Peter Bell. Address: 1726 18th St., NW, Washington, DC 20009. Phone: (202) 328-9171. Fax: (202) 265-4435.

—Laurel Phoenix ◄

► National Housing Conference

Founded in 1931, the National Housing Conference (NHC) promotes public awareness of the nation's housing needs and advocates national housing policies designed to make decent housing available, affordable, and accessible to all citizens. It holds educational forums and annual conferences and represents the housing industry before Congress, the administration, and federal agencies. The conference publishes a quarterly magazine and weekly legislative fax, free with membership. NHC's membership includes housing authority officials, community development specialists, builders, lawyers, civic leaders, financiers, architects and planners, religious organizations, labor groups, and national housing and housing-related organizations. Former name: National Public Housing Conference. Current Membership: 700. Staff size: 9. Executive Director: Robert Reid. Address: 815 15th St. NW, Suite 538, Washington, DC, 20005. Phone: (202) 393-5772. Fax: (202) 393-5656.

—Laurel Phoenix ◄

► National Housing Institute

The National Housing Institute (NHI) was established in 1975 to provide information on affordable housing and community building to state and local policymakers, nonprofit groups, and practitioners in the low-income housing and community development field. The institute publishes the bimonthly *Shelterforce* magazine as well as research reports and working papers on affordable housing. Contact person: Harold Simon, Executive Director. Address: 439 Main Street, Orange, NJ 07050. Phone: (973) 678-90060. Fax: (973) 678-8437. E-mail: hs@nhi.org.

—*Caroline Nagel* ◀

▶ National Housing Law Project

The National Housing Law Project (NHLP) is a legal advocacy group dedicated to helping poor people gain access to decent, affordable housing. It works to protect and expand the existing affordable housing stock, and it monitors the efficacy of government housing programs. NHLP works with attorneys, members of Congress, and local communities interested in affordable housing to promote housing rights and related policies that create stable, well-served neighborhoods. In addition, it combats segregation, works to improve minority communities, and promotes fair operation practices of publicly assisted housing. NHLP oversees the Legal Services Homelessness Task Force, which works in housing, public assistance, and child welfare to fight homelessness. Workshops and training events are sponsored for low-income housing advocates and housing organizations. Publications include the *Housing Law Bulletin* (bimonthly), *HUD Housing Programs: Tenants' Rights, FmHA Housing Programs: Tenants' and Purchasers' Rights,* and special reports on various topics. Director: Manuel Romero. Address: 614 Grand Avenue, Suite 306, Oakland, CA 94610. Phone: (510) 251-9400. Fax: (510) 451-2300.

—*Laurel Phoenix* ◀

▶ National Housing Trust

The National Housing Trust (NHT) is a national nonprofit organization established in 1986, dedicated to the preservation of affordable housing. The trust has its national headquarters and a small staff located in Washington, D.C. It had its original support from, and continues to be supported by, major foundations, including the Ford Foundation and Fannie Mae.

The board of directors includes representatives from major interests associated with housing and preservation, including tenant advocates, owners and managers, state agencies, urban scholars, and housing professionals.

The organization plays a number of roles, including advocacy and public policy analysis in support of preservation; research and clearinghouse functions on program development for public and nonprofit organizations, and technical assistance in the consulting of owners and tenants in connection with specific preservation projects. The project-based and policy research undertaken by the NHT has been central to the development of the original low-income housing preservation policies and to the continuing efforts to revise and implement the policy.

Finally, as a result of a contract with the Department of Housing and Urban Development, the trust has held a series of public education events and training sessions throughout

the United States to introduce various stakeholders to the intricacies of preservation project work.

The trust also does technical and project-based consulting work with private, nonprofit, and public agency clients. The technical assistance work performed by the trust is especially essential in those parts of the United States that do not have statewide or regional nonprofits or intermediaries with extensive development experience and the flexibility to develop capacity in the area of preservation. Such capacity exists in states such as Massachusetts and California but is not available in many other states. For more information, contact the current president, Michael Bodaken, at 1101 30th Street, NW, Suite 400, Washington, DC 20007. Phone: (202) 333-8931.

—*Phillip L. Clay* ◀

▶ National Institute of Senior Housing

Founded in 1981, the National Institute of Senior Housing (NISH) is a constituent unit of the National Council on the Aging (NCOA). It serves as a network of housing providers and advocates promoting quality housing options for older people. It represents the concerns of both residents and providers of housing before state, national, and local decision makers. NCOA offers an annual conference, training series, networking, and information dissemination. Publications included in membership fee are *NCOA Networks* (bimonthly) and *Innovations in Aging* (quarterly). Funded by and operated under the NCOA. Current members: 700+. Program Assistant: Carol McClendon. Address: 409 Third St. SW, Suite 200, Washington, DC 20024. Phone: (202) 479-1200. Fax: (202) 479-0735. TDD: (202) 479-6674. (SEE ALSO: **Elderly**)

—*Laurel Phoenix* ◀

▶ National Leased Housing Association

The National Leased Housing Association (NLHA) is dedicated to expanding the supply of low- and moderate-income housing and to strengthening the role of the government in providing such housing. Representing developers, lenders, public housing authorities, and nonprofit agencies, the NLHA lobbies Congress and government agencies for the development of housing for elderly, handicapped, and low-income people. The association has had an active role in formulating a national policy to preserve the low-income housing inventory (particularly Section 8 rental housing), including tax credits for low-income rental housing development. The NLHA publishes a newsletter that reports legislative and administrative activity in the leased-housing field and holds two meetings per year for members. Founded in 1972, the NLHA has 550 organizational members and operates on an annual budget of $375,000 (1995). Staff size: 4. Executive Director: Denise B. Muha. Address: 1300 19th St. NW, Suite 410, Washington, DC 20036. Phone: (202) 785-8888. Fax: (202) 785-2008.

—*Caroline Nagel* ◀

► National Low Income Housing Coalition

The National Low Income Housing Coalition (NLIHC) is a nonprofit educational organization that conducts research and provides information on the U.S. affordable housing crisis to Congress, the executive branch, the media, and the public. NLIHC publishes a weekly newsletter, occasional "roundups" on community topics, research publications, and advocacy-related outreach materials. NLIHC is the nation's foremost organization in support of low-income housing and helps to foster the development of a growing number of state-based housing coalitions throughout the United States. The coalition was founded in 1975. A national conference is held annually in Washington, D.C., usually during early April. A publications list is available. Membership fees are scaled. President: Helen Dunlap. Address: 1012 14th St., NW, No. 600, Washington, DC 20005. Phone: (202) 662-1530. Fax: (202) 393-1973. E-mail: info@nlihc.org. (SEE ALSO: **Affordability; Homeownership**)

—*Patty Vrabel* ◄

► National Low Income Housing Preservation Commission

The National Low Income Housing Preservation Commission was established in 1987 as a nonprofit bipartisan group with a fourfold mission:

1. To determine the possible magnitude of loss of low-cost, federally subsidized housing and the causes of the loss
2. To examine alternative ways to minimize the loss of subsidized housing stock
3. To the extent possible, to recommend ways to offset the negative effect of any losses on low-income households
4. To analyze the cost of alternative solutions to the U.S. Treasury

The commission was created with the support of committees of both the U.S. House of Representatives and the Senate. It had financial support from the National Corporation of Housing Partnerships and the Ford Foundation. The commission's report was issued in 1988.

The specific focus of the commission's work was the 645,000 federally assisted rental housing units constructed since 1961 with subsidy contracts and restrictions that would begin to expire in the early 1990s. Just over half of those units had contracts with exposure to contract terminations between 1991 and 1995. The commission did not address other sources of low-cost housing loss such as the expiration of Section 8 contracts, demolition of public housing units, or the shrinkage of the stock of unsubsidized private low-rent housing.

At the end of 1995, approximately 115,000 units were in some preservation status. This means there had been a notice of intent filed by the owner. In most cases, no action had been taken on the expression of intent because of pending legislation. In 68% of the units, the owner had given an intention to extend low-income housing status implicitly on the condition that support for avoiding default is available. Owners of 31% indicated an intention to sell. Such sales are contingent on identifying purchasers who qualify under the government's rules.

Preservation is a national issue. In 13 states, at least 25 projects have a preservation status. Part of the current debate has to do with whether Congress will revise the preservation legislation to bring it in line with current realities of rents, budget restrictions, and other factors. Another part of the debate is whether political judgment favoring vouchers should be substituted for the original commitment of Congress to preserve units.

The commission sponsored the development of a computer model—the Preservation Analysis Model—to simulate the behavior of project owners under a variety of conditions. Through the model and other data, the commission assessed the behavior of nonprofit owners obligated to maintain subsidized housing over the 40-year life of the mortgage, private owners obligated for various reasons over a 40-year period, and owners eligible to prepay at the end of 20 years.

The model found that if the government failed to develop and support a preservation policy, the overwhelming majority of assisted units would be lost to the low-rent stock. This included 43% of at-risk units to default (because with a lack of subsidy, they would be unable to operate) and 38% of units whose owners would likely prepay and no longer make the units available as low-rent housing. In short, the commission found that the stock of assisted housing was in serious trouble and required additional subsidies or other assistance to remain viable. This set of needs would frame what has come to be known as the preservation policy.

The commission then sought to identify what actions the government might take, both to maintain owner commitment to the low-income housing units and to address the economic forces that owners faced at both ends of the market. To this end, the commission analyzed three types of government intervention and the associated costs. It reviewed remedies that would address the default possibilities, remedies that would discourage prepayment, and broad programmatic remedies designed to address problems facing this sector. Each of these remedies was framed so that tenants would benefit similarly under each.

On the basis of the findings from this analysis, the commission made a series of recommendations. The most significant was that the federal government commit to preserving the housing stock at risk. Although some exceptions were made with respect to units in very high cost areas, where cost-effective preservation may be difficult to demonstrate, and for units in poor condition, the commission was otherwise very strong in advocating the preservation of units.

The commission's priority on preserving the stock was not universally accepted. Many in Congress proposed that it is more cost-effective for the government to provide vouchers or housing certificates to those low-income families who might be displaced if units were in poor condition, too costly to preserve, or were taken from the affordable stock.

On the basis of the commission's recommendations and other events, Congress passed emergency legislation in 1988 (the Low Income Housing Preservation and Residential Homeownership Act [LIHPRHA]). Permanent legislation followed in 1990 to implement a commitment to preserving the nation's low-income housing. The act provided for financial relief and for the maintenance of affordable rents for low-income households. The act also preserved housing, authorized funds to meet the cost of urgent repairs, and made other adjustments needed to stabilize and preserve the federally assisted housing at risk of loss.

As budget reductions and political forces changed in Washington, the commitment to an expensive and comprehensive approach to preservation, and to preservation as a goal, began to wane. In 1995, the Department of Housing and Urban Development (HUD) began to back off from its support, and Congress moved toward amendments or revisions to the legislation and budget reductions that relied more heavily on voucher and certificate support for families affected by the loss of housing. Market conditions have also changed so that the initial estimates of loss from prepayment are now in question. The high costs of preservation are an issue as well, given HUD's other commitments. (SEE ALSO: **Affordability;** *Low-Income Housing Preservation and Residential Homeownership Act of 1990*)

—*Phillip L. Clay* ◄

► National Multi Housing Council

The National Multi Housing Council (NMHC) is a national association of corporations in the rental housing industry that promotes governmental policies favorable to the preservation of rental housing; campaigns to combat rent control legislation; disseminates information on environmental, tax, finance, and fair housing legislation; disseminates information on building codes and other issues affecting the rental housing industry; maintains a comprehensive library and information clearinghouse; and hosts quarterly and annual industry meetings. The council also distributes several publications to members, including a biweekly journal, *Washington Update,* which reports developments and trends in rental property legislation; four quarterly newsletters, *Environmental Update, Tax Update, Building Codes Update,* and *Market Trends;* two periodic newsletters, *Research Notes,* and *Technology Update;* and a bimonthly newsletter, *Seniors Housing Update.* In addition, the council publishes periodic reports and advocacy pieces on issues of concern to the industry. It operates on an annual budget of more than $4,000,000 and is funded from membership dues and investment income. It was founded in 1978 as the National Rental Housing Council. Membership: 765 corporations nationwide. Staff size: 27. President: Jonathan Kempner. Address: 1850 M Street, NW, Suite 540, Washington, DC 20036. Phone: (202) 974-2300. Fax: (202) 775-0112.

—*Caroline Nagel* ◄

► National Resource and Policy Center on Housing and Long Term Care

Established in 1994, the National Resource and Policy Center on Housing and Long Term Care conducts research on supportive housing options and linking of housing with services for older persons with low to moderate incomes, who are frail, live in rural areas, experience homelessness, are members of minority groups, or fit some combination of these characteristics. It also provides training and technical assistance to local and state governments, area agencies on aging, and state agencies for housing and aging. Publications include books, monographs, reports, and journal articles. The center has been funded by the U.S. Administration on Aging. Staff size: 14. Coordinator: Julie Overton. Address: Andrus Gerontology Center, University of Southern California, Los Angeles, CA. 90089-0191. Phone: (213) 740-1364. Fax: (213) 740-8241. E-mail: natresctr@ usc.edu. (SEE ALSO: **Elderly**)

—*Jon Pynoos* ◄

► National Resource Center on Homelessness and Mental Illness

The National Resource Center on Homelessness and Mental Illness (NRCHMI) focuses on the needs of homeless persons with serious mental illnesses. Founded in 1988, the center operates under contract to the Federal Center for Mental Health Services (CMHS) within the Department of Health and Human Services. It disseminates information about housing and support services to homeless persons with mental illnesses, their families, and service providers through workshops, publications, and an extensive database of published and unpublished materials on homelessness and mental illness. The center also provides on-site technical assistance to CMHS grantees on financing, development, management, and operation of housing and services for homeless persons with mental illnesses. NRCHMI publications include *Access,* a periodic bulletin featuring reports on research, funding, public and private initiatives, and program development in the field; a housing-related assistance resource list; a resource list of national organizations dealing with homelessness, housing, and mental illness; and various commissioned papers on service issues. All services, products, and publications are free of charge. It is funded by the CMHS. Staff size: 6. Project Director: Deborah L. Dennis. Address: 262 Delaware Avenue, Delmar, NY 12054. Phone: (800) 444-7415. Fax: (518) 439-7612. (SEE ALSO: **Homelessness;** *Psychiatric Disabilities, Housing of Persons with*)

—*Laurel Phoenix* ◄

► National Rural Housing Coalition

The National Rural Housing Coalition was established in 1969 to lobby Congress for programs and policies to improve low-income housing in rural areas. The coalition publishes *Legislative Update* and *FmHA Notes*. Membership: 250 individuals and organizations nationwide. Annual budget: $80,000 (1993). Contact person: Robert A. Rapoza. Address: 601 Pennsylvania Ave., NW, Suite 850, Washington, DC 20004. Phone: (202) 393-5229. Fax: (202) 393-3034. E-mail: NRHC@Rapoza.org. (SEE ALSO: *Rural Housing*)

—*Caroline Nagel* ◄

► National Shared Housing Resource Center

The National Shared Housing Resource Center (NSHRC) was established in 1981 by Gray Panther activist Maggie Kuhn to support shared-housing programs for the elderly, disabled, homeless, and single parents. NSHRC volunteers offer training and technical assistance for governmental and nonprofit shared-housing sponsors and aid them in the planning, financing, and management of their programs. The center also serves as an information clearinghouse, collecting and disseminating shared-housing information to government officials, housing professionals, sponsors, researchers, and consumers. In addition, NSHRC publishes a quarterly newsletter, which features research findings and legislative updates, and a variety of manuals on group residence programs, fund-raising and financing, marketing and education, zoning, and coalition building. The center is funded through donations and membership and publication fees. Membership: 300 nationwide. Address: 321 East 25th Street, Baltimore, MD 21218. Phone: (410) 235-4454. Fax: (410) 235-7141. (SEE ALSO: *Shared Group Housing*)

—*Caroline Nagel* ◄

► National Tenant Union

Tenant activism in the United States developed steadily during the 1970s and 1980s. Tenant groups existed in almost every city and many suburbs. These included building-level, citywide, and several statewide tenant groups. In 1975, tenant leaders founded the national magazine *Shelterforce* to report on and encourage tenant activism. The publication has helped to give the movement a sense of identity and coordination, and its editors (along with New Jersey and New York tenant leaders) took the first steps to form the National Tenant Union (NTU), established in 1980 to help coordinate the growing number of tenant activities.

The NTU played a key role in fighting attempts by organized landlord groups during the 1980s to pass national legislation that would prohibit cities with rent control from obtaining federal housing assistance. It was dissolved by 1991 because of the difficulty of raising money to support a staff. (SEE ALSO: *Tenant Organizing in the United States, History of*)

—*John D. Atlas* ◄

► Native Americans

There are more than 550 federally recognized Native American reservations, communities, and Alaskan Native villages throughout the United States. Each community has its own political, economic, and social institutions, shaped by the environment and the history of the tribe. The diversity of tribal communities poses unique opportunities and challenges to the provision of housing services to this population. Developing and designing housing policies and programs that meet local needs and resources is often a difficult task. This entry presents an overview of the housing conditions of Native American households, federal housing policy and programs, and the issues affecting the provision of housing for Native Americans in tribal areas. It will not touch on issues affecting Native Americans in urban areas.

To fully appreciate the housing situation of Native American reservation populations, a general understanding of the political and legal status of tribes and their relationship to the federal government is useful. A thorough understanding of tribal housing issues requires knowledge of the larger policy context of Native American communities. Therefore, a brief historical overview of federal Indian policy is presented first.

The Political Status of Tribal Governments and Federal Indian Policy

As "domestic dependent nations," tribes are not municipal arms of state or federal governments but are independent sovereigns with the authority to exercise control over their affairs. As such, tribes hold a unique political and legal status within the United States. The federal government has recognized tribal sovereignty through treaties, in the U.S. Constitution, and in many congressional acts, and tribal sovereignty has been upheld in the U.S. Supreme Court (*United States v. Wheeler*, 1978, 435 US 313, 98 S. Ct 1079).

As sovereign nations, tribal governments have the authority to control their lands and resources and to regulate their internal affairs. This authority comes from the inherent rights of sovereignty and is not granted by Congress. Included among the powers that tribal governments can exercise are the abilities to define membership, provide tribal services, make laws, develop codes and adjudicate legal disputes, and regulate the physical environment and political, social, and economic activities on tribal lands. As such, the provision of housing is a responsibility of tribal governments.

The ability of tribes to exercise their sovereign powers, however, is limited by the actions of Congress. Congress holds absolute (plenary) authority to regulate tribal affairs. Over the past two centuries, Congress has used its authority to enact policy that has both reaffirmed and diminished tribal sovereignty. There are five major periods in federal Indian policy.

Removal and the Establishment of Reservations (1830-1880)

During this period, many tribal communities were uprooted and relocated to "reservations"—small land areas set aside for tribes, sometimes located hundreds of miles from their original homes. By 1891, a majority of the Indian population was confined to reservations, villages, or rancherias.

Assimilation and the Reservation Period (1880-1934)

Through the administration and regulation of reservation areas, Congress initiated a series of policies designed to incorporate tribal people into mainstream society. New institutions, such as the Indian agency, the Indian police, the Court of Indian Offenses, and Indian boarding schools, were established to replace traditional tribal structures. The administrative responsibilities of the Bureau of Indian Affairs (BIA) increased greatly during this time.

During this period, the federal government also initiated the breakup of tribal land holdings. Under the General Allotment Act (or the Dawes) of 1887 tribal lands were divided into individual allotments and distributed to tribal members. Remaining lands were sold to non-Indians. By 1934, more than 90 million acres of tribal land—two-thirds of tribal land holdings of 1887—had been lost. Sixty million acres of these lands were sold as "surplus."

Indian New Deal (1934-1953)

The passage of the Indian Reorganization Act (IRA) in 1934 signaled a major reversal of federal Indian policy. Through this act and subsequent legislation, the federal government attempted to strengthen the role of tribal institutions that they had previously undermined. Under the IRA, tribes were encouraged to organize politically and establish tribal courts. Funds were made available to purchase land, provide educational loans, and establish a revolving credit fund. Many of these benefits were short-lived.

Termination and Relocation (1953-1970)

During the 1950s, a number of measures were passed to reduce the status of tribal governments and extinguish the federal relationship with tribes. In 1953, the House of Representatives passed Resolution 108, whose purpose was to remove federal responsibility to tribal communities, thereby terminating their legal and political status. More than 100 tribes were targeted for termination. Many tribes that were legally terminated have been able to reverse this action.

During this period, Congress also sought to encourage Native Americans to move off tribal lands and relocate to major cities by providing employment and training opportunities for those willing to move. State governments were granted the authority to exercise criminal and civil jurisdiction over reservations for the first time.

Self-Determination (1970-)

In 1970, President Nixon articulated a new federal policy toward Native Americans. The purpose of "self-determination without termination" was to "strengthen the Indian's sense of autonomy without threatening his sense of community." Over the past 25 years, a great deal of federal legislation has been passed to affirm tribal rights. For instance, the Indian Self-Determination and Educational Assistance Act of 1975 and the Tribal Self-Governance Demonstration Project of 1988 have allowed tribes to regain control over many services provided by federal agencies.

The unique relationship between tribes and the federal government means that congressional actions have an enormous impact on tribal institutions. With each legislative session, federal Indian policy supports or challenges tribal sovereignty and the ability of tribes to provide community services. The effects of past federal policy are equally as important. One of the most damaging has been the legacy of the 19th-century allotment policy on tribal land ownership and status. As a result of the breakup of tribal lands, land within a tribal area may be owned by an individual, by the tribe, or by federal, state, or local government, creating a highly fragmented pattern of landownership. In addition, land may be held in trust or restricted status or may be fee simple. Each type of property ownership is subject to different laws and jurisdictions. Trust lands, for instance, cannot be alienated or encumbered without approval from the BIA. These complex land arrangements hamper a tribe's ability to conduct economic activities and provide infrastructure and support services. This has a significant impact on housing development. Issues surrounding land status and ownership affect site acquisition and housing development. Restrictions on land have also meant that conventional financial institutions have been reluctant to provide capital for projects in tribal communities.

Housing in Tribal Areas

In 1996, the Department of Housing and Urban Development (HUD) published a major statistical study of the housing conditions of Native American households. The *Assessment of American Indian Housing Needs and Programs,* the most comprehensive national report on Indian housing yet completed, attempted to quantify what is often visibly evident—that for a large percentage of Native Americans, particularly those living in tribal areas, housing conditions remain substandard and lag behind the rest of the United States.

Using standard measures of housing quality, quantity, and price, the report found that American Indian and Alaskan Native (AIAN) households face a larger share of housing problems than the national population. Not only are they more likely to live in substandard housing, they face severe overcrowding and a lack of affordable housing opportunities.

Analyzing data from the 1990 census, the study found that 40% of all AIAN households had one or more housing problems, compared with 27% of households nationally. Not surprisingly, low-income households had more severe housing problems. Whereas 3.8% of the national population lacked plumbing or kitchen facilities, the comparable figure for AIAN households was 5.5%, indicative of severe physical housing deficiencies. Of all AIANs households, 12% were living in overcrowded units compared with 2.7% of the national population. These figures imply a serious shortage of decent housing 29% of all AIAN households paid more than 30% of their income for housing, compared with 24% for the national population, suggesting a lack of affordable housing opportunities. Although 48% of all

AIAN households have incomes of 80% or more than the national median, they have much lower rates of homeownership than comparable populations.

The housing environment for American Indians and Alaskan Natives varies greatly by geographical area. AIANs living in tribal areas suffer most from housing shortages and physical inadequacies. About 14% of households lack plumbing or kitchen facilities, and 21% of AIAN households living in tribal areas are overcrowded. Affordability is a greater problem for AIANs living in metropolitan areas and areas just outside reservation lands, as are homelessness and access to housing assistance.

For AIAN households in tribal areas, housing problems are more severe in Alaska and the Arizona/New Mexico region. For instance, the HUD survey found that 71% of AIAN households in Alaskan villages experienced one or more housing problems, compared with 30% in Oklahoma.

The type of housing problem also varies by geographical region. Affordability is a much greater problem in the South-Central region than in other areas, whereas overcrowding is considerably higher in Arizona/New Mexico, affecting about 45% of all households.

Federal Housing Assistance: Policy and Programs

Approximately one-fourth of American Indians and Alaskan Native households in tribal communities live in publicly assisted housing units. Federal housing assistance has been a major component in the development of Indian housing for the past 30 years and has shaped how housing services are provided.

In contrast to other policy areas, such as education, health, and economic development, there has been very little federal legislative policy regarding Indian housing. Rather, tribes have been allowed to participate in existing mainstream publicly assisted housing programs. In 1961, tribal governments first became eligible for federal housing assistance under the 1937 U.S. Housing Act, through an administrative rather than a legislative decision. To access these programs, tribes were required, through tribal ordinances, to establish Indian housing authorities (IHAs)—agencies responsible for the administration of public housing funds and management of federally assisted housing units. The first IHA was established in Pine Ridge, South Dakota, in 1961. In 1998, there were more than 190 IHAs, serving more than 450 tribes.

Indian housing authorities are equivalent to public housing authorities (PHAs) operating in many cities. Although created under tribal law, an IHA is as an independent corporation with a separate board of directors and corporate bylaws. Until 1995, when the regulations governing them were substantially revised, IHAs operated largely under the same regulations as their urban and rural counterparts. In many tribal areas, the lack of alternative housing stock has meant that IHAs have been the primary housing provider for reservation residents.

As with public housing, in 1965 the administrative oversight for IHA-managed housing was placed under HUD. During the early 1980s, a separate Office of Indian Housing (OIH) was created with six regional offices. Currently, the Office of Native American Programs (ONAP), within the Office of Public and Indian Housing (PIH), and six regional field offices are responsible for the oversight of HUD-funded housing units.

Although a range of federal housing assistance programs are available, tribes and Indian housing authorities have relied largely on HUD programs. Five types of housing programs have been available.

Indian Housing Programs

Through a competitive grant process, Indian housing authorities have received funding for the acquisition and construction of rental or homeownership units. Two major programs were available (technically, the programs no longer exist): the Rental Program, and the Mutual Help Homeownership Opportunity Program. Similar to the low-rent program in public housing, IHAs acquire or construct rental units and receive operating subsidy for management of the units. Mutual Help is a program unique to Indian housing. It is a lease-purchase program, whereby the tenant makes a monthly payment, over a period of up to 25 years, toward the purchase of the unit. Funding for the modernization and rehabilitation of housing units has also been available through the Comprehensive Improvement Assistance Program (CIAP) and the Comprehensive Grant Program (CGP).

In tribal areas, there are approximately 80,000 HUD-funded units in management, with an additional 10,000 under construction. In 1996, these programs were funded at a little more than $400 million dollars, of which 41% was spent for new housing development, 39% for modernization programs, and 20% for operating subsidy.

Tenant-Based Assistance

Indian housing authorities can access tenant-based rental assistance, such as Section 8 certificate and voucher programs.

Resident Support Services

HUD offers a number of competitive grant programs that support resident services. The Public Housing Drug Elimination Program, Tenant Opportunity Program, and Family Investment Centers are designed to assist public housing residents to become more self-sufficient.

Tribal Block Grants

Both the Community Development Block Grant Program and the HOME Program have statutorily authorized set-asides for tribal governments. These programs provide funding for housing-related activities, including land acquisition, infrastructure development, and economic development activities.

Private Financing Assistance Programs

In an effort to encourage private sector investment, a number of housing finance programs have been targeted to tribal communities. Section 184 offers guarantees for loans made to Native American individuals, IHAs, or tribes on restricted or trust lands. Until recently, the Federal Housing Administration (FHA) offered a mortgage insurance program, Section 248, for single-family properties on trust lands.

Although HUD assistance predominates, other federal agencies offer additional housing resources. A limited number of housing grants, under the BIA Housing Improvement Program (HIP), are available for housing rehabilitation to very low income households. The Rural Housing Services, within the Department of Agriculture provides direct and guaranteed loans targeted toward low-income rural populations. The Department of Veterans Affairs has also developed a special pilot program to provide housing loans for veterans living on trust lands. However, use of these federal resources has been limited. The low participation rates for these programs has been attributed to several factors, including unworkable or unacceptable regulatory requirements, lack of awareness regarding the availability of assistance, and lack of outreach by federal agencies.

Impediments to the Provision of Housing in Tribal Areas

The reliance on federal assistance and its inability to meet the housing demands of tribal communities have raised the question as to what barriers exist in providing housing to tribal residents. In 1989, Congress established the National Commission on American Indian, Alaska Native, and Native Hawaiian Housing. The purpose of the commission, which began its work in 1991, was to explore the barriers to the development of decent, safe, and affordable tribal housing. Through field hearings, interviews, and meetings with tribal leaders, housing program participants, federal officials, and many others, in its 1992 report, the commission identified a number of impediments. Among them were (a) a chronic shortage of federal housing funding and lack of political support, on the part of both Congress and tribes, for housing appropriations and legislation; (b) the limited management capabilities of tribal housing agencies because of high turnover among staff and board members (the commission recommended additional technical assistance and training to address this issue); (c) regulatory constraints caused by burdensome and inappropriate housing regulations within Indian programs; and (d) a lack of cultural sensitivity to tribal housing traditions and a lack of tribal input in housing design and construction. Many of the issues raised by the commission were subsequently addressed in the 1995 revised HUD program regulations that sought to increase tribal participation and provide additional flexibility for federal Indian housing programs.

Other barriers cited by the commission have been much less tractable. Two issues in particular stand out. The first is the lack of a comprehensive and integrated approach to housing, and the second is the lack of access to conventional housing financing.

Development of housing is a costly and complicated process. Not only does it involve construction of the housing unit, but it is intimately tied to land and site concerns, as well as to infrastructure services, such as utilities and roads. In tribal communities, these factors are particularly complex, confounded by competing jurisdictions and regulations. In addition to HUD, two other federal agencies are important in the provision of tribal housing. The BIA, as part of its trust responsibilities, has administrative oversight of all tribal and individual trust property and resources. It is responsible for issuing title status reports for land-ownership verification on individual trust properties. In this capacity, the BIA must review and approve all land leases and sites involving trust lands. The BIA is also responsible for road and highway maintenance and construction. The Indian Health Service, within the Department of Health and Human Services, is another key agency because it has oversight of the development and maintenance of sanitation facilities, and construction of water and sewer lines—infrastructure necessary for housing developments. In addition, tribes may have their own laws and regulations governing land and economic infrastructure development. Integration and coordination of services across agencies is essential to the successful completion of housing projects.

The issue of access to conventional housing finance has also been critical. The inability to attract private investment confronts both tribal governments and residents. Anecdotal evidence suggests that the lack of access may, in part, be based on lender's perceptions that lending in tribal areas presents too many risks. Land status, tribal sovereignty, and legal jurisdiction issues create an uncertain investment environment. Individual borrowers, because of low income levels, lack of credit histories, and seasonal or unstable employment, are often seen as credit risks. Remoteness, high construction costs, the lack of a housing appreciation market, and infrastructure constraints, as well as small loan volume, raise the risks of lending. As a result, financial institutions are unwilling to invest in tribal projects or make available mortgage loans. Furthermore, a lack of knowledge about private financing opportunities on the part of tribal housing providers has also limited private sector participation.

The Native American Housing Assistance and Self-Determination Act of 1996

In October 1996, President Clinton signed the Native American Housing Assistance and Self-Determination Act (NAHASDA). It went into effect on October 1, 1997. This legislation is historic for Indian housing in several respects. First, the provision of "affordable homes in safe and healthy environments" for Native Americans living in tribal areas is recognized as a trust responsibility of the federal government and an "essential element in the special role of the U.S. in helping tribes and their members to improve their housing conditions and socioeconomic status." This recognition finally defined housing as an important policy issue affecting tribal communities. Second, the act repeals the 1937 U.S. Housing Act with respect to Indian housing and terminates tribal participation from many existing HUD programs. Tribes are no longer eligible for programs such as Rental and Mutual Help, McKinney Homeless Assistance Act, Youthbuild, and HOME. In consequence, Indian housing is no longer linked to public housing programs.

Third, the act allocates funding directly to the tribes, or a designated agency, as a block grant rather than as distinct program funding channeled to Indian housing authorities. Federal funds can be used for development, rehabilitation, and acquisition of housing; support services; and model housing activities. In addition, tribes can restructure how they deliver housing services to residents, which allows them to integrate HUD funding with other financial resources. This provides tribes an opportunity to design housing services that better meet local housing needs. Fourth, the act encourages strategic planning by requiring tribes to submit one- and five-year housing plans. The development of a housing plan provides tribes the opportunity to coordinate housing and related services comprehensively. Finally, the act contains provisions to encourage the leveraging of public resources in an attempt to bring more private financing into tribal areas.

In implementing the legislation, HUD was required to develop regulations for the new Indian Housing Block Grant (IHBG) program through a negotiated rule-making process with tribal governments. These negotiations took place in 1997.

Conclusion

The effective provision of housing and the delivery of services requires many resources, including land, capital, and infrastructure. It also requires a supportive economic, legal, and political environment. In certain tribal communities, these elements may be fragmented. As a result, tribal residents continue to face serious housing problems. For them, physical housing deficiencies, overcrowded conditions, and a lack of affordable housing opportunities are among the most severe in the nation.

The passage of the NAHASDA provided a new framework for Indian housing. Most important, the Act strengthened the role of tribal governments in the provision of housing and granted them greater control as local decision makers. It thus provides an opportunity for better integration of housing with other tribal activities and services.

—*Juliet King*

Further Reading:

Cohen, Felix S. 1982. *Handbook of Federal Indian Law.* Rev. ed. Charlottesville, VA: Michie.

Cooley, Martha, ed. 1992. *Building the Future: A Blueprint for Change.* Final Report. Washington, DC: National Housing Commission on American Indian, Alaska Native and Native Hawaiian Housing.

Housing Assistance Council. 1986. *Indian Housing in the U.S.: A History.* Washington, DC: Author.

———. 1996. *Case Studies of Lending in Indian Country.* Washington, DC: Author.

National American Indian Housing Council. 1997. *Expanding Home-ownership Opportunities in Native American Communities: The Role of the Private Sector.* Washington, DC: Author.

O'Brien, Sharon. 1989. *American Indian Tribal Governments.* Norman: University of Oklahoma Press.

Prucha, Francis. 1990. *Documents of the United States Indian Policy.* Lincoln: University of Nebraska Press.

U.S. Department of Housing and Urban Development, Office of Indian Housing. 1993. *The Indian Housing Program: History and Current Status of Indian Housing in the United States.* Washington, DC: Author.

U.S. Department of Housing and Urban Development, Office of Policy, Development and Research. 1996. *Assessment of American Indian Housing Needs and Programs: Final Report.* Washington, DC: Author. ◄

► Negative Amortization

With the development of alternative mortgage instruments (AMIs), it is possible (in some cases, required) to have negative amortization, or the increasing of the outstanding balance of the mortgage after making debt service payments. For example, whenever the debt service payment is less than the interest owed for the period, the failure to cover the interest obligation results not only in no principal payment but, in fact, adds to the outstanding balance of the mortgage. This "negative amortization" has important implications for lenders, borrowers, housing markets, and the choice of mortgage instruments.

In the typical mortgage repayment schedule, each debt service payment reduces the outstanding balance of the mortgage. This is because under the fixed-rate mortgage (FRM), the debt service payment exceeds the amount of interest owed each period. The difference between the debt service payment and the interest payment is called *principal* or *amortization*. For an FRM, there can only be positive amortization.

Certain instruments promise negative amortization; others permit negative amortization under certain conditions. For example, the use of the graduated-payment mortgage (GPM) *requires* that negative amortization occur for a fixed period of time. For an adjustable-rate mortgage (ARM), negative amortization can take place under some conditions.

For lenders, negative amortization increases the riskiness of the loan. This "negative sinking fund" means that not only is the lender's money not being systematically repaid but that the lender is expected to make additional loans each period. Furthermore, depending on the uncertainty of the repayment schedule, it may be difficult to measure the risk for the lender. It is not surprising that lenders view vehicles with negative amortization very cautiously.

For borrowers, a mortgage instrument with negative amortization implicitly means that the future outstanding balance of the mortgage will be larger than the current mortgage balance. Uncertainty about the future affects the decisions of borrowers as well. Borrowers might like the current benefits of reduced payments that are known with certainty (i.e., are calculable). However, with uncertainty about the borrower's future needs, negative amortization increases the debt claim for the borrower in the future, and the borrower cannot calculate the burden of the claim for the future.

For housing markets, instruments with negative amortization might expand participants' choices in the mortgage market. However, some instruments (such as GPMs) have been unsuccessful mechanisms in most mortgage markets. At the same time, the existence of instruments with negative amortization is not a bad thing per se. Negative amortization is a risky proposition, but this is insufficient to conclude that it is welfare reducing.

Regarding the choice of mortgage instruments, negative amortization may make certain types of mortgages unattractive, especially to lenders and, increasingly in the future, the secondary mortgage market. In cases in which negative amortization results in serious pricing problems for lenders, these mortgages will probably not survive. In other cases, negative amortization might result under specialized conditions and might not be as harmful as in other instruments. Over time, there may be new mortgage instruments that avoid some of the problems associated with negative amortization and achieve the same benefits for lenders and borrowers as the current instruments. In this manner, the best instruments are likely to prevail because competition will remove inferior tools and replace them with superior ones. (SEE ALSO: **Mortgage Finance**)

—*Austin J. Jaffe*

Further Reading

Blankenship, Frank J. 1989. *The Prentice Hall Real Estate Investor's Encyclopedia.* Englewood Cliffs, NJ: Prentice Hall.

Fabozzi, Frank J. and Franco Modigliani. 1992. *Mortgage and Mortgage-Backed Securities Markets.* Boston: Harvard Business School Press.

Jacobus, Charles J. 1996. *Real Estate Principles.* Upper Saddle River, NJ: Prentice Hall.

McKenzie, Dennis J. 1996. *Essentials of Estate Economics.* Upper Saddle River, NJ: Prentice Hall.

Peiser, Richard B. 1992. *Professional Real Estate Development: The ULI Guide to the Business.* Washington, DC: Urban Land Institute.

Shim, Jae K. 1996. *Dictionary of Real Estate.* New York: John Wiley.

Tosh, Dennis S. 1990. *Handbook of Real Estate Terms.* Englewood Cliffs, NJ: Prentice Hall. ◀

▶ Neighborhood Housing Services of America

Neighborhood Housing Services of America (NHSA) purchases below-market-rate, "high-risk" loans originated by local Neighborhood Housing Services (NHS) organizations. NHSA creates a secondary market for home purchase mortgages, home rehabilitation and multifamily loans, and real estate development financing made available through NHS revolving loan funds and other lending sponsored by the Neighborhood Reinvestment Corporation (NRC). Using capital from insurance companies, financial institutions, pension funds, and foundations, NHSA has invested $236 million to provide liquidity to NHS lenders who otherwise would be underserved or ignored by traditional institutional investors.

The Urban Reinvestment Task Force, formed as the predecessor to NRC by the U.S. Department of Housing and Urban Development (HUD) and the Federal Home Loan Bank Board, capitalized NHSA in 1974 with an initial $250,000 HUD grant as part of a national NHS demonstration program of neighborhood preservation and low-cost home financing. Since then, mortgage-backed securities issued by NHSA have been placed through purchase agreements with 21 different social investors, totaling $244 million. Several large insurance companies are among the leading financial supporters of NHSA, including Prudential, Allstate, USAA, Equitable, Metropolitan Life, Aetna, State Farm, and Nationwide.

NHSA also provides technical assistance to NRC affiliates and has created special credit programs with private mortgage insurers and lending institutions, Fannie Mae,

Freddie Mac, the National Cooperative Bank, the Federal Home Loan Bank system, and state and local governments to expand NHS lending capacity. Since 1990, NHSA has offered assistance and support to the Social Compact, a network of 200 of the nation's leading financial services institutions formed to publicize and raise awareness about local community reinvestment programs. In 1993, NHSA committed $200 million to help finance the NRC-NeighborWorks® Campaign for Home Ownership, a five-year, $650 million national initiative to increase the number of low-income homeowners. Three years later, NHSA launched a five-year, $250 million capital campaign to strengthen its financial reserves and expand its investment capacity. NHSA receives operating and capital grants from NRC and several foundations.

According to a 1994 estimate, the average size of NHSA home purchase loans was $66,000, compared with $97,000 to $98,000 for conventional mortgages sold to Fannie Mae (formerly, the Federal National Mortgage Association) and the Federal Home Loan Mortgage Corporation (Freddie Mac). The average size of NHSA home rehabilitation loans was $15,000. The willingness of NHSA to purchase these loans permits greater underwriting flexibility for NHS revolving loan funds. NHSA is instrumental to the success and longevity of the NHS concept by replenishing the supply of capital available for making targeted local investments, and forging secondary market relationships with national private sector investors. Contact: Mary Lee Widener, President. Address: 1970 Broadway, Suite 470, Oakland, CA 94612. Phone: (510) 832-5542. (SEE ALSO: *Fannie Mae; Federal Home Loan Bank System; Federal Home Loan Mortgage Corporation;* **Homeownership;** *Lending Institutions; Neighborhood Reinvestment Corporation; Redlining; Secondary Mortgage Market*)

—*John T. Metzger*

Further Reading

Clay, Philip L. 1981. *Neighborhood Partnerships in Action: An Assessment of the Neighborhood Housing Services Program and Other Selected Programs of Neighborhood Reinvestment.* Washington, DC: Neighborhood Reinvestment Corporation.

NeighborWorks® Network and Neighborhood Reinvestment Corporation. 1996. *1996 Annual Report.* Washington, DC: Neighborhood Reinvestment Corporation.

Urban Systems Research and Engineering, Inc. 1981. *Evaluation of the Urban Reinvestment Task Force.* Washington, DC: U.S. Department of Housing and Urban Development. ◀

▶ Neighborhood Reinvestment Corporation

The Neighborhood Reinvestment Corporation (NRC) is a national nonprofit agency created by Congress in 1978 to sponsor and support Neighborhood Housing Services (NHS) organizations and related efforts in large and medium-sized cities across the United States. This network of organizations, known as NeighborWorks®, now totals 171 groups (and 1,500 staff members) working in 43 states, the District of Columbia, and Puerto Rico. The low-

and moderate-income neighborhoods served by these groups contain 2% of the total housing stock in the United States and 4.8 million residents, of whom half are racial minorities and one-fifth are Hispanic. From 1991 to 1996, NRC affiliates produced $1.6 billion in public-private investment, 45% by financial institutions.

The first NHS was formed on Pittsburgh's Central North Side in 1968 by local residents, foundations, government officials, and 13 banks and savings institutions. The NHS program consisted of a "high-risk," below-market-rate, pooled revolving loan fund for low-income homeownership and home repairs; financial counseling and technical assistance; and municipal code enforcement. NHS succeeded in stabilizing and revitalizing the Central North Side after the area had been "redlined," and the Federal Home Loan Bank Board (FHLBB) decided to replicate the program as a model for encouraging investment by savings and loans in older urban neighborhoods. NHS expanded to Washington D.C., Cincinnati, Oakland, and Dallas, and in 1974, the U.S. Department of Housing and Urban Development (HUD) made a $3 million grant to launch a national demonstration program with FHLBB through the Urban Reinvestment Task Force. The task force created Neighborhood Housing Services of America to purchase NHS loans and funded the Apartment Improvement Program in Yonkers, New York, as a model strategy for rehabilitating multifamily housing.

NRC replaced the task force in 1978, was awarded an annual federal grant, and has since attracted commercial banks and insurance companies to invest in local NHS organizations. Its board of directors includes all of the federal financial regulators and the secretary of HUD. NRC distributes grants to NeighborWorks® groups for loan funds, program administration, and various community services and assists them in raising money from local government, lenders, and foundations. NRC also provides technical assistance, staff training, and planning and evaluation support. In many large cities (such as Chicago), NHS has expanded into multiple neighborhoods. NRC has also advanced the European concept of mutual housing in the United States, using Section 316(a) of the Housing Act of 1980 to start mutual housing associations (MHAs) in Baltimore, New York, and 10 other areas. These MHAs have developed nearly 5,000 permanently affordable units of community-based, resident-managed multifamily housing.

NRC enjoys strong bipartisan political support. Its annual federal appropriation grew from $12 million in 1980 to $19 million in 1988, when most federal housing and community development programs suffered cutbacks. It is now funded at four times the original amount and in 1992 received an additional grant of $5 million to capitalize local homeownership revolving loan funds and equity investments in rental and mutual housing. During the 1990s, NRC has used new federal housing initiatives such as the HOME program and the Affordable Housing Program of the Federal Home Loan Bank system, along with continued local spending through the federal Community Development Block Grant program, to expand its activities. NRC uses public funds to leverage private investment and restore the tax base in the targeted communities. In 1993, NRC and NeighborWorks® launched a five-year national Campaign for Home Ownership that has received support from several large financial institutions and HUD. By mid-1997, the campaign had exceeded its goals, generating $858 million in housing investment and 12,500 new homeowners. NRC has also convened a National Insurance Task Force to develop collaborative partnerships between insurance companies, industry regulators, and NeighborWorks® groups to expand the availability of property and casualty insurance. In conclusion, NRC's work has increased the ranks of low-income and minority homebuyers, spawned new innovations in urban housing policy, and, in many cities, helped to reverse the decline of poor and working-class neighborhoods by facilitating private sector reinvestment. Contact: George Knight, Executive Director. Address: 1325 G Street NW, Suite 800, Washington, DC 20005. Phone: (202) 376-2400. Fax: (202) 376-2600. (SEE ALSO: *Code Enforcement; Federal Home Loan Bank System;* **Homeownership;** *Local Government; Mutual Housing; Neighborhood Housing Services of America; Redlining*)

—*John T. Metzger*

Further Reading

ACTION-Housing, Inc. 1975. *The Neighborhood Housing Services Model: A Progress Assessment of the Related Activities of the Urban Reinvestment Task Force.* Washington, DC: U.S. Department of Housing and Urban Development.

Ahlbrandt, Roger S., Jr. and Paul C. Brophy. 1975. *Neighborhood Revitalization: Theory and Practice.* Lexington, MA: D. C. Heath.

Bratt, Rachel G. 1990. *Neighborhood Reinvestment Corporation-Sponsored Mutual Housing Associations: Experiences in Baltimore and New York.* Washington, DC: Neighborhood Reinvestment Corporation.

Clay, Philip L. 1981. *Neighborhood Partnerships in Action: An Assessment of the Neighborhood Housing Services Program and Other Selected Programs of Neighborhood Reinvestment.* Washington, DC: Neighborhood Reinvestment Corporation.

Knight, George. 1997. "What's Working: Insurance as a Link to Neighborhood Revitalization." In *Insurance Redlining: Disinvestment, Reinvestment, and the Evolving Role of Financial Institutions,* edited by Gregory D. Squires. Washington, DC: Urban Institute Press.

NeighborWorks® Network and Neighborhood Reinvestment Corporation. 1996. *1996 Annual Report.* Washington, DC: Neighborhood Reinvestment Corporation.

Research Triangle Institute. 1991. *Evaluation of Neighborhood Housing Services: Final Report.* Washington, DC: Neighborhood Reinvestment Corporation.

Urban Systems Research and Engineering, Inc. 1981. *Evaluation of the Urban Reinvestment Task Force.* Washington, DC: U.S. Department of Housing and Urban Development. ◄

► Netherlands Journal of Housing and the Built Environment

The *Netherlands Journal of Housing and the Built Environment* is a quarterly, English-language, refereed international journal sponsored by the Netherlands Institute for

Physical Planning and Housing (NIROV), the Faculty of Architecture, Urban Planning and Housing (Delft University of Technology), the Netherlands Graduate School for Housing and Urban Research (NETHUR), and OTB Research Institute for Policy Sciences and Technology (Delft University of Technology). It focuses on the interests of researchers, policymakers, architects, and others in the fields of housing and urban studies. Articles cover a broad range of topics, including cooperative, public, low-income, and seniors' housing; real estate and building markets; and urban development. It was originally published in 1985 as the *Netherlands Journal of Housing and Environmental Research* and is circulated internationally. Price: About $70 (special rates available). Publisher: Delft University Press, Stevinweg 1, NL 2628 CN Delft. Phone: 31 15 278 32 54. Fax: 31 15 278 16 61. Editorial Office: OTB Research Institute of Policy Sciences and Technology, P.O. Box 5030, NL—2600 GA Delft. Phone: 31 15 278 35 23. Fax: 31 15 278 44 22. E-mail: Haffner@otb.tudelft.nl.

—*Caroline Nagel* ◀

▶ New Towns

Although the history of self-contained new settlements dates to the fortified villages of the early Roman Empire, the modern idea of the new town is generally traced to the invention of the garden city in England as an alternative to urban congestion and unhealthy conditions produced by rapid industrialization in the late 19th century. In its most complete form, the new town is an independent urban complex, planned in advance, built on virgin soil, and containing the necessary residential, employment, and cultural opportunities to provide a full measure of economic and social life. In practice, however, the more prevalent form of new town is the satellite city composed primarily of residential areas dependent for employment on a nearby urban center to which it is frequently linked by rail. As such, it shares certain characteristics with the broader category of planned suburbs, notably with regard to street networks and open space development. But the new town is distinguished by its identity as an independent social and political unit whose civic infrastructure extends beyond the residential neighborhood and school district.

Planned communities have been constructed for a variety of purposes other than the goal of orderly urban growth and regional development. Utopian societies such as the Shakers built new communities to support sectarian religious or philosophical beliefs. Industrialists such as railroad car manufacturer George Pullman and soap magnate W. H. Lever built company towns to attract and control a stable workforce. Towns built for political and military purposes include the *bantustans* and African townships built to enforce racial segregation in South Africa under the Urban Areas Act of 1923 and the Group Areas Act of 1950, strategic hamlets built by the United States to contain and neutralize the native population during the Vietnam War, and new settlements built in Israel to establish the new nation in the 1950s and to consolidate the annexed West Bank territory in the 1980s. The resumption of this practice in the 1990s has exacerbated tensions between the Israeli and Palestinian peoples. New capital cities such as Brasilia (Brazil), by Lucio Costa and Oscar Niemeyer, and Chandigargh (E. Punjab), by Le Corbusier, celebrate political power with imposing new construction.

The new town is the most complex form of planned community. Its intention is to rationalize land use, transportation, and building location by incorporating residential, industrial, commercial, cultural, and recreational facilities in a single new development. The construction of a new town is a lengthy and expensive undertaking. It requires "patient money," sponsors who are willing and able to make a substantial up-front investment in land acquisition and infrastructure (roads, utilities, water systems, etc.) and to wait years before the investment shows a return. For this reason, new towns are more common in countries with strong traditions of direct government involvement in development, notably Western Europe, Scandinavia, and the state socialist economies of the former Soviet bloc, than in economies such as the United States, dominated by the ideology of private enterprise.

The garden city idea, progenitor of the modern new town, was first put forward by Ebenezer Howard, a British court stenographer and "social inventor," in his book *Tomorrow: A Peaceful Path to Real Reform* (1898), republished in 1902 as *Garden Cities of Tomorrow*. Howard's plan was not to reform the 19th-century industrial city but to replace it with a polycentric regional network of "social cities" consisting of a ring of garden cities of 32,000 people on 6,000 acres of land surrounding a central city of 58,000, all linked by intermunicipal railways. The garden city was designed for living and working, combining the economic and cultural opportunities of the city with the salubrious natural environment of the countryside. The designation of more than 80% of the land as a permanent greenbelt would both restrict the growth of the city and ensure a natural preserve for agricultural and recreational purposes.

Howard's idea went beyond the physical order of the region to its social structure as well. Just as the physical garden city would synthesize town and country, the social contract would blend individual self-interest and collective enterprise under the banner of "freedom and cooperation." The vehicle for this would be municipal ownership of the land, which would be leased to individuals or companies for private construction of dwellings or factories. The purpose was to capture the increase in value of the urbanized land for the collective benefit of the town rather than for private gain.

Howard himself organized the limited-dividend corporation that built the first garden city at Letchworth in 1903, based on plans by architects Raymond Unwin and Barry Parker. After a slow start, the town eventually attracted a number of industries. Living accommodation for a wide range of income groups was made possible through a series of Cheap Cottage exhibitions, cooperative apartments, and modest row houses in addition to privately built single-family homes. Howard resided in Letchworth until moving to the second garden city at Welwyn, which he initiated in 1921.

Figure 19. Town Center in Columbia, Maryland
Photograph by Tony Schuman.

His work engendered a spate of garden city experiments in lands as distant as Japan and Australia, with the most significant activity in France and Germany. Henri Sellier, director of social housing for the Department of the Seine, planned a series of *cités jardins* on the outskirts of Paris between 1916 and 1939. These districts, such as Suresnes, took the form of garden suburbs, comprising primarily residential structures. In Germany, architects Ernst May and Martin Wagner built a series of satellite settlements in and around Frankfurt and Berlin, respectively. These developments, composed of attached single-family row houses and flats, were notable for their extensive landscaping and modern architecture. Built by a combination of municipal authorities and cooperative building societies, these housing estates, known as *Siedlungen,* benefited from an earlier far-sighted program of municipal land purchases.

In Great Britain, the garden city program was institutionalized through the New Towns Act of 1946 to guide reconstruction following the destruction of World War II. Starting with Stevenage in 1949, more than 30 new towns have been built, each with populations upward of 60,000. Although some have faulted the design quality of these new cities built from scratch, and many residents still commute to jobs outside the new towns, the program is regarded as highly successful in controlling the effects of urban sprawl and preserving a significant greenbelt. The largest of the new towns, Milton Keynes, with a population in excess of 200,000, is a series of villages linked by a road network to a central commercial area, like a mixture of Letchworth and Los Angeles. The British new towns were built by nonprofit development corporations financed by the national government, with land publicly owned and leased to private builders.

Other countries adopting new town programs include Sweden, where the satellite towns of Vällingby and Farsta were built on land purchased earlier by the national government and linked to nearby Stockholm by a subway system. The most comprehensive Western new town effort was the French *villes nouvelles* program initiated in the 1970s. The core of the program is a series of five new towns built on the outskirts of Paris linked to the center by a regional express train. These new cities incorporate existing small villages within their boundaries, consolidating them into a continuous urban region. Perhaps the most extensive new towns program was carried out in the former Soviet Union, with the objective of opening up new regions to development. Facilitated by a centralized economy that controlled all construction investment, thousands of new towns were developed around industrial plants, such as Togliattigrad in Russia, a city of more than 500,000 built around a giant auto factory. Soviet housing construction was primarily based on heavy prefabricated components.

In the United States, the first important advocate of the garden city idea was architect Clarence Stein. As head of the New York State Commission on Housing and Regional Planning, Stein traveled to England in 1919 to visit Letchworth and Welwyn and returned a disciple of Howard and Unwin. In 1924, together with landscape architect Henry Wright and Stein, realtor Alexander Bing, Stein founded the limited-dividend City Housing Corporation to build a U.S. garden city. Their first project, Sunnyside Gardens, in Queens, New York (1924-1928), was not a garden city but a residential neighborhood built within the confines of the existing street grid. Consisting mostly of traditional two-story row houses, the site plan is distinguished by shared-landscaped gardens in the block interiors.

In 1928, the corporation purchased agricultural land in Fair Lawn, New Jersey, and commenced construction of Radburn, planned as a garden city of 25,000 people. The site plan was divided into three sectors, each surrounding an elementary school, following Clarence Perry's concept of the neighborhood unit. The City Housing Corporation went bankrupt during the Great Depression, and only one of the three neighborhoods was built, with a total of 640 dwellings. The planned industry never materialized. Although falling short of the garden city ideal, Radburn introduced several other concepts that influenced the course of U.S. suburban development, notably separation of vehicular from pedestrian traffic through a system of cul-de-sacs (dead-end streets) and tunnels, and reduced private yards in favor of a large shared central greenway. As at Sunnyside, efforts were made to keep housing affordable by tight space planning and construction of a variety of semidetached and row house types in addition to rental apartments.

Stein, Wright, and Bing, together with a handful of visionaries, including author and social critic Lewis Mumford and geotechnic planner Benton MacKaye, constituted an informal group known as the Regional Planning Association of America (RPAA). For 10 years, starting in 1923, this group was the foremost national voice for balanced communities within balanced regions. President Roosevelt's New Deal initiatives, such as the Tennessee Valley Authority and the Appalachian Trail, designed by MacKaye, were heavily influenced by RPAA polemics. The greenbelt towns built by the Resettlement Administration under Rexford Guy Tugwell, the federal government's most successful effort at regional planning, stem directly from RPAA thinking, although massive unemployment rather than national planning policy was the decisive impetus. The need to put people to work in construction during the great Great Depression caused Congress to override its distaste for direct government involvement in housing. Congress appropriated funds under the Emergency Relief Appropriations Act to build five new towns, of which three were actually constructed: Greenbelt, Maryland, outside Washington, D.C.; Greenhills, Ohio, outside Cincinnati; and Greendale, Wisconsin, outside Milwaukee. Conceived as satellite residential communities within an easy commute of nearby urban centers, each was a self-governing entity with a full complement of educational, recreational, and residential options, including private row houses and rental apartments. Under pressure from Congress, the housing was sold in the 1950s, with preference given to veterans and existing residents.

Notwithstanding the success of these early models, it was 40 years before the federal government again sponsored new town construction. The model came from two successful private initiatives launched in the early 1960s: Columbia, Maryland, founded by mortgage banker and shopping mall developer James W. Rouse, and Reston, Virginia, named for its founder Robert E. Simon. Although each encountered early financial difficulties (Reston was forced into bankruptcy), both survived to become successful new towns with populations in excess of 75,000. Both eventually attracted significant industrial and office developments and are known for the quality of their schools, services, and recreational opportunities made possible by the use of cluster planning. They pioneered in clustering higher-density row housing to preserve large areas of the natural landscape, a concept later codified in many local zoning ordinances as planned unit developments (PUDs). Because both Virginia and Maryland are states with strong county government systems, neither new town has municipal self-government; both have residents' associations to maintain public green spaces and recreational areas. Although most housing is privately built and owned, both towns have a sprinkling of subsidized units. Columbia has been notably successful in achieving a racially integrated community.

As private ventures, these new towns meshed better with the U.S. private enterprise ideology than did the government-financed European model. Consequently, when the federal government established a program to guide regional development through the construction of new towns, it did so by offering loan guarantees and grant assistance to private developers through the Housing and Urban Development Act of 1968 (Title IV) and the Urban Growth and New Communities Act of 1970 (Title VII). For a variety of reasons, including understaffing at the Department of Housing and Urban Development (HUD) and the inadequacy of proposal evaluation criteria, the program was a failure: Only 1 of the 13 new communities funded under this program survived. The government had far better success when it built new communities directly, under the greenbelt program in the 1930s and earlier housing for shipbuilders and munitions workers during World War I. These efforts were undertaken during clearly defined national emergencies, and the government invested in high-

Figure 20. Cooperative Residence in Letchworth, England
Photograph by Tony Schuman.

quality construction and engaged nationally prominent architects and planners to direct the programs.

A variant of the new towns program, the new-town-in-town, aims at applying the benefits of comprehensive planning to redeveloping the central city. One successful project is New York City's Roosevelt Island, located in the East River just off midtown Manhattan, begun in 1968. The Urban Development Corporation (UDC), a nonprofit agency chartered by the State of New York, used proceeds from the sale of tax-free bonds to finance the necessary site preparation and infrastructural development. Housing construction was assisted by state and federal low-interest mortgage programs and rent subsidies. With construction two-thirds complete in 1995, the island had a residential population of 7,500, linked to Manhattan by subway and aerial tramway. The housing stock, a mixture of rental and cooperative apartments, offers a wide range of unit sizes. All new construction on Roosevelt Island is barrier free.

With the withdrawal of federal backing and a shrinking supply of inexpensive land for development, there was little new town development activity after 1970. Notable exceptions have been communities such as Las Colinas, outside Dallas, Texas, and Irvine, in Orange County, California, incorporated in 1971, where unified land holdings in large family-owned ranches obviated the need for costly site acquisition.

For the most part, suburban development in the United States continues to be dominated by speculative tract developers on land made accessible by private automobiles traveling on a federally subsidized highway network. The postwar period has seen the decentralization not only of the residential population but of corporate offices and industrial plants, entertainment complexes, and service facilities. The concentration of each of these activities in separate enclaves clustered near major highway intersections has produced a phenomenon described as "edge city," a spontaneous settlement form driven by the individual investment decisions of private developers that is in many ways the antithesis of new town planning practice and principles.

Although some celebrate this regional landscape as the natural and proper working of the private market, growing concern about the environmental consequences of unplanned sprawl has produced a resurgence of interest in planned communities generally. The neotraditional development (NTD) promoted by architects Andres Duany and Elizabeth Plater-Zyberk, following the success of their resort community of Seaside, Florida, champions the virtues of urban design guidelines. The Transit-Oriented Development (TOD) concept developed by Peter Calthorpe is based on high-density, mixed-use "pedestrian pockets" organized around surface light-rail links to urban centers. These two current models are not complete new towns but fragments, a shift in scale that may improve their chances of implementation. By the mid-1990s a growing movement coalesced around these pioneering efforts under the banner of "the new urbanism."

Perhaps ironically, a movement conceived as an antidote to urban congestion is being reborn as a response to suburban sprawl. The inability to accomplish decentralization in an orderly fashion has led to a rediscovery of the merits of concentration. The new town, an idea that began a century ago by detaching itself from the central city, is in the process of restoring the umbilical connection. In the intervening years, however, suburban migration has increased regional segregation and stratification by race and income, thereby exacerbating urban economic and social tensions. As a result, the present challenge for advocates of planned regional development is to address issues of racial and social balance as well as environmental conservation. A hundred years of new town practice has demonstrated the validity of Howard's conviction that the goal of creating vibrant human communities is a matter of social reform as much as of physical development. (SEE ALSO: *Federal New Communities [Title IV] 1968 and [Title VII] 1970; New Urbanism; Planned Unit Development*)

—*Tony Schuman*

Further Reading

Burby, Raymond J. and Shirley Weiss. 1976. *New Communities, U.S.A.* Lexington, MA: Lexington Books.

Calthorpe, Peter. 1993. *The Next American Metropolis.* New York: Princeton Architectural Press.

Christensen, Carol A. 1986. *The American Garden City and the New Towns Movement.* Ann Arbor: UMI Research Press.

Howard, Ebenezer. 1965. *Garden Cities of Tomorrow.* Cambridge: MIT Press.

Katz, Peter. 1993. *The New Urbanism: Toward an Architecture of Community.* New York: McGraw-Hill.

Peel, Mark. 1995. "The Rise and Fall of Social Mix in an Australian New Town." *Journal of Urban History* 22(1):108-40.

Scott, Mel. 1969. *American City Planning.* Berkeley: University of California Press.

Stein, C. S. 1978. *Toward New Towns for America.* Cambridge: MIT Press.

Wang, L. H. and Anthony G. O. Yeh. 1987. "Public Housing-Led New Town Development." *Town Planning Review* 9(1):41-63.

Ward, Stephen V., ed. 1992. *The Garden City: Past, Present and Future.* London: E & FN Sponm. ◄

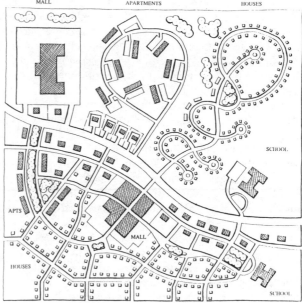

Figure 21. Comparison of Low-Density Sprawl with Traditional Development
Andres Duany and Elizabeth Plater-Zyberk, Town Planners.

► New Urbanism

In the late 1980s, a new approach to the creation of communities began to emerge in the United States, Canada, and Australia. Based on the walkable neighborhoods, villages, and small towns built prior to World War II, the new urbanism seeks to reintegrate the components of modern life—housing, workplace, shopping, and recreation—into compact, pedestrian-friendly, mixed-use neighborhoods set in a larger regional framework providing open space and transit. The new urbanism has been promoted as an alternative to suburban sprawl, a form of low-density development that consists of large, single-use "pods"—office parks, housing subdivisions, apartment complexes, shopping centers—all of which must be accessed by private automobile (see Figure 21).

Initially dubbed "neotraditional planning," the new urbanism is best known for projects built in new growth areas such as Seaside (Walton County, Florida, 1981; Duany and Plater-Zyberk, Town Planners), Kentlands (Gaithersburg, Maryland, 1988; Duany and Plater-Zyberk, Town Planners) and Laguna West (Sacramento County, California, 1990; Calthorpe Associates). The principles that define new urbanism can also be applied successfully to infill sites within existing urbanized areas. The leading proponents of new urbanism believe that such infill development should be given priority over new development. However, many of the social, political, and economic realities in North America favor development at the metropolitan edge.

The new urbanism comprises four major principles. First, all development should be in the form of compact, walkable neighborhoods or districts. Such places should have clearly defined centers and edges. The center includes a public space such as a square, a green, or an important street intersection. The center may also include public buildings such as a library, church, or community center; a transit stop; and retail businesses.

Neighborhoods and districts should also be compact (typically, no more than one-quarter mile from center to edge) and detailed to encourage pedestrian activity without excluding automobiles altogether. Streets should be laid out as an interconnected network (usually in a grid or modified grid pattern) forming coherent blocks where building entrances front the street rather than parking lots. Public transit should connect neighborhoods to each other and to the surrounding region.

Third, a diverse mix of activities (residences, shops, schools, workplaces, and parks, etc.) should occur in proximity. Also, a wide spectrum of housing options should enable people of a broad range of incomes and ages to live within a single neighborhood or district. Large developments featuring a single use or serving a single market segment should be avoided. For example, rental outbuildings attached to larger individually owned residences and units over shops can contribute to the provision of affordable housing in a community, obviating the need for large-scale apartment blocks.

Finally, new urbanist planning gives priority to public buildings and spaces. Civic buildings, such as such as government offices, churches, and libraries, should be sited in prominent locations. Open spaces, such as parks, playgrounds, squares, and greenbelts, should be provided in convenient locations throughout a neighborhood. (Because of the size of their playing fields and the need to serve multiple neighborhoods, schools are often sited in greenbelts between neighborhoods.)

New urbanism serves as an umbrella term that encompasses two major design schemes. The first of these is the *traditional neighborhood development* (TND) (see Figure 22) conceived by the Miami-based urban design firm of Andres Duany and Elizabeth Plater-Zyberk with the participation of transportation engineer Rick Chellman, planner Dan Cary, and developer Arnold B. Chace. This approach uses a site-specific master plan in conjunction with detailed codes to prescriptively regulate the buildout of the community. Unlike conventional zoning that deals primarily with density and use, new urbanist codes define the physical form of a community. Such codes specify individual block, building, street, and open space types for every parcel and lot within a community. A neighborhood, or village (as a neighborhood is called when it is freestanding in the countryside), is of finite dimension, allowing one to walk from any location in the neighborhood to the center in about five minutes (consistent with the one-quarter mile distance mentioned in the principles of new urbanism).

The compact settlement pattern of the TND supports a fine-grained mix of uses and activities within proximity. It also promotes walking as a viable alternative to driving for many daily errands. Neighborhood centers include public

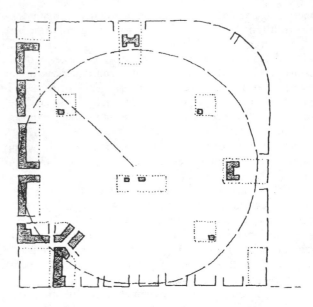

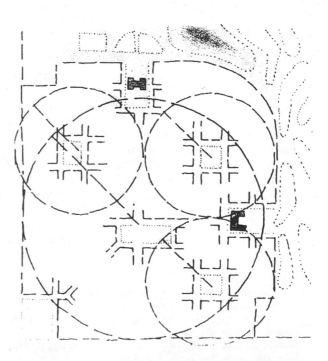

Figure 22. The TND model proposes a five-minute walk (top)—no more than one-quarter mile—for one's daily needs, and a three-minute walk (bottom) to a neighborhood park.
Andres Duany and Elizabeth Plater-Zyberk, Town Planners.

open space, such as a commons or square, retail and/or civic uses, and a transit stop. The community's highest-density housing is typically located near the center. A complete mix of housing types, suitable for a range of ages and income groups, is found within the neighborhood, with lower-density, single-family homes near the periphery. Neighbor-

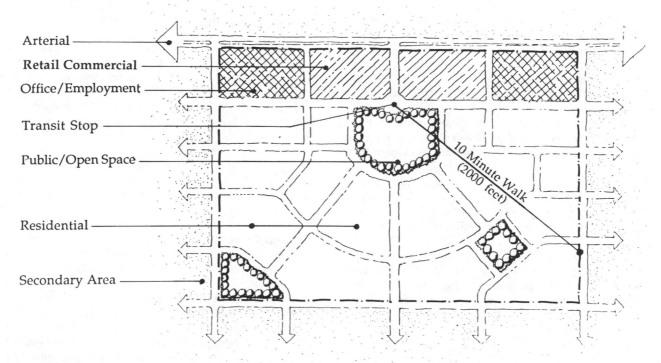

Arterial

Retail Commercial

Office/Employment

Transit Stop

Public/Open Space

Residential

Secondary Area

10 Minute Walk (2000 feet)

Figure 23. TOD Model of Transportation and Land-Use Strategies. The TOD model integrates transportation and land-use strategies at both the neighborhood and regional scale.
Calthorpe Associates.

hood edges may be defined by man-made elements such as boulevards, highways, or railroad tracks, or by natural boundaries, such as a greenbelt or river.

According to Duany and Plater-Zyberk, all growth within a region, whether in the form of infill or "green-field" development, should be directed into TNDs, configured either as neighborhoods, villages, or where some special use is dominant, districts (e.g., university campus, airport, harbor). This strategy, they have reasoned, is the best way to ensure that most citizens will be within convenient walking distance of public transit and most of the needs of daily life.

The second major new urbanism scheme, the *transit-oriented development* (TOD) (see Figure 23) was developed by Peter Calthorpe in conjunction with Daniel Solomon, Douglas Kelbaugh, and others. It is similar to the TND in many ways (one-quarter mile walking radius, open space at the center), but it differs in several key respects.

True to its name, the TOD presumes a major transit (rail or bus) connection at the heart of each community. This feature also occurs in some TNDs, but TODs are designed to work within the structure of regional road (and sometimes rail) networks that have been laid out for optimal transit service. This structure usually includes long, straight multi-lane arterials that are too large to be "tamed" for pedestrian use within a TOD. Thus, such roads remain outside the local street network of the TOD. The TND approach attempts to accommodate transit with a loop of smaller, less direct, local streets that connect the centers of neighborhoods to each other and to a major rail or bus connection in a "town center TND."

Another key difference is that TODs are designed to function in a context of conventional low-density development. In fact, Calthorpe has designated a "secondary area," consisting of single-family detached housing, low-intensity employment-generating uses, and/or large community parks, that extends up to a further quarter mile *beyond* the core area of the TOD. Although residents of the secondary area may be more likely to drive rather than walk to the community's center, Calthorpe has reasoned that the population base of this larger area is necessary to support transit and retail uses, particularly in the early years of a community's development.

According to Calthorpe, the presence of the secondary area also creates an overall "product" mix more suited to present-day market preferences for single-family homes. The secondary area plays an important political role as well: It uses single-family housing as a transition zone around the relatively high-density core of the TOD. This approach may be the best strategy for inducing suburban communities to accept density and a mix of uses near their homes.

By 1998, relatively few examples of each scheme had been completed. A much higher number of projects were still in the planning stages (up to 150 by some estimates). Projects completed after the recession of the early 1990s showed strong market acceptance. Projects such as Newpoint in Beaufort, South Carolina; Celebration in Orlando, Florida; and Seaside in Santa Rosa Beach, Florida, have enjoyed dramatic price appreciation relative to market comparables.

Developers, planners, local government officials, and citizens have shown great interest in new urbanist design approaches, particularly in regions experiencing major conflicts related to growth. Many see the new urbanism as a "win-win" approach that enables a community's growth to be channeled into a physical form that is more consistent with the scale of existing neighborhoods, discourages auto use, is less costly to service, and uses less land and natural resources.

Despite such benefits, the new urbanism has yet to be broadly embraced as a model for urban growth. One reason for this is that its physical design standards and implementation practices are not fully compatible with the regulatory framework in most regions of the United States and Canada. For example, many fire departments require streets wider than those proposed by new urbanists. Zoning laws also often discourage accessory living units within established residential areas.

Another reason for the slow adoption of new urbanism has been that the real estate profession is highly segmented by land-use category (for example, single-family housing, multifamily housing, hotels, elder care facilities, retail, recreation, offices, and warehouses). Professionals rarely work outside the boundaries of their own specialty. Each category has its own practices, markets, trade associations, and financing sources. The highly integrated development strategy advocated by the new urbanists requires a more holistic approach to community building than the real estate industry is currently structured to deliver.

Despite such barriers, public opposition to conventional suburban development is creating greater demand for alternatives. To address this need, in 1993, a coalition of architects, urban designers, developers, government officials, and others formed the Congress for the New Urbanism (CNU) to advance the principles of new urbanism and promote their broad application. Since then, the organization has hosted a series of annual meetings and drafted a *Charter of the New Urbanism* (ratified in May 1996). For additional information, CNU can be contacted in San Francisco, California, at (415) 495-2255. (SEE ALSO: *Commuting; Floor Area Ratio; Mixed-Income Housing; New Towns; Planned Unit Development; Residential Development; Setback Requirement; Social Mix; Subdivisions; Suburbanization; Zoning*)

—*Peter Katz*

Further Reading

Barnett, Jonathan. 1995. *The Fractured Metropolis: Improving the New City, Restoring the Old City, Reshaping the Region.* New York: Icon Editions.

Calthorpe, Peter. 1993. *The Next American Metropolis: Ecology, Community, and the American Dream.* New York: Princeton Architectural Press.

Duany, Andres and Elizabeth Plater-Zyberk. 1991. *Town and Town-Making Principles.* New York: Rizzoli.

Katz, Peter. 1995. *The New Urbanism: Toward an Architecture of Community.* New York: McGraw-Hill.

Kelbaugh, Doug, ed. 1989. *The Pedestrian Pocket Book.* New York: Princeton Architectural Press.

Kunstler, James H. 1996. *Home from Nowhere: Remaking Our Everyday World for the 21st Century.* New York: Simon & Schuster.

Mohney, David and Keller Easterling, eds. 1991. *Seaside: Making a Town in America.* New York: Princeton Architectural Press. ◀

▶ Nonprofit Housing

Since the early 1980s, the federal government has drastically reduced the amount it spends subsidizing the construction and renovation of housing for low- and moderate-income households. In many U.S. cities, nonprofit organizations have increasingly filled the void, producing more than 300,000 housing units. Tremendous diversity exists in the nonprofit or "voluntary" sector, both in terms of the types of housing provided and in the activities engaged in by individual organizations. Despite this diversity, all nonprofit housing shares one distinguishing characteristic— the housing is not operated to maximize financial returns for its owners.

The Legal Framework

Nonprofit housing is usually owned by entities incorporated under state nonprofit corporation law. One of the core requirements of these laws is the "nondistribution constraint." The nondistribution constraint forbids directors and managers of a nonprofit corporation from earning a profit from the activities of the organization. In most states, property owned by certain categories of nonprofit organizations is exempt from real property taxation. Most nonprofit housing organizations also qualify for an exemption from the federal income tax under Section 501(c) of the Internal Revenue Code. Those organizations that qualify as public charities under Section 501(c)(3) may also receive tax-deductible donations from individuals and foundations. Like state nonprofit corporation law, federal law requires tax-exempt entities to ensure that no part of their net earnings inures to the benefit of any private shareholder or individual.

First- and Second-Generation Nonprofit Housing

Although the prominence of nonprofit housing has grown since the early 1980s, the concept of nonprofit housing has been around for a long time. For example, churches and labor unions sponsored nonprofit housing cooperatives for their members after World War I. These units were typically not targeted to low-income households and received limited public subsidy, mostly in the form of advantageous financing and reduced land prices.

The first federal program to make extensive use of nonprofit organizations in subsidized housing production was the Section 202 program, begun in 1959. Under this program, nonprofit organizations received below-market-interest-rate loans from the federal government to construct and operate housing for elderly persons. In the 1960s and particularly the early 1970s, the activities of nonprofit housing providers expanded with the enactment by the federal government of the Section 221(d)(3) and 236 programs. Under these two initiatives, participating nonprofit organizations and limited-dividend investors received below-market-interest-rate mortgage loans to finance the construction of housing for low- and moderate-income households. Typically, the nonprofit sponsoring entities of Section 202, 221(d)(3), and 236 housing were churches, labor unions, and fraternal organizations. Usually, these organizations built only one housing project and viewed housing development as only an adjunct to their other activities. By 1977, nonprofit organizations had constructed more than 2,000 projects throughout the nation.

Modern Community-Based Nonprofit Housing

Most housing built by nonprofit organizations since the early 1980s has little in common with first- and second-generation nonprofit housing. Nonprofit housing is now typically built and owned by community development corporations (CDCs) that operate in small geographic areas such as a single neighborhood. The CDC, at least in theory, represents the interests of neighborhood residents and encourages their participation in the management of the organization. A recent survey of CDCs reports that 44% of the directors of CDCs were neighborhood residents. Approximately 2,000 CDCs existed in 1994 and had built approximately 400,000 units of affordable housing.

Nonprofit housing providers function in a number of different roles. Although many CDCs engage in activities that are not strictly housing related, such as economic development and social service provision, their most important activity is usually housing production and management. Some organizations build, own, and operate housing for low- and moderate-income residents. Others specialize in management or the provision of technical assistance. Similarly, among nonprofit housing developers there exists tremendous diversity in the types of housing produced. Some build houses that are sold to individual homebuyers. CDCs have also been instrumental in building housing that is subsequently transferred to cooperative corporations or mutual housing associations. The main difference between housing cooperatives and mutual housing corporations is that the former typically owns only one building, whereas the latter owns several properties. Some CDCs also create community land trusts, in which they retain title to the land and lease homes to residents on a long-term basis. Several nonprofit housing organizations provide housing for populations with special needs, such as the homeless, battered women, and people with physical and mental disabilities.

The most common activity of nonprofit housing organizations is the construction or renovation of multifamily rental housing for low- and moderate-income households. Unlike their second-generation predecessors, modern nonprofit housing providers do not participate in a well-defined subsidized housing program. Instead, CDCs typically finance their projects by putting together seven or more different sources of equity, subsidy, and debt. Most nonprofit housing developers receive money from their local governments' Community Development Block Grant allocation to underwrite organizational costs and project equity. In addition, charitable foundations and networks of foundations, such as the Local Initiatives Support Corporation (LISC), often provide financial assistance in the form of grants or low-interest loans. Additional debt finance is typically obtained through tax-exempt bonds, loans from state and local housing finance agencies, and mortgage loans from local financial institutions. In some cities, such as Boston and Chicago, public and private funding sources have formed formalized networks or "housing partner-

ships" with nonprofit housing providers to finance and plan housing production.

An important source of funds for many nonprofit housing projects is the Low Income Housing Tax Credit (LIHTC). Under the Internal Revenue Code, private investors in low-income housing may qualify for a tax credit equal to a large proportion of the cost of construction or renovation. Although nonprofit organizations cannot use these credits themselves because they have no tax liability, they can form partnerships with profit-motivated entities and individuals and, in effect, sell tax credits for equity investments. Typical investors include local banks or corporations. To facilitate private sector investment in LIHTC transactions, LISC and the Enterprise Foundation have established pooled investment funds. Congress has granted nonprofit housing developers preferential status with respect to the LIHTC by requiring that states earmark at least 10% of their credit allocations to projects in which nonprofit housing organization's own an interest.

Level of Production

Data limitations make it impossible to arrive at a precise estimate of the number of housing units produced by the nonprofit sector in the United States. A survey of CDCs estimates that as of 1991, 320,000 housing units had been produced. According to another (1993) estimate, from 1960 to 1990, nonprofit housing organizations produced more than 736,000 housing units with federal subsidies. If this estimate is correct, nonprofit organizations own considerably less than 1% of all housing units in the nation. Nonprofit housing constitutes approximately one-third of the social housing stock (i.e., subsidized housing not owned by profit-motivated investors) in the United States, the remainder being supplied by local public housing authorities.

Compared with most Western capitalist countries, the United States is an anomaly. For example, in France and Germany, 16% and 25%, respectively, of the housing stock is composed of social housing, virtually all of it owned by nonprofit organizations. Even in Great Britain, a nation with the largest publicly owned housing stock in Western Europe, 3% of all housing is owned by nonprofit housing associations.

Evaluating the Performance of Nonprofit Housing Providers

Nonprofit housing development is not spread evenly throughout the nation. Cities in the Northeast, Midwest, and Pacific states typically have more CDCs than do those in the South and in Plains states. As might be expected, nonprofit housing production is greatest in large cities. The extent to which CDCs operate in low- and moderate-income neighborhoods is also subject to tremendous variance. Some cities, such as New York and Chicago, have more than 100 CDCs, whereas other large cities, such as Dallas and Los Angeles, have significantly fewer.

Although Congress has increasingly reserved a preferential role for CDCs in delivering low- and moderate-income housing, there are little systematic data to evaluate the relative performance of CDCs in building and managing housing. Anecdotal accounts reported in the popular media frequently recount the successful completion of a project and its positive impact on the surrounding community. The only comprehensive studies of nonprofit housing are evaluations of second-generation nonprofit housing organizations that participated in the Section 202 and 236 programs. In general, the record of nonprofit organizations in providing elderly housing is good, although these types of developments likely present considerably fewer management challenges compared with low- and moderate-income housing. In contrast, evaluations of the performance of nonprofit sponsors of Section 236 low- and moderate-income housing projects suggest the need for closer evaluation of nonprofit housing providers. A 1977 report to Congress by HUD reported that 27% of the projects owned by nonprofit organizations had defaulted on their loans, had a loan assigned to HUD, or had a building taken over by HUD compared with only 12% of those owned by limited-dividend corporations. A study completed by the General Accounting Office one year later indicated rates of failure that were four times higher for nonprofit organizations compared with profit-motivated sponsors.

There is little doubt that several of the conditions identified as causing relatively high defaults among second-generation nonprofit housing providers do not apply to modern nonprofit housing organizations. Most of the organizations that sponsored housing during the 1960s and 1970s did so on a "one-shot" basis. Housing was not the main focus of the organization, and members had little or no experience in the complicated financial and managerial issues that surround development. Furthermore, few sources of technical assistance were available to provide needed expertise. Today, CDCs typically view housing production as their most important activity and intend to be repeat players. In addition to amassing knowledge by experience, CDCs can also turn to an array of consultants and intermediaries to provide technical assistance. Indeed, a recent report by the New York Equity Fund on the performance of LIHTC projects in New York City indicates that with respect to some performance indicators, nonprofit housing compares favorably with housing owned and operated for profit.

Despite their widely publicized achievements, the performance of modern nonprofit housing organizations needs to be carefully monitored. Although the level of technical expertise available to these organizations has increased since the 1970s, so has the level of complexity of housing programs. Rather than dealing with only one or two sources of capital, most nonprofit housing today involves combinations of seven or more funding sources. Organizations are frequently thinly capitalized and undiversified, making them vulnerable to unanticipated economic changes and management miscalculations. In addition, because of their emphases on developing individual communities and obtaining neighborhood participation, CDCs typically produce relatively small quantities of housing, thereby failing to exhaust scale economies. Finally, the absence of a profit motive, although no doubt salutary in helping nonprofit organizations gain the trust of neighborhood residents and government agencies, may also diminish their incentive to minimize costs.

In an era of scarce housing resources, evaluating the performance of the nonprofit sector is vital. Perhaps one of the reasons for the absence of careful analysis comparing nonprofit housing providers with profit-motivated developers of subsidized housing is the methodological difficulty such a study would entail. Unlike most profit-motivated housing developers, CDCs typically view the construction of housing as only one part of an integrated program of community development. In many U.S. inner-city neighborhoods, social order has broken down as working-class families and the social institutions they supported fled increasing concentrations of poverty. CDCs, by involving neighborhood residents in community-based development, seek to empower residents and restore neighborhood social structure. Therefore, any evaluation of the success or failure of nonprofit housing providers that fails to account for the external effects of housing development on the surrounding community would underestimate the benefits generated.

Conclusion

As government subsidies for the construction of low- and moderate-income housing dwindled in the early 1980s, nonprofit organizations in many cities attempted to fill the void by patching together available funds to build low-cost housing. Despite the absence of any systematic evaluation of the performance of nonprofit organizations as providers of low-cost housing, their ingenuity, tenacity, and accomplishments were rewarded in the latter half of the decade by Congress's granting them preferential treatment under a number of new housing programs. The desirability of further public investment in nonprofit housing hinges on the objectives that government seeks to accomplish. Providing housing through nonprofit developers probably costs more than subsidizing individuals directly through demand-oriented programs such as Section 8 housing certificates and vouchers. Community-based nonprofit housing will also do little to facilitate the movement of poor households from inner cities to more economically and racially mixed neighborhoods. On the other hand, community-based nonprofit housing may be the most effective and, perhaps, the only method available to achieve concentrated neighborhood redevelopment objectives that seek to go beyond bricks and mortar and rebuild local institutions. (SEE ALSO: **Community-Based Housing;** *Community Development Corporations*)

— *Michael H. Schill*

Further Reading

Bratt, Rachel G., Langley C. Keyes, Alex Schwartz, and Avis C. Vidal. 1994. *Confronting the Management Challenge: Housing Management in the Nonprofit Sector.* New York: Community Development Research Center.

Goetz, Edward G. 1992. "Local Government Support for Nonprofit Housing: A Survey of U.S. Cities." *Urban Affairs Quarterly* 27: 420-35.

Hansmann, Henry B. 1990. "The Role of Nonprofit Enterprise." *Yale Law Journal* 89:835-901.

National Congress for Community Resource Development. 1995. *Tying It All Together.* Washington, DC: Author.

New York Equity Fund. 1997. *Building a Future: A Report on the Performance of Low-Income Housing Tax Credit Projects in New York City.* New York, NY: Author.

Stoecker, R. 1997. "The CDC Model of Urban Redevelopment: A Critique and an Alternative." *Journal of Urban Affairs* 19:1-22.

Schill, Michael H. 1994. "The Role of the Nonprofit Sector in Low Income Housing Production: A Comparative Perspective." *Urban Affairs Quarterly* 30:74-101.

U.S. Department of Labor. 1947. *Nonprofit Housing Projects in the United States.* Bulletin No. 896. Washington, DC: Government Printing Office.

Vidal, Avis C. 1992. *Rebuilding Communities: A National Study of Urban Community Development Corporations.* New York: New School for Social Research.

Walker, Christopher. 1993. "Nonprofit Housing Development: Status, Trends, and Prospects." *Housing Policy Debate.* ◄

► Nontraditional Households

The traditional household in the United States is a married couple or a married couple with children who live separately from all other households. In the most traditional form, the husband earns wages by working outside the home and the wife is responsible for housework and child care. Most housing is designed with only these traditional households in mind, the assumption being that other kinds of households will act like traditional ones and adapt themselves to the housing provided. Some exceptions are made for the elderly; retirement communities and other forms of housing for the elderly provide social spaces and services that supplement private dwelling units.

The image of the traditional household is not a reflection of what is most common but rather an idealization of what households should be. Several problems arise. First, a smaller and smaller percentage of all households match the image. So the idealization becomes farther and farther removed from people's actual living arrangements, depriving these other arrangements of legitimacy and value as an acceptable, even desirable, way of living and working. Second, the housing provided for this imagined traditional household makes the lives of those who do not meet the norm unnecessarily difficult. In some cases, designing only for the norm deprives households of any housing whatsoever because conventional housing is not affordable.

Households composed exclusively of married couples and two parents with children never reflected the true variety of how people lived in the United States. At the very least, people lived alone, with others to whom they were not related, and in family situations other than two parents with children. Some people earned wages by working in the home, and women as well as men have been wage earners. The "traditional" has always served as a norm, but in many ways, the gap between the norm and reality has grown. From 1970 to 1991, the proportion of all households composed of a married couple with children decreased from 40% to 26%, whereas the proportion of other kinds of households with children increased from 5% to 9% and the proportion of households composed of individuals living alone increased from 17% to 25%. The largest change in the form of family households is the dramatic increase in single-parent families; the proportion of such families rose from 13% in 1970 to 29% in 1991.

It is also much more common today for all women to work outside the home, even when they have young children; more than half of mothers with children under six are employed either part- or full-time. Also, it is becoming increasingly common for both men and women to do wage work in the home, either on a full-time or part-time basis. Although men are gradually contributing more time to child care and housework, women still bear primary responsibility for these tasks.

Architects, community and housing organizations, policymakers, and residents themselves have recognized the need for creating housing and communities that more fully meet the needs of nontraditional households, and so a variety of new types of housing have emerged. The renovation and new construction of single-room occupancy (SRO) housing offer affordable and convenient accommodations for working and retired people who are willing to share a kitchen or are satisfied with small pullman kitchens. With the integration of social services, SRO housing can also accommodate people with special needs. Sharing kitchens and other social spaces is also a feature of smaller-scale residential settings and of much of the transitional housing for single-parent families. One consequence of recognizing the particular characteristics and needs of contemporary households has been to combine the provision of support services with the provision of housing; this is true for housing for single parents and SRO housing.

Separations between different households and different activities are also being reduced in intergenerational housing, in houses and apartments designed for sharing by two or three independent adults, in houses and apartments that intentionally accommodate wage work, and in housing that incorporates child care facilities. The sharing of meal preparation and its associated tasks and the creation of closer relationships between households are key objectives of cohousing, a form of housing adopted from Denmark where complete, private dwellings are supplemented by a range of shared spaces.

The creation of housing designed for nontraditional households is severely constrained both by conventional attitudes that idealize the traditional family and by building codes and zoning ordinances that require conventional dwelling units—that is, for traditional households. Many zoning ordinances in single-family neighborhoods exclude certain nontraditional households—namely, more than a certain number of unrelated individuals living together or the performance of certain forms of wage work in the home. These same laws often forbid the creation of accessory apartments that might accommodate older or younger relatives of the homeowner. Housing for single-parent families, group homes, or other housing for those with special needs may be excluded by law or by the antagonism of neighborhood residents. Despite these obstacles, the increasing variety of housing types is beginning to respond to the growing numbers of nontraditional households. However, there is still room for more innovation and more exploration of alternatives, and there is a pressing need for greater acceptance of those alternatives and the households who may occupy them. (SEE ALSO: *Household; Single-Parent Households*)

—*Karen A. Franck*

Further Reading

Franck, Karen and Sherry Ahrentzen, eds. 1989. *New Households, New Housing*. New York: Van Nostrand Reinhold.

Goldscheider, Frances K. and Linda J. Waite. 1991. *New Families, No Families? The Transformation of the American Home*. Berkeley: University of California Press.

Hayden, Dolores. 1981. *The Grand Domestic Revolution: A History of Feminist Designs for American Houses, Neighborhoods and Cities*. Cambridge: MIT Press.

———. 1984. *Redesigning the American Dream*. New York: Norton.

Hemmens, George C., Charles J. Hoch, and Jana Carp, eds. 1996. *Under One Roof: Issues and Innovations in Shared Housing*. Albany: State University of New York Press.

McCamant, Kathryn and Charles Durrett. 1994. *Cohousing: A Contemporary Approach to Housing Ourselves*. Berkeley, CA: Ten Speed.

Perin, Constance. 1977. *Everything in Its Place: Social Order and Land Use in America*. Princeton, NJ: Princeton University Press.

Skolnick, Arlene. 1991. *Embattled Paradise: The American Family in an Age of Uncertainty*. New York: Basic Books.

Sprague, Joan Forrester. 1991. *More Than Housing: Lifeboats for Women and Children*. Boston: Butterworth Architecture. ◄

► Not in My Back Yard

NIMBY is an acronym for "Not in My Back Yard." It appeared in the 1980s as a battle cry against environmentally or locally objectionable development projects—for instance, low-income housing, hazardous waste facilities, nuclear sites, and polluting factories.

NIMBY objectors use several basic arguments. They may contend, for example, that the NIMBY is unneeded anywhere, as in the case of nuclear power plants or strip mines. Alternatively, they may argue that the NIMBY is needed, but not where it is proposed—say, in a wetland. Then again, they may contend that its siting or operating features—for instance, its citizen participation mechanisms—are inadequate. Or perhaps its effects, such as pollution, will be harmful. In many NIMBY disputes, these arguments appear simultaneously.

NIMBY projects create the problem of having in society's midst large numbers of big projects that no one wants nearby. The United States routinely resolves the dilemma through land use regulations, pollution controls, environmental impact statements, citizen participation, eminent domain, preservation areas, and lawsuits. These procedures are used to help justify or eliminate NIMBY projects.

NIMBY objectors have succeeded in blocking many NIMBY projects. Few large new freestanding hazardous waste facilities have been sited anywhere in the United States since the late 1970s. Only two large metropolitan airports, in Denver and Austin, have appeared since the early 1960s. The lack of locations for new prisons has caused such overcrowding in many existing jails that some systems—for instance, in New York City, Chicago, Pittsburgh, and in Texas, Florida, and Connecticut state systems—have had to release convicted criminals. Most big cities, even relatively wealthy ones, have not begun a major low-income housing, mass transit, or highway project in 15 years. In many suburbs, even middle-income housing has become hard to site. NIMBY is prominent among a number of contributing factors.

Some environmentalists are now retreating from their previous hard-line position against NIMBY projects, lest they be seen as elitist. Instead of resisting every NIMBY project, environmentalists are increasingly seeking middle ground—new practical working relationships between local (backyard) goals and regional and national ones or innovative ways to mesh environmental and economic imperatives.

Thus, environmentalists are increasingly willing to allow expansion of existing NIMBY projects in exchange for agreements prohibiting new ones. Environmentalists are also paying more attention to the racial, ethnic, and class patterns underlying the siting of projects, as they are reevaluating the equity of whose backyards in fact end up with NIMBYs.

Other recent acronyms have become associated with NIMBY—for instance, LULUs (Locally Unwanted Land Uses), TOADS (Temporarily Obsolete Abandoned Derelict Sites), BANANA (Build Absolutely Nothing Anywhere Near Anyone), NIABY (Not in Anybody's Back Yard), NIMTOO (Not in My Term of Office), and PIBBY (Placed in Black Back Yards). (SEE ALSO: *Brownfields; Environmental Contamination: Toxic Waste; Locally Unwanted Land Use; Temporarily Obsolete Abandoned Derelict Sites*)

—*Frank J. Popper*

Further Reading

Brion, Denis J. 1991. *Essential Industry and the NIMBY Phenomenon*. New York: Quorum.

DiMento, Joseph and LeRoy Graymer, eds. 1991. *Confronting Regional Challenges: Approaches to LULUs, Growth, and Other Vexing Governance Problems*. Cambridge, MA: Lincoln Institute of Land Policy.

Morris, J. A. 1994. *Not in My Back Yard: The Handbook*. San Diego: Silvercat.

Piller, Charles. 1991. *The Fail-Safe Society: Community Defiance and the End of American Technological Optimism*. New York: Basic Books. ◄

► Nursing Homes

Despite the widespread perception that the United States relies more heavily on nursing homes to provide care for the elderly and others with chronic disabilities, most long-term care is provided by families. This remains as true today as it was decades ago, despite the challenges entailed in the aging of the population and the chronic diseases and disabilities associated with this demographic trend. Moreover, the percentage of elderly in nursing homes or other residential long-term care settings in the United States is very similar to that found in most other industrialized nations, such as Sweden, the Netherlands, Denmark, and Great Britain, although facilities providing long-term care in other countries may be known by a variety of different names and may include long-stay chronic care or rehabilitative hospitals. In addition, most older persons will not enter a nursing home or will be there for only a relatively short stay, usually posthospital. But for the very old and those with serious, chronic, or long-term disabilities, nursing homes remain an important part of the health care system. On any given day, about 5% of the elderly over the age of 65 are in a nursing home in the United States, as in most of Europe.

Nursing homes of the late 1990s look very different from the "mom-and-pop" and denominational homes for the aged that characterized the industry prior to the passage of Medicare and Medicaid in the mid-1960s. Facilities often are larger, more likely to be operated as a for-profit business, and are more likely to provide a greater number of services. These range from intensive rehabilitative care to daily nursing care and concentrate on meeting residents' need for assistance with basic activities of daily living (ADLs), such as eating, bathing, dressing, toileting, locomotion, and transfer (e.g., from bed to chair). Moreover, facilities increasingly provide more specialized or distinctive services, such as hospice care, subacute care, and specialized care for persons with Alzheimer's disease. More important from a resident's point of view, the scandals of the 1970s and early 1980s, involving serious and widespread poor quality have given way to a more pervasive model in which nursing homes focus on helping residents attain and maintain their maximum possible functioning in an environment that addresses quality of life and respects residents' rights. These developments have been a result of widespread changes in the dominant philosophy guiding nursing homes, state and federal regulations governing nursing homes, and the ways in which nursing homes are reimbursed, as well as of the emergence of a more fully realized group of advocates for residents and their families.

The Nursing Home Industry

Nursing homes are licensed by states to provide daily nursing care and other supportive services in a residential setting. The federal government has also certified the vast majority of nursing homes, so the Medicare or Medicaid programs can reimburse them for providing care to persons eligible for coverage. Facilities certified for reimbursement from the Medicaid program are called *nursing facilities*. Facilities certified for reimbursement from Medicare are called *skilled nursing homes*. Facilities certified for both Medicare and Medicaid are referred to as *dually certified*. According to data from the Health Care Financing Administration (HCFA), as of 1995, there were more than 17,000 nursing homes certified so that they could be reimbursed from public funds for providing nursing care to Medicare or Medicaid beneficiaries, providing care to about 1.7 million persons in 1996.

Before the passage of Medicare and Medicaid, the nursing home industry was characterized by small mom-and-pop homes or denominational facilities. With the advent of Medicare and Medicaid, both the number and nature of nursing homes changed, giving rise to an industry that in 1997 accounted for the third largest segment of health care spending. Three-quarters of the homes were owned by proprietary enterprises, and many of them, as well as some nonprofit homes, were part of multifacility systems or nursing home chains, which owned and operated multiple facilities, often in several states.

In 1997, the average nursing home was a freestanding, for-profit enterprise that had a high probability of being affiliated with a "chain." It contained more than 100 beds and had a high overall occupancy rate, usually above 90%. However, it is important to note that the characteristics of nursing homes vary considerably by state. Although almost 90% of Texas nursing facilities operate under for-profit arrangements and almost 80% are chain affiliated, other states, such as Minnesota and Pennsylvania, have a not-for-profit sector that constitutes more than half of all facilities, and still other states (e.g., New York and Maine) discourage interstate chains from operating in the state.

The nursing home industry varies from state to state, but the industry is also more generally influenced by a number of national trends. Across the United States, facility services and functions are, over time, becoming more diverse. More facilities have begun to "specialize" or engage in "product differentiation." A number of hospital-based facilities operate as "step-down" units or subacute rehabilitation facilities that care for post-acute-care Medicare patients, requiring relatively complex nursing care or rehabilitative services. Other facilities have started Alzheimer's special care units for treating residents with fairly serious problems in cognitive functioning (i.e., memory, orientation, decision making) but who are mobile and need only moderate assistance with ADLs. Between 10% to 15% of facilities have at least one specialized unit, and the presence of such units seems to vary by state, ownership, and chain affiliation. Still other nursing facilities operate within a setting that includes multiple-care modalities or levels of care. Such nursing facilities might be integrated into continuing care retirement communities (CCRCs) or into settings that provide a range of residential care, from board and care to assisted living to congregate apartments to nursing home care.

Nursing Home Residents

Just as we see changes over time in the nature of the nursing home industry and in the source and amount paid for nursing home care, we also see changes in the characteristics of nursing home residents. The general trend has been toward more impaired persons residing in nursing homes as a result of a variety of factors, including the aging of the population, increasing morbidity (i.e., health care problems and disability) as the population ages, and public policies aimed at reducing the use of both hospitals and nursing homes. As a result, the typical nursing home resident in 1997 was older and more impaired than in the past.

The average nursing home resident in 1996 was an 83-year-old widow who had a number of chronic medical conditions and serious functional limitations. Of all nursing home residents, 90% were elderly, and more than 45% were among the oldest old (i.e., 85 and over). Of residents, 75% were women, and only 12% were married. Less than 10% of residents were members of a racial or ethnic minority. More than 60% of all residents were cognitively impaired (i.e., had memory, orientation, or decision-making problems), and more than half had urinary incontinence. Data suggest that more than 80% needed daily assistance with more than three ADLs, including bathing, dressing, locomotion (walking or using a wheelchair), and toileting, as well as transferring, eating, and bed mobility. Thus, although nationally, a small percentage of residents need assistance with only bathing, dressing, and managing their medications and might be candidates for a lower level of care, most nursing home residents need substantial hands-on care and monitoring by a nurse on a daily basis.

Nursing Homes and Quality of Care

There have also been significant changes in the quality of care and life offered by nursing homes over the last three decades. One factor in this has been changing professional training and standards or mores among nursing home operators and staff. For example, a group of nonprofit nursing homes (Kendall-Crossland) in Pennsylvania "jump-started" the move among like-minded caregivers in the United States to "untie" the elderly. Another significant factor has been the most far-reaching set of reforms in the regulation of nursing homes since the passage of Medicare and Medicaid. In 1987, Congress enacted a broad set of nursing home reforms in the Omnibus Budget Reconciliation Act (OBRA), following the recommendations of an Institute of Medicine (IOM) study (done in 1986) on nursing home regulation and quality of care. Most of the OBRA provisions went into effect in October 1990, but even before this date, many nursing homes on their own adopted the restraint reduction provisions of the law, as well as reducing the inappropriate use of psychotropic medications. Together with the impetus of federal nursing home regulations and support from consumer advocacy groups and long-term care ombudsmen, this led to a dramatic reduction in the use of physical restraints across the nation since 1990. Other OBRA provisions required more training of nursing assistants, who are the primary caregivers in most nursing homes; the presence of a registered nurse (RN) on duty at least one shift a day; a greater emphasis on quality of life and residents' rights; the use of a uniform, comprehensive assessment system that focuses on maintaining maximum practicable functioning to guide the residents' plans of care; a survey system that focuses on what residents actually experience on a daily basis (rather than on the home's records or "paper compliance" with standards); and a range of enforcement sanctions to encourage homes to stay in compliance with federal standards of care. Findings from an evaluation of the new resident assessment system suggest that since the implementation of the OBRA 1987 reforms, there have been significant improvements in key elements of the process of care in nursing homes and in resident outcomes, particularly in terms of preventing decline and helping residents maintain functional independence as long as possible.

Expenditures and Payment for Nursing Home Care

Although growth in the number of nursing homes and in the number of persons receiving nursing home care was dramatic since the passage of Medicare and Medicaid, the growth in expenditures, public and private, for nursing home care has been even more dramatic. In 1960, total expenditures for nursing home care in the United States stood at $1 billion. By 1980, that figure had increased to $20 billion. In 1991, expenditures for nursing home care totaled $59.9 billion. By 1994, spending on nursing home care topped $72.3 billion. In addition, the nursing home sector experienced price inflation equal to or greater than that seen in other segments of the health care sector. In fact, from 1989 through 1991, expenditures for nursing home care grew more quickly than expenditures for hospital care, physician services, and dental services. From 1993 to 1994, the rate of increase in spending on nursing homes exceeded the rate of growth in overall national health care expenditure.

Prior to Medicare and Medicaid, most nursing home care, as much as 80%, was paid for by individuals and their families "out-of-pocket," although some public monies did flow to nursing homes through programs such as "old age assistance." In 1997, private, out-of-pocket spending on nursing homes was still substantial. Indeed, over the last decade, between 40% and 50% of all dollars spent on nursing home care came from individual nursing home residents and their families. Even when public programs, such as Medicare or Medicaid, cover the nursing home stay, individuals still contribute through copayments and deductibles, to paying for their care. The remaining 50% to 60% of expenditures on nursing home care comes from third-party payors—that is, public and private insurers or programs, such as private long-term care insurance, Medicare, Medicaid, and the Department of Veterans Affairs. However, the bulk of these third-party payments has come from Medicare and Medicaid, mostly Medicaid. For example, in 1994, the total of all spending on nursing home care was $72.3 billion. Of that total, 36% or approximately 26 billion came from the individual out-of-pocket payments from individuals. More than $44 billion came from third-party payors, and slightly more than $41.8 billion, or almost 58% of the expenditures, came from federal and state governments through Medicaid and, to a much lesser extent, Medicare programs.

Thus, there are only two primary payors for nursing home care: individuals and their families paying out-of-pocket and the Medicaid program. Although most people think of Medicare as the providing coverage for the health care of the elderly, that is true only for acute (hospital) and ambulatory (e.g., physician, therapies) care. Medicare's coverage for long-term care, although growing in the area of home health, has for the most part been trivial. Indeed, Medicare has paid only between 2% and 3% of all the expenditures for nursing home care except for the brief period when the expansion of Medicare, known as coverage for "catastrophic care," was in place (1990–1991) and Medicare coverage for nursing home care expanded dramatically. This is because Medicare coverage has been restricted to a limited number of days of nursing home care and effectively limited to those persons who benefit from rehabilitative care. Most nursing home care actually needed by the elderly does not meet this definition or does so only for a relatively short time. Similarly, private long-term care insurance, although generating much publicity, has also paid relatively little of the nation's bill for nursing home care. As studies for the American Association of Retired Persons (AARP) and the United Seniors Health Cooperative have found, such policies have tended either (a) to be unaffordable, (b) not to cover the services needed by the elderly, or (c) to have such restrictive eligibility criteria that most nursing home care is not actually covered.

Medicaid, then, has been the primary public program that pays for nursing home care, covering all or part of the nursing home bill for between 60% and 70% of all nursing home residents. Medicaid is a program that combines federal and state (and sometimes county) funds to pay for the nursing home care for persons who are poor or have become poor (or medically needy) in paying for nursing home care. Medicaid also pays for acute and ambulatory care, as well as for some other services and for poor persons who are aged, blind, disabled, or are members of families with dependent children, but a substantial portion of all Medicaid dollars has gone for nursing home care for the elderly. In 1989, almost 30% of Medicaid funds went into payments for nursing home care, a figure that has remained fairly constant throughout the 1990s. In some states, the percentage of Medicaid funds going to nursing homes has been even higher. In Wisconsin in 1991, for example, 40% of Medicaid assistance funds went to the nursing home industry, and many states were having trouble meeting their long-term care costs. Thus, one concern among advocates for the elderly has been the effect that continued federal and state budget reductions and pressures on the Medicaid program will have on the availability and coverage of nursing home care.

Because Medicaid is a combined federal and state program, states have had primary responsibility for determining key elements of nursing home policy. As a result, Medicaid's policies have varied considerably from state to state in terms of beneficiary eligibility and facility payment policies. Eligibility determinations have varied by the definition states use for *poverty* (that is, the income and assets a person may have and whether he or she is eligible for nursing home coverage), as well as by the level of impairment a person must have before becoming eligible for Medicaid coverage of the nursing home stay (usually impairment in two or more ADLs). Changes in eligibility can result in thousands of elderly nursing home residents becoming ineligible for Medicaid coverage at the same time that they cannot afford to pay privately.

States have varied even more in *how* they pay for nursing home care. This variation covers both the system under which they have paid nursing homes for care of a Medicaid beneficiary and the amount they have paid. These systems have varied from retrospective systems, in which nursing homes are paid based on their reasonable and allowable costs, usually up to a ceiling amount, to prospective systems, in which a rate is set in advance, although most systems have been a mix of the two in terms of how they actually function. Some states have adopted a system that pays facilities based on the characteristics and care needs of their residents, paying a higher rate for very impaired residents, whereas other states have based payment on the services the residents need and actually receive. Importantly, systems have had different incentives in terms of quality of care and access for heavy-care residents. Furthermore, no matter what reimbursement system they use, poorer states usually pay less and richer states usually pay more. In 1993, the average Medicaid per diem for nursing home care varied from $49 to more than $200 per resident day. Private-pay rates charged by nursing homes have been consistently higher than the Medicaid rate in all states but Minnesota, which has had a "rate equalization" law.

Nursing Homes and the Future

Regardless of improvements in the quality of care offered by nursing homes, both the elderly and policymakers have preferred that long-term care be delivered in another setting, preferably the person's home, whenever possible. This is because the elderly and their families continue to prefer the environment and autonomy often associated with living independently in one's own home, even in the presence of chronic disabilities. Policymakers, faced with the aging of the population and the already high cost of nursing home care, have been seeking other alternatives in an effort to contain costs. Thus, there is likely to be continued expansion of home care, domiciliary or board and care, and assisted living. However, as a practical matter, nursing homes will remain an important component of the nation's health care system and a vital option for the elderly and their families for both short-stay rehabilitative episodes of care and for longer-term care of persons with multiple disabilities and the need for daily nursing care and supervision.

The proliferation of special care units, particularly special care units for persons with Alzheimer's disease, is likely to continue. Ideally, the effective and life-enhancing practices of such units will become part of the standard of care in all nursing homes. In addition, there is likely to be growth in special rehabilitation units and subacute care units, as well as in the provision of more clinically complex and rehabilitative care in all nursing homes, as policymakers, third-party payors, and hospitals continue to minimize the use of and length of stay in acute care settings. However, given the demographics and the prevalence of chronic diseases, including dementias, that require daily nursing care and supervision, as well as substantial assistance with ADLs, nursing homes will continue to play a critical role in providing long-term care. Nursing homes are likely to concentrate even more on enhancing the quality of life and autonomy of residents in their facilities, with greater attention to activities and residents' psychosocial needs; to promoting a wider range of choices for residents in daily activities, meals, and rooms or apartments; and to developing residential settings that are less institutional in nature. (SEE ALSO: **Elderly;** *Section 232*)

—*Catherine Hawes and Charles D. Phillips*

Further Reading

Buchanan, R. J., R. P. Madel, and D. Persons. 1991. "Medicaid Payment Policies for Nursing Home Care: A National Survey." *Health Care Financing Review* 13:55-72.

Harrington, D., C. L. Estes, A. del la Torre, and J. H. Swan. 1993. "Nursing Homes under Prospective Payment." Pp. 113-31 in *The Long-Term Care Crisis: Elders Trapped in the No-Care Zone*, edited by C. L. Estes, J. H. Swan, & Associates. Newbury Park, CA: Sage.

Hing, E., E. Sekscenski, and E. Strahan. 1989. *The National Nursing Home Survey: 1985 Summary for the United States*. Vital and health Statistics, Series 13, No. 97, DHHS Publication No. (PHS)

89-1758, Public Health Service, Washington, DC: Government Printing Office.

Letsch, S. W., H. C. Lazenby, K. R. Levit, and C. A. Cowan. 1992. "National Health Expenditures, 1991." *Health Care Financing Review* 14:1-2.

Shaughnessy, P. W. and A. M. Kramer. 1990. "The Increased Needs of Patients in Nursing Homes and Patients Receiving Home Health Care." *New England Journal of Medicine* 322:21-27.

Weissert, W. G., C. M. Cready, and J. E. Pawelak. 1988. "The Past and Future of Home and Community Based Long-Term Care." *Milbank Memorial Fund Quarterly* 66:309.

▶ Obsolescence

Housing is long-lived. Some estimates have put its life as much as 300 years for the United States, and of course, there are castles in England and Europe hundreds of years older than that. But most housing will last only 40 to 120 years. It's not that it falls down. Few houses do. Nor does it physically wear out from use and "break down" in the way a piece of machinery does. Really old housing is more apt to merely get more decrepit than to break down, more apt to be increasingly undesirable than to become absolutely useless. It becomes obsolete and shunned while it still stands.

When housing fails to remain standing, it is usually because it was destroyed by fire or storm or, more likely, was deliberately demolished. In other words, society builds houses because it needs them and later deliberately removes them when they have outlived their usefulness rather than allowing them to deteriorate and ultimately disintegrate from old age.

Housing's usefulness is bound up in two concerns: physical deterioration and economic obsolescence. These two different aspects of housing are frequently confused, for they are inevitably intertwined.

Physical deterioration is the simpler concept. As structures grow old, they suffer wear and tear by their inhabitants. They also suffer insults from nature: storms, winds, fires, floods, and earthquakes. All of these physical transgressions on the structure weaken it and in weakening it make it even more susceptible to further weakening. A small, easy-to-patch hole in the roof can let rain penetrate into a unit's interior where it can cause considerably more damage—and at much greater cost to repair. This type of physical deterioration is to be expected, and most owners promptly repair such damage before it grows worse. Thus, a rigorous and conscientious physical repair and maintenance program can ensure that a structure will last a long time.

Why then do owners not keep a structure intact for centuries or until storm or fire inadvertently takes it away? The answer is *obsolescence*. The structure might be as physically sound as the day it was built, but over time, it will come to be considered to have been built in the wrong place or to be built to the wrong style or to be so old-fashioned in the way that it was laid out and the conveniences and amenities installed that it is no longer "worth" making physical repairs. That is, the physical condition of the structure is beside the point because the structure is obsolete on one of more of these other dimensions. Thus, the owner will decide to demolish the building, despite its relatively good physical condition, because it is more profitable to build something else in its place.

Typically, decisions on physical maintenance are made with an eye to the unit's degree of obsolescence as well. Often, owners will reduce the quality of physical repairs for housing that is becoming economically obsolete so that the time path or "trajectory" of physical deterioration begins to match—and sometimes reinforce—that of economic obsolescence. A summary indicator that includes both physical deterioration and economic obsolescence is the depreciation in value of the housing as it ages.

Because housing has changed so much in its basic construction, amenities, and configuration during the 20th century, we cannot impute longevity for, for example, units built under Federal Housing Administration (FHA) standards since the mid-1930s from harder evidence on longevity for houses built in the 1880s and 1890s. Nor should we assume that units inexorably follow a continuous decline. Maintenance and repair, and especially periodic remodeling to update a unit, mean that in practice the decline is interspersed with periodic upgrades. Indeed, because over its life span so many of the unit's appliances, water heaters, space heaters, and sometimes the electrical service and wiring and plumbing have been changed—and perhaps a room or two added—it becomes problematic to assess precisely how "old" a unit really is. Appraisers speak of "effective age" rather than age based on original year built to take these factors into account.

Changes in fashion can have much the same effect, especially when coupled with historic preservation efforts or gentrification where older, deteriorating and obsolescing parts of town become popular places to live once again. Some experts have suggested a rule of thumb about these changes. For one- to four-unit structures, the first major decision by the owner as to whether to rehabilitate or tear down is made between 30 and 60 years; if rehabilitated, then another 20 years will pass before a major decision must be made once again; but in between these decision points, the structure is relatively immune from deliberate demolition. Another rule of thumb is that owner-occupants tend to invest more in their units and prolong the life of housing more than will landlords who provide rental housing.

The issue of housing deterioration and obsolescence will play a larger part in public policy over the next 20 years. The nation's housing stock is growing older, as is its population. The median age of housing units was 25 years in 1940, dropped to 21.5 in 1973, but aged again to 26 years by 1989, and rose further to 28 years by 1993, stabilizing at 28 years again in 1995. Nor is a major housing construction boom apt to occur over the next 10 years sufficient to reverse this aging trend.

Given this trend, public policy will be primarily concerned not with older individual housing units but, rather, with groups of older housing units—older neighborhoods. Efforts to deal with this problem in the 1950s and 1960s applied a mistaken analogy—simple physical deterioration and obsolescence was branded as "blight" and said to be spreading like a cancer. Radical surgery was thought to be the necessary cure. Thus, urban renewal programs engaged in large-scale bulldozing and demolition of older units and, later, when that policy failed, in large-scale rehabilitation. Both programs were relatively insensitive to demographic change, migration patterns, job relocation, and discrimination. Yet all these forces help account for economic obsolescence and will affect chances for success of any future large-scale rehabilitation and preservation efforts that must inevitably come to the fore once again with the nation's aging stock. (SEE ALSO: **Abandonment;** *Depreciation of Property; Maintenance*)

—*William C. Baer*

Further Reading

Arnott, Richard, Russell Davidson, and David Pines. 1983. "Housing Quality, Maintenance and Rehabilitation." *Review of Economic Studies* 50(3):467-94.

Baer, William C. 1991. "Housing Obsolescence and Depreciation." *Journal of Planning Literature* 5(4):323-32.

Corgel, John B. and Halbert C. Smith. 1982. "The Concept and Estimation of Economic Life in the Residential Appraisal Process." *The Real Estate Appraiser and Analyst* 48(4):4-11. ◄

► *Open House International*

First published in 1976 at the Stichting Architecten Research Institute in the Netherlands, *Open House International,* with a circulation of 650, is published quarterly in the United Kingdom. Its audience includes university libraries, students, and faculty as well as public and private architecture and planning and urban development organizations. Its subscribers and authors are in the Third World, Latin America, the Far and Middle East, Europe, and North America.

It is a refereed journal of an association of institutes and individuals concerned with housing, design, and development in the built environment. This association, Open House International Association, focuses on exchanges that enable the various professional disciplines dealing with the built environment to understand the dynamics of housing and so contribute more effectively to it. Its more general aim is to improve the quality of the built environment through encouraging a greater sharing of decision making and to help develop the necessary institutional frameworks that will support the local initiatives of people in the housing process.

Articles deal with theories, ideas, tools, and methods concerning alternative housing and development processes. Particularly, they cover user participation and community roles in government housing policy in both developed countries and the Third World. Articles by educators and practitioners address issues of policy, training and education, technology, professional roles, design practice, rehabilitation and upgrading, and a broad range of other topics incorporating increased attention to urban regeneration.

The journal has an international editorial board and international correspondents. Each issue, typically contained in the range of 65 pages, has an editorial, main articles, and book reviews as well as announcements of major housing publications, conferences, university and other training programs, and events. Its graphics, photographs, and text are printed in black and white. On occasion, articles are organized in theme issues on a special topic or geographical area. Once each year, an entire issue is dedicated to the results of an international competition covering approaches that may be transferable and interchangeable in evolutionary planning and neighborhood and housing design. An international panel of judges selects the top submissions, which subsequently appear in the journal. Subscription price for universities and libraries, 85.00 per year, for companies and businesses, 55.00, and for indviduals and students, 35.00. Editor: Nicholas Wilkinson. Editorial and subscription information: University College London, The Bartlett, Development Planning Unit (DPU), 9 Endsleigh Gardens, London WC1H OED. (SEE ALSO: **Industrialization in Housing Construction**)

—*Stephen Kendall* ◄

► **Operation Breakthrough**

In the United States, the 1968 Housing Act defined the need for 26 million housing units over the next decade, 6 million for low- and moderate-income families. Operation Breakthrough was initiated in 1969 by the secretary of the Department of Housing and Urban Development (HUD)— George Romney. It was intended that Operation Breakthrough would result in the much wider use of factory-based industrialized methods of prefabricated house construction. The favorite analogy was the mass production of automobiles. Mass production of factory-built houses was seen as necessary to quickly produce sufficient, good-quality housing at low cost. At that time, it appeared that Europe had achieved great success following such an approach.

Complementing the need for faster development of housing production technology itself, was the need to remove obstacles in the way of an aggregated, large, and continuous market. The lack of such a market was seen as the greatest obstacle to accelerating the application of industrialized methods to house construction. Capital-intensive methods of house construction were also needed because of the shortage of skilled labor in the traditional building process. For these various reasons, Operation Breakthrough was directed not only at technological advancement of housing but at breaking through the various nonhardware constraints to more efficient production of housing.

Thus, Operation Breakthrough was a multifaceted program aimed at the following:

► Encouraging industry to propose ideas available for volume production methods
► Encouraging conventional housing system advancement through support of applied research and development
► Aggregating the market for housing, including both the demand for housing and the available land for such housing
► Encouraging state and local to adjust building codes and zoning codes that contributed to fragmentation, high cost, and the difficulty of producing for a broad market
► Improving methods of financing
► Making more effective use of the total workforce
► Conducting tests and evaluations with authoritative test validation that would serve as a basis for approval of advanced housing concepts and for development of performance standards for housing systems

The program's primary objective was to include the production of at least 1,000 dwelling units per year for five years, using up to five different technologies as called for by Section 108 of the Housing and Urban Development Act of 1968. To assist in reaching the primary objective, the program was also to stimulate the modernization and broadening of the housing industry through increased emphasis on better design and greater use of improved techniques within the housing industry and through increased participation by other organizations that possess the necessary talents, interest, and capabilities for such a commitment.

The reference to "other organizations" included companies that had a large-scale manufacturing capability. Operation Breakthrough was designed to help industrial companies develop new building materials and new construction techniques. The assumption was that cooperation among leading industrial corporations in mass production of housing was essential and would prove sufficiently profitable to attract investment capital on the scale required to launch a new high-technology industry. Although the attention to size of company and the need for new construction methods had often been referred to in Europe, this reference to the potential role of organizations outside the conventional construction industry had not been seriously considered. This marked a major difference in the U.S. approach compared with that adopted in Europe, although the method of encouraging the policy—demonstration projects—was similar. To encourage big companies that were engaged in other fields to enter the housing market, HUD provided the support for design, testing, and prototype construction of these concepts and then gave companies an opportunity through those sites and their marketing capability to market these concepts for volume construction. More than 2,000 requests for proposals were sent out to companies. It was then planned to choose prototypes for construction from between 10 and 20 contractors. These prototypes were to be built on eight regional sites, following which various prospective sponsors (private and public) would be invited "to examine these various concepts and select those they want to have produced in volume in their areas to satisfy their needs." It was intended that the operation should be "largely privately financed." This was the second major departure from the way industrialized building was developed in Europe.

On March 17, 1970, in a statement on combating inflation in the housing industry, President Nixon said,

> In the longer run, new materials, new techniques, improved designs and innovations in marketing are needed to improve the efficiency of the building industry. In order to encourage these necessary advances in housing, the Department of Housing and Urban Development has sponsored Operation Breakthrough. I strongly endorse these experimental techniques for housing construction. (HUD 1970, p. 2)

At the same time, HUD Secretary Romney predicted that "by 1980 two-thirds of all the housing built in this country will be basically factory made—either assembled components or completed systems" (HUD 1970, p. 3).

The rhetoric is called in question by the small scale of actual activity in Operation Breakthrough. There were difficulties in even realizing its relatively small objectives. Beginning in 1969, Operation Breakthrough subsidized the erection of 2,795 units by a variety of building systems on nine demonstration sites. Paralleling Operation Breakthrough, the Department of Defense conducted, independently by its constituent services, industrialized housing demonstration projects totaling 3,160 units at 12 sites. Operation Breakthrough itself was not able to achieve its "primary objective" of 1,000 units per year over five years. Furthermore, by 1971, two and a half years after the inception of the experiment, no units had actually been produced. Later, evaluation showed that Operation Breakthrough failed to produce housing units at lower cost than traditional methods or to aggregate markets that were large enough to absorb truly mass-produced factory housing.

Operation Breakthrough was overtaken by events that questioned the necessity of such programs. Major manufacturing industries did not pursue the path indicated by the theory of industrialized building. The Committee on Industrial Housing of the National Academy of Engineering, in a study of Operation Breakthrough, concluded that it was not necessary to involve major manufacturing companies to increase the quality and amount of housing or decrease the cost. Furthermore, it deemed fallacious a comparison between the production of housing and the mass production of automobiles.

Furthermore, it was difficult to provide housing for low-income groups. Romney stated before the House Committee on Appropriations on April 10, 1972, "These conditions cannot be corrected by the housing or other programs of our Department no matter how well they are administered. Housing needs for the poor in the central city cannot be met without cures for the social conditions" (U.S. House 1972, p. 36)

Operation Breakthrough failed to make any impact on the market, which suggests that industrialized building was unable to compete with either traditional methods or the mobile home industry. The nature of the technology involved required government intervention to institute the necessary degrees of coordination. The fate of Operation Breakthrough reinforces the suggestion that industrialized production of housing—along European lines—is not possible in the United States unless there is a significant public housing program. (SEE ALSO: **Industrialization in Housing Construction**)

—*Robert McCutcheon*

Further Reading

Committee on Industrialized Housing of the National Academy of Engineering. 1972. *Industrialized Housing: An Inquiry into Factors Influencing Entry Decisions by Major Manufacturing Corporations.* Washington, DC: National Academy of Engineering.

Downs, A. 1974. "The Successes and Failures of Federal Housing Policy." *The Public Interest* 34(Winter):132-37.

Fuerst, J. 1974. "Public Housing in the United States." In *Public Housing in Europe and America,* edited by J. Fuerst. London: Croom Helm.

McCutcheon, R. T. 1990a. "Modern Construction Technology in Low-Income Housing Policy: The Case of Industrialized Building and the Manifold Links between Technology and Society in an Established Industry." *History of Technology* 12:136-76.

———. 1990b. "The Role of Industrialized Building in Low-Income Housing Policy in the USA." *Habitat International* 14(1):161-76.

Nelkin, D. 1971. *The Politics of Housing Innovation: The Fate of the Civilian Industrial Technology Program.* Ithaca, NY: Cornell University Press.

Patman, P. F. et al. 1968. *Industrialized Building: A Comparative Analysis of European Experience.* Department of Housing and Urban Development, Division of International Affairs, Special Report. Washington, DC: Department of Housing and Urban Development.

U.S. Congress, Joint Economic Committee. 1969. *Industrialized Housing.* Hearings before the Subcommittee on Urban Affairs of the Joint Economic Committee, Congress of the United States, Part I. Washington, DC: Government Printing Office.

U.S. Department of Housing and Urban Development. 1970. *Operation Breakthrough.* Washington, DC: Author.

U.S. House of Representatives. 1972. Subcommittee of the Committee on Appropriations. *Testimony of the Secretary of HUD, George Romney.* Hearing. 92d Cong. 2d sess. p. 36. ◀

▶ Owner Building

Throughout the world, most dwellings are built by their eventual occupants. The same was true in North America until about the turn of the century, although in urban areas, building for profit rather than for use had become the norm before then. Even so, as late as 1949, a survey by the U.S. Bureau of Labor Statistics revealed that one-quarter of new nonfarm single-family dwellings were owner built. The figure in the late 1990s was probably below 10%.

Building for use and profit are often combined, usually in one of two ways. In the first, a prospective homeowner purchases a lot and initiates construction. How much of the work he or she then contracts out depends, chiefly, on what resources—time, manual skills, and finances—he or she has to work with. (Most owner-builders, even today, are men.) The importance of skills is implied by the fact that owner-builders rely much more on contractors for some jobs than for others. A recent survey of owner-builders in New Brunswick, Canada, for example, found that three-quarters hired out the electrical work but that more than half put up their own drywall and more than three-quarters did their own interior painting and wallpapering. At one time, it was very common for owner-builders to do almost everything themselves, often with the help of friends, family, or neighbors. Today, this is less common. The New Brunswick survey found that some contract labor was employed in every case.

The alternative method of combining building for profit and use is for a professional builder to erect an unfinished "shell." Here, there is an almost infinite range of possibilities. Toward one extreme, the builder lays a basement, frames the exterior walls, and puts on a roof, leaving everything else to the buyer. Toward the other, and more commonly, the builder might leave only an attic or basement to be finished by the owner at his or her leisure. This shades imperceptibly into home improvement. Many older homes, originally built and sold in "finished" form, have subsequently been improved with finished basements and, with skylights or raised roofs, attics as well. Although the finishing of shells can involve a great deal of sweat equity, because the initiative comes from a professional builder, these cannot properly be regarded as "owner building."

People build homes for themselves for economic or ideological reasons: to save money or to make a social statement. In the past, the economic motive was dominant. Most owner-builders had modest incomes, and a disproportionate number were blue-collar workers. Many had no savings and built in stages, whenever they could save enough to purchase materials. At various times, however, a significant minority of owner-builders have sought to "get back to the land." During the 1910s, there was a minor middle-class fad for owner building. This underwent a resurgence during the 1930s and 1940s and then again in the late 1960s. In each period, owner building was associated with a skepticism, and sometimes hostility, toward the consumer society of urban capitalism.

Because the economic motives for owner building have been dominant, its incidence has reflected conditions in the housing market. The circumstance that has most favored owner building is rapid urban growth. The best-documented cases of owner building between the 1880s and the 1950s occurred where recent migrations of low-income workers created a severe shortage of housing. Thus, there was a good deal of owner building by Polish immigrants in Milwaukee in the 1890s, by British immigrants in Toronto after 1900, by African American migrants outside Cincinnati in the 1920s, and by southern whites in the Los Angeles area in the 1930s. A national surge of owner building came after World War II, when returning veterans created a demand for housing that professional builders were at first incapable of meeting.

Because owner building has been associated with migrations of specific peoples, individual owner-builders have often received help from others in the same group. Poles in Milwaukee organized their own building and loan societies. The British in Toronto relied more on mutual aid, sometimes organized through local churches. The same was true of African Americans in Cincinnati. Many post-World War II veterans received training and other assistance from local Legion posts. Even so, in North America much more than in Europe, owner building has reflected the efforts of individuals and their immediate kin. In Europe, notably Sweden, co-ops have coordinated self-help building by substantial groups of people. In addition, municipal programs of aided self-help—comparable to modern "site-and-service" schemes in the Third World—date back to the interwar years. In Vienna and Stockholm, for example, local municipalities provided basic services, technical assistance, and building materials at cost. The first municipal scheme in North America, however, appears to have been that of Bartonville, a suburb of Peoria, Illinois, in the late 1940s. It was not widely copied.

In the long run, owner building has declined for three main reasons. First, a growing number of people have acquired homes by purchasing them, albeit on credit. This reflects the growth of real incomes, especially from the 1950s to the 1970s, and the refinement of long-term amortized mortgages, coupled with mortgage insurance. The latter depended on federal initiatives during the 1930s. Second, the introduction and enforcement of building regulations of growing complexity have been a deterrent. They have raised the minimum cost of the home, pushing it beyond the reach of those who might once have been able to begin by building only a shack. Also, the very complexity of most modern building codes is daunting to the amateur. Third, land has come to make up a large proportion of the total cost of housing. At the turn of the century, about 10% of the cost of a finished dwelling was required for the lot. Today, the figure is closer to 30%. Today, no matter how much of their own labor they invest, owner-builders cannot reduce their shelter costs as much as was once possible. To some extent, this reflects the fact that the relative cost of raw land has risen. Just as important are the steadily rising standards for servicing, which have added both to the cost of land and to municipal taxes. One way or another, then, the state has played a significant role in the long-term decline of owner building.

Much the same considerations help to explain why, within urban areas, owner building has been concentrated toward the fringe. Land is cheapest there because it is less accessible and because servicing requirements are lower. Again, building codes are likely to be more lax—or even nonexistent—in fringe areas. All major U.S. cities had introduced building regulations by the turn of the century, but as late as the 1950s, many suburbs, and especially unincorporated districts, had not. Cheap, unregulated land played a critical role in drawing thousands of owner-builders from the city of Toronto into its suburbs in the first quarter of this century. In different degrees, and over a growing geographical range, the same process has been at work in other metropolitan areas throughout the 20th century.

The merits of owner building, and of policies to encourage it, have usually been debated in a Third World context but are relevant to the North American scene. Proponents have argued that working on one's own home has economic and intrinsic benefits: that it fosters pride and thrift and helps low-income families who might otherwise be overcrowded or even homeless. Often, they argue for aided self-help, where municipalities provide serviced sites and technical advice. Critics contend that building by amateurs is inefficient and that it helps to depress wage rates, contributing to the oppression of already exploited people. The latter argument is likely to be relevant only in situations in which very large numbers of owner-builders are active. Although most critics of owner building have tended to lie on the left of the ideological spectrum, support has crossed political boundaries. There are good reasons for believing that aided self-help could play a greater role in helping to address the low-income housing problem in the United States and other advanced capitalist societies. (SEE ALSO: *Do-It-Yourself;* **Homeownership;** *Participatory Design and Planning; Self-Help Housing*)

—*Richard Harris*

Further Reading

Ashton, B. and David Bruce. 1993. "Affordability and Self-Built Housing: The Role of Financing, Skills and Quality in a New Brunswick Case Study." Research report of the Rural and Small Town Research and Studies Programme, Department of Geography, Mount Allison University, Sackville, New Brunswick.

Duncan, S. S. and A. Rowe, 1993. "Self-Provided Housing: The First World's Hidden Housing Arm." *Urban Studies* 30(8):1331-54.

Harris, Richard. 1996. *Unplanned Suburbs: Toronto's American Tragedy, 1900-1950.* Baltimore: Johns Hopkins University Press.

Harris, R. Forthcoming. "Slipping through the Cracks: The Origins of Aided Self-Help Housing, 1918–1953." *Housing Studies.*

Kern, Ken. 1975. *The Owner-Built Home.* New York: Scribner's.

Murphy, K. 1958. "Builders of New One-Family Homes, 1955-56." *Construction Review* 4(8-9):5-15. ◀

▶ Owner Take-Back Financing

During periods when nominal interest rates are historically high, the use of the fixed-rate mortgage (FRM) becomes impaired. When inflation is expected in the future, the financial burden of repayment shifts heavily to the borrower because of a single, fixed rate for the entire loan. During these periods, the financial system is strained because affordability is reduced, traditional finance instruments are unattractive, and capital flows are reduced from lenders to borrowers.

One result of high interest rates is the movement away from FRMs to alternative mortgage instruments (AMIs). Demand for adjustable-rate mortgages (ARMs) has risen as interest rates have risen in recent years. This is due to affordability concerns as well as the ability of the ARM to be priced without the burden of the tilt effect. Another outcome is the rise of owner take-back financing for home sales. This phenomenon is best viewed as an attempt to augment the financial system products when economic conditions make the traditional instruments relatively unattractive.

Sometimes called seller or creative financing, owner take-back financing creates real estate finance concurrent with the sales agreement of the real property. Instead of relying on third-party lenders for financing, the gap between the sale price and the first mortgage available from the financial institution or mortgage company is supplied by the seller of the real property. In this sense, the seller "takes back" a mortgage as part of the agreement.

Of course, the seller would rather sell the rights to the real property for cash, but during periods of high interest rates, cash settlements may not always be possible. In addition, high interest rate periods are associated with slowdowns in real estate markets, and seller financing can assist buyers in closing the deal.

Owner take-back financing is generally short term in nature, with or without amortized repayment provisions. This type of financing is a second mortgage and thus, places the lender (former owner) in a riskier debt position than the primary mortgage issuer. Generally, but not always, these types of mortgages carry higher interest rates than do first mortgages.

One issue is the extent to which selling prices capitalize the financing value of these mortgages that are written at terms inconsistent with other mortgages in the market. For example, imagine a seller-financed mortgage at a very low interest rate offered to the buyer to assist in closing the sale by the seller. The low interest rate on the second mortgage represents a dollar value to the borrower (buyer) and an opportunity loss to the lender (seller). It is an interesting empirical question whether such a "giveaway" can be captured into a higher sales price. The evidence suggests that during the late 1980s, roughly half of such "creative" financing value was passed through in higher sales prices.

Although sometimes controversial, especially among inexperienced buyers and sellers, this type of financing has served as one of the most flexible features of the U.S. financial system since the 1980s. With continuing innovation in U.S. financial markets and a greater reliance on the secondary mortgage market for funds, the use of owner take-back financing may not be as prevalent in future periods with high interest rates. On the other hand, this characteristic is one of the more useful elements available. (SEE ALSO: **Homeownership; Mortgage Finance**)

—*Austin J. Jaffe*

Further Reading

Brueggeman, William B. and Jeffrey D. Fisher. 1993. *Real Estate Finance and Investments*. 9th ed. Homewood, IL: Irwin.

Clauretie, Terrence M. and James R. Webb. 1993. *The Theory and Practice of Real Estate Finance*. Fort Worth, TX: Dryden.

Fabozzi, Frank J. and Franco Modigliani. 1992. *Mortgage and Mortgage-Backed Securities Markets*. Boston: Harvard Business School Press.

P

▶ Panelized Housing

Panelized housing is a kind of prefabrication involving the production of wall, floor, or roof panels and their erection on-site. Following a house design, planar parts—walls, floors, and sometimes roofs—are made in panel sections in an off-site plant or, in large projects, in an on-site facility. At the site, panels are assembled with other parts, which come from other suppliers, the panel producer, or both. Panelized methods can be incorporated into site-built and modular house production.

Production of wall and floor panels has been used in house building from the 1880s, including the shipment of building elements internationally. During the 1920s and 1930s, U.S. companies began using panelized construction, influenced in part by the need to increase efficiency, reduce costs, and improve quality. A number of well-known architects, including Walter Gropius, Konrad Wachsmann, and Carl Koch, undertook panelized projects in the years before and after World War II.

Experiments have been conducted with concrete, wood, steel, and aluminum for panel framing and with wood products, steel, aluminum, copper, and numerous cementitious and synthetic materials for cladding. Some of the architects who participated in these efforts succeeded in the market when they used technical systems and processes compatible with and used in site-built construction. However, most of the experiments have not succeeded in competing with conventional methods.

Beginning in the 1960s and ending in the early 1990s, when the last major concrete panel company ended production, some U.S. companies produced concrete panels for large residential building projects, based on pioneering European projects dating from the 1940s.

Most companies currently making panelized houses use the same materials, trade skills, and sequences used in site-built detached and low-rise housing: light wood frame technology. Wooden wall studs and floor and roof members can be replaced with equivalent parts made of steel or composite materials. The cross-section dimensions of framing members, normally determined by structural calculations, may, in the case of exterior panels, be sized also by the insulation requirements.

Panels thus made from wooden or substitute "sticks" fastened together and clad with other materials are "framed" panels. Because the cavities within framed panels are useful as a place to install insulation and to route resource distribution systems such as ducts, wiring, and piping, panels may arrive at the site as "open panels" ready for these elements to be installed, or as "closed panels" in which distribution system elements and insulation are placed in the panel before delivery to the site, where final connections and inspections are made.

Panels can also be made of solid insulating cores with a variety of skins on both sides or with normal framing members with plywood glued and nailed to each side to create added structural performance. These are often called "stress-skin" panels. Roof panels can be made in the same way. Sizes depend on the house design and panel-handling capability in the plant and on the site. Entire walls can be brought to the site, including windows, doors, and other specified parts, and are normally one or two stories high and from 4 to 40 feet long.

The panelized housing approach is a growing part of homebuilding in the United States, accounting in 1997 for approximately 750,000 units; 13% of all panelized homes made in the United States are exported. (SEE ALSO: **Industrialization in Housing Construction**)

—*Stephen Kendall*

Further Reading

Bender, Richard. 1973. *A Crack in the Rear View Mirror: A View of Industrialized Building.* New York: Van Nostrand.

Bruce, Alfred and Harold Sandbank. [1944] 1972. *A History of Prefabrication.* New York: Arno.

Hallahan Associates. 1989. *Factory Built Housing in the 1990s: A Study on Modular and Panelized Housing from 1984 to 1994.* Washington, DC: Building Systems Council of the National Association of Home Builders.

Herbert, Gilbert. 1984. *The Dream of the Factory-Made House: Walter Gropius and Konrad Wachsmann.* Cambridge: MIT Press.

Kelly, Burnham. 1959. *Design and the Production of Houses.* New York: McGraw-Hill. ◄

► Participatory Design and Planning

All design professions seek to do good as they define good, in ways that also increase the power, status, and income of their members through what has been described as a benevolent professional imperialism. However, it must also be observed that this approach has not always been benevolent for the ultimate recipients of those professional services. This role model is characterized as design and planning solutions stemming from the professional's subjective ideas and values. Professionals make planning and design decisions, alone, denying exposure to other opinions. Participatory design, on the other hand, is an alternative approach that involves users directly in the decision-making process that determines the quality and direction of their lives. It is an open, broad-based process of collaboration between citizens, professionals, and public administrators.

Who decides what for whom is a central issue in housing and human settlement. John Turner, an advocate for self-determination, believes that when people are in control of decisions about the design, construction, and management of their housing, the process and product will affect their social well-being. When people have no control over the housing process, the housing produced may instead become a barrier to achieving personal fulfillment and a burden on the economy.

People become invisible in the housing process to the extent that housing providers either do not see them at all or see them as stereotyped individuals. This blindness can be the result of a genuine desire to improve the living conditions of as many people as possible. Providers tend to have a fixed idea of what is good housing and, consequently, discount the input of the dwellers. This contention is generally based on assumptions that public participation is inefficient and time-consuming, that people do not know what they want, or that people trained in housing know better about users needs than users do. On the basis of these beliefs, the housing needs of many people in the world have been reduced to specifications of codes and standards, however well intentioned they may be.

Planner Paul Davidoff pointed out the effects of participation by stating that it reduces the feeling of anonymity and communicates to the users a greater degree of concern on the part of management or administration. Furthermore, participation is an inclusive process in which the planner or designer represents the special-interest group, just as a lawyer represent a client's interest in court. In this respect, participation is a matter of control over decisions. Urban renewal, for example, was more than an ineffectual solution to the problems of inner cities: It was the conscious uprooting and destruction of some neighborhoods for the benefit of others. Professionals became the instruments of the establishment. Their expertise was not sufficient to resolve urban problems. People need to participate in the creation of their environment; they need the feeling of control. It is the only way that their needs and values can be taken into consideration. Participation is a means of protecting the interests of groups of people as well as of individuals, because it satisfies needs that are very often totally or partially ignored by organizations and by expert planners and architects. Participation involves face-to-face interaction of individuals who share a number of values important to all. Participation will be a major aspect in a society in which freedom of all citizens, in all aspects of social life, is well ensured.

Issues in Participation

Participation is a general concept covering different forms of decision making. Conceptualizing it means asking simple questions of who, what, where, how, and when?

- ► *Who* are the parties to be involved in participation? Generally, people who will be affected by design and planning decisions should be involved in the process of making those decisions.
- ► *What* do we wish to have performed by the participation program? For example, is the participation intended to generate ideas, identify attitudes, disseminate information, resolve some identified conflict, or review a proposal, or is it merely to serve as a safety valve for pent-up emotions?
- ► *Where* do we wish participation to lead? What are the goals to be accomplished?
- ► *How* should people be involved? Appropriate participation methods need to be identified to achieve desired objectives. Methods need to be matched to purposes. Participation in policy-making offers a degree of citizen power. Methods such as community workshops and charrettes (intensive seminars) allow for diverse interest and promote human resource development. They may afford the opportunity for participants to have control over decisions. Public hearings, on the other hand, may survey citizens for input, but decision makers are still free to make decisions without regard to that input.
- ► *When* in the planning process is participation needed or desired? It is necessary to decide in what part of the planning process the participants should be involved—development, implementation, evaluation, or some combination thereof.

The scope of participation has been described by some to include information exchange, the resolution of conflicts, and supplementing planning and design. With participation, residents are actively involved in the development process. There will be a better maintained physical environment, greater public spirit, more user satisfaction and significant financial savings. The main purposes of participation are these:

- ► To involve people in design decision-making processes and, as a result, increase their trust and confidence in organizations, making it more likely that they will accept decisions and plans and work within the systems when seeking solutions to problems
- ► To provide people with a voice in design and decision making to improve plans, decisions, and service delivery
- ► To promote a sense of community by bringing people together who share common goals

Stages of Participation

The types and degrees of participation depend on several factors and vary in accord with the circumstances. In the most modest kind of participation, the user helps to shape a building by acting as the client of an architect. The fullest kind of participation is the kind in which users do the building themselves. Participation can be viewed in four categories that can lead to agreement about what the future should bring:

► *Goal setting*. This stage involves discovering or redis-covering the realities of a given environment of a situation so that everyone in the process is speaking the same language based on experiences in the field where change is proposed.

► *Programming*. This entails going from awareness of the situation to understanding its physical, social, cul-tural, and economic ramifications. It means sharing with each other so that the understanding, objectives, and expectations of all participants become resources for planning, not hidden agendas that could disrupt the project later on.

► *Design*. This phase concentrates on working from awareness and perception to a program that fits the situation under consideration. In it, participants make actual physical designs based on their priorities for professionals to use as resources to synthesize alter-native and final plans.

► *Implementation*. Many community-based planning processes stop with goal setting, programming, and design, often with fatal results to a project because it ends people's responsibilities just when they could be of most value: when the how-to, where-to, when-to, and who-will-do-it must be added to what people want and how it will look. People must stay involved throughout the process and take responsibility with professionals to see that there are results.

The planning that accompanies the design of any par-ticipation program should first include a determination of goals. Participation objectives will differ from time to time and from issue to issue. Once they are stated, it becomes clear that participation is perceived differently depending on the type of issue and people involved. If differences in perception and expectations are not identified at the outset and realistic objectives are not made clear, the expectations of those involved in the participation program will likely not be met and people will become disenchanted.

Planning for participation needs considerable time. When sufficient time is allowed to analyze issues, partici-pants, resources, and objectives prior to the choosing of participation methods, the chance of success is greatly en-hanced.

Conclusion

Participation can be seen as direct public involvement in decision-making processes. Citizens share in decisions that help shape the quality and direction of their lives. People will come together if change can and will occur. Participa-tion can function if it is active, directed, and if those involved experience a sense of achievement. At the same time, it requires a reexamination of traditional design and planning procedures to ensure that participation becomes more than affirmation of the designer or planner's inten-tions. Ensuring community participation in the design process requires effective tools. Experiences in design par-ticipation show that the main source of user satisfaction is not only the degree to which design needs have been met but also the feeling of having influenced the decisions.

The theories and practices of participation can be syn-thesized into the following five statements:

1. *There is no best solution to design problems.* Each problem has a number of solutions. Solutions to design and planning problems are traditionally based on two sets of criteria: (a) *facts*—the empirical data concerning material strengths, economics, building codes, and so forth—and (b) *attitudes*—interpretation of the facts, the state of the art in any particular area, traditional and customary ap-proaches, and value judgments. Thus, design and planning decisions are by nature biased and depend on the values of the decision maker(s).

2. *Expert decisions are not necessarily better than lay decisions.* Given the facts with which to make decisions, the users can examine the available alternatives and choose among them. The designer or planner involved in such an approach should be considered a participant who is ex-pected to identify possible alternatives and discuss conse-quences of various alternatives and state an opinion, just as the users state opinions and contribute their expertise, not to decide among alternatives.

3. *A design or planning task can be made transparent.* Alternatives considered by professionals are frameworks in their own minds and can be brought to the surface for the users to discuss. After understanding the components of design decisions and exploring alternatives, the users in effect can generate their own plan rather than react to one provided for them. The product is more likely to succeed because it is more responsive to the needs of the people who will use it.

4. *All individuals and interest groups should come to-gether in an open forum.* In this way, people can openly express their opinions, make necessary compromises, and arrive at decisions that are acceptable to all concerned. By involving as many interests as possible, not only is the product strengthened by the wealth of input, but the user group is strengthened by the wealth of input and by learning more about itself.

5. *The process is continuous and ever changing.* The product is not the end of the process. It must be managed, reevaluated, and adapted to changing needs. Those most directly involved with the product, the users, are best able to assume those tasks. (SEE ALSO: **Community-Based Hous-ing**; *Owner Building; Self-Help Housing*)

—*Henry Sanoff*

Further Reading

Goodman, Robert. 1971. *After the Planners.* New York: Simon & Schuster.

Hatch, Richard C. 1984. *The Scope of Social Architecture*. New York: Van Nostrand Reinhold.

Hester, Randolph T. 1990. *Community Design Primer*. Mendocino, CA: Ridge Times Press.

Marcus, Clare C. and Wendy Sarkissian. 1986. *Housing as if People Mattered*. Berkeley: University of California Press.

Sanoff, Henry. 1990. *Participatory Design*. Raleigh, NC: Sanoff.

————. 1992. *Integrating Programming, Evaluation and Participation in Design*. Brookfield, VT: Ashgate.

Turner, John F. C. 1977. *Housing by People: Towards Autonomy in Building Environments*. New York: Pantheon. ◀

▶ Partnership for the Homeless

Founded in 1982, The Partnership for the Homeless is the nation's largest service-providing agency dealing exclusively with the issues surrounding homelessness. Serving New York City, it seeks to motivate various units of government, religious denominations, private industry, charitable foundations, and the general public to formulate programs and provide funding for the housing and care of homeless people. The partnership relies on more than 11,000 volunteers and more than 600 churches and synagogues to help provide shelter, transitional and permanent housing, employment training, and health, mental health, and other services. Its 1997 revenue of $9,211,886 was funded 50% from government agencies, 50% from private support. Over 85% of revenue goes directly to program services for homeless people. Programs include Emergency Services, providing overnight shelters; Project Domicile, a permanent rehousing program; HIV/AIDs Services, providing permanent housing and casework services for clients suffering from HIV/AIDS; Employment Services, providing assistance in moving clients away from government benefits and toward financial independence; Furnish a Future, a furniture distribution program to furnish newly acquired apartments; and Peter's Place, an emergency drop-in center serving frail, elderly homeless people 24 hours a day. The partnership publishes a quarterly newsletter for its volunteers and others concerned with helping homeless people. Contact person: Pat Hildebrand. Address: 305 Seventh Ave., 13th Floor, New York, NY 10001-6008. National toll free number: (800) 438-0005. New York toll free number: (800) 235-3444. Phone: (212) 645-3444. Fax: (212) 477-4663. (SEE ALSO: **Homelessness**)

—*Laurel Phoenix* ◀

▶ *People and Physical Environment Research*

People and Physical Environment Research (PAPER) is a periodic journal focusing on human-environment interaction primarily in the Southwest Pacific region. Its primary objective is to facilitate communication between researchers to improve the physical environment. *PAPER*'s audience consists of academics. Circulation: 120, mostly Australia and New Zealand. First published in 1988. Price: $35, Australian currency for four issues. Submission editor: Ross Thorne, c/o the Department of Architecture,

Building GO4, University of Sydney, NSW Australia, 2006. Phone: 61 2 351-2826. Fax: 61 2 351-3855.

—*Laurel Phoenix* ◀

▶ Physical Disabilities, Housing of Persons With

Persons with physical disabilities include those with motion or mobility impairments. They may use a variety of mobility aids, including artificial limbs, walking aids, or wheelchairs, or they may rely on a caregiver. Unless housing is designed to accommodate persons with such disabilities, they cannot live independently nor can they be cared for by caregivers without difficulty. Dwellings designed to be usable by physically disabled persons are described as "accessible" or as "barrier-free housing."

Such housing is designed to permit persons with physical disabilities to move about the dwelling and to comfortably use the spaces and features of the home. For example, dwellings for persons in wheelchairs are generally on one level, with wider doorways and corridors. The rooms have sufficient space to allow the person to maneuver the wheelchair between furniture and fixed features such as closets. Cabinets and fixtures are designed to allow use from a seated position or with limited hand mobility. Design adaptations are necessary for the entire home, including location, access to the building, garage or carport, ramps or chairlifts, entrances, lighting, electrical switches and outlets doors and windows, heating and cooling systems, floors, security and fire safety, vestibules, corridors, level changes and stairs, living and dining areas, kitchen, bathroom, and bedrooms. Technical standards and manuals for barrier-free design exist in most developed countries.

As the numbers of persons with physical disabilities increase due to improvements in medical care and due to an aging population, there is a need to increase the stock of accessible dwellings. There are two major challenges: (a) appropriate design of such housing and (b) policies to increase the number of such housing units in the existing stock.

Until the last few decades, barrier-free housing was either specially built or an existing house was adapted to the specific needs of each person with physical disabilities. Design manuals and standards provided design details. However, the design challenges are considerably different if such houses are to be available on the market. Such dwellings must accommodate a wide variety of disabilities rather than being custom designed for a particular disabled person or even a particular disability. These dwellings must be comfortable for use by other occupants as well. Barrier-free housing may be more expensive than other housing if it is larger, but it need not be if it is carefully designed. Special equipment, such as elevators or hoists, that is institutional rather than residential may not only be expensive, but it also may not be domestic in scale or attractiveness.

Although the demand for barrier-free housing is rising, the market has been slow to respond. Therefore, increasing such stock is a challenge for public policy. The challenges include the lack of information and the choice of policy instruments. Policy development is hindered because there are no reliable statistics on the number of barrier-free dwellings in the housing stock or how accessible they are. There continues to be debate regarding the level of government responsibility and the appropriate policy instruments. Legislation at the national level, such as the Americans with Disabilities Act in the United States or the Charter of Rights and Freedoms in Canada, generally proscribes discrimination in public buildings and spaces. Private property, including dwellings, is exempt. Local governments that apply building codes and bylaws may also require barrier-free public buildings and spaces, but the requirements for private property tend to be limited to issues of safety and health. Therefore, rental housing and social housing units are more likely than private dwellings to be barrier free. Furthermore, in most developed countries, most of the housing stock required for the next half century has already been built.

Several forces require an increase in barrier-free housing stock for rent and ownership. First, in many Western countries, disabled persons have the right to equal opportunities in their communities and their workplaces. Furthermore, in roughly 20 years, one in five persons will be over the age of 65, and disabilities correlate strongly with older ages. In addition, households prefer to age in their own homes rather than moving to specially designed homes. Because the life span of housing is at least 50 years, it is highly likely that during its life span a dwelling will be occupied by a person with some disability. It is becoming evident that retrofitting housing is not cost-effective. It is more economical to build barrier-free housing in the first place. This has led to the search for supply-side policies for barrier-free housing rather than demand-side policies for housing specific disabled persons. Thus, a shift is occurring from specially adapted housing to "adaptable housing" that can be easily modified according to the changing needs of the occupants.

These developments have given rise to "universal design," recognizing that good design will be valuable for many users under different conditions. The notion of universal design grew from the idea that the costs and benefits of barrier-free design must be considered for all users rather than only those who are disabled. Choice and flexibility can be enhanced for all consumers at lower cost if the general housing stock is universally designed. (SEE ALSO: *American National Standards Institute (and Other) Accessibility Standards and Housing; Barrier-Free Design; Center for Universal Design; Smart House*)

— *Satya Brink*

Further Reading

Brink, Satya. 1994. "Housing for an Aging Society: The Meaning of Barrier Free Design." *Man Environment Relations Association (MERA) Journal* 2(2):17-23.

Canestro, N. and R. Null, eds. 1995. "Universal Design." *Housing and Society* 22(1&2, Special issue).

Housing for Everyone: Universal Design Principles in Practice. Washington, DC: U.S. Department of Housing and Urban Development.

Imrie, R. 1996. *Disability and the City: International Perspectives.* New York: St. Martin's.

Liebrock, C. 1992. *Beautiful Barrier Free: A Visual Guide to Accessibility.* New York: Van Nostrand Reinhold.

Lifchez, Raymond and Barbara Winslow. 1979. *Design for Independent Living: The Environment and Physically Disabled People.* Berkeley: University of California Press.

Peloquin, Albert A. 1994. *Barrier Free Residential Design.* New York: McGraw-Hill.

Rostron, J. 1984. "The Physically Disabled: An International Audit of Housing Policies." *Public Health, London* 98:247-55. ◀

▶ Planned Unit Development

A planned unit development (PUD) is an alternative zoning plan for mixed land use that is often primarily residential but that incorporates several land uses into one area, development project, or neighborhood. It differs from traditional zoning because several different land uses are provided for in one area rather than the traditional segregated land uses. PUDs incorporate several of the key concepts of the British new town movement, with planning conducted on an area (or unit) within a traditionally zoned locality, usually comprising from a few hundred acres up to about one thousand acres. Begun on the West Coast of the United States, PUDs attempt to deal with the many problems of land use and housing—urban and suburban sprawl, affordable housing, and allocations of public and private space.

Although a PUD requires that developers include mixed use and diversity in their development plan, the specific details are often negotiated between a local planning board and a developer. In some cases, some light industry and high-density apartments might be included along with single-family residences, in addition to parks or public recreation areas. In other cases, the PUD may require the developer to meet certain standards regarding streets or schools. In addition, a PUD creates more efficient land use by pooling property that traditionally constituted a homeowner's private space into public spaces for use by the entire community.

A PUD is meant to offer a more interesting and socially useful land use pattern than typically would be found in an exclusively residential neighborhood or area. PUD zoning patterns, also called "cluster developments," have been created in U.S. localities such as New Hope, Pennsylvania; Reston, Virginia; Twin Rivers and Teaneck, New Jersey; and Columbia, Maryland. A number of states, including New Jersey, Pennsylvania, Maryland, and Missouri, have enacted enabling legislation for PUDs. In the mid-1960s, the Urban Land Institute proposed model PUD legislation, followed by the American Society of Planning Officials, which in the mid-1970s also proposed a model of enabling legislation.

At its best, some benefits accrue to all parties. Housing may be more affordable for prospective buyers. A developer may find larger profits because the PUD provides for permission to develop some commercial areas or build to a higher density than had been allowed under traditional zoning plans. The locality may find savings in infrastructure costs (fewer roads to maintain, fewer utility needs) because of higher density; in addition, more public space or un-spoiled areas might be kept free from development.

In some instances, PUDs have not been deemed successful because localities have been unable or unwilling to negotiate enough "public" benefits from the agreement with the developer. Few public spaces were created and few amenities otherwise were available to the public, but the developer still obtained substantial profits because of the relaxed requirements on density or mixed use. (SEE ALSO: *New Towns; New Urbanism; Subdivision*)

—*John S. Klemanski and C. Michelle Piskulich*

Further Reading

Burchell, Robert W., ed. 1973. *Frontiers of PUD: A Synthesis of Expert Opinion.* New Brunswick: Rutgers, State University of New Jersey, Center for Urban Policy Research.

Johnson, William C. 1989. *The Politics of Urban Planning.* New York: Paragon.

Keller, Suzanne. 1986. *Creating Community: The Role of Land, Space, and Place.* Cambridge, MA: Lincoln Institute of Land Policy.

Moore, Colleen Grogan. 1985. *PUDs in Practice.* Washington, DC: Urban Land Institute. ◀

▶ Points

One of the most widespread practices associated with residential mortgages in the United States is the use of discount points (or, more simply, "points"). These charges are attached to most mortgages and are paid at closing. Points are additional charges by the lender and paid by the borrower when the mortgage and note are originated.

Each point is defined as a percentage of the loan. Thus, for a $100,000 mortgage at a certain interest rate with two points, the borrower agrees to all of the terms in the mortgage, including the promise to repay the note according to the prescribed schedule and to pay $2,000 (2% of $100,000) at the closing of the loan. In effect, the borrower agrees to repay a $100,000 loan but effectively receives only $98,000. Obviously, the use of points can dramatically affect the affordability of the mortgage package and can have a major impact on the cost of the loan.

It can easily be shown that points increase the cost of borrowing in residential mortgage markets. Therefore, it is often advised that borrowers need to worry about the *effective* cost of borrowing (i.e., the real cost of borrowing after the effects of points, closing costs, and other provisions are taken into account) rather than the nominal cost, generally measured by the interest rate on the loan. Points frequently add 25 to 50 basis points (or ¼ to ½ percentage points) to the effective cost of borrowing.

Why do residential mortgage lenders use points in their contracts? If lenders wanted to charge 50 basis points more, why not simply increase the nominal interest rate on the loan? Traditionally, it was argued that mortgage lenders charged points because usury laws prohibited them from charging higher mortgage interest rates in many states. However, usury laws in mortgage markets have been held to be unenforceable since 1980 as against public policy, yet the use of points continues.

Another explanation has been that points were used because they were induced by the income tax code (i.e., points were deductible from personal taxable income in the year they were paid and, thus, the after-tax cost to the borrower was reduced by the tax law). However, since the Tax Reform Act of 1986, points can no longer be deducted in the year paid (the new law permits them to be amortized over the life of the loan). This law thus reduced the tax advantage to the borrower paying points. However, the use of points in mortgage markets has remained unchanged after 1986.

Currently, observers think that points are used as a more effective device rather than prepayment penalties for long-term mortgages. Because the economic burden of points is higher the shorter the repayment period, borrowers who choose to prepay their mortgage notes face a higher effective cost of borrowing when points are used. Given fixed costs in originating mortgages, points serve as deterrents to quick refinancing and to helping to cover the fixed costs at financial institutions.

Most financial institutions maintain a menu of mortgage options with combinations of interest rates and points. Although it used to be very difficult to compare mortgages with different interest rates and points, sophisticated calculators and computer spreadsheet programs have dramatically increased the ability to perform the necessary calculations to make informed decisions. Points used to be thought of as a major deterrent to affordability as well, but over the past decade, lenders have often permitted the financing of points as well. Points have become one of the institutionalized features in residential (and some commercial) mortgage markets but are no longer as important as in the past. (SEE ALSO: *Closing Costs*)

—*Austin J. Jaffe*

Further Reading

ABA Standing Committee on Lawyers' Title Guaranty Funds. 1991. *Buying or Selling Your Home: Your Guide to: Contracts, Titles, Brokers, Financing, Closings.* Chicago: American Bar Association.

Blankenship, Frank J. 1989. *The Prentice Hall Real Estate Investor's Encyclopedia.* Englewood Cliffs, NJ: Prentice Hall.

Clauretie, Terrence M. and James R. Webb. 1993. *The Theory and Practice of Real Estate Finance.* Fort Worth, TX: Dryden.

Jacobus, Charles J. 1996. *Real Estate Principles.* Upper Saddle River, NJ: Prentice Hall.

Jaffe, Austin J. and C. F. Sirmans. 1995. *Fundamentals of Real Estate Investment.* 3d ed. Englewood Cliffs, NJ: Prentice Hall.

McKenzie, Dennis J. 1996. *Essentials of Estate Economics.* Upper Saddle River, NJ: Prentice Hall.

Peiser, Richard B. 1992. *Professional Real Estate Development: The ULI Guide to the Business.* Washington, DC: Urban Land Institute. ◀

▶ Polarization

Polarization is a pattern of structured social and spatial inequality, characterized by a declining middle and growing extremes in the distribution of valued goods and services in advanced capitalist societies.

Origins and Uses

The concept was popularized in Great Britain and the United States in the 1980s to capture a variety of changes observed beginning in the 1970s indicating a reversal of previous trends toward decreasing inequalities and a growing middle-class society. In addition to its broader conceptualization in terms of social class, polarization may refer to a variety of social and spatial divisions: between the employed and the unemployed, between professional and skilled workers and the unskilled, between high-income and low-wage earners, between citizens and immigrants, between blacks and whites, between North and South (in Britain), between inner cities and suburbs, between gentrified areas within cities and nongentrified areas, and between homeowners and renters of public housing.

Causes and Issues

Analysts attribute polarization to global economic changes, including tertiarization and the decline in manufacturing; changes in economic and social policy, especially the new right politics; and the contraction of the welfare state. Although the term *polarization* draws attention to and expresses concern about rising inequalities, the concept raises some conceptual and empirical questions and is not universally accepted. For example, it is not clear whether the new inequalities are systematic and cumulative, and analysts disagree on where to draw the new lines of cleavage. There is a tendency to focus on opposing categories in all areas of economic, social, and political life and to disregard the possibility that, although there may be a new trend toward polarization in one domain, there may be countertrends in other domains.

Housing Polarization

Although one can observe a trend of government withdrawal from housing in most advanced industrial societies in the 1980s, such withdrawal has been most consequential in Great Britain. In the United States, where government involvement in housing for low-income households was residual to begin with, the term polarization is used less frequently. This is not to suggest that the situation of low-income households has not worsened but that the changes have been less dramatic.

In Great Britain, the Thatcher government's policy to sell public housing (known as council housing) stock to sitting tenants at a fraction of the assessed value (Housing Act 1980 with the "right to buy") led British analysts to introduce the concept *sociotenurial polarization* in the mid-1980s. It refers to the changing distribution of housing tenure and income, where households with limited resources are increasingly concentrated in council housing, in particular in undesirable housing estates located in the inner city or the periphery, whereas the more skilled and affluent households able to purchase the council housing they previously occupied join the ranks of homeowners. Council housing, which represented 30% of housing tenure in 1980 and included a variety of social classes, was becoming residual, as better-off (middle-aged, skilled working-class) tenants purchased their homes at discounts ranging from 35% to 50%. Although the tendency for the more affluent to become homeowners began in the mid-1960s, the "right to buy" accelerated this trend, and those remaining in council housing are increasingly the long-term unemployed, recipients of social assistance, female-headed households, Afro-Caribbean immigrants (Asian immigrants are predominantly homeowners), and the elderly. Between 1963 and 1985, the percentage of households in the bottom one-third of income who rented council housing increased from 26% to 57%.

Council housing is thus becoming residualized, as tenants are increasingly characterized by low incomes. The council sector is also becoming more distinct in terms of types of dwellings, because the dwellings sold have been predominantly row houses or semi-detached houses, not apartments. Yet at the same time, the category of homeowners has become more heterogeneous, raising questions concerning the polarization thesis.

The ranks of homeowners in Britain increased throughout the 1980s and 1990s, and in 1995, the rate of homeownership had risen to 66.9%. After a high of 65.6% in 1980, private homeownership in the United States declined in the following decade and then again increased to 65.7% in 1997. However, homeownership in the United States declined from a high in 1980 of 65.6% to stabilize at 64% in the early 1990s. In contrast to Britain, where the percentage of young people who become homeowners has been increasing, the most dramatic decline of homeownership in the United States has been in this age category. Because homeownership is the major source of wealth accumulation, a decline of homeownership among younger age groups contributes to polarization of wealth.

Unlike in Britain, public housing in the United States has always been residual, representing only 1.5% of the housing stock today. Affordability has become a problem for many households, from moderate-income families unable to qualify for a mortgage to the poor unable to get adequate rental housing or receive housing assistance (even if they meet the eligibility requirements). Private sector rents also have risen rapidly, and the supply of subsidized housing has been shrinking.

Despite these polarizing trends, some analysts have pointed out that local governments have stepped in with an array of new programs. Although these local programs have not offset federal cuts, a variety of community-based housing movements are providing alternative housing to low-income households. These new actors produce "third-sector" housing that serves public needs while being privately owned. (SEE ALSO: *Privatization*)

—*Barbara Schmitter Heisler*

Further Reading

Bentham, G. 1986. "Socio-Tenurial Polarisation in the UK 1953-83: The Income Evidence." *Urban Studies* 23:157-62.

Davis, John Emmeus. 1994. *The Affordable City: Toward a Third Sector Housing Policy*. Philadelphia: Temple University Press.

Forrest, Ray and Alan Murie. 1988. *Selling the Welfare State: Privatisation of Public Housing*. London: Routledge.

Forrest, Ray and Alan Murie, eds. 1995. *Housing and Family Wealth: Comparative International Perspectives*. New York: Routledge.

Goetz, Edward G. 1993. *Shelter Burden: Local Politics and Progressive Housing Policy*. Philadelphia: Temple University Press.

Hamnett, C. 1984. "Housing the Two Nations: Socio-Tenurial Polarisation in England and Wales, 1961-81." *Urban Studies* 21:389-405.

Henwood, Doug. 1996. "America: Still World Capital of Inequality." *Left Business Observer,* January 22, pp. 4-5.

Hudson, R. and A. Williams. 1990. *Divided Britain*. London: Belhaven. ◄

► Postoccupancy Evaluation

The measurement of housing occupants' behavioral reactions to their social and physical surroundings is the primary purpose of postoccupancy evaluations (POEs). Since the early 1960s, systematic building evaluations have been conducted to determine the ways in which program and design objectives have been realized in residential settings. POEs are distinct from less formal methods of building or site assessments in that the evaluator seeks to place the occupied housing environment in the context of well-defined building owner, occupant, and designer goals and to measure the degree to which these goals are enhanced or hindered by the physical setting. Behavioral measures collected in these evaluations are concerned with privacy, security, use of internal and external space, building image, resident satisfaction, and personalization of space. The ultimate goal of accumulating information from POEs is the creation of design guidelines that will assist housing authorities and designers to plan effectively for the needs of the occupants. To date, a large number of POEs have been conducted in a variety of housing environments and in a number of countries, but few standardized or generalizable design guides have been established.

The term *postoccupancy evaluation* has been formalized within the design and behavioral science professions to define systematic assessments of physical environments and the effects these environments have on the people who inhabit and use them. In this respect, POEs are often collaborative research efforts that draw on the theories and methods of sociology, anthropology, psychology, interior design, architecture, landscape architecture, and human factors design. A useful compendium of POE methods and applications, compiled by Preiser, places POEs in the larger context of building diagnostics. This form of environmental analysis encompasses a wide range of environmental assessments that, in addition to behavioral measurements, includes building component performance and economic analyses. The essential elements of a POE are (a) clearly defined client and occupant goals that are used to inform the design process, (b) standardized instruments (questionnaires, interviews) that record and measure behavioral and attitudinal responses to the occupied environment, (c) theoretical models that show how those responses are linked to the client and occupant goals and physical conditions, and (d) standards of comparison that place data from any given POE in the context of similar occupant groups and physical settings. Recent advances in POE theory and applications have been focused in work and health care environments,

primarily because client groups that own and manage these facilities have vested interests in ensuring human responsiveness in process-oriented behavioral settings.

POEs in housing began in Great Britain during the 1960s. Governmental housing authorities became concerned that the rapid expansion of the housing stock built after World War II should be tested against the specific needs of the British public. The Ministry of Housing and Local Government, the Department of the Environment, and a number of government-sponsored building research units produced a series of policy and design recommendations that followed from a variety of empirical and demographic studies during the 1960s and early 1970s. In North America, social scientists and environmental researchers who developed specific theories of human behavior in the residential setting initiated the foundations for POE studies. Typical of this period is work by Gans (*Levittowners*), Hall (*Hidden Dimension*), and Jeffery (*Crime Prevention through Environmental Design*). Similar to governmental agencies in Britain, the U.S. Department of Housing and Urban Development (HUD) and state and local housing authorities began research in public housing during the mid-1960s. These efforts, executed primarily by academic researchers, resulted in methods and evaluation protocols that remain current today. Sources detailing this research include studies by Lansing, Marans, and Zehner, who undertook behavioral analyses of planned residential communities; Cooper, who describes the use of structured occupant interviews to test basic design assumptions about housing needs; and Francescato, Weidemann, Anderson, and Chenoweth, who outline methods for measuring occupant perceptions of specific housing attributes. Following the models of evaluation established by researchers in Britain and the United States, governmental housing authorities in Canada, New Zealand, and Australia executed a series of POEs and established ongoing programs of housing assessments during the 1970s.

The 1980s saw a decline in POEs in the residential setting in both Britain and North America, owing primarily to shifts in governmental housing policies toward privatization of housing stocks and reductions in housing research budgets. Systematic housing evaluations and behavioral research increased during the 1980s in Scandinavia and The Netherlands, in part owing to the growing interest in cohousing and cooperative-based housing construction projects. Detailed case study analyses and basic research findings are reported in the journals *Scandinavian Housing and Planning Research* and *The Netherlands Journal of Housing and the Built Environment,* and articles outlining technical and economic studies in housing can be found in the journal *Synopses of Swedish Building Research*. In the early 1990s, a number of housing authorities in various countries initiated pilot POE programs in concert with large-scale housing projects. An example of these efforts is the in-house evaluation agency established by the Korean National Housing Corporation, which has been charged with monitoring the effectiveness of design processes in meeting occupant needs. In general, the history of successful POEs of housing since the 1960s can be characterized by high degrees of direct governmental support and intervention in the establishment of clear client and occupant goals, the first and primary element of POEs described above. Whereas systematic building evaluations in commercial and health care settings can place human behavior in the context of instrumental (worker productivity, workplace safety, and increases in health indices) and economic (reduction in absenteeism, office efficiency, and facility use rates) measures of building performance, the creation of clearly defined owner and user goals in housing is problematic in the absence of central planning controls or strong residents' organizations. Ultimately, this lack of clearly defined residential goals makes it difficult to translate POE findings into objective design guidelines. To overcome this inherent difficulty in housing evaluations, the evaluator must rely on methods that address a diverse range of occupant needs and design variations.

The primary methods of data collection employed in POEs are (a) structured interviews or written-response questionnaires administered to the client organization, designers, and inhabitants of the occupied environment; (b) systematic observation and recording of behavior and physical conditions in the housing environment; and (c) analyses of data and comparison of results with similar evaluation studies. Because the purpose of these evaluations is to collect information that will be useful in the improvement of the housing environment, the interview and analysis processes must be carefully designed and executed to ensure the accuracy and representative nature of the behavioral data. References that outline applicable survey techniques are found in books by Bechtel, Marans, and Michelson (*Methods in Environmental and Behavioral Research*) and Zeisel (*Inquiry by Design*). These research methods are often supplemented by ethnographic techniques of analysis that search out the cultural dimensions of housing settings and enhance the evaluator's understanding of the individual characteristics of the residents and the collective nature of the residential community. Altman and Werner outline a number of such research methods in their book *Home Environments*.

In comparison to POEs in the workplace, housing evaluations tend to require more time to execute and greater interaction with the occupants of the environment. This is due in part to the complex nature of the residential setting, in which the perceptions of many environmental users, living in separate spaces, often with diverse cultural points of view must be measured in relation to a common set of design principles and user needs. To communicate this complexity, anecdotal quotes from the housing occupants, children's drawings of their houses and neighborhoods, and photographic documentation of significant features of the environment are important supplements to the more quantitative forms of data typically associated with survey research.

POEs have the potential to provide behavioral and environmental information that can be used to formulate objective design guidelines and housing standards. To date, such guides to housing policymakers and designers are rudimentary and can be found only in fragmented and isolated references. One of the most comprehensive and concise sources is *Housing as if People Mattered* by Cooper Marcus

and Sarkissian. This study draws together data from approximately 100 POEs—primarily British and North American—and sets out a series of recommendations for those making decisions about housing programs and designs. Information is provided about privacy, building image, use and personalization of spaces within and around dwellings, security, accessibility, and management of housing environments. The guidelines, although based on systematic evaluations of housing, are qualitative outlines of design action rather than prescriptive physical standards. Another reference that provides a useful synopsis of public housing quality in the United States is the HUD document *Case Study and Site Examination Reports of the National Commission on Severely Distressed Public Housing*. Although these reports rely only on the perceptions of designers and managers of housing projects, they do outline a series of factors that affect housing quality and inventory environmental conditions in public housing that can lead to improved living conditions.

A review of results from the POE literature as a whole reinforces the importance of several critical behavioral issues in housing designs. One is the need of housing occupants to find expression and control within their own home environment. The ability of housing residents to modify and personalize their immediate surroundings is a significant factor in promoting individual satisfaction and well-being.

A second issue is the degree of security and levels of privacy that the physical setting provides within the individual dwelling and throughout the housing development. A sense of personal ownership and the perception of collective responsibility are fostered in places that allow control of the environment without creating impressions of confinement.

A third issue is the provision of rich and varied physical settings that can accommodate a wide range of both structured and serendipitous activities. This factor is especially important in the ways that various age groups within a housing development interact with or separate themselves from each other at various times and seasons.

A fourth issue is the image created by the environment that captures and translates in form the nature and character of the housing community as well as the individual residents. A sense of place in housing is an essential factor in communicating meaning to both the occupants and the outside community the social values and cultural norms of the housing occupants. Finally, factors that deal with the pragmatics of maintaining the facilities and site—trash collection, cleaning and repairing, traffic control—are critical aspects of housing designs that often determine the eventual success or failure of the environment. In many instances, these factors serve as environmental thresholds that, if poorly designed, inhibit the ways in which housing residents are able to carry out basic activities.

A number of research organizations have been instrumental in disseminating the results of housing POEs. These are the Environmental Design Research Association (EDRA), the International Association for People-Environment Studies (IAPS), and People and Physical Environment Research (PAPER). The annual proceedings of these organizations provide current developments in POE theories and report the findings of ongoing evaluation studies. (SEE ALSO: **Behavioral Aspects**)

—Kent F. Spreckelmeyer

Further Reading

Altman, Irwin and Carol M. Werner, eds. 1985. *Home Environments.* Human Behavior and Environment Series, Vol. 8. New York: Plenum.

Bechtel, Robert B., Robert W. Marans, and William Michelson. 1987. *Methods in Environmental and Behavioral Research.* New York: Van Nostrand Reinhold.

Cooper, Clare C. 1975. *Easter Hill Village: Some Social Implications of Design.* New York: Free Press.

Cooper Marcus, Clare and Wendy Sarkissian. 1986. *Housing as if People Mattered.* Berkeley: University of California Press.

Francescato, Guido, Sue Weidemann, James R. Anderson, and Richard Chenoweth. 1979. *Residents' Satisfaction in HUD-Assisted Housing: Design and Management Factors.* Washington, DC: U.S. Department of Housing and Urban Development.

Gans, Herbert. 1982. *The Levittowners: Ways of Life and Politics in a New Suburban Community.* 2d ed. New York: Columbia University Press.

Hall, Edward. 1966. *The Hidden Dimension.* Garden City, NY: Doubleday.

Jeffery, C. R. 1971. *Crime Prevention through Environmental Design.* Beverly Hills, CA: Sage.

Lansing, John B., Robert W. Marans, and Robert B. Zehner. 1970. *Planned Residential Environments.* Ann Arbor, MI: Institute for Social Research.

Preiser, Wolfgang F. E., ed. 1989. *Building Evaluation.* New York: Plenum.

U.S. Department of Housing and Urban Development. 1992. *Case Study and Site Examination Reports of the National Commission on Severely Distressed Public Housing.* Washington, DC: Author.

Zeisel, John. 1981. *Inquiry by Design.* Monterey, CA: Brooks/Cole. ◀

▶ Power of Eminent Domain

The power of eminent domain is the authority of the state to acquire property rights. Private ownership does not preclude the state from obtaining title to assets under certain conditions. This authority is granted by express language in the Fifth Amendment of the U.S. Constitution when it says, "Nor shall private property be taken for public use, without just compensation." Although there is equivalent language in almost all state constitutions, no individual can claim the power of eminent domain; it is an authority granted to the government.

Eminent domain has been granted to the state and federal governments since before the American Revolution. However, in the Colonial period, it was generally employed unevenly. Thus, at the 1789 Constitutional Convention, the need arose to include the "takings clause" as one of the first 10 amendments by requiring that government acquisitions of private property be limited to the extent that "just compensation" would be required. Courts have traditionally held that just compensation should be equal to the market value of the property taken.

In a famous Supreme Court case in 1897, the court held that the just compensation requirement was part of the "due process" requirement of the Fourteenth Amendment. As a result, both the Fifth and Fourteenth Amendments are frequently cited in condemnation suits.

Disputes have also arisen over the meaning of the term *public use*. Although it is reasonably clear what *public* means, public use has been expanded in some jurisdictions. Traditionally, public use referred to the building of highways, parks, railroads, and utilities. However in some cases, public use seems to be stretched. For example, a controversial case in Michigan during the early 1980s satisfied the Michigan public use requirement by meeting the "public purpose" test, and the assemblage of a large tract involving the demolition of numerous of privately owned houses, churches, and businesses was upheld to build an automobile plant by a large private company. The claim was that Michigan met the public purpose test because economic opportunities would be expanded as a result of the taking.

The power of eminent domain is exercised through a condemnation procedure. The condemnee often disputes the valuation of the property taken, and often, experts are called to court by the state, the property owner, and sometimes the court. Because real estate valuation is not an exact science, valuation estimates often differ. Studies have shown that valuation estimates sometimes vary with the size of the property taken, a result taken to mean that just compensation awards may be partially determined by the quality of legal representation within the legal procedure.

Eminent domain is sometimes conceived of as a limitation on private property, because a private owner cannot stop the government from acquiring the property if needed for public use and just compensation is paid. In effect, the owner lacks the right to stop the government from converting the private property to public property. However, the just compensation requirement goes far to protect private property owners. The requirement that just compensation be paid ensures that if the government wants the property, it must pay the market price for the bundle of rights. In this way, governments are precluded from taking whatever they wish; they, too, must pay for what they want. The just compensation requirement is frequently absent in countries with weaker concepts of private property. In former socialist countries, any private property that did exist could be taken at any time, without any payment of just compensation.

Controversy has plagued eminent domain proceeds for much of the history of the United States. With the rise of statutory land use controls in the 20th century, governments have sought to control land through regulation (under the authority of police power) rather than by taking property under the power of eminent domain. In recent years, several Supreme Court cases have been decided on behalf of private owners against governments that had passed regulations limiting private property. Some have argued that many of the regulations ought to have been treated as takings; if so, just compensation is due the owners. It is often complicated to decide whether land use restrictions should be viewed as police power regulations or as takings under eminent domain. (SEE ALSO: **Abandonment;** *Property Rights;* **Urban Redevelopment**)

—Austin J. Jaffe

Further Reading

Epstein, Richard A. 1985. *Takings: Private Property and the Power of Eminent Domain.* Cambridge, MA: Harvard University Press.

Munch, Patricia. 1976. "An Economic Analysis of Eminent Domain." *Journal of Political Economy* 84:473-97.

Zarembka, Arlene. 1990. *The Urban Housing Crisis: Social, Economic, and Legal Issues and Proposals.* Westport, CT: Greenwood. ◀

▶ Prefabrication

Prefabrication is a method of production in which a part intended to fit in a specific assembly is made at a distance from the site of its use. However, in large projects, prefabrication work can occur adjacent to the site of assembly to simplify delivery. Prefabrication takes place on the order of its user, rather than on the initiative of its producer. Prefabrication can be done by hand or by sophisticated machines. Examples of prefabricated parts include many kitchen cabinets, roof trusses, many windows, and panelized houses.

Prefabricated housing products are usually made of dimensioned lumber, fasteners, sheet materials, coatings, hardware, electrical or mechanical parts, and/or special products made to the order of the prefabricator by another producer.

Prefabrication has long been employed to make parts for buildings. This has been useful when there was an advantage to making parts away from the building site for reasons of improved working conditions, quality assurance, and protecting sophisticated equipment and materials in a more efficient way than could be done on-site, and when the site was too small for the efficient storage of products and final production of the parts.

In modern, advanced plants, fabrication processes and materials stocks are normally in place, or suppliers are ready for quick delivery to supply processes ready to be set in motion when an order is received. The parts used in making these items are almost always industrially produced. For example, the parts for windows may be made by mass production (glass, aluminum hinges, insect screen, weatherstripping) but assembled into complete units by franchised dealers only when orders are placed. The window in that instance is prefabricated from industrialized parts. Some of these parts are made by the window company itself for use in all its window lines, and some are made by other manufacturers, also for the wider market.

Prefabrication and industrialization represent two fundamentally different but entirely compatible production modes, both of which are necessary today for housing production. The important distinction is the locus of initiative to produce. Prefabrication produces for use, whereas industrialized production is for trade. Prefabrication begins where industrial production stops. (SEE ALSO: **Industrialization in Housing Construction**)

—*Stephen Kendall*

Further Reading

Bender, Richard. 1973. *A Crack in the Rear View Mirror: A View of Industrialized Building.* New York: Van Nostrand.

Bruce, Alfred and Harold Sandbank. [1944] 1972. *A History of Prefabrication.* New York: Arno.

Fitchen, John. 1986. *Building Construction before Mechanization.* Cambridge: MIT Press.

Habraken, N. John. 1988. *Transformations of the Site.* Cambridge, MA: Awater Press and MIT Department of Architecture.

Herbert, Gilbert. 1984. *The Dream of the Factory-Made House: Walter Gropius and Konrad Wachsmann.* Cambridge: MIT Press.

Kaiser, Edgar F., Chairman. 1968. *A Decent Home: The Report of the President's Committee on Urban Housing, Technical Studies.* Vol. 2. Washington, DC: Government Printing Office.

Kelly, Burnham. 1959. *Design and the Production of Houses.* New York: McGraw-Hill. ◀

▶ President's Committee on Urban Housing (Kaiser Committee)

The President's Committee on Urban Housing, popularly known as the Kaiser Committee, was appointed by President Lyndon Johnson and made its report to him on December 11, 1968. It was named for its chairman, Henry F. Kaiser, head of the Kaiser Industries Corporation. The committee's tenure paralleled those of the Douglas Commission (National Commission on Urban Problems) and the Kerner Commission (National Advisory Commission on Civil Disorders). Unlike these two national commissions, however, the Kaiser Committee had a relatively narrow focus, concentrating its attention on the housing industry and the development of both subsidized and market rate housing. Consequently, the report of the Kaiser Committee lacks the breadth of scope of the reports of the other two committees, each of which dealt broadly with the still continuing problems associated with urban poverty and racial isolation.

That the Kaiser Committee focused its efforts on the housing industry was no accident. Of its 18 members, 8, including Kaiser, were from the building industry, 3 were from financial institutions, and 3 were from labor unions. Consequently, the committee's emphasis on the housing construction industry, on housing finance, and on labor was no surprise. Consistent with this focus, the committee's primary historical significance was its emphasis on greater use of the private sector to develop subsidized housing. Reflecting on the history of federal housing subsidy programs, the committee concluded:

> The nation has been slow to realize that private industry in many cases is an efficient vehicle for achieving social goals. The Federal housing subsidy programs of the 1930s assumed that the initiation and ownership of subsidized housing were the direct and full responsibilities of government. . . . Not until 1961 did subsidy programs permit participation by profit-motivated entrepreneurs. Since then reliance on the private sector has expanded rapidly. . . . Nevertheless, some programs still make too little use of the talents of private entrepreneurs. . . . One of the basic lessons of the history of Federal housing programs seems to be that the programs which work best—such

as the FHA mortgage insurance programs—are those that channel the forces of existing economic institutions into productive areas (p. 54).

Emphasis on using the private sector to produce subsidized housing was reflected in the Housing and Urban Development Act of 1968, developed in consultation with the Kaiser Committee, which launched the Section 235 and 236 programs for the subsidization of privately developed owner and rental-occupied dwelling units. The 1968 act was consistent with the Kaiser Committee's most significant recommendation: that the nation commit to a national housing goal, in which 26 million new and rehabilitated housing units be produced by 1978, of which 6 to 8 million would be federally subsidized units. However, only about 2.7 million subsidized units were constructed or rehabilitated during the period of 1968 to 1978, although about 18.5 million unsubsidized units were produced during this period. Moreover, after 1973, when the Nixon Administration began to dismantle existing federal housing programs (culminating in the Housing and Community Development Act of 1974), the national housing goal failed to have any impact on national housing policy.

The Kaiser Committee's other recommendations focused on enhancing the efficiency of the building industry and ensuring that adequate financing, land, materials, and labor would be available to achieve the national housing goal. Nevertheless, the committee did foresee some role for the nation's existing housing stock, recommending that portions of monies appropriated for the Section 235, 236, and rent supplement programs be used for existing housing. The committee also called for creation of an experimental housing allowance program, which was adopted by Congress in 1970.

Some of the committee's most historically significant recommendations pertained to housing finance and the availability of credit. The committee recommended that the Government National Mortgage Association (GNMA) be permitted to issue securities, backed by a federal guarantee of principal and interest, that would be used to finance home mortgages. This recommendation has been realized in the GNMA mortgage-backed securities program, in which Federal Housing Administration (FHA) and Department of Veterans Affairs (VA) mortgages are pooled, with timely payment of principal and interest guaranteed by GNMA, and made available to investors not traditionally involved in the mortgage market. The Kaiser Committee also recommended that GNMA undertake purchase of mortgages for privately owned subsidized housing. This program, known as the GNMA Tandem Program, was used primarily in the 1970s, to finance Section 8 new construction units at below-market interest rates. (SEE ALSO: *Government National Mortgage Association; National Commission on Urban Problems; Privatization*)

—*Charles E. Connerly*

Further Reading

Fish, Gertrude S., ed. 1979. *The Story of Housing*. New York: Macmillan.

Hartman, Chester W. 1975. *Housing and Social Policy*. Englewood Cliffs, NJ: Prentice Hall.

Hays, R. Allen. 1995. *The Federal Government and Urban Housing: Ideology and Change in Public Policy*. Albany: State University of New York Press.

President's Committee on Urban Housing. 1967. *The Report of the President's Committee on Urban Housing: Technical Studies*. Washington, DC: Government Printing Office.

———. 1968. *The Report of the President's Committee on Urban Housing: Technical Studies*. Washington, DC: Government Printing Office.

———. 1969. *A Decent Home: The Report of the President's Committee on Urban Housing*. Washington, DC: Government Printing Office.

Weicher, John C. 1980. *Housing: Federal Policies and Programs*. Washington, DC: American Enterprise Institute. ◀

▶ President's Conference on Home Building

The President's Conference on Home Building and Home Ownership was called by President Herbert Hoover and met in Washington, D.C., December 2 to 4, 1931. Some 3,700 leaders and experts in housing and planning at that time attended the conference. President Hoover charged the participants with focusing on homeownership and hinted that the delegates should endorse his proposal for creating a system of home loan banks. That proposal was later adopted in 1932 as the Federal Home Loan Bank Act.

The conference is primarily significant for the expression by the cochairperson of the conference, Interior Secretary Ray Lyman Wilbur, that the housing industry in the United States was facing a crisis in which, if the industry failed to provide adequate housing at moderate prices, government intervention in the housing market would be necessary. Various conference committees also envisioned increased government action in the eradication of slums and the provision of replacement housing, as well as the reform of mortgage lending. The conference did not address the issue of federal government-induced development of housing. The conference, therefore, is viewed as a way station in the development of a national housing policy, a transition between the pre-Depression view that the federal government had no business in the housing sector and the Roosevelt-era programs, such as Federal Housing Administration mortgage insurance and the public housing program, in which the federal government played a direct role in the housing market.

—*Charles E. Connerly*

Further Reading

Fish, Gertrude S., ed. 1979. *The Story of Housing*. New York: Macmillan.

Mitchell, J. Paul. 1985. *Federal Housing Policy and Programs: Past and Present*. New Brunswick: Rutgers, State University of New Jersey, Center for Urban Policy Research.

President's Conference on Home Building and Home Ownership. 1932. *Housing Objectives and Program*. Washington, DC: Government Printing Office. ◀

▶ Price-Level Adjusted Mortgage

A price-level adjusted mortgage (PLAM) is a lending instrument in which an overall price index (such as the gross national product deflator or Consumer Price Index) is used periodically to adjust subsequent mortgage payments for the effect of inflation during the period. It is one among several alternatives to a conventional fixed-rate mortgage (FRM); others include the adjustable-rate mortgage (ARM), graduated-payment mortgage (GPM), and the shared-appreciation mortgage (SAM). When inflation is persistently high (as it was in the late 1970s and early 1980s), such alternatives become attractive to lenders or borrowers. In a conventional FRM, lender and borrower each gamble on the rate of inflation. If consumer prices rise faster than expected, lenders risk a loss in the purchasing power of their capital over the life of an FRM. Lenders, on the other hand, count on inflation to make FRM payments more affordable with the passing of time. In other words, a proportion of the mortgage interest rate can be thought of as an inflation premium. For lenders, the attraction of ARMs or PLAMs is that they offer protection against changes in mortgage rates; the borrower assumes the risk of increased inflation. In this, a PLAM is a special version of an ARM: one that protects lenders only against interest rate changes that arise because of unanticipated inflation. For borrowers, the attraction of these mortgage alternatives is that they reduce either the cost of the mortgage in its earlier years (the so-called tilt problem) or the cost of borrowing. (SEE ALSO: *Alternative Mortgage Instruments*)

—*John R. Miron*

Further Reading

Downs, A. 1985. *The Revolution in Real Estate Finance*. Washington, DC: Brookings Institution.
Fabozzi, F. J. and F. Modigliani. 1992. *Mortgage and Mortgage-Backed Securities Markets*. Boston: Harvard Business School Press.
Pozdena, R. J. 1988. *The Modern Economics of Housing: A Guide to Theory and Policy for Finance and Real Estate Professionals*. New York: Quorum (see chap. 8). ◀

▶ Principal

Using the standard fixed-rate mortgage (FRM), each debt service payment is divided into two portions: the interest payment (or the payment *on* the mortgage balance) and the principal payment (or the payment *of* the mortgage balance). The principal payment is the portion of the debt service used to repay the mortgage note.

Many consumers and housing analysts have often misunderstood the payment of principal. Many novice borrowers think that interest and principal payments are made in equal proportions when a mortgage is being repaid. However, it should be emphasized that each mortgage payment (called debt service) can be divided into an interest payment and a sinking-fund payment. The sinking fund is calculated so that the accumulation of all payments over the life of the mortgage will sum to the amount of the mortgage owed. Thus, the standard FRM is said to be a "self-amortizing loan" because the sum of the sinking-fund payments will completely repay the amount borrowed.

Another misunderstanding is the accumulation of principal payments under the amortization procedure. Because all interest must be paid first out of the debt service payment and because the outstanding balance owed at the beginning of the loan is large, a high percentage of the payment goes to paying interest. Any remaining debt service payment greater than the amount of interest owed each period goes to the principal payment (this is the sinking fund). As time goes on, the percentage allocated to interest of the debt service falls and the percentage allocated to principal rises. But for 25- or 30-year mortgages, at most interest rates, the so-called equity buildup proceeds very slowly.

Perhaps the greatest misunderstanding is the mistaken value placed on the equity buildup when making the "rent versus buy" decision. It is frequently argued that if a household rents, it ends up only with "rent receipts." However, if it purchases and finances with debt, the household acquires an equity (i.e., ownership) interest by systematic repayment of the mortgage through principal repayments. At the time of sale, it is argued, that instead of rent receipts, the household will acquire wealth through equity buildup. The problem with this strategy is that equity buildup consists of a series of payments to repay the mortgage, dollar-for-dollar right out of the household's checking account. In other words, the principal repayments are extra payments (over and above interest) specifically set aside (in a sinking fund) to repay the mortgage note. Therefore, principal payments are required under the FRM and have nothing to do with the decision to rent or buy.

Principal payments are also important because as equity buildup occurs, the owner acquires a larger and larger ownership interest in the property. With such a large interest, the owner is likely to maintain the property better, try harder than ever not to default on the mortgage, and consider the equity in the property as an increasingly important part of the household's portfolio. Therefore, as a form of household savings, principal plays an essential role in the accumulation of wealth for households. (SEE ALSO: **Mortgage Finance**)

—*Austin J. Jaffe*

Further Reading

ABA Standing Committee on Lawyers' Title Guaranty Funds. 1991. *Buying or Selling Your Home: Your Guide to: Contracts, Titles, Brokers, Financing, Closings*. Chicago: American Bar Association.
Jacobus, Charles J. 1996. *Real Estate Principles*. Upper Saddle River, NJ: Prentice Hall.
Jaffe, Austin J. and C. F. Sirmans. 1995. *Fundamentals of Real Estate Investment*. 3d ed. Englewood Cliffs, NJ: Prentice Hall.
Shim, Jae K. 1996. *Dictionary of Real Estate*. New York: John Wiley.
Tosh, Dennis S. 1990. *Handbook of Real Estate Terms*. Englewood Cliffs, NJ: Prentice Hall. ◀

▶ Private Rental Sector

Renting versus Homeownership: Housing "Preference" in the United States

Private rental housing can be defined as privately owned single-family and multifamily housing units leased for rental by the owner. In 1995, about 34 million households lived in such units, only 16.5% of whom (mainly the very poorest) reported receiving any form of government subsidy or assistance.

"Housing" in the United States is symbolized by the freestanding single-family home, in which approximately two of every three people live. In a country where homeownership is virtually regarded as a birthright, private rental housing is seen as the illegitimate stepchild. Public policy and tax laws have long favored homeownership. In particular, federally guaranteed mortgages result in lowered interest rates, while at the same time private ownership confers significant tax advantages over renting: All mortgage debt service and local property taxes are fully deductible on state and federal tax returns.

As a consequence of consumer preferences and public policy, the proportion of people in the United States who own their own homes has risen steadily throughout much of this century, reaching a high of 65.6% in 1980 (see Table 22). During the 1980s, this process was briefly reversed, with the proportion of owners declining to 63.8% in 1986—the result of rising mortgage interest rates and home costs, which froze many people out of the dream of homeownership. In the late 1990s, the proportion of owners was again rising, having reached 64.5% by 1993 and increasing to 65.7% in 1997.

With few exceptions (primarily singles and young professionals, as well as residents of a handful of large cities such as New York), private rental housing is regarded as a fate to be endured until one can afford to make a down payment on a house. Permanent renting, in this prevailing view, is the unhappy destiny of those too economically marginal to buy. Renters are on the average considerably poorer than homeowners in the United States—and becoming more so. Whereas average renter income stood at 64.5% of average homeowner income in 1972, by 1995, this ratio had dropped to 53.9%.

It is worth noting that although the United States has had rates of homeownership comparable with a number of other industrialized nations, its rates of private rental have been among the highest in the industrial world (Table 23). That is because many other industrial nations have extensive public housing programs or other forms of social (i.e., nonprofit) ownership. In Britain, for example, roughly the same percentage of households own (64%), but private

TABLE 22 Percentage of Households Owning Their Own Housing, 1973 to 1997

1973	1976	1980	1983	1986	1990	1993	1997
64.4	65.2	65.6	64.9	63.8	64.1	64.5	65.7

TABLE 23 Housing Tenures as a Percentage of Occupied Dwellings (Selected Industrialized Nations)

Country	Privately Owned	Private Rental	Public Rental	Other Social Rental	Other Tenures
United States (1988)	64.0	31.0	2.0	—	3.0
Canada (1986)	63.0	33.0	2.0	2.0	—
Great Britain (1987)	64.1	7.5	25.9	2.5	—
Federal German Republic (1984)	42.0	38.0	—	20.0	—
Netherlands (1987)	43.0	12.0	7.0	37.0	1.0
Italy (1981)	65.9	29.7	5.6	—	5.6
Japan (1983)	62.3	24.9	7.6	—	5.2

SOURCE: Van Vliet-- (1990, Table 1.4).

rental rates have historically been far lower, because council housing (or public housing) was commonplace. Although some of the council housing stock was sold off during the Thatcher–Major privatization schemes, public ownership is still commonplace. Similar patterns are found in Germany and the Netherlands.

Private Rental Housing and Affordability

Private rental housing has thus confronted a restricted market composed largely of those households for whom homeownership is either undesirable or economically unviable. This has contributed to an increasing gap between renters' incomes and the cost of rental housing. Between 1972 and 1982, average renters' incomes declined in every year except one, falling from $19,270 to $15,305 over the decade, a decline of 21% (all figures are in 1989 dollars). Although renters' incomes increased slightly during the remainder of the decade (to $17,300 in 1989), they have dropped sharply since that time, falling to $15,618 in 1993. During the same period, average rents have risen, from $372 in 1972 to $403 in 1993. As a consequence, renters who in 1978 were averaging about 22% of their income on rent, in 1994 were averaging closer to 27%, and if housing-related expenditures such as heating and utilities are included, the figure rises to 31%.

The affordability gap in rental housing is largest among low-income renters. One of every 4 renters lives in poverty, compared with only 1 of 13 homeowners. Federal standards assume that most households cannot afford to spend more than 30% of their income on rent; yet 8.6 million impoverished renter households in the United States average close to half. The incidence of poverty among renters is especially acute among minorities, accounting for 46.3% of all Hispanic renter households and 54.6% of all black renter households in 1995. Such households truly can be said to suffer "shelter poverty": Their substantial housing costs leave virtually nothing for food, clothing, health care, and other necessities. For example, the median monthly gross rent (including utilities and heating) spent by poverty-level families in 1995 was $391. In that same year, the poverty level for a three-person family was defined as $12,158.

After paying the median rent, such a family would have less than $20 a day left over for all other expenses.

The fact that private rental housing is primarily destined for lower-income people results in downward pressures on rents, resulting in a long-term pattern of private disinvestment in the rental housing market. In 1970, the number of low-rent units exceeded the number of low-income households by 2.4 million units, nearly a third of all rental units. By the mid-1980s, this surplus of low-rent units had turned into a deficit of some 3.7 million units. Between 1978 and 1985, the number of low-rent units declined by a half million; the number of low-income renters grew by 3.6 million. The problem has become even more acute in recent years. In 1972, for example, there were approximately 1 million multifamily housing starts. (The vast majority of multifamily units are destined for the private rental market, although a small number are owner-occupied condominiums.) Annual starts dropped steeply during the next 3 years, falling to 268,000 in 1975. For the next 10 years, starts rose and fell in roughly 3-year cycles, never again approaching the 1 million mark; the peak year—1985—saw only 669,000 units constructed. Since 1985, production has dropped every year, reaching a low of only 162,000 units in 1993 but then rose again to 270,800 in 1997.

Interestingly, rental housing vacancy rates have steadily increased during the same period, from 5.3% in 1972 to 7.7% in 1997. This suggests that as rents have increased faster then renters' incomes, some renters have responded by doubling up (or becoming homeless). At the same time, the dwindling profitability of rental housing relative to other forms of investment has led to the virtual cessation of rental housing construction, especially for the lower-income part of the market.

Federal, State, and Local Regulation of Rental Housing

There is virtually no direct federal regulation of privately owned, unsubsidized rental housing. There are no federally mandated rent controls, protections against evictions, building code standards, or warranties of habitability. Only in response to the civil rights movement of the 1960s did the federal government begin to regulate private rental housing by guaranteeing access to minority groups. In 1962, the Kennedy Administration issued the first fair housing orders, containing limited prohibitions against discrimination in federally assisted housing projects. Title VIII of the landmark 1968 Civil Rights Act contained broad prohibitions against discrimination in both private and public housing. Most recently, the Fair Housing Amendments Act of 1988 extended protections to the handicapped and families with children.

The principal impact of the federal government on private rental housing is not to be found in direct regulation, however, but in the indirect effects of credit and tax policies. Virtually all housing construction and purchase in the United States is financed through private credit, rendering housing costs and starts highly susceptible to fluctuations in credit availability and cost. Fluctuations in interest rates regulate the rental housing market in several ways. First, and most obviously, they directly affect rent levels through their effect on mortgage payments. Second, interest rates also add directly to the cost of construction, sales prices, and other consumer credit. Finally, interest rates have a strongly countercyclical effect on the housing economy, producing alternate periods of boom and bust. Federal income tax policies also serve to indirectly regulate rental housing, because they strongly affect profitability. Income property owners are able to reduce their taxable income not only by deducting actual expenses but also by depreciating physical assets. This provides paper losses and therefore higher profits while at the same time encouraging speculation, which drives up sales prices and thereby, rents.

State and local housing regulation dates to New York City's Tenement House Act of 1867, which prescribed minimum standards for fire, safety, ventilation, and sanitation of multifamily housing, and the Tenement House Act of 1901, which created more adequate enforcement mechanisms. The right of state and local governments to regulate housing has been repeatedly upheld by the courts; the right of localities to zone for land use dates to the landmark 1926 *Euclid v. Ambler* decision. The Supreme Court's ruling in that decision reflects the central principle that a key function of local regulation is to maintain property values. The Court argued that zoning restrictions enhance constitutional property rights by providing a predictable environment for one's investment. Regarding rental housing, the *Ambler* decision held that apartments could be legally restricted to prescribed zones because they posed a threat to the value of single-family homes.

Inasmuch as zoning proved to be a crude mechanism for controlling land use, cities and counties have over the years devised a host of diverse and often contradictory means for regulating housing, including health and safety requirements and minimal standards for building siting, design, and construction. In addition, development in some communities is governed by locally enacted general plans to which particular projects must conform. Regulations managing growth and development (permits, impact fees, growth controls, linkage programs) have increased markedly in the past quarter century. These include mandatory environmental review, down-zonings, service hookup moratoria, and building permit allocation systems.

Growth-controls are found primarily in the faster growing regions of California, suburban New York and New Jersey, and Florida. They are frequently accompanied by regulations governing landscaping, environmental preservation, and architectural aesthetics, as well as special restrictions on land use in areas deemed to be of special historical, architectural, or environmental value. Finally, where municipal finances are limited, efforts have been made to shift the costs of infrastructure such as roads and sewer lines from the general taxpayer to the developer and thereby eventually to the final owner or renter. In recent years, some communities have also adopted regulations aimed at increasing the supply of affordable housing. Among these are "inclusionary" zoning requirements, whereby developers of large-scale projects must include a designated number of units for specified low-income households or other designated categories and "linkage" programs requiring commercial developers to help provide housing for the workers who will find eventual employment in their projects. The most visible efforts to address rental housing costs consist of local rent controls.

The Effects of Regulation

In this climate, the rental housing crisis of the late 1970s led conservative policymakers in both parties to identify "excessive" government interventions as the chief culprit for rising rental housing costs, calling for an end to housing regulations altogether. Although the call for privatization in the rental housing industry originated during the Democratic Carter presidency, it flourished during the fervently antiregulatory climate of the Republican Reagan and Bush Administrations and showed no sign of abating during the first two years of the Democratic Clinton Administration. The Republican Party's 1994 Congressional triumph ensures that such antiregulatory pressures will continue at least into the near future.

Many forms of housing regulation have come under attack. These include health, safety, and environmental regulations, growth controls, and other limitations on development and—most especially—rent controls. The National Association of Home Builders, for example, has faulted "overregulation" for as much as 25% of home sales prices, and former President Reagan's Commission on Housing concluded that states should prohibit localities from limiting development unless "a vital and pressing governmental interest" can be demonstrated. An important study commissioned for the bipartisan National Housing Task Force identified a large number of ways in which local land use regulations are responsible for high building costs; impact fees and outdated building codes alone were held to contribute as much as 30%. The National Multi-Housing Council, a lobbying and information group that represents the largest owners and developers of rental housing in the United States, has claimed (without providing evidence) that regulations currently add from $50 to $100 to monthly rents, thereby threatening the rental housing industry with extinction. Local rent controls have been held responsible for everything from rental housing shortages to homelessness. Although the courts have repeatedly upheld the legality of local rent regulations, their supposedly adverse effects have been used to justify federal and state efforts to ban or weaken local controls.

There has been a great deal of research on the impact of regulations on housing costs, much of it consisting of anecdotes about particular projects or statistical analyses that fail to compare locales with varying degrees of regulation. Regulations that add to developers' costs have been estimated to add anywhere from 2% to 30%, with the higher figures generally coming from industry reports. Despite the belief that the U.S. housing economy is "overregulated," most research indicates that local regulations are relatively minor determinants of housing production and costs. In most housing markets, new construction typically accounts for only a small percentage of the total housing stock, and restrictions on construction are likely to have only a marginal effect on overall price levels. An extensive review of a large number of studies concerned with the relationship between various forms of regulation and housing prices concluded that *supply restrictions* typically add

under 10% to housing, generally in the 5% range. One econometric analysis concluded that "the combined effect of increasing development densities by one unit per acre, reducing development fees by 50%, and doubling supplies of vacant land—all drastic steps—would be to lower the sales price of a new home by . . . roughly 6%." Direct regulatory costs have been characterized as especially onerous; yet most research indicates that they generally add at most a few percent to total costs. The effect of rent controls on investment in rental housing has been extensively studied; prevailing economic theory notwithstanding, virtually all studies to date have concluded that moderate rent controls have no adverse effects on either new construction or maintenance of controlled units.

The Future of Private Rental Housing

During the past two decades, economic globalization has contributed to the conversion of the U.S. economy from one based on high-wage industries, to one increasingly reliant on low-wage services. This process has accelerated in recent years, producing growing inequality. These changes do not auger well for the private rental housing market nor for those who rely on it for housing. Lower-income households in particular are likely to be caught in a continuing squeeze between their limited incomes on the one hand and rising rents and diminished supply on the other.

It would appear that in light of such long-term structural changes, some form of government support is necessary to make low-cost housing economically viable. The United States, more than virtually any other industrial society, relies on market forces to deliver housing. For lower-income households, government enters the picture only as a last resort, primarily through subsidizing the very bottom of the private rental market in hopes of restoring its profitability. In contrast with many European and some Asian countries, publicly assisted housing is thoroughly disdained as the sole recourse of those who are too poor to afford even the least desirable private rental units. This is indeed unfortunate, because it is questionable whether the current limited mix of public housing, private rental subsidies, housing allowances and vouchers, tax credits, and public-private partnerships will be adequate to the task of housing the bottom third of the U.S. population in the years to come. (SEE ALSO: *Contract Rent; Flat-Rent Concept; Rent Burden; Rent Control; Rent Decontrol; Rent Gap; Rent Strikes; Section 515; Security Deposit;* **Tenure Sectors**)

—*Richard P. Appelbaum*

Further Reading

Appelbaum, Richard P. and John I. Gilderbloom. 1988. *Rethinking Rental Housing*. Philadelphia: Temple University Press.

Dolbeare, Cushing. 1992. *The Widening Gap: Housing Needs of Low Income Families*. Washington, DC: Low Income Housing Information Service.

Downs, Anthony. 1988. *Residential Rent Controls: An Evaluation*. Washington, DC: Urban Land Institute.

Joint Center for Housing Studies. Annually. *The State of the Nation's Housing, 1994*. Cambridge, MA: Harvard University, Joint Center for Housing Studies. Website: http://builder.hw.net/monthly/archives/harvard/exec.htx

Keating, D. (ed.). 1998. *Rent Control: Regulation and the Rental Housing Market.* New Brunswick: Center for Urban Policy Research.

Kemeny, J. 1994. *From Public Housing to Social Housing: Rental Policy Strategies in Comparative Perspective.* New York: Routledge.

Leonard, Paul A., Cushing N. Dolbeare, and Edward B. Lazere. 1989. *A Place to Call Home: The Crisis in Housing for the Poor.* Washington, DC: Center of Budget and Policy Priorities, Low Income Housing Information Service.

Stone, Michael. 1993, *Shelter Poverty: New Ideas on Housing Affordability.* Philadelphia: Temple University Press.

Van Vliet--, Willem. 1990. *International Handbook of Housing Policies and Practices.* New York: Greenwood.

Varady, David P. and Barbara J. Lipman. 1994. "What Are Renters Really Like? Results from a National Survey." *Housing Policy Debate* 5(2):491-532. ◀

▶ Privatization

Privatization in the housing sector is the shift in the ownership of dwelling units, firms providing maintenance, and firms constructing housing from state organizations to private individuals and firms. Privatization on a significant scale is a phenomenon of the 1980s and 1990s, beginning in Britain and developing later in the United States and the planned economies of Eastern Europe, the former Soviet Union, and China. Privatization of housing finance—the shift from state funding, often at below-market interest rates, to private bank funding—has also been occurring in these countries and certain developing nations. The primary argument for privatization has been to achieve an increase in the economic efficiency with which housing services are produced and allocated.

Different mechanisms have been involved for the transfer of different types of assets. Regarding the ownership of dwelling units, the transfer is accomplished through (a) restitution (restoring ownership of the property to the former owner from whom the state took it); (b) sales by the state to private entities, usually individuals, at nearly universally deep price discounts; and (c) free-of-charge transfers of title.

State construction enterprises have been transferred to private ownership through (a) restitution (although rarely), (b) negotiated sale or free transfer to workers and management of the firm, and (c) sales, either as part of national privatization "voucher" programs (in which citizens have been able to use vouchers freely provided to them by the government to purchase shares in companies) or completely open sales conducted on a tender or negotiated basis. Often, construction companies have been broken into smaller components as part of the privatization and rationalization process.

The privatization of maintenance companies occurred in much the same way as construction enterprises. However, the experience has seemed to be for replacement of such firms rather than their privatization and restructuring.

Finally, privatization of housing finance, particularly mortgage finance, has been enabled by government action making private lending potentially profitable. The key here has been the elimination of interest rate controls on private lenders and of the origination of deeply subsidized loans by government-sponsored banks, which made it impossible for private banks to compete. Of course, in the formerly planned economies, creation of private banks also had to be permitted.

The primary gain from privatization was expected to be improved efficiency. Privatization accompanied by the introduction of genuine competition carried the realistic expectation of enhanced performance by construction and maintenance firms. In countries with a large share of the total stock under state ownership, the transfer of ownership of dwellings was the sine qua non for the creation of a housing market by making a critical mass of units tradeable on freely negotiated terms. Private ownership was expected to encourage investment and, in multifamily buildings, make tenants more demanding for quality maintenance services. Substantial budget savings were typically expected as the new owners paid for maintenance, replacing the former system of low rents and large maintenance subsidies.

The privatization decision process has differed sharply between firms and dwellings. Privatization of construction enterprises and maintenance firms, of which a substantial share of the total assets have been government owned, has generally been mandated by national policies. Although the pace of privatization has differed markedly among countries, most firms realized its inevitability. They therefore chose to privatize when management believed it could strike the best deal, although in some countries management did not even have much to say about the timing of the transaction.

In contrast, the tenants of a government-owned unit were offered the opportunity to privatize; they were not compelled to take the step. Important considerations include (a) the value of the unit on the market; (b) the amount of rehabilitation required by the unit and the building in which it is located and who will pay for it; (c) the price being charged by the state for the unit, including processing fees; (d) the strength of the tenant's rights "without privatization" to remain in the unit and to pass it on to his or her children; and (e) the extent of the increase in operating costs (maintenance fees, utilities, and property taxes) expected if the unit is privatized. Evidence available for the United Kingdom, Hungary, and Russia suggests that tenants took these factors into account when making their decisions, with unit value being the dominant determinant.

Housing Privatization in the United States. Within the United States, privatization has dealt largely with the transfer of ownership of publicly owned units to sitting tenants and secondarily with transfer of management to private entities. Quantitatively, ownership transfer has been unimportant: titles to only a few thousand units have been transferred to sitting tenants from a public housing stock of about 1.3 million units (constituting only about 3% the rental stock). In 1997, there was little expectation that the program would ever transfer large numbers of units.

The low volume of transfers has been directly related to a key characteristic of the environment in which the program has operated: Typically, the sitting tenants in public housing are poor, making questionable the prospects of most for paying even operating costs. Another defining feature of the U.S. program is that it has been motivated by

TABLE 24 Policy Environment for Privatizing Social Housing in Eastern Europe as of Mid-1993

	Poland	Hungary	Bulgaria	Russia
Government-owned housing				
Number of housing units (thousands)[a]	2,078[b]	703	275	32,696
Percentage of total stock	19	20	9	67
Transfer to local government (m/y)	5/90[c]	10/90[d]	always 12/91	—
Rent determination				
Controlled by	Central government	Local government	Central government	Local government
Rents being adjusted to move to market[e]	No	No	No	Yes[f]
Modification of tenant rights				
Extent	Major	Major	Major	Modest
Status	Pending	Enacted	Enacted	Enacted
Laws permitting sales of units	1970	1982	1991[g]	1988[h]
Percentage managed by				
Local government	Almost 100	Almost 100	Almost 100	40[i]
Private firms	Negligible	Unknown	Unknown	Negligible

a. Number of units at the start of the privatization process.
b. Excludes enterprise housing.
c. Some questions about legal status due to registration problems.
d. Further clarification of law passed in July 1991.
e. Rents on state units increasing faster than inflation with intention of at least reaching full-cost recovery.
f. Provided 1992 law passed by the Supreme Soviet is implemented.
g. Although sales were permitted earlier, only as of 1991 were price restrictions removed.
h. Significantly modified in July 1991 and December 1992.
i. Balance managed by enterprises and government agencies for their employees.

the proposition that making tenants into owners will have a significant impact on their interest in participating in the country's mainstream economic life. Each local program has contained job training and other components designed to foster and exploit the expected newly awakened interest of tenants to participate in the labor market. Thus, unlike initiatives in other countries, privatization in the United States has been motivated less by housing sector reform (although reducing the size of the public housing stock has been an objective) and more by poverty amelioration.

The U.S. federal government has actively promoted the hiring of private management firms by public housing authorities (PHAs) for at least two decades. For about the same period, tenant management has been the subject of experiments, and in a few cities, it has been successful. An unknown number of PHAs have hired private firms to manage at least some of their projects. Tenant management has been active at about 10% of the country's 3,000 PHAs, typically with tenants responsible for some management tasks at one or two of a PHA's projects. Large-scale attempts at evaluating the efficacy of private and tenant management compared with PHA management have been inconclusive.

Housing Privatization in Western Europe. Within Western Europe, privatization has really meant the transfer of unit ownership to tenants in the United Kingdom, where the Thatcher government originated the policy on a significant scale. The dominance in housing privatization of the United Kingdom among Western European countries corresponds to its much higher share of publicly owned housing—22% in 1989, compared with 6% in the Netherlands and traces in other countries. (Social housing, normally operated by nonprofit sponsors, is much more important in the other nations.)

Having sold more than 1.2 million public housing units during the 1979 to 1992 period—an amount equivalent to about a fifth of 1979 council (public) housing stock—the United Kingdom has often been cited as an example of a country successfully implementing a public housing sales program. The large volume of units sold resulted from a concerted policy to make purchase attractive to sitting tenants: increasing rents coupled with sales prices set at substantial discounts from the market price (averaging about 45% over the period). The easy availability of housing finance has been important in permitting tenants to purchase.

The United Kingdom also has tried to initiate the large-scale transfer of ownership of whole projects from public authorities to nonprofit "housing associations." Under the scheme, the tenants must approve the transfer. By 1995, however, tenants proved reluctant to approve such transfers, and private landlords have found it difficult to borrow sufficient funds to finance purchase of the projects.

Housing Privatization in Eastern Europe and the Republics of the former Soviet Union. In Eastern Europe and the former Soviet Union, privatization differs dramatically from the experiences just recounted. At the beginning of the transition to market economies, privatization of unit ownership was emphasized, but privatization of firms constructing and maintaining housing also has been significant.

The transfer of unit ownership has garnered the most attention in the West, in part because of the view that increasing homeownership would increase the population's stake in market-oriented reforms. Tables 24 and 25 give summary information on the unit ownership aspect of the

TABLE 25 Progress of Privatization in Selected Eastern European Countries as of Mid-1993

	Poland	Hungary	Bulgaria	Russia
Terms and conditions of sale				
Price basis	Appraised value	Net appraised market[a]	Fixed tariff[b]	Free
Price discount (%)	14[c]	50 to 85	80 to 90[d]	—[e]
Source of financing	State savings bank	Seller	State savings bank	—
Loan term (years)	20	25	30	—
Mortgage interest rate (%)	38.5	0[f]	49[g]	—
Market interest rate (%)	40 to 50	29.0	44.0	170
Housing units sold				
Period of sale	1986 to 1992	1990 to 1992	1958 to 1993[i]	1988 to June 1993[h]
Estimated units sold (thousands)	189	210	Exact figure unknown	5,850
Units sold as a percentage of state housing stock at start of privatization	10	30	Over 90	18

a. Appraised value less value of tenant investments.
b. Last updated in 1991.
c. These are current rules. Earlier discounts were greater. Also, it is believed that local governments in fact offer discounts greater than the maximum permitted nder the law.
d. This is a rough average; because the tariff is set on a national basis and unit prices vary by market, the size of the discount also varies.
e. — = not applicable.
f. No interest for first 6 years; market rate thereafter.
g. 2% loans before January 1991 converted to 10%; now floating rate (44% if housing savings account).
h. Significant sales began only in 1992.
i. Since 1958, it has been possible for tenants to purchase their units. Over 90% of all municipal units developed have been sold.

privatization process in four countries. Remarkable is the difference in the share of units owned by the state at the beginning of the privatization process: Among these countries, the range is from 9% in Bulgaria (much less than in the United Kingdom) to 67% in the Russian Federation. One constant among these countries was that state rental housing involved enormous subsidies for both development and operation. Thus, privatization was highly rational for these governments if pricing policies were not to be changed.

The general policy environment for the state rental sector during the 1990s was also highly variable among the countries in this group. Central government control of rents remained, for example, in Poland where they have increased substantially less than inflation. In Russia, in contrast, the power to raise rents was shifted to local Soviets in April 1992; but in December 1992, the Supreme Soviet passed a law requiring rents to be increased over a five-year period to cover full operating costs. Hungary and Bulgaria have taken action to substantially reduce tenants' rights (by, for example, making it easier to evict tenants who do not pay their rent), although little change in this area has occurred elsewhere. On the other hand, in all these countries, ownership of and responsibility for the state housing stock has been transferred to local governments.

The record of the share of units actually privatized differs importantly among these countries. Although several countries had laws from the 1960s and 1970s permitting privatization, a significant volume of privatization began only in 1989 or later everywhere except Bulgaria. As shown in Table 25, these countries adopted alternative policies on the terms of sale, with Russia adopting a free-of-charge policy at the end of 1992 (after 18 months of a low-charge or no-charge policy determined by local governments) and Poland and Hungary having a comparatively tough policy of modest discounts (in the initial stages of privatization, discounts were deeper in both countries).

In terms of the volume of units transferred, Bulgaria's long-standing policy of encouraging homeownership puts it in a separate class. Among the others, Poland has made slow progress, whereas Russia and Hungary put programs in operation that seem likely to transfer the majority of units to tenants.

Privatization of unit ownership on the scale being undertaken in the formerly centrally planned economies has not been without its costs. First, because of the idiosyncrasies of the allocation of units under the old regime, members of the Party and *nomenklatura* had the best units and therefore have received the largest windfall gains from privatization. Second, there have been other equity issues. The clearest concerned those on the waiting list for housing, who may never have received any benefit from the state, whereas those who privatized may have enjoyed decades of low-cost housing and a large capital gain. In some countries, where those who constructed their own housing received little, if any, state support, there has been the further equity issue between them and privatizing renters. No country in the region has yet acted to reform its waiting list procedures or to make ex post facto compensatory payments.

Third, low- or no-cost privatization has denied local governments of a very large source of potential revenue. How much revenue has actually been lost is far from clear, given the apparent sensitivity of the tenant's decision to the size of the net capital gain from privatizing. Moreover, few tenants have possessed significant financial assets after the surge of inflation and low interest rates on savings deposits that has generally characterized the initial stages of economic reform. Hence, long-term housing finance would have had to be available to have generated much in the way of effective demand. Because few transitional economies have been in a position to undertake responsible mortgage lending under the prevalent inflationary conditions, this has not been a realistic prospect.

Privatization of maintenance of current or former state rental units has begun in most formerly planned economies. Typically, these have involved experiments or demonstrations, with a strong mayor or city council taking difficult decisions. Change on a significant scale has yet to be achieved. One common pattern, however, seems to be the replacement of the public companies with private firms rather than the transformation of the state companies into more efficient firms.

The record on the privatization of residential construction firms is less well documented. There is general agreement that the transformation process had, by fall 1992, progressed most in Czechoslovakia, Hungary, and Poland and was probably proceeding at a faster pace in these countries. One hallmark of the process has been the down-sizing of the huge construction *kombinants,* through both breaking up the firms into more specialized entities and reducing overall capacity. Capacity reduction has been necessitated by the universal massive decline in total production during the transition period. As in the maintenance field, the formation of numerous new construction companies has been an essential component of the full privatization process.

In 15 years, housing privatization mushroomed from a pioneering policy of a conservative government in the United Kingdom to a cornerstone policy in a dozen countries or more. Although in good measure, the spread of the policy has resulted from the discrediting of the central planning model of economic management, it also reflects a more general international search for housing policies engendering lower subsidies, greater consumer choice, and improved efficiency. (SEE ALSO: *Polarization; Public Housing Homeownership Demonstration Program; Public/Private Private Housing Partnership; Turnkey Public Housing*)

—*Raymond J. Struyk and Nadezhda B. Kosareva*

Further Reading

Clapham, David. 1995. "Privatisation and the East European Housing Model." *Urban Studies* 32(4-5):679-94.

Forrest, Ray and Alan Murie. 1988. *Selling the Welfare State.* London: Routledge.

———. 1990. *Moving the Housing Market: Council Estates, Social Change and Privatization.* Aldershot, UK: Avebury.

Hegedus, Jozsef, Katharine Mark, Raymond Struyk, and Ivan Tosics. 1993. "Local Options for Transforming the Public Rental Sector: Empirical Results for Two Cities in Hungary." *Cities* 10(3):257-71.

Katsura, Harold, and Raymond Struyk. 1991. "Privatizing Eastern Europe's State Rental Housing: Proceed with Caution." *Housing Policy Debate* 4(2):1251-74.

Legrand, Julian and Ray Robinson, eds. 1984. *Privatization and the Welfare State.* London: Allen & Unwin.

Schill, Michael H. 1990. "Privatizing Federal Low Income Housing Assistance: The Case of Public Housing." *Cornell Law Review* 75(4):878-948.

Whitehead, Christine. 1991. "Privatizing Housing: An Assessment of the U.K. Experience." *Housing Policy Debate* 4(1):101-39. ◀

▶ Progressive Architecture

First published in 1920 under the name *Pencil Points, Progressive Architecture (PA)* is a monthly publication serving professional architects and planners. *PA* explores the processes behind architecture; the people, policies, and producers that make it happen. It examines typical problems in architecture, featuring articles on technology and practice as well as design. *PA* evaluates buildings after they have been in use to see what has worked and what has not. *PA* occasionally accepts articles for submission two months prior to publication. Articles should be sent with an SASE to Mr. John Morris Dixon, Editor, *Progressive Architecture,* 600 Summer St., P.O. Box 1361, Stamford, CT 06904. Circulation: U.S., 46,500; worldwide, 56,000. Price: U.S., $48 per year; Canada, $65 per year; Foreign, $130 per year. Administrative Editor: Mary B. Coan. Address: 600 Summer St., P.O. Box 1361, Stamford, CT. 06904. Phone: (203) 348-7531. Fax: (203) 348-4023.

—*Laurel Phoenix* ◀

▶ Property Rights

One of the fundamental considerations for all societies is an individual's relationships to objects and other human beings. Thus, the concept and specification of property rights is an essential aspect of society. Property rights means different things to different people. In one sense, property rights may be viewed narrowly and somewhat legalistically. In another sense, property rights can be viewed philosophically and quite broadly. No matter which path is chosen, property rights refer to an area that is often hotly contested and certain to remain controversial.

There can be no question that the issue of property rights was important to the founders of the United States. Evidence abounds about the appreciation of and importance attached to property ownership during the 17th century in the North American colonies. Indeed, property right protections extend back to the Magna Carta of 1215.

John Locke's *Second Treatise on Government* (1689) had enormous influence on the Whigs who founded the United States during the middle of the 18th century. Locke's emphasis on "natural rights" to property affected the thinking and values of the framers of the U.S. Constitution, especially portions of the Bill of Rights, at the Constitutional Convention in Philadelphia in 1787. However, language in the Declaration of Independence in 1776 shows Locke's influence as well: Thomas Jefferson's expression of "life, liberty, and the pursuit of happiness" was taken directly from Locke's concern that government needed to protect the "Lives, Liberties and Estates" of its citizens.

In addition, the Whig philosophy also emphasized the economic importance of private property; property ownership was viewed as essential for the development of the new nation. In fact, the added importance given to the federal government at the Constitutional Convention stemmed from the belief that individuals had a right to property and that private rights were critical for economic growth. In the absence of federal protections (such as the Fifth Amendment), it was feared that state legislatures would overregulate individuals and businesses and that the United States would not develop very rapidly.

In addition to the forbidding of expropriation of property by the government, property rights extended to the prohibition against confiscation by taxation, the protection of property rights in slaves, the forbidding of interference by the state in contracts, and the protection of individuals under "due process."

By the end of the Civil War, slavery was no longer a legal form of property and industrialization had replaced agriculture as the major form of economic production. This revolutionary change in economic structure adumbrated a future when the Supreme Court would view so-called laissez-faire constitutionalism as the proper approach to the protection of property rights.

By the turn of the 20th century, property rights became increasingly constrained by legislative directives and court decisions that sought to address issues such as urban congestion, reliance on child labor, overcrowded housing, and public health. In addition, pressure was growing to regulate monopolistic practices common to some industries in the 19th century. In cities, zoning and land use restrictions passed an important constitutional challenge in the 1926 case *Euclid v. Ambler Realty Co.* (372 U.S. 365), and the regulation of urban land became a public rather than a private matter. Individual property rights in land have been subject to significant public regulation throughout most of the 20th century.

With the Great Depression of the 1930s, the social agenda favored economic reform, management of the national economy, and government involvement in solving social ills and in the drive to promote the general social welfare. The New Deal programs of President Roosevelt moved away from private property and often conflicted with them. Civil liberties and civil rights were deemed to be more important than property rights, and property issues received considerably less attention.

In the post-World War II period, economic growth and technological change characterized life in the United States. Property rights protections did not receive much attention as the legal agenda concentrated on other social issues. With the so-called Reagan Revolution in 1980, there was a return to property rights. Several land use cases characterize this period as well, although most cases were narrow in scope. With the changing players on the Supreme Court, property rights cases received renewed attention.

It is fair to say that property rights have always been one of the elements on the U.S. agenda. During some periods, legislation and court decisions protecting property rights have appeared; in others, regulation of property rights was deemed to be important. Property rights is an area in which there will probably always be tension between the constitutional guarantees to individuals and the need for social change in society. Trying to balance the rights of individuals and the democratic demands for reform in a modern society is a microcosm for the historical development of property rights throughout the history of the United States. (SEE ALSO: *Eviction; Power of Eminent Domain; Right to Housing*)

—*Austin J. Jaffe*

Further Reading

Becker, Lawrence C. 1977. *Property Rights*. London: Routledge & Kegan Paul.

Brueggeman, William B. and Jeffrey D. Fisher. 1993. *Real Estate Finance and Investments*. 9th ed. Homewood, IL: Irwin.

de Alessi, Louis. 1980. "The Economics of Property Rights: A Review of the Evidence." *Research in Law and Economics* 2:1-47.

Jaffe, Austin J. 1996. "On the Role of Transaction Costs and Property Rights in Housing Markets." *Housing Studies* 11(3):425-34.

Jaffe, Austin J. and C. F. Sirmans. 1995. *Fundamentals of Real Estate Investment*. 3d ed. Englewood Cliffs, NJ: Prentice Hall.

MacPherson, C. B., ed. 1978. *Property*. Toronto: University of Toronto Press. ◄

► Property Tax

One of the tax levies most heavily used to finance local public services is the property tax. This assessment differs from income and other taxes in several important ways.

First, the property tax is an ad valorem (or "according to value") tax. In other words, it is a levy whose basis for taxation is the assessed value of the real estate. Most jurisdictions define a specific relationship between assessed value and market value (e.g., 25%, 50%, 80%, 100%, or others). So the tax burden is based on the market value of the property, and the property tax is a wealth rather than an income tax.

Second, property tax receipts almost always fund local needs and local public services. Income, excise, estate, and specialized tax collections are used elsewhere; property taxes fund local public schools, police and fire departments, park districts, libraries, senior citizen centers, and so on. In the last half century, the tax burden associated with paying property taxes has grown as demand for local public services has expanded. City and municipal governments continue to search for alternative revenue sources because property taxes can no longer be expected to fund local government's growth in expenditures.

Third, the property tax is consistently found to be the most unpopular tax in taxpayer surveys, despite the fact that in the United States, the federal income tax takes a larger percentage of household income on average. With the expansion of state (and sometimes local) income taxes, the income tax burden widens even more. The unpopularity of the property tax stems from its "shock value" in collection: It is frequently billed once or twice a year. Even though property taxes are often escrowed for disbursement on receipt of the annual charge, this displeasure with paying the property tax bill may be viewed as a measure of how successful the policy of income tax withholding has proven to be.

Fourth, because property taxes are largely used for local government programs, taxpayers probably have more to say about the use of their property tax funds than with any other source of taxation. The California tax revolt in the early 1970s known as Proposition 13 was led by local activists who viewed the level of property tax burdens and the use of these revenues as unacceptable. Income taxes have probably funded more wasteful spending relative to property taxes, but it is much more difficult to have the same kind of accountability from the federal government.

There are numerous issues associated with property taxes. One is the assessment process necessary to assign tax liabilities. Tax assessment varies considerably from jurisdiction to jurisdiction; small towns may require less tax revenue than major cities, but the fairness associated with tax assessment is also important. This is not intended to suggest that small town tax assessors do a poor job. However in recent years, the use of sophisticated statistical assessment procedures (so-called mass appraisal methods) have dramatically cut down of the likely errors of taxing similar properties differently. Led by the International Association of Assessing Officers (IAAO), these methods have led valuation practice to higher standards of performance. Smaller assessment operations might not be able to achieve the same levels of performance as larger jurisdictions, given their staffs and budgets.

Another issue is the relationship between property taxes, local government services, and the market value of property. The evidence shows that in general, homeowners tend to perform a cost-benefit calculus between local services provided and local tax burdens when deciding whether to locate in one jurisdiction or another. This "voting with your feet" phenomenon is sometimes known as the "Tiebout hypothesis" after the author who first proposed it in 1954. It predicts that local expenditures and receipts will be important enough to lead to a sorting out of households into remarkably homogeneous neighborhoods. Calls for regionalization of property tax receipts are in response to these findings.

Another issue deals with the effects of property taxes on real estate development. Following the Henry George tradition, critics charge that taxing improvements (buildings) discourages economic development but that taxing sites (land) would have the opposite effect. Land taxes would reduce speculation and encourage productive usage of land because the cost of holding undeveloped land would be increased. Some cities (e.g., Pittsburgh) and some countries (e.g., Australia) have had very good success with land valuation taxes. The issue of who bears the burden of property taxes, including land taxes, is an important one.

The use of property taxes is likely to continue as a primary source of local government finance. However, unlike earlier in the post–World War II period, additional growth in government services is not likely to be paid for by higher property tax receipts. The challenge for government officials is to find new sources of public finance and, in some cases, figure out ways to manage the demand for high-quality government services with a tax-paying public that is very sensitive to the burden of additional property taxes. (SEE ALSO: *Land Value Taxation; Property Tax Abatement; Property Tax Delinquency; Proposition 13*)

—*Austin J. Jaffe*

Further Reading

DiPasquale, Denise and William C. Wheaton. 1996. *Urban Economics and Real Estate Markets*. Englewood Cliffs, NJ: Prentice Hall.

Jacobus, Charles J. 1996. *Real Estate Principles*. Upper Saddle River, NJ: Prentice Hall.

O'Sullivan, Arthur. 1993. *Urban Economics*. 2d ed. Homewood, IL: Irwin.

Pryke, M., ed. 1994. "Property and Finance, Theme Issue." *Environment and Planning A* 26:167-264.

Shim, Jae K. 1996. *Dictionary of Real Estate*. New York: John Wiley.

Tosh, Dennis S. 1990. *Handbook of Real Estate Terms*. Englewood Cliffs, NJ: Prentice Hall. ◄

► Property Tax Abatement

Real estate tax abatement is one of a number of incentives available to states and their local governmental units to encourage residential, commercial, and industrial development. Tax abatement is a temporary nonrecognition of increased real estate taxes that result from improvements to real property. For example, a vacant piece of land valued for real estate tax purposes at $100,000 will pay $2,000 per year in real estate taxes, assuming a 2% tax rate on the value of that land. The construction of a $1,000,000 office building on that land will result in a real estate tax obligation of $22,000 per year. A tax abatement will ordinarily freeze real estate taxes at their predevelopment level, $2,000 in the example, for some period of years, usually between 10 and 30.

Tax abatement must be distinguished from tax exemption. The latter negates *any* obligation for real estate taxes and is permanent rather than temporary. Moreover, tax exemptions are not development incentives. Rather, they are granted in recognition that the potential taxpayer is undertaking a function such as charity that the state would otherwise be required to perform. In the case of religious organizations, tax exemptions are granted, at least in part, because the Constitution may prohibit the imposition of a tax liability on property used for religious purposes.

Virtually all states permit their local political subdivisions to abate real estate taxes. Most often these abatements are permitted for local revitalization projects: to restore downtown business and commercial areas, to increase industrial and employment opportunities, and to provide housing for low- and moderate-income families. There are nearly as many forms of tax abatements as there are states that permit them. New York, for example, permits a 30-year tax abatement for up to 50% of the full assessed value of newly constructed or rehabilitated housing units for low- and moderate-income families. Missouri permits real estate tax abatements on all improvements to land owned by an "urban redevelopment corporation" for 10 years, provided the property is developed by that corporation and is consistent with the local community's redevelopment plan. Ohio, along 38 other states, permits tax abatements in "enterprise zones" for up to 10 years on 100% of the value of real estate that has been purchased by any person or corporation desiring to operate a "large manufacturing facility."

The theory of tax abatement is that if the costs of landownership are reduced sufficiently, persons will be encouraged to increase, even maximize, the use of their land in ways that are socially productive, creating new jobs and more affordable housing and arresting the decay caused by a weak, unsubsidized market. This increased social utility is said to justify the public's investment in the form of decreased tax revenues. For example, real estate tax scholars such as Dick Netzer have estimated that real estate taxes approximate 25% of the gross rents that an owner might expect from the use of his or her land. Assuming a full tax abatement on the value of any improvements to the land, an owner of an apartment building could reduce the rent charged for any unit by 25% below that which would have to be charged without the abatement and still retain the same level of profitability. Similarly, a businessperson with a choice of sites for a new plant could sustain a 25% difference in costs between operating in an "inner-city" redevelopment area and operating in a less costly suburban or rural environment.

It is difficult to assess how much new development tax abatements have spurred. It is even more difficult to assess (a) the costs of tax abatement and (b) whether the social utility of the development activity has been worth the cost. A priori, tax abatements are likely to have had some influence. Empirical data, however, suggest that taxes have very little impact on investment decisions, despite popular perceptions to the contrary. With respect to housing, the impact may be even more problematic. First, despite the fact that taxes constitute approximately 25% of gross rents, the cost savings to each potential tenant is relatively insignificant, less than $1,000 per year of income per family. Additional subsidies are thus necessary to produce affordable new or substantially rehabilitated housing for low- and moderate-income families. This is confirmed by empirical studies of housing tax abatements in both New York City and St. Louis, which indicate that the benefit of new or newly rehabilitated, tax-abated housing units has been largely to middle- and upper-middle-income persons. To the extent this is true, tax abatement tends to be a regressive form of subsidy, reducing money for public services such as public education, welfare, and other social services used primarily by lower-income residents and increasing the dollars available to those in higher-income brackets.

More important, because the need for tax revenues makes tax abatements a limited resource, it is politically more profitable to funnel what tax abatement subsidies are available into higher-visibility projects such as office buildings and industrial facilities bringing higher-income persons to an area and increasing a community's employment, giving a city the advantages of economic multiplier effects and increased payroll taxes. Because housing for lower- (as opposed to upper-) income families generally gives a community neither of those two advantages, the political impetus for housing tax abatements is often lacking.

In conclusion, real estate tax abatements are viewed by land developers as a commodity: If one community pro-

vides them, all communities that compete with that community for development must also make them available. As applied to housing, real estate tax abatements have been of limited value in creating a favorable market for lower-income housing. They can, however, be one tool to gentrify a community's housing stock if coupled with enhanced public services such as public safety and education. (SEE ALSO: *Property Tax; Property Tax Delinquency*)

—*Melvyn R. Durchslag*

Further Reading

Aaron, Henry. 1972. *Shelters and Subsidies*. Washington DC: Brookings Institution.
Heilbrun, James. 1996. *Real Estate Taxes and Urban Housing*. New York: Columbia University Press.
Hellerstein, Walter. 1986. "Selected Issues in State Business Taxation." *Vanderbilt University Law Review* 39:1036.
Krumholz, Norman. 1991. "Equity and Local Economic Development." *Economic Development Quarterly* 5:291-300.
Mandelker, Daniel R., Gary Feder, and Margaret Collins. 1980. *Reviving Cities with Tax Abatement*. New Brunswick: Rutgers, State University of New Jersey, Center for Urban Policy Research.
Netzer, Dick. 1967. *The Economics of the Property Tax*. Washington, DC: Brookings Institution.
Rosen, Harvey S. 1985. "Housing Subsidies: Effects on Housing Decisions, Efficiency and Equity." Pp. 375-420 in *Handbook of Public Economics*. Vol. 1, edited by Alan J. Auerbach and Martin Feldstein. The Netherlands: Elsevier.
Sternlieb, George, Elizabeth Roistacher, and James W. Hughes. 1976. *Tax Subsidies and Housing Investment: A Fiscal Cost Benefit Analysis*. New Brunswick: Rutgers, State University of New Jersey, Center for Policy Research. ◄

► Property Tax Delinquency

Ad valorem taxes are property taxes charged according to the value of the real estate that generally became a senior or superior lien on property in the event they are not paid. Ad valorem tax values as found on the local tax rolls or tax assessment rolls are generally based on the current market value (or some percentage of market value) of the property without regard to the original cost or cost to replace the subject property.

Although property taxes, the collection of delinquent taxes, and the penalties for nonpayment vary from state to state, the following is a generalized overview of the failure to pay any taxes that may be due from any authorized public entity. Authorized taxing authorities include but are not limited to counties, cities, school districts, hospital districts, flood districts, transportation districts, or any "special district" sometimes referred to as muncipal utility districts (MUDs). Failure to pay any taxes due from any of the aforementioned entities at the appropriate time can result in a number of problems for property owners, ranging from penalties to the loss of the property at a public sale for "back taxes."

A variety of circumstances can result in the nonpayment of property taxes without owners' full knowledge that their taxes are in fact delinquent. An example of this is the common budget mortgage that includes monthly payments of principal, interest, taxes, and insurance (PITI). The property owner pays mortgage payments to the lender and one-twelfth of the anticipated taxes due at the end of the year. The mortgage company "escrows" these monthly budgeted tax payments and generally pays the property taxes on behalf of the property owner. Because loans are bought and sold and loan servicing companies change from time to time, sometimes a property owner's taxes are not paid or paid late as an oversight. In this case, the lender or mortgage servicing firm would most likely be held responsible for any penalties or interest due for late taxes paid. However, the property owner is ultimately responsible for taxes' being paid.

Other common situations that result in taxes becoming delinquent include changes in owners' addresses without notification of the tax assessor office, death, and bankruptcy. If, for whatever reason, the owners did not receive a tax notice on their property, it is generally not the taxing entity's responsibility to find the property owner but, rather, the property owner's responsibility to find the tax office and pay the taxes. Property tax protests or strikes are not common because failure to pay taxes can lead to a loss of the property.

Delinquent property taxes are superior to preexisting liens and mortgages and can cause loss of collateral held by banks or other lenders if they are not paid. This is the primary reason that most mortgages require owners of properties to stay current in their property taxes owed and insist on collecting the taxes with the mortgage payments. Property taxes are in fact superior to Internal Revenue Service (IRS) liens placed on properties.

The definition of a property tax delinquency is being late one day past the due date as found on an owner's tax statement, whether the tax statement was actually received or not. Most tax assessors and tax collectors have penalties and late fees that progressively increase and compound as time passes from the date taxes are due. Sometime, property owners fail to realize that delinquent taxes for a particular year may have been inadvertently missed and subsequent years' taxes paid on time. The compounding effect escalates substantially in a relatively short period of time.

If any one of the many taxing authorities demands payment, it is possible that a property sale or auction could be held. The steps of tax delinquency and loss of property are generally as follows:

1. Default on taxes due occurs.
2. Penalties and interest accrue.
3. A law firm is hired to collect the taxes plus legal fees.
4. Formal notices are sent to the property owner (if whereabouts known), and classified notices are "posted" in the courthouse and in local newspaper.
5. A tax sale or sheriff's sale is held on the courthouse steps at which time the taxes, interest, penalties, and legal fees are paid from the proceeds along with judgments or personal liens filed on the owners for any balance owing.
6. The buyer at the tax sale or sheriff's sale receives a tax deed or sheriff's deed that may or may not allow the previous owner the right of redemption for some period of time (varies by state).
7. The buyer at the tax sale or sheriff's sale may earn very high interest or premiums (25%-50%) if the property is redeemed (varies by state) with stated time periods but generally not longer than two years.
8. The tax deed or a sheriff's deed is filed on record, the property passes to the highest bidder at the sale that occurred previously, and possession procedures begin according to law.

Delinquent taxes are a very serious situation for property owners and can result not only in the loss of property but also in the continued liability of tax penalties due, repayment of mortgages on property no longer owned, and personal judgment, liens, and bad credit reports for the *former* owners. Buyers are generally protected from tax liens and back taxes through the purchase of title insurance or tax certificates that the taxing authorities acquire through their real estate lawyer. (SEE ALSO: *Property Tax; Property Tax Abatement*)

—*John S. Baen*

Further Reading

Darling, Richard. 1984. "Property Taxation: A Complicated Process." *Real Estate Today* (November/December):19.

Jacobus, Charles and Bruce Harwood. 1990. *Texas Real Estate*. 5th ed. Englewood Cliffs, NJ: Prentice Hall. ◀

▶ Proposition 13

During the mid-1970s, California residents experienced rapidly increasing property values. With stabilized tax rates, this resulted in rapidly increasing property tax levies. When the state legislature failed to recognize these rapidly increasing payments, an initiative was sponsored to limit these taxes. The secretary of the State of California assigned the number 13 to the initiative for the June 1978 California ballot, and so it became known as Proposition 13.

Proposition 13 was among the initial wave of successful property tax limitations that swept across the United States during the late 1970s and early 1980s. When it passed in June 1978, it did several things:

▶ It limited the property tax rate to 1% of market value except to pay for previously incurred voter-approved debt.

▶ It rolled back property assessments to their March 1975 levels.

▶ It limited future value increases to change of ownership, new construction, and inflation (a maximum of 2% per year).

▶ It prohibited any new sales or transaction taxes on the sale of real property.

▶ It required a two-thirds voter approval for new "special taxes."

Although subject to much litigation, the basic features of Proposition 13 have been upheld by the California and U.S. Supreme Courts.

The state legislature responded to Proposition 13 by passing implementing legislation to do the following:

► Shift some school districts' property tax revenues to other local governments. The state general fund replaced the schools' losses, centralizing control over school finance at the state level.
► Allocate property tax revenues that resulted from new development to the jurisdictions where the increase occurred (called the situs provision).
► Assume financing of much of the local health and education programs.

Despite frequent but marginal amendments, Proposition 13 and the implementing legislation still govern property tax allocation in California.

The combination of Proposition 13 and its implementing legislation affected land use and housing in at least four important ways: development financing, municipal incorporations, residential mobility, and rent controls.

Because Proposition 13 initially eliminated the ability of local government to pledge its full faith and credit for debt repayment, the property tax could not support local general obligation debt. A later amendment to Proposition 13 allowed local officials to issue general obligation bonds with two-thirds voter approval. Nevertheless, officials invented new financial instruments to pay for the infrastructure necessary for land development. For example, a "community facilities district," instituted by owners of undeveloped land, can issue debt and also pay for services. These both would be financed by property liens on the development that would generate a tax stream. This financial arrangement assumed the ability of the developer to sell the property. If the property did not sell, the debt could go into default. This non-ad valorem lien appeared on the property tax bill alongside the regular property tax liability, and there is some anecdotal evidence that it is also capitalized into the price of the house. In addition, local officials expanded their use of traditional assessment districts to finance land improvements.

Developer fees and exactions became increasingly important. Jurisdictions imposed these fees not only to recover their processing costs (e.g., planning, staff, and services) but for schools, parks, and other general facilities required by the development. Total fees on the developer exceeded $15,000 per dwelling unit in some jurisdictions. Although developers often argued that the fee was merely shifted forward to the home purchaser, their ability to do so depended on market conditions. During the long duration in California of the recession of the early 1990s, market conditions probably left the burden of these fees on the developer.

Because new housing rarely generated enough property tax revenues to pay for the service demands of the new residents, jurisdictions were tempted to engage in fiscal zoning (sometimes called the "fiscalization of land use"). Local officials encouraged commercial development, which generated both property and sales tax revenues, rather than residential development. In particular, shopping malls became even more important, and competition for these malls was intense. Many communities became addicted to commercial growth to finance their activities, and some were greatly stretched when the 1990 recession struck the state.

Until 1986, school districts were also unable to issue general obligation bonds. They gained new authority to impose developer fees on new construction. These fees were on top of the other local fees, charges, and liens. Although the state restricted school district's fees, schools often found ways to circumvent these limits.

An additional land use phenomenon attributed to Proposition 13 was the increased number of municipal incorporations. In the decade after the proposition's passage, the rate of incorporation doubled. One reason for this increase was that there was no longer a potential property tax increase awaiting the residents of the new city. Proposition 13 had fixed the rate. However, with California's population increases exceeding the national average for long periods, many local areas desired the ability to regulate how local land was to be developed. With no additional costs and potential control over land use decisions as a benefit, it made sense for more jurisdictions to make this move.

A residential lock-in effect may also occur because of the property tax increases that residents would face if they moved to a new dwelling. Some evidence indicates that residents remain in their homes longer since the passage of Proposition 13: For example, in Los Angeles County, over 47% of the homeowners did not move between 1978 and 1997.

During the campaign for its passage, the proponents of Proposition 13 urged renters to vote in its favor, asserting that landlords would pass through the tax reduction to tenants and rents would be lowered. When this did not occur, a number of rent control ordinances were passed, ranging from the quite draconian to those allowing vacancy decontrol. (For example, in at least one California jurisdiction, units could not be withdrawn from the rental stock regardless of reason.) Most studies of the less draconian types seem to indicate only marginal effects on new multifamily construction. During the 1995 legislative session, the state mandated that rental units be decontrolled upon vacancy, with the proviso that the new rent be again subject to controls. (SEE ALSO: *Property Tax*)

—*Jeffrey I. Chapman*

Further Reading

Amador Valley Joint Union High School Dist. v. State Board of Equalization [22 Cal. 3d 208 (1978)]

Chapman, Jeffrey I. 1981. *Proposition 13 and Land Use.* Lexington, MA: Lexington Books.

Misczynski, Dean. 1986. "The Fiscalization of Land Use." Pp. 73-106 in *California Policy Choices*, Vol. 3, edited by John J. Kirlin and Donald R. Winkler. Sacramento: University of Southern California.

Nagy, J. 1977. "Did Proposition 13 Affect the Mobility of California Homeowners?" *Public Finance Review* 25(1):102-16.

Nordlinger v. Hahn. [112 S.Ct. 2326 (1992)].

Stocker, Frederick D. 1991. *Proposition 13: A Ten Year Retrospective.* Cambridge, MA: Lincoln Institute of Land Policy. ◄

► Pruitt-Igoe

The Pruitt-Igoe public housing development in St. Louis occupies a unique position in the history of U.S. public housing. Constructed in 1954 and dramatically demolished in 1972, the project has come to symbolize all the alleged inadequacies of the federal public housing program. Architects have been particularly vocal in their criticism of the project, taking the position that its flawed design was the major cause of so-called social pathologies among project residents. Yet other researchers interpret the Pruitt-Igoe story as a demonstration of the racial discrimination and economic injustice embedded in U.S. society.

Pruitt-Igoe was created under the Housing and Redevelopment Act of 1949, which made federal funds directly available to cities for slum clearance, urban renewal, and public housing. Like many U.S. cities in the postwar era, St. Louis planned to use redevelopment funding to raze the economically blighted neighborhoods north and south of the central city and replace them with revenue-generating commercial development. The city constructed public housing projects at the periphery of the redevelopment areas to rehouse those displaced by the slum clearance activities.

Pruitt-Igoe was among the projects slated to receive the displacees. Planning for the project began in 1950, when the St. Louis Housing Authority acquired a 57-acre site in the north side African American ghetto and engaged the architecture firm of Leinweber, Hellmuth and Yamasaki. The architects' final design consisted of 33 identical 11-story slab buildings arranged in roughly parallel rows and separated by open plazas. Each building incorporated a number of features aimed at minimizing the inconvenience of living in elevator buildings. Glazed internal galleries located on every third floor were intended to substitute for conventional yards and gathering spaces. Skip-stop elevators transported residents to the gallery levels, from which they would walk up or down a flight of stairs to their units.

Although the St. Louis Housing Authority originally conceived Pruitt-Igoe as two segregated sections (Pruitt for African Americans and Igoe for caucasians), a 1954 Supreme Court decision forced desegregation just as the project was being completed. Caucasian families in St. Louis tended to oppose living in an integrated environment, so the majority of the project's first tenants were African Americans.

Pruitt-Igoe attracted little negative attention during the first few years of its existence. Its residents expressed satisfaction with their new living environment, which represented a substantial improvement over the unsanitary and overcrowded housing they had vacated. Yet by the late 1950s, tenants began to complain of deterioration. A burgeoning supply of low-cost housing in the private market drew potential tenants away from the project and pushed up vacancy rates. Faced with lower than anticipated income from rents, the housing authority was unable to adequately maintain and repair the buildings. Tenants reported broken elevators that went unrepaired for days. As the years passed, the project was increasingly plagued by incidents of van-

Figure 24. Demolition of Pruitt-Igoe Public Housing
Photograph courtesy of St. Louis Post-Dispatch.

dalism and violent crime. In 1969, Pruitt-Igoe tenants joined residents of other St. Louis projects in a massive rent strike to protest unbearable living conditions.

By the early 1970s, the St. Louis Housing Authority and the Department of Housing and Urban Development (HUD) officials agreed on a plan to demolish Pruitt-Igoe. In addition to its dilapidated condition and the incidents of violence, the project represented a tremendous financial drain on the housing authority. Demolition began in March 1972 and continued until the project was completely razed in 1976. The dramatic photograph of one of the buildings crashing to the ground has come to symbolize the project's demise and serves as the centerpiece of a debate that continues to this day about the conclusions to be drawn from Pruitt-Igoe's history.

Since the project's demolition, a body of literature has emerged that links Pruitt-Igoe's demise to an intrinsically flawed design. These studies cited Pruitt-Igoe as an example of how environmental design in high-rise public housing can cause residents to engage in destructive and antisocial behavior. For example, designers have suggested that because the glazed galleries were not visible from apartments, the residents made no effort to police them. A number of architecture critics have taken the design argument a step further, suggesting that Pruitt-Igoe's architectural failings are the direct result of the inadequacy of modernist design ideology. Pruitt-Igoe has become the most commonly cited U.S. example of the alienating effects of modernist design principles such as the "tower in the park," exemplified by LeCorbusier's iconic postwar housing project, the Unité d'Habitation in Marseilles.

A number of researchers have interpreted the Pruitt-Igoe story as a demonstration of the consequences of racially discriminatory housing and redevelopment policy and disagree with the view that the project's architecture was central to its demise. Political scientist Eugene Meehan has argued that Congress's political ambivalence toward public housing produced a token program burdened with impossible fiscal constraints. In his view, restrictive federal regulations forced the St. Louis Housing Authority to construct poor-quality housing and then prevented the housing authority from adequately maintaining the buildings. After conducting extensive fieldwork at Pruitt-Igoe, sociologist Lee Rainwater argued that the vandalism and violence reflected not poor design but, rather, the residents' collective response to institutionalized poverty and discrimination.

Pruitt-Igoe was the product of an urban redevelopment strategy that allowed cities to displace poor, African American families and rehouse them in poorly constructed public housing projects on undesirable sites at high densities. Design flaws played only a secondary role in Pruitt-Igoe's fate when compared with the city's redevelopment strategy, crises in the local economy, and financially unrealistic federal regulations. Those who point only to the architectural problems divert attention from the racial discrimination and economic injustice that underlie the history of public housing in the United States. (SEE ALSO: *High-Rise Housing*)

—*Kate Bristol*

Further Reading

Bristol, Katharine G. 1991. "The Pruitt-Igoe Myth." *Journal of Architectural Education* 44(3):163-71.

Jeffery, C. R. 1971. *Crime Prevention through Environmental Design.* Beverly Hills: Sage.

Meehan, Eugene. 1979. *The Quality of Federal Policymaking: Programmed Failure in Public Housing.* Columbia: University of Missouri Press.

Montgomery, Roger. 1985. "Pruitt-Igoe: Policy Failure or Societal Symptom." Pp. 229-43 in *The Metropolitan Midwest: Policy Problems and Prospects for Change,* edited by Barry Checkoway and Carl V. Patton. Urbana and Chicago: University of Illinois Press.

Montgomery, Roger and Kate Bristol. 1987. *Pruitt-Igoe: An Annotated Bibliography.* Council of Planning Librarians Bibliography Series, No. 205. Chicago: Council of Planning Librarians.

Rainwater, Lee. 1970. *Behind Ghetto Walls: Black Family Life in a Federal Slum.* Chicago: Aldine. ◀

▶ Psychiatric Disabilities, Housing of Persons With

During the 1990s, the critical need for stable housing linked to effective supports has emerged as a major policy issue within the fields of mental health and rehabilitation. Factors contributing to this emergence include significant numbers of people with psychiatric disabilities among the nation's homeless, a continuing decline of affordable housing, and the increasing articulation of people with psychiatric disabilities themselves about a desire for further community integration through normal housing. These trends have significantly challenged mental health systems, which traditionally have focused on the provision of treatment, and the development of congregate, segregated facilities rather than supporting people in regular, community housing and, in the process, have stimulated substantial community opposition to such mental health facilities.

The nation's affordable housing crisis has had a particularly negative impact on people with psychiatric disabilities, not because of their mental illness but because they generally reside in the lowest 20% of income groups in the country, making their access to even "affordable" housing provided by state or federal governments difficult. A recent study concluded that there is not a single community in the United States in which a person on Supplemental Security Income (the funding that most people with serious mental illnesses receive) can afford an efficiency or one-bedroom apartment, using the federal standard of affordability.

Over 60 recent studies of the kinds of housing and supports that people with psychiatric disabilities want conclude emphatically that the great majority want integrated housing in which they live side by side with people without disabilities. These studies conclude that (a) people value privacy and autonomy more than anything else; (b) people prefer not to live with other ex-psychiatric patients but, rather, to live with either friends or romantic partners; (c) they would like to have services available on a 24-hour basis but do not want service professionals to live with them; and (d) the greatest barrier to their housing aspirations is limitations in income.

These studies of consumer preferences, along with the affordable housing crisis, as well as the development of much more effective case management and crisis service programs, have combined to develop a new approach to housing in mental health called supported housing. The basic tenets of this approach involve the primacy of choice among consumers about their housing, the use of normal integrated housing, the provision of flexible supports tailored to the individual, and a focus on community membership and integration. Mental health systems throughout the United States have adopted this approach in their policies and have begun implementing a variety of strategies to promote improved housing outcomes.

These strategies include (a) using demonstrably effective community support services, such as expanded assertive community treatment in case management programs, which provide a variety of clinical and support services on a 24-hour basis; (b) outreach crisis services oriented toward both preventing and resolving crises wherever they may occur; (c) the provision of support to landlords, neighbors, employers, and other key community members; (d) education and support programs for families; and (e) expanded employment programs. In addition, state and local mental health authorities have become involved in moving away from funding group homes and other congregate-care facilities and toward developing instead funding resources more relevant to integrated housing development. Examples include the widespread establishment of local housing development corporations, which form public and private partnerships to access rental housing broadly in the community and to develop new housing, in partnership with other neighborhood or statewide organizations, in which people with psychiatric disabilities are integrated with other members of the community, including people of mixed incomes.

Finally, state and local mental health agencies are establishing rental subsidy programs that frequently serve as a bridge funding mechanism so that consumers who are on long waiting lists for Section 8 (federal rental subsidies) can have their rents supplemented until they qualify for these subsidies or vouchers.

Studies from communities in which this supported housing approach has been implemented indicate reasonable costs compared with group homes or hospital settings, much more positive life outcomes, and much higher levels of consumer satisfaction. Therefore, it is likely that this approach will be dominant in the future.

Continued research is needed on strategies for facilitating meaningful choices for consumers, as well as for developing housing and supports. The costs and benefits of housing and support initiatives need to be documented, and clinical interventions best suited to normal housing need to be identified. More broadly, research is needed on successful strategies for community integration, membership, and participation, including employment-related studies, as well as those related to social integration.

In summary, the mental health field made significant progress in reconceptualizing its roles and responsibilities for housing and supports. New knowledge about what consumers want, and about the needs of families, are emerging, as are new policies, funding mechanisms, and programs through which agencies and systems can respond to these needs. It is both significant and hopeful that these changes have resulted in an increased focus on the basic need for a home, and a deepening commitment to pursue strategies that will help people with psychiatric disabilities have their own homes with the supports that they want. (SEE ALSO: *Halfway House; Health; National Resource Center on Homelessness and Mental Illness*)

—*Paul J. Carling*

Further Reading

Anthony, W. A. and A. Blanch. 1989. "Research on Community Support Services: What Have We Learned?" *Psychosocial Rehabilitation Journal* 12(3):55-81.

Carling, P. J. 1995. *Return to Community: Building Support Systems for People with Psychiatric Disabilities.* New York: Guilford.

Knisley, M. B. and M. Fleming. 1993. "Implementing Supported Housing in State and Local Mental Health Systems." *Hospital and Community Psychiatry* 44:456-61.

Livingston, J. A., D. Srebnik, D. A. King, et al. 1992. "Approaches to Providing Housing and Flexible Supports for People with Psychiatric Disabilities." *Psychosocial Rehabilitation Journal* 16(1):27-43.

Morse, G. and R. J. Calsyn. 1992. "Mental Health and Other Human Service Needs of Homeless People." In *Homelessness: The National Perspective,* edited by M. J. Robertson and M. Greenblatt. New York: Plenum.

Racino, Julie Ann et al. 1993. *Housing, Support, and Community: Choices and Strategies for Adults with Disabilities.* Community Participation Series, Vol. 2. Baltimore, MD: Paul H. Brookes.

Ridgway P. 1987. *Avoiding Zoning Battles.* Washington, DC: Intergovernmental Health Policy Project.

Ridgway P. and A. M. Zipple. 1990. "The Paradigm Shift in Residential Services: From the Linear Continuum to Supported Housing Approaches." *Psychosocial Rehabilitation Journal* 13(4):11-31.

Tanzman, B. 1993. "An Overview of Surveys of Mental Health Consumers' Preferences for Housing and Support Services." *Hospital and Community Psychiatry* 44:450-55. ◄

► Public Housing

The public housing program in the United States was authorized by the U.S. Housing Act of 1937, also known as the Wagner-Steagall Act. It was the first major federal program aimed at providing low-rent housing to low-income households. Although housing problems had been acknowledged for decades, not until the Great Depression in the 1930s did the federal government become involved with housing on a wide scale. The housing initiatives that were enacted during this period, including public housing, had several goals, only one of which involved the provision of improved housing. Other key objectives involved creating employment opportunities, stimulating the economy, and removing slums.

Passage of the 1937 housing act followed several other key New Deal measures that related to housing, notably the creation of the Home Owners' Loan Corporation, the Federal Home Loan Bank System, and the Federal Housing Administration. One of the major reasons the enactment was relatively long in coming was that from the outset, public housing was a controversial idea and even President Franklin Roosevelt had to be coaxed. There was strong opposition from several key private interest organizations, including the National Association of Real Estate Boards, the Chamber of Commerce of the United States, the U.S. Savings and Loan League, and later, from the National Association of Home Builders. Together, these organizations launched scathing attacks on public housing, accusing it of being socialistic and representing unfair government competition with private enterprise. In addition to opposing the 1937 legislation and the eventual reactivation of public housing (as part of the Housing Act of 1949), they played major roles in organizing local communities to oppose the siting of public housing. As a result, many public housing developments were built in poor locations, where abutters (owners of contiguous property) were scarce.

Public housing has never been a broadly popular program. Although it provides low-rent housing to some 1.3 million households, it has never managed to shed its institutional image, with phrases such as "vertical ghettos" and "government-supported slums" frequently used in the media to describe public housing.

Much of the recent scholarly literature on public housing has argued that these overall negative images are unfair and inappropriate. Although public housing has certainly had its share of problems, it provides affordable housing for millions of people who do not have other options for attaining housing.

How Public Housing Works

The first stage of public housing development involves a state's enacting legislation that allows local governments to create their own public housing authorities (PHAs). Until 1987, when the financing method of public housing was changed to a direct grant for development, PHAs, also known as local housing authorites (LHAs), borrowed money for the construction of public housing by floating government-backed, tax-exempt bonds. The federal government guaranteed the payment of interest and principal on these notes through an agreement with the PHA known as an annual contributions contract. Maintenance and renovation costs were to be covered through the rentals paid by tenants. This financing formula worked well until maintenance and utility costs rose in the 1960s and 1970s. This coincided with a change in the demographics of the public housing population, with the original occupants being replaced by a new lower-income group.

In response to the lack of revenues to maintain the buildings, which were requiring more funds due to aging, and to a tenantry that was decreasingly able to meet the mounting costs, a series of amendments to the public housing statutes were enacted between 1969 and 1971. Known as the Brooke amendments (in honor of Senator Edward Brooke of Massachusetts, who was the key sponsor of the legislation), these initiatives provided additional operating subsidies to public housing developments and capped the rental contribution of public housing tenants at 25% of family income. By 1972, operating subsidies had already reached over $102 million; 10 years later, the amount had increased more than tenfold, to $1.3 billion. For fiscal year 1998 authorizations for operating subsidies were about $2.9 billion. Since 1975, operating subsidies have been provided through the Performance Funding System (PFS).

Another way in which the federal government supports the operation of public housing is through its modernization program. From 1980 to 1992, modernization funds were provided through the Comprehensive Improvement Assistance Program (CIAP). The Cranston-Gonzalez National Affordable Housing Act of 1990 created a new program for all medium and large PHAs. Known as the Comprehensive Grant Program, funds are now provided to these PHAs on a block grant basis, using a formula. Small PHAs will continue to receive modernization funds under a modified CIAP program. In FY 1998, about $2.5 billion in modernization funds were provided to PHAs.

As part of the Housing and Community Development Act of 1992, Congress strengthened a relatively new program aimed specifically at severely distressed public housing developments. The Major Reconstruction of Obsolete Public Housing Program (MROP) authorized the Department of Housing and Urban Development (HUD) to set-aside up to 20% of the funds appropriated for public housing development for major repairs on severely distressed, or obsolete, projects. According to a 1988 study conducted by Abt Associates, a private consulting firm, about $22.2 billion will be needed to adequately upgrade and modernize the public housing stock.

In addition to the conventional public housing program described above, in 1965 two variants on the public housing program were introduced—the Leased Housing program and the Turnkey Program. Both of these programs involved increased roles for the private sector. In the Leased Housing program, also known as the Section 23 program (which served as a prototype for the Section 8 Existing Housing and Voucher housing allowance-type programs), the housing authority entered into long-term contracts with land-lords and paid the difference between the unit's market rent and a proportion of the tenant's income.

In the Turnkey program, a developer enters into a contract with an LHA to construct a project at a specified price. On completion, the developer "turns the key over" to the PHA, which then assumes ownership and management of the development.

Some public housing is designated "Indian Public Housing," specifically earmarked for Native Americans.

Development of public housing has had many ups and downs. In the decade 1964 to 1974, about 600,000 units, over 40% of the public housing stock, was constructed. During the Reagan and Bush administrations, congressional appropriations were sufficient to construct only about 5,000 or fewer new public housing units per year. Under President Clinton, starting in FY 1996, no funds were appropriated for new public housing development. This reduction in new funding was accompanied by a major effort to reform and restructure public housing, which included demolition of severely distressed developments, without replacement of units, and targeting units to somewhat higher-income households.

Numbers and Types of Units and Characteristics of Residents

There are about 1.3 million public housing units in the United States, providing homes to some 3.3 million people. A majority of residents are nonwhite. White residents tend to be much older than nonwhites, with the latter generally having at least two children. Over three-quarters of all public housing households are headed by single adults, usually an elderly person living alone or a single parent with children. More than half of all public housing tenants depend on welfare for their incomes; another 25% depend on social security or disability payments. Families admitted to public housing before 1981 had to have incomes below 80% of the area median income and were considered "lower income." Since then, however, most families admitted to public housing have been "very low income," with incomes below 50% of area median income. In 1995, the median income of public housing familes with children was $6,190, and for elderly residents, median income was $7,010. Incomes of public housing residents may rise somewhat over the next several years, because federal preference targeting very low income households for public housing have been eliminated. As of 1995, the average monthly rent was $169.

The public housing stock includes about 13,000 public housing developments administered by some 3,400 PHAs. About 100,000 public units are considered "severely distressed." This means that they are deteriorated, that they have high vacancy rates, and that unemployment rates among a low-income population are high. In addition, the developments are saddled with high incidences of crime, are expensive and difficult to manage, and are viewed as unstable communities. The majority of PHAs are small, managing less than 200 units, but most public housing units are located in developments owned by large housing authorities. Of all public housing units, 66% are administered by 134 PHAs—about 4% of the total number of housing authorities. Despite the popular image of the public housing stock as consisting exclusively of large, high-rise buildings, this housing type is not, in fact, dominant. Only 28% of public housing developments have four or more stories; 38% of developments have two- or three-story buildings, and about 34% contain single-story and single-family detached structures. Thus, the conventional image of public housing as being primarily large, unattractive developments is not valid. Moreover, to the extent that public housing is not aesthetically pleasing, at least some of the reason is significant opposition by private real estate interests, which were opposed to any housing being built that would directly compete with the existing privately owned stock of rental housing.

Costs of Public Housing

One of the key criticisms of public housing has been that it is too expensive. Critics have argued, particularly since the 1970s, that far more families can be served through direct housing allowance approaches, such as the Section 8 Existing Housing program, than by subsidizing public housing units. Although there are a variety of ways to compare the costs of public housing with other federal subsidy programs, it does not appear to be more costly to subsidize an existing public housing unit than to subsidize a household in another unit of privately owned existing housing (e.g., through the Section 8 program). The cost of constructing a new unit of public housing is greater than the cost of subsidizing a household in a privately owned existing unit. However, the public housing approach creates a new public resource—a unit of affordable housing—whereas the Section 8 approach does not.

Other Criticisms of Public Housing

Many criticisms have been lodged against the public housing program. Some have been justified; others have not been. In many cases, where criticism is warranted, there are reasonable explanations underlying the problems. For example, public housing has been criticized for poor management. Although there have been many instances of developments being run poorly, sometimes the reason has been inadequate funds available to run the buildings properly. In other cases, mismanagement has, indeed, been due to shortcomings or, in rare instances, fraud, on the part of housing authority personnel. The negative image of the design and siting of public housing, which consists of many buildings with few amenities and that are inconveniently located, also has truth to it. On the one hand, many buildings are unattractive and locations are sometimes poor. On the other hand, opposition on the part of the private real estate industry often was the reason that the physical aspects of public housing emerged the way they did. To sum up, there have been many criticisms of the public housing program and, although some are justified, many have served to unfairly undermine the program.

Major Dilemmas, Debates, and Problems

Public housing is likely to undergo some sweeping changes. In December 1994, HUD issued its *Reinvention Blueprint*, which among other proposals, called for the transformation of public housing. Over the past several years, public housing has been the subject of much discusion and scrutiny. The following provides an overview of some of the key debates that have surrounded public housing since the 1980s.

Should public housing be provided to the neediest households or to a broader spectrum of low-income households? Federal housing policy has shifted a number of times over who should have access to public housing units. In the 1960s and 1970s, for example, attempts were made to make public housing available to both low-income and very low income households. In the 1980s, HUD policy focused on serving very low income families. In the Housing and Community Development Act of 1987, for example, Congress directed that housing assistance be provided on a preferential basis to households (a) on waiting lists who were living in substandard housing, homeless, or living in shelters; (b) who had been involuntarily displaced from their previous dwellings; and (c) paying more than 50% of their income for rent. Under President Clinton, these preferences were phased out, although local housing authorities have the option of retaining them.

Should public housing residents manage their own housing? As conditions in many public housing developments worsened, some tenants began to organize themselves and won the right from their local housing authorities to manage their developments, under contract to their respective PHAs. Some of the most prominent examples of resident management of public housing are found in St. Louis; Washington, D.C.; and Boston.

In 1976, HUD, with the assistance of the Ford Foundation, launched a three-year resident management demonstration project, modeled after the successful tenant-initiated efforts. The results of the National Tenant Management Demonstration were mixed. Although tenant management was found to be a "feasible alternative to conventional public housing management" and, in a number of areas, "tenants were able to manage their developments as well as prior management," tenant management was not found to be significantly better in areas such as rent collections, vacancy rates, and speed of responding to maintenance requests.

The Housing and Community Development Act of 1987 authorized HUD to provide funding to PHAs to assist the development of new resident management corporations. With a maximum of $100,000 expended on each initiative, resident management corporations did not become a dominant feature of most local housing authorities' operations. Funding for technical assistance and training for tenant management was reauthorized as a part of the Cranston-Gonzalez National Affordable Housing Act of 1990. During the Bush administration, HUD Secretary Jack Kemp strongly supported resident management and advocated that it be a precursor to resident ownership of public housing.

Should public housing be sold to residents? Since 1974, HUD has had the power to sell public housing units to tenants, although through the end of 1983, HUD had authorized only 1,731 such sales. During the mid-1980s, under the Reagan administration, interest in selling public housing units to tenants was fueled by the expansion of Britain's public housing sales program under Prime Minister Margaret Thatcher. In 1985, HUD launched the Public Housing Ownership Demonstration program, which authorized 17 PHAs to transfer 1,315 public housing units to tenants. As of late 1989, only 25% of the total had been transferred to resident owners. Evaluators of the program cautioned that a comprehensive sales program would be very costly and that additional cost-benefit analyses should be done to assess the relative merits of providing home-ownership opportunities through public housing sales, as opposed to other approaches.

Even before the results of the demonstration program were made available, the Housing and Community Development Act of 1987 established a new program that promoted sales of public housing units. Public housing tenants were given the ability to purchase units from resident management corporations that had acquired the development from the PHA, provided that a new unit of public housing was provided for each unit removed from the public housing stock. The requirement that public housing units sold or otherwise disposed be replaced on a one-to-one basis was restated as part of the Cranston-Gonzalez National Affordable Housing Act of 1990. Title IV, Homeownership and Opportunity for People Everywhere (HOPE), authorized the sale of public and Indian housing, as well as FHA distressed and foreclosed multifamily buildings and single-family homes owned by a public agency. However, in addition to actual replacement of the disposed public housing, the 1990 act stipulated that 5-year Section 8 certificates or vouchers also could satisfy the replacement requirement. Nevertheless, no specific funding for replacement of public housing units was authorized. Also, as of FY 1995, Congress suspended the replacement requirement for that fiscal year and it is unlikely that it will be reinstated in the foreseeable future.

Opponents of public housing sales argued that this was a means to privatize public housing and enable the government to absolve itself of the responsibility of maintaining and funding public housing. The issue, they argued, is that the government should fund the public housing program so that developments can provide good living environments for low-income people. Proponents, who rallied around former HUD Secretary Jack Kemp's persuasive appeals to providing homeownership opportunities, saw public housing sales as a means of providing an entry to the middle class for the poor residents of public housing. Under the Clinton administration, public housing sales have received little support.

How can PHAs address crime and drugs in their developments? Although crime and drugs are certainly not unique to public housing, the concentration of a low-income population has often proved easy prey for criminals. This, combined with the frequent neglect by municipal police departments and an unwillingness to enforce leases and evict tenants found breaking the law, has often resulted in public housing developments becoming havens for a host of illegal activities. During the 1970s, both HUD and the newly formed Law Enforcement Assistance Administration (LEAA) were actively engaged in an anticrime agenda, with many of the demonstration programs focused on public housing.

In 1978, Congress enacted the Public Housing Security Demonstration Act, aimed at developing, evaluating, and improving innovative anti-crime and security methods to reduce crime in public housing developments and their surrounding neighborhoods.

During President Carter's administration, the Public Housing Urban Initiatives Anti-Crime Program was created, which focused on major physical rehabilitation, management assistance, and anti-crime programs for seriously distressed projects. In addition, it emphasized coordination between local housing authorities, city government, and other relevant federal departments. Although an initial evaluation of the program was encouraging, the final report, produced during the Reagan administration, signaled the end of a series of policy initiatives started in the early 1970s in the management and anti-crime research and demonstration areas.

In 1985, however, HUD again turned to the issues of crime and security in public housing developments. The "Oasis Technique," modestly supported by HUD, emphasized the development of partnerships between residents, housing authority personnel, police, housing managers, and social service providers. Three years later, Congress enacted The Anti-Drug Abuse Act of 1988, which authorized $8 million to assist PHAs enhance security personnel and make physical changes in the developments that would promote resident security.

Overall, the problems of crime and drugs in public housing, particularly in large, inner-city developments, are still presenting daunting challenges. In fairness, however, private housing, both subsidized and unsubsidized, that attempts to service a similar population is likely to encounter the identical problems.

How can public housing be used to promote family self-sufficiency and to enhance the quality of life for residents? The large number of low-income households living in public housing makes these developments attractive locales for providing various types of services that can promote economic independence for residents. Family Investment Centers and the Family Self-Sufficiency Program were authorized by the Cranston-Gonzalez National Affordable Housing Act of 1990; both are aimed at assisting low-income households living in subsidized housing to achieve self-sufficiency. Family Investment Centers are limited to public housing developments and encourage the creation of special facilities for service providers in or near public housing developments. Lafayette Courts, a public housing development in Baltimore, served as the protype for Family Investment Centers. In the Family Self-Sufficiency Program, localities develop strategies to coordinate federal housing assistance with public and private services. Participants include current Section 8 certificate and voucher holders, as well as public housing residents.

In 1992, Congress created a new approach to enhancing the quality of life for public housing residents. Following the recommendations of the National Commission on Severely Distressed Public Housing, the HOPE VI-Urban Revitalization Demonstration Program is aimed at developing new initiatives to change the physical condition of public housing, promoting resident self-sufficiency, providing comprehensive services, lessening concentrations of poverty by reducing the density of public housing developments, and encouraging development of mixed-income communities. This is to be accomplished by encouraging PHAs, residents, and local communities to work with HUD and by forging partnerships with public, private, and nonprofit organizations. According to HUD, the mission of the HOPE VI program is to "end the physical, social and economic isolation of obsolete and severely distressed public housing by recreating and supporting sustainable communities and lifting residents from dependence and persistent poverty."

Additional important dilemmas that continue to surround the public housing program include whether public housing should be viewed as "housing of last resort" or whether stringent eviction policies, for households who are in violation of their leases, should be implemented. On the one hand, social service providers ask, If problem clients cannot turn to public housing, where can they go? On the other hand, public housing advocates claim that tenants who are unable to obey rules or who are engaged in illegal activities can create problems for the overall public housing community. Another lively debate concerns whether tenant assignments to public housing should be race neutral or whether they should strive to foster racial integration. Finally, HUD is actively considering deregulating the best-managed public housing authorities.

In conclusion, despite the ongoing problems facing the public housing program, this housing stock represents a critical national resource. Although it has been much maligned, it provides decent-quality affordable housing for millions of low-income people for whom other options are lacking. One of the clearest indications of the ongoing need for public housing is the number of households currently on waiting lists for public housing. According to the National Association of Housing and Redevelopment Officials, as of 1991, it was estimated that 1 million households were waiting for a public housing unit. If these households had other, better options, this demand for public housing would disappear. In the meantime, therefore, public housing remains a critical component of the U.S. social "safety net." (SEE ALSO: *Council of Large Public Housing Authorities; Drugs and Public Housing; Family Self-Sufficiency; Leased Housing; Military-Related Housing; National Commission on Severely Distressed Public Housing; Public Housing Authorities Directors Association; Public Housing Authority; Public Housing Homeownership Demonstration Program; Public Housing Program, Admission to and Occupancy of; Resident Management; Scattered-Site Public Housing; Turnkey Public Housing; Wagner-Steagall Housing Act*)

—*Rachel G. Bratt*

Further Reading

Bratt, Rachel G. 1989. *Rebuilding a Low-Income Housing Policy*. Philadelphia: Temple University Press.
Connerly, Charles E. 1986. "What Should Be Done with the Public Housing Program?" *Journal of the American Planning Association* 52(2):142-55.
Council of Large Public Housing Authorities and Public Housing Authorities Directors Association. 1992. *Public Housing: Open Your Eyes, Open Your Mind*. Washington, DC: Author.
Freedman, Leonard. 1969. *Public Housing: The Politics of Poverty*. New York: Holt, Rinehart & Winston.
Meehan, Eugene J. 1979. *The Quality of Federal Policy Making: Programmed Failure in Public Housing*. Columbia: University of Missouri Press.
National Association of Housing and Redevelopment Officials. 1990. *The Many Faces of Public Housing*. Washington, DC: Author.
Spain, Daphne. 1995. "Direct and Default Policies in the Transformation of Public Housing." *Journal of Urban Affairs* 17(4):357-76.
"Theme Issue: Public Housing Transformations—New Thinking About Old Projects." 1995. *Journal of Architectural and Planning Research* 12(3). ◀

▶ Public Housing Authorities Directors Association

The Public Housing Authorities Directors Association is a national, nonprofit association representing the directors of public housing agencies before Congress and the Department of Housing and Urban Development. It conducts workshops, seminars, and trade shows; provides members with informational resources; and publishes a biweekly newsletter titled *The Advocate*. The association operates on an annual budget of $1.5 million and is funded by membership dues, publication sales, and fees from conferences and workshops. Founded in 1979. Membership: 1,500 public housing directors nationwide. Staff size: 10. Executive Director: Wallace Johnson. Address: 511 Capitol Court, N.E., Washington, DC 20002-4937. Phone: (202) 546-5445. Fax: (202) 546-2280. (SEE ALSO: **Public Housing**)

—*Caroline Nagel* ◀

▶ Public Housing Authority

Public housing authorities, or PHAs, are public entities created by local levels of government to implement the U.S. federal public housing program, created in 1937. The public housing program provides direct federal subsidies to PHAs to defray both the capital and operating costs of running the housing, which is occupied by low-income households.

State enabling legislation authorizes local governments to create PHAs. The decision to create a PHA is voluntary. Only if a locale wants to operate public housing is it required to establish a PHA. After the creation of the public housing program in 1937, larger urban areas generally were the first to create PHAs, with smaller cities and suburban areas following, sometimes much later. Many local communities still do not have PHAs.

The public housing stock includes about 13,000 developments administered by some 3,400 PHAs. The majority of PHAs are small, managing less than 200 units, but most public housing units are located in developments owned by large housing authorities. In 1992, of all public housing units, 66% were administered by 134 PHAs—about 4% of the total number of housing authorities.

PHAs are responsible for a complex set of tasks, requiring various kinds of expertise: developing and acquiring developments, maintaining and modernizing them, selecting and evicting tenants, and providing social services to residents. Thus, the tasks of PHAs involve those of developer, landlord, agent of the federal government, and social service provider.

PHAs often have been the target of criticism, particularly for alleged poor management. Although there have been many instances of developments being run poorly, sometimes the reason has been inadequate funds available to operate the buildings properly. In other cases, however, mismanagement has, indeed, been due to shortcomings or, in rare instances, fraud, on the part of housing authority personnel. In some locales, PHAs have had the reputation of providing patronage jobs, with inadequate or unqualified staff.

PHAs have also been criticized for maintaining racially segregated developments. In the earliest years of the public housing program, developments were built either for black or white occupancy. However, even after overt segregation was abandoned, many PHA personnel continued to assign tenants to developments according to racial characteristics, thereby maintaining racially homogeneous developments.

PHAs are constantly challenged by having to operate and maintain complex physical structures; deal with myriad social problems, as often manifested by their low-income clientele; and mediate the multiplicity of problems associated with poor, inner-city neighborhoods, the location of many public housing developments. (SEE ALSO: **Public Housing**)

—*Rachel G. Bratt*

Further Reading

Council of Large Public Housing Authorities and Public Housing Authorities Directors Association. 1992. *Public Housing: Open Your Eyes, Open Your Mind.* Washington, DC: Author.

Kolodny, Robert. 1979. *Exploring New Strategies for Improving Public Housing Management.* Report prepared for the U.S. Department of Housing and Urban Development, Office of Policy Development and Research. Washington, DC: Department of Housing and Urban Development.

National Association of Housing and Redevelopment Officials. 1990. *The Many Faces of Public Housing.* Washington, DC: Author.

Struyk, Raymond J. 1980. *A New System for Public Housing: Salvaging a National Resource.* Washington, DC: Urban Institute. ◄

► Public Housing Homeownership Demonstration Program

In 1985, HUD selected 17 public housing authorities (PHAs) to participate in the Public Housing Homeownership Demonstration. The participating PHAs proposed selling 1,315 units of public housing to their tenants. An evaluation found many unforeseen obstacles to the sale of public housing units. Four years after the sales program began, only 25% of the units selected for sale had actually been transferred to residents. The results of this evaluation were used in the design of the HOPE 1 program introduced in 1990 (see Homeownership and Opportunity for People Everywhere).

Impressed by the success in selling public housing in Britain, the Reagan Administration proposed and Congress approved the Public Housing Homeownership demonstration in 1985. That demonstration was designed to "find practical ways to enable lower-income public housing tenants to own their own homes through the sale of Public Housing Authority units." Legislative authority for these sales already existed in Section 5(h) of the National Housing Act of 1974. HUD solicited applications to participate in the demonstration from local housing authorities, who were given great latitude in the design of the sales programs. Of the 36 applications received, HUD selected 17 PHAs to participate in the demonstration (Baltimore, MD; Chicago, IL; Denver, CO; Los Angeles County, CA; McKeesport, PA; Muskegon Heights, MI; Nashville, TN; Newport News, VA; Paterson, NJ; Philadelphia, PA; Reading, PA; St. Mary's County, MD; Tulsa, OK; St. Thomas, VI; Washington, DC; Wichita, KS; and Wyoming, MI). They proposed selling a total of 1,315 units of public housing over a 36-month period.

About one-half of the units selected for sale were detached, single-family units, and the rest were either duplex, triplex, town house, or low-rise apartment units. The sponsoring PHAs had to ensure that the units were in good condition at the time of sale. HUD, however, did not provide funds for the needed repairs, leading most housing authorities to select for sale their newer units in good repair.

Of the 17 sales programs, 12 involved the fee simple sale of units to tenant, 4 involved the sale of units as limited-equity cooperatives, and 1 involved condominium sales. The sponsoring PHAs determined how the sale prices were to be established. Even though they had the option of selling the units for as little as $1, all the program participants decided to charge higher prices and set minimum incomes for participating in the program. Most authorities set unit prices at their appraised values and then individually discounted them to what the sitting tenants could afford.

The demonstration guidelines required the participating PHAs to include safeguards against buyers reaping windfall profits from the quick sale of the units. The most frequent way of accomplishing this was for PHAs to hold a deferred payment, silent-second mortgage on the difference between the appraised value of the units and the discounted sale price. This silent-second mortgage is repayable if the buyer moves within a specified period of time, ranging from 5 to 28 years.

Only 320 of the 1,315 units selected for sale were actually transferred to tenants within four years of the start of the program. The major reasons for the difficulty in transferring units were (a) a lack of commitment to the program on the part of the sponsoring authorities or their governing boards, (b) difficulty in finding public housing tenants who had both the means and the desire to participate in the program, and (c) difficulty relocating tenants who did not want to participate in the demonstration.

A comparison of the characteristics of those who purchased units with those of all public housing authority tenants showed that participants had substantially higher incomes, were more likely to have at least one full-time wage earner, and were more likely to be two-parent rather than single-parent households. Those who did purchase homes, however, were generally found to be very satisfied with the units they purchased.

The results of the evaluation led to several recommendations for the design of future public housing homeownership programs. First, to interest more housing authorities in participating in the program, HUD would have to offer to replace the units sold. Second, more technical assistance should be provided to the sponsoring PHAs in designing and executing their sales programs. Third, HUD would need to provide the participating PHAs with funds to make repairs to the units to be sold. Fourth, special allocations of vouchers need to be provided to assist the sponsoring authorities in relocating families who are either not qualified to buy or are not interested in becoming homeowners. (SEE ALSO: *Privatization;* **Public Housing**)

—*William M. Rohe*

Further Reading

Rohe, W. and M. Stegman. 1990. *Public Housing Homeownership Assessment.* Washington, DC: Department of Housing and Urban Development.

———. 1991. "Public Housing Homeownership: Will It Work and for Whom?" *Journal of the American Planning Association* 58(2):144-57.

———. 1993. "Converting Multifamily Housing to Cooperatives: A Tale of Two Cities." In *Ownership, Control and the Future of Housing Policy,* edited by R. A. Hays. Westport, CT: Greenwood. ◀

▶ Public/Private Housing Partnership

Public/private housing partnerships, as defined by the National Association of Housing Partnerships (NAHP), are organizations that have substantial leadership and participation by the private sector, have the support of government, and see themselves as catalysts for programs under which affordable housing is produced, owned, and well managed—all necessary elements in the creation of desirable and successful communities. They act as intermediaries, expand housing opportunities, and help revitalize inner-city, suburban, and rural communities. Their goals reflect the political landscape of their locale. They are coalitions that can achieve objectives beyond the capacity of government or individual organizations acting alone.

In 1998, there were more than 100 formally organized public/private housing partnerships in states and cities across the United States with new ones formed each year.

A partnership is a voluntary association between two or more persons or entities who agree to carry on a business together, with mutual participation in profits and benefits. A public/private partnership is one in which private persons or entities carry out specific programs or projects in conjunction with public agencies, sharing control and using both public and private resources.

Public/private partnerships are not a new idea. For decades, there have been collaborative efforts by private civic groups to address community concerns. These endeavors have been in many cases at the behest of government or at least with its support. Business leadership was a prime mover in drafting the Chicago Plan in 1909. A variety of similar public/private efforts over the last 100 years have been devoted to building civic centers, rebuilding commercial areas, improving educational systems, building recreational facilities, and since the Housing Act of 1949, redeveloping urban sites and neighborhoods.

In its 1982 report, the Committee for Economic Development stated, "To make full use of the private sector's potential, local government will need to adopt an entrepreneurial approach that anticipates needs, seeks out opportunities, and encourages an effective coalition of public and private efforts. The private sector, in turn, needs to determine what it realistically can contribute." The report suggested that there were both policy and operational dimensions for public/private partnerships efforts. Policies need to include "a process that produces consensus on community goals, agreement on institutional roles, and sustained support for action."

Significant participation in partnerships by the business community occurs when business views the improvement of social and economic conditions in their communities of sufficient self-interest to warrant direct participation and support. In partnerships, risks are shared among the private and public players, enabling them to put their reputations on the line without jeopardizing the individual organization or agency they represent.

The term *public/private partnership* is now used to cover all kinds of organizations and programs in which there is collaboration between government and the private sector, with significant leadership and participation by the private sector. These partnerships are becoming increasingly involved with social problems facing local government such as providing jobs, education, and affordable housing.

Successful implementation of programs and sustained commitments of support are best achieved if the partnership is formally organized and viewed as a permanent institution. Most public/private partnerships are in fact organized as nonprofit corporations. Because public/private partnerships must respect the various interests of all those participating, they are sometimes complex and can be difficult to manage. Their success depends to a considerable measure on their corporate leadership and the degree to which support from government is explicit.

Factors that have led to the creation of public/private housing partnership and that influence their development include these:

▶ The reduction in the federal government's role as the primary initiator and funder of low-income housing
▶ Acceptance by state and local government of greater responsibilities for solving social problems, such as the lack of affordable housing
▶ Governmental requirements that force lending institutions to be more active and increase investment in low-income urban and rural areas
▶ Increased support for nonprofits and other community-based organizations as housing producers and owners

▶ Interest in privatizing services now performed by government

Roles played by public/private housing partnerships include the following:

▶ Advocating and initiating housing programs
▶ Serving as developer and owner
▶ Aggregating public and private resources
▶ Administrating loan, equity, and grant pools and reserve funds
▶ Administrating government subsidies
▶ Providing technical assistance
▶ Implementing homeownership counseling and other programs
▶ Coordinating programs beyond the capabilities of individual organizations or neighborhoods
▶ Supporting delivery of related social services
▶ Overseeing asset management

Accomplishments of local public/private housing partnerships have varied. The New York City Housing Partnership has facilitated the construction and sale of over 10,000 housing units. The Metropolitan Boston Housing Partnership has assisted community development corporations (CDCs) develop 2,000 housing units, administers rental assistance subsidies, and is undertaking a range of homeownership and resident empowerment programs. The Charlotte-Mecklenburg Housing Partnership has aggregated a $17.5 million loan pool for single-family mortgages, provided homebuyer counseling, and served as its own developer. The Indianapolis Neighborhood Housing Partnership administers local government funds, provides funding for CDC housing projects, and has helped over 2,000 low- and moderate-income households purchase homes.

Public/private housing partnerships are more than project developers. They have broad, multipurpose agendas in addition to being initiators and catalysts for action by others. Local partnerships have played a key role in support of community-based development organizations. They are uniquely qualified to achieve consensus about objectives, bring the necessary parties to the table, and successfully complete programs of significant scale. New local public/private housing partnerships are being established each year, and new directions in federal housing policy suggest that local public/private housing partnerships will be important players in future government-assisted housing. (SEE ALSO: *National Association of Housing Partnerships, Inc; Privatization; Turnkey Public Housing*)

—*Robert B. Whittlesey*

Further Reading

Committee on Economic Development. 1982. *Public-Private Partnership: An Opportunity for Urban Communities.* Special Summary. New York: Author.

National Association of Housing Partnerships. 1994. *A Catalogue of Local Housing Partnerships.* Boston: Author.

Suchman, Diane R. 1990. *Public/Private Housing Partnerships.* Washington, DC: Urban Land Institute. ◀

▶ Pueblos

Pueblo is the term given by the 16th-century Spanish explorers for the multistoried, terraced structures of stone or adobe they encountered in what is now northern New Mexico and Arizona. These tall, apartmentlike structures housed communities of agriculturalists who farmed the lands surrounding their village. The historic pueblos were built around one or more square, rectangular, or linear plazas. Today, *Pueblo Peoples* (a cultural term applied after the 1848 annexation of the Southwest by the United States) reside along the Rio Grande in north-central New Mexico; at Acoma, Laguna, and Zuni Pueblos in western New Mexico; and atop the Hopi Mesas in northeastern Arizona. Many of these pueblos (especially along the Rio Grande) have lost their traditional tightly nucleated, multistoried form, although almost all continue to use sacred plaza space. A few pueblos (such as Acoma) have retained their traditional form but are now occupied only part of the year (primarily during ceremonial cycles) by owners who maintain full-time residences in modern, more convenient dwellings elsewhere (often nearby). Pueblo peoples today continue to build and repair their homes, but the use of modern materials (such as cinderblock) is increasingly common.

Development of the Pueblo Form

Until about A.D. 700, in the northern Southwest of the United States, people lived in below-ground pithouses. The first crude above-ground structures were of "wattle and daub" (a woven stick and twig framework coated with mud) and provided extra storage space for pithouse dwellers. During the next 200 years, this form developed into the "unit pueblo," a crescent-shaped suite of 5 to 10 rooms (usually masonry) used for both domestic and storage purposes. In front of the rooms were one or two pit structures, probably used for domestic as well as some ceremonial purposes (a prototype for the "kiva," described below), and beyond that, there was a "trash midden" that contained domestic trash but that was also a sacred space where the things of everyday life went back to the earth.

By A.D. 900, a new form of pueblo structure developed and was eventually found over much of the northern Southwest: the Chacoan Great House. Massively and elaborately built with thick walls, stout roofs, and a carefully planned layout that was strikingly different from the accretional, unplanned roomblocks characteristic of historic pueblos. Encompassing 100 to 800 rooms up to 5 stories tall, Chacoan Great Houses were first built in Chaco Canyon, a remote valley in northwestern New Mexico. Although similar in underlying form, Great Houses differed dramatically in scale from the unit pueblos that continued to be built, usually in a community surrounding a Great House. Great Houses were apparently used primarily for ceremonial purposes by the surrounding community that lived in unit pueblos. By A.D. 1100, Great Houses could be found across the northern Southwest and were connected to Chaco Canyon by a series of wide, straight roads, demonstrating that this remote canyon was the center of a large regional system.

The nature of this regional system and the reason for 250 years of Great House construction and use remains uncertain, although it almost certainly was both a religious and economic system. When the Chacoan regional system collapsed about A.D. 1150, many Great Houses were remodeled and used for domestic purposes, likely serving as a model for the subsequent traditional pueblo style.

By the mid-1200s, most of the population of the northern Southwest lived in large, multistoried pueblos like those the Spanish encountered when they first entered the Southwest. Between 1150 and 1250, people in the Southwest had shifted from a dispersed settlement pattern of small unit pueblos around a Chacoan Great House to aggregated, high-density, apartment living that characterized the remainder of the prehistoric period and the early historic period. These highly aggregated pueblos have been explained partly as a response to the need for defense (historically, nomadic tribes raided pueblos) and also as the result of the development of the kachina religion, still found among Pueblo people today. Kachina ceremonies are usually conducted in plazas and viewed from surrounding rooftops.

Traditional Pueblo Architecture

Traditional pueblo buildings consisted of a set of attached dwellings (in archaeological terms, a "roomblock"), each dwelling occupied by a nuclear or extended family. Buildings tended to face southeast, allowing the house to be warmed by the sun on cold winter mornings. Southeast, where the winter sun rises, is also an important ritual direction to Pueblo people. Typically, Pueblo families inhabited a suite of rooms that ran from front to back of the roomblock, and these rooms were passed from generation to generation within the family. Rooms and entire houses were built and modified as needed for expanding families or for young couples becoming established, but over centuries, the plaza orientation was maintained.

Lower-story rooms, especially those in the back of the suite were used for storage, whereas the upper-story, front-facing rooms were living rooms. Pueblos were built of stone masonry or, at some Eastern Pueblos, of mud built up in courses by hand. In these multistoried structures, roofs had to be sturdy. The roof of a lower-story room served as the floor of an upper-story room, and terraced roofs served as open-air space for domestic tasks, especially food preparation. The flat roofs were constructed of stout beams (*vigas*), perpendicular smaller poles (*latillas*), a layer of twig and brush matting, and a tight cap of earth. Doors were usually not found in lower-story rooms; these rooms were entered through a ceiling hatch. Prior to Spanish contact, Pueblo people did not use chimneys but vented smoke from interior hearths through doors, ceiling hatches, or small vents in the wall.

Located in or adjacent to the plaza were *kivas* or, as the first Spanish explorers termed them, *estufas* (ovens—because they were warm inside). These were subterranean or semisubterranean religious structures that served as gathering places for sacred societies, places where rituals were performed, places where ceremonial dancers prepared for

Figure 25. Main Plaza at the Hopi Pueblo of Oraibi
Photograph by J. Hillers, 1879. Courtesy Smithsonian Institution, National Anthropoligical Aarchives, Photo No. 56394.

performances in plazas, or simply as meeting places for male members of the sacred societies who owned the kivas. Kivas were generally rectangular in shape in Western Pueblos, such as Hopi, and round among the Eastern Pueblos along the Rio Grande in New Mexico. Four wooden posts usually supported a roof consisting of a lattice-work of beams, poles, and brush, which was covered with a cap of earth to seal out moisture. They were entered through a roof hatch. Inside, a bench often encircled the kiva and a hearth was often placed beneath the roof hatch. Among other features found on the floor, many Pueblo kivas contained a small hole, called a *sipapu,* that represented the traditional place of emergence from the underworld for the community.

Each pueblo community saw itself as the center of a natural universe whose focus was often a rock shrine in the plaza that represented the spiritual "essence" of the village. The terraced structures were built up around and defined that sacred space. Beyond the dwellings were the fields, and beyond that, places where village people hunted, gathered wild plants, or visited sacred springs or shrines. High distant mountains defined the edge of the pueblo world.

Pueblo Evolution and the "Pueblo Style"

When the need for defense from nomadic tribes diminished in the mid-19th century, many of the pueblos began to reverse the use of multiple stories, and today, only a few have more than one or two stories. Lower stories were opened up with doors and windows and used for habitation, and upper-story rooms were torn down. A decline in the cohesiveness of the Pueblo community has been blamed for the construction of individual family houses farther and farther from the center of some Pueblos, changing the aggregated character of the traditional style.

Historic accounts of traditional pueblos and studies of modern pueblos are used by archaeologists to help them interpret abandoned prehistoric pueblos found across the northern Southwest. The flat-roof, open-beam, multistory, terraced, earth-toned Pueblo style of building continues in some modern pueblos and has been copied by contemporary architects and used for buildings as diverse as hotels and shopping malls. It is a dominant style in New Mexico; in Santa Fe, building codes prohibit the use of other styles. The "Southwestern" style of interior decorating, with softly rounded, thickly plastered spaces, muted lighting, and the use of pinks, light browns, and turquoise, also owes much to the traditional pueblos. This style can be found far beyond the Southwest in domestic and commercial architecture. (SEE ALSO: *Cultural Aspects; Hogan; Native Americans*)

—*Catherine M. Cameron*

Further Reading

Adams, E. Charles. 1983. "The Architectural Analogue to Hopi Social Organization and Room Use and Implications for Prehistoric Southwestern Culture." *American Antiquity* 43:44-61.

Markovich, Nicholas C., Wolfgang F. E. Preise, and Fred G. Strum. 1990. *Pueblo Style and Regional Architecture*. New York: Van Nostrand Reinhold.

Mindeleff, Victor. [1891] 1989. "A Study of Pueblos Architecture, Tusayan and Cibola." *Eighth Annual Report of the Bureau of American Ethnology 1886-1887*. Washington, DC: Smithsonian Institution Press.

Morgan, William N. 1994. *Ancient Architecture of the Southwest*. Austin: University of Texas Press.

Nabokov, Peter and Robert Easton. 1989. *Native American Architecture*. Oxford, UK: Oxford University Press.

R

► Racial Integration

There has been less research on racial integration than on segregation. Racial integration is a process whereby two or more racial groups move into and out of the same neighborhoods so as to maintain approximately proportional representation among the racial groups. Unlike segregation, integration is rare in most U.S. metropolitan areas.

Extent of Racial Integration

Few recent studies have been done to assess the extent of racial integration. In 1967, the National Opinion Research Center conducted a national study that revealed that 36 million people in the United States, or 19% of the population, lived in racially integrated neighborhoods. The study defined an integrated neighborhood as one in which the proportion of black residents in the neighborhood was approximately equal to the proportion of black families in the total city or metropolitan area. In this sense, integration can be viewed as the opposite of segregation.

The total number of households in integrated neighborhoods in 1967 was estimated at 11 million. Furthermore, the number of blacks living in racially integrated neighborhoods tended to be small. Of the 11 million households found in integrated neighborhoods, only about 7% were black. More than 20 years later, racially integrated neighborhoods were still rare. It was estimated that in 1990 only 20% of people in the United States lived in racially integrated neighborhoods.

Some progress toward racial integration has occurred through movement of the black middle class to suburban communities. But racial discrimination has continued to restrict a large percentage of the black middle class from participation in racial integration, even though income is not a barrier.

Barriers to Racial Integration

Certain barriers restrict movement into a neighborhood by members of racial minority groups. The first barrier involves illegal acts of racial discrimination—acts that continue at very high rates. Whites are often steered away from housing in predominantly black areas, and blacks are steered away from housing in predominantly white areas. Such practices undermine racial integration by preventing the open participation of all races in the competition for housing in all neighborhoods.

The second barrier to racial integration is the legacy of segregation. The cumulative effects of private and public acts of discrimination, carried on throughout history, have created negative myths about racial integration in housing. For many, if not most whites, a prevalent myth is that any racial integration in housing inevitably leads to racial transition, resegregation, declining property values, and neighborhood deterioration.

Unlike segregation in the public schools, where the courts have mandated racial integration, segregation in housing has continued with few examples of mandatory racial integration. The earliest example of mandatory integration began in Chicago in 1966 with a lawsuit called the *Gautreaux* case. Ten years later, in 1976, the court ruled in favor of a metropolitan-wide remedy involving the relocation of several thousand low-income residents (most of whom were black) to the suburbs of Chicago.

The most recent case was Yonkers, New York, where in 1986 Yonkers officials were found guilty of intentional discrimination in the location of federally subsidized housing. The officials were charged with promoting racial segregation by systematically locating subsidized housing in predominantly black areas. The city officials were ordered to construct public housing outside of black areas and to develop long-term plans for integration of the subsidized housing.

Factors That Facilitate Racial Integration

Research suggests that stable, racially integrated neighborhoods are most likely to be found when the following circumstances are present: (a) neighborhoods have many amenities, (b) integration is supported by city government, (c) public housing is deconcentrated, (d) a school desegregation program is in place, and (e) there is an effective regional fair housing program.

Neighborhood amenities include the quality and type of housing stock that will make it possible to affirmatively market the neighborhood, because affirmative marketing is an important tool for both achieving and maintaining racial integration. However, any fairly adequate neighborhood can still be successfully marketed to both blacks and whites, provided certain other factors are also present. One such factor that facilitates racial integration is the support of city government for the integration efforts. Integration can occur without this support, but it can be more difficult to accomplish.

Two factors seem to be most critical to the success of racial integration: (a) systemwide school desegregation and (b) the deconcentration of public housing. A regional school desegregation program eliminates racially identifiable schools, thus making the racially integrated neighborhood competitive with surrounding communities in attracting young families of both racial groups. Unless the public neighborhood schools are racially integrated, a majority of white families with children will not perceive the neighborhood as desirable and will not move into it.

Racially segregated public housing operates in much the same way as racially segregated public schools. Thus, public housing must be deconcentrated—that is, dispersed throughout the metropolitan region—if racial integration of neighborhoods is to succeed. Many people perceive public housing as undesirable because of poverty and, to some extent, racial composition (i.e., public housing is often disproportionately black, Hispanic, or both). Thus, a concentration of public housing in a neighborhood will make it more difficult to market. Finally, there should be an effective regionally based fair housing program to carry out affirmative marketing or other programs that promote racial integration.

Racial Integration Efforts

Voluntary organized efforts to bring about racial integration in housing began in 1950, following the 1948 *Shelley v. Kraemer* (334 U.S. 1 [1948]) Supreme Court decision. This decision declared that restrictive covenants, designed to prevent the sale or resale of housing to persons of specified race, color, religion, national origin, or ancestry were unenforceable by either state or federal courts because such enforcement would constitute governmental action in furtherance of racial discrimination.

One of the first organizations to work toward racial integration was the National Committee against Discrimination in Housing, founded in 1950. The committee has several local chapters around the United States. These chapters assist individual victims of discrimination in filing fair housing suits, systematically test for evidence of racial discrimination, and provide legal counsel to victims of discrimination in trial proceedings.

Since 1966, the Leadership Council for Metropolitan Open Communities, located in Chicago, has been actively working to bring about racial integration in the metropolitan Chicago area. Despite hard work and persistence, little racial integration has occurred. Such lack of impact has caused many to question whether private enforcement efforts in other metropolitan areas can succeed in bringing about racial integration.

Efforts to achieve racially integrated neighborhoods have been hampered by a lack of coordinated metropolitanwide efforts. The success of any single municipality's efforts to promote racial integration depends in large measure on the degree to which many municipalities in the same metropolitan area are involved in similar efforts. When only one municipality is involved in promoting racial integration, those whites who do not want integration often move to a segregated municipality nearby.

Racial Integration Methods

One method designed to increase racial integration is reducing racial discrimination through a technique called the fair housing audit. The fair housing audit is a field survey method designed to determine the extent of minority access to housing throughout a metropolitan area. In a fair housing audit, two families—one black and one white—are matched in terms of social and economic characteristics and trained to pose as home seekers. Both families separately contact a real estate agent or landlord and attempt to purchase a home or rent an apartment. The treatment the black and white families receive is compared for any evidence of differences. The extent of differential treatment indicates whether racial discrimination has occurred.

A second method designed to promote racial integration is the guaranteed home equity program. This method was first introduced in Oak Park, Illinois, a suburb of Chicago. The objective of the program was to guarantee homeowners in racially changing neighborhoods that their property values would not fall below fair market value determined at the time of entry into the program. In other words, the program's goal was to promote racial integration by stabilizing housing values so that a family's property would not decline below its appraised value. The ultimate purpose of the program was to reduce the out-migration of whites and aid neighborhood stability and racial integration.

A third method designed to increase racial integration is integration maintenance. Integration maintenance programs employ a variety of race-conscious techniques to maintain racially balanced populations. These programs, used most often in municipalities located near existing black areas, are likely to attract more blacks into the area. White demand for housing in the area is usually declining. The general belief is that without intervention in the housing market, white demand will continue to decrease while black demand increases—resulting in segregation. Integration maintenance programs are designed to prevent the resegregation process and to maintain stable racial integration. The goal of most integration maintenance programs is to maintain blacks as a minority population. Thus, realtors may be encouraged by neighborhood stabilization organizations to steer blacks to white areas and whites to predominantly black or integrated ones. Integration maintenance programs started in the mid-1950s and became widespread in the 1960s. These programs have had both successes and failures. To be successful, the programs must be implemented early when there is the first indication of racial integration. Integration maintenance programs rely heavily on affirmative marketing involving aggressive advertising of the neighborhood in an attempt to recruit home seekers of a certain race to a neighborhood where their race is underrepresented.

Integration maintenance programs, although designed with good intent, usually restrict the choice of blacks by applying quotas to control the rate of black entry into white areas. Such programs may restrict blacks rather than provide them with equal access to housing. Defendants have attempted to justify integration maintenance programs on the grounds that they are necessary to preserve racial integration. They are enacted from altruistic motives and are free of discriminatory interest.

Nevertheless, integration maintenance has been perceived by some as not integration or fair housing at all but as discrimination—a program designed by whites to maintain white dominance and continue white supremacy. Most integration maintenance programs do not depend on the consent of blacks, individually or as a group. The governing bodies of predominantly white municipalities usually impose integration maintenance to prevent certain neighborhoods or the entire municipality from becoming predominantly black.

Traditionally, blacks have not participated significantly in the political processes that produced integration maintenance plans. Therefore, mistrust has occurred on the part of some minority organizations and minority home seekers as to the intent of such predominantly white communities. Such groups have raised the question, "Is the intent to expand the access of minorities to all communities, or is it to restrict minority access to some communities by maintaining a certain racial quota?"

Thus, integration maintenance programs pose a dilemma for supporters of fair housing. The issue centers around the intent of national fair housing legislation, passed to eliminate discrimination by assuring that all housing be accessible to all racial and ethnic groups. Because the intent was accessibility, through elimination of discrimination, the law did not address voluntary segregation. This is the passive interpretation of the act.

Another interpretation by some fair housing groups is that the law was intended to promote racial integration. The conflict between the more activist position and the passive one has resulted in continued controversy as supporters of the activist position continue to implement integration maintenance programs.

One method that is less controversial and probably more effective is the *affirmative-integration incentive method*. This method is similar to affirmative marketing and integration maintenance, both of which provide information about available housing. However, the affirmative-integration incentive method goes beyond affirmative marketing and differs from integration maintenance. It not only promotes racial integration by providing information about available housing to both whites and nonwhites where they are underrepresented, but the method provides an economic incentive that makes it advantageous to purchase or rent such housing. The emphasis, then, is not on the "maintenance" of racial groups to achieve racial integration but on using economic incentive as the primary tool. The economic incentive should be strong enough to (a) counteract current racial steering and (b) encourage black and white home seekers to make housing choices unconstrained by the persisting (segregative) effects of past discrimination.

An affirmative-integration incentive loan program is designed to serve both black and white home seekers. The program does not set aside a quota of loans for particular racial groups. Participation in the program is voluntary. The program merely gives home seekers, with the resources to buy a house, an incentive to purchase in an area where their own race is underrepresented. The homebuyers themselves possess the freedom to participate or not. They are not restricted as to where they can live if they choose not to participate. Such incentives may include low-interest loans to cover part of the down payment or any needed maintenance.

Unlike integration maintenance, which is perceived by some to restrict choices on the basis of race, the use of affirmative-integration incentive programs expands the housing choices of both blacks and whites. Such incentive programs may be initiated by local, county, state, or federal governments or by institutions in the private sector. The Oakland County Center for Open Housing in Bingham Farms, Michigan, has been operating an affirmative-integration incentive program. These programs, which got their start in 1986 in Shaker Heights, Ohio, are still in their infancy and are limited in scope. They have had some degree of success in stabilizing integration. However, more time and more research are needed to assess their true potential.

To sum up, unlike racial segregation, racial integration is rare in U.S. metropolitan areas. It is estimated that in 1990, only 20% of people in the United States lived in racially integrated neighborhoods. The major barrier to racial integration in housing continues to be racial discrimination. Most methods to reduce racial discrimination and increase racial integration have not been effective. Affirmative-integration incentive programs may hold the most promise to reverse the persistent pattern of racial residential segregation in large metropolitan areas. (SEE ALSO: **Discrimination;** *Military-Related Housing; Segregation; Tipping Point*)

—*Joe T. Darden*

Further Reading

Bradburn, Norman, Seymour Sudman, and Galen L. Gockel. 1970. *Racial Integration in American Neighborhoods.* Chicago: University of Chicago, National Opinion Research Center.

Bullard, Robert D., ed. 1994. *Residential Apartheid: The American Legacy.* Los Angeles: University of California, CAAS Publications.

Cromwell, Brian A. 1990. *Pro-Integrative Subsidies and Their Effect on Housing Markets.* Cleveland, OH: Federal Reserve Bank of Cleveland.

Galster, George and Edward W. Hill, eds. 1992. *The Metropolis in Black and White.* New Brunswick: Rutgers, State University of New Jersey, Center for Urban Policy Research.

Massey, Douglas, S. and Nancy Denton. 1993. *American Apartheid: Segregation and the Making of the Underclass.* Cambridge, MA: Harvard University Press.

Saltman, Juliet. 1990. *A Fragile Movement: The Struggle for Neighborhood Stabilization.* New York: Greenwood.

Thomas, Suja. 1991. "Efforts to Integrate Housing: The Legality of Mortgage Incentive Programs." *New York Law Review* 66(3):940-978.

Tobin, Gary, ed. 1987. *Divided Neighborhoods: Changing Patterns of Racial Segregation.* Newbury Park, CA: Sage. ◄

▶ Real Estate Agency

Generally, an agent is anyone who represents the interests of another person (a principal) in dealing with third parties. When an agent has general authority to represent a principal, the relationship is called a *general agency*. When the agent's duties are specific and the agent cannot bind the principal in contract, the relationship is called a *limited*, or *special agency*. Because of their limited function (procuring a willing buyer, seller, landlord, or tenant), real estate brokers and salespersons are characterized as special agents.

Historically, U.S. courts have deemed the broker-client relationship to be a fiduciary relationship requiring trust and competence on the part of the broker or affiliated salesperson. Most states have two classes of real estate licenses: broker and salesperson. A broker can engage independently in real estate activities, but a salesperson can engage in such activities only if his or her license is affiliated with a broker. The salesperson is an agent of the broker, who in turn is an agent of the client. The salesperson who serves the client is the client's agent, through the salesperson's affiliation with the broker.

With some variance from state to state, an agent owes a principal seven fiduciary duties:

1. *Loyalty:* The agent must always act in the best interests of the principal and not engage in self-dealing.
2. *Obedience:* The agent must follow all lawful directions of the principal.
3. *Disclosure:* The agent must disclose all relevant information to the principal that could have a material effect on the principal's business.
4. *Confidentiality:* The agent must keep secret any information that the principal wants to remain confidential or that could weaken the principal's bargaining position.
5. *Accounting:* The agent must account for all the principal's funds or property coming into the agent's hands.
6. *Reasonable skill and care:* The agent must exercise reasonable skill and care in carrying out the principal's business.
7. *Diligence:* The agent must actively and diligently pursue the best interests of the principal in carrying out the purpose of the agency.

Fiduciary duties are a part of the substantive laws of each state, although the Real Estate Commission or Real Estate Department of a state may also promulgate related agency rules and regulations that subject real estate licensees to discipline for violations.

The residential real estate agency relationship is usually first created when the seller of a house enters into a listing contract with a broker (the listing broker). In return for a commission, the broker agrees to procure a ready, willing, and able buyer for the house. With some variations, there are four basic types of listing contracts.

1. *Open listing:* The seller nominates a broker as agent but reserves the right to nominate other sales agents. The first agent to procure a buyer (unless the seller procures one first) receives the commission. This type of listing is common in commercial transactions and some subdivision sales but is generally not otherwise used in residential sales.
2. *Exclusive agency listing:* The seller nominates a broker as the exclusive sales agent, but the seller reserves the right to sell the property without the agent's assistance and thereby avoid paying a commission.
3. *Exclusive right to sell listing:* The seller nominates a broker as the exclusive sales agent and agrees that any sale, even by the seller's efforts, will result in a commission to the broker. This is the predominant form of listing in residential transactions.
4. *Net listing:* The seller contracts with the broker to procure a buyer at a net cash price to the seller. Any sales proceeds in excess of the required net to the seller are retained by the broker as a commission. This type of listing is illegal in most states due to the inherent conflict of interest on the part of the broker.

The choice of contract is a matter of negotiation between seller and broker.

Often, a cooperating broker (the selling broker) procures a buyer and the two brokers split the sales commission. Until recently, such a cooperative sale was typically structured as a subagency relationship. The selling broker derived its authority, compensation, and responsibilities from the seller by carrying out the listing broker's task of finding a buyer. Thus, the selling broker was legally deemed to be a subagent of the seller. This legal structure for cooperative brokerage was frequently used because most residential cooperative sales originated through multiple-listing services (MLSs) and the MLSs typically required brokers to use a standard subagency contract provided by the MLS. An MLS is a cooperative arrangement among participating brokers to pool the listings of its members to gain greater exposure for the listing and thus increase the possibility of a sale. Placing a listing with an MLS is an invitation to other brokers to procure a buyer for the property. As MLSs proliferated after World War II, so did subagency sales of residential real estate.

Despite the duty of loyalty owed to the seller, selling brokers often acted as if they represented the buyer. A 1983 study by the Federal Trade Commission showed that in cooperative residential sales, most buyers thought that the selling broker represented them. In such a case, the broker may be characterized by a court as an undisclosed dual agent, with possible adverse consequences. The selling broker owes a duty to the seller but somehow gives the impression that he or she represents the buyer. Two conflicting fiduciary duties result from this so-called accidental or unintended agency. Through the 1970s and 1980s, brokers began to face a rapidly increasing number of lawsuits for conduct alleged to be in breach of fiduciary duties. Also, unhappy homebuyers sought to expand the scope of fiduciary duties owed by brokers into areas such as the duty of the broker to inspect the property and inform the purchaser (who may not be the broker's client) of any potential defects. Despite the increased liability, MLSs and professional organizations resisted change to the subagency relationship. Some brokers, realizing the problems with subagency, began to practice buyer's brokerage, representing buyers in real estate negotiations and rejecting subagency.

In 1993, the National Association of Realtors® dropped the requirement for mandatory subagency in its locally affiliated MLSs, paving the way for further development of buyer's brokerage and other alternative forms of broker-client relationships. One such alternative that began to receive attention in the 1990s is the concept of facilitator, which is similar to the form of brokerage in many European and Middle Eastern countries. The facilitator acts as a neutral mediator to the sale whose sole function is to bring the parties together and facilitate a sale. The broker represents neither party.

In response to the confusion that attends many residential agency relationships, many states now require that brokers provide some form of written disclosure of their agency status to their client or to all parties in the transaction. One continuing area of problems is the degree of care owed by a broker to a party he or she does not represent in the transaction. The broker, as agent, owes some degree of fiduciary duties to the party he or she represents, but the duties owed to the other party continue to pose problems. Brokers make a distinction between clients (the parties they represent) and customers (the parties they do not represent), but the trend in the courts is to require at least a standard of fair dealing and, in some states, affirmative duties of inspection and disclosure, even to customers.

Due to the constraints of antitrust laws, there are no fixed rates of commission for real estate brokerage services. The traditional compensation has been a percentage of the sales price, contingent on the broker's procuring a ready, willing, and able purchaser. The recent emergence of "buyer's brokerage" has brought some changes in brokerage compensation, including retainer fees, noncontingent fees, and reduced percentage fees. Commissions for commercial properties and acreage have historically shown wide variances depending on the size and difficulty of the transaction, usually in the 2% to 10% range. Residential brokerage commissions have traditionally remained around 6% with recent lower fees for nontraditional brokerage firms.

Between 1968 and 1995, existing home sales in the United States more than doubled from 1,569,000 to 3,802,000. As the volume of existing home sales increases, the use of real estate brokers will likely increase apace. (SEE ALSO: *Appraisal; Carrying Costs; Real Estate Licensing*)

—*Roy T. Black*

Further Reading

Black, Roy T. 1994. "Proposed Alternatives to Traditional Real Property Agency: Restructuring the Brokerage Relationship." *Real Estate Law Journal* 22(3):201-13.

Collette, Matthew M. 1988. "Sub-Agency in Residential Real Estate Brokerage: A Proposal to End the Struggle with Reality." *University of California Law Review* 61(2):399-457.

Federal Trade Commission. 1983. *The Residential Real Estate Brokerage Industry*. Washington, DC: Government Printing Office.

National Association of Realtors. 1996. *Annual Volume of Existing Home Sales*. Washington, DC: Author.

Nelson, Theron R. and Thomas A. Potter. 1994. *Real Estate Law: Concepts and Applications*. St. Paul, MN: West.

Rohan, Patrick J., Bernard H. Goldstein, and Charles S. Bobis. 1989. *Real Estate Brokerage Law and Practice*. New York: Matthew Bender. ◄

► Real Estate Developers and Housing

More than in most other countries, the private ownership of land and buildings is widespread in the United States; millions of people own, buy, and sell real property. Mass participation in the real estate market has been a fundamental characteristic of the economic life of the United States since its origin. Real estate developers have made an important contribution to the progress achieved during the past half century in working toward the goal declared by Congress in 1949 of "a decent home and a suitable living environment for every American family." The homeownership rate has increased from 43% of households in 1940 to 66% nearly 60 years later. Substandard housing has been reduced from nearly 50% of the dwelling units in 1940 to less than 5% in 1998.

Early History

Throughout the 18th and 19th centuries, people in the United States fought for greater legal rights and opportunities to become property owners. No institution was left untouched by this sweeping movement. Legislatures and the courts established and enforced new laws and definitions of the rights inherent in property and contracts, land was physically surveyed and real estate market mechanisms organized to facilitate sales, a vast array of subsidies was granted to prospective settlers to enable them to afford to own land, and enormous public investments in improved transportation and infrastructure helped make the land accessible and productive. As cities grew in the 19th century, more and more land was subdivided into building lots within existing urban areas and in the open countryside to establish new cities. The biggest single use of land in these metropolitan communities was for housing the steadily growing population. Land (and in some cases, buildings) was continually carved up to provide dwelling units for new residents.

An enormous amount of urban housing in the United States in the 19th century consisted of single-family detached dwellings because of the abundance of cheap land, inexpensive construction materials, and a constant stream of innovations in transportation technology that made residential dispersion possible. In the older and more crowded cities of the early 19th century, attached row houses (typically constructed in block groups by speculative builders) and multifamily dwellings converted from spacious mansions accommodated a higher-density population that walked to work in areas where available space was limited. Later in the century, a number of other dwelling types made their debut, including luxury apartment buildings, squalid tenements, and structures housing two to four families with modest-income owners often living in one of the units and in some cases constructing the buildings themselves.

Not only was it possible for the first time for millions of people to become urban property owners, but many were also actively engaged in the real estate business. Selling one's own or someone else's property as an agent was a completely unregulated activity in the 19th century and occupied the time and energy of a substantial segment of the population, especially during boom times. Some of these vendors indulged in unethical, fraudulent, fly-by-night practices that at times lent sales agents, developers, and landlords an unsavory image and later led to calls for reform by angry private citizens, concerned industry leaders, and progressive public officials.

In addition to ownership, sales, and property management, building construction was also a widespread endeavor. Most contractors and subcontractors, particularly in the residential field, were small-scale operators, often shuttling back and forth between contractor and laborer. The vast majority of houses were built under contract to the owner-user, with a significant segment actually built by the owner-user with the help of family and friends. Many large and small builders also constructed houses as speculative investments, although, generally, just one or two and seldom more than five such houses per year. *Merchant homebuilding,* as this method came to be known, did not really begin to dominate the housing industry until the 1950s. The standard approach for even sophisticated real estate developers was primarily to sell finished building lots, not completed houses.

The 1920s and the Great Depression

One of the most notable trends of the early 20th century, particularly during the 1920s, was the tremendous increase in the construction of apartment buildings. Earlier waves of urbanization in the United States, outside of New York and a handful of other major cities, had been based on a relatively low-density pattern of small, detached single-family houses, attached row houses, or duplexes. Some cities, such as Boston, had triple-deckers, and in many cities, large older houses were subdivided into multiple units. This pattern began to change dramatically during the 1920s. Real estate investors, developers, lenders, and contractors all became active participants in the production of new apartment buildings. The apartments were built primarily as rental units, although in a few cities, some of the buildings were sold to occupants for cooperative ownership. These new structures, built mainly with brick or stucco exteriors, ranged from very fashionable luxury residences with doormen and other services to more modest housing, and from individual six-unit buildings to high-rises to large complexes equipped with schools, parks, and community centers.

New single-family houses were also built in record numbers during the 1920s. The peak year, 1925, established an all-time high for starts of new housing that was not surpassed until 1950. At the height of the boom, new suburban subdivisions entered the market daily along the U.S. "crab-grass frontier." Although most of the houses were only modestly improved with basic infrastructure and amenities, a small but significant group of community builders was increasingly developing large-scale, well-planned, fully improved subdivisions complete with extensive landscaping, parks and parkways, and shopping centers. This pattern of development, with roots in the 19th century, was becoming more common and growing in scale of operations and degree of capital investment during the 1920s. The most eloquent representative of this trend was Jesse Clyde Nichols of Kansas City, Missouri, developer of the world-famous Country Club District and a founder of the Urban Land Institute. Part of what inspired Nichols and others to build the ideal of a stable, family-oriented, and beautifully landscaped community was their exposure to the European Garden City movement, launched by Ebenezer Howard in England at the turn of the century.

The garden city movement was a response to the rapid growth and overcrowding of the grimy, unsanitary, and crime-ridden industrial cities of the West. Howard envisioned balanced, self-contained, and modestly sized communities, each with an adequate economic base for manufacturing employment near workers' housing; each democratically self-governing with public ownership of land and community facilities; each physically well planned with plenty of greenery, open space, and easy transportation; and all part of a regional system of small cities separated by a permanent greenbelt of agricultural land. The philosophy of garden cities as it spread around the world contained four elements: environmental reform, social reform, town planning, and regional planning. Many of the actual development efforts, including J. C. Nichols's Country Club District, were motivated primarily by interests in environmental reform and town planning, with far less concern for the other two elements. The most ambitious attempt to translate the full expression of Howard's ideas in the United States was the City Housing Corporation (CHC) of New York, headed by Alexander Bing. Linking up with a group of visionaries called the Regional Planning Association of America headed by critic Lewis Mumford and architects and planners such as Clarence Stein, Henry Wright, and Catherine Bauer, Bing attracted sufficient investment capital to establish the CHC with the intention of building a garden city in the United States. After developing one successful preliminary project called Sunnyside Gardens in New York, the CHC bought a large parcel of land in Fair Lawn, New Jersey, within commuting range of Manhattan, and began in 1928 developing Radburn, "a town for the motor age."

The vigorous spirit of reform and modernization that characterized the early 20th century was married with the tremendous growth and institutional development of the real estate industry through the movement for "professionalization." Many elements of the flourishing business organized trade associations to upgrade standards of practice; to isolate, ostracize, and where possible eliminate unsavory activities; and to cooperate with the public sector and other segments of the business world and the general public to protect the interests of real estate and enhance its political stature and economic viability. The National Association of

Realtors, for example, was established in 1908 to seek government licensing of the brokerage business.

By the late 1920s, the speculative craze for subdivision lots was abating, and many of the people who had bought on credit in anticipation of rapid and profitable resales were defaulting on their loans and property tax assessments. Through 1931, new investment, development, sales, and leasing continued in many markets, and real estate entrepreneurs kept hopes alive; in the following year, however, everything began grinding to a halt, and bankruptcy became the normal state of affairs. Financing was unavailable and real estate plummeted in value. Much of the market was frozen, flooded with properties for sale or lease that no one wanted, even at heavily discounted prices and rents. By 1933, nearly half of all home mortgages were in default and 1,000 properties a day foreclosed. Annual starts of new housing had dropped by more than 90%, from the record-breaking peak of 937,000 units in 1925 to the dismal trough of 93,000 units in 1933.

For real estate developers, the most important federal agency created by the New Deal was the Federal Housing Administration (FHA). The FHA mutual mortgage insurance system promoted cost-efficient production of small houses and affordable homeownership for middle-income families. The FHA's conditional commitment enabled developers and merchant builders of subdivisions to obtain debt financing for large-scale construction of new residential neighborhoods and communities, complete with finished houses and full installation of improvements, ready for immediate occupancy by people who were able to buy with modest savings because they qualified for FHA-insured mortgages. The FHA model of real estate development represented a dramatic advance over previous methods of subdividing and selling unimproved lots that had been fairly common in the 1920s. The FHA's property standards and neighborhood standards helped to improve the minimum levels of quality in the design, engineering, materials, equipment, and methods of land development and housing construction. The FHA also introduced new techniques for analyzing market demand and using underwriting to limit overbuilding and excessive subdividing.

The FHA's underwriting guidelines strongly favored new housing over used, suburban locations over central-city sites, whole subdivisions over scattered units, single-family houses over apartments, and Caucasians over African Americans. For older cities and racial minorities, these policies were inequitable, discriminatory, and disastrous. But for the growth of white, middle-class suburbia, they were of crucial importance. Although the FHA did insure mortgages on suburban garden apartments, its overall thrust helped reverse the trend of the late 1920s toward increased construction of apartments and significantly boosted large-scale homebuilders and suburban home-ownership.

Two new organizations, both spin-offs from the National Association of Realtors, also emerged from the crucible of economic crisis and political reform that characterized the 1930s. The Urban Land Institute started as a small, elite organization, primarily of big commercial and residential developers. The National Association of Home Builders was formed in 1943 to lobby the federal government during wartime to allow the continuation of private development of housing for sale or rent financed with the aid of FHA mortgage insurance.

The Modern Era

The FHA and the new Veterans Administration home loan guarantee program, created in 1944, stimulated large-scale homebuilding and the availability of mass financing through life insurance companies, savings and loans, mutual savings banks, and other sources. Residential developers grew rapidly in size, and the entire industry dramatically increased its production. By 1949, 10% of the builders constructed 70% of the houses, 4% of the builders constructed 45% of the houses, and just 720 firms built 24% of the houses. These figures reveal the growing concentration of housing production after World War II among a small group of well-capitalized "community builders" and represent a drastic change. This trend continued during the 1950s with increasing expansion in size, scale, and volume by the big homebuilders. Part of this change in the real estate industry after the war can be attributed to the experiences during the war, when the federal government encouraged and subsidized residential developers to quickly mass-produce private housing for war workers. The biggest of all the homebuilders immediately after the war was Levitt & Sons, developers of Levittown, New York. Levittown was the largest private housing project in the United States at the time. The first houses were completed in 1947, and by the early 1950s, Levitt & Sons had built 17,500 houses on 4,000 acres of potato fields in Hempstead on central Long Island, about 30 miles east of New York City.

By the late 1950s, the incredible postwar demand for new suburban single-family houses had largely been satisfied, and builders and developers began searching for new products and markets. Developers such as William Zeckendorf built middle-income and luxury housing in central-city locations through local urban renewal programs and the provisions of Title I of the 1949 Housing Act. Under Title I, the federal government paid anywhere from two-thirds to three-fourths or more of the "write-down," equal to the total direct public cost of land assembly, demolition, site replanning, and improvements, minus the revenue from the sale of the land to the private redevelopers. These projects were usually built with tax exemption and other financial subsidies and were frequently criticized for lengthy delays and the displacement of small businesses and low- or moderate-income residents.

Growing dissatisfaction with the federal public housing program led low-income housing advocates to favor a subsidized public-private approach. New federal housing programs were created during the 1960s, including the Section 221(d)(3) program (below-market interest rates), the Section 202 program (housing for the elderly), and a variety of others. These programs generally served a somewhat higher-income market than public housing and by the 1970s were producing a much greater volume of new units. Further initiatives occurred in 1968 with the passage of the landmark National Housing Act, which included a program to encourage production of rental housing (Section 236) and a subsidy to cover mortgage interest for low-income homeowners (Section 235). In 1974, the Section 236 program was replaced by yet another variant, Section 8. Particularly during the 1960s and 1970s, these programs helped produce hundreds of thousands of units of new housing, many of good quality. The federal government drastically cut back these programs during the 1980s, eliminating some entirely and reducing others by as much as three-fourths of their annual budget, compared with the late 1970s. State and local governments as well as philanthropic institutions and nonprofit organizations have also contributed resources to this complex system of producing housing, and some for-profit builders have made this activity a major portion of their business.

The massive population influx of the postwar baby boomers who reached adulthood and formed separate households, the shift in population growth from the Frostbelt states to the Sunbelt states, the substantial increase in single and divorced households, and the rise in the numbers, income, and wealth of senior citizens all had a major impact on housing development. The housing industry responded by building and rehabilitating a record volume of units in the 1970s and maintaining high production through much of the 1980s. Condominiums as a new form of owning an individual apartment burst onto the scene in the early 1970s, accounting for a significant portion of new and converted multifamily units.

Prices, especially of single-family houses, rose rapidly in many markets as demand outran supply, with the costs of new and existing housing and developable land all outpacing the increase in median household income during the past two decades. The gap in wealth between homeowners and renters widened, and both long-standing tenants and newly formed households strained their resources to rush into homeownership before prices rose higher and to take advantage of anticipated appreciation in equity and tax benefits. Mostly on the West and East Coasts at various times from the mid-1970s to the early 1990s, sales and prices of housing rose and fell in successive waves of speculative frenzy followed by recessionary panic. Construction of multifamily housing was boosted substantially in the early and mid-1980s by the use of syndications, accelerated depreciation, passive loss, and other income tax benefits created through the Economic Recovery Tax Act of 1981. The reduction of these benefits under the 1986 tax reform law immediately set off a significant reduction in new investment and development and a rapid decline of the syndication industry. Homebuilding was also slowed by the collapse of many savings and loans, particularly in the Sunbelt, and by a "credit crunch" in commercial banking.

In the early 1990s, developers were having a much harder time financing new projects. Although the "credit crunch" for homebuilding eased somewhat by the mid 1990s, annual housing starts remained below the boom years of the 1970s and 1980s. In addition, many new federal, state, and local laws and practices—such as environmental impact reviews, historic preservation requirements, growth controls, sewer moratoriums, impact fees, linkage payments, and regional "fair share" agreements for inclusionary housing—all served to slow the process and add to the costs of real estate development in many communities. Developers find themselves increasingly involved in public relations campaigns and public policy initiatives, working with local residents, business and civic groups, community leaders, and government officials to have projects approved by agreeing to pay a greater share for public facilities and amenities and finding new ways to address neighborhood concerns and mitigate the perceived negative effects of proposed development. (SEE ALSO: *American Real Estate and Urban Economics Association; European Real Estate Society; Residential Development; Section 608*)

—*Marc A. Weiss and John T. Metzger*

Further Reading

Abrams, Charles. 1939. *Revolution in Land.* New York: Harper.
Dowall, David. 1990. "The Public Real Estate Development Process." *Journal of the American Planning Association* 56:504-12.
Downs, Anthony. 1985. *The Revolution in Real Estate Finance.* Washington, DC: Brookings Institution.
Eichler, Ned. 1982. *The Merchant Builders.* Cambridge: MIT Press.
Hoyt, Homer. 1993. *One Hundred Years of Land Values: The Relationship of the Growth of Chicago to the Rise in Its Land Values.* Chicago: University of Chicago Press.
Jackson, Kenneth T. 1985. *Crabgrass Frontier: The Suburbanization of the United States.* New York: Oxford University Press.
Mayer, Martin. 1978. *The Builders: Houses, People, Neighborhoods, Governments, Money.* New York: Norton.
Miles, Mike E., Richard L. Haney, Jr., and Gayle L. Berens. 1996. *Real Estate Development: Principles and Process.* Washington, DC: Urban Land Institute.
Weiss, Marc A. 1987. *The Rise of the Community Builders: The American Real Estate Industry and Urban Land Planning.* New York: Columbia University Press. ◄

► Real Estate Law Journal

The *Real Estate Law Journal* is a quarterly journal focusing on current issues in real estate law. First published in 1971, this journal currently has a circulation of 3,500 copies nationwide. Submissions Editor: Jayne Allen. Address: One Penn Plaza, 40th Floor, New York, NY 10119. Phone: (212) 971-5241. Fax: (212) 971-5025.

—*Caroline Nagel* ◄

► Real Estate Licensing

Real estate licensing laws became common in the early 20th century in response to apparent widespread improper real estate practices. State licensing laws protect the public by ensuring that real estate licensees have met minimum standards of ability and that the licensees follow professional rules of conduct. Most states have a real estate commission that administers the state licensing law. A real estate commission typically consists of appointed officials. Some states administer the licensing laws through state licensing boards.

Real estate licensing covers brokerage activities. Generally, a real estate broker is any person who, for a fee or commission, negotiates or assists in procuring prospects for the listing, sale, purchase, exchange, renting, lease, or option for any real estate or its improvements.

All 50 of the United States have licensing statutes for real estate brokers and salespersons, although the requirements for obtaining a license vary from state to state. For example, the requirement for classroom hours of prelicensing education varies from no requirement at all to a high of 200 hours. Sixty hours is a common requirement. Salespersons and brokers must pass written exams before receiving their licenses. Usually, a salesperson must hold a salesperson's license for some period before qualifying for a broker's license. Two years is a common requirement. A salesperson must affiliate his or her license with a broker and work as an agent under the broker's direction.

After obtaining a license, a broker or salesperson must complete the state's continuing education requirements to keep the license valid. A common requirement is six classroom hours per year. Usually, real estate licenses are valid only within the state that issues them. A broker who wishes to do business in more than one state must meet the licensing requirements for every state in which he or she wishes to do business. Most states will issue nonresident licenses.

State laws require real estate brokers to have financial responsibility. Most states require that brokers maintain escrow checking accounts for clients' money. Many states have Real Estate Recovery Funds to protect the public from broker misconduct. Some states require that brokers post bonds for public protection.

Licensing laws provide exceptions for certain activities. A licensed practicing attorney acting incident to the practice of law would not need a real estate license. An employee of a property owner (such as an apartment resident manager) who leases property or collects rents or a person acting as an attorney-in-fact under a duly executed power of attorney to convey real estate would also typically be exempt.

Real estate licensees are subject to private lawsuits for misconduct, such as fraud or breach of contract. In addition, each state has rules and regulations subjecting real estate licensees to disciplinary action. Examples of violations could include falsifying contracts or racial discrimination. If a complaint is made, the licensee is notified of the charges. The licensee has the right to a hearing and, if found guilty, may be disciplined by the licensing authority, including having his or her license revoked. (SEE ALSO: *Real Estate Agency*)

—Roy T. Black

Further Reading

Association of Real Estate License Law Officials. 1994. *1994 Digest of Real Estate License Laws*. Bountiful, UT: Author.

Lindeman, Bruce J. 1994. *Real Estate Brokerage Management*. 3d ed. Englewood Cliffs, NJ: Regents/Prentice Hall.

Nelson, Theron R. and Thomas A. Potter. 1994. *Real Estate Law: Concepts and Applications*. St. Paul, MN: West. ◀

▶ Real Estate Mortgage Investment Conduits

Despite the success of the market for mortgage-backed securities (MBSs), there has been a demand for more specialized instruments. Two examples of such instruments are collateralized mortgage obligations (CMOs) and real estate mortgage investment conduits (REMICs).

REMICs were introduced as part of the Tax Reform Act of 1986 in response to the significant tax limitation placed on CMOs by the federal tax code. CMOs had grown in importance in the previous years as an attractive vehicle for institutional real estate investors. The Internal Revenue Service had ruled that organizations that divided claims to various classes of interests could not receive trust status. CMOs were forced to be classified as collateralized debt and thus introduced accounting and taxation problems. The Tax Reform Act of 1986 permitted an alternative instrument, REMICs, which eliminated the tax problems for issuers, because REMICs qualified as sales of assets for tax purposes.

Technically, the issuer of these securities can elect whether or not a CMO should be treated as a REMIC for tax purposes. If so, the REMIC does not affect the liability side of the issuer's balance sheet and avoids the tax problem. Therefore, a REMIC can be viewed as a special type of CMO in response to the federal tax regulations.

Given the added flexibility of REMICs, many observers predict that REMICs will continue to develop and expand the options available for institutional investors. For example, the frequency-of-payments option (whether cash flows are paid monthly, quarterly, semiannually, annually, or at some other interval) may be expanded with REMICs. Also, various types of mortgage strips could develop for different types of interest rate instruments. (SEE ALSO: *Collateral Mortgage Obligations*)

—Austin J. Jaffe

Further Reading

Fabozzi, Frank J. and Franco Modigliani. 1992. *Mortgage and Mortgage-Backed Securities Markets*. Boston: Harvard Business School Press.

Goodman, John L., Jr. 1992. "The Future Supply of Mortgage Credit." *Real Estate Review* 22(Spring):35-44.

Konstas, Panos. 1987. "REMICS: Their Role in Mortgage Finance and the Securities Market." Federal Deposit Insurance Corporation *Banking and Economic Review* (May/June):1-9.

Winger, Alan R. 1987. "The Securitization of Real Estate Finance." Federal Home Loan Bank of Cincinnati *Review* 1(August):2-5. ◀

▶ Real Estate Review

Real Estate Review is a quarterly national publication providing comprehensive coverage of real estate topics and issues. It contains articles on (a) real estate financial and investment issues, (b) discounted cash flow analysis, (c) property market analysis, (d) real estate portfolio analysis, (e) lease negotiation issues and concerns, (f) income tax effects on property acquisition and financial rates of return, (g) legal and environmental factors effecting real property, (h) property development concerns, (i) real property valuation issues, (j) real estate securities and real estate investment trusts, and (k) a wide range of articles on other topics of interest to the real estate professional. Each issue also contains a letters to the editor section and special sections on the topics of law and taxation, appraisal, and executive compensation. As of winter, 1998, the price is $109.00 per year in the United States and Canada and $144 per year elsewhere. *Real Estate Review* is published by The West Group, 610 Opperman Drive, Eagen, MN 55123-1396. The editor is Norman Weinberg, and the managing editor is Naomi Miller Weinberg. Editorial offices are located at 24 Oak Street, Woodmere, NY 11598. Phone: (516) 295-2179. Fax: (516) 569-2448. E-mail: rer@ix.netcom.com.

—*Joseph S. Rabianski* ◄

► Real Estate Settlement Procedures Act of 1974

Settlement is the process by which ownership in a home passes from seller to buyer. It usually requires the services of professionals. The Real Estate Settlement Procedures Act of 1974 (RESPA) regulates the conduct of service providers when a single-family home is bought with a loan from a federally insured depository institution or a federally regulated lender, is insured by a federal agency, or will be sold to Fannie Mae (formerly, the Federal National Mortgage Association), Ginnie Mae (GNMA), or Freddie Mac (FHLMC). In practice, nearly all single-family mortgages are covered.

The purposes of RESPA are (a) "more effective advance disclosure to homebuyers and sellers of settlement costs," (b) "elimination of kickbacks or referral fees" among service providers "that tend to increase unnecessarily the costs of certain settlement services," (c) "reduction in the amounts home buyers are required to place in escrow accounts," and (d) "significant reform . . . of local record keeping of land title information."

The main substantive provisions are contained in the following:

► *Section 4:* HUD must prescribe a standard form for the statement of settlement costs. The person conducting the settlement (usually an attorney or escrow agent) must give it to the borrower at settlement or, on request, one day before.

► *Section 5:* The lender must provide a good faith estimate of the settlement costs the buyer is likely to incur.

► *Section 6:* The lender must tell the borrower at application if the lender will service the loan or sell it to another firm. The borrower must receive certain written notice in the event of such transfers and has a right to written answers about the loan account within 20 days. If the lender demands escrows for taxes and insurance premiums, the lender must pay those bills as they come due.

► *Section 8:* No person may give or receive a kickback, fee, or any other thing of value in return for referring business to a settlement provider. The penalty for violating this provision is a fine of up to $10,000 or a prison term of not more than one year. The prohibition does not apply to payments for services actually performed. For example, a lender may pay a mortgage broker for qualifying a loan applicant, and real estate agents may split fees among themselves. Controlled business arrangements may or may not be permitted; it matters, for instance, whether the buyer knew of the relationship between the parties or could choose other service providers and whether the owners' profit represented normal return on investment.

► *Section 9:* No property may be sold with a requirement that title insurance must be purchased from any particular company.

► *Section 10:* Lenders may not require excessive escrow payments. If a balance represents more than two months' taxes and insurance premiums, payments are excessive. Lenders must give borrowers itemized annual statements of their escrow accounts.

Evaluations of RESPA must be tentative. Research is emerging on the following aspects:

1. *Advance disclosure:* Good-faith advance estimates and actual settlement cost disclosures seem to be routinely provided. It is not clear that these measures have changed buyer behavior or market outcomes; there have been no systematic studies.

2. *Elimination of kickbacks and referral fees:* The classic kickback is from a title insurance company to a real estate agent or broker. In 1994, the Department of Housing and Urban Development (HUD) collected more than a million dollars in fines in the Philadelphia area alone from title insurance and real estate firms charged with giving and receiving referral fees. RESPA has probably reduced the prevalence of grosser forms of kickback but has not affected the economic incentives for service providers to pay them. It may therefore have stimulated more subtle forms of kickback arrangements.

3). *Reduction of escrow payments:* HUD regulations implementing the ban on escrow balances in excess of two months of taxes and premiums are new at the time of writing, and it is too early to discuss their effectiveness.

4. *Reform of land title recordation:* No progress has been made toward this goal.

One common complaint against RESPA is that it adds more paperwork to settlement. Many consumers do not understand the forms, so they do little good, and providers have to fill them out, which raises settlement costs. Another criticism is that the referral fee prohibition inhibits innovative market arrangements and joint ventures that would reduce cost to consumers. Finally, some argue that RESPA is not necessary because lenders and real estate agents depend heavily on their reputations for business, especially repeat business. To date, there is no adequate empirical basis for evaluating these criticisms.

In 1974, the main legislative alternative was "lender-pay," under which the lender was responsible for all settlement costs. Although these costs would then be passed on to the borrower, competition among lenders would give each lender a strong incentive to control settlement provider fees. A second alternative is reform of local government property record systems, which are often blamed for the high cost of title search, evaluation, and insurance. (SEE ALSO: *Closing Costs; Escrow Account*)

—Mark Shroder

Further Reading

Bourdon, Richard. 1994. *The Real Estate Settlement Procedures Act: Is It Working?* Report No. 94-841 E. Washington, DC: Library of Congress, Congressional Research Service.

Crowe, David A., John C. Simonson, and Kevin E. Villani. 1981. *Report to Congress on the Need for Further Legislation in the Area of Real Estate Settlements.* Washington, DC: U.S. Department of Housing and Urban Development, Office of Policy Development and Research. ◄

► Redlining

Access to credit to purchase and improve housing is essential for the health of any neighborhood. The unavailability of credit for residents in many urban neighborhoods has undermined community development efforts, particularly in older and predominantly nonwhite neighborhoods. The practice commonly known as *redlining*—refusing to make credit available or the willingness to do so only under restricted terms, due to geographic location and unrelated to any objective determination of risk—is a long-standing reality of U.S. urban areas (This word is derived from the practice of literally drawing red lines on maps to mark areas where residents would not be eligible for certain credit products. Although much more common among financial institutions two or three decades ago, such maps still exist in the late 1990s.) Redlining denies the opportunity for many individuals to purchase a home and maintain their neighborhoods. More problematic have been the racially discriminatory consequences that reinforce the dual housing markets and segregated living patterns so characteristic of urban communities in the United States.

As with other components of the housing industry, mortgage lenders have long provided services along racial lines. Until 1950, the National Association of Real Estate Boards told its members, "A realtor should never be instrumental in introducing into a neighborhood . . . members of any race or nationality, or any individual whose presence will clearly be detrimental to property values in the neighborhood" (Judd 1984, p. 284). Frederick Babcock (1932), a leading theoretician of real estate principles and property appraisal practices, wrote more than 60 years ago that "there is one difference in people, namely race, which can result in very rapid decline. Usually such declines can be avoided by segregation" (p. 91). The Home Owners Loan Corporation (HOLC), created by President Roosevelt in 1933 to provide financial support for homeownership, developed a rating system in which neighborhood desirability was determined largely by the racial composition of its residents. "Residential security maps" were then developed in which a rating was assigned to clearly label the attractiveness of identifiable neighborhoods. HOLC's appraisal methods were adopted by others in the housing industry,

including the Federal Housing Administration (FHA), which during its early years warned in its underwriting manuals that "if a neighborhood is to retain stability, it is necessary that properties shall continue to be occupied by the same social and racial classes" (FHA 1938, par. 937). Lending has long been influenced by that admonition from the early years of the FHA.

The FHA was instrumental in nurturing the post World War II suburbanization of U.S. cities, the racial exclusivity of suburban communities, and the segregation of metropolitan areas. Through 1959, less than 2% of FHA-insured loans went to black homeowners. During the 1960s, the agency did a turnabout and, working with unscrupulous real estate agents and mortgage lenders, flooded central cities with FHA-insured loans. Unfortunately, many of these borrowers, a disproportionate share of whom were black, in fact could not afford to maintain the homes they were sold. When they defaulted on the loans, the properties were taken and sold again, often to similarly unwary buyers. Blockbusting and other scare tactics were used, making many white urban homeowners anxious to sell and resulting in the rapid transition of several neighborhoods from predominantly white to virtually all-black communities. With the closing costs paid up front and the mortgages insured by the federal government, realtors and lenders had every incentive to exploit these buyers and turn the properties over several times. In 1997, approximately 80% of black and white people in the United States live in virtually all-white or all-black neighborhoods and such mortgage-lending practices constitute one major direct cause.

If the more explicitly racist practices have faded, continued problems with credit availability throughout the 1960s and 1970s in many urban neighborhoods stimulated widespread organizing by community groups. Public protests of redlining on the part of lenders and lax law enforcement by federal financial regulatory agencies resulted in the passage of two key federal laws, the Home Mortgage Disclosure Act (HMDA) in 1975 and the Community Reinvestment Act (CRA) in 1977. HMDA requires most mortgage lenders to disclose the census tract, race, gender, and income for each application they receive and the disposition of all applications. CRA requires federally regulated mortgage lenders to continuously and affirmatively assess and respond to the credit needs of their entire service areas, including- low and moderate-income communities.

Subsequent research using HMDA data and other sources of information has confirmed suspicions of community groups that their neighborhoods were redlined and that racial minorities were not receiving the same treatment as white borrowers. Data on lending practices through the 1980s and mid-1990s reveal that racial minorities are rejected twice as often as whites even among applicants with similar incomes. Fewer loans are made and applications are more likely to be rejected in central-city neighborhoods, even after controlling for the number and condition of housing units, median income, population turnover, and other demographic characteristics.

Test audits by fair housing groups (in which pairs of individuals posing as mortgage loan applicants, matched on every conceivable characteristic, except one—their race—inquire about the possibility of home loans at selected institutions) reveal that race frequently affects the treatment by the loan officers. For example, a white tester is often encouraged to apply for a conventional loan, whereas the black tester is advised to go to another institution to apply for an FHA loan. In the case of marginally qualified applicants, white testers will often be counseled on how they can improve their applications, whereas the black testers are informed that they do not qualify. Mortgage-lending testing is in its infancy today, but this tool, which has long been used to monitor racial steering by real estate agents, will no doubt be used more extensively to assess lending practices in the future.

The Boston Federal Reserve Bank provided the most compelling evidence of racial discrimination when it examined 3,100 loan files at 300 area banks. Controlling on 38 variables from the loan files, researchers found that blacks were rejected 60% more often than comparable white applicants.

Many factors contribute to the redlining problems confronting the nation's cities. Financial characteristics of borrowers, the condition of property, and the degree of risk do vary across neighborhoods. Some of the underwriting rules that are neutral on their face have adverse racial effects. For example, some lenders will not provide mortgage loans on lower valued or older properties, homes that racial minorities are more likely to inhabit. Use of phrases such as "pride of ownership" or "declining neighborhood" in underwriting guidelines and practices often still reflect racial stereotypes and negative images of older urban communities. Discrimination by real estate agents and property insurers can discourage a potential borrower from coming to the bank in the first place. The underwriting guidelines of private mortgage insurers (private mortgage insurance is often required for low-income borrowers) and secondary mortgage market institutions (to whom lenders sell an increasing share of their mortgages over time) also restrict the availability of credit.

In the 1990s, redlining problems received more widespread publicity and attention from lenders, regulators, and policymakers. Using leverage provided by the CRA, lenders have entered into reinvestment agreements with community groups in more than 100 cities, totaling over $210 billion in lending commitments. The U.S. Department of Housing and Urban Development has begun funding private nonprofit fair housing groups to conduct mortgage-lending testing programs. Since 1992, the U.S. Department of Justice settled several pattern-and-practice lawsuits brought by the federal government against mortgage lenders for racial discrimination. Fannie Mae and Freddie Mac, two major secondary mortgage market institutions, have developed more flexible underwriting guidelines to facilitate greater access to credit for low-income homebuyers. Fannie Mae's chairman and chief executive officer observed that "discrimination continues to limit access to housing and mortgage credit for many citizens. The challenge now facing the housing community is to fashion solutions that remedy these disturbing findings."

Redlining persists, although it may be more subtle today than in earlier years. Although debates and controversies continue, there is evidence that progress is being made to increase the availability of credit in older urban communities and to begin the process of dismantling the dual housing market. (SEE ALSO: **Abandonment;** *Community Reinvestment Act; Home Mortgage Disclosure Act*)

—*Gregory D. Squires*

Further Reading

Bradbury, Katharine L., Karl E. Case, and Constance R. Dunham. 1989. "Geographic Patterns of Mortgage Lending in Boston, 1982–1987." *New England Economic Review* (Sept./Oct.):3-30.

Cloud, Cathy and George Galster. 1993. "What Do We Know about Racial Discrimination in Mortgage Markets?" *Review of Black Political Economy* 22(1):101-20.

Hunter, William C. and Mary Beth Walker. 1995. "The Cultural Affinity Hypothesis and Mortgage Lending Decisions." *Journal of Real Estate Finance and Economics* 13:57-70.

Jackson, Kenneth T. 1985. *Crabgrass Frontier: The Suburbanization of the United States.* New York: Oxford University Press.

Massey, Douglas S. and Nancy A. Denton. 1993. *American Apartheid: Segregation and the Making of the Underclass.* Cambridge, MA: Harvard University Press.

Munnell, Alicia H., Geoffrey M. B. Tootell, Lynn E. Browne, and James McEneaney. 1996. "Mortgage Lending in Boston: Interpreting HMDA Data." *American Economic Review* 86(1):25-35.

Schafer, Robert and Helen F. Ladd. 1981. *Discrimination in Mortgage Lending.* Cambridge: MIT Press.

Squires, Gregory D., ed. 1992. *From Redlining to Reinvestment: Community Responses to Urban Disinvestment.* Philadelphia: Temple University Press. ◀

▶ Refinancing

Refinancing is the creation of a new mortgage on a property prior to the maturity of the existing mortgage. Typically, it involves increasing the loan amount, obtaining better terms for the sum owed, or both. Four basic reasons underlie the motivation to refinance. First, lower interest rates can result in lower mortgage payments and, hence, more disposable income or cash flow. Second, lower interest rates can enable a borrower to replace an adjustable-rate mortgage with a fixed-rate mortgage to "lock in" a lower and more predictable payment that will not fluctuate over time; this practice frequently occurs at the bottom of an interest rate cycle. Third, irrespective of interest rates, the borrower can increase the size of the mortgage to tap into the property's equity and thereby generate tax-free cash for a variety of purposes. In contrast, if the property was sold, the proceeds would bear tax consequences. And fourth, increasing the mortgage amount (if the loan is assumable) can make the property more attractive to a buyer because the size of the required down payment can be reduced. In addition, a larger mortgage carries the additional tax advantage of higher interest payments, which for some individuals can reduce taxable income. (SEE ALSO: **Mortgage Finance**)

—*Daniel J. Garr*

Further Reading

ABA Standing Committee on Lawyers' Title Guaranty Funds. 1991. *Buying or Selling Your Home: Your Guide to: Contracts, Titles, Brokers, Financing, Closings.* Chicago: American Bar Association.

Jacobus, Charles J. 1996. *Real Estate Principles.* Upper Saddle River, NJ: Prentice Hall.

Shim, Jae K. 1996. *Dictionary of Real Estate.* New York: John Wiley.

Tosh, Dennis S. 1990. *Handbook of Real Estate Terms.* Englewood Cliffs, NJ: Prentice Hall. ◀

▶ Religion and Housing

Of the many factors that can affect the design, construction, occupation, and use of housing, an important one is religion. Religion is defined here as a set of institutionalized social beliefs, values, attitudes, practices, and rituals that indicate what is appropriate and inappropriate, what is good and evil. It includes notions of an ideal or good life and prescribes rules, activities, and behavior the faithful observance of which will lead to salvation and ultimate good for the followers. Religion and related beliefs help people deal with complex questions concerning, for example, the origin of the world, humans, natural and supernatural phenomena, the existence of an omnipotent force, the meaning of life, and death and beyond. Religion thus provides prescriptions for a specific lifestyle.

Religion affects housing in a number of ways, directly through prescriptions and proscriptions regarding building and indirectly through values, preferences, and requirements for lifestyle, and activities and artifacts for the conduct of religious activities, which in turn affect the designs of the spaces housing these activities.

Direct prescriptions affecting housing can specify the location, nature, and design of housing. The Hindu *Vastu Shastra,* for example, specifies where, what, how, and when one should build. It describes principles for location, orientation, dimensions, layout, timing of construction, landscaping, materials, colors, and iconography. Similarly, feng-shui principles, which originated in China and are followed in modified form by Vietnamese, Koreans, and others, also provide directives for location, orientation, dimensions, layout, landscaping, colors, and so on. Feng-shui attempts to achieve harmony between the occupant and the house by channeling through design positive and negative energy in beneficent ways, which lead to safety, longevity, and prosperity of the occupants.

Religion may require a certain orientation for the house or may express preferences for particular orientations and auspicious or inauspicious directions. For example, Zoroastrians believe that north is an inauspicious direction because Ahriman (who represents evil) is believed to reside there. Ahura Mazda (who represents good) is believed to reside in the south, which is seen as an auspicious direction. Zoroastrians prefer to not face their houses and the primary rooms northward; south is the preferred direction. For Muslims, orientation toward Mecca during sacred and religious activities is important. In Iran, this is approximated to facing south. Conversely, inauspicious, impure activities should not be conducted facing the holy city. In Iran, toilets are oriented so that one does not face south. For Hindus, auspicious activities are not to be conducted facing south, which is believed to be the direction of death. Rather, for Hindus, east, the direction of the rising sun, has the "greatest cosmological significance" and is the most auspicious direction. The Navajo hogan entrance faces east to welcome the morning sun.

Requirements for the practice of religion in the home can also affect the design of the house. In Hinduism, for example, the emphasis is on individual communion with god, not on congregational prayer. The home is the place to establish a location for the image of the god, worship, and meditation. Zoroastrians similarly have a space in the home for their sacred fire, the *Atesh Dadgah,* and for rituals that need to be conducted at home. Sikhs have a special space for the *Guru Granth Sahib,* their holy text. Vietnamese Buddhist houses have an altar for religious icons, artifacts, pictures of ancestors, and ritual objects and offerings. Greek Orthodox Christian houses have a small shrine, called the *iconostassi.* Hopis have sacred *kiva*s, circular sacred chambers for fertility rites, often underground.

Religion affects the interior arrangement of houses in two ways. First, it establishes the need for certain spaces for religious purposes or for keeping religious objects and specifies characteristics for each of those spaces. For example, the space for domestic prayer in the Hindu house is the *pooja* (worship) room. In the Greek house, the *iconostassi* is located near the ceiling on a wall near the east. Second, religion sometimes specifies the relationship between spaces. The *pooja* space is located in the purest or highest space in the house, following the definition of sacred and profane spaces and activities. Separative strategies may be called for. In the Hindu house, the sacred is expected to be separated and distanced from profane spaces, such as toilets.

Religions may have additional requirements that influence the design of houses. For example, orthodox Jews have strict separation of one category of food items from another. This led some followers of Orthodox Judaism to have two sets of refrigerators, ovens, sinks, dishwashers, and storage for two sets of dishes to ensure the separation of milk products from meat products. Separation of the sexes is important to Muslims. Muslim houses in Iran, Iraq, India, and elsewhere have separate quarters for men and women.

In some congregational religions, most religious activities are performed in the church. For Catholic Christians, only one of the six sacraments, extreme unction, can be performed at home in a consecrated area. Catholic homes in the 19th century had very few religiously oriented design features or artifacts, although it was common practice to have homes consecrated by priests. By the late 19th century, Catholic homes in Victorian America had oratories or chapels for prayers and places for several artifacts, such as St. Bridget's cross, crucifixes, a statue of madonna and child, hymnals, altars, candles, tapers, holy water, sometimes arched stained glass windows, crimson draperies, flowers in vases, a Holy Bible (as home Bible reading became common), and Angelus clocks to remind them of prayers three times a day.

Religion has also affected the layout of housing in towns. In the past, the church, synagogue, temple, or mosque was often the most significant building in the skyline of a town. It was the focal point around which the town grew.

Differences in the ways that specific religions affect housing, as illustrated in the examples, are important because they contribute to the diversity and richness found in the architecture of the different religious groups.

Housing, in turn, can affect the practice and continuance of religious practices, particularly for noncongregational religions, by presenting or lacking opportunities for keeping religious objects or conducting certain rituals. The residents can reject designs that overlook these factors. For example, government-provided housing proved to be inappropriate for the Tonto Apache and Navajo. If people are compelled to live in inappropriately designed houses, this may require adjustment of religious practices or negatively affect adherence to religion. (SEE ALSO: *Cultural Aspects*)

—*Sanjoy Mazumdar*

Further Reading

Boyce, Mary. 1971. "Zoroastrian Houses of Yazd." Pp. 125-46 in *Iran and Islam: Papers in Memory of V. Minorsky,* edited by C. E. Bosworth. New York: Bibliographic Distributors.

Chua, Beng Huat. 1998. "Adjusting Religious Practices to Different House Forms in Singapore." *Architecture and Behaviour* 4(1):3-25.

Mazumdar, Sanjoy & Shampa Mazumdar. 1997. "Religious Traditions and Domestic Architecture: A Comparative Analysis of Zoroastrian and Islamic Houses in Iran." *Journal of Architectural and Planning Research* 14(3):181-208.

Mazumdar, Shampa & Sanjoy Mazumdar. 1994. "Of Gods and Homes: Sacred Space in the Hindu House." *Environments* 22(2):41-49 (Special issue on "Environments of Special Places").

McDanneil, Colleen. 1986. *The Christian Home in Victorian America, 1840-1900.* Bloomington: Indiana University Press.

Oliver, Paul. 1997. *The Encyclopedia of Vernacular Architecture of the World.* 3 vols. Cambridge, UK: Cambridge University Press.

Pavlides, Eleftherios and Jana Hesser. 1989. "Sacred Space, Ritual and the Traditional Greek House." Pp. 275-94 in *Dwellings, Settlements and Tradition: Cross Cultural Perspectives,* edited by J. P. Bourdier and N. AlSayyad. Lanham, MD: University Press of America.

Rapoport, Amos. 1969. *House Form and Culture.* Englewood Cliffs, NJ: Prentice Hall.

———. 1982. "Sacred Places, Sacred Occasions and Sacred Environments." *Architectural Design* 9(10):75-82.

Saile, D. G. 1985. "The Ritual Establishment of Home." Pp. 87-111 in *Home Environments,* edited by I. Altman and C. Werner. New York: Plenum. ◀

▶ Rent Burden

Rent burden, generally stated as a percentage of household income, is an indicator of the cost of rental housing in the context of socioeconomic factors. When rent burdens are high and incomes are low, there is very little left to pay for other necessities. Unexpected expenses, failure to manage scarce resources, or both also make low-income renters vulnerable to eviction for nonpayment of rent. It is not uncommon to find lower-income renters in the United States paying 50%, 60%, and even 70% of income for private market housing. Such high rent burdens are predictors of impending crises such as homelessness.

Federal housing programs target assistance to lower-income renters in consideration of the following:

- ▶ *Cost burden > 30%:* The extent to which gross housing costs, including utility costs, exceed 30% of gross income, based on data published by the U.S. Bureau of the Census
- ▶ *Severe cost burden > 50%:* The extent to which gross housing costs, including utility costs, exceed 50% of gross income
- ▶ *Rent burden > 50% (severe cost burden):* The extent to which gross rents, including utility costs, exceed 50% of gross income

These criteria are used for tenant selection preferences in federally assisted public or Section 8 housing. Preference is to be given first to applicants whose rent burdens are determined to be severe. Once residents gain access to HUD-sponsored housing, they are required to pay 30% of their adjusted household income toward rent.

In the early days of public housing, under the Wagner-Steagall Housing Act of 1937, residents paid 20% of their income for rent. Later, rents varied according to what was needed for operating costs. In 1969, the Brooke Amendment limited rents charged by local housing authorities to 25% of a tenant's income, which became the rent burden standard for most federal programs over the next decade. In 1981, under the Reagan administration, Congress approved an increase to a 30% rate.

In both public housing and Section 8 programs, the federal government pays the balance of housing costs not covered by residents' shares. For example, the Section 8 certificate program operated by the U.S. Department of Housing and Urban Development (HUD), covers the difference between the HUD fair market rent (FMR) for a comparable unit within the local market and 30% of adjusted gross household income. Adjusted income is determined according to household size after deducting certain additional expenses (e.g., standard allowances for the elderly, each dependent, child care, and medical expenses). A personal benefit expense or utility allowance also enters into the housing contributions made by eligible households to ensure that, after payment of utilities, the federal statutory standard is not exceeded.

Rent burden standards serve the purpose of allocating scarce housing subsidies according to measurable criteria. Although the federal rules seem to suggest that rents not in excess of rent burden standards are affordable to lower-income households, this is not necessarily true. In any analysis of rent burdens, it is necessary also to consider contextual factors, such as the cost of nonhousing essentials, noncash forms of public assistance, the supply of available low-cost rental units, waiting lists for subsidized housing, overcrowding, housing quality, government control, race, discrimination, family size and composition, welfare status, and so on. (SEE ALSO: **Affordability;** *Affordability Indicators; Affordable Housing Indexes; Housing-Income Ratios;* **Private Rental Sector;** *Rent Control; Rent Decontrol*)

—*Marjorie E. Jensen*

Further Reading

Daye, Charles et al. 1989. *Housing and Community Development: Cases and Materials.* 2d ed. Durham, NC: Carolina Academic Press.

Feins, Judith D. and Terry Saunders Lane. 1983. "Defining the Affordability Issue." In *Housing Supply and Affordability,* edited by Frank Schnidman and Jane A. Silverman. Washington: Urban Land Institute.

Stone, Michael E. 1993. *Shelter Poverty: New Ideas on Housing Affordability.* Philadelphia: Temple University Press.

Wright, Gwendolyn. 1989. *Building the Dream: A Social History of Housing in America.* 4th ed. New York: Pantheon. ◀

▶ Rent Control

Roughly 40% of the world's urban dwellers are renters; in many cities in developing countries, two-thirds or more of the housing stock is rental. A majority of countries have some form of price control on some of or all their rental housing stock.

Rent control is usually thought of as a policy applied to private markets, but publicly provided housing is also subject to controls and to some of the attendant problems, such as reduced revenue and maintenance. For example, most urban housing in Russia and in China is or was owned by the government or by state enterprises. Rents are based on historical costs and are extraordinarily low in real terms. As a consequence, housing subsidies are a huge share of government budgets. Many units are undermaintained because of lack of financing. Severely controlled prices can cause problems for public as well as private housing.

History

Rent controls are often instituted in response to a major economic or political shock that limits the responsiveness of the housing market. Most European nations introduced rent control during World War I, only to liberalize in the interwar years. Controls were reintroduced in World War II in Europe, North America, and under European colonial influence, much of the developing world. Most jurisdictions in the United States and Canada removed controls in the postwar years; however, controls were maintained in much of Europe and the developing world.

Features

There are many different kinds of rent control regimes. For example, one key feature is whether regulations set the level of rents or control increases in rent. Others include (a) *how* controlled rents are adjusted for changes in costs (with cost pass-through provisions or adjustments for inflation), (b) how close the adjustment is to changes in market conditions, (c) how it is applied to different classes of units, or (d) whether rents are effectively frozen over time. Other key provisions that vary from place to place include (a) breadth of coverage, (b) how initial rent levels are set, (c) treatment of new construction, (d) whether rents are reset for new tenants, and (e) tenure security provisions. Rent control's effects can vary markedly depending on these specifics, market conditions, and enforcement practices.

For example, in New York City there are three main classes of rental housing. Controlled rental housing comprises mainly pre-1947 apartments with rents set on what is roughly a (financial) cost plus basis. Since 1969, units built after 1947 (and some pre-1947 units that had been decontrolled) have been subject to "rent stabilization," under which a board comprising landlord, tenant, and "general public" representatives sets annual guidelines for percentage increases. Since 1971, both controlled and stabilized units are removed from the system whenever tenants turn over, but since 1974, once new tenants negotiate rents the units come under stabilization once again.

As another example, Los Angeles has had rent control in place since 1978. But compared with New York its regulatory framework is fairly flexible. New construction, "luxury" units, and detached units are exempt; about 63% of rental units fall under controls. Rents are reset for new tenants by negotiation, and increases thereafter are limited to the Consumer Price Index. Evictions are permitted for "just cause." There is a "sunset provision" that provides for the end of controls if the rental vacancy rate reaches 5%.

Other U.S. markets with controls include Santa Monica, Boston, Cambridge, and several jurisdictions in northern New Jersey. Compared with most controlled U.S. and Western European markets, developing country and recently socialist markets tend to have stricter regimes (more setting of levels rather than regulating increases, fewer exemptions for small landlords or new units, etc.). Generally, comparative research finds that although qualitative outcomes of controls are often similar, magnitudes differ.

Analysis

Rent control can be analyzed as an implicit tax on housing capital. In the simplest case, where imposition of controls reduces the price of an existing stock of rental housing, the tax is borne by landlords for the benefit of tenants. Over time, as the market adjusts to controls, the incidence of the "tax" becomes more complicated.

At least nine alternative adjustment mechanisms can equilibrate a notionally controlled market. The maintained hypothesis is that markets must adjust in some fashion in the long run, given alternative opportunities for landlords and a housing stock of limited durability. Four of the adjustments can be embodied in rent control laws:

1. Indexing (keeping real rents constant)
2. Reassessment for new tenants
3. Differential pricing of new and existing units
4. Differential pricing for upgraded units

Three are market responses that many would generally consider undesirable outcomes:

5. Outright evasion
6. Side payments such as key money
7. Adoption by tenants of maintenance expenditures

Other adjustments include the following:

8. Accelerated depreciation and abandonment

9. Distortions in consumption, not only in the composite housing services but in also crowding, length of stay, mobility, and tenure choice

Although it is commonly accepted that rent control can reduce the efficiency of the rental market, the magnitude of such effects is debated, and proponents of controls usually justify them on distributional grounds. Three key questions are, What are the efficiency losses from controls? Do they redistribute income as intended? Are the benefits to some tenants worth the costs?

Static Analysis

Static analysis takes the current stock of units as given and considers how costs and benefits are distributed. Studies that calculate the static cost borne by owners of existing rental units show that the reductions from market rent can be substantial but that tenants, in general, value the implicit subsidy of controls less than it costs.

The "static" cost of controls to landlords is the difference between the controlled rent they are allowed to charge for a particular unit and the market rent for that unit in the absence of controls. Obviously, this cost is also the implicit subsidy to the tenant from controls. But the tenant may value this implicit transfer less than an equivalent amount of money income.

For example, suppose a unit would rent for $300 in the absence of controls, and the controlled rent is $100. The landlord loses $200 and the tenant gains a $300 unit for $100. If the tenant would happen to demand a more or less identical unit in the absence of controls, his or her benefit is also $200, and we say the "transfer efficiency" of controls is 100%; that is, the benefit to the tenant is equal to the landlord's cost. We may or may not like the idea of transferring income from landlord to tenant in this fashion, but that is a separate issue.

Now consider the case in which the tenant would, in the absence of controls, choose a larger unit (which would rent for, say, $500). Assume the tenant is in our original controlled unit; he or she receives $300 worth of housing services for $100 and so saves $200. But to receive this transfer, the renter has had to reduce his or her housing consumption, and our net benefit measure should adjust for that. In the jargon of economics, the household has lost "consumer's surplus" from having to move to a smaller unit to get the rent control discount. The concept is discussed in more detail in any introductory economics text, and the magnitude of such changes can be estimated using econometric techniques (see Further Reading at end of this entry). Suppose the lost consumer's surplus was estimated using such methods and found to be $55; the household's *net* benefit would be $200 minus $55, or $145. In this example, the landlord incurs a cost of $200 to benefit the tenant only $145. The "transfer efficiency" is only 72%. Landlords incur a cost of $1.38 for every $1 of tenant benefit.

Rent control in New York City in 1968 appeared to redistribute income, but very weakly, and in no way proportional to its cost. A number of other studies have been carried out along these lines. For example, in Cairo, Egypt,

monthly rents for a typical unit are less than 40% of estimated market rents. "Key money" (illegal up-front payments to landlords) and other side payments make up about a third of the difference. In Amman, Jordan, the static cost of controls is about 30% of estimated market rent; the benefit to the typical tenant is only 65% of cost.

Dynamic Analysis

Rent control can also impose dynamic costs (i.e., undesirable changes in the stock of housing over time). Controls can reduce dwelling maintenance, reduce the useful life of dwellings, and inhibit new construction. Controls provide strong incentives to convert rental units to other uses. These market responses shift the incidence of rent control's costs forward to tenants, over time. It is theoretically possible to design a rent control regime that does not discourage maintenance and starts with a pricing scheme that rewards maintenance and new construction. In practice, revaluation and maintenance inspections are expensive and difficult to organize, and new construction can still be adversely affected by the expectation of future controls. Research in Los Angeles found that, generally, tenant benefits were substantially less than landlord costs; the transfer efficiency in three representative cases ranged from 65% to 83%.

Given their apparent importance, dynamic effects of controls are understudied. For example, no one has yet credibly analyzed the effects of controls on the aggregate supply of housing. Despite many studies that imply controls qualitatively reduce returns to rental investors, given the myriad ways real-world regimes work and ways around controls (legal and illegal), the size of the net aggregate effect on supply remains unknown. Countries with stricter rent control regimes have been found to invest less in housing, in the aggregate, but although they controlled for demand (income and demographics), they were not able to control for other constraints on housing markets (e.g., land use constraints, financial constraints). Because these may well be correlated with the strength of controls, the jury is still out.

Distributional Issues

One reason controls appeal to many is that they are believed to transfer income from supposedly wealthy landlords to poor tenants. But much remains to be learned about the actual distributional effects of controls. There are three main questions of interest, related to (a) the distributional effects between landlords and controlled tenants, (b) the distributional effects between controlled tenants and uncontrolled households (homeowners and uncontrolled tenants), and (c) distributional outcomes within the class of controlled tenants.

Such evidence as exists casts doubt on controls' effectiveness as income transfer mechanisms. In Cairo and Bangalore, India, for example, no relationship was found between the benefits gained from reduced rent and household income, because rent control is not well targeted to low-income groups. In Kumasi and Rio de Janeiro, benefits were found to be somewhat progressive.

Another questionable assumption behind redistribution as a rationale for controls is the notion that landlords are rich and tenants are poor. In Cairo, Kumasi, and Bangalore, the income of tenants and landlords was compared, and although the landlords' median income was higher in all three, there was significant overlap. In Cairo, for example, about 25% of tenants had incomes higher than the landlord median, and about 25% of landlords had incomes lower than the tenant median. There is no guarantee that the transfers will occur only from high-income landlords to low-income tenants.

Some tenants are, on balance, worse off under controls because of constraints on housing consumption. And in markets with significant uncontrolled sectors, rent controls can drive up the price of uncontrolled housing, an important unintended consequence further complicating the incidence of its costs.

Concluding Comments

Consensus of empirical research is that rent controls usually fail to meet the goals sought by their advocates. But the dire consequences of simple models are also not born out. Although a stringent system may reduce rents, authorities could achieve the same benefit if rent supplements were paid to low-income households—without distorting the supply of housing. (Of course, rent control subsidies are "off-budget," which accounts for some of their political popularity.) In many situations, reform of other housing policies might be more important than reform of rent control. Because weak control may create only minor distortions, other policies may be the cause of more serious problems. For example, overly restrictive land use regulation or an inadequate housing finance system may be the real constraint on the production of housing. In such cases, a focus on rent controls may divert attention from the real problems and, at the same time, fail to achieve the desired effect. (SEE ALSO: *Rent Burden; Rent Decontrol*)

—*Stephen Malpezzi*

Further Reading

Appelbaum, R. P. 1990. "The Redistributional Impact of Modern Rent Control." *Environment and Planning A* 22:601-14.

Arnott, Richard, ed. 1988. "Rent Control: The International Experience." *The Journal of Real Estate Finance and Economics on Rent Controls* 1(3)[Special issue].

Gilderbloom, John I. and Richard P. Appelbaum. 1988. *Rethinking Rental Housing*. Philadelphia: Temple University Press.

Keating, W. Dennis, M. B. Teitz, and A. Skaburskis, eds. 1998. *Rent Control: Regulation and the Rental Housing Market*. New Brunswick: Rutgers, State University of New Jersey, Center for Urban Policy Research.

Kutty, Nandinee K. 1996. "The Impact of Rent Control on Housing Maintenance: A Dynamic Analysis Incorporating European and North American Rent Regulations." *Housing Studies* 11(1):69-88.

Lowry, Ira S. 1992. "Rent Control and Homelessness: The Statistical Evidence." *Journal of the American Planning Association* 58(2):224.

Malpezzi, Stephen and Gwendolyn Ball. 1991. *Rent Control in Developing Countries*. World Bank Discussion Paper No. 129 (84 pp.). Washington, DC: World Bank.

Murray, Michael P., C. Peter Rydell, C. Lance Barnett, Carol E. Hillstead, and Kevin Neels. 1991. "Analyzing Rent Control: The Case of Los Angeles." *Economic Inquiry* 29:601-25.

Navarro, Peter. 1985. "Rent Control in Cambridge, Mass." *The Public Interest* 78(Winter):83-100.

Olsen, Edgar O. 1972. "An Econometric Analysis of Rent Control." *Journal of Political Economy* 80:1081-1100. ◀

▶ Rent Decontrol

Rent controls have been studied more than any other housing market regulation, with the possible exception of zoning. Because debate rages about the effects of different kinds of controls, it is not surprising that little is known about different ways of eliminating or relaxing controls. The comparative static models, discussed in the companion entry on controls, provide little guidance on how best to get there from here. By their *static* nature, these models seem to imply that immediate, complete "blanket decontrol" works as well as any other possible strategy. But given the lack of knowledge of housing market dynamics, this conclusion should be treated skeptically.

The first, and still one of the best, discussions of alternative methods of decontrol (or relaxation of controls) remains Arnott's (1981) monograph. His study was carried out as part of an evaluation of reform options for Ontario's rent regulation. He lists seven forms of decontrol:

1. *Vacancy decontrol.* Units are decontrolled as they become vacant.
2. *Vacancy rate decontrol.* Particular housing submarkets (defined on the basis of the location or type of unit) with a vacancy rate above some statutory level are decontrolled.
3. *Rent-level control.* Rent-level decontrol could be more appropriately called "decontrol from the top down," because it involves decontrolling the most expensive units first and the lease expensive last. The rent level above which units are decontrolled can depend on the location or the type of unit.
4. *Floating up and out.* This designation covers any gradual relaxation of controls that applies uniformly across housing submarkets. When controls means restrictions on the rate of rent increase, floating up and out means gradually raising the guideline increase above the underlying rate of inflation. When the control program contains a rate-of-return provision, this kind of decontrol could entail raising the rate of return.
5. *Contracting out.* This is a form of vacancy decontrol; the landlord and tenant negotiate a sum that the landlord pays the tenant if he or she vacates.
6. *Local option.* A higher jurisdiction that currently administers controls allows lower jurisdictions to choose whether or not to retain them. Usually, the higher jurisdiction requires the lower to administer the controls if the latter decides to retain them.
7. *Blanket lifting.* All rent control provisions are suddenly and completely lifted.

Lack of quantitative estimates of relative effects precludes choosing a clear-cut winner for Ontario's rent control regime, much less one method best for all regimes and all market conditions. The method implied by comparative static analysis, blanket decontrol, would have high social costs, in particular because at the time of writing, Arnott judged Ontario's rental market to be in a state of excess demand.

The general conclusion about methods of decontrol from the limited research so far is that general conclusions are difficult to draw. Nevertheless, in the absence of market and regime-specific analysis, a policy package focused on "floating up and out" has much to recommend it.

Encouraging new investment, while at the same time protecting low-income renters, may best be accomplished by exempting new construction, indexing rent increases for existing units, and floating up and out of controls (i.e., a gradual transition from controlled rents to market rents over a period of years). To maintain landlord confidence in the decontrol program, it is preferable to have an end date when controls are withdrawn completely. Rents could be increased annually by, say, the Consumer Price Index plus a percentage of the previous year's rent until a set date when the final increase to market levels would be implemented. Any units reaching their market level before this date would, of course, remain there. This phasing would smooth the path of adjustment, giving tenants who could not afford their current housing at the market rent time to find suitable alternatives. This would improve the "fairness" of the decontrol program and also help ensure its political viability. Countries that have attempted rapid decontrol under conditions of excess demand have seen rapid initial runups in rents lead to the reimposition of controls. Brazil, in particular, has experienced a virulent "stop-go" cycle over time.

In the final analysis, despite the arguments for floating up and out, it is impossible to provide a blanket prescription as to which method is best under all conditions. Each market must be analyzed individually to assess the need to remove controls and the most appropriate method to do so. Whatever method is adopted, it is important that the reform be *credible*. Sufficient political consensus must exist behind the decontrol plan for suppliers to believe in the new "rules of the game" and, in particular, to believe that they will not change again in a few years' time. It is also important to package reforms in rent control regulations with reforms in other areas, such as land and finance, to ensure a positive supply response to removal or relaxation of controls. Housing subsidies targeted to low-income households can also be brought to bear as a collateral action. (SEE ALSO: *Rent Burden; Rent Control*)

—Stephen Malpezzi

Further Reading

Arnott, Richard (with Nigel Johnston). 1981. *Rent Control and Options for Decontrol in Ontario*. Toronto: Ontario Economic Council.

Malpezzi, Stephen and Gwendolyn Ball. 1991. *Rent Control in Developing Countries*. World Bank Discussion Paper No. 129. Washington, DC: World Bank.

Rydell, C. Peter, C. Lance Barnett, C. E. Hillstead, Michael P. Murray, Kevin Neels, and R. H. Sims. 1981. *The Impact of Rent Control on the Los Angeles Housing Market*. Santa Monica, CA: RAND. ◄

► Rent Gap

Rent gap is the difference between how much a household can afford to spend on shelter and what the market requires it to allocate. Because the monthly expense for housing is the single largest budget item for most households, lower-income tenants may be compelled to spend so much of their available resources on housing that they cannot afford other necessities. Unless a subsidy is provided to bridge the rent gap, food, clothing, and medical care needs will go unmet. In addition, if households are unable to pay a sufficient economic rent, the impact of this shortfall can have negative consequences for the housing stock itself.

What percentage of income can a household afford to spend on shelter while at the same time having enough resources to purchase other necessities? In the United States, a percentage of 20% to 30% has been traditionally accepted as a rule of thumb. This dates back to the 1880s, when "a week's wages for a month's rent" was a widely accepted benchmark. This housing-income ratio still prevails and is accepted in the U.S. Section 8 Existing Housing Subsidy Program in which a tenant is required to expend 30% of adjusted income for rent.

However, analyses of the demand for housing have suggested that affordability is a function of more than just income. The influences of age, household size, sex of household head, and race are all likely to be significant variables in determining a level of affordability. As a result, standards of affordability in Europe's advanced social democracies reflect these factors, and accordingly, the proportion of income appropriate for shelter allocation is typically much lower. The lower the accepted threshold, the greater the resulting public policy challenge to span the rent gap. This can be accomplished by providing a housing allowance to address all or part of the rent gap.

The concept of rent gap may also be applied to the maintenance and preservation of the housing stock. If rents are constrained by inadequate tenant incomes, they may be too low to permit the maintenance of the housing stock. As a result, owners of rental units will defer or ultimately eliminate needed repairs. This eventually leads to abandonment of the affected units and to the potential destruction of an entire neighborhood if this scenario is repeated in other structures. A housing allowance program can be implemented to bridge all or part of the rent gap. Another and more complex solution involves the coordination of a housing allowance with a general income maintenance policy. (SEE ALSO: *Housing Allowanc; Shelter Poverty*)

—Daniel J. Garr

Further Reading

Feins, Judith D. and Terry Saunders Lane. 1981. *How Much for Housing? New Perspectives on Affordability and Risk*. Cambridge, MA: Abt.

Howenstine, E. Jay. 1986. *Housing Vouchers: A Comparative International Analysis*. New Brunswick: Rutgers, State University of New Jersey, Center for Urban Policy Research.

Salins, Peter D. 1980. *The Ecology of Housing Destruction*. New York: New York University Press.

▶ Rent Strikes

Despite the long and colorful history of tenant/landlord relations in U.S. society, scholars have written surprisingly little on tenants in the United States. Moreover, little reliable research has been conducted on rent strikes.

The history of rent strikes in the United States can be viewed as a tale of how the underdogs in society can *sometimes* use the democratic process to improve their position. Tenants have successfully organized for both short-term (e.g., avoiding eviction and rent hikes) and long-term (e.g., rent control) gains. They have also had their share of defeats.

The history of rent strikes in New York can be traced back to April 1, 1904. Over the course of the next nine decades, both issues and actors changed as the struggle between tenants and landlords unfolded. As New Yorkers attempted to protect their places of residence or acquire livable conditions, several distinctive trends emerged. Early struggles frequently evolved from concerted efforts by renters to even gain a seat at the bargaining table. This period of the tenants' movement has its parallels in early labor/management disputes. Having negotiated this first obstacle, the tenants' movement could employ a wider spectrum of tactics, but it also encountered new barriers, ranging from "grassroots militancy to bureaucratizing institutionalization" (Lawson 1986). As the focus of tenants' groups became more clearly defined, the organization of support systems changed from radical political factions to special-interest groups. The demographics of the leadership of the tenants' movement also varied from one time period to the next. In the initial phases, Jews and Italians provided the direction for the cause; later, African and Hispanic Americans assumed leadership positions. The one perpetual presence throughout has been that of women.

The 1904 experience of Jewish immigrants residing on New York's Lower East side is a benchmark for unraveling the historical web of rent strikes in the United States. These tenants organized to accomplish several major objectives, including warding off evictions, gaining "rent rollbacks," and attaining contractual protection in the form of leases. In so doing they blazed a trail. However, in 1907 the tide seemed to turn, as a coalition of Jewish merchants, Socialists, and thousands of concerned residents in Brooklyn and a stronger opposition composed of organized landlords and an unsympathetic media defeated Manhattan. This failed rent strike rendered many of the participants homeless.

The Great Depression witnessed a revival of rent strikes. The rent strike of 1932 to 1933 brought in new actors and new methods. Chief among them were Communists and African Americans. To counter the upsurge in the tenants' movement, both landlords and government representatives resorted to strong-arm tactics, previously reserved for labor strikes. However, as the New Deal brought legitimacy to

advocacy groups, organizers were able to couple "direct action with litigation and lobbying" (Lawson 1986). New gains made during this era included the creation and enforcement of rent controls. New Deal legislation also broadened the scope of housing policies. In the early 1970s, renters began demanding (a) controls on rents, (b) adequate living conditions, (c) eviction protections, and (d) political participation. The relative success or failure of this more recent movement has been a matter of considerable debate.

Although laws governing rent strikes vary from state to state, a California Supreme Court case (*Green v. Superior Court*, 10 Cal. 3d. 616 [Jan. 15, 1974]) provides an example of how this issue is governed, ruling that "a rent strike is . . . legal in California, if (a) the landlord has materially breached his implied warranty of habitability by failing to correct serious housing code violations, (b) tenants did not cause the violations, and (c) the landlord was given notice of the violations and a reasonable time to correct them" (Copek and Gilderbloom 1992).

Of all actions available to tenants, rent control generates the greatest involvement. Other actions do not generally create sufficient support. Rent strikes involve a high degree of personal risk—arrest, loss of possessions, legal costs, fines, and perhaps even imprisonment. Few tenant organizations used rent strikes during the 1970s, 1980s and early 1990s. Campaigns to lower interest rates or increase housing subsidies are too remote in their targets, too long term in their potential results, and too indirect and diffuse in their impact.

Tenant consciousness is an ideological outcome of rent strikes. Tenant consciousness develops when tenants (a) view themselves as a group sharing similar problems, (b) have a common understanding of the causes of their problems, and (c) have a collective political purpose that responds effectively to these problems. Sources of support for rent strikes and other reforms include political orientation, relative deprivation, class-consciousness, willingness to act, perception of system blame or inefficiency, and landlord/tenant conflict. Other important factors concern resources, leadership, technology, and ideology in the framing of political issues.

Rent strikes have generally been unsuccessful in the United States. Nevertheless, when used effectively, they can be influential instruments of social change. Like most entrepreneurs, landlords cannot afford to have their assets frozen for even a short period of time. Rent strikes can seriously erode their financial interests. Hence, withholding rent monies may coerce property owners to quickly recognize the demands of their tenants. Rent strikes, however, are not risk-free, and the costs associated with losing can be severe. (SEE ALSO: *Eviction; Tenant Organizing in the United States, History of*)

—*John I. Gilderbloom and Russell N. Sims*

Further Reading

Capek, Stella M. and John I. Gilderbloom. 1992. *Community Versus Commodity: Tenants and the American City*. Albany: State University of New York Press.

Dreier, Peter. 1982. "The Status of Tenants in the United States." *Social Problems* 30(2):179-98.

Dreier, Peter and John Atlas. 1980. "The Housing Crisis and the Tenants' Revolt." *Social Policy* (January):13-24.

Gilderbloom, John I. 1988. "Tenants Movements in the United States." Pp. 269-82 in *Handbook of Housing and the Built Environment in the United States,* edited by Elizabeth Huttman and Willem van Vliet--. New York: Greenwood.

Heskin, Allan D. 1981. "A History of Tenants in the United States: Struggle and Ideology." *International Journal of Urban and Regional Research* [Special issue on housing] 5(2):178-204.

Lawson, Ronald, ed. 1986. *The Tenant Movement in New York City, 1904-1984.* New Brunswick, NJ: Rutgers University Press.

Lipsky, Michael. 1970. *Protest in City Politics: Rent Strikes, Housing and the Power of the Poor.* Chicago: Rand McNally.

Marcuse, Peter. 1981. "The Strategic Potential of Rent Control." Pp. 86-94 in *Rent Control: A Source Book,* edited by John Gilderbloom. Santa Barbara, CA: Foundation for National Progress. ◄

► Resident Management

As an alternative to conventional housing management, resident management is a form of multifamily management in which the tenants form a corporation for the purpose of providing management services. Known also as tenant management, resident management has been associated in the United States primarily with public housing, although it has also been implemented in other types of low-income and subsidized housing. Proponents of resident management argue that it leads to improved housing quality, more efficient and cost-effective housing management, and tenant empowerment. Critics contend that it is an abrogation by landlords of their responsibility to provide safe and decent housing and is part of an overall scheme by government to do away with public housing.

Although a few private developments were tenant managed in New York during the 1960s, the beginnings of resident management are commonly traced to the early 1970s when it arose almost simultaneously in public housing in Boston and St. Louis. In both instances, it happened because tenants could no longer endure the inadequate level of management being provided by their housing authorities.

In Boston, residents at the 1,100-unit town house and high-rise Bromley-Heath development became managers after assuming responsibility for health and social service delivery and developing a drug treatment center. In St. Louis, five developments became resident managed between 1973 and 1975 after a 1969 rent strike resulted in a massive reorganization of the St. Louis Housing Authority. Funds to initiate resident management in St. Louis were provided by the Ford Foundation.

The Ford Foundation's experience in St. Louis led to its joint sponsorship, with the Department of Housing and Urban Development (HUD), of a National Tenant Management Demonstration Program involving seven public housing sites in six cities between 1976 and 1979. Although a 1981 evaluation of the demonstration program found some positive outcomes, it also concluded that the effort had been costly and the benefits were not sufficient to warrant expanding resident management to other public housing sites. Some of the demonstration sites initially continued with tenant management, but by the mid-1980s, only one resident-managed site remained. Also, of the five tenant-managed St. Louis sites, three eventually returned to conventional housing authority management.

Seemingly dead as a viable concept, resident management was revived in the late 1980s primarily because of reported successes at two developments, Cochran Gardens in St. Louis and Kenilworth-Parkside in Washington D.C. A leader in promoting resident management at this time was Robert Woodson, a conservative policy analyst and head of the Washington-based National Center for Neighborhood Enterprise (NCNE). A *60-Minutes* television segment featuring Cochran Gardens thrust resident management into the public consciousness and provided support for claims that it was a potentially powerful tool for revitalizing troubled public housing.

Resident management was supported by the Reagan administration and became a key component of the Bush administration's public housing policy under HUD secretary Jack Kemp. Kemp and other conservatives held that resident management is an intermediate stage in a process of empowerment that ultimately leads to homeownership. Federal housing legislation in 1987 and 1990 established procedures for creating resident management corporations (RMCs) and funding mechanisms for converting from conventional to resident management. Also included were procedures to convert resident-managed developments to resident ownership.

In an evaluation of resident management, done at the close of the Bush administration, the consulting firm, ICF, reported to HUD that there were 11 "mature" resident-managed developments, dating back to before 1988. ICF also reported that over 300 resident organizations in public housing had received some assistance from the federal government regarding resident management and that 80 "emerging" RMCs were in process of taking over management at their developments. However, the report noted that only about one-third of the emerging RMCs had actually become involved in housing management and that resident management apparently takes a long time to implement.

The Clinton administration has not supported resident management with as much vigor as did Jack Kemp and the Bush administration. Although resident management is still a viable tenant option, the Clinton administration is more occupied with the restructuring and demolition of existing public housing units than with management issues.

Researcher Daniel Monti has identified four conditions for the success of resident management:

1. Adequate funds for operating subsidies, modernization, and technical assistance
2. Grassroots organizing efforts that build resident support from the bottom up
3. A professional, neither too hostile nor too friendly, relationship between an RMC and a housing authority
4. Ties between the residents and institutions other than the housing authority

Resident management efforts that lack one of the four conditions do not seem to succeed over time.

The transition from conventional to resident management is a slow and potentially costly process, during which a resident board of directors must learn how to oversee an organization and staff must learn how to be housing managers. After the not-for-profit RMC is formed, its board established and trained, and staff hired, a period of what is commonly called "dual management" can begin. Dual management takes one of two forms. With the first, dual management is a transitional period of from 6 months to a year during which the resident managers work alongside housing authority managers learning their jobs and gradually assuming more and more management responsibility. By the final day of dual management, the resident managers have assumed all the work. With the second form of dual management, residents choose to take on some of the tasks of managing, leaving others to the housing authority. As time passes, residents may take on more tasks, but this second form of dual management can continue indefinitely even into the period in which residents are said to be "fully" managing their development.

Similarly, resident management itself can be of two types—full and partial management. Under full management, the resident corporation acts as an independent contractor hired to perform all management functions at the development. A full-management RMC develops its own budget process and is paid for the work it does. Under partial management, the RMC behaves more like a partner of the housing authority and management is actually shared. Residents usually choose to assume tasks associated with on-site management, such as rent collection, screening of new tenants, maintenance, and security. The housing authority, in turn, usually handles tasks associated with central management such as planning and implementing major renovations and purchasing equipment and supplies. With partial management, budgeting responsibility usually remains with the housing authority.

Although there is considerable support for resident management among researchers, policy analysts, and public housing tenants, not everyone believes it is an appropriate model for public housing. The failure rate of RMCs is high, and they need to be continually nurtured and provided with technical support and money by a sympathetic housing authority and external supporters. Not everyone agrees that the end result of resident management is or should be resident ownership. Recent studies of HUD attempts to sell public housing units to residents have shown that sales are difficult to accomplish and that residents who become homeowners often encounter difficulties in meeting mortgage payments.

Many proponents claim that resident management empowers tenants, but a review of the history of resident-managed developments suggests that resident management is an outcome of empowerment rather than a cause of it. It can even be argued that by concentrating too much on becoming managers, residents can be distracted from other serious social and economic needs of their communities. In such cases, resident managing may subvert a process of community organization and empowerment.

Some researchers have also suggested that it is unfair to expect residents to shoulder the burdens of management, especially in situations in which the housing authority is seemingly unable to provide adequate management. These critics argue that the poor in public housing are faced with enough obstacles to living without being given the additional burden of managing their developments. They point out that middle- and upper-income renters expect management services to be provided, so why should not public housing residents be afforded the same services.

Resident management would appear to be a response to the general failure of public housing in the United States. Resident management has been successful at several locations, and under certain conditions, it can be an appropriate alternative to conventional housing management. It must, however, be viewed in a broader context of community organization, development, and empowerment. Although some public housing developments will continue to be resident managed and others will become resident managed in the future, it is unlikely that resident management will replace conventional public housing management. Instead, resident management should be viewed as one of several strategies needed for maintaining and revitalizing public housing. (SEE ALSO: **Community-Based Housing;** *Resident Management Corporation; Tenant Organizing in the United States, History of*)

—*William Peterman*

Further Reading

Chandler, Mittie Olion. 1991. "What Have We Learned from Public Housing Resident Management?" *Journal of Planning Literature* 6:136-43.

ICF. 1992. *Evaluation of Resident Management in Public Housing.* Washington, DC: U.S. Department of Housing and Urban Development.

———. 1993. *Report on Emerging Resident Management Corporations in Public Housing.* Washington, DC: U.S. Department of Housing and Urban Development.

Lane, Vincent. 1995. "Best Management Practices in U.S. Public Housing." *Housing Policy Debate* 6(4):867-904.

Manpower Demonstration Research Corporation. 1981. *Tenant Management: Findings from a Three Year Experiment in Public Housing.* Cambridge, MA: Ballinger.

Monti, Daniel J. 1993. "People in Control: A Comparison of Residents in Two U.S. Housing Developments." Pp. 177-94 in *Ownership, Control, and the Future of Public Housing,* edited by R. Allen Hays. Westport, CT: Greenwood.

Peterman, William. 1993. "Resident Management and Other Approaches to Tenant Control of Public Housing." Pp. 161-76 in *Ownership, Control, and the Future of Public Housing,* edited by R. Allen Hays. Westport, CT: Greenwood.

———. 1996. "The Meanings of Resident Empowerment: Why Just About Everybody Thinks It's a Good Idea and What It Has to Do with Resident Management." *Housing Policy Debate* 7(3): 473-90. ◄

► Resident Management Corporation

Resident management—that is, public housing residents taking over management services at a development—is a popular alternative to conventional housing authority management. Resident management is carried out by a not-for-profit tenant organization known as a resident management corporation (RMC).

Most resident management corporations evolve out of residents' organizing efforts and tenants' struggles to improve the quality of their developments. Often, tenants conclude that existing housing authority management practices lie at the heart of a development's problems and see taking over management as one or possibly the only way to bring about improvements. They then form an RMC with the intent of assuming management responsibilities.

RMCs are similar in structure to other not-for-profit organizations, particularly those that are community based. The corporation's board of directors is made up of tenants, and membership in the corporation is usually defined as being all adults living in the development. Staff of the RMC may or may not consist of tenants, although most RMCs attempt to provide as many employment opportunities for residents as possible. Many RMCs also have advisory boards or committees made up of housing experts, local officials, planners, and other nonresidents interested in seeing the resident management effort succeed.

Because residents chosen to be board members of an RMC usually have little or no experience in the operation of an organization as complex as an RMC, the initial phase of resident management is commonly a one- to two-year period of board training in preparation for beginning management. This is followed by the hiring and training of staff, including a director, a manager, an assistant manager, maintenance and security supervisors and staff, and others needed to perform the necessary management tasks.

The period during which the staff is trained is called "dual management"—a time during which an RMC and the housing authority staff share management responsibilities. Following dual management, which can last a year or more, the RMC and housing authority sign a formal management contract. According to the terms negotiated, the RMC may assume full management responsibilities (full management) or may elect to do some management tasks while leaving others to the housing authority (shared management).

Evaluative research of RMCs, performed by Daniel Monti, finds four conditions that are basic to resident management success: (a) adequate and continuing funds for operation, renovation, and technical assistance; (b) ongoing support and involvement of tenants; (c) a professional relationship with the housing authority that is neither too friendly nor too confrontational; and (d) links with community organizations and officials other than the housing authority. The absence of any one or more of these conditions can result in a weak and ineffective RMC and failure of the resident management effort.

According to a study by ICF, at the beginning of 1993, 11 "mature" public housing RMCs had been doing management for five years or more and about 25 new or "emerging" RMCs undertaking at least one or more management tasks. More recently, "resident management" has come to mean a number of things, making it difficult to say how many RMCs exist. (SEE ALSO: *Resident Management*)

—*William Peterman*

Further Reading

ICF. 1992. *Evaluation of Resident Management in Public Housing.* Washington, DC: U.S. Department of Housing and Urban Development.

———. 1993. *Report on Emerging Resident Management Corporations in Public Housing.* Washington, DC: U.S. Department of Housing and Urban Development.

Monti, Daniel J. 1993. "People in Control: A Comparison of Residents in Two U.S. Housing Developments." Pp. 177-94 in *Ownership, Control, and the Future of Public Housing,* edited by R. Allen Hays. Westport, CT: Greenwood.

Peterman, William. 1993. "Resident Management and Other Approaches to Tenant Control of Public Housing." Pp. 161-76 in *Ownership, Control, and the Future of Public Housing,* edited by R. Allen Hays. Westport, CT: Greenwood. ◄

► Residential Autobiographies

Residential autobiographies are people's personal narrative histories of their relationships with the micro- to macro-scale environments in which they have lived. They describe the sounds and smells of home, school, neighborhood, and landscape, the places that were loved and hated, their implicit and explicit meanings. Most important, residential autobiographies are chronicles of how the physical environment affects people's emotions, cognition or attitudes, and behavior.

The technique of residential autobiographies was originally developed in the late 1970s by landscape architects Cooper-Marcus and Helphand to elucidate the origin and nature of personal values for place. Over the next 15 years, it was adapted by a variety of phenomenologically oriented professionals in psychology, geography, architecture, design, and other disciplines to examine the archetypal character of environmental experience. Among the spaces that researchers identified as significant were kitchens, bedrooms, bathrooms, streets, recreational facilities, playgrounds, backyards, and natural landscapes, such as beaches and mountains. There is also general agreement on the remembered qualities of experience, including seclusion, enclosure, privacy, control, challenge, and the transcendence provided by places that encourage reverie and imagination. Other research suggests that individual characteristics such as gender, age, culture, and class have an effect on the character of the places identified and types of experience remembered. Although most research has elicited positive associations with place, some studies have identified negative and traumatic images of abuse or violence in association with the same spaces that are identified by others as sanctuaries.

Residential autobiographies take many forms, traditionally combining written narratives with graphic images, but they may also be auditory, olfactory, and or tactile. Cooper Marcus has used guided visualizations to assist in evoking the remembered passage from front door to the interior to the dining table and the bedroom. Such visualizations have been followed by sketches of the remembered images. Others have used a variety of verbal and written techniques to evoke the memories and have imposed less structure on the generated form of the autobiography. These techniques have elicited collages of photographs, tape recordings of prom and ritual music, or boxes filled with objects as emblems of pleasant or painful memories. Autobiographies used in a classroom setting are often shared in a process that permits long forgotten memories to surface, as each group member triggers associations for others. It is crucial, however, that the residential autobiography be used carefully and with sensitivity to the fact that eliciting environmental memories may be extremely painful, even traumatic for some.

The value of residential autobiographies to housing professionals is threefold. First, they are useful tools with which to explore the *individual*'s experience of spaces, enabling practitioners and students of housing to examine the origin and nature of environmental aesthetics and preferences. This may help professionals to avoid imposing their own design values on clients whose attitudes and needs may differ in significant ways from their own. Second, residential autobiographies may provide useful information on *group* values for spaces. Knowledge of the relationships between demographic factors, culture, religion and epistemology, and the design of spaces may be critical to those responsible for planning, maintaining, and evaluating housing at the micro- to macrolevel—from interior design to community planning.

Finally, residential autobiographies elucidate environmental archetypes that reflect both positive and negative associations with places. Such knowledge may permit the development of structures and the planning of landscapes that naturally and unconsciously evoke positive or negative emotions and that may thereby influence behavior, supporting or discouraging the use of particular places. For example, Michael Brill has proposed use of negative archetypal symbols to discourage the use of the landscape over a nuclear waste facility, by including grids of metal spikes simulating painful thorns, whereas at the other end of the spectrum, designers of health care facilities have used gardens and natural landscapes to promote healing. Furthermore, Cooper-Marcus has applied the archetypal desire for security to housing design for clients through the use of environmental autobiographies. Using interviews and other techniques, she has noted the continuing influence of a childhood setting on current choices of dwelling, location, dwelling form, garden design, and interior decoration.

Use of feng shui philosophy and other symbols in housing environments can similarly suggest the existence of familiar or alien territory, and influence people's attitudes, emotions, and behaviors. (SEE ALSO: **Behavioral Aspects**)

—*Nora J. Rubinstein*

Further Reading

Bachelard, G. 1969. *The Poetics of Space*. Boston: Beacon.
Boschetti, M. 1987. "Memories of Childhood Homes." *Journal of Interior Design Education Research* 13(2):27-36.
Chawla, Louise. 1992. "Childhood Place Attachments." Pp. 63-86 in *Place Attachment*, edited by Irwin Altman and Setha M. Low. New York: Plenum.
Marcus, Clare Cooper. 1995. *House as a Mirror of Self: Exploring the Deeper Meanings of Home*. Berkeley, CA: Conari.
Rubenstein, N. 1994. "There's No Place Like Home." In *Equitable and Sustainable Habitats*, edited by E. Arias and M. Gross. Oklahoma: Environmental Research Associates. ◀

▶ Residential Development

The development of residential areas involves the subdivision of land followed, sometimes after a lapse of time, by the construction of dwellings. The processes involved at each stage have changed over the past century, and vertical integration within the development industry has brought the two stages into closer relation.

A century ago, land subdividers sold individual lots, either to speculators or to builders. They were not usually involved in the building process and often made little attempt to influence its character. Over time, they came to realize that they could make more profit by imposing a variety of controls that, in effect, guaranteed minimum subdivision standards for the eventual homebuyer. This they accomplished directly, by installing services and restricting building methods and patterns of occupancy, as well as indirectly, through the agency of the local state. Even in the 19th century, a few entrepreneurs had combined land subdivision with homebuilding, Samual Gross in Chicago being one of the most prominent examples. Gross, however, was very much the exception. Only around the time of World War II, encouraged by new federal mortgage insurance guidelines, did a significant number of "community builders" begin to integrate their operations in this fashion. The activities of the Levitts and Fritz Burns signaled the emergence in the postwar years of the vertically integrated land developer, who was involved in all stages of the development process.

Even in the postwar years, small builders have usually dominated the construction scene. Residential construction has often been described as backward: "the industry that capitalism forgot." There is some truth to this. Technological change has been steady but slow. Nevertheless, over the past century or so, cumulative changes have been significant. For the most part, dwellings are assembled on-site from parts that have been manufactured elsewhere. This distinguishes houses from most other goods and helps disguise the fact that the production of construction materials is as streamlined as, say, that of auto parts. Bricks, cut lumber, pipes, nails, wire, drywall, and shingles, not to mention bathroom fittings, kitchen appliances, windows, and doors, are mill- or factory-made in large quantities. Some of these materials—bricks, for example, and lumber—may be regarded as traditional, but their production has steadily been standardized and made more efficient. The introduction of grade marking in lumber between the wars, for example, had an enormous impact on its marketing, distribution and use. Other materials are comparatively recent innovations: concrete became widely used in residential construction only around the turn of the century; drywall after World War II.

If there was ever a revolution in construction technology in North America, however, it occurred in the mid-19th century with the adoption of the technique of balloon framing—lightweight construction using milled lumber held together with nails. This simultaneously reduced the cost of construction and the skills required. By the early 20th century, it had become widespread throughout most of North America, whereas in Europe, bricks-and-mortar were still the more costly norm. In the 20th century, some on-site experiments have been made with industrial assembly techniques, as workers put together preassembled wall units that incorporate doors, windows, and wiring. Such techniques have never been as widely used in North America as in Europe, however, mainly because they offer few cost advantages over balloon framing. Instead, the comparative cheapness of wood encouraged companies to manufacture entire house kits. Manufacturers can ship these anywhere, with detailed instructions included. In the early 20th century, kits became a common and distinctive feature of the North American construction scene. Since World War II, the manufacturers of mobile homes have met some of the same market. In 1998, these accounted for about one-fifth of the U.S. annual production of new dwelling units.

The builders of the remaining four-fifths of all homes usually operate in one of three ways: on speculation, on contract, or for themselves. Speculative builders erect dwellings without having particular buyers in mind. Like the manufacturers of most consumer goods, they read "the market" and target particular segments that are always defined by income and sometimes also by age, family characteristics, ethnicity, and—less obtrusively—race. They pitch their advertising accordingly. Using an innovation developed in the 1920s, they build "model" homes to impress potential buyers. In contrast, contractors are employed to do particular jobs, for a specified sum. The employer may be a speculative builder who "contracts out" instead of doing the work in-house or a governmental agency wishing to produce publicly subsidized housing. The use of contractors is necessary for small builders whose scale of operation cannot support the profitable use of expensive equipment, such as backhoes. It is also widely used by larger builders, who maximize flexibility by minimizing their permanent workforce.

Contractors are influenced by the same considerations, and the large ones will subcontract particular jobs. These in turn may be further distributed in a complex social division of labor. Some contractors are employed directly by the eventual homeowner or the owner's agent, frequently an architect. In this manner, the owner is able to exert direct control over the construction process and to get a house that suits his or her particular needs. Recognizing that "control" can be a selling point, and in an effort to reduce the risks of speculative construction, large builders have begun to offer customized versions of otherwise standard designs. Typically, buyers sign a contract and pay a deposit before the foundation is dug. In return, they are able to specify particular features and materials, as well as being able to monitor the construction process. Together, speculative builders and contractors make up the building industry. In addition, quite a number of people build their own homes. This is especially true in the fringe areas of cities, in smaller towns, and rural areas, where building regulations are less stringent. Some owner-builders do

everything themselves. More typically, they seek advice and contract out particular jobs. The line between owner building and contractor building, then, easily becomes blurred.

These three methods of construction are associated with different segments of the housing market. Customized contractor building is most common at the top end of the market, where prospective homeowners can afford the inefficiencies of having a unique home built for them, often under the supervision of an architect. In contrast, most of those who have built homes for themselves have done so because they could not afford to acquire a home any other way. Erecting dwellings that began as shacks, many had very low incomes indeed. Speculative builders have always been most active in the middle segment of the market, where households can afford not to have to work on their own home but where limited budgets have compelled them to accept fairly standard designs.

Over time, speculative building has grown at the expense of both alternatives. Owner building was common even in cities until the 1920s and made a brief but strong resurgence in the late 1940s and 1950s. This was one of the most distinctive features of residential construction in Canadian and U.S. cities. Rising incomes and land prices, low-ratio amortized mortgages, and the introduction of building regulations have eroded the practice. The expanding operations of some speculative builders have made it possible to realize economies of scale. In developments such as Levittown, builders make a small margin on each unit but thrive by building whole subdivisions at a time. Combined with the changes in residential financing that were sponsored by the federal government during the 1930s, this speculative activity brought mortgaged homeownership within the reach of many moderate-income households after World War II. Vertical integration and the greater efficiencies of larger-scale production have also allowed speculative builders to move upmarket, especially by offering limited custom features to more affluent clients. One of the benefits that large builders have been able to offer higher-income households is the planned, controlled residential subdivision. The growth of speculative construction has depended in part on the integration of homebuilding with land development, and many of the larger builders are more properly regarded as land developers.

Although speculative building has become the norm, sweat equity and contracting will remain very important. One reason the latter has persisted in the face of competition from developers and the manufacturers of kits and mobile homes has to do with the services provided by building supply dealers. The construction industry is unusual in that, except for the largest companies, most producers of housing rely on retailers for their raw materials. In the early 20th century, competition from mail-order companies compelled lumber dealers to diversify, in effect providing a one-stop service to the small builder. By the late 1940s, their services included not only a wide range of building supplies but also credit, job and product information, estimations on contracts, and house plans. They helped small builders compete with larger competitors.

Another strength of small builders in general is their greater flexibility. Construction is highly cyclical. More than usual, it depends on the general state of the economy, mainly because houses are durable and flexible in their use. In any year, construction adds only a small amount to the total stock. For months, and even years, an increase in the demand for housing can be met by using the existing stock more intensively: by taking in lodgers or by dwelling subdivision. During boom periods, speculative activity expands in absolute and relative terms, as erstwhile contractors become ambitious. During the sometimes severe downturns, all builders are hurt, but those with substantial capital investments are usually affected most of all. The share of speculative construction drops as builders go bankrupt or shed their workforce and return to contract work. This cyclical pattern places a premium on flexibility and in particular on subcontracting. As long as severe economic downturns continue to occur, large builders are never likely to dominate the construction industry for very long.

This is especially true because so many people in the United States own their own homes. This virtually guarantees that some form of owner labor, even if only in the reduced form of do-it-yourself repairs and improvement, will remain common. Owners, as opposed to tenants, can more easily repair their own homes. They are also better able to extend, or adapt, their dwelling to changing family needs, creating new space for a small business, elderly relatives, or paying tenants. For demographic reasons, and because of a growing interest in historic preservation, an increasing proportion of all construction activity is being devoted to renovation rather than the production of new dwellings. Owners are in a position to do much of this themselves; pride in being handy and the prospect of saving money provide many with the incentive. For the remainder, the only alternative is to hire a contractor. In active renovation markets, there have been local attempts to franchise residential contracting services, and speculative renovations have been undertaken of multi-unit dwellings. Even here, however, it is usually the owner of the apartment building who takes the risk, not the company that actually undertakes the renovations. In general, the growth of home renovation favors the owner-builder and small contractor. Speculative activity may sometimes become the norm for homebuilding, but it will never dominate residential construction activity as a whole. (SEE ALSO: *Construction Industry; New Urbanism; Real Estate Developers and Housing; Subdivision; Subdivision Controls*)

—*Richard Harris*

Further Reading

Checkoway, B. 1986. "Large Builders, Federal Housing Programs, and Postwar Suburbanization." In *Critical Perspectives on Housing*, edited by Rachel G. Bratt, Chester Hartman, and Ann Meyerson. Philadelphia: Temple University Press.

Doucet, M. and J. Weaver. 1991. *Housing the North American City*. Montreal and Kingston: McGill-Queen's University Press.

Dowall, David E. 1990. "The Public Real Estate Development Process." *Journal of the American Planning Association* 56:504-12.

Harris, R. 1996. *Unplanned Suburbs: Toronto's American Tragedy, 1900-1950*. Baltimore: Johns Hopkins University Press.

Schlesinger, T. and M. Erlich. 1986. "Housing: The Industry Capitalism Didn't Forget." In *Critical Perspectives on Housing*, edited by Rachel G. Bratt, Chester Hartman, and Ann Meyerson. Philadelphia: Temple University Press.

Warner, S. B. 1962. *Streetcar Suburbs: The Process of Growth in Boston, 1870-1900*. Cambridge, MA: Harvard University Press.

Weiss, M. 1987. *The Rise of the Community Builders: The American Real Estate Industry and Urban Land Planning*. New York: Columbia University Press. ◀

▶ Residential Lead-Based Paint Hazard Reduction Act of 1992

The Residential Lead-Based Paint Hazard Reduction Act of 1992, Title X of the Housing and Community Development Act of 1992 (P.L. 102-550), was enacted to do the following:

- ▶ Encourage the development of an infrastructure to eliminate lead-based paint hazards
- ▶ Develop a broad program to evaluate and reduce lead-based paint hazards in the housing stock
- ▶ Establish reasonable standards of care for those engaged in lead-based paint hazard reduction activity
- ▶ Develop federal policy regarding the sale, rental, and renovation of homes that takes lead-based paint hazards into account
- ▶ Develop cost-effective methods for evaluating and reducing lead-based paint hazards
- ▶ Reduce the threat of childhood lead poisoning in federally assisted housing
- ▶ Educate the public concerning lead-based paint hazards

In Title X, the government enacted major changes in federal law pertaining to the control of lead-based paint hazards and the reduction of lead exposure. It mandates coordinated action by several federal agencies, including Housing and Urban Development (HUD), the Environmental Protection Agency (EPA), the Department of Labor (DoL), the Department of Health and Human Services (DHHS), and the General Accounting Office (GAO). The following describes its major provisions.

Development of Infrastructure

Title X authorized a grant program for state and local governments for the evaluation and reduction of lead-based paint hazards in privately owned housing built before 1978 and occupied by families of low and moderate income. Title X also authorized a related program of grants to states to establish state training, certification, or accreditation programs for contractors.

Evaluation and Reduction of Lead Hazards

Title X prescribed hazard evaluations and reductions for federally assisted housing built before 1978. Abatement of lead-based paint hazards is required in rehabilitation projects receiving more than $25,000 per unit in federal funds. Title X also sets requirements for lead-based paint hazard reduction and control in residential units sold by federal agencies. All target housing—that is, units built prior to

1978—must be evaluated for lead-based paint hazards prior to sale. Target units constructed prior to 1960 must have lead-based paint hazards abated. For units built between 1960 and 1977, an inspection report is to be provided to prospective purchasers.

Under the Cranston-Gonzalez National Affordable Housing Act, state and local governments must develop a comprehensive housing affordability strategy (CHAS) as a prerequisite to receiving federal housing or community development funds. Title X requires that each CHAS include an estimate of the number of housing units with lead-based paint hazards and occupied by low-income families. Actions taken or proposed to remedy the problem must also be included.

Worker Protection

The act stipulated that the EPA establish a regulatory framework governing lead-based paint activities to ensure that individuals engaged in such activities are properly trained, that contractors are certified, and that training programs are accredited. Lead-based paint activities include abatement, inspection, and risk assessment. These regulations must also include standards for performing lead-based paint activities and require that all activities be performed by certified contractors.

The EPA will approve state programs for training and certification of lead-based paint contractors. State programs must be at least as protective as the federal program and must provide adequate enforcement. In states without a program, the EPA must establish, administer, and enforce federal programs.

The EPA is also required to study the extent to which persons engaged in various types of renovation and remodeling activities are exposed to lead or create a lead-based paint hazard. Contractors are required to conduct an exposure assessment for each job classification where there is potential exposure to lead. If exposures are above the permissible limit, they must reduce the exposure of workers to lead dust and fumes, develop a written compliance plan, designate a competent person to oversee worker protection efforts, and conduct medical surveillance. Workers with high levels of lead must be removed from exposure.

Public Education

Title X requires that before selling or leasing all pre-1978 housing, sellers and lessors must do the following:

- ► Disclose all known lead-based paint and lead-based paint hazards in the dwelling
- ► Provide EPA's lead hazard information pamphlet, prepared pursuant to Section 406 of the Toxic Substances Control Act, to the purchaser or lessee
- ► Allow purchasers a 10-day opportunity to inspect the dwelling for lead-based paint hazards
- ► Include a lead warning statement, with specific statutory language, in each contract to sell the residential property

These requirements were implemented in 1996.

Federal public education and outreach activities include a clearinghouse, established in April 1993, and a hotline, established in November 1992. The hotline phone number is 1-800-LEAD-FYI. The clearinghouse phone number is 1-800-424-LEAD. (SEE ALSO: Environmental Contamination: Lead-Based Paint Hazards)

—*Barbara A. Haley*

Further Reading

U.S. Department of Housing and Urban Development. 1990. *Comprehensive and Workable Plan for Abatement of Lead-Based Paint in Privately Owned Housing: Report to Congress.* HUD-PDR-1295. Washington, DC: Author.

———. 1997. *Moving Toward a Safe America: A Report to the Congress of the United States.* Washington, DC: Author. ◄

► Residential Location

Households do not distribute themselves randomly across the metropolitan landscape. For example, childless households constitute the vast majority of apartment and condominium dwellers. Many elderly households congregate in retirement communities. And all households with one or more members who work are limited in their residential choice by where these members are employed.

Two dominant factors that sort households spatially are income and race. Households having similar incomes or of the same race are more often found clustered together than interspersed with households who differ from them in these two important respects. This clustering becomes more and more pronounced as urban areas increase in size. In larger metropolitan areas, not only neighborhoods but entire communities differ significantly from one another with respect to income and racial characteristics. In large and small metropolitan areas alike, neighborhood characteristics associated with these differences are usually more important than dwelling unit characteristics in the residential choices that households make.

Unlike the residential sorting that occurs as a consequence of differences in age or family type, the spatial separation of racial and income groups is widely viewed as not entirely benign. Racial separation reflects present and past discriminatory practices in the housing market, not simply voluntary association. Income separation, especially if at a large geographic scale, exacerbates and perpetuates societal differences between the haves and have-nots. Public schooling in lower-income areas is usually inferior to that which children living in other areas receive. Also, the spatial isolation of lower-income households serves to distance residents from dominant cultural norms, aggravating the conditions that restricted their locational opportunities in the first place. To the extent that market forces and public policies serve to bar certain racial minorities and lower-income groups from living in places that would afford them more equal opportunity to develop and use their talents, the achievement of broadly accepted societal objectives is obstructed.

For approximately half a century, and against considerable political resistance, efforts to deal with perceived artificial barriers to residential locational choice have been a major focus of national housing policy. The 1960s civil rights legislation and its sequelae eliminated de jure racial discrimination. The Housing Act of 1974 took aim at income isolation, requiring that federally subsidized residential construction not be placed in the midst of low-income neighborhoods. At the state level, legislation and judicial decisions removed the most egregious manifestations of exclusionary zoning being practiced by communities seeking to prevent the construction of low-priced housing in their jurisdictions. A few states adopted "inclusionary" zoning under which municipalities have an obligation to provide subsidized new homes within their borders.

Despite all these efforts, several indicators measuring the degree of income and racial mixing in residential areas show that spatial separation persists. Most blacks, for example, still reside in almost entirely black neighborhoods, and a larger proportion of low-income households live in predominantly low-income neighborhoods than was the case several decades ago.

The residential concentrations of blacks are due, in varying degrees depending on the metropolitan area, to their lower-than-average incomes, their reluctance to become the only black households in otherwise all-white neighborhoods, and to continuing racial discrimination and intimidation in many parts of the housing market. When black households do move from traditional black neighborhoods to less racially homogeneous environments, their destination is frequently an already existing black enclave or a home near the edge of their former community. In these situations, the apparent racial integration of their new residential environment is usually either temporary or misleading.

Despite the persistence of racial separation, surveys addressing racial attitudes of blacks and whites indicate growing amenability to interracial neighborhoods. Some observers believe these changing attitudes may begin to translate into shifts in residential behavior in the coming years, particularly among black families in the top half of the income distribution where discrimination has become less frequent.

Spatial separation of income groups is a more complex phenomenon. Although exclusionary zoning is commonly thought to be the major cause, most of the separation that exists would occur in the absence of any zoning at all. There are two reasons for this. The first, and paramount, reason relates to the fact that about half of the households in the United States cannot afford to buy even the most inexpensive new homes. As a consequence, as urban areas grow, the new homes built to accommodate this growth are purchased or rented entirely by upper-income households. Over time, then, entire neighborhoods and communities of newer, hence more expensive, homes become the location of choice for most upper-income households, whereas households of lesser means live in older areas that have dropped in value either in absolute terms or relative to the new neighborhoods. Although there are notable exceptions to this income separation/neighborhood change market dynamic (e.g., Beverly Hills, California, and Chestnut Hill, Pennsylvania), it fairly accurately explains spatial patterns of income groups in U.S. cities and how these patterns change over time.

The second reason income separation occurs is that within the new-construction market, builders target their developments to particular groups of buyers or renters. They do not try to cater to several different income or taste groups in the same residential complex, unless it is quite large. This marketing practice does not generally create income separation on such a large geographical scale as to adversely affect access to good schools and employment opportunities on the part of the households acquiring homes in the lower prices ranges of the new-construction market.

For both political and financial reasons, a significant reduction in the extent of income separation is unlikely to be achieved through public initiatives. Subsidizing low-income households to live in even a modestly priced new home is very expensive, so only a tiny fraction of these households could be assisted in a residential dispersal program. In addition, the political resistance to subsidized income mixing in residential neighborhoods is severe. There are worries, correctly or incorrectly, about the effect of such mixing on property values and the quality of the neighborhood living environment. Also, questions about which low-income households should be assisted in a dispersal effort and the geographic scale at which income mixing should occur to achieve its objectives have not been addressed. Much of the political resistance comes, moreover, from central cities, which would lose housing subsidy dollars to the suburbs if an income-mixing housing strategy were ever to be seriously pursued. As blacks move into predominantly white communities at a gradually increasing rate while income segregation remains or intensifies, cities housing the bulk of the low-income population will continue to experience population loss and related financial problems. Where disadvantaged groups in the population live will, likewise, continue to be a critical but intractable public policy problem. (SEE ALSO: *Residential Preferences; Segregation*)

—*William G. Grigsby*

Further Reading

Downs, Anthony. 1973. *Opening Up the Suburbs: An Urban Strategy for America*. New Haven, CT: Yale University Press.

Farley, Reynolds, Charlotte Steeh, Tara Jackson, Maria Krysan, and Keith Reeves. 1993. "Continued Racial Segregation in Detroit: 'Chocolate City, Vanilla Suburbs' Revisited." *Journal of Housing Research* 4(1):1-38.

Galster, George and Steven Hornburg. 1995. *Housing Policy Debate* 6(1[Special issue]).

Grigsby, William, Morton Baratz, George Galster, and Duncan Maclennan. 1987. "The Dynamics of Neighborhood Change and Decline." *Progress in Planning* 28(1):10-19.

Massey, Douglas S. and Nancy A. Denton. 1993. *American Apartheid: Segregation in the Making of the Underclass*. Cambridge, MA: Harvard University Press. ◄

► Residential Mobility

Residential mobility customarily refers to change of a person's usual residence from one housing unit to another. The concept can also apply to the mobility of households if all members of a household exit or enter a dwelling unit together. In the past, the concept of residential mobility has at times been restricted to mean relatively short-distance moves in contrast to longer-distance moves that represent "migration," and although this distinction still has heuristic value, attempts to establish—theoretically or empirically—a distance where residential mobility ends and migration begins have been fruitless. Increasingly, residential mobility refers simply to change of usual residence (housing unit) without regard to distance moved or the crossing of thresholds, such as city or county boundaries.

Critical to the concept is the notion of what is a person's or a household's usual or customary dwelling. In more developed societies, most persons maintain a single residence where they receive mail, where they spend most of their time, and that they regard as home; for these persons, a change in usual residence is unambiguous. Persons who maintain more than one residence, engage in extensive travel or commuting, or experience seasonal movements may strain the notion of what is usual residence and hence what is change in usual residence. In the United States, increased attention has been given to "snowbirds," retired or semiretired persons who spend summers in the cooler northern parts of the country and winters in the south. Other persons pursue occupations that require travel away from home for extended periods or may spend most of a normal workweek away from a home base. Some persons maintain loose ties to more than one household (housing unit) and may even be homeless for varying periods. These groups illustrate why some researchers conceptualize residential mobility as part of a larger set of spatial movements differentiated by duration, purpose, and periodicity. How they are treated in surveys strongly affects measurement of the amount of residential mobility and characteristics of movers.

Residential mobility is most often measured in surveys and censuses that have fixed rules to determine for each household a single "usual residence." Several approaches may then be used to retrospectively measure change in residence. The Current Population Survey, conducted monthly by the U.S. Bureau of the Census, is a large national sample that illustrates this practice. The March questionnaire asks, "Did you live in this [your current] housing unit on this date one year ago?" for each person 16 years old and over (persons 1 through 15 years of age are assigned a parent's mobility status). Persons who answer "no" are asked additional questions to identify place of residence one year earlier so as to show origins and destinations of persons' changing residence.

Another important source of data on residential mobility is the American Housing Survey, conducted nationally every other year by the U.S. Bureau of the Census for the Department of Housing and Urban Development. This survey asks householders (the person in whose name the housing unit is owned or rented), "In what month and year did you move into this unit?" Householders who report that they have lived in their current housing unit for less than 12 months are asked additional questions on place of previous residence. The U.S. censuses of 1960, 1970, 1980, and 1990 asked, "Did you live in this residence five years ago?" and also asked questions to identify place of residence five years

earlier. Censuses in several other countries have used these and similar practices to measure residential mobility.

Many ad hoc or local surveys focus on varied facets of residential mobility. Surveys often employ different intervals of time for ascertaining change of usual residence and have included questions on number of times moved, distance of last move, reason or reasons for moving (often distinguishing between the decision to depart the old residence and how the new one was chosen), expectations regarding future moving, and other aspects relevant to the decision to move or stay. A few surveys have sought to obtain residence histories by asking how long a person had lived in the current residence, where he or she lived before that, and so forth. Other approaches to obtaining residential mobility histories have sought to identify residence at significant life cycle events, such as the place of entering school, graduating from high school, taking a first post-school job, and so on.

Longitudinal surveys are an increasingly important source of data on residential mobility. These surveys seek to obtain characteristics of the same individuals at successive points in time and follow persons who change residence, thereby generating data on residential mobility. In the United States, the longitudinal surveys most often used for these purposes are the Panel Study of Income Dynamics conducted by the University of Michigan, the National Longitudinal Surveys conducted by Ohio State University, and the U.S. Bureau of the Census Survey of Income and Program Participation. In other cases, longitudinal surveys have followed different cohorts, such as persons who graduate from high school or college in a given year. Typically, longitudinal surveys have fixed intervals for following persons (e.g., one-year follow-ups). They offer exciting opportunities for analyzing a wide range of antecedents, concomitants, and consequences of residential mobility, but they are complex to work with, and they have chronic problems—and differing degrees of success—in following persons who move. They sometimes have different practices regarding persons who exit households to enter group quarters, such as college dormitories, the military, prisons, long-term hospitals, or who leave the United States; the measured amount of mobility can be strongly influenced by these persons.

Research using a life course conceptualization—that is a conceptualization that focuses on the progress of an individual through life as a series of events, many of which involve household relocation—has strengthened earlier research findings and provided new interpretations of individual behavior. Longitudinal research has reiterated the role of the demand for space as part of the logic of moving, but it has also shown the importance of household composition change in stimulating mobility and housing purchase. Perhaps even more important, the same research has shown that changing economic conditions influence the rate of movement. Thus, the longitudinal models are capable of embedding individual behavior in the wider societal and economic contexts.

Other approaches to studying residential mobility have been used less frequently in recent years. One of these is to study "vacancy chains." The idea behind this approach is that when a new housing unit is added to the housing stock, it is usually occupied by an existing household whose departure from another housing unit creates a vacancy. That vacant unit, in turn, is typically occupied by a household whose move creates a vacancy, and so on. Hence, addition of a new housing unit creates a succession of moves in a "multiplier effect." Knowing the size of the multiplier effect in different housing markets and the characteristics of the successions of movers helps planners assess the effects of new housing construction on a community and its residents. Following the chains of moves to their end identifies units as they pass out of the housing inventory.

Underlying the vacancy chains approach is the familiar concept of housing "filtering"—the theory that over time, housing units tend to filter down to less affluent segments of the population. Residential mobility is thus the key to the pace of the process and the actors involved. Collecting such data is costly, and this approach to residential mobility is rarely used.

A more limited version of the filtering or vacancy chains approach is simply to examine individual housing transitions. Administrative records can be used for this purpose, and a few surveys return to the same housing units to collect data. When different households are present at the two dates, analysts focus on the turnover by comparing who moved in with who moved out. Although the result shows only a single housing transition rather than the chain of moves involved in filtering, results of such analyses still illuminate critical events, such as the racial or ethnic successions that over time can transform entire cities. Often, the goal is to compare and contrast who moves out with who moves into housing units of different, sizes, values, characteristics, and locations. This approach looks at residential mobility from the fixed perspective of a housing unit.

By almost any set of criteria, most changes of residence are voluntary, but obviously, some are not. Evictions and foreclosures clearly produce involuntary moves. Other involuntary moves occur when rents (a) are raised and a unit becomes unaffordable to current occupants, (b) a unit is converted from rental to condominium status or for other owner-occupiers and is unaffordable to current renters, or (c) a unit is lost to demolition or conversion to nonresidential uses; in these cases, "reverse filtering" can occur as units pass from less affluent to more affluent occupants. For some movers, change of residence may be considered involuntary if it is in response to undesirable changes in neighborhood composition or conditions. For the very old, deteriorating health can decrease the ability to live independently and result in unintended and unwanted moves to live near or with other relatives or in assisted-living settings. Reliable methods for clearly distinguishing the voluntary and involuntary motives for change of residence have not been found.

The best known attempt to develop a theory of the level of residential movement at the societal level is Zelinsky's "hypothesis of the mobility transition," presented in the early 1970s. Zelinsky asserted that over time, residential mobility follows an S-shaped curve, being low in preindustrial agrarian societies, rising with industrialization, and attaining a plateau at a high level of industrial development. In highly advanced societies, residential mobility might fluctuate with economic cycles or might tend to fall if widening commuting radii allow commuting to substitute for changes of residence. Recent speculation suggests that technological and organizational changes that put modems, fax machines, and personal computers in more homes may allow more persons to work at home and reduce the need to change residence or engage in other forms of spatial mobility.

Confirming the general thrust of the Zelinsky hypothesis, internal migration has been found to increase with societal transformation from predominantly rural and agricultural to urban and industrial. In more developed Western nations, the level of residential mobility appears to have peaked.

The Zelinsky model does little to explain why developed nations vary substantially in levels of residential mobility. The annual rate of residential mobility (the percentage of the population changing usual residence) is about 17% to 19% in the United States, Canada, Australia, and New Zealand. In most of Western Europe and in Japan, annual rates of moving are generally 6% to 11%. Differences among countries have not been fully accounted for but are thought to derive from (a) long-term historical patterns (high rates of moving tend to persist), (b) the degree of government intervention and regulation of housing markets (less regulation, more mobility), and (c) even the nature of the urban hierarchy and settlement patterns (mobility may be decreased if all roads lead to Rome and a single metropolis is dominant).

Rates of residential mobility vary greatly within countries, but even the low-mobility regions of North America, Australia, and New Zealand generally exhibit annual rates of mobility as high or higher than high-mobility regions in Japan and Western Europe. Areal variation in rates of residential mobility within countries partly reflects demographic variables (e.g., younger populations have higher rates of moving) and various "structural" variables that represent the degree of openness of housing markets and the absence of restrictive practices such as rent control.

Most theories of residential mobility have been directed not at why countries or areas have different rates of moving but at why individuals or households move. Characteristics that differentiate highly mobile from more sedentary groups have been studied extensively. High-mobility groups tend to include persons in their 20s, renters, low-income persons, the unemployed, and single persons, especially single men. Low-mobility groups include persons at midlife (rates of moving increase at very advanced ages) and racial or ethnic groups that face discrimination in housing markets. Moves are often linked to specific life cycle transitions such as completion of education, setting up a new household, the birth of a child, divorce, and retirement from the labor force (although most persons do not move when they retire but, rather, "age in place").

Otherwise, the presence of overcrowded housing or other mismatch between housing attributes and needs and expectations raises the odds of moving. Dissatisfaction with neighborhood features or conditions raises the likelihood of moving, but many individuals, especially middle-class households, appear to move in anticipation of future needs—before levels of dissatisfaction become very high. For this reason, levels of neighborhood satisfaction or conditions may have less than their expected effect in predicting who will move. Having recently made extensive repairs to a home lowers homeowners' odds of moving, suggesting a fix-it-or-leave-it decision-making process.

In the United States, rates of residential mobility were at their peak in the 1950s and 1960s. Rates of moving were then often 20% to 21% per year, based on today's age distribution (to take out the effect of aging baby boomers, many of whom are now in their 50s, an age when rates of residential mobility are low). In the early 1990s, annual rates of moving have averaged close to 17%, representing a decline of at least 15% in annual rates of residential mobility.

Since around 1970, mobility rates have tended to drift downward for most demographic categories, but especially for persons in their 20s—the age at which rates of residential mobility are at their highest. Declining mobility at this age reflects postponement of marriage and household formation, because more persons at this age group have stayed in their parents' homes in response to prices of housing relative to perceived opportunities for long-term employment. Stagnant family incomes and declining housing affordability are other factors that appear to have lowered the nation's rate of residential mobility since around 1970. (SEE ALSO: *Housing Norms; Market Segmentation; Migration; Residential Preferences*)

—*Larry Long*

Further Reading

Clark, W. A. V. 1992. "Comparing Cross-Sectional and Longitudinal Analyses of Residential Mobility and Migration." *Environment and Planning A* 24:1291-1302.

Gober, Pat. 1993. *Americans on the Move.* Washington, DC: Population Reference Bureau.

Lansing, John B., Charles Wade Clifton, and James N. Morgan. 1969. *New Homes and Poor People: A Study of Chains of Moves.* Ann Arbor, MI: Institute for Social Research.

Long, Larry. 1988. *Migration and Residential Mobility in the United States.* New York: Russell Sage.

Rogers, Andrei. 1992. *Elderly Migration and Population Redistribution.* London: Belhaven.

Rossi, Peter H. 1980. *Why Families Move.* 2d ed. Beverly Hills, CA: Sage Publications.

U.S. Bureau of the Census. 1993. "Geographical Mobility: March 1991 to March 1992." *Current Population Reports.* Series P20-473. Washington, DC: U.S. Government Printing Office.

Zelinsky, Wilbur. 1971. "The Hypothesis of the Mobility Transition." *The Geographical Review* 61:219-49. ◀

▶ Residential Preferences

Residential preferences reflect desired types of housing situations and encompass many dimensions of housing, including structural type, tenure, location, neighborhood housing and population composition, and political jurisdiction.

Determining the type of housing that people want is important for several reasons. First, residential preferences are economically important. Housing expenditures are large, representing a significant proportion of household income. In the United States, housing's economic impact is considered so great that the number of housing starts per month is used as a barometer for the health of the general economy. Moreover, the housing industry is a major source of employment, including that for construction workers, landlords, real estate brokers, appraisers, lenders, and more. Housing "stimulates" other types of consumption as well, including purchase of furniture, appliances, and other household items. The economic centrality of housing at a variety of levels, from the household to the nation-state, makes the study of residential preferences an important economic enterprise. As a commodity that is largely produced in the private sector, housing is developed to appeal to people's tastes and preferences. Hence, there is a need to determine whether there is a correspondence between the types of housing supplied on the market and the types of housing that people want.

Second, residential preferences are relevant to policy. Certain types of housing (especially suburban, owner-occupied, single-family housing) are supported and even promoted by public policy (e.g., through tax policy, land use planning techniques, and direct and indirect subsidies). The bias of public policy toward supporting some types of housing and not others makes it important to determine whether policy *reflects* widespread preferences or instead helps shape residential desires.

Third, residential preferences are socially important. Through housing, people gain varying levels of access to important life-sustaining amenities, including shelter, comfort and satisfaction, education and employment, transportation, recreational activities, safety, and other people. Housing is a locus of family life and a critical linchpin for members' opportunities for socioeconomic mobility. Therefore, a crucial question is whether people's desires for housing situations correspond to those that will enable them to optimize the quality of their lives. This is a particularly important question for single parents, and minority and poor households.

Finally, residential preferences are politically important. Housing is alleged to be a major source of political stability through engendering satisfaction with the political and economic system undergirding the production of housing. Therefore, whether people's housing situations reflect what they want is regarded as a fundamental political issue.

What Is Housing?

Determining people's residential preferences is complicated because housing is a multidimensional phenomena, including structural type (e.g., single-family home), tenure (own or rent), location, political jurisdiction, and neighborhood population and housing composition. People's preferences for housing reflect their desires for different aspects of each residential situation—distance from employment, homogeneous racial composition, mixed land use, presence or absence of public transportation, and so on. Residential preferences are typically referred to as desires for particular characteristics of the "housing bundle." It is the preferred bundle of housing characteristics that makes up people's residential preferences, not simply desires for a particular type of housing unit.

But it is also the case that housing bundles are packaged in predictable ways, making it difficult to discern preferences for each bundle characteristic. There is a large correspondence between housing's locational attributes and its structural type. For example, suburban housing is more likely to be owner-occupied, single-family housing located in homogeneous communities. City housing is less likely to be detached single-family units and more likely to provide greater access to racial diversity. Therefore, it is difficult to discern people's preferences for some residential attributes (e.g., structural type) independent of others (e.g., location).

Preferences for one set of characteristics may proxy for others. For example, expressed preferences for single-family housing may proxy for desires for white, upper-income neighborhoods in suburbs.

The predictability of packaged housing bundles may require households to make trade-offs among different sets of characteristics. For example, households may prefer the bundle of attributes associated with housing in suburbs but prefer not to live in suburbs per se. Yet obtaining that preferred bundle may require trading away the city location for the less desired suburban one.

Housing Norms

Some believe that residential preferences are guided by a set of normative principles that are the socially prescribed mix of housing bundle characteristics. These alleged housing norms are argued to be social laws governing the types of housing situations that people in the United States ought to live in. These norms represent structural type (single-family, detached dwelling), amenities (private outdoor space), tenure (ownership), and construction (conventionally built). According to this line of thought, the most preferred residential situation will "satisfy" the most norms. In this sense, housing norms are regarded as explanations for housing preferences. People prefer particular housing bundles because the norms prescribe these preferences. People *ought* to live in single-family, owner-occupied housing (the norm) and, therefore, *want* to live in this type of residential situation (the preference). People ought not to be renting multifamily housing (the norm) and therefore find this type of housing situation less desirable (the preference).

The housing norm explanation for housing preferences cannot be empirically proven because it is tautological. One cannot document the presence of this social law other than through the patterns of activities it is alleged to govern. Of course, many people live in single-family, owner-occupied homes and say that they want them and like them. But significant numbers (almost half the U.S. population) live in nonnormatively prescribed housing situations (e.g., rental, multifamily housing). Moreover, the U.S. political system has not experienced any challenges because of the alleged dissatisfaction associated with violating housing norms (i.e., renting). Therefore, housing norms may be neither universal nor normative. The presumption that universal housing norms govern U.S. housing preferences is more aptly regarded as a characteristic of a housing ideology or culture than as part of U.S. society. The admonition that single-family homeownership is the "American dream" is the clearest expression of this ideology.

Decomposing Residential Preferences: Wants, Needs, and Constraints

Although the housing bundle represents many different attributes, some are more important than others. The most important dimensions of housing are structural type, tenure, location, and neighborhood population composition. Each dimension can be examined separately. Yet in reality, they are rarely independent of each other; variations in housing bundle characteristics tend to be highly correlated.

Structural Type

Discussions of residential preferences typically begin with desires for particular types of structures. Independent of other housing bundle attributes, the most preferred structural type varies inversely with the density of housing. All else being equal, the most desired type of housing structure, the detached single-family dwelling, is one with the lowest levels of housing and population density.

Tenure

To own one's housing is considered to be a primary housing desire for U.S. families. Underlying this preference are several rationales, including satisfaction through the greater control associated with homeownership, investment opportunities through housing value appreciation, the ability to save through building equity in housing, and savings through tax benefits that accompany homeownership (e.g., deduction of mortgage interest from taxable income and deferred capital gains). In addition, homeownership has been alleged to be a tangible desire in its own right because of the positive feelings engendered through the act of owning property.

Yet the positive benefits underlying preferences for homeownership need to be evaluated relative to the costs and benefits associated with existing tenure alternatives in the U.S. housing system. Although some choices other than conventional homeownership are available to a limited degree (e.g., land trusts and cooperatives), the main alternative to homeownership is to rent one's housing. Therefore, assessing why people may prefer homeownership has to take into account the housing bundle characteristics associated with the rental housing market. People may want to own their housing because renting provides an undesirable residential situation, not because homeownership is intrinsically valued in its own right.

Location

Location, is a multidimensional housing bundle characteristic in its own right. Location provides varying levels of accessibility to other people, opportunities, amenities, and resources. A location also brings with it varying levels of costs (e.g., auto insurance, property taxes, etc.) and benefits (public schools, garbage collection, libraries, etc). Hence, a critical issue is the relationship of location on residential preferences.

Yet little is known about how location affects preferences for different housing situations, in part, because of the high correspondence between housing structural characteristics, tenure and location. For example, single-family, owner-occupied housing tends to be provided in particular locations (e.g., white suburbs) so that people who want single-family homeownership may get certain locational attributes whether they want them or not. Therefore, it is not clear what people desire from location independent of their desires for a particular structural type and tenure situation.

Neighborhood Composition

Neighborhood population composition is a characteristic of location. But its centrality in U.S. housing market dynamics makes it an important point of emphasis in evaluating residential preferences. The most salient population characteristics are family type, age, income, race, and ethnicity. Due to local zoning and other economic factors, neighborhoods tend to be homogeneous in terms of land use. Neighborhood population characteristics tend to vary with the types of housing contained in the community. Therefore, neighborhoods tend to be homogeneous in terms of population composition. For example, higher-income, married-couple families with children tend to live in communities dominated by owner-occupied, single-family homes, whereas younger single people with lower incomes tend to live in communities with more rental multifamily housing. The homogeneity of neighborhood population composition by age, income, and family type is, in part, a product of the housing contained in the neighborhood, whether people prefer to live in homogeneous communities or not. Homogeneous neighborhoods by income and family type appear to be the outcome of the intermingling of residential preferences, public policy, and housing market dynamics.

Although neighborhood land use may help to create and maintain homogeneous neighborhoods by income and family type, it cannot explain consistently high levels of racial and ethnic segregation in U.S. cities. Moreover, research on different groups' definitions of racial integration suggests that it is white preferences that drive high levels of racial segregation, not the preferences of either African Americans or Hispanics. Whites express desires to live among very few racial and ethnic minorities. Apparently, they view the presence of any minorities as integration. Minorities desire parity and view neighborhoods composed of about 50% minority as integration.

Yet it is wrong to suggest that white preferences for racial homogeneity are the cause of high levels of segregation. Housing audits consistently document that both African American and Hispanic households face discrimination in urban housing markets. Widespread discrimination is not an expression of white residential preferences but a reflection of white racist power. White racist power is manifested in the discriminatory activities of housing market institutions, activities that violate fair housing laws.

Residential Preferences and Gender

Typically, residential preferences have been approached as household preferences, not the residential desires of indi-

vidual members within these households. A reawakening of feminist analyses of housing now suggests that contemporary housing patterns may reflect the preferences of men, not women. Low densities, homogeneous land use, and the segregation of housing from place of employment put constraints on women's ability to combine employment and family responsibilities. From this perspective, the growth and expansion of homogeneous suburban communities and the decline of more diverse central cities is an expression, in part, of male dominance or patriarchy.

Residential Preferences and Public Policy

Does public policy reflect residential preferences, or does it shape housing desires? Since the New Deal, federal and local policies have consistently supported particular types of housing situations over others. Homeownership has been heavily subsidized through both federal loan guarantees and tax incentives. Suburbanization has been encouraged through both federal policy on transportation and housing finance. Local zoning and land use policies have either excluded multifamily land use from suburban environments or segregated it as an offending land use pariah. The lack of enforcement of housing codes combined with low levels of funding for housing subsidies for the poor have worked to escalate the physical deterioration of low-income housing in cities. The failure to enforce fair housing laws has maintained high levels of racial segregation. Fiscal mechanisms to finance public education maintain huge disparities in educational resources across metropolitan regions.

Public policy has supported suburban single-family homeownership. As a result, it has made this type of housing more attractive, particularly compared with the housing alternatives that policy has helped to make less attractive. Public policy may not shape residential preferences, but it certainly makes some housing preferences more rational than others and, as a result, may limit the public discourse on the viability of alternative housing situations to the alleged American dream. (SEE ALSO: *Norms; Residential Location; Residential Mobility*)

—*Anne B. Shlay*

Further Reading

Hayden, Dolores. 1984. *Redesigning the American Dream: The Future of Housing, Work and Family Life.* New York: Norton.

Jackson, Kenneth T. 1985. *Crabgrass Frontier: The Suburbanization of the United States.* New York: Oxford University Press.

LaGory, Mark and John Pipkin. 1981. *Urban Social Space.* Belmont, CA: Wadsworth.

Massey, Douglas S. and Nancy A. Denton. 1993. *American Apartheid: Segregation and the Making of the Underclass.* Cambridge, MA: Harvard University Press.

Michelson, William. 1977. *Environmental Choice, Human Behavior and Residential Satisfaction.* New York: Oxford University Press.

Morris, Earl W. and Mary Winter. 1978. *Housing, Family and Society.* New York: John Wiley.

Perin, Constance. 1977. *Everything in Its Place: Social Order and Land Use in America.* Princeton, NJ: Princeton University Press.

Tremblay, Kenneth R. Jr. and Don A. Dillman. 1983. *Beyond the American Dream: Accommodation to the 1980s.* Lanham, MD: University Press of America. ◀

▶ Residential Satisfaction

There is often a need to assess how well a residential environment meets the requirements, goals, and expectations of its inhabitants—that is, how satisfied they are with it. In broad terms, any such assessment may be viewed as an indicator of residential satisfaction. More specifically, residential satisfaction indicates people's response to the environment in which they live. In this definition, the term *environment* refers not only to physical aspects of residential settings, such as dwellings, housing developments, and neighborhoods, but also to the social and economic aspects (and organizational or institutional aspects, if any) of such settings. By using appropriate techniques of data collection and analysis, one can measure the degree of residential satisfaction in a specific setting and determine which physical, social, and organizational factors contribute to it. This information is useful because it adds to our understanding of how people respond to certain residential types or conditions and may suggest ways to improve the design and construction or rehabilitation of residential settings.

Satisfaction research has been used extensively in market studies to ascertain which features of products and services best meet consumers' needs and thus are likely to cause consumers to repeat their purchase or to recommend the product or service to others. To a degree then, residential satisfaction studies are part of the broader field of market research. However, although certain facets of residential environments can be reasonably viewed as products or services, others are unique to housing. Housing usually has a degree of capital investment and permanence that distinguishes it from most other products and services. Moreover, a person's place of residence, one's house, is bound up with connotations such as refuge from life's daily vicissitudes, status in socioeconomic structures, symbols of self, and many other psychological and social aspects having to do with people's affective relationships with their living environment.

Because of housing's singular characteristics, researchers of residential satisfaction have developed their own conceptual frameworks and methods, distinct from market research. Indeed, the initial impetus for and interest in this field of study did not come from the market, but from the recognition that the needs, expectations, and interests of low- and moderate-income people (who have inadequate or no access to the housing market) were generally neglected. This acknowledgment occurred at different times in different countries, beginning in the late 1950s in the Scandinavian countries and Great Britain and eliciting increasing interest during the following decades as public housing for low- and moderate-income households began to show clear signs of not being well received by its inhabitants. For instance, a number of studies by the Swedish Institute for Building Research reported high vacancy rates, particularly in high-rise housing, despite a general housing shortage in that country. The Economic Commission for Europe reported similar findings in a variety of other countries. In the United States, Pruitt-Igoe, a large public housing complex in St. Louis, became notorious, among others, for numerous vacancies, vandalism, and high rates of crime.

Thus, the aim of residential satisfaction assessments has been primarily to provide information that can be used to improve the residential conditions of those people who cannot make their preferences, requirements, and expectations known through the market mechanisms that are, at least in theory, available to the more affluent populations.

Residential satisfaction research is also a form of post-occupancy evaluation (POE), a field that focuses on evaluating the performance of buildings and other environments by assessing it against generally accepted criteria or by comparing an environment's intended use with its actual use. In this perspective, residential satisfaction becomes a criterion against which to evaluate the performance of housing from the inhabitants' point of view. However, although POE paradigms tend to privilege the physical environment and its design, the concept of residential satisfaction, as mentioned earlier, also includes economic, social, and organizational aspects.

The most commonly used model of residential satisfaction acknowledges that any explanation of person-environment relations must account for the interaction between people's characteristics and the social and physical components of environments. In housing research, this has led to viewing residential satisfaction as an attitude, a concept that has been extensively studied in social psychology.

Attitudes are usually considered to reflect beliefs, feelings, and behavioral tendencies. A conceptual model of residential satisfaction can be constructed, in which people's evaluations of the places where they live (residential satisfaction) can be linked to their beliefs, feelings, and behavioral tendencies, as shown in Figure 26.

In such a model, the category of beliefs typically includes people's perceptions about various physical features of the place where they live, about their neighbors, and about management. It also includes people's experiences with housing, such as comparisons with prior places of residence, and their expectations for places where they might live in the future. The category of feelings usually encompasses memories and connotations associated with housing, symbolic content attributed to it, and aesthetic judgments about it. The category of behavioral intentions generally entails plans for moving or staying, participation with other residents in activities related to the place where they live, and attempts to personalize their surroundings—that is, to use devices such as colors of paint, landscaping, or decorative elements to imprint a personal mark on the environment.

All these aspects of satisfaction are subjective; they reflect each inhabitant's personal views, perceptions, experiences, behavioral norms and values, and emotions. But the attitude model of residential satisfaction also accounts for objective variables—that is, conditions other than residents' opinions. Objective variables include environmental, demographic, and personal characteristics. Objective environmental aspects include physical characteristics, such as space per dwelling; density of dwelling units; layout, type, and number of parking spaces; and the like. Environmental characteristics also include social and organizational features, such as interaction among residents, children's play and teenagers' activities, activities of and participation in residents' councils or other tenants' organizations, other

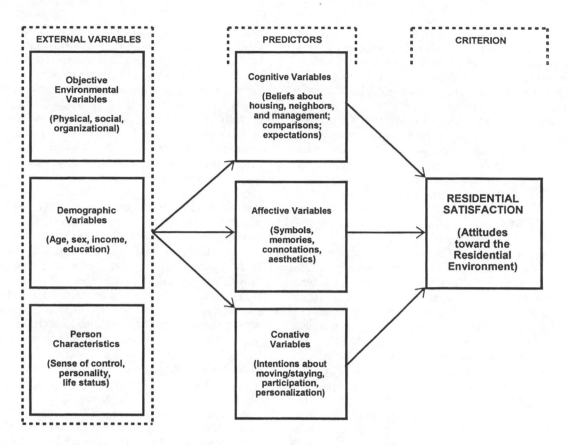

Figure 26. Conceptual Model of Residential Satisfaction: Beliefs, Feelings, and Behaviors

formal or informal social activities, and policies and practices of managers and maintenance personnel. Demographic characteristics consist of variables such as the age, sex, income, and educational attainment of the residents. Personal characteristics that can be objectively measured with standard psychological scales include traits such as sense of control over events in one's life, type of personality, and status in one's life cycle.

Work in the psychology of attitudes suggests that beliefs, feelings, and behavioral tendencies predict satisfaction but that external objective factors (environmental, demographic, and personal) may underlie or influence subjective attitudes. This is but another way of saying that residents' subjective views, feelings, and intentions mediate between the objective physical, social, and organizational environment and residential satisfaction. This helps explain, for example, the often-reported phenomenon that housing that appears satisfactory to "experts" may not necessarily satisfy the inhabitants.

Given this model, it is not surprising that survey research methods—methods of data gathering that seek information from subjects by means of questionnaires and interviews—play a major role in the design of studies aimed at understanding and predicting residential satisfaction. It is often useful to compare the inhabitants' responses with those of others. Management and maintenance personnel, planners and architects, and residents in the surrounding community can be surveyed for this purpose. Observations of the en-

vironment and of behavior (often with photographic documentation) and archival research are used to collect environmental and demographic data.

Data analysis is complex and is typically accomplished by using descriptive statistics, such as calculations of means and standard deviations, and multivariate statistical methods. Descriptive statistics yield results about levels of residential satisfaction. Knowing such levels, especially if it is possible to compare them with levels obtained from other housing environments, is of obvious value. However, it is even more valuable to ascertain which components of a housing environment are most strongly related to residential satisfaction and how satisfied residents are with these components. This information can guide interventions aimed at improving unsatisfactory housing conditions, especially when resources are limited and therefore must be targeted to items that will maximize satisfaction.

Some researchers have also attempted to determine whether residential satisfaction influences residential behavior. If such a link could be established, it might be possible to answer questions such as, Is it likely that the more satisfied the residents, the less likely they will move? Are satisfied residents less likely to vandalize their housing than unsatisfied ones?

Residential satisfaction studies have focused on assisted housing, but market-oriented housing providers may also profit from a more refined and sophisticated understanding of residential satisfaction than that which market demand-

and-supply data are able to provide. This may become especially important as the market increasingly diversifies. The "traditional family" of a married couple with children that the housing industry still considers its mainstay customer in fact represents a declining type in the United States and other industrial countries.

Finally, an important aspect of residential satisfaction research to be emphasized is that people's responses to the places where they live change over time. Because of this, studies conducted at one time may be of debatable value if not repeated at appropriate intervals and supplemented by methods of inquiry that consider time and change.

In residential satisfaction research, as in other fields, further study is likely to sharpen conceptual models and research tools. A number of pending issues exist. For example, the viewpoint of the residents, paramount as it should be, needs to be integrated into a framework in which that of other actors (policymakers, architects, managers, lenders) is also carefully delineated. Given the importance of people's feelings and emotions about the places where they live, more sophisticated methods of studying people's responses to housing should be developed. And because robust knowledge generally emerges from cumulative findings, more residential satisfaction studies should be conducted. Repeated application of the methodology and comparisons of findings from different studies would also strengthen our confidence in residential satisfaction as a reliable and valid criterion for evaluating housing. (SEE ALSO: **Behavioral Aspects;** *Housing Norms*)

—*Guido Francescato*

Further Reading

Cooper Marcus, Clare and Wendy Sarkissian. 1986. *Housing as if People Mattered.* Berkeley: University of California Press.

Francescato, Guido, Sue Weidemann, James R. Anderson, and Richard Chenoweth. 1979. *Residents' Satisfaction in HUD-Assisted Housing: Design and Management Factors.* Washington, DC: U.S. Department of Housing and Urban Development.

Francescato, Guido, Sue Weidemann, and James R. Anderson. 1989. "Evaluating the Built Environment from the Users' Point of View: An Attitudinal Model of Residential Satisfaction." In *Building Evaluation,* edited by Wolfgang F. E. Preiser. New York: Plenum.

Franck, Karen and Sherry Ahrentzen. 1989. *New Households, New Housing.* New York: Van Nostrand Reinhold.

Stokols, Daniel. 1986. "Transformational Perspectives on Environment and Behavior." In *Cross-Cultural Research in Environment and Behavior,* edited by William H. Ittelson, Masaaki Asai, and Mary Ker. Tucson: University of Arizona.

Weidemann, Sue and James R. Anderson. 1985. "A Conceptual Framework for Residential Satisfaction." In *Home Environments,* edited by Irwin Altman and Carol Werner. New York: Plenum. ◄

► Resolution Trust Corporation

The Financial Institutions Reform, Recovery, and Enforcement Act of 1989 (FIRREA) established the Resolution Trust Corporation (RTC) as a temporary government agency charged with the responsibility of disposing of bankrupt thrifts and their assets. The agency was established with staff reassigned from the Federal Deposit Insurance Corporation (FDIC). When the RTC was dissolved in December 1995, its responsibilities were transferred to the FDIC, which plays a parallel role in resolving bankrupt banks. The Resolution Finance Corporation was established to organize funding for the bailout. Interest payments are covered by Treasury funds (authorized incrementally by Congress), and principal payments are covered by assessments on the Federal Home Loan Bank system and the money raised through the sale of assets. The RTC's purpose was to take control of bankrupt thrifts and either sell them to a solvent institution (a bank, bank holding company, or another savings and loan) or, if no buyer could be found, dissolve the firm, sell off its assets, and use the money raised (supplemented by RTC funds) to pay off depositors and creditors. This proved an immensely complex task, because bankrupt thrifts held a wide range of assets, including mortgages, real estate, financial instruments, and junk bonds. By March 1994, the RTC had closed or sold 680 of 743 failed institutions and liquidated assets valued at more than $393 billion at approximately 90% of book value. The remaining assets have been poorer quality and much harder to sell. In addition to auctions and bulk sales, the RTC pioneered securitization as a method of disposing of large volumes of assets. That is, a range of financial instruments (or commercial properties or mortgages) from a number of insolvent thrifts and with varying levels of risk and value were packaged together, and investors were sold shares (securities) in the ownership of the pool of assets. (SEE ALSO: *Lending Institutions; Savings and Loan Industry*)

—*Heather MacDonald* ◄

► Resolution Trust Corporation's Affordable Housing Disposition Program

The Resolution Trust Corporation (RTC) was established in August 1989 (as part of the savings and loan bailout bill) to dispose of bankrupt thrifts or, if no buyer could be found for a thrift, to dissolve the business, sell off its assets, and pay off its depositors and creditors. The RTC's operations were funded by the Resolution Funding Corporation, through budget allocations, issues of Treasury bonds, and payments from the thrift industry. The RTC ceased to exist December 1995, when its operations and remaining assets were transferred to the Federal Deposit Insurance Corporation (FDIC), which operates the insurance funds that guarantee deposits in banks and thrifts. Real estate made up about 12.8% of bankrupt thrift assets in 1991, and approximately 11.6% of these were single-family homes or rental properties. Most thrift failures (and thus, most thrift real estate assets) were concentrated in the "oil patch" states—Texas, Oklahoma, Louisiana, Arizona, and Colorado.

In an attempt to use the tremendous cost of the bailout to achieve some public good, the RTC was required to establish an affordable housing disposition plan (AHDP). Single-family homes appraised at less than $67,500 would be reserved for sale to low- or moderate-income households (earning less than 115% of area median income) for a period of 97 days after the RTC placed the property on the market. Rental property with units appraised at below specific values depending on apartment size (for instance, three-bedroom apartments valued at less than $52,195) would be reserved for a similar period for sale to organizations that would reserve at least 35% of units for rent to a low-income or very low income renter household (earning less than 60% of area median income). Other property (e.g., warehouses) that could not be sold was to be donated to organizations providing housing for homeless people.

Affordable housing advocates and state and local housing agencies presented many arguments in support of the AHDP. Although housing prices in the states where most RTC assets were concentrated had declined (in part, this had helped cause the widespread bankruptcies), many families in those states still could not afford decent housing. Existing housing could be used far more cheaply to satisfy needs than building new housing. Opponents of the program argued that the RTC's first responsibility was to reduce the cost to taxpayers by selling properties as fast as possible and for as high a price as possible and that the AHDP would increase the cost of the bailout. Supporters replied that the AHDP could in practice reduce taxpayer costs and minimize damage to local property markets. First, every day the RTC owned a single-family home, it paid out $17.50 in property taxes and management and maintenance fees to private contractors. If homes could be sold at a discount to needy households, holding costs would be reduced. Second, property owners and realtors in areas with concentrations of RTC-owned property were concerned that the longer the oversupply of vacant homes remained on the market, the lower property values would fall. If RTC-owned homes could be sold soon, and to buyers who would not normally be in the market for homes, property values would return to normal sooner.

The 1989 legislation authorized (but did not require) the RTC to employ a number of strategies to make the AHDP work. Homes could be sold at a discount to income-qualified buyers, and the RTC could offer seller financing, at below-market rates if necessary. However, dissension over the AHDP within the RTC and its oversight board (composed of the secretaries of the Treasury and of Housing and Urban Development and the chairs of the Office of Thrift Supervision and FDIC) raised a number of roadblocks during the first three years of the program.

Significant ground was lost during those first three years. By June 1992, 4,112 single-family homes had passed through the AHDP program unsold (14.3% of the total 28,819 available), and 9,529 had been sold. Until March 1991, qualified homes held in conservatorships (the stage before the RTC took ownership of a thrift and its assets) were not included in the program; it is estimated that a total of 3,998 houses appraised at less than $67,500 were sold to unqualified buyers during this time. The multifamily (rental) sales program began more slowly than the single-family program. By June 1991, 13 multifamily properties had been sold under the program, just 2.8% of the 471 available, and 226 properties had passed through the AHDP unsold. One year later, this record had improved somewhat

to 184 sold, 21.3% of those available, but sales were poor compared with the 33% of single-family properties sold by that time.

The problems with the program over this initial period fall into four groups:

1. *Information and marketing difficulties:* The system of information clearinghouses and private contractors used to provide information and market homes in the program was inadequately funded at first and did not work effectively in some regions. The RTC began holding auctions and housing fairs in June 1991 to speed up the process, but housing advocacy groups argued that low-income buyers were disadvantaged by the sales methods used, the lack of information on properties available, and the RTC's failure to advertise discounts.

2. *Reluctance to use seller financing:* Seller financing for single-family homes was not made available until May 1991, but many income-qualified buyers were still unable to take advantage of it because they did not meet the requirements of loan underwriters. Seller financing for multifamily properties was not made available until January 1992; this was a serious problem for many potential buyers because it was very difficult to obtain financing for rental housing from banks over this period.

3. *Sales procedures that did not maximize the number of rental units reserved for low-income households:* The RTC's emphasis (in some regions) on bulk rather than individual sales disadvantaged many smaller purchasers (such as nonprofit agencies) and favored large corporations. Because nonprofit agencies would be more likely to reserve all rental units for income-qualified households rather than the minimum 35%, this practice may have minimized the impact of multifamily sales on the affordable housing supply. Purchasers were also allowed (until May 1992) to aggregate the 35% low-income occupancy requirement in bulk purchases so that low-income units would be concentrated in some properties, whereas more desirable units were rented at market rates. Nonprofits and public agencies were disadvantaged by the RTC's assessment of "substantially similar bids"; many nonprofit and public agencies reserving 100% of units for low-income occupants lost out to marginally higher bids from private firms reserving only the minimum 35% of units for low-income tenants.

4. *Program monitoring and enforcement difficulties:* Information on buyers' income was not collected in some regions (in particular, the Southwest and West, where 58% of sales occurred), so it is possible that many homes in the AHDP program were sold to wealthier households. By December 1991, only 42% of purchasers had reported their incomes to any agency, and not all these reports were verified.

These problems were gradually resolved, and by 1993, the consensus was that the AHDP was working more effectively. By February 1993, 16,780 single-family homes had been sold, at an average price of $28,000; over the following year, 4,447 more homes were sold. A total of 431 multifamily properties (containing 37,000 units) had been sold under the program by 1993, and a further 242 were sold the following year, at an average price one-seventh that of newly built public housing. The RTC had established an information clearinghouse network of 12 federal home loan banks and 38 state housing finance agencies as well as a network of technical assistance providers, including 68 local nonprofit organizations and 8 national organizations. Approximately $100 million in seller financing had been provided to homebuyers (including financing for repairs), and both temporary and permanent financing were available for nonprofit and public purchasers of multifamily housing. A direct sales program had been established to make it easier for public agencies and nonprofits to buy properties, and a substantial number of properties were sold to these organizations subsequently. Finally, internal controls on program monitoring and enforcement had been improved, with the assistance of state housing finance agencies. Over this period, the RTC learned many valuable lessons about the effective management of a property disposition program, and many low- and moderate-income households that would not have been able to buy a home (estimated to have been half of all RTC homebuyers) benefited from the program. (SEE ALSO: **Affordability**; *Resolution Trust Corporation*)

—*Heather MacDonald*

Further Reading

General Accounting Office. 1992a. *Resolution Trust Corporation: Affordable Multifamily Housing Program Has Improved But More Can Be Done*. Report to the Chairman, Subcommittee on Housing and Community Development, Committee on Banking, Finance and Urban Affairs, House of Representatives. GAO/GGD-92-137. Washington, DC: Author.

———. 1992b. *Resolution Trust Corporation: More Actions Needed to Improve Single Family Affordable Housing Program*. Report to the Chairman, Subcommittee on Housing and Community Development, Committee on Banking, Finance and Urban Affairs, House of Representatives. GAO/GGD-92-136. Washington, DC: Author.

Johnson, Sara. E. 1990. "RTC's Affordable Housing Program: Reconciling Competing Goals." *Housing Policy Debate* 1(1):87-130.

Kettl, Donald F. 1991. "Accountability Issues of the Resolution Trust Corporation." *Housing Policy Debate* 2(1):93-114.

MacDonald, Heather I. 1995. "The Resolution Trust Corporation's Affordable-Housing Mandate: Diluting FIRREA's Redistributive Goals." *Urban Affairs Review* 30(4):558-79.

U.S. Congress. House. Subcommittee on Financial Institution Supervision, Regulation and Insurance, of the Committee on Banking, Finance and Urban Affairs. 1993. *Resolution Trust Corporation's Affordable Housing Program*. 103d Cong., 1st Sess. Washington, DC: Government Printing Office.

———. Senate. Committee on Banking, Housing and Urban Affairs. 1992. *RTC's Operations and the Affordable Housing Program*. 102d Cong., 2d Sess. Washington, DC: Government Printing Office. ◀

▶ Restrictive Covenant

A restrictive covenant is a contractual agreement to restrict the development, use, or occupancy of real property. Often used in the absence of or in addition to public zoning ordinances, restrictive covenants are private agreements written into deeds or leases of real property and are enforceable by law as any normal contract provision. These provisions are not required to protect the safety, health, or welfare of community residents because the agreements are between individuals rather than the result of public policy. In addition, because these covenants are written into the deed or lease, they remain in force even as property changes hands regardless of whether or not the restriction is included in the new agreement.

Restrictive covenants are used by developers and planners to influence the character of neighborhoods. In the late 19th and early 20th centuries, visionary developers used these restrictions to create model communities. Restrictions included in the deeds might cover the minimum lot and home size, building heights and setbacks, and future land uses. In some cases, developers even interviewed prospective tenants to ensure that they met their vision for the community.

Covenants may be required in communities where no public zoning ordinances have been adopted. Houston, Texas, for example, requires restrictive covenants prior to issuing building permits. They also may be desirable in unincorporated areas to protect the investment of developers and, ultimately, landowners by preventing undesirable businesses from locating in a residential development and by creating a homogeneous community by controlling the style, size, and character of homes.

The use of restrictive covenants expanded to cover racial integration, particularly after the first black migration northward following World War I. Although generally viewed as a means to prevent blacks from moving into white neighborhoods, restrictive clauses were also used to block other minority groups. These restrictions meant that blacks were allowed to compete for only a small sector of the housing market. Because the demand for housing by minorities exceeded the supply available to them, the cost of housing was significantly inflated for minority group members. This reality might mean that property values declined as the residents were unable to make repairs or improvements and that neighborhoods in which minorities resided were unstable as families who failed to meet financial obligations were evicted. Moreover, housing segregation was reflected in school segregation.

Challenges to the practice were brought under the 14th Amendment to the Constitution, which precluded the denial to any citizen of the United States of life, liberty, or property without due process of law. Developers believed that because the covenants were written into private contracts between individuals, and not the result of government policy, they were legal.

In *Shelley v. Kraemer* (334 U.S. 1 [1948]), the U.S. Supreme Court ruled that restrictive covenants themselves could not be outlawed because they were private agreements between individuals and as such did not violate the 14th amendment. Equal protection under the law would be violated if the state acted to enforce the agreement. Thus, individuals are free to include restrictions based on race, but they cannot expect the state to use its police powers to enforce the contract.

Following the *Shelley* case, racial discrimination based on restrictive covenants was seriously impaired. Some communities were able to maintain the practice by requiring that potential residents be accepted as members of a home-owners association wherein membership was not granted to members of "undesirable" minority groups. Other forms of discrimination, such as racial steering and redlining, are still used to limit access by minority groups to all housing markets. (SEE ALSO: **Discrimination**)

—C. Michelle Piskulich and John S. Klemanski

Further Reading

Levine, Hillel and Lawrence Harmon. 1992. *The Death of an American Jewish Community*. New York: Free Press.

Stach, Patricia Burgess. 1988. "Deed Restrictions and Subdivision Development in Columbus, Ohio, 1900-1970." *Journal of Urban History* 8(1):42-68.

———. 1989. "Real Estate Development and Urban Form: Roadblocks in the Path to Residential Exclusivity." *Business History Review* 63(2):356-83. ◄

► Retirement Communities

Throughout much of the Western world, increasing numbers of older people are retiring or modifying their employment status. Whereas most healthy retirees remain in the homes they occupied while part of the labor force, others will move. Of those who change residence, some retirees will move to smaller homes requiring less time, energy, and money; others will move in with willing children, relatives, and friends. Still others will opt for an entirely new housing arrangement, either within the same community or elsewhere. These latter moves are most often made by people less than 70 years of age, and for many, they are likely to be to destinations offering amenities such as a mild climate and recreational and scenic attractions.

Choosing a place to move involves two closely related decisions; one dealing with the selection of a dwelling and the other involving the setting or community in which that dwelling is located. In choosing a community, those who move after retirement must consider whether they wish to remain in the setting where they lived before retirement or move to another setting. For retirees and other older people who decide to live elsewhere, the question becomes one of first choosing a region and then selecting the particular community within that region offering the greatest number of attributes capable of fulfilling individual needs. In making these decisions, older people must consider the attractions and limitations of living in an age-integrated setting or one that is age-segregated, attracting a predominantly older and relatively healthy population. Age-segregated or retirement communities can either be planned or they can evolve naturally.

Planned Retirement Communities

Planned retirement communities are aggregations of housing units with at least a minimal level of service and modest community facilities (i.e., clubhouse or meeting room) that are designed for at least 100 residents who are 50 years of age or older, relatively healthy and active, and retired. Using this definition, continuing care retirement communities, retirement residences, granny flats, congregate housing, and other forms of assisted living are not considered retirement communities. Planned retirement communities, as defined, are likely to be found in the United States and can be categorized as new towns, retirement villages, and retirement subdivisions. Each type can be differentiated by its scale, the characteristics of its residents, and the level of services it offers.

The scale of the retirement community refers to the size of its population, whereas resident characteristics focus mainly on the age of the population and its health status. Services include the type and quantity of recreational and leisure, social, commercial, and health facilities and care available to the residents.

Retirement New Towns

The largest of the planned retirement communities are new towns, designed for relatively young, healthy retirees interested in a leisurely but highly active lifestyle within in a totally self-contained community setting. As privately built developments, they contain a wide range of commercial and health services and an extensive network of outdoor and indoor recreational facilities and leisure programs. New towns are most commonly located in the Sunbelt and western United States because the climate is conducive to year-round outdoor activity. They are large in size and contain at least 5,000 residents. For example, the population of Sun City and of Sun City West, Arizona (near Phoenix), is more than 50,000 people. Although new towns represent a small proportion of retirement communities in the United States, they house approximately one third of the entire population living in planned retirement communities.

Retirement new towns contain numerous supportive services and facilities for older people. In fact, when compared with other types of retirement communities, new towns offer the most extensive network of recreational, medical, commercial, and financial services and programs. Opportunities are available for outdoor and indoor recreational pursuits ranging from golf and swimming to drama, arts and crafts, billiards, and bridge tournaments. Retirement new towns also contain a full range of health care and medical services. Hospitals are often present, and clinics, doctors' offices, and laboratories are almost always available. The communities may even include nursing homes or a continuing care retirement community. For example, Sun City has several continuing care centers within its jurisdiction, and others are situated in adjacent communities. Such health care facilities and services enable residents of the new towns to feel secure about remaining there as they age and as their health status changes. Finally, retirement new towns have an abundance of shopping, banking, and opportunities for dining out. Residents are able to deal with their personal affairs without having to leave the community.

The housing available to prospective new town residents varies from single-family homes, duplexes, and town houses to high-rise buildings designed for residents seeking security and continuing health care. In addition to homeownership, there are rental units, condominiums, and cooperative living arrangements.

Residents of retirement new towns vary in age but are likely to be relatively young and active retirement age people. In the early 1980s, the average age was about 70 years old, although the population of some places was considerably younger. For example, the typical resident in Sun City Center, Florida, was in his or her mid-60s. Households, for the most part, consisted of couples with a moderately comfortable income. Many maintained another home or apartment in a northern state and used it during the summer months. Others traveled extensively, although fewer than a third of the residents worked full- or part-time. Several engaged in volunteer work in the retirement community or in the surrounding communities. This resident profile has not dramatically changed since an aging population has been offset by an increase in the number of young retirees moving in.

As large-scale, privately sponsored communities offering a variety of services, retirement new towns contribute substantially to the local economy through tax revenues and the consumption and savings of their residents. The introduction of a retirement new town into a local governmental unit is likely to improve its tax base. Tax base improvement comes directly from new housing construction and other forms of development adding to the tax roles and indirectly through adjacent land development, attracted in part by the presence of the retirement community. New town residents, many of whom are affluent, are consumers of various goods and services that use sizable amounts of retail, institutional, and office space. These spaces, in turn, are taxed by local government. Retirement new towns are often characterized by their extensive security systems, and many police themselves; others have voluntary fire departments, and all contribute tax dollars to the local school system without using it.

Retirement Villages

These medium-sized retirement communities are designed to house a young, healthy retirement and preretirement population in a secure setting offering an array of outdoor recreational facilities and programs. Unlike the retirement new towns, they contain limited commercial facilities and, if health care is available, it is sparse and unobtrusive. Most retirement villages are privately developed, although some have been built under union or church sponsorship. Villages tend to be smaller than retirement new towns, ranging in size from 1,000 to 5,000 people. Unlike new towns, they are not planned as full-service, self-sufficient developments built within the fabric of local governmental units. And unlike most new towns where nonresidents have easy access to businesses and services located within them, villages tend to be self-contained physical entities, often secured by gatehouses and enclosed walls. An example is Leisure World at Laguna Hills, California. These retirement communities are designed in part to provide security for residents and separation for recreational and other leisure amenities located within. Although they are commonly found in the Sunbelt states, retirement villages have also been developed in a number of northern states.

Retirement villages offer an extensive network of recreational and communal facilities, including a clubhouse, crafts and game rooms, swimming pools, golf courses, and tennis courts. Recreational and leisure facilities often rival those found in retirement new towns. Because they are not planned to be autonomous, full-service communities, the extent to which villages provide shopping facilities on-site varies greatly. Some contain a few private businesses, and others have commercial establishments adjacent to their properties. Still others, such as Hawthorne near Leesburg, Florida, have neither on-site nor nearby commercial facilities. Similarly, the extent to which health care is available varies greatly among villages. Most offer only emergency medical service, although others, such as Leisure World in Maryland, have a fully staffed medical clinic and a pharmacy.

The mixture of housing in retirement villages varies extensively. Options include single, detached homes; town houses; low-rise and high-rise apartment buildings; and manufactured homes. There is variety in housing tenure as well. In addition to traditional homeownership arrangements, cooperative and condominium units, rentals, and manufactured homeownership combined with lot rental are available. Regardless of tenure arrangements, the residents typically pay a monthly fee, covering operating costs of recreational facilities and other community services.

As in the case of retirement new towns, retirement villages tend to attract relatively young, active, and healthy people. Occupancy is typically limited to adults who are at least 50 years old, although some communities house younger adults. For the most part, the population consists of retired couples who are college educated and financially independent. Not all retirement villages are designed to house an affluent population. Over time, however, such villages tend to attract wealthier retirees as the communities mature. In part, the changing financial status of the residents is associated with rising service costs. As costs rise, monthly fees increase or a service charge is instituted for services and activities previously included as part of the basic monthly fee. As a result, retirement villages become increasingly expensive for retirees on limited and fixed incomes. Many move out and are replaced by more affluent residents.

Retirement Subdivisions

Subdivisions tend to be privately developed residential areas planned predominantly for independent and healthy older populations. But unlike the retirement new towns and villages, retirement subdivisions have limited outdoor recreational facilities and support services. They vary in size and are intended to be an integral part of the fabric of the surrounding community, which is usually rich in services and other amenities. The larger setting is viewed as the major attraction for prospective residents who want to live in a homogeneous elderly setting and who seek this type of living arrangement that is unencumbered by a costly infrastructure. Retirement subdivisions, therefore, tend to be located within urban areas of Florida and other Sunbelt states so as to take advantage of the favorable climate, the proximity to other older people, and the abundance of health care and recreational services. Despite their being characterized as retirement communities that house predominantly an over-50 population, subdivisions are often indistinguishable in appearance from surrounding residential environments. That is, young adults may live in close proximity to the older residents.

Services and facilities in subdivisions tend to be limited and may include shuffleboard courts and a small community building combining a card room, a meeting area, or laundry facilities. Commercial, medical, and nursing facilities are unlikely to be incorporated into the subdivision infrastructure because they are readily available in the surrounding area.

Housing in the retirement subdivision is usually limited to conventionally built, single-family homes or to manufactured homes. Although most residents own their dwellings, tenure arrangements vary in the manufactured-housing developments. In some places, the occupant owns both the home and lot, whereas at other sites, the manufactured home is purchased and the lot is leased from the developer.

Although many retirement subdivisions pose limitations on the age of prospective residents, others do not or restrict occupancy to persons 18 years of age or older. In some, at least one resident must be 50 years old; still others have sections of the development set aside for younger couples. Most, however, house a predominantly older population, a majority of whom are retired and were attracted to the community by its relatively homogeneous elderly population and modest living costs.

Households typically consist of married couples in their 70s and in good health; however, residents tend to be less affluent than their counterparts in retirement new towns and villages. Subdivisions contain fewer services and lower housing costs and, consequently, are more affordable than other types of retirement communities.

Naturally Occurring Retirement Communities

Naturally occurring retirement communities (NORCs) are housing developments that are neither planned nor marketed for older people but, rather, attract a predominance of residents who are 60 years of age and older. Such communities offer a supportive social environment and access to services and facilities that prolong independent living for older people. NORCs can vary in size from an entire neighborhood housing several thousand residents, most of whom are elderly, to a single apartment complex or building. Because they occur naturally rather than being planned, they are not specifically designed for older people and therefore may not be totally age segregated. The original residents are most likely to be young, and as they grow older, others of similar age move in. In time, the older residents outnumber the younger ones. Because they are not marketed or promoted as retirement communities, NORCs are often not recognized as such even by their own residents, managers, and developers.

NORCs differ from the planned retirement communities in that housing and services within them are not originally designed to serve an older population. In fact, naturally occurring retirement communities may offer few, if any, services to their residents. Large-size NORCs approximate retirement subdivisions and are typically found in older neighborhoods.

In sum, retirement communities, whether planned or naturally occurring, offer independent older people an opportunity to live with their contemporaries who share a common background and relatively good health. Residents of retirement communities can essentially create their own social worlds and surround themselves with others who value their worth and are more sensitive to what it means to grow old. In both planned retirement communities and NORCs, older people can organize themselves to protect against criminal activities, ranging from keeping watch or serving as buddies for their peers to the formation of posses. Planned retirement communities offer occupants a relatively unchanging, ordered, predictable, and protective setting and a predictable lifestyle. In contrast, NORCs are more likely to be subject to changes in population makeup and surrounding land uses. And in contrast to many planned retirement new towns and villages, NORCs are integratively linked to the broader community within which the older people live.

In the United States and elsewhere, most older people live in age-integrated settings. Indeed, planned retirement communities are not the preferred place of residence for many independent and healthy retirees. For some, retirement communities are geriatric ghettos, and to be identified with them implies impending senility and old age. Others agree with urban planners and sociologists who suggest that age-segregated communities do not reflect the true diversity of society and promote isolation and social conflict between their residents and those living outside the community. Such segregated living arrangements, particularly in gated communities, are prone to exclusivity and disengagement from the surroundings. This is likely to occur when retirement community services, including recreation programs, police protection, and health services have been privatized and supported by local residents, and the attitude persists that paying taxes to a local government to support the broader public is inequitable.

Another issue faced by retirement villages and subdivisions catering to independent and active older people is the growing need for health care services as the population ages. Whereas some retirement communities provide modest services to meet the health care needs of their residents, others have no plans to introduce such services in response to the growing health care demands of an aging population. In such instances, the increasing cost burden rests with the residents themselves, and the burden of meeting these demands falls on the surrounding community. Prospective retirement community residents, retirement community builders, and governmental units responsible for development decisions need to be cognizant of these potential problems. Nonetheless, new retirement communities catering to an active lifestyle are expected to play an increasingly important role in housing the growing elderly population in the United States. (SEE ALSO: *Continuing Care Retirement Communities;* **Elderly**)

—Robert W. Marans

Further Reading

Brecht, Susan B. 1991. *Retirement Housing Markets: Project Planning and Feasibility Analysis.* New York: John Wiley.

Golant, Stephen M. 1992. *Housing America's Elderly: Many Possibilities/Few Choices.* Newbury Park, CA: Sage.

Hunt, Michael E., Allan Feldt, Robert W. Marans, Leon Pastalan, and Kathlyn Vakalo. 1984. *Retirement Communities: An American Original.* New York: Haworth.

Hunt, Michael E. and Gail Gunter-Hunt. "Naturally Occurring Retirement Communities." *Journal of Housing for the Elderly* 3(3/4):3-21.

Marans, Robert E., Michael E. Hunt, and Kathlyn L. Vakalo. 1984. "Retirement Communities." In *Elderly People and the Environment,* edited by I. Altman, M. P. Lawton, and J. Wohlwill. New York: Plenum.

McKenzie, Brian. 1994. *Privatopia: Homeowner Associations and the Rise of Residential Private Government.* New Haven, CT: Yale University Press.

Rosow, Irving. 1987. *Social Integration of the Aged.* New York: Free Press.

Streib, Gordon, Anthony J. LaGreca, and William E. Folts. 1986. "Retirement Communities: People, Playing, Prospects." In *Housing An Aging Society,* edited by R. Newcomer, M. P. Lawton, and T. Byert. New York: Van Nostrand Reinhold.

Stroud, H. B. 1995. *The Promise of Paradise: Recreational and Retirement Communities in the United States since 1950.* Baltimore, MD: Johns Hopkins Press. ◄

► Reverse-Equity Mortgage

During the 1980s, the U.S. financial system was deregulated in many ways. One of the fundamental changes occurred in the freedom that financial institutions were given to develop new and innovative products. Many of these products were financial instruments for consumers. One of the more unusual instruments was the reverse-equity mortgage (or as it is sometimes known, the reverse-annuity mortgage [RAM]).

Like all financial instruments, reverse-equity mortgages are designed to solve a financial problem. In this case, elderly homeowners may have reduced income but growing consumption needs. At the same time, they have a large portion of their wealth as an equity interest in their paid-up home. A reverse-equity mortgage provides cash flow payments to the owner in exchange for a claim on the asset at the end of the mortgage. In this sense, the flows are reversed from the typical mortgage note.

It has been estimated that the elderly in the United States hold more than two-thirds of their assets in paid-up equity interests in the houses in which they reside. Reverse-equity mortgages enable such households to convert this wealth into cash without selling their residence or taking out a second mortgage. In this regard, this instrument solves the same type of financial problem for the elderly as a conventional mortgage does for the first-time buyer: The mortgage better matches payments and flows with consumption preferences and needs. It trades present and future consumption with the interest rate as the agreed-on price for this privilege. The equity portion of the homeowner declines with each payment received. This is because the homeowner has collateralized the equity and the value of the homeowners' interest is declining as payments are received.

The lender faces a number of risks when issuing a reverse-equity mortgage. There is the chance of prepayment, a risk inherent in all modern U.S. mortgages. If interest rates fall, the homeowner will have incentive to prepay the outstanding balance of the mortgage. There is also uncertainty about the life of the loan. Most reverse mortgages have specific maturities. However, to recover the lender's interest, the property must be sold, and real estate sales do not occur immediately. Finally, there is exposure to risk because, effectively, the date when the lender's interest is satisfied is unclear. Changes in the real property market can weaken the lender's collateralized position.

Borrowers also face some risks in that the maturity of the mortgage may not match well with their expected planning horizons. Should they live longer than the mortgage maturity date, homeowners will need to seek additional funding to pay off the claim.

It is widely expected that with an aging and increasingly wealthy population, reverse-equity mortgages will grow in acceptance and usefulness. Government agencies such as the Federal Housing Administration and the U.S. Department of Housing and Urban Development have become involved with this instrument. In the future, we should anticipate a growth in its use, especially as the general population becomes more familiar with alternative mortgage instruments and their usefulness in household finance. (SEE ALSO: **Mortgage Finance;** *National Center for Home Equity Conversion; Section 255*)

—*Austin J. Jaffe*

Further Reading

Bridewell, D. A. 1997. *Reverse Mortgages and Other Senior Housing Options.* Chicago: American Bar Association.

Clauretie, Terrence M. and James R. Webb. 1993. *The Theory and Practice of Real Estate Finance.* Fort Worth, TX: Dryden.

Fabozzi, Frank J. and Franco Modigliani. 1992. *Mortgage and Mortgage-Backed Securities Markets.* Boston: Harvard Business School Press.

Friedman, Joseph and Jane Sjogren. 1980. *The Assets of the Elderly as They Retire.* Cambridge, MA: Abt Associates.

May, Judith V. and Edward J. Szymanoski, Jr. 1989. "Reverse Mortgages: Risk Reduction Strategies." *Secondary Mortgage Markets* (Spring)6:8-10.

National Center for Home Equity Conversion. 1991. *Home Equity Conversion in the United States.* Washington, DC: American Association of Retired Persons.

Rasmussen, D., I. Megbolugbe, and B. Morgan. 1997. "The Reverse Mortgage as an Asset Management Tool." *Housing Policy Debate* 8(1):173-94.

Weintrobe, Maurice D. 1989. *Reverse Mortgages: Problems and Prospects for a Secondary Market and an Examination of Mortgage Guarantee Insurance.* Madison, WI: National Center for Home Equity Conversion.

► Right to Housing

The transformation by the international community during the past five decades of the concept of housing *needs* into housing *rights*—rights equal in stature and implication to other widely accepted human rights—has been a significant formula for more genuinely ensuring equal access for everyone to a safe and secure place to live. Every government has voluntarily accepted legal obligations to respect, protect, and fulfill housing rights. These rights are enshrined throughout dozens of international treaties, declarations, resolutions, and other agreements. Likewise, more than 50 national constitutions, representing all regions and all legal systems, contain housing rights provisions.

Judicial and quasi-judicial bodies at the international, regional, national, and local levels, although often reluctant to apply housing rights norms as vigorously as they may other rights, have nonetheless begun producing judicial decisions and other legal pronouncements giving housing rights greatly enhanced clarity and precision. This is particularly true at the international level, where United Nations and other human rights bodies have begun applying housing rights standards more vigorously than ever before.

Likewise, governmental housing rights obligations under international law have, in fact, reached levels of exactitude easily sufficient to determine whether or not States have acted in full conformity with these important legal duties toward their citizenry. At the same time, the inherent individual, family, and group entitlements to the full enjoyment of housing rights are by now largely accepted components of human rights law.

As advantageous as these developments may appear on the surface, however, the legal recognition and subsequent normative accuracy accorded the obligations and guarantees inherent within housing rights remain comparatively meager steps, as exemplified so clearly by the ease with which many governments have simply pretended housing rights do not even exist. When questioned by U.N. bodies about their record on housing rights, no government will deny they recognize such rights, but few governments can be justifiably proud of the domestic stances they have taken on these basic standards.

Human rights standards in their timeless, universal, and fundamental nature establish clear minimum standards as to how governments should treat their citizens and respect their inherent dignity as human beings. Human rights law similarly establishes (or should establish) a series of enforcement mechanisms designed to ensure vindication when these rights are infringed. In many respects, however, national courts have been far more hesitant to apply housing rights standards than their international equivalents. Enforcing housing rights—judicially or otherwise—remains a difficult undertaking, although such obstacles are more clearly based on perception than on the legalistic capabilities of housing rights to enforcement. Yet the need for much-enhanced measures of enforcement are crucial, particularly in view of the massive violations to which this human right is routinely subjected.

Ongoing Violations of Housing Rights

Although most governments perhaps fear the diplomatic ostracism that would surely emerge were they to publicly oppose the principle of housing rights, few have taken the required steps at the domestic level to ensure the widespread enjoyment of this right. This is readily visible in any large city or impoverished rural area in nearly every country. Instead, many States have consciously undertaken precisely the opposite route of action, leading in many instances to

nothing less than gross and systematic violations of internationally recognized housing rights and, consequently, of human rights.

Although rarely invoked with the passion associated with infringements of more commonly assumed notions of human rights, distinct and identifiable violations of housing rights are extremely widespread. Of all such violations, the most blatant and total violation of housing rights is the still common practice of eviction.

Although international law has repeatedly, and in no uncertain terms, reaffirmed that "forced evictions constitute a gross violation of human rights, in particular the right to adequate housing," this act of governmental complicity continues to destroy the homes and communities of millions of dwellers year after year. According to the Centre on Housing Rights and Evictions, as of June 1994 more than 5.6 million persons were threatened with pending forced evictions in some 38 countries.

International legal bodies such as the U.N. Committee on Economic, Social and Cultural Rights have sought to strenuously enforce the right not to be evicted. Since 1991, both this committee and the U.N. Commission on Human Rights have held the governments of the Dominican Republic, Panama, Nicaragua, Italy, the Philippines, the Sudan, Zaire, Israel, and others liable for such violations under international law. In several such cases, these decisions have had a major bearing on decreasing the practice of eviction within these countries. Many more countries have been told by the United Nations to confer security of tenure on all dwellers lacking legal protection against eviction.

Forced evictions are perhaps the most prominent breach of housing rights obligations, are readily identifiable, and are of such a nature that governmental accountability is relatively easy to establish. Acts of forced eviction, both past and pending, tend additionally to receive extensive judicial attention in many countries, and few courts would refuse to consider appeals based on this practice, whether perpetuated by the private or public sectors.

Stemming from the stance taken on forced evictions, it might consequently appear that a significantly large and growing homeless population within a given country or the existence of sprawling slums are evidence of additional housing rights violations. It is precisely situations such as these and others in which some of the difficulties in housing rights lie.

International human rights law does not oblige States to provide a home to anyone on demand. However, this legal regime does require countries to use the "maximum of their available resources" toward the "progressive realization" of housing rights. These qualifications, although in many respects necessary and reasonable, have often led to restrictive interpretations by States of housing rights provisions and to States' taking these rights far less seriously than the law actually dictates.

Nonetheless, the U.N. Committee on Economic, Social and Cultural Rights responsible for monitoring State compliance with the International Covenant on Economic, Social and Cultural Rights, has declared that "a general decline in living and housing conditions, directly attributable to policy and legislative decisions by States, and in the absence of accompanying compensatory measures, would be inconsistent with the obligations found in the Covenant," and thus a violation of international law.

What is often forgotten in the housing rights debate are the numerous negative duties of States vis-à-vis the housing sector, which effectively require neither resource allocation nor time to achieve these progressively. In effect, these duties of restraint to (a) allow self-determined housing processes to take place, (b) allow the free expression of opinion concerning housing, (c) allow the free functioning of housing advocacy groups, (d) ensure that people and communities are not arbitrarily evicted from their homes and land, and (e) prevent acts of racial or other forms of harmful discrimination from occurring within the housing domain are each attainable by all governments, notwithstanding levels of development or societal wealth. Even agreements not of a human rights nature, such as Agenda 21 adopted during the UN Conference on Environment and Development (UNCED) in 1992 and the U.N.-sponsored Global Strategy for Shelter to the Year 2000 recognize this point.

Similarly, the provision of security of tenure for everyone, coupled with the conscious removal by governments of legislative, regulatory, and other constraints to housing can be forceful steps toward ensuring substantially greater access to adequate housing for all sectors of society. Such measures, to be effective in strengthening housing rights and the prospects of this right achieving practical effect, must be combined with promotional, policy, and legislative steps consciously designed toward taking housing rights as seriously, for instance, as property rights.

Enforcing Housing Rights

The enforcement of housing rights or related legal provisions, such as nondiscrimination or equality of treatment clauses, remain some of the most difficult stumbling blocks preventing these rights from achieving full fruition. Of course, many housing rights attributes are capable of judicial scrutiny and daily are the basis of court decisions. Yet not only are the judicial path and concomitant legal remedies effectively the final resort of housing struggles, but still most governments would argue that housing rights are not even justiciable.

The United Nations has stated, for instance, that many component elements of the right to adequate housing are consistent with the provision of domestic legal remedies and include, but are not limited to, the following:

▶ Legal appeals aimed at preventing planned evictions or demolitions through the issuance of court-ordered injunctions
▶ Legal procedures seeking compensation following an illegal eviction
▶ Complaints against illegal actions carried out or supported by landlords (whether public or private) in relation to rent levels, dwelling maintenance, and racial or other forms of discrimination
▶ Allegations of any form of discrimination in the allocation and availability of access to housing
▶ Complaints against landlords concerning unhealthy or inadequate housing conditions.

In some legal systems, it would also be appropriate to explore the possibility of facilitating class action suits in situations involving significantly increased levels of homelessness.

Obviously, this list is nonexhaustive, but it does make amply clear that courts can entertain housing rights complaints without any legal difficulties, if they are so willing. Indeed, efforts by nongovernmental organizations and lawyers toward enforcing housing rights are expected to receive an added boost through the International Housing Rights Lawyers Alliance. One of the central ambitions of the alliance will be to systematically pursue all available international, regional, and national legal avenues with a view to building what it labels as an "impermeable wall of globally applicable housing rights jurisprudence" whereby the rights of dwellers receive more and more legal substance.

Recent Headway on Housing Rights

Despite the persistence of severe political and other constraints, it is indisputable that housing rights have gained substantially in stature and attention during the past decade. Unthinkable in the past, for instance, the right to adequate housing was a key issue at the 1996 Habitat II Conference in Istanbul, and global debate from the grassroots to the governmental level is increasing as to the feasibility of the possible adoption of an International Convention on Housing Rights.

The United Nations Sub-Commission on Human Rights now has a special rapporteur on housing rights, and governments are regularly chastised in legal fora for the practice of forced evictions. Likewise, the contents, obligations, and entitlements of the right to housing are now generally accepted under international law, and grassroots housing rights movements and campaigns are emerging throughout the world. Each of these laudable achievements can by no means be discarded as meaningless, albeit well-intentioned, bits of progress in a tremendously complicated social arena.

Not surprisingly, attention to and reliance on the protection afforded by housing rights has blossomed in the 1990s within the independent sector, civil society, and nongovernmental and community-based organizations. Without this renewed grassroots emphasis on the promise of housing rights, in fact, few if any of the recent strides on the housing rights front would have taken place.

Until the present, no government has shown a willingness or overt commitment toward pushing housing rights anywhere beyond the vague rhetoric of a few words in relatively obscure treaties. The entire housing rights issue remains frequently marginalized, if not ignored by most governments, international agencies, judges, housing researchers, and others who might be expected to be natural allies in the perpetual struggle for this right.

Indeed, activities surrounding the right to adequate housing, despite major advances, are beginning to encounter the proverbial wall of resistance by the very powers who have gladly accepted this right in principle yet actively prevent this right from attaining the legal and practical stature it deserves.

Conclusions

Although much has been achieved on housing rights since the early 1980s, the structurally transformative nature of these rights have only rarely taken on the required scope. Housing rights continue to face governmental resistance generally associated with official views on economic, social, and cultural rights.

Part of the reason that governments have been so averse to housing rights stems from the simplistic and ill-informed perceptions of what housing rights actually imply in terms of law, policy, and practice. The right to housing has been tragically and very wrongly reduced in the minds of many politicians to duty to provide an adequate house to all who demand it directly from the State.

Adequate, affordable, and accessible housing is now widely recognized as one of the most fundamental attributes to living a life in which the inviolable rights of every individual to a place to live in peace, dignity, and security can be attained and sustained. Such indisputable truths may be accepted at least rhetorically on all sides of the political spectrum, and most governments continue to devote substantial public resources to the housing sector. However, few governments have comprehensively enacted the requisite legislative, economic, and social measures required to attain this right for all throughout society.

Although practice has shown the inherent imperfections of exclusively market-driven or State-controlled approaches toward ensuring everyone a decent home, it is clear that the former has become in many respects the ideology of the present. As the unbridled free-market policies advocated by the World Bank, the International Monetary Fund and General Agreement on Tariffs and Trade continue to trickle down into the domestic legislative and policy domains of nearly all States (both developed and developing), the very notion of housing rights has been placed under severe threat.

Despite the universal failures of market-based solutions to housing crises anywhere in providing housing for everyone, especially the poor, the new "enabling" approach seems in many respects to run the risk of being destined to repeat history's many eras of housing deprivation. Even in the richer nations that during the past half century have devoted huge sums to the housing sector, literally millions of homeless persons now roam the streets unable to pay for and without access to the hugely insufficient number of affordable housing units. Were housing rights accorded the stature in practice that they are given under law, perhaps more reasonable, democratic, human-oriented, and practical solutions could be found to humanity's growing underclass. (SEE ALSO: *Eviction; Habitat International Coalition; History of Housing; Housing Courts; Property Rights*)

—*Scott Leckie*

Further Reading

Centre on Housing Rights and Evictions. 1993a. *Bibliography on Housing Rights & Evictions*. Utrecht, The Netherlands: Author.

———. 1993. *Forced Evictions and Human Rights: A Manual for Action*. Utrecht, The Netherlands: Author.

General Comment No. 4 on the Right to Adequate Housing. 1991. Article 11(1) of the Covenant on Economic, Social and Cultural Rights, adopted by the U.N. Committee on Economic, Social and Cultural Rights (U.N. doc. E/1992/23, pp. 114-120).

Leckie, Scott. 1992. *Housing as a Need, Housing as a Right: International Human Rights Law and the Right to Adequate Housing*. London: International Institute for Environment and Development, Human Settlements Programme.

Sachar, Justice Rajindar. 1992. *Working Paper on Promoting the Realization of the Right to Adequate Housing*. Submitted to the 44th Session of the U.N. Sub-Commission on Prevention of Discrimination and Protection of Minorities (U.N. doc. E/CN.4/Sub.2/1992/15).

Steinhagen, Renee and Andrew Scherer. 1990. "United States Violations of the Right to Shelter under International Law." *IADL Journal of Contemporary Law*(1):69-86. ◄

► Rollover Mortgage

One of the oldest examples of a mortgage other than the fixed-rate mortgage (FRM) is called the Canadian rollover (or, simply, the rollover mortgage). As traditionally employed in most mortgage markets in Canada, the rollover mortgage is like an FRM, except that after a short period (in Canada, 5 years), the interest rate to be applied to the outstanding balance of the loan is "rolled over" to the current interest rate to be applied to long-term loans. In effect, the rollover mortgage fixes the mortgage interest rate for a short period and then adjusts the interest rate on the remaining balance.

In some ways, one could view a rollover mortgage as a hybrid between an FRM and an adjustable-rate mortgage (ARM). The rollover mortgage is fixed for 5, 7, or 10 years; subsequently, the interest rate adjusts according to current inflationary expectations, as measured by an agreed-on index. In virtually all cases, the term of the original loan is not extended.

As a result of the structure of the mortgage, the debt service payment schedule resembles a series of multiyear steps. If interest rates increase over the life of the loan, the payments will be a series of upward steps. If interest rates continue to decline over the life of the mortgage, the pattern will be a series of downward steps.

The rollover mortgage has had limited success in the United States. However, hybrid mortgages that resemble the Canadian rollover mortgage are becoming increasingly popular in the United States. For example, a new and popular hybrid in the United States fixes the interest rate for seven years. Subsequently, the mortgage balance converts to an ARM. With refinancing as an option, a borrower can largely replicate the effects of a rollover mortgage.

In the future, it is likely that there will be an expanded menu of FRM/ARM hybrids. As markets continue to develop, it should be possible for borrowers to find the type of hybrid that closely matches their needs. In such a market, the pricing of options to repay at fixed mortgage rates will be very competitive. (SEE ALSO: **Mortgage Finance**)

—*Austin J. Jaffe*

Further Reading

Fabozzi, Frank J. and Franco Modigliani. 1992. *Mortgage and Mortgage-Backed Securities Markets*. Boston: Harvard Business School Press.

Jacobus, Charles J. 1996. *Real Estate Principles*. Upper Saddle River, NJ: Prentice Hall.

Shim, Jae K. 1996. *Dictionary of Real Estate Terms*. New York: John Wiley.

Tosh, Dennis S. 1990. *Handbook of Real Estate Terms*. Englewood Cliffs, NJ: Prentice Hall. ◀

▶ ROOF

ROOF Magazine contains summaries of important research, analysis of new policies, essential facts and figures, legal updates, and reviews, both in Great Britain and the rest of Europe. Published by Shelter, articles focus on the interests of housing practioners and policymakers, housing authorities, and housing associations. Price: £48 for organizations and institutions; £26 for individuals; non-UK subscriptions, £64. Editor: Julian Blake, 88 Old Street, London EC1V 9HU. Phone: 0171-505-2161. Fax: 0171-505-2167. Email: ROOF@compuserve.com.

—*Caroline Nagel* ◀

▶ Rural Housing

A clear, simple definition of what constitutes rural housing does not exist. Based on the U.S. Bureau of the Census definition, rural housing is located in a geographic area not classified as urban. Urban housing comprises all housing units in urbanized areas and in places of 2,500 or more inhabitants outside urbanized areas. An urbanized area comprises an incorporated place and adjacent densely settled (1.6 or more people per acre) surrounding area that together have a minimum population of 50,000. A metropolitan area must include a city of 50,000 population or an urbanized land area of at least 50,000 population with a total metropolitan population of at least 100,000. To be classified as a farm unit, occupied housing units must report sales of agricultural products of at least $1,000 during the prior 12-month period.

Rural areas have some unique characteristics that require consideration in the development of programs and policies to address housing needs. Rural areas are remote from urban centers, have a low population density, and usually depend on one particular industry. This industry is often based on the use of natural resources, such as agriculture or mining, or on low-wage, low-skill manufacturing. Although a majority of people in the United States are well housed, housing for many rural residents is inadequate or imposes severe financial burdens. Furthermore, rural United States is diverse in its needs and problems. Concerns of groups such as the homeless, Native Americans, and migrant farmworkers require special attention.

In 1995, approximately 28.6% of the U.S. population lived in rural areas (73 million persons). Of these, about 5.6% were farm households. That same year, 26.8% of rural housing units were outside urbanized areas. Using a broader definition of *rural,* 38% of all occupied housing units are in rural areas. Close to half of those 37 million rural units are located in the rural portions of metropolitan areas.

The rural population as a percentage of the total U.S. population has been declining since the first census was taken in 1790. In 1790, 95% of the population lived in rural areas, compared with 26% in 1991. The urban population first exceeded the rural population in 1920, although more than 80% of all land is still designated as nonmetropolitan. Selected rural areas continue to experience high growth rates. These are primarily areas of scenic beauty, such as the Rocky Mountains, the Ozarks, the Blue Ridge Mountains, Upper Michigan, and the Appalachian Mountains.

Social and Economic Characteristics

Demographic and income trends help shape housing markets, problems, and policies. A 1995 comparison of urban and rural households shows that urban households are younger. The percentage of the population 65 years or older comprises 27.6% of the population in nonmetropolitan areas, compared with 16% in metropolitan areas. Rural householders are not as well educated as urban householders. Approximately 22.3% of rural householders did not complete high school, compared with 18% of urban householders. Rural households are larger than urban households: 2.4 persons compared with 2.2. Rural households are more likely than urban households to have single children under 18 years of age.

They also have lower incomes. In 1995, the median family income of rural owners was $35,600, compared with $42,400 for urban owners. Rural renters had a median family income of $21,700, similar to that of urban renters. About 13% of rural nonmetropolitan owners are low-income households, compared with 9% of rural metropolitan owners. Nonmetropolitan per capita income has been consistently lower than metropolitan per capita income, and the gap is increasing. Unemployment is also higher in nonmetropolitan areas.

Housing Characteristics

According to the 1995 American Housing Survey, about 81% of all rural households owned homes, compared with only 59% of urban households. About 88% of rural farm households owned homes. Thus, ownership policies are comparatively more important to rural residents.

The predominant housing structural type in rural areas is single family detached. About 17% of rural housing units are mobile homes, compared with 2% of urban housing units. However, more urban owners lived in single-family homes than did rural owners—89% versus 83%. Thus, rural renters are more likely than urban renters to live in single-family units—60% versus 29%.

Rural housing is newer than urban housing. The median age of urban housing was 30 years, compared with 21 years for rural housing. Rural owners occupied newer units than did urban owners. However, rural renters occupied older units than urban did renters.

In 1995, the median value of rural homes was $80,318—81.5% of the median value of urban homes ($98,503). The ratio of home value to income is similar: In rural areas, it is 2.1; in urban areas, it is 2.4. Rural owners with mortgage payments had a median monthly housing cost of $459, compared with $623 for urban owners. A similar situation exists for rural renters. The median rent for rural households was $450, compared with $534 for urban renters.

Quality of Housing

Physical conditions of housing have been worse in rural than in urban areas. Agricultural laborers, especially migrant farmworkers, live in some of the worst housing conditions in the United States. The rural-urban differences in quality have been greatly reduced, although 2% of households were still living in physically inadequate housing in 1996.

Rural households are more likely than urban owners to have severe plumbing problems as well as moderate heating and upkeep problems. Rural renters are also more likely than urban renters to have severe plumbing problems and moderate heating problems. Irrespective of location, renters are more likely than owners to have these physical housing problems.

Rural households are less likely to have access to public sewers than urban households and more likely to use septic tanks, cesspools, or chemical toilets than urban households. Whereas 94% of urban households use public sewers, only 32% of rural households do so. A public system or private company provides water to 98% of urban households, compared with only 58% or rural households. Wells are the primary source of water for 40% of rural households. Despite the problems of adequacy, rural households rated their neighborhoods higher than urban households. Forty-two percent ranked it as best, compared with 29% of urban households.

Homelessness

Homelessness is substantial in nonmetropolitan areas, even when only those people seeking shelter and services are counted. It has been estimated that about 12.3% of the national homeless population is located in nonmetropolitan areas. The rural homeless are more difficult to count than the urban homeless because they are hidden in places such as vacation campsites and shacks. Native Americans and migrant farmworkers represent a significant portion of the homeless population in nonmetropolitan areas. Family conflict, especially domestic violence, is a major factor in homelessness in rural areas. However, characteristics of the homeless population vary greatly from location to location.

Financing

The provision of housing depends on the availability and cost of capital. There is a long history of special concern about rural financial markets. Research has indicated that residential financing is less available and more costly in rural areas than in urban areas, which may contribute to the major role that manufactured housing plays in rural areas. Most manufactured homes are placed through dealers who offer financing.

Special programs have been created to finance rural housing, owing to the concern of Congress and others that rural households face a shortage of private sector loan funds. One of the major programs influencing the quality and availability of housing in rural areas is Rural Housing and Community Development Service (RHCDS), formerly the Farmers Home Administration (FmHA). RHCDS home loan programs are directed to very low income, low-income, and moderate-income borrowers. Families can obtain financing for modest homes in nonmetropolitan areas. RHCDS also provides funding for some rural rental housing. Another source of funding for residential mortgages is the Farm Credit System, which makes loans for farms and modestly priced rural housing.

Other federal financing assistance includes Federal Housing Administration (FHA) and Department of Veterans Affairs (VA) mortgages. Of all owned mortgaged housing units in nonmetropolitan areas, 2.3% received RHCDS assistance, 7.2% received FHA assistance, and 4.1% received VA assistance. This compares with metropolitan areas in which .6% received RHCDS assistance, 14.0% received FHA assistance and 6.5% received VA assistance. Thus, the federal programs for rural areas played only a small role in residential financing of single-family homes. However, government-sponsored enterprises, such as Fannie Mae (formerly, the Federal National Mortgage Association) and the Federal Home Loan Mortgage Corporation, may have more prominent roles in the future.

Affordability

A comparison of monthly housing costs as a percentage of current income indicates that urban housing residents have slightly higher housing cost burdens than do rural households. In 1995, the median urban owner with a mortgage spent 23% of income on housing, compared with 19% for the median rural owner. For owners without mortgage payments, the percentages are identical (13%). Rural renters pay about 29% of their income on housing.

Characteristics of New Housing

New housing units account for approximately 10% of the housing stock each year. Although rural housing is newer than urban housing, 82.3% of all new units in 1996 were built in metropolitan areas. However, for the last five years, the percentage of new units in rural areas has been increasing. In 1992, 78% of all new units were in metropolitan areas. The average and median sales prices of new homes are higher in metropolitan areas than in nonmetropolitan areas. In 1996, the average new home price in nonmetropolitan areas was $129,400 and the median new home price was $119,000, whereas the average new home price in metropolitan areas was $170,600 and the median price was $143,700. New homes located outside metropolitan areas are smaller, with an average square footage of 1,915 (compared with 2,170), are less likely to have central air-conditioning, and have fewer bathrooms. The average price per square foot was $60.25 in nonmetropolitan areas, compared with $62.95 in metropolitan areas.

Financing methods for new housing varies. Nonmetropolitan housing was purchased with cash 24% of the time, compared with 8% for metropolitan purchases. Conventional financing was used for 70% of nonmetropolitan purchases and for 75% of metropolitan purchases. Government-sponsored financing (FHA, VA, RHCDS) was used for 14% of metropolitan housing mortgages compared with 5% in nonmetropolitan areas. (SEE ALSO: *Council for Affordable and Rural Housing; Farmers Home Administration; Farmworker Housing; Housing Assistance Council; National Rural Housing Coalition; Section 502; Section 515*)

—*Carol B. Meeks*

Further Reading

Dacquel, Laami T. and Donald C. Dahmann. 1993. "Residents of Farms and Rural Areas: 1991." *Current Population Reports*. Bureau of the Census, P20-472. Washington, DC: Government Printing Office.

Fannie Mae Office of Housing Research. 1995. *A Home in the Country: The Housing Challenges Facing Rural America*. Washington, DC: Author.

Housing Assistance Council. 1991. *Rural Homelessness: A Review of the Literature*. Washington, DC: Author.

Meeks, Carol B. 1988. *Rural Housing: Status and Issues*. HP #19. Cambridge: MIT Center for Real Estate Development.

Tin, Jan S. 1993. "Housing Characteristics of Rural Households: 1991." *Current Housing Reports*. Bureau of the Census, Series H121/93-5. Washington, DC: Government Printing Office.

U.S. Bureau of the Census, 1996. *Current Construction Reports*. Series C25. Washington, DC: Government Printing Office.

S

▶ Sage Urban Studies Abstracts

Sage Urban Studies Abstracts is a quarterly publication of cross-indexed abstracts covering the spectrum of urban studies, including urban history, architecture, planning, housing, urban real estate, economic and land use issues, public services, local government administration, politics, law, and comparative urban analysis. Books, articles, pamphlets, government publications, significant speeches, legislative research studies, theses and dissertations, as well as other fugitive material are included. The audience is primarily academic libraries, urban planners, and those concerned with urban studies. Circulation: approximately 500 internationally. First published in 1973. Price: Institutions $375 per year, Individuals $114 per year. Submissions of materials to be considered for inclusion are welcome, including extended summaries of dissertations or theses, but cannot be returned. Documentation Editor: Paul McDowell. Address: 2455 Teller Rd., Thousand Oaks, CA 91320. Phone: (805) 499-0721, ext. 7145. Fax: (805) 499-0871.

—*Laurel Phoenix* ◀

▶ Sale-Leaseback

Given the emphasis placed on property ownership in many markets, a strategy that combines leasing and owning appears to be complicated. One such strategy is a well-known financing method called sale-leaseback, which is widely used in many commercial and industrial real estate markets.

A sale-leaseback arrangement transfers a title to a piece of real property and at the same time, leases the same property back from the new owner. Transference of title means legal ownership of the rights to the asset has been exchanged. The creation of the new lease means that the new owner conveys back some of the rights to the former owner via the terms of the lease. Also, frequently, there is an option exchanged to the former owner to repurchase the asset if the former owner wishes to do so at some time in the future, at the end of the lease.

For commercial transactions, the cash flows associated with ownership or leasing of the property may be quite similar. For example, the owner is given use of the property and can deduct interest payments as tax expenses. Similarly, the tenant (lessor) is given use of the property for the duration of the lease, and the lease payments are tax deductible as business expenses for the duration of the lease. Because the cash flow payments are very similar, the sale-leaseback strategy can be seen as a financing method rather than as a decision driven by tenure choice.

There are of course, differences between tenures. The risk associated with ownership is greater than the financial risk associated with leasing. There is a chance that the asset may lose its usefulness as time goes on, and because leasing will not require ownership of the entire bundle in perpetuity, this choice would be preferred in such a case. On the other hand, tenants can lose their lease, even after long periods of tenancies, if economic conditions for the site change. Ownership can ensure against such surprises.

Most sale-leaseback arrangements are structured to reduce taxes or to alter the balance sheet positions of firms with too much or too little real property. An entire section of accounting practice is devoted to evaluating the legality of some sale-leaseback arrangements if driven entirely by the purpose of avoiding taxes. Sale-leasebacks are financing methods for gaining control of real property without transferring title yet aiding the cash flow of the investor. In this sense, owning property is not essential for a real estate investor. (SEE ALSO: **Mortgage Finance**)

—*Austin J. Jaffe*

Further Reading

Brueggeman, William B. and Jeffrey D. Fisher. 1993. *Real Estate Finance and Investments.* 9th ed. Homewood, IL: Irwin.

Clauretie, Terrence M. and James R. Webb. 1993. *The Theory and Practice of Real Estate Finance.* Fort Worth, TX: Dryden.

Jaffe, Austin J. and C. F. Sirmans. 1995. *Fundamentals of Real Estate Investment.* 3d ed. Englewood Cliffs, NJ: Prentice Hall.

Sirmans, C. F. 1989. *Real Estate Finance.* 2d ed. New York: McGraw-Hill. ◀

▶ Savings and Loan Industry

During the past 10 years, the federal government's bailout of the savings and loan (S&L) industry cost U.S. taxpayers $160 billion. In 1997, there were just 1,600 S&Ls—compared with 12,000 back in the 1920s. Their chief mission is to take deposits and make home mortgage loans to individual housing consumers. In the mid-1980s, S&Ls (or "thrifts" as they are also called) had combined assets of $1.5 trillion dollars. In 1991, S&Ls earned $1.2 billion. In 1992 and 1993, S&Ls earned a whopping $10.1 billion. At midyear 1994, the S&L industry posted combined earnings of $2 billion. However, the thrift industry then had combined assets of just $775 billion—a decrease of almost 50%.

By almost all measures, the S&L industry is a business in decline. Although in 1995 these institutions made large profits, many were being bought up by commercial banks or were choosing to liquidate themselves by converting to independent mortgage banking firms. By the end of the 20th century, it is very possible that only 100 or fewer traditional S&Ls will remain.

History

Until the early 1980s, the typical S&L had only one single task: It took deposits from individuals in the community and lent out those funds in the form of home mortgages. Unlike a commercial bank, an S&L did not make large business loans, did not lend on large land projects, and did not speculate in the equities market. By law, S&Ls were not allowed to do these things. And for 50 years, the S&L industry thrived. Then inflation struck.

In the 1960s, Congress, worried about the increasing cost of homes, placed a cap on the amount of interest that S&Ls could pay to their borrowers. The reasoning was that if S&Ls did not have to pay much for deposits, they would not charge homebuyers much for mortgages. But this financial logic proved to be fallacious when the annual inflation rate soared to 14% in the late 1970s and continued unchecked into the next decade.

As S&Ls were limited to paying just 5.5% for deposits, customers pulled their savings out of thrifts in search of higher yields elsewhere. This posed a major problem for thrifts, because they had many fewer available deposits to lend out in the form of mortgages.

In 1980, Congress acted again to help the industry deal with the devastating effects of inflation. It passed the Depository Institutions Deregulation and Monetary Control Act, which removed the interest rate ceiling on deposits and allowed thrifts to offer money market accounts. Now, for the first time, thrifts could offer market rates on their deposits. But this too posed a problem. Thrifts were designed to borrow money from depositors at 3% and lend it out at 6%. This allowed for a 3% profit as the mortgages they made were kept on their books. In other words, thrifts held onto these assets. With the new money market accounts, thrifts now were paying 10% for deposits and, for example, were making home mortgages, charging 13%. Unfortunately, these thrifts still had old mortgages on their books that were yielding 6% or even less. (The contract life of most mortgages is 30 years, although the average life is about seven years.)

Because thrifts were losing much money by paying depositors 10% while earning just 6% on their old mortgages, the S&L industry asked Congress for help. S&Ls generously donated money to elected politicians that served on committees that oversaw their industry, so Congress was very receptive. The result was the most significant financial legislation in 50 years: The Garn-St Germain Depository Institutions Act of 1982.

In essence, the Garn-St. Germain Act deregulated the S&L industry. It freed S&Ls from the constraint of making only home loans. It allowed S&Ls to engage in other types of lending—commercial, land development loans, and even large business loans. (It also allowed federally chartered S&Ls to make adjustable-rate mortgages for the first time, something only state-chartered institutions had been allowed to do.) Thrifts also were allowed to invest in securities, especially junk bonds, something that even banks were not allowed to do. The Reagan administration, Congress, and many financial experts on Wall Street believed these new lending powers would allow S&Ls to enter new lines of business. Conversely, these new powers would translate into new lending products that would make up for the large losses being suffered by S&Ls during this time.

Worsening of the Problems

After Congress acted, several individual states passed parity laws that deregulated S&Ls even further. In California, for example, the Nolan Act was signed into law. The bill allowed state-chartered California S&Ls to invest all of their deposits in almost any business. This free reign soon resulted in catastrophe because deposits at S&Ls were (and still are) insured by the federal government. If a savings and loan failed, the federal government (the nation's taxpayers) would pick up the loss. (By law, the federal government insures deposits up to $100,000 per account.)

Between 1982 and 1987, hundreds of S&L operators lent money on large land projects, without giving sufficient thought to the financial soundness of these real estate developments. The industry also attracted outside entrepreneurs who sometimes viewed S&Ls as personal piggy banks. Almost overnight, thrifts were transferred into money-lending conglomerates that lent billions on projects such as hotels, casinos, shopping malls, mushroom and windmill farms, fast-food restaurants, commodities and brokerage firms, and junk bonds. There were no restraints, because it could all be done using federally insured deposits.

Many of the new "entrepreneurs," attracted to the S&L industry in the early 1980s, turned out to be con men and hustlers—swindlers intent on draining as much money from the system as they could before moving on—men such as Charles Keating of Arizona, Charles Bazarian of Oklahoma, and mafia figure Michael Rapp of New York.

As Congress and the Reagan administration deregulated the S&L industry, they also cut the number of examiners overseeing these institutions. The examiners who remained during the "deregulatory" atmosphere of the 1980s were underpaid (entry-level examiners made just $14,000 in

1984) but also had little experience analyzing the new and exotic investments thrifts were getting into.

It was not long before S&L failures rippled across the nation. Not only were swindlers draining S&Ls of money, but old line S&L executives who were basically honest were entering into business transactions they did not fully understand. To make matters worse, thrifts were so plush with new deposits during this time they were flooding commercial developers with money, which caused certain markets—Texas, Florida, Oklahoma, and Southern California—to become overbuilt. With overbuilding came a recession in commercial real estate values that caused these loans to go into default. When the loans went bad it was discovered that the value of the underlying properties had been inflated and that sometimes "kickbacks" had been paid to loan officers and even management. Other times, S&L management had actually sold land to the S&L at inflated prices and cashed out at a huge profit.

Rescue Efforts

By 1987, hundreds of savings and loan institutions were insolvent, having made billions of dollars worth of bad loans. By this time, the Federal Bureau of Investigation was beginning to take notice and began a probe of several large S&L failures. The criminal investigations started in Dallas and Houston and spread to other major metropolitan areas, including Miami, Oklahoma, Los Angeles, Phoenix, San Diego, and eventually, Boston. It also was revealed that many elected officials had received large campaign donations from corrupt S&L operators in return for passing laws that benefited these owners, including new rules that allowed them to stay in business for several more years. The most celebrated case of S&L political corruption was the "Keating Five" affair in which Arizona businessman Charles Keating Jr. donated more than $1 million to Senators Alan Cranston (California), Dennis DeConcini (Arizona), John Glenn (Ohio), John McCain (Arizona), and Don Riegle (Michigan) in an attempt to block federal S&L regulators from closing down his S&L, Lincoln Savings. After it was discovered that Keating was stealing money from his own S&L, he was convicted and sent to prison. By 1989, with land values declining rapidly and with many crooked S&L deals being unwound by regulators and the FBI, the nation's taxpayers were called on to bail out (pay off) depositors of hundreds of failed S&Ls. In August 1989, newly elected President George Bush signed into a law the Financial Institutions Reform, Recovery and Enforcement Act (FIRREA). Among other things, the bill provided $50 billion in taxpayer money to restore the nation's thrift industry. (In time, the $50 billion would prove to be just a down payment and at least another $90 billion would be needed; previously, Congress had already voted to give the industry a $20 billion shot in the arm.)

Even more important, FIRREA reregulated the S&L industry, ending all of the liberal lending laws passed during the Reagan administration. The bill also created a new government agency, the Resolution Trust Corporation whose sole mission was to liquidate all insolvent S&Ls. That task was completed until 1995.

Current and Prospective Developments

In the early 1980s, many financial experts on Wall Street argued that S&Ls could not make a profit by issuing just home mortgages and that they needed to make other types of loans to subsidize their mortgage business. However, in time, it would be proven that the S&Ls that survived the carnage of the 1980s were those institutions that stuck to home mortgage lending. The interest rate mismatch whereby S&Ls had low-yielding loans and high-rate deposits was a major problem. But as inflation waned and S&Ls began to make more adjustable-rate loans (which allowed them to increase the yield on their loans as inflation grew), the industry began to see the light of day again. More important, the S&L industry—and all mortgage lenders, for that matter—was helped greatly by the creation of a secondary market for mortgages. The secondary market involves the resale of loans to major institutional buyers, such as pension funds, insurance companies, and other large investors. Often, these loans are packaged into securities and sold through Wall Street. By selling the loans right away—which many S&Ls do now—the risk of higher interest rates is shifted to the buyer. The S&L removes the loan from its balance sheet and receives cash. The money can then be used for new loans or can be invested in safer, government Treasury bonds until the institution decides what to do with the cash.

The huge taxpayer bailout has hurt the image of the S&L industry. There were only 1,600 S&Ls left in 1997, and many consumers are reluctant to deal with S&Ls. Ironically, with interest rates falling in the early 1990s, S&Ls became quite profitable by sticking to home mortgage lending. However, by this time, the damage was done. Many S&Ls themselves no longer liked the name "S&L" and began changing it to "savings bank," even though, legally, they were not banks. In the 1990s, with the boom in mortgage lending, S&Ls began to do well again but faced strong competition from many independent mortgage banking firms. These independent companies do not take deposits and instead borrow their mortgage money from Wall Street and large commercial banks. With the image of the S&L industry tarnished, these "nonbank" competitors began to steal large market shares away from S&Ls. In 1993 and 1994, mortgage banking firms accounted for 50% of all home mortgages made in the United States. With their market share falling, many S&Ls began selling their franchises to commercial banks. This trend of selling out likely will continue unabated through the rest of the decade, and by the end of the century, the only S&Ls left will be a handful of large multibillion dollar thrifts. It is anticipated that by the year 2000 the S&L industry, as it was once known, could be extinct, although its chief product, the mortgage loan, will continue to live on through other types of lenders. (SEE ALSO: *Lending Institutions; Resolution Trust Corporation*)

—*Paul Muolo*

Further Reading

Bird, Anat. 1992. *Can S&Ls Survive? The Emerging Recovery, Restructuring & Repositioning of America's Savings & Loans*. Burr Ridge, IL: Irwin.

Brumbaugh, Robert D. Jr. 1992. *The Collapse of Federally Insured Depositories: The Savings & Loans as Precursor.* New York: Garland.

Calavita, K., R. Tillman, and H. N. Pontell. 1997. "The Savings and Loan Debacle, Financial Crime, and the State." *Annual Review of Sociology* 23:19-38.

Davis, L. J. 1990. "Chronicle of a Debacle Foretold: How Deregulation Begat the S&L Scandal." *Harper's,* September.

Ewalt, Josephine Hedges. 1962. *A Business Reborn: The Savings and Loan Story, 1930-1960.* Chicago: American Savings and Loan Institute Press.

Kane, Edward J. 1989. *The S&L Insurance Mess: How Did It Happen?* Washington, DC: Urban Institute Press.

Meyerson, Ann. 1986. "Deregulation and the Restructuring of the Housing Finance System." Pp. 68-98 in *Critical Perspectives on Housing,* edited by Rachel G. Bratt, Chester Hartman, and Ann Meyerson. Philadelphia: Temple University Press.

Mitchell, J. Paul. 1985. *Federal Housing Policy and Programs: Past and Present.* New Brunswick: Rutgers, State University of New Jersey, Center for Urban Policy Research.

Pryke, M., ed. 1994. "Property and Finance." *Environment and Planning A* 26(Special issue). ◀

▶ Scandinavian Housing and Planning Research

Scandinavian Housing and Planning Research is a quarterly publication of the Swedish Institute for Housing Research, the Finnish Ministry of Environment, the Norwegian Institute of Urban and Regional Research, and the Danish Building Research Institute. It focuses on the interests of architects, planners, and policymakers in the housing field. Articles present original research on the physical, economic, and social aspects of housing and planning. First published in 1983. Major changes are contemplated for 1998, including an international expansion of the editorial board and a name change reflecting a broad focus. Price: NDK 715 for institutions; NDK 330 for individuals. Contact: Jim Kemeny, Managing Editor, Institute for Housing Research, Box 785, S-801 Gävle, Sweden. Phone: 46 26 147700. Fax: 46 26 147802. World Wide Web: http://www.soc.uu.se/staff/jim_k.html.

—*Caroline Nagel* ◀

▶ Scattered-Site Public Housing

During the 1950s, 1960s, and early 1970s, when the bulk of the public housing in the U.S. stock was constructed, much of the housing was built on single sites, with one or more buildings per site. As early as the mid-1950s some large-scale public housing developments encountered serious problems, primarily in the management of these large buildings with a burgeoning population of very low income households with many children.

In response, public policy slowly shifted away from the construction of large projects on single sites for families and toward the development of public housing on sites scattered through a city's geographic area or scattered across a neighborhood. This approach also supported at-tempts at desegregation. Scattered-site public housing usually consists of a number of single-family homes built or purchased in a community by the public housing authority. Generally, this type of housing is indistinguishable from privately owned units and is better integrated into the community than many conventional public housing developments. A 1990 report by the National Association of Housing and Redevelopment Officials estimated that scattered-site developments represent 18% of all public housing developments nationwide. (SEE ALSO: **Public Housing**)

—*Rachel G. Bratt*

Further Reading

Kolodny, Robert. 1979. *Exploring New Strategies for Improving Public Housing Management.* Report prepared for the U.S. Department of Housing and Urban Development, Office of Policy Development and Research. Washington, DC: Department of Housing and Urban Development.

Lord, J. Dennis and George S. Rent. 1987. "Residential Satisfaction Scattered-Site Public Housing Projects." *Social Science Journal* 24(3):287-302.

National Association of Housing and Redevelopment Officials. 1990. *The Many Faces of Public Housing.* Washington, DC: Author.

Struyk, Raymond J. 1980. *A New System for Public Housing: Salvaging a National Resource.* Washington, DC: Urban Institute.

U.S. Department of Housing and Urban Development. 1996. *Scattered-Site Housing: Characteristics and Consequences.* Washington, DC: Author. ◀

▶ Second Homes

In 1990, just more than 3 million housing units, about 3% of all housing units in the United States, were designated by the U.S. Bureau of the Census as second homes. This number includes those vacant housing units that were "reserved for seasonal, recreational, or occasional use." Because some second homes are used by more than one family and because some families visit other housing units on a periodic basis, it is likely that the percentage of all households that spend parts of the year living in a place other than their primary residence is greater than 3%.

The incidence of second-home use in the United States is less than in many countries, especially several countries in Europe where higher proportions of the primary residences are flats or apartments. More than 20% of all households in Sweden, Norway, and France have second homes, and in some urban areas such as Vienna, the percentages are even higher. In Australia and New Zealand, about 5% of households have second homes. Canada has a percentage just higher than the United States and has significant concentrations of second homes in the Great Lakes region, especially within a few hours drive of Toronto.

In the United States, second homes take on a variety of forms. Some are individual cottages or cabins in lake, mountain, or seashore environments in various parts of the United States—the Upper Peninsula of Michigan, the Jersey shore, rural New England, and the Colorado Rockies, for example. Other second homes are parts of planned subdivisions, often built on lakes. Some of these homes are used seasonally, others are used on a shorter-term basis, such as for

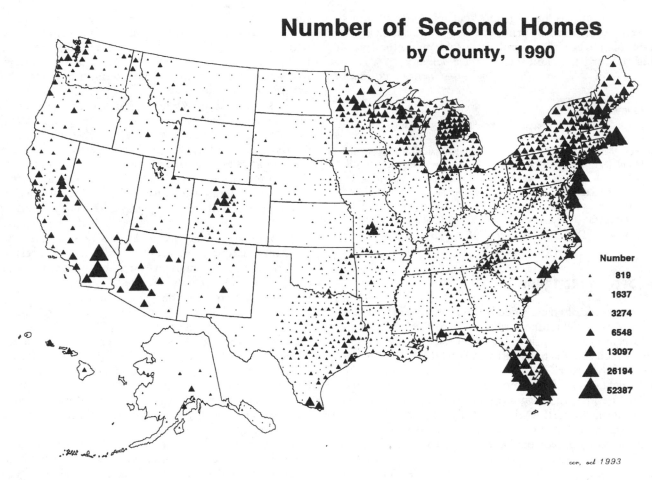

Figure 27. Number of Second Homes, by County, 1990
SOURCE: U.S. Bureau of the Census.

weekend getaways. Another large group of second homes reflects long-distance seasonal movements to and from the Sunbelt, including homes used in the winter by elderly snowbirds who flock to states such as Florida, Arizona, and California from a variety of more northerly locations in both the United States and Canada.

Traditionally, most second homes have been in the Northeast, especially New England and upstate New York, and in the Great Lakes region of the Upper Midwest. In 1970, 61% of second homes were in the North Central and Northeast census regions. By 1990, that percentage had decreased to 47%, a majority of second homes now being in the South and West regions. Among states, both the greatest number and the most rapid growth of second homes has been in Florida, where the number rose from 84,000 in 1970 to 418,000 in 1990. As a percentage of total housing units, however, the concentration is much less in Florida (6.0%) than in Vermont (16.6%) and Maine (15.0%).

Most very high income groups have had high rates of ownership for second homes. Certain communities—Palm Beach, Florida; Newport, Rhode Island; and Palm Springs, California among them—are widely recognized as seasonal playgrounds for the upper classes. Today, a broader range of people have second homes, especially upper- and upper-middle-class households. Income tax laws in the United States allow the deduction of mortgage interest for a second home, in addition to one primary residence, provided that the second residence is used for at lease 14 days during the tax year. If the second residence is rented to others during the year, nonrental personal use is required for at least 10% of rental days.

Within individual family life cycles, the use of second homes peaks in the preretirement stage, when adults are more than 40 years of age and children are teenagers or have left the household. For many families, having a second home in a particular area is a precursor to a retirement move to that same area. Their repeated visits to the second-home location establish strong familiarity with and ties to that place. When the decision to migrate at or near retirement age is made, the second-home location becomes a most logical destination. In some areas of the United States, including the Great Lakes and New England regions, large numbers of second homes are converted to year-round residences and occupied by retirees. Many retirees, on the other hand, continue to move to and from second homes on a seasonal or more frequent basis. As a result, the populations of many "retirement" areas fluctuate significantly with the seasons as the number of occupied second homes increases dramatically during one season. This undulation of population can cause significant problems for local planning and for health and social service delivery.

—*Curtis C. Roseman*

Further Reading

Coppock, J. T. 1977. *Second Homes: Curse or Blessing?* London: Pergamon.

Gartner, William C. 1987. "Environmental Impacts of Recreational Home Developments." *Annals of Tourism Research* 14(1):38-57.

Longino, C. F., Jr. and V. W. Marshall. 1990. "North American Research on Seasonal Migration." *Aging and Society* 10:229-35.

Roseman, Curtis C. 1985. "Living in More Than One Place: Second Homes in the United States—1970 and 1980." *Sociology and Social Research,* 70(1):63-67. ◄

► Second Mortgage

Mortgages are often classified by their priority in the event of borrower's default and, subsequently, the legal process of foreclosure. The priority is important because it indicates which claim is to be satisfied first. It is especially important when there are insufficient resources to cover all of the debt claims, as is often the case when default occurs.

The permanent, long-term mortgage, traditionally a fixed-rate mortgage (FRM) is generally the first (or primary) mortgage in most real estate financing arrangements. This is achieved by agreement between borrower and lender within the mortgage document (i.e., the parties agree that subsequent to a default by the borrower, the lender has the first claim to the liquidation value of the asset to recover the lender's outstanding debt balance). For the lender, the establishment of a priority claim against the property helps to secure collateral for the loan, which lowers the riskiness of the agreement. For the borrower, the acceptance of the claim against the asset enables the borrower to repay the mortgage note over a very long period of time (as much as 30 years in most states) despite numerous changes expected in the economy and, in some cases, in the borrower's life.

However, sometimes, a first mortgage may be insufficient. Additional financing may be desirable or even necessary by the borrower, but because of lending policies at the first mortgage's institution, a higher amount is not available. Alternatively, a buyer in a real estate transaction might not be able to produce a sufficient down payment to complete the financing arrangements of the transaction. However, the willingness of the seller to supply some of the financing to fill the gap could keep the deal alive. A borrower might want to use some of the equity in the house by refinancing a portion of the equity without disturbing the first mortgage. Finally, a high-leverage strategy might be to supplement the first mortgage with one or more additional mortgages so as to reduce the down payment and increase the returns from financial leverage. All of these situations employ second mortgages.

Second mortgages are supplementary financing instruments used to achieve one or more of the financing objectives of the borrower. Some second mortgages originate in financial institutions, but many do not. Some are marketable in secondary mortgage markets, but many are not (although in the 1950s and 1960s, some small investors used to buy and sell second mortgages as investments). Many second mortgages are short term in duration, such as three to five years.

All second mortgages have certain characteristics. By definition, they have a lower priority in the event of default than do first mortgages. Thus, they are always thought to be riskier than first mortgages. This has usually meant that the interest rates on second mortgages were higher than on first mortgages (but in the 1980s, this was not always true). Second mortgages always run a greater risk of being worthless compared with first mortgages in the event of a default. Second mortgages are less standardized than first mortgages so that the marketability of these loans is questionable. Also, the name "second mortgages" refers to a class of claims (i.e., several second mortgages may be used at the same time). If several "seconds" are used, holders of the mortgages will want to know the priority of their claims.

Second mortgages vary in their repayment schedules. Although short term in nature, they often employ balloon payments at the end of their terms. Thus, they often take the form of balloon mortgages. If so, borrowers need to worry about financing the outstanding balance due at the end of the balloon mortgage. Second mortgages often make it easier to finance projects when cash flows are tight. "Interest-only" financing is also a frequent choice for second mortgages.

In the 1990s, financial institutions have offered "home equity loans" as second mortgages. Homeowners with sufficient equity in their homes could create second mortgages similar to lines of credit based on the balance of the primary mortgage and the existing equity in the property. Because mortgage interest remains tax deductible in the U.S. tax system and other interest is no longer deductible, there has been considerable growth in home equity loans as second mortgages. In the event of financial difficulties in repaying the home equity loan, foreclosure procedures could force the homeowner to lose the property as a result of the second mortgage.

Second mortgages are best viewed as supplementary financing vehicles in residential housing markets. They also provide flexibility and permit buyers and sellers to try to structure deals that could otherwise possibly not be completed. Although second mortgages are riskier than first mortgages for borrowers and lenders, they are a valuable part of the U.S. mortgage finance system. (SEE ALSO: **Mortgage Finance**)

—*Austin J. Jaffe*

Further Reading

Jacobus, Charles J. 1996. *Real Estate Principles.* Upper Saddle River, NJ: Prentice Hall.

Shim, Jae K. 1996. *Dictionary of Real Estate.* New York: John Wiley.

Sirmans, C. F. and Austin J. Jaffe. 1988. *The Complete Real Estate Investment Handbook.* 4th ed. New York: Prentice Hall.

Tosh, Dennis S. 1990. *Handbook of Real Estate Terms.* Englewood Cliffs, NJ: Prentice Hall. ◄

► Secondary Mortgage Market

The secondary mortgage market in the United States provides a market for mortgage loans to be bought and sold. The principal agencies involved are Fannie Mae (formerly, the Federal National Mortgage Association) a government-regulated but privately held company; the Government National Mortgage Association (GNMA or Ginnie Mae) under the U.S. Department of Housing and Urban Development; and the Federal Home Loan Mortgage Corporation (FHLMC or Freddie Mac), also a federally sponsored organization. In addition to buying and selling individual loans and packages of loans, secondary mortgage market activities include the packaging and sale of mortgage-backed investments such as pass-through securities and mortgage-backed bonds. The effect of the secondary mortgage market is to increase the amount of capital available for housing and to increase the geographic mobility of that capital.

Most real estate in the United States is bought with a mortgage—that is, a loan secured by the property. If the borrower defaults, the lending institution can foreclose and obtain possession of the property. The institution then sells the property to recoup the remaining loan principal. This primary mortgage market concerns the process of making mortgage loans to borrowers. However, the mortgage loan itself is also tradable on the secondary mortgage market. The secondary mortgage market involves what the lending institutions do with mortgages once the loans have been made to borrowers.

Although mortgages can be bought and sold like any other good, they have some special characteristics. Every mortgage is given to a specific borrower for a specific property. Individual borrower characteristics can make an enormous difference to the quality of a mortgage: Two households each borrowing $100,000 may have vastly different chances of paying it back successfully because of differences in income, occupations, educational levels, other assets, and other liabilities even though both qualified for the loan. Perhaps even more important, each mortgage is secured by a specific property with attributes that can make it a better or worse investment (e.g., region of the country, condition, neighborhood, age, amenities, etc.). To trade individual mortgages would thus require an investor to be as aware of all aspects of the borrowers and properties as the original lender. Not much trading of mortgages would get done under those circumstances.

However, an active secondary mortgage market benefits the economy and housing provision because it allows capital to flow more easily between parts of the country and taps different sources of capital for housing. If Texas is growing and needs more mortgage capital, an investor in Massachusetts could get a better rate of return investing in Texas mortgages than in Massachusetts mortgages and the Texas borrower would benefit from having more mortgage money available because rates would be somewhat lower. If the Massachusetts investor could be relieved of worries about the soundness of individual mortgage transactions, the number of potential investors could be expanded. Increased liquidity, mobility of capital, and expanding housing opportunities through increased mortgage availability

are seen as acceptable policy objectives. Consequently, the federal government established and supports a secondary mortgage market in the United States through a variety of agencies.

First, the Federal Housing Administration's (FHA's) mortgage insurance program provides a group of standardized loans. These loans are made at similar interest rates with similar terms, and similar down payments. The FHA's standards are applied to the properties and borrowers so there is some quality control, and should the borrower default, the FHA pays off the loan. Therefore, an investor can purchase FHA-insured loans with relatively little risk and far less detailed knowledge of the local markets. Loans guaranteed by the Department of Veterans Affairs (VA) act in a similar fashion.

The FHA began operations in 1934, and in 1938 the Federal National Mortgage Association (now known as Fannie Mae) was chartered to provide a market for FHA-insured (and after World War II, VA-guaranteed) loans. Lenders could originate an FHA-insured loan to a borrower in their local area; the loan could then be sold to Fannie Mae so that the lender would immediately have the capital to loan out again (rather than receiving it in a long, slow stream more than 30 years). The lender, in most cases, would continue to service the loan. This means that the lender would continue to collect the monthly payments, keep track of the escrow accounts and pay the insurance premiums and the property taxes and so on, for a fee. The financial institution thus continued to make money originating loans and servicing them but did not tie up capital for long periods of time.

Fannie Mae was rechartered in 1954, with the intent that it would gradually become a private company, a goal achieved by 1968. Fannie Mae no longer receives financial support from the federal government (although it has a line of credit with the U.S. Treasury, and it is doubtful that it would be allowed to fail), but it is regulated by the federal government. It continues to be a primary buyer of federally insured and guaranteed mortgages (mainly from mortgage companies) and finances its operations through the sale of stocks and bonds. Ginnie Mae took over Fannie Mae's roles of managing direct loans made in an earlier phase of Fannie Mae's existence and performing special assistance functions. In addition, Ginnie Mae has the assignment of providing a guarantee for pools of FHA and VA mortgages (discussed further below). Finally, a third agency, Freddie Mac, was added in 1970 to purchase conventional loans (i.e., loans that are not FHA insured or VA guaranteed), mostly from savings and loan associations.

Fannie Mae and Freddie Mac buy and sell loans. Theoretically, either can deal in insured, guaranteed, or conventional loans, but in practice, Fannie Mae mainly works with insured and guaranteed mortgages and Freddie Mac with conventional mortgages. When lenders want to lend more money quickly, they sell loans to Fannie Mae and Freddie Mac. When there is more mortgage money in the market than is needed, Fannie and Freddie Mac sell loans to take up some of that capital.

In addition to providing a market for the sale and purchase of mortgage loans, the secondary mortgage market includes a variety of mortgage-backed investments. Ginnie Mae's role, mentioned above, of guaranteeing FHA-insured and VA-guaranteed loan pools is a good example. The loans themselves are already insured or guaranteed, so if an investor buys a share in a group or pool of the loans, the investor is safe from loss of capital in case of default on the part of mortgage borrowers. However, if a borrower defaults, it could take time to settle the debt. The investor would eventually receive the loan repayment, but it would take time, and there would be some loss of returns because of that delay. Ginnie Mae guarantees "timely payment of principal and interest" by pools of these mortgages. Thus, the investor will be paid immediately, and Ginnie Mae will bear the cost of the time to settle things. This is an example of a pass-through security in which an investor buys shares in a pool of mortgages and the principal and interest are "passed through" to the investor on a regular basis.

Many other kinds of mortgage-backed securities are available, and the number is continually growing. Mortgage-backed bonds in which an investor buys bonds backed by mortgages constitute a second type of mortgage-related security. In this case, the mortgage originator continues to own the mortgage and receives the principal and interest payments; the bond buyer receives payment on the bond at a specified rate over a specified term. Mortgage pay-through bonds, a third example, combine characteristics of mortgage-backed bonds and mortgage pass-through securities. They are bonds with fixed rates, and the mortgage originator continues to own the mortgages, but the principal and interest repayments are passed through to the investor as in a pass-through security. (SEE ALSO: *Fannie Mae; Mortgage Finance*)

—*Hazel A. Morrow-Jones*

Further Reading

Brueggeman, William B. and Jeffrey D. Fisher. 1993. *Real Estate Finance and Investments*. 9th ed. Homewood, IL: Irwin.

Daye, Charles E., Daniel R. Mandelker, Otto J. Hetzel, James A. Kushner, Henry W. McGee, Jr., Robert M. Washburn, Peter W. Salsich, Jr., and W. Dennis Keating, eds. 1989. *Housing and Community Development: Cases and Materials*. 2d ed. Durham, NC: Carolina Academic Press.

Lederman, Jess, ed. 1992. *The Secondary Mortgage Market: Strategies for Surviving and Thriving in Today's Challenging Markets*. Chicago: Procus.

Miles, Mike E., Emil E. Malizia, Marc A. Weiss, Gayle L. Berens, and Ginger Travis. 1991. *Real Estate Development: Principles and Process*. Washington, DC: Urban Land Institute. ◄

▶ Section 8: Lower Income Rental Assistance

Section 8 supports rental housing subsidies provided by the U.S. Department of Housing and Urban Development (HUD) to private housing owners on behalf of low-income households to allow them to live in decent and affordable housing. Generally, subsidies from HUD equal the difference between the rent charged for the housing unit and 30% of the household's income, adjusted for the household's size. "Section 8" refers to the program's location in housing law—namely, Section 8 of the U.S. Housing Act of 1937, as amended.

Section 8 housing assistance takes a number of different forms but falls into two basic categories: tenant-based assistance and project-based assistance. Generally, under tenant-based assistance, low-income households may choose to live in any housing unit meeting HUD's rent and quality standards, if the housing owner agrees to participate. Under project-based assistance, assisted households must live in specified properties for which HUD has provided Section 8 assistance.

Evolution

Prior to the 1970s, HUD programs had two approaches for providing rental housing assistance to low-income households residing in privately owned housing. Under the first approach, HUD offered subsidies that reduced the interest rate, sometimes to as low as 1%, on mortgages used to finance low-income housing. However, as rents increased to cover escalating operating costs over the years, this housing became increasingly less affordable to low-income households. Under the second approach, HUD contracted with public housing authorities to lease units in the private market and then sublease the units to low-income households.

Congress made a major shift in federal housing assistance when it enacted the Section 8 program as part of the Housing and Community Development Act of 1974. One of the major goals of the Section 8 program was to refocus federal housing policy away from locating large numbers of low-income households in properties concentrated in specific areas of the community and toward economically integrating these households throughout the community.

The Section 8 program is flexible; it can assist households down to any income level. Generally, HUD pays the difference between the rent charged by a housing owner (called contract rent) and the assisted household's rental contribution, which is generally 30% of the household's income, adjusted for the household's size. With some exceptions, the rent charged by the owner cannot exceed a "fair market rent" (shelter plus utilities) established by HUD for a unit with a given number of bedrooms in different housing markets. Fair market rents are set at the 40th percentile of an area's rental housing—that is, the level at which about 40% of a market area's rental housing can be obtained. The rental charge must also meet a "rent reasonableness test" to ensure that it is not out of line with comparable unsubsidized housing in the community. Generally, households are eligible for Section 8 assistance if their income does not exceed 80% of the median income for the area.

As it has evolved over the past two decades, several types of assistance are authorized under Section 8. Tenant-based assistance comprises the certificate and housing voucher programs. Households that use certificates and vouchers rent from owners of their choice, as long as the housing meets HUD's rent and quality standards and the housing owner agrees to participate in the Section 8 program. A major difference between a certificate and a voucher is that a household assisted through certificates must rent a unit that is within the HUD-established fair market rent and pay 30% of its income for rent, whereas a household with a voucher may elect to pay more than the fair market rent if it pays the difference itself. Another difference is that the voucher program has a so-called shopper's incentive, which allows a household to rent units below the fair market rent and keep the difference. Currently, the certificate and voucher programs are the primary means by which Congress is adding to the size of the assisted-housing stock each year.

Project-based assistance includes rental assistance (a) for newly built or rehabilitated housing (which falls under HUD's new-construction, substantial-rehabilitation, and moderate-rehabilitation programs), (b) to facilitate the sale of HUD-insured properties undergoing default or foreclosure (the property disposition program), (c) to provide a steady stream of income for financially troubled properties receiving other HUD assistance (the loan management set-aside program), and (d) to help preserve other HUD-assisted low-income housing when housing owners' contractual obligations with HUD expire (the preservation program). Because their rent subsidies are attached to particular units, households who move lose their Section 8 assistance unless they move to another subsidized unit.

Increasingly, federal housing law is requiring that Section 8 subsidies be used as a means to help move households toward economic independence rather than allowing the housing subsidy to be used as an end. For example, in 1988, legislation added "portability" to the certificate and voucher programs to increase housing choices for households and facilitated their movement to other areas of the United States, where, for example, employment opportunities may be greater. In 1990, Congress enacted the Family Self-Sufficiency Program to encourage a small but growing number of households assisted through certificates and vouchers to move toward economic independence. Also, in 1992 Congress enacted (a) the Moving to Opportunity for Fair Housing demonstration program, which permits households receiving project-based assistance to move from areas with a high concentration of poverty to areas with a low concentration of poverty and (b) a provision that allows Section 8 rental subsidies, under certain conditions, to be used for homeownership opportunities. More recently, HUD has proposed using tenant-based rental assistance to help link welfare recipients to jobs as a means of beginning to implement 1996 welfare reform legislation.

Administration

Generally, for tenant-based subsidies, HUD contracts with public housing agencies (instrumentalities of local or state governments) to administer its programs. HUD enters into an annual contributions contract with the housing authorities and provides funds to the agency to administer the programs. In return for a fee, the housing agency is to ensure that low-income households applying for assistance are eligible and that housing units selected by assisted households meet HUD's rent and quality standards. The housing agency enters into a housing assistance payment contract with owners to pay the housing subsidy. For most project-based assistance, HUD enters into housing assistance payment contracts directly with the housing owners and pays subsidies directly to them. For both the tenant-based and project-based programs, assisted households pay their rental contribution directly to the housing owner.

Congress appropriates budget authority (authority to make financial commitments) to cover the expected costs of the next fiscal year's Section 8 contracts with housing authorities and owners. For most kinds of Section 8 assistance, if congressional budget authority is insufficient to cover costs, Congress supplies additional budget authority (called contract amendments) to continue the subsidies.

The Section 8 program is not an entitlement program. Under entitlement programs, such as Medicaid, anyone who meets the eligibility requirements can receive assistance. But under the Section 8 program, assistance is limited to the number of housing units already supported under congressional appropriations for Section 8 housing and the number of additional units (called "incremental units") that Congress has added each year to the assisted-housing stock. HUD and congressional policy has been to renew all Section 8 contracts that are due to expire to ensure that low-income families will continue to receive assistance.

Current Status

As of 1996, about 3.2 million low-income households were receiving Section 8 housing assistance. Of those assisted, about 1.4 million households were receiving certificate and voucher assistance, and about 1.8 million households were receiving project-based assistance. Federal outlays for Section 8 assistance in fiscal year 1997 are expected to be in the neighborhood of $16 billion. (SEE ALSO: *Fair Market Rent*)

—*James Ratzenberger*

Further Reading

Congressional Research Service. 1991. *Housing Assistance in the United States*. Washington, DC: Author

Congressional Research Service. 1993. *HUD Housing Assistance Programs: Their Current Status*. Washington, DC: Author.

Rosenthal, Donald B. 1984. "Joining Housing Rehabilitation to Neighborhood Revitalization: The Section 8 Neighborhood Strategy Area Program." In *Public Policy Formation,* edited by R. Eyeston. Greenwich, CT: JAI. ◄

► Section 101: Rent Supplements

The Department of Housing and Urban Development (HUD) provides federal payments to reduce rents for certain disadvantaged low-income persons. HUD may pay rent supplements on behalf of eligible tenants to certain private owners of multifamily housing insured by the Federal Housing Administration or to owners of certain projects financed under a state or local program of loans, loan insurance, or tax abatement. The payment makes up the difference between 30% of the tenant's adjusted income and the fair market rent determined by HUD. However, the subsidy may not exceed 70% of the HUD-approved rent for the specific unit. HUD may pay the supplements for a maximum term of 40 years.

Legislation establishing this program was enacted as part of the 1965 Housing and Urban Development Act, but new rent supplement contracts are no longer available. The program was suspended under the housing subsidy moratorium of January 5, 1973. Private, nonprofit, limited-dividend, cooperative, or public agency sponsors carrying mortgages insured under HUD programs were able to apply for rent supplements. Eligible tenants were limited to low-income households that qualified for public housing and were either elderly, handicapped, displaced by government action, victims of national disaster, occupying substandard housing, or headed by a person serving on active military duty. (SEE ALSO: *Moratorium on Federally Assisted Housing Programs*)

—*Mara Sidney* ◄

► Section 106: Counseling for Homebuyers, Homeowners, and Tenants

The Department of Housing and Urban Development (HUD) offers housing counseling to homebuyers, homeowners, and tenants under HUD programs and for homeowners with conventional mortgages or mortgages insured or guaranteed by other government agencies, including the Department of Veterans Affairs (VA) and the Farmers Home Administration (FmHA). HUD provides the service through HUD-approved counseling agencies. For fiscal year 1996, HUD funded 183 counseling agencies and 4 intermediaries. These agencies are public and private nonprofit organizations with housing counseling skills and knowledge of HUD, VA, and conventional housing programs. HUD awards housing counseling grants on a competitive basis to its approved agencies when Congress appropriates funds for this purpose. The funding helps the approved agencies partially meet their operating expenses.

Counseling consists of (a) housing information, (b) purchase and rental of housing, (c) home equity conversion mortgage application guidance, (d) money management, (e) budgeting and credit counseling to avoid mortgage default and rent delinquencies that led to foreclosure or eviction, and (f) home maintenance. The objective of the counseling is to help homebuyers, homeowners, and tenants improve their housing conditions and meet their responsibilities.

Legislation establishing this program was enacted in 1968 and subsequent years. Homebuyers, homeowners, and tenants under programs indicated are eligible to receive the counseling services from HUD-approved housing counseling agencies. Public and private nonprofit entities may apply through a local HUD office for HUD approval as a counseling agency.

—*Mara Sidney* ◄

► Section 108: Loan Guarantees

Section 108 is the loan guarantee provision of the Community Development Block Grant (CDBG) program adminis-

tered by the Department of Housing and Urban Development (HUD). Section 108 provides communities with a source of financing for economic development, housing rehabilitation, public facilities, and large-scale physical development projects.

Eligible applicants include the following public entities: (a) metropolitan cities and urban counties (i.e., CDBG entitlement recipients), (b) nonentitlement communities assisted in the submission of applications by states that administer the CDBG program, and (c) nonentitlement communities eligible to receive CDBG funds under the HUD-administered Small Cities CDBG program. The public entity may be the borrower, or it may designate a public agency to be the borrower.

Activities eligible for Section 108 financing include the following:

- ► Economic development activities eligible under CDBG
- ► Acquisition of real property
- ► Rehabilitation of publicly owned real property
- ► Housing rehabilitation eligible under CDBG
- ► Construction, reconstruction, or installation of public facilities (including street, sidewalk, and other site improvements)
- ► Related relocation, clearance, and site improvements
- ► Payment of interest on the guaranteed loan and issuance costs of public offerings
- ► Debt service reserves
- ► Public works and site improvements in colonias
- ► In limited circumstances, housing construction as part of community economic development, Housing Development Grant, or Nehemiah Housing Opportunity Grant programs

As with the CDBG program, all projects and activities must either principally benefit low- and moderate-income persons, aid in the elimination or prevention of slums and blight, or meet urgent needs of the community.

The Section 108 program has undergone several major changes since its establishment in 1974. In 1987, HUD was directed by Congress to use a private sector financing mechanism to fund the loan guarantees as opposed to using federal funds. In 1990, legislative changes increased public entities' borrowing authority to five times the CDBG allocation, extended the maximum repayment period to 20 years, and made units of general local government in nonentitlement areas eligible to apply for loan guarantee assistance. Beginning in FY 1993, Congress increased the guarantee authority for the Section 108 program to more than $2 billion and authorized similar amounts in FY 1994 and FY 1995. The 1996 authorization was $433 million. (SEE ALSO: *Community Development Block Grant*)

—*Mara Sidney* ◄

► Section 201: Flexible Subsidy

The Department of Housing and Urban Development (HUD) provides federal aid for troubled multifamily housing projects as well as capital improvement funds for both troubled and stable subsidized projects. The Flexible Subsidy program has two components—operating assistance for troubled projects and capital improvements loans. The Operating Assistance Program provides funds that will assist in restoring the financial and physical soundness of privately owned, federally assisted multifamily housing. Operating Assistance focuses on correcting deferred maintenance, financial deficiencies, replacement reserve, and operating deficits.

The Capital Improvement Loan Program (CILP) provides funds for subsidized multifamily housing projects to implement major capital improvements. CILP assists both stable and troubled projects that require major capital improvements to sustain a project's future viability. Both programs are designed to maintain the use of the property for low- and moderate-income people and are conditioned on the project owner's ability to provide management satisfactory to HUD.

Legislation establishing this program was enacted in 1978. Projects with mortgages insured or held by HUD and subsidized under Section 23, Section 236, or Section 221(d)(3) are eligible, along with below-market-interest-rate or the rent supplement program and projects that were constructed more than 15 years before the request for assistance under the Section 202 program. Also eligible are noninsured projects developed by state agencies and receiving HUD financial assistance under one of the above subsidy programs. In fiscal year 1996, HUD used Flexible Subsidy funds only for Section 202 projects for the elderly or disabled. (SEE ALSO: *Multifamily Housing; National Commission on Severely Distressed Public Housing*)

—*Mara Sidney* ◄

► Section 202: Housing for the Elderly and the Handicapped

For the past 36 years, the Section 202 housing program has been known as the flagship of the Department of Housing and Urban Development (HUD) housing production programs because of its longevity, its effectiveness in targeting needy elderly and disabled persons, and its history of caring and efficient management through nonprofit local owners and sponsors. The program was originally authorized by Congress in 1959 and has been modified several times by different administrations to emerge as the primary federal financing vehicle for constructing subsidized rental housing for older adults and disabled persons. The current program provides direct loans at below-market interest rates to private, nonprofit sponsors to develop housing designed to meet the unique needs of the occupants. Since 1974, the Section 202 program has been paired up with the Section 8 certificate subsidy program, which ensures that tenants do not pay more than 30% of their income on rent.

As of 1996, 7,547 Section 202 facilities were in service, housing more than 387,000 persons. Unlike other federal housing subsidy programs, there have been very few defaults by 202 sponsors, and national surveys of sponsors and facility managers, conducted in 1983 and 1988, indicated that the Section 202 program has been a model of cooperation between the federal government and private, not-for-profit sponsors to produce secure, barrier-free, and supportive independent-living accommodations. Most of the nonprofit sponsors are religious organizations; community-based development corporations, unions, fraternal organizations, and cooperatives are examples of other nonprofit sponsors. Most sponsors own and manage just one project.

The Section 202 program can best be analyzed by dividing its history into four distinct phases, initiated by changes in program goals and tenant eligibility as congressional and executive policy evolved.

The Moderate-Income Eligibility Phase

The first phase ran from 1959 to 1972 and can be labeled as the period of "moderate-income eligibility." During this period, the Section 202 program provided an interest subsidy on building loans aimed at lowering the cost of housing production (3% interest for up to 50 years). The below-market interest rates and nonprofit ownership meant that rents were affordable to persons unable to afford market rate apartments but whose income was too high to qualify for public housing.

More than 45,000 units in 335 projects were built during this phase. Individual projects for elderly persons were relatively large, averaging 153 units. Most of the units were efficiency apartments, and the projects tended to be located in large cities and near community-based service facilities to accommodate resident support needs. After the Housing Act of 1968, the Nixon administration decided to phase out the Section 202 program and replace it with Section 236, which provided interest subsidies to both nonprofit and for-profit sponsors who obtained private financing to build elderly and multifamily housing.

By the early 1970s, no new Section 202 commitments were being made and the program was terminated. Critics of the early years of the program felt it was aimed at too "wealthy" a clientele. This was dictated, however, by the relatively shallow subsidy consisting of only a low-interest loan for building construction. On a per-unit basis, the early phase of the program was also far less costly than subsequent phases. To house "poorer" clients, the later phases of the program had to increase the subsidies, and per-unit costs increased significantly.

The Low-Income Eligibility Phase

The period between 1974 and 1983 saw a revival of the Section 202 program and has been titled the "low-income phase" because the 1974 Housing Act established a new mission to serve low-income persons—those with incomes at or below 80% of the median income. Interest rates were tied to the U.S. Treasury borrowing rate (later capped at 9.25%), and units were made eligible for Section 8 rental assistance. The addition of the rental assistance on top of the low-interest subsidy made units affordable to a lower-income population. Under Section 8 rental assistance as initially conceived, tenants paid 25% of their adjusted yearly income and the rental assistance paid the difference between the tenant contribution and a locally determined "fair market rent" (FMR). The FMR set a ceiling on what sponsors received, and this could be no higher than comparable housing in the local market.

The 1974 act also targeted the Section 202 program, setting aside 20% to 25% of loans for rural areas and requiring increased participation by minority and frail older adults. Frail persons were to be assisted by new provisions that encouraged support services and larger staff.

The largest number of projects and units were built during this middle period, with more than 20,000 units per year approved in the latter part of this phase. In addition, the unpopular efficiency units were replaced. More than 90% of all units built were one-bedroom apartments. Average size of the projects declined from the 153-unit average in the first phase to an average of 92 units in this phase.

The Very Low Income Eligibility Phase

The third, or "very low income" phase, covers the period of 1984 to 1990, following regulatory changes implemented by the Reagan administration. Income eligibility was lowered to below 50% of the local median income. At the same time, tenant rent contributions increased from 25% to 30% of adjusted yearly income. Because relatively little had been done to address the low participation rate of minorities in the previous phase, HUD also introduced priority selection criteria for minority sponsors. The results of these changes were significant. Minority sponsorship rose from 7.5% in the first phase to 17.3% in this phase as of 1988. One third of the residents in the very low income phase had incomes below $5,000 by 1988 compared with just 17.9% of the residents of projects built before 1974.

The focus on poorer people and minorities was not the only change introduced during the Reagan and Bush administrations. This period marked a significant decline in the funding of the Section 202 program, in line with federal cuts in most housing and social welfare programs. The funding cuts resulted in a decrease in both the number and quality of projects built. By FY 1989, there were fewer than 7,200 new senior projects and fewer than 2,300 new projects for handicapped persons. Project size declined again to an average of 56 units per project between 1985 and 1988 and to just 47 units per project in FY 1988-89. In addition, a series of regulations in the name of "cost containment" were introduced in 1982 that had negative effects on facility development and design because of severe limitations on construction funding and the elimination of many amenities.

In 1988, the average waiting time for senior applicants to Section 202 housing was 8 years; 11 years in the large urban centers. During these later years of the Reagan administration and throughout the Bush administration, Section 202 projects that had been awarded funding also began to lapse into protracted bargaining sessions either because of prolonged HUD field office reviews of every detail in the construction plans or because HUD set unrealistically low FMR scales, which made debt servicing infeasible for many project sponsors. At the very time when more low-income seniors than ever before were applying for Section

202 housing, the funding was being cut and the construction of the few new facilities available was becoming bogged down in bureaucratic delays.

The Current Phase of the Program

The National Affordable Housing Act (NAHA) became the relevant housing legislation starting in 1990. The act separated elderly and disabled housing. The Section 202 program remains for elderly persons, and Section 811 has become the new housing program for the disabled. Most of the Section 202 sponsors want facilities designed to house independent elderly persons. This has been true throughout the program and remains so today. Nevertheless, the NAHA recognized the growing number of very old and frail seniors applying to and living in Section 202 facilities and makes provisions for facilities that include support services. Sponsors can apply for funding of new Section 202 with support services, or apply for funding to add support services to an existing Section 202 facility designed for independent living where residents have aged in place and require support to avoid a move to more dependent living accommodations. However, in 1977, only about 10% of the sponsors of new facilities built between 1990 and 1996 used this support service option.

Demand continues to remain very high for Section 202 housing, but funding continues to remain low. More sponsors are applying than can be funded, and more seniors are waiting for units than can be accommodated. In 1993, about 9,000 total units were funded. This dropped to 7,800 in 1994, dropped again to 7,700 in 1995, and rose only slightly to 8,700 units in 1996.

Projects for the Disabled

The above demographic statistics primarily address Section 202 facilities for elderly persons. The typical Section 202 project for younger disabled persons is a group home that houses 18 tenants with an average age of 35 years. More than two-thirds of these tenants have very low or no income. Almost half of the tenants come directly from institutional care, so these facilities have a major role in deinstitutionalization. These facilities have more extensive barrier-free design and sufficient communal space to provide many support services on-site. The majority of sponsors have always been community-based groups rather than religious organizations. Under the new Section 811 program, public bodies with 501(c)3 nonprofit status are eligible to apply and are becoming a major sponsor type. In 1996, 1,900 of these small Section 811 facilities were funded.

Conclusion

The most important finding when reviewing the entire history of the Section 202 and 811 programs is that these have been very popular and successful programs. They have fewer defaults, higher-quality construction, more sensitive management, and as a result, no social stigma attached to them, as well as highly satisfied clients.

Residents of Section 202 housing are clearly aging in place with greater and greater support and management needs. In 1997, projects built during the first phase housed a population with an average age of age 77, and more than 35% were over the age of 80. A similar profile fits about 25% of the applicants on waiting lists. The biggest change over the coming years is going to be the growing number of residents and applicants who need congregate support services made up of at least five hot meals per week and on-site housekeeping and personal care services. Only 28% of Section 202 facilities provided even one support service at the start of the 1990s, and this percentage had not changed appreciably through 1996. The older and larger facilities appear to be best prepared to provide support for their tenants. These projects have the building capacity to accommodate congregate services, a tenant population wealthy enough to afford services in the absence of adequate support subsidies, and facilities large enough to provide economies of scale. Despite the overall impression of high-quality construction in Section 202 facilities, evidence is mounting that many of the program changes made in the name of cost containment in the 1980s are resulting in facilities that are inadequate to serve the growing frail senior population.

Program changes that provide for support services and mechanisms to facilitate timely construction approval have strengthened the program. Although the goals of providing facilities uniquely designed and managed to serve the needs of low-income and frail seniors and younger persons with chronic disabilities is largely being met by the Section 202 and Section 811 programs, the total need for such subsidized housing is clearly not being met. The low yearly funding levels remain the biggest concern, severely limiting these programs in reaching most of the needy clients and worthy sponsors throughout the United States. (SEE ALSO: Elderly; *Senior Citizens Housing Act of 1962*)

—Leonard F. Heumann

Further Reading

Gayda, Kathy S. and Leonard F. Heumann. 1989. *The 1988 National Survey of Section 202 Housing for the Elderly and Handicapped.* Report to the U.S. Congress, Subcommittee on Housing and Consumer Interests of the Select Committee on Aging. Washington DC: Government Printing Office.

Heumann, Leonard F. 1993. "Aging in Place in Housing Designed for Independent Living: The Case of the U.S. Section 202 Program." In *Aging in Place with Dignity: International Solutions Relating to the Low-Income and Frail Elderly,* edited by Leonard F. Heumann and Duncan P. Boldy. Westport, CT: Praeger.

U.S. Senate, Special Committee on Aging. 1984. *Section 202 Housing for the Elderly and Handicapped: A National Survey.* Washington, DC: Government Printing Office. ◀

▶ Section 203(b) and (i): One- to Four-Family Home Mortgage Insurance

The Department of Housing and Urban Development (HUD) provides federal mortgage insurance to finance homeownership and the construction and financing of housing. By insuring commercial lenders against loss, HUD encourages them to invest capital in the home mortgage market. HUD insures loans made by private financial institutions

for up to 97% of the first $25,000 of value, 95% of the value between $25,000 and $125,000, and 90% of the amount above $125,000 up to the high-cost limit ceiling for the area. The terms for the loan can be for up to 30 years. The loan may finance homes in both urban and rural areas. Maximum insurable loans range from $81,548 for a one-family home to $156,731 for a four-family home. The secretary of HUD may increase these amounts on an area-by-area basis in cases of high prevailing housing costs. Less rigid construction standards are permitted in rural areas.

HUD and FHA-insured homeowners threatened with foreclosure owing to circumstances beyond their control, such as job loss, death, or illness in the family, may apply for assignment of the mortgage to HUD, which, if it accepts assignment, takes over their mortgages and adjusts the mortgage payments for a period of time until the homeowners can resume their financial obligations.

Legislation establishing this program was enacted in 1934. Any person able to meet the cash investment, the mortgage payments, and credit requirements is eligible. The program is generally limited to owner-occupants. (SEE ALSO: *Mortgage Insurance; National Housing Act of 1934*)

—Mara Sidney ◀

▶ Section 203(k): Rehabilitation Mortgage Insurance

The Department of Housing and Urban Development (HUD) offers mortgage insurance to finance the rehabilitation of one- to four-family properties. HUD insures loans to finance rehabilitation of an existing property, to finance rehabilitation and refinancing of the outstanding indebtedness of a property, and to finance purchase and rehabilitation of a property. An eligible rehabilitation loan must involve a principal obligation not exceeding the amount allowed under Section 203(b) home mortgage insurance.

Legislation establishing this program was enacted in 1961. Any person able to make the cash investment and the mortgage payments is eligible. (SEE ALSO: *Mortgage Insurance*)

—Mara Sidney ◀

▶ Section 207: Multifamily Rental Housing

The Department of Housing and Urban Development (HUD) offers federal mortgage insurance to finance construction or rehabilitation of a broad cross-section of rental housing. HUD insures mortgages made by private lending institutions to finance the construction or rehabilitation of multifamily rental housing by private or public developers. The project must contain at least five dwelling units. Housing financed under this program, whether in urban or suburban areas, should be able to accommodate families (with or without children) at reasonable rents.

Legislation establishing this program was enacted in 1934. Investors, builders, developers, and others who met HUD requirements could apply for funds to a lending institution approved by the Federal Housing Administration (FHA) after conferring with their local HUD office. The housing project had to be located in an area approved by HUD for rental housing and in which market conditions showed a need for such housing. By the 1990s, however, the program was not being used. Instead, new construction and substantial rehabilitation multifamily rental projects were insured under Section 221(d)(3) and (4) because those programs are more advantageous to the developer and lender. (SEE ALSO: *Housing Act of 1954; Mortgage Insurance; Section 221[d][3] and [4]*)

—*Mara Sidney* ◄

▶ Section 213: Cooperative Housing

The Department of Housing and Urban Development (HUD) provides federal mortgage insurance to finance cooperative housing projects. HUD insures mortgages made by private lending institutions on cooperative housing projects of five or more dwelling units to be occupied by members of nonprofit cooperative ownership housing corporations. These loans may finance (a) new construction; (b) rehabilitation, acquisition, improvement, or repair of a project already owned; (c) resale of individual memberships; (d) construction of projects composed of individual family dwellings to be bought by individual members with separate insured mortgages; and (e) construction or rehabilitation of projects that the owners intend to sell to nonprofit cooperatives.

Legislation establishing this program was enacted in 1950. Nonprofit corporations or trusts organized to construct homes for members of the corporation or beneficiaries of the trust are eligible, as are qualified sponsors who intend to sell the project to a nonprofit corporation or trust. By the 1990s, however, the program was not being used. Instead, new construction and substantial rehabilitation cooperative projects are currently insured under Section 221(d)(3) because it is more advantageous to the cooperative. (SEE ALSO: *Cooperative Housing; Mortgage Insurance; Section 221[d][3]*)

—*Mara Sidney* ◄

▶ Section 220: Mortgage and Major Home Improvement Loan Insurance for Urban Renewal Areas

The Department of Housing and Urban Development (HUD) offers federally insured loans used to finance mortgages for housing in urban renewal areas, areas in which concentrated revitalization activities have been undertaken by local government, or to alter, repair, or improve housing in those areas. HUD insures mortgages on new or rehabilitated homes or multifamily structures located in designated urban renewal areas and areas with concentrated programs of code enforcement and neighborhood development. HUD insures supplemental loans to finance improvements that will enhance and preserve salvageable homes and apartments in designated urban renewal areas.

Although investors, builders, developers, individual homeowners, and apartment owners are eligible, the program is not used frequently. (SEE ALSO: *Housing Act of 1954*)

—*Mara Sidney* ◄

▶ Section 221(d) pursuant to Section 223(g): Mortgage Insurance for Single-Room Occupancy

The Department of Housing and Urban Development (HUD) offers mortgage insurance for the new construction and substantial rehabilitation of single-room occupancy (SRO) facilities. The insured SRO program is designed to expand the availability of affordable housing for low- and moderate-income persons, thereby helping to prevent homelessness. An SRO project must be an unsubsidized project with five or more units. Units may contain kitchen or bathroom facilities, but the provision of those facilities within the units is not required. Sanitary facilities and kitchen facilities may be shared among tenants.

Legislation providing general authorization for this program was enacted in 1974. Nonprofit, public body, limited distribution, and general mortgagors are eligible under HUD regulations. Cooperative and investor-sponsor mortgagors are not eligible. (SEE ALSO: *Housing Act of 1954; Single-Room Occupancy Housing*)

—*Mara Sidney* ◄

▶ Section 221(d)(2): Homeownership Assistance for Low- and Moderate-Income Families

The Department of Housing and Urban Development (HUD) offers mortgage insurance to increase homeownership opportunities for low- and moderate-income families, especially those displaced by urban renewal. HUD insures lenders against loss on mortgage loans to finance the purchase, construction, or rehabilitation of low-cost, one- to four-family housing. Maximum insurable loans for an owner-occupant are $31,000 for a single-family home (up to $36,000 in high-cost areas). For a larger family (five or more persons), the limits are $36,000 or up to $42,000 in high-cost areas. Higher mortgage limits apply to two- to four-family housing.

Legislation establishing this program was enacted in 1954. Anyone may apply; displaced households qualify for special terms. (SEE ALSO: *Displacement; Housing Act of 1954; Mortgage Insurance*)

—*Mara Sidney* ◄

► Section 221(d)(3) and (4): Multifamily Rental Housing for Moderate-Income Families

The Department of Housing and Urban Development (HUD) provides mortgage insurance to finance rental or cooperative multifamily housing for moderate-income households, including projects designated for the elderly. Single-room occupancy (SRO) projects are also eligible for mortgage insurance. The department insures mortgages made by private lending institutions to help finance construction or substantial rehabilitation of multifamily (five or more units) rental or cooperative housing for moderate-income or displaced families. Projects in both cases may consist of detached, semidetached, row house, walk-up, or elevator structures. SRO projects may consist of units that do not contain a complete kitchen or bath.

The principal difference between the programs is that HUD may insure up to 100% of replacement cost under Section 221(d)(3) for public nonprofit and cooperative mortgagors but only up to 90% under Section 221(d)(4), irrespective of the type of mortgagor. Legislation establishing Section 221(d)(3) was enacted in 1954, and legislation establishing Section 221(d)(4) was enacted in 1959. Sections 221(d)(3) and 221(d)(4) mortgages may be obtained by public agencies; nonprofit, limited-dividend, or cooperative organizations; private builders; or investors who sell completed projects to such organizations. In addition, profit-motivated sponsors may obtain Section 221(d)(4) mortgages. Tenant occupancy is not restricted by income limits. (SEE ALSO: *Housing Act of 1954; Mortgage Insurance; Multifamily Housing*)

—Mara Sidney ◄

► Section 222: Homes for Service Members

The Department of Housing and Urban Development (HUD) provides federal mortgage insurance enabling members of the armed services on active duty to purchase a home partially subsidized by the service. HUD allows the Departments of Transportation and Commerce to pay the HUD mortgage insurance premium on behalf of service members on active duty under their jurisdiction. The mortgages may finance single-family dwellings and condominiums insured under standard HUD home mortgage insurance programs.

Legislation establishing this program was enacted in 1954. Service personnel on active duty in the U.S. Coast Guard or employees of the National Oceanic and Atmospheric Administration who have served on active duty for two years are eligible for the program. (SEE ALSO: *Military-Related Housing; Mortgage Insurance*)

—Mara Sidney ◄

► Section 223(e): Housing in Declining Neighborhoods

The Department of Housing and Urban Development (HUD) offers mortgage insurance to purchase or rehabilitate housing in older, declining urban areas. In consideration of the need for adequate housing for low- and moderate-income families, HUD insures lenders against loss on mortgage loans to finance the purchase, rehabilitation, or construction of housing in older, declining but still viable urban areas where conditions are such that normal requirements for mortgage insurance cannot be met. The property must be in a reasonably viable neighborhood and must be an acceptable risk under the mortgage insurance rules. The terms of the loans vary according to the HUD or Federal Housing Administration (FHA) program under which the mortgage is insured. HUD determines if a project should be insured under Section 223(e) and become an obligation of the Special Risk Insurance Fund, which supplements other HUD mortgage insurance programs.

Legislation establishing this program was enacted in 1968. Home or project owners ineligible for FHA mortgage insurance because property is located in an older, declining urban area are eligible for the program. (SEE ALSO: *Mortgage Insurance*)

—Mara Sidney ◄

► Section 223(f): Existing Multifamily Rental Housing

The Department of Housing and Urban Development (HUD) offers federal mortgage insurance under Section 207 pursuant to Section 223(f) for the purchase or refinancing of existing apartment projects; to refinance an existing cooperative housing project; or for the purchase and conversion of an existing rental project to cooperative housing. HUD insures mortgages under Section 207 pursuant to Section 223(f) to purchase or refinance existing multifamily projects originally financed with or without federal mortgage insurance. HUD may insure mortgages on existing multifamily projects under this program that do not require substantial rehabilitation. A project must contain at lease five units, and construction or substantial rehabilitation must have been completed for three years.

Legislation establishing this program was enacted in 1974. Investors, builders, developers, and others who meet HUD requirements are eligible. (SEE ALSO: *Mortgage Insurance*)

—Mara Sidney ◄

► Section 231: Mortgage Insurance for Housing for the Elderly

The Department of Housing and Urban Development (HUD) offers federal mortgage insurance to finance the construction or rehabilitation of rental housing for the elderly or handicapped. To ensure a supply of rental housing suited to the needs of the elderly or handicapped, HUD insures mortgages made by private lending institutions to build or rehabilitate multifamily projects consisting of five or more units. HUD may insure up to 100% of the estimated replacement cost for nonprofit and public mortgagors but only up to 90% for private mortgagors. Congregate care projects with central kitchens providing food services are not eligible. All elderly (62 or older) or handicapped persons are eligible to occupy units in a project insured under this program.

Legislation establishing this program was enacted in 1959, although multifamily housing for the elderly is now financed under the Section 221(d)(3) and (4) programs. (SEE ALSO: **Elderly**; *Mortgage Insurance*)

—Mara Sidney ◀

▶ Section 232: Nursing Homes, Intermediate Care Facilities, and Board-and-Care Homes

The Department of Housing and Urban Development (HUD) offers federal mortgage insurance to finance or rehabilitate nursing, intermediate care, or board-and-care facilities. HUD insures mortgages made by private lending institutions to finance construction or renovation of facilities to accommodate 20 or more patients requiring skilled nursing care and related medical services or those in need of minimum but continuous care provided by licensed or trained personnel. Board-and-care facilities may contain no fewer than five one-bedroom or efficiency units. Nursing home, intermediate care, and board-and-care services may be combined in the same facility covered by an insured mortgage or may be in separate facilities. Major equipment needed to operate the facility may be included in the mortgage. Facilities for day care may be included. As of October 1988, existing projects already insured by HUD are also eligible for purchase or refinancing with or without repairs under Section 232.

Investors, builders, developers, private nonprofit corporations or associations, public agencies (nursing homes only), or public entities licensed or regulated by the state to accommodate convalescents and persons requiring skilled nursing care or intermediate care may qualify for mortgage insurance. Patients requiring skilled nursing, intermediate care, and/or board-and-care are eligible to live in these facilities. (SEE ALSO: *Assisted Living; Board-and-Care Homes; Mortgage Insurance; Nursing Homes*)

—Mara Sidney ◀

▶ Section 234: Condominium Housing

The Department of Housing and Urban Development (HUD) provides federal mortgage insurance to finance the construction or rehabilitation of multifamily housing by sponsors who intend to sell individual units and to finance acquisition costs of individual units in proposed or existing condominiums. HUD insures mortgages made by private lending institutions for the purchase of individual family units in multifamily housing projects under Section 234(c). Sponsors may also obtain mortgages insured by the Federal Housing Administration (FHA) to finance the construction or rehabilitation of housing projects that they intend to sell as individual condominium units under Section 234(d). A project must contain at least four dwelling units; they may be in detached, semidetached, row house, walk-up, or elevator structures. The maximum mortgage amount for a unit mortgage insured under Section 234(c) is the same as the limit for a Section 203(b) mortgage in the same area. A condominium is defined as joint ownership of common areas and facilities by the separate owners of single dwelling units in the project.

Legislation establishing this program was enacted in 1961. Qualified profit-motivated or nonprofit sponsors may apply for a blanket mortgage covering the project after conferring with their local HUD-FHA field office. Any creditworthy person may apply for a mortgage on individual units in a project; however, it is generally limited to owner-occupants. (SEE ALSO: *Condominium; Mortgage Insurance; Multifamily Housing*)

—Mara Sidney ◀

▶ Section 235: Home Mortgage Interest Reduction

Section 235 was the first major U.S. subsidy program that permitted "lower-income" households to purchase rather than rent housing. It was enacted as part of the Housing and Urban Development Act of 1968.

Under the program, most of the housing was new and developed by private builders. The homes were financed through private lending institutions, and the mortgages were insured by the Federal Housing Administration (FHA) at a market interest rate. The Department of Housing and Urban Development (HUD) made interest subsidy payments to the mortgagee in an amount sufficient to cover the difference between the total costs of debt service, taxes, insurance, mortgage insurance, and an amount equal to 20% of the household's adjusted income. The maximum amount of the subsidy payment was the difference between the cost of debt service (plus the mortgage insurance premium) at the market rate of interest and at 1%. An eligible lower-income household was one whose income was not in excess of 135% of the local public housing limit for initial occupancy (taking into account the size of the household).

The Section 235 program was a success compared with previous federal efforts. Between 1968 and 1972, 400,000 units were produced under the program, compared with a total of 200,000 units built under public housing programs between 1935 and 1949 and with 410,000 public housing units built in the 1950s.

However, the program had a number of serious flaws. The program was inoperable in many metropolitan areas of the North and West. The mortgage limits of the program were $18,000 (1968 dollars) for a three-bedroom unit and $21,000 for larger houses—or $21,000 and $24,000, respectively, in high-cost areas. The median cost of a new home purchased under Section 235 was approximately $18,500. The median value of existing homes in the areas in which the program was popular was lower. More than 50% of those insured were in two of HUD's 10 administrative regions, the South and Southwest.

The program's incentives pushed toward a higher-priced house. Because the subsidy amount is based on debt service, the highest subsidies were paid for families that chose the most expensive houses. Although the program was advertised as a program designed for poor households, it missed its mark.

The segment of the program with respect to existing units in inner-city areas was racked with problems. Many of the households defaulted when they could not meet both mortgage payments and unexpected major maintenance costs. In marginal houses and neighborhoods, the program had deficiencies that could not be overcome. The subsidy was insufficient to avoid exposing the homeowner to the danger of unexpected catastrophes without pricing the buyer out of the market.

The Section 235 program also had a threefold equity problem. The first source of discontent came from families who were the intended beneficiaries: the households with incomes between $3,500 and $5,000. Fewer than 4% of the families entering the program in 1972 had incomes under $4,000. The National Tenant's Organization called the program a "perversion of subsidies from the neediest families." The second group that found the program unfair were the economic peers of the beneficiaries. As HUD Secretary George Romney noted at the Real Estate Boards Conference in Honolulu in 1972,

I don't blame people who are writing in to us and Congressmen, and saying, look, Joe Blow next door makes the same amount of money, he's got the same number of kids. Why should he get his mortgage interest costs reduced to 1 percent and I still have to pay 7 and 8 percent? . . .

I don't think you can avoid all inequity, but when it gets to that magnitude . . . you've got a real problem, and a growing problem, because even if we achieved the national housing goals by 1978, only one fourth of the families eligible for these housing programs will be getting assistance. There will be three complaining like Joe Blow, and only one that's getting help, and I'm telling you that isn't politically feasible.

The third source of discontent were households with a higher income, which did not understand why they were taxed to provide a better house, at a lower price, to households of perceived lesser worth next door.

In January of 1973, the program was suspended. After a brief revival in the mid-1970s with a lower subsidy and a higher income beneficiary, it was terminated. (SEE ALSO: *Housing and Urban Development Act of 1968; Section 236*)

—*Irving H. Welfeld* ◄

► Section 236: Rental and Cooperative Housing Interest Reduction

This program was designed to produce new and substantially rehabilitated rental units for "lower-income" households. Like the Section 235 program, it was an interest subsidy program enacted as part of the Housing and Urban Development Act of 1968. Under Section 236, housing was developed by private nonprofit and limited-profit (6% to 11% of the mortgage amount) developers. The maximum mortgage was 100% of the replacement cost of the project for nonprofits and 90% for limited profits. Projects were financed through private lending institutions, and the mortgages were insured by the Federal Housing Administration (FHA) at a market rate. Mortgage interest payments were made to the lenders as if the interest rate of the mortgage was equal to 1%, and HUD paid the mortgagee the difference between 1% and the market rate. Each tenant paid the greater of a basic rental (calculated on the assumption of a mortgage with a 1% rate) or 25% of adjusted income. To be eligible for a subsidy, a tenant had to have an income that was not in excess of 135% of the local public housing for initial occupancy (taking into account the size of the household). Rent collected by the mortgagor in excess of basic charges was returned to HUD. Ownership was restricted to nonprofit and limited-profit sponsors.

Section 236 was a success in producing housing. Between 1968 and 1972, 365,000 units were produced. The program was operable in all areas of the United States. It was also very profitable even for limited-profit developers because the project's equity was syndicated to high-income individuals who were not interested in the meager returns of the projects but, rather, in the sizeable "losses" resulting from the accelerated depreciation of the project. Section 236's production success was based on the key elements considered in calculating a mortgage based on debt service: the interest rate, the income limits, the rent-income ratio, and the estimate of operating expenses.

The 1% rate meant that a small amount of net income could support a large debt. In the Section 236 program, which provided for 40-year loans, every $1 supported $27 of mortgage principal.

In the case of estimating rental income, the lower-income limits were surprisingly high—often quite close to the median in the locality. The rent-income ratio was higher than in previous programs. Instead of projecting gross income of the project by that assuming a family at the income limit was paying 20% of its income for rent, FHA used 25% as the multiplier. This raised gross income projections by 25%.

To move from gross income to the amount available for debt service, it is necessary to estimate future expenses.

FHA underwriters were willing to accept the developers' low projections of expenses to further the national production goals for subsidized housing. The result was high mortgages that permitted developers to build throughout the United States.

The same elements that made Section 236 a production success also made it a management failure. Loans were processed with the assumption that all the residents would have incomes of close to $10,000, whereas the actual income of tenants was $6,000. This meant that tenants were paying far more than 25% of their income. To add to the problem, the operating guestimates proved to be far too low.

Tenants with lower incomes had to bear a high rent burden, and that threatened the stability of the project. In 1972, to protect the projects from poor people, HUD issued a regulation (later overturned by a federal court) that would have barred entry to households who would have to pay 35% or more of their income.

The legal and financial structure of the program resulted in a basic dilemma: If the projects were to be built, optimistic assumptions would have to be made, but these assumptions would price the poor out of the program; however, if realistic assumptions were to be made, the mortgages would be insufficient to get the project built.

The program was suspended in January of 1973, and no new commitments were issued. Faced with foreclosure of many of these projects, during the 1970s HUD applied a patchwork of programs to keep the projects afloat, including subsidies to cover a portion of tax and utility increases, flexible subsidies, and the use of Section 8 funds to provide higher rents to landlords and higher subsidies to tenants. A substantial amount of Section 8 funds (that were to be freely used to help tenants pay the rent in private unsubsidized housing) were set aside to cure the "loan management" problems of subsidized projects. There was hitch: The subsidy was tied to the project. If the tenant moved out, he or she could not take the subsidy along. (SEE ALSO: *Housing and Urban Development Act of 1968; Section 235*)

—*Irving H. Welfeld* ◄

► Section 237: Special Credit Risks

The Department of Housing and Urban Development (HUD) offers mortgage insurance and homeownership counseling for low- and moderate-income families with a credit history that does not qualify them for insurance under normal underwriting standards. HUD is also authorized to provide (through local HUD-approved organizations) budget, debt management, and related counseling services to these families when needed. Applicants may seek credit assistance under most Federal Housing Administration (FHA) home mortgage insurance programs. The insured mortgage limit is $18,000 ($21,000 in high-cost areas).

Legislation establishing this program was enacted in 1968. Low- and moderate-income households with credit records indicating ability to manage their financial and other affairs successfully if given budget, debt management,

and related counseling are eligible. One does not formally apply for this program. Applicants rejected for other programs are automatically screened for eligibility. (SEE ALSO: *Housing Credit Access; Mortgage Insurance*)

—*Mara Sidney* ◄

► Section 241: Supplemental Loans for Multifamily Projects and Health Care Facilities

The Department of Housing and Urban Development (HUD) offers federal loan insurance to finance improvements and additions to and equipment for multifamily rental housing and health care facilities. The program provides owners of eligible low-income housing with an adequate return on their investment and the ability to finance the acquisition of eligible low-income housing.

HUD insures loans made by private lending institutions to pay for improvements or additions to apartment projects, nursing homes, hospitals, or group practice facilities that already carry HUD-insured or HUD-held mortgages. Projects may also obtain FHA insurance on loans to preserve, expand, or improve housing opportunities, to provide fire and safety equipment, or to finance energy conservation improvements to conventionally financed projects. Major movable equipment for nursing homes, group practice facilities, or hospitals also may be covered by a mortgage under this program.

Owners of eligible low-income housing who have filed a plan of action under HUD's mortgage prepayment programs may receive an equity take-out loan insured by HUD to enable them to receive an adequate return on their investment, and purchasers of eligible low-income housing may receive HUD-insured acquisition financing.

Legislation establishing this program was enacted in 1968. (SEE ALSO: *Mortgage Insurance; Multifamily Insurance*)

—*Mara Sidney* ◄

► Section 245: Graduated-Payment Mortgages

The Department of Housing and Urban Development (HUD) offers federal mortgage insurance for graduated-payment-mortgages. HUD insures mortgages to finance early homeownership for households that expect their incomes to rise substantially. These mortgages allow homeowners to make smaller monthly payments initially and to increase their size gradually over time. Five plans are available, varying in length and rate of increase. Larger than usual down payments are required to prevent the total amount of the loan from exceeding the statutory loan-to-value ratios. In all other ways, the graduated-payment mortgage is subject to the rules governing ordinary HUD-insured home loans. Growing-equity-mortgages (GEMs) are insured under the same statutory authority.

Legislation establishing this program was enacted in 1974. All lenders approved by the Federal Housing Administration (FHA) may make graduated-payment mortgages; creditworthy owner-applicants with reasonable expectations of increasing income may qualify for such loans. (SEE ALSO: *Graduated-Payment Mortgage; Mortgage Insurance*)

—*Mara Sidney* ◄

► Section 251: Adjustable-Rate Mortgages

The Department of Housing and Urban Development (HUD) offers federal mortgage insurance for adjustable-rate mortgages (ARMs). Under this HUD-insured mortgage, the interest rate and monthly payment may change during the life of the loan. The initial interest rate, discount points, and margin are negotiable between the buyer and lender.

The one-year Treasury Constant Maturities Index is used for determining the interest rate changes. One percentage point is the maximum amount the interest rate may increase or decrease in any one year. Over the life of the loan, the maximum interest rate change is five percentage points from the initial rate of the mortgage. Lenders are required to disclose to the borrower the nature of the ARM loan at the time of loan application. In addition, borrowers must be informed at least 25 days in advance of any adjustment to the monthly payment.

Legislation establishing this program was enacted in 1983. All lenders approved by the Federal Housing Administration may make ARMs; creditworthy applicants, who will be owner-occupants, may qualify for such loans. (SEE ALSO: *Adjustable-Rate Mortgages; Mortgage Insurance*)

—*Mara Sidney* ◄

► Section 255: Home Equity Conversion Mortgage Insurance Demonstration

The Department of Housing and Urban Development (HUD) provides federal mortgage insurance to allow borrowers who are 62 years of age and older to convert the equity in their homes into a monthly stream of income or a line of credit.

Under the Home Equity Conversion Mortgage (HECM) Insurance Demonstration, the Federal Housing Administration (FHA) insures reverse mortgages that allow older homeowners to convert their home equity into spendable dollars. Reverse mortgages provide a valuable financing alternative for older homeowners who wish to remain in their homes but have become "house-rich and cash-poor." Any lender authorized to make HUD-insured loans may originate reverse mortgages.

Borrowers may choose from among five payment options:

1. *Tenure:* The borrower receives monthly payments from the lender for as long as the borrower lives and continues to occupy the home as a principal residence.

2. *Term:* The borrower receives monthly payments for a fixed period selected by the borrower.
3. *Line of credit:* The borrower can make withdrawals up to a maximum amount, at times and in amounts of the borrower's choosing.
4. *Modified tenure:* The tenure option is combined with a line of credit.
5. *Modified term:* The term option is combined with a line of credit.

The borrower retains ownership of the property and may sell the home and move at any time, keeping the sales proceeds in excess of the mortgage balance. A borrower cannot be forced to sell the home to pay off the mortgage, even if the mortgage grows to exceed the value of the property. An FHA-insured reverse mortgage need not be repaid until the borrower moves, sells, or dies. When the loan is due and payable, if the loan exceeds the value of the property, FHA insurance will cover any balance due the lender.

Legislation establishing this program was enacted in 1987. All borrowers must be at least 62 years of age. Any existing lien on the property must be small enough to be paid off at settlement of the reverse mortgage. (SEE ALSO: **Elderly**; *Mortgage Insurance; Reverse-Equity Mortgage*)

—*Mara Sidney* ◄

► Section 502: Single-Family Housing Program

Since 1949, the Section 502 Single-Family Housing Program has been the mainstay of the housing program component of the Farmers Home Administration (FmHA—now the Rural Housing Service [RHS]) of the Department of Agriculture. Authorized by the Housing Act of 1949, it is the largest and most popular of FmHA programs, resulting in more than 1.5 million loans to rural farm and non-farm households, totaling nearly $25 million. The number of loans peaked in the mid-1970s with a single-year commitment of 132,000 units but has decreased to less than 50,000 units annually in recent years.

There are two types of Section 502 single-family loans: insured and guaranteed. Insured loans are made directly by the FmHA (RHS) for the construction or purchase of new, modest-design housing and the purchase, rehabilitation, and refinancing of existing housing. To be eligible, borrowers must have incomes at or below 80% of the area median income; reside in housing that is not decent, safe, and sanitary; be unable to obtain a loan from private lenders with reasonable rates and conditions; and have sufficient incomes to repay the loan. Eligible borrowers can receive 33-year, no-down-payment loans for up to 100% of the cost of the unit's purchase, construction, or rehabilitation. Loans with smaller terms are made on existing structures with useful lives of less than 33 years.

The loans have market rates of interest based on the federal cost of borrowing. To be affordable to lower-income borrowers, however, the interest rates are written down by the FmHA (RHS) through an interest credit mechanism to as low as 1%, depending on the family's income and the loan amount. The total payment for interest, principal, taxes, and insurance must be within 20% of adjusted family income. Borrowers pay the greater of the loan payment at 1% interest rate, or 20% of adjusted income, minus real estate taxes and insurance. Effective in fiscal year 1996, RHS reduced subsidy levels and required new borrowers to make payments on a graduated, sliding-scale basis. For example, a family at 50% of median income pays the greater of the loan payment at 1% interest or 22% of income, while a family at 80% of median pays the greater of the loan payment at 6.5% interest or 26% of income. Efforts by the Clinton administration in 1994 to increase the maximum family contribution to the federal rental payment standard—30% of income—were defeated by low-income-housing advocates who argued that the cost of utilities and maintenance would force many low-income borrowers to pay well in excess of 30%.

Even with interest credit, many very low income borrowers with incomes of 60% or less of the area median income could not afford Section 502 loans and were not being served. Consequently, in 1983, Congress earmarked 40% of all insured loan dollars for very low income families and authorized the FmHA (RHS) to make loans to these borrowers with terms up to 38 years and permit up to 25% of the monthly payment to be deferred. In the more expensive markets, however, many such families still could not show repayment ability, and this set-aside of Section 502 funds was undersubscribed.

Typically, a builder can obtain a conditional commitment of Section 502 funds for the permanent, take-out financing and use this commitment to obtain short-term construction financing from a commercial lender. A builder can also use FmHA (RHS) funds for construction financing to be rolled over into permanent financing if the builder has a contract with the borrower.

The insured program is often used in combination with the FmHA's (RHS's) self-help housing program. Under the self-help method, nonprofit development corporations usually enter into a contract with the FmHA (RHS) under its Section 523 program to organize a certain number of low-income and very low income families and provide technical and supervisory construction assistance within the contract time, usually one to two years. These builders may also receive funds for site acquisition and land improvements under the Section 524 program. The families save money on construction costs by joining with other families and undertaking a major portion of the construction themselves, with the more highly skilled work contracted out. The value of their labor, called "sweat equity," is deducted from the amount of the Section 502 permanent loan that the FmHA (RHS) provides once the units are conveyed to the families. Guaranteed Section 502 loans are provided to higher-income families whose incomes do not exceed the area median income. The loans are made by commercial lenders at the market rate of interest to borrowers perceived as high risk and guaranteed by the FmHA (RHS). The guaranteed program is a fairly recent addition to the FmHA's (RHS's) housing component and has been undersubscribed in some states. Nonetheless, in the 1990s, Congress has tended to favor the guaranteed over the direct approach as reflected in increasing appropriations relative to the direct loan program. (SEE ALSO: *Farmers Home Administration [Rural Housing Service]; Housing Credit Access; Rural Housing*)

—Robert Wiener

Further Reading

California Coalition for Rural Housing. 1992. *Affordable Housing Handbook*. Sacramento: Author.

Collings, Art and Linda Kravitz. 1986. "Rural Housing Policy in America: Problems and Solutions." In *Critical Perspectives on Housing*, edited by Rachel Bratt, Chester Hartman, and Ann Meyerson. Philadelphia: Temple University Press.

Housing Assistance Council. 1994. *A Guide to Federal Housing and Community Development Programs for Small Towns and Rural Areas*. Washington, DC: Author. ◄

► Section 515: Rural Rental Housing Program

Since the early 1960s, the Section 515 Rural Rental Housing Program has been the main federal program for producing rental housing in rural areas and small towns. Administered by the Farmers Home Administration (FmHA—now the Rural Housing Service [RHS]) of the Department of Agriculture, about 400,000 low- and moderate-income units (as of FY 1993) have been built. In combination with FmHA's (RHS's) deep-subsidy rental assistance program, which functions like HUD Section 8 project-based subsidies, Section 515 currently serves a majority of households with very low incomes. It is one of the few 1960s/1970s-era rental housing production programs still remaining and the only one that places the government in the role of direct lender.

The program was created by the Senior Citizens Housing Act of 1962 (P.L. 87-723; 76 Stat. 670), which amended Title V of the Housing Act of 1949 by adding a new Section 515. FmHA (RHS) was authorized to provide government loans for the construction, rehabilitation, acquisition, and operation of rental and cooperative housing specifically for elderly persons and families of low or moderate income living in rural farm and nonfarm areas. In 1966, the program was expanded to include any rural low- or moderate-income household.

According to the report of the Senate Committee on Banking and Currency, Section 515 was intended to address the needs of a large group of previously unserved rural elderly—those who were unable to afford homeownership or who no longer had the incentive or capacity to maintain ownership of their homes. Because of the lack of appropriate alternatives, they remained in farm and nonfarm homes or moved to cities. The elderly programs of the Housing and Home Finance Agency (HHFA)—the predecessor to the Department of Housing and Urban Development (HUD)—operated mostly in urban areas and had failed to meet the physical, social, and financial needs of the rural elderly.

FmHA (RHS) was directed to make two kinds of loans. A new revolving loan fund was established from which FmHA (RHS) was to make direct, below-market-interest-rate (BMIR) loans, similar to HHFA's Section 202 program, for low- and moderate-income elderly. Eligible borrowers were private nonprofit corporations and consumer cooperatives. Loans were to be for 100% of development cost with maturities up to 50 years and subsidized rates of interest not to exceed the maximum rate in the Section 202 program, then 3.5%.

FmHA (RHS) was further authorized to insure loans made by private lenders to individuals, trusts, partnerships, associations, or corporations, similar to HHFA's Section 231 elderly insured program. Loans were for 95% of development cost, with interest rates linked to the federal cost of borrowing. The purpose was to provide multifamily housing for elderly households whose moderate incomes enabled them to afford insured but unsubsidized private financing.

In 1968, the landmark Housing and Urban Development Act introduced a new and innovative interest credit approach. In lieu of making BMIR loans, FmHA (RHS) was authorized to issue notes at the federal borrowing rate. The borrower's account was then credited with an amount that effectively reduced the interest rate to as low as 1%, depending on the tenants' incomes. This approach yielded lower rents than attainable under the BMIR method.

Section 515 remained a relatively low-production program until 1972 when FmHA (RHS) administratively modified it to permit limited-profit syndicates to receive subsidized loans. Within years, these builders and operators, motivated primarily by the tax-sheltered depreciation benefits derived from investment in such properties, eclipsed other entities as the main borrowers, more than doubling previous production in the program.

Although dramatically increasing production, FmHA's (RHS's) administrative expansion of the program resulted in unanticipated consequences for the tenants of the housing. A unique feature of FmHA (RHS) financing was a "graduation" provision that arose from its mandate to serve as lender of last resort. Borrowers were required, at the first opportunity, to prepay their loans and graduate to non-FmHA financing. Thus, from the late 1970s to the mid-1980s, thousands of units were prepaid by their limited-profit owners when tax shelters expired. Many converted to market rate housing with economic hardship and displacement of tenants.

In 1979, Congress moved to retroactively restrict prepayments on all current projects and require a 20-year "lock-in" on all future projects. One year later, Congress repealed the retroactive provision, allowing all pre-1979 projects to continue prepaying. A temporary moratorium on prepayments was imposed in emergency legislation in 1986, and prepayments were permanently restricted a year later in the Emergency Low-Income Housing Preservation Act (ELIHPA). In 1989, legislation required all future projects to remain affordable for the full 50 years of the mortgage.

In recent years, the Section 515 program has undergone several important changes. With passage of the federal low-income housing tax credit in 1986 and elimination of attractive depreciation offsets against other income, many Section 515 projects are now being developed by partnerships of nonprofits and private investors drawn by the tax credit. In addition, legislation in 1990 created a nonprofit set-aside requirement, 7% of appropriated funds, initially, rising to 9% in 1992. Since 1994, reported abuses and inefficiencies in the program, including the high cost of tax credit-financed transactions, have exposed the program to lengthy Congressional review, delays in reauthorization, and major cost-saving reforms, including a change in loan terms to 30 years. From fiscal year 1994 to fiscal year 1998, production activity slowed as appropriations declined from $512 million to $69 million. (SEE ALSO: *Farmers Home Administration;* **Private Rental Sector;** *Rural Housing*)

—*Robert Wiener*

Further Reading

Collings, Art and Linda Kravitz. 1986. "Rural Housing Policy in America: Problems and Solutions." In *Critical Perspectives on Housing,* edited by Rachel G. Bratt, Chester Hartman, and Ann Meyerson. Philadelphia: Temple University Press.

Housing Assistance Council. 1994. *A Guide to Federal Housing and Community Development Programs for Small Towns and Rural Areas.* Washington, DC: Author.

Wiener, Robert. 1990. "Prepayment Hits Countryside Hard." *Shelterforce* (March/April):18-21. ◄

► Section 608: Title VI, National Housing Act

In 1942, six months after the United States entered World War II, Section 608 was added to Title VI of the 1934 National Housing Act to stimulate private production of rental housing for war workers through federally insured mortgages. Between 1942 and the termination of the program in 1950, 465,683 units in more than 7,045 developments were built, to meet the housing needs of war workers and returning veterans during the postwar housing shortage. But rumors of fraud and malfeasance on the part of both private developers and Federal Housing Administration (FHA) staff resulted in a full-scale investigation of Section 608 and other FHA programs by the Senate Committee on Banking and Currency in 1954. Leading real estate developers, builders, mortgage bankers, and housing professionals were called to testify. The committee's final report found striking patterns of abuse and corruption, yet few were prosecuted because the statute of limitations had run out.

In the original 1942 legislation, Section 608-insured mortgages were calculated as $1,350 per room, with a cap of $5 million; mortgages were not to exceed 90% of the "reasonable replacement cost" of the completed development; and the occupants of the development were to be war workers. After the war, Section 608 was extended and revised by a series of amendments drafted to meet new emergency conditions: (a) the acute postwar housing crisis,

(b) the struggling building industry, and (c) the claims of the private sector that rent control and material shortages discouraged investment. These revisions included more flexible definitions of "project cost" and new formulas for calculating the maximum mortgage amount.

In response, the "608" program expanded dramatically. In 1947 alone, more than $350 million in mortgage commitments were processed, twice the total of those between 1942 and 1946. By the end of the program, $3,439,678,928 in mortgages had been insured. But a recession in the late 1940s caused building material costs to drop, and this increased the potential for smart builders to complete their projects at less than the amount of their original mortgage, a practice called "mortgaging out." Amid growing rumors of windfalls, Section 608 was allowed to expire on March 1, 1950.

To qualify for mortgage insurance, developers and builders had to follow the FHA's "minimum requirements," which dictated both the design and the location of multi-family moderate income rental developments. More than 50% of the Section 608 projects were located in six states: New York, New Jersey, Maryland, Virginia, Pennsylvania, and California, often near the growing defense industry. The FHA also published design guidelines in pamphlets and special inserts in architectural journals. Because project approval depended on following these design programs as closely as possible, many of the Section 608 projects (and those in Section 207) are remarkably similar. Most were built as two-story structures arranged around courtyards, a familiar apartment building typology. Variations in materials, roof forms, and decorative elements reflected regional preferences, but site plans almost always included segregation of pedestrian and auto traffic and some form of courtyard.

The FHA investigation began on April 12, 1954, when President Eisenhower asked the administrator of the Housing and Home Finance Agency (HHFA) to take protective custody of all FHA files. This action was in response to a report by the Internal Revenue Service that showed windfall profits for a number of Section 608 projects. At the same time, the Federal Bureau of Investigation (FBI) found widespread fraud in the Title I home improvement program. Both of these sections of the National Housing Act were investigated at the same time—internally by the HHFA and by the FBI and the Senate Committee on Banking and Currency (which was also considering the Housing Act of 1954). Authorized and funded under Senate Resolution 229, the Senate committee, chaired by Indiana Republican Homer Capehart, heard 43 days of public testimony between June and early October in Washington, New York, Los Angeles, New Orleans, Chicago, Indianapolis, and Detroit. A team of special investigators reviewed records for 543 projects. Public testimony on the FHA administration and financing of housing production often led to other topics, including slum clearance, public housing, racial integration, and cooperative developments. Many of those who were subpoenaed were members of the National Association of Home Builders, the National Association of Real Estate Boards, or the Mortgage Bankers Association.

Instead of testifying against "socialistic" programs such as public housing, as many had in the past, they now found themselves defending their own business practices.

The report on the FHA investigation was issued in January 1955. After noting the important contribution of Section 608 housing during the housing shortage, the committee stated that it was "not prepared to accept the premise that adequate rental housing cannot be made available to the American people except when unconscionable profits are realized through abuses and irregularities in the program." It found that many of the developers of Section 608 projects received mortgages far above their actual land and construction costs. Other findings included inflated architect fees, fraudulent financial statements, and acts of corruption, including "substantial entertaining and wining and dining of FHA people by builders," as well as employment of FHA staff by the builders whose projects they were approving. Despite the work of the investigative team, led by special counsel William Simon, the only criminal charges resulting from the investigation were brought against one FHA employee, Clyde Powell. There was, however, an informal "grey list" of builders and developers, preventing their participation on subsequent FHA insured projects.

Section 608 most often appears in reviews of federal housing policy and programs as just a footnote, an example of fraud and scandal best forgotten. A closer examination of both the housing produced under this program and its subsequent investigation suggests that the history of Section 608 can provide a broader context for our understanding of federal housing policy in succeeding decades. The FHA investigation identified many of the dangers inherent in a system of private sector production of affordable rental housing. Although some changes were made in the administration and underwriting of FHA mortgage insurance programs, "abuses and irregularities" continued to plague federal housing programs, as the HUD scandals of the early 1970s and late 1980s remind us. (SEE ALSO: *History of Housing; Mortgage Insurance; Real Estate Developers and Housing*)

—*Gail Sansbury*

Further Reading

"Apartment Boom." 1950. *Architectural Forum* 92(1):97-106.

"FHA's Five-Year-Old Scandal." 1954.*House and Home*, May, Special insert.

Sansbury, Gail. 1997. "Dear Senator Capehart: Letters Sent to the U.S. Senate's FHA Investigation." *Planning History Studies* 11(2):19-46.

U.S. Congress. 1950. Senate Committee on Banking and Currency. *Conference on Section 608, Title VI, National Housing Act.* 81st Cong., 2d sess.

———. 1954. Senate Committee on Banking and Currency. *Hearings on the FHA Investigation.* 83rd Cong., 2d sess.

———. 1955. Senate Committee on Banking and Currency. *The Report on the FHA Investigation, Pursuant to S. Res. 229.* 84th Cong., 1st sess.

U.S. Federal Housing Administration. 1942. *Minimum Requirements for Rental Housing Projects.* Submitted under Title VI, Section 608, National Housing Act. Washington, DC: Government Printing Office.

———. 1947. *Planning Rental Housing Projects.* Washington, DC: Government Printing Office.

Welfeld, Irving. 1992. *HUD Scandals: Howling Headlines and Silent Fiascoes.* New Brunswick, NJ: Transaction Publishing. ◀

▶ Section 810: Urban Homesteading

Section 810 of the Housing and Community Development Act of 1974, as amended, marked the entry of the federal government into urban homesteading. Section 810 authorized the federal Urban Homesteading Demonstration project. Program operations began under the Urban Homesteading Demonstration Program in 23 cities in the fall of 1975. HUD selected an additional 16 demonstration cities in May 1977. Results from the evaluation of the demonstration cities were encouraging enough for HUD to announce its promotion to operating status in September 1977. Section 810 urban homesteading became a fully operational program when final guidelines were published in 1978. The National Affordable Housing Act terminated the Section 810 Urban Homesteading Program effective October 1, 1991.

The broad Section 810 guidelines established by Congress aimed to ensure the fairness, timely completion, and coordination of urban homesteading with other neighborhood improvement efforts—not to dictate specific procedures for satisfying these requirements. The legislation required a coordinated approach toward neighborhood improvement that prohibited scattered-site or citywide homesteading. Local and state governments participating in urban homesteading were to specify their own goals and design programs based on local considerations within this framework. However, by 1988, HUD was encouraging local urban homesteading agencies (LUHAs) to plan on homesteading a minimum of five properties per year for their programs to be cost-effective and have discernible neighborhood impact.

Section 810 legislation imposed guidelines in three areas: (a) how long homesteaders were given to make repairs, (b) how long they must reside in the property to receive final title, and (c) homesteader eligibility criteria. Program regulations regarding homesteader selection changed over time. Homesteaders were initially given 18 months to make repairs sufficient enough to permit occupancy and three to four years to complete work conforming to local building code standards. These figures were later changed to three and five years, respectively.

The flexibility of Section 810 criteria allowed localities to develop and pursue their own goals. Localities were also permitted discretion in carrying out functions required in the urban homesteading process: selection of neighborhoods, properties, and homesteaders; conditional conveyance of properties; and arrangement of rehabilitation financing. Cities used the Section 810 program to provide low- and moderate-income homeownership opportunities and stabilize neighborhoods or to target a higher-income population.

The legislation provided for the transfer of unoccupied one- to four-family properties owned by the U.S. Department of Housing and Urban Development (HUD), the Department of Veterans Affairs, the Farmers Home Administration, and the Resolution Trust Corporation to states and local governments with homesteading programs approved by HUD. Until 1979, only HUD-owned properties were eligible for Section 810 reimbursement.

Under the final rules for Section 810, properties were eligible for acquisition if the appraised as-is fair market value of the property did not exceed $25,000 for a one-unit, single-family residence or an additional $8,000 for each unit of a two- to four-family structure. Earlier, the maximum value had been $20,000 with an additional $5,000 allowed per unit. HUD had the ability to authorize acquisitions when the value of a one-unit property was as much as $35,000.

Participating states and localities conditionally conveyed the properties at a nominal cost or no cost to homesteaders who agreed to repair them as required. The repairs had to entail all defects that posed a threat to health and safety and meet local code standards. When these conditions were met, the homesteader acquired fee simple title to the property. This stipulation encouraged individual homeownership and discouraged speculation.

Funding

Funds authorized for Section 810 were used to reimburse the Federal Housing Administration Fund or the Rehabilitation Loan Fund for losses incurred under the urban homesteading program and to reimburse the secretary of Veterans Affairs, the secretary of Agriculture, and the Resolution Trust Corporation for properties conveyed for use in connection with Section 810 urban homesteading programs. In 1984, city-acquired properties became eligible for purchase using Section 810 funds within the Local Urban Property Homesteading Demonstration Program. This demonstration program concluded in July 1987 with 11 cities as participants. The majority of Section 810 funds have been used to acquire HUD-owned properties.

Section 810 did not provide funds for the rehabilitation of properties. Those funds had to be secured from other sources. The sources, in descending order, were the Section 312 Rehabilitation Loan Program, Community Development Block Grant, other public funds, and private funds (including commercial loans and personal savings).

Fiscal support of urban homesteading was initially inconsistent. Between 1975 and 1980, Congress appropriated $55 million for the acquisition of urban homesteading properties. During fiscal years 1980 and 1981, no appropriations were made. After that time, the annual appropriations ranged between $12 million and $14.4 million dollars. When the program was ended, Congress had appropriated a total of $168.2 million to support the acquisition of properties for the program.

Selection of Homesteaders

The issue of who the homesteaders should be was debated prior to the passage of Section 810. At the heart of the debate was whether lower-income individuals in need of housing or upper-income individuals, with more disposable income and access to credit, should benefit from the program. Congress evaded the issue and gave local program implementers the responsibility of choosing homesteaders. The implementers were to consider two factors in the selection process—the need for housing and the capacity to make the necessary repairs and improvements. The tension between the "housing need" criterion, which favored lower-income applicants, and the "capacity to repair" criterion, which favored higher-income applicants, was left unresolved.

In 1983, regulations were changed to exclude owners of other residential properties. In addition, the applicant's ability to contribute toward repairs was taken into account. Priority consideration was required for those meeting three criteria: (a) rent payments in excess of 30% of adjusted income, (b) residence in substandard housing, and (c) little or no prospect of obtaining housing in the near future. Finally, the Housing and Community Development Act of 1987 replaced the three-part priority requirement with a single mandate to give preference to lower-income households (with incomes below 80% of the area median) before other applicants.

Program Variations

The Housing and Urban-Rural Recovery Act of 1983 authorized HUD to undertake a Local Property Urban Homesteading Demonstration Program under Section 810(i). The purpose of the program was to provide funds to test the feasibility of local properties early in the tax foreclosure process. The premise of the demonstration was that tax-delinquent properties often lose most of their economic value during the lengthy local foreclosure process. The expectation was that timely acquisition would reduce the likelihood of vandalism and neighborhood deterioration and increase the feasibility of rehabilitation.

The 1983 Housing and Urban-Rural Recovery Act also authorized HUD to conduct a Multifamily Urban Homesteading Demonstration Program to determine whether it is both practical and cost-effective for localities to help lower-income tenants acquire and rehabilitate multifamily projects for ownership. The multifamily demonstration would support only the use of HUD-owned properties. The response to the program announcement revealed that potential participants were primarily interested in federal technical assistance and did not intend to use HUD properties. The funds set aside for this demonstration were then reallocated into the regular Section 810 program.

Section 810 was a relatively small initiative compared with other housing and community development programs. It provided funds to secure properties but none for rehabilitation or administration. In effect, localities that chose to participate in it were required to divert some of their Community Development and Section 312 allocations to that program. Section 810 urban homesteading did not fare well in cities where it was in tight competition for rehabilitation and administrative dollars. Some communities used local revenues or Community Development Block Grant

funds to operate their own homesteading programs. (SEE ALSO: *Urban Homesteading*)

—*Mittie Olion Chandler*

Further Reading

Borgos, Seth. 1986. "Low Income Homeownership and the ACORN Campaign." In *Critical Perspectives in Housing*, edited by Rachel G. Bratt, Chester Hartman, and Ann Meyerson. Philadelphia: Temple University Press.

Chandler, Mittie Olion. 1988. *Urban Homesteading: Programs and Policies*. Westport, CT: Greenwood.

———. 1991. "The Evolution of Urban Homesteading: Planning for Lower-Income Participation." *Journal of Planning Education and Research* 10(2):145-154.

U.S. Department of Housing and Urban Development. 1981. *Evaluation of the Urban Homesteading Demonstration Program: Final Report*. Vol. 2. Washington, DC: Government Printing Office.

———. 1986. *Consolidated Annual Report to Congress on Community Development Programs*. Washington, DC: Government Printing Office.

———. 1987. *The Local Property Urban Homesteading Demonstration*, Washington, DC: Government Printing Office.

———. 1988. *Consolidated Annual Report to Congress on Community Development Programs*. Washington, DC: Government Printing Office.

———. 1992. *Consolidated Annual Report to Congress on Community Development Programs*. Washington, DC: Government Printing Office. ◀

▶ Section 811: Supportive Housing for Persons with Disabilities

The Department of Housing and Urban Development (HUD) provides assistance to expand the supply of housing with supportive services for persons with disabilities. Capital advances are made to eligible nonprofit sponsors to finance the development of rental housing with supportive services for the disabled. The advance is interest free and does not have to be repaid so long as the housing remains available for 40 years to very low income persons with disabilities. Project rental assistance covers the difference between the HUD-approved operating cost per unit and the amount the resident pays.

Legislation establishing this program was enacted in 1990. Nonprofit organizations may qualify for assistance. Occupancy is open to very low income persons with disabilities between the ages of 18 and 62.

—*Mara Sidney* ◀

▶ Security Deposit

A security deposit is any money, property, or right given by a tenant to a landlord to be held by, or on behalf of, the landlord as security for the performance of an obligation or payment of a liability by the tenant and to be returned to the tenant on the happening of some event (e.g., lease termination). As with any other civil contract, the landlord can sue a tenant for breach of a lease condition and seek a court-imposed settlement. However, some landlords may find it difficult to even locate, let alone collect a court award from, former tenants. The security deposit is simply a convenience for landlords, who can then impose, and be reimbursed for, claims to damages up to the amount of the security deposit and put the onus on a dissatisfied tenant to apply for a court settlement.

Whatever their merits in principle, security deposits are problematic in practice. Some tenants can ill afford this lump sum payment. Others find that the landlord, acting in bad faith, refuses to return the deposit or overestimates damage and that applying to the court to redress the situation is too costly. Furthermore, landlords typically do not pay interest on the security deposits that they hold. There is also a question of who is responsible for repaying a security deposit to the tenant when a rental property is sold. Although the security deposit is part of a "covenant" between landlord and tenant, this has not been seen by U.S. courts to be a covenant that obliges the new landlord to assume the responsibilities of the former landlord. For their part, landlords find that some tenants simply reduce their rent payment at the end of a lease by the amount of the security deposit.

In light of these problems, many jurisdictions across North America and elsewhere now regulate the taking, holding, and disposition of security deposits as an element of landlord-tenant legislation. Most jurisdictions have a strict upper limit on the amount of a security deposit (generally related to the monthly rent paid). Some require that the landlord pay interest on the security deposit. In some jurisdictions, the status of security deposits has been changed from a debt (an amount owed by the landlord to the tenant) to a trust with a corresponding change in performance expected of the deposit holder. Some jurisdictions even have a public office (e.g., rentalsman) that holds security deposits and mediates disputes between landlord and tenant. (SEE ALSO: *Lease; Private Rental Sector*)

—*John R. Miron*

Further Reading

Jacobus, Charles J. 1966. *Real Estate Principles.* Upper Saddle River, NJ: Prentice Hall.

Miceli, T. J. 1992. "Habitability Laws for Rental Housing." *Urban Studies* 29(1):15-24.

Miron, J. R. 1990. "Security of Tenure, Costly Tenants and Rent Regulation." *Urban Studies* 27(2):167-83.

Shim, Jae K. 1996. *Dictionary of Real Estate.* New York: John Wiley.

Tosh, Dennis S. 1990. *Handbook of Real Estate Terms.* Englewood Cliffs, NJ: Prentice Hall. ◄

► Segregation

The overall pattern of housing in the United States is one of segregation. Segregation in housing exists wherever the residential distribution of one population group differs from that of another population group. Segregated housing leads to segregation in other areas of life—schooling, religion, recreation, and employment, for example. Housing segregation is related to inequality and subordination;

it limits the options for social mobility by consigning the segregated group to inferior life chances.

There is a continuing debate about the causes of residential separation. The debate is about the relative role of housing affordability, residential preferences, job location, and private acts of discrimination in creating the patterns of separation in the residential fabric and how to assess the relative weight of discrimination in creating residential separation. Current thinking recognizes that the creation of separation in the residential mosaic is a complex process and the explanation is multifaceted. HUD studies show that as many as half of black renters and buyers may experience discrimination. Further research is needed to assess the ways in which discrimination is translated into separation in the residential structure.

Measuring Segregation

The most common measure of segregation is the index of dissimilarity, which assesses the degree to which two groups are unevenly distributed over neighborhoods in metropolitan areas. The basis on which unevenness is determined is the percentage composition of the metropolitan area as a whole. If a metropolitan area is 30% black and 70% white, for example, then the expected criterion for nonsegregation is that every neighborhood would be 30% black and 70% white. As each neighborhood deviates from the percentage racial composition of the metropolitan area as a whole, the segregation index increases. The index ranges from zero, reflecting no segregation (i.e., complete evenness of racial groups over neighborhoods), to 100, indicating complete unevenness. The higher the index, the greater the amount of segregation. Other, complementary indices of segregation exist as well, for example, Lieberson's Index of Isolation.

Segregation by Race

The most prevalent form of segregation in the United States is segregation by race in general and segregation between blacks and whites in particular. Before 1910, the level of segregation of blacks did not differ significantly from that of European immigrant groups. This period has been described by some authors as the period before the ghetto—that is, a time when most blacks were not severely segregated from whites. Historical investigations of several cities, however, reveal sharp increases in black residential segregation between 1910 and 1930. For most northern cities, this was the first period of large-scale black migration from the South. It resulted in the formation of black ghettos, as blacks were restricted to certain sections of cities and became the majority racial group in those sections. Until the 1970s, southern cities were less segregated than those in the North.

From a historical standpoint, residential segregation between blacks and whites increased to high levels and has remained there. The level of segregation of blacks in the 1990s is higher than the levels for any other group (European ethnic groups, Hispanics, Asians, etc.).

The reason relates to the issue of color and more specifically black color. Unlike Europeans, black migrants to cities in the North faced impenetrable barriers to housing outside of predominantly black areas. European immigrants who were not restricted because of race were often able to overcome discrimination based on ethnicity and to reside outside of ethnic areas. Black migrants to Northern cities, although native-born Americans, were usually restricted, both by law and social custom, from moving outside of predominately black areas, regardless of income.

Thus, the factor of color sharply distinguished the residential segregation of blacks from that of European immigrants. Although the native-born white population viewed European immigrant groups as inferior, there was clearly a discrimination continuum based on skin color. Thus, segregation between native-born whites and white European immigrant groups has been lowest. On the other hand, segregation has been highest between native-born whites and native-born blacks. The levels of segregation between whites and Asians have usually been between these two extremes.

In short, residential segregation between native whites and European ethnic groups declined as those groups increased their social and economic status through better education, better jobs, and higher levels of income. This same relationship between residential segregation and socioeconomic status has also been true for Hispanics and Asian Americans.

However, because of deeper prejudice and discrimination against blacks, no such relationship exists between black residential segregation and socioeconomic status. The implications are clearly different for other minority groups compared with blacks. For other minority groups, socioeconomic mobility leads to significantly reduced levels of residential segregation and ultimately to greater assimilation. For blacks, socioeconomic mobility is no guarantee of freedom of spatial mobility—that is, freedom to move into the residential area of one's choice subject only to ability to pay. Black suburbs are an indication of this. Chicago is an example of a metropolitan area with highly segregated city *and* suburban neighborhoods. Thus, the opportunity for assimilation by blacks is less.

More recently, a new segregation has emerged that is manifested in public housing where there is both economic and racial segregation. This segregation has been shaped and influenced by action of the federal and local governments.

Segregation in Public Housing

Before 1954, when the U.S. Supreme Court declared racially separate schools unconstitutional, most public housing projects were designed to specifically serve black residents or white residents, but not both. Although the policy after the 1960s was not to discriminate on the basis of race, the old segregated pattern has largely remained. Thus, public housing that serves nonelderly families is usually overrepresented with blacks and located in central cities, whereas public housing that serves elderly individuals is usually disproportionately white and located in the suburbs.

Public housing authorities have traditionally based site selection on racial characteristics of neighborhoods. Because neighborhoods were usually racially segregated, public housing resulted in reinforcing the existing segregation. Local politics played an important role. Neighborhoods with strong political organizations often prevented public housing construction in their neighborhoods.

Race is thus an important factor related to segregated public housing. Projects have generally been located in areas that were already black or rapidly becoming black. In 1993, the last year in which racial data on public housing are available, the segregation between blacks and whites in the largest public housing projects was very high (i.e., 70%) and with few exceptions closely corresponded with the level of segregation in private housing.

As in private housing, blacks are the most segregated group in public housing. In allocating assignments to public housing, federal and local governments took an aggressive role in reinforcing existing patterns of racial segregation.

The Causes of Segregation

At least three different theories have been presented to explain neighborhood segregation. First, class theory states that racial groups are distributed unequally by income, education, and occupation. Because neighborhoods are located in different parts of metropolitan areas by cost of housing, racial segregation of neighborhoods occurs.

Second, voluntary segregation or preference theory suggests that whites and nonwhites prefer to live in separate neighborhoods with members of their own race. As a result, blacks and whites voluntarily segregate themselves into racially homogeneous neighborhoods.

Third, discrimination theory states that nonwhite racial groups are denied equal access to housing in white neighborhoods by a combination of practices such as white refusal to sell or rent, racial steering by white real estate brokers, and other prohibitive actions by financial institutions, builders, and state, local, and federal governmental agencies.

Most empirical evidence has shown the invalidity of the class theory. Census data have shown that, given the same occupation, education, and income, most blacks and whites still do not live in the same neighborhoods. Blacks and whites in poverty usually live in separate neighborhoods and so do affluent blacks and whites. If blacks and whites were distributed over neighborhoods on the basis of class, most neighborhoods would contain numerous blacks and whites, and residential segregation in cities and suburbs would be low. Thus, neighborhood segregation between blacks and whites does not occur merely because blacks are poorer, less educated, or in lower-status jobs.

The invalidity of the voluntary or preference theory has been shown by the results of surveys of black attitudes toward housing and race. Most surveys suggest that a majority of blacks prefer racially mixed as opposed to all-black neighborhoods. Most blacks prefer mixed neighborhoods for positive reasons of racial harmony. The minority of blacks who prefer all black neighborhoods prefer them for reasons related to a desire to avoid interracial tension and strife. In other words, their preference is strongly influenced by the perceived reactions of whites to black residency in the neighborhood.

The strongest evidence identifies racial discrimination as the primary cause for segregated housing in the United States. Although blatant discriminatory practices in housing are not as common today as they were prior to the passage of the Fair Housing Act of 1968, a great deal of research documents the validity of the discrimination theory. In general terms, racial discrimination in housing exists wherever individuals of a certain racial group are prevented for racial reasons from obtaining the housing they want in the location they prefer.

Black individuals are prevented from exercising a purely economic choice through the use of techniques, policies, or both designed to avoid selling or renting housing to blacks in a given location. Such discrimination by race hampers the development of integrated neighborhoods, despite improved white attitudes toward blacks and increases in the socioeconomic status of blacks.

Black residential segregation is the result of a cumulative process involving past and present discriminatory practices by real estate brokers, homebuilders, financial institutions, the federal government, and state and local governments. Like other urban social problems that have deep roots in history, black residential segregation cannot be totally understood without comprehending the roles that each actor has played in causing these patterns.

Real estate brokers have played the role of primary "gatekeeper" through whom most information and activity in the housing market must pass. Traditionally, real estate brokers have operated on the assumption that residential segregation is a business necessity and a moral absolute.

Prior to the 1968 Fair Housing Act, white real estate brokers simply refused to show or sell blacks homes in predominantly white areas. The Civil Rights Act of 1968 and the Supreme Court decision of the same year have made discriminatory behavior by white real estate brokers more difficult. Although black home seekers still occasionally encounter outright refusal of brokers' services, it is no longer common practice for white brokers simply to refuse to sell a particular house to a black home seeker. Discriminatory behavior, however, has not disappeared.

The most effective, subtle, and widespread discriminatory technique used by white real estate brokers presently is "racial steering." This is a practice by which a real estate broker directs buyers toward or away from particular houses or neighborhoods according to the buyer's race. Black home seekers are steered away from white areas, whereas whites are directed to them. Real estate brokers have used a variety of reasons to justify racial steering. One reason is based on the long-standing assumption of the housing industry that black and white groups will not willingly share the same residential areas. This assumption results from traditional housing industry values.

Although discrimination is illegal, recent national studies conducted by the U.S. Department of Housing and Urban Development indicate that widespread discrimination against black home seekers and renters continues. The resistance by whites to housing desegregation has far more explanatory power for continued racial segregation of neighborhoods than either the economic status of blacks or the theory that blacks prefer to live in black neighborhoods.

Given the severity and the persistence of discrimination in housing, it is clear that solutions to the problem of housing segregation must go beyond current weakly enforced antidiscrimination policies. (SEE ALSO: *African Americans; Asian Americans;* **Discrimination;** *Index of Isolation; Racial Integration; Residential Location; Segregation Index; Tipping Point; Yonkers*)

—*Joe T. Darden*

Further Reading

Bickford, Adam and Massey, Douglas S. 1991. "Segregation in the Second Ghetto: Racial and Ethnic Segregation in American Public Housing, 1977." *Social Forces* 69(4):1011-36.

Bullard, Robert D., ed. 1994. *Residential Apartheid: The American Legacy.* Los Angeles: University of California, Los Angeles, CAAS Publications.

Darden, Joe T. 1973. *Afro-Americans in Pittsburgh: The Residential Segregation of a People.* Lexington, KY: D. C. Heath.

Frey, William H. and Farley, Reynolds. 1996. "Latino, Asian, and Black Segregation in U.S. Metropolitan Areas: Are Multiethnic Metros Different?" *Demography* 33(1):35-50.

Massey, Douglas S. and Nancy Denton. 1993. *American Apartheid: Segregation and the Making of the Underclass.* Cambridge, MA: Harvard University Press.

Momeni, Jamshid, A., ed. 1986. *Race, Ethnicity and Minority Housing in the United States.* Westport, CT: Greenwood.

Taeuber, Karl E. and Alma F. Taeuber. 1965. *Negroes in Cities: Residential Segregation and Neighborhood Change.* Chicago: Aldine.

Tobin, Gary, ed. 1987. *Divided Neighborhoods: Changing Patterns of Racial Segregation.* Newbury Park, CA: Sage. ◄

► Segregation Index

Segregation indices are statistical measures of levels of racial residential segregation. These indices facilitate comparisons among cities and other areas, comparisons among racial and ethnic groups, analyses of economic and other contributory causes, and studies of consequences. Two of the most commonly used segregation indices have to do with *dissimilarity* and *exposure.* The uses and interpretation of each index depend on its specific characteristics; multiple indices are required because segregation is a complex social pattern.

The most common use of segregation indices in housing studies is to assess the pervasiveness and persistence of residential segregation among racial and ethnic groups in urban areas. These indices are also used for measuring the segregation of economic and social groups, such as low-income households and single-parent families. The indices have many applications beyond housing segregation, notably as measures of school segregation, occupational segregation, industrial location, and income inequality.

The underlying statistical concepts and many of the measures were developed by economists and geographers, beginning early in the century and continuing to the present. Applications to racial segregation began in the late 1940s as sociologists sought to use newly available census data to quantify the racial and ethnic organization of urban life that was central to urban studies and urban social issues.

The measurement of racial residential segregation may be illustrated by describing the most frequently used measure, the index of dissimilarity. It measures the unevenness in the distributions of two groups among subareas. Imagine a map of any U.S. metropolitan area (or city), with subareas shaded according to the percentage of residents who are African Americans. (To simplify, assume that every resident is either African American or white.) Segregation is obvious. There are clusters of subareas where few whites live, other clusters where few African Americans live, and only a scattering of subareas with substantial presence of both groups.

The dissimilarity index (D) assigns a score ranging from 0 to 1 (often expressed as a percentage ranging from 0 to 100) based on the amount of racial diversity among subareas. If all subareas on the map have the same racial composition and the same shading, there is no segregation and D = 0. A score of 1 indicates complete segregation: every subarea has only African American residents or only white residents.

The data needed to calculate the index are as follows:

- ▶ The number of subareas (N)
- ▶ The number of residents in each subarea (t_i) and in the total area (T)
- ▶ The number of minority residents in each subarea (m_i) and the total minority population (M)
- ▶ The proportion of minority residents in each subarea ($p_i = m_i \div t_i$) and in the total area ($P = M \div T$).

$$D = \Sigma\, t_i(p_i - P) \div 2TP\,(1 - P).$$

In words, the index is a weighted average of the deviations between subarea minority proportions and the total minority proportion. It is also the minimum proportion of minority (or nonminority) residents who would have to be moved to produce zero segregation (complete evenness).

The term *minority group* is used in a very flexible way for purposes of calculation. The terms *m* and *M* may designate any group. To measure the segregation between the designated minority group and the aggregate nonminority population, include the total population in the terms *t* and *T*. To measure segregation between two specific groups, such as African Americans and Asian Americans, use *t* and *T* as the totals of the two groups, excluding all other residents.

The dissimilarity index is calculated using the total area's minority proportion as a standard. In a metropolitan area that is 75% minority, the index tallies the deviations of subareas from the 75% figure. In a metropolitan area that is 6% minority, the index tallies the deviations of subareas from the 6% figure. Differences in scores on the dissimilarity index are not directly affected by such differences in aggregate racial composition. This is one reason the dissimilarity index is widely used for comparing segregation among metropolitan areas (and cities) and for tracking changes in segregation through time.

Differences in aggregate racial composition are of great interest for some purposes, and another segregation index may be useful. Consider the two metropolitan areas with minority proportions of 75% and 6% and assume that neither area has much segregation. In one metropolitan area, the typical subarea is approximately 75% minority. When minority or nonminority persons venture around their residential neighborhoods (subareas), about 75% of the persons they contact (or encounter or are exposed to) are minority persons and about 25% are nonminority persons. In the other metropolitan area, the neighbors to whom a resident is exposed are about 6% minority persons and 94% nonminority persons. The exposure index captures this difference. (The exposure index is also referred to as the contact index, P^* index, P-star index, and isolation index.)

The formula for the exposure index uses symbols that identify the two subgroups for which the index is calculated. For example, let W = total white population and w_i = subarea white population, and let M and m represent the minority population (everyone who isn't white):

$$E_{wm} = \Sigma\, (w_i)\, (m_i \div t_i) \div W.$$

This is the exposure of whites to minorities. In words, it is an average of the minority proportions that whites encounter in their residential neighborhoods.

For each metropolitan area, there are four versions of this basic index: two "interaction" indices, E*wm* and E*mw*, and two "isolation" indices, E*ww* and E*mm*. These indicate the average exposure in their neighborhoods of whites to minorities, minorities to whites, whites to other whites, and minorities to other minorities.

A surprisingly large array of other indices can be calculated from the simple data set used to calculate dissimilarity indices and exposure indices. There has been much debate about the advantages and disadvantages of various measures, sometimes with an emphasis on methodological aspects and sometimes with more attention to conceptual issues. One criticism of the index of dissimilarity is that the zero point, perfect evenness, is unrealistic; an alternative index defines the zero point as a random distribution. Another criticism is that the dissimilarity and exposure indices can be calculated only from data for two groups. In a multi-ethnic city there are many pairs of groups: whites versus nonwhites, African Americans versus Asian Americans, Japanese Americans versus Korean Americans, and so on. Some indices allow simultaneous consideration of more than two groups.

All segregation indices are sensitive to how subareas are defined. Census data are usually used, so the choice is census tracts, block groups, or blocks. The smaller the subarea, the greater the amount of segregation that can be detected.

For dissimilarity, exposure, and some of the other segregation indices, the only information used about a subarea is its racial composition. This gives rise to what is called the *checkerboard problem*. If there are 32 all-minority subareas and 32 all-white subareas, the dissimilarity score is 1 whether the subareas are arranged into one minority ghetto and one white ghetto, or into a checkerboard of 64 interspersed mini-ghettos. Indices have been developed to take account of a variety of features of the location of each subarea. Location may be defined (a) by the racial or other attributes of the adjoining subareas, (b) by distance from one or more "ghettos," (c) by distance from the city center or from workplaces employing minorities, and (d) in other ways that capture specific features of the geographic pattern.

Segregation indices have proved a useful tool for social science and policy analysis. Their use has facilitated exploration of connections among spatial arrangements, social organization, the processes of assimilation, and emerging patterns of racial and ethnic differentiation. In response to the urban racial conflicts of the mid-1960s and later urban racial issues, policy analysts became more interested in quantitative studies of residential segregation. Indices documented the prevalence of racial segregation in all regions of the United States at all census dates through 1990. Analyses of indices demonstrated that racial economic differences did not account for much of the segregation, that segregation between blacks and whites was more extreme and more persistent than segregation between other racial and ethnic groups, and that housing discrimination continued to be influential long after it was outlawed.

Dozens of statistical measures are available for use as segregation indices. Each index summarizes a particular feature of a residential pattern. The availability of a wide array of measures is a valuable resource. Residential patterns are extraordinarily complex. Analysts can now select from a well-stocked toolbox those specific measures that best fit each topic. Studies in several nations use multiple indices for several racial and ethnic groups to assess the evolving nature of the residential mosaics. (SEE ALSO: **Discrimination**; *Index of Isolation; Segregation*)

—*Karl Taeuber*

Further Reading

Duncan, Otis D. and Beverly Duncan. 1955. "A Methodological Analysis of Segregation Indexes." *American Sociological Review* 20:210-17.

James, David R. and Karl E. Taeuber. 1985. "Measures of Segregation." Pp. 1-32 in *Sociological Methodology*, edited by Nancy Brandon Tuma. San Francisco: Jossey-Bass.

Massey, Douglas S. and Nancy A. Denton. 1988. "The Dimensions of Residential Segregation." *Social Forces* 67:281-315.

Massey, Douglas S. and Nancy Denton. 1993. *American Apartheid: Segregation and the Making of the Underclass*. Cambridge, MA: Harvard University Press.

Momeni, Jamshid, A., ed. 1986. *Race, Ethnicity and Minority Housing in the United States*. Westport, CT: Greenwood.

White, Michael J. 1986. "Segregation and Diversity Measures in Population Distribution." *Population Index* 52:198-221. ◄

► Self-Help Housing

In the past, *self-help housing* has referred to new home-building in which some or all of the building work depends on labor by the occupiers, usually in cooperating groups. Overlaps with neighborhood improvements such as utilities and community facilities with self-management and with maintenance have broadened the meaning consider-

ably. In this newer and wider sense, almost all preindustrial housing and most low- and moderate-income housing in low-income, rapidly urbanizing countries are "self-built." Hence, most current literature focuses on low-income, rapidly urbanizing areas.

Until recent times, a great majority of the world's population housed themselves. When the *enclosing*—that is, privatizing—of common land took place with the modernization of agriculture, significantly large numbers of wage-earning people were housed by their employers, a tradition continued in Europe by the early industrialists but abandoned to speculative tenement builders as overhead costs escalated. With rapid industrialization and the explosive growth of cities, poverty and unsanitary conditions threatened public health and order. Charitable institutions responded, initiating the institutionalization of housing. Subsequently, "public" or "social" housing became the norm for households with incomes below the mortgage market thresholds in many mixed economies and for a majority in command economies with no mortgage markets. The many different forms of "self-help housing" can be seen as survivals of traditional, vernacular modes: homemade, homespun, homegrown goods for home use only, not meant to be sold to others. In light of the newly emerging development paradigm, it seems not only that these survivors have a future but that "sustainable development" depends on a recovery of vernacular values and community-based production.

Most who undertake the generally burdensome task of self-building whether as laborer or manager, individually or cooperatively, do so because it is the only way to get a habitable home or one that meets their priority needs, whether material, economic, social, creative, or aesthetic. The wide variety of motives for self-building are evident only when "housing" is understood as a process—when it is seen that what matters most about the activity and its products is what they do for the users, not their material standards alone. In addition to shelter and material comfort, a dwelling has to provide access to what the dwellers need to maximize future opportunities as well as for present livelihood; and of course, there must be sufficient security of tenure to make the effort worthwhile. If housing is assessed by its material standards alone, as in the old development paradigm, the meaning of self-help (or any other way of building) cannot be properly understood, problems are misstated, and policies fail.

A vast majority of permanent, self-built homes are in rapidly urbanizing and newly industrializing countries where most households cannot afford "formal" market prices and where governments can afford to subsidize only privileged minorities. Between three-quarters and one-half of most cities in these countries have been developed by the residents themselves—either on land they have squatted or on undeveloped land that they have bought in illegal subdivisions. Because most such self-builders live in shacks on their plots as soon as they are acquired and in dwellings while under construction, observers often confuse "progressively developing" settlements with unimproving or deteriorating shanty-towns. In many respects, the former are similar to self-help housing in industrialized countries; the latter are not.

A great majority of First World self-builders are individual owner-builders acting as their own general contractors and carrying out varying proportions of the building work themselves—in the United States, for instance, from none at all to 100%. Most U.S. self-builders actually carry out between one-quarter and one-third of the work themselves, contracting out heavy work, such as excavation, and skilled work, such as plastering. Most squatter builders in contemporary Peru also employ workers—bricklayers, for instance, with whom available household members, including women and children, work as assistants. As recently as the 1960s, 20% of all new homes in the United States were built in this way, and proportions were much higher, of course, in low-income countries. U.S. owner-builders' median income was the same as that for the U.S. population as a whole throughout the 1960s. There were, and probably still are as many U.S. owner-builders and as many with above median incomes as below. They gain much the same material and social benefits as squatter-builders in low-income countries: Both have higher or much higher ratios of equity to income than do conventional mortgage buyers; relative to the investments made and, when squatters also have secure tenure, both achieve higher space and building standards. Many self-builders gain self-confidence, and family and community relations are strengthened. On the down side, risks may be greater, especially for low-income squatter-builders, whose health can be damaged by overwork and living in incomplete dwellings for long periods. They also suffer frequent material losses because of inadequate models and the lack of technical assistance.

More recently, since mortgage credit systems became widely available, speculative commercial provision of housing has provided alternatives to self-help for higher-income earners, and (generally, subsequent) state welfare provision has helped to close the affordability gap. The technical, financial, procedural, and legal norms established by corporate market and state housing providers tend to restrict community-based, local, and personal initiative. Rising land and other resource costs, increasing technical complexity, and regulations, such as those prohibiting occupation during construction, also inhibit cooperative or individual self-building.

Much of the literature on self-help housing refers to agency-assisted and agency-sponsored projects and programs. Pioneering governmental and private voluntary rehabilitation and development programs developed aided and mutual self-help systems. Eleanor Roosevelt promoted one of the first: an aided and mutual self-help project in rural Pennsylvania, carried out by her husband's administration in 1931 and 1932, at the height of the Great Depression. Federally assisted self-help housing projects for low-income farmers and Native American people in the United States have been carried out on a small scale since the 1950s. Private voluntary organizations such as Self-Help Enterprises of Visalia, California and, more recently, Habitat for Humanity, championed by Jimmy Carter, have concentrated similar efforts on low-income African Americans and Mexican immigrants. These aided self-help programs, mostly for groups and generally organized by the sponsoring agencies, were and are quantitatively insignificant in the United States, however, compared with individual owner-builders.

Agency-directed and self-managed self-help housing procedures have very different potentials. When management is in agency hands, cost savings are limited to "sweat equity"—the manual labor provided by the self-helpers. "Management equity," however, can be equally significant. The sensitivity and commitment demanded of the field staff and backup management in directed programs generally exceed realistic expectations of public administration. And the number of dedicated voluntary bodies is limited. These limitations impose standardized procedures and supervised, disciplined, and scheduled group work. However, sponsored and technically assisted self-help programs can be highly successful and cost-effective especially when carried out by nongovernmental organizations, when group work is minimized, and when lightweight technologies are used—as demonstrated by British, Swedish, U.S., and Third World experience. (SEE ALSO: **Community-Based Housing**; *Owner Building; Participatory Design and Planning; Sweat Equity;* **Third World Housing**; *Urban Homesteading*)

—*John F. C. Turner*

Further Reading

Broome, Jon and Brian Richardson. 1991. *The Self-Build Book.* Bideford, Devon, UK: Green Books.

Grindley, William. 1972. "Survivors with a Future." In *Freedom to Build, Dweller Control of the Housing Process,* edited by John F. C. Turner and Robert Fichter. New York: Macmillan.

Kolodny, Robert. 1986. "The Emergence of Self-Help as a Housing Strategy for the Urban Poor." Pp. 447-62 in *Critical Perspectives on Housing,* edited by Rachel Bratt, Chester Hartman, and Ann Meyerson. Philadelphia: Temple University Press.

Mathéy, Kosta, ed. 1992. *Beyond Self-Help Housing.* London: Mansell.

Turner, Bertha, ed. 1988. *Building Community: A Third World Case Book.* Building Communities Books and Habitat Forum Berlin for Habitat International Coalition. (Available from Habitat International Coalition, Cordobanes 24, Colonia San Jose Insurgentes, 03900. Mexico D.F.)

Turner, John F. C. 1976. *Housing by People.* New York: Pantheon.

Ward, Peter. 1982. *Self-Help Housing: A Critique.* London: Mansell. ◀

▶ Senate Committee on Banking, Housing and Urban Affairs

The Committee on Banking, Housing and Urban Affairs has chief responsibility within the U.S. Senate for overseeing the nation's housing policy, urban affairs, and banking industry. Agencies under its purview include the U.S. Department of Housing and Urban Development, the Federal Reserve System, the Federal Deposit Insurance Corporation, and the Resolution Trust Corporation. Subcommittees relevant to housing issues include the Subcommittee on Housing Opportunity and Community Development (named Housing and Urban Affairs until 1995) and the

Subcommittee on HUD Oversight and Structure, created at the start of the 104th Congress.

Legislation proposed in the Senate is routed to the appropriate committee, where members hold hearings on the proposal, debate its merits, and bargain among themselves, with the executive branch, and with interest groups. The committee chair and the relevant subcommittee chair decide whether and when the committee will consider a bill. Eventually, the committee either produces a resolution ready for debate and vote in the Senate or decides to abandon the proposal altogether.

The committee system evolved as a way for Congress to handle its diverse and growing workload. The Senate's 17 standing committees, permanent committees with fixed jurisdictions, allow members to specialize in two or three policy areas. The Senate committee structure has fewer subcommittees and is more centralized than that of the House. Although hearings may be held at the subcommittee level, legislation is more often marked up in full committee meetings.

Committees have research staffs to aid their information gathering and administrative tasks. Usually, the staff is organized on a partisan basis, with the committee chair overseeing hiring decisions and staff activities; the minority party typically receives at least one-third of the committee staff. In addition, under Senate rules, senators may hire one personal aide for each of their committee assignments.

The committee with jurisdiction over housing policy, originally named the Committee on Banking and Currency, was established in 1913. Senate reforms in 1946 gave the committee jurisdiction over public and private housing, and in 1977, urban development and mass transit were officially added to its domain. Thirteen senators served on the committee in 1947, and by 1989, membership had reached 21. In 1995, at the start of the 104th Congress, 16 senators were appointed as members.

At the start of each Congress, senators request committee assignments from the Committee on Committees for their political party. Current committee members may hold their assignments or request a transfer. Several criteria guide committee assignments, from member interest and expertise to geographical balance to electoral considerations, such as how members might best serve their constituencies or improve reelection prospects. The Senate majority party will hold a majority on every committee.

Committee chairpersons are chosen based on seniority, meaning that the ranking member of the majority party will become the chair. Unlike the House, the Senate has not diverged from the norm of seniority rule. Committee chairs appoint subcommittee members and chairs. Observers have described Senators as having more individual influence than House representatives, leading to a different sort of committee politics. In particular, committee chairs have had less autonomous influence over their committees and have worked more closely with ranking minority members and fellow majority party members to reach agreement regarding committee action.

The committee's housing agenda has shifted over time as the president, interest groups, committee members, and other legislators developed the direction of federal housing policy. Thus, the committee has worked on legislation ranging from the federal government's first major intervention in the U.S. housing market during the Great Depression to the urban renewal and community development programs of the 1960s era to retrenchment and privatization under Reagan. Typically, groups lobbying the committee have wanted either more or less intervention from government, no matter which particular housing issue is under examination. In general, proponents of an expansive federal housing policy have included urban interests, such as the National League of Cities and the Conference of Mayors, the National Association of Housing and Redevelopment Officials, and advocacy groups for the poor, such as the National Low Income Housing Coalition. Opponents of government intervention typically include housing industry trade associations representing builders and realtors, along with the banking industry and business interests such as Chambers of Commerce.

Committee members themselves divide along party lines, where the split also occurs according to degree of support for government intervention, with Democrats typically pressing for continued or increased intervention and Republicans seeking government restraint. Although Democrats have controlled the Senate far more frequently than the Republicans, the long-standing strength of Southern Democrat senators, traditionally more conservative than other Democrats, has tended to temper the Senate's liberalism. An urban-rural split also has emerged over the years among committee members.

The Senate and House housing committees often pass different versions of housing bills, and compromises are usually reached in conference committees with members from the House and Senate. But even when a compromise has been reached, committee proposals can be altered on the Senate floor through the introduction of amendments. In addition, passage of a bill does not guarantee that the funding it authorizes will be appropriated. The Appropriations Committee has become an increasingly significant policymaker, able to blunt the effect of a program authorization bill by funding it at a low level. A presidential veto can also prevent a bill from becoming a law.

The committee's party orientation changed in 1995 when Republicans won the Senate majority after 10 years in the minority. As part of a general effort by Republicans to reduce the size of the bureaucracy, staff was cut for the Subcommittee on Housing Opportunity and Community Development. A new subcommittee was formed to oversee HUD, an agency that Republicans had been criticizing for overregulation and inefficiency. Indeed, at the start of the 104th Congress, the new chair of the Subcommittee on Housing and Community Opportunity demanded a reduction in federal housing regulations.

Over time, the committee's actions and relations with its House counterpart and with the president have fluctuated. During the early years of Reagan's presidency, when the Senate was controlled by Republicans and the House by Democrats, the Senate Banking committee was friendlier to presidential housing policy proposals than was the House committee. Although housing subcommittee members did not support as extreme a retreat from government intervention in housing as did Reagan, they were willing to decrease funding levels of housing programs and tended to support more privatized assistance mechanisms, such as vouchers. Strong ties to urban interest groups and constituencies, however, made these senators less willing than Reagan to eliminate Community Development Block Grants and Urban Development Action Grants.

At other times, however, the Senate committee has been less supportive of presidential policy than its House counterpart. When Nixon placed a moratorium on housing programs in 1973, Senate Banking Committee members fought for their revival, leading the Senate to pass a bill restoring homeownership and rental subsidy programs that focused on housing construction. At the same time, the House committee developed legislation more in line with Nixon's block grant proposals and directed funds toward suburbs as well as large cities.

In addition to drafting and passing legislation, Congress oversees the executive branch bureaucracy, and so, in theory, the Banking Committee oversees implementation of housing legislation. Committee oversight techniques include informal communication between committee and agency staff members, program evaluations by congressional support agencies such as the General Accounting Office, oversight and program reauthorization hearings, and review of proposed agency rules. In practice, because of committees' heavy legislative workloads, congressional oversight occurs sporadically, often when scandals or crises gain public attention. For example, in 1954, the Senate committee investigated corruption in Federal Housing Administration programs, holding hearings in Washington, D.C., and in cities across the United States. Eventually, the committee issued a report, many FHA employees were fired or resigned, FHA regulations were tightened, and safeguards were written into the 1954 Housing Act.

Committee hearings, reports (which include the text of proposed legislation), and summaries of activity for each congressional session are published by the Government Printing Office. They are available in libraries that are depositories of federal government publications. Committee staff members can be reached at SD-534 Dirksen Senate Office Building, Washington, DC, 20510-6075. Phone: (202) 224-7391. For hearings and schedule information, call (202) 224-0791. (SEE ALSO: **Federal Government;** *House of Representatives, Committee on Banking and Financial Services*)

—Mara Sidney

Further Reading

Congress and the Nation. 1965-1993. Vols. 1-7. Washington, DC: Congressional Quarterly Service.

Congressional Quarterly Almanac. Washington, DC: Congressional Quarterly Service. Published annually from 1948.

Congressional Quarterly Weekly Report. Weekly news magazine. Washington, DC: Congressional Quarterly Service.

Rieselbach, Leroy N. 1995. *Congressional Politics: The Evolving Legislative System.* Boulder, CO: Westview. ◀

▶ Senior Citizens Housing Act of 1962

The Senior Citizens Housing Act of 1962 (P.L. 87-723; 76 Stat. 670) was one of a series of congressional initiatives, beginning in the late 1950s, intended to meet the unique housing needs and requirements of the elderly through the provision of favorable loans and grants. It also represented one of the first acts passed in the 1960s and 1970s that departed from the earlier public housing effort by encouraging private builders and operators of low-cost apartments.

The act recognized that senior citizens, increasing numbers of whom were living well beyond retirement age, were confronted with special housing obstacles. First, their age made it difficult for them to secure liberal mortgage financing. Second, because the great majority had fixed, low to moderate incomes, they were unable to afford housing better adapted to their physical, psychological, and economic needs.

The act singled out for special attention the rural elderly living on farms and in nonfarm areas where the problems of mortgage credit and availability of suitable housing were even more acute. According to the report of the Senate Committee on Banking and Currency, the legislation was designed, especially, to serve a large group of previously unserved rural elderly—those who were unable to accumulate sufficient assets to afford decent housing or who no longer had the incentive or capacity to maintain ownership of their homes. They remained in farm and nonfarm homes, not by choice but because of the absence of appropriate alternatives. Others were compelled to move to cities to find alternatives. The elderly programs of the Housing and Home Finance Agency (HHFA)—predecessor to the Department of Housing and Urban Development (HUD)—operated mostly in urban areas and had failed to fill this need.

Assistance was to be administered by both HHFA and the Rural Housing Service (RHS)—formerly the Farmers Home Administration (FmHA)—of the Department of Agriculture. First, with respect to HHFA, $100 million in additional funds were provided for the highly successful Section 202 program for the production of affordable rental and cooperative housing for the elderly. This program, adopted in the Housing Act of 1959, offered below-market-interest-rate loans directly from the government to private nonprofit corporations, consumer cooperatives, and certain public bodies. The loans had 50-year maturities and interest rates at that time of 3.5%.

Second, the centerpiece of the legislation was a new program for the elderly to be administered by FmHA (RHS) in rural areas. Amending Title V of the Housing Act of 1949, the act created the Section 515 elderly housing program. Modeled after HHFA's Section 202 program, it provided direct loans for the construction, rehabilitation, acquisition, and operation of rental and cooperative housing for elderly households of low or moderate income. The eligible borrowers and terms of the loans were substantially the same as those for the Section 202 program.

The act also authorized FmHA (RHS) to insure loans made by private lenders to individuals, trusts, partnerships, associations, or corporations, similar to HHFA's Section 231 insured program for the elderly. The purpose was to stimulate the provision of multifamily housing for elderly persons and families whose incomes enabled them to afford the terms of insured but unsubsidized private financing, primarily households of moderate income and above.

Finally, the act made two other changes in FmHA (RHS) programs benefiting the elderly. The existing Section 502 homeownership program was broadened to make it easier for the elderly to receive low-interest loans to purchase and rehabilitate existing housing and to purchase sites for new housing. Moreover, the maximum grant amount under the existing Section 504 rehabilitation program was increased, in part, to enable elderly owner-occupants to better improve the health and safety conditions of their dwellings. (SEE ALSO: **Elderly**; *Section 202*)

—*Robert Wiener*

Further Reading

Housing Assistance Council. 1994. *A Guide to Federal Housing and Community Development Programs for Small Towns and Rural Areas.* Washington, DC: Author.

Collings, Art and Linda Kravitz. 1986. "Rural Housing Policy in America: Problems and Solutions." In *Critical Perspectives on Housing,* edited by Rachel G. Bratt, Chester Hartman, and Ann Meyerson. Philadelphia: Temple University Press.

Wiener, Robert. 1987. "Rural Housing: The Prepayment Debate." *Economic Development and Law Center Report* 17(3-4):1-11.

Wiener, Robert. 1990. "Prepayment Hits Countryside Hard." *Shelterforce* (March/April):18-21. ◄

► Setback Requirement

Setback requirement, or yard requirement, is the required minimum distance between a property line and the nearest wall of a building on that property. Commonly used to refer to the setback from the front property line, the term can also be used with respect to side and rear setbacks. Setback requirements normally form part of a zoning ordinance, although they can also be established by restrictive covenants on title. Proponents argue that setbacks provide privacy and quiet through distancing of neighbors, better access to sunlight, and an exterior route for residents to travel from front yard to rear yard. Side setbacks also help in containing a fire that starts in a neighboring building as well as providing access for firefighters.

Essentially, setback requirements use vacant land to solve housing design problems. However, land is costly, both initially and in terms of the consequences of low-density development for the provision of transportation and other networks and services. Other ways of solving these housing design problems are through the use of different building materials (for example, better fireproofing and sound insulation) or imaginative designs. In terms of the latter, zero lot line developments have been proposed to spur more land-intensive suburban development. (SEE ALSO: *Floor-Area Ratio; New Urbanism; Zoning*)

—*John R. Miron*

Further Reading

Chapin, F. S. and E. J. Kaiser. 1994. *Urban Land Use Planning.* 4th ed. Urbana: University of Illinois Press.

Schultz, M. S. and V. L. Kasen. 1984. *Encyclopedia of Community Planning and Environmental Management.* New York: Facts on File Publications (see pp. 354, 436). ◄

► Shared-Appreciation Mortgage

A shared-appreciation mortgage (SAM) is a loan that grants the lender a fraction of the amount that the property securing the mortgage increases in value between the closing of the loan and the earlier of (a) loan maturity or repayment and (b) sale or transfer of the property. In return for this lump sum payment (the "contingent interest"), the lender discounts the mortgage interest rate set below the prevailing market rate for a conventional fixed-rate mortgage.

A SAM is special kind of shared-equity mortgage (SEM), wherein lender and borrower partner in ownership of the property; in a SAM, the partners share only the increase in property value over the life of the mortgage. With SAMs, the interest rate discount depends on the lender's expectations about property price inflation in that local market. In a volatile housing market, SAM lenders incur considerable risk. The downside risk to lenders is that they overestimate the appreciation in property prices; the attraction to lenders is that they can protect themselves against unexpectedly high inflation. The advantage to SAM borrowers is that they can rearrange their lifetime stream of payments: less in monthly interest charges now in exchange for a balloon payment (the contingent interest) at a future date. One difficulty with SAMs is that they reduce the net financial incentive (that is, the capital gain that must now be shared with the lender) for homeowners to fully maintain or actively renovate or add to a dwelling over the life of the SAM. (SEE ALSO: *Alternative Mortgage Instruments*)

—*John R. Miron*

Further Reading

Downs, A. 1985. *The Revolution in Real Estate Finance.* Washington, DC: Brookings Institution.

Fabozzi, F. J. and F. Modigliani. 1992. *Mortgage and Mortgage-Backed Securities Markets.* Boston: Harvard Business School Press.

Pozdena, R. J. 1988. *The Modern Economics of Housing: A Guide to Theory and Policy for Finance and Real Estate Professionals.* New York: Quorum Books (see chap. 8). ◄

► Shared Group Housing

Shared group housing is a living arrangement in which three or more unrelated adult persons (but usually less than 20) operate as a single housekeeping unit in a conventional residential structure. Typically, occupants have their own bedrooms and sometimes their own bathrooms, but they share the kitchen, living room, and other common living areas. They also share responsibility for the usual residential upkeep and homemaking tasks, although one or more hired persons may do the housekeeping, laundry, home repairs, and meal preparation and provide transportation assistance. A full-time manager overseeing the running of the household, sometimes, but not always, will reside full-time in the house. Among the benefits associated with this housing option are its rent savings (monthly costs may be as much as 50% less than conventional accommodations), its attractive and comfortable surroundings, the security derived from the presence of others, the reduced burden of household chores, and the benefits of companionship, friendship, and social interaction. Residents in turn give up some degree of personal autonomy and must practice cooperative and coordinating behaviors.

Shared group housing alternatives date from at least as early as the late 19th century in the United States. They were occupied by a variety of users, including members of religious groups, single female workers, and single elderly persons. Interest in this option declined after World War II, but in the 1980s, various groups (elderly consumers, state and local governments, community organizations) showed renewed interest in shared group housing as a housing alternative. By the early 1990s, more than 140 organizations, mostly nonprofits, were involved in shared-housing programs, up from about 100 in 1981. Eclectic groups of persons occupy these facilities, including students, elderly persons, single parents, people with mental and physical disabilities, and people with AIDS. An increasing number of state governments are financing and administering shared-housing programs in response to the need for affordable housing for their lower-income and more dependent populations. Most state programs limit eligibility for shared-housing sponsorship to nonprofit and public organizations, although a few encourage joint ventures between nonprofit and for-profit developers. The remainder of this entry focuses on shared group housing in the United States that is primarily occupied by elderly persons.

Characteristics of Shared Group Housing Occupied by Elderly Persons

Usually, a nonprofit organization (often with a religious affiliation), and less frequently a government agency or proprietary (for-profit) group, develops and manages a shared group housing facility. Typically, the organization purchases or leases a house that originally housed a single household or converts what was once a nonresidential building (an old school, a hotel, a nursing home, or a hospital) into a multiroom residential structure. The "house" is sometimes received as a gift from a donor benefiting from tax advantages. The renovation, furnishing, and administrative costs may be partly subsidized with government funding and private charitable donations. Shared-housing renovation is considered very cost-effective compared with new construction. The nonprofit organizations that operate shared group residence programs have small budgets and a median staff size of five (with at least one full-time staff person, along with part-time and volunteer staff). Support services provided in shared group housing are funded in part by the residents (from their monthly rents) and in part from charitable and religious organizations, foundations, and government programs (addressing aging and human service needs).

Along with securing the residential structure, the sponsoring organization often screens the capabilities of prospective older residents. It may also arrange for a live-in housekeeper to take care of home management (e.g., housekeeping, laundry) and meal activities. At the very least, it may provide some "resource person" to advise, monitor, and resolve residents' problems and offer 24-hour, seven days a week, on-call service in case of emergencies. The older occupants may also have access to a variety of community-based services sponsored or administered by the organization responsible for the shared group residence. Examples include emergency hotline contacts, home-delivered meals, housekeeping and repair services, counseling, legal assistance, health assessment, and recreational programs.

An average of eight residents occupies this housing option, mostly women between the ages of 70 and 90. Some shared group residences are intergenerational, however, and are occupied by students, displaced homemakers, and working couples. On average, the older occupants are more frail than elderly persons living in conventional dwellings, but they are ambulatory and sufficiently independent to conduct their everyday activities without assistance. Although most occupants enjoy the family-oriented, communal living, this is not usually why they select this alternative. Rather, because of their own or other people's changed circumstances, they recognize that it is becoming harder for them to live completely on their own. Among the more important precipitating circumstances are these:

► They were living with someone who recently died or became too disabled to care for them.
► They were unable or did not want to rely on their grown children.
► They felt insecure about living alone.
► They were experiencing chronic health problems or physical disabilities.
► They were living in declining neighborhoods with higher crime rates.
► They were having difficulty affording their current accommodations.

In exchange for the benefits of group living, residents must adapt to a smaller amount of private space, give up some of their privacy, lose some decision-making control over the everyday operation of their living arrangements, and adjust to the sharing of living space with strangers. The financial and sharing benefits of these accommodations may also have the effect of reducing their Supplemental Security Income and food stamp benefits in some states. No accurate estimates exist about how many shared group living facilities are found in the United States. There have been no national surveys, and state governments do not regulate many of these facilities.

This will not be the last home for many of the residents in these facilities. They will be confronted with new physical and mental disabilities requiring them to receive assistance with their activities of daily living and a more supportive housing situation. Sponsoring organizations and residents of shared group projects will differ as to how and on what basis they decide that this tipping point is reached. The extent that these organizations allow such disabled residents to remain even as they should be receiving more assistance is undocumented. On average, older persons occupy these accommodations for fewer than two years, but average stays range from one month to five years.

Shared Group Housing Differs from Other Elderly Housing Options

Shared group housing is among the most affordable of housing arrangements available to more vulnerable elderly persons seeking to maintain their autonomy. Along with its generally satisfying extended family ambience, it also offers an alternative for poor, single, elderly persons who might otherwise end up in nursing homes and thus cost the state Medicaid dollars. Not all are in favor of this option, however, and potential neighbors often try to block the presence of these residential arrangements (the NIMBY— "not in my backyard"—syndrome). This residential alternative is often excluded from single-family neighborhoods by local zoning ordinances and building regulations that employ restrictive (that is, nuclear) definitions of *family*, by the imposition of building regulations (such as fire codes) that are applied to multifamily buildings and that demand prohibitively high renovation costs or by excessively high licensing fees.

Shared group living differs in other significant ways from several other housing options. It should be distinguished from home-matching arrangements in which an elderly homeowner shares his or her own home (whether house or apartment) with an unrelated person (a housemate), who pays rent or barters rent for his or her help. Organizations that provide shared group housing also often arrange these matchups. Shared group accommodations may also be physically similar to board-and-care or foster care shelter. A fundamental difference, however, is that occupants in shared group living accommodations are not viewed as boarders or residents needing protection but as more self-reliant residents who participate equally in an extended family, communal-like living situation. Thus, they have much more say than board-and-care residents do over the management and operations of the facility. Shared group living also distinguishes itself from congregate housing options, where residents occupy their own self-contained apartment units, are more correctly characterized as tenants, and where the building facility and its services are usually professionally managed. Finally, a smaller and as yet unstudied group of older people occupy shared group residences that they themselves have organized and developed without any major assistance (other than perhaps of an advisory nature) from an outside organization.

The largest and most prominent organization that promotes home matching for elderly persons is the National Shared Housing Resource Center (NSHRC) in Baltimore, Maryland, established in 1981 by Gray Panther activist Maggie Kuhn. NSHRC is a volunteer-led organization directed by 10 regional coordinators. Staff members offer training and technical assistance to shared-housing sponsors, and the organization serves as an information clearinghouse on shared-housing trends and issues. (SEE ALSO: **Elderly**; *National Shared Housing Resource Center*)

—*Stephen M. Golant*

Further Reading

Franck, Karen A. and Sherry Ahrentzen. 1991. *New Households, New Housing*. New York: Van Nostrand Reinhold.

Golant, Stephen M. 1992. *Housing America's Elderly: Many Possibilities, Few Choices*. Newbury Park, CA: Sage.

Hemmens, George C., Charles J. Hoch, and Jana Carp, eds. 1996. *Under One Roof: Issues and Innovations in Shared Housing*. New York: State University of New York Press.

Horne, Jo and Leo Baldwin. 1988. *Home-Sharing and Other Lifestyle Options*. Washington, DC: Scott, Foresman.

Jaffe, Dale J. 1989. *Shared Housing for the Elderly*. New York: Greenwood.

Streib, Gordon F., Edward Folts, and Mary Anne Hilker. 1984. *Old Homes—New Families: Shared Living for the Elderly*. New York: Columbia University Press. ◄

► Shared Ownership

Property may be owned or "held" in a variety of ownership forms that range from real estate owned by two persons who share the same real estate at the same time (such as a traditional home) to strangers sharing the same place at different times (such as time-share vacation condominiums). Co-owners or investors need to consider many multiple-ownership forms prior to taking title to real property.

Perhaps the most common form of joint ownership is the "traditional" family home in which both husband's and wife's names appear on the deed at the time the property is acquired. Variations on the co-ownership forms vary widely by states according to where the property is located, not where the owners reside.

Multiple ownership in real property may also occur as a result of property being "willed" in total to multiple parties or heirs rather than in separate parcels or tracts of land. Other co-ownership opportunities occur in investment properties, shared residential housing, joint ventures, and various types of partnerships and trusts. The following briefly describes various forms of shared ownership.

Tenants in Common

This form of shared ownership may occur when two or more persons' individual names appear on a deed, with each owning an individual interest in the whole property. Whether by will (by devise to heirs) or at the time of acquisition, this form of ownership presents special problems in the event of disagreement on the management, disposition, or use of the property. Even more problematic is the event of an existing mortgage with multiple parties liable when a default occurs by any one of the parties, leaving one making payments and the other retaining a "free ride" in the title ownership. The challenge of undivided ownership is how to resolve problems when owners do not agree. In the case of a home, the conflicting parties may agree to sell or a court may order the property sold. Land, however, may be ordered partitioned or divided by the parties. A co-ownership or operating agreement prior to taking title is an important document that can be signed by all owners to address present and future ownership questions and issues that will arise.

Joint Tenancy

Joint tenancy is similar to tenants in common with one clear distinction, and this is the right of survivorship. On the death of a "joint tenant" owner, his or her interest does not descend to the heirs or pass by will, but the entire ownership remains and passes to the surviving joint tenant(s). Married couples often use this form to acquire property because it avoids probate court and legal expenses after the death of one spouse. This form of shared ownership is not legal in many states.

Community Property

This form of ownership by married couples is based on old Spanish heritage as found in Arizona, California, New Mexico, and Texas, which operates on the premise that all property acquired during marriage with community funds (income, rents, interest, and royalties) is jointly owned by each spouse. The sale of these properties (homestead or investment properties) generally requires consent and signatures of both parties on a deed, regardless of whose name(s) appears on the title. It is very difficult to retain separate property in community property states unless by will or separate agreement between the partners.

General Partnership and Limited Partnership

Shared ownerships by general partnership or limited partnership interests in real estate are similar in that each has a designated general partner(s) who has (have) or share mortgage and general liability. Limited partners have limited liability to the extent that their initial investment is all that is at risk. Partnership agreements vary widely as to the management and percentage of ownership shares required to make decisions. These are both private and public partnership ownership organizational structures.

Condominium Ownership

Condominium ownership is "shared" to the extent that although the units are individually owned, the land is co-owned and maintained by the collective total of unit owners in a project. Condominium declarations generally outline the co-ownership rules and regulations and require unit owners to join a homeowner or condominium association to manage and operate the commonly owned property.

Other forms of co-ownership or shared ownership of property can border on the ownership of securities or "shares" such as *corporation* (ownership of stock—no tax benefits), *Sub Chapter* S Corporations (ownership of stock—flow-through tax benefits), and various types of *trust ownership forms* (land trusts, inter vivos and testamentary trusts). A very popular form of trust in recent years has been the real estate investment trust (REIT), which offers co-ownership opportunities with liquidity through Wall Street on the sale or resale of the shares. (SEE ALSO: **Homeownership**)

—*John S. Baen*

Further Reading

Peiser, Richard B. 1992. *Professional Real Estate Development: The ULI Guide to the Business.* Washington, DC: Urban Land Institute.
Shim, Jae K. 1996. *Dictionary of Real Estate.* New York: John Wiley.
Tosh, Dennis S. 1990. *Handbook of Real Estate Terms.* Englewood Cliffs, NJ: Prentice Hall. ◀

▶ Shelter Poverty

Shelter poverty is a sliding scale of housing affordability that challenges the conventional percentage-of-income standards. The traditional notion of how much people reasonably can be expected to pay for housing is the familiar 25% of income rule-of-thumb. Yet even this standard has been subjected to various adjustment and interpretations. For a time, U.S. analysts and policymakers thought 20% was appropriate, but since the early 1980s, 30% of income has become the rule generally applied. The adoption of any universal percentage of income is, however, conceptually flawed, whether it be 10%, 20%, or 30%, rather than 25%.

Because housing costs generally make the first claim on a household's disposable income, with nonhousing expenditures having to adjust to what is left, saying that a household is paying more than it can afford for housing presumably means that after paying for housing, it is unable to meet its nonshelter needs at a minimum level of adequacy. Consider, for example, two households of comparable disposable (i.e., after-tax) incomes—one small, the other large. Obviously, the large household would need substantially more for its nonshelter necessities than would the small household to achieve a comparable material quality of life. This implies that a larger household can afford to spend less for housing—if it is to meet its nonshelter needs at the given level of adequacy—than can the small household of the same income. Next, compare two households of the same size but different after-tax incomes. Both would need to spend about the same amount to achieve the same standard of living in terms of their nonshelter items. The higher-income household thus could afford to spend more for housing, as a percentage of income as well as in dollars. This suggests that a logical standard of affordability for housing is a sliding scale, that explicitly takes into account the cost of a minimum standard for nonshelter necessities, with income and household size as the principal parameters.

Households paying more than they can afford on this standard are *shelter poor*—the squeeze between their housing costs and incomes leaving them unable to meet their nonshelter needs at a minimum level of adequacy. That is, shelter poverty is a form of poverty that results from the burden of housing costs rather than just limited incomes. On this basis, only if a household would still be too poor to meet its nonshelter needs even if its shelter cost was reduced to zero should its condition be regarded as absolute poverty rather than shelter poverty.

Although every household has its own unique conditions of life, there do exist historically and socially determined notions of what constitutes a minimum or decent standard of living. They represent norms or standards around which a range of individual variations can be recognized and about which there is philosophical debate. The Lower Standard Budgets developed by the U.S. Bureau of Labor Statistics (BLS) represent one such standard. Using cost components from the BLS Lower Standard Budgets (other than shelter costs and taxes), updated (because the federal government has not computed the budgets since 1981) with components of the Consumer Price Index, scaled for households of various sizes and types, and taking into account taxes as a function of income and household composition, it is possible to operationalize the shelter poverty concept.

On the shelter poverty standard, for each size and type of household there is a level of income below which such households cannot afford to pay anything for shelter and still maintain the BLS lower budget standard for nonshelter items and pay their taxes. Households below the zero point of affordability would not have enough income to meet their nonshelter needs at the BLS Lower Standard Budgets level even if their housing were free. These zero points of affordability increase dramatically with household size. The differences by household size are primarily the result of larger households' needing to spend more to achieve the same basic level for nonshelter necessities, but these differences are compounded by the effects of taxes.

Social Security taxes do not take into account household size at all, whereas the personal exemption in the federal income tax provides only a modest offset for the higher level of nonshelter costs incurred by larger households.

On the other hand, for each type and size household, there is a level of income above which such households can afford to pay more than the traditional 25% of income for shelter and a slightly higher income above which they can afford more than the current Department of Housing and Urban Development (HUD) standard of 30%. Not surprisingly, the minimum income required to afford more than the conventionally determined amount is much greater for larger households than smaller: Households of six persons need five times as much as one-person households to be able to afford over 25% and over 30% of income for housing.

The shelter poverty concept has been applied to decennial census and American Housing Survey data to determine the extent and distribution of shelter poverty in the United States. In 1993, 29.6 million households—31.3% of all households were shelter poor. This included 14.8 million renter households (44.2%) and 14.9 million homeowner households (24.3%). The number of households that were shelter poor grew by 57% between 1970 and 1993. However, the total number of households in the United States has grown at about the same rate so that the incidence of shelter poverty remained relatively stable—fluctuating between about 30% and 35% as the overall economy as gone up and down.

The number of households that are shelter poor is nearly identical to the number paying over 30% of income for housing. The two approaches do, however, suggest rather different distributions of affordability problems. Shelter poverty is considerably more severe among larger and lower-income households and less severe among smaller and higher-income households than suggested by the conventional approach to affordability.

This difference has substantial policy implications. For example, rental assistance programs would provide payments on a sliding scale rather than on the basis of 30% of income if the shelter poverty approach were adopted. Policymakers have been reluctant to use the concept of shelter poverty, in part because it is not widely known and familiar, and in part because it would involve acknowledging both the inadequancy and the inequity in existing approaches to housing assistance. (SEE ALSO: **Affordability**; *Affordability Indicators*)

—*Michael E. Stone*

Further Reading

Stone, Michael E. 1993. *Shelter Poverty: New Ideas on Housing Affordability*. Philadelphia: Temple University Press. ◄

► *Shelterforce Magazine*

Shelterforce is the nation's oldest continually published housing and community development publication. For more than two decades, it has been a primary forum for organizers, activists, and advocates in the affordable housing and neighborhood revitalization movements.

Shelterforce began in 1975 as a "how-to" tenants' newspaper, founded by tenant and grassroots activists. *Shelterforce* helped tenants, tenant organizers, and tenant advocates (e.g., legal aid lawyers) learn more effective ways of securing tenants' rights to safe, decent, and affordable homes. Over the years, as inner-city neighborhoods experienced disinvestment, crime, and family disintegration, *Shelterforce's* focus changed. It began to examine a wider range of housing and community issues—but always with the goal of empowering individuals and groups to take control of their communities and effect real change. As a reflection of this more expansive view, *Shelterforce* also shifted to a magazine format in the 1980s.

In 1983, the National Housing Institute (NHI) incorporated as an independent nonprofit research organization and publisher of *Shelterforce*. NHI has performed original research on topics such as saving subsidized housing, preventing homelessness, and creating jobs. NHI explores the issues causing the crisis in housing and community in the United States, including poverty and racism, disinvestment and lack of employment, breakdown of the social fabric, crime, and lack of education. NHI studies how these and other factors affect people as they try to build decent, affordable housing in safe and viable neighborhoods.

Shelterforce is driven by NHI's vision of once-devastated communities "rebuilt" by empowered residents. *Shelterforce* has increasingly reported not only on physical redevelopment but also on the equally important factors of economic development and efforts to repair the social fabric in distressed communities. The magazine reflects the view that successful community revitalization requires both local, bottom-up community organizing and a national agenda that supports urban redevelopment and grassroots revitalization efforts. *Shelterforce* provides activists with clear analysis of important policy issues and concise descriptions of rules and regulations, while providing local, state, and national policymakers with input from the grassroots. Although *Shelterforce* highlights innovative strategies and partnerships created to organize low-income communities, it also examines why some efforts have failed. By exploring its subjects in a critical, journalistic style, *Shelterforce* helps activists more effectively rebuild and organize communities.

Shelterforce is a bimonthly journal. Price: $30 per year for organizations, $18 per year for individuals, and $5 for a single copy. Contact information: National Housing Institute; Executive Director/Editor: Harold Simon; 439 Main Street, Orange, NJ 07050-1523. Phone: (973) 678-9060. Fax: (973) 678-8437. E-mail: hs@nhi.org or kc@nhi.org. World Wide Web: http://www.nhi.org. (SEE ALSO: **Affordability**)

—Karen Ceraso ◀

▶ Shimberg Center for Affordable Housing

The Shimberg Center for Affordable Housing (SCAH) is a research, information, and referral center for issues relating to housing and community development. Founded in 1988 at the University of Florida, the Shimberg Center serves to facilitate the provision of safe, decent, and affordable housing and community development and to establish Florida as the national and international model for successful affordable housing delivery. The SCAH publishes a bimonthly newsletter titled *Affordable Housing ISSUES,* as well as periodic research reports and technical notes. It is an original member of the National Consortium of Housing Research Centers and the coordinator for Working Commission W63: Affordable Housing, a subgroup of the Netherlands-based International Council for Building Research and Documentation. In addition, it conducts research for government agencies and private sector organizations on a contractual basis. The Shimberg Center operates on an annual budget of $300,000 (1993) plus contract research with three full-time staff members. Contact person: Dr. Robert C. Stroh, Director. Address: University of Florida, P.O. Box 115703, Gainesville, FL 32611-5703. Phone: (352) 392-7697. Fax: (352) 392-4364. E-mail: affhsng@nervm.nerdc.ufl.edu. World Wide Web: http://www.bcn.ufl.edu/shimberg. (SEE ALSO: **Affordability**)

—Robert C. Stroh ◀

▶ Single-Parent Households

Single-parent households were not a new phenomenon when increases were observed in the 1970s, but such households were forming for different reasons. Unlike earlier decades, when solo parenting was primarily the result of widowhood, the rise of single-parent households resulted from increases in divorces, out-of-wedlock births, and never-marrieds with children. The rising numbers, and the predominance of women as single parents, were also identified with increases in poverty. By the 1990s, one child in four was living with only one parent, primarily in the central city and in deteriorating or substandard housing conditions.

Census Definitions of Single Parents
In 1970, the U.S. Bureau of the Census distinguished single parents from two-parent families for the first time. Most single parents maintain their own household; some are living with someone else, related or not. Other defining characteristics include gender and race and ethnicity, each of which affects income, tenure status, and location of residence. Because data are unavailable for some racial and ethnic groups, such as Asian Americans and the subgroups in this category, the researcher should be mindful of drawing stereotypes about "model minorities," extended families, and women's roles.

Defining Single Parents by Gender, Income, and Race and Ethnicity
More women than men head single-parent households. In 1991, almost 87% of single-parent households were women. Overall, more single parents are white but, relatively, single parenting is more prevalent among African Americans and Hispanics. The high rates of single parents among African American families (in 1991, about 63% compared with about 33% for Hispanics and 23% for whites) have been attributed to declining marriage rates and increasing divorces. Divorce and separation are more likely to be the reasons that single-parent households form among white households, whether male or female.

Comparisons of income for single parents reveal wide disparities by gender, type of household, and by race and ethnicity. Women earn less than men. The 1990 average income for a mother who was a single parent was $13,100, compared with $26,000 for a father and $41,300 for households with two parents present. Whites earn more than African Americans and Hispanics. White single mothers earned on average $14,900, compared with $10,300 for African American single mothers and $10,100 for Hispanic single mothers. Distinct differences appear in subgroups of Hispanics. For example, poverty levels have been found to be higher for Hispanic single-mother families who are Puerto Ricans. Although poverty rates for similar household types among whites and African Americans decreased between 1980 and 1990, rates increased for Puerto Ricans. The disparity in incomes reflects lower levels of educational attainment among single parents as well as among their parents. In 1991, dropout rates in high school for young Hispanic women between the ages of 16 and 24 were higher than for other groups. About 31% of Hispanic women dropped out, compared with about 17% of African American women and 9% of white women.

Characteristics That Define the Quality of Life for Single Parents
These differences affect choices about housing location and type of housing: In 1990, 27.2% of single-parent households with children under the age of 18 were homeowners compared with 64.7% who were renters. When single mothers earn enough, and despite passage of amendments to the 1988 Fair Housing Act, many female-headed house-

holds who want to become homeowners experience discrimination by financial institutions. Among single mothers who are renters, public housing has been a resource. A 1989 survey of more than 200 public housing authorities found that more than 60% of households were families with children; a little more than three-quarters of these families were headed by single parents. Case studies of limited-equity cooperatives in Canada and the United States have documented the advantages for single parents in co-ops where they experienced a better standard of living at affordable prices. Using 1987 data, Karen Fox Folk analyzed variations in economic status according to types of living arrangements. Both African American and white single-parent women who lived in their parents' homes were less likely to receive public assistance, but lesser demands on household work applied only to white single mothers, who were more likely to be living with two parents.

Among women and children of the middle class, studies have shown that after divorce, multiple moves will lead to increasingly inferior housing in contrast to the single fathers and their children. Displacement from the marriage may result in moving in with parents or friends on a transitional or semipermanent basis until stability is achieved and a move can be made from the suburbs to the central city where housing may be more affordable. The number of single parents who are homeless has been increasing. Homeless women with children are more likely to be younger African Americans. The U.S. Conference of Mayors has estimated that single parents made up about two-thirds of all homeless families in the 1990s. The lucky ones find refuge in battered women's shelters and transitional housing, but permanent affordable housing is scarce. Public housing has long waiting lists, and its future as a low-income housing alternative looks bleak.

Widespread discrimination on the basis of gender, income, race and ethnicity, and the presence of children reduces the housing choices of single parents. In addition, incomes of single mothers suffer because of the government's failure to enforce child support awards. About 50% of all single mothers received child support awards in 1989. Among those who received support, about half were paid all they were owed, but 25% received no payment at all. The pattern was similar in 1991 with 51% receiving the full amount, 25% receiving nothing, and 24% receiving partial amounts. Of all custodial parents, about 94% are mothers who live below the poverty level. African American and Hispanic women are less likely to be awarded child support given the higher unemployment among African American and Hispanic males than White males. Single mothers, perceived as vulnerable, are also subject to sexual harassment by landlords at the home base and by bosses or other workers at the workplace. Efforts to alleviate this type of discrimination may be at the cost of missing a day's work or result in loss of a subsidy.

Low incomes also affect other aspects of daily life that together constitute a housing package. No or limited access to services, particularly child care and transportation, limits job options. Working at part-time or seasonal jobs usually means lack of health insurance. Fear for personal safety contributes to isolation within the home. Other barriers to jobs include racial discrimination in home seeking and mortgage lending practices.

Guidelines for Houses and Neighborhoods to Serve Single Parents in the United States

Although some historical material identifies needs of single parents, their greater visibility among designers and planners began to emerge only in the 1980s. Built and proposed projects were sometimes referred to as "shelter plus" or "lifeboats." In 1983, an advocacy project of the Bergen County Chapter of the League of Women Voters in New Jersey identified problems that single parents faced in finding and affording housing. The League held some of the first workshops in the United States that brought together single parents and professionals from the housing, finance, and government sectors. A report was issued with guidelines, references to working models and programs, and a bibliography. In Minnesota, guidelines were published in 1988 for designing and retrofitting new and existing housing and neighborhoods to meet the needs of single-parent families. Community forums, panels, and workshops helped to conceptualize workable models.

During the 1980s, design competitions broadened awareness about the increased special needs of single parents. In 1984, the New American House design competition was the first to incorporate an innovative program aimed at nontraditional households through requiring separate space for wage work within the house. By early 1991, an inventory existed of built projects explicitly directed toward serving the needs of single parents, including housing that combined community facilities and job training, new purpose-built housing, and rehabilitated and redesigned facilities. Canadian case studies and European examples, particularly in Sweden, Holland, and England, have become more widely known in the United States.

Equally significant to the recommended floor plans in individual houses was the growing attention to design guidelines for the neighborhood environment. Issues of scale, land uses other than housing, and location of services are of particular importance to single parents with low incomes.

Public Opinion about Single Parents and Policy

Perhaps the most contentious underlying issue that influences public opinion about housing and services policy toward single parents has been the debate about morality regarding work and family. In this debate, African American single parents are frequently singled out as "problem" families, deviant or pathological, who pass on a culture of poverty that is wedded to welfare. This linkage is then used to argue that government support through AFDC (Aid to Families with Dependent Children, replaced in 1996 with Temporary Assistance to Needy Families, or TANF) reduces the desire for marriage and that welfare payments discourage marriage and job searches. Social histories of family life in the United States offer another view: AFDC has permitted unmarried women with children greater freedom to set up independent homes. That this occurs at all is surprising, and welfare studies document episodic patterns of reliance on welfare because of problems with child care, low wages, and loss of jobs. Analyses of the welfare grant substantiate that the benefit (an average of $373 per month or $4,476 per year in 1993) is well below the federal definition for the poverty line. The Institute for Women's Policy Research (IWPR) examination of census data from 1984 to 1991 pointed to welfare as a safety net for single mothers when their labor force participation brought in insufficient earnings. One expert who analyzed the different sides on issues of increasing nonmarital child-bearing concluded that the findings were ambiguous. Other positions state that the cause of the growth in single-parent households and out-of-wedlock births lies in the lack of African American marriageable men, which in turn is tied to their high unemployment rates and low potential of earning power. This then prevents them from supporting wives and children. A related perspective attributes growth to the disproportionately high numbers of African American men in prisons.

Perspectives based on analyzing the effects of capitalism, the dual labor market, and racial oppression also bring to light the critical role that African American women play in nurturing and passing on a cultural legacy. Carol Stack's *All Our Kin* is widely cited as an example of ways poor women weave together survival strategies that include bartering goods and providing room for others, even under conditions of overcrowding. The literature on needs and roles of single mothers in development planning in countries of Central America, South America, the Caribbean, India, and Africa similarly focuses on strategies to overcome poverty. Poor women are assisted with training in local economic development projects such as credit unions, construction companies, and cooperative housing. These provide a supportive structure and opportunities for women to acquire leadership skills.

Awareness of differences among single-parent households rather than generalized statements should inform government policy. The prevalence of single parenting among poor women in central cities needs to be analyzed in relation to the kinds of short- and long-term opportunities in local labor markets. Even for many single parents with jobs, affordable housing is a major expense. The lives of single parents in the United States, particularly mothers and children, will suffer from government withdrawal of income assistance, the lack of permanent housing subsidies, and the closing of child care facilities. Welfare-to-work programs may serve some but are highly dependent on the availability of jobs at the local level. The de facto result may mean a return to the ideology of the two-parent household as the model family and women's proper role as within the house. The irony is that two-parent households can no longer enjoy a better quality of life through one paycheck. Whether because of this or as a result of single parenting, the intervening supports are similar. (SEE ALSO: *Family Self-Sufficiency; Nontraditional Households*)

—*Jacqueline Leavitt*

Further Reading

De Bell, Meagan, Hsiao-Ye-Hi, and Heidi Hartmann, with Jill Braunstein. December 1997. "Single Mothers, Jobs and Welfare: What the Data Tell Us." *Research-in-Brief*, Institute for Women's Policy Research, Washington, DC, pp. 1-8.

Folk, Karen Fox. November 1995. "Single Mothers in Various Living Arrangements: Differences in Economic and Time Resources." Institute for Research on Poverty, University of Wisconsin–Madison, Discussion Paper No. 1075.

Leavitt, Jacqueline. 1985. "The Shelter-Service Crisis and Single Parents." Pp. 153-76 in *The Unsheltered Woman*, edited by Eugenie L. Birch. New Brunswick: Rutgers, State University of New Jersey, Center for Urban Policy Research.

———. 1989. "Two Prototypical Designs for Single Parents: The Congregate House and the New American House." Pp. 161-86 in *New Households, New Housing*, edited by Karen A. Franck and Sherry Ahrentzen. New York: Van Nostrand Reinhold.

Mulroy, Elizabeth. 1988. *Women as Single Parents: Confronting Institutional Barriers in the Courts, the Workplace, and the Housing Market*. Dover, MA: Auburn House.

———. 1990. "Single-Parent Families and the Housing Crisis: Implications for Macropractice." *Social Work* 35(6):542-46.

———. 1995. *The New Uprooted: Single Mothers in Urban Life*. Westport, CT: Auburn House.

Rawlings, Steve W. 1989. "Single Parents and Their Children." *Current Population Reports*. U.S. Bureau of the Census Special Series P-23, No. 162. Washington, DC: Government Printing Office.

Rosenbaum, James E. 1991. "Black Pioneers: Do Their Moves to the Suburbs Increase Economic Opportunity for Mothers and Children?" *Housing Policy Debate* 2(4):1179-1213.

Sprague, Joan Forrester. 1991. *More Than Housing: Lifeboats for Women and Children*. Boston: Butterworth Architecture.

Stack, Carol. 1975. *All Our Kin: Strategies for Survival in a Black Community*. New York: Harper Colophone.

Winkler, Anne E. 1992. "The Impact of Housing Costs on the Living Arrangements of Single Mothers." *Journal of Urban Economics* 32:388-403. ◀

▶ Single-Room Occupancy Housing

In the United States, there has been a long but largely unrecognized tradition of living permanently or for long periods of time in single, furnished rooms without private kitchens. Although intended primarily for single persons, such accommodations have also provided housing to couples and families with children in the downtown sections of cities since the early 19th century. Single-room housing in private homes and apartments, commercial establishments, and charitable institutions was common up through World War II. Then urban renewal, urban redevelopment, and the public and private market's emphasis on family housing in apartments and houses all contributed to the widespread decline, conversion, and destruction of single-room accommodations. Between 1970 and 1980 alone, an estimated 1 million single-room occupancy (SRO) rooms were converted or destroyed nationwide. When the occupants could not find other SRO rooms, they became homeless; the loss of SRO housing is a major cause of urban homelessness. The large-scale loss of this form of convenient and affordable housing was first noted in the 1970s, but not until the 1980s were significant efforts made to stem the loss and to improve the conditions and the status of SRO housing. In the 1990s, SRO housing was viewed both as a way to provide affordable housing for independent single persons in downtown locations and, with a variety of social service arrangements, as a way to house populations with special needs who might otherwise be homeless or in institutions.

In the late 19th and early 20th centuries, a great variety of single-room accommodations serviced an equally great variety of people. Luxury and middle-class hotels with restaurants, libraries, parlors, valet, and many other services provided permanent or long-term homes, in rooms and suites, to those who chose the conveniences of hotel living and could afford the luxury of this type of housing. The Ansonia Hotel in New York is one well-known example; so is the Sheldon where Georgia O'Keefe and Alfred Steiglitz lived in the 1920s, at the height of hotel living. Private residential clubs, particularly for men, offered similar conveniences and amenities as did commercially run residences for single women, such as the Barbizon Hotel, also in New York. More modest accommodations predated these hotels in the form of boarding houses in private homes where residents ate home-cooked meals and in private or commercial lodging or rooming houses that offered rooms but no meals. Small, sparsely furnished rooms in cheap hotels and exceedingly small, unfurnished cubicles in "flophouses" were also available to men. Single men and women with more money could find safer and more comfortable rooms (and meals) in residences built and managed by philanthropic organizations such as the YMCA, the YWCA, and the Salvation Army.

Despite the number and the variety of these SRO-type accommodations and the fact that many people lived in them their entire adult lives, housing reformers and housing law did not recognize them as housing at all. In the 1920 Model Tenement Housing Law in New York, Lawrence Veiller, one of the most renown housing reformers in U.S. history, referred to all hotel residents as homeless men and women. The property standards adopted by the Federal Housing Administration in 1937 considered SRO buildings to be substandard because of the size of the rooms and the absence of private kitchens and baths. This meant that after World War II, no mortgage funds or loans restricted for housing could be used for SROs, and until the 1980s, no housing subsidies or other housing programs could be used for SRO buildings or SRO residents.

The social and legal denigration of a very reasonable, and affordable, form of housing for single people was due partly to the idealization of the private house with its own kitchen and bathrooms and partly to the idealization of the nuclear family. Until very recently, living in a single state beyond some period that might be considered transitional (as in youth, widowhood, or old age) was widely considered an aberration. Those who did so, particularly when they lived in single rooms downtown, were believed at best to be socially isolated; at worst a threat to society—that is, if such people were recognized at all. Often, occupants of residential hotels were not even acknowledged to be residents of that city or neighborhood. So while urban renewal and urban redevelopment were converting or destroying SRO hotels, virtually no effort was made to help their occupants find other accommodations. In building public housing, all attention was given to families or to the elderly, but not to single people whose homes might well have been destroyed in building that housing.

By the 1970s, many SRO hotels had become seriously deteriorated, poorly maintained, and very unsafe places to live and thus were easy targets for urban redevelopment. With the vast deinstitutionalization of mental patients, SRO hotels also became home to an increasingly vulnerable population. However, even when living conditions are very poor, SROs continue to provide affordable, convenient housing for low-income households. Furthermore, many SRO residents are neither transient nor isolated but are active participants in various social networks. At the same time, SRO housing offers residents a high degree of independence and allows for a diversity of lifestyles and habits.

By the late 1970s, housing activists and others began to recognize that SRO housing, the residents, and the lifestyles deserved attention and support. The City of Portland, Oregon, pioneered in the preservation, renovation, and quality management of SRO hotels, persuading the U.S. Congress and the Department of Housing and Urban Development (HUD) to allow SROs to be eligible for federal housing assistance. In 1982, under a federally funded demonstration program, Eugene, Oregon, began the physical and social improvement of SRO hotels with the participation of profit and nonprofit developers and with HUD subsidizing the rents of low-income residents. High-quality maintenance and management were key to the program. Similar efforts, under public or private ownership with a variety of funding sources and for various populations, followed in other cities. Cases of renovation and new construction multiplied. For a period of time, the City of San Diego encouraged private developers to build new SRO hotels such as the Baltic Inn, completed in 1987. Newly constructed SRO buildings may also house formerly homeless people, as in several examples designed by Skidmore, Owings and Merrill in New York. The number and variety of cases continues to increase.

As a type of dwelling, the single room is extremely versatile. A pullman kitchen and tiny bath can be added, providing a high degree of privacy, or several rooms, with or without private baths, can be grouped together to share a single kitchen and living room, possibly in the form of an apartment with its own locked door. The single rooms may also be arranged along corridors according to a hotel model. The size, location, and function of shared spaces and management spaces show great variation. Depending on the number and arrangement of private and public rooms, the building may create the atmosphere of a large hotel, a small hotel, a boarding house, or even a shared private house. Single, furnished rooms with shared kitchens may also be included in housing that has full apartments. All of these models are being adopted in the 1990s.

The SRO has been rediscovered as an acceptable housing type that costs significantly less to build or renovate and to rent than apartments. For the most part, SROs are again recognized as housing, which makes them eligible for housing programs and subsidies. With attentive management and the provision of social services, or a program of referrals to services, the SRO offers a way to house people with special needs, including the recently homeless, substance abusers, former mental patients, and frail elderly persons. Thus, SRO housing provides a variety of possible living arrange-

ments to meet a variety of needs. (SEE ALSO: *Efficiency Apartment;* **Homelessness;** *Lodging Accommodation*)
—*Karen A. Franck*

Further Reading

Franck, Karen A. and Sherry Ahrentzen, eds. 1989. *New Households, New Housing.* New York: Van Nostrand Reinhold.

Garside, P. L., R. W. Grimshaw, and F. J. Ward. 1990. *No Place Like Home: The Hostels Experience.* London: Her Majesty's Stationery Office.

Greer, Nora Richter. 1988. *The Creation of Shelter.* Washington, DC: American Institute for Architects Press.

Groth, Paul. 1994. *Living Downtown: The History of Residential Hotels in the United States.* Berkeley: University of California Press.

Hoch, Charles and Robert A. Slayton. 1989. *New Homeless and Old: Community and the Skid Row Hotel.* Philadelphia: Temple University Press.

Meyerowitz, Joanne J. 1988. *Women Adrift: Independent Wage Earners in Chicago, 1880-1930.* Chicago: University of Chicago Press.

Paul, Bradford. 1981. *Rehabilitating Residential Hotels.* Washington, DC: National Trust for Historic Preservation. ◀

▶ Sites-and-Services Schemes

The central concept of the sites-and-services scheme is a shift of focus from the provision of a complete house to the provision of a serviced plot and the delegation of the responsibility for the construction of the house to the allottee.

Initial attempts by governments of developing countries to reduce the housing problems of the urban poor focused on the direct construction of walk-up apartments and single-family housing units by a public agency with considerable government subsidy. Such housing proved too expensive for low-income households, and the large subsidies quickly depleted public resources. To save costs, housing agencies reduced plot sizes and located housing schemes in the urban fringe where land values were lower. The cost of infrastructure was brought down by lowering standards (unpaved roads, on-site sewage disposal) and providing shared facilities (public water taps, community toilets). Introducing low-cost building materials and simple construction technologies reduced the cost of the house. The next step was to provide serviced plots only, ask the allottees to use their own labor to construct the houses (sweat equity), and encourage the formation of building groups of families to achieve greater efficiency (mutually aided self-help). To ensure standards, the project authorities designed the houses and controlled all other aspects of the project.

In the late 1960s and early 1970s, students of housing such as Turner and Mangin pointed out that the tenure status, housing standards, regular loan repayments, and location of the low-cost housing projects did not match the needs and resources of the urban poor. They argued that squatter settlements form a much more suitable living environment for the urban poor in Third World cities, because they offer freedom to build—that is, squatters can build (or have built) what, how, and when they want. Turner suggested regularizing and upgrading rather than demolishing squatter settlements by providing what squatters cannot

acquire by themselves: secure land tenure (sites) and basic infrastructure (services). This would serve as an incentive to the squatters to improve their housing. However, squatter settlement regularization and upgrading is a reactive form of planning and rewards those who have taken illegal possession of someone else's land. To forestall the development of more squatter settlements and to provide alternative accommodation for the population of squatter settlements that cannot be regularized, Turner urged governments to develop simultaneously quasi-squatter settlements under planned conditions with vacant plots (sites) and basic infrastructure (services).

So site(s)-and-service(s) schemes evolved out of two quite different lines of thought. The first one emphasizes the labor input of the allottee—participation in the form of self-help construction. The allottee builds the house according to a design prepared by the housing agency so that it meets all the standards, but it is more affordable for a low-income family. The second line of thought emphasizes freedom to build: The dweller is in full control of the construction process and can construct the house according to his or her needs and priorities, resources, and abilities without constraint from externally imposed standards (progressive development).

Definition

The purest form of a sites-and-services scheme is an urban residential land subdivision where the developer sells or leases serviced plots—that is, land with basic infrastructure (roads, water supply, drainage, a human waste disposal system, and electricity). Whereas other public or private sector land subdivisions target middle- and higher-income groups, sites-and-services schemes usually target low-income households.

The infrastructure in the scheme may be available on the plot or be shared by a number of households. The allottee is responsible for the construction of the house, for which the project may extend loans. The beneficiaries usually have to meet definite criteria: They should fall in a particular income bracket and have a regular income or fixed employment to ensure loan recovery; they cannot own any other urban property, and they must have lived in the city for an extended period to prevent the schemes from becoming magnets for rural migrants. Sites-and services schemes are usually developed by the public sector, because the profit margin on serviced land subdivision projects for low-income groups is too small to interest private developers.

Beyond the above, there is little consensus among writers and researchers about what the concept of sites-and-services precisely stands for. The differences of opinion center around two questions: Is freedom to build a critical component of a sites-and-services scheme? Is a low-income housing project still a sites-and-services scheme if the housing agency provides some development on the plot? Both questions relate to the issue of standards.

Standards

The Turner school of thought argues that the urban poor have little flexibility in the way they can spend their income, and dwellers' control over the housing process is essential to meet their needs and priorities. Standards are oppressive, because they reduce dwellers' control and may force a household to live beyond its means or to abandon its dwelling and return to the squatter settlement. In this view, a low-income housing project cannot succeed unless the authorities give the household freedom to build—that is, control over the building process. These arguments are difficult to accept for public housing agencies, the public health bureaucracy, and political officials. Administrators hesitate to waive their own building regulations set to protect the public health and safety in order to make housing affordable for low-income groups. In their eyes, freedom to build would make the government responsible for the development of new slum areas. Political officials expect a visible impact from their housing policies and reject projects where the (initial) housing conditions are not much different from those of squatter settlements. Most governments would prefer sites-and-services schemes based on mutually aided self-help rather than schemes based on freedom to build. Some writers have indeed criticized sites-and-services schemes for enabling the government to evade its responsibility to provide adequate housing to the urban poor.

In view of these objections, housing agencies usually introduce some standards in sites-and-services schemes as well as measures to assist the allottees to construct the best possible house, and they closely monitor and guide the construction process to ensure that the allottees do not turn the scheme into a slum. They build model houses that allottees can visit, and they prepare type plans for the allottees to follow. There is often a fixed period for the completion of a basic house to deter the allottees from erecting a hut on the plot without any plan for a permanent house. Some projects require the construction of a latrine before any other development on the plot to ensure basic sanitary standards. Strict zoning regulations have been imposed in some projects: The allottees can use the plot for residential purposes only and are not allowed to open a shop or a workshop. Even the rental of rooms may be prohibited, because allottees are not supposed to make a profit from a government-subsidized scheme.

To enable the allottees to build the basic house within the set period, many sites-and-services projects extend loans that have to be paid back over an extended period of time (e.g., 20 years). Rather than giving a cash loan, the housing agency may provide the loan in the form of building materials. Such a loan allows for cost reductions through bulk purchases and ensures that the quality of the materials meets the standards and that no secondhand materials are used. The housing agency may provide training in construction skills to the allottees to ensure good-quality housing; if building groups are promoted, the project may also organize community leadership training. Sites-and-services schemes are often located in the urban fringe, and allottees could lose their sources of income in the city as a result of the move. Therefore, schemes sometimes have funds for income and employment generation and skill training.

Types of Sites-and-Services Schemes

A typical sites-and-services scheme provides vacant serviced plots, but many housing agencies have introduced some initial contractor-built development on the plot to ensure standards, thereby deviating from the basic concept of sites-and-services. Because the connection of the house to the utility networks requires technical skills, the plots in some sites-and-services schemes have a wet wall (one or more walls for the toilet and the kitchen with connections for water and sewerage) or a sanitary core (a toilet and kitchen). Because the roof is the most difficult and the most expensive part of the house to construct, some schemes supply the frame for a house plus a roof; the allottee has to construct the walls and complete the dwelling. Because allottees complained that they had to occupy (and some times pay for) one house while building another, housing agencies introduced a contractor-built room on the plot (core house), thereby abandoning the sites-and-services concept. With each addition to the vacant site and the services, the cost of the serviced plot obviously increases, making it less affordable for the lowest-income group.

In Hyderabad, Pakistan, authorities have tried to reduce the costs for the allottees by providing unserviced plots and water supply by tankers. The allottees pay the full cost of the plot on approval of their application. Because the government owns the land, this is a relatively small amount that is affordable to the lowest-income groups. The residents have to save for the infrastructure services, which are provided once the neighborhood has collected sufficient funds to finance it (incremental development scheme). This makes cost recovery superfluous. Allottees are obliged to occupy their plot and to start building their house as soon as their application has been approved. Given the initially harsh conditions in the scheme, this is expected to deter those who would buy the plot only for speculative purposes.

Issues

Sites-and-services schemes try to imitate the development process of squatter settlements under planned conditions, but the planning principles create conditions different from those in a squatter settlement. Moreover, some assumptions about squatter settlement have proven to be incorrect.

When promoting the concept of sites-and-services schemes, the World Bank insisted that such schemes must be self-financing and replicable, and it stressed the importance of cost recovery. However, the need to make regular payments, to have a minimum and regular income, and to have fixed employment renders a sites-and-services scheme less accessible to the lowest-income groups. Only the more established members of the working class may move to such schemes, leaving others behind to fend for themselves in squatter settlements. This is undesirable, because persons who make a living in the informal sector are in large part dependent on trickle-down income from the wage sector by selling their goods and services to their more established neighbors.

Because of the focus on a particular target group, the selection of applicants poses problems to both applicants and authorities. The urban poor have problems producing evidence of their monthly household income and its source, the absence of urban property, and the time spent in the city. It is equally difficult for the project authorities to verify the information supplied by the applicants. Evaluations of sites-and-services schemes show that many plots in a scheme fall into the hands of middle-income groups that may not be in immediate need of housing or that buy the plot for speculative purposes.

Community spirit is generally stronger in squatter settlements than in sites-and-services schemes. Squatter settlements develop more gradually, and squatters often face the common threat of eviction, which is absent in sites-and-services schemes. Squatter settlements are often populated by families bound by ethnicity, religion, occupation, caste, or region of origin, whereas allottees in sites-and-services schemes come from different parts of the city and are selected on an individual basis. This is an obstacle for the creation of building groups and neighborhood communities. A considerable effort by community organizers is required to form groups able and willing to jointly work on the construction of houses for the group members.

The idea that squatters construct their houses through self-help labor is only partially true. Squatters may build their first, temporary shelter themselves, but many tend to leave a consolidation of the house to hired craftsmen. These can produce better-quality work than the household members, who usually lack construction skills and whose earnings from their normal work often exceed the monetary value of participation in construction work as unskilled laborers. They are better off doing their usual job and using the money earned to pay the craftsmen to work on their house, providing, at most, unskilled labor. If project authorities promote self-help construction and the formation of building groups, the costs of training and supervision may be considerable.

Land forms an important part of the cost of a plot in a sites-and-services scheme, so most projects are located in the urban fringe, where land values are low. The distance between the scheme and the existing service networks makes off-site infrastructure expensive. Such costs may be only be partially charged to the allottees, because the infrastructure also benefits other city dwellers, but its construction is sometimes delayed until the area is further developed, making the sites-and-services scheme practically unlivable for an extended period of time. Transportation costs to the centers of employment tend to be high, which forms another obstacle for the urban poor to move to a sites-and-services scheme.

Sites-and-services schemes are often inappropriate for women-headed households. Women who are the head of a household may not meet such eligibility criteria as a minimum household income and a regular employment; in some countries, women may not even have the right to own property. Because illiteracy levels are often higher among women than among men, women have less access to information about a new sites-and-services scheme and face more difficulties applying for a plot. Conditions for self-help construction and speed of consolidation often do not take into account the multiple roles of women, and the

house and settlement designs are often ill suited for the home-based, income-generating activities on which many women depend. The remote location of many sites-and-services schemes places a particularly heavy burden on women; they often work part-time and need rapid and cheap access to the place of employment. (SEE ALSO: **Third World Housing**)

—*Kioe Sheng Yap*

Further Reading

Aliani, Adnan Hameed and Kioe Sheng Yap. 1990. "The Incremental Development Scheme in Hyderabad: An Innovative Approach to Low Income Housing." *Cities* 7(2):133-48.

Laquian, Aprodicio A. 1983. *Basic Housing: Policies for Urban Sites, Services and Shelter in Developing Countries.* Ottawa, Ontario, Canada: IDRC.

Mangin, W. 1967. "Latin American Squatter Settlements: A Problem and a Solution." *Latin American Research Review* 2:65-98.

Moser, Caroline O. N. and Linda Peake, eds. 1987. *Women, Human Settlements and Housing.* London: Tavistock.

Peattie, Lisa R. 1982. "Some Second Thoughts on Sites-and-Services." *Habitat International* 6(1/2):131-39.

Rakodi, Carole and Penny Withers. 1995. "Sites and Services: Home Ownership for the Poor?" *Habitat International* 19(3):371-89.

Swan, Peter J., Emiel A. Wegelin, and Komol Panchee. 1983. *Management of Sites and Services Housing Schemes: The Asian Experience.* Chichester, UK: Wiley.

Turner, John F. C. and Fichter, Robert, eds. 1972. *Freedom to Build: Dweller Control of the Building Process.* New York: Macmillan.

Van der Linden, Jan. 1986. *The Sites and Services Approach Reviewed: Solution or Stopgap to the Third World Housing Shortage?* Aldershot, UK: Gower. ◀

▶ Size of Unit

Size of unit is any measure of the amount of private, indoor space accessible to household members. Analysts sometimes use unit size as a measure of quality of life. Hence, size can be measured along any dimension thought to give rise to utility: for example, floor area, design and layout, or facilities and features. To illustrate, consider four measures incorporated into the 1991 American Housing Survey (AHS; see Table 26).

One measure is the number of rooms. In general, a room is a livable, interior space separated from other rooms by walls. The AHS counts bedrooms, living rooms, dining rooms, kitchens, recreation rooms, permanently enclosed porches suitable for year-round use, lodgers' rooms, and other finished and unfinished rooms. Also included are rooms used for offices by a person living in the unit. An L-shaped space or other multipurpose space (e.g., a living-dining area or kitchen-den area) is counted as one room only, unless a space is separated from adjoining rooms by built-in floor-to-ceiling walls extending at least a few inches from the intersecting walls. Movable or collapsible partitions or partitions consisting solely of shelves or cabinets are not considered built-in walls. By convention, bathrooms are not counted as rooms.

A second measure of dwelling size is number of bedrooms. The AHS includes any rooms used mainly for sleep-

TABLE 26 Size of Dwelling, by Tenure of Household, Occupied Private Dwellings, United States, 1993

	Owner Occupied	Rented
All year-round occupied dwellings	63,544	34,150
By number of rooms (% of all dwellings)		
1 room	0	2
2 rooms	0	3
3 rooms	1	22
4 rooms	10	32
5 rooms	22	23
6 rooms	26	12
7 rooms	19	5
8 rooms	12	2
9 rooms	6	1
10 or more	4	0
Median rooms	6.2	4.2
By number of bedrooms (% of all dwellings)		
None	0	0
1 bedroom	3	29
2 bedrooms	23	43
3 bedrooms	52	20
4 or more	22	4
Median bedrooms per dwelling	3.0	1.9
By complete bathrooms (% of all dwellings)		
None	0	1
1 bathroom	30	72
1 and a half	18	10
2 bathrooms or more	52	17
Single detached and mobile homes (% of all dwellings reporting)	52,480	8,627
Less than 500 square feet	1	3
500–749 square feet	3	11
750–999 square feet	8	18
1,000–1,499 square feet	24	32
1,500–1,999 square feet	23	18
2,000–2,499 square feet	18	8
2,500–2,999 square feet	10	4
3,000–3,999 square feet	9	3
4,000 square feet or more	5	2
Median square footage	1,814	1,270

SOURCE: *American Housing Survey for the United States in 1993, Table 1A-3.*
NOTE: The American Housing Survey does not include households in group quarters.

ing, even if also used for other purposes. Rooms reserved for sleeping, even if used infrequently (e.g., guest rooms), are also counted as bedrooms. On the other hand, rooms used mainly for other purposes, even though also used for sleeping, such as a living room with a hideaway bed, are not considered bedrooms. A housing unit consisting of only one room, such as a one-room efficiency apartment, is classified by definition as having no bedroom.

A third measure of dwelling size is the number of bathrooms. The AHS classifies a housing unit as having a com-

plete bathroom if it has a room with a flush toilet, bathtub or shower, a sink, and hot and cold piped water. All facilities must be in the same room to be a complete bathroom. A half bathroom has either a flush toilet or a bathtub or shower but does not have all the facilities for a complete bathroom.

The fourth measure of dwelling size is floor area. The AHS measures square footage of floor area for single, detached dwellings and mobile homes. Their measure excludes unfinished attics, carports, attached garages, porches that are not protected from the elements, and mobile home hitches. Both finished and unfinished basement areas are included.

All of the above measures of dwelling size are more appropriate to single, detached dwellings than to other kinds of dwellings. Such measures are less satisfactory in the case of large rental projects, condominiums, and housing cooperatives, where communal spaces such as a laundry room, party room, gym, studio, or child care facility are used by residents in lieu of space within their private dwelling. (SEE ALSO: *Crowding; Space Standards*)

—*John R. Miron*

Further Reading

Holdsworth, D. W. and J. Simon. 1993. "Housing Form and Use of Domestic Space." Pp. 188-202 in *House, Home, and Community: Progress in Housing Canadians, 1945-1986,* edited by J. R. Miron. Montreal: McGill-Queen's University Press.

U.S. Department of Commerce, Bureau of the Census. 1995. *Current Housing Reports: American Housing Survey for the United States in 1995.* Washington, DC: Author. ◄

► Slaves, Housing of

The type and quality of slave housing varied greatly according to region, size of the slave labor operation, and the type of slave labor enterprise. However, during the slave era in the United States (1619-1863), most slaves who worked in the fields on plantations lived in groups of as many as a dozen in crudely constructed, sparsely furnished, one- or two-room log cabins.

The first Africans came ashore in Virginia in 1619, where both Africans and Indians were enslaved during the 17th and early 18th centuries. Although one-third of the 5,500 slaves recorded in South Carolina in 1710 were Indians, after 1720, because of disease and deaths associated with Indian wars, Africans constituted the overwhelming majority of slaves in that state. The housing of slaves in both Virginia and the Carolinas, however, reflected structural influences of both Africans and Indians (e.g., open temporary cabins, clay walls, cribbed log walls, stick and clay chimneys, and hearths in the middle of the floors), as well as of Europeans.

Slaves were one of the most rapidly growing segments of the population, increasing from less than 700,000 in 1790 to more than 2 million by 1830 and to 3.9 million by 1860. Most slaves lived on plantations that harvested tobacco (Maryland, Virginia, and North Carolina), rice

(South Carolina and Georgia), or sugar (Louisiana). In 1860, there were more than 46,000 plantations in the South, with 11,000 of them housing more than 800,000 slaves in groups of 50 or more.

The increasing slave population in the colonies in the 19th century (because of births and continued importation) was housed according to varying patterns. For example, in late 17th and early 18th century Maryland, males substantially outnumbered females and usually lived with other unrelated slaves in rooms in the main farmhouse or in outbuildings. By the 1750s when the number of slaves on plantations with more than 20 slaves had increased and the sex ratio had become more balanced, tax records revealed separate "Negro dwellings," presumably constructed as separate living quarters for slave couples.

In South Carolina, however, where a strong plantation economy had emerged in the 1690s and there were large slave populations per plantation, the slaves built their own segregated quarters (or separate villages) shortly after arriving from Africa. These quarters were small, African-style, rectangular huts constructed with mud walls and thatched roofs and with a hearth in the middle of the earthen floor. The windowless slave cabins found in South Carolina bear a striking resemblance to houses in Central Africa, an area from which many slaves came. If slave blacksmiths, carpenters, and bricklayers built the quarters, even if the design was simple, the workmanship was good.

In 1850, of the 3.2 million slaves in the United States, 400,000 lived in urban communities, sharing the homes of their owners (in lofts, attics, or rooms behind the kitchen), where they were domestic servants, porters, common laborers, or skilled craftsmen. Slaves who were skilled craftsmen (such as carpenters, masons, and blacksmiths) were sometimes hired out to townspeople by plantation owners and were allowed to live in the houses or shops of their employers. If townspeople had large numbers of slaves, they often constructed two-story dormitory-like structures with kitchens, laundries, and other workrooms on the first

floor and slave bedrooms on the second. These quarters were generally furnished in a manner compatible with the main house.

Records from 1860 reported 411,000 mulatto slaves, the offspring from the sexual union of slaves and whites, who were housed in a variety of ways. Some received favored status and residence in the master's house, either as a house slave or as a family member. Other mulatto slaves lived in the slave quarters with their mothers or with older slave women no longer capable of working in the fields. Still others were sold by their masters or fathers to other plantation owners as either house or field slaves.

Functions of Slaves

Plantation slaves were housed in cabins or quarters close to their worksites—the barn or stables and the fields or pastures. Slaves who worked in the master's house, also known as the Big House, lived there or in nearby quarters. Slaves living in the Big House were sometimes children whom the master's wife had decided to raise (mulattos and others) and who waited on the master's children. House servants and coachmen slept in the attic, and slave children slept on the floors or on trundle beds in the rooms with their young masters or mistresses. Slaves who lived in the Big House had better accommodations than slaves housed elsewhere on a plantation.

The location of the kitchen was symbolic of the social order between slave owners and slaves. Although originally part of the Big House (through the 17th century), in the 18th century, it became detached from the house along with the servants who used it. On some plantations, it became a separate two-room structure, with one room as quarters for the cook and her family. Other mansions retained the kitchen in the house but moved the unpleasant kitchen work (such as cleaning and scaling fish or putting up pork) to a separate kitchen quarter near the house. House servants lived and ate in the kitchen quarters, and laundry for the master's family was done there as well.

Quarters for the field slaves were placed far enough from the Big House to keep their sights, sounds, and smells from offending the master but close enough to allow surveillance of slaves. Although black overseers were housed in the quarters with field slaves, their houses were larger than the houses of other slaves but smaller than the houses of the white overseers, who lived elsewhere on the plantation.

Construction Styles

The most common design of cabins for field slaves had square or rectangular rooms and contained either one or two units. Slave cabins were often whitewashed to brighten their interiors because windows were few and tiny. Slave quarters were built out of the same construction material and in the same style as the Big House, with log construction the most common (see Figure 28), and brick, stone, and frame housing the "top of the line." Some slave housing was built of tabby, a primitive sort of concrete derived from oyster shells (North Florida), and "ragged huts made out of poles" were reported in Alabama.

Figure 28. Example of Single-Unit Log Cabin Used as Slave Quarters
Photograph by Wilhelmina Leigh.

Figure 29. "Post-and-Beam" Construction Used in Slave Quarters in Virginia
Photograph by Wilhelmina Leigh.

In southern Maryland, by the late 18th century, the single-unit log cabin had become the typical slave dwelling—one story high (with one room down and a loft), a gable roof, and a chimney exterior to one gabled end. There was usually one door centered in the long wall, with one or two windows on either side of the door, one in the gable end, or no windows at all. The downstairs room had a wooden or log floor and served as the kitchen, dining room, bedroom, bathing room, and sitting room for the family.

In Virginia, slave quarters were the so-called earthfast houses built by British pioneers. These "post-and-beam" houses had posts extended into the ground as their foundations (see Figure 29). The posts were topped with beams, and gabled roofs and perhaps chimneys completed the structures. Earthfast slave houses were distinguished from 18th century earthfast houses built by whites only by their earthen root cellars. Wood-lined and used to store food, these cellars were dug near the chimneys.

In another example, from French Louisiana in the 19th century, slave housing built according to a design from northwestern France had replaced the prototypical log cabin. It was one-story tall, two rooms wide and two rooms deep, with a central chimney between the two front rooms and a steeply pitched roof.

Condition of Slave Housing

Slave quarters seem cramped and primitive when compared with the housing of large slave owners, although the rural white poor and small slave owners lived in one- and two-room log cabins, as did most slaves. However, most non-slave inhabitants of log cabins, selected the type, style, and location of their residences, whereas slaves had the one-room log house in a village setting thrust on them with little, if any, choice.

Slave housing was usually small, dilapidated, without windows, and practically without furnishings. Fully finished dwellings with floors, doors, and windows were the exception for plantation slaves. In addition, slave housing was often crowded and drafty, being labeled "laboratories for disease," by some agrarian reformers.

Slave owners felt that a 16 × 18 foot square was not too large for a man, a woman, and three or four small children. Most cabins contained at least two families, and the average number of slaves per one-room cabin ranged from 3.7 to 8.8 in the 18th and 19th centuries. In 18th-century Virginia, on some plantations, groups of 8 to 24 slaves were communally housed in one- or two-room structures.

For many slaves, beds were a wooden plank or a collection of straw and old rags thrown on the floor. The overseer and the colored carpenter on some plantations provided furnishings such as tables and benches. On other plantations, slaves made their simple furnishings during their limited spare time, using whatever materials they could acquire. (SEE ALSO: *African Americans*)

—*Wilhelmina A. Leigh*

Further Reading

Ferguson, Leland. 1992. *Uncommon Ground: Archaeology and Early African America, 1650-1800*. Washington, DC: Smithsonian Institution Press.

Franklin, John Hope and Alfred A. Moss, Jr. 1988. *From Slavery to Freedom: A History of Negro Americans*. 6th ed. New York: Knopf.

McDaniel, George W. 1982. *Hearth and Home: Preserving a People's Culture*. Philadelphia, PA: Temple University Press.

Sobel, Mechal. 1987. *The World They Made Together: Black and White Values in Eighteenth-Century Virginia*. Princeton, NJ: Princeton University Press.

Vlach, John Michael. 1993. *Back of the Big House: The Architecture of Plantation Slavery*. Chapel Hill: University of North Carolina Press. ◄

► Slums

From about the middle of the 19th century to the present, city neighborhoods inhabited by low-income populations and usually characterized by physical dilapidation have been called slums. The word *slum* has never been defined officially, even during the decades when "slum clearance" was an essential element of the federal Urban Renewal Program. Rather, the term has been used in a plastic manner, with local political forces specifying which urban places should be so labeled.

Slums are undesirable places, in part because of the attributes of their residents, and, in part, it is often claimed, because slums contribute to those attributes. In the latter sense, slums have been identified as causing, or at least contributing to, disease, crime, family deviance, unemployment, and poverty. Although the term slum has fallen into disuse in recent decades in the United States, synonyms and euphemisms are prevalent. These include *poverty, inner-city,* or *underclass neighborhoods; ghettos* or *black ghettos;* or simply *the central city*.

Slums need to be understood on two levels—objective and ideological. Objectively, slums represent a spatial expression of social and economic inequality: People who are poor and different from dominant groups in race, ethnicity, or religion are physically concentrated in particular city neighborhoods. The concentration usually involves the interaction of market-driven processes (households with low incomes can afford only the cheapest accommodations in the most dilapidated areas) and sociopolitical processes ("undesirable" populations are prevented from living with dominant groups by means of discrimination and, often, violence).

The variability of the process of spatial concentration over time is noteworthy. Thus, Italian and Jewish immigrants, for example, were concentrated in "slums" early in the 20th century but later were dispersed throughout the urban landscape. African Americans, in contrast, were relatively dispersed in the 19th century and became highly segregated only over the course of the 20th century. Today, poor whites are spatially dispersed and tend to live on the periphery of urban areas, in contrast to blacks who are forced into the centers.

On the ideological level, the characteristics of people and places are blurred. People are labeled as slum or ghetto residents and thereby stigmatized. Moreover, the processes that concentrate stigmatized populations and simultaneously create "desirable" places (the slums versus the suburbs) are obfuscated so that responsibility is shifted from dominant social groups to the subordinate "slum dwellers" ("ghetto poor" or the like).

The consequences of the concentration of poor people, now usually people of color, in urban slums are significant. On the one hand, low-income populations weaken the fiscal base of municipal governments. In the U.S. system, where municipalities assume prime responsibility for providing public services, the result is that low-income slum populations receive relatively inferior services, including inferior schooling. Their ability to escape poverty is accordingly reduced. On the other hand, advocates of slum clearance from the 19th century to the 1960s believed that the physical crowding and dilapidation of the slum affected poverty in a causal manner.

More recently, the causal argument has shifted to the social side. Spatial concentration, in this view, undermines the economic prospects of poor people, other things being equal. In other words, there is a cost to growing up in a poor neighborhood that results from the culture and dynamics of the social environment. This argument has attained particular focus in the debates over the existence and consequences of a so-called underclass. Scholars such as William Julius Wilson and John Kasarda have claimed that the concentration and isolation of the poor have contributed both to the development of dysfunctional patterns of behavior and to a spatial "mismatch" between the neighborhoods where the poor live and those with expanding job opportunities. Although the "underclass debate" has been couched in a universalistic language, its subject matter has been primarily African Americans and their ghetto neighborhoods.

So far, the evidence supporting this most recent slum theory of poverty is wanting. A great majority of poor people do not live in concentrated city neighborhoods—the slums of old. Although research has documented the liabilities of a poor environment with regard to schooling, crime,

and services, it has been extremely difficult to demonstrate the existence of "neighborhood" effects that systematically reduce the prospects of social mobility. Nonetheless, it is—and always has been—easier to indict the slum for producing poverty rather than the economic inequality and the politics that support it. (SEE ALSO: *Slums;* **Urban Redevelopment**)

—*Norman Fainstein*

Further Reading

Fried, Mark and Peggy Gleicher. 1961. "Some Sources of Residential Satisfaction in an Urban Slum." *Journal of the American Institute of Planners* 27:305-15.
Gaskell, S. Martin, ed. 1990. *Slums.* Leicester, UK: Leicester University Press.
Gelfand, Mark I. 1975. *A Nation of Cities: The Federal Government and Urban America, 1933-65.* New York:, Oxford University Press.
Riis, Jacob A. 1902. *The Battle with the Slum.* New York: Macmillan.
Sternlieb, George. 1973. *Residential Abandonment: The Tenement Landlord Revisited.* New Brunswick: Rutgers, State University of New Jersey, Center for Urban Policy Research.
Suttles, Gerald D. 1968. *The Social Order of the Slum: Ethnicity and Territory in the Inner City.* Chicago: University of Chicago Press.
Walker, Mabel L. 1938. *Urban Blight and Slums: Economic and Legal Factors in Their Origin, Reclamation, and Prevention.* Cambridge, MA: Harvard University Press (with special chapters by Henry Wright, Ira S. Robbins, and others).
Ward, David. 1989. *Poverty, Ethnicity, and the American city, 1840-1925: Changing Conceptions of the Slum and the Ghetto.* Cambridge, UK: Cambridge University Press.
Yelling, J. A. 1992. *Slums and Redevelopment: Policy and Practive in England, 1918-1945.* New York: St. Martin's. ◀

▶ Smart House and Home Automation Technologies

A "smart house" that responds to the dwellers' needs and desires by adjusting lighting, temperature, even ambient music, has appeared in science fiction for much of the 20th century. From LeCorbusier's vision of the house as a machine for living, to Negroponte's architecture machine, home automation technologies are the latest extension of a century-long fascination with housing and mechanism. However, with the development of new electronic technologies and their integration with older, traditional building technologies, the intelligent home is at last becoming a real possibility.

The basic idea of home automation is to employ sensors and control systems to monitor a dwelling, and accordingly adjust the various mechanisms that provide heat, ventilation, lighting, and other services. By more closely tuning the dwelling's mechanical systems to the dwellers' needs, the automated "intelligent" home can provide a safer, more comfortable, and more economical dwelling. For example, the electronic controller of an automated home can determine when the dwellers have gone to bed and turn off the lights and lower the thermostat; it can monitor burglar and fire alarms; it can anticipate hot water usage and optimize the operation of the water heater.

The Smart House Project was initiated in the early 1980s as a project of the National Research Center of the National Association of Home Builders (NAHB) with the cooperation of a collection of major industrial partners. The smart house technology is one realization of home automation ideals using a specific set of technologies. In smart house technology, the dwelling is wired with a single multiconductor cable that includes electric power wires, communications cables for telephone and video, and other conductors that connect appliances and lamps with electronic devices that control the supply and switching of power.

A principal benefit claimed for smart house technology is safety. With current technology, electric power is provided to all appliances that are plugged into a wall outlet. Cutting a wire, inserting a screwdriver into a wall outlet, or a fault inside the appliance can result in a severe electric shock. The smart house, in contrast, provides power only to outlets that have appliances plugged in and turned on, and the smart house controllers monitor the circuit, disconnecting power at the first indication of a short circuit or other failure. In addition, when sensors detect gas and water leaks, smoke, and other abnormal conditions, the electronic controllers can shut down the appropriate devices and trigger the alarm.

A second benefit is convenience. Traditional wiring in North America provides only 110-volt power outlets, with occasional 220 volt service for heavy appliances. However, many of today's consumer electronic appliances such as radios, personal computers, and even power tools step the line voltage from 110 down to 6, 9, 12, or 24 volts. Smart house technology includes provision for power at several voltage levels, eliminating the need for numerous small power adapters at each appliance. The smart house cabling provides a single outlet for power and communications; gas is provided through flexible tubing and its own, quick-connect outlets. In addition, smart house technology can automatically control the temperature, humidity, and lighting in the dwelling on a room-by-room basis.

A third benefit is economy. Smart house technology can adjust the power supplied to each appliance according to need. In the traditional scheme, each appliance is provided with sufficient power to provide for its peak use. In addition, the smart house controllers can schedule the operation of heavy-power-consuming appliances (such as dishwashers, electric water heaters, and air conditioners) to take maximum advantage of off-peak electric rates. These adjustments could result in lower utility costs.

A great deal of groundwork has been done to coordinate the building industry infrastructure in preparation for bringing smart house technology to market. Whether smart house and other home automation technologies succeed at changing standard residential construction, it is clear that at least some of the innovations are already entering the marketplace, albeit in fragmentary forms. For example, ground fault interrupted circuits, which detect an electric shock incident (a "ground fault") and shut down immediately, are now commonly used in bathrooms, kitchens, and in outside outlets. Likewise, telephone, television, and other communication services are undergoing tremendous changes, and this will no doubt affect the way these services

are delivered inside the dwelling. In some instances, third-party vendors sell the sensors and controller units to connect the security system, lights, telephone, and other devices to a personal computer. For example, these devices allow the homeowner to call home and instruct the computer to "turn on the oven at 4 p.m. and heat up the hot tub."

Smart house and other home automation technologies require widespread changes in the way buildings are made, changes that call for cooperation among the manufacturers of construction components, utility suppliers, and regulatory agencies that oversee the building industry. With the rapid changes in electronic and materials technology, it may well be that new standard building technologies will become obsolete by the time it can gain a footing. Providers for cable television, telephone, and new communications services are struggling to define technological standards, and the ownership of copper wire and fiber-optic cable networks to each dwelling has been a valuable resource held by local telephone and electric companies. These organizational changes in the urban infrastructure may well have an impact on the way housing is wired and cabled for power and communications. The new information infrastructure being developed offers the possibility of broader participation in civic and community activities, access to educational resources, as well as to work, shopping, and entertainment. Already the fax and the modem enable people to work at home and browse information databases over conventional telephone lines. Higher bandwidth communication technologies are being developed to provide electronic community town meetings, distance learning, home shopping, and video on demand. However, it remains to be seen whether the technology will enhance and enrich the lives of citizens.

Finally, even after technical and organizational challenges are met, many will find a fine line between an intelligent house that maintains comfort levels and an overbearing house that monitors the inhabitants too closely. Few people object to using a thermostat to control the temperature in a house, but most cherish the power to set and reset the thermostat. As the hardware and software to control home automation systems become increasingly complex, human interface designers must make it easy for dwellers to program the house and to override preprogrammed settings. (SEE ALSO: *Construction Technology; Physical Disabilities, Housing of Persons With*)

—*Mark D. Gross*

Further Reading

Allen, Edward, ed. 1975. *The Responsive House*. Cambridge: MIT Press.

Arno, P. S., K. A. Bonuck, and R. Padgug. 1994. "The Economic Impact of High-Technology Home Care." *Hastings Center Report* 24(5):S15.

Arras, J. D. 1994. "The Technological Tether—An Introduction to Ethical and Social Issues in High-Tech Home Care." *Hastings Center Report* 24(5):S1.

Negroponte, Nicholas. 1975. *Soft Architecture Machines*. Cambridge: MIT Press.

Smith, Ralph L. 1988. *Smart House: The Coming Revolution in Housing*. Columbia, MD: GP Publishing. ◄

▶ Social Housing

Social housing is a vague concept, which some observers may see as an advantage, but generally, it refers to housing subsidized by the state. Its historical roots are also rather obscure. Among the first explicit references to public subsidies for housing for lower-income families was the Labouring Classes Dwelling Act of 1866 in Britain. The more specific term *social housing,* however, has been in regular use in the European housing policy literature only late in this century and usually with respect to the public rental sector. In the social-democratic context of European politics, this sector has gradually become labeled as "social rented housing."

In its contemporary setting, social housing has two broader connotations. The first and most inclusive meaning encompasses all types of housing that receive some form of public subsidy or social assistance, directly or indirectly. In this sense, much of the private sector stock would qualify as social housing, at least to the degree that it receives a public subsidy. That subsidy might include tax relief on mortgage interest for owner-occupiers, tax shelters for homeownership savings, subsidies to builders, depreciation allowances for investors in residential properties, and the below-cost provision of collective public services, such as roads, schools, water, and sewers for housing in new suburban developments.

The second and more common application of the term combines traditional public housing with newer forms of publicly supported and nonmarket housing, such as cooperatives, rent-geared-to-income, limited-dividend, and nonprofit housing typically provided by social agencies, community groups, nonprofit private firms, and political organizations other than governments. The latter, often referred to as "third-sector" housing, includes housing for the elderly, the handicapped, and other special-needs groups.

All of these forms, unlike market housing, share similar attributes. They are collectively managed; operated on a not-for-profit basis; and have rents, for a proportion of the units, set according to ability to pay. The subsidies for such projects may serve to reduce initial capital costs (e.g., low-interest mortgages) or downstream operating costs (e.g., rent supplements), or both. Most of these new forms of social housing are also designed to attract a wider mix of income, household types, and social groups than traditional public housing now serves.

Why has the use of the social housing label become so widespread? One obvious reason is the desire on the part of policymakers and planners to avoid (or at least reduce) the stigma traditionally attached to earlier public housing projects and programs. Negative reaction to such projects, based on their size and homogeneity, and on fears that they would reduce property values, increase crime, and become ghettoes for the very poor, was especially strong in the United States and Canada. As a planning strategy, it was assumed that mixed social housing, on a smaller scale, was more likely than large and homogeneous public housing projects to be acceptable to local governments and neighborhood associations. Local conflicts and resistance (the

NIMBY—"not in my backyard"—syndrome) were not, however, significantly reduced.

A second reason is more pragmatic, reflecting the reality of a broadening base of housing provision in Western economies during the 1970s and 1980s. Investment in traditional public housing has declined almost everywhere in relation to investment in other forms of socially assisted housing, including that provided by not-for-profit organizations supported by, but administered at arm's length from, governments. A new and more neutral term was needed to embrace these increasingly diverse forms of housing provision. At the same time, there has been a shift away from subsidies for particular buildings (the old public housing model) to subsidies for people in the private market (e.g., rent supplements, allowances, and vouchers). The latter have the advantage that they create temporary social housing units within the private rental sector as the need arises, while avoiding the continuation of large bureaucracies.

Both as a concept and as a strategy for housing policy, however, social housing has not been without its critics. One such criticism is that the term masks a substantial shift in government housing policy and investment priorities away from housing for the poorest members in society to housing for those less difficult and costly to house—such as lower- and low-middle-income households and the elderly. Governments can argue that the level of investment in the entire social housing envelope has expanded, even if far fewer resources are going directly to those most in need through the public housing sector.

On the other hand, in its broadest interpretation, the concept is useful in that it stresses that there is not a clear division between public and private housing in contemporary Western societies. Many possible forms of housing provision lie between the polar extremes of government-managed public housing and individually owned market housing. All housing is socially produced, the vast majority by private entrepreneurs, some by government, and a small but increasingly significant proportion by a host of other largely nonprofit organizations, such as charities, housing associations, foundations, community trusts, unions, churches, and ethnic institutions. Yet all housing receives some form of public subsidy and is therefore, at least in part, social housing. The difference in subsidy is one of degree and type. Other important contrasts between social and market housing rest in the differing equity and tax treatment and the differing nature of control that each offers over one's home environment. (SEE ALSO: *European Social Housing Observation Unit; Welfare State Housing*)

—*Larry S. Bourne*

Further Reading

Ball, M., M. Harloe, and M. Martens. 1988. *Housing and Social Change in Europe and the USA.* New York: Routledge.

Bourne, L. S. 1981. *The Geography of Housing.* London: E. Arnold.

Burns, L. and L. Grebler. 1986. *The Future of Housing Markets.* New York: Plenum.

Danemark, B. and I. Elander, eds. 1993. *Social Rented Housing in Europe: Policy, Tenure and Design.* Delft, The Netherlands: Delft University Press.

Karn, V. and H. Wolman. 1992. *Comparing Housing Systems: Housing Performance and Policy in the US and Britain.* Oxford, UK: Oxford University Press.

Kemeny, Jim. 1995. *From Public Housing to the Social Market: Rental Policy Strategies in Comparative Perspective.* London: Routledge.

Miron, J. 1993. *House, Home and Community: Progress in Housing Canadians.* Montreal: McGill-Queen's University Press.

Van Vliet--, Willem, ed. 1993. *International Handbook of Housing Policies and Practices.* New York: Greenwood. ◄

► Social Mix

Social mix is a policy in housing intended to achieve a blend or balance of a population's social and economic characteristics and mainly aimed at improving the viability of communities. As a social policy, it has been guided more by ideology than by research.

Social Mix: Housing Programs in Search of a Policy

Social mix may be described as the distribution of households in a given housing project, neighborhood, community, or larger geographic area defined according to some social or economic characteristic—for example, income, education, occupation, disability, race, ethnicity, culture, age, gender, household composition, stage in the life cycle, lifestyle, or political preference. Communities that are homogeneous in regard to any of the above characteristics might be considered less socially mixed, or more segregated. The "grain" of mix can be defined at varying geographic scales, such as block levels, housing projects, neighborhoods, census tracts, and census metropolitan areas.

The social welfare implications of a social-mix strategy involve relationships between social mix and some outcome measure, such as employment or use of services, often mediated by intervening variables such as housing location and design.

There is more opinion than fact on the subject of social mix. The evidence from the social sciences and planning literature is incomplete. Both positive and negative consequences have been found. For example, social disorder is often attributed to relative deprivation effects of social mix. Not poverty in isolation, but poverty in the midst of plenty has been seen to foster violence and property crime. The tendency for private consumer services to follow income, reflecting free-market forces, presents a problem of access to services when there is a lower socioeconomic mix. A social mix in housing also has human aspects. When housing management, real estate industries and financial institutions act as gatekeepers by steering certain groups away from some areas to protect property values and powerful community interests, the rights of citizens to choose their location of residence is constrained. The mixing of the elderly and nonelderly with mental disabilities in public housing has posed problems in dealing with disruptive behaviors frightening the elderly residents.

The mixing of able-bodied residents with the physically challenged in nonprofit cooperative housing has been associated with some positive social interaction between the two groups. Ethnic minorities will frequently self-select to live in communities with similar ethnolinguistic characteristics for ease of communication, increased political solidarity, and the availability of formal and informal support systems.

There is conflicting evidence on the desirability of similar socioeconomic groups living together. Some social scientists have made the case for less socially mixed communities on the basis of evidence that lower-status people have closer neighborhood ties and social support networks in more homogeneous communities. In contrast, others have found no evidence to support the view that individuals of comparable social rank living together would be more satisfied with their community.

Although the concept is often poorly defined and the results of social-mix studies are frequently inconclusive, social mix has been preferred by many policymakers. The relationship of social mix to factors of design, management, and maintenance of housing and neighborhood and community services is complex. Therefore, more research is needed to evaluate the quality of life effects of social-mix policies. In the absence of solid research to inform housing and planning policy, program decisions tend to be guided more by ideology than by support from scientific inquiry.

Social Mix: The Ideological Debate

Social policy debate on social mix generally revolves around the principles of universality, selectivity, and functionality. A universalist frequently argues for a more mixed approach to planning housing and communities. A selectivist is inclined to take a more targeted approach that often results in segregation effects. In Canada, nonprofit cooperative housing was intended to promote a universal policy applied to social housing in which most of the housing units were to be made available to all members of society. Means-tested public housing, based on need, is more selectivist. The universalist model tends not to separate disadvantaged members of society from the mainstream population. Rather, the universality principle treats all categories of individuals similarly. Implementing a universal policy in social housing means that public sector housing would be equally available to any member of society. This equal treatment tends not to segregate or stigmatize those in receipt of the public benefit of social housing. Selectivists on the other hand argue that it serves the aim of redistributive justice if governments supply housing directly to those in greatest need.

A third ideological approach to social mix is the functionalist perspective. This approach suggests that the selectivist stance is used to compensate for the "inefficiencies" of the housing market by government's playing economic handmaiden to the housing industry through the minimal provision of public housing to accommodate those who cannot afford to pay for housing in the marketplace. Beneficiaries must qualify by submitting themselves to often intrusive and demeaning needs and means tests. Functionalists argue that the poor thereby serve as scapegoats for market inefficiencies with the public "blaming the victims" of these inefficient market forces through the provision of stigmatized and segregated public housing. Concentrations of the poor also help to create higher-paying jobs in public services and volunteer organizations for middle-class professionals to administer the institutions of welfare, including social services and social housing. (SEE ALSO: *Segregation*)

—Morris Saldov

Further Reading

General Accounting Office. 1992. *Housing Persons with Mental Disabilities with the Elderly.* Washington, DC: Author.

Ginsberg, Yona and Robert W. Marans. 1979. "Social Mix in Housing: Does Ethnicity Make a Difference?" *Journal of Ethnic Studies* 7(3):101-11.

Heumann, Leonard F. 1996. "Assisted Living in Public Housing: A Case Study of Mixing Frail Elderly and Younger Persons with Chronic Mental Illness and Substance Abuse Histories." *Housing Policy Debate* 7(3):447-72.

Hjärne, Lars. 1994. "Experiences from Mixed Housing in Sweden." *Scandinavian Housing & Planning Research* 11:253-57.

Lazerwitz, Bernard. 1985. "Class, Ethnicity, and Site as Planning Factors in Israeli Residential Integration." *International Review of Modern Sociology* 15(Spring-Autumn):23-44.

Peel, Mark. 1995. "The Rise and Fall of Social Mix in an Australian New Town." *Journal of Urban History* 22(1):108-40.

Sarkissian, Wendy. 1976. "The Idea of Social Mix in Town Planning: An Historical Review." *Urban Studies* 13:231-46.

U.S. Department of Housing and Urban Development and Department of Health and Human Services. 1993. *Creating Community: Integrating Elderly and Severely Mentally Ill Persons in Public Housing.* Prepared by the National Resource Center on Homelessness and Mental Illness. Delmar, NY: Policy Research Associates. ◄

▶ Solar Housing

The use of solar energy in building forms has challenged designers from ancient times and continues to do so today. Housing located in temperate and cold climate regions is an appropriate building type for solar energy use because of the need for both daylight and heating for most interior spaces. Many types of solar systems fall into categories of either passive or active technologies. There is a broad range of use for solar systems, from space and hot-water heating to natural lighting and electricity production. An enormous benefit of a solar system in housing is in the use of a pollution-free, renewable energy source—the sun.

The use of solar energy in housing is not a new innovation. Historically, it dates back to the fourth century B.C. in Greece and probably much earlier than that. Ancient builders learned to design houses to take advantage of the sun's energy during the moderately cool winters and to avoid the sun's heat during the hot summers. Thus, solar housing came into being—designing buildings to make optimal use of the sun by responding to its changing positions during different seasons. The Greeks knew that in winter the sun path in the Northern Hemisphere was in a low arc across the southern sky and, therefore, openings could capture much needed heat. In summer, the sun path was much higher overhead so roof overhangs provided shading. That most early buildings were made of stone enabled the storage of the solar energy. Early evidence also shows that solar principles were used not only for single, isolated villas but

Figure 30. 19th Street Solar Housing in Boulder, Colorado. This project received a HUD Demonstration Award (construction costs were $250,000 in 1975).
Photograph by Phillip Tabb.

for groups of houses within an urban context as well. Villages and small towns were planned for the benefits of the sun with optimal community and building plan shapes, east/west street orientations, and good solar access to most buildings and outdoor public places.

Today, solar technology occupies a less important role in the housing delivery industry because contemporary housing form is derivative of many varied and complex determinants and environmental technologies. However, for many projects, typically in more remote locations or within energy conscious communities, solar housing continues to evolve because of cost savings, quality of space, and an environmental ethic related to the use of renewable energy sources. The principles of solar geometry and the greenhouse effect are the same as those used by the Greeks; the building construction technologies have changed the most. For housing projects that consider solar heating in a more primary way, two inextricably bound principles are necessary to render solar housing practical and efficient: (a) the incorporation of energy conservation measures designed to keep the energy demand as low as possible and (b) the complete integration of the solar technology components, whether they are actual technologies or parts of the building. The conservation measures typically include improved envelope insulation, double glazing, weather-stripping, and attention to the thermal details of construction. The solar technologies include the components of collector, overnight storage, and some system of heat distribution. There are two basic types of solar technologies—passive and active systems.

Passive Systems

Passive solar systems have no mechanical or electrical components, which gives them their *passive* reference. Passive systems are characterized by fluctuations in interior temperatures offset by daily solar gains and nightly thermal storage. The primary principle behind a passive solar system is the greenhouse effect. Shortwave solar radiation passes through glass and is trapped within the building, thereby creating heat or long-wave radiation. The energy either heats the interior air directly or is stored in the building mass for night use. There are three generic passive solar space heating systems: direct gain, thermal wall, and sunspace. Typically, solar housing concepts integrate a number of these generic solar systems types with energy savings of between 20% and 70%, depending on climatic context and type of the solar design. The direct-gain system is the simplest and most cost-effective of the passive solar systems. It consists of an energy-conserving envelope with south-facing glass exposed to internally distributed thermal mass. The south-facing glass can take the form of discrete windows or banks of glass. Thermal storage is distributed throughout the interior space and is usually an integral part of the floor system and partition walls. The direct-gain system can easily be integrated into conventional housing forms and aesthetics because of the architectural nature of its components.

French engineer, Felix Trombe, originated the thermal-wall system. It is effective because it combines solar collection directly to a concentrated thermal mass storage and, therefore, can achieve more controlled passive space heating than is possible with direct-gain systems. The south-facing glass is coupled to a concentrated thermal mass layer. The thermal mass is either masonry (poured concrete, concrete block, adobe, or brick), 8 to 12 inches thick, or vertical water containers 12 inches in diameter or larger. The advantage of this system is found in the storage efficiency. Sometimes these systems employ movable insulation, positioned between the glass and thermal mass, to reduce night energy losses, or a mechanical heat recovery subsystem. The thermal wall is particularly appropriate for rooms that need privacy because of the opaque nature of the collection and storage system.

The sunspace system is the most architecturally interesting passive system because it combines the principles of solar collection and storage with usable or functional space. The sunspace can either be an integral part of the house or it can be isolated from the primary functions. Sunspace glazing is either vertical glass with an insulated roof or a combination of vertical and sloped glass. The principal advantage of the isolated gain of sunspace is its buffering ability to withstand greater temperature fluctuations or swings—lower in winter and higher in summer. The functional activities of a sunspace are governed by the southern orientation with high solar intensity, and, therefore, they are limited to activities that include solar collection, greenhouse growing, circulation, dining, sitting, or children's play.

Active Systems

Two types of active solar systems are appropriate for housing projects. The first is the solar domestic hot-water heating system that has been used since the 1920s, and the second is the photovoltaic system for electricity production, a product of the early U.S. space program. Both systems employ mechanical or electrical components such as pumps, fans, inverters, and battery storage. Domestic hot-water systems simply consist of one or two 4- × 8-foot solar collectors and an insulated water storage tank along with a distribution system and electronic controls. The solar participation for these systems can vary between 40% and 80% depending on the size of the system. The photovoltaic system consists of a photocell collector array made up of collector modules, an inverter (converting direct current [DC] to alternating current [AC]), battery storage for approximately three days, and conventional electrical distribution throughout the house. For photovoltaic systems to remain cost-effective, appliances need to be efficient and electricity use needs to be monitored and coordinated with the solar gain and battery storage capacity. As much as 100% systems are becoming increasingly more desirable for off-grid or energy-independent housing in more remote locations because of the high costs of existing utility infrastructure connections.

The architectural integration of solar systems into housing must respond to certain physical parameters to function effectively. Building and solar collector orientation, building aspect or plan shape, solar collector tilt angles and the physical integration of systems' components into the building form, thermal mass location within the interior, and thermal zone coupling all need to be addressed. To maintain thermal comfort indoors, there must be clear access to the solar energy, collector orientation must be approximately south, and there must be adequate amounts of thermal mass for heat storage overnight, properly located for effective heat distribution. Very often, solar technologies, particularly passive ones, are not additive technological appendages but, rather, are integral architectural parts of a building—windows, exterior and interior walls, roofs, floors, and so on. User interfaces with solar systems can be important. Measurable improvements in efficiency can be made with appropriate behaviors, such as use of night insulation and daytime window management and use of lower thermostat set point temperatures when the house is unoccupied.

The limitations required by solar technologies very often cause developers to ignore energy as a formative consideration. Solar energy is a dispersed form of energy and, consequently, solar systems inherently have collector area intensity. Regardless of system type, the collector requirement means fairly large amounts of glass or solar panels integrated into the south roof or face of the building. Site slope, density of development, orientation of roads, building plots and buildings, self-shadowing, complexity of program and form, need for privacy, and cost control all can discourage the realization of solar housing. For example, housing development projects of medium to high densities—in excess of 40 to 50 dwelling units per acre—may be difficult to construct for a high solar participation to the total building energy needs. In climatic regions where solar radiation levels are poor, conventional technologies are most likely to be used. Although using solar energy in housing continues to intrigue many homeowners and developers, it is likely to be fully commercialized only in the face of more severe shortages of conventional energy sources, increased evidence of atmospheric pollution, and a marked increase in energy costs.

Building energy consumption amounts to approximately one-third of the total U.S. energy need—heating, cooling, and electricity. It is conservatively estimated that the use of solar technologies in housing could reduce the demand for conventional fossil fuels by 20% nationally. The use of both solar technologies and energy conservation measures in the housing market could effectively improve this estimate to at least 50%. Conservative uses of solar energy for daylighting and sun tempering will occur in most populated temperate climate regions. More advanced solar housing projects will continue to evolve in more climatically advantageous regions where greater efficiencies can be realized. (SEE ALSO: *Energy Conservation*)

—*Phillip Tabb*

Further Reading

Balcomb, J. D., ed., 1992. *Passive Solar Buildings.* Cambridge: MIT Press.

Butti, K. and J. Perlin. 1980. *A Golden Thread: 2500 Years of Solar Architecture and Technology.* Palo Alto, CA: Cheshire Books.

Knowles, Ralph. 1981. *Sun, Rhythm, Form.* Cambridge: MIT Press.

Kreider, Jan. 1982. *The Solar Heating Design Process.* New York: McGraw-Hill.

Mazria, Edward. 1979. *The Passive Solar Energy Book.* Emmaus, PA: Rodale.

Olgyay, Victor. 1963. *Design with Climate.* Princeton, NJ: Princeton University Press.

Tabb, Phillip. 1984. *Solar Energy Planning.* New York: McGraw-Hill. ◄

▶ Space Standards

Space standards is a term with two connotations: positive and normative. The positive connotation refers to the amount or nature of space occupied by a typical household. Common measures of the amount of space include number of rooms (or bedrooms) and floor area per dwelling. Measures of the nature of space include dwelling fittings, household equipment, car parking, usable outdoor space, and suitability of the dwelling with respect to the needs of occupants.

The normative connotation takes the form of statements about the minimum amount, nature, or quality of housing that consumers ought to be able to enjoy. In its simplest form, the latter specifies the living area, number of rooms, or facilities required for the decent accommodation of a household of given characteristics. Standards have varied over the years and differ from country to country. In the United Kingdom, for example, new council housing (public housing) was built to an average standard of around 800 square feet in the 1930s, 1,000 square feet during the 1940s, and about 800 square feet during the 1950s.

Normative space standards are widely used by governments, planners, and housing advocates. Commonly, they use space standards to identify households that are inadequately housed. Governments and public planners also use space standards in designing public housing and in determining the eligibility of private sector housing units for public subsidies. Some communities have also used space standards, as a kind of exclusionary zoning, to keep out inexpensive, small, or low-quality housing.

How do normative space standards get defined? On the one hand, standards articulate community goals, and hence reflect community affluence; richer societies can afford (or want) a more costly space standard. Standards also reflect prevailing attitudes toward the poverty and deprivation of others. In addition, standards reflect the particular social norms that give rise to the notion of "decent" housing. Finally, standards evolve because of information and understanding that arises about the causes and consequences of inadequate housing.

An early normative standard used internationally calls for at least one room per person in a dwelling. Households with more than one person per room are deemed to be "crowded." Built into this notion of crowdedness are assumptions about the desirability of privacy, the definition of a "room," and the ability of residents to obtain privacy given enough rooms. The definition of household is problematic here, to the extent that consumers share accommodation to reduce the cost of housing. In recent decades, more sophisticated crowding standards have emerged that link number of rooms to household composition as well as to household size. In Canada, for example, the National Occupancy Standard specifies the minimum number of bedrooms that a household should have by assigning (a) a bedroom for the parent, or parents, separate from their children; (b) a separate bedroom for other singles aged 18 or older; and (c) bedrooms for children at no more than two per bedroom and wherein children aged five or more do not share a bedroom with persons of the opposite sex. This example raises the broader question of how space needs differ with the characteristics of the household. The need for play space, for example, has led some jurisdictions to promote ground-oriented housing for households with children.

However, normative space standards can be cast more broadly than simply in terms of crowdedness. Another category of standard in common use looks to facilities or features present in the dwelling: for instance, source of water, toilet and bath facilities, kitchen facilities, sanitation and refuse disposal, central heating, electricity and wiring, thermal and sound insulation, access to sunlight, and children's play space. Still other categories of standard look to the frequency of breakdown of dwelling equipment, safety and repair of common areas (e.g., elevators), the cost of repairs, and the condition of the dwelling (e.g., state of repair, presence of vermin). (SEE ALSO: *Crowding; Health Codes; Housing Occupancy Codes; Size of Unit; Substandard Housing*)

—*John R. Miron*

Further Reading

DeChiara, Joseph, Julius Panero, and Martin Zelnick, eds. 1995. *Time Saving Standards for Housing and Residential Development.* New York: McGraw-Hill.

U.S. Department of Commerce, Bureau of the Census. 1995. *Current Housing Reports: American Housing Survey for the United States in 1995.* Washington, DC: Author. ◄

► Speculation

Concern with housing affordability accents a "use value" orientation toward urban space. Yet under capitalism, "exchange value" orientations dominate. Thus, the goals of speculators oriented toward exchange value conflict with individuals and families concerned with the use value of their housing. The conflict between these two competing orientations has important implications for housing. Speculation is intricately tied to housing and neighborhood processes, including urban renewal, gentrification, and suburban land sprawl, that, in turn, influence housing prices.

Among real estate capitalists, speculators are distinctive. They seek unearned profits from rising property values, not from developing land. Sometimes, though, it is difficult to tell where conventional land investment ends and speculation begins. Buyers of property are generally considered speculators if transactions are executed expressly in anticipation of future financial gains—and without value-adding activity on the buyers' part.

The types of speculators range widely. They include investors who buy converted apartments, "flipping" them to quickly make significant profits. This type is pronounced in areas where strong pressures exist for available housing. Others are serendipitous speculators. For example, when in 1986, 36,000 New York tenants were offered their recently converted apartments at substantial discounts, some got windfall profits in excess of $1,000,000 for buying and immediately reselling converted units. These individuals do not fit the popular image of speculators.

Today, corporate speculators have grown in significance. Since the 1970s, large corporations have increasingly become involved in real estate speculation and development. Until the 1950s, relatively small capitalists dominated the land and housing industries. But corporate investment has advanced rapidly. In the 1960 to 1975 period, 300 of the top 1,000 U.S. corporations developed real estate departments. Moreover, until recently, most lenders were reluctant to finance speculative ventures. But the situation changed with financial deregulation under the Reagan and Bush administrations in the 1980s. With a relaxed regulatory environment, financial institutions invested an increasing share of funds in speculation-driven projects. A classic example is the $300 million Phoenician Hotel developed by Charles Keating and his former Lincoln Savings and Loan near Scottsdale, Arizona. In the heyday of deregulation, the institution also speculated $100 million in Austin, Texas, real estate before the S&L was seized by the government in 1989.

The connection between speculation and housing can be demonstrated in several areas, including urban renewal programs, gentrification, and suburban sprawl.

Urban Renewal Programs

Urban renewal can be illustrated by the 1949 Housing Act, which authorized a massive program of urban redevelopment. The act envisioned "slum" clearance and the provision of decent, low-rent housing for the poor by relying on free-market mechanisms. Yet a report for the National Commission on Urban Problems characterized it as a "fed-

erally financed gimmick to provide relatively cheap land for a miscellany of profitable or prestigious enterprises" (Feagin and Parker 1990, p. 259). Federal funds were used to bulldoze large central-city areas considered "blighted," with the improved land frequently sold to speculators. In this way, many central cities have been redeveloped since World War II. But heavy social costs have been exacted. Those forced out of moderate-rent housing have been disproportionately poor or elderly.

A related result of unrestrained speculation is housing deterioration in central cities. Housing stock quality declines for several reasons. One is that speculators often buy property to "bank" it until a more profitable use can be constructed, in the process foregoing all maintenance and upkeep. In many cases, building code violations accumulate to the point that the structure must be condemned. More dramatically, housing quality declines in areas with speculative activity because speculators frequently employ arson and other heavy-handed techniques to clear tenants from a property bought for speculative purposes. For example, when speculation was at a fever pitch in Atlantic City after legalized gambling was established, there were more than 175 fires in one four-month period in the once fashionable Inlet area, the vast majority of which were determined by fire officials to be arson related. The physical condition of these older, often inner-city, areas are often characterized as the product of poor people without the means or cultural values to maintain property or in terms of some inevitable invasion and succession process. An alternative view stressed here accentuates the profitability sought by speculators in slum real estate.

Gentrification

Gentrification, too, has been fueled by speculative activity. Gentrification can be defined as the emergence of middle- and upper-middle-class enclaves in formerly deteriorated, inner-city neighborhoods. It frequently makes housing unaffordable for many with the displacement of lower-income urbanites by higher-income earners. Speculators buy property in areas potentially attractive to better-off white-collar families. Existing moderate-income tenants are forced out by eviction or rising rents—and the affordable housing stock declines. Absentee speculators play a pivotal role in gentrification, buying up low-income housing, rebuilding or replacing it, and then selling it to young, white, affluent families with professional workers. Meanwhile, poorer, indigenous residents are pushed out. Displacement by this exchange-value-driven speculative development has been documented in many U.S. cities.

Urban regeneration in the form of gentrification continues in many U.S. cities, with new hotels, exclusive neighborhoods, and specialty shops signaling the return of well-off people. But it also signals the demise of lower- and moderate-income households and communities. Whether private or public, urban revitalization tends to hurt those who are least able to defend themselves—the elderly, the nonwhite, and female-headed households. Those displaced must move and compete with other families of modest incomes for a declining number of decent housing units in the central city.

Surburban Sprawl

A typical explanation of suburban sprawl emphasizes rising affluence leading to demand for single-family housing. In reality, suburban growth has involved considerably more planning and purpose. Although the specifics vary, real estate speculators have radically shaped suburban development. In urban fringes, land will often go through a series of speculators before it becomes a suburban development. Studies have illustrated some of the negative social effects of speculation. Residential land prices, particularly in suburban (formerly farm) areas, have risen 200% to 400% faster than the price of housing in other areas. In some suburban areas, such as those surrounding Los Angeles, the price of residential land escalated 40% a year in the late 1970s, largely the result of speculative activity.

Conclusion

Speculative activity shapes urban development in a way that discourages the construction and preservation of affordable housing and stable moderate-income neighborhoods. Real estate speculators are powerful actors, as individual entrepreneurs or as corporations, who buy and sell land in search of profit. Cities are centers of conflict; exchange-value-oriented speculators gain in this struggle, and use-value-oriented citizens needing inexpensive housing lose. New projects in central cities and suburbs may give the appearance of progress, but the availability of decent housing grows increasingly problematic. (SEE ALSO: *Gentrification; Growth Machines; Land Value Taxation*)

—*Robert E. Parker*

Further Reading

Feagin, J. R. and R. Parker. 1990. *Building American Cities: The Urban Real Estate Game.* Englewood Cliffs, NJ: Prentice Hall.

George, H. 1962. *Progress and Poverty.* New York: Robert Schalkenbach Foundation.

Judd, D. and M. Parkinson, eds. 1990. *Leadership and Urban Regeneration.* Urban Affairs Annual Reviews, Vol. 37. Newbury Park, CA: Sage.

Lindeman, B. 1976. "Anatomy of Land Speculation." *Journal of the American Institute of Planners* (April):142-52.

London, B., B. Lee, and S. Lipton. 1986. "The Determinants of Gentrification in the United States." *Urban Affairs Quarterly* 21(3):369-87.

Sakolski, A. M. 1932. *The Great American Land Bubble.* New York: Harper & Brothers.

Skaburskis, A. 1988. "Speculation and Housing Prices: A Study of Vancouver's Boom-bust Cycle." *Urban Affairs Quarterly* 23(4):556-80.

Wade, R. C. 1959. *The Urban Frontier.* Chicago: University of Chicago Press. ◀

▶ Squatter Settlements

Because of a lack of affordable housing, millions of families in cities and towns of Third World countries are forced to live in what is commonly referred to as squatter settlements. *Squatter settlement* is a generic term loosely used for a wide range of low-income settlements developed (a) on vacant land by low-income families and informal-sector entrepreneurs without permission by the landowner, (b) independently of the authorities charged with the external or institutional control of local building and planning, or (c) both. Because the occupation is unauthorized, security of land tenure is low and the squatters are reluctant to invest much money in their housing. As a result, housing is often of a low quality. Because the squatters are unable to construct roads, drainage, and sewerage networks and the authorities consider the settlements illegal, unsanitary conditions often prevail.

Accurate statistics on the size of the population of squatter settlements are difficult to obtain because of definitional problems and inadequate methods of data collection. Moreover, authorities tend to underestimate the extent of inadequate housing because they ignore communities outside city boundaries or do not enumerate them correctly. Even so, U.N. estimates indicate that in many cities of developing countries, 40% to 50% of the inhabitants live in slums and informal settlements.

Definition

Squatter settlements can be narrowly defined as aggregates of houses built on lands not belonging to the housebuilders but invaded by them, sometimes in individual household groups, sometimes as a result of organized collective action. However, such a definition obscures all sorts of subtleties of possession, such as partial recognition of tenure and indirect acceptance of possession or tenure by the landowner and the authorities. It also emphasizes only the legal aspects of the settlements, ignoring many social aspects.

To take such considerations into account and to avoid the pejorative word *squatter,* other terms are used: *autonomous settlement, spontaneous settlement, extralegal settlement, popular settlement, unauthorized settlement, uncontrolled settlement, informal settlement, unplanned settlement, irregular settlement, shanty settlement,* or *marginal settlement.* For similar reasons, many writers prefer the local names for such settlements: *favelas* (Brazil); *barriadas* (Colombia, Panama, Peru); *barrios piratas* or *clandestinos* (Colombia); *bustees* or *jhuggis* (India); *katchi abadis* (Pakistan); *kampung* (Indonesia); *bidonvilles* (Morocco); *gourbivilles* (Tunisia); *barong-barong* (Philippines); and *gecekondu* (Turkey).

Sometimes the term *slum* is used to refer to squatter settlements. However, there is an important difference between (inner-city) slums and squatter settlements. Slums can be defined as legally constructed permanent buildings where the housing conditions are substandard due to age, neglect, subdivision, and consequent overcrowding; over time, living conditions in slums tend to deteriorate. Housing in squatter settlements is substandard, because the settlements lack security of tenure and basic infrastructure services. If the perceived level of security of tenure increases (and other conditions are favorable), housing in squatter settlements tends to improve over time.

Origin

Because the opportunities to develop a squatter settlement depend on the legal, political, and sociocultural system and the economic and physical conditions of the city, squatter settlements differ from city to city and from country to country. Some squatter settlements have a very haphazard road pattern, whereas other settlements are well planned. Squatter settlements are often also internally heterogeneous: Within one settlement the quality of the houses can range from simple huts to two-story permanent buildings. Not all residents of squatter settlements are poor (and not all urban poor live in squatter settlements). Sizable squatter settlements attract shops and small-scale industries, schools, and clinics. Squatter settlements can be classified according to the way they come about as unorganized invasions, organized invasions, and illegal subdivisions.

Pure squatter settlements are gradual encroachments or unorganized invasions of vacant land by low-income families who would otherwise not have a place to live. In some cases, the land occupation may be free of any charge; in many other cases, however, the squatters have to pay protection money to middlemen or representatives of the law. The simplest form of such settlement is represented by the pavement dwellers who build their shelters of temporary materials on roads and footpaths, often against the walls and fences of adjacent buildings. Most squatter settlements are built on land that is vacant because it is not easily accessible, is disputed, is part of a road or railway reserve, or has no designation as yet. A family occupies a small piece of land and builds a simple structure, usually not more than a hut of temporary materials. If the landowner or the authorities do not demolish the structure and evict the occupants, the family will gradually improve its shelter and other families will settle next to them. A growing population and a gradual consolidation of the houses make it increasingly difficult for the authorities to remove the settlement, and this encourages the squatters to further improve their houses.

The poor have learned through experience that certain types of land provide a higher level of tenure security than others. Private landowners tend to be more alert to attempts to occupy their land, and squatters may, therefore, prefer to occupy government-owned land. They mobilize local politicians to exert pressure on the authorities not to demolish the settlement. Squatters have also realized that the authorities are more likely to demolish a single house than a large settlement. This has led to organized invasions. Some invasions involve relatively small groups of families who join together on an informal basis shortly before the occupation of the land; others are highly organized and involve hundreds of families. The leaders carefully plan the invasion and meet frequently to recruit members and select a site. They may choose a public holiday or the visit of some foreign dignitary as the invasion date so that the authorities will be reluctant to use force to evict the squatters. They establish contacts with political or religious leaders and sympathetic journalists to secure their support.

In some cities, squatting on government land has developed into a sophisticated system of illegal subdivision of public land that houses millions of people. Illegal subdividers procure the protection from local politicians, municipal administrators, and police officials and contact city planners to ascertain that the intended site has not been designated for any important function. They know that a well-

planned settlement is less likely to be demolished than an unorganized one; so the new settlement has a regular layout, a proper road network, and spacious rectangular plots. The first plots are sold at nominal prices or even given free of charge, and the first families have to occupy their plots immediately. Once several hundred families have settled, the subdivider organizes water supply and transportation, and shops open for daily necessities. This makes the area attractive for other settlers, and plot prices rise steadily. At that moment, the illegal subdivider starts making profit. Unorganized invasions are rare nowadays, and squatting in the form of illegal subdivisions has become a big business in some cities. However, because land in good locations has become scarce and valuable, public and private landowners have become more alert to squatting.

Illegal subdivisions should be distinguished from what may be called informal subdivisions. The informal subdividers are landowners or leaders with the customary authority to allocate land. They subdivide the land in their possession or under their authority without taking into account subdivision regulations and sell the plots without permission from the authorities. Some landowners allow families to settle on their land against a nominal rent with the understanding that they can cancel the agreement at any time. The landowner does not permit the families to consolidate their houses, and the residents are reluctant to do so in view of the low security of tenure. The houses are built without a building permit and do not meet any building standards.

Characteristics

The location of squatter settlements depends on the availability of land and the location of employment opportunities. Urban poor who rely on the informal sector for their livelihood, will try to live as close as possible to the city center, but vacant land in the center is scarce. Some cities have small pockets of squatter houses in the heart of the city (minisquatters). Squatter settlements are often found along railway lines and on the banks of rivers that cross the city; these strips are not under any development pressure and provide easy access the city center. Squatter settlements can also be found on steep hills that remain unoccupied because of the danger of landslides. More established urban poor who can afford the transportation cost prefer to live in the urban fringe where security of tenure is higher and more space is available.

It is often assumed that houses in squatter settlements are built entirely by the occupants through self-help, but this is rarely the case. If squatters consolidate their houses, hired craftsmen and small-scale contractors usually build the semipermanent and permanent structures. There is also a market for squatter housing, and the occupant of a house is often not the original squatter of the land. Although housing in squatter settlements is usually owner-occupied, rental housing is common. In Nairobi, Kenya, the protection of the squatter settlements by powerful politicians against demolition by the authorities resulted in the commercialization of squatter housing and the development of a rental squatter housing market.

Policy

Many city administrators see squatters settlements primarily as eyesores, and until the early 1970s, it was the stated policy of most governments to demolish squatter settlements. In the late 1960s and early 1970s, John Turner and others drew the attention to the positive aspects of squatter settlements and pointed out that squatter settlements are not a problem but, rather, a solution to the housing problem. Squatters have the resources, skills, and personal motivation to secure adequate shelter for themselves and are willing to use these to improve their housing conditions, provided they have security of tenure. So to improve the housing conditions of the urban poor, the authorities should regularize the settlements and provide basic infrastructure services. During Habitat, the United Nations Conference on Human Settlements in Vancouver (Canada) in 1976, governments endorsed these ideas and adopted the Vancouver Plan of Action, which recommended that the informal sector should be supported in its efforts to provide shelter, infrastructure, and services, especially for the less advantaged, and that governments should concentrate on the provision of services and on the physical and spatial reorganization of spontaneous settlements in ways that encourage community initiative and link "marginal" groups to the national development process.

The 1975 Housing Sector Policy Paper of the World Bank advocated the improvement of squatter housing as a means to retain and improve the existing housing stock that might otherwise be demolished, while maintaining access to employment and social services in relatively central locations for low-income residents. The bank started funding squatter settlement regularization and upgrading projects that legalized land tenure and provided basic infrastructure, because most governments could afford neither the financial cost of conventional housing solutions nor the political cost of bulldozing existing squatter settlements.

Squatter settlement regularization and upgrading projects usually have three components: (a) the regularization of the layout of the settlement (reblocking), (b) the issuing of freehold or leasehold titles to the residents (legalization), and (c) the provision of basic infrastructure (upgrading). Some projects also provide housing loans, but the actual improvement of the houses is the responsibility of the residents. Some projects try to recover the cost of the infrastructure construction but charge no or only a nominal amount for the cost of the land to keep housing affordable for the residents. A World Bank evaluation revealed that upgrading projects reached a larger percentage of the poorest urban residents than sites-and-services schemes.

A few governments (e.g., Sri Lanka, Indonesia) have actually undertaken large-scale squatter settlement regularization and upgrading programs. Regularization is not only opposed by city administrators and politicians who see squatter settlements as eyesores in the city but also by representatives of the landed class who fear massive land occupations by squatters. It is, on the other hand, unlikely that the urban poor can occupy land without some sort of consent (open or concealed) of the authorities. Politicians often see low-income families in search of shelter as a potential vote bank and support their efforts to squat on vacant land. Once settled, squatters are vulnerable to extortion and eviction because of their unauthorized land occupation, and they tend to seek the patronage of politicians who can also be helpful to obtain municipal services and eventually regularization. However, politicians tend to prefer piecemeal infrastructure improvements that they can dole out in exchange for political support rather than a blanket improvement of entire settlements in a planned manner. Furthermore, with the increasing emphasis on an unhindered functioning of the market, many squatter settlements face a renewed threat of eviction, while their regularization is becoming more unlikely. (SEE ALSO: **Third World Housing**)

—*Kioe Sheng Yap*

Further Reading

Gilbert, Alan and Josef Gugler. 1992. *Cities, Poverty and Development: Urbanization in the Third World.* 2d ed. Oxford, UK: Oxford University Press.

Hardoy, Jorge E. and David Satterthwaite. 1989. *Squatter Citizens: Life in the Urban Third World.* London: Earthscan Publications.

Keare, Douglas H. and Scott Parris. 1982. *Evaluation of Shelter Programs for the Urban Poor: Principal Findings.* World Bank Staff Working Paper No. 547. Washington DC: World Bank.

Mangin, W. 1967. "Latin American Squatter Settlements: A Problem and a Solution." In *Latin American Research Review,* 2:65-98.

Sandhu, R. S. and B. Aldrich, eds. 1995. *Housing the Poor, Policy and Practice in Developing Countries.* London: Zed Books.

Skinner, Reinhard J., John L. Taylor, and Emiel A. Wegelin, eds. 1987. *Shelter Upgrading for the Urban Poor: Evaluation of Third World Experience.* Manila, Philippines: Island Publishing.

Turner, John F. C. 1976. *Housing by People: Towards Autonomy in Building Environments.* London: Marion Boyers. ◀

▶ State Governments

Traditionally, state governments in the United States have played minor roles in programs of direct activity in housing and urban development activity but very important roles in providing the enabling powers for local governments to undertake such functions and in establishing the authority and standards for locally administered indirect activities such as building codes, zoning and land use planning, minimum housing standards, and real estate taxation. States began to confer comprehensive zoning authorization on municipalities early in the 1920s. States became particularly aggressive in the area of mechanical elements of building codes. In 1966, the U.S. Advisory Commission on Intergovernmental Relations reported that four-fifths of the states had one of the following: statewide plumbing, electrical, boiler, and elevator codes. Almost half had a statewide fire code or regulations. However, only five states had enacted statewide building construction codes, and none was mandatory for all construction. By 1973, however, a survey by the U.S. Department of Housing and Urban Development documented that 15 states had statewide building codes, 28 had preemptive laws governing factory-built housing, and 38 had preemptive regulations for mobile home construction.

The first direct involvement of states in slum clearance and low-income housing came in the 1930s when five states (New York, Massachusetts, Connecticut, New Jersey, and Pennsylvania) established state-assisted public housing programs, parallel in concept to the federally assisted public housing program authorized in 1933. About 127,000 low-income housing units were built under these programs prior to 1960, but new development activity stopped after this date.

A major expansion of the direct role of states in housing and urban development did not come until the 1960s with the establishment of state housing finance agencies. By 1982, agencies in 47 states had assisted in the financing of single-family homes, and 17 states had assisted in financing multifamily housing programs serving low- and moderate-income families; by 1988, the cumulative volume of this activity reached more than 1.5 million assisted-housing units. In addition, 13 states had established housing rehabilitation grant or loan programs, and 15 states had authorized tax incentives for lower-income housing development.

Under the federal Community Development Block Grant program enacted in 1974, federal assistance was made available to states to administer the program for small communities.

Program Expansion in the 1980s

Since 1980, there has been a new upsurge in state activity in housing and urban development. This has been widely attributed to the significant cutbacks in federal housing assistance under the Reagan administration. But it also reflects changing conditions and new state capacity in developing and administering these programs. In terms of changing conditions, shifts in the economy were important factors, as states became more aggressive in trying to rebuild their economic bases. Also, there was an increased state sensitivity to (a) the lack of available rental housing for lower-income households (particularly the elderly, the homeless, the physically and mentally disabled, and farmworkers), (b) the lack of homeownership opportunities for first-time homebuyers, (c) the extent of substandard housing and deteriorating neighborhoods, and (d) the lack of affordable housing to accommodate workers related to business development and job initiatives. A 1988 study documented that more than 300 new state housing programs had been enacted since 1980. This "third generation" of state housing action has included a broad range of new legislation covering low-income housing trust funds, new rental and homeownership assistance programs, special needs housing (including frail elderly, single-parent households, developmentally disabled, and homeless persons), and low-income housing requirements for regions within states. A 1992 survey by the Council of State Housing Agencies cataloged more than 600 affordable housing programs in all 50 states: homeownership (50 states, 225 programs), rental housing (49 states, 245 programs), special needs housing (34 states, 100 programs), economic development (14 states, 31 programs), and housing finance and technical assistance (40 states, 169 programs). These new programs have been accompanied by changes in state administrative structures to accommodate the new activity. States are also assuming new responsibilities for state comprehensive planning and growth policies.

One of the most important resources for financing state-assisted housing developments has been the availability of federal Low-Income Housing Tax Credits (LIHTC), authorized under the Tax Reform Act of 1986 (P.L. 99-154). Almost all states have used such credits to stimulate private investment and to reduce rent levels for low-income families.

Assistance Mechanisms for Expanded Activity

As in the case of expansion of housing and urban development activity at the local level, state governments have initiated a number of new mechanisms to carry out their new programs. These include state housing partnerships, housing trust funds, fair housing requirements, and enterprise zones.

State Housing Partnerships

Up to 1990, four states (Massachusetts, Wisconsin, California, and Ohio) had adopted housing partnerships, providing unique possibilities for involving the private sector in a new partnership arrangement as well as new state-local relationships around the partnerships concept. The Massachusetts Housing Partnership established in 1985 created a new delivery system for state housing and urban development assistance based on increasing the ongoing capacity of local governments to carry out comprehensive, cohesive housing and urban development strategies. Localities were required to move through three stages of demonstrating capacity before being fully qualified as partners with the state. Local public-private partnerships were required as well as a development strategy agreed on by the partners. These early efforts became a bridge to the HOME investment partnerships established under federal legislation in 1990. The HOME program is a federal housing block grant providing funds to states and localities to undertake flexible, wide-ranging housing activities through partnerships among states, localities, private industry, and nonprofit corporations. In 1996, state housing finance agencies awarded $174 million in HOME funds (50% of the total they received) to nonprofit organizations.

State Housing Trust Funds

In 1997, the Center for Community Change identified 34 active state housing trust funds with annual revenues ranging from $1 million or less to $10 million (Indiana) to $52 million (Maryland). Revenue sources for these funds vary widely. Fifteen of the funds are funded in whole or in part from real estate-related sources, including real estate transfer taxes. In 1996, the Council of State Housing Agencies reported that 8 of the state trust funds received support from state appropriations or general revenue. Of the total trust funds, almost all are administered entirely or partly by state government agencies. Some funds involve a separate body established by law to set policy, and 20 have advisory commissions.

Fair Housing Programs

The New Jersey Fair Housing Act of 1985, enacted as a response to the *Mt. Laurel* court decision, put this state in the forefront of states willing to take positive action to pursue low-income housing goals and counter exclusionary practices. New Jersey's municipalities have a constitutional obligation to zone for low-income housing and promote its development. The 1985 act created the Council on Affordable Housing to provide an administrative process and the Balanced Housing Fund as a funding source to promote the achievement of the act's goals. A 1993 evaluation of the record of the Fair Share program documents 13,600 total housing units built, rehabilitated, or under construction located in 280 developments in 125 localities in New Jersey's 21 counties. The largest contributor among five major sources of subsidy was "inclusionary development" (builder set-asides—usually 20% of market rate developments for lower-income occupancy).

Another state that deals directly with the fair housing distribution issue is California, which has a required "housing element" under the comprehensive planning provisions of state legislation. In recent legislation in Florida, all localities are required to prepare a comprehensive plan that contains a housing element that includes standards, plans, and principles to be followed. Over the last five years, there appears to be a focus away from "fair share" housing plans and "housing elements" in comprehensive plans to broader-based planning and development strategies as discussed in the following section.

Enterprise Zones

Using tax incentives to spur business investment in seriously depressed urban areas in the form of "enterprise zones" was first proposed in the United States by the federal government in 1981. Although federal legislation was not passed, 26 states had created such zones by 1985, at least partially in response to the prospect of federal legislation.

An assessment of state enterprise zones in 1993 raised questions about their focus and productivity. However, the enterprise zone concept took on new life in 1993 with the congressional passage of legislation establishing nine "empowerment zones" and 95 "enterprise communities." Although these new federal programs have differences with the earlier proposals, there are important relationships with the original enterprise zone concept. In addition, some states have adapted the concept of "enterprise zones" to their particular needs. In 1996, Michigan adopted a "Renaissance Zone" program establishing tax-free zones to revitalize inner cities and poorer rural communities. Localities that apply for and receive zone designation have authority to grant almost total relief from state and local taxes to businesses and families within the zone for up to 15 years. Of the 20 localities that applied for designation, 12 were chosen. The stated objective is to revitalize disadvantaged communities by reducing financial burdens to investment in industry, business, and families.

Future Issues: Structures and Strategies

Unlike the broad array of housing and urban development structures that have recently evolved at the local level, states have tended to broaden existing agencies to assume new responsibilities: basically, these are the state housing finance agencies and the cabinet-level Departments of Housing and Community Development. As indicated above, all of the 34 state housing trust funds are administered by these agencies, often with officially established advisory committees or commissions. All of the four state housing partnerships are separate corporations. Originally, the Massachusetts Partnership was an entity under the Governor's Office of Communities and Development but was replaced by state legislation in 1990 that created the Massachusetts Housing Partnership Fund, a quasi-public corporation financed by the state's banking industry. In 1988, every state had an administrative entity incorporating the functions of housing and urban development. Maryland created a comprehensive Department of Housing and Community Development in 1987, and Kansas established a Division of Housing in the Department of Commerce in 1992. A diversity of titles reflects the evolutionary status of housing and community development functions in most states.

Similar to the trends in organizational structure described above, state housing and urban development strategies are still largely unformed. Since the termination of the federally assisted 701 comprehensive planning program in 1981, states and local governments have been left to their independent initiatives, and there is an uneven pattern and practice across the United States. The promising development of comprehensive "urban strategies" that took place in the late 1970s as a counterpiece to national urban policy initiatives has largely faded out. These strategies were initially defined to cover growth management, economic development and employment, community revitalization, and fiscal reform. At least eight to nine states developed urban strategy approaches. As the momentum creating comprehensive state urban strategies faded, it was replaced with a new movement but with a different origin and focus: state urban growth management. In the 20 years ending in 1992, eight state governments enacted laws encouraging local governments to prepare comprehensive growth management programs consistent with specific state criteria: Oregon, Florida, New Jersey, Maine, Vermont, Rhode Island, Georgia, and Washington. It has been noted that all of these states have substantial seacoast or lakeside regions, indicating the influence of environmental protection in establishing these programs. The newest innovation in state growth planning and development is the creation of so-called Smart Growth programs, a comprehensive approach to deal with suburban sprawl and economic segregation. Representative of this movement is the program adopted by Maryland in 1997 called "Neighborhood Conservation and Smart Growth Initiatives." Smart growth is defined as a balance between preservation of land, open spaces, and unique character and economic growth. The program targets most state infrastructure funding and economic development, housing, and other program monies to those places that local governments have identified as "Priority Funding Areas." Giving further momentum to this movement is the multiyear initiative of the American Planning Association called the *Growing Smart Legislative Guidebook* released in 1996 and 1997. The *Guidebook* contains model state legislation for planning agencies and plans at state, regional, and local levels, as well as elements of affordable housing, transportation, and economic development. An entire chapter is devoted to regions that are increasingly important components of planning and development strategies. (SEE ALSO: **Federal Government;** *Housing Finance Agencies; Local Government; National Council of State Housing Agencies*)

—Mary Nenno

Further Reading

Council of State Housing Agencies. 1997. *Fact Book: State HFA Fact Book, 1996 NCSHA Annual Survey Results.* Washington, DC: Author.

Council of State Community Development Agencies. 1997. *THE STATELINE: State Housing and Development News.* "The Michigan Renaissance" (January–February):1-5; "Smart Growth in Maryland" (May–June):1-3; "An Overview of State Housing Trust Funds" (July–August):7-9. Washington, DC.

Gale, Dennis E. 1992. "Eight State-Sponsored Urban Growth Programs." *Journal of the American Planning Association* 58(4):425-39.

"The Michigan Renaissance." 1997. *THE STATELINE: State Housing and Development News* (January–February):1-5.

Nenno, Mary K. 1989. *Housing and Community Development: Maturing Functions of State and Local Government.* Washington, DC: National Association of Housing and Redevelopment Officials.

"An Overview of State Housing Trust Funds." 1997. *THE STATELINE: State Housing and Development News* (August):7-9.

Petherick, Glenn D. 1992. *State HFA Housing Catalog and (1993) State Housing Trust Funds: Innovative Sources of Financing for Affordable Housing.* Washington, DC: Council of State Housing Agencies.

"Smart Growth in Maryland." 1997. *THE STATELINE: State Housing and Development News* (May–June):1-3.

Terner, Ian Donald and Thomas B. Cook. 1988. *New Directions for Federal Housing Policy: The Role of the States.* Cambridge: MIT, Center for Real Estate Development. ◀

▶ Steering

Ever since the passage of Title VIII of the Civil Rights Act of 1968, it has been illegal for real estate agents to "steer" those seeking homes or apartments because of the client's race or that of the area in which the suggested dwellings were located. Specifically prohibited practices include the following:

- ▶ Directing people to a particular community, neighborhood, development, or section of a building because of race
- ▶ Discouraging people from occupying any dwelling because of the race of persons in the neighborhood or building
- ▶ Exaggerating drawbacks or failing to inform people of desirable features of a dwelling or neighborhood to perpetrate such discouragement
- ▶ Communicating to prospective purchasers or renters that they would not be comfortable or compatible with existing residents of a neighborhood or building because of race

As summarized in the regulations implementing the Fair Housing Amendments Act of 1988, "It shall be unlawful because of race . . . to restrict or attempt to restrict the choices of a person by word or conduct in connection with seeking, negotiating for, buying or renting a dwelling so as to perpetuate, or tend to perpetuate, segregated housing patterns . . ."

The questions to be addressed here are these: How frequently does steering occur in contemporary U.S. metropolitan housing markets? If it occurs, what specific practices are employed? To what extent are the housing choices of minority and majority race home seekers thereby differentiated and limited?

The way that steering has been most often investigated is through a research technique known as the "fair housing audit." In an audit, two teammates pose as home seekers. They are matched on both actual and some fictitious characteristics so that both in person and "on paper," teammates appear virtually identical from the perspective of a housing agent. Of course, the teammates differ by race or whatever characteristic is being audited. After their individual contacts with housing agents, they independently record the number and location of units suggested and shown, the remarks made about certain neighborhoods, and so on. The audit coordinator compares the teammates' records in an attempt to detect any systematic differential treatment in the racial or other attributes of the neighborhoods shown or commentary provided by the agent.

Since 1974, more than three dozen audit studies of steering have been completed in metropolitan areas across the United States. In addition, the U.S. Department of Housing and Urban Development commissioned nationwide audit studies in 1977 and 1989, portions of which investigated steering. They have consistently revealed that steering is a widespread phenomenon that occurs in at least 20% of the dealings with real estate agents and sometimes in as often as 60% of the cases. It also appears that selective commentary editorializing on the pros or cons of an area has been practiced as much if not more than differential patterns of home showings.

The consequence of this steering has rarely been to limit the number or concentration of geographic alternatives available to minority auditors, nor have all their options in predominantly white communities typically been precluded. Rather, steering in the sales sector most often has constituted a failure to show white auditors options in areas (and school districts) with nontrivial proportions of minority residents and a propensity to show minority auditors disproportionate numbers of homes in areas currently possessing or expected to possess significant proportions of minority residents. These conclusions hold regardless of whether geography is defined by block, tract, municipality, or school district.

Real estate advertising practices can also abet steering. The selective use of photographs of real estate agents and models posing as residents has been found illegal in several court cases because it can signal the racial composition of the neighborhood in question and who might feel "welcome" there. Statistical studies of print media real estate advertising have shown that the likelihood of any given home for sale being advertised is substantially less in racially mixed and predominantly minority-occupied neighborhoods than in white-occupied ones and that the latter areas are more often described in the advertisements with favorable adjectives, even when they have the same median incomes and property values as those with higher fractions of minority residents. Thus, print advertising not only can identify the racial composition of neighborhoods to prospective home seekers, but it also tends to render minority-occupied neighborhoods less visible and as inferior.

In theory, there are five nonmutually exclusive reasons that motivate real estate agents to steer. First, agents may have a personal aversion to the notion of racial integration. Under this "segregationist" explanation, agents steer so as to give themselves the satisfaction of perpetuating the type of society they deem desirable or even to protect a neighborhood. Second, agents may steer because they believe they are fulfilling the wishes to the client and thus do not waste time by showing options that would not be attractive. Under this "anticipated client preference" explanation, the agent believes that virtually all whites would be reluctant to live in racially mixed areas and that most minorities prefer to live in mixed areas and would be reluctant to be the first in an all-white area. Third, agents may steer because they fear loss of white clientele. Under this "customer prejudice" explanation, agents presumably would suffer a loss of potential home listings by prejudiced white home sellers if these agents acquired a reputation of introducing minorities to erstwhile all-white neighborhoods. Fourth, agents may steer minority clients into predominantly minority-occupied areas if they believe that a sizeable fraction of white home sellers with whom they hold listing contracts will refuse to sell to a minority person. This is the "anticipated discrimination" motive. A variant on this theme is that an agent may develop close ties with mortgage lenders, and if the agent believes that these lenders will not make loans to applicants of a certain race in particular neighborhoods, the agent will steer to avoid such eventualities. Fifth, agents may believe that perpetuating segregation, and its corollary—unstable racially mixed areas—is more profitable to them in the long run because it maximizes the amount of housing turnover and, hence, their aggregate sales commission income. Under this "turnover maximizing" explanation, agents steer so as to abet white fears in racially mixed areas that the area will eventually resegregate, thus speeding their out-migration and increasing the potential for real estate sales commissions to be earned. Existing evidence is mixed on which motivator, if any, predominates.

Regardless of cause, it is clear that steering spatially distorts the demands of white and minority home seekers in such a way that residential segregation is perpetuated. If whites are disproportionately made aware of and shown homes in preponderantly white-occupied areas and verbally encouraged to select them, if whites are encouraged to believe in the inferiority of minority-occupied areas, and if minorities are shown relatively few homes in predominantly white-occupied areas and given no encouragement to select them, predominantly white residential areas will persist. It is thus not surprising that a majority of whites live in census tracts with 1% minority households or less.

A corollary implication is that racially integrated areas will remain rare and transitory. By steering away white home seekers and steering in minority home seekers to such areas, the potential for racial transition and eventual resegregation is abetted. Clearly, steering is partly to blame for the perpetuation of a racially segregated society. (SEE ALSO: Discrimination)

—George C. Galster

Further Reading

Galster, George. 1990a. "Racial Steering by Real Estate Agents: Mechanisms and Motivations." *Review of Black Political Economy* 19:39-62.

——. 1990b. "Racial Steering in Urban Housing Markets: A Review of the Audit Evidence." *Review of Black Political Economy* 18:105-29.

Helper, Rose. 1969. *Racial Policies and Practices of Real Estate Brokers.* Minneapolis: University of Minnesota Press.

Newburger, Harriett. 1989. "Discrimination by a Profit-Maximizing Real Estate Broker in Response to White Prejudice." *Journal of Urban Economics* 26:1-19.

Pearce, Diana. 1979. "Gatekeepers and Homeseekers: Institutional Patterns in Racial Steering." *Social Problems* 26:325-42.

Turner, Margery. 1992. "Discrimination in Urban Housing Markets: Lessons from Fair Housing Audits." *Housing Policy Debate* 3:185-216. ◄

► Stewart B. McKinney Homeless Assistance Act

The Stewart B. McKinney Act, passed in 1987, was the U.S. government's first concerted effort to deal with homelessness in the United States. The act treated homelessness as a crisis and created 21 programs, operated through seven different agencies. The funding amounts were never sufficient to meet the needs seen by service providers. Some activism was centered on moving the federal government from treating homelessness as a crisis to treating it as endemic to the society and attempting to end it permanently. However, the change in congressional leadership and the budget battles of 1995 created more immediate concerns for saving the programs and their funding.

Homelessness in the United States became a major public issue during the 1980s. Many states and communities attempted to ameliorate (or at least hide) the problem but found that local funds were insufficient and that the causes of homelessness were often outside of local control. The U.S. government created a few scattered programs early in the decade but made no concerted effort to assist until the passage of the Stewart B. McKinney Act of 1987. Given the climate of the Reagan administration, it was something of a coup that the act passed at all. Even so, it reflected the belief that homelessness was a temporary problem and that only emergency measures were necessary. The act contained a potpourri of programs with a wide range of types and funding. In addition, an Interagency Council on the Homeless was established to coordinate efforts. The agencies involved include the Department of Housing and Urban Development (HUD), the Federal Emergency Management Agency (FEMA), the Department of Health and Human

Services (DHHS), the Department of Education, the Department of Agriculture, the Department of Veterans Affairs, and the Department of Labor.

Each department undertook somewhat different activities. HUD's efforts incorporated aspects of the Section 8 program, as well as the Emergency Shelter Grants and Supportive Housing Demonstration program (emergency and transitional and permanent housing, respectively), among others. FEMA worked through the Emergency Food and Shelter National Board. The DHHS focused heavily on mental health and substance abuse programs and so on. Some of the funds were automatically distributed to large cities and urban counties under formulas; others were competitive grants. Programs changed with subsequent amendments to and reauthorizations of the act. To be eligible for funds, communities and states had to complete comprehensive homeless assistance plans (CHAPs) and have them approved by HUD. Smaller communities and rural areas had the option to completely ignore the McKinney programs, and many did—in part because of the complexity of the paperwork and in part because of the (mistaken) sense that homelessness is purely an urban problem. The McKinney Act supplied some much needed funding, but it did not begin to replace the amounts that had been cut from domestic programs under the Reagan administration—cuts that probably helped to exacerbate the homelessness problem in the first place. In addition, it was a piecemeal set of programs. Local providers either did not use the programs or became "nonprofit entrepreneurs" in their efforts to mix and match the various money sources most effectively. This undoubtedly took efforts away from other worthwhile activities and may have directed funds to areas other than those of most need; for example, some funds could not be used for operating expenses, so buildings may have been overrehabilitated when increased staff was the actual need. There is evidence that in some states the McKinney Act increased the state's funding for homeless assistance over and above the additional federal funds.

In the 1980s, the McKinney Act was important as a symbol of federal recognition of the homelessness issue. In the early 1990s, there was a growing acknowledgment that long-term prevention and transitional housing were the areas on which policy needed to focus. Homeless advocates celebrated the 10th anniversary of the McKinney Act in 1997. In that year, Congress kept funding levels constant for the four programs administered by HUD (no increase for inflation) and increased funding modestly in the area of education for homeless children. The homeless veterans reintegration project received funding after a year's hiatus. The McKinney Act continues to provide a range of programs operated through different federal departments. Unfortunately, advocates remain in the position of battling for McKinney's continuance in every congressional session rather than developing permanent solutions to the problems that underlie homelessness. (SEE ALSO: **Homelessness**)

—*Hazel A. Morrow-Jones*

Further Reading

Blau, Joel. 1992. *The Visible Poor. Homelessness in the United States.* New York: Oxford University Press.

Burt, Martha R., Lynn Burbridge, Barbara Cohen, Pam Holcomb, Janet Kahn, Eric Patashnik, Therese Van Houten, and Regina Yudd. 1988. *State Activities and Programs for the Homeless: A Review of Six States.* Washington, DC: Urban Institute Press.

Hopper, Kim and Jill Hamberg. 1986. "The Making of America's Homeless: From Skid Row to New Poor, 1945-1984." In *Critical Perspectives on Housing,* edited by Rachel G. Bratt, Chester Hartman, and Ann Meyerson. Philadelphia: Temple University Press. ◄

► Student Housing

In countries influenced by the tradition of English universities, housing authorities and social policy professionals usually consider student housing as an educational rather than a housing issue. This carries over to when lack of student housing causes students to occupy low-income private rental housing. Although many universities consider student housing an educational issue, particularly in the United States where there is an emphasis on out-of-class student development, university-supplied accommodation ranges from below 10% to above 75% of the student enrollment at individual universities. The remaining students live in a number of alternatives, from boarding with strangers or in hostels to sharing houses and apartments to living with parents or other relatives.

Research into student housing for postsecondary students is mostly published in the educational or educational administration literature and is generally concerned with on-campus housing, variously termed *colleges, residence halls, dormitories,* or *houses* and, occasionally, *hostels.* Some research also comes from social psychology and environmental psychology, also known as environment-behavior or person-environment studies. The architects' guide to design, *Time-Saver Standards for Building Types* (De Chiara, Panero, and Zelnik 1995), provides a sensible appraisal of research with recommendations for design of "dormitories." For the uninitiated, however, the literature communicates confusing signals. An overall picture of all types of housing supplied to and used by students is generally lacking, except for studies (often unpublished) commissioned by individual universities, cities, or states.

Paradoxes in student housing are associated with how students, not being seen as productive workers like their peers might be, are discriminated against in their not being accepted as autonomous adults. The second paradox concerns the philosophy of university culture: Most English-speaking countries and former colonies of England established universities along the lines of those originally set up in England in medieval times. These were referred to as communities of scholars and had some aspects carried over from upper-class secondary colleges, whereas others were quasi-monastic. The university college was both residence and educator, not so much through classes, lecture halls, and laboratories but through individual tutors. Such communities were by nature highly elitist. The students depended on parents to pay the fees and the concept "in loco parentis"—in place of parents—came to describe the philosophy of such colleges in both Britain and the United

States. Whereas this concept was waning in the late 1960s and through the 1970s, some welcomed its apparent return in the United States in the more conservative 1980s and 1990s. They extol the virtues of living in residence halls, citing that students in them achieve higher grade point averages (GPAs) than students who commute. However, other research has found that residents both in students apartments adjacent to campus and in dormitories on campus had lower GPAs than students who commute to the campus. Most of these educational or college-administration-initiated studies do not specify who the commuters are, how long they have to travel, their type of accommodation, or the quality of their work-study environment.

Environmental psychologists have criticized campus living in the United States. These criticisms relate to sharing rooms, long corridors with many rooms, communal bathrooms, and the large populations within one building. Sometimes, residents have felt crowded and shown signs of social withdrawal because of loss of control and privacy. Crime, vandalism, and victimization became such a "hidden" problem on U.S. campuses that in 1990, Congress required universities to make public all reported criminal activity.

Off-campus housing and its adequacy for study has not been well researched. Britain has had a tradition of students' lodging or boarding with nonfamily households, however unsatisfactory the arrangements. Like Australia, it is now finding a growing preference among students for apartments or town houses to share with two or three other students, rather than the traditional residence halls. From economic necessity, many students live with their parents.

In the early 1990s, in Britain, Europe, and Australia, but less so in United States, university enrollments have increased considerably, creating a housing shortage for students both on campus and in the private rental market. The latter disadvantages both students and other groups, such as families competing for scarce housing resources. Yet the problem is still treated as one of education, not of housing policy. (SEE ALSO: *Behaviorial Aspects*)

—Ross Thorne

Further Reading

Baum, Andrew, Glenn E. Davis, and Stuart Valins. 1979. "Generating Behavioral Data for the Design Process." Pp. 175-96 in *Residential Crowding and Design*, edited by John R. Aiello and Andrew Baum. New York: Plenum.

Brothers, Joan and Stephen Hatch, eds. 1971. *Residence and Student Life: A Sociological Inquiry into Residence in Higher Education*. London: Tavistock.

De Chiara, Joseph, Julius Panero, and Martin Zelnik, eds. 1995. *Time-Saver Standards for Housing and Residential Development*. 2d ed. New York: McGraw-Hill.

Delucchi, Michael. 1993. "Academic Performance in College Town." *Education* 114(1):96-100.

Palmer, Carolyn J. 1993. *Violent Crimes and Other Forms of Victimization in Residence Halls*. Asheville, NC: College Administration Publications.

Scherer, Jacqueline. 1969. *Students in Residence: A Survey of American Studies*. London: National Foundation for Educational Research in England and Wales.

Winston, Roger B., Scott Anchors, and Associates, eds. 1993. *Student Housing and Residential Life: A Handbook for Professionals Committed to Student Development Goals*. San Francisco: Jossey-Bass. ◀

▶ Subdivision

A subdivision is a tract of land that has been divided into lots and developed for residential purposes. Usually, it is identified as a separate plan in the local land registry office.

Today, the typical subdivision is the product of a single company that has been responsible for all stages in the development process. Developers often buy tracts of land ahead of their needs. When they consider that part of their holdings are ripe for development, they register a plan description that specifies lot and street boundaries. Typically, they then provide hard services (water, sewer, roads, and sidewalks) and begin construction of dwellings. At this time, the developer builds a model home and begins marketing, usually by employing a name intended to create an attractive image. For decades, bucolic names were preferred (e.g., "Forest Glen") but those with historic or snob associations (e.g., "Colonial Estate") have recently become more popular. The name also signals the fact that modern subdivisions are usually quite homogeneous in terms of architectural styles, dwelling types, and most important, price.

The present meaning of *subdivision* has evolved steadily over the past century. In the 19th century, subdividers registered plans and then sold undeveloped lots to individual buyers. Many people speculated in land, and lots often passed through several hands before being built on. The original subdivider exercised some influence over the way in which an area grew. Lot sizes and street widths affected what types of dwelling were likely to be built. The influence was not strong, however, and many areas soon contained dwellings that differed considerably in terms of style, type, and price. In working-class districts especially, many also contained commercial and small industrial businesses.

Over time, a series of controls were introduced. The first were imposed by subdividers themselves, who came to realize that by introducing building and deed restrictions they could enhance the marketability of lots in their subdivisions. Building regulations typically specified a certain quality of dwelling construction and a minimum price. Until they were struck down as unconstitutional after World War II, deed restrictions prevented buyers from selling lots or dwellings to members of specified ethnic or racial groups. African Americans and Jews were often singled out. Used together, these two forms of regulation were widely used to target subdivisions at specific segments of the housing market.

A developer could control an individual subdivision but not neighboring districts. Especially during the 1920s, some developers began to realize they stood to gain from encouraging local governments to impose subdivision and zoning controls. These could ensure, for example, that the street layout of adjacent subdivisions was compatible, with no inconvenient jogs. They could also ensure that the viability of a subdivision of expensive homes would not be threatened by the subsequent growth of neighboring industry. For this reason, U.S. developers were prominent in pushing for public regulation of land use and land subdivision. Developers and local governments also came to realize that there was a need to control the absolute amount of subdivision activity. During real estate booms, such as those of the 1900s and 1920s, a great many plans were registered. Far too many building lots, many of them with hard services, were put onto the market. This glut led to scattered development and inefficient service provision. The experience of the Great Depression, when construction virtually ceased and many serviced vacant lots grew weeds, underlined the need for this type of control. Between the wars, municipal and county governments introduced subdivision approval procedures that attempted to control not merely the form of new subdivisions but also their absolute number. To this day, however, such efforts have been hampered by the increasing fragmentation of government in most metropolitan areas.

Public and private agencies exert a variety of controls over land subdivision. Increasing efforts have been made to evaluate the environmental impact of proposed development. Although the number of regulations continues to grow, land use planning remains weaker in the United States than in Canada or in Europe where the rights of the land developer have less priority than does the public good. (SEE ALSO: *Planned Unit Development; Residential Development; Subdivision Controls; Zoning*)

—*Richard Harris*

Further Reading

Cullingworth, J. B. 1993. *The Political Culture of Planning. American Land Use Planning in Comparative Perspective.* New York: Routledge.

Doucet, M. and J. Weaver. 1991. *Housing the North American City.* Montreal and Kingston: McGill-Queen's University Press.

Weiss, M. 1987. *The Rise of the Community Builders: The American Real Estate Industry and Urban Land Planning.* New York: Columbia University Press. ◄

► Subdivision Controls

Subdivision controls are the regulations controlling partitioning of a larger parcel of land into two or more individual properties for the purpose of selling them to separate owners. The nature and restrictiveness of subdivision control in practice depends on how many lots are to be created and on the change in land use envisaged after subdivision. A minor subdivision typically involves no new streets and relatively few lots. Usually, it is subject to a much simpler set of subdivision controls and requirements. Subdivision control is practiced largely in the urban fringe, where it regulates the conversion of agricultural and woodland plots into suburban neighborhoods, industrial parks, and commercial areas. However, subdivision control can also be defined broadly to include the registration of condominiums, and these are found throughout the urban area.

In general, subdivision control takes the form of permit approval. In other words, a developer cannot subdivide a parcel of land without first gaining permission from a state or local authority. Proponents argue that subdivision control permits planners to ensure that (a) all lots are of an appropriate size and shape and have access to a public right of way; (b) road and other public utilities are adequately and efficiently developed; (c) sufficient land is set aside for appurtenant uses, including public open space; (d) attention is paid to floodplains, site contours, and environmental preservation; and (e) development will be compatible among neighboring land uses. In addition, subdivision control provides an opportunity for other affected parties in the community to make their views known about the impacts of the proposed subdivision.

Subdivision controls also generally provide for statutory dedications and exactions. Statutory dedications are transfers of land or an interest in land (e.g., an easement) to public ownership required by subdivision ordinances as a condition of subdivision approval. Exactions, also known as development charges, development impact fees, and lot levies, are cash contributions (or in-kind transfers) required of a developer as a condition of subdivision approval. Exactions can take many forms: for example, paying for construction of public roads and other infrastructure within the subdivision. Exactions can also include cash payments in lieu of statutory dedications. (SEE ALSO: *Residential Development; Subdivisions; Zoning*)

—*John R. Miron*

Further Reading

Bossons, J. 1993. "Regulation and the Cost of Housing." Pp. 110-35 in *House, Home, and Community: Progress in Housing Canadians, 1945-1986,* edited by J. R. Miron. Montreal: McGill-Queen's University Press.

Bramley, G. 1993. "The Impact of Land Use Planning and Tax Subsidies on the Supply and Price of Housing in Britain." *Urban Studies* 30(1):5-30.

Delaney, C. J. and M. T. Smith. 1989. "Pricing Implications of Development Exactions on Existing Housing Stock." *Growth and Change* 20(4):1-12.

Schultz, M. S. and V. L. Kasen. 1984. *Encyclopedia of Community Planning and Environmental Management.* New York: Facts on File (see pp. 100, 134).

Slack, E. and R. Bird. 1991. "Financing Urban Growth through Development Charges." *Canadian Tax Journal* 39(5):1288-1304. ◄

► Subsidy Approaches and Programs

When governments in capitalist economies have chosen to intervene to alter private market allocations of housing during the 20th century, they have aimed at a number of targets and employed a number of strategies. These strategies and targets have not, at any given time, been the result of conscious rational selection among the full range of available alternatives, but, rather, the result of prevailing political and economic pressures filtered through the ideologies and perceptions of the political elites making the choices.

The first fundamental choice is the choice of the target(s) for housing subsidies. Although it is common to think of housing assistance as aimed primarily at the poor or working class, in fact, the middle class has been the beneficiary of some of the largest government subsidies in a number of countries. This targeting of the middle class has resulted from (a) a deliberate policy of encouraging homeownership among those already in a good position to make that choice or (b) housing assistance aimed at broad, society-wide housing shortages, which affect substantial elements of the middle class as well as the less advantaged. In sum, housing subsidies have been targeted at a broad range of incomes, and programs must be discussed according to the nature of their primary beneficiaries.

The second fundamental choice is the tenure to be favored by a housing subsidy strategy. In some developed societies, such as Switzerland and Germany, rental housing is either the prevailing tenure, or is, at the least, regarded as acceptable for the middle class and poor alike. This has shaped the type of programs undertaken by these governments. In others, such as the United States and Great Britain, homeownership has typically been regarded as the most desirable form of tenure for the middle class and as a status to be encouraged for the poor as well. To homeowners are attributed desirable traits such as stability, responsibility, and civic-mindedness.

A third fundamental choice is the mode of production: new construction, rehabilitation, or reliance on an existing, standard housing stock. Many programs rely on a mix of these modes, but the emphasis shifts over time, based on the adequacy and condition of the housing stock. Acute housing shortages encourage new construction; large numbers of deteriorated but sound units encourage rehabilitation, whereas an adequate stock of physically sound units encourages strategies that make them more widely affordable.

Over these fundamental choices may be superimposed a classification of the mechanisms of government involvement. Bearing in mind that some housing programs represent complex and unique mixtures of strategies, the following broad categories seem most useful in encompassing the programs used in the United States and other developed capitalist countries. Each is presented with examples of specific programs.

Tax System Strategies

One of the most common modes of subsidizing housing for the middle class is through selective tax exemptions for certain elements of housing expenditures. In the United States and Great Britain, these benefits have been made available largely to individual homeowners. In the United States, for example, homeowners can deduct their mortgage interest and local property taxes from their federal taxable income, and they are not taxed on the imputed rent they earn through owning their dwelling. This subsidy primarily benefits upper-income taxpayers and cost the U.S. Treasury about $70 billion in 1997.

Other selective tax advantages have been used as incentives for private developers to invest in housing for low- to moderate-income households. During the 1970s, the sheltering of income through accelerated depreciation was a major tax benefit to investors in the United States. This has since been phased out and replaced with the Low-Income Housing Tax Credit.

State and local governments in the United States have also reduced the cost of housing by financing it with public bond issues. The interest on such bonds is exempt from federal taxation, thus lowering the interest rates that must be offered to investors. The Tax Reform Act of 1986 limited the total per capita indebtedness that states and localities could incur in this way, thus reducing the availability of this subsidy mechanism.

Housing Finance Strategies

This category includes an almost infinite variety of devices for lowering borrowing costs, making more credit available for housing construction, or both. It can involve the creation of government-owned credit institutions, the provision of government-protected secondary mortgage markets, the provision of interest subsidies to private investors, the provision of mortgage insurance to guarantee the security of private loans, or many combinations of these and other devices.

During the Great Depression of the 1930s, the U.S. government sought to rescue collapsing financial institutions and assist the submerged middle class by creating the Federal Housing Administration (FHA) mortgage insurance program, by regulating savings and loan associations, and by creating a secondary mortgage market under the auspices of the Federal National Mortgage Association (now known as Fannie Mae.) After World War II, many European countries faced acute housing shortages brought on by war destruction, the postwar population boom, or both. Decent, affordable housing was also seen as a part of the basic "social contract" between the state and its citizens. Sweden and France provide examples of strongly "socialized" housing finance systems designed to meet these goals.

In the United States, the importance of federal mortgage insurance has declined somewhat in recent years, as the private mortgage market has become more regularized, and private mortgage insurance is readily available. FHA loans now serve primarily the low- to moderate-price segment of the homeownership market. Also, the savings and loan system has undergone drastic shrinkage during the 1980s, leaving banks and national mortgage companies as the primary lenders.

Another form of financial assistance is the direct interest subsidy, in which the public entity "buys down" the interest costs of the private developer and the savings are then reflected in reduced rents to low- or moderate-income tenants. The Section 236 program, which existed in the United States from 1969 to the mid-1970s, was an example of this type of subsidy, applied to rental housing. The Section 235 program applied the same type of subsidy to the purchase of homes by low-income households.

The Section 236 program was criticized for the frequent financial difficulties of participants, for high construction costs, and for the minimal rent reductions it produced.

Unlike housing allowances, this form of subsidy is relatively inflexible in that it does not increase to reflect increases in operating or maintenance costs. The Section 235 program also ran into many difficulties, Because of poor management and poor screening and counseling of recipients. However, the subsidy mechanism itself was not central to these difficulties.

Public Housing Strategies

From the 1930s through the 1960s, the most common housing assistance strategy for lower-income persons in the United States and in many European countries was low-rent housing units owned and operated by national or local governments. In many countries (e.g., Great Britain), these units also housed a large segment of the working and middle class. Many were flats in multistory structures, creating high-density land usage. Problems with the physical condition and the social environment in these units led to intense criticism of this mode of assistance. Efforts to "deconcentrate" public housing and to improve its management were made in the late 1960s. However, criticism of publicly owned housing also led to the adoption of the alternative strategies described here.

In the 1980s, Great Britain instituted a massive program of selling these units to their tenants, which was emulated in the United States on a much smaller scale. Public housing sales tend to benefit the economically better-off tenants and to remove the most attractive units from the public stock, leaving the remainder even more "residualized," as housing of last resort for the very poor. Nevertheless, in the formerly state socialist countries of Eastern Europe in the 1990s, thousands of units of government-owned housing were sold off to their tenants as part of an effort to "recommodify" housing.

Housing Allowance Strategies

Housing allowances are vouchers or cash assistance given to households (primarily renters) to reduce their out-of-pocket housing expenses to some level considered affordable by policymakers. Growing disillusionment with publicly owned housing and with subsidies to private developers to produce low-cost housing led many governments to embrace housing allowances in the 1970s.

A major benefit attributed to housing allowances is the deconcentration of lower-income persons from large, stigmatized government projects; however, programs vary in the degree of flexibility granted to recipients in their choice of housing. The Section 8 New Construction, Moderate Rehabilitation, and Substantial Rehabilitation programs tied the receipt of the housing allowance to residence in specific projects, by signing long-term contracts with the developers of these projects.

In the 1980s, the emphasis shifted to certificates or vouchers that could be used in any *existing* housing unit meeting minimum physical standards. This mode of assistance proved less costly than allowances tied to new construction, and it gave the recipients greater choice of where to live. These programs were made workable by the fact that growing production in the private housing sector in-

creased the availability of standard units. The problem was widely perceived to have shifted from providing enough units to making units affordable to lower-income households. Such programs can, therefore, be problematical in areas with extremely tight rental housing markets. Housing allowances now provide most of the federally assisted units in the United States and in a number of Western European countries as well. Recipients of housing vouchers are not always as widely dispersed as might be expected because of racial and ethnic barriers to moving to some locations and because the rent levels set by governments as the basis on which these subsidies are calculated do not keep up with rent increases in the private sector.

Nonprofit Housing Development Strategies

Not-for-profit corporations have taken on increasing importance as developers and managers of assisted housing since 1990, partly because of disillusionment with both publicly owned and private, for-profit housing developments. Although their activities may cover multiple neighborhoods or an entire city, frequently, these corporations are based in a particular neighborhood. To reduce rents to affordable levels for low-income households, they rely on special housing finance mechanisms (including grants and low-interest loans from governments, private financial institutions, or charitable foundations). In addition, nonprofits often rely on vouchers to subsidize the rents of their lowest-income residents. These organizations may develop either rental or owner-occupied housing, depending on circumstances in their community and are considered superior to alternative forms of ownership by many observers because of their focal concern on the preservation of low- and moderate-income housing and neighborhoods. Their decision-making processes typically involve residents of their units or of the surrounding neighborhood. These nonprofit corporations may also provide additional services to residents, such as job training or assistance with family crises. (SEE ALSO: *Affordability; Demand-Side Subsidies; Housing Allowances; Housing Subsidies in Industrialized Nations: History and Issues; Specific Programs and Legislation; Supply-Side Subsidies*)

—R. Allen Hays

Further Reading

Bratt, Rachel. 1989. *Rebuilding a Low Income Housing Policy*. Philadelphia, PA: Temple University Press.

Hays, R. Allen. 1995. *The Federal Government and Urban Housing: Ideology and Change in Public Policy*. Albany: State University of New York.

Karn, Valerie and Harold Wolman. 1992. *Comparing Housing Systems*. Oxford, UK: Clarendon.

Keyes, Langley C. and Denise Dipasquale. 1990. *Building Foundations: Housing and Federal Policy*. Philadelphia: University of Pennsylvania Press.

Mitchell, J. Paul. 1985. *Federal Housing Policy and Programs: Past and Present*. New Brunswick: Rutgers, State University of New Jersey, Center for Urban Policy Research.

Van Vliet--, Willem, ed. 1990. *International Handbook of Housing Policies and Practices*. New York: Greenwood. ◄

► Substandard Housing

By several measures, citizens of the United States are among the best housed in the world. According to the 1991 American Housing Survey, the median size of occupied units is 1,697 square feet with 674 square feet per person. more than half of the units have three bedrooms or more; 97.6% have complete plumbing. Only 2.7% are considered to be overcrowded with 1.01 or more persons per room.

From today's vantage point, it is difficult to fathom the squalid housing conditions that were commonplace in the 1800s and the first half of the 20th century. Although the relationship between bad housing and frequent epidemics of smallpox, typhus, yellow fever, and cholera was established as early as 1842 by New York's city sanitary inspector, the publication in 1890 of *How the Other Half Lives* by police reporter Jacob A. Riis was instrumental in raising public awareness of the horrors of the tenements. Riis documented 40 families living in accommodations for five, and in a room 13 feet square he found 12 men and women. Privies, where they existed, were often not maintained. Garbage and human waste accumulated in cellars and public hallways. Sloppy construction practices and the absence of adequate fire escapes created deathtraps.

The work of Lawrence Veiller, Robert W. DeForest, and other reformers led to the establishment of enforceable building and housing maintenance standards. These regulations ensured that new construction would be of higher quality.

With time, there were significant improvements in the overall housing stock, but problems persisted. A survey of 1,500 communities in 1925 and 1926 found that 17.8% of the homes lacked flush toilets and 31.7% did not have bathtubs. At that time, more than three-fourths of all farm dwellings in 1930 lacked modern improvements or conveniences. The U.S. Bureau of the Census found by 1940 that more than 44% of all housing units lacked complete plumbing; this dropped to 5.5% by 1970.

Gains were most pronounced in cities because of urban renewal initiatives, code enforcement programs, and the availability of public water and sewer services. Overcrowding was ultimately alleviated because of a significant decline in immigration along with the development of new housing in suburban areas. The persistence of substandard housing in rural areas can be attributed largely to rural poverty.

Substandard housing is created in several ways:

► *Poor workmanship.* This may be due to lack of skills, inadequate supervision of construction personnel, and dishonesty.

► *Compromises in the interest of profit.* Tenements, for example, were a result of efforts to increase revenues by maximizing the density on a given lot. Even the common practice in the 1950s to locate trailer parks in flood-prone areas is an example of profit-oriented compromises. In the cases cited, amenities, services, and safety features were often not provided.

► *Obsolescence.* As a structure ages, it requires basic maintenance, including periodic painting, reroofing,

updating of wiring and plumbing, and other repairs. Deferred maintenance can lead to significant damage that ultimately may be too expensive to repair.

► *Disinvestment.* Sometimes, property owners choose not to maintain a housing unit. This may be due to a concern that they will not realize a sufficient return on their investment or they may be "milking" the property for its rental revenues.

► *Changing standards.* Unincorporated areas often allow the construction of housing without the provision of sidewalks, paved streets, and other amenities such as parks. If subsequently annexed by cities that have a history of requiring urban infrastructure, these areas will be considered to be substandard and in some cases referred to as slums.

► *Changing values.* Householders' expectations have changed dramatically over time. The house of 40 years ago may be considered insufficient for want of space, a second (if not a third bathroom), or a two- or three-car garage. Small houses (both old and new) may be in compliance with building regulations, but they are considered to be substandard as defined by the housing market.

Local governments have various alternatives available to them for maintaining or improving housing quality. Problems of poor workmanship and compromises in the interest of profit can be addressed through building codes. Enforcement of a housing code is an appropriate tool for dealing with obsolescence and disinvestment. A change in standards, as in the case cited, is likely to require the expenditure of public funds, although where household incomes are high enough, a local improvement district can be established with landowners paying for improvements through a special assessment of their property. Changing values becomes a problem when the construction industry begins to cater exclusively to the high-end market and as a consequence fails to provide more affordable housing alternatives. In this case, local governments need to resist demands for minimum house and lot size requirements that would price some households out of the market.

Researchers and policymakers have long struggled with the development of an operational definition for substandard housing. Although the 1949 Housing Act established a "decent home and suitable living environment for all Americans," it failed to define any of the terms used. Analysts defined substandard housing as lacking complete plumbing (in urban areas), needing major structural repairs, or both. In 1940, the first decennial Census of Housing included a question about complete plumbing facilities; enumerators also recorded their evaluation of overall structural condition. By 1970, the U.S. Bureau of the Census decided to drop the structural condition question because it was determined that there was a high incidence of classification error and the census was to be conducted by mail. The question was included in a small sample surveyed in 1971 and subsequently dropped. More detailed data were collected in the Annual Housing Survey, which was started in 1973. Although respondents were asked to rate the overall quality of their unit, it was determined that they gave higher ratings than the enumerators. The traditional measures of quality—lack of plumbing and the need for major repairs—are no longer useful. Many of the problems encountered today represent only temporary deficiencies. Further research has focused on identifying single attributes or clusters that would serve as a measure of housing quality, with little success. (SEE ALSO: *Colonias; History of Housing; Space Standards*)

—*Deborah A. Howe*

Further Reading

DeForest, Robert W. and Lawrence Veiller. 1903. *The Tenement House Problem.* Vols. 1 and 2. New York: Macmillan.

Riis, Jacob A. 1971. *How the Other Half Lives.* New York: Dover.

Weicher, John C. 1989. "Housing Quality: Measurement and Progress." In *Housing Issues of the 1990s,* edited by Sara Rosenberry and Chester Hartman. New York: Praeger. ◀

▶ Suburbanization

U.S. suburbs are a product of the last 150 years. Although the existence of peripheral suburban places can be traced back to antiquity, before the mid-19th-century, U.S. "suburban" places were largely self-contained villages that were not bound socially or economically to the nearby city. A dramatic mid-19th-century breakthrough in transportation technology, the commuter railroad, made possible a selective suburban migration of businessmen and professionals who could afford the costs, both temporal and economic, of commuting. By 1850, half of Boston's 400 lawyers were already commuting from suburban locations. Post-Civil War industrialization of central cities, with the resulting pollution from steam engines and congestion of factories and tenements increased the advantages of a peripheral residence. However, the suburban option was available only to the well-to-do who could afford the time and expense of commuting by rail. It took the development of a practical electric streetcar in 1888 in Richmond, Virginia, to make middle-class suburbanization a reality. The streetcar meant that it was no longer necessary for middle-class workers to live within walking distance of the place they worked. For a five-cent fare, a middle-class worker could live as far as 10 miles from the central business district and commute to work. The turn-of-the-century middle-class housing developments built along the streetcar right-of-way meant that the middle class could divorce place of work from place of residence and live in socially, ethnically, and economically homogeneous suburban neighborhoods. Streetcars fueled the suburban real estate growth machine.

Although streetcars made substantial suburbanization possible, the automobile made it the dominant residential pattern. In 1920, suburbs housed only 15% of the nation's population, but there were already 9 million automobiles, up from 2.5 million only five years earlier. The 1920s saw middle- and upper-middle-class suburbanization such that by the early 1930s, more than half the commuters in all but the largest cities such as New York and Chicago were already driving to work. The common image of suburbia as a place

of substantial freestanding, single-family homes populated mainly by upper-middle-class, white, Anglo-Saxon Protestants (WASPs) who voted Republican was substantially set at this time.

Postwar Mass Suburbanization

At the end of World War II in 1945, suburbs housed 20% of the U.S. population; the central city was still the hub of the metropolitan area. In the following decades, the United States transformed from a nation where cities dominated to one of increasing suburban ascendancy. By 1970, the U.S. Bureau of the Census reported a suburban population larger than that of central cities. The 1990 U.S. census recorded some 115 million Americans living in suburbs, whereas 78 million lived in central cities, and 56 million in nonmetropolitan places. In 14 states, including populous ones such as California, Michigan, New Jersey, Ohio, and Pennsylvania, an absolute majority of the residents are suburbanites. Perhaps more practically, more than half the nation's voters now live in the suburbs; a fact that has serious consequences for financially stressed central cities.

A number of factors account for postwar mass suburbanization. First and greatest weight has to be given to the role of federal government policies. Significant among them was the liberalization of mortgage-lending programs, which provided potential home purchasers a massive subsidy to suburbanize. Veterans Administration (VA) loan guarantees made loans available to ex-GIs with no money down and a 25- or 30-year repayment schedule. Moreover, VA loans, and newly liberalized Federal Housing Administration (FHA) loans were made at interest rates below those of conventional mortgages. Prior to World War II, mortgages were commonly given only for a five-year period with a balloon payment at the end. To get a mortgage, it was common to have put down a front-end payment of at least a third of the value of the property. Postwar VA loans meant that for the first time, young middle- and working-class families just getting economically established could purchase a home. As of 1954, in the second Levittown north of Philadelphia, a veteran could buy a two-bedroom rancher for no money down, total closing costs of $10 and monthly payments of $59 for principal and interest. The suburbs were no longer the exclusive domain of upper-middle-class WASPs.

The government provided an additional subsidy to suburbanize by constructing a federally financed freeway system. The freeways, which were strongly supported by downtown business interests in the expectation that they would bring more shoppers and business downtown, radically cut the time required to commute to outer suburbs. For the new suburbanites, distance in miles became less important than time in minutes.

A second reason for postwar suburbanization was the private profit motive. The federal policies, just mentioned, were not formulated in a vacuum. Rather, they reflected intensive lobbying and strong pressures from lenders, developers, builders, automobile manufacturers, oil companies, the real estate industry, and other parties for whom large-scale (suburban) development was much more profitable than infill and urban growth.

The third reason for suburbanization was related to the first. As a consequence of the government programs, it was often cheaper for a young family to purchase in the suburbs than to rent in the city. Besides, because of the tight postwar housing market, there were few landlords in the city eager to rent apartments to young people with growing numbers of children. The fourth factor was the postwar baby boom, which lasted until 1964. The baby boom created some 10 million new households after the war. There was no room for these families in the already overcrowded city housing market. Economics and demographics, not a search for "togetherness" or a desire to escape city ills, fueled the first two decades of postwar mass suburbanization.

Fifth, much of the suburbanization following the war was simply a reflection of the fact that by the 1950s, most large cities, particularly in the Northeast and Midwest, had already developed virtually all the land within their boundaries. This meant that without substantial annexation, new housing built on open land would by definition be suburban. Thus, if one were seeking one of the lower-priced new homes, one automatically became a suburbanite.

Finally, there was the element of choice. Survey data consistently demonstrate that U.S. residents express a strong preference for freestanding, single-family homes. Regardless, VA and FHA loans did not cover rental apartments, and town houses were not being built.

Noticeably absent from the above list are most of the "popular" explanations given for postwar suburbanization, such as urban crime and racist "white flight."

The mass suburbanization of middle-class whites in the decades following World War II was not flight from perceived city ills but, rather, the movement of young families with children toward the advantages of new single-family housing subsidized by government loan polices. Young couples moved to where the new homes were being built, and they were being constructed in the suburbs.

U.S. style suburban sprawl, however, is not inevitable. Following World War II, other nations, especially those in Europe with tight land use control policies such as Sweden and the Netherlands, built high-density, high-rise suburban housing. Great Britain, by contrast, built many low-rise, but high-density, row houses. Government policy, as much or more than consumer choices, set suburban housing patterns.

Suburban Variation

Although suburbia is often spoken of as being relatively homogeneous, there is considerable variation. There are rich suburbs, poor suburbs, garden apartment suburbs, retirement suburbs, and industrial park suburbs. Income levels for individual suburbs reported in the 1990 census figures ranged from $62,000 per capita per person in Kenilworth on Chicago's North Shore to $4,900 at Ford Heights on the other side of Chicago. In the East and Midwest, established high-income suburbs often date back to the exclusive romantic suburbs of the 19th century with large homes on landscaped lots. Upper-status suburbs are often screened off from casual observation by extensive shrubbery and trees. Such suburbs tend to have an older age structure and few women in the labor force. Population turnover is low and many of them are socially closed WASP communities. Newer upper-income suburbs, particularly in the West and Southwest, place less emphasis on ethnic and religious background and more on cash. Luxury high-rise condominiums are increasingly found in newer suburbs for the well-to-do in retirement locations such as Florida.

Prior to World War II, working-class suburbs were largely older industrial or factory suburbs—for example, the working-class suburbs of Cicero west of Chicago. The suburb attained notoriety during the 1920s when Al Capone temporarily made it his headquarters. Residentially, such suburbs were dominated by streets of simple but well-maintained working-class homes on small lots. Following the war, working-class suburbs tended to follow the Levittown pattern insofar as the subdivisions were built on cleared outlying farmland and housing styles were usually limited to one or two models. Workers moving to such suburbs were often following their jobs as industrial plants suburbanized. Many of these plants have now closed, leaving working-class suburbs with aging structures and an eroding commercial and residential tax base. Suburbs, compared with city neighborhoods, are more likely to have status persistence over time. For example, Chevy Chase, northwest of Washington, D.C., and Berwyn west of Chicago remain respectively upper-middle-class and ethnic working class just as they were 75 years ago. Some high-status suburbs in the East and Midwest have maintained their positions for more than a century. Affluent suburbs have been especially successful in holding their positions by using their resources to control entry. This is done through exclusionary zoning to maintain certain home and lot sizes, through maintaining strict building codes, or by keeping taxes for schools and services at a high level. Suburbs enjoying a high status attract new high-status residents, which reinforces the reputation.

Decentralization to Suburban Outer Cities

The Burgess zonal hypothesis, which assumed that urban growth radiated from a single central busines district (CBD) through concentric rings of decreasing density, lost empirical validity as a growth model following post-World War II decentralization of offices and industry as well as of housing. The traditional pattern of growth has been turned inside out. Today, there is a multinucleated pattern of suburban centers. Many suburbs have been transformed into new economic and commercial cores. Industrial parks have supplanted central-city factories. In 1997, there were twice as many manufacturing jobs in suburbs as in central cities. Nationwide, downtown office space vacancy rates exceed those of surrounding suburbs by a third. Growing outer suburbs increasingly find themselves sandwiched between demographically and economically stressed central cities and declining rural areas.

Minority Suburbanization

The 1990 census reported that more than 8 million of the 30 million African Americans in the United States were suburbanites. Although media stories often give the impression that African Americans reside overwhelmingly in inner-city neighborhoods, one of every three African Americans living within a metropolitan area is a suburban resident. There are some 40 metropolitan areas where the African American suburban population exceeds 50,000. Generally, black growth occurs where white growth occurs. As of 1990, Washington, D.C., had 620,000 black suburbanites, Atlanta had 463,000, and Los Angeles 401,000.

African American suburbanization is a fact. The question is whether black suburbanization reflects increasing housing integration or merely the growth of black enclaves across city boundaries into suburban areas. Are African Americans living in contemporary suburbs living a suburban lifestyle or are they only displaced poorer urbanites who have been moved farther out? The pattern until the 1980s was substantially one of black spillover from central cities into older inner-ring suburbs. Suburban blacks were more likely to live in suburban communities with lower per capita income, less adequate housing, and strained local finances. The 1980s and 1990s show less of the invasions-succession model and more parallel growth. Overall, suburbs are becoming more diverse racially. Spillover is not seen in major metropolitan areas having the largest black populations. The pattern is more of a "leapfrog" effect with blacks moving into newer subdivisions on the periphery. Stable multiracial suburbs are more common. The old patterns of suburban racial segregation are increasingly becoming history, but racial steering by real estate agents and discrimination by banks, lending institutions, and insurance companies still persists. Race remains a core issue in U.S. society. African American suburbanites increasingly resemble white suburbanites insofar as they are predominantly middle-class, married-couple families. As a group, African America suburbanites have income levels slightly below those of white suburbanites but above those of black and white city dwellers.

Within a little more than a decade, the largest minority population in the United States will be Hispanics. Approximately 43% of the 23 million Hispanics in the United States were suburbanites according to the 1990 census. Hispanic suburban growth is greatest in the areas of the greatest economic growth. The eight metropolitan areas having the greatest number of Hispanic suburbanites all are in the Sunbelt; five in California, two in Texas, and one in Florida. The Los Angeles metropolitan area counts more than 1.7 million suburban Hispanics. Hispanics, unlike African Americans, have faced wide variations in patterns of physical segregation. Historically, segregation was most common in border states such as California and, especially, Texas. For Hispanics, suburban residence is associated with lower levels of segregation and higher levels of association with non-Hispanic whites.

Asians are the most suburban of the minority groups in the United States. This is particularly true of new legal and more affluent immigrants, who are most likely to bypass the old central-city ethnic enclaves and settle directly in the suburbs, especially in the South and West. It is also true of some eastern cities. In Washington, D.C., there is no central-city Koreatown, only 800 of the 44,000 metro area Koreans actually live in the District of Columbia. Legal Asian immigrants with educational skills tend to move directly into those suburban neighboods where employment and schooling opportunities for children are greatest.

Although U.S. suburbs have existed for 150 years, mass suburbia is a post-World War II phenomenon. Federal government policies heavily subsidized suburban growth. Suburbs have changed dramatically over the past quarter of a century, from overwhelmingly residential areas to the nation's primary shopping, office, and manufacturing location. Outer cities are not only newer, they also differ in being private rather than public places. Suburbs have long been criticized for being all-white communities, but they are now becoming more multiracial and multi-ethnic. (SEE ALSO: *Levittowns; New Urbanism*)

—*J. John Palen*

Further Reading

Binford, Henry C. 1985. *The First Suburbs*. Chicago: University of Chicago Press.
Fishman, Robert. 1987. *Bourgeois Utopias*. New York: Basic Books.
Garreau, Joel. 1991. *Edge Cities*. New York: Doubleday.
Jackson, Kenneth T. 1985. *Crabgrass Frontier: The Suburbanization of the United States*. New York: Oxford University Press.
Kelly, Barbara M., ed. 1989. *Suburbia Reexamined*. New York: Greenwood.
Kling, Rob, Spencer Olin, and Mark Poster, eds. 1991. *Postsuburban California*. Berkeley: University of California Press.
Palen, J. John. 1995. *The Suburbs*. New York: McGraw-Hill.
Teaford, Jon. 1996. *Post-Suburbia; Government and Politics in the Edge Cities*. Baltimore: Johns Hopkins University Press.
Warner, Sam Bass. 1962. *Streetcar Suburbs: The Process of Growth in Boston, 1870-1900*. Cambridge, MA: Harvard University Press. ◄

► Supply-Side Subsidies

Supply-side subsidies go to suppliers of housing and reduce the costs of provision. Supply-side housing subsidies make it cheaper for builders and landlords to provide housing. The subsidy can take many forms. It involves an explicit or implicit flow of funds initiated by government activity. Thus, it can be a cash payment to providers of housing, or it can be in the form of a fiscal concession that reduces the tax payments that otherwise would have to be made.

The subsidy can be in the form of a measure that reduces the payments on a loan used to provide housing. The loan might be provided by government at submarket rates, or the interest payments on a privately raised loan may be reduced by a variety of measures, such as government's contributing to the payments or underwriting the loan and thus reducing the interest required. The subsidy could involve initiatives by government, which means that land for housing production is supplied at a low cost. When the subsidy results from explicit expenditure by government, it may be said to be *direct*. When it does not, it may be called an *indirect* subsidy.

A supply-side subsidy may assist the building of new housing, the improvement of existing housing or simply the continuing provision of housing. The dwellings involved could be for homeownership or renting. When linked to improvement, the subsidy may be tied to goals of neighborhood and environmental improvements as well as an increase in housing quality.

In practice, one of the most significant uses of supply-side subsidies has been to promote the production of rented housing for lower-income households. Where such housing has been provided by public sector landlords or housing associations, it has usually been referred to as *public housing* or *social housing*. The large-scale building of such housing in Europe has been subsidized by a range of construction grants, low-cost loans from government sources, and the provision of low-cost land either owned or procured by government. Public or social rented housing has also been supported by continuing supply-side subsidies in the form of assistance with meeting ongoing expenses, such as loan repayments and management and maintenance costs.

Because they support buildings or objects, supply-side subsidies have been referred to as *object* subsidies in contrast to housing allowances, which support demand and have been called *subject* subsidies. There has been much debate worldwide about the relative merits of object and subject subsidies.

If the principal purpose of housing policy is to reduce shortages, object or supply-side subsidies have the advantage that they address the problem directly by seeking to increase housing supply. The large and highly perceptible burden of such subsidies on government budgets, coupled with a belief that major shortages had been eliminated, led many European governments to cut these subsidies from the 1970s onward.

The cuts were often supported by arguments about the inequitable distributional effects of supply-side subsidies. Although, usually, the subsidies have been given to housing associations or other social landlords, on the condition that they favor low-income tenants and keep rents down, the principal allegation has been that they do not, in fact, help households in the greatest need and that they give disproportionate benefits to tenants who can afford to pay more. If rent reductions affect all tenants equally, irrespective of income, the subsidies do not target help where it is most needed.

If the production problem is separated from the distributional or affordability problem, it can be argued that supply-side subsidies are well suited to tackling the former but not the latter problem. In practice, they have been used to tackle both problems. They have sometimes had a third objective: reducing unemployment. Thus, for example, the public housing subsidies in the United States in the 1930s were supported partly on the grounds of providing work for the construction industry. Housing production subsidies in Europe have also been used in the past to promote the growth of the economy. Their reduced priority in recent decades is linked to changes in economic policy that favor fiscal restraint.

Some of the arguments against supply-side subsidies are really arguments against the design and management of the housing they have been used to produce rather than the principle of such subsidies. Thus, large-scale poorly planned housing developments in the public sector have given ammunition to those who want to reduce supply-side support.

Supply-side subsidies go to the private as well as to the public sector. Private owners can receive support on the condition that they give access to low-income tenants. This support might be in the form of a tax concession or tax expenditure such as an accelerated depreciation allowance. Some nonprofit housing associations in Europe are treated for tax purposes as charities and receive all the fiscal privileges that go with this status. In many countries, government provides incentives for the building of houses to purchase. The subsidies involved are often conditional on the prospective homeowners' being in lower-income groups.

The subsidy need not be given by central government. In the United States, a range of federal programs as well as several state-sponsored subsidies have supported production. In some European countries, local government may supplement funding that is provided centrally.

There is thus much diversity in the arrangements that surround supply-side subsidization. This diversity extends to a range of conditions attached to the granting of subsidy. Supply-side subsidies with strong conditions attached give government the capacity to influence the design, location, occupancy, and rent levels of housing. This degree of control will be seen by some supporters of this approach to be beneficial, but critics contrast this with the degree of decentralized control and individual choice that they see attached to demand-side subsidies. (SEE ALSO: *Demand-Side Subsidies;* **Subsidy Approaches and Programs**)

—*Michael Oxley*

Further Reading

Ball, Michael, Michael Harloe, and Maartje Martens. 1988. *Housing and Social Change in Europe and the USA.* London: Routledge.

Boelhouwer, Peter and Harry van der Heijden. 1992. *Housing Systems in Europe: Part 1. A Comparative Study of Housing Policy.* Delft, The Netherlands: Delft University Press.

Bratt, Rachel G. 1986. "Public Housing: The Controversy and Contribution." Pp. 335-61 in *Critical Perspectives on Housing,* edited by Rachel G. Bratt, Chester Hartman, and Ann Meyerson. Philadelphia: Temple University Press.

Hallett, Graham, ed. 1993. *The New Housing Shortage: Housing Affordability in Europe and the USA.* New York: Routledge.

Oxley, Michael. 1991. "Housing subsidies in Western Europe." Pp. 1783-92 in *Management, Quality and Economics in Building,* edited by Artur Bezelga and Peter Brandon. London: Spon.

Oxley, Michael and Jacqueline Smith. 1996. *Housing Policy and Rented Housing in Europe.* London: Spon.

Stegman, Michael, ed. 1970. *Housing and Economics: The American Dilemma.* Cambridge: MIT Press (see Section 3 "Evaluating Housing Programs"). ◄

► Support-Infill Housing

Support-Infill housing is an approach to the design and construction of residential buildings. When applied to detached houses, the approach distinguishes between a more durable, long-lasting "base building" or shell and a more changeable "interior fit-out." When applied to multifamily buildings, a distinction is made between (a) the group (an apartment building owner, condominium association, or cooperative), (b) the individual (household, single person, or family), and (c) the physical elements under the control of each. In such housing, the common infrastructure of services, structure, walls, and spaces is clearly separated from interior partitions, fixtures, and all the piping, wiring, and air-conditioning equipment determined for each individual dwelling. Thus, it is a scheme to balance individual freedom with a coherent community in complex dwelling environments. One of the principal purposes of this approach is to enable a variety of dwellings to come about initially and over the long term, corresponding to the variety and changing characteristics of households.

Between 1920 and 1970, governments responsible for the production of large numbers of housing units adopted a strategy known as "mass housing." An international phenomenon, advocated by government officials and large bureaucratic private organizations, mass housing was employed in both capitalist and socialist societies. Its principal characteristic was the design and construction at one time, under central control, of large projects of uniform dwellings. Professionals saw this as a necessary and appropriate means of achieving control, efficiency, and improved housing standards.

In the late 1950s, observers of mass housing began to see the deleterious effects of this practice. They saw that mass housing produced senseless uniformity, which, with bureaucratic control, led to a loss of individual freedom and negative social and economic consequences.

Among those observers, N. John Habraken suggested that the problems associated with mass housing stemmed from the elimination of the household as a force in the housing process, causing a disruption of the "natural process" wherein households take direct action and responsibility to establish dwellings. He suggested that the mass housing "solution" was based on a lack of understanding of the balance of forces—community and household—needed to support socially, economically, and physically healthy dwelling environments.

To ensure that decisions about individual dwellings could be made individually, "supports" would have to be designed with the capacity to hold a variety of dwelling unit plans. He proposed that supports would be architecturally specific to the locality, use appropriate technology, and accommodate a variety of dwelling plans. "Infill" products for creating a variety of dwellings would be made by manufacturers. His concept distinguished between what belonged to the community sphere (the *support,* containing the building structure, shared circulation and utility systems, and most of the facade) and the individual sphere (the *infill,* including interior nonload bearing walls, cabinets, fixtures, and the piping, wiring, and mechanical systems for each dwelling). By adopting this approach, the fine-grained qualities of a viable neighborhood, lost in mass housing, could then be regenerated.

This concept constituted a breakthrough combining increased adaptability with more efficient production, as demonstrated by recent projects in the Netherlands, Japan, China, Britain, Finland, Sweden, and the United States. Developers, housing associations, and contractors have begun to understand that large projects can be realized efficiently and cost-effectively even though each dwelling is different. This approach is known as open building. An Open Building Working Commission in CIB (International Council for Building Research Studies and Documentation) was formed in 1996 as a platform for further implementation of support-infill projects and products. (SEE ALSO: *Flexible Housing;* **Industrialization in Housing Construction**)

—*Stephen Kendall*

Further Reading

Habraken, N. J. (with Boekholt, Thyssen, and Dinjens). 1976. *Variations: The Systematic Design of Supports.* Cambridge: MIT Press.

———. 1972. *Supports. An Alternative to Mass Housing.* New York: Praeger; London Architectural Press.

———. 1998. *The Structure of the Ordinary.* Cambridge: MIT Press.

Kendall, Stephen H. 1988. "Management Lessons in Housing Variety." *Journal of Property Management* September/October): 22-27.

———. 1993. "Open Building: Technology Serving Households." *Progressive Architecture* (November):95-98.

van der Werf, Frans. 1980. "Molenvliet-Wilgendonk: Experimental Housing Project, Papendrecht, The Netherlands." *Harvard Architecture Review* 1(Spring):161-69. ◄

► Sweat Equity

Sweat equity is a self-help technique employed by persons whose participation in the physical rehabilitation and construction work contributes to their acquisition of housing. Sweat equity describes the means by which those wishing to own property contribute their skills toward its renovation for full or partial ownership of the property. Self-help can be the product of individual efforts as well as those of family members and friends. The crucial issue is the exchange of work on a property for some dimension of ownership. The usual mode of ownership is that of a conventional single-family home; however, with the investment of sweat equity in a multifamily structure, residents can become owner-occupants through cooperative ownership. Sweat equity beneficiaries are not necessarily persons with low incomes.

The earliest urban self-help housing efforts arose in New York City in the early 1970s. In buildings abandoned by the landlords, tenants began to join forces to make repairs, maintain, and secure the premises where they resided. The courts sometimes sanctioned these actions, but in other cases, the actions were without legal authorization. Sweat equity occurred where the self-help aspects went beyond tenant ownership and cooperative management to direct participation in reconstruction. Self-help can be viewed as an end in itself by promoting self-reliance, inexpensive rehabilitated housing and mutual ownership. Sweat equity includes the use of voluntary labor to reduce the cash costs of renovation while representing the down payment for each participant.

Self-Help and Homesteading

The historical precedents for the Section 810 Urban Homesteading Demonstration Program, extending back to the 1862 Homestead Act and continuing through the subsistence homesteads of the 1930s to the early urban homesteading efforts in Wilmington, Delaware, and elsewhere, have all viewed the homesteader's own efforts at property improvement as a key ingredient. The purported benefits of self-help in housing are that homesteaders form an enduring attachment to their land and property and reduce the costs to acquire it. There has been a strong association in the public mind between self-help or sweat equity and urban homesteading. This association has been only partially realized in urban homesteading programs, whether federal or local in origin. The contribution of sweat equity to Section 810 urban homesteading repairs was not acknowledged in the regulations until 1983. Sweat equity has been limited by local rules that require homesteaders to be certified prior to undertaking technical work, such as wiring, plumbing, and heating, and by local provisions that restrict sweat equity contributions to cosmetic improvements.

The Baltimore local homesteading program was operated in neighborhoods and on scattered sites. The scattered-site structures were all located within the central-city area. Some lower-income households participated in the scattered-site component of the program where instances of hands-on rehabilitation or sweat equity and homesteaders acting as their own general contractors were commonplace. The Philadelphia urban homesteading program sought individuals who, beyond meeting formal requirements, had proven construction skills or other ability to undertake sweat equity.

Theoretically, sweat equity encourages cost control and fosters homesteader attachment to the property. There is often a concern that sweat equity implies a trade-off between quality and speed of rehabilitation, although some research refutes this point. Management of the locally operated homesteading program rather than the agent of repair was found to be the most important determinant of the quality of workmanship in the Section 810 Urban Homesteading Demonstration. One government report found that if self-help rehabilitation by homesteaders formed a substantial portion of the work, it generally took longer to complete but produced work that was equal in quality to that performed by professional contractors. The report also found that sweat equity is compatible with quality repairs and cost reduction.

Cities have varied in their allowance of sweat equity repairs in homesteading programs. Typically, the amount of sweat equity varied with the application of rehabilitation standards. Where rehabilitation standards were higher, sweat equity was restricted or controlled; where rehabilitation standards were less stringent, greater involvement of the homesteader in work planning, contractor selection, and repair work was encouraged.

Sweat equity raises the question of who is best prepared to apply the concept of self-help to housing revitalization. Local homesteading programs have tended to favor middle-income applicants partly on the assumption that they have more skills and more cash to apply to home improvement and rehabilitation. Low-income squatters who take over abandoned properties without title or authority have been among those who have challenged this contention.

Self-help skills, meaning the ability to perform some of the necessary rehabilitation work, were not required in any of the 23 original Section 810 Urban Homesteading Demonstration cities. In 12 cases, however, self-help was considered a factor in defining financial capability. In these instances, a homesteader who could perform some of the rehabilitation reduced the size of the loan needed to finance repairs to the home. Conceivably, a program would qualify persons with a relatively low income if they possessed a high level of construction skills.

Use or nonuse of sweat equity has been guided by varied viewpoints toward necessary rehabilitation. These viewpoints, noted during the Section 810 Urban Homesteading Demonstration, included the following:

- ► Self-help rehabilitation is highly unreliable. Some homesteaders will simply fail to complete the work within a reasonable amount of time or to meet minimum workmanship standards.
- ► Sweat equity is unnecessary if repairs can be performed by contractors for significantly less than their aftermarket repair value.
- ► Monitoring and providing technical assistance are expensive for homesteading agencies.
- ► Most homesteaders cannot be expected to undertake the repair of major code violations that usually require licensed trades persons.
- ► Sweat equity is necessary if the cost of repairs places a financial burden on lower-income homesteaders.
- ► Homesteaders should be given the opportunity to undertake their own repairs as long as this does not lead to continued existence of hazards.

Reliance on sweat equity versus contracted rehabilitation has varied widely. Factors that may determine the policy on sweat equity include the philosophical inclinations of program administrators, the condition of the properties, and the financial capacity of the homestead applicants. In cases in which the repair costs are high relative to after-repair market values, sweat equity may be a matter of necessity rather than of choice. The four basic approaches to self-help rehabilitation of urban homesteads include (a) no self-help, (b) self-help limited to cosmetic items (such as painting, limited carpentry, cleanup, and yard tasks), (c) medium-scale rehabilitation (significant amount of work that may require demonstrated knowledge of construction), and (d) significant reliance on self-help.

In some cities where sweat equity is permitted for significant tasks, proof that the homesteader can perform the work in a satisfactory manner is required. In some instances, homesteaders have to show that they have undertaken construction work for other homeowners, and occasionally, they must pass skills tests. A contract is frequently required to commit the homesteader to perform specified work requirements.

Effects of Sweat Equity

Reviews of how effectively sweat equity has advanced homeownership among lower-income persons differ. Sweat equity has been suggested as a way for tenant organizations to respond to the threat of disinvestment by landlords. Sweat equity may be one of a few ways that these organizations can acquire ownership or control at prices they can afford. A risk is that, having acquired the property, the owners may be unable to manage the long-term costs of ownership.

The Urban Homesteading Assistance Board (UHAB) is a nonprofit housing service established in 1973 to advocate and provide technical assistance for self-help housing groups. The approach tried by UHAB was to allow the use of voluntary sweat labor to perform most of the construction work under the supervision of paid, skilled (and licensed) professionals.

UHAB helped the New York City Department of Housing Preservation and Development to set up an urban homesteading program in 1981, when the city held possession of 112,000 in rem units. City policy had been to resist long-term responsibility for these units, and it used a number of means to dispose of them, including auctioning them off, sometimes to gentrifiers, sometimes to private landlords. Small portions of these properties were given to low-income tenants as sweat-equity ventures. The city performed work requiring licensed professionals, and the homesteaders did demolition, roofing, carpentry, and interior work. (SEE ALSO: *In Rem Housing; Self-Help Housing; Urban Homesteading*)

—*Mittie Olion Chandler*

Further Reading

Atlas, John and Peter Dreier. 1986. "The Tenants' Movement and American Politics." In *Critical Perspectives on Housing*, edited by Rachel G. Bratt, Chester Hartman, and Ann Meyerson. Philadelphia: Temple University Press.

Chandler, Mittie Olion. 1988. *Urban Homesteading: Programs and Policies*. Westport, CT: Greenwood.

Kolodny, Robert. 1986. "The Emergence of Self-Help as a Housing Strategy for the Urban Poor." In *Critical Perspectives on Housing*, edited by Rachel G. Bratt, Chester Hartman, and Ann Meyerson. Philadelphia: Temple University Press.

Sanchez, Jose Ramon. 1986. "Residential Work and Residual Shelter: Housing Puerto Rican Labor in New York City from World War II to 1983." In *Critical Perspectives on Housing*, edited by Rachel G. Bratt, Chester Hartman, and Ann Meyerson. Philadelphia: Temple University Press.

Schuman, Tony. 1986. "The Agony and the Equity: Self-Help Housing." In *Critical Perspectives on Housing*, edited by Rachel G. Bratt, Chester Hartman, and Ann Meyerson. Philadelphia: Temple University Press.

U.S. Department of Housing and Urban Development. 1977. *The Urban Homesteading Catalogue*. Vol. 2. Washington, DC: Government Printing Office.

———. 1980. *Neighborhood Self-Help Case Studies*. Washington, DC: Government Printing Office.

———. 1981. *Evaluation of the Urban Homesteading Demonstration Program: Final Report*. Vol. 2. Washington, DC: Government Printing Office. ◄

► Syndication

When a housing development is being purchased or constructed, financing is needed to pay for this development. Much of the financing will be in the form of debt financing, money loaned in return for a mortgage promising to surrender the property if the money is not repaid. To the extent that this debt financing does not cover the total cost of the development, other financing must be found, and this financing is usually equity financing. Equity financing means investments made by individuals, corporations, or partnerships who become owners of the development. As owners, they will receive a share of any profits that the development generates during its operation and on sale of the development in the future. Investor owners may also invest to obtain a share of any tax benefits that the development may generate. The process of finding investors in real estate who will accept tax benefits as some or all of their return on investment is referred to as *syndication*.

Syndication is performed either through a fee-for-service company or through a public agency or other nonprofit entity interested in finding investors to build housing. The fee-for-service companies, called syndicators, prepare an investment opportunity statement used to solicit interest among individuals or corporations with capital to invest. Once the investors have been found, the syndicators arrange for the exchange of investment dollars, the cash flow, and the tax benefits between the development and the investors. The public sector and nonprofit syndicators perform similar duties, but they tend to solicit specific investors located in close proximity to the development. These investors usually have a particular interest in assisting a development, such as a development designed to revitalize a neighborhood in which the investor is located or a development that serves employees of the investor's firm.

The syndication process has been active for many years. Over time, investors have made equity contributions to developments for a variety of types of tax benefits. These tax benefits include Historic Rehabilitation Tax Credits, surplus depreciation, and most recently, Low-Income Housing Tax Credits. These investments are generally made to develop newly constructed or substantially rehabilitated buildings, especially housing, but syndication can, under very limited circumstances, even be applied to properties that are already in operation and have been syndicated previously.

Because the tax laws dictate the amount of tax benefits that a property generates, syndication is integrally tied to the tax laws. To syndicate a property, a set of conditions, stemming from the tax laws, must be satisfied. These include a set of disclosure rules that govern the flow of information about an investment to the investors, ensuring that investors are fully informed as to the risks and consequences of their participation in the development. The tax laws also limit the form of ownership of the development, which is usually a limited partnership, restricting the extent to which each investor-partner may share in the risks and rewards of the development's returns over time. Finally, the tax laws regulate the amount of the tax benefits themselves and set rules as to when and how they may be claimed by the investors.

These tax laws are subject to periodic change by Congress, and as such, the interest of investors in these syndicated investments changes in response. Prior to 1986, syndication was a large industry working in all areas of real estate. In housing, syndication was a common way to secure equity capital for all types of income earning residential development, such as apartment complexes. However, after the Tax Reform of Act of 1986, this activity was sharply curtailed. The act severely reduced the tax benefits available to investors by reducing the amount of depreciation that could be claimed by a development, by limiting the amount of tax benefits that a taxpayer could claim in any one year, and by reducing the value of tax benefits through lowering overall tax rates. The act did create a new program, the Low-Income Housing Tax Credit, which depends heavily on syndication for the its implementation.

Syndication brings together investors and developers to help finance housing and other forms of real estate development. The process depends heavily on the tax laws but has managed to adjust to periodic changes in the tax laws, bringing needed investment dollars into the housing development process. (SEE ALSO: **Housing Finance**; *Low-Income Housing Tax Credits; Tax Reform Act of 1986*)

—*Kirk McClure*

Further Reading

Dowall, David E. 1990. "The Public Real Estate Development Process." *Journal of the American Planning Association* 56:504-12.

Howell, Joseph T. 1983. *Real Estate Development Syndication.* New York: Praeger.

Jacobus, Charles J. 1996. *Real Estate Principles.* Upper Saddle River, NJ: Prentice Hall.

Pryke, M., ed. 1994. "Property and Finance." *Environment and Planning A* 26(Theme issue):167-264.

Shim, Jae K. 1996. *Dictionary of Real Estate.* New York: John Wiley.

Tosh, Dennis S. 1990. *Handbook of Real Estate Terms.* Englewood Cliffs, NJ: Prentice Hall.

T

▶ Tax Credit Advisor

The Tax Credit Advisor is a nationwide, monthly publication offering news, ideas, and information for federal low-income housing tax credit program participants. Its price is $269 per year, $199 per year for nonprofits. Editor: Glenn Petherick. Address: 1726 18th St., NW, Washington, DC, 20009. Phone: (202) 328-9171. Fax: (202) 265-4435.

—Laurel Phoenix ◀

▶ Tax Expenditures

The term *tax expenditure* refers to the cost to the U.S. Treasury of a special provision of the tax code that reduces individual or corporate tax liability. The impact of a tax expenditure on the federal budget is thus equivalent to that of a direct outlay. The major types of tax expenditures are exclusions from income, credits, deductions, preferential tax rates, and deferral of tax liability. Tax expenditures are most similar to those direct spending programs that have no spending limits and that are available as entitlements to those who meet the statutory criteria established for the programs. The Congressional Budget and Impoundment Control Act of 1974 required inclusion of an analysis of the cost of tax expenditures in the proposed federal budget submitted annually to Congress by the administration. These estimates, prepared by the Treasury Department, are based on the amounts that would be collected if the special tax expenditure were not available. In recent years, the proposed federal budget has also included a series of "outlay estimates," or the estimated cost of achieving a comparable benefit by direct payments instead of reduced taxes.

Housing-related tax expenditures dwarf direct federal expenditures for housing. The total cost of housing tax expenditures for FY 1997 was an estimated $105 billion. In contrast, direct federal outlays for housing assistance were estimated at $28 billion. A major reason for this difference is that homeowner deductions, the bulk of the housing tax expenditures, are unrestricted entitlements, available to all who qualify for them, and because they are well-known and easy to deduct, a substantial portion of homeowners take advantage of them. In contrast, federal housing assistance for low-income people is not only not an entitlement, but HUD has estimated that there are an estimated 5.3 million unserved very low income renter households with "worst case" housing needs, for whom housing assistance is unavailable. This compares with approximately 5 million renter households currently receiving assistance under federal low-income housing programs. The National Low Income Housing Coalition has estimated that the nation's low-income housing needs could be met in a relatively short period for a fraction of the housing tax expenditures that now benefit upper-income people.

The Mortgage Interest Deduction

The deductibility of mortgage interest on owner-occupied homes, including second homes, is the largest of the housing tax expenditures and the third largest of all tax expenditures. Treasury figures show that the FY 1997 cost of this expenditure, $49.1 billion, was exceeded only by the cost of employer contributions for health insurance and care and the exclusion of employer pension plan contributions and savings.

Contrary to popular belief, the mortgage interest deduction was not enacted because of a policy decision to promote homeownership. Rather, it resulted from a general provision on interest deductibility carried over from an emergency Civil War income tax into the federal income tax law enacted after adoption of a constitutional amendment in 1913. The deductibility of mortgage interest had little impact either on tax expenditure levels or on the housing market until the broadening of the tax base and the rise in homeownership that followed World War II.

Until 1987, when it was capped at the owner's basis in the residence up to a maximum of $1 million, there were no limits on the amount of mortgage debt on which interest could be deducted.

The Joint Tax Committee, an arm of Congress, estimated that 27.8 million of the nation's 84.1 million taxpayers would claim this deduction in 1996. Those *not* making use of the deduction were either renters or owners who did not have mortgages or who used the standard deduction.

TABLE 27 Distribution, by Income Class, of Tax Returns and Benefits of Homeowner Deductions, 1995 Rates, Law, and Income Levels

Income Class (in 1,000s)	Mortgage Interest Deductions						Real Estate Tax Deduction				
		Returns		Amount		Average Benefit per Taxpayer	Returns		Amount		Average Benefit per Taxpayer
	Thousands of Taxable Returns	Thousands of Returns	Percentage of Taxable Returns	Millions of Dollars	Percentage of Tax Expenditures		Thousands of Returns	Percentage of Taxable Returns	Millions of Dollars	Percentage of Tax Expenditures	
Below $10	2,324	29	1.2	47	0.1	1,621	14	0.6	1	0.0	71
$10 to $20	9,538	420	4.4	173	0.3	412	358	3.8	45	0.3	126
Subtotal	11,862	449	3.8	220	0.4	490	372	3.1	46	0.3	124
$20 to $30	13,669	1,364	10.0	685	1.2	502	1,339	9.8	207	1.5	155
$30 to $40	14,202	2,661	18.7	1,919	3.3	721	2,628	18.5	526	3.7	200
$40 to $50	11,674	3,436	29.4	3,270	5.6	952	3,486	29.9	868	6.1	249
$50 to $75	17,566	8,516	48.5	11,005	18.9	1,292	8,712	49.6	2,895	20.3	332
Subtotal	57,111	15,977	28.0	16,879	28.9	1,056	16,165	28.3	4,496	31.6	278
$75 to $100	7,790	5,590	71.8	12,253	21.0	2,192	5,668	72.8	3,112	21.9	549
$100 to $200	5,817	4,540	78.0	16,359	28.0	3,603	4,812	82.7	4,032	28.3	838
$200 and over	1,565	1,293	82.6	12,624	21.6	9,763	1,275	81.5	2,543	17.9	1,995
Subtotal	15,172	11,423	75.3	41,236	70.7	3,610	11,755	77.5	9,687	68.1	824
Total	84,145	27,849	33.1	58,335	100.0	2,095	28,292	33.6	14,229	100.0	503

SOURCE: Derived from data in Joint Committee on Taxation, *Estimates of Federal Tax Expenditures for Fiscal Years 1996 to 2000*, U.S. Government Printing Office, 1995.

The deduction is clearly regressive. A taxpayer in the 15% bracket would benefit by a $150 tax reduction for each $1,000 in mortgage interest paid. However, fewer than a half million of the 12 million taxpayers with incomes below $20,000 (4%) even claim the deduction, and the average benefit of the deduction to these taxpayers is $490. A taxpayer in the 31% bracket would benefit by $310 for each $1,000 in interest and is far more likely to have a bigger mortgage. For example, the average benefit for taxpayers with incomes above $200,000 is $9,763, and 83% of them claim the deduction (see Table 27).

Taxpayers with incomes between $20,000 and $75,000 compose a majority (57.4%) of the beneficiaries of the mortgage interest deduction. However, the 16 million middle-income taxpayers taking this deduction receive only 30% of the benefits, whereas households with incomes above $75,000—the top 11% of taxpayers—receive 70% of the benefits.

State and Local Property Tax Deduction

The tax expenditure for allowing owners to deduct state and local property taxes on owner-occupied housing, including second homes, is the 10th largest of all federal tax expenditures. Like the mortgage interest deduction, this provision is a carryover from the Civil War tax law that was reenacted in 1913.

Although the amount is far smaller, the pattern of beneficiaries of the property tax deduction is similar to that of the mortgage interest deduction. Of middle-income taxpayers, 57% benefit from the deduction, receiving 32% of the benefits. Of the total expenditures, 68% benefit households with incomes above $75,000. According to the Joint Tax Committee estimates, using 1995 incomes and tax lev-els, some 39.7 million taxpayers had incomes below $40,000 and received $0.8 million of the benefits of the property tax deduction, whereas 1.6 million had incomes above $200,000 and received $2.5 billion in benefits. Beneficiaries with incomes between $40,000 and $200,000 received a total of $10.9 billion in benefits.

Deferral and Exclusion of Capital Gains on Sale of Home

Until the Internal Revenue Service Code was amended in 1990, taxpayers who sold their primary residence and used the proceeds to buy or build another home within two years could defer payment of a capital gains tax on the sale for as long as they continued to own their primary home. Taxpayers aged 55 or over at the time of sale could exclude up to $125,000 of gain from primary home sales from their taxes. These capital gains provisions carried a hefty cost to the Treasury, estimated for FY 1997 at $12.2 billion for the deferral and $12.5 billion for the exclusion.

Investor Deductions

The foregoing homeowner deductions were estimated by the Treasury to account for 79% of the $104.5 billion cost of housing tax expenditures in 1997. Yet neither Congress nor the executive branch has seriously considered changes in these deductions. Tax reform, as applied to housing tax expenditures, has instead focused on the less important provisions, totaling $23.6 billion, aimed at stimulating investment in new or rehabilitated housing or broadening opportunities for homeownership for families with incomes too low to benefit from the homeowner deductions.

The major investor provisions are the tax exemption for bonds issued by state and local agencies to finance low- and moderate-income housing construction and for the Low-

Income Housing Tax Credit, enacted in 1986 to replace less efficient provisions for accelerated depreciation of privately financed low-income housing. Unlike the homeowner deductions, which are open-ended entitlements, the major investor deductions available within a state are covered by per capita limits imposed by tax reform legislation in 1986. The volume cap for mortgage revenue bonds and rental housing bonds is combined with student loans and industrial development bonds and set at $50 per capita or a minimum of $150 million per state.

Mortgage revenue bonds began to be used for housing in the 1970s, but their use did not become widespread until the cuts in direct federal expenditures for subsidized housing in the 1980s. In 1976, the tax expenditure for housing bonds was $75 million. By 1980, it reached $445 million, jumped to $1.1 billion in 1981, and rose to a peak of $3.4 billion in 1987 when the volume cap took effect. The estimated cost of mortgage revenue bonds to the Treasury for fiscal 1997 was $2.7 billion, of which about one-third was for rental housing bonds.

Single-family mortgage revenue bonds may be issued to finance purchases by low- and moderate-income first-time homebuyers. States may also issue mortgage credit certificates to individual homebuyers for up to 25% of their annual allocations. A credit is much more valuable (and costly) than a deduction. A deduction reduces the amount of income that is taxed; a credit is subtracted from the amount of the tax itself. Housing prices are limited to 90% of the average area purchase price. Rental housing bonds may finance multifamily projects in which at least 20% of the units are reserved for families with incomes below 50% of area median income or in which at least 40% have incomes below 60% of median.

The Low-Income Housing Tax Credit, enacted in 1986 to replace special depreciation provisions for privately financed low-income rental housing, has become a major source of rental housing production. The credit to investors is structured to have a present value of 70% of the capital cost of the project over a 10-year period.

Developers or owners using the credit must, for at least 15 years, rent either 20% of the units to households with incomes below 50% of area median or 40% to households with incomes below 60% of area median. There is an additional set-aside of 15% of the low-income units to be available, at reduced rents, to households with incomes below 40% of median. Because the tax credit is available only for units earmarked for low-income occupancy, there is a significant incentive in the program for owners to go beyond the minimum targeting requirements. On the other hand, it is difficult in many areas to provide housing within the 30% of income rent restrictions without additional subsidies beyond the tax credit itself. Tax credits are allocated to state housing finance agencies at the rate of $1.25 per capita per year, and each state agency is responsible for administering the program under fairly broad guidelines set forth in the law. The 1997 tax expenditure for the Low-Income Housing Tax Credit was an estimated $2.3 billion.

Almost two decades ago, when their cost was far smaller than at present, an analysis of homeowner tax preferences made the following points: Because they provide tax savings for owners over tenants with comparable incomes and differential savings for owners with comparable incomes, they create horizontal inequities, and because they disproportionately benefit upper-income households, they create vertical inequities. Finally, they skew the housing market by increasing the demand for single-family homes and more expensive housing at the expense of rental housing and less expensive single-family housing for first-time buyers.

It appears that the cost and benefits of the housing tax expenditures, particularly the homeowner deductions, will be central to any future discussions of a flat tax. Some of the current flat tax proposals would eliminate them; others would preserve them. (SEE ALSO: *Capital Gain; Low-Income Housing Tax Credits; Tax Incentives; Taxation of Owner-Occupied Housing*)

—*Cushing N. Dolbeare*

Further Reading

Dolbeare, Cushing N. 1986. "How the Income Tax System Subsidizes Housing for the Affluent." In *Critical Perspectives on Housing*, edited by R. Bratt, C. Hartman, and A. Meyerson. Philadelphia: Temple University Press.

Hecht, Bennett L. 1994. *Developing Affordable Housing: A Practical Guide for Nonprofit Organizations*. New York: John Wiley.

Hellmuth, William. 1977. "Homeowner Preferences." In *Comprehensive Income Taxation*, edited by Joseph A. Pechman. Washington, DC: Brookings Institution.

Joint Committee on Taxation. 1995. *Estimates of Federal Tax Expenditures for Fiscal Years 1996-2000*. Washington, DC: Government Printing Office.

National Low Income Housing Coalition. 1995. *Working Paper on Federal Housing Trust Fund Proposal*. Washington, DC: Author.

U.S. Government Printing Office. 1998. *The Budget of the United States Government: Fiscal Year 1999*. CD-ROM. Washington, DC: Author.

Wood, Gavin A. 1988. "Housing Tax Expenditures in OECD Countries: Economic Impacts and Prospects for Reform." *Policy and Politics* 16(4):235-50. ◄

► Tax Incentives

Government, at any level, sometimes seeks to enhance the amount of investment being applied to some area of the economy. If the government sees that insufficient amounts of housing are being built, it may feel the need to cause more investment dollars to flow into housing development. The means through which the government causes these investment dollars to flow into the areas of need may vary. Government may provide a direct subsidy, or it may offer indirect incentives through the use of tax benefits. Direct subsidy is the expenditure of public funds to address a specific need, such as when the federal government gives funds to local housing authorities to run public housing. Tax incentives are elements of the tax laws that reduce the taxes owed by a taxpayer. This reduction comes as compensation for the taxpayer investing in certain types of projects deemed to be in the public interest. For example, a taxpayer who invests in affordable rental housing for low-income families may be able to claim the Low-Income-

Housing tax credit, reducing the amount of taxes otherwise owed to the government.

Many types of tax incentives are given in regard to housing, ranging from the commonplace to the esoteric. For example, a well-known and popular federal tax incentive is the one permitting each owner-occupant of a home to deduct, for income tax purposes, the interest payments on home purchase loans as well as property taxes paid on the home. A more esoteric form of tax incentive has been granted to savings and loan banks, provided that these banks make a certain percentage of their loans for home purchases.

However, despite the broad range of tax incentives that exist in housing at all levels of government, federal income tax incentives command the most attention within the housing investment community. These fall into three categories: deductions, capital gains treatment, and credits.

Deductions are among the most commonly recognized forms of tax incentives granted, especially by the federal government. It is a principle of federal taxation that taxes are not levied against the gross income of a taxpayer but against the income net of the costs of raising that income. This is true for income from housing as with all other forms of income. If a developer owns a rental property, taxes are paid on that amount of rental income that remains after deducting the costs of operating the property.

In calculating these costs that may be deducted, the developer may deduct all interest paid on a loan taken out to develop or purchase the property, plus any property taxes paid on the property, as well as other maintenance and operating expenses. In addition, for purposes of taxation, the property owner is assumed to have suffered some loss in the property's value. This loss, called depreciation, will frequently reduce a property's taxable income to zero. In fact, the property may generate more depreciation than is necessary to eliminate all federal income tax liability, making it possible to sell the surplus depreciation through syndication.

The federal government frequently alters the method through which the amount of depreciation is calculated. When the government feels it needs to encourage greater investment in housing, it may alter the calculation process, permitting greater depreciation deductions. An increase in the deductions lowers the taxes owed on such investment, thereby raising the return on investment and making that investment more attractive.

Of course, the government may do the reverse, lowering the amount of deductions that can be claimed so as to discourage further investment. Overly generous depreciation deductions were blamed for much of the excessive investment in real estate that occurred during the late 1980s. It has been claimed that these tax incentives encouraged investors to develop housing that did not provide a good return on investment, but given the tax incentives, the housing returned a good after-tax profit. As a result, too much housing was built, depressing the value of all housing in the marketplace. A dramatic reduction in these deductions was implemented through the Tax Reform Act of 1986. This act brought to an end the excessive flows of capital into real estate and, with it, the overbuilding.

Homeowners benefit from a special version of these deductions. A household that owns and occupies its own home is allowed to deduct property taxes and interest paid on loans to purchase the home similar to the deductions granted to owners of rental property. Unlike the owners of rental property, however, a homeowner is not taxed on the rental income value of its home. As such, homeowners are allowed to claim deductions against income that they do not have to claim, which has the effect of lowering the taxes paid by some homeowners. This deduction has been criticized extensively by economists as being a poorly designed tax incentive. These special deductions are of value only to those households who have enough total deductions to exceed the amount of the standard deduction granted to all individual taxpayers. Furthermore, those taxpayers in higher-tax brackets will enjoy a greater reduction in taxes for each dollar of deduction than will a household in a lower tax bracket. As such, these special deductions tend to be available only to middle- and upper-income households and to favor the rich over the less well-off. These tax incentives have also been blamed for causing households to invest more of their money into housing than they really need, draining valuable investment resources from other sectors of the economy. Despite these criticisms, these special homeowner deductions have retained their political popularity over time and in 1997 alone were estimated to cost the federal government more than $65 billion in foregone tax revenue.

A second major form of tax incentives given to housing is found in the preferential treatment of capital gains for purposes of taxation. Capital gains are simply the profits made from the acquisition and later sale of a capital good, such as stocks or a piece of real estate. This profit is generally defined as the difference between the selling price of the capital good (after deducting selling costs such as brokerage fees) and the purchase price.

Investment analysts argue that taxation of this profit in the same manner as other income is unfair given that it may take many years to generate this profit. As such, the profit obtained from buying and selling a home or an apartment complex may reflect the inflationary change in its price, rather than a real gain in its value. To compensate for this inflation effect, the government frequently exempts some fraction of capital gains from taxation and applies the capital gains tax rate only against the remainder. When the fraction that is exempt from taxation is large, capital gains are given preferential treatment relative to other investments that generate ordinary income that is taxable each year. This causes investment to flow into capital goods such as housing. When the fraction that is exempt is low or zero, investment tends to flow away from capital goods such as housing and into other investments generating ordinary income. The Tax Reform Act of 1986 reduced the exemption to zero, discouraging investment in housing and other capital goods. However in 1991, the federal government set the maximum capital gains tax rate at 28% even if the taxpayer is in a higher tax bracket, and proposals for further adjustments to the taxation of capital gains are frequently debated in Congress. In 1997, the capital gains tax rate was further reduced to 20%.

Here again, homeowners enjoy a special benefit. If a homeowner buys a home and later sells that home, any profit on that home is a capital gain and would normally be subject to taxation. Tax law allows taxpayers to exclude up to $250,000 of gain from the sale of their home ($500,000 for married couples) provided that the taxpayer used the home as a principal residence for a period of at least two of the five years prior to the sale. This tax incentive has been supported as a means to encourage households to save for retirement. It has also been criticized for encouraging excessive investment in housing and for benefiting those areas of the nation that enjoy rapid house price appreciation over those that do not. As with the homeowner deductions, this capital gains treatment of owner-occupied housing is politically popular, and the exclusions and deferred provisions were estimated to cost the federal government about $23 billion in 1997.

Tax credits are yet another way that the federal government uses the tax codes to encourage investment. Tax credits are an especially simple method to encourage investment. Unlike deductions and capital gains provisions of the tax law, credits permit recipients to reduce their taxes by the exact amount of the credit. Two credits have been especially important to housing in recent years, the Historic Rehabilitation Tax Credit and the Low-Income Housing Tax Credit.

The Historic Rehabilitation Tax Credit is granted to an investor who rehabilitates income-earning properties of significant historical importance. The credit amount has varied in the past, but it is generally 20% of the rehabilitation costs. The Low-Income Housing Tax Credit is similar. It grants tax credits in varying amounts against the costs of acquiring and building apartments for low- or moderate-income occupancy. The credits are granted for the first 10 years of what must be at least 15 years of occupancy by low- or moderate-income households. Both of these credits can be claimed only after the construction is completed and the occupancy begun. However, developers can translate the credits into money to help finance the project through the process of syndication. The two credits have been instrumental in increasing the flow of investment dollars to these two types of housing development. During its peak years of the early 1980s, the Historic Rehabilitation tax credit leveraged about $800 million of investment in more than 1,300 projects per year. During the late 1980s and early 1990s, the Low-Income Housing Tax Credit brought about the development of about 100,000 apartments per year.

Tax incentives have generally been recognized has an important tool used by government to stimulate investment in various types of housing. These tax incentives have been the subject of much debate by economists and policy analysts as to both the efficiency and the equity issues involved. Government prefers tax incentives to direct subsidies because they are generally easier to administer than direct subsidies. Over time, tax incentives have proven to be a relatively simple means to encourage investment in housing. However, the beneficiaries of this investment have not always been those who were intended to benefit, the dollars foregone by the government have not always been in the amounts expected, and the need for these tax incentives has not always been clear. However, these incentives endure even though the Tax Reform Act of 1986 reduced many of them in scale. It seems likely that tax incentives will continue to play a major role in guiding the amount of investment that goes into housing. (SEE ALSO: *Depreciation Allowance for Landlords; Low-Income Housing Tax Credits; Property Tax Abatement; Tax Expenditures; Taxation of Owner-Occupied Housing*)

—*Kirk McClure*

Further Reading

Dowall, David E. 1990. "The Public Real Estate Development Process." *Journal of the American Planning Association* 56:504-12.

Downs, Anthony. 1985. *The Revolution in Real Estate Finance.* Washington, DC: Brookings Institution.

"Financing." 1992. *Economic Development Review* 10(Spring):4-73.

Major, John B. and Fung-Shine Pan, eds. 1996. *Contemporary Real Estate Finance: Selected Readings.* Upper Saddle River, NJ: Prentice Hall.

McClure, Kirk. 1990. "Low and Moderate Income Housing Tax Credits: Calculating Their Value." *Journal of the American Planning Association* 56:363-69

National Council of State Housing Agencies. 1989. *The Low-Income Housing Tax Credit in the 1990s.* Washington, DC: Author.

Peiser, Richard B. 1992. *Professional Real Estate Development: The ULI Guide to the Business.* Washington, DC: Urban Land Institute.

Phyrr, Stephen A., James R. Cooper, Larry E. Wofford, Steven D. Kaplin, and Paul D. Lapides. 1989. *Real Estate Investment: Strategy, Analysis, Decisions.* New York: John Wiley.

Pryke, M., ed. 1994. "Property and Finance." *Environment and Planning A* 26(Special issue):167-264.

Wood, Gavin A. 1990. "The Tax Treatment of Housing: Economic Issues and Reform Measures." *Urban Studies* 27:809-30. ◄

► Tax Increment Financing

In more than 30 states, local governments are permitted to establish tax increment financing (TIF) districts in which the *increment* in property values associated with revitalization activity may be recycled into the district rather than going into the general funds of municipal government and county government. TIF funds are frequently used as a source of payback for revenue bonds used to finance basic infrastructure improvements in a revitalization area. TIF has been generally used as a substitute for the federal Urban Renewal Program, which existed from 1949 to 1974. Consistent with the Urban Renewal Program, TIF districts are typically established in areas defined by state statute as being slums or blighted, and TIF funds are generally used to fund activities having to do with urban renewal. TIF funds have been used for affordable housing and in some instances have been used to make direct loans or grants for affordable housing rather than used as the source of payment for revenue bonds.

In California, for example, state statute requires that community redevelopment agencies set aside 20% of tax increment funds for low- and moderate-income housing. Los Angeles has used TIF funds for the development of affordable housing and for the operating expenses associated with providing social services in single-room occupancy housing. Other jurisdictions have also used TIF funds for funding affordable housing for low-income households. This works particularly well when the TIF district includes commercial areas, such as a downtown, that undergo revitalization, thereby increasing property values and the associated tax increment.

The chief advantage of TIF is that it allows governments to raise money for affordable housing and other redevelopment purposes without having to allocate funds from existing taxes and without having to raise taxes. Instead, funds come from the increase in property values associated with the initial stages of revitalization activity. Of course, if such revitalization activity does not lead to increased property values, there is no increment and no new funds are generated. But if successful revitalization does take place, the associated increase in property values can be recycled back into the TIF district and used to finance further revitalization. In this way, TIF can be a self-reinforcing process.

The chief disadvantages of TIF are twofold: First, if revitalization does not result in higher property values, no increment is created. This can be especially harmful if local government borrows against the expected flow of TIF funds. The second disadvantage is that other units of government, aside from the municipal jurisdiction that enacts the TIF, may be bothered that the increase in property tax revenues does not flow into their treasuries. Although it can be argued that if no revitalization occurred, these jurisdictions would also not receive any more funds, school districts and county governments looking at a successful TIF will wonder why they cannot enjoy the benefits associated with the increased property tax revenues. (SEE ALSO: **Housing Finance**)

—*Charles E. Connerly*

Further Reading

Bloch, Susan. 1988. *Tax Increment Financing: A Tool for Community Development.* Washington, DC: Neighborhood Reinvestment Corporation.

Huddleston, Jack R. 1986. "Distribution of Development Costs under Tax Increment Financing." *Journal of the American Planning Association* 52:194-98.

Klemanski, John S. 1989. "Tax Increment Financing: Public Funding for Private Economic Development Projects." *Policy Studies Journal* 17:656-71.

"Financing." 1992. *Economic Development Review* 10(Spring):4-73.

Pryke, M., ed. 1994. "Property and Finance." *Environment and Planning A* 26(Special issue):167-264. ◀

▶ Tax Reform Act of 1986

The Tax Reform Act of 1986 (P.L. 99-514) instituted a new tax credit for owners of low-income rental housing to encourage the creation of housing for low-income people. The tax credit was contained in Title II of the landmark tax legislation adopted that year, which is best known for reducing tax rates and curtailing many special tax breaks.

The tax credit is 9% a year of the value of the units occupied by low-income tenants (not receiving other fed-

eral subsidies) for a period of 10 years in projects in which a sufficient number of people have qualifying incomes (either 20% of the units are occupied by people with incomes under area median income or 40% of the units are occupied by people earning below 60% of the area median income). If tenants receive other federal subsidies, the tax credit would be 4%. Newly constructed, purchased, or rehabilitated units would be eligible for this credit. If the use of the units is changed within 15 years, the credit has to be repaid along with penalties.

Making the new tax credit even more attractive was a provision in the Omnibus Budget Reconciliation Act of 1986 (P.L. 99-509) that allowed investors to claim as part of their investment the loans they had received to buy low-income housing—instead of being able to claim only those funds of their own that they had invested. This provision increased the tax credit that investors received. (SEE ALSO: *Tax Incentives*)

—*Nathan H. Schwartz*

Further Reading

Congressional Information Service. 1987. *CIS Annual 1986: Legislative Histories.* Washington, DC: Author.

Congressional Quarterly. 1987. *Congressional Quarterly Almanac: 99th Congress, 2nd Session, 1986.* Vol. 42. Washington, DC: Congressional Quarterly Service. ◀

▶ Taxation of Owner-Occupied Housing

Owner-occupied housing is accorded unusual and special treatment under the tax laws of many nations. Except in a few countries (e.g., Scandinavian countries), no attempt is made to tax the annual return on investment in owner occupancy. In many countries (e.g., Australia), the capital gains arising from housing investment are untaxed. In other countries (e.g., Germany), the depreciation of owner-occupied housing is considered a deductible personal expense in computing tax liability. The tax benefits given to homeowners in the United States are among the most generous in the developed world.

Consider an individual who chooses between an investment in owner-occupied housing and an equivalent investment of some other form, say, in common stocks. The investment in owner-occupied housing offers three kinds of tax advantages. First, under the U.S. Internal Revenue Code, the returns on investment in owner-occupied housing are untaxed. In contrast, the dividends yielded by common stock are reported as income and taxed in the year accrued. Second, any capital gains arising from the investment can be deferred. Moreover, a large capital gains exclusion is available to those over the age of 55. In contrast, capital gains in the stock market are taxed in the year they are realized. Third, some of the expenses associated with homeownership, notably property taxes and interest payments, can be itemized as deductions in computing federal tax liability under the personal income tax. No other interest payments are deductible as personal expenses under the Internal Revenue Code. This favorable treatment also extends to personal income taxation under the laws of all of the 50 states.

The net effect of these provisions of the U.S. tax law is to reduce the price of homeownership, relative to renting, by a sizable amount. Moreover, as a result of these policies, the relative price of homeownership varies by income level and the level of inflation.

It is useful to think of the annual cost of homeownership as the cost of using the stock of residential capital. Under standard simplified conditions, the user cost of residential capital (in equilibrium, the annual rent $R1$ of the stock) will be related to the value of the stock, V, according to

$$R1 = iV, \qquad [1]$$

where i is the rate of interest. In equilibrium, the annual rent generated will be equal to the annual cost of owning the capital stock. Housing is subject to an annual property tax, at an effective rate t. Annual expenditures of $100d\%$ of house value are required for maintenance and to offset depreciation. In addition, the owner can expect capital gains of $100g\%$.

This means that in the absence of income tax considerations, the user cost relationship is

$$R2 = (i + t + d - g)V, \qquad [2]$$

where the term in parentheses is the user cost of residential capital. This is the cost of holding the stock for one year. Note that, in the absence of income taxes, the user cost does not vary with the level of inflation. Suppose, for example, interest rates are $i^* + a$ and capital gains rates are $g^* + a$, where the asterisks represent real values and a is the level of inflation. On substitution into Equation 2,

$$R3 = ([i^* + a] + t - d - [g^* - a])V, \qquad [3]$$

the a's simply cancel each other out, and $R3 = R2$.

Now suppose nominal capital gains are untaxed, mortgage interest payments are deductible from gross income, and property taxes are similarly deductible. Suppose net income is taxed at a marginal rate of $T\%$. Under these circumstances, the expression relating equilibrium rents to values is

$$R4 = ([i^* + a] [1 - T] + t[1 - T] + d - g^* - a)V. \qquad [4]$$

The expression in parentheses is the net after-tax user cost of residential capital. First, note that inflation is no longer neutral. The asymmetry between the tax treatment of interest payments and capital gains means that the after-tax cost of homeownership varies inversely with the level of inflation in the economy. Note also that the net cost of homeownership declines with the tax rate on income and the value of the house:

$$R4 = R3 - T(i^* + a + t)V. \qquad [5]$$

If federal tax rates increase with income or if higher-income households live in jurisdictions with higher prop-erty tax rates, the cost of homeownership declines with income. More important, as long as housing is a normal good with a positive income elasticity, the net cost of homeownership declines with income. Furthermore, a given level of inflation in the economy reduces the user cost more for higher-income owners than for lower-income homeowners.

More generally, the analysis shows that the costs of homeownership are sensitive to macroeconomic stabilization policies and to the structure of income tax rates. As the marginal tax rates of the highest-income U.S. households fell from 70% to 30% and then rose to 40% during the course of the decade 1983 to 1993, although the inflation rate plummeted from 15% to 5%, the implicit policy toward housing and homeownership varied substantially.

For example, at reasonable values of the variables in Equation 4 (say, $i^* = g = 3\%$, $t = d = 2\%$, $T = 30\%$), then as inflation goes from 6% to 1%, the after-tax user cost of residential capital roughly doubles. Similarly, at reasonable values of the variables (for example, $a = 3\%$ and, as before, $i^* = g = 3\%$, $t = d = 2\%$), then as income tax rates go from 40% to 20%, the after-tax cost of owner occupancy increases by more than one-third. These are substantial price changes induced entirely by taxation and macroeconomic considerations. There are a number of predictable effects of this subsidy to owner-occupants.

First, the fluctuations in tax rates can be expected to induce changes in the quantity of housing consumed by households. Reductions in the user cost of owner occupancy can be expected to increase consumption. Second, changes in the relative price of owning versus renting can be expected to affect the homeownership propensities of households. Third, the magnitude of the implicit subsidy, and its distribution across income classes, is sensitive to tax policy. Finally, the social loss to society from mispricing owner-occupied housing is sensitive to the extent of mispricing.

Econometric research suggests that the demand for housing is moderately price inelastic. A price elasticity of −0.6 or −0.7 is reasonable. Based on these numbers, a 15% change in user cost arising from tax rate policy or stabilization policy (well within the changes in the past decade) could lead to a change of roughly 10% in housing consumption.

In the United States at least, it appears that homeownership probabilities are more responsive to income and household demographics than to the relative price of owning versus renting. In fact, it has been estimated that the entire subsidy to owner-occupied housing that arises through the tax treatment of housing increases homeownership rates by only a couple of percentage points. The magnitude of the implicit subsidy arising through the personal income tax code is large and extremely regressive. The subsidy is available only to owners, who are usually more affluent than renters, and only to those who find it advantageous to itemize their deductions. The propensity to itemize deductions increases with income. Finally, as noted above, for those owners who do itemize deductions, the magnitude of the subsidy increases with income.

The absolute value of the revenues foregone by the federal treasury as a result of the tax treatment of owner-occupied housing are routinely estimated by the Congressional Budget Office and the Joint Committee on Taxation of the Congress. The revenue costs of these tax subsidies are large, and more than half of the revenue losses accrue to the top 15% of the income distribution. A more relevant benchmark of the costs of the tax treatment of homeownership may be in comparison with other government housing programs (whose principal beneficiaries are lower-income households). When compared with other housing assistance programs (principally public housing, low- and moderate-income subsidies, housing rehabilitation, etc.), the value of homeowner subsidies is large—more than the entire budget of the Department of Housing and Urban Development, for example. The social cost of the disparities introduced by favorable tax treatment of owner-occupied housing is even larger than the budgetary costs to the government—because the mispricing of housing assets entails an additional deadweight loss to the economy.

Proposals to change the combination of deductions and exclusions that characterize the tax treatment of housing are routinely advanced, but they do not receive strong political support. Some changes have been wrought by imposing a cap on interest deductions (for second homes). Such a cap could easily be extended to principal residences. Similarly, it would be rather easy to include imputed rent as taxable income, using rules of thumb in much the same way that depreciation is estimated for business property. Moreover, it would be easy, as an administrative matter, to structure the cap or the imputed rental adjustment so as to affect tax liability only at the high end of the income distribution. (SEE ALSO: **Homeownership;** *Imputed Rental Income; Tax Expenditures; Tax Incentives*)

—*John M. Quigley*

Further Reading

Hanushek, Eric and John M. Quigley. 1980. "What Is the Price Elasticity of Housing Demand?" *Review of Economics and Statistics* 42:449-54.

Jorgenson, Dale. 1971. "Econometric Studies of Investment Behavior: A Survey." *Journal of Economic Literature* 9:1111-47.

Rosen, Harvey S. 1985. "Housing Subsidies: Effects on Housing Decisions, Efficiency, and Equity." Pp. 375-420 in *Handbook of Public Economics,* edited by Alan J. Auerbach and Martin Feddstein. Amsterdam: North Holland.

Rosen, Harvey S., Kenneth T. Rosen, and Douglas Holtz-Eakin. 1984. "Housing Tenure, Uncertainty, and Taxation." *Review of Economics and Statistics* 46:102-19. ◄

▶ Temporarily Obsolete Abandoned Derelict Sites

The name temporarily obsolete abandoned derelict sites (TOADS) appeared in 1990 to describe closed, boarded-up, or decaying housing projects, factories, warehouses, schools, dump sites, mines, railroad lines, canals, waterfronts, and tracts of overgrown undeveloped land. TOADS are usually superannuated (locally unwanted land uses, or LULUs), often literally burned-out sites. TOADS have hit bottom on the land use cycle. Without some form of treatment or reclamation, they will not rise on that cycle.

Meanwhile, they become eyesores or worse—the sites of fire hazards, drug trades, illegal garbage and toxic dumping, and other dangers to public safety and the environment.

TOADS show up in cities, suburbs, small towns, and rural areas but especially in inner-city neighborhoods. Studies of the 15 largest U.S. cities and of medium-sized cities in New Jersey demonstrate that TOADS are economic pariahs: They produce no revenues, lower nearby property values, create public costs, and are expensive to secure and clean up.

TOADS are often so repellent that frightened individuals, families, and business owners leave their vicinity, producing further abandonment and more TOADS and LULUs. Those who remain in TOADS neighborhoods are almost always the poor and elderly who cannot afford to leave. These marginalized people suffer from extraordinarily high rates of family and street violence, drug abuse, infant mortality, AIDS, and other illnesses and injuries.

Historically, TOADS have appeared in the wake of economic changes that resulted in bankruptcies and closings—for example, of tanning plants, slaughterhouses, shipyards, or railroad lines. More recently, large numbers of big-city companies, such as those in the textile, steel, and pharmaceutical industries, have moved to the suburbs, outside the region, or abroad. The resulting closings and abandonment have created large numbers of TOADS. Urban renewal programs, gentrification, and the ongoing siting of LULUs also have often produced abandonment in surrounding areas—and thus, more TOADS.

The TOADS problem appears to be worsening in the United States and other industrialized societies. Economic stress means that resources to redevelop TOADS are often not available. TOADS are unwanted, except as sites to dump pariah land uses and people, and they do not have a powerful public to promote their rehabilitation. In the United States in particular, TOADS neighborhoods exemplify the urban malaise of the 1990s. (SEE ALSO: *Brownfields; Environmental Contamination: Toxic Waste; Infill Housing; Locally Unwanted Land Uses; Not in My Back Yard*)

—*Michael R. Greenberg and Frank J. Popper*

Further Reading

Greenberg, Michael R., Frank J. Popper, and Bernadette M. West. 1990. "The TOADS: A New American Urban Epidemic." *Urban Affairs Quarterly* 3:435-54.

Greenberg, Michael R., Frank J. Popper, Bernadette M. West, and Dona Schneider. 1992. "TOADS Go to New Jersey: Implications for Land Use and Public Health in Mid-Sized and Large U.S. Cities." *Urban Studies* 1:117-25.

Greenberg, Michael R., Frank J. Popper, Dona Schneider, and Bernadette M. West. 1993. "Community Organizing to Prevent TOADS in the United States." *Community Development Journal* 1(January):55-65.

Greenberg, Michael R. and Frank J. Popper. 1994. "Finding Treasure in TOADS." *Planning* 4(April):24-28.

Wallace, Roderick. 1988. "A Synergism of Plagues: Planned Shrinkage, Contagious Housing Destruction and AIDS in the Bronx." *Environmental Research* 47:1-33. ◄

► Tenant Organizing in the United States, History of

If there are a significant number of landlords and tenants in a society, there will be conflict between the two, leading to the formation of tenant organizations. This is certainly true for the United States. Early settlers came to this country in an attempt to escape the institution of landlordism in Europe. They did not succeed. Landlords came too. As a result, landlord and tenant disputes are documented as far back as the mid-18th century in the United States. In the colonial period, this resistance took the form of groups of tenants attempting to forcibly prevent evictions. If unsuccessful on the spot, the groups returned at a later day, ejected whomever had taken over the property, and returned it to the original tenant. Those who resisted often later found their homes dismantled and strewn about the countryside in pieces.

One can find articles in colonial papers before the American Revolution complaining about rent-gouging landlords, and by the time of the Revolution, tenants took their stand in the Sons of Liberty. Tenants in Philadelphia especially looked forward to the day when they would be able to elect their own to office to prevent landlords from passing laws that made it so easy to evict a tenant. Tenants had assumed that after the Revolution they would have the right to vote. (Before that, one had to be a landowner to vote.) They responded by organizing to gain the right to vote, demanding universal male suffrage.

The fight for suffrage came to head in New York in 1821. Landlords argued that among landowners one could "always expect to find moderation, frugality, order, honesty, and a due sense of independence, liberty, and justice." Among the tenants were a "motley and undefinable population, the idle and profligate." The forces supporting tenant suffrage responded that the root problems of society could be found not in the masses but in the "wealth of aristocracy bearing down on the people." They found no proof that ownership of land led to "elevation of mind," gave "stability to independence," or added "wisdom to virtue" (Carter and Stone 1821).

The debate in New York was peaceful. This was not true everywhere. In one of the more extraordinary chapters in United States history, a tenant army fought for the right to suffrage in what is known as the Dorr War. Tenants denied the vote staged their own election of candidates and passed a constitutional amendment to give them the vote. They outpolled the voting property owners who voted at the official state election and sought to establish a government in the state. Federal troops put down the uprising.

White male tenants obtained the vote in all the states in this period, but this did not end the rest of the landlord-tenant conflicts. One of the most colorful outbreaks of this inevitable struggle was in the Hudson River Valley in the middle of the 19th century. The 20-year long effort is known as the Anti-Rent Movement. A battle raged on between the landlords (most of whom had gained their 2 million acres of land by grant from the King of England) and some 300,000 tenants. To the tenants, many of whom were descendants of veterans of the Revolutionary War, they were fighting the second revolution. The tenant families had been paying rent on this land for 60 years, and they thought that was enough. They were champions of equal rights or "down rent" fighting the aristocratic landowners. As they stated, "Honor, justice and humility forbid that we should any longer tamely surrender that freedom which we have so freely inherited from our gallant ancestors . . . We will take up the ball of the Revolution where our fathers stopped it and roll it to the final consummation of freedom and independence of the masses" (Christman 1945, p. 20). The tenants formed "Indian" bands modeled after their forefathers who put on the Boston Tea Party. They wore calico costumes and masks to hide their identity when they appeared in public. Their leader was a country doctor known as "Big Thunder" who organized as he made his rounds on the back roads. The anti-rent Indians tarred and feathered sheriffs who sought to evict rent striking tenants and removed them from the county.

As in any movement worth its salt, they made up parodies to popular songs of the day to express their position. A song based on "Oh Dear, What Can the Matter Be" was titled "The Landlord's Lament." One verse went as follows:

I used to get rich through the poor toiling tenant,
And I spent all their earnings in pleasures satanic,
But now I confess I'm in a great panic,
Because I can get no more rent!

The antirenters held political conventions to endorse candidates for office who would bring about the reforms to break up the old manorial estates and force the conveyance of the land to the tenants. Although the Indian bands were eventually broken by force, the movement was successful. Taxes were levied that took much of the profit out of this landlording, and a great deal of the property was sold off to the tenants.

As the United States shifted from an agrarian to an urban society, landlord-tenant struggles became more common in the cities. The Panic of 1837 brought on demonstrations for which the cry was, "Bread, Meat, Rent, Fuel! Their Prices Must Come Down!" In 1848, during the Anti-Rent Movement, tenants in New York City formed the "Tenant League." They called for rent regulation, eviction protection, and the construction of better-quality affordable housing. The demands of the Tenant League reappeared periodically in the later half of the century and eventually led to the adoption of the first housing codes in the state in 1865.

Periods of economic instability were common in the 19th century. During each panic or depression, the pattern of 1837 was repeated. The depression of 1874, for example, resulted in mass evictions, 90,000 homeless people, and protest demonstrations. Tenant organizing came to a climax in New York when the labor party ran Henry George for Mayor on a pro-tenant platform. George lost.

The story continues in the 20th century with peaks after the turn of the century, during both world wars, the Great Depression, the 1960s, and continuing into the current period. Mass rent strikes took place in New York City in

1904 and 1908. In 1904, the New York Rent Protective Association was formed with 1,000 members, but it soon foundered on issues of politics. Some of the group felt the organization could not succeed with "an avowedly political base," whereas others stated, "We are not here as Socialists nor as labor unionists but as tenants" (Weissman 1986, p. 43). The 1908 tenant rebellion was thought to be "better and more tightly organized" (p. 45). However, although victories were won, as in 1904, the renters' organizing lasted only a short time and "disappeared as quickly as it had first appeared" (p. 47).

The "rent" problem grew only worse with the onslaught of World War I. The "lack of construction, increased demand for apartments, and rampant speculation in New York's tenement neighborhoods led to increasing landlord-tenant bitterness during the last years of the war" (Spenser 1986, p. 51). As maintenance problems, rent gouging, and evictions grew, so did the ranks of New York tenant leagues (estimated at 25,000 members). Similar conditions in many other parts of the country also spurred tenant activism. With rent strikes growing in number and size, elected officials in New York moved to cut off the agitation. The result, in 1920, was New York's first experience with rent control (the Emergency Rent Laws). In the rest of the United States, rent mediation was employed. With this legislation in New York came tenant organizations that focused on the administration and maintenance of the rent laws rather than on further organizing or agitation.

By 1929, much of the influence of tenant organizations had waned. In 1929, the Emergency Rent Laws expired. There was little protest. It did not take long, however, for tenant activism to return with more ferocity than ever before. The Great Depression re-created intolerable conditions for tenants. Tenant incomes dropped faster than rents. The inevitable result was a growing number of evictions. This time, the Communist Unemployed Councils came to many tenants' aid. In the early 1930s, they first physically resisted evictions with crowds sometimes reaching 3,000 to 5,000 people. They also organized mass rent strikes. The Communist tactics met with significant police resistance, and by the mid-1930s, their influence had declined to be replaced by large citywide coalitions and organizations more akin to those in the tenant movement today that seek legislative reform as the outcome of their organizing. The goal of these groups was to bring back rent control, enforce the codes, and have the government begin to provide housing for its citizens where the market would or could not.

The drive for local rent control was suspended when, during World War II, the federal government brought rent control to the entire United States. It began again after the war when the federal government withdrew. Although tenants in many parts of the country resisted the end of rent control, they succeeded only in New York. The success in New York did not resolve all housing problems for tenants there, or in the rest of the country, and the struggle between landlords and tenants continued. After the war, however, the government became a more active player with the mass dislocations caused by urban renewal and the issues associated with public housing. Anti-displacement and public housing tenant organizing followed.

The issue of housing conditions was central to the major early Harlem rent strikes of the 1960s that led to anti-slum reforms in New York. The issue of then-unregulated rents that remained outside New York spawned movements across the country later in the decade. In the late 1960s and early 1970s, these movements brought local rent control to many communities. None of these movements was more significant than that which brought rent control to more than 110 cities in New Jersey, leading to the formation of the New Jersey Tenants Organization and the creation of a national tenant magazine, *Shelterforce,* which gave voice to tenant concerns and helped increase awareness of tenant issues across the United States. These efforts also helped spread the warranty of habitability to many cities nationwide.

In the late 1970s and early 1980s, tenant activism and rent control played a major part in progressive reform in a number of smaller cities across the country, including the city of Santa Monica in California. A tenant-led coalition, spurred by the passage of statewide Proposition 13, became a majority in the early 1980s with the mission of preserving the rent control law previously passed by the voters. The proponents of Proposition 13, a tax limitation initiative, had promised its passage would lead to rent relief. When it did not, tenants across the state revolted, causing spontaneous organizing and making tenants a political force in the state. Local rent control and landlord tenant law reform were major beneficiaries of the movement.

In New York, fights to maintain rent control and attacks on the slums continued. The tenant movement was also heavily affected by an increasing housing abandonment problem. Tenants in several places in New York took steps to take over the operation of their housing and form tenant cooperatives, a movement that has spread since its New York beginnings. Jack Kemp, secretary of the Department of Housing and Urban Devlopment at the time, supported the organization of tenants into resident management corporations and sought to sell the stock to the tenants. The interest in transferring ownership has dwindled, but resident organizing and an interest in greater participation in management continue. (SEE ALSO: *Eviction; National Tenant Union; Private Rental Section; Rent Strikes; Resident Management*)

—*Allen David Heskin*

Further Reading

Baar, K. K. 1977. "Rent Control in the 1970s: The Case of the New Jersey Tenants' Movement." *Hastings Law Journal* 28:631.

Carter, N. H. and W. L. Stone. 1821. *Reports of the Proceedings and Debates of the Convention of 1821.* New York: E. & E. Hosford.

Christman, H. 1945. *Tin Horns and Calico.* Cornwallville, NY: Hope Farm Press.

Heskin, A. D. 1983. *Tenants and the American Dream: Ideology and the Tenant Movement.* New York: Praeger.

———. 1991. *The Struggle for Community.* Boulder, CO: Westview.

Lawson, R. with M. Naison, eds. 1986. *The Tenant Movement in New York City, 1904-1984.* New Brunswick, NJ: Rutgers University Press.

Leavitt, J. and S. Saegert. 1990. *From Abandonment to Hope.* New York: Columbia University Press.

Naison, M. 1972. "Rent Strikes in New York City." In *Tenants and the Urban Housing Crisis,* edited by S. Burghardt. Dexter, MI: New Press.

Spenser, J. 1986. "New York City Organizations and the Post-World War I Housing Crisis." In *The Tenant Movement in New York City, 1904-1984,* edited by R. Lawson with M. Naison. New Brunswick, NJ: Rutgers University Press.

Weissman, J. J. 1986. "The Landlord as Czar: Pre-World War I Tenant Activism." In *The Tenant Movement in New York City, 1904-1984,* edited by R. Lawson with M. Naison. New Brunswick, NJ: Rutgers University Press.

Williamson, C. 1960. *American Suffrage from Property to Democracy 1760-1860.* Princeton, NJ: Princeton University Press. ◄

► Tenement House Law of 1867

The first comprehensive housing law in the United States, this act, officially titled "An Act for the Regulation of Tenement and Lodging Houses in the Cities of New York and Brooklyn," is significant because it offered the first legislative definition of a tenement, set minimum standards for acceptable habitation, and established a means of enforcing its provisions. This law worked in conjunction with the cities' building codes to regulate the construction and maintenance standards for multiple dwellings.

State legislators passed the law in direct response to the prevalence of unsanitary and unsafe housing in New York City. Since the beginning of the 19th century, New York's population had grown exponentially, exploding from about 60,000 in 1800 to just under 1 million 60 years later. The city's built-up area, defined largely by the limits of contemporary transportation, also expanded but not nearly enough to accommodate the population growth without severe congestion. By midcentury, the city's northern boundary ran to 14th Street with less dense development extending to 42nd Street. About half the city's population lived in multiple dwellings. Either newly constructed buildings that covered up to 90% of the lot or converted single-family houses, commercial, or industrial structures, most tenements were characterized by extreme overcrowding, poor sanitation, and an absence of light and air. Often, landowners maximized their land value by constructing multistory front and back tenements, leaving a narrow strip between for privies, tap water, and garbage. Health reports of the time are filled with accounts of large families (and their boarders) squeezed into one or two rooms totaling under 200 square feet where they cooked, slept, and worked. At the passage of the Tenement Act of 1867, the city contained more than 15,000 tenements housing about a half million people, many in the 20,000-unit inventory of damp, dark, and airless cellar dwellings.

Reformist groups, such as the New York Association for Improving the Conditions of the Poor (founded in 1843) and the Citizen's Association Council of Hygiene and Public Health whose 1865 report documented the execrable conditions of tenement housing, pushed the lawmakers into action. The elected officials were also concerned about health—the city had suffered two cholera epidemics with more than 7,500 deaths in the previous decade—and public

safety—the draft riot of 1863 originated in the slum districts.

The provisions of the law contained minimum standards for new and existing dwellings. For example, all bedrooms had to have a window, all units had to have access to a fire escape, roofs were to be in good repair, stairwells were to have banisters, and there was to be one privy for every 20 tenants. Cellar dwellings were by permit only, and these units had to meet specific standards for light and ventilation. All infectious disease was to be reported to the Metropolitan Board of Health, and certain levels of cleanliness were to be maintained.

Overseeing the enforcement of these rules was the Metropolitan Board of Health, created the year before passage of the Tenement House Act. With jurisdiction over five counties, including New York and Kings (Brooklyn), the board's nine commissioners appointed a sanitary superintendent and employed 15 inspectors.

Although this legislation was clearly an important pioneering effort in its time, in actuality, it was quite weak, filled with loopholes and unenforceable by a small staff. First, the definition of a tenement included only those buildings with more than three families. A significant number of substandard dwellings were in tenements of three or fewer families. Lot coverage of these buildings was also not regulated other than specifying a 10-foot minimum space between front and back coverage. Furthermore, the standards were not so strict as they appeared. For example, the fire escape provision could be met by placing a wooden ladder at the back of the building, and in ventilating bedrooms, small interior windows were acceptable. There was no real regulation of the privies because the law was silent regarding their size, building materials, or proximity to the clean water supply. Finally, the number of inspectors was entirely too few to accomplish the assigned tasks. (SEE ALSO: *Substandard Housing*)

—*Eugenie Birch*

Further Reading

DeForest, Robert W. and Veiller, Lawrence. 1903. *The Tenement House Problem.* Vols. 1 and 2. New York: Macmillan. ◄

► Tent City

Tent City in Boston is the result of nearly 20 years of successful grassroots community activism that stands as an example of how "people power" can move from protest to large-scale development. In 1968, shortly after the assassination of Martin Luther King, Jr., a group of about 100 activists occupied a three-acre parking lot. The site is on a key block, immediately in Boston's South End, an ethnically diverse neighborhood of historic row houses close to the heart of downtown (see Figure 31). They were protesting the early stages of gentrification and the large-scale demolition of older housing being carried out in the name of progress and urban renewal. The protesters pitched tents and built improvised shelters to dramatize the need for affordable housing.

A national award-winning 269-unit mixed-income residential development complete with a child care center, retail stores, and a large underground parking facility opened in 1988 on the same parking lot. The site is immediately next door to the upscale Copley Place shopping, office, and hotel complex and Boston's ritzy Back Bay. The name of the development, "Tent City," ensures that history will not be forgotten. Most remarkably, the development remains true to the original vision of the protesters—a rent structure that ensures a residential community with a full range of family incomes, a design that fits in well with the rest of the historic neighborhood, and upkeep and management services that are first rate.

Tent City is also unusual in its ownership structure, which places legal control in the hands of tenants and the surrounding community through the Tent City Corporation, a nonprofit community development corporation that is a direct descendant of the original protesters. To obtain the development and financial expertise it needed, it contracted with The Community Builders, a nonprofit housing development firm that provides a full range of real estate development, finance, management, and social service skills to community-based development organizations. Under this arrangement, Tent City Corporation provided more than community vision; it retained legal control over all basic decisions affecting the final development—rent levels, design quality, income diversity, and unit sizes. As a result, Tent City is one of the premier examples of a local community group with negligible real estate experience maintaining full ownership control of its development without giving up or sharing the type of decisions usually required in joint ventures with for-profit developers.

First-quality design and construction are central ingredients in Tent City's success. The design consists of four-story walk-up row houses—the typical building style in the neighborhood—with 93 large-family units on three sides of the 3.3-acre site. In the corner adjacent to the taller, denser uses of the Back Bay, the row houses merge into a midrise building that steps from 5 to 12 stories. The midrise building contains 176 one- and two-bedroom units, as well as a day care center and 6,500 square feet of retail space. Architects Goody, Clancy and Associates created the design to knit Tent City into the neighborhood architectural context and to provide a natural transition from one neighborhood to another. Below Tent City is a two-story, 700-car underground parking facility paid for by adjacent Copley Place. The parking was central to the compromise that made the development possible because Copley's owners, JMB/Urban, controlled about half of the Tent City site. The City of Boston, through the Boston Redevelopment Authority (BRA), and a handful of small private owners controlled the balance. Without community support, there would be no parking; without JMB/Urban's land, assembly could be tied up in a long, costly eminent domain procedure.

Financing to meet the multifaceted objectives of the Tent City Corporation was challenging and required a high degree of creativity. Some 13 separate sources of capital and operating funding were blended to cover the $38 million development cost and meet the mixed-income tenancy

Figure 31. Tent City in Boston, Massachusetts, is a leading example of nonprofit, community-owned quality housing financed creatively to be affordable to a full range of income groups.
Photograph by Steve Rosenthal, Architectural Photography, Auburndale, Massachusetts.

goals. Financing came from Massachusetts Housing Finance Agency (first mortgage and "shallow" mortgage interest subsidy), Commonwealth of Massachusetts (backup rent subsidy if HUD Section 8 tenant-based subsidies are interrupted), private equity (through the sale of tax benefits to investors), and various smaller loans and grants from the City, BRA, foundations, and other government agencies.

The creation of a mixed-income community was one of the basic elements of Tent City Corporation's vision for the development. This was accomplished by reserving 25% of the housing units for low-income families, 50% for moderate-income families, and 25% at market rate. All unit types are scattered throughout the development with no differences in amenities. To cover ongoing costs, in a development where rents are low, an unusual source of ongoing support was used—the stream of funds being repaid to the city from an Urban Development Action Grant (UDAG) that helped build the neighboring Copley Place commercial complex. Tent City targets these UDAG repayment funds to the moderate-income units, guaranteeing that half of its housing units are reserved for moderate-income families, a group that has been left out of most affordable rental housing. In most "mixed-income" housing developments, there is often a large gap between the low-income tenants—typically limited to 20% to 25% of the units—and the higher-income "market" tenants who often make up 75% to 80% of the tenant population. By making moderate-income families the centerpiece of the Tent City community, the artificiality, awkwardness, and community tensions of 80:20 mixed-income housing is avoided.

There are many measures of Tent City's success. Vacancies are virtually nonexistent even in the market rate units. The development has continued to be financially sound even through large swings in the real estate market. People involved in housing from around the United States and the world visit in a steady stream. Tent City won the Urban Land Institute Award of Excellence in 1990 and the 1995 World Habitat Award given by the United Nations, as well as national recognition for design from *Architectural Record, Progressive Architecture,* and the *New York Times.* But perhaps the most important measure of success is that people who know Tent City best, those who live there and those in the surrounding neighborhood, are proud of it. It is a beacon of hope that collective community effort combined with vision, persistence, creativity, and competence can take on significant problems and bring about positive changes that make a lasting difference in the life of a city. (SEE ALSO: *Mixed-Income Housing;* **Urban Redevelopment**)

—*Joshua Posner*

Further Reading

Posner, Joshua. 1989. "Tent City: Creative Financing for Affordable Housing." *Urban Land,* (November):6-11. ◀

▶ Tenure Sectors

Residential property can be divided into three sectors: (a) housing that is privately owned and conveyed through market means, (b) housing that is publicly owned and conveyed through nonmarket means, and (c) housing that is privately owned and conveyed through nonmarket means. Different forms of tenure are to be found within each of these three housing sectors. Different public policies have been developed to support each sector. Historically, however, there has been a blurring of the boundaries between "public" and "private" and between "market" and "nonmarket" housing. Some forms of housing straddle the line dividing one sector from another (see Figure 32).

Housing Tenure

Housing tenure (from the Latin *tenere,* "to hold") refers to any number of legal arrangements for securing, permanently or temporarily, a possessory interest in land and buildings used for human shelter. Residential real estate may be held in many different ways. There exist, therefore, many different forms of tenure.

These multiple forms of housing tenure can be classified and clustered by sector according to (a) who holds the possessory interest and (b) how that interest is priced and conveyed. Thus, housing may be owned by individuals or by corporations of the *private* sector or by governmental (or quasi-governmental) entities of the *public* sector. Housing may be priced and conveyed through the *market* or may be priced and conveyed through *nonmarket* means. Three tenure sectors are differentiated when housing is classified by ownership and conveyance: (a) privately owned market housing, (b) publicly owned nonmarket housing, and (c) privately owned nonmarket housing. A fourth sector of publicly owned market-priced housing is too rare to warrant more than a passing mention.

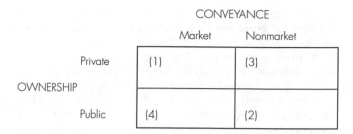

Figure 32. Classification of Tenure Sectors

Contained within each of these tenure sectors are multiple variations on each sector's distinctive approach to owning and conveying residential real estate. To employ a traditional metaphor, each sector contains a variety of ways in which a property's "bundle of rights" may be parceled out. Many "sticks," including the following, are to be found in this bundle:

- ▶ The right to exclude others
- ▶ The right to sell to others
- ▶ The right to profit from a property's sale
- ▶ The right to improve or demolish existing buildings
- ▶ The right to use the ground or mine the minerals underneath these buildings
- ▶ The right to enjoy the air and light overhead

All of these rights may be permanently held by a single owner or temporarily held by a single renter. The popular perception of tenure tends to divide residential property into precisely this dichotomy of *owning* versus *renting.*

The reality is much more complex. Any property's bundle of rights may be held in total by more than one owner— or by more than one renter. The bundle of rights may itself be untied and taken apart, with its individual sticks being separately apportioned among several owners or renters.

Private, Market Housing

The largest tenure sector in the United States, containing more than 90% of the nation's residential units, is made up of market-oriented housing, either owner occupied or renter occupied, having the following characteristics:

- ▶ *It is privately owned.* Title to residential real estate is held by an individual or by a private corporation, not by an instrumentality of the state.
- ▶ *It is market priced.* Prices for the sale or rent of residential real estate are established by the market. Housing is conveyed, except in cases of gift, inheritance, or foreclosure, through market transactions between a willing seller (or lessor) and a willing buyer (or renter). Access to housing is available only to those prospective occupants who can pay the market price.
- ▶ *It is profit oriented.* The function of residential real estate is not only to meet the residential needs of the property's occupants but to accumulate wealth for the property's owners.

The housing of this sector, variously known as *market housing, for-profit housing, commodity housing,* or *private sector housing,* comes in many different types and tenures. Such housing can be found in single-family houses and in multi-unit towers. It can be found in buildings where a single owner-occupant possesses all of the sticks in the property's bundle of rights. It can be found in buildings organized as a condominium or a cooperative where homeowners possess a common interest in residential property, where the bundle of rights is divided up among multiple owner-occupants. And it can be found in buildings leased out by absentee owners, buildings in which the occupants possess only the right to remain in residence for a specified period of time. Housing policy in the United States has been heavily biased toward this single sector.

On the *rental* side, governmental support for market housing has focused on providing tenant-based rental assistance for low-income households, enabling them to pay market-priced rents to private, profit-oriented landlords.

On the *homeowner* side, support for this sector has focused on subsidizing (i.e., lowering) either the initial down payment or the continuing monthly payments required of homebuyers who are borrowing long-term money for the purchase of market-priced homes. The most successful public policy promoting private homeownership has been the creation of the long-term, fixed-rate, high loan-to-value mortgage, made possible by federal insurance and the federal backing of secondary mortgage markets. The most substantial homeowner subsidies, however, have come in the form of tax expenditures, the largest being the federal deductibility from individual income of mortgage interest and real estate taxes. By 1997, this single subsidy was costing the U.S. government $67.9 billion a year.

Public, Nonmarket Housing

The housing contained within this governmental sector, known as "public housing" in the United States and in Canada, shares these defining characteristics:

▶ *It is publicly owned.* Title to residential real estate is held by an instrumentality of the state, typically a municipal corporation.
▶ *It is price restricted.* A limit is placed on the future price at which the property's units may be rented, preserving their affordability for a targeted class of low-income residents. Rents are established by public policy, not by the market.
▶ *It is means tested.* Access to rental housing is available only to prospective tenants whose incomes and other social characteristics make them eligible for admission.

Aside from the special case of military housing, the only form of publicly owned, nonmarket housing that has received regular—although somewhat grudging—governmental support in the United States has been that which is owned and managed by locally chartered, municipal corporations known as public housing authorities (PHAs). There are 3,060 of these local authorities. Together, they own 1.36 million units of housing.

The ownership of housing by other governmental or quasi-governmental bodies has been rare, although not unknown. At the height of the New Deal, for instance, several innovative experiments in public, nonmarket housing were launched by the Resettlement Administration, the Farm Security Administration, and the Public Works Administration. With the passage of the Wagner-Steagall Housing Act of 1937, however, most of the responsibility for owning and operating public housing shifted from the federal government to local authorities.

Except for the interruption of World War II, when the federal government became briefly involved in planning settlements and providing housing for defense workers, this trend toward a decentralized system of public housing continued. Federal funding for publicly owned housing ebbed and flowed in the years that followed, federal regulations came and went, but the ownership and management of public, nonmarket housing remained almost the exclusive purview of local authorities appointed by municipal governments.

Private, Nonmarket Housing

The numerous models of housing contained within this third sector represent a clear alternative to the more familiar models of both the market and the state. Numbering an estimated 700,000 units in the United States, such housing has these defining characteristics:

▶ *It is privately owned.* Title to residential real estate is held by an individual or by a private corporation, not by an instrumentality of the state.
▶ *It is price restricted.* A contractual limit is placed on the future price at which the property's units may be rented or resold. Prices are established by a predetermined formula, not by the market. These price restrictions do not lapse when the housing changes hands, nor do they expire after a short duration to allow owners to cash in on the housing's appreciated value. By design and intent, the housing is to remain affordable in perpetuity.
▶ *It is socially oriented.* The property's primary function is to meet the social needs of current, and future, occupants, not to accumulate wealth for the property's owners. Although the need for safe, decent, and affordable housing is paramount here, the property's "social orientation" often includes a collaborative component as well, whereby individual households are linked together in a residential network of pooled risk, mutual aid, and operational support.

Various names have been attached to such private, nonmarket housing: *social housing, nonspeculative housing, decommodified housing, nonprofit housing,* or when applied solely to homeownership arrangements, *limited-equity housing.* A more precise label is *third-sector housing,* denoting a nongovernmental domain within which the preeminence of social needs over private accumulation is institutionalized—and perpetuated.

Many different models of private, nonmarket housing have been developed, although six may serve as points of reference for all the rest. Arrayed along a tenurial contin-uum between owner-occupied housing and renter-occupied housing, they are the following:

1. The deed-restricted, owner-occupied house
2. The community land trust
3. The limited-equity condominium
4. The limited-equity (or zero-equity) cooperative
5. The mutual housing association
6. Various forms of nonprofit rental housing—some of it resident managed, some not

Although governmental support for private, nonmarket housing has been common in many Western countries, es-pecially in Sweden, Denmark, and Canada, such housing has seldom been a favored recipient of public largess in the United States—with two exceptions. Federal support for nonprofit rental housing, targeted to the elderly, has been provided through the Section 202 program since 1959. Federal support for limited-equity and zero-equity housing cooperatives was once available through the Section 213, Section 236, and Section 221(d)(3) programs.

The proliferation of private, nonmarket housing during the 1980s and early 1990s attracted new governmental support to this sector. At the federal level, set-asides and priority funding for the types of nonmarket housing being developed by community-based nonprofits were built into the Financial Institutions Reform, Recovery, and Enforce-ment Act of 1989 and the Cranston-Gonzales National Affordable Housing Act of 1990. At the state level, 23 states established housing trust funds, most of which provided priority support for private, nonmarket housing and for the nonprofit developers of such housing.

It is at the municipal level, however, that third-sector housing has had its widest acceptance. By the 1990s, public support for nonmarket models and nonprofit organizations had become a significant ingredient in the general mix of housing programs in a number of U.S. cities. In a few, it had become the touchstone for nearly every policy and every program promoting the construction or rehabilita-tion of affordable housing.

Overlapping Sectors

All forms of housing tenure and all types of housing policy do not fit neatly or completely into three sectors. There is a blurring of the boundary between public and private, for example, in the case of publicly subsidized, privately owned housing with project-based rental assistance. Developed by profit-oriented investors, using a variety of federal pro-grams such as Section 236, Section 221(d)(3), and Section 8, this housing bears all of the ownership characteristics of private, market housing; yet the units are priced and conveyed through the same sort of price-restricted, means-tested mechanisms characteristic of public, nonmarket housing.

The boundary surrounding private, nonmarket housing has proven, in practice, to be just as permeable. The use of Low-Income Housing Tax Credits by nonprofit housing developers, for example, has attracted profit-oriented in-vestors into publicly regulated, price-restricted projects that

are privately owned by a for-profit partnership in which a nonprofit corporation has a financial and managerial stake. Such an arrangement not only muddies the clear distinction between "public" and "private" but between "market" and "nonmarket" housing as well.

Most of the ways in which residential property is owned and conveyed can be contained within three tenure sectors. There will always be a degree of spillover, however. Some forms of housing straddle the line dividing one tenure sector from another. (SEE ALSO: **Community-Based Housing;** *Community Land Trust; Cooperative Housing;* **Homeown-ership;** *Joint Tenancy;* **Private Rental Sector**)

—*John Emmeus Davis*

Further Reading

Barton, Steve. 1996. "Social Housing versus Housing Allowances: Choosing between Two Forms of Housing Subsidy at the Local Level." *Journal of the American Planning association* 62(Winter): 108-19.

Davis, John Emmeus. 1994. *The Affordable City: Toward a Third Sector Housing Policy.* Philadelphia, PA: Temple University Press.

Kaufman, Phyllis and Arnold Corrigan. 1987. *Understanding Condo-miniums and Co-ops.* Stamford, CT: Longmeadow.

Kemeny, Jim. 1981. *The Myth of Home-Ownership: Private Versus Public Choices in Housing Tenure.* London: Routledge & Kegan Paul.

Krinsky, John and Sarah Hovde. 1996. *Balancing Acts: The Experience of Mutual Housing Associations and Community Land Trusts in Urban Neighborhoods.* New York: Community Service Society of New York.

Stone, Michael E. 1993. *Shelter Poverty: New Ideas on Housing Af-fordability.* Philadelphia, PA: Temple University Press. ◄

► Third World Housing

Housing problems in the Third World involve location, quality, durability, high cost, long-term finance, tenure, turnover, and neighborhood externalities, just as they do in industrialized countries. But the context is quite differ-ent because of much greater poverty, inequality, and popu-lation growth. Housing conditions vary from sub-Sahara Africa to Latin America and from the Middle East to Mon-soon Asia, but similarities are nevertheless sufficient to group them together as "Third World Housing."

Until the 1950s, affluent Third World households lived in mansions, while a majority of the overwhelmingly poor population lived in rural traditional villages and com-pounds or in rooms above urban shops, in tenements, or in the subdivided former dwellings of the rich. That housing system resembled the one of Europe before its Industrial Revolution. The resemblance ended as the Third World resolutely tried to develop with plans that initially copied either the post-World War II European recovery or the Stalinist transformation of the USSR. Modest economic success in most countries, however, was accompanied by an unprecedented acceleration of population growth and a torrent of rural-urban migration. Inner-city slums dete-riorated rapidly with overcrowding, and squatter settle-ments sprawled about the urban fringes.

As late as the mid-1960s, economists generally believed that attention to the housing shortage would be a diversion from the real task of raising productivity. Housing investment was not promoted. Housing was considered welfare spending with an excessively high "capital-output ratio" that poor countries could not possibly afford until enough "productive" industrial capital had been installed. No finance should be diverted to housing, and workers should somehow find shelter through doubling up or makeshift accommodations.

Lack of professional economic concern meant that decisions about building codes, zoning, and the layout of streets and water pipes were made in an extravagant manner by urban planners and architects trained in Western countries and eager to appear "modern." They rarely modified their proposals to match the poverty of households or governments, and their plans could not be carried out. The inevitable response of the poor was crowding into back alleys and courtyards, squatting in parks and along railroad tracks, building on poles in lagoons, and buying minute unserviced lots on clandestinely subdivided private farms. Illegality and substandard dwellings led governments to evict, to demolish, and to eradicate millions of huts and shacks, especially if land was needed for other purposes. Worse than inner-city slums and peripheral squatter settlements, however, would have been the greater housing shortage without them.

In large Latin American cities, one-fourth of housing has typically been built without permits, and two-thirds or more lacks authorization in Sub-Sahara Africa and some large South Asian cities. As a result, about 10% to 20% of urban housing in the Third World has been built of cardboard, cloth, refuse, and other temporary materials. Unauthorized housing in Africa generally lacks indoor piped water. Elsewhere, water was eventually provided so that access ranges from one-third or less (Malawi, Cote d'Ivoire) to nearly all (Colombia, Korea). Not only is housing quality less in the Third World, but so is quantity in the form of floor space. Urban housing in the poorest countries of South Asia and Sub-Sahara Africa has only 5 to 10 square meters per person, or about one-fourth of the space typical in European cities. East Asian and Latin American cities are about halfway between. Little floor space per person reflects not just the prevalence of small one- and two-room dwellings but overcrowding and doubling up of households—homelessness in disguise.

In African cities, only three or four dwellings per thousand inhabitants are built annually, far short of needs. The rest of the Third World averages six or seven units, which is about that of industrialized countries, where population growth is much less. Much of this construction has been financed informally because mortgage lending has been a very small proportion of the portfolios of Third World public and private financial institutions. Nevertheless, compared with gross urban product, the value of newly built dwellings was generally higher in the Third World (5-7%), except for Africa (2%), compared with that of rich countries (4%). The effect of such building on the housing stock has also been such that (again, except for Africa) housing prices and rents compared with household incomes are not much

different in the Third World, compared with their levels in industrialized countries. Prices were four to six times annual incomes, and for tenants, rents averaged 15% to 20% of incomes. Rents of 10% and prices twice the annual income were more typical of Africa. Because of the absence of developed real estate markets and data collection, such figures must be regarded with caution. In general, housing conditions in the Third World seem to be worse than could be expected from household income levels and willingness to pay for shelter.

Three solutions appeared in the 1960s, based not on sophisticated economic analysis but on the notion of a "housing gap." The part of the dwelling stock that was officially assessed as adequate was compared with the rate of growth of households, and the difference was a housing gap that had to be closed with new construction. The first solution was subsidizing mortgages and savings-and-loan associations for new lower-middle-class "social interest" housing so that vacated dwellings could "filter" to the poor. The number of units that could be provided in this manner proved to be minuscule compared with accelerating migration, however, and in Latin America inflation rapidly decapitalized the lending intermediaries.

A second solution was prefabrication, renamed "industrial systems building." Dwelling construction, it was thought, was at last being transformed by mass production technology in France, Scandinavia, and Eastern Europe, so why not adapt these methods to the overcrowded Third World? Experiments were made with concrete panels, stacked boxes, spun fiberglass, extruded plastics, sprayed cement, bamboo reinforcing, and several other inventions. Projects with experimental technology earned priority for loans. International competitions and meetings were held on the subject. After two decades of trials, however, only stabilized soil blocks, fiber-reinforced roofing tiles, and a machine for handpressing concrete blocks survived as "intermediate technology." Conventional bricks, blocks, and reinforced concrete remained cheapest as materials, and low wages allowed them to be handled and put into structures economically.

The third solution was "organized self-help." If building was labor-intensive and if spare time was abundant in the Third World, why not organize the unemployed poor to build their own houses with loans just for land and for the materials? Professionals would manage the building process to ensure that dwellings were safe and that layouts would leave space for later roads, water pipes, and sewers. Experience in Puerto Rico, Colombia, and elsewhere, however, showed that, considering administrative expenses and subsidies, these dwellings still cost more than nations could afford to shelter everyone. Besides, labor from creditworthy families had a high opportunity cost, making it wise to subcontract "self-help" building work to professional artisans. Thus, even intervention to organize self-help proved to be unrealistic.

In the 1970s, the perspective changed. Large international conferences on the environment and human settlements were held in Stockholm and in Vancouver, Canada, and the United Nations Centre for Human Settlements (Habitat) was set up in Nairobi, Kenya. Housing was elevated to the status of a "basic need," requiring an appropriate delivery system. Thus, the World Bank and others began to finance sites-and-services projects. Most important was redefining squatter slums from marginal urban blights to settlements of "hope" with "informal housing," built independently of regulations or large-scale assistance. If households were left alone and feared no eviction, they could locate their dwellings and use them in a way that would generate more benefits than simple shelter: Besides, "capitalist market values or state socialist productivity values . . . both inevitably inhibit the investment of personal and local resources on which the housing supply ultimately depends" (Turner 1976-77, p. 70). In Lima, Peru, where John F. C. Turner studied conditions with José Matos Mar and William Mangin, this awareness caught on first. In the Orangi Project of Karachi, under the guidance of Akhter Hameed Khan, even an extensive sewer system was installed through informal autonomous neighborhood collaboration.

Meanwhile, housing finance for the Third World middle class also came to be better understood and mobilized. Households who reduce consumption to acquire assets are the ultimate source of savings. Any number of financial intermediaries may channel the decision to release resources from consumer goods production to their use for making capital goods. The usual assumption is that saving, or the release of resources, depends ultimately on the security and investment yield of the capital goods. The financial intermediaries channel resources to highest-yield options, partly by providing liquidity to the savers and long-term security to the borrowers.

In the case of much housing, however, the savings are generated by the expectation of owning the specific capital good as a dwelling and a hedge against inflation. Families who wish to buy housing would not equally reduce their consumption to acquire some other type of asset. They will repay mortgages and gradually acquire their dwelling but will not alternatively set aside comparable amounts for savings accounts or portfolios of stocks and bonds. In Third World countries, paper claims are viewed with little confidence. Beyond a certain point, therefore, less housing construction does not mean that more factories and offices can be built and equipped. Consumption will rise instead. Some financial institutions are involved in both housing and non-housing transactions with complex intermediation so that this constraint, or segmentation, is easily overlooked. Because housing is built with local resources, little international finance is needed for direct assistance. If housing finance expands the domestic money supply, however, and raises employment and incomes, imports will be stimulated indirectly, and a foreign loan will help with that. Some countries have borrowed foreign exchange nominally for housing but actually to import equipment and materials for other purposes.

Filtering usually stopped as new households, vacating nothing but nevertheless needing loans, occupied dwellings. Finance had to be available for old as well as newly constructed dwellings, contrary to the self-interested pressure of contractors and of policymakers with a naive fixation on direct employment. Moreover, repayments of mortgages had to cover inflation, either through high nominal interest rates or through indexing the principal to a price index, as was the policy for some years in Israel, Chile, Brazil, and Colombia. Mexico and Turkey solved an obvious problem by indexing deposits to retail prices and repayments to wage increases, with differences added to (or subtracted from) the mortgage maturity. Robert Buckley of the World Bank (1988) then stressed that unrealistic controls on housing finance and impossible use of dwellings as collateral (because of forbidden foreclosures and evictions) blocked proper functioning of the entire financial system and therefore impeded all development. Thus, housing had moved from being an ignored sector to being on center stage.

Research has supported the removal of constraints on the construction and use of dwellings. Because the poor need income more than space, even loss of living space through locating enterprises in dwellings has come to be tolerated, especially the subletting of rooms, first recorded with unnecessary alarm in the World Bank's Dandora project in Nairobi. Gradually, the taboo against private landlords and against a free rental market has weakened with the awareness that private letting of dwellings is going on anyway, even in informal settlements, serves a need, and is seriously impaired by rent controls. Large-scale rental housing under public auspices has been a success only in the extraordinary city-states of Hong Kong and Singapore, and even here, many units were eventually converted to condominiums.

By the 1990s, a consensus emerged that unavailable finance and unaffordable serviced land, above all, blocked the improvement of housing. The willingness to make high rent or mortgage payments (as a share of income) indicated a demand elasticity about as high in poor as in rich countries, but rigid yet shaky financial intermediaries were ill designed for mobilizing this willingness to pay. A confused urge to subsidize led to earmarked finance and arbitrary discrimination. Needed instead was a way of subsidizing only the most needy and to do that in an open "transparent" fashion. Chile led the world here with a novel system of one-time cash grants linked to smaller mortgages on market terms that would be attractive to savers. Even this system needed years to gain widespread acceptance, meaning willingness to allow foreclosure and eviction of defaulters.

Perhaps the biggest hurdle to better housing in the Third World is the slow and costly expansion of urban land, owing to the political power of large-scale landowners and of numerous small land speculators—anyone buying a few lots as an inflation hedge. From the Philippines to Brazil, these groups conspire to enhance their investment by keeping taxes negligible on idle land. Such low taxes then raise the demand for idle land as an asset, while lowering supply of serviced land by depriving government of revenues for extending roads and infrastructure to new sites. The scarcer that empty urbanized land is, the more its price will rise, making it an ever-better investment. A recent World Bank-UNCHS (U.N. Centre for Human Settlements) survey of more than 50 countries found that, compared with raw land, newly serviced plots for sale on the urban fringe had twice the price increase in Third World countries as did rezoned and newly equipped land on the fringes of large cities of high-income nations. Monopolistic scarcity premiums for land were fueled by the rapid urban expansion in poor countries.

"Enabling markets to function" has become the central theme of the "Global Shelter Strategy to the Year 2000," adopted by the U.N. General Assembly in 1988. The strategy involves lowering undue government regulations of all types, as well as removing monopolistic private constraints in the land market. One must not, however, expect that in the course of development, poor countries will move relentlessly toward ever more private markets, copying the affluent nations. Free markets may cause unsightly building configurations, inadequate green space, inconvenient separation of homes from work, wasteful traffic congestion, and dangerous segregation of social groups. Coping with these externalities means costs that poor countries may be less able and willing to pay. Experiments with intervention, nonintervention, controls, taxes, subsidies, conflict, and compromise are therefore likely to come and go. Interventions are most likely after wars, migrations, revolutions, earthquakes, and storms that compel reallocation of remaining resources. (SEE ALSO: *Colonias;* **Cross-National Housing Research;** *Environment and Urbanization; Global Strategy for Shelter;* Habitat International; *Habitat: U.N. Conference on Human Settlements; Self-Help Housing; Sites-and-Service Schemes; Squatter Settlements; U.S. Agency for International Development*)

—*W. Paul Strassmann*

Further Reading

Abrams, Charles. 1964. *Man's Struggle for Shelter in an Urbanizing World.* Cambridge: MIT Press.
Buckley, Robert. 1994. Household Finance in Developing Countries: The Role of Credible Contracts. *Economic Development and Cultural Changes,* 42(2):317-32.
Burns, Leland and Leo Grebler. 1977. *The Housing of Nations: Analysis and Policy in a Comparative Framework.* London: Macmillan.
Currie, Lauchlin. 1966. *Accelerating Development: The Necessity and the Means.* New York: McGraw-Hill.
Gilbert, Alan. 1992. "Third World Cities: Housing, Infrastructure, and Servicing." *Urban Studies* 29:435-60.
Rodwin, Lloyd, ed. 1987. *Shelter, Settlement, and Development.* Boston: Allen & Unwin.
Turner, John F. C. 1976. *Housing by People: Towards Autonomy in Building Environments.* New York: Pantheon.
World Bank. 1992. *Housing: Enabling Markets to Work: A World Bank Policy Paper.* Washington, DC: Author. ◀

▶ Tipping Point

Researchers investigating the changing racial composition of urban neighborhoods have often focused on residential succession, which is the changing of a predominantly white neighborhood to one that eventually becomes almost exclusively African American. Of key significance is the presumption of the irreversibility or inevitability of the succession process. This irreversibility is widely associated with the idea of a "tipping point." As defined by sociologist Morton Grodzins, this is the point at which the percentage of African Americans in an area exceeds the limits of a neighborhood's tolerance for interracial living.

The commonly held convention is that once the percentage of African Americans passes the tipping point, it is assumed that whites will leave these neighborhoods at accelerated rates and be replaced by blacks until the neighborhood has become entirely African American. It is further assumed that predominantly black residential neighborhoods are undesirable and should be avoided.

The broader context in which tipping point has been applied is housing segregation. It has been assumed, for example, that an integrated residential neighborhood can be maintained if the tipping point is not exceeded. Conversely, it is assumed that once past the tipping point, previously racially integrated neighborhoods will become predominately black segregated enclaves.

The key difficulty with the idea of the tipping point is that it implies a hard, quantifiable yardstick that can be applied to any locale where there is some question about the status of neighborhood racial succession. In reality, no such measure has been validated. Instead, individuals desirous of maintaining integrated neighborhoods have held strongly to the belief that passing the tipping point converts a desirable arrangement into an undesirable situation.

Measuring the Tipping Point

Although Grodzins first introduced the term in 1958, he did not attempt to quantify when neighborhoods reached the tipping point. Indeed, Grodzins believed that the tipping point would vary depending on many different circumstances. Initial attempts at defining the tipping point came more from anecdotal evidence than from empirical studies, suggesting that 30% is the upper limit.

In an attempt to develop more rigorous quantitative measures of the concept, Thomas Schelling employed a series of mathematical models to examine under what circumstances it might be possible to calculate a tipping point. In the process, he refined the idea by suggesting that tipping could function in two ways. The first involves whites leaving an area once it is perceived that there are too many blacks. This, Schelling called "tipping out." When blacks move into a previously all-white neighborhood, the phenomenon is labeled "tipping in." Schilling concluded that blacks tip in when a neighborhood reaches about 10% black. Whites tip out when the black population reaches about 18%. According to Schelling, a range from 10% to 18% black tends to stabilize in-and-out movement for both groups. He suggests that a white range of 82% to 90% is a natural equilibrium.

But not all scholars are convinced that tipping necessarily occurs when blacks move into white residential neighborhoods. Goering, for example, notes that time-series data collected in Philadelphia and Detroit and in Chicago in the mid-1960s reveal no evidence of a tipping point. Several factors need to be identified before an answer can be given to the question of whether and when neighborhoods would change racially.

Emerging evidence also calls into question the earlier assumptions regarding the inevitability and irreversibility of neighborhood succession from white to black. Lee's 1986 analysis of sample data collected from 58 large cities suggests that certain cities may be experiencing what he calls "reverse" racial change—that is, a change from black to white or more broadly black to nonblack. More recent studies have suggested that this idea of residential succession may not be the same for Latinos and Puerto Ricans as it is for African Americans. In 1984, Massey and Mullan found that nationwide residential succession was much less prevalent in Hispanic areas than in black areas. In New York, residential succession for Puerto Ricans is not the "inevitable" process described for blacks. In Los Angeles, from 1970 to 1990, predominantly African American neighborhoods have undergone varying degrees of racial succession, but the rate of black out-migration varies tremendously, and some new neighborhoods are experiencing simultaneous Latino and black immigration with no accelerated rates of white flight.

Despite the lack of significant empirical evidence supporting the existence of a specific tipping point, there is a widely held belief that a tipping point exists and that once a neighborhood has tipped, the resulting neighborhood pattern is undesirable. The strength of this belief has influenced the way lenders, realtors, real estate appraisers, and landlords conduct their business. Even the courts have cited existence of this "fact." For example, in *Otero v. New York City Housing Authority* (484, F. 2d, 1122, 1973), a court of appeals ruled that the housing authority had the duty to prevent the creation of a segregated black neighborhood. The court accepted the argument that the housing authority had to consider whether increased concentrations of non-whites would act as a "tipping factor" that would hasten an increase in the nonwhite population in the surrounding neighborhoods, leading to a steady loss of total white population.

Is the Concept of Tipping Point Useful?

Perhaps the most functional aspect of the tipping point concept is that it has helped to focus the attention of advocates, researchers, and policymakers on the fact that neighborhoods do undergo racial succession. As a result, there is widespread belief, at least among most whites, that policies and actions need to be in place to prevent the tipping point from being reached, thereby protecting property values and ensuring greater neighborhood stability. For many African Americans and other minority group members and for those concerned with equity in the housing market, barriers enacted to prevent tipping from occurring are little more than discriminatory actions designed specifically to limit access to housing markets by minorities.

But the fact that the concept of tipping point has become conventional wisdom does not mean that it is theoretically sound. There is little empirical evidence to demonstrate that tipping is irreversible, that it primarily accounts for African Americans moving into white neighborhoods, or that neighborhoods with a majority African American population are undesirable residential locations for whites. Indeed, studies on gentrification have documented that certain predominantly inner-city African American neighborhoods are now attracting whites in relatively large numbers. Results from the 1990 census have also shown that African Americans are suburbanizing faster than they ever have in the past. Neither of these phenomena is accounted

for by the idea of a tipping point. Perhaps more important is that the concept is not very useful in explaining how or why residential succession occurs. Indeed, focusing on the tipping point diverts attention from the larger, more pertinent issues—namely, continued housing segregation in the United States.

Implications for Research and Public Policy

Two implications can be readily drawn regarding research and the idea of a tipping point. The first is that a significant amount of empirical work needs to be undertaken before the theoretical legitimacy of the term can be validated. Schelling's work was an attempt to do this. But little replication of his efforts has been attempted, perhaps because researchers have concluded that greater clarity about the tipping point will not explain why residential succession occurs, what its significance is, and how its outcomes relate to the continuation or diminution of residential segregation. Thus, the second implication regarding tipping point is that a more productive focus of research inquiry should be on the arena of housing equity—that is, what perpetuates residential segregation and what steps can be done to afford minority group members of all income levels the opportunity to reside in neighborhoods of their choice. Research on segregation indices provides empirical evidence on the intractability of housing segregation but has not as yet provided meaningful policy solutions for reducing racial isolation.

Research is fairly conclusive concerning the cost that segregated neighborhoods impose on African Americans. African Americans in predominantly black residential neighborhoods pay a disproportionate share of their income for housing, occupy the oldest residential units most in need of repair, and live in areas where the quality of local schools has declined, where basic services are lacking, and where they have the highest probability of becoming victims of violent crime.

For policymakers and those seeking to influence the decision-making process, studying residential succession has become important because residential mobility has usually been viewed as a way of improving one's quality of life. That African Americans have systematically had a difficult time entering residential neighborhoods of choice has meant that their ability to obtain better-quality housing, better educational opportunities for their children, and access to better and higher-paying jobs and to avoid crime-ridden neighborhoods is much less compared with their white counterparts.

Although researchers may disagree about whether there really is a tipping point at which housing units in formerly all-white neighborhoods begin to convert to all-black residential units, there is little disagreement that housing segregation in the United States persists. Most minorities, and particularly African Americans, have not had the opportunity to find housing of their choice, irrespective of the ability to pay. For these individuals, eliminating barriers to continued racial isolation is more meaningful than refining or empirically validating the existence of the tipping point. (SEE ALSO: *Racial Integration; Segregation*)

—*J. Eugene Grigsby, III*

Further Reading

Goering, John M. 1978. "Neighborhood Tipping and Racial Transition: A Review of Social Science Evidence." *Journal of the American Institute of Planners* 44:68-78.

Grodzins, Morton. 1958. *The Metropolitan Area as a Racial Problem.* Pittsburgh: University of Pittsburgh.

Lee, Barrett A. 1986. "From Black to White." *Journal of the American Planning Association* (Summer).

Massey, Douglas S. and Brendan P. Mullan. 1984. "Process of Hispanic and Black Spatial Assimilation." *American Journal of Sociology* 89:836-73.

Offensman, John R. 1995. "Requiem for the Tipping-Point Hypothesis." *Journal of Planning Literature* 10(2):131-41.

Rapkin, Chester and William Grigsby. 1960. *The Demand for Housing in Racially Mixed Areas.* Berkeley: University of California Press.

Rosenberg, Terry J. and Lake, Robert W. 1976. "Toward a Revised Model of Residential Segregation and Succession: Puerto Ricans in New York." *American Journal of Sociology* 8(March):142-50.

Schelling, Thomas C. 1972. "A Process of Residential Segregation: Neighborhood Tipping." In *Racial Discrimination in Economic Life,* edited by Anthony H. Pascal. Lexington: Lexington Books.

Taeuber, Karl E. and Alma F. Taeuber. 1965. *Negroes in Cities, Residential Segregation and Neighborhood Change.* Chicago: Aldine.

Wolf, Eleanor P. 1963. "The Tipping Point in Racially Changing Neighborhoods." *Journal of the American Institute of Planners* 29:217-22. ◄

▶ Title

A *title* represents an ownership interest in real property. A *deed* is a legal instrument that conveys title and evidences ownership. The best possible title is called a *marketable title,* which means that the ownership interest can be sold to a reasonably prudent buyer at market value.

Title can be transferred voluntarily or involuntarily. Voluntary transfers can be accomplished by deed, will, or patent. Involuntary transfers of title may occur by public sale, intestate succession, bankruptcy, erosion, adverse possession, or eminent domain. That is, title can be transferred by a public grant by the government to individuals, by a private grant between individuals, or by action of law.

A title examination is an investigation of public records to evidence ownership by the use of either a *chain of title* or an *abstract of title.* The chain of title is a list of all previous owners, whereas the abstract is a summary of all transactions that affect the ownership of the property. Any defects or clouds on the title may be protected by the use of title insurance, a policy that insures the purchaser, lender, or both from any defect on the title. (SEE ALSO: *Title Insurance; Title Search*)

—Deborah Pozsonyi

Further Reading

ABA Standing Committee on Lawyers' Title Guaranty Funds. 1991. *Buying or Selling Your Home: Your Guide to: Contracts, Titles, Brokers, Financing, Closings.* Chicago: American Bar Association.

Jacobus, Charles J. 1996. *Real Estate Principles.* Upper Saddle River, NJ: Prentice Hall.

Smith, H. and J. Corgel. 1987. *Real Estate Perspectives.* Homewood, IL: Irwin. ◄

▶ Title Insurance

Title insurance is a comprehensive indemnity contract in which a title insurance company agrees to compensate the insured for losses arising from title defects. Title insurance covers losses caused by past events as opposed to hazard insurance, which covers losses created by events yet to occur. A title insurance policy is not written for a fixed or specific period of time. It continues in effect as long as the insured is subject to the risks that the policy covers.

The potential risks covered by title insurance include (a) errors made in the examination of the public records, (b) mistakes in interpreting the legal effect of instruments appearing in the records, and (c) certain facts that could not be discovered by a search of the public records (known as "off-record risks"). Examples of the types of losses covered by title insurance include forged documents, misfiled documents, undisclosed heirs, misinformation regarding marital status, improper interpretation of wills, mistakes related to persons with similar names, boundary disputes, and unrecorded liens. Should there be a challenger to the insured's title based on one of the categories covered by the policy, the title insurance company is obligated to defend the title in all related litigation.

Not all risks are covered by title insurance policies. Risks commonly excluded from coverage are (a) defects that come into existence after the owner takes title to the property, (b) misrepresentation and concealment by the insured, (c) unrecorded mechanics' liens at the time of the title search, (d) survey defects (unless the insurer has access to the survey results), (e) zoning and governmental police power rights, (f) unrecorded leases and conditional sales contracts, and (g) defects known to the insured at the date of the policy but not recorded in the public records or disclosed to the insurer.

Two primary types of title insurance policies are written—owner's policies and lender's policies. An owner's policy covers the title interests of the owner and the owner's heirs, devisees, or successors in interest. The insured is protected until the property is sold or otherwise transferred. The coverage terminates when the owner divests himself or herself of title. An owner's policy is not transferable from an insured seller to a buyer. A lender's policy is usually issued for the benefit of a mortgage lender and its assignees. This policy protects the lender in the same fashion as an owner's policy but is limited in coverage to the outstanding loan balance. The policy terminates when the borrower's debt is paid and the mortgage that secures the debt is released. Lenders' policies can be transferred to assignees who purchase mortgage loans from lenders in the secondary mortgage market. (SEE ALSO: *Title*)

—Jeffery M. Sharp

Further Reading

Floyd, Charles F. and Marcus T. Allen. 1994. *Real Estate Principles.* 4th ed. Chicago: Real Estate Education Co.

French, William B. and Harold F. Lusk. 1979. *The Law of the Real Estate Business.* 4th ed. Homewood, IL: Irwin.

Hinkel, Daniel F. 1995. *Practical Real Estate Law.* 2d ed. Minneapolis/St. Paul, MN: West.

Seidel, George J. 1993. *Real Estate Law.* 3d ed. Minneapolis/St. Paul, MN: West. ◄

► Title One Home Improvement Lenders Association

Title One Home Improvement Lenders Association (TOHILA) provides education, advocacy, and referral services to financial institutions that make federally insured home improvement loans under the Department of Housing and Urban Development/Federal Housing Authority (HUD/FHA) Title I program. The association hosts four educational conferences per year, featuring workshops on topics such as marketing, loan origination and underwriting, processing and servicing, filing claims, flood insurance, quality control, staffing and personnel management, warehouse lines of credit, and secondary market opportunities. TOHILA represents its members' interests in federal legislative and regulatory actions that affect the Title I program. It also works with investment banks and other secondary market agents to encourage their interest in Title I and disseminates information to the trade and to consumers about Title I home improvement loans. Publications include *The Title I Advisor,* a quarterly newsletter for lenders covering all aspects of Title I operations and industry news (free for members, $140 per year for nonmembers); *The Title I Deskbook;* and *The Complete Guide to Getting Your Title I Claims Paid.* Founded in 1988, TOHILA has a membership of approximately 350 members nationwide. Staff: 7. Executive Director: Peter Bell. Address: 1625 Massachusetts Ave., NW, #601, Washington, DC. 20036. Phone: (202) 328-9171. Fax: (202) 265-4435. (SEE ALSO: *FHA Title I Home Improvement Loan Program)*

—*Laurel Phoenix* ◄

► Title Search

A title search is an examination of the public records to determine if there are any defects in the owner's chain of title or potential claims by third parties. Representatives of title companies or abstractors usually conduct title searches. Most of the records examined will be located in the courthouse or records building in the county in which the subject parcel of land is located. Each state designates a county official as the recorder and custodian of the land records, such as the court clerk or the registrar of deeds. Parcels extending beyond one county require a search in each county.

The extent of the records search will vary by state and purpose. A full search requires the examiner to prove the chain of title from the present back to a predetermined date in the past. In most states, chain of title must be proven for a 50- or 60-year period. Therefore, defects older than the stated period, subject to limited exceptions, will not cloud the title. Common examples are unrecorded instruments older than the period of limitation. In a few states, a full examination will require searching the records back to the original source of title, such as a government patent grant or award of title.

A limited title search examines a period shorter than that required for a full title search. Often, limited searches are conducted when a parcel is protected by a title insurance policy covering the period of a full search. Limited searches are common in loan assumption and second mortgage transactions. Limited searches usually cover the period from the time the title insurance was issued to the present.

The mechanics of a title search involve an examiner's finding documentation to prove each transaction in the chain of title. This process is followed by records searches for liens or other claims that could affect the owner's title status. The process commences with an examination of the grantor-grantee index. Most counties will index each recorded document by which an ownership interest in land is conveyed by the names of both the grantor and the grantee. The grantee index is an alphabetized list of those persons or entities who had interests in land conveyed to them. Documents recorded in the index include deeds, mortgages, trust instruments, easements, leases, liens, and any other document purporting to grant an interest in land. The grantee index is the most useful tool for tracing the chain of title. The examiner begins with the name of the current owner of the property, who should be the most recent grantee. Locating the grantee's name in the index will reveal the name of the grantor. That grantor was the grantee of the previous transaction. The grantee index will reveal the recorded document proving that conveyance. The examiner continues this process successively back to the document originating title or for the time period designated by statute. With the chain of title created from the grantee index, the examiner cross-checks the grantor index to verify the conveyances and to ensure that the same interest in land has not been conveyed by the same grantor more than once. The process begins from the oldest transaction and continues to the present. This process will also reveal the presence of mortgaged interests, easements, liens, and other potential title claims by third parties.

Title examiners also review plat or tract indices to verify the accuracy of the parcel's legal description. Recorded plats may also reveal restrictive covenants. If so, the examiner tests the subject parcel for compliance with those covenants.

After establishing a chain of title and confirming the appropriate information against the plat or tract index, the examiner must review each document in the chain of title for errors and potential third-party claims. These potential problems include (a) discrepancies in the legal description, (b) questions about the identities of the parties to the transactions, (c) compliance with legal formalities (such as signature and witnessing standards), (d) conveyances of less than full title (such as life estates), (e) divorces and other marital estate claims, and (f) other unresolved claims or covenants set forth in the recorded documents.

Finally, the examiner must search for claims that exist outside the recorded chain of title. These claims may originate from a number of sources. The more common sources are the judgment index (unsatisfied money judgments recorded against a title holder), tax lien indices (unpaid federal, state, local, or property taxes in the name of a title holder), recorded *lis pendens* and civil court dockets (evidence in the land or court records that a lawsuit is pending against a title holder or an interest in the subject parcel), probate court records (to ensure that heirs of a deceased title holder do not have claims), the Uniform Commercial Code secured transaction index (verifying that no security interests exist against the property's fixtures), and the index of Mechanics' and Materialmen's Liens (verifying that all obligations to mechanics and materialmen providing labor or goods to the property have been satisfied). After completing this process, the examiner issues a written report containing the examiner's conclusions regarding the status of title, including any exceptions. (SEE ALSO: *Title*)

—*Jeffery M. Sharp*

Further Reading

Floyd, Charles F. and Marcus T. Allen. 1994. *Real Estate Principles.* 4th ed. Chicago: Real Estate Education Co.

French, William B. and Harold F. Lusk. 1979. *The Law of the Real Estate Business.* 4th ed. Homewood, IL: Irwin.

Henszey, Benjamin N. and Ronald Friedman. 1984. *Real Estate Principles.* 2d ed. New York: John Wiley.

Hinkel, Daniel F. 1995. *Practical Real Estate Law.* 2d ed. Minneapolis/St. Paul, MN: West.

Seidel, George J. 1993. *Real Estate Law.* 3d ed. Minneapolis/St. Paul, MN: West. ◄

► Transitional Housing

Transitional housing typically serves a particular population: for example, single mothers and children, single parents, women, men, two-parent families, or people infected with AIDS. More study and many more examples of this housing type exist for single mothers and children, now one-fourth of all U.S. families and the focus for the discussion here. Transitional housing is sometimes identified as second-stage or bridge housing. Programs focus on helping residents toward self-sufficiency before helping them find permanent housing. The term of residency is temporary, but this period may be as long as two years, longer than any permanent period a family may have experienced. Transitional support services can also be provided for those who have been settled in permanent housing, expanding services beyond a transitional building to a whole neighborhood. This approach has also been called *service-enriched housing.*

Yet definitions of transitional housing are blurred. For example, commonly defined emergency housing generally serves residents for days, weeks, or a few months. But if transitional or permanent housing is not available, the time can last for a year or more. In some cases, residents move from short-term emergency to permanent housing with no transitional period. Particularly in New York City, emergency housing funds are used for stays as long as a year to support a program that may be called *transitional,* a definition used by some for any housing between homelessness and a permanent residence. Affecting the length of this kind of transitional stay is the intensity and effectiveness of hous-

ing relocation assistance and social services. The U.S. General Accounting Office has reported to the U.S. Senate that both more time in transitional housing and more supportive services appears to ensure greater resident success in maintaining stable posttransition lives.

What makes housing permanent? The answer is simple: having the choice to live in a location for as long as one wishes without unwanted displacement. There are many transitional residents in what is ordinarily defined as permanent housing. College housing is transitional, as are trial partnerships and shared houses. Some argue that without a substantial increase in affordable permanent housing, a shelter system will institutionalize homelessness. The "American dream" approach to family and housing leads some to believe that an increase in affordable housing would solve problems of homelessness. However, although more affordable housing is desperately needed, it would not end the impact of domestic violence, substance abuse, or teen motherhood on the lives women and children. Affordable housing in and of itself is not enough to help children who have led traumatic lives or mothers who need child care to work.

Life stabilization, therefore, is necessary during a transitional period before poor women who head households can take advantage of affordable, permanent housing. Both short- and long-term goals ensure stability. Particularly for short-term programs, many transitional sites, especially domestic violence shelters, exclude men to eliminate any possibility of violence. Others create a women's community to enhance bonding and self-help because conditioning of both men and women reinforces the dependence of women in a mixed setting, particularly for those from strong patriarchal traditions. Within a community of others with similar life experiences, women see positive reflections and models for themselves on a day-to-day basis.

Separatism for women and children is not an end in itself, but it is particularly helpful during emergency and transitional housing periods. For some women, it is an important long-term alternative. Separatism ordinarily acknowledges contemporary facts of life, including the involvement of men in the lives of women and children. Help for single mothers has been labeled by extremists as discrimination against men, driving families apart. Yet often, women and children who remain in families because they have no other alternative are abused.

Emergency housing generally has less private household space than transitional or permanent housing—one or, at most, two rooms—and residency may be limited from several weeks to several months. But it can include functions similar to transitional housing: program and community spaces for group cooking and dining, socializing, children's play, one-to-one counseling, group sessions, program administration offices, in-house volunteer staff training space, and private space for residents to meet with volunteers who provide important support. Residents (a) balance family needs with scant resources, (b) attempt to obtain housing vouchers, (c) secure welfare payments and food stamps, (d) apply for and secure transitional housing or a permanent rent subsidy certificate or voucher, and (e) organize children's education, all of which can involve travel. An unusual emergency model, the Shelter for Victims of Domestic Violence in Albuquerque, New Mexico, has its own one-room schoolhouse located at the shelter site staffed by the public school system.

Transitional housing is perceived in two ways: (a) as a stopgap for homelessness and (b) as an impetus toward life improvement. This stage typically lasts from six months to two years. As a stopgap measure, the transitional period primarily fills the need between homelessness and permanent housing. For life improvement, the transitional period is over when a single mother has achieved her goal: a foundation for long-term family stabilization and self-sufficiency. Transitional services augment those provided in emergency housing, giving residents access to counseling, skill development, and a support network.

Housing design, child care, and other service spaces are defined differently by different sponsors. Previous building uses influence the choice to some extent. Reclaimed large houses are more likely to be shared because subdivision into apartments requires many architectural changes. Buildings that were once temporary lodgings, such as hotels or hospitals that have a transitional function, ordinarily have less private household and, therefore, more shared space. For clusters of houses, former apartment buildings, and new construction, private apartments are more typical. Buildings of all types, however, have been designed with a variety of shared and private spaces. Most transitional housing has been partially funded by the Supportive Housing Demonstration Program under the Stewart B. McKinney Act of 1987. This funding has encouraged innovation and diversity.

Sharing

Single mothers who have been victims of domestic violence, who are in their teens, who are recovering from substance abuse, or who have been imprisoned, benefit from close peer support and sharing. Shared spaces include bathrooms, kitchens, dining rooms, and living rooms. Sharing encourages spontaneous cooperation in baby-sitting and pooling resources. Each mother need not shop and cook every day. These tasks can be traded, giving each more time for job development and life improvement tasks. But it can also be difficult if the life and parenting styles of apartment mates are dissimilar, if counseling help is limited, and if housemates change often. Sharing encourages a buddy system and can be used to teach interpersonal skills as part of a program's goals. Sharing can also be part of a staged program approach as at the Visions Teen Parent Home in Massachusetts, where four teen parents share an apartment, each with private bedroom space. Later in the program, they move at the same site to their own private apartments.

Sharing can be attractive to some program planners as a romantic notion of togetherness. It can be less attractive to a household for whom privacy is an unattained goal. Some see sharing as integral to the definition of transition, connecting permanence with being on one's own. Although some planners of transitional housing are concerned that incentive to move onward could be diminished by a private comfortable apartment similar to typical permanent housing, no evidence is available to support this disincentive

theory. Some agencies argue for sharing because it appears to have space economies. Experience with congregate housing, however, shows that shared spaces are more successful when adequate private space is also provided.

Private Apartments

Independent living in private apartments is enhanced by additional shared space to promote peer support. In the redesign of college dormitories for single mothers and children, separate apartments were created, giving mothers private study space and access to shared laundry-playroom space. Grouping apartment doorways and clusters of apartments can also strengthen peer connectionss, similar to tenements where neighbor bonding is strengthened by shared stair landings for front and back doors. With separate apartments, more service outreach to families and nutritional counseling may be necessary. More emphasis through formal programming brings residents together in shared community space. Yet homeless households value an apartment of their own, even if it is small. Research shows that only sharing that is chosen is truly appreciated. With limited on-site services, and particularly for single-parent or family housing that includes male residents, private apartments are typical.

Services

Depending on program goals and individual needs, services can include counseling in self-esteem, parenting, budgeting, nutrition, job training, and career planning. A narrow or broad range of these on-site services may be offered to residents and also to the surrounding community. Some social service providers recommend some services in the wider neighborhood for transitional residents to establish ties that can be sustained after they move to permanent housing. Alternatively, if services are primarily provided at the site, transitional housing can remain a center of support for those who have moved to permanent housing and for the neighborhood. Peer support advantages coupled with service provision convenience suggest a minimum number of 5 or 6 families living in proximity. Some service providers recommend an optimal subgroup size of 8 to 10 households within large sites. This subgroup is small enough to allow households to know each other and large enough for diversity. Subgroups at large sites can be created by physical proximity and clustering for group services to those with similar backgrounds.

On-site child care offers the greatest convenience and similarity to a home environment. It can be the focus for a parenting program for teens who must devote time to finishing high school while they are learning parenting skills. A child care center in transitional housing can encourage its use after the family moves to permanent housing, becoming a point of continuity in family life. Those who have moved on and return for services become role models for new residents. If continuity of child care service is not offered to families leaving transitional housing, the typically long wait for a new center and potential interruption of services can be yet another obstacle in a household's progress. This disruption can be prevented by having children move from on- to off-site child care during the time the family lives in transitional housing.

Basic space components in transitional housing typically include the following:

- ▶ Furnished single rooms or suites of rooms
- ▶ Furnished or partly furnished shared or private apartments
- ▶ Private and/or community kitchens and dining space
- ▶ Office, counseling, and community space
- ▶ Child care space, both indoors and outdoors
- ▶ Storage for a family's possessions
- ▶ Adjunct functions, such as job training

Although some families move directly from emergency to permanent housing, transitional housing assists as the stage between. Home roots and more stable lives are established when all three stages are available in the same or nearby neighborhoods. Emergency and transitional housing may be located at the same site, as in Los Angeles, where residents from the Chernow House emergency shelter can move to the adjacent Triangle House for transitional housing. Others are helped to find nearby permanent housing. In other cases, transitional and permanent housing are combined. Some transitional sites integrate with permanent housing by continuing to provide child care and counseling, as at Rainbow House in Pennsylvania and Virginia Place in Kentucky. The Greyston Family Inn in New York and Project Family Independence in Boston take an ambitious approach, planning to assist households in transition to stabilize their lives and become permanent homeowners at the same site.

Unless the transitional period has brought a single mother to an income level of market rate rents, subsidies are generally necessary. A progression from rental to ownership is the preferred permanence option for most, but renting is typical for most leaving transitional housing. Without child care, services, a community of support, safety, and nearby job opportunities providing more than subsistence wages, affordable housing does not necessarily end homelessness. Without an adequate economic and social support, sickness or other crises may start the homeless cycle again. A network of other single-mother families contributes to household stability and quality of life. (SEE ALSO: *Family Self-Sufficiency;* **Homelessness;** *Single-Parent Families*)

—*Joan Forrester Sprague*

Further Reading

Adkins, Laura, ed. 1989. *The Search for Shelter Workbook.* Washington, DC: American Institute of Architects.

Dandekar, Hemalata, ed. 1993. *Shelter, Women and Development: First and Third World Perspectives.* Ann Arbor, MI: George Wahr.

Ford Foundation. 1985. *Women, Children, and Poverty in America.* New York: Author.

Franck, Karen and Sherry Ahrentzen, eds. 1989. *New Households, New Housing.* New York: Van Nostrand Reinhold.

Greer, Nora Richter. 1988. *The Creation of Shelter.* Washington, DC: American Institute of Architects.

McClain, Cassie and Janet Doyle. 1984. *Women and Housing: Changing Needs and the Failure of Policy.* Toronto: Canadian Council on Social Development/Lorimer.

Sprague, Joan Forrester. 1991. *More Than Housing: Lifeboats for Women and Children.* Boston: Butterworth. ◀

▶ Turnkey Public Housing

Created in 1965 in the United States, the Turnkey program was a variant on the public housing program. Along with the Leased Housing program, it was a response to criticisms about the conventional public housing program, and it provided increased opportunities for the private for-profit sector to participate.

Under the Turnkey program, a developer entered into a contract with a local housing authority to construct a project. The developer then sold the project (or "turned the key" over) to the housing authority at the stipulated price. According to data collected in the late 1970s, about one-third of all public housing developments had been built in this way, but it was unclear whether it had managed to reduce either the time or costs of development or whether it improved the quality.

Although the private for-profit development community had always been staunchly opposed to the conventional public housing program, because it essentially bypassed their constituents, the turnkey program was popular among developers. Testifying before the U.S. Senate in 1968, for example, the president of the National Association of Home Builders advocated that all new public housing authorizations be directed primarily to the Turnkey program. (SEE ALSO: *Privatization;* **Public Housing;** *Public/ Private Housing Partnership*)

—*Rachel G. Bratt*

Further Reading

Kolodny, Robert. 1979. *Exploring New Strategies for Improving Public Housing Management.* Report prepared for the U.S. Department of Housing and Urban Development, Office of Policy Development and Research. Washington, DC: Department of Housing and Urban Development.

Meehan, Eugene J. 1979. *The Quality of Federal Policy Making: Programmed Failure in Public Housing.* Columbia: University of Missouri Press.

U.S. Senate. 1966. *Hearings before the Committee on Banking and Currency.* 89th Cong., 2d sess., March 5-20, p. 293.

U

▶ UNCHS Habitat News

UNCHS Habitat News is a former, triannual publication of the U.N. Centre for Human Settlements (UNCHS) that focused on the interests of housing professionals, officials, and scholars. Published since 1983, it covered the activities of UNCHS, as well as a variety of human settlement issues on the international, regional, and national levels with special reference to questions of affordability and sustainable development. Circulation: 10,500 worldwide. (SEE ALSO: *U.N. Centre for Human Settlements*)

—*Caroline Nagel* ◀

▶ Uniform Relocation Assistance and Real Property Acquisition Policies Act of 1970

The Uniform Relocation Assistance and Real Property Acquisition Policies Act of 1970 was the federal government's attempt to introduce order and fairness into a fairly chaotic situation of eminent domain land takings and population displacement because of urban renewal and highway construction programs.

Accurate data on eminent domain activity have never been easy to obtain, and anticipatory activities (moveouts, sales) before actual taking actions mean that official data severely understate the displacement impact of such government actions. But clearly, the effects of government projects on land, structures, and people are immense and not evenly distributed across demographic characteristics. The National Commission on Urban Problems, in its 1968 report, noted that the urban renewal, highway, and public housing programs, through 1967, had collectively demolished 1,054,000 dwelling units, directly displacing 2 to 3 million persons, and "it has been primarily the poor, the near poor and the lower middle-class whose houses have been demolished" (p. 82). Although that conclusion was silent on the race issue, the urban renewal program so heavily concentrated on "slum clearance" of inner-city minority neighborhoods ripe for "higher and better uses" that the popular sobriquet for the program was "Negro removal."

A National Association of Home Builders study (Hartman 1971) gave an even higher estimate: that total housing demolition by all public programs (federal, state, and local) in the 1950 to 1968 period amounted to 2.38 million units.

Scattered studies of the effects of these land takings reported consistently disturbing findings. Increased housing costs were extremely widespread and of considerable proportions, often irrespective of improvements in housing conditions or the family's ability to absorb these increased costs. Relocatees tended to cluster in the immediate vicinity of the displacement project, a pattern suggesting at best marginal improvement in living conditions and often giving rise within a few years to a repeat of the displacement experience. Patterns of racial segregation became more pronounced. Relocation assistance was inadequate and used by only a minority of displacees. Overcrowding did not decrease and in some instances increased. Displacement and relocation often occasioned severe social and personal disruption. The degree of housing improvement was far less than displacees and the general community had the right to expect from programs ostensibly devoted to the public welfare and that resulted in the costs associated with displacement and relocation.

Earlier statutory provisions—mainly in the 1949 Housing Act, which introduced urban renewal—in theory, guaranteed decent, affordable, appropriately located relocation housing but offered few programs to carry out this mandate. Litigation on behalf of displacees was abundant, and court decisions on occasion provided needed protection and resources. The 1970 Uniform Relocation Act represented an attempt by Congress not only to ensure adequate treatment but to create uniform treatment among the many federal agencies with taking powers. (The act provided no assistance for those displaced by state and local governments or by private action.)

Major Features

Expanded eligibility for benefits. Persons no longer had to wait until actual project execution to obtain benefits but became eligible on official notice to vacate. Left uncovered were people who, knowing forced displacement was inevitable, left before officially required to do so.

Moving payments. There would be a payment of actual moving expenses or a moving allowance (up to $300) plus a $200 "dislocation allowance," higher amounts than previously available.

Relocation housing for homeowners. There would be a payment of up to $15,000 in excess of the condemnation award or negotiated price, in recognition that comparable replacement housing and financing costs might be higher than the "fair market value" of the previous home.

Relocation housing for tenants. Displaced tenants and displaced homeowners who become tenants became eligible for up to $4,000 over a four-year period, in recognition of the virtual inevitability of higher rents for the new quarters (but limiting this compensation to a four-year period); the payment also could be used as down payment for a home purchase, providing the displacees matched any payment more than $2,000 with their own funds. These payments were higher than what was previously offered, and for the first time, single householders who were neither elderly nor handicapped became eligible to receive this aid. (Nearly a third of all relocated households were individuals, and their incomes were generally lower than family incomes.)

Relocation assistance/replacement housing. Improved information, planning, and home-finding services were mandated. In addition, the act provided that the federal funding agency could become a "houser of last resort," using project funds directly to provide replacement housing if no other sources were available.

The Uniform Relocation Act represented a significant advance in protection for displacees. However, because each federal agency (Department of Housing and Urban Development, Department of Transportation, General Services Administration, Department of Defense, Department of Justice, etc.) was left to promulgate its own regulations, some of the uniformity goal was adulterated. The major advance was increased monetary payments to displaced families and individuals, although for tenants, these are only of short-term duration. The underlying problem of inadequate housing resources for lower-income households and the rent inflation this shortage induces—which these taking programs, of course, exacerbate—was left virtually untouched. (SEE ALSO: *Displacement*)

—Chester Hartman

Further Reading

Hartman, Chester. 1964. "The Housing of Relocated Families." *Journal of the American Institute of Planners* 30:266-86.

Hartman, Chester. 1971. "Relocation: Illusory Promises and No Relief." *Virginia Law Review* 57:745-817.

LeGates, Richard T. and Chester Hartman. 1981. "Displacement." *Clearinghouse Review* 15:207-49.

National Commission on Urban Problems. 1968. *Building the American City.* Washington, DC: Author. ◄

▶ Uniform Residential Landlord and Tenant Act

U.S. law regulating the landlord-tenant relationship underwent a major transformation during the 1960s and the 1970s. The early decisions that changed the law were decisions by state courts recognizing an implied warranty of habitability flowing from the landlord to the tenant.

From medieval times in England, the landlord-tenant relationship had been recognized as founded on both property law and contract law, although the possessory concepts of property law predominated. Once possession was transferred from the landlord to the tenant, the responsibilities for the conditions of the premises passed to the tenant.

After World War II, attorneys representing tenants in urban multifamily apartment buildings mounted major efforts to change the law concerning responsibilities for conditions of the premises. In a series of well-known cases in the late 1960s and early 1970s, a majority of state courts that considered the matter agreed with the argument that the realities of urban residential living in the 20th-century United States necessitated a change in the conceptual framework of landlord-tenant law. Changed expectations of the parties, together with a relative lack of bargaining power for urban tenants and increasing concern about the poor housing conditions of low-income tenants led courts to conclude that urban residential landlords were impliedly promising to provide "a package of shelter" meeting basic standards of health and safety reflected in local housing codes. In 1972, the Conference of Commissioners on Uniform State Laws, responding to increasing requests that state legislatures define the responsibilities of both landlords and tenants, approved the Uniform Residential Landlord and Tenant Act (URLTA).

URLTA is a comprehensive treatment of the landlord-tenant relationship. It adopts a contract law rather than a property law mode of analysis. Contract law principles such as the requirement to act in good faith and the prohibition against unconscionable terms are integral parts of the statute.

In addition to adopting a contract law approach, the statute also imposes certain regulatory standards on the relationship. For example, the statute prohibits tenants from agreeing to waive or forego rights or remedies under the act, to confess judgment, to pay landlord's attorney fees, or to allow exculpation, limitation, or indemnification of any liability of the landlord under the law.

URLTA spells out obligations of both the landlord and the tenant. Landlords may not demand excessive security and must return security deposits or account for them within a specified period of time. Landlords are required to maintain the premises, including complying with requirements of applicable building and housing codes materially affecting health and safety, making all repairs to put and keep the premises in a fit and habitable condition, keeping all common areas clean and safe, maintaining in good and safe working order electrical, plumbing, sanitary, heating, ventilating, and other facilities supplied or required to be supplied by the landlord. Landlords must also provide and maintain appropriate receptacles for trash and garbage removal and must supply running water and reasonable amounts of hot water and reasonable heat within specific periods of time.

These duties are not waivable except that a landlord and a tenant of a single-family residence may agree in writing that the tenant will perform landlord duties associated with trash removal, providing water and heat, and specified repairs and remodeling, provided that the transaction is "entered into in good faith and not for the purpose of evading the obligations of the landlord." Also, in limited situations, tenants of multifamily dwellings may agree to perform specified repairs, maintenance, alterations, and remodeling.

URLTA requires tenants to comply with building and housing codes materially affecting health and safety, to keep the part of the premises that the tenant occupies and uses as clean and safe "as the condition of the premises permit," to dispose of all trash and other waste in a clean and safe manner, to keep all plumbing fixtures "as clear as the condition permits," and to use in a "reasonable manner" all electrical plumbing, sanitary, heating, and other facilities and appliances, including elevators. Tenants must refrain from deliberately or negligently destroying, defacing, or damaging any part of the premises and must conduct themselves in a manner that will not disturb their neighbors' peaceful enjoyment of the premises.

The act authorizes landlords to adopt rules and regulations from time to time concerning use and occupancy and specifies that such rules and regulations are enforceable against tenants only if the regulations (a) are designed for the convenience, safety, or welfare of the tenants; (b) are reasonably related to the purpose for which they are adopted; (c) apply to all tenants in a fair manner; (d) are sufficiently explicit to fairly inform tenants of what they must and must not do; and (e) are not designed to evade obligations of the landlord. Tenants must receive notice of applicable rules and regulations when entering into the rental agreement or when the rules and regulations are adopted.

The act also spells out remedies for noncompliance, authorizing among other remedies, tenants' use of self-help to correct minor defects and assertion of landlord's noncompliance as a defense to an action for possession for nonpayment of rent. Landlords' remedies include traditional eviction for nonpayment of rent and assessing tenants for the costs of repairing damages caused by tenants.

At least 15 states had enacted legislation based on URLTA as of 1997. Thirteen of those states adopted it between 1972 and 1978. The other two states, Rhode Island and South Carolina, adopted the act in 1986. Other states have approached the question of responsibility for conditions of the premises through court decision or by individual pieces of legislation focusing on specific parts of the relationship but not treating it comprehensively. (SEE ALSO: *Displacement; Housing Credit Access; Tenant Organizing in the U.S., History of;* **Urban Redevelopment**)

—*Peter W. Salsich, Jr.*

Further Reading

Backman, J. 1980. "Tenant as Consumer: Comparison of Developments in Consumer Law and in Landlord/Tenant Law." *Oklahoma Law Review* 1(33):17-22, 35-39

Blumberg, R. and B. Robbins. 1976. "Beyond URLTA: A Program for Achieving Real Tenant Goals." *Harvard Civil Rights-Civil Liberties Law Review* 1(11):3-22

Cunningham, R. 1979. "The New Implied and Statutory Warranties of Habitability in Residential Leases: From Contract to Status." *Urban Law Annual* 3(16):65-74, 127-29.

Cunningham, R., W. Stoebuck, and D. Whitman. 1993. *The Law of Property.* 2d ed. St. Paul, MN: West (see pp. 312-16).

Glendon, M. 1982. "The Transformation of American Landlord-Tenant Law." *Boston College Law Review* 502(23):528-45.

National Conference of Commissioners on Uniform State Laws. 1985. "Uniform Residential Landlord and Tenant Act." Pp. 427-508 in *Uniform Laws Annotated.* St. Paul, MN: West (also see 1995 supplement).

Rabin, E. 1984. "The Revolution in Residential Landlord-Tenant Law: Causes and Consequences." *Cornell Law Review* 69(517):520-40.

Schoshinski, R. 1980. *American Law of Landlord and Tenant.* Rochester, NY: Lawyers Co-operative Publishing (also see 1995 supplement). ◄

► United Nations Centre for Human Settlements

It has been estimated that one-fourth of the world's population lacks adequate shelter. It is the task of the U.N. Centre for Human Settlements, also known as "Habitat," to focus international attention on shelter and associated human settlement needs. Even though most of Habitat's programs deal with developing countries, its scope encompasses the entire U.N. system. In 1997, Habitat was undergoing an organizational reappraisal that may put it into closer administrative relationship with other U.N. agencies such as the U.N. Environment Programme.

The U.N. Centre for Human Settlements was established in 1978 by the U.N. General Assembly on the foundation of the U.N. Centre for Housing, Building and Planning, which it replaced. This was two years after the Vancouver, Canada, meeting of governments and nongovernmental organizations titled, Habitat: U.N. Conference on Human Settlements. In addition to the U.N. Secretariat, four individuals played important roles in organizing this conference: Helena Benitez (educator from the Philippines), Constantinos Doxiadis (Greek architect-planner), Margaret Mead (U.S. anthropologist), and Barbara Ward (British historian). The resulting "Vancouver Declaration" called on all U.N. member states to adopt "spatial strategy plans" and "human settlement policies" to "meet progressive minimum standards for an acceptable quality of life." It should be noted that "housing" was even then considered but one element in human settlements. (A second worldwide U.N. conference, known as "Habitat II," was held in 1996 in Istanbul, Turkey.)

In 1988, the U.N. General Assembly enacted another major declaration in its Resolution 43/181, Global Strategy for Shelter to the Year 2000. It calls for "adequate shelter for all by the year 2000," with the strategy's major focus on "improving the situation of the disadvantaged and the poor" and on honoring "sustainable development." The resolution asks the U.N. Centre for Human Settlements to monitor and coordinate the strategy.

With such ambitious assignments, it is perhaps surprising that Habitat's "regular" budget is rather modest, approximately $5 million per year. Some 120 professional and 150 support staff discharge a variety of functions from Habitat's Nairobi, Kenya, headquarters and from several offices spanning the globe. Habitat conducts research and training, provides information to policymakers as well as to grassroots leaders, makes available methods for technical cooperation, and provides technical assistance to communities seeking appropriate sewer facilities and to bankers concerned about innovative mortgage systems. It has published a wide range of technical reports (from advisories on how to increase citizen participation to a bibliography on cooperative housing), the periodical *Habitat News,* and the multilingual newsletter *Shelter Bulletin.*

However, Habitat does not have the resources to fund, by itself, the actual construction of projects. The U.N. Centre for Human Settlements depends on partners within and outside the U.N. system to plan and implement most of the approximately 300 projects with which it is involved. Some of the major collaborators are the U.N. Development Programme (UNDP), the World Bank, and other donor organizations, such as the Danish International Development Agency.

Nongovernmental organizations (NGOs) that support Habitat's programs and try to influence its policies include the NGO Committee on Shelter and Community, the National Committee for HABITAT (a U.S. NGO), the Habitat International Coalition, and ECO-HAB, a New York-based international NGO originally formed to highlight settlement concerns at the 1972 Stockholm U.N. Conference on the Environment.

The human settlements program has grown considerably since the original Habitat Conference. It has adapted to the new emphases and concerns, for example, by expanding its activities in the areas of human resource development, institution strengthening, participatory approaches to human settlements planning and management issues, defining and supporting the role of women in human settlements development, the production and dissemination of analyses and statistical data on human settlements, and the environment. The collaboration with international agencies, both within and outside the U.N. system, is stronger than ever before. The Commission on Human Settlements, in particular, has taken the lead in developing new perspectives on human settlements issues that also have a bearing on wider development issues: Among them are the following:

▶ The New Agenda for Human Settlements, which launched an "enabling approach" to human settlements development and identified urbanization as a challenge and an opportunity for development rather than simply the result of rural underdevelopment and uncontrolled migration

▶ The 1987 International Year of Shelter for the Homeless and the Global Strategy for Shelter to the Year 2000, which translated into an action-oriented, national strategic context the enabling approach pioneered by the New Agenda

▶ The Urban Management Programme, in cooperation with UNDP and the World Bank

▶ The evolution of research, training, and technical cooperation efforts from a project-by-project basis to an integrated program approach

Solutions to human settlements problems have taken the form of (a) policy recommendations across the whole spectrum of human settlements; (b) specific technical guidelines covering building materials, the construction sector, water supply and sanitation, appropriate modes of transport, environmental planning and management, energy, and the assessment of training needs; and (c) field-testing of appropriate and replicable approaches in practically all developing countries. Achievements can be measured in terms of the many countries that, through technical cooperation and the application of policy guidelines and technical research, have succeeded in modernizing their policies, strengthening their human settlements institutions, promoting effective decentralization, and mobilizing the resources of many new actors for human settlements development and thus improving the settlements conditions of relatively large numbers of their population. But there have also been failures: (a) At the policy level, many countries still perceive human settlements, and shelter in particular, as a top-down responsibility resting under the sole responsibility of one ministry. (b) Decentralization policies rarely accompany the transfer of functions and responsibilities with resources, staff, and legislation enabling local authorities to expand their sources of revenue. (c) The poor, their settlements, and their informal activities are still seen as disturbing problems rather than resources to be included as legitimate actors in the mainstream of development. (d) More generally, human settlements are still perceived by too many policymakers as a social expenditure sector rather than what they are and what they can be—one of the indispensable foundations of social and economic progress, economic growth, human development, and environmental improvement.

The U.N. Conference on Environment and Development, the Earth Summit, held in Rio de Janeiro in 1992 also adopted important recommendations for sustainable development of human settlements. These are contained in a separate chapter of Agenda 21, the ambitious action plan adopted by the conference and endorsed by the U.N. General Assembly.

A new U.N. body, the Commission on Sustainable Development, has been established to monitor and report on the implementation of Agenda 21 by both national and international organizations. The vast scope and implications of human settlements programs will be given due consideration for sustainable development by the commission in its efforts to provide cohesive direction for sustainable development by all sectors. Contact address: P.O. Box 30030, Nairobi, Kenya. Phone: 2542 62 1234. Fax: 2542 62 4266. (SEE ALSO: *Global Strategy for Shelter;* **Third World Housing;** *U.S. Agency for International Development; World Bank*)

—*Hans B. C. Spiegel*

Further Reading

Okpala, D. C. I. 1996. "ViewPoint: The Second United Nations Conference on Human Settlements (Habitat II)." *Town Planning Review* 18(2):iii-xii.

U.N. Centre for Human Settlements. 1991. *Global Strategy for Shelter to the Year 2000.* Nairobi, Kenya: Author.

———. 1991. *Operational Activities Report 1991*. Nairobi, Kenya: Author.

———. n.d. *UNCHS (Habitat) Profile*. Nairobi, Kenya: Author.

———. 1996. *An Urbanizing World: Global Report on Human Settlements, 1996*. Oxford, UK: Oxford University Press.

United Nations. 1976. *Report of Habitat: United Nations Conference on Human Settlements, Vancouver, 31 May—11 June 1976*. New York: Author.

———. 1993. *Agenda 21: Programme of Action for Sustainable Development*. Earth Summit, U.N. Conference on Environment and Development, Rio de Janeiro, Brazil, 1992. New York: Author. ◄

► United States Agency for International Development

Mission and Origin

The U.S. Agency for International Development (USAID) was created in 1961 under the Kennedy administration and designed to function as a semiautonomous agency under the State Department. An integral part of USAID is the Office of Environment and Urban Programs, formerly the Office of Housing and Urban Programs. For more than 30 years, the Office of Housing and Urban Programs has aimed to respond to the housing needs of low-income urban families in developing nations by focusing on the municipality as a method of decentralizing housing opportunities. The office's Urban Environmental Credit Program (formerly the Housing Guaranty Program) is USAID's primary capital resource for urban environmental infrastructure, shelter, and related programs. The office also promotes policy reform in the areas of shelter and infrastructure finance and of urban environmental management and municipal management. These three foci make up the agency's current philosophy: To meet the challenges of rapid growth, a "properly managed urbanization" must be in place to stimulate the economy and provide basic urban services and housing for rapidly multiplying populations in the developing world. Arguably, housing contributes directly to economic growth, which in turn affects the local and national economy.

The current mission of the USAID Office with regard to housing is to promote reliance on market forces, individual initiative, and the private sector to finance and produce shelter. This mission is carried out through the Urban Environmental Credit Program, formerly the Housing Guaranty Program. This program is USAID's primary capital resource for shelter and provides long-term financing for low-income shelter and urban development to developing countries deemed creditworthy.

The Budget

Since its inception, USAID has authorized more than $2.9 billion in guarantees to finance housing programs and projects. More than 40 countries and 210 projects have received support in the form of $2.4 billion under contract. To support related urban technical assistance, research, and training in the mid 1990s, USAID received approximately $4 million annually in grant funds. Budget and personnel cuts, which began in 1985, have created challenges for the office and have frustrated an agency decision to gradually increase its urban capabilities. These cuts have also reduced efforts to enhance the expertise and resources of the regional offices in developing countries. However, training activities and priority research have been able to survive these financial constraints.

Organization and Personnel

With approximately 50 U.S. professionals and an equal number of full-time consultants, the Office of Housing and Urban Programs is quite small, relative to other government agencies. The office provides capital assistance as well as training and technical assistance through a network of regional housing and urban development offices (RHUDOs) and AID mission-based housing and urban development advisers, as well as its AID Washington, D.C., office. As one of three offices in the Center for the Environment, the office's programs address the impact of urbanization on the environment and focus on the contribution that sound urban management can make to environmental quality.

The Washington, D.C., office is divided into four divisions: Field Operations, Urban Environment, Policy and Municipal Development, and Shelter and Infrastructure Finance. The office supports regional programs, provides leadership in research and training related to USAID initiatives, and promotes opportunities for economic growth.

Relationships and Collaboration

Although a theme of USAID assistance focuses on decentralization, the agency does see a particular role for the governments in housing provision. The office recognizes that a government can help to make private initiatives more efficient through its ability to establish "realistic building standards," make credit available to low-income families, and provide serviced land at a reasonable cost, as well as deal with land tenure. The USAID objective is for governments to capitalize on the informal sector responsible for the majority of new housing for low-income families in less developed countries.

The office actively participates in international meetings that focus on urbanization questions. Once each year, each RHUDO sponsors a meeting on the policy issues of the region.

USAID has strong connections to several organizations involved with shelter. In 1990, USAID cosponsored the Third International Shelter Conference with the World Bank, the U.N. Centre for Human Settlements (Habitat), and the U.S. National Association of Realtors. Collaborative work with these organizations and USAID has also taken place in the areas of research and training. The ties with the World Bank are particularly strong. These close connections have resulted in projects financed jointly as well as significant agreements made on the policy agenda.

Approaches and Changes

The approach taken by USAID in responding to housing worldwide in the 1990s is the Urban Environmental Credit Program. This program involves collaboration between USAID and a host country housing institution. These institutions, such as a national housing bank, a housing development corporation, a government ministry, a central savings and loan, or a similar entity in the private sector, act as a borrower in the collaboration.

In the process, the host country makes a request to USAID. In turn, USAID works with host country officials to evaluate the institutional context and establish the type of shelter program to be supported. Once the project has been mutually agreed on, it is further developed, then authorized by USAID. At this point, the parties enter into an implementation agreement that delineates the use of the proceeds of the loan and sets conditions that must be fulfilled for disbursements under the loan to take place. Simultaneously, the borrower evaluates the U.S. capital markets to determine the most favorable terms for a U.S. government-guaranteed loan. A negotiation process ensues in which the borrower and the U.S. lender agree on the terms of financing. These terms must reflect the prevailing interest rates for U.S. securities of comparable maturity and are formalized in a loan agreement. This agreement is then subject to USAID approval and must include provisions regarding the payment and transfer agent, prepayment rights, lenders' fees, terms, amortization, and other charges. Any deviation from the inclusion of these additional provisions must be approved by the Office of Housing and Urban Programs. Repayment of these loans is guaranteed by the full faith and credit of the U.S. government. To ensure this guaranty, after the signing of the loan agreement, USAID signs a contract and charges a fee for its guaranty. In addition, USAID requires that the government of the borrowing country sign a full-faith-and-credit guaranty of repayment of the loan and outstanding interest.

This financial philosophy of the Urban Program has shifted significantly from the sites-and-services programs of the earlier years of USAID. In sites-and-services housing programs, the philosophy (a) embraced a grassroots level of identifying the nature of the housing site, the financial capacity of the people, the local materials cost, and reasonable targets for the type of residents to be housed and (b) concentrated on the sanitation of the site as its first priority. The second investment made in a sites-and-services program was to distribute the housing sites to a cooperative of 50 to 100 people, which took responsibility for the site with the cost of the investment distributed in terms of liability per household. The homeowners cooperative became responsible to collect individual homeowners' payments and collectively submit them to the host country housing institution, thus placing the burden of default on the cooperatives. Homeowners were able to spread payments out over a period of time, with the cooperatives recognizing the sporadic payment schedule typical of a peasant community. Thus, with the flexible payment schedule, the cooperatives made it possible for any citizen to participate in homeownership.

However, with the change of philosophy of the Housing Guaranty Program, responsibility for default is placed indirectly on the individual homeowner because cooperatives are not involved in the lending process. The loan is secured vis-à-vis a steady paycheck, which essentially limits lending to those individuals on the government or a corporate payroll. Thus, housing is now more readily available to government and corporate employees, whereas the average, lowest-income peasant does not have the credit for the Housing Guaranty loan.

Accomplishments

Undoubtedly, the greatest accomplishments of the USAID housing programs has been the ability of the program to multiply lending capacities of developing countries across the world. The Housing Guaranty Program targets housing as a bankable loan, primarily because the rate of default has dropped significantly, and housing is now considered a "risk-free" loan. Because the Housing Guarantee Program lends its money primarily to government and corporate employees, the benefits are particularly great for individuals in developing countries who are considered "bankable" and less beneficial for those who are not.

Challenges Ahead

The USAID Office of Environment and Urban Programs will continue to face difficult challenges in the future as populations are estimated to rise exponentially in the developing world. These challenges include dealing with increased problems of urbanization as delineated by USAID in its recommendations to the U.S. Congress to revise foreign assistance legislation. The recommendations include issues that the agency feels more adequately address the urban policy and urban-related program authority issues.

In areas less recognized by USAID, future challenges include providing shelter at a price and loan structure that allows even the citizens at the lowest income levels to participate in homeownership. In addition, challenges remain to offer housing that is built using indigenous materials, which is preferable for many reasons: (a) The materials are biodegradable; (b) the architecture is culturally sensitive; (c) it makes the broadest use of local skills; (d) it allows for continued upkeep by the homeowners; and (e) it provides housing that is comfortably heated and cooled throughout the year. With continued refocusing of the USAID housing initiatives to embrace a future aim to fulfill the greatest need rather than search for the lowest risk, USAID programs could then provide comfortable, pleasing housing to even more of the world's people. (SEE ALSO: **Third World Housing;** *United Nations Centre for Human Settlements; World Bank*)

—*Carol Wiechman Maybach*

Further Reading

Gunn, Angus M. 1978. *Habitat: Human Settlements in an Urban Age.* Oxford, UK: Pergamon.

Laguian, Aprodicio A. 1983. *Basic Housing: Policies for Urban Sites, Services, and Shelter in Developing Countries.* Ottawa, Ontario, Canada: International Development Research Centre.

Stren, Richard et al. 1992. *An Urban Problematique: The Challenge of Urbanization for Development Assistance.* Toronto: Centre for Urban and Community Studies.

U.S. Agency for International Development. 1992. *Office of Housing and Urban Programs.* Washington DC: Author.

Yeh, Stephen H. 1979. *Housing Asia's Millions: Problems, Policies, & Prospects for Low-Cost Housing in Southeast Asia.* Ottawa, Ontario, Canada: International Development Research Centre.

Yeung, Y. M. 1983. *A Place to Live: More Effective Low-Cost Housing in Asia.* Ottawa, Ontario, Canada: International Development Research Centre. ◄

► United States Bureau of the Census

The Early Years

The U.S. Bureau of the Census collects information about the nation's people and its institutions, producing more than 2,000 reports each year. The data collected by the census bureau reflect contemporary social and economic concerns. The first decennial census in 1790 counted the population to fulfill the constitutional mandate of apportioning the House of Representatives; it also counted the number of "free, white, males, sixteen years and older," apparently to assess potential military manpower.

Subsequently, the census of 1810 included inquiries on manufacturing; in 1820, inquiries on agriculture and commerce were added. The seventh decennial census in 1850 was the first to collect housing statistics when a count of "dwelling houses" was obtained. More housing data were added when the 1860 census counted dwelling houses, slave houses, and included an inquiry on value of real estate. The 1870 and 1880 censuses contained several questions on Indian housing. The 1880 Census Act contained several important milestones in the evolution of census taking. The act established a census office and a superintendent for the duration of the census period and also provided for specially appointed enumerators to replace the U.S. marshals who had conducted censuses since 1790. In addition, the confidentiality of the information collected by census employees was codified.

Housing content was expanded for the 1890 census when the Single Tax League persuaded Congress to direct the superintendent of the census to ask a series of questions on home and farm ownership, indebtedness, the value of the property, the amount of the mortgage, if any, and the reason the mortgage was placed.

The Census of Housing

During the decade of the 1930s, the effect of the Great Depression on all aspects of U.S. life, including housing, was severe and far-reaching. There was virtually no new construction, foreclosures on homes occurred at the rate of 20,000 per month, the quality of existing structures deteriorated because there was little money for repairs and upkeep, and families were forced to double up with relatives or seek makeshift shelters in hopes of coping with the crisis. The poor condition of the nation's housing and the need for corrective action was voiced by President Franklin D. Roosevelt in his second inaugural address (1937) when he stated, "I see a third of the nation ill housed." In 1939, Congress passed a law authorizing the collection of housing information and income and appropriated funds to accomplish the task.

A committee principally composed of federal government agencies chaired by Dr. Ernest Fisher of the Housing and Home Finance Agency developed a set of questions. That first census of housing laid a comprehensive foundation for the content of future censuses. Many of the same inquiries were asked in the 1990 census. Those 1940 inquiries may be classified into three broad groups:

1. *Facilities and equipment,* which included toilet facilities, bathtub or shower, electric light, refrigeration, radio, heating equipment, and heating and cooking fuel
2. *Physical characteristics,* which included the size and type of structure, exterior material, whether in need of major repairs, year built, rooms, and water supply
3. *Financial characteristics,* including value, rent, utility cost, mortgage status, present debt, mortgage payments, taxes included, interest rate, and type of mortgage holder

In subsequent censuses, many inquiries were dropped, and others were added as the data needs of the United States changed. Thus, the 1950 census dropped questions on estimated rent for owned homes and original purpose of structures because the data were little used. The question on exterior building material was dropped because interviewers could not accurately identify the variety of building materials. The questions on mortgages were shifted to a separate residential finance survey because the detailed nature of the data required collection from the lender. New questions included whether the housing unit was a mobile home and whether the household had a television set.

The 1960 housing census incorporated substantial changes in content and procedures. The scope of the inquiries was broadened, but the extended use of sampling held the total time required by the average household to answer the census to 1950 levels. Items on electric lighting, refrigeration, and the kitchen sink were dropped because of near total saturation. Added items included bathrooms, bedrooms, source of water, sewage disposal, automobiles available, and several appliance inquiries. The 1960 census was the first to have a substantial mail-out and mail-back component.

The 1970 housing census showed modest changes in content. Questions on radio and enumerator rating of the structural conditions of the housing unit were dropped. New inquiries included whether the household had a UHF-equipped television set (the Federal Communications Commission was allocating the UHF channels at the time) and condominium and cooperative ownership.

The 1980 housing census had a few, albeit important, changes in content. One dealt with a change in the definition of a housing unit. Participants at local public meetings (LPMs), particularly in the South, pointed out that mobile homes were becoming an increasingly important part of their housing supply. The bureau's practice since 1940 of excluding vacant mobile homes would seriously understate the housing supply in certain areas.

Therefore, the housing unit definition was modified to include vacant mobile homes, and about 800,000 units were added to the inventory as a result. Inquiries about trucks and vans were added because LPM participants from the South and West told the census bureau that a significant proportion of households in their areas had pickup trucks or vans for family transportation. Perhaps because of the high inflation in the 1970s, the bureau began to shift content emphasis away from structural or equipment inquiries and toward economic or financial aspects. Consequently, the questions on appliances were dropped, and a new group of questions was added to measure the out-of-pocket expenses of single-family homeowners. These questions included whether the unit was mortgaged or not mortgaged and questions about the mortgage payment, real estate taxes, fire and hazard insurance, and utility costs (previously obtained only from renters.)

The selection of items for the 1990 census was guided by several basic criteria:

▶ Only essential items were considered; essential was defined as having application to broad public policy issues or needed to meet federal, state, or local government statutory requirements or to administer programs.
▶ There would be no significant increase in the number of questions.
▶ No controversial subjects would be included.
▶ For each subject, the census bureau had to be able to formulate a clear and concise question that would yield accurate data.

The 1990 census content items are shown in Table 28.

Current Housing Programs

In addition to the decennial census, a number of surveys provide information on housing:

The American Housing Survey (AHS). The AHS collects a broad range of housing, household, and neighborhood characteristics for the United States, its regions, and for selected metropolitan areas.

The Housing Vacancy Survey (HVS). This is a survey of units identified as vacant in the monthly Current Population Survey. The HVS is the only source of quarterly and annual statistics on rental vacancy rates, homeowner vacancy rates, and homeownership rates. Vacancy rates are a component of the index of leading economic indicators published by the Department of Commerce. The census bureau publishes vacancy rates cross-classified by selected characteristics, such as rent or value, number of rooms, and units in structure, as well as selected characteristics of the vacant-for-rent and vacant-for-sale universes. The data are shown for the United States, the four census regions, inside or outside metropolitan areas, the 50 states and District of Columbia, and the 61 largest metropolitan areas.

The Survey of Market Absorption (SOMA). The census bureau monitors the demand for newly built rental and

TABLE 28 1990 Census Content Related to Housing

100% Component

Population
    Household relationship
    Sex
    Race
    Age
    Marital status
    Hispanic origin

Housing
    Number of units in structure
    Number of rooms in unit
    Tenure—owned or rented
    Value of home or monthly rent
    Congregate housing (meals included in rent)
    Vacancy characteristics

Sample Component

Population
    Social characteristics
    Education—enrollment and attainment
    Place of birth, citizenship, and year of entry to U.S.
    Ancestry
    Language spoken at home
    Migration (residence in 1985)
    Disability
    Fertility
    Veteran status

Housing
    Year moved into residence
    Number of bedrooms
    Plumbing and kitchen facilities
    Telephone in unit
    Vehicles available
    Heating fuel
    Source of water and method of sewage disposal
    Year structure built
    Condominium status
    Farm residence
    Shelter costs, including utilities

Economic characteristics
    Labor force
    Occupation, industry, and class of worker
    Place of work and journey to work
    Work experience in 1989
    Income in 1989
    Year last worked

NOTE: Questions dealing with the subjects covered in the 100% component were asked of all persons and housing units. Those covered by the sample component were asked of a portion or sample of the population and housing units.

condominium apartments through information collected in the monthly SOMA. This survey measures the rate at which privately financed, nonsubsidized apartments are absorbed—that is, taken off the market by being rented or sold. The bureau publishes absorption rates by rent classes, price classes, and number of bedrooms in four quarterly and two annual reports, for the United States, the four census regions, and by inside or outside metropolitan areas. The information is of value to builders, bankers, market analysts, land planners, and government officials interested in the multi-unit housing market.

Construction Statistics

In 1959, the Bureau of the Census assumed responsibility for current construction survey programs from the Bureau of Labor Statistics. Although there are annual statistics on housing starts from 1889 and on the value of new construction since 1915 and there are monthly data on building permits since 1945, the bureau's assumption of the program marked a substantial increase in the comprehensiveness and improvement in the timeliness of construction data. The surveys noted above were expanded and new ones introduced, several of them sponsored and published by the Department of Housing and Urban Development (see list under Current Construction Surveys below). In 1967, construction industries were added to the census bureau's economic censuses and have been conducted in years ending in 2 and 7 since then. The Census of Construction Industries includes all establishments that operate as building contractors, heavy construction contractors, and special trade contractors or as land subdividers and developers. The 1987 and 1992 census included the following topics:

- ▶ Number of employees
- ▶ Payrolls
- ▶ Payments for subcontract work and for materials
- ▶ Rent for structures and equipment
- ▶ Cost of power, fuel, and so on
- ▶ Fringe benefits
- ▶ Selected purchase of services
- ▶ Value of construction work done
- ▶ Capital expenditures, assets, and depreciation
- ▶ Inventories
- ▶ Value of construction work by type, location, and ownership

The 1987 data were tabulated in three principal series:

- ▶ Industry Series, CC87-I, presents data for the nation and for states for establishments with payroll in each of the 27 construction industries.
- ▶ Geographic Area Series, CC-87-A, presents data for census divisions, states, and large metropolitan areas.
- ▶ Subject Series, CC87-S, presents data in a single report for the United States only, classified by industry and legal form of organization.

Current Survey Series

Most current construction data are published under the general title *Current Construction Reports*. Major series are shown below:

- ▶ Housing Starts, Series C20, is a monthly report showing data for the United States and regions on new housing units started under private ownership.
- ▶ Housing Completions, Series C22, is a monthly report on the number of privately owned housing units completed for the United States and regions.
- ▶ New One-Family Housing Sold, Series C25, is a monthly report for the United States showing totals for new, privately owned one-family houses sold and for sale. This series includes quarterly supplements that provide additional information on financing and sales prices. It also includes an annual report, *Characteristics of New Housing*, that provides selected physical and financial data for new units.
- ▶ Value of New Construction Put in Place, Series C30, is a monthly report on the total value of new private and public construction.
- ▶ Housing Units Authorized by Building Permits, Series C40, presents monthly data on the number and value of new housing units authorized by permits. Data are shown for the United States, regions, states, selected metropolitan statistical Areas, and about 5,000 permit-issuing places. The annual report includes data for approximately 17,000 permit-issuing places.
- ▶ Expenditures for Residential Improvements and Repairs, Series C50, issues quarterly reports covering residential property owners expenditures for the United States.

Products and Services

Printed Reports. The most widely used and readily available product continues to be printed reports. Nearly all of the bureau's censuses and surveys issue results in the form of statistical tables in such reports. The amount of data that can be efficiently put into a printed report is limited, so for many data series, far more information is available on microfiche or magnetic tape.

Microfiche. Nearly all census bureau reports are available on 4″ × 6″ microfiche.

Computer Tapes. These are particularly well suited to users who need to handle very large amounts of data or data for many small geographic areas (blocks and census tracts, for example.) Several types include the following: (a) Summary Tape Files from the 1990 census, with considerable detailed data and geographic summaries not published elsewhere; (b) Public Use Microdata Files, which allow users to design their own tabulations; and (c) Geographic Reference Files.

CD-ROM Files. These are 4¾″ disks with very large data storage capacity. They are used with a CD-ROM reader connected to a microcomputer. Several of the 1990 census files and some current survey data are available on CD-ROM.

CENDATA. This is an on-line service available commercially and used for access from remote terminals or microcomputers. It carries selected current data, press releases, and other information from census bureau programs.

Computer-assisted telephone interviewing and handheld data capture devices are now being used in bureau surveys along with computer-designed maps and other geographic products and innovative statistical techniques. The bureau's commitment to state-of-the-art technology is also reflected in on-line access to census data and issuing data on laser disks. (SEE ALSO: *American Housing Survey;* **Federal Government;** *HUD Statistical Yearbook; U.S. Department of Housing and Urban Development;* U.S. Housing Market Conditions)

—*William A. Downs* ◀

▶ United States Department of Housing and Urban Development

The U.S. Department of Housing and Urban Development, known as HUD, is one of the newest cabinet-level departments of the federal government. Created in 1965 under the Johnson administration, its mandate is

> to achieve the best administration of the principal programs of the Federal government which provide assistance for housing and for the development of the Nation's communities; to assist the President in achieving maximum coordination of the various Federal activities which have a major effect on urban community, suburban, or metropolitan development; to encourage the solution of problems of housing, urban development, and mass transportation through state, county, town, village, or other local and private action, including promotion of interstate, regional, and metropolitan cooperation; to encourage the maximum contributions that may be made by vigorous private homebuilding and mortgage-lending industries to housing, urban development, and the national economy; and to provide for full and appropriate consideration at the national level of the needs and interests of the Nation's communities and of the people who live and work in them. (P.L. 89-174, 42 U.S.C.A. 3537a [1965])

HUD's goals are lofty and its mission is broad; generally, it has not been viewed as successful in meeting its objectives.

Forerunners of HUD and Political Context of Its Formation
Not until the Great Depression did the federal government become directly involved with housing on a wide scale. But this intervention was largely to shore up the faltering banking industry and to stimulate the construction industry, not primarily to promote affordable housing. The creation of the Federal Housing Administration (FHA) in 1934 (which provided mortgage insurance on privately originated loans) and the United States Housing Authority in 1937 (which administered the public housing program) were two of the major first components of a federal housing strategy.

During World War II, the National Housing Agency (NHA) was created to coordinate the various housing-related activities of the federal government, particularly related to production of war-worker housing. Although the possibility existed for the NHA to become a permanent

agency following the war, it got caught in the political cross fire over the future of the public housing program and the debate over the federal government's postwar role in slum clearance efforts. In a major setback for advocates of a strong federal role in housing and urban redevelopment, in 1946 a conservative Republican Congress rejected permanent status for the NHA. Two years later, however, a central, albeit weak, coordinating agency was formed, the Housing and Home Finance Agency (HHFA), which was responsible for administering the public housing program, as well as the urban renewal program, which was created in 1949.

The Democratic platform of the 1960 presidential campaign advocated the creation of a new cabinet-level department of urban affairs. Although both President Kennedy and Robert Weaver, his appointee as HHFA administrator, were in favor of such a move, they were unsuccessful in persuading Congress to support the new department. As with many other pieces of unfinished business of the New Frontier, it was the task of the President Johnson and the so-called Great Society to implement much of the Kennedy agenda.

In the aftermath of the Watts riots in Los Angeles in the summer of 1965, Congress overwhelmingly passed legislation creating HUD. President Johnson named Robert Weaver as the first secretary of HUD, making him the first African American to head a cabinet-level department.

Historical Problems Confronting HUD
From the outset, HUD was given a broad mandate but limited powers to address the myriad housing and urban problems facing the nation. Many key activities affecting urban areas (such as transportation and economic development) were administered by other entities, such as the Bureau of Public Roads (which later became the Department of Transportation) and the Office of Economic Opportunity. Merged into HUD were the FHA, the Public Housing Administration, the Urban Renewal Authority, and Fannie Mae (formerly, the Federal National Mortgage Association). (Fannie Mae was spun off from HUD in 1968 and transformed into a private corporation with public responsibilities; the Government National Mortgage Association, or Ginnie Mae, a secondary mortgage market entity empowered to purchase subsidized loans, was also created and placed under HUD control.) However, other key housing-related agencies, notably the Federal Home Loan Bank Board, the Department of Veterans Affairs, and the Farmers Home Administration (now the Rural Housing and Community Development Service), with authority over the thrift-housing financing system and the housing programs for veterans and farmers, respectively, continued to function independently. Also important is that the array of subsidies for housing that operate through the federal income tax system are under the jurisdiction of the Internal Revenue Service, not HUD. Furthermore, the housing portion of welfare payments constitute another large amount of funding going into housing over which HUD has no control.

A second key problem facing HUD has been inconsistent support by both the president and Congress. HUD was created with all the high hopes surrounding President Johnson's domestic agenda, including a congressional commitment in 1968 to produce or substantially rehabilitate 26 million units, including 6 million for low- and moderate-income households. Although there was some support during President Nixon's first term, Nixon's 1973 moratorium on HUD's subsidized housing programs signaled a nearly uninterrupted two-decade decline in federal support for housing and urban programs; the only exception being the early years of the Carter administration, 1976 to 1980. At best, President Reagan was hostile to large-scale federal spending for housing and urban activities and, at worst, HUD, under his secretary, Samuel Pierce, was laden with corruption and abuse of programs. With the appointment of President Bush's HUD Secretary, Jack Kemp, the housing/urban agenda was given more attention. However, the primary focus of HUD during Secretary Kemp's tenure was on resident management of public housing and, ultimately, privatizing the stock of public and subsidized housing through the HOPE program (Homeownership and Opportunities for People Everywhere), which was enacted as part of the Cranston-Gonzalez National Affordable Housing Act of 1990. By the time President Clinton took office in early 1993, HUD was widely perceived as an agency teetering on collapse. A major evaluation of HUD, authorized by Congress in 1992 to be carried out by the National Academy of Public Administration, opened its final report with the message: "The Clinton administration and the secretary of HUD offer what may be the last best chance to create an accountable, effective department" (National Academy of Public Administration, 1994, p. ix).

Lack of consistent presidential support for HUD is reflected in HUD's budget. With the advent of the Reagan administration, HUD's new budget authority steadily declined from its peak of $33 billion to only $15 billion by 1989, the most severe cutback of any domestic, cabinet-level federal department.

Another long-standing problem facing HUD is that its programs often have had conflicting components. The Urban Renewal Program, which fostered "slum clearance" was in direct conflict with another of HUD's key goals, to promote affordable housing, since removing the slums often meant dramatically reducing the supply of low-rent housing. Also, there has been an ongoing debate over the extent to which FHA should be viewed as a business or primarily as a socially motivated agency.

Some of HUD's problems have been because of the structure of the programs that Congress has created—a structure that depends on the private sector, in large part, to implement nearly all of the federal government's housing and urban development initiatives. As would be expected, the private homebuilding industry is geared toward maximizing profit, whereas the ostensible goal of many of HUD's programs has been to promote social welfare. It is hardly surprising that these two sets of goals often conflict.

A final source of HUD's problems lies in the very nature of the issues with which it is concerned. HUD has been asked to address a host of serious and complex problems, the origins of which generally lie outside the locale in which the problems manifest themselves. For example, poverty and unemployment are largely the outcomes of international, national, and regional economic decisions rather than local actions or inactions. Also, the very complexity of the problems makes it difficult, if not impossible, to "get it right the first time." Instead, even at best, one might expect a process of trial and error. However, the very visible nature of many of HUD's blunders makes the agency susceptible to criticism rather than a sense that mistakes are part of a learning process.

For example, the Section 235 program, which was enacted in 1968 and was the nation's first subsidized home-ownership program, was beset by many serious problems, primarily because of lack of adequate oversight by HUD in implementing the program. However, rather than improving the program and correcting its defects, the bad press and visibility of the problems signaled the end of the program and effectively signaled a policy shift away from federal support for low-income homeownership. (The Nehemiah program, a relatively small-scale subsidized homeownership program, since 1987, received federal support, but it took nearly 15 years for such a program to recapture federal interest.) Overall, HUD is saddled with a large number of programs to administer. According to the National Academy of Public Administration, between 1980 and 1992, HUD's statutory mandates increased from 54 to just more than 200 programs. Although many of these programs are inactive, HUD has responsibility for a large number of separate initiatives. To make matters even more complex, Congress is constantly changing or creating new programs, the administration of which is in HUD's domain.

Structure of HUD

HUD's organization has undergone numerous changes since its inception. In early 1998, there were 9,697 staff members, representing one of the smallest cabinet-level departments. A secretary, who is a presidential appointee, heads the department. An administrative staff consisting of a deputy secretary and eight assistant secretaries assists this individual. The major divisions of HUD are Housing/FHA, Community Planning and Development, Fair Housing and Equal Opportunity, Public and Indian Housing, Congressional and Intergovernmental Relations, Policy Development and Research, Public Affairs, and Administration. Prior to the Clinton administration and the appointment of his first HUD Secretary, Henry Cisneros, HUD's field staff involved some 81 offices, composed of both regional and area offices. Under Cisneros, HUD's field operation was changed to include 52 state offices (50 states plus the District of Columbia and Puerto Rico) and 29 area offices, servicing major metropolitan areas.

In December 1994, in the aftermath of the election of the 104th Republican-controlled Congress, HUD issued a new *Reinvention Blueprint*. In a harsh self-condemnation, the document called on HUD to "do a better job" than it had done in the past. Under HUD Secretary Andrew Cuomo, the department continued its self-examination and reform initiatives, which include consolidation and privatization of programs and a series of efforts to improve management of the agency. Downsizing the scope of federal housing programs and placing more responsibility on state and local governments are key philosophical components of HUD as it is evolving. HUD's budget for FY 1998 is more than $24 billion.

It is unclear whether HUD's new organization will be better equipped to address and solve the housing and urban ills facing the nation. It is certain, however, that the essence of the problems will continue to present enormous challenges for the agency. (SEE ALSO: *Department of Housing and Urban Development Act of 1965;* **Federal Government;** U.S. Bureau of the Census)

—*Rachel G. Bratt*

Further Reading

Bratt, Rachel G. and W. Dennis Keating. 1993. "Federal Housing Policy and HUD: Past Problems and Future Prospects of a Beleaguered Bureaucracy." *Urban Affairs Quarterly* 29(1):3-27.

Cityscape. 1995. 1(3[Special issue commemorating HUD's 30th Anniversary]).

Gelfand, M. 1975. *A Nation of Cities: The Federal Government and Urban America 1933-1965.* New York: Oxford University Press.

McFarland, M. Carter. 1978. *Federal Government and Urban Problems: HUD: Successes, Failures, and the Fate of our Cities.* Boulder, CO: Westview.

National Academy of Public Administration. 1994. *Renewing HUD: A Long-Term Agenda for Effective Performance.* Washington, DC: Author.

Weaver, Robert C. 1985. "The First Twenty Years of HUD." *Journal of the American Planning Association* 51:463-74.

Welfeld, Irving H. 1992. *HUD Scandals: Howling Headlines and Silent Fiascoes.* New Brunswick, NJ: Transaction Publishers.

Wood, Robert and Beverly M. Klimkowsky. 1990. "HUD in the Nineties: Doubt-Ability and Do-Ability." In *The Future of National Urban Policy.* edited by Marshall Kaplan and Franklin James. Durham, NC: Duke University Press. ◄

▶ Urban Affairs Review

The *Urban Affairs Review* (formerly the *Urban Affairs Quarterly*) is published six times per year, presenting the empirical and theoretical research of urban scholars in a variety of disciplines. First published in 1965, its articles pertain to all urban issues, including urban policy, urban economic development, metropolitan governance and service delivery, housing, residential and community development, and social and cultural dynamics. Circulation: 1,925 internationally. Price: $66 per year for individuals; $225 per year for institutions. Editor: Dennis R. Judd, Public Policy Research Centers, University of Missouri—St Louis, 8001 Natural Bridge Road, St. Louis, MO 63121-4499.

—*Caroline Nagel* ◄

▶ Urban Age

Formerly called "The Urban Edge," *Urban Age* was first published in 1992 with the aim of stimulating debate and interaction on various urban topics in developed and developing countries, including housing. Each issue focuses on a theme. Recent topics include international migration, transportation, environmental problems, and violence.

Published quarterly, *Urban Age* is $20.00 for developed country subscribers and free of charge for developing country subscribers. Editor: Margaret Bergen. Address: Room 4K - 256, The World Bank, 1818 H Street NW, Washington DC 20433. Fax: (202) 522-3224.

—*Laurel Phoenix* ◄

▶ Urban Geography

Urban Geography is a semiquarterly journal focusing on the research interests of geographers, urban planners, policymakers, and scholars in related social science fields. First published in 1980. Articles are problem oriented and cover issues of race, ethnicity, and poverty in the city; international differences in urban form and function; historical preservation; the urban housing market; service provision; and urban economics. Price: $349 per year, North America; $389 per year, elsewhere; $78 per year, individuals. Submissions Editor: Dr. James O. Wheeler, Coeditor, Department of Geography, University of Georgia, Athens, GA 30602-2502.

—*Caroline Nagel* ◄

▶ Urban Homesteading

Homesteading is the term applied to offers of land or property by units of government for development and occupancy by homestead settlers. Urban homesteading was first proposed in the United States in the late 1960s as a way to address the pervasive problem of housing abandonment. In many central cities, large quantities of locally owned, privately owned, and federally held properties were vacant and deteriorating. These deteriorating structures threatened neighborhood stability and contributed to blighting conditions. By 1970, mention of the term *urban homesteading* as a partial solution had become widespread. Few labored under the delusion that urban homesteading was a cure-all for the problem of housing abandonment. Public statements and analyses in the press emphasized that urban homesteading could be useful if carefully applied under special circumstances.

Origin

The federal government has periodically used homesteading to redistribute population to areas unable to attract a significant number of residents—either because of remote location or other disincentives. The legislative origin of homesteading can be traced back to the Armed Occupation Act passed in 1842. It provided for the armed occupation of the unsettled part of the peninsula of East Florida. Under this law, any head of household more than 18 years of age who was able to bear arms could own a parcel of land in southeast Florida after a five-year residency period. Only those persons who did not own land or were not Florida residents were eligible. Occupants were expected to begin construction of a house within one year and to cultivate five acres of the land. Even though the Rural

Homestead Act is usually cited as the earliest homesteading legislation, the Armed Occupation Act was the first statute to enable the government to give land to settlers.

The Rural Homestead Act was signed into law in 1862. This law allowed any U.S. citizen or naturalized citizen more than 21 years of age who had never borne arms against the United States to receive not more than 160 acres of land in specified areas of the western and midwestern United States. With the exception of the $10 application fee, the land was given free, provided that the homesteader build a structure on the land, cultivate part of it, and live on it for five years. The act further specified that if the land was abandoned for more than six months, it would revert to the government. No lands acquired under this act could become liable for the satisfaction of debts incurred prior to the issuing of the homestead patent.

A parallel act, the Morrill Act, was passed in 1862 creating land grant colleges and assigning them the task of advising and counseling rural homesteaders. It provided that, as their primary focus, the original land grant colleges would bridge the gap between the advanced state of the agricultural art and the realities of the new homesteaders.

Many of the program elements and legal provisions of modern urban homesteading are related directly to the Rural Homestead Act. These include the eligibility requirements, the stipulation of self-occupancy, and the period of occupancy before gaining clear title to the homestead. Provisions for support services can be traced to the Morrill Act. The provision that protected rural homesteaders from the loss of their property because of prior debts is presently translated into the need for financial counseling and support. Penalties that were imposed for failure to comply with the regulations are also stipulated today in many urban homesteading ordinances.

Local Programs

Some local urban homesteading programs predated the establishment of the federal program, and many were established later. Local urban homesteading programs in Baltimore, Philadelphia, and Wilmington, Delaware, preceded the federal program. Philadelphia Councilman Joseph Coleman prepared reports on the local program and testified before Congress to encourage the legislation to enable the federal support of urban homesteading. The reported success of the local programs led to the support of urban homesteading among federal legislators. With the promulgation of Section 810 of the Housing and Community Development Act of 1974, the federal homesteading legislation, several cities operated both local and federal programs. Baltimore, Philadelphia, and Wilmington were among the cities that operated both locally and federally enacted urban homesteading programs. Each of these cities applied for and was selected as an Urban Homesteading Demonstration location in 1975.

Baltimore officials attained renown for their well-directed homesteader program. This was accomplished through the use of more city surplus housing inventory than federally held properties. In the 1970s, vacant city properties in Baltimore reached a high of 2,500, and the Department of Housing and Urban Development (HUD) inven-

tory peaked at less than 200. The city program was operated in neighborhoods and on scattered sites. The structures found most suitable for homesteading were small- to moderate-sized row houses that could be rehabilitated at modest cost. The objective of the local program was to retain and attract middle-class residents. Properties were transferred for the sum of $1; the average rehabilitation cost in 1978-1979 was $45,000. The assessed house values had doubled by 1987.

The Baltimore case illustrates how city homestead officials were able to exercise more flexibility with the local program than with the federal program. For example, the federal program required the identification of target neighborhoods, whereas the city program used properties in scattered locations.

The rules governing local programs could differ in other ways from those imposed by the federal program. For example, the Baltimore program mandated that homesteaders satisfy all fire and safety requirements within 6 months after the loan was settled and move into the property within 24 months. Rehabilitation work was to meet all applicable code standards, and the required length of residence prior to transferal of title was 18 months. The Section 810 regulations allowed a longer time for repair of the property and required a longer period of residency before final title was granted. The Baltimore city program was criticized for not being used in the right neighborhoods and not serving the most needy citizens.

Local guidelines in Philadelphia required homesteaders to begin rehabilitation no later than 60 days after title was acquired, to complete the rehabilitation within two years, and to live in the structures at least five years. In this case, the local regulations were closer the federal rules.

The federal government attempted to support local urban homesteading before the enactment of the Section 810 federal program. In early 1974, under the Property Release Option Program (PROP), HUD made 4,100 houses available to 43 cities for use in urban homesteading or other public purposes. Houses made available under PROP could have no remaining property value; hence, some cities refused to accept them and others demolished the structures for parks and open space. Nonetheless, PROP was an incremental step toward greater involvement by the federal government in homesteading.

Program Dimensions

There are a number of similarities among urban homesteading programs regardless of their origin. In most cases, the number of homestead properties available always exceeded the number of interested applicants. A lottery system was widely used to select recipients of the homestead properties. Typically, the urban homestead program administrators determined the ability of the applicant to complete the necessary rehabilitation work prior to the lottery. The abundance of prospective homesteaders has meant that communities maintain waiting lists of applicants who can step in if a homesteader drops out for some reason or who can be considered when additional properties are available. The means of completing the rehabilitation work vary from sweat equity, in which the homesteader completes some portion of the work, to contractual arrangements with licensed contractors. In most cases, some financial investment is required regardless of the labor contributed by the homesteader. Low-interest rate, subsidized loans were made available in some cities irrespective of homesteader income. In other instances, the low-interest loans were available only to lower-income persons. Section 312 federal loans were the primary source of low-interest loans, and their use was limited to lower-income persons.

Urban homesteading efforts brought other issues to the forefront. For example, in Philadelphia, the city controlled very few of the abandoned residential properties, and few of those were structurally suitable for homesteading because substantial deterioration had occurred during lengthy periods of vacancy. This issue highlighted the need for action to obtain repairable properties. The Pennsylvania state legislature passed a special fast-take law that enabled the city to take title to vacant, tax delinquent properties in six months. The city sped up the acquisition of full title to tax-foreclosed properties by six to nine months.

Gift properties, donated by owners in exchange for absolution from delinquent taxes, provided another source of properties for Philadelphia. In addition, the city developed a donor-taker provision that allowed delinquent taxpayers to donate their houses to the homesteading pool and deduct the assessed value from back taxes owed. The city also experimented with the use of spot condemnation powers derived from urban renewal legislation to obtain properties designated by homesteaders to expand the pool of homestead properties.

Program Variation

Urban homesteading can take on many forms as a federal, city-operated program. Units of state and local government or nonprofit quasi-public entities have administered urban homesteading programs. The Philadelphia Housing Development Corporation (PHDC) was established to develop and promote housing opportunities for low- and moderate-income people. Philadelphia city officials turned to PHDC to operate the urban homesteading programs and to target them to their low- and moderate-income clientele.

Provisions for lower-income homesteaders differ from those in place for others. Financing options in Philadelphia included a two-tiered process to provide small start-up loans to begin reconstruction and to start accumulation of equity, typically for six months at 6% interest rate from a loan fund—the Urban Homesteading Finance Corporation. Long-term financing was provided by the Pennsylvania Housing Finance Agency at variable rates ranging from 3% to 7% for terms of 5 to 15 years. This two-part process emerged from the necessity to address the difficulty of making a long-term financial commitment before building equity through property rehabilitation.

The objectives of urban homesteading programs have varied among localities. They have been used to promote lower-income homeownership or to encourage upper-income households to locate in designated areas. The architectural style of homesteaded properties varies from small row houses to large, elegant historic structures. Urban homesteading has also been attempted with multifamily structures, and it has been a response by tenant organizations to the threat of disinvestment by landlords.

Cooperative conversion programs in New York City during the 1970s emerged as a result of severe tenant frustrations with apartment buildings abandoned by the private sector and developed as part of the tenants' efforts to salvage their homes. Tenants and community organizers attempted to rehabilitate and maintain buildings while providing housing and tenant ownership opportunities. Although very few buildings were converted to low-income-cooperatives, the New York programs stimulated a federal demonstration program.

The federal demonstration derived its legal authority from Section 510 of the 1978 amendments to the 1970 Housing Act to determine the feasibility of expanding homeownership in urban areas, giving special attention to multifamily housing. Under Section 510, HUD selected seven cities to participate in multifamily rehabilitation projects. Buildings rehabilitated by community groups and private developers were turned over to cooperative or condominium ownership. No sweat equity was allowed, unlike in the New York programs. An evaluation of the multifamily demonstration projects found that low-income cooperative ownership could be achieved only with long-term Section 8 subsidies for a majority of the tenants.

The limited federal effort and the continued existence of vacant, abandoned properties led to the passage of local homesteading legislation in cities such as Detroit, Pittsburgh, and St. Louis. These efforts corresponded with actions of squatters and other protestors (associated with ACORN—Association of Community Organizations for Reform Now) to assert the continued need for low-income housing in some cities despite the federal homesteading program. Concerns raised by some of these protests were incorporated in the 1983 Housing and Urban-Rural Recovery Act, which gave homesteaders more time to meet local housing standards, gave priority access to Section 312 low-interest loans, required an emphasis on need in selection of homesteaders, and established the demonstration program to acquire privately owned vacant houses for homesteading purposes.

Evaluations of urban homesteading programs have found that they contribute to the improvement of nearby properties and to neighborhood stabilization efforts. Urban homesteading programs continue in various forms. The term has been applied to any situation in which lower-income persons—public housing residents, for example—secure ownership of properties through self-help initiatives. With the discontinuation of the federally funded program in 1992, the emphasis on local programs is likely to grow. Local and state programs continue to use outstanding Section 810 properties and funds as well. (SEE ALSO: *Local Urban Homesteading Agency; Section 810; Self-Help Housing; Sweat Equity; Urban Homesteading Assistance Board;* **Urban Redevelopment**)

—*Mittie Olion Chandler*

Further Reading

Atlas, John and Peter Dreier. 1986. "The Tenants' Movement and American Politics." In *Critical Perspectives on Housing*, edited by

Rachel G. Bratt, Chester Hartman, and Ann Meyerson. Philadelphia: Temple University Press.

Borgos, Seth. 1986. "Low Income Homeownership and the ACORN Campaign." In *Critical Perspectives on Housing,* edited by Rachel G. Bratt, Chester Hartman, and Ann Meyerson. Philadelphia: Temple University Press.

Bratt, Rachel G. 1989. *Rebuilding a Low-Income Housing Policy.* Philadelphia: Temple University Press.

Chandler, Mittie Olion. 1991. "The Evolution of Urban Homesteading: Planning for Lower-Income Participation." *Journal of Planning Education and Research* 10(2):140-54.

Hughes, James W. and Kenneth D. Bleakly, Jr. 1975. *Urban Homesteading.* New Brunswick, NJ: Rutgers University Press.

Rohe, William M. 1991. "Expanding Urban Homesteading: Lessons from the Local Property Demonstration." *Journal of the American Planning Association* 57(4):444-56.

U.S. Department of Housing and Urban Development. 1981. *Evaluation of the Urban Homesteading Demonstration Program.* Washington, DC: Government Printing Office.

————. 1987. *The Local Property Urban Homesteading Demonstration.* Washington, DC: Government Printing Office.

Von Hassel, M. 1996. *Homesteading in New York City: 1978-1993.* Westport, CT: Bergin & Garvey. ◄

► Urban Homesteading Assistance Board

The Urban Homesteading Assistance Board (UHAB) assists local residents throughout New York City to develop and administer their own low-income housing. UHAB is a not-for-profit organization that has trained people living in some 20,000 dwelling units to acquire, finance, repair, and redesign their own apartments and to manage their own housing associations.

When UHAB was founded in 1973, an alarming number of apartment buildings in New York City were being abandoned by their landlords. A serious shortfall of low-income housing ensued while the city's own resources were stretched by a municipal financial crisis bringing it to the brink of bankruptcy. In a number of neighborhoods, residents joined together to save their homes through self-help efforts. UHAB provided key assistance in making tenant-controlled building management a reality.

Since then, UHAB has helped residents create and maintain their own homesteading groups, low-income housing cooperatives, and tenant associations. UHAB trains residents in the considerable technical understandings and skills required in such endeavors. This assistance takes various forms: informing residents of federal, state, and city housing programs and conducting training sessions for building management, cooperative conversion, repair and bookkeeping services, fire and liability insurance programs, discounted architectural services, mediation services, and so forth.

Development and management manuals, many available in both English and Spanish, are available. They include titles such as *A Guide for Tenants Who Manage Their Own Buildings, A Guide to Payroll Bookkeeping, A Guide to Cooperative Ownership, A Guide to Maintenance and Repair,* and, significantly, *Managing Your Lawyer.* UHAB is one of four sponsors of the monthly *City Limits,* New York's urban affairs news magazine.

UHAB's $2 million yearly budget includes significant contracts with governmental housing agencies for specific training activities, other fees-for-services from participating housing organizations, and foundation grants and other contributions.

Because of increasing national interest in self-help housing, UHAB has also provided assistance to homeownership projects in many other cities. But UHAB's major focus remains New York City.

In sum, UHAB provides an indispensable service to any homesteading and other self-help housing endeavor: technical assistance that is not only technically competent but also sensitive to and skilled in dealing with residents living in poverty and working-class neighborhoods. UHAB recognizes that this constituency already possesses profound urban knowledge and experiences and builds its assistance on these assets. Because of the low financial resources of this constituency, however, UHAB depends largely on outside funding, especially city contracts. Staff size: 25. For further information, contact UHAB, 120 Wall St., New York, NY 10005. Phone: (212) 479-3311; Fax: (212) 479-6457. E-mail: reicher@uhab.org. (SEE ALSO: *In Rem Housing; Urban Homesteading*)

—*Hans B. C. Spiegel*

Further Reading

Leavitt, Jacqueline and Susan Saegert. 1990. *From Abandonment to Hope: Community-Households in Harlem.* New York: Columbia University Press.

Peirce, Neal R. and Robert Guskind. 1993. *Breakthroughs: Re-Creating the American City.* New Brunswick: Rutgers, State University of New Jersey, Center for Urban Policy Research.

Urban Homesteading Assistance Board. 1984. *UHAB's Tenth Year Report and Retrospective (1974-1984).* New York: Author.

————. 1988. *Self-Help: In Our Own Words (1974-1988): UHAB's Fifteen Year Report.* New York: Author. ◄

► Urban Institute

The Urban Institute is a nonprofit, nonpartisan public policy research organization established in Washington, D.C., in 1968. The purpose of the institute is to investigate the social and economic problems confronting the United States and to evaluate governmental programs designed to alleviate those problems. In recent years, this mission has expanded to include the analysis of similar problems and policies in developing countries and in Eastern Europe. The organization is divided into seven independent centers, each with a specific research focus: Health Policy, Human Resources Policy, Income and Benefits Policy, International Activities, Population Studies, Public Finance and Housing, and State Policy. The Urban Institute conducts public policy seminars and interdisciplinary research workshops throughout the year and publishes *Policy and Research Report,* a triannual journal highlighting institute research. It also maintains a research library of 38,000 books and 620 periodicals, available to the public by appointment. The institute operates with an administrative staff of 60 and a research-related staff of 150 on an annual budget of $30 million (1994). It is funded by federal and state governments and by private sector foundations and corporations. Contact person: Susan Brown, Director of Public Affairs Office. Address: 2100 M Street, NW, Washington, DC 20037. Phone: (202) 857-8709. Fax: (202) 429-0687.

—*Caroline Nagel* ◄

► Urban Land Institute

For more than 50 years, the Urban Land Institute (ULI) has been widely recognized as the preeminent voice in the United States for encouraging and fostering high standards of land use planning and real estate development.

The Urban Land Institute was founded in 1936, when many U.S. cities were experiencing both suburban expansion and urban decay, with limited public sector planning and no guidance available to the private sector. No central planning or research group existed in the United States to research, analyze, and encourage responsible patterns for long-term urban growth or to conduct inquiries into what constitutes sound real estate development projects and practices. These circumstances led Cincinnati real estate entrepreneur Walter Schmidt and six prominent real estate leaders to petition the National Association of Real Estate Boards (NAREB, the forerunner of the National Association of Realtors) to establish a separate research institute within NAREB. This proved to be too limiting, and in 1940, the ULI became an independent institute. Today, it is a nonprofit, nonpartisan research and education institute under Section 501(c)(3) of the Internal Revenue Code.

Established during the Great Depression, the ULI had original objectives very similar to its current guiding principles. These early objective were (a) to study and interpret real estate trends, (b) to examine principles through which private enterprise could effectively develop real estate, (c) to develop a body of knowledge in real estate and allied subjects, (d) to publish text and technical journals based on that knowledge, and (e) to act as a statistical clearinghouse for the dissemination of real estate data. The institute's continuing focus on nonpartisan research and education has made it a widely respected and often-quoted organization in urban planning, land use, and development.

The ULI membership has grown from 230 members at the start to some 13,000 professionals in 50 states and 40 countries today. One-third of the membership consists of developers and owners, approximately one-third is public officials and academics, and the remaining one-third includes professionals such as attorneys, architects, planners, appraisers, and engineers. The ULI (a) sponsors educational programs and forums to encourage an international exchange of ideas and sharing of experiences, (b) initiates research that anticipates emerging land use trends and issues and proposes creative solutions based on this research, (c) provides advisory services on particular land use problems, and (d) publishes a wide variety of materials to disseminate information on land use and development.

The ULI provides information on an array of subjects affecting the real estate community in books, papers, seminars, and workshops. Research covers residential, hotel and recreation, office and commercial, and retail development; inner-city issues; transportation and growth management; financing; and the environment. The ULI's residential research program addresses issues and trends affecting both market rate and affordable housing, development of inner-city housing, and multifamily housing.

During the 1990s, the program was expanded substantially with regard to the multifamily housing industry. As a result, the ULI has published several books of interest to developers, investors, financiers, and managers of rental apartment communities. For example, the annual publication of *Managing Environmental Mandates of Multifamily Housing: A Compendium of Federal Regulations* provides useful assistance to those concerned with how environmental regulations affect apartment owners and managers. *A Reevaluation of Residential Rent Controls* examines the pros and cons of rent controls. In 1997, the ULI offered an addition to its *Community Builders Handbook Series* with the *Multifamily Housing Development Handbook*.

ULI has expanded its role as a provider of primary data to the real estate industry. In 1996, it formed an affiliation with the Multifamily Housing Institute to launch a major new database for use by capital markets, developers, investors, and managers of rental apartment communities. Data are collected on more than 29,000 properties and 3 million units in four primary categories:

1. Static property characteristics
2. Affordable housing characteristics
3. Property (income and expense) characteristics
4. Loan characteristics

Data are available through the Internet so users can customize reports electronically. To complement this state-of-the-art database, ULI copublishes with the National Apartment Association an income and expense book for apartments titled *Dollars and Cents of Multifamily Housing*. For more information on the Urban Land Institute, call 1-800-321-5011 or 202-624-7000. Address: 1025 Thomas Jefferson Street, NW, Washington, DC 2007-5201. Fax: 202-624-7140.

—*Lloyd W. Bookout* ◄

► Urban Redevelopment

The passage of Title 1 (Slum Clearance and Urban Development) of the Housing Act of 1949 inaugurated a series of national, state, and local measures aimed at redeveloping central cities in the United States. The original rationale for these programs was the elimination of slums; more recently, redevelopment efforts have had economic development as their explicit goal. U.S. urban redevelopment efforts have differed from European approaches to what the British term *urban regeneration* in being less state directed. Thus, U.S. programs have placed greater emphasis on private sector actors as both initiators of planning and implementers of redevelopment. European policymakers, however, have increasingly emulated U.S. models, especially in the United Kingdom since the election of the national Conservative government in 1979.

The first U.S. redevelopment programs, which were conducted by local renewal authorities receiving federal funding, involved the clearance of large sites occupied by dilapidated structures and constructing new buildings. Federal funding later also became available for rehabilitation of existing structures, but demolition remained the dominant method for dealing with deteriorated structures. Beginning in the 1970s, the federal role in urban redevelopment diminished considerably, and municipal governments increasingly relied on local public-private partnerships to spur redevelopment activities. Although many redevelopment efforts have targeted residential neighborhoods, the bulk of spending on redevelopment has been directed at reviving central business districts (CBDs) and their peripheries. Early programs were criticized for ineffectiveness in stimulating new development and for displacing low-income minority residents from their neighborhoods to make way for businesses and upper-income residential occupation. More recent schemes have succeeded in creating "new downtowns" of office buildings and retail centers but have been blamed for creating "two cities," one for the rich and one for the poor.

Title 1 of the 1949 act was the legislative mandate for the federal urban renewal program until 1974. Employing federal funds appropriated under this act, local authorities used the power of eminent domain to acquire privately held land deemed "blighted"; once a site had been aggregated and prepared, they sold the land to a private for-profit or nonprofit developer at a lower price or else turned it over to a public agency. Public agencies that made use of urban renewal land included public housing authorities, educational institutions, and hospitals. Localities were not required to participate in the program, and many cities, particularly in the South, chose not to do so.

Federal urban renewal grants were discretionary, and to receive support, participating municipalities had to conform to a detailed and onerous set of guidelines. The federal government paid for between two-thirds and three-fourths of the net project cost (the difference between the purchase price and selling price of the land plus costs of demolition and improvements). The local share could be either a cash payment or an in-kind contribution, such as the construction of a school or fire station.

The act declared that the redevelopment area should be predominantly residential in character, but it did not require the replacement of low-income housing that was destroyed, nor did it provide funds that would make it feasible for private builders to construct housing for low-income occupants. Although public housing was constructed on urban renewal land in many cities, the federal program demolished far more units of low-income housing than were built.

Business leaders frequently took the initiative in planning urban renewal programs. The most famous example of a business-led planning group was in Pittsburgh, where the Allegheny Conference, made up of corporate chief executives, spearheaded the conversion of the city from a manufacturing to a service center. Corporate elites in other cities emulated the Pittsburgh example, and business groups such as the San Francisco Planning and Urban Research Association (SPUR) in San Francisco, "the Vault" in Boston, and the Greater Baltimore Committee formulated similar redevelopment programs. These involved the construction of large office, tourism, and retail projects in old CBDs and the production of luxury housing on centrally located sites through new construction, rehabilitation of deteriorated dwellings, and adaptive reuse of old manufacturing and warehouse facilities.

Because city governments were seeking to attract and retain middle-class residents, they encouraged the development of upper-income housing on centrally located properties. The Boston Renewal Authority, for instance, razed apartment buildings housing a multi-ethnic working-class community located on a large tract adjacent to downtown and affording excellent views of the Charles River and replaced them with a luxury high-rise project. Similarly, in San Francisco's Western Addition a neighborhood occupied by Asian and African Americans was cleared for the construction of a Japanese business center, new retail establishments, and housing for people of higher income. A sizable portion of New York's Greenwich Village was torn down to make way for high-rise apartments intended to house the faculty and students of New York University. Critiques of these projects by Herbert Gans, Chester Hartman, and Jane Jacobs, respectively, made these three cases notorious; attacks by these and other intellectuals on the federal urban renewal program did much to discredit it.

Because federal urban renewal funds were available for land purchase and clearance rather than for new construction, redevelopment schemes required the participation of private investors to come to fruition. Private investors, however, could bid on property only after it was selected for renewal and the land was cleared. Frequently, no investors were interested in the site, resulting in the decades-long existence of large, centrally located, cleared areas in many cities. Many of these original urban renewal tracts were not filled until the downtown real estate boom of the 1980s.

During the 1960s, community-based movements strongly resisted redevelopment schemes that would displace neighborhood residents. Although the federal government did not discontinue its urban renewal program, which continued to fund programs for downtown development, it responded to protest with the passage of the Demonstration Cities and Metropolitan Development Act of 1966 (later called Model Cities). This act redirected the thrust of programs within poor neighborhoods to emphasize community preservation, coordination of physical and social planning, rehabilitation rather than demolition, community participation, and a focus on target areas.

Criticisms of the federal urban renewal program for destroying usable structures within project areas, reducing the stock of affordable housing, and creating "bombed out" districts in urban centers led to modifications of the 1949 legislation. In 1954, incentives for rehabilitation rather than wholesale clearance were added, and the requirement of a redevelopment plan was imposed. In 1968, local urban renewal authorities were mandated to set up project area committees to represent affected community residents. The Uniform Relocation Assistance Act of 1970 ensured that displaced residents would receive adequate relocation payments. Despite these reforms, clearance continued to be the predominant approach to urban renewal throughout the history of the federal program.

In 1974, both the federal Urban Renewal and Model Cities Programs ended with the passage of the Housing and Community Development Act, which introduced the Community Development Block Grant (CDBG). CDBG funds were distributed on a formula rather than a project basis. Local elected officials set community priorities for using the funds. The grant could be used for a broad variety of redevelopment activities, ranging from housing rehabilitation to road and sewer repairs to downtown improvements. Although initially the Model Cities' intent of combining physical and social services was retained in the CDBG program, the social service component was gradually dropped in most cities. During the history of CDBG, the amount of local discretion concerning its use has varied, depending on the policies of the national administration in power. Thus, during the Carter years, municipalities were required to target most of their grants to low-income areas and to devise a scheme for citizen participation in determining expenditures. Under the Reagan and Bush administrations, targeting was loosened, a broader range of activities became permissible under the program, and citizen participation requirements were dropped.

The CDBG program remains the principal federal program subsidizing urban redevelopment. Its funding level, however, has shrunk substantially in constant dollars, and in 1997 it constituted only a small proportion of the funds devoted to redevelopment in most cities. For many cities, the CDBG represented a substantial decline in federal support. The small size of the grant and remaining targeting regulations mean that the CDBG does not usually contribute to major renewal schemes, which therefore are not constrained by federal guidelines.

In 1977, to provide more funding for specific projects, Congress enacted the Urban Development Action Grant (UDAG) program. UDAGs were offered on a discretionary basis for the sole purpose of fostering economic development and could be spent only if the private component of the project was committed in advance. They thereby overcame the problem of the old urban renewal program whereby land sat vacant until a developer could be found. The UDAG program, although it did stimulate a considerable amount of new downtown investment, was gradually eliminated during the 1980s in conformity with President Reagan's intention of reducing governmental intrusion in the economy.

Municipalities needing large amounts of capital for major projects such as stadiums, convention centers, or industrial parks must rely on revenue bonds. Until the 1980s, investors purchasing these bonds enjoyed federal tax exemptions on the interest even when the funds were re-lent to private developers. The 1986 federal tax act, however, ended exemptions for interest on public authority bonds if the revenues did not directly fund construction of public facilities.

Municipal and state governments, increasingly forced to rely on their own sources of revenue to promote redevelopment, devised a group of mechanisms to entice private developers to invest in central cities. Among the devices most frequently used are various kinds of tax inducements. Developers, building owners, and businesses moving into redevelopment areas receive property and sales tax exemptions and abatements for a period of years after new construction; developers are exempted from paying sales taxes on construction materials; and various tax district programs return portions of the property tax or special assessments to the area that generates the funds rather than to the municipality's general fund. Revenues raised in tax districts may be used for further capital expenditure within the area, as backing for loans, or for purchasing services such as special security forces or sanitation crews. Tax increment financing employs any increase in property taxes resulting from a redevelopment project to pay back the bonds used to finance the public portion of the investment. These various tax expenditure programs spare public officials from raising funds to subsidize development. Their disadvantage is that they cut into the municipal tax base, thereby increasing the rate of taxation on homeowners and unaffected businesses.

To attract development, municipal governments have also agreed to relax zoning and environmental regulations. In return, they have typically asked developers to contribute public amenities. For example, in New York City, developers are allowed to build extra floors on their buildings in return for refurbishing a subway station or providing spaces for public use.

Because central city governments suffering from fiscal stress have so much difficulty raising money to finance public works and low-income housing, they have sought to make deals in which developers trade contributions in cash or kind for permission to build. This strategy could be successfully adopted only during the periods of real estate boom like the 1980s and was limited to cities where developers were competing for the right to build. In some cities (e.g., Boston and San Francisco), these arrangements became formalized in linkage requirements, whereby office developers must contribute to a housing fund based on the number of square feet that they construct. Another similar scheme is "80-20" housing, whereby developers of market rate housing receive zoning bonuses or tax abatements in return for constructing 20% of their units for low- or moderate-income residents.

As redevelopment has proceeded on an ad hoc basis, with each project involving a customized package of financial arrangements, regulatory relief, and developer contributions, the function of redevelopment planners has increasingly become bargaining with investors rather than charting the future of their municipalities. The municipal agencies that carry out redevelopment functions tend to be semiautonomous institutions with business-dominated boards. They work on locating and marketing sites that will be attractive to investors, then put together financial packages and extract contributions from them in return.

Neighborhood groups have also sought new sources of funding in the absence of federal support. Neighborhood commercial and housing revitalization has been carried out under the auspices of community development corporations (CDCs). These entities raise funds from a variety of public and private sources, including churches, foundations, governments, banks, and financial intermediaries such as the Enterprise Foundation. They have become the principal redevelopment actors in low-income urban neighborhoods, and some have produced housing on a large scale. Their commercial successes, however, have been limited, and the overall number of housing units they have created has been quite small relative to need and to the number of units once built by public housing authorities.

Enterprise zones are districts in which governments promote redevelopment by offering businesses a variety of tax and regulatory relief programs. Originally proposed, but not adopted, as federal legislation, a majority of states enacted enterprise zone statutes during the 1980s. State-mandated enterprise zones provide exemptions from state and local taxes and regulations only. Under the Clinton administration, a modified enterprise zone bill was adopted in 1993, providing for the establishment of nine empowerment zones (seven of which are urban) and 50 enterprise communities. Aimed at stimulating redevelopment in poor neighborhoods, the program provides businesses in these districts with federal tax and regulatory relief as well as some grants.

Since the termination of the federal urban renewal program, large-scale displacement as a consequence of demolition has mainly halted. Cities have typically used still-vacant old urban renewal sites, filled wetlands, or derelict railyards and ports as the location for major projects. Rehabilitation of existing structures rather than new construction has been the principal approach to housing improvement, and rehabilitation subsidies have been a principal use of CDBGs. In a number of cases, rehabilitation loans have contributed to gentrification. In other instances, redevelopment projects have produced "secondary displacement"; that is, they have stimulated the privately funded gentrification of nearby neighborhoods.

Throughout the history of federal sponsorship of local redevelopment programs, the size and emphases of the federal program fluctuated. Regulations concerning demolition versus rehabilitation, minimum size of projects, citizen participation, low-income targeting, and relocation changed constantly. Funding levels oscillated, never exceeding 1.5% of the federal budget and rarely topping 1%.

During the last 20 years, with the decline in federal involvement and oversight, the differences among cities in the character of their redevelopment programs have increased. A number of variables account for these differences:

▸ The electoral coalition that is in power
▸ The amount of entrepreneurship shown by public officials
▸ The competitive advantages the city possesses
▸ The racial composition of the population

In general, urban redevelopment programs have focused on CBD-centered, office- and retail-led development, with benefits to the general population dependent on the trickle down from these enterprises. Some cities, however, have had programs that emphasized neighborhood improvements or focused on manufacturing revival, high-tech, or health strategies. (SEE ALSO: **Abandonment;** *Blight; Brownfields; Columbia Point; Community Development Corporations; Condominium Conversion; Demonstration Cities and Metropolitan Development Act of 1966; Enterprise Foundation; Historic Preservation; Incumbent Upgrading; Infill Housing; In Rem Housing; Model Cities Program; National Association of Housing and Redevelopment Officials; National Commission on Urban Problems; Power of Eminent Domain; Slums; Tent City; Urban Homesteading; Urban Renewal Agency*)

—*Susan S. Fainstein*

Further Reading

Fainstein, Susan S., Norman I. Fainstein, Richard Child Hill, Dennis R. Judd, and Michael Peter Smith. 1986. *Restructuring the City.* Rev. ed. New York: Longman.

Frieden, Bernard J. and Lynne B. Sagalyn. 1990. *Downtown, Inc.* Cambridge: MIT Press.

Gelfand, Mark. 1975. *A Nation of Cities: The Federal Government and Urban America, 1933-1965.* New York: Oxford University Press.

Judd, Dennis and Michael Parkinson, eds. 1990. *Leadership and Urban Regeneration.* Newbury Park, CA: Sage.

Mollenkopf, John. 1983. *The Contested City.* Princeton, NJ: Princeton University Press.

Squires, Gregory, ed. 1989. *Unequal Partnerships.* New Brunswick, NJ: Rutgers University Press.

Stone, Clarence and Heywood Sanders, eds. 1987. *The Politics of Urban Development.* Lawrence: University Press of Kansas.

Wilson, James Q., ed. 1966. *Urban Renewal: The Record and the Controversy.* Cambridge: MIT Press. ◄

▸ Urban Renewal Agency

In 1949, Congress enacted urban redevelopment legislation (Title I of the Housing Act). Planning for the clearance and redevelopment of urban slum areas had begun during World War II. Under Title I, a new Urban Renewal Agency (URA), reporting to the Housing and Home Finance Agency (HHFA), provided federal funding to localities willing to match these funds to redevelop blighted areas.

The URA had to approve all local redevelopment plans and then supervise the disbursal of federal funds and ensure the completion of the projects. Over the 25-year life of urban renewal, the URA, because of its central administrative role, became embroiled in the many controversies that surrounded the Urban Renewal Program, even though the primary responsibility for planning and implementation lay with local governments. These included selection of project areas and developers, the cost of redevelopment, undue red tape and long delays in the completion of projects, the failure to relocate displaced households (especially poor minorities), and the loss of below-market-rate housing that was not replaced. The URA itself became the target of critics of urban renewal, because it generally rubber-stamped local plans and projects.

As urban renewal came under increasing attack in the 1960s, so did the URA. In 1965, the URA was absorbed into the newly created U.S. Department of Housing and Urban Development (HUD). By 1968, with reforms instituted in reaction to protests and lawsuits, urban renewal began to decline, even as its emphasis shifted from large-scale clearance projects to smaller-scale rehabilitation projects. The ill-fated Model Cities Program created in 1966 was seen as an alternative to urban renewal. In 1974, urban renewal was among the HUD programs (including also Model Cities) terminated to create the Community Development Block Grant (CDBG) program. (SEE ALSO: **Urban Redevelopment**)

—*W. Dennis Keating*

Further Reading

Bellush, Jewel and Murray Hausknecht, eds. 1967. *Urban Renewal: People, Politics and Planning.* New York: Anchor.

Gelfand, Mark I. 1975. *A Nation of Cities: The Federal Government and Urban America, 1933-1965.* New York: Oxford University Press.

Wilson, James Q. 1966. *Urban Renewal: The Record and the Controversy.* Cambridge: MIT Press. ◄

▶ Urban Studies

Urban Studies is an international journal published 11 times per year by the University of Glasgow. Contributions to the journal are drawn from the fields of economics, planning, statistics, sociology, demography, and public administration. Articles cover issues salient to the study of urban regions, including housing finance, transportation, rent control, urban renewal, land price theories, economic development theories, and environmental behavior. First published in 1964. Circulation: 2,500 worldwide. Price: $548 for institutions; $150 for individuals (1998). Managing Editors: Professor W. F. Lever, Professor R. Paddison, and Mr. J. Bannister. Address: Adam Smith Building, University of Glasgow, Glasgow G12 8RT, United Kingdom.

—*Caroline Nagel* ◄

▶ U.S. Housing Market Conditions

U.S. Housing Market Conditions is a quarterly publication containing detailed information on national and regional housing markets in the United States. Each issue includes current and historical data on diverse topics, such as housing starts and completions, delinquencies and foreclosures, home prices, housing investment, homeownership rates, affordability, interest rates, mortgage originations, home sales, low-income housing tax credits, housing permits, and vacancy rates. Brief narratives describe the characteristics of new databases that become available, and summaries spotlight national trends and local developments. Subscription: $30 per year. Available from HUD USER,

P.O. Box 6091, Rockville, MD 20849. Phone: 1-800-245-2691. World Wide Web: http://www.huduser.org/pubs.html. (SEE ALSO: *HUD User; U.S. Bureau of the Census; Housing Survey; Housing Markets*)

Laurel Phoenix ◄

▶ Usury Laws

Usury laws are regulations that prohibit charging interest on loans above a certain level, often a specified ceiling. Usury laws were used in Biblical times as well as in ancient Greece and Rome as indicated in historical documents. These laws have long had great appeal throughout the ages and, until recently, applied to mortgage markets for residential loans in the United States.

Usury laws are intended to protect consumers of credit (borrowers) from unscrupulous or unethical sellers of credit (lenders) by making charging very high rates of interest against public policy. "Very high rates of interest" are usually defined as effective interest rate charges above a certain level. Usurious interest rates need not be two or three times market rates; they can also be five basis points above the specified usury ceiling.

Mortgage markets are only one type of market for which usury laws have been passed. Consumer credit, automobile loans, and credit card revolving charge cards are also frequently subjected to usury laws in the United States. Often, these statutes are part of the consumer protection legislation that has become prominent in most states during the past decades. However, it is important to note that usury laws are always state regulations, so there are differences in the legal environments on this issue between various states.

Extensive research has been conducted on the economic effects of usury laws. Much of the research has provided empirical evidence about the operation of the usury laws in specific lending markets. Usury laws were in operation for mortgage markets in many states until 1980, and there is a body of knowledge about the effects of usury ceilings for housing and mortgage finance as well.

With respect to mortgage markets, the weight of the evidence is overwhelmingly clear: Usury laws often failed to assist low- and middle-income borrowers contract with lenders in financial institutions and often had the perverse effect of reducing the supply of mortgage funds for low-income borrowers. Thus, it is generally agreed, at least in mortgage markets, that usury laws, if anything, tended to hurt the consumers they were intended to protect.

One experiment was to use a moving interest rate ceiling (i.e., prohibit mortgage interest rates above a specified limit and when interest rate levels change, adjust the usury ceiling accordingly). The idea was to try to mitigate the adverse effects of restricting the supply of mortgage funds as found by the empirical research but to continue to guard against gross violations of normal interest rate charges. However, the research evidence failed to show that this experiment had any effect on economic behavior whenever the usury ceiling was above the market rate of interest.

After an abundance of evidence accumulated during the 1960s and 1970s and with an important governmental study of the U.S. financial system called the Hunt Commission Report in 1972, federal legislation in 1980 prohibited the enforcement of any usury laws in mortgage markets for federally insured mortgages (a majority of those made in the United States). Effectively, usury laws were made null and void for mortgages beginning in 1980s in the United States. (SEE ALSO: *Housing Credit Access; Interest Rates*)

—*Austin J. Jaffe*

Further Reading

ABA Standing Committee on Lawyers' Title Guaranty Funds. 1991. *Buying or Selling Your Home: Your Guide to: Contracts, Titles, Brokers, Financing, Closings.* Chicago: American Bar Association.

Brown, Kevin W. 1987. *Usury and Consumer Credit Regulation.* Boston: National Consumer Law Center.

Jacobsohn, David B. 1980. *Federal Regulation of Banking: The Depository Institutions Act.* Boston: Warren, Gorham & Lamont.

Pryke, M., ed. 1994. "Property and Finance." *Environment and Planning A* 26(Special issue).

Stegman, Michael A. 1986. *Housing Finance and Public Policy: Case and Supplementary Readings.* New York: Van Nostrand Reinhold.

V

▶ Vacancy Chains

The move of one family or individual into a house or apartment often sets off a chain reaction of additional moves: A second family or individual moves into the house or apartment vacated by the first, a third into the housing unit vacated by the second, and so on. These chain reactions in housing moves are known more formally as *vacancy chains,* and their study has supplied a variety of important facts about housing markets, such as the number, ethnic status, and income levels of families and individuals getting housing through typical vacancy chains. Their study has also raised questions about the possibility of stimulating moves through vacancy chains to achieve practical housing goals. Because the study of vacancy chains looks directly at the microlevel processes through which housing units move from one household to another, it can provide information about the allocation of housing that cannot be gained through other approaches.

A vacancy chain begins when a vacancy initially comes into the housing market through either the construction of a new house or apartment or the departure of a household from the housing market as a consequence of death, retirement, and so on. This initial vacant unit is taken by the first household in a vacancy chain, and the chain continues as one household replaces another in its recently vacated dwelling. Eventually, a vacancy chain has to end, and it does so either when a household takes a dwelling without leaving behind a vacant unit (e.g., when the occupants move from their parents' houses), or when the last unit in a chain remains permanently vacant or is removed from the housing stock by being destroyed.

By convention, each vacancy chain is assumed to have just one vacancy. The chain is composed of the moves that this vacancy makes, starting with its move from the initial vacant dwelling and ending with its move out of the housing market (see Figure 33). The multiplier effect is defined as the length of a chain: how many times the vacancy moves after its entrance into the housing market, including the final move outside the system. For example, the multiplier effect of the chain in Figure 33 is 3.0. If a chain ends with a household moving in from outside the housing market (from parents' housing, college dormitories, etc.), the number of mobile households is the same as the multiplier effect, but if a chain ends with the last dwelling being destroyed or abandoned, the number of families is 1.0 less than the multiplier effect. In studies of the national housing market in the United States, the average chain has a multiplier effect of around 4.0 or so with between three and four households moving in a typical chain. In studies of more restricted, local housing markets in the United States and of housing markets in Eastern Europe, the average number of households per chain is often considerably smaller—from a little more than 1.0 to about 2.5.

Usually, chains begin in larger and more expensive housing units and work their way through the housing stock to smaller and less expensive dwellings. In parallel form, families and individuals getting a dwelling at the beginnings of chains often have greater incomes and social status than those at the ends of chains. So dwelling units often trickle down from more to less affluent groups. Over the course of many years, as a dwelling is involved in a series of different vacancy chains and as its occupants come and go, it may deteriorate in quality and filter down from one ethnic or status group to another.

Data on vacancy chains can be collected from a series of linked interviews following households moving into new housing or from cross-sectional studies of turnover in housing units. These data can be used in standard Markov chain (renewal) models to predict things such as (a) average numbers of households involved in chains; (b) estimates of the numbers of households moving from old housing units of specific types to new housing units of specific types, given a distribution of vacancy chains starting in various classes of housing units; and (c) the numbers of houses destroyed, abandoned, or taken by first-time buyers or renters. Although not yet employed in housing, Markov techniques developed for analyzing vacancy chains in jobs, could describe the housing "careers" of households—that is, the types of units and the lengths of time that households occupy them over the course of their lives.

A variety of recent research offers further information about vacancy chains. For example, Hua suggests ways to link vacancy chain models with other models of the behav-

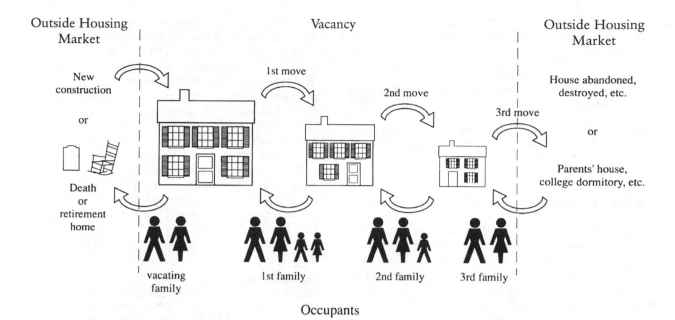

Figure 33: Schematic Representation of a Vacancy Chain. A chain starts when a new house or apartment is constructed or an existing one becomes vacant as a result of, for example, death or relocation to a nursing home. The chain continues when a first household moves into that initial unit, its (old) home is subsequently taken by a second household, and so forth. Eventually, the chain ends when the last unit is removed from the housing stock (for example, by demolition) or is taken by a household that leaves no old unit behind (for example, new households that form when students leave a dormitory or children leave the house of their parents).

ior of housing buyers and sellers; Williams et al. develop mathematical programming models to predict the flow of vacancy chains in housing markets given constraints such as mortgage rates, housing costs, population growth, and numbers of first-time buyers; and Chase outlines an as yet unstudied aspect of housing vacancy chains: the impact of chains on individuals and organizations such as real estate agents, lawyers, moving companies, banks, and state and local governments connected to the people getting new housing.

Researchers have indicated divergent opinions concerning the practical use of vacancy chains started in more expensive housing to provide trickle-down opportunities for disadvantaged populations. The studies of Lansing et al. and Marullo address this controversy by providing detailed information on the ethnic status and incomes of groups getting housing at various links in chains in the national housing market. Making informed decisions about the effect of vacancy chains on the provision of housing for disadvantaged groups would appear to depend on detailed empirical knowledge of specific housing markets and how vacancy chains operate in them, local political constraints, and the potential funds available for housing subsidies or land development.

Although vacancy chains trace a deceptively simple process, their study involves different conceptual formulations and methodological techniques than many more standard approaches to the allocation of housing. In many economic situations, a form of simple competition prevails: When one group or individual obtains a good, the other groups or individuals interested in that good are denied its use, and the process goes no further. But when goods, like housing, are distributed through vacancy chains, an earlier success by one family or individual in obtaining a good sets in motion a process through which other families or individuals can then also obtain goods in which they are interested. So rather than concentrating on the economic behavior of independent individuals, studies of vacancy chains look directly at the small-scale processes through which interdependent sequences of individuals actually get new housing and leave old housing. Because it traces these interrelated sequences of movement, the study of vacancy chains can provide information about the impact of housing starts and departures from the housing markets on the numbers, locations, and types of families and individuals subsequently moving and on the organizations and other individuals connected to those moving that is not possible with other approaches to the study of housing. (SEE ALSO: *Filtering; Residential Mobility; Vacancy Rate*)

—*Ivan D. Chase*

Further Reading

Chase, Ivan D. 1991. "Vacancy Chains." *Annual Review of Sociology* 17:133-54.

Hegedüs, J. and I. Tosics. 1991. "Filtering in Socialist Housing Systems: Results of Vacancy Chain Surveys in Hungary." *Urban Geography* 12:19-34.

Hua, Chang-i. 1989. "Linking a Housing Vacancy Chain Model and a Behavioral Choice Model for Improved Housing Policy Evaluation." *Annals of Regional Science* 23:203-11.

Lansing, J. B., C. W. Clifton, and J. N. Morgan. 1969. *New Homes and Poor People*. Ann Arbor: University of Michigan Institute for Social Research.

Marullo, Sam. 1985. "Housing Opportunities and Vacancy Chains." *Urban Affairs Quarterly* 20:364-88.

Sands, G. and L. L. Bower. 1976. *Housing Turnover and Housing Policy: Case Studies of Vacancy Chains in New York State*. New York: Prager.

Williams, H. C. W. L., P. Keys, and M. Clark. 1986. "Vacancy Chain Models for Housing and Employment Systems." *Environment and Planning A* 18:89-105. ◀

▶ Vacancy Rate

Housing market participants, tenants, landlords, and property managers need to understand the nature and significance of the vacancy level and the vacancy rates for dwelling units in both the single family and the multifamily housing markets. Several important aspects of the vacancy level and rate need to be recognized.

The vacancy rate can be applied to both a structure and to a market. In general, the vacancy rate is the amount of vacant space divided by the total amount of space available for rent at a point in time. In a residential context for a single structure, the vacancy rate in a multifamily structure is the number of vacant units divided by the total number of dwelling units in the apartment building.

The vacancy rate in a residential rental market is the number of vacant units in the market divided by the total number of units in the market. The market vacancy rate changes as demand and supply conditions in the market change. If demand increases while supply remains constant or if demand increases by more than the supply increases, the vacancy level and the vacancy rate will decline. If on the other hand, supply increases while demand remains constant or if supply increases by more than demand increases, the vacancy level and the vacancy rate will increase.

In an economic context for a market, the level of vacant units and thereby the vacancy rate is the measure of excess supply in the market at a rent level above the equilibrium rent level. The vacancy level is the measure of the number of dwelling units that do not clear the market.

The vacancy rate for a multifamily dwelling structure can and most typically does change as the market vacancy rate changes. However, the vacancy rate for a specific multifamily building can be greater than the market rent if (a) potential renters view it as inferior to the other available comparable units at the same rent level or (b) the marketing and the leasing activities of the landlord or property manager are inferior to the actions of the other owners and property managers supplying dwelling units to the market.

Specific leasing arrangements and turnkey operations of the owner or property manager also affect the vacancy rate for a single multifamily dwelling structure. An apartment building can be fully leased during a time period and still exhibit vacancy in its financial accounts. For example, a tenant moves out of an apartment unit on the last day of June and a tenant is found for July. However, it will take five days to prepare the apartment for the new tenant; the walls need to be repainted, the kitchen and bathroom need to be cleaned, and the carpeting needs to be steam cleaned. The new tenant moves into the apartment on July 6 and starts paying rent on that day. Even though the apartment building appears to be fully occupied for the entire year, this single apartment was not generating rent for 100% of the year. It lost one-sixth of a month's rent. The vacancy rate from a financial perspective for this unit is 1.67%. This figure is calculated by using a 360-day year representing 12 months of 30 days each. The unit receives rent for 354 days, so the vacancy rate is $6 \div 360$ or $1 - 354 \div 360$.

The vacancy rate is conceptually the reverse of the occupancy rate. The vacancy rate (V) and the occupancy rate (O) for a structure sum to unity: $V + O = 1$. This relationship is true for both a single property and for the market. (SEE ALSO: *Vacancy Chains*)

—*Joseph S. Rabianski*

Further Reading

Carn, Neil G., Joseph S. Rabianski, Ronald Racster, and Maury Seldin. 1988. *Real Estate Market Analysis: Techniques and Applications*. Englewood Cliff, NJ: Prentice Hall (see pp.136-37).

Fanning, Stephen F., Terry V. Grissom, and Thomas Pearson. 1994. *Market Analysis for Valuation Appraisals*. Chicago: Appraisal Institute (see pp. 284, 304-05, 319).

Friedman, Jack P., Jack C. Harris, and Bruce J. Lindeman. 1993. *Dictionary of Real Estate Terms*. Hauppauge, NY: Barron's Educational Series (see p. 363). ◀

▶ Vernacular Housing

Long before governments and housing agencies took to supplying housing, people were constructing houses for themselves. Such housing of, by, and for the people, designed and constructed without the help of specialized professionals, such as formally trained designers, architects, planners, and engineers, is called *traditional,* or more broadly, *vernacular* housing.

Designers of vernacular housing usually remain anonymous. Design of the building is often seen as routine rather than a major creative achievement. Vernacular housing calls on available knowledge and tradition. The mode of construction and level of quality of workmanship may be very high but usually is found in the local population.

An important quality of vernacular housing is that it usually draws on locally available materials. Vernacular housing uses technology, materials, designs, and forms that follow accepted local procedure or tradition, although they may appear innovative to outsiders. Most vernacular housing ameliorates prevalent environmental problems, such as extremes of climate.

Cultural values, beliefs, preferences, customs, traditions, and practices heavily influence vernacular housing. The layout of housing and the community can be affected by cultural values, including selection of location of the community, where each group can live, and the relationship between public and private space. Cultures help define design problems, and cultural values influence the choice of design solutions. Vernacular housing may represent cultural conceptions of the world. Cultural values and ideas also influence preference for a particular lifestyle, which in turn affects the choice of architectural solutions. One of these cultural factors, religion, affects vernacular housing by suggesting appropriate and inappropriate designs.

Environmental factors, such as climate conditions and the presence of animals or plants, also affect the design of vernacular housing. Examples are frequent flooding, droughts, and heavy snowfall. To prevent rising floodwaters from entering, houses have been raised on stilts. To avoid desert winds from bringing sand and dust into the house, a vernacular house may have few and small openings. To reduce the discomfort associated with high humidity, vernacular designs may use bamboo shutters that permit airflow and facilitate cross ventilation. Conversely, in hot dry climates, some houses are designed to be underground, as in China, Libya, and Iran, or built of mud above ground as in the southwestern United States, Iran, and Africa. To cool rooms, vernacular housing in hot dry areas may use *badgirs* (wind catchers), or *khas-khas* screens periodically doused with water. Often, vernacular designs are effective low-tech solutions that ameliorate climatic discomforts.

Building materials also affect the design of vernacular housing. Easily available local materials are typically used. Examples are mud, adobe, sun-dried brick, wood, bamboo, leaves, ice blocks, and stone. The nature and characteristics of building materials influence the design. For example, mud and mud brick require thick walls and relatively small openings and small spans and therefore small rooms. Use of bamboo, a material with good tensile and seismic characteristics, enables larger spans and openings.

Building technology influences the design of vernacular housing as well as the sizes of rooms, building height, and the choice of building materials. For example, the size of a span depends on whether there is a load-bearing wall, simple columns and beams, or an arch, dome, or truss. Hay or straw is added to mud to increase its strength, cow dung is applied to a mud building to increase its resilience to water, and mud is formed into bricks and dried in the sun to create a modular building block. Technology extends the choices available from naturally available materials to manufactured materials, such as kiln-fired brick, machine-cut stone, and machine-cut lumber.

There is much variety in the vernacular housing in different nations and regions. This variety is testimony to the richness in both the nature of problems and ingenuity of solutions.

The vernacular housing of each region has characteristic features making it distinct from other regions, although there are many similarities in the materials and technology used and in climatic conditions. Some researchers study vernacular architecture from these perspectives, focusing on, for example, housing in hot dry and desert areas or the architecture of mud. These studies tend to look at vernacular housing across cultures and concentrate on similarities.

Vernacular housing is typically unregulated by authorities, such as government or appointed bodies. Because professionals are not involved, the architecture is thought to be a form of self-expression of the local people, an expression of identity of an individual or culture. Even though within any given culture there is often uniformity, when carefully examined, individual units tend to be distinct; they are a form of individual self-expression within an overall framework.

Vernacular housing is often seen as socially, culturally, and environmentally appropriate. It is believed by many to solve design problems creatively and as being good at developing innovative forms. Conversely, studies also illustrate that many modern housing may not be culturally appropriate. Examples are designs and regulations at variance with preferred lifestyles among the Motilone, Tonto Apache, Navajo, and Mexican Americans. But vernacular housing can also have negative dimensions. Its construction may not be permanent and require ongoing maintenance. Furthermore, it can perpetuate social inequities, status differentials, and economic disparities.

Studies of vernacular housing reveal at least two points of view. On the one hand, the architecture has been viewed with admiration for its innovativeness, beauty, and simplicity, and many modern-day architects see it as an inspiration for their designs. On the other hand, there is a critical view that sees vernacular housing as inadequate according to health and safety criteria. (SEE ALSO: *Cultural Aspects; Earth-Sheltered Housing; Hogan; Igloo; Pueblo; Yanomami Shapono*)

—*Sanjoy Mazumdar*

Further Reading

Bourdier, Jean-Paul and Nezar AlSayyad, eds. 1989. *Dwellings, Settlements and Tradition: Cross-Cultural Perspectives.* Lanham, MD: University Press of America.

Mazumdar, Sanjoy and Shampa Mazumdar. 1994. "Societal Values and Architecture: A Socio-Physical Model of the Interrelationships." *Journal of Architectural and Planning Research* 11(1):66-90.

Oliver, Paul. 1987. *Dwellings: The House across the World.* Austin: University of Texas Press.

———. 1997. *Encyclopedia of Vernacular Architecture of the World.* Cambridge, UK: Cambridge University Press.

Pader, Ellen-J. 1994. "Spatial Relations and Housing Policy: Regulations That Discriminate against Mexican-Origin Households." *Journal of Planning Education and Research* 13:119-35.

Rapoport, Amos. 1969. *House Form and Culture.* Englewood Cliffs, NJ: Prentice Hall.

Rudofsky, Paul. 1964. *Architecture without Architects.* New York: Doubleday.

Turan, Mete, ed. 1990. *Vernacular Architecture: Paradigms of Environmental Response.* Aldershot, UK: Avebury.

Upton, Dell and John Michael Vlach, ed. 1986. *Common Places: Readings in American Vernacular Architecture.* Athens: University of Georgia Press.

Wagner-Steagall Housing Act

Also known as the U.S. Housing Act of 1937, the Wagner-Steagall Housing Act created the public housing program, the first major federal housing initiative aimed at assisting low-income households. Although housing problems facing the poor had been acknowledged for decades, federal intervention was not forthcoming until the Great Depression of the 1930s. At that time, the need for improved housing was coupled with another key goal: to stimulate the economy and create jobs. The 1937 act reflected these dual objectives: "to alleviate present and recurring unemployment and to remedy the unsafe and insanitary housing conditions and the acute shortage of decent, safe and sanitary dwellings for families of low income."

Senator Robert F. Wagner, Sr., of New York was the key legislator responsible for passage of the act. On the House side, the chairman of the Banking and Currency Committee, Representative Henry Steagall of Alabama, was also instrumental in its enactment, despite his personal opposition on the grounds that it was socialistic.

The bill was vigorously opposed by private interest groups, such as the National Association of Real Estate Boards, the Chamber of Commerce of the United States, and the U.S. Savings and Loan League, which viewed direct government provision of housing as inappropriate government competition with private enterprise.

The public housing program, which was authorized by the Wagner-Steagall Act, has had a stormy history. It has, however, succeeded in providing housing to millions of households who otherwise would have found it difficult or impossible to locate affordable housing on the private rental market. (SEE ALSO: **Public Housing**)

—*Rachel G. Bratt*

Further Reading

Freedman, Leonard. 1969. *Public Housing: The Politics of Poverty.* New York: Holt, Rinehart & Winston.

Keith, Nathaniel S. 1973. *Politics and the Housing Crisis since 1930.* New York: Universe Books.

McConnell, T. L. 1957. *The Wagner Housing Act.* Chicago: Loyola University Press.

Meehan, Eugene, J. 1975. *Public Housing Policy: Myth versus Reality.* New Brunswick: Rutgers, State University of New Jersey, Center for Urban Policy Research. ◄

Welfare State Housing

Welfare state housing aimed primarily at low-income households is provided on a nonprofit basis within capitalist economic systems. Forms of subsidy, allocation, and ownership vary. Welfare state housing may refer to dwellings directly provided and managed by local or central government or to state-subsidized housing managed and owned by a diverse range of housing associations and cooperatives. This could include housing provided by employers, religious bodies, or other voluntary organizations.

Forms of Intervention

Throughout this century, governments have intervened in the housing market in a variety of ways. These interventions have taken the form of (a) grants and subsidies to prospective renters or owners to enable households to purchase housing on the private market, (b) grants and subsidies to private landlords to encourage them to rent to lower-income households, (c) financial and other assistance to encourage the growth of voluntary and cooperative forms of housing, and (d) direct provision with government, usually at the local level, as developer, manager, and owner of the dwellings. The diversity of forms of housing intervention and institutional arrangement complicates definitions. Two criteria are important, however, in differentiating welfare state housing from any other housing provided with state assistance. The key institutions should operate on a nonprofit basis, and allocation should be on the basis of some assessment of need rather than on ability to pay. This excludes, for example, general assistance to homeowners and more privatized forms of social housing based on subsidies to private landlords.

Welfare state housing (more usually referred to as social housing) is, as the name implies, associated with welfare state capitalism. It is to be differentiated therefore from direct housing provision in former state socialist societies and is linked with a particular period of development in particular countries. In general terms, we can refer to a period extending from around 1920 to the present day and to the core capitalist countries of Europe, North America, and Australia.

Welfare state housing developed initially in the period following the end of World War I. A number of factors contributed to its emergence as an important form of intervention in housing. These varied in significance from country to country, but key factors were (a) a growing housing shortage through industrialization and urbanization; (b) recognition of the links between bad housing and ill health; (c) shortages exacerbated by wartime damage and reduced building; (d) changed social expectations, the growth of working-class radicalism, and perceived threats to the existing social order; and (e) the evident inability of the private sector and charitable bodies to provide affordable housing in sufficient numbers and quality to cope with absolute housing shortages.

Welfare state housing as a key element in housing provision became more firmly established in many countries in the second period of social change and disruption following World War II. Changes in the political complexions of government combined with prewar shortages exacerbated by war damage required extensive building programs. New forms of intervention in housing to complement, and compensate for, market-based provision became part of a broader strategy of planned intervention in social and economic life—associated with Keynesian/New Deal welfarism and political commitments to full employment. Direct state provision via local municipalities was to varying degrees one of the chosen methods of boosting housing output.

Substantial welfare housing sectors are associated with countries such as Great Britain and The Netherlands. These two examples provide sharply contrasting forms of large public housing sectors. In Great Britain, by 1979, one-third of households lived in housing owned and managed by local councils. In The Netherlands, at the same point, more than 40% of households were in a public sector made up of a variety of housing associations with only a small percentage of the public sector owned and managed by municipalities. At the other extreme is the United States with only 3% of housing in the welfare sector and Australia with around 5%.

Policy Change

Enthusiasm and support for welfare state housing has tended to be greatest at times of social disruption and unrest—primarily during the periods immediately following the two world wars. In these periods, social housing was aimed primarily at the better-off sections of the working class and the lower middle classes. In both economic and political terms, these groups were seen as strategically important by governments, associated, for example, with the large-scale relocation of industry. Nevertheless, such interventions have generally been viewed as temporary measures to cope with pressing housing problems. Both the quality and quantity of provision in the welfare state housing sector have tended to decline as markets and economies have revived. This has led to higher-density building and to shifts

from providing family housing to providing apartments. The conception of welfare state housing as housing for general needs has been transformed into a narrower conception of smaller-scale provision for those cleared from the urban slums and those deemed to have special needs associated with age or disability.

The general economic recession of the mid-1970s is associated particularly with changed policies toward housing provision for low-income households. Public expenditure cuts, an apparent disillusionment with mass housing solutions, and the general resurgence of neoliberal ideologies led to policies of privatization and deregulation in housing. This has involved sharp cuts in building programs for low-income households, greater emphasis on means testing, stricter eligibility rules, and a shift from directly subsidizing the building of dwellings to various forms of housing allowances. It has also involved policies of selling individual publicly owned dwellings to the tenants themselves and transfers of ownership or management to private or quasi-private landlords. These policies have been pursued to varying degrees depending on past patterns of provision in different countries. The sale of welfare housing has been most actively pursued in countries with large, directly state-owned and -managed welfare housing sectors. Welfare housing sectors dominated by state landlordism (e.g., Ireland, Great Britain) have been most vulnerable to policy change. In these situations, central government has been able to impose its will directly on municipal or other state organizations through legislation and control over capital resources. In Great Britain, for example, more than 2 million dwellings were sold in the 1980s, representing about one-third of the stock. In situations in which the relationship between the welfare housing sector has been less direct and ownership more diverse, it has been more difficult for central government to develop privatization policies.

Problems

Two particular areas have been the focus of critical attention in welfare housing sectors. The first concerns management style that has in many cases been seen as bureaucratic, moralistic, paternalistic, and insensitive to the diversity of needs of poorer households. The second concerns the mass, high-rise, modernist housing estates, typically located on the periphery of cities. These low-quality, high-density developments, often housing slum clearance families, have come to represent some of the most concentrated housing problems facing governments.

Social Role

A key issue in social housing and one that could be seen as a defining feature is the extent to which it is the exclusive domain of low-income households. Such issues are often referred to in terms of social mix, meaning the degree to which different social classes or income groups are represented in the public housing sector. Public housing sectors that are most exclusively targeted on lower-income households could be seen as the purest versions of welfare state housing—as housing for those households unable to gain accommodation through the market. This is a feature of more residual forms of provision. The United States and Australia are typically cited as prime examples of countries with small, lower-quality social housing sectors accommodating poor people in the context of housing systems that are generally based on private enterprise. At the other end of the spectrum are countries such as Denmark and Sweden with a strong social democratic tradition that have developed public housing sectors that accommodate a wider cross section of the population. In this context, housing sits within a broader conception of a social market, with the welfare system in general serving a broad base of the population. There are also exceptional examples in the capitalist enclaves of Singapore and, to a lesser extent, Hong Kong. In both these cases, there are large public sectors of directly provided and owned state housing. In Hong Kong, by the late 1980s, about 45% of the population was in public housing. In Singapore, in 1987, 86% of the population was in public housing.

Much debate in recent years has focused on the proper role of welfare state housing and the degree to which this type of housing should be reserved for the poor. One perspective is that those who can afford to gain access through the market should do so. Means-tested access to welfare state housing and pricing according to ability to pay are ways of encouraging this outcome. Particular policy difficulties arise when household circumstances change. Poor households allocated to welfare state housing may become better off. Policies of selling to individual tenants, rent differences linked to household incomes, and grants and other measures to encourage better-off households to seek accommodation elsewhere are all methods that have been used to refocus welfare housing on poor people. Better-off households vacate a public sector, which is then reallocated to a poorer household.

There are, however, those who argue from a different perspective that housing exclusively for, and associated with, poor people will inevitably be stigmatized, second-class housing. If better-off households are encouraged to leave a welfare sector, this sets in train a spiral of decline that accentuates the differences in social status between the welfare and other housing sectors. In some countries, it may also lead to welfare housing sectors associated with particular ethnic minorities. From this viewpoint, the greater the social mix within the social sector, the less stigmatized will be the housing situations of the poor.

The Future

In general, the policy drift throughout the 1980s was away from public sector housing provision amid a desire by governments to withdraw from the sector and for housing supply and demand to be regulated by market forces. A greater proportion of available public resources is now spent on renovating poor-quality dwellings from earlier eras. In many countries, there have been sharp cuts in investment in welfare housing, greater selectivity in housing expenditures, and various privatization policies. These strategies continue to dominate in many countries, and the general trend is toward more residual forms of welfare housing sectors. There is, however, equal recognition that in some countries—particularly those with relatively un-

developed public housing sectors—some new forms of intervention are required to cope with affordability problems for lower-income households.

The future of welfare housing provision will also be affected by developments in other housing tenure forms, particularly the pattern of recruitment into homeownership and the nature of any problems that arise within that tenure. Few commentators, however, envisage a scenario of renewed mass housing provision for lower-income households by large state landlords. More likely is a greater diversity of approaches using a variety of agencies and institutions, less centralized forms of provision, and a further blurring of distinctions between public and private sectors in welfare housing. (SEE ALSO: *Centrally Planned Housing Systems;* **Cross-National Housing Research;** *Social Housing; Tenure Sectors*)

—*Ray Forrest*

Further Reading

Boelhouwer, P. and H. van der Heijden. 1992. *Housing Systems in Europe. Part 1: A Comparative Study of Housing Policy.* Delft, The Netherlands: Delft University Press.

Bratt, R. G., C. Hartmann, and A. Meyerson, eds. 1986. *Critical Perspectives on Housing.* Philadelphia: Temple University Press.

Danermark, Bertha and Ingemar Elander. 1994. *Social Rented Housing in Europe: Policy, Tenure and Design.* Delft, The Netherlands: Delft University Press.

Dunleavy, P. 1981. *The Politics of Mass Housing in Britain 1945-75.* Oxford, UK: Clarendon.

Emms, Peter. 1990. *Social Housing: A European Dilemma?* Bristol, UK: SAUS.

Harloe, M. 1994. *The People's Home? Social Rented Housing in Europe and America.* London: Blackwell.

Lundqvist, Lennart J. 1992. *Dislodging the Welfare State? Housing and Privatization in Four European Nations.* Delft, The Netherlands: Delft University Press.

Power, A. 1994. *Hovels to High Rise.* New York: Routledge. ◀

▶ Women and Environments

Women and Environments is an annual international journal presenting feminist perspectives on women's relationships to their built, natural, and social environments. Articles cover major themes of interest to architects, designers, and feminist scholars and activists, including housing and community planning. First published in 1976. Circulation: 1,500. Price: $22 for four issues; businesses and institutions add $10; add $8 if outside Canada. Contact person: Lisa Dale, Magazine Manager. Address: 736 Bathurst Street, Toronto, Ontario, Canada M5S 2R4. Phone: (416) 516-2600. Fax: (416) 531-6214. E-mail: weed@web.net. World Wide Web: http://www.web.net/weed

—*Judith Kjellberg Bell* ◀

▶ Women and Shelter Network

A coalition of nongovernmental organizations working with women to improve their shelter situation at the community level exchanges information and mutual support through the Women and Shelter Network. Initiative centers at the country level and reference centers at the regional level promote and organize activities and keep the secretariat supplied with information that is shared in a global newsletter produced once a year in English and Spanish. Other organizations and individuals can also receive the newsletter.

As of 1998, there were centers in 25 countries in Africa, Asia, Latin America and the Caribbean, North America, Europe, and Australia. Several regional network meetings and numerous national meetings have been organized, and the group of reference centers has held its eighth international meeting to exchange experience and guide network activities.

Women rarely have the same rights or access as men to land and housing. They suffer extra burdens when services are lacking because of their traditional role in looking after the household. In poor urban neighbourhoods, this role extends to the provision of services in the community, and in rural areas, women are often the caretakers of the environment because of their traditional role in managing their families' subsistence. Therefore, womens' concerns are the survival concerns of the whole community.

Women's role in construction is often unrecognized and unpaid, and their skills and job opportunities are fewer than men's. Furthermore, they are insufficiently represented in organizations at community, local, national, and international levels, despite the active role they play in their neighborhoods.

The network enables women at the community level to share their experiences and search for strategies and methods that will change their situations, and the situations of their communities, for the better. It also aims at making their concerns more widely known by allowing their voices to be heard and bringing what they say to the attention of organizations at the local, national, and international level.

The network was formed in 1988 as an initiative of Habitat International Coalition (HIC), the nongovernmental organization alliance on human settlements, of which it is an active part. It has three representatives on the HIC board to ensure that women and shelter issues are addressed by the coalition.

The plan of action formulated in 1988 called for the group of reference centers to be accountable to and representative of women at the grassroots level, through the network structure, and for this structure to be built up over time through outreach in all regions. It also called for this group to participate in U.N. Centre for Human Settlements (Habitat) meetings and to influence the Habitat program.

As a result of an interregional seminar held in 1989, Habitat responded by strengthening its Women and Habitat Program and endorsing collaboration with the network in the U.N. Global Strategy for Shelter to the Year 2000. Later, Resolution 13/13 was passed by the Human Settlements Commission, inviting "Governments and the UN Centre for Human Settlements (Habitat) to develop a closer cooperation with the HIC Women & Shelter Network and similar NGOs at national, regional and international levels." This was later endorsed by the U.N. General Assembly.

This gives initiative centers in different countries the opportunity to contact and influence government authorities on issues of concern. It has been useful in the campaign against forced evictions, initiated by HIC. Women and children are most affected by forced evictions because they are the ones most often at home. They, therefore, suffer most when they have no place to live. The secretariat can also act to influence governments or international agencies involved in cases of forced evictions, based on information provided through the network.

The newsletter contains reports from the different regions and stories from women living in poor communities, told in their own words and using their own concepts. Women's own perceptions of their shelter situation; their priorities and concerns; their strategies, problems, and successes are thus disseminated and help set new agendas for housing and human settlements planning.

Action at community level is, however, the most important part of the network's activities. It enables women in poor neighborhoods to meet each other, analyze their shelter problems, and take action on them. Membership: approximately 1,000 worldwide. Budget: $30,000, funded by Canadian International Development Agency (CIDA) and MISEREOR and membership dues ($25 per year). (SEE ALSO: *Habitat International Coalition*)

—*Diana Lee-Smith* ◀

▶ Women as Housing Producers

Housing production is a complex process involving builders, developers, designers, landscape architects, contractors, lawyers, salespeople, real estate brokers, bankers, planners, managers, and people in an array of government agencies, including building departments and public works. Women as housing producers participate in all of the above and also have primary responsibility for transforming houses into homes. Because women's visibility is still emerging in many areas, these roles are difficult to quantify, but patterns emerge.

Unidentified Women as Housing Producers

Many Indian women in North America are known as the architects of their communities, designing, producing, and constructing dwellings. For those who lived on the Great Plains, their responsibilities also included choosing camp sites and ensuring that entire villages could be easily dismantled and re-erected. We do not know their individual identities, nor those of many middle- and upper-class women in colonial America who studied domestic architecture as part of etiquette training. Similarly unidentified are the names of African American women who were part of the production process on plantations in addition to caring for their own households. Domestic labor was largely engaged in spinning and quilting, activities essential to making houses livable, and domestic skills were rarely included among skills of an architect.

Naming Women as Housing Producers

Lady Deborah Moody laid out the streets and town center for the Village of Gravesend, the first English settlement in Dutch New Amsterdam, in 1646. Biddy Mason, an African American slave in California, gained freedom in Los Angeles in 1856 and invested her earnings in that city's real estate in 1866. Women producers who merged house production and homemaking in the 19th and 20th centuries include the sisters Catherine Beecher and Harriet Beecher Stowe and Charlotte Perkins Gilman. Catherine Beecher's *Treatise on Domestic Economy* showcased her early designs for improving the private suburban home. With her sister, she suggested improvements in household organization in a later book, *The American's Woman's Home*. Gilman positioned the single-family house within the broader community through her depiction of utopia in *Herland*. Dolores Hayden, in *The Grand Domestic Revolution,* discusses a range of proposals by less-known women who focused on transforming economic and domestic work, through producer and consumer cooperatives; starting commercial enterprises such as apartment hotels; and organizing nonprofits for collective housekeeping and cooperative dining.

A Profile Takes Shape: When Women Are Counted

When statistics on women in the construction trades became available in 1982, female representation was minimal. Painting was the only specialty in which women reached 5.5%. By 1991, the number of women in trades had grown by 23%, but the percentage had slipped in carpentry and masonry, whereas painting had stayed the same. When figures are compared for an 11-year period between 1983 and 1994, women in construction trades (except for supervisors) increased from 1.9% to 2.3%, but women as carpenters decreased from 1.4% to 1.0%. In 1995, only 1% of carpenters were women.

A complete picture of the status of women-owned businesses does not yet exist. The census has been providing data for every five years since 1972 for women-owned businesses (defined as those in which 50% or more of the owners are women). Women-owned construction firms, general builders, and land developers have increased their share over time; however, this number may be deceptively low because many male-headed construction firms are family businesses, and wives, whether or not on the payroll, are involved in decision making. Census data in 1982 and 1987 provided trend information by major industry groups. In construction, about 94,300 women-owned firms existed in 1987, including general building, heavy construction, special trade contractors, and subdividers and developers, reflecting a substantial rise from almost 59,000 in 1982. By 1992, the number of women-owned firms had increased to 183,695. A majority in construction were special trade contractors, and this category had almost doubled in numbers of firms. Yet most self-employed contractors face the stress of sustaining their business in the first year. Discrimination against women in construction still exists even on federally funded projects that mandate hirings. In finance, insurance, and real estate—including banking, insurance carriers, and real estate—the increase went from about 246,400 in 1982 to 437,360 in 1987. Real estate accounted for the majority of the total.

In 1992, the census of women- and minority-owned business expanded to include firms whose stock is more than one-half owned by women. As of 1996, nearly 8 million business were women-owned, with one in eight owned by a woman of color. Employment in women-owned firms exceeded the national average in nearly every region in the United States and in nearly every major industry. Between 1987 and 1996, the top growth industries for women-owned business included construction (up 171%), wholesale trade (up 157%), transportation/communication (up 140%), agribusiness (up 130%), and manufacturing (up 112%). Aggregrate figures mask differences in employment by gender, although the proportion of females in some occupations has increased from 1975 to 1995. Female financial managers, for example, increased from 24% to 50% during this 20-year period. The National Foundation for Women Business Owners reported that women-owned firms are highest in such nontraditional, goods-producing sectors, although more than half of women-owned firms remained in services, with 19% in retail trade and 10% in finance, insurance, and real estate.

In 1988, the Women's Council of Realtors traced its 50-year history. A woman's name first appeared in 1882 when Cora Bacon Foster, a builder and realtor, formed the Real Estate Board in Houston, Texas. The next known identification of women in real estate was in 1911 and 1912 when the National Association of Realtors (later Real Estate Boards) admitted 3 women. In 1924, 8 women formed a Women's Division in California; by 1944, the roster had grown to 19 active and 25 "salesman" members. The Department of Labor reported the steady increase of women real estate agents from 2,927 in 1910 to 9,208 in 1920. By 1935, the Bureau of the Census reported that 44% of the 22,000 practicing real estate agents were women. Between 1985 and 1995, women as a percentage of total employment in the category that includes real estate increased from 41% to 49.8%. Today, women dominate residential sales, a trend noticeable by the late 1970s. In 1994, a group of Wisconsin women formed Women in Real Estate Development to draw attention to their role in commercial real estate.

Becoming Housing Producers through Design and Urban Planning

Schools of architecture emerged after 1862. Apprenticeship remained the common path for men to take. In 1882, Minerva Parker Nichols was one of the few women who trained through apprenticing to a male architect. In 1903, Hazel Wood Waterman went to work for Hebbard and Gill, the leading San Diego firm for apprenticing architects. Three years later, at age 41, Waterman received her first commission and continued to practice until she was 64. She retired to northern California where she resided at the Berkeley City Women's Club, designed by Julia Morgan. Morgan is the best known of the early women architects in the United States, and her many commissions in California included YWCAs and the Hearst Castle.

Architecture schools were still not admitting women in 1915 when the Cambridge School was begun in Boston, Massachusetts. At first, operating out of the architectural office of Henry Frost and Bremer Pond, two Harvard University instructors in the Graduate School of Design, and later, under Smith College's auspices, the Cambridge School opportunity lasted until 1942. Even as schools increasingly admitted women, female students across the United States expressed dissatisfaction with training that ignored women's special needs, such as child care. In the 1970s and 1980s, as a way to compensate for the absence of gender in the typical curriculum, the Women's School of Planning and Architecture ran weeklong summer courses. The lack of role models in schools is still reflected by low percentages of tenured women professors: In 1997, for example, the Association of Collegiate Schools of Architecture reported that 12% of all architecture faculty members were women.

The 100th anniversary of the American Institute of Architecture (AIA) in 1988 coincided with celebrating the election to membership 100 years earlier of the first woman, Louise Blanchard Bethune. It wasn't until 1992 that a woman, Susan Maxman, became the first president of the AIA. By the early 1990s, 18% of all architects were women. Not everyone joins the professional organization. As of June 1995, about 10.5% of AIA members were women (5,622 of a total membership of almost 53,706) and women architects of color were 1.06% (574). Women as a percentage of total employment in architecture in 1995 were 19.8%. Women's presence in architecture is clearly not equal to men's, as evidenced by more than percentages or membership in AIA. Women are less likely to be owners of firms with greater control over projects. When in partnerships with well-known architects, prestigious prizes are more likely to acknowledge the man, as in the case of Robert Venturi and Denise Scott Brown. Women are less likely to receive major commissions for museums, university campuses, and entertainment and civic centers. Within landscape architecture, women are also reported to be fewer in numbers, to be paid less, and to face unfriendly practices in maternity leave and work schedules. As architectural and planning critics, Ada Louise Huxtable, Sybil Moholy-Nagy, and Jane Jacobs have achieved prominence.

Middle-class women's organizations have provided a training ground in various aspects of housing production, particularly as it relates to urban planning. The late 19th and early 20th centuries were a period when middle- and upper-class women were deemed the civilizers of society, whose authority was based in the house. Their talents at housekeeping and their absence from the paid labor force reflected favorably on their husbands' ability to support a household. Some women organized literary or study gatherings that evolved during the first quarter of the 20th century into clubs that focused on social issues. Their work became known as municipal housekeeping. Charles Mulford Robinson, an early city planner, attributed the presence of women's clubs to the existence of children's playgrounds in cities.

By 1912, thousands of women's clubs were studying civic conditions, including cleanliness, sanitation, government, and welfare. They sponsored construction of buildings while developing community institutions such as maternity homes, orphan asylums, residences for working women, and old-age homes. Women's clubs acted as informal planning authorities. The Women's Civic League and the Woman's Club of Colorado Springs obtained $2,500 that they used to hire Robinson to produce a comprehensive plan for the city. In Minnesota, club women recommended town planning commissions with the objective of village and city beautification. The "City Mothers" of Idaho Falls, Idaho, started a Village Improvement Society and planted trees, bought and developed land for a town park, and secured a tax levy for the support of a library. The Women's Improvement Association in Westport, Connecticut, laid 2,000 feet of sidewalk. In Corte Madera, California, women were responsible for installing and maintaining streetlights until the town assumed management. In 1923, the General Federation of Women's Clubs conducted a survey in 33 states that documented the availability of public water, sewer, and garbage systems, as well as the provisions in homes for electricity, telephone, central heat, and so on. The Federation also lobbied to change the census definition of women from "nonproductive" to "housewife."

In the African American community, separate institutions were a necessity, given discrimination. African American women of the middle class organized the National Association of Colored Women's Clubs and played a leading role in forming orphanages, old folks homes, and girls' shelters. Interest in this history has led to identification of women, in and out of the club movement, who made a difference in housing production. In Los Angeles, Charlotta A. Bass was editor and publisher of The California Eagle from 1912 to 1951, a period that witnessed increasing segregation of African Americans. In 1914, when opponents threatened her and boarded the house that Mrs. Mary Johnson bought on an all-white block, Charlotta A. Bass called on club women to march on the home, pressure the Sheriff's Department to remove the boards, and allow Mrs. Johnson to reclaim her house. Through such actions to lift restrictive covenants and continuous newspaper coverage of lawsuits against campaigns to keep neighborhoods white, Bass helped influence real estate development in Los Angeles and other parts of the segregated United States.

Degree-granting urban planning programs became more prevalent after World War II. With the general increase of women in college enrollment, in 1997, the American Planning Association's (APA) annual member survey (of almost 27,000 respondents) reported an increase in women from 18.1% in 1981 to 30%. Responses from planners who are men and women of color fluctuated over the 10 years. Men were more predominant than women, and among women, the representation was far greater for whites. Of the almost 6,000 who responded to questions about interruptions in their careers, women's primary reason was child care, followed by return to school. In comparison, men's primary reasons were return to school and travel. Women planners were practicing in all areas, but far fewer women than men were directors of planning and related agencies or principals of firms. The data suggested that women were underrepresented in development firms. Salary differences between men and women persisted, and despite a greater percentage increase for women, the median salary in dollar amounts had widened by $90. By 1995, despite similar years of employment, only those women and men with less than 5 years' experience earned the same $30,000; for those with 5 to 10 years' experience, men earned $40,000 compared with women's $38,900; and for those with more than 10 years' experience, men's salaries were $56,000 compared with women's $50,000.

Housing Production and Community Development: The Role of Women

Community development was associated with women decades before becoming part of the curricula in programs of urban planning. Housewives in women's clubs became involved in investigations of alleys and houses. In Indiana, one housewife, Albion Fellows Bacon, dedicated years of volunteer work in the successful effort for legislation to alleviate slum conditions. The work of Mary Simkhovitch, Catherine Bauer, and Edith Elmer Woods illustrates different paths in becoming housing producers. Social work, critical writing, and political economy informed their lobbying for national legislation. Each recognized that code enforcement was an insufficient response to the housing crisis and that new construction was necessary. These women advocated a strong role for federal intervention in housing, and in 1937, Congress adopted the U.S. Housing Act and established public housing. Other women took staff positions in the new bureaucracies. Catherine F. Lansing was a mainstay within the New York City Public Housing Authority, where she tirelessly advocated for community facilities as part of the early reform vision for public housing.

On-the-job training for women in community development, particularly management, has dated back for quite some time. The history of women as managers of private and public housing can be traced to an approach Octavia Hill pioneered in England in the 19th century. There, women housing managers organized in the years before World War I. The Octavia Hill system went beyond women's collecting rent to their setting up tenants' groups, creating meeting space, and returning investment to the building, such as whitewashing public hallways. Although Hill has been accused of imposing middle-class values on the poor, others have interpreted her work differently. In the mid-1930s, May Lumsden and Catherine Lansing were able to introduce some of these ideas to Langley Post, the first general director of the New York City Public Housing Authority. Drawing on Octavia Hill principles, women were hired to be rent collectors, but the approach fell victim to struggles over the scope of civil service requirements and union protests about workloads for office staff. Today, in the United States, women, who are the majority of residents in public housing and frequently de facto managers, are leading figures in resident management programs. Women have also continued to play a leading role in code enforcement through reclaiming landlord-abandoned buildings, as in New York City, using their networking skills to organize other tenants.

Women's Involvement in the Housing Production Process

Women's multiple roles and the relationship to housing production are better recognized in the literature of developing nations. A training module of the United Nations, for use by planning authorities in producing low-income housing projects, begins by acknowledging women's relative exclusion from public life and identifying three reasons for the importance of their participation: (a) as an end in itself that provides the right to participate in and implement projects that affect their roles as wives and mothers; (b) as a means to improve projects, given women's primary use of housing; and (c) as a self-generating activity that encourages them to participate more in the life of the community and that builds leadership. The training modules aim to include women at all stages of housing production, from project formulation, including forming eligibility criteria, recruitment of residents, site and house planning, and financing; to project implementation, including construction and obtaining community services; to project management, including maintenance and cost recovery. At the U.N. Habitat II Conference in Istanbul in June 1996, the women's caucus ensured that similar principles were included.

The Women's Institute for Housing and Economic Development, in Boston, Massachusetts, has its roots in an all-woman's architectural office. Responding in the 1970s to issues such as the multiple needs of single parents and the pulls between work and the home, the Institute filled a gap through providing technical expertise in both housing and economics. The Institute pointed out women's absence as developers because of their lack of access to credit and credibility to attain financing. Recognizing this, in 1980, a group of women entrepreneurs and bankers launched Women's World Banking (WWB). By 1990, WWB was a worldwide network of local affiliates that provided assistance to low-income women entrepreneurs. They have served more than 500,000 micro-enterprise clients and made, brokered, or guaranteed more than 200,000 loans averaging $300 each, with a repayment rate of more than 95%. Services include counseling and training in marketing, costing, pricing, production, and general management principles. Affiliates have developed loan guarantee agreements and revolving funds, helped prepare and screen loan proposals, and supervised loan proposals.

Other trends in community development in the 1990s concern building and retrofitting cities to be safe, as part of a redefinition of healthy communities. Women activists have joined forces with women in community development and city planning agencies to develop and implement guidelines for security. Housing cooperatives are being organized, as a response to the lack of affordable shelter; they have proven also to be effective training grounds for developing women's leadership skills. Through these and other efforts, women produce housing but also link housing to the wider community. In so doing, women carry on a tradition of linking homemaking, in which they continue to spend more hours than men, to the many facets of house production. (SEE ALSO: *Feminist Housing Design; Women as Users of Housing*)

—Jacqueline Leavitt

Further Reading

Birch, Eugenie Ladner. 1983. "From Civic Worker to City Planner: Women and Planning, 1890-1980." Pp. 396-427 in *The American Planner: Biographies and Recollections,* edited by Donald Krueckeberg. New York: Methuen.

Beecher, Catherine and Harriet Beecher Stowe. [1869] 1972. *The American Woman's Home.* Salem, NH: Ayer.

Cole, Doris. 1973. *From Tipi to Skyscraper: A History of Women in Architecture.* Boston: MIT Press.

Dubrow, Gail Lee. 1992. "Women and Community." Pp. 83-118 in *Reclaiming the Past,* edited by Page Putnam Miller. Bloomington: Indiana University Press.

Gilman, Charlotte P. 1979. *Herland.* New York: Pantheon.

Goldfrank, Janice. 1995. *Women Builders and Designers: Making Ourselves at Home.* Watsonville, CA: Papier-Mache Press.

Hayden, Dolores. 1989. *The Grand Domestic Revolution: A History of Feminist Designs for American Homes, Neighborhoods, and Cities.* Cambridge: MIT Press.

Leavitt, Jacqueline and Susan Saegert. 1990. *From Abandonment to Hope: Community-Households in Harlem.* New York: Columbia University Press.

Moser, Caroline O. N. and Linda Peake. 1987. *Women, Human Settlements and Housing.* New York: Tavistock.

Torre, Susana. 1977. *Women in Architecture: A Historic and Contemporary Perspective.* New York: Whitney Library of Design.

U.N. Development Programme. (various dates). *Gender in Development.* Monograph series. New York: United Nations.

Wooton, Barbara, H. 1997. "Gender Differences in Occupational Employment." *Monthly Labor Review* 120(4):15-24. ◄

► Women as Users of Housing

Women and men experience housing differently in several ways: access to decent, affordable housing; the meaning of home in their lives; activities and labor performed in the home; spatial and temporal use; and security.

There are many different types of women, and it is impossible to categorize "women" in any singular manner. Women experience the home through their culture, ethnicity, race, class, region, religion, age, and sexuality and through their own temperaments, life history backgrounds, and personal concerns with maintaining or challenging social mores. The research documenting women's use of housing and home covers several countries—Iran, Egypt, India, Mexico, Tunisia, Israel, Greece, India, and Japan, for example—and covers different segments of U.S. cultures, classes, and regions, although Caucasian middle-class women dominate North American studies. Women's use of the home has been analyzed by their marital and parental status, their orientations toward familism or individualism, and their acceptance and demonstration of different sex, occupational, and family roles.

Following is a summary of some of the more salient issues concerning women as users of housing, with special reference to North America.

Access to Decent Housing

In many countries, women predominate in certain types of low-income households (for example, in the United States and Great Britain, women are overrepresented in single-parent and single elderly households). As such, their access to adequate housing is greatly reduced. Female householders in the United States, for example, are the largest subgroup of the poorly sheltered population. According to housing and census statistics, female householders, both homeowners and renters, have a higher incidence of housing quality problems (i.e., physical adequacy of a dwelling, the extent of crowding, and the level of affordability) than the general population. Female householders occupy more than 40% of problem-ridden dwellings in the United States Women are the fastest-growing subgroup among the homeless, constituting between 15% and 20% of the total homeless population in 1997.

Inadequate housing conditions for women are even more striking in the cities and large urban centers of developing countries because of the high proportion of inadequate housing in general and because of women's central and complex roles in the household. Because of increased female migration from rural to urban areas in many developing countries, women's numbers are growing at a more rapid rate in urban centers than is the overall population. Women's role in housing is central because of their substantial responsibilities for the economic well-being of the household, the relevance of household location to women's employment opportunities, and the opportunity costs of women's time spent in obtaining urban services and participating in mutual and self-help schemes. Once again, urban women in developing countries are generally the poorest segment of the population. They are less well educated than their male counterparts, responsible for both managing the household and contributing to its income, disproportionately represented in informal-sector occupations, and have a propensity to be heads of households. Consequently, women rarely have equal rights or access to land and dwellings. They suffer additional burdens when water and fuel services are absent because of their traditional role in providing such services to their families.

Few reliable data are available on women and urban housing. The difficulties encountered in collecting such data are compounded by the situation of these women. They are often not at home during working hours, and when they are at home, social mores often dictate that they not speak to surveyors or strangers.

Meaning of Home

Women's involvement in the home, as well as the spatial organization of the home itself, affects the evaluation of the dwelling, both as a practical place in which to live and work and as a vehicle for personal expression.

Although both sexes attach positive meanings to the concept of home, they do not necessarily perceive the home in the same manner. Interviews with Chicago suburban households in the late 1970s found that men prefer home objects reflecting or portraying action (e.g., sports trophies), whereas women prefer objects reflecting contemplation (e.g., photographs). The research also found that women, compared with men, hold special attachments to the more communal areas of the home, such as the kitchen, dining room, and living room. Other research on differences between U.S. middle-class women and men in the meaning of home found that married women were more likely to view the home as an expression of identity, of family and personal relationships, of personalization, and of refuge. Married women who were primarily homemakers felt more strongly about their homes as reflections of their identities compared with employed women. For married men, home more likely refers to a physical place.

For Israeli women, regardless of employment status, home was found to reflect a place for family, self-expression, and security. However, employed women and men were more likely to view home as a sense of control, compared with the perceptions of full-time homemakers. Employment status or the amount of time spent outside the home perhaps results in a different sense of control, or power, within the residence.

Women other than middle-class, Caucasian, and married are frequently neglected by researchers examining the meaning of home. In an exception, a study of low-income single women in Australia found various housing types in which these women reside: hostels, rooming and boarding houses, rental apartments, with parents, in prostitution houses, and in temporary shared quarters with men. For many, security and independence were core concepts in the meaningfulness of their homes in their lives. Security involved financial and emotional stability as well as the freedom from fear of being evicted. Independence meant financial independence from family or partner and control over one's environment. In another study, women agoraphobics saw the home as a defense, through isolation, against the unknown threats of the public world. Another study of Muslim communities in Greece found the women residents in the multifamily *Organi* enclave experiencing meaning within the collective use of space.

Labor and Activity in the Home

For the last four decades, research has shown that U.S. wives shoulder the responsibility for and complete the vast majority of household tasks and that husbands and children provide supplementary help. Husbands are more likely to undertake those household tasks that have clear and identifiable boundaries (e.g., mowing the lawn), tasks that have greater discretion in both how and when to complete them (e.g., minor household repairs), and tasks with greater leisure components (e.g., playing with children). Advances in domestic technology have not reduced women's work in domestic labor in industrialized countries. Rather, household technology brought with it higher standards of cleanliness and larger homes to clean, confounding the opportunity for work reduction.

However, the sex stratification patterns of domestic labor change somewhat when women work outside the home. Husbands' household work and child care can be responsive to increased demand for their labors, fostered by wives' employment or the presence of young children. Nevertheless, findings from numerous large-scale, time budget studies show that even when working, fully employed women are more responsible than men for a majority of child care and housework duties in the home, not only in North American countries but also in European countries and Japan.

Studies of poor and working-class households reveal that women cope with economic and housing adversity by intertwining their domestic activities with those of the larger community. The "work of kinship," in particular, links many working-class households to the larger community. The domestic skills and responsibilities of women, particularly of poor women, can be seen as hidden resources that can extend to opportunities and crises that occur in the neighborhood or community.

Time and Space in the Home

Does this gender difference of labor in the home translate to different space use in the home? In Michelson's time budget study of Toronto households, fully employed married men and women spend the same amount of time in core rooms of the home, except for the kitchen. However, although the time spent in the core rooms is essentially the same, the experiences within those spaces differ. In the living room, men compared with women are more often involved in passive leisure activities, such as reading or watching television. In the kitchen, women are more likely than men to be involved in domestic work. In bedrooms and bathrooms married women are more likely than men to be involved in taking care of children. Women, compared with men, spend their time in the home in activities with others, often their children. Thus, although full-time employment diminishes the difference between men and women in the distribution of time spent in particular rooms of the home, it does not similarly adjust for the activities and interactions with others in these rooms. Differences also existed among women along the lines of employment and marital status. Fully employed women spend a greater proportion of time in the bedroom and less time in the kitchen, for example.

Such spatial usage has likely consequences for the design of housing. Empirical work and historical analyses demonstrate that as women participate increasingly in labor outside the home, they seek more open and flexible spaces in the home rather than several sole-purpose rooms.

But access to various spaces in the home is also culturally structured, as evidenced in several countries and communities where there are strong social norms dictating time and space requirements for women. History reveals many examples of sexually segregated dwellings associated with women's lower social status. For example, in ancient Greece, where women's status was similar to that of slaves, houses consisted of a series of apartment-like *megarons* surrounding separate women's and men's courtyards. Several contemporary domestic situations couple sex segregation in the dwelling with women's lowly social status: the black tent (*ghezhdi*) of the nomadic Ghilzai, the *Haveli* of India, the Bedouin tents of the Sudan, the domed *indlu* of the African Zulu, the Tibetan tent, the reed houses of coastal Yeman, and the Turkish *yurt*.

Security

An estimated 28 million U.S. wives are battered by their husbands each year, almost half of all married women in the country. The FBI claims that an incident of such abuse occurs every 15 seconds, making it the most common crime in the United States. Historically, police have been reluctant to intervene in domestic violence, a situation partially conditioned by social mores that a home is a "man's castle" and public interference in private family "spats" is wrong.

Domestic violence occurs most frequently in the kitchen, followed by the bedroom and living room. Sometimes battles progress from one room to another. Other studies of domestic homicide show that the bedroom is the deadliest room in the house and that the victims there are most often women. When family murders occur in the kitchen, traditionally women's "territory," the victims are most often men.

Ironically, although men are most likely to be victims of personal violence, women are more fearful than men of interpersonal crime because they perceive a unique and severe threat not felt by men—sexual violence. Their fear is not aspatial. Women feel more at risk in certain places and at certain times. Ironically, the geography of violence against women suggests that they should be more fearful at home and of men they know; yet they perceive themselves to be in greater danger from strange men and in public space. Former U.S. Surgeon General Dr. Antonia C. Novello announced in 1991, "the home is actually a more dangerous place for American women than the city streets" ("Physicians Begin," 1991).

Thus, despite the crime statistics that show that women are more at risk at home and from men they know, women are still socially encouraged to perceive the home as a haven of safety and to associate the public world where the behavior of strangers is unpredictable with male violence. The social construction of masculinity has defined the male stranger as potentially aggressive and powerful and the male partner as family provider and protector. This belief and imagery is so strong that women who are attacked by male partners still hold an overpowering image of potential aggressors as strangers and of violent places as public spaces. Women's frequent economic and social dependence on men encourages them to suppress or deny violence committed by men they know in the private sphere. Their perceptions of space are also influenced by how the public media interpret where violent assaults are committed. The media rarely display the stories of women abused by their partners at home—certainly nowhere near the number of stories of violence to women in public space.

For most women, escape from domestic violence depends on the existence of places of safety outside the home. The feminist community has reacted accordingly by advocating and creating thousands of shelters for battered women and children. (SEE ALSO: *Feminist Housing Design; Women as Housing Producers*)

—*Sherry Ahrentzen*

Further Reading

Birch, Eugenie Lander. 1985. *The Unsheltered Woman: Women and Housing in the 1980s.* New Brunswick: Rutgers, State University of New Jersey, Center for Urban Policy Research.
Hayden, Dolores. 1984. *Redesigning the American Dream: The Future of Housing, Work, and Family Life.* New York: Norton.
Leavitt, Jacqueline and Susan Saegert. 1990. *From Abandonment to Hope: Community-Households in Harlem.* New York: Columbia University Press.
Michelson, William. 1984. *From Sun to Sun: Daily Obligations and Community Structure in the Lives of Employed Women and Their Families.* Totowa, NJ: Rowman & Allanheld.
"Physicians Begin a Program to Combat Family Violence." October 17, 1991. *New York Times,* p. A16.
Spain, Daphne. 1993. *Gendered Spaces.* Chapel Hill: University of North Carolina Press.
Weisman, Leslie Kanes. 1992. *Discrimination by Design.* Urbana: University of Illinois Press. ◄

► Women's Institute for Housing and Economic Development

Founded in 1981, the institute's mission is to create housing and economic security for low-income women and their families in Boston. It provides technical assistance to community and women's groups on housing and business development and develops innovative affordable housing. Workshops and training are offered in the areas of strategic planning, financial management, women's economic development, development of affordable housing, alternative housing models, and cooperative housing development. Since its founding, the Women's Institute has helped to create more than 275 units of housing for homeless families, battered women, pregnant and parenting teens, women with AIDS, and the first permanent housing for grandparents raising their grandchildren. Publications include *Making It Ourselves: A Primer on Women's Housing and Business Development* and *More Than Shelter: A Manual on Transitional Housing.* Staff: 8. Executive Director: Felice Mendel. Address: 14 Beacon St., Suite 608, Boston, MA 02108. Telephone: (617) 367-0520. Fax: (617) 367-1676.

—*Laurel Phoenix* ◄

► World Bank

Since the middle of the 1980s, there has been a growing awareness about the desirability of better management of urban development in less developed and relatively poor societies. At all levels—local municipal administration and state and federal government—the message is one of how (a) to achieve better value for capital expenditures in providing urban services, utilities, land, and housing; (b) cut wastage through removing inefficiencies and "leakages"; (c) improve cost recovery; and (d) ensure a greater level of financial replicability for such programs so that these are ongoing rather than one-time efforts that end when the money runs out. This changing awareness reflects a broader paradigm shift away from large-scale urban projects in which government expects to be the principal provider, toward a position in which the role of public administration is to facilitate equitable and replicable urban development processes, in part by offering conditions conducive for privately raised capital to become involved. Increasingly, a major concern is to ensure self-sustaining environmental management. Often, too, removing subsidies involves a complete revision of taxation and consumption charges.

The World Bank has provided an important leadership role in setting the agenda for housing and land market analyses, for identifying policy approaches, and in "pump-priming" the implementation of new urban policy approaches. Since the early 1970s, the World Bank has been very influential: initially, in effectively recasting housing policies in developing countries; later, during the 1980s, at developing lines of urban development strategy; and during the 1990s, in promoting the New Urban Management Programme. These policy shifts have been achieved through the dissemination of its publications and technical expertise, through demonstration projects, through lines of credit, and through structural adjustment policies.

This has not been without criticism: The Word Bank and the International Monetary Fund have tended to apply similar remedies, as doctrine, irrespective of a country's circumstances, with the result that programs continue to be supported and promoted even after it is clear that they do not work. Contrary to the broad aims of both institutions to foster development and to improve conditions in which the poor live, the result is often the application of austerity measures that are antigrowth and that harm the poor.

The Shifting Paradigm: From "Urban Projects" to the New Urban Management Programme

The structure of World Bank urban policy was initially laid out in the 1972 Urbanization Sector Working Paper. The document was one of the most lucid evocations of the urban crisis in developing countries and the potential to be played by supportive self-help policies. Moreover, it outlined the bank's philosophy of the *urban problem,* and although in other areas of activity the philosophy may have changed radically, in terms of urban policy, the bank has held fast to its fundamental views during the following decades. It incorporated phrases such as "urban efficiency," a "harnessing of market forces," the "improvement of urban management," and so on. Urban policies of the 1990s have involved more of the same rather than a radically new departure. The greater emphasis on poverty alleviation was "pasted onto the prevalent ideology without, however, altering its fundamental slant" (Ayers 1983). Thus, although the bank appeared to have become more skeptical during the 1970s of the ability of market forces alone to provide urban solutions without guided state intervention, cost recovery was still the driving force behind all World Bank projects.

The 1972 Urbanization Working Paper identified four key areas in which bank policies could improve the urban environment. First, it advocated low-cost solutions so as to make shelter affordable to more households. In essence, this meant moving away from completed housing "packages" toward actions and "elements" that would support self-help and so-called sweat equity approaches. As a result, there was an upsurge in programs such as sites-and-services schemes, core units, and upgrading. Institutionalization of these programs was often accompanied by the development of major new housing finance institutions, some of which later developed into second-line financing for worker housing of different forms. Second, services were to be extended

without, so the bank hoped, a need for subsidy. Rather, financial aid from the bank was to prime the pump for provision and greater capture of the installation costs and consumption charges by local authorities. In fact, service installation often remained heavily subsidized. Third, it promoted the growing desirability of urban planning and investment procedures to improve the technocratic nature of policy formulation and application in developing countries. Fourth, programs were required to be self-financing and, therefore, to be replicable.

Thus, the World Bank played an important part in reforming urban policy in the less developed countries, producing a sea change in attitude among governments and other agencies and inducing a shift toward self-help as more than a "politically rhetorical pledge." This turnaround, however, did not lead to the bank's adopting the policy recommendations made by Turner and his followers. Although Turner's ideas were seen as compatible with the bank's own perception of the problem, the specific arguments and policy recommendations won less attention. The principal planks in the policy agenda were urban development projects such as sites-and-services and upgrading programs. To provide what was hoped would be a "demonstration effect," the bank promoted some 13 sites-and-service programs in various less developed countries. By the late 1970s, however, it became apparent that even minimally or unserviced plots were beyond the means of the lowest 20% to 30% of many of the urban poor. These structural constraints demanded policy options that were difficult for governments to accept: a reduction in the already minimum standards, smaller plots sizes, trade-offs of "consolidation" and dwelling improvements through self-help, or heavier land and servicing subsidies. From this time onward date the earlier critiques of self-help housing philosophy and of the World Bank policy prescriptions (Ward 1982; Mathey 1992).

Partly in response, the World Bank began to look to the marketplace as the principal provider. In 1983, bank economist Johannes Linn published a monograph titled *Cities in the Developing World: Policies for Their Equitable and Efficient Growth.* It closely reflects World Bank thinking during the 1980s. Although the housing problem was identified as a major impediment requiring government intervention, Linn maintains an approach consistent with standard bank ideology—as revealed by the subtitle. Land and housing markets presented an impediment because they were inefficient, defined in terms of supply and demand. Thus, although there is a concern with equity throughout the book, there also remains the emphasis on efficiency. For Linn, the problem was largely one of inefficiency and bottlenecks in the supply of land—particularly fully serviced land. The policy prescriptions that arise therefore are those that will facilitate the more smooth operation of the (largely) privately produced land market.

The market supply mechanisms have become "blocked," and the role of public sector intervention is to clear those blockages. Specifically, Linn proposes a range of actions that include (a) the "regularization" of illegal ("clouded") land titles, ostensibly to provide tenure security to de facto owners; (b) limitations to be imposed on the monopoly control of land by a single group or landowner and that may result in artificially created scarcity and high land prices; and (c) the development of effective land registration and cadaster, from which clear property ownership and taxation responsibilities may be identified. The bank no longer considered the public supply of land through low-cost residential subdivisions to be an appropriate response.

This policy approach was very influential, especially proposals (a) and (c) above. These became viewed as a key means to extend the property register as a basis for the application of property taxes, valorization charges, consumption costs, and the many other levies that the local state is empowered to adopt. These two policy development areas are tied to the notion of cost recovery but have lately begun to offer an important mechanism for generating resources necessary for city development programs and for reducing dependence on state and federal lines of credit. "Indeed, from a position during the 1980s of encouraging the withdrawal and downsizing of the state along with privatization and deregulation, since the late 1990s, the World Bank has begun to seek ways to strengthen the state's capacity to intervene, especially at the local level" (World Bank 1997).

This brings up the third and latest promotional "orthodoxy"—the New Urban Management Program (NUMP). This began in 1986 as a 10-year program; in 1998, it continues to operate. Phase 1 (1986-1990) aimed to concentrate on urban land and infrastructure management and municipal finance; Phase 2 (1991-present) seeks to promote the sustainability and adoption of these new approaches at the local level and to improve the quality of research. During Phase 2 and beyond, the primary concerns are to improve the effectiveness of public administration within the urban sector. In the bank's own words, it focuses "urban operations on citywide policy reform, institutional development, and high priority investments and to put the development assistance in the urban sector in the context of broader objectives of economic development and macroeconomic performance." The new focus accelerates an already existing trend within the bank for policy emphasis to concentrate on support for national housing finance systems, urban management, and local government revenue generation and away from urban "projects" and the "urban product."

Some researchers have argued that this represents a much more limited and less ambitious approach than either of the two earlier phases. Yet given that NUMP seeks to address total management and government practices, not just individual projects or lines of funding, and does so against the backdrop of more neoliberal macroeconomic management, it is actually very radical. The emerging priority issues related to different areas of NUMP are relatively straightforward and embrace many of the activities that were often incipient during the "urban projects" phase and that became more visible during the 1980's emphasis on "market efficiency." Municipal finance is at the top of the agenda. Under this heading, three categories deal directly with finance and institution building, whereas the fourth category identifies community participation and the need

for urban managers to respond to public needs. The second principal agenda item sets out the broad categories of bank thinking on infrastructure improvement and again presents these in terms of financial management. The third agenda issue advocates the removal of "constraints" from the land market. Here, the bank argues for "management" over "administration." Land markets are to be freed from constraints, supported, and formalized. Effective land management also requires institution building (planning offices, land registries, etc.) as well as the development of administrative capacities and regulatory measures that will allow effective implementation of those institution's programs. The fourth agenda item concerns urban environmental issues for which the bank sees the need for *more* regulation for environmental control. Although, the bank seems willing to assert the need to improve and increase the amount of legal controls, there is a consistent emphasis on management, operational efficiency, and property rights.

Urban Productivity

Underlying the application of NUMP is the rather nebulous concept of "urban productivity." Neither the loans associated with direct intervention in urban projects nor the broader macroeconomic structural adjustment loans have proven adequate to overcome urban poverty. Thus, the bank began to perceive the desirability in market economies of raising local productivity as a key mechanism to bring about growth. The bank's emphasis on the economic dimension of the city, its contribution to the wider economy, has been an element of policy since at least 1985. Prior to that, the bank tended to see the city as a net drain on resources—as a consumer of investment and subsidy rather than as a producer. Indeed, the bank appears to have adopted much of the thinking behind recent work on informal urban, labor, and transport markets in Peru. This argues that the poor, and more specifically the informal sector, have enormous economic potential and are contributing ever-larger sums to the national economy. Moreover, the urban poor have done most to keep developing economies afloat during the 1980s and 1990s. The rise of the informal sector has taken place, however, in the context of institutional and political constraints: bribery, corruption, legal restrictions, and the arbitrary use of power. Thus, the bank places "paramount importance" on reducing the constraints to urban productivity. It advocates governments' (a) strengthening the management of urban infrastructure and reinforcing the institutional capacity for the state to operate and maintain this system; (b) improving the urban regulatory framework to increase market efficiency and enhance the role of the private sector in shelter and infrastructure provision; (c) improving the financial capacity of municipal institutions through the more effective division of resources and responsibility between central and local government; and (d) strengthening financial services for urban development.

Like the previous two rounds of World Bank orthodoxy, this latest initiative has been undergoing careful scrutiny, and criticism has begun to settle on the dangers associated with leaving housing supply and production solely to market forces, making growth paramount, and ignoring issues of welfare and equity. Another potential problem is the apparent contradiction between acknowledging that the informal sector (or the "unregulated" sector) may be highly productive and sensitive to low-income needs while at the same time seeking to make more efficient and effective systems of city management and administration predicated on appropriate regulation and controls to integrate the informal within the formal sectors. At the time of writing (1998), it still remains unclear if attempts to raise levels of urban productivity will ultimately be successful or whether the NUMP will work, particularly among poor and very poor populations. What is clear, however, is that in the future there will be intensifying demands for housing and land markets to be more effectively managed within a wider context of efficient and replicable structures of city management and financing. It is also clear that in both the housing and the city administrative spheres, the World Bank will continue to exercise a significant role of leadership. (SEE ALSO: *United Nations Centre for Human Settlements; U.S. Agency for International Development*)

—*Peter M. Ward*

Further Reading

Ayers, R. L. 1983. *Banking on the Poor: The World Bank and World Poverty.* Boston: MIT Press.

Baken, R-J. & J. Van der Linden. 1993. "Getting the incentives right: Banking on the Formal Private Sector." *Third World Planning Review* 15(1):1-27.

Campbell, J. 1990. "World Bank Urban Shelter Projects in East Africa: Matching Needs with Appropriate responses?" Pp. 205-23 in *Housing Africa's Urban Poor,* edited by P. Amis and P. Lloyd. Manchester, UK: Manchester University Press.

Jones, G. & P. M. Ward. 1994. "Tilting at Windmills: Paradigm Shifts in World bank Orthodoxy." In *Methodology for Land and Housing Market Analysis,* edited by G. Jones and P. M. Ward. London: UCL Press.

Linn, J. 1983. *Cities in the Developing World: Policies for Their Equitable and Efficient Growth.* World Bank research publication. Oxford, UK: Oxford University Press.

Mathey, Kosta, ed. 1992. *Beyond Self-Help Housing.* London: Mansell.

Pugh, C. 1989. "Housing Policy Reform in Madras and the World Bank." *Third World Planning Review* 11(3):249-73.

Pugh, C. 1994. "Housing Policy Development in Developing Countries: The World Bank and Internationalization, 1972-93." *Cities* 11(3):159-80.

Sisters in the Wood: A survey of the IMF and the World Bank. 1991. *The Economist* (October [Special suppl.]).

Toye, J. 1989. "Can the World Bank Resolve the Crisis of Developing Countries?" *Journal of International Development* 1(2):261-72.

Turner, John F. C. 1976. *Housing by People.* New York: Pantheon.

Ward, Peter M., ed. 1982. *Self-Help Housing: A Critique.* London: Mansell.

World Bank. 1972. *Urbanization Sector Working Paper.* Washington, DC: Author.

———. 1991. *Urban Policy and Economic Development: An Agenda for the 1990s.* Washington DC: Author. ◀

▶ World Health Organization and Housing

The houses that people inhabit for shelter, comfort, warmth, and safety all too frequently cause harm, harbor pests, create health risks, and encourage sickness. Most countries of the world are faced with serious and complex housing problems that adversely affect health. The World Health Organization (WHO) has long been concerned that in some regions, vast populations subsist in rudimentary and overcrowded shelters with less than elementary facilities for healthy living.

It should be noted that the various WHO publications and guidelines on health and housing do not set any international standard for housing; standards are a national responsibility for each country, and WHO's responsibility is in relation to developing health criteria that set out the relationship between housing conditions and health status.

The urbanization of the world is occurring rapidly, bringing new housing challenges. Within 15 years, more than 20 to 30 cities will have more than 20 million people. More important, human-made environments will account for the living space of most of the world's population. By 1998, at least 600 million people in the urban areas of developing countries were living under life- and health-threatening housing conditions.

Many technical, social, planning, and policy factors relating to housing may affect physical and mental health and social well-being. These factors can be expressed in terms of basic human requirements that can accordingly be incorporated into housing standards, policies, and targets relevant to a country's needs, resources, and priorities.

In contrast to absolute environmental standards that prescribe optimal housing conditions, in 1990, WHO defined the health implications of various environmental conditions in terms of three "levels of environmental conditions":

1. "Desirable levels of environmental conditions which promote human health and well-being"
2. "Permissible levels of environmental conditions which are not ideal, but which are broadly neutral in their impact on health and well-being"
3. "Incompatible levels, which, if maintained, would adversely affect health and well-being"

Lawrence has developed a "checklist of housing indicators," comprising eight classes of indicators and their subcomponents, that may serve as a reference model for understanding and analysis of health in housing relationships. The classes of indicators concern the following:

1. *Architecture and urban design,* including not only buildings but also neighborhood characteristics, such as access to public and private services and facilities, sanitary and solid waste disposal, site drainage, and so on
2. *Housing administration,* including building costs, financing of housing, housing allocations policy, and housing maintenance
3. *Societal factors,* including social values related to housing, aspirations and needs of specific groups (e.g., children), leisure activities, and so on
4. *Household demography,* including size and composition of households, male and female domestic roles, and so on

5. *External and internal environmental conditions,* including external air pollution, noise, density, sunlight, ventilation, and internal pollution because of human activities and construction materials
6. *Ergonomics and safety,* including hygienic services, food storage and preparation facilities, design of domestic fittings and equipment, and special needs of residents (e.g., the aged or disabled)
7. *Individual human factors,* including perception and appraisal of the residential environment, sense of security in and around the house, visual and aural privacy, and personal characteristics of residents
8. *Residential mobility and well-being,* including issues of housing choice and varied composition of housing stock as well as vacancies in housing stock within household budget

Programs to promote health in housing continue to pose a major challenge to public health authorities in all countries and continue to take second place to the better-understood health programs related to provision of medical and hospital facilities and specific programs against well-known diseases. However, in recent years, several comprehensive approaches to public health and health in housing, such as the WHO Healthy Cities Programme, have succeeded in defining a practical methodology to address housing and health. It should be clear that no single agency can address all the preceding eight issues, and this has led to the "partnership approach" described next.

Key "Health in Housing" Concepts
Two key concepts inform WHO efforts in health and housing: intersectoral collaboration for health and supportive environments.

Intersectoral Collaboration for Health
A focus of the Health in Housing Programme is the development of policies and management practices that attach importance to health as a goal of development at the level of city and local government or rural district. It requires all agencies concerned with energy, food, agriculture, macroeconomic planning, housing, land use, transportation, and other areas to examine the health implications of their policies and programs and to adjust them to better promote health. This activity is often termed "intersectoral collaboration for health," or ISC. A key weakness of earlier efforts on ISC was the failure to define "who is to do what," to define whether (and how) the health sector was to be involved, and to develop political support for ISC. To succeed, ISC requires at least three elements:

1. The measurement of the health impacts of various development activities, an effort that requires carefully designed and executed activities to link health problems with environmental and social conditions, which may take place by health authorities in collaboration with universities
2. An analysis of both the adverse impacts on health of various development activities and of the potential

opportunities to enhance health that almost every development activity presents

As a result of this analysis, the health sector may, for example, develop a policy on health in housing, health in the workplace, health in schools, and so on.

3. Advocacy by the health sector in relation to each implementing agency to implement the health-related policies and an appropriate program of health promotion

ISC is enhanced by political support. Municipalities may commit themselves to a Healthy City or Healthy Village process that will involve formulating and adopting a municipal health plan and developing solutions to problems on a communitywide basis. Such solutions involve partnerships of municipal government agencies (health, water, sanitation, housing, social welfare, etc.), universities, nongovernmental organizations, private companies, and community organizations and groups.

Supportive Environments
The idea of a supportive environment turned out to be valuable for the promotion of health in many countries. The home, the school, the village, the workplace, and the city are the places where people live and work. Health status is determined more by the conditions in these settings than by the health care facilities that we can provide. These are settings in which health may be maintained and indeed created, by engaging all participants and authorities that operate in the settings, in activities to create a supportive environment for health.

In relation to the settings of neighborhoods and cities, the local government may examine the health implications of its work and develop a municipal health plan, with participation of community organizations, local institutions, and so on. In relation to the school setting, parents, school principals, and education authorities must together develop a plan to install in the school adequate water and sanitation facilities, and a good safe playground—and perhaps allow the children to participate in these activities. WHO has documented examples of these comprehensive health development approaches (which usually have titles such as Healthy Villages, Healthy Schools, Healthy Workplaces). (SEE ALSO: **Health**)

—*Greg Goldstein*

Further Reading
Lawrence, R. 1993. "An Ecological Blueprint for Healthy Housing." In *Unhealthy Housing: Research Remedies and Reform,* pp. 338-60. Edited by R. Burridge and D. Ormandy. London: E & FN Spon.
Ranson, R. 1988. *Guidelines for Healthy Housing.* Environmental Health Series, Vol. 31. Geneva: World Health Organization.
World Health Organization. 1990. *Indoor Environment: Health Aspects of Air Quality, Thermal Environment, Light and Noise.* Geneva: Author.
———. 1989. *Health Principles of Housing.* Geneva: Author.
———. 1995. *Twenty Steps for Developing a Healthy Cities Project* (2nd edition). Geneva: Author.

Y

▶ Yanomami *Shapono*

Shapono is the name of the communal house of the Yanomami Indians of the Venezuelan and Brazilian Amazon Forest. It serves not only as a shelter for the entire community but also as a setting for the social organization. The average size of a Yanomami community is 60 people with a range from 35 to 150. Smaller and larger communities sometimes are recorded, but they are almost always temporary conditions.

The shapono can be described as a large donutlike structure with a central plaza that is cleared to bare ground. The inner ring is constructed by erecting 20- to 30-foot poles about six inches in diameter, and the outer wall is of shorter poles about four feet high. Roof rafters about 30 feet in length connect the front and rear poles to give a slanted roof with a 60-degree incline. The roof is then covered with any of several varieties of broad leaves from the surrounding forest. The life span of the structure depends on how well the roof is constructed, primarily a function of the density of the leaves. A well-constructed roof can withstand the tropical sun and rains for up to five years, at which time it must be replaced or abandoned. Most communities move before this time, about every two to three years.

When a community decides where it will begin clearing a plantain garden, it constructs temporary shelters similar to those used in their nightly camps while on a trek. These flimsy frames with leaves laid on top shelter just one nuclear family. After about a year when the crops begin to produce, the residents begin construction of the shapono.

While in residence at the shapono, the community periodically goes on treks to hunt and gather wild foods. This becomes necessary when the plantains in the gardens become exhausted. The entire community packs up all of its belongings and abandons the shapono for a period of several weeks to several months. During this time, the structure remains empty. In the community's absence, the plantains mature in the village gardens. Thus, the Yanomami are only partially adapted to a settled agricultural way of life and rely on their old hunting and gathering ways for as much as half of the year. This also limits the time and effort they will invest in making their communal structure.

Most of a Yanomami's life is spent within the shapono. It is therefore important not only as a shelter but serves as the stage on which Yanomami play out their lives. Members of each family hang their hammocks in a triangular fashion around a hearth under a section of the leafed roof. Brothers and their families live next to each other, and members of another lineage occupy more distinct sections of the leafed ring. As they look onto the central plaza and beyond, they are in constant contact with the entire community. Without walls or partitions, there is little privacy within the shapono. For sex, giving birth, bathing, and other bathroom matters, they go out to the surrounding forest.

The spacing of the hammocks is important for reciprocal exchanges between households. Food is most often shared with the closest neighbors. The community is integrated also by exchanging food with in-laws on the far side of the shapono.

Within the confines of the shapono, the shamans perform curing rituals, and the leaders or headmen give their daily addresses to the community. Periodic feasts are held when another community is invited to consume large quantities of ripe plantain drink. When someone dies, the deceased is cremated in the center plaza, and the ashes are

Figure 34: Yanomami *Shapono*
Photograph by Kenneth R. Good.

638

mixed with plantain drink and consumed by relatives in a funeral ceremony held usually with another village.

When unexpected visitors arrive from another village, the adult men engage in an all-night ritual chant in the central plaza for purposes of affirming friendship and displaying articles for the subsequent trading session held also in the shapono.

In short, all the family, social, and political activities occur within the confines of the communal shapono. It is a place of security and camaraderie. Outside the confines of the shapono is the forest where men and women make their living by hunting and gathering. Only the plantain and banana gardens adjacent to the shapono offer a buffer to the great forest beyond. (SEE ALSO: *Cultural Aspects; Vernacular Housing*)

—*Kenneth R. Good* ◄

► Yonkers

Yonkers, New York, is the site of two fiercely contested housing segregation cases. For two years, city officials steadfastly resisted a federal court order calling for construction of public housing units in white neighborhoods, giving Yonkers a national reputation for racial hostility. Construction of 200 scattered-site public housing units did not begin until 1991, almost five years after the court order was issued. A second case led to the establishment of an enhanced Section 8 outreach program, which enables poor residents of Yonkers' minority-concentrated southwest area to obtain rental subsidies for private housing throughout Westchester County. In part because of the Yonkers desegregation orders, policymakers and researchers nationwide have begun to pay increased attention to housing mobility strategies that move poor households from concentrations of poverty and minority isolation to more affluent—and often largely white—neighborhoods.

Background

Once a prosperous industrial suburb of Greater New York, Yonkers is an ethnically diverse and highly segregated city (pop. 190,000 in 1990) in Westchester County. In 1980, the U.S. Department of Justice, joined by a local chapter of the National Association for the Advancement of Colored People (NAACP), filed suit against the city of Yonkers, charging deliberate segregation of public housing and schools. On November 11, 1985, Federal District Judge Leonard B. Sand ruled that there were long-standing efforts by city leaders to appease white homeowners by concentrating all new public housing in poor, minority neighborhoods (*United States v. City of Yonkers, et al.* Civil Action 80 C1V. 6761 [LBS] 1985). In 1982, he noted, 97% of the city's 6,800 subsidized housing units were located in or adjacent to the southwest quadrant. Furthermore, although the minority proportion of Yonkers grew from 2.8% in 1940 to 18.8% in 1980, this increase was from 3.5% to 40.4% in the southwest section during the same period and only 2.0% to 5.8% elsewhere in the city. Judge Sand also found that discriminatory siting of public housing had created a dual system of neighborhood schools; minority children were denied the educational opportunities available to their white peers.

A second housing segregation case involved the provision of Section 8 rental subsidies to minority families. In 1991, Westchester/Putnam Legal Services and the law firm of Sullivan and Cromwell filed suit against the U.S. Department of Housing and Urban Development, the Yonkers Municipal Housing Authority, and several state and county agencies. The suit claimed that the federal rights of minority Section 8 recipients were being violated, because approximately 1,300 such families were restricted to Southwest Yonkers and many of the apartments rented to minority Section 8 recipients there were grossly substandard and unsafe.

Court-Ordered Desegregation

In May 1986, Judge Sand ordered the city of Yonkers to immediately integrate its public schools and to address the long-standing segregation of its public housing in two ways: (a) by requiring the construction of mixed-income housing by private developers and (b) by constructing 200 units of public housing in neighborhoods outside the city's minority-concentrated southwest area.

The school order was immediately implemented through busing and the creation of theme-oriented magnet schools. But Yonkers' city council fiercely resisted the housing order, declaring its fears—and those of local neighborhoods associations—that low-income housing in middle-income white neighborhoods would lead to high crime, a decline of property values, and racial turnover. As city officials defied the federal court and openly slurred public housing residents, Yonkers was vilified in the national press—becoming a codeword among housing advocates for racial hostility and "not-in-my-backyard" (NIMBY) resistance to fair housing initiatives. In addition, the city of Yonkers incurred $460,000 in contempt-of-court fines before agreeing, in September 1988, to implement Judge Sand's order.

In April 1991, construction began on 200 low-rise, town house-style public housing units, arranged in enclaves of 14 to 48 units each across seven sites. The first tenants arrived in June 1992; the final units were occupied in November 1994. No progress has been made on the order to build mixed-income housing—a proposal criticized by some city officials and private developers as infeasible in Yonkers' weak housing market.

Section 8 Case

In 1993, a consent decree established an enhanced Section 8 outreach program, designed to help Section 8 participants obtain rental housing outside minority neighborhoods if they wish to do so (U.S. District Court 91 C1V.7181 [RPF] 1993). Here, because the court found discriminatory effects owing to county as well as city action, Section 8 recipient families can access neighborhoods throughout Westchester County, not just in the city of Yonkers.

Housing Mobility Programs

Yonkers' court-ordered desegregation programs include two types of housing mobility strategies: *unit mobility* (wherein government builds or purchases housing) and *tenant mobility* (wherein government provides rental subsidies but units are privately owned). Housing mobility is receiving increased attention from policymakers nationwide who are looking for cost-effective, politically feasible strategies for reducing racial segregation and persistent poverty in inner-city neighborhoods in the United States. It is also of great interest to researchers who seek to understand more about the relative importance of family, neighborhood, school, and other effects on social attainment.

Much of the interest in housing mobility programs owes to the compelling evidence that poverty in many U.S. cities has become more concentrated, and racial segregation worse, since the civil rights era. The suburbanization of high-wage jobs and out-migration of the black middle class has concentrated chronic poverty in socially isolated inner-city neighborhoods, creating an "urban underclass."

A 1993 study of Chicago's Gautreaux program compared black residents of central-city public housing who used rental subsidies to enter white middle-income neighborhoods with counterparts who moved to city neighborhoods. It found significant positive effects of suburban residence on job holding and educational attainment. Inspired in part by these findings, housing advocates in the Clinton administration launched Moving to Opportunity (MTO), a voucher-based initiative in five cities patterned after Gautreaux. (SEE ALSO: **Discrimination;** *Segregation*)

—*Joe T. Darden and Xavier de Souza Briggs*

Further Reading

Gallagher, Mary Lou. 1994. "HUD's Geography of Opportunity." *Planning* (July):12-13.

Massey, Douglas S. and Nancy A. Denton. 1993. *American Apartheid: Segregation and the Making of the Underclass.* Cambridge, MA: Harvard University Press.

Rosenbaum, James and Susan Popkin. 1993. "Employment and Earnings of Low-Income Blacks Who Move to Middle-Income Suburbs." Pp. 342-65 in *The Urban Underclass,* edited by Christopher Jencks and Paul Peterson. Washington, DC: Brookings Institution.

Wilson, William Julius. 1987. *The Truly Disadvantaged: The Inner City, the Underclass, and Public Policy.* Chicago: University of Chicago Press.

Z

► Zoning

Zoning emerged in the United States as part of the progressive reform movement at the turn of the century. Its roots were contradictory. City planners and housing advocates saw zoning as the means to provide civilized restraints to urban development, whereas zoning's more powerful real estate and business sponsors saw it as a way to predict property values, protect real estate values, and maximize profits. Zoning's remarkable spread through the nation's towns and cities and its development as a sometimes inconsistent body of law—now progressive, now regressive—run back in time and thought to these contradictory themes.

Zoning refers to the legislative method of controlling the use of all the land in a municipality by regulating use, lot size, bulk, height, density of population, setbacks, and yards of buildings. This is accomplished by dividing all land into certain districts (residential, industrial, commercial, etc.), with each district having specific rules as to how land and buildings can be developed and used. Unlike most local regulations, such as building and housing codes or subdivision regulations, zoning differs from district to district rather than being uniform throughout the city.

Zoning is an exercise of the police power: the inherent power of government to legislate for the health, welfare, and safety of the community. Operationally, zoning matters are considered by a community's planning commission, which makes its recommendations to city council for action. A board of zoning appeals normally handles variances and appeals from the city council's zoning decisions.

Zoning allows the community to control density of population so that new development has adequate light and air and serviced properly. It helps predict and stabilize real estate investments, helps separate differing and possibly conflicting land uses, and is the most commonly used legal device available for implementing the community's comprehensive land use plan. Since its inception, zoning has also been used to exclude most low-income households from suburban communities and, indirectly, to exclude most members of minority groups. An early use of zoning for this purpose was the racially based zoning system prepared for Atlanta by the prominent planning consultant Robert Whitten. The plan proposed to divide all land in the city into three districts: white, colored, and undetermined. Whitten argued that "race zoning . . . is simply a common sense method of dealing with facts as they are."

Early Developments

Before comprehensive zoning, the use of land and buildings was controlled only by the doctrine of common-law nuisance; "using your property so as not to injure another's." This approach seemed appropriate to a land-rich country committed to laissez-faire economic and social policies. In fact, it proved inadequate to the task of maintaining a decent living environment in the explosion of development that characterized the U.S. city in the late 19th and early 20th centuries when more than 18 million immigrants poured into cities. These cities were plagued by crowded and disorderly development, with high-density tenements everywhere, boiler shops next to hospitals, and steel mills next to residential neighborhoods.

U.S. cities reacted to the crowding and disagreeable conditions of vast immigration and industrialization by limiting the areas where the most noxious industries could operate and by legally prohibiting the worst overcrowding through regulations governing the height of buildings. Boston adopted the first height restrictions in the United States in 1892, and in 1920, Congress adopted similar regulations for Washington, D.C. New York City adopted the first comprehensive zoning ordinance in 1916 after hearings that typically blended real estate interests and reform. The need to protect the security of real estate investments took equal precedence with the need to provide the public with more air and light and lessen unhealthy congestion in housing. Indeed, Hubbard and Hubbard in their national survey of zoning and planning in 1929 noted, "The protection afforded property values by zoning has been quickly recognized by financial institutions everywhere" (p. 189).

The growth of zoning was truly phenomenal following passage of the New York City ordinance. It was as if the urban United States was running a fever and was sure it would perish if it did not have zoning immediately. In 1924, only eight years after the New York City ordinance, the U.S.

Department of Commerce, under Herbert Hoover, with the assistance of the U.S. Chamber of Commerce, published the Standard State Zoning Enabling Act. It sold more than 55,000 copies and was adopted almost verbatim in almost half the states. Lawyers Alfred Bettman and Edward M. Bassett, with their long interest in city planning and zoning, and consultants such as Robert H. Whitten, helped local communities frame their own zoning ordinances—often using the 1916 New York City ordinance as a model. By 1930, 768 municipalities with 80% of the nation's urban population had adopted zoning controls.

Whereas zoning boomed, urban planning languished. Although zoning is ostensibly the means to implement a community's land use plan and although the Standard State Zoning Enabling Act states that "zoning regulations shall be made in accordance with a comprehensive plan," most of the communities rushing to adopt a zoning ordinance did so without preparing such a plan. In many cases, plans were prepared decades after the zoning ordinance was adopted. And in most cases, zoning ordinances were adopted as law by city councils, whereas urban land use plans were adopted, not by councils, but only by planning agencies.

Zoning continues to receive widespread public support up to the present. No aspect of local government, with the exception of taxation, has generated as much intense public interest in the post-World War II era. Its administration is carried out by thousands of mostly unpaid citizens who serve on planning commissions, zoning boards, and boards of zoning appeals. In 1970, an estimated 25,000 people served in this capacity in the metropolitan areas of the United States. Perhaps part of its popularity is based on the fact that in zoning, the lay citizen is "vested with the heady power of direct participation in decision making."

Significant Zoning Cases

The constitutionality of zoning was not tested until the celebrated test case of the *Village of Euclid vs. Ambler Realty Co.* was brought before the U.S. Supreme Court in 1926 (*Village of Euclid v. Ambler Realty Co.* 272 US 365 1926). In deciding the Euclid case, the Supreme Court held that comprehensive zoning was a legitimate and constitutional exercise of the police power.

The Ambler Realty Co. had purchased property in the small village of Euclid, Ohio, which it intended for factory sites. Subsequently, in 1922, Euclid adopted a comprehensive zoning ordinance in which part of Ambler's land was zoned for residential use. Ambler brought suit, charging that the residential zoning substantially reduced the value of its industrial land. In so doing, Ambler leveled a frontal attack on the very concept of zoning as outside the ambit of the police power and therefore a "taking" of property forbidden by the due process clause of the Fourteenth Amendment.

In rejecting Ambler's claim, the Supreme Court established the framework within which all subsequent constitutional challenges to zoning have been conducted. The Court's opinion contained three principles that have shaped all subsequent zoning litigation. First, the Court emphasized that the scope of the police power is not rigid but

elastic and capable of expansion to meet the complex needs of an urbanizing society. Second, "taking" challenges based on alleged dollar loss in property values would not be judged solely on the basis of diminution in value. Rather, any such diminution would be considered as only one factor in a calculus that weighed the community's interest in orderly development against the landowner's claim to unrestricted property use. Third, the Court extended to zoning enactments a legislative presumption of validity that it had not formerly received. The *Euclid* decision was of such importance that from 1926, when the case was decided, until late 1973, the Supreme Court accepted only one other zoning case.

As zoning matured from a novelty to an accepted institution and as the pace of suburban development accelerated after World War II, more and more critics began to raise questions about the use of zoning to classify and segregate the general population according to income, race, or station in life. These critics pointed to the conflict between the legitimate desire of a community for orderly growth, the separation of inappropriate land uses, and the "preservation of community values" as opposed to the need across the metropolitan region to provide affordable housing and opportunities for racial integration. Noting the extreme racial and economic segregation of most U.S. metropolitan areas, these critics suggested that large-lot or "snob" zoning, and the manipulation of other land use regulations simply protected existing interests in property by making new development more expensive and played a significant role in economic and racial exclusion. They pointed to common exclusionary practices such as large minimum lot size requirements, restrictions on apartments and multifamily housing, and imposition of fees and exactions on new housing. The two cases that follow illustrate the problem.

In June of 1980, more than seven years after the U.S. Justice Department brought suit, the U.S. District Court of Northern Ohio found the city of Parma liable for violating sections of the Federal Fair Housing Act (*United States vs. City of Parma Ohio*, 494 F. Supp. 1049 C.N.D. Ohio 1980). The court found that Parma, through a systematic pattern of actions and inaction, had followed a long-standing practice of excluding blacks from residing in the city. Included among these actions was the charge that Parma's city council had (a) denied a building permit to a subsidized apartment development, (b) mandated excessive parking requirements to raise housing costs for apartments, and (c) passed a zoning ordinance sharply restricting height limits to block any subsidized apartments. In the view of the court, zoning and other land use regulations were being used for exclusionary and discriminatory purposes. The court-ordered remedy nullified several of these regulations and required Parma to promote the development of racially integrated subsidized housing.

A zoning case that raises similar issues (or, more accurately, a collection of cases) began in the 1960s and continues to unfold. The cases are broadly known as *Mt Laurel I & II* (*Southern Burlington County NAACP v. Township of Mount Laurel*, 67 N.J., 151, 336 A 2nd 713 1975).

Mt. Laurel is a pleasant, growing suburb of Philadelphia in the southwestern part of New Jersey. In common with most of the municipalities in the state, Mt. Laurel had imposed large minimum lot zoning and other land use restrictions to exclude low-income families. A grouping of poor residents of the township, joined by various other public interest groups and led by the local National Association for the Advancement of Colored People (NAACP) brought a case against the township's blatantly exclusionary zoning policy. This case eventually found its way to the New Jersey Supreme Court.

The court's finding was a landmark. For the first time, a court proclaimed the idea that a community's zoning and other land use regulations had to provide for the construction of a regional "fair share" of low- and moderate-income housing. The court held that exclusionary zoning violated the equal protection and due process guarantees in the New Jersey state constitution. A subsequent group of cases known as *Mt. Laurel II* established a court-administered remedial system designed to achieve the fair share goal. The court's remedial system challenged the state legislature to address fair share issue. In 1985, the New Jersey legislature responded with a Fair Housing Act (New Jersey Stat. Ann. Secs 52:27D—301 to 334), which replaced court supervision of municipal fair share obligations with an administrative agency, the Council on Affordable Housing.

The *Mt. Laurel* litigation was a landmark, but unfortunately for Mt. Laurel advocates, the ruling was based on the New Jersey state constitution, and its effects have been largely restricted to that state. Even in New Jersey, its effects have been compromised by the weakness of government in the face of powerful socioeconomic forces and the judiciary's lack of powers of implementation.

Current Zoning Problems

Under sound planning principles, zoning should implement a previously prepared long-range master plan. Indeed, it is difficult to defend zoning not based on a comprehensive plan that includes estimates of a community's present problems and future needs. However, most communities adopted their zoning ordinance before the preparation of a master plan. In 1927, when 569 cities throughout the United States reported adoption of a zoning ordinance, only 181 of them had also completed master plans. Most authorities agree that one of the most critical defects in zoning has been the failure of many communities to integrate zoning with land use planning.

Another problem is the extent to which zoning tends to preserve the status quo. Most zoning ordinances permitted land uses prior to passage of the ordinance to be continued as "nonconforming." It was expected that the passage of time would eliminate such nonconformities. But natural elimination has been slow, especially in areas of weak market demand. As a result, zoning has not succeeded in improving conditions in areas already deteriorating when the zoning was enacted. Some authorities have called for zoning to be retroactive so that nonconformities would be eliminated at the end of a given time period.

A crucial problem raised in the *Mount Laurel* litigation and numerous other cases around the United States has to do with local zoning patterns that bear little relationship to the needs of the metropolitan area as a whole. All U.S.

metropolitan areas are economically interdependent, but zoning controls are a matter of local discretion with powerful local interest and support. Instead of locating industry, housing, or commercial developments from the perspective of what is best for a region, zoning decisions become politicized by efforts to generate net local fiscal gains, avoid deleterious uses, or protect the status quo. Aside from the discriminatory aspects of such procedures, the very ability of some local governments to be financially viable is threatened by such fiscal zoning. Many authorities emphasize that a minimum level of metropolitan zoning coordination is essential if we are to avoid purely fiscal zoning and the hardening of economic segregation and racial exclusion.

Such home rule power has also to do with the difficulty of siting facilities with unpleasant side effects but that are essential for the functioning of modern metropolitan areas. These locally unwanted land uses (LULUs), such as landfills, sludge disposal plants, sewage treatment facilities, airports, and the like, are necessary, but they evoke a "not-in-my-backyard" (NIMBY) response. Like low-income housing and halfway houses, too often they are barred by local zoning and other land use regulations.

Finally, a 1994 case involving conditional regulations may call into question important aspects of zoning and other land use control practices. In *Dolan v. Tigard* the city required an uncompensated dedication of land as a condition for the issuance of a building permit; the U.S. Supreme Court found this to be a "taking." In general terms, a regulatory taking results when a governmental regulation places such burdensome restrictions on a landowner's use of property that the government has for all intents and purposes denied the owner the use of the property.

The Court ruled that conditions requiring developers to dedicate portions of their property to the government without receiving compensation can be justified only if the government "makes some sort of individualized determination that the required dedication is related both in nature and extent to the impact of the proposed development." The *Dolan* case continues the Supreme Court's explanation of how dedications and exactions, often used in zoning for environmental and infrastructure mitigation purposes, must establish a "rough proportionality" or "nexus" between the nature of the dedication requirement and a legitimate public interest. This continues a trend in which a more conservative court has made it more difficult for government to regulate land use. In light of this decision, local government, planners, and regulators must exercise greater caution in imposing conditions on development proposals.

Conclusion

Zoning is a very popular function in modern local government. It has been the target of numerous changes and challenges since it appeared in the United States in the early part of the 20th century. Typically, three groups have mounted challenges:

1. Landowners or developers who normally resist limitations imposed by the regulations and placed on their development options

2. Present residents of communities who usually seek to impose their own vision of desirable community standards through land use regulations and who often seek to screen out land uses and racial and economic groups not in accord with this vision
3. Residents of surrounding communities in the metropolitan area who are concerned with the adverse effects of restrictive local land use policies on the region as a whole

These residents would like to trade their older environments and public facilities for newer ones; their options are limited by zoning regulations that artificially raise housing prices and screen them out.

The interests of the third group seem likely to present the United States with more frequent challenges in the future. Zoning, which represents a compromise between private rights and public interests, must be continually modified and broadened to meet these challenges. (SEE ALSO: *Exclusionary Zoning; History of Housing; Inclusionary Zoning; New Urbanism; Restrictive Covenant; Setback Requirement; Subdivision Controls; Subdivisions*)

—*Norman Krumholz*

Further Reading

Babcock, R. F. 1961. *The Zoning Game*. Madison: University of Wisconsin Press.

Cullingworth, J. Barry. 1993. *The Political Culture of Planning*. New York: Routledge.

Haar, C. M. 1955. "The Master Plan: An Impermanent Constitution." *Law and Contemporary Problems* 20:353-418.

Hubbard, T. K. and H. V. Hubbard. 1929. *Our Cities of Today and Tomorrow: A Survey of Planning and Zoning Progress in the United States*. Cambridge, MA: Harvard University Press.

Randle, Wm. L. 1993. "Professors, Reformers, Bureaucrats, and Cronies: The Players in *Euclid vs. Ambler*." In *Zoning and the American Dream*, edited by Charles M. Haar and Jerold S. Kayden. Chicago: American Planning Association Press.

Rosenberg, G. W. 1991. *The Hollow Hope: Can Courts Bring About Social Change?* Chicago: University of Chicago Press.

Scott, M. 1969. *American City Planning since 1890*. Berkeley: University of California Press.

Toll, S. I. 1969. *Zoned America*. New York: Grossman.

Whitten, R. 1922. "Social Aspects of Zoning." *Survey*, June 15, 1922, pp. 418-19.

Appendix A

List of Nodal Entries

Each nodal entry contains cross-references to satellite entries that cover other aspects of the same topic. Each satellite entry, in turn, provides additional cross-references to a smaller number of related entries. (Nodal entries in the list of cross-references are set in bold roman type.) Each satellite entry also leads back to the nodal entry and, via it, to other satellite entries. Nodal entries thus serve as gateways or central points in thematically organized clusters of entries. Not every entry in this volume is part of such a cluster.

Abandonment

Affordability

Behavioral Aspects

Community-Based Housing

Cross-National Housing Research

Discrimination

Elderly

Fair Housing Amendments Act of 1988

Federal Government

Health

Homeownership

Homelessness

Industrialization in Housing Construction

Mortgage Finance

Private Rental Sector

Public Housing

Subsidy Approaches and Programs

Tenure Sectors

Third World Housing

Urban Redevelopment

Appendix B

List of Organizations

American Affordable Housing Institute
American Association of Homes and Services for the Aging (AAHSA)
American Association of Housing Educators (AAHE)
The American Institute of Architects (AIA)
American Land Title Association (ALTA)
American Planning Association (APA)
American Real Estate and Urban Economics Association
American Seniors Housing Association (ASHA)
Appraisal Institute (AI)
Assisted Living Federation of America (ALFA)
Association of Community Organizations for Reform Now (ACORN)
Association of Local Housing Finance Agencies (ALHFA)
Building Systems Councils of the National Association of Home Builders (BSC)
Center for Universal Design
Center for Urban Policy Research, Rutgers University
Centre for Urban and Community Studies
Community Associations Institute (CAI)
Consortium for Housing and Asset Management (CHAM)
Cooperative Housing Foundation (CHF)
Council for Affordable and Rural Housing (CARH)
Council of Large Public Housing Authorities (CLPHA)
Council on Tall Buildings and Urban Habitat
Elderly Housing Coalition
Enterprise Foundation
Environmental Design Research Association (EDRA)
European Network for Housing Research (ENHR)
European Real Estate Society (ERES)
European Social Housing Observation Unit
Federal Housing Finance Board
Fannie Mae
Fannie Mae Foundation
Foundation for Hospice and Homecare (FHH)
Government National Mortgage Association (GNMA)
Habitat for Humanity International
Habitat International Coalition

Hospice Foundation of America (HFA)
Housing and Development Law Institute (HDLI)
Housing Assistance Council (HAC)
Institute of Real Estate Management (IREM)
Interagency Council on the Homeless
International Association for Housing Science (IAHS)
International Association for People-Environment Studies (IAPS)
International Federation for Housing and Planning (IFHP)
Joint Center for Housing Studies
Mortgage Bankers Association of America (MBA)
National Alliance to End Homelessness
National American Indian Housing Council (NAIHC)
National Apartment Association (NAA)
National Association for Home Care (NAHC)
National Association of Affordable Housing Lenders (NAAHL)
National Association of Housing and Redevelopment Officials
National Association of Housing Cooperatives (NAHC)
National Association of Housing Partnerships, Inc. (NAHP)
National Association of Realtors (NAR)
National Center for Home Equity Conversion
National Center for Housing Management, Inc. (NCHM)
National Center for Lead-Safe Housing
National Coalition for the Homeless (NCH)
National Commission on Severely Distressed Public Housing
National Commission on Urban Problems (Douglas Commission)
National Community Reinvestment Coalition
National Consortium of Housing Research Centers (NCHRC)
National Council of State Housing Agencies (NCSHA)
National Fair Housing Alliance (NFHA)
National Foundation for Affordable Housing Solutions

National Homebuyers and Homeowners Association
National Housing and Rehabilitation Association (NH&RA)
National Housing Conference
National Housing Institute (NHI)
National Housing Law Project (NHLP)
National Housing Trust
National Institute of Senior Housing (NISH)
National Leased Housing Association (NLHA)
National Low Income Housing Coalition (NLIHC)
National Low Income Housing Preservation Commission
National Multi Housing Council (NMHC)
National Resource and Policy Center on Housing and Long Term Care
National Resource Center on Homelessness and Mental Illness (NRCHMI)

National Rural Housing Coalition
National Shared Housing Resource Center (NSHRC)
National Tenant Union
Neighborhood Housing Services of America
Neighborhood Reinvestment Corporation
Partnership for the Homeless
Public Housing Authorities Directors Association
Shimberg Center for Affordable Housing
Title One Home Improvement Lenders Association (TOHILA)
Urban Homesteading Assistance Board
Urban Institute
Urban Land Institute (ULI)
Women's Institute for Housing and Economic Development

Appendix C

List of Periodicals

Architecture and Behavior/Architecture et
    Comportement
Built Environment
Canadian Housing/Habitation Canadienne
Cityscape
City Limits
EKISTICS: The Problems and Science of Human
    Settlements
Environment and Behavior
Environment and Planning A: Urban and Regional
    Research
Environment and Planning B: Planning and Design
Environment and Planning C: Government and Policy
Environment and Planning D: Society and Space
Environment and Urbanization
Habitat International
Habitat World
Housing Affairs Letter
Housing and Development Reporter
Housing and Society
Housing Finance
Housing Finance International
Housing Finance Review
Housing Policy Debate
Housing Studies
International Journal of Urban and Regional Research
Journal of Housing and Community Development
Journal of Housing Economics

Journal of Housing for the Elderly
Journal of Housing Research
Journal of Property Management
Journal of Real Estate Research
Journal of Social Distress and the Homeless
Journal of the American Real Estate and Urban
    Economics Association
Journal of Urban Economics
Netherlands Journal of Housing and the Built
    Environment
Open House International
People and Physical Environment Research (PAPER)
Progressive Architecture (P/A)
Real Estate Law Journal
Real Estate Review
Roof
Sage Urban Studies Abstracts
Scandinavian Housing and Planning Research
Shelterforce Magazine
Tax Credit Advisor
UNCHS Habitat News
Urban Affairs Review
Urban Age
Urban Geography
Urban Studies
U.S. Housing Market Conditions
Women and Environments

Appendix D

Major Federal Legislation and Executive Orders Authorizing HUD Programs

National Housing Act, 1934 (Public Law 73-479)

Title I: Property Improvements

Section 2: Manufactured Housing (Loan Insurance) Property Improvement (Loan Insurance)

Title II:

Section 203: Homes (One-to Four Family) (Mortgage Insurance)

Section 203(h): Disaster Housing (Mortgage Insurance)

Section 203(i): Suburban and Outlying Areas or Small Communities (Mortgage Insurance)

Section 203(k): Major Home Improvements (Loan Insurance)

Section 207: Multifamily Housing (Mortgage Insurance)

Section 213: Cooperative Housing (Mortgage Insurance)

Section 221(d)(2): Homeownership Assistance for Low- and Moderate-Income Families (Mortgage Insurance)

Section 221(h): Major Home Improvements (Loan Insurance)

Section 222: Homes for Service Members (Mortgage Insurance)

Section 223(a)(7): Refinancing of Existing Insured Multifamily Rental Housing (Mortgage Insurance)

Section 223(e): Housing in Declining Neighborhoods (Mortgage Insurance)

Section 223(f): Purchase or Refinance: Existing Multifamily Rental Housing (Mortgage Insurance)

Section 231: Mortgage Insurance for the Elderly

Section 232: Nursing Homes, Intermediate Care Facilities, and Board and Care Homes (Mortgage Insurance)

Section 233: Experimental Housing (Mortgage Insurance)

Section 234: Condominium Housing (Mortgage Insurance)

Section 235: Interest Supplements on Home Mortgages

Section 236: Interest Supplements on Rental and Cooperative Housing Mortgages

Section 237: Mortgage Credit Assistance for Homeownership Counseling Assistance for Low- and Moderate-Income Families

Section 240: Purchase of Fee Simple Title from Lessors (Mortgage Insurance)

Section 241: Supplemental Loans for Multifamily Projects

Section 242: Nonprofit and Public Hospitals (Mortgage Insurance)

Section 245: Graduated Payment Mortgages

Section 247: Single Family Mortgage Insurance on Hawaiian Home Lands

Section 248: Single Family Mortgage Insurance on Indian Reservations

Section 249: Reinsurance Contracts

Section 251: Adjustable Rate Single Family Mortgages

Section 252: Shared Appreciation Mortgages for Single Family Housing

Section 253: Shared Appreciation Mortgages for Multifamily Housing

Section 255: Home Equity Conversions Mortgages (Demonstration)

Title III: Government National Mortgage Association

Title VIII:

Section 809: Armed Services Housing for Civilian Employees (Mortgage Insurance)

Section 810: Armed Services Housing in Impacted Areas (Mortgage Insurance)

Title X: Land Development (Mortgage Insurance)

Title XI: Group Practices Facilities (Mortgage Insurance)

U.S. Housing Act of 1937 (P.L. 93-383, which replaced P.L. 75-412)

Housing Act of 1949 (P.L. 81-560)

Title I: Urban Renewal Projects

Housing Act of 1954 (P.L. 83-560)

Title VII: Section 701: Comprehensive Planning Assistance

Housing Act of 1959 (P.L. 86-372)

Title II: Section 202: Senior Citizen Housing (Direct Loans)

Housing Act of 1964 (P.L. 88-560)

Title III: Section 312: Rehabilitation Loans

Title VIII: Part 1: Federal-State Training Programs

Housing and Urban Development Act of 1965 (P.L. 89-117)

Title I: Rent Supplements

Title VII: Community Facilities

Section 702: Grants for Basic Water and Sewer Facilities
Section 703: Grants for Neighborhood Facilities

Department of Housing and Urban Development Act (P.L. 9-174)

Demonstration Cities and Metropolitan Development Act of 1966 (P.L. 89-754)

Title I: Model Cities

Title X: Sections 1010 and 1011: Urban Research and Technology

Civil Rights Act of 1968 (P.L. 90-284)

Title VIII: Fair Housing

Housing and Urban Development Act of 1968 (P.L. 90-448)

Title I: Homeownership for Lower-Income Families

Title IV: New Communities

Title VIII: Government National Mortgage Association

Title XI: Urban Property Protection and Reinsurance

Title XIV: Interstate Land Sales

Housing and Urban Development Act of 1969 (P.L. 91-152)

Housing and Urban Development Act of 1970 (P.L. 91-609)

Title V: Research and Technology

Title VII: National Urban Policy and New Communities

Housing and Community Development Act of 1974 (P.L. 93-383)

Title I: Community Development Block Grants

Title II: Assisted Housing

Section 8: Lower-Income Rental Assistance

Title III: Mortgage Credit Assistance

Section 306: Compensation for Substantial Defects
Section 307: Coinsurance
Section 308: Experimental Financing

Title VI: Mobile Home Construction and Safety Standards

Title VIII: Miscellaneous

Section 802: State Housing Finance Agency Coinsurance
Section 809: National Institute of Building Science (NIBS)
Section 810: Urban Homesteading
Section 811: Counseling and Technical Assistance

Emergency Home Purchase Assistance Act of 1974 (P.L. 93-449)

Emergency Housing Act of 1975 (P.L. 94-50)

Title I: Emergency Homeowner's Mortgage Relief

Housing Authorization Act of 1976 (P.L. 94-375)

Housing and Community Development Act of 1977 (P.L. 95-128)

Title I: Community Development

Title II: Housing Assistance and Related Programs

Title III: Federal Housing Administration Mortgage Insurance and Related Programs

Title IV: Lending Powers of Federal Savings and Loan Associations; Secondary Market Authorities

Title V: Rural Housing

Title VI: National Urban Policy

Title VIII: Community Reinvestment

Title IX: Miscellaneous Provisions

Housing and Community Development Amendments of 1978 (P.L. 95-557)

Title I: Community and Neighborhood Development and Conversion

Title II: Housing Assistance Programs

Title III: Program Amendments and Extensions

Title IV: Congregate Services

Title V: Rural Housing

Title VI: Neighborhood Reinvestment Corporation

Title VII: Neighborhood Self-Help Development

Title VIII: Livable Cities

Title IX: Miscellaneous

Housing and Community Development Amendments of 1979 (P.L. 96-153)

Title VI: National Commission on Native American, Alaska Native, and Native Hawaiian Housing

Title VII: Miscellaneous

Title VIII: Section 8 Rent Adjustments

Cranston-Gonzalez National Affordable Housing Act (P.L. 101-625)

Title I: General Provisions and Policies

Title II: Investment in Affordable Housing

Subtitle A: HOME Investment Partnerships
Subtitle B: Community Housing Partnerships
Subtitle C: Other Support for State and Local Housing
    Strategies
Subtitle D: Specified Model Programs
Subtitle E: Mortgage Credit Enhancement
Subtitle F: General Provisions

Title III: Homeownership

Subtitle A: National Homeownership Trust
    Demonstration
Subtitle B: FHA and Secondary Mortgage Market
Subtitle C: Effective Date

Title IV: Homeownership and Opportunity for People Everywhere Programs

Subtitle A: HOPE for Public and Indian Housing
    Homeownership
Subtitle B: HOPE for Homeownership of Multifamily
    Units
Subtitle C: HOPE for Homeownership of Single Family
    Homes

Title V: Housing Assistance

Subtitle A: Public and Indian Housing
Subtitle B: Low-Income Rental Assistance
Subtitle C: General Provisions and Other Assistance
    Programs

Title VI: Preservation of Affordable Rental Housing

Subtitle A: Prepayment of Mortgages Insured Under
    National Housing Act
Subtitle B: Other Preservation Provisions

Title VIII: Housing for Persons with Special Needs

Subtitle A: Supportive Housing for the Elderly
Subtitle B: Supportive Housing for Persons with
    Disabilities
Subtitle C: Supporting Housing for the Homeless
Subtitle D: Housing Opportunities for Persons with AIDS

Title IX: Community Development and Miscellaneous Programs

Subtitle A: Community and Neighborhood Development
    and Preservation
Subtitle B: Disaster Relief
Subtitle C: Regulatory Programs
Subtitle D: Miscellaneous Programs

Housing and Community Development Act of 1992 (P.L. 102-550)

Multifamily Housing Property Disposition Reform Act of 1994 (P.L. 103-233)

Housing Opportunity Program Extension Act of 1996 (P.L. 104-120)

Index of Contributors

Index of Authors Cited

Subject Index

Entries in this volume are in bold; nodal entries are in bold italic and are preceded by an asterisk. See Appendix A for a list and explanation of nodal entries.

In this index, cross-references have been kept to a minimum. Readers interested in coverage of related aspects of a certain topic are strongly encouraged to consult the cross-references provided under the "See also" headings at the end of entries in the *Encyclopedia*.

About the Contributors

Sherry Ahrentzen is Professor of Architecture at the University of Wisconsin—Milwaukee. Her research, focusing on new forms of housing to better address the social and economic diversity of the United States, has been published extensively in journals and magazines, such as *Journal of Architectural and Planning Research, Environment and Behavior,* and *Progressive Architecture.* She also coedited the book *New Households, New Housing.*

Fred Andreas, a practicing architect, has worked on various projects in architecture, urban design, and planning throughout the country. He has 18 years of experience in architecture, planning, and environmentally sustainable design. He teaches a variety of courses at the University of Colorado's College of Architecture and Planning, including Sustainable Design/Solar Technology, Architectural Drawing, and Design Studio.

Richard P. Appelbaum is Professor of Sociology and Director of the Institute for Social, Behavioral, and Economic Research at the University of California at Santa Barbara, where he is also Codirector of the Center for Global Studies. He is a founder and the editor of *Competition and Change: The Journal of Global Business and Political Economy.* His most recent books include *States and Economic Development in the Asian Pacific Rim* (Sage, 1992) and *Sociology* (3rd edition, 1998) His research focuses on the social, cultural, and political ramifications of economic globalization.

Wayne R. Archer is a Professor in the Department of Real Estate and Finance at the University of Florida.

John D. Atlas is a public interest attorney specializing in housing and president of the National Housing Institute, a think tank based in Orange, New Jersey. He served on the Advisory Board of the Resolution Trust Corporation from 1992 to 1995.

Diana E. Axelsen is a senior production editor in the Books Divison at Sage Publications. She previously taught philosophy and women's studies at Spelman College in Atlanta, Georgia, and at California Lutheran University in Thousand Oaks. She helped develop bioethics courses at the Morehouse College School of Medicine and has published in this area.

John S. Baen is Associate Professor of Real Estate at the University of North Texas. He has written four books and more than 70 articles in the areas of real estate appraisal, investments, and marketing.

William C. Baer teaches housing and community development as well as the history of property rights at the University of Southern California. He is currently applying market analysis to the 17th-century London housing market.

David W. Bartelt is Professor of Geography and Urban Studies at Temple University in Philadelphia. He works with neighborhood groups and community organizations on issues of physical and economic revitalization in the Philadelphia region and is the coauthor of several monographs on the city's neighborhoods and on housing issues. He is also coauthor of *Philadelphia: Neighborhoods, Division and Conflict in a Postindustrial City* (1991).

Robert B. Bechtel is Professor of Psychology at the University of Arizona in Tucson. His housing research began in Kansas City in the 1960s; branched to Project Arrowhead in Cleveland, Ohio, in 1970; and has since extended to Alaska, Australia, Canada, Iran, Israel, and Saudi Arabia. His previous works include *Enclosing Behavior* (1979), *Methods in Environmental and Behavioral Research* (coedited with Robert Marans and William Michelson, 1989), and *Environment and Behavior: An Introduction* (1997). He is also editor of *Environment and Behavior.*

Judith Kjellberg Bell is editor of publications for the Centre for Urban and Community Studies, University of Toronto, and was for several years editor of the magazine *Women & Environments*.

Peter Bell is president of Dworbell, Inc., a Washington, D.C.-based association management and communications firm that specializes in housing and banking issues. He serves as Executive Director of the National Housing & Rehabilitation Association, the Title One Home Improvement Lenders Association, and the National Rehabilitation Lenders Association, and is president of the National Reverse Mortage Lenders Association. He is also publisher of *Tax Credit Advisor,* a monthly newsletter covering the development and financing of affordable housing built with federal low-income housing tax credits.

Philip R. Berke is Associate Professor of Land Use and Environmental Planning at the Department of City and Regional Planning, University of North Carolina. His research interests focus on state and local development management, sustainable development, and natural hazard mitigation in developed and developing countries. He has also worked as a practicing planner and consultant to state and local governments.

Christine Benglia Bevington is a New York architect who has worked in the architectural firms of Candilis/Josic/Woods and Edouard Albert in Paris, and of Marcel Breuer and Edward Larrabee Barnes in New York City. She has also taught housing design at New York Institute of Technology, Columbia University, and the Pratt Institute. In 1979, she founded ACWM (Architecture for Child Woman and Man), which focuses on architectural issues affecting children.

Eugenie Birch is Professor of City and Regional Planning at the University of Pennsylvania and past President of the Association of Collegiate Schools of Planning.

A. Kermit Black is Director of the Center for Housing and Urban Development at Texas A&M University. He has been centrally involved in its Colonias Program, which concerns the delivery of education, health, human services, and community development in Texas' colonias through community resource centers.

Roy T. Black teaches real estate at Georgia State University. He is the author of *The Georgia Real Estate Guide to License Law, Brokerage, and Related Topics,* as well as numerous journal articles on real estate.

Leslie Black-Plumeau is a senior program evaluator for housing and community development issues at the U.S. General Accounting Office in Washington, D.C.

Lloyd W. Bookout is Director of Housing and Community Development Research/Education at the Urban Land Institute, a nonprofit research and education institute based in Washington, D.C. He is the author of numerous books, research papers, and articles on issues affecting real estate development.

Larry S. Bourne is Professor of Geography and Planning and Director of the Planning Program at the University of Toronto and past Director of the University's Centre for Urban and Community Studies. He is the author or editor of several basic texts in urban geography and planning, including *The Changing Social Geography of Canadian Cities* (1993).

Rachel G. Bratt is Professor and Chair of the Department of Urban and Environmental Policy at Tufts University. She is the author of *Rebuilding a Low-Income Housing Policy* and coeditor of *Critical Perspectives on Housing.* Her recent work involves studying models of self-sufficiency programs operated by nonprofit housing development organizations.

Xavier de Souza Briggs is a planner and sociologist at Harvard University. He is coauthor of *The Social Effects of Community Development* (Community Development Research Center), a major report on the effects of community development corporations on their target neighborhoods, and is active in community planning and research around the country.

Satya Brink, an environmental sociologist, works for the Government of Canada, managing social research. Her clients for housing research projects were international organizations such as the Organization for Economic Cooperation and Development (OECD) and UNCHS-Habitat, the governments of Sweden and France, universities, and the private sector. She served on the Standing Committee for Barrier-Free Design for the Canadian Building Code for 10 years and has published widely on accessible housing and communities.

Kate Bristol is a Senior Program Manager for the Housing Authority of the County of Marin in California. She has taught architecture and city planning at the University of California, Berkeley, and has written extensively on public housing in the United States.

Mary E. Brooks is the Director of the Housing Trust Fund Project, a special project of the Center for Community Change. The Project is a clearinghouse of information on housing trust funds throughout the country that provides technical assistance to organizations and agencies working to create these funds. She created the project some 12 years ago and has provided assistance to dozens of jurisdictions. She is the author of numerous publications on low-income housing and housing trust funds.

Ronald Brooks is a postdoctoral Fellow in Public Health at University of California—Los Angeles. His recent research is on AIDS health education among Latino populations. He also teaches on housing issues at California State College at Northridge and at UCLA.

Patricia Burgess Director of Planning and Urban Design at Cleveland State University's Maxine Goodman Levin College of Urban Affairs. An urban planner and historian who researches the history of planning, zoning, real estate, and urban development, she is author of *Planning for the Private Interest: Land Use Controls and Residential Patterns in Columbus, Ohio, 1900-1970* (1994).

Richard J. Buttimer, Jr. is Assistant Professor of Finance and Real Estate in the Department of Finance and Real Estate at the University of Texas at Arlington.

Margaret P. Calkins is President of IDEAS, Inc., Innovative Designs in Environments for an Aging Society, a research and consulting firm focusing on the therapeutic potential of the environment—organizational and social, as well as physical—particularly as it relates to frail and impaired elderly. She is author of *Design for Dementia,* the first comprehensive design guide for people with dementia.

Charles Cambridge is a full-blood and enrolled member of the Navaho Tribe and holds a doctorate in anthropology from the University of Colorado, Boulder. He has taught at several universities and is currenting continuing his research in appropriate technology and traditional architectural designs and on Acquired Immune Deficiency Syndrome among American Indian populations. He initiated the Solar Hogan Project at the University of Colorado and is affiliated with Kimochi, Inc., and Medicine Bow, Inc.

Catherine M. Cameron is Assistant Professor of Anthropology at the University of Colorado, Boulder. She specializes in the archaeology of the American Southwest and was involved in a long-term research project in Chaco Canyon, New Mexico. She has also participated in archaeological projects in Alaska, Mexico, Europe, and the Near East.

Ayse Can is Director of Community Development Research and Training at the Fannie Mae Foundation. She oversees research, technology, professional education, and training programs targeted at underserved populations and communities. Her recent research has focused on the spatial modeling of house price and neighborhood dynamics; development of geographically based housing affordability indicators; and development of spatial analytical research and business tools for housing and the mortgage finance industry.

Paul J. Carling is Executive Director of the Center for Community Change, an international research and training center focused on full community integration for people with major mental illnesses, located at Trinity College of Vermont, in Burlington, where he also directs the Program in Community Mental Health. He has consulted in nearly all of the United States and Canadian provinces, as well as in Europe, Australia, and New Zealand, on issues of housing, community support, and community integration. His most recent work is *Return to Community: Building Support Systems for People with Psychiatric Disabilities.*

James H. Carr is Senior Vice President of the Fannie Mae Foundation, where he is responsible for research, policy development, program evaluation, and professional educa-

tion training. He has been published in *Vital Speeches of the Day* and has written extensively on housing and urban policy, housing finance, community reinvestment, and state and local finance. He is an instructor for the Neighborhood Reinvestment Corporation and an adviser to the Organisation for Economic Cooperation and Development (OECD).

Roger W. Caves is Professor and Coordinator of the Graduate City Planning Program and Undergraduate Urban Studies Program at San Diego State University. He is the author of *Land Use Planning: The Ballot Box Revolution* (1992), editor of *Exploring Urban America: An Introductory Reader* (1995), and coeditor of the American Planning Association's Housing and Human Services Division's *Housing and Human Services Quarterly.*

Karen Ceraso is Senior Editor of the magazine *Shelterforce,* published by the National Housing Institute in Orange, New Jersey.

Mittie Olion Chandler is Associate Professor and Director of the Master of Urban Planning, Design, and Development and the Master of Science in Urban Studies Programs in the Maxine Goodman Levin College of Urban Affairs at Cleveland State University. Her research interests include low-income housing, public housing resident management, fair housing, and urban politics. She is the author of *Urban Homesteading: Programs and Policies.*

Jeffrey I. Chapman teaches public finance and public policy classes at the Sacramento Center, School of Public Administration, University of Southern California. He has done research on Proposition 13 since 1978.

Ivan D. Chase is Director of the Laboratory for the Experimental Study of Social Organization at the State University of New York at Stony Brook. His area of research is social organization in humans and animals, and he has written about vacancy chains, dominance hierarchies, and the division of labor in a variety of species.

Louise Chawla is Associate Professor in an interdisciplinary honors program at Kentucky State University in Frankfort. She is a developmental and environmental psychologist whose research focuses on children's environmental experience, the development of environmental concern and activism, and design and planning to meet children's housing and community needs. She is the author of *In the First Country of Places.*

Phillip L. Clay is Professor of City Planning at the Massachusetts Institute of Technology. His special expertise includes housing and community development. He was a member of the Low Income Housing Preservation Commission and served as Assistant Director of the MIT-Harvard Joint Center for Urban Studies.

Robert W. Collin is the first faculty member of the Department of Environmental Studies at the University of Oregon.

He is a lawyer, planner, and social worker who has published extensively on homelessness, environmental equity, and municipal finance. He provides legal research for civil rights advocates in the South, to environmental justice advocates at the U.S. Environmental Protection Agency, and to the native peoples of Alaska.

E. Raedene Combs is Professor in the Department of Family and Consumer Sciences at the University of Nebraska—Lincoln. She has published on housing affordability and housing for the elderly, particularly older women.

Michael K. Conn is Director of Research for Girl Scouts of the U.S.A. He is also a member of the Children's Environments Research Group in the Environmental Psychology Program at the City University of New York Graduate Center.

Charles E. Connerly is Professor of Urban and Regional Planning at Florida State University. He was coeditor of the *Journal of Planning Education and Research* from 1991 to 1996.

Daniel M. Cress is Assistant Professor of Sociology in the Department of Sociology at the University of Colorado at Boulder. His research interests include social movements and social problems. He is working on a project examining protest activity by homeless people in the United States.

Joe T. Darden Professor of Geography and Urban Affairs at Michigan State University. His research interests are urban and social geography, with an emphasis on minority groups. He is the author of more than 100 publications, including the coauthored book *Detroit: Race and Uneven Development*. He was Vice President of Lansing's Fair Housing Center from 1984 to 1987 and is currently evaluating the long-term impact of scattered-site housing in Yonkers, New York.

John Emmeus Davis is a cofounder and principal of Burlington Associates in Community Development. He was formerly Housing Director and Enterprise Community Coordinator for the City of Burlington, Vermont. He is the editor and coauthor of *The Affordable City: Toward a Third Sector Housing Policy*, the author of *Contested Ground: Collective Action and the Urban Neighborhood*, and coauthor of *The Community Land Trust Handbook*.

Charles J. Delaney is Associate Professor of Real Estate at Baylor University in Waco, Texas, and has been Visiting Research Scholar at the Centre for Urban and Social Research at Swinburne University of Technology in Melbourne, Australia. He has published widely on the topic of impact fees as an alternative financing mechanism for the provision of physical and social infrastructure necessitated by growth and development.

Cheryl P. Derricotte is Director, Professional Practice, Housing and Regional/Urban Design at The American Institute of Architects in consultant and housing policy analyst. She has been a member of the American Institute of Certified Planners (AICP) since 1995.

Cushing N. Dolbeare has been a Consultant on Housing and Public Policy since 1971. She is founder of the National Low Income Housing Coalition and the Low Income Housing Information Service and was executive of those organizations from 1977 to 1984 and 1993 to 1994, respectively. She has also served as Executive Director of the National Rural Housing Coalition and as interim Executive Director of the National Coalition for the Homeless. She began her housing career in 1952 as Assistant Director of the Baltimore Citizens Planning and Housing Association and then spent 15 years as Assistant Director and Managing Director with the former Philadelphia Housing Association.

David E. Dowall is Professor of City and Regional Planning at the University of California at Berkeley. He teaches courses on housing, infrastructure, and urban land economics. He has worked in over 30 countries, conducting research and providing advice on land management, housing, and infrastructure planning.

William A. Downs is retired Chief of the former Physical Characteristics Branch of the Division of Housing Statistics at the U.S. Bureau of the Census, Washington, D.C.

Peter Dreier is the E. P. Clapp Distinguished Professor of Politics and director of the Public Policy Program at Occidental College in Los Angeles. He has written extensively on issues of affordable housing.

Melvyn R. Durchslag is Professor of Law at Case Western Reserve University School of Law where he teaches and writes in the areas of state and local government law and constitutional Law.

Michael R. Edelstein is Professor of Environmental Psychology at Ramapo College in Mahwah, New Jersey. His research has focused on psychosocial impacts of contamination and environmental hazard issues and on ecological literacy and ecopsychology. His publications include *Contaminated Communities: The Social and Psychological Impacts of Residential Toxic Exposure* (1988) and *Radon's Deadly Daughters: Science, Environmental Policy and the Politics of Risk* (1998).

David R. Ellis is Associate Director of the Center for Housing and Urban Development for Planning and Economic Development and Visiting Professor of Urban Planning in the Department of Landscape Architecture and Urban Planning at Texas A&M University.

Harold A. Ellis is an Assistant County Counselor with St. Charles County in Missouri. He holds a J.D. degree from Washington University in St. Louis.

Leo F. Estrada is Associate Professor of Urban Planning in the School of Public Policy and Social Research at U.C.L.A.

As a social demographer, he has specialized in ethnic and racial trends, particularly of the Latino population of the United States. He has conducted research on public housing and is a consultant to the Housing Authority of the City of Los Angeles.

Norman Fainstein is Dean of the Faculty and Professor of Sociology at Vassar College. He has previously held teaching or administrative posts at City University of New York, the New School for Social Research, and Columbia University. He has published several books and numerous articles concerned with urban political economy, race relations, social movements, and public policy.

Susan S. Fainstein is Professor of Urban Planning and Policy Development at Rutgers University. She is coeditor of *Divided Cities: New York and London in the Contemporary World* and author of *The City Builders: Property, Politics, and Planning in London and New York.* The author of many articles on urban politics and policy, she has written extensively on urban redevelopment and comparative urban policy. She recently completed a three-year evaluation of the Minneapolis Neighborhood Revitalization Program.

Roberta M. Feldman is Codirector of the City Design Center at the University of Illinois at Chicago. She has written and lectured widely on resident activism in public housing. She was a founding coeditor of the *Journal of Architectural and Planning Research* and is past chair of the Board of Directors of the Environmental Research Association. She is a member of the Board of Trustees of the Graham Foundation for Advanced Studies in the Fine Arts.

Carol Fennelly is with the Community for Creative Non-Violence, Washington, D.C.

Price V. Fishback is Professor of Economics at the University of Arizona and Research Associate with the National Bureau of Economic Research. He is the author of *Soft Coal, Hard Choices: The Economic Welfare of Bituminous Coal Miners, 1890-1930* (1992).

Will Fleissig is an architect, urban designer, and developer of affordable housing. He is a partner and cofounder of continuum Partners, LLC, a Colorado-based real estate group focused on developing pedestrian-oriented and mixed-use projects throughout the western United States. He is Senior Associate at the University of Colorado's Real Estate Center and has taught urban design and development courses at University of California—Berkeley, University of California—Los Angeles, University of Southern California, and Harvard.

Ray Forrest is Professor of Urban Studies and Director of Research in the School for Policy Studies, University of Bristol, England.

Guido Francescato is Professor of Architecture at the University of Maryland, College Park. He coauthored *Resi-*

dents' Satisfaction in HUD-Assisted Housing: Design and Management Factors.

Karen A. Franck is Associate Professor in the School of Architecture and the Department of Humanities and Social Sciences at the New Jersey Institute of Technology. She is coeditor of *New Households, New Housing* (1989) and *Ordering Space: Types in Architecture and Design* (1994) and author of the monograph *Nancy Wolf: Hidden Cities, Hidden Longings* (1996).

Dorit Fromm is a practicing architect. She is the author of *Collaborative Communities: Cohousing, Central Living, and Other New Forms of Housing with Shared Facilities* and of numerous articles in the housing field.

Dennis E. Gale is Department Chair and Henry D. Epstein Professor of Urban and Regional Planning, College of Urban and Public Affairs, Florida Atlantic University. Formerly, he was Director of Planning and Management Research at the Urban Institute in Washington, D.C. and taught for 14 years at George Washington University. He is the author of *Neighborhood Revitalization and the Postindustrial City* (1984), *Washington, D.C.: Inner City Revitalization and Minority Suburbanization,* (1987) and *Understanding Urban Unrest: From Reverend King to Rodney King* (Sage, 1996)

George C. Galster is the Clarence Hilberry Professor of Urban Affairs at Wayne State University in Detroit, Michigan. Previously, he was Principal Research Associate and Director of Housing Research at the Urban Institute. He has published extensively on the topics of metropolitan housing markets, racial discrimination and segregation, neighborhood dynamics, residential reinvestment, community lending and insurance patterns, and urban poverty. His latest books are *The Maze of Urban Housing Markets* (1991), *The Metropolis in Black and White,* (1992), and *Reality and Research* (1995).

Daniel J. Garr is Professor of Urban and Regional Planning at San Jose State University in San Jose, California. He has developed subsidized housing units in northern California.

John I. Gilderbloom is Associate Professor of Urban Policy and Economics in the College of Business and Public Administration at the University of Louisville. He also directs three research centers there: HANDS (Housing and Neighborhood Development Strategies), SUN (Center for Sustainable Neighborhoods), and the Urban Center on Aging. He is the author of numerous journal articles, chapters, monographs, and opinion pieces on housing.

Patricia Gober is Professor of Geography at Arizona State University where she teaches courses in population and urban geography. She has written extensively on migration and housing demography. She is the author of a recent Population Reference Bureau monograph, *Americans on the Move.*

John M. Goering is Supervisor of Social Science Analysis in the Office of Policy Development and Research at the U.S. Department of Housing and Urban Development. He manages civil rights studies and conducts research on the racial composition of public housing He has taught at the University of Leicester (England), Washington University (St. Louis), and the Graduate Center of the City University of New York and has conducted research on legal and undocumented immigration for the U.S. Senate. His publications include many articles and books concerned with segregation in public housing and discrimination in mortgage lending.

Edward G. Goetz is Associate Professor in the Housing Program at the University of Minnesota. His research interests include local housing policy and politics, community-based housing development, local economic development policy, and homelessness. He is the author of *Shelter Burden: Local Politics and Progressive Housing Policy* (1993) and coeditor of *The New Localism: Local Politics in a Global Era* (1993). He has published articles on local housing policy and is working on the neighborhood-based politics of subsidized housing.

Stephen M. Golant is Professor of Geography at the University of Florida. He is currently directing the CASERA PROJECT—Creating Affordable and Supportive Elder Renter Accommodations —funded by the Retirement Research Foundation of Chicago. He is the author of *Housing America's Elderly: Many Possibilities, Few Choices.*

Gideon S. Golany is Distinguished Professor of Urban Design in the Department of Architecture at Pennsylvania State University. He specializes in geospace design, urban design with climate, and new-town planning. Recipient of several Fulbright Research Awards and Honorary Professor of the China Academy of Sciences, he has authored or edited more than 25 books, including *Geo-space Urban Design* (1996); *Ethics and Urban Design* (1995); *Vernacular House Design and the Jewish Quarter in Baghdad* (1994); and *Chinese Earth-Sheltered Dwellings: Indigenous Lessons for Modern Urban Design* (1992).

Greg Goldstein works in the Urban Environmental Health Unit of the World Health Organization Programme on Environmental Health, Geneva, where he is Coordinator for Healthy Cities (Asia, Africa, Latin America). His interests include urban health development and city health planning; the development and application of indicators on health and environment; social marginalization of peri-urban dwellers; and health issues of small-scale industries. In 1986 and 1987, he was Project Manager for the Pakistan Health Financing Study.

Kenneth R. Good is Associate Professor of Anthropology at Jersey City College, New Jersey. He has studied the Yanomami for 23 years and is author of the book *Into the Heart* (1996) and of several articles in *Natural History* and other journals. With the National Geographic Society, he did a documentary on the Yanomami, titled "Yanomami Homecoming."

Michael R. Greenberg is Codirector of the New Jersey Graduate Program in Public Health. His research focuses on environmental health policy. His book *Environmentally Devastated Neighborhoods: Perceptions, Realities, and Policies,* coauthored with Dona Schneider, describes the dilemmas of living in neighborhoods with TOADS.

J. Eugene Grigsby III is Director of the Advanced Policy Institute and Professor at UCLA's School of Public Policy and Social Research. He is also the Los Angeles Coordinator of the Mega-Cities Project, a consortium of 20 of the world's largest cities, and President of the Mega-Cities Coordinating Council Executive Committee. As a member of the *Los Angeles Times* Board of Advisors, he writes a regular column on urban economic issues. His most recent books are *Shaping a National Urban Agenda* and *Residential Apartheid.*

William G. Grigsby is Professor Emeritus of City and Regional Planning and former director of the Center for Energy and the Environment at the University of Pennsylvania.

Mark D. Gross is Associate Professor of Planning and Design at the University of Colorado, where he codirects the Sundance Laboratory for Computing in Design and planning and teaches design computing. His interests include artificial intelligence in design, computational models and methods of architectural design, and systematic approaches to architecture.

Barbara A. Haley is an applied sociologist in the Office of Lead Based Hazard Control and Poisoning Prevention at the U.S. Department of Housing and Urban Development (HUD). She is working on the evaluation of the HUD grant program for lead-based paint hazard reduction and analyzing the results of HUD's national assessment of lead hazard awareness. She is the author of *American Health Care in Transition: A Guide to the Literature* (1997).

Cassandra M. Hanley is a graduate student at Duke University.

Patrick H. Hare is an independent consultant in Washington, D.C.

Richard Harris teaches urban historical geography at McMaster University, Hamilton, Canada. He has written about residential segregation, home ownership, owner building and suburban development. He is the author of *Unplanned Suburbs: Toronto's American Tragedy, 1900-1950* and is writing a book about owner building in North America from the 1890s to the 1950s.

Chester Hartman is President and Executive Director of the Poverty & Race Research Action Council in Washington, D.C.

Catherine Hawes is a Senior Policy Analyst and Codirector of the Program in Aging and Long-Term Care at Research Triangle Institute (RTI), a nonprofit research organization in North Carolina.

R. Allen Hays is Director of the Graduate Program in Public Policy and Professor of Political Science at the University of Northern Iowa. He is the author of *The Federal Government and Urban Housing: Ideology and Change in Public Policy* (1995) and the editor of *Ownership, Control and the Future of Housing Policy* (1993). He has also authored articles on a variety of housing policy issues and other topics in U.S. local government.

Jozsef Hegedüs, an economist and sociologist, has been working on housing and urban questions in Hungary and Central and Eastern Europe since the late 1970s. He has published several articles and is coeditor of books on these issues. He is one of the principals of the Metropolitan Research Institute in Budapest, founded in 1989. His current research is in housing finance, subsidy policy, and local government finance.

Barbara Schmitter Heisler is Professor in the Department of Sociology and Anthropology at Gettysburg College. Her research interests are comparative with a focus on Western Europe. She has published numerous articles on immigration, citizenship, poverty, the welfare state, and housing and is working on a book on poverty and social exclusion in Germany, the Netherlands, and Britain.

Allen David Heskin is Professor in the Department of Urban Planning, School of Public Policy and Social Research at the University of California—Los Angeles. He is author of *Struggle for Community*, and coauthor of *The Hidden History of Housing Cooperatives*.

Leonard F. Heumann is Professor of Urban and Regional Planning and Psychology at the University of Illinois at Urbana-Champaign. He is coauthor of *Housing for the Elderly: Planning and Policy Formulation in Western Europe and North America*, coeditor of *Aging in Place with Dignity: International Solutions Relating to the Low Income and Frail Elderly*, and author of numerous other publications.

Charles J. Hoch teaches planning and housing courses in the College of Urban Planning and Public Affairs at the University of Illinois at Chicago. He coauthored the book *New Homeless and Old: Community and the Skid Row Hotel* and is coeditor of *Under One Roof: Issues and Innovation in Shared Housing* (1996).

Daniel Hoffman is Policy Director for the Pennsylvania Low Income Housing Coalition and former Research Director of the American Affordable Housing Institute, Rutgers, the State University of New Jersey.

Lily M. Hoffman is Associate Professor of Sociology at City College and the Graduate Center, City University of New York. She directed the Rosenberg/Humphrey Program in Public Policy and served as consultant for a national community development program involved in housing. She is the author of *The Politics of Knowledge: Activist Movements in Medicine and numerous Planning* and articles on urban economic development, housing, and urban planning.

C. Scott Holupka is a Research Associate a the Washington, D.C., Office of Vanderbilt University's Institute for Public Policy Studies. His recent research includes an evaluation of the Homeless Families Program, developed and jointly sponsored by The Robert Wood Johnson Foundation and the U.S. Department of Housing and Urban Development, as well as several studies of services-enriched, supportive housing.

Deborah A. Howe is Professor of Urban Studies and Planning at Portland State University in Portland, Oregon. Her professional work has focused on community planning and the development of low- and moderate-income housing.

E. Jay Howenstine is an independent consultant and a member of the Board of Editors of the international journal on urban policy, *Cities*. He was Coordinator on Workers' Housing Policy at the International Labour Office in Geneva, Switzerland, from 1948 to 1967 until returning to the U.S. Department of Housing and Urban Development, where he was the International Research Coordinator until retirement. He authored *Housing Vouchers: A Comparative International Analysis* and contributed to the international symposium, *The New Housing Shortage: Housing Affordability in Europe and the USA.*

J. David Hulchanski is the Dr. Chow Yei Ching Professor of Housing in the University of Toronto's Faculty of Social Work, where he teaches and supervises research in community development, housing, and social policy. His research focuses on housing policy issues, including rental housing problems, discrimination in housing markets, social housing programs, homelessness, and housing and human services issues in relation to land use planning.

Elizabeth Huttman is founding president of the Research Committee of Housing and the Built Environment of the International Sociological Association and professor emerita of California State University. She is coeditor of *Handbook on Housing and the Built Environment* and *Urban Housing Segregation of Minorities in Western Europe and the U.S.* and the author of several book chapters on homelessness and on the housing of immigrants.

Austin J. Jaffe is the Philip H. Sieg Professor of Business Administration and the Research Director of the Institute for Real Estate Studies at Pennsylvania State University. He is author or coauthor of several books, including *Real Estate Investment Decision Making, Fundamentals of Real Estate Development,* and *Property Rights and Privatisation in the Baltic Countries,* as well as numerous articles. He teaches in the areas of real estate financial analysis, the economics of urban property rights, and international property and housing issues.

Marjorie E. Jensen teaches community-based housing at the University of Rhode Island and is a coauthor of *Developing Community Housing Needs Assessment and strategies: A Self-Help Guidebook for Nonmetropolitan Communities* and of *Home Buyers Guide: Financing and Evaluating Prospective Homes.* She has developed training curricula for novice home buyers and landlords and for HOPE 3, Family Self-Sufficiency, and Habitat programs.

James A. Johnson is chairman and CEO of Fannie Mae. He has served as a faculty member at Princeton University, as a managing director in corporate finance at Lehman Brothers, as Vice-President Walter Mondale's executive assistant and chair of his presidential campaign in 1984, and as president of the Washington consulting firm Public Strategies. He is also chairman of the board of trustees of the Brookings Institution and was recently appointed chairman of The John F. Kennedy Center for the Performing Aarts.

Peter Katz is principal of Urban Advantage, an Oakland, California-based strategic consulting firm that promotes the inherent benefits of urban places. He was founding executive director of the Congress for the New Urbanism and is the author of *The New Urbanism: Toward an Architecture of Community.*

W. Dennis Keating is Professor of Law and Urban Planning and Associate Dean at the College of Urban Affairs at Cleveland State University, Cleveland, Ohio. He teaches housing policy and law and has authored numerous books, articles, essays, and reports on housing issues. He has been a housing consultant to the U.S. Department of Housing and Urban Development, U.N. Habitat, state and local governments, and national and local foundations in the United States.

Barbara M. Kelly is the Curator of the Long Island Studies Institute at Hofstra University where she also teaches U.S. history. She is the author of *Expanding the American Dream; Building and Rebuilding Levittown* and the editor of *Suburbia Reexamined* and *Long Island: The Suburban Experience.*

Stephen Kendall is a research architect and design educator and has written extensively on Open Building at the international level. He has also conducted major investigations into the introduction of this practice into the U.S. housing industry. His work includes studies of advanced methods for constructing sustainable multi-tenant buildings of all kinds, based on Open Building principles. He is joint coordinator of the CIB Task Group 26 Open Building Implementation.

Leslie Kilmartin is Professor of Regional and Urban Studies at the Bendigo campus of La Trobe University. He is also Pro-Vice-Chancellor in charge of that campus and immediate Past President of the Research Committee on Housing

and the Built Environment of the International Sociological Association. He has written on planning and housing issues in Australia and developing societies.

Juliet King is working on her doctoral dissertation in sociology at the University of Wisconsin—Madison. She is former research associate at the National American Indian Housing Council, a nonprofit organization that provides technical assistance, training, research, and informational services to tribes, Indian housing authorities, and tribally designated housing entities (TDHEs) on Indian housing and related issues.

John S. Klemanski is MPA Director at Oakland University, Rochester, Michigan. His major research interest is in urban economic development politics and policy. He is coauthor of *The Urban Politics Dictionary.*

Nadezhda B. Kosareva is President of the Institute for Urban Economics in Moscow, a private nonprofit public policy research institute. She has been one of the primary architects of housing reform in Russia.

Lauren J. Krivo is Associate Professor of Sociology at Ohio State University. She studies housing inequality, residential segregation, and race-specific crime in cities and urban neighborhoods in the United States. Recently, she has published articles on immigrant characteristics and Hispanic-Anglo housing inequality and on crime in extremely disadvantaged neighborhoods.

Norman Krumholz is professor at Cleveland State University aand has had a 20-year career as a practicing urban planner. He served as President of the American Planning Association in 1987. His book (with John Forester) *Making Equity Planning Work* won an Outstanding Publication award in 1991 from the American Collegiate Schools of Planning.

Joseph Laquatra is Associate Professor and Extension Housing Specialist in the Department of Design and Environmental Analysis at Cornell University. He conducts research and develops educational programs for Cornell Cooperative Extension and other groups that focus on technical and socioeconomic issues related to housing. His programs have concentrated on energy efficiency, environmental issues, affordability, and housing policy.

Jacqueline Leavitt teaches Urban Planning at UCLA's School of Public Policy and Social Research. She is coauthor of two books, *From Abandonment to Hope: Community-Households in Harlem* and *The Hidden History of Housing Cooperatives,* and is author of *Defining Cultural Differences in Space: Public Housing as a Microcosm.*

Scott Leckie is Codirector of the Centre on Housing Rights and Evictions (COHRE), based in the Netherlands, and is legal advisor to Habitat International Coalition (HIC), a global coalition for housing rights composed of members from over 90 countries. He has written extensively on a range of human rights issues, with a focus on the human right to adequate housing. His most recent books include *Destruction by Design: Housing Rights Violations in Tibet* (1994) and *When Push Comes to Shove: Forced Evictions and Human Rights* (1995).

Barrett A. Lee is Professor of Sociology and Senior Scientist in the Population Research Institute at Pennsylvania State University. His research interests include urban homelessness, neighborhood change, residential mobility, and racial and ethnic settlement patterns.

Diana Lee-Smith is a founding member of Mazingira Institute, Nairobi. She is also the founder of Habitat International Coalition Women and Shelter (HIC WAS) and was its secretary for eight years. She is Editor of *Settlements Information Network Africa (SINA)* and does research on gender, urbanization, and environment.

Wilhelmina A. Leigh, Senior Research Associate at the Joint Center for Political and Economic Studies, does policy research in the areas of housing and health. She is author of "U.S. Housing Policy in 1996: The Outlook for Black Americans" (in *The State of Black America,* edited by Audrey Rowe and John M. Jeffries) and coeditor of *The Housing Status of Black Americans.*

Stephen J. Lepore is Associate Professor at Carnegie Mellon University, where he teaches and conducts research on the influence of the social and built environment on human health and psychological well-being.

Stanley Lieberson is the Abbott Lawrence Lowell Professor of Sociology at Harvard University. He is a former president of the American Sociological Association.

Jane H. Lillydahl is Professor of Economics and Associate Dean of the College of Arts and Sciences at the University of Colorado, Boulder. She has published in the areas of urban and labor economics.

Y. Thomas Liou is Associate Professor in the Institute of Architecture and Urban Planning at Seng Chia University in Taiwan.

Larry Long is a demographer with the U. S. Census Bureau's Housing and Household Economic Statistics Division. He is author of *Migration and Residential Mobility in the United States* and numerous studies of U.S. population distribution.

Stella Lowry is Medical Assistant Editor of the *British Medical Journal.* Previously, she worked in various hospital specialties. She is the author of *Housing and Health* and *Medical Education* and serves on the Education and Psychological Medicine Committee of the Cancer Research Campaign, the BMA's Working Group on Medical Education and the Steering Group on Curricula Reform at King's College Hospital School of Medicine, London.

Heather MacDonald teaches Housing and Community Development in the Graduate Program in Urban and Regional Planning at the University of Iowa. She has published a number of articles on the restructuring of the housing finance system since 1989.

Ronald L. Mace is the founder and Program Director of the Center for Universal Design at the School of Design at North Carolina State University in Raleigh, where he is also Research Professor in Architecture. He is president of Barrier Free Environments-Architecture P.A. He is author of *The Accessible Housing Design File, Mobile Homes: Alternative Housing for the Handicapped, The Planners Guide to Barrier Free Meetings,* and has written extensively on accessible and universal design.

Stephen Malpezzi is on the faculty of the Department of Real Estate and Urban Land Economics at the University of Wisconsin—Madison. An economist, he was at the Urban Institute from 1977 to 1981 and at the World Bank from 1981 to 1990.

Daniel R. Mandelker is Stamper Professor of Law at Washington University in St. Louis. He is the author of a legal treatise, *Land Use Law,* and coauthor of a law school textbook, *Planning and Control of Land Development.*

Robert W. Marans is Professor of Architecture and Urban Planning at the University of Michigan's College of Architecture and Urban Planning and a research scientist at the University of Michigan's Institute of Social Research. His research has dealt with retirement housing, new towns, and public housing and the contributions of physical settings to the quality of community life and individual well-being. His current work deals with planning for parks and recreation, international tourism, and neighborhood revitalization in distressed cities.

Peter Marcuse, a lawyer and urban planner, teaches housing and comparative planning at Columbia University. Now at work on a history of the New York City Housing Authority, he has written widely on housing policy, redlining, racial segregation, urban divisions, New York City's planning history, property rights and privatization, and the history of housing.

Carol Wiechman Maybach is a doctoral student in international comparative education at the University of Colorado, Boulder. She has been involved extensively with nonprofit organizations and service projects in the United States and abroad for the past 15 years.

Sanjoy Mazumdar is Associate Professor in the School of Social Ecology at the University of California, Irvine. His research focuses on vernacular architecture, housing, religion, and work environments.

Kirk McClure is Associate Professor in the Graduate Program in Urban Planning at the University of Kansas in Lawrence. His teaching and research interests cover housing policy and housing finance, especially in the area of provision of affordable housing. He has also worked in the design and implementation of affordable housing programs with the Boston Redevelopment Authority and with the Commonwealth of Massachusetts through its housing finance agency.

Kathy McCormick is Housing Director, Housing and Human Services, City of Boulder, Boulder, Colorado.

Robert McCutcheon is Professor and Head of the Department of Civil Engineering at the University of the Witwatersrand, Johannesburg, South Africa. He is particularly interested in employment creation in the construction industry.

Judith McDonnell is Associate Professor of Sociology at Bryant College, Smithfield, Rhode Island. She is coeditor of *European Immigrant Women in the United States* (1994). Her research interests include race and ethnicity and housing policy.

Michael McDougall is a Professor of City and Regional Planning at California Polytechnic State University, San Luis Obispo, California.

Carol B. Meeks is Dean of the College of Family and Consumer Sciences and Professor of Human Development and Family Studies at Iowa State University. Previously, she was Professor and Head of the Department of Housing and Consumer Economics at the University of Georgia. Her primary research interests including rural housing, adoption of technology, and regulation.

Isaac F. Megbolugbe is the Practice Leader for Price Waterhouse Housing Finance Group. He previously served as Senior Director of Housing Finance Research at Fannie Mae and as a senior economist for the National Association of Home Builders. He has written and spoken extensively on housing and mortgage finance issues.

Heather C. Melton is a graduate student in the Department of Sociology at the University of Colorado, Boulder.

John T. Metzger is Assistant Professor of Urban and Regional Planning at Michigan State University. He has developed innovative public-private housing and community reinvestment programs and has contributed articles on housing and urban policy to several books and journals.

William Michelson is Professor of Sociology at University of Toronto, where he is also an associate of the Centre for Urban and Community Studies. His research has focused on how people's surroundings affect their everyday lives. He is the author of *Man and His Urban Environment: A Sociological Approach*, and *Environmental Choice, Human Behavior and Residential Satisfaction*, as well as recent articles and chapters on the behavioral effects of experimental housing in Sweden.

John R. Miron is Professor of Geography and Planning, Centre for Urban and Community Studies, University of Toronto. He is the author of *Housing in Postwar Canada: Demographic Change, Household Formation and Housing Demand* (1988), and editor-in-chief of *House, Home, and Community: Progress in Housing Canadians, 1945-1986* (1993).

Iouri Moisseev specializes in research of global human settlements conditions and trends. Before joining the United Nations Centre for Human Settlements, he was Associate Professor in Theory of Town Planning at the Moscow Institute of Architecture.

Harvey Molotch is Professor of Sociology at University of California, Santa Barbara, where he does research in the areas of urban development, environmental problems, and news media studies. He is coauthor of *Urban Fortunes*.

Edward D. Montfort is Chief of the AHS Branch of the Division of Housing Statistics at the U.S. Bureau of the Census, Washington, D.C.

Burrell E. Montz is Director of the Environmental Studies Program and Professor of Geography and Environmental Studies at Binghamton University, in Binghamton, New York. She is coauthor of *The Great Midwestern Floods of 1993* and *Natural Hazards: Explanation and Integration* (1997)

Earl W. Morris is Professor Emeritus of Housing at the University of Minnesota. He has authored numerous journal articles, books, and monographs on the sociology of housing, household well-being, and gerontology. He is currently engaged in comparative research in the United States, Poland, Korea, and Mexico that focuses on the connections between major macroeconomic factors and household well-being.

Shirley A. Morris is a Research Associate in the Research and Training Institute at the Hebrew Rehabilitation Center in Boston, where she has been project manager for several national studies.

Hazel A. Morrow-Jones is Associate Professor of City and Regional Planning in the Austin E. Knowlton School of Architecture at the Ohio State University. She has published on policy to assist the homeless, homeownership, the impact of federal policies on housing patterns, and the movements of homebuyers in U.S. cities. Her most recent research with the Ohio Housing Research Network focuses on sprawl and the loss of city homeowners to suburban jurisdictions.

Paul Muolo is the coauthor of *Inside Job: The Looting of America Savings and Loans*, a New York Times best-seller. He currently serves as senior editor of *National Mortgage News*, an independent trade publication. His freelance work has appeared in the *New York Times*, the Washington Post, *Barrons*, and numerous other publications. He has been a guest lecturer at several universities and business groups.

Alan Murie is Professor of Urban and Regional Studies and Director of the Centre for Urban and Regional Studies at the University of Birmingham, England. He is coeditor of *Housing Studies* and coauthor *Housing Policy and Practice, Selling the Welfare State, Home Ownership,* and *Housing and Family Wealth.*.

Dowell Myers teaches urban planning and demography at the University of Southern California. He is author of *Analysis with Local Census Data and Housing Demography.*

Caroline Nagel is a doctoral student in Geography at the University of Colorado, Boulder. She is currently doing dissertation research on Arab immigrant communities in Europe.

Mary K. Nenno was Visiting Fellow at the Urban Institute, Washington, D.C., from 1992 to 1996. She is the author of *Ending the Stalemate: Moving Housing and Urban Development into the Mainstream of America's Future* (1996). She has been involved in the housing and urban development field for over 40 years, beginning as a staff member of the Buffalo Municipal Housing Authority and then serving in various positions on the staff of the National Association of Housing and Redevelopment Officials in Washington, D.C.; her last position was Associate Director for Policy Development.

Thomas S. Nesslein teaches public policy economics at the University of Wisconsin, Green Bay. His research focuses on urban economics, in particular, housing market analysis and public policy. His article "Housing: The Market Versus the Welfare State Model Revisited" (1988) received the Donald Robert Memorial Prize for best article in *Urban Studies* and a *Scandinavian Housing and Planning Research* award.

Lawrence Q. Newton is a senior researcher/writer in the Department of Communications at Fannie Mae. He does housing policy research and investigates trends and developments that may affect the housing and finance industries. He previously was a research fellow ?at the Rutgers University Center for Urban Policy Research and is coauthor of *Development Impact-Assessment Handbook.*

Michael Oxley is Professor of Housing and Director of the Centre for Comparative Housing at De Montfort University, Leicester, United Kingdom. He is coauthor of *Housing Policy and Rented Housing in Europe.* He has published several articles relating to economic, and internationally comparative, aspects of housing policy.

Ellen-J. Pader, is Professor of Regional Planning at the University of Massachusetts–Amherst. An anthropologist by training, her publications include analyses of housing policies in the United States, particularly in relation to housing discrimination based on the cultural construction of law and housing policy.

J. John Palen is Professor of Sociology at Virginia Commonwealth University. He is the author of a dozen books of which the most recent are *The Suburbs* (1995) and *The Urban World* (1996). His urban interests include gentrification, suburbanization, and Southeast Asian urbanization.

Risa Palm is Dean of the College of Arts and Sciences and Professor of Geography at the University of North Carolina–Chapel Hill. She is author of 12 books, including *Geography of American Cities* and *Natural Hazards: An Integrative Framework for Research and Planning.*

Jelena Pantelic is Urban Development Specialist at the World Bank, South Asia Region, Infrastructure Unit. She is managing two disaster recovery projects in India, following the Maharashtra earthquake in 1993 and the Andra Pradesh cyclone of 1966, and is preparing an urban rehabilitation project for Bombay.

Robert E. Parker is Associate Professor in the Department of Sociology at the University of Nevada, Las Vegas. He is the coauthor with Joe R. Feagin of *Building American Cities* (1990) and the author of *Flesh Peddlers and Warm Bodies: The Temporary Help Industry and Its Workers*(1994). He is working on a book on Las Vegas.

Susan Peck is the Director of the Housing Assistance Council (HAC) Western Regional Office in Mill Valley, California. She also administers a Native American and Colonias housing capacity-building project to increase the use of Rural Housing Services (RHS, formerly Farmers Home Administration, FmHA) housing programs in traditionally underserved areas.

Richard B. Peiser directs the Lusk Center for Real Estate Development at the University of Southern California. He is author of *Professional Real Estate Development: The ULI Guide to the Business* and editor of *The Lusk Review for Real Estate Development and Urban Transformation.* He is actively engaged in affordable housing provision and master-planned community development.

William Peterman is Professor of Geography at Chicago State University, where he is the director of the department's Neighborhood Assistance Center and Calumet Environmental Resource Center.

Charles D. Phillips is the Director of the Myers Research Institute at the Menorah Park Center for the Aging in Beachwood, Ohio.

Laurel Phoenix is a doctoral student in Watershed Management and Hydrology at the College of Environmental Science and Forestry at SUNY in Syracuse, NY. Her interests include housing for special populations, stormwater modeling, and regional water quality planning.

Christopher G. Pickvance is Professor of Urban Studies at the University of Kent, Canterbury, England. He coedited *State Restructuring and Local Power* and *Place, Policy and Politics,* and has written many articles on housing, urban protest, and urban theory.

C. Michelle Piskulich is Associate Professor of Political Science at Oakland University in Rochester, Michigan. Her research interests include welfare policy and nonprofit management and leadership.

Sarah E. Polster is Public Information Coordinator of the American Planning Association in Washington, D.C.

Frank J. Popper teaches in the Urban Studies Department at Rutgers University. He first suggested the concept of LULUs and with Michael Greenberg that of TOADS.

Joshua Posner is a real estate developer and consultant in affordable housing and community revitalization. From 1983 through 1996, he was Director of Housing Development at the Community Builders, a leading nonprofit affordable housing development, management, and human services firm in the northeastern United States. He has managed and directed the development of housing developments in urban, suburban, and rural locations, for families, senior citizens, and special needs populations.

Deborah Pozsonyi is Assistant Professor at Georgia State University. She specializes in real estate investments, real estate finance, arbitrage pricing models, and the formation of mimicking portfolios.

Hugo Priemus is Managing Director of OTB Research Institute for Housing, Urban and Mobility Studies at Delft University of Technology in the Netherlands. He has been Visiting Professor at the De Montfort University in Leicester and the European Faculty in Strasbourg and is Vice Chairman of the European Network for Housing Research.

Cedric Pugh is a faculty member in the School of Urban and Regional Studies at Sheffield Hallam University, England. He writes on urban and housing issues in developing countries. Some of his housing research has been comparative, including OEDC countries, Kenya, Mexico, Singapore, Sri Lanka, India, People's Republic of China, and Russia.

Jon Pynoos is Director of the National Resource and Policy Center on Housing and Long Term Care, Director of the Division of Policy and Services Research of the Andrus Gerontology Center at the University of Southern California, and UPS Foundation Professor of Gerontology, Public Policy and Urban Planning at USC. He is the author of several books on elderly housing policy.

John M. Quigley is Chancellor's Professor of Economics and Public Policy at the University of California, Berkeley. He is also Research Associate at Berkeley's Fisher Center for Real Estate and Urban Economics and president of the American Real Estate and Urban Economic Association.

Joseph S. Rabianski is Professor and Chair of the Department of Real Estate, College of Business Administration, Georgia State University in Atlanta. He is a coauthor of *Principles of Real Estate Decisions, Shopping Center Appraisal and Analysis, Corporate Real Estate, Real Estate Market Analysis,* and *Georgia License Law, Brokerage and Real Estate.* He has also published extensively in real estate-oriented journals.

Nancy L. Randall is interested in low-income housing policy, nonprofit real estate development, and the effects of large urban institutions on their local communities.

Amos Rapoport is Distinguished Professor of Architecture at the University of Wisconsin—Milwaukee. He is one of the founders of environment-behavior studies. He has edited and coedited several books and is the author of over 200 papers, chapters, and articles as well as of five books and several monographs. His work has been translated into 13 languages.

James Ratzenberger reviews governmental activities for the U.S. General Accounting Office in Washington, D.C. For the last 10 years, he has specialized in low-income rental housing issues.

Linda Redmond is Communications Manager at the National Council of State Housing Agencies.

William M. Rohe is the Dean E. Smith Professor of City and Regional Planning and the Director of the Center for Urban and Regional Studies at the University of North Carolina at Chapel Hill. He is the coauthor of *Planning with Neighborhoods* (1985) and author of numerous articles on housing and community development policy and practice. He has conducted evaluations of federal housing programs and studied the impacts of community policing in communities throughout the country. His recent research has focused on the efficacy of public housing programs in assisting residents to achieve economic independence.

Elizabeth A. Roistacher is Professor of Economics at Queens College of the City University of New York. She is coauthor of *Tax Subsidies and Housing Investment* and has written numerous articles on housing and urban policy. She previously served as deputy assistant secretary for economic affairs in the U.S. Department of Housing and Urban Development.

Curtis C. Roseman is Professor of Geography at the University of Southern California. His primary research interests include cyclical migration and migration of ethnic groups. He is editor, with H. D. Laux and G. Thieme, of *EthniCity: Geographical Perspectives on Ethnic Change in Modern Cities* (1996).

James E. Rosenbaum is Professor of Sociology, Education, and Social Policy at Northwestern University. He has published three books and many articles on work, education, and housing opportunities. His studies of the Gautreaux program contributed to the federal Moving to Opportunity program, implemented by the U.S. Department of Housing and Urban Development.

Jeffrey Rubin is Professor of Economics at Rutgers, the State University of New Jersey. He is also on the faculty of the Institute for Health, Health Care Policy and Aging Research at Rutgers.

Nora J. Rubinstein teaches at Rutgers University in the Psychology Department and at the Pratt Institute in the Interior Design Program. She has conducted research on the impacts of trauma on the meaning of *home* and the effects of idealized environments on attitudes toward place. Her research has appeared in various publications, and she has consulted with both the for-profit and nonprofit sectors.

Ronald C. Rutherford is Professor of Finance and Real Estate and Elmo James Burke, Jr., Chair in the Management of Building/Development, in Division of Economics and Finance at the University of Texas at San Antonio.

Walter Rybeck is director of the Center for Public Dialogue, a land economics research group. Formerly, he was Washington bureau chief for Cox Newspapers, assistant director for the National Commission on Urban Problems, and editorial director at the Urban Institute. He has written extensively on housing, poverty, local finance, and social justice issues.

Susan Saegert is Professor of Environmental Psychology and Director of the Center for Human Environments at the City University of New York Graduate School and University Center. She directs the work of the Housing Environments Research Group, focused on the relationship of housing to human development and the social fabric of communities.

Morris Saldov has studied social policy and housing and worked as a consultant for Canada Mortgage and Housing Corporation, Ontario Ministry of Housing, and City of Toronto Non-Profit Housing.

Peter W. Salsich, Jr. is McDonnell Professor of Justice in American Society at Saint Louis University School of Law. He has served as Chair of the American Bar Association's Commission on Homelessness and Poverty and as the first Chair of the Missouri Housing Development Commission. He teaches property, land use, real estate transactions, state and local government, and housing law courses.

Fahriye Hazer Sancar is Professor of Planning in the College of Architecture and Planning at University of Colorado. Her research has been in the area of planning and design processes and includes basic research on the psycho-logical aspects of design, field research on vernacular environments, and land use planning and design review processes and regulations.

Henry Sanoff is Distinguished Professor of Architecture at the School of Design, North Carolina State University. A founder of the Environmental Design Research Association, he is the author of *School Design, Integrating Programming, Evaluation and Participation in Design, Participatory Design: Theory and Practice,* and many other works on design. His research-based design consulting includes community projects throughout Japan, Korea, Australia, Brazil, and Slovenia.

Gail Sansbury is a doctoral candidate in Urban Planning at the University of California—Los Angeles School of Public Policy and Social Research.

Paul Anthony Saporito is an architect who has worked extensively in planning, designing, and constructing small-scale, affordable housing. He is a member of the Congress for the New Urbanism.

Gerald W. Sazama is Associate Professor of Economics at the University of Connecticut, Storrs, and a member of the Board of Directors of the National Association of Housing Cooperatives. His research interests are in the areas of the economics of housing and higher education. He has published articles in *Housing Policy Debate* and the *Economics of Education Review.*

Andrew Scherer is Director of the Legal Support Unit of Legal Services for New York City. He teaches housing law and policy at the City University of New York Law School and planning law at the Columbia University Graduate School of Architecture, Planning, and Preservation. He is the author of *Residential Landlord-Tenant Law in New York.*

Michael H. Schill is Professor of Law and Urban Planning at New York University, where he directs the Center for Real Estate and Urban Policy. He has published two books and several articles on housing policy, finance, and discrimination. His current research projects include analyses of the determinants of housing abandonment in New York City, the impact of public housing on neighborhood poverty rates, enforcement of laws prohibiting discrimination in the housing market, and the housing conditions of immigrants.

Tony Schuman is a Registered Architect and Associate Professor of Architecture at the New Jersey Institute of Technology. He has written about housing design and policy in France, the United States, and Nicaragua, and is working on a book about planned communities in the United States.

Nathan H. Schwartz teaches politics at the University of Louisville. He is working on a comparative study of the politics of housing in the United States and Western Europe.

Robert G. Schwemm is the Ashland Professor at the University of Kentucky College of Law. He has had a broad range of experiences dealing with fair housing law, including serving as private counsel in several landmark housing discrimination cases and working with the Civil Rights Division of the U.S. Department of Justice. He is the author of *Housing Discrimination: Law and Litigation.*

Joseph J. Seneca is University Vice President for Academic Affairs at Rutgers University. He is chairman of the New Jersey Council of Economic Advisors, which provides economic data, analysis, and advice for state government.

Jeffery M. Sharp is Professor of Business Law in the Smeal College of Business Administration at The Pennsylvania State University. He has authored and coauthored numerous articles and books on property and environmental law.

Sylvia Sherwood Founding Director of the Social Gerontological Research Deparment, is Senior Fellow Emeritus of the Research and Training Institute of the Hebrew Rehabilitation Center for Aged in Boston and is also Clinical Professor of Community Health at the Brown University Medical School. She was principal investigator/director of studies evaluating congregate housing and other intervention programs in terms of quality of life, reducing institutional days, and cost-effectiveness.

Anne B. Shlay is Associate Director of the Institute for Public Policy Studies and Associate Professor of Geography and Urban Studies at Temple University. Her work has focused on urban development, lending discrimination, welfare reform, homelessness, and child care.

Mark Shroder is an economist with the U.S. Department of Housing and Urban Development. He is coauthor of *The Digest of the Social Experiments.*

Mara Sidney is a Ph.D. candidate in political science at the University of Colorado, Boulder. Her research has focused on the local politics of affordable housing and on local and national policies to address residential segregation.

Hilary Silver is Associate Professor of Sociology and Urban Studies at Brown University. Her recent publications include two government reports, *Firing Federal Employees: Does Race Make a Difference?* and *Rhode Island State Government Contracting With Minority- and Women-Owned Businesses, and several articles on social exclusion prepared for the International Labour Office.*

Russell N. Sims is a doctoral student in Urban and Public Affairs at the University of Louisville.

John W. Smith teaches political science at Henry Ford Community College, Dearborn, Michigan, and is coauthor of *The Urban Politics Dictionary.*

Marc T. Smith is Associate Director of the Shimberg Center for Affordable Housing at the University of Florida. He teaches courses on housing economics, housing policy, and land development. His research focuses on housing needs analysis, housing program evaluation, and housing markets.

Neil Smith is Professor of Geography and Senior Fellow at the Center for the Critical Analysis of Contemporary Culture, Rutgers University. He is author of several books, including *Uneven Development* (1991) and *The New Urban Frontier: Gentrification and the Revanchist City* (1996).

Jeff Sovern is Professor of Law at St. John's University School of Law.

Hans B. C. Spiegel is Emeritus Professor of Urban Affairs and Planning at Hunter College, City University of New York. Formerly, he was Deputy Assistant Commissioner of the U.S. Urban Renewal Administration, Associate Professor of Urban Planning at Columbia, and visiting professor and researcher in India, Germany, South Korea, Philippines, and Kenya. He edited the three-volume *Citizen Participation in Urban Development* and is Contributing Editor of *Nonprofit & Voluntary Sector Quarterly.*

Joan Forrester Sprague is an architect and planner who founded housing and economic development organizations in several states. She has been a consultant with a number of nonprofit and public agencies and is the author of *More than Housing; Lifeboats for Women and Children.*

Kent Spreckelmeyer is Professor of Architecture in the School of Architecture and Urban Design at the University of Kansas. He is the coauthor of *Evaluating Built Environments and Creative Design Decisions.*

Gregory D. Squires, Professor of Sociology and member of the Urban Studies Programs Faculty at the University of Wisconsin—Milwaukee, recently served two years as a consultant to the Department of Housing and Urban Development's Office for Fair Housing and Equal Opportunity on discriminatory practices by financial institutions. He is the author of *Capital and Communities in Black and White.*

Daniel Stokols is Professor of Social Ecology and Dean of the School of Social Ecology at the University of California, Irvine. His current research examines the effects of physical and social conditions within work environments on employees' health, performance, and social behavior. He is past President of the Division of Population and Environmental Psychology of the American Psychological Association (APA) and a Fellow of Divisions 9, 27, and 34 within the APA and the American Psychological Society. He serves on the editorial boards of the *Journal of Environmental Psychology* and the *Journal of Architectural Planning and Research,* and received the Annual Career Award of the Environmental Design Research Association in 1991.

Michael E. Stone is Professor of Community Planning at the University of Massachusetts–Boston. For nearly 30 years, he has been involved in advocacy, research, and policy development around issues of income adequacy and housing affordability.

W. Paul Strassmann teaches economics at Michigan State University. He has written a number of books, including *Housing and Building Technology in Developing Countries; The Experience of Upgrading in Cartagena, Colombia* and is coauthor of *The Global Construction Industry: Strategies for Entry, Growth, and Survival.*

Robert C. Stroh is Director of the Shimberg Center for Affordable Housing at the University of Florida. He also serves as international coordinator for Working Commission W63: Affordable Housing of the Netherlands-based International Council for Building Research Studies and Documentation. He is coeditor of *Wood Frame House Construction,* among other publications.

Raymond J. Struyk is a Senior Fellow at the Urban Institute. During 1992-98, he was the resident director in Moscow of the Housing Sector Reform Program in the Russian Federation.

Phillip Tabb is the founding principal of the Vesica Group, a consulting firm specializing in sustainable planning and solar design. He has been a practicing architect and planner for 25 years and has won numerous Department of Housing and Urban Development and Department of Energy solar demonstration grants. He is a founding director of the Academy for Sacred and Sustainable Architectural Studies.

Karl Taeuber is a sociologist who pioneered use of demographic methods to investigate trends in residential and school segregation. He recently retired after 31 years on the faculty of the University of Wisconsin—Madison.

Ross Thorne researches person-environment interaction in housing and social history of movie and "live" theatres at the University of Sydney, Australia and is editor of the journal *People and Physical Environment Research.*

Ivan Tosics, a mathematician and sociologist, has been working on housing and urban questions of Hungary and Central and Eastern Europe since the late 1970s. He has published several articles and is coeditor of books on these issues. He is one of the principals of the Metropolitan Research Institute in Budapest, founded in 1989. His primary focus is on rental policy and local government finance.

John F. C. Turner is the coauthor of *Freedom to Build: Dweller Control of the Housing Process* (1972) and of *Building Community: A Third World Case Book* (1988) and author of *Housing by People: Towards Autonomy in Building environments* (1976). He is currently writing on neigh-

borhood regeneration as a mirror and a directive agent of paradigm change.

Lawrence J. Vale is Associate Professor of Urban Studies and Planning at the Massachusetts Institute of Technology. He is the author of *Architecture, Power, and National Identity* as well as numerous articles and two forthcoming books about American public housing reform.

Jan van Weesep is Professor of Urban Geography and Urban Policy at Utrecht University in the Netherlands and was Director of its Urban Research Program for ten years. His research interests cover a broad array of housing issues on which he has published widely. Recently, he has focused on the contribution of housing provision to equity policies and how this can affect the position of cities in international competition for economic growth. He is coeditor of *European Cities in Competition.*

David P. Varady is Professor in the School of Planning, University of Cincinnati. During 1995-1997, he was a Distinguished Senior Scholar at the Center for Urban Policy Research, Rutgers University, where he directed several U.S. Department of Housing and Urban Development research projects. He is the author of *Selling Cities: Attracting Homebuyers through Schools and Housing Programs* and *Neighborhood Upgrading: A Realistic Assessment.*

Patty Vrabel is with the National Low Income Housing Coalition, Washington, D.C.

Allan D. Wallis is Associate Professor of Public Policy at the University of Colorado, Denver, and director of research for the National Civic League. He is author of *Wheel Estate: The Rise and Decline of Mobile Homes* (1991).

Peter M. Ward is Professor of Sociology and a member of the faculty at the LBJ School of Public Affairs at the University of Texas at Austin. He has served as adviser to the Mexican government and to several international development agencies. His recent books include *Methodology for Land and Housing Market Analysis* and *Urbanization by Stealth: Colonias and Public Policy in the Texas and Mexico Border Region.*

Leslie Kanes Weisman is Associate Professor and former Associate Dean of the School of Architecture at New Jersey Institute of Technology. She is a cofounder of the Women's School of Planning and Architecture and of Sheltering Ourselves: A Women's Learning Exchange, an international education forum on women, housing, and economic development. She is the author of *Discrimination by Design: A Feminist Critique of the Man-Made Environment* (1992) and a coeditor of *The Sex of Architecture* (1996).

Marc A. Weiss is senior adviser to the Secretary of the U.S. Department of Housing and Urban Development and Associate Professor and Director of the Real Estate Development Research Center at Columbia University. He is the author of *The Rise of the Community Builders: The Ameri-*

can Real Estate Industry and Urban Land Planning (1987) and coauthor of *Real Estate Development Principles and Process* (1991).

Irving H. Welfeld is a senior analyst at the U.S. Department of Housing and Urban Development. He has taught at the University of Maryland and served as a Visiting Distinguished Robert La Follette Professor at the University of Wisconsin—Madison. He has published extensively in law and housing journals and is the author of *Where We Live* (1988) and *HUD Scandals—Howling Headlines and Silent Fiascoes* (1992).

Andrew White is Editor of *City Limits* magazine, an urban affairs monthly in New York City, and executive director of City Limits Community Information Service, a nonprofit organization dedicated to the dissemination of information about the revitalization New York's low- and moderate-income communities. His work has appeared in *The Village Voice, Metropolis,* the *San Francisco Chronicle,* and elsewhere.

Jill L. White is a Director at Schwab Washington Research Group in Washington, D.C., and is a former Director of Policy Analysis at the National Association for Home Care in Washington, D.C.

Robert B. Whittlesey is Chairman of the Board of the National Association of Housing Partnerships, Inc. and former Executive Director and President of the Metropolitan Boston Housing Partnership, Inc. He was the founding Executive Director of Greater Boston Community Development, Inc. (now the Community Builders) and also served as the Court-Appointed Master in the *Perez et al. vs. Boston Housing Authority* case. He previously taught at the University of Massachusetts and the Kennedy School of Government at Harvard University.

Robert Wiener is Executive Director of the California Coalition of Rural Housing Project, a statewide coalition of builders, government officials, community activists, and low-income families that advocate affordable housing for rural and low-income Californians. He is also a lecturer in the Department of Human and Community Development, University of California, Davis, where he teaches courses on housing policy and community research methods. He

has written extensively in housing journals and is coeditor of and contributor to *Housing in Rural America: Building Affordable and Inclusive Communities* (Sage, forthcoming).

Mary Winter is Professor of Human Development and Family Studies at Iowa State University. Her teaching and research is in the field of family resource management. In recent years, she has examined family responses to unusual conditions, studying families in the United States in which someone earns income working at home, and the responses of families in Poland and Mexico to changing macroeconomic conditions.

Robert C. Wood is John F. Andrus and Henry Luce Professor Emeritus at Wesleyan University. He is former Secretary of the U.S. Department of Housing and Urban Development, former Chair of the Massachusetts Bay Transportation Authority, and former superintendent of the Boston public schools. He currently lives in Boston.

Ping Xu is Associate Professor of Architecture and Landscape Architecture at the University of Colorado in Denver. She worked as an architect in China for eight years. She is also a feng-shui consultant, practicing in the United States, Brazil, China, and Italy.

Yap Kioe Sheng teaches housing policy and planning theory at the Asian Institute of Technology in Bangkok. He worked as a researcher and consultant in Pakistan, Sri Lanka, Thailand, Laos, and Vietnam. He is the author of *Leases, Land and Local Leaders* and edited *Low-Income Housing in Bangkok: A Review of Some Housing Sub-Markets.*

John Yinger is Professor of Economics and Public Administration at the Maxwell School, Syracuse University. He is the author of *Closed Doors, Opportunities Lost: The Continuing Cost of Housing Discrimination* and coauthor of *America's Ailing Cities: Fiscal Health and the Design of Urban Policy.*

Sharon Zukin, Professor of Sociology at Brooklyn College and the Graduate School, City University of New York, has written books and articles about cultural gentrification, housing investment, and urban redevelopment. She is the author of *Loft Living: Cultural and Capital in Urban Change, Landscapes of Power* and *The Cultures of Cities.*